CLINICAL DECISION SUPPORT AND BEYOND

CLINICAL DECISION SUPPORT AND BEYOND

Progress and Opportunities in Knowledge-Enhanced Health and Healthcare

---

THIRD EDITION

Edited by

ROBERT A. GREENES
Biomedical Informatics, Arizona State University, Phoenix, AZ, United States

GUILHERME DEL FIOL
Department of Biomedical Informatics, University of Utah, Salt Lake City, UT, United States

Academic Press is an imprint of Elsevier
125 London Wall, London EC2Y 5AS, United Kingdom
525 B Street, Suite 1650, San Diego, CA 92101, United States
50 Hampshire Street, 5th Floor, Cambridge, MA 02139, United States
The Boulevard, Langford Lane, Kidlington, Oxford OX5 1GB, United Kingdom

ISBN: 978-0-323-91200-6

For information on all Academic Press publications
visit our website at https://www.elsevier.com/books-and-journals

Publisher: Stacy Masucci
Acquisitions Editor: Linda Versteeg-Buschman
Editorial Project Manager: Barbara L. Makinster
Production Project Manager: Sajana Devasi P K
Cover Designer: Matthew Limbert

Typeset by STRAIVE, India

Contents

11. Terminologies, ontologies and data models

Thomas A. Oniki, Roberto A. Rocha, Lee Min Lau, Davide Sottara, and Stanley M. Huff

12. Grouped knowledge elements

Claude Nanjo and Aziz A. Boxwala

13. Infobuttons and point of care access to knowledge

Guilherme Del Fiol, Hong Yu, and James J. Cimino

14. Information visualization and integration

Melanie C. Wright

15. The role of standards: What we can expect and when

Kensaku Kawamoto, Guilherme Del Fiol, Bryn Rhodes, and Robert A. Greenes

16. Population analytics and decision support

John Halamka and Paul Cerrato

17. Expanded sources for precision medicine

Darren K. Johnson and Marc S. Williams

18. Knowledge resources

Guilherme Del Fiol and David A. Cook

IV

Adoption of clinical decision support and other modes of knowledge enhancement

19. Cognitive considerations for health information technology in clinical team environments

Amy Franklin and Jiajie Zhang

20. Governance and implementation

Richard Schreiber and John D. McGreevey III

21. Managing the investment in clinical decision support

Tonya Hongsermeier and John Glaser

29. Integration of knowledge resources into applications to enable CDS: Architectural considerations

Preston Lee, Robert A. Greenes, Kensaku Kawamoto, and Emory A. Fry

30. Getting to knowledge-enhanced health and healthcare

Robert A. Greenes and Guilherme Del Fiol

Contributors

Gaelen P. Adam Health Services, Policy, and Practice, Brown University School of Public Health, Providence, RI, United States

Jessica S. Ancker Department of Biomedical Informatics, Vanderbilt University Medical Center, Nashville, TN, United States

David W. Bates Brigham and Women's Hospital, Boston, MA, United States

Nicole M. Benson McLean Hospital, Belmont, MA, United States

Aziz A. Boxwala Elimu Informatics, Inc., La Jolla, CA, United States

Steven Brown Department of Biomedical Informatics, Vanderbilt University Medical Center, Nashville, TN; Office of Knowledge Based Systems, Department of Veterans Affairs/VHA/OHI/CIDMO, Washington, DC, United States

Paul Cerrato Mayo Clinic Platform, Mayo Clinic, Rochester, MN, United States

James J. Cimino Informatics Institute, The University of Alabama at Birmingham, Birmingham, AL, United States

David A. Cook Office of Applied Scholarship and Education Science, Mayo Clinic College of Medicine and Science; and Division of General Internal Medicine, Mayo Clinic, Rochester, MN, United States

Kathrin Cresswell Usher Institute, The University of Edinburgh, Edinburgh, United Kingdom

Sharon E. Davis Biomedical Informatics, Vanderbilt University Medical Center, Nashville, TN, United States

Guilherme Del Fiol Department of Biomedical Informatics, University of Utah Health, University of Utah, Salt Lake City, UT, United States

Apurva Desai Office of Knowledge Based Systems, Department of Veterans Affairs/VHA/OHI/CIDMO, Washington, DC, United States

Floyd Eisenberg iParsimony, LLC, Washington, DC, United States

Amy Franklin School of Biomedical Informatics, Center for Digital Health and Analytics, University of Texas Health Science Center at Houston, Houston, TX, United States

Hamish S.F. Fraser Brown Center for Biomedical Informatics, Brown University, Providence, RI, United States

Emory A. Fry Cognitive Medical Systems, San Diego, CA, United States

John Glaser Harvard Medical School, Acton, MA, United States

Robert A. Greenes Biomedical Informatics, Arizona State University, Phoenix, AZ, United States

John Halamka Mayo Clinic Platform, Mayo Clinic, Rochester, MN, United States

Peter Haug Department of Biomedical Informatics, University of Utah; Advance Decision Support Group, Intermountain Healthcare, Salt Lake City, UT, United States

Tonya Hongsermeier Elimu Informatics, El Cerrito, CA, United States

Stanley M. Huff Graphite Health, Murray, UT, United States

Robert A. Jenders Center for Biomedical Informatics; Medicine, Charles Drew University; Clinical and Translational Science Institute, University of California; Department of Medicine, University of California, Los Angeles, CA, United States

Darren K. Johnson Department of Genomic Health, Geisinger, Danville, PA, United States

Kensaku Kawamoto Department of Biomedical Informatics, University of Utah Health, University of Utah, Salt Lake City, UT, United States

Kristin J. Konnyu Health Services, Policy, and Practice, Brown University School of Public Health, Providence, RI, United States

Lee Min Lau 3M Health Information Systems, Murray, UT, United States

Preston Lee BDR Solutions, Silver Spring, MD, United States

Leslie A. Lenert Medical University of South Carolina, Charleston, SC, United States

Farah Magrabi Australian Institute of Health Innovation, Macquarie University, Sydney, NSW, Australia

Michael E. Matheny Biomedical Informatics, Vanderbilt University Medical Center; Geriatrics, Research, Education, and Clinical Care Service, Tennessee Valley Healthcare System VA, Nashville, TN, United States

Saverio M. Maviglia Semedy Inc., Needham; Division of General Internal Medicine and Primary Care, Department of Medicine, Brigham and Women's Hospital, Harvard Medical School, Boston, MA, United States

John D. McGreevey III University of Pennsylvania Health System; Perelman School of Medicine, University of Pennsylvania, Philadelphia, PA, United States

Timothy Miksch Biomedical Informatics, Arizona State University, Phoenix, AZ; Information Technology, Mayo Clinic, Rochester, MN, United States

Claude Nanjo Department of Biomedical Informatics, University of Utah, Salt Lake City, UT, United States

Shamim Nemati Department of Medicine, University of California San Diego, San Diego, CA, United States

Lucila Ohno-Machado Section on Biomedical Informatics and Data Science, Yale School of Medicine, New Haven, CT, United States

Thomas A. Oniki 3M Health Information Systems, Murray, UT, United States

Vimla L. Patel Cognitive Studies in Medicine and Public Health, The New York Academy of Medicine; Department of Biomedical Informatics, Columbia University, New York, NY; Biomedical Informatics, Arizona State University, Phoenix, AZ, United States

Mor Peleg Department of Information Systems, University of Haifa, Haifa, Israel

Bryn Rhodes Alphora, Orem, UT, United States

Beatriz H. Rocha Division of General Internal Medicine and Primary Care, Department of Medicine, Brigham and Women's Hospital, Harvard Medical School, Boston; Wolters Kluwer Health, Waltham, MA, United States

Roberto A. Rocha Semedy Inc., Needham; Division of General Internal Medicine and Primary Care, Department of Medicine, Brigham and Women's Hospital, Harvard Medical School, Boston, MA, United States

Jorge A. Rodriguez Division of General Internal Medicine and Primary Care, Brigham and Women's Hospital; Harvard Medical School, Boston, MA, United States

Ian Jude Saldanha Health Services, Policy, and Practice; Epidemiology, Brown University School of Public Health, Providence, RI, United States

Hojjat Salmasian Children's Hospital of Philadelphia, Philadelphia, PA, United States

Lipika Samal Division of General Internal Medicine and Primary Care, Brigham and Women's Hospital; Harvard Medical School, Boston, MA, United States

Christopher H. Schmid Biostatistics, Brown University School of Public Health, Providence, RI, United States

Richard Schreiber Penn State Health Holy Spirit Medical Center, Camp Hill; Geisinger Commonwealth School of Medicine, Scranton, PA; Department of Health Policy and Management, Johns Hopkins Bloomberg School of Public Health, Baltimore, MD, United States

Jane Shellum Information Technology, Mayo Clinic, Rochester, MN, United States

Edward H. Shortliffe Department of Biomedical Informatics, Columbia University, New York, NY; Biomedical Informatics, Arizona State University, Phoenix, AZ; Department of Population Health Sciences, Weill Cornell Medical College, New York, NY, United States

Davide Sottara Mayo Clinic, Rochester, MN, United States

Thomas A. Trikalinos Health Services, Policy, and Practice; Biostatistics, Brown University School of Public Health, Providence, RI, United States

Meghan Reading Turchioe Columbia University School of Nursing, New York, NY, United States

Marc S. Williams Department of Genomic Health, Geisinger, Danville, PA, United States

Melanie C. Wright College of Pharmacy, Idaho State University, Meridian, ID, United States

Hong Yu Center for Biomedical and Health Research in Data Sciences and Miner School of Computer & Information Sciences, University of Massachusetts, Lowell, MA, United States

Jiajie Zhang School of Biomedical Informatics, University of Texas Health Science Center at Houston, Houston, TX, United States

Preface to the third edition, 2023

What a difference a decade makes! If you consider the more than six decades of work in the realm of computer-based decision support and guidance, you would not have been able to say the same for the first four+ decades, when progresses in terms of methods, adoption, and impact were more-or-less glacial. Foci were on both (a) theory in terms of models of decision making and cognition, including probabilistic reasoning, decision analysis, and artificial intelligence and (b) support for simple cognitive processes such as identification of abnormal test results, reminders for interventions such as preventive care, drug dose and drug interaction checks, and order sets for particular indications. Main issues included the limited tools for integration into actual clinical systems, lack of interoperability and portability of approaches, and low adoption of electronic health record (EHR) systems, which limited the implementation of effective CDS to a few vanguard academic medical centers, using home-grown EHR systems. The state of the art over that period was summarized in the first edition of this book, published in 2007. We include the Preface for that edition in the pages that follow.

The second edition, published in 2014, was a substantial revision, recognizing that progress was occurring on many fronts. While not yet resulting in a huge uptick in use of clinical decision support, the technological, political, social, and economic drivers that then existed were beginning to align, warranting a reexamination of the state of the art. Progress was occurring in expansion of data sources, standards and interoperability, evaluation methods, knowledge management, governance and management of health systems, and accelerated adoption of commercial EHR systems with embedded clinical decision support capabilities. The Preface for the second edition is also included.

This third edition, published in 2023, represents a fresh look at the realm of computer aids to decision making, with an expanded focus on the broader context in which knowledge and guidance are able to shape our activities in many realms, for all the participants in the health and health care "collective." In this (almost) decade since the previous edition, we note that not only is progress occurring on many fronts, but the *name of the game is changing*! We believe that health care is ripe for a drastic transformation, partially driven by technologies such as clinical decision support and digital health.

Consider the following:

1. Universal adoption of EHR systems with embedded clinical decision support capabilities and increasing support for standards-based integration with external clinical decision support tools and services.
2. A growing and expanding set of data sources are now available for characterizing health and disease as well as current activities, context, and direction of change, including sensors and app data, genomics, population databases, narrative text, images, and the environment,

creating a rich environment for improving the ability to provide ubiquitous advice, guidance, and workflow support.

3. Progress in distributed system architectures, linkage/communication capabilities, and methods for capturing and interpreting data in the real world in computable form are expanding the kinds and numbers of knowledge resources available and the ability to integrate them into a variety of care processes.
4. Increasing interoperability, with standardized approaches advancing for representing data, knowledge and health system processes, methods to access them, and approaches to integrating them into the health care workflow.
5. Our perspectives on health and health care have expanded over the past decade, to encompass the goals of wellness, health maintenance, early proactive management of disease, and an overall more patient-centered approach. This has involved increased appreciation of the social determinants of disease, the importance of personal choice, and the role of the environment and other factors on a patient's health. The COVID-19 pandemic of the early 2020s brought increased attention to public health and the importance of population-wide databases for optimal health management. The goal of a continuously improving Learning Health System is advocated, with a virtuous cycle of practice generating data, data yielding knowledge, and knowledge then being applied to practice. The Quintuple Aim is posited as a goal—the concurrent pursuit of improving health, enhancing the care experience, reducing costs, ensuring health equity, and promoting workforce well-being.
6. Also sparked by the COVID-19 pandemic, a substantial increase in the use of telehealth and digital health approaches (e.g., at-home testing, home monitoring through sensors and self-management apps) has led to an explosion in innovations that allow patients to receive care at their homes.
7. Increasing use of artificial intelligence approaches in a broad range of health care domains.

With these expanded perspectives and capabilities, the kinds of opportunities for use of knowledge to aid health and health care have broadened—hence our title change to *Clinical Decision Support* ***and Beyond***. The *Beyond* portion of this title refers to all the additional ways in which knowledge can be generated and applied to health and health care, both by individuals regarding their own health by the health system practitioners and by health systems themselves.

It is fair to say that the first five decades of interest in computer aids to decision making have been largely defined by technology limitations, data incompatibility minimal interoperability/sharing, and reliance on local initiatives, despite broad efforts at standardization, cooperation, and calls for improved infrastructure. These limitations have resulted in high costs, lowered expectations, and tepid enthusiasm. With the reduction in technology limitations and the appreciation of the broader context of health and health care, the emphasis has shifted more to the promotion of broad principles, and organizational efforts, both at governmental and public-private consortium levels.

In previous editions, case histories of home-grown EHR systems at three large medical centers, some academic, were detailed. These are now less relevant, given that commercial EHR systems, although they need to evolve over time to more open and layered architectures, are pretty much ubiquitous. The main challenges now are how to adapt these systems and other infrastructure to provide a cohesive environment for optimizing the health system, including individual care, patient self-care, population health, and public health. Thus, these issues are examined more fully in this edition with new and expanded chapters.

As a result of the expanded perspective, we intend for this book to serve a broader audience of practitioners and managers of health care systems generally and the technological methods used by those systems to facilitate optimal decision making and action.

We are also pleased to have Guilherme Del Fiol as the new Co-Editor of this book. With an MD and PhD in biomedical informatics, Guilherme brings a wealth of experience in the investigation of standards-based clinical decision support interventions as well as adapting these interventions to advance health equity.

Robert A. Greenes and Guilherme Del Fiol

Preface to the second edition, 2014

When we published our first edition of this book, in 2007, the effort to incorporate computer-based clinical decision support (CDS) into health care had already been pursued for more than four decades. Most of the progress had occurred in academic medical centers in the United States, and in scattered other locales worldwide. Some of the successes had been adapted and incorporated in a variety of proprietary electronic health record (EHR) systems, but by and large, this was only scratching the surface in terms of what the potential of CDS could achieve, to ensure the safety, quality, and cost-effectiveness of health care practice. It also had provoked some backlash, based on instances in which it was not done well, was not sufficiently patient-specific or useful, and caused interruptions of workflow or increased effort to deal with. Further, although standards existed for exchanging some forms of CDS or delivering CDS as a service, most implementations in EHRs were in proprietary formats, integrated into applications in proprietary ways, and virtually no sharing was occurring. As a result, larger health care organizations had to develop mechanisms for handling their own (a) knowledge generation and validation; (b) knowledge management, curation, and update; and (c) adaptation and integration into applications and workflow settings—three time-consuming and difficult lifecycle processes. Smaller health care entities—community hospitals and practices, for example—basically had no resources to do any of this.

As a result of this situation, stakeholders were beginning to recognize the need for larger-scale collaborative or national initiatives to promote adoption, and develop both shared knowledge and tools to make it easier to do so. In our first edition, we proposed that efforts along those lines be developed.

What we did not foresee is that in the 7 years since that preceding edition, there has been tremendous change on a number other fronts—some of which were already occurring but the impact of which had not yet been felt—that would bring about the situation we have today: namely, that there are many new drivers for CDS adoption and many initiatives that are stimulating the process. Drivers include advances in genomics and precision medicine, increasingly fine-tuning the knowledge needed for care of individuals; increased use of home monitors, sensors, fitness trackers, and personal health records by individuals to record their own health status; new standards and interoperability as well as natural language processing and image feature extraction methods, enabling us to integrate data from multiple sources to create "big data" repositories, that can be used to provide new capabilities for population-based prediction and management of cohorts of patients; new understanding of cognitive processes and development of methods for visualization and management of complex data; and an increasing "app culture" in which lightweight applications can be developed that pull data from various sources and provide new opportunities to integrate, organize, analyze, and display data on top or in conjunction with existing EHR systems. Initiatives include national policies in a number of countries fostering EHR adoption and convergence on features that enhance integration of data, measurement and reporting of quality, and use of CDS; new health care finance models for reimbursing care based on pay-for-value rather than fee-for-service; and new care delivery models incentivizing increased levels of care coordination, patient engagement, and emphasis on wellness, disease prevention, and early detection of problems.

As a result of all these changes, we have added a number of new chapters, and substantially revised other chapters to capture the multiple forces now aligning for increased CDS adoption. This includes looking at the technical, cultural, organizational, policy, legal, and financial factors that are now at play, or beginning to come into play. We also identify some key requirements that are starting to take shape but don't have organized effort behind them yet to take CDS to scale. Nonetheless, perhaps over the next 7 years or so, we will see that many of these ideas also become reality!

This book is intended to familiarize stakeholders having the various backgrounds and perspectives we just mentioned with the factors involved, and to equip them to be effective in this fast-moving field.

We are very pleased to have expanded the contributed list of chapters and to have, overall, the participation of a fantastic set of authors that are truly experts in their respective areas. I am extraordinarily grateful to this wonderful group of colleagues for sharing their knowledge and passion for this important activity.

Robert A. Greenes

Preface to the first edition, 2007

In looking for a way to introduce this book, I came across a speech by Admiral Hyman G. Rickover, delivered at Columbia University in 1982, in which he said,

> *Good ideas are not adopted automatically. They must be driven into practice with courageous impatience. Once implemented they can be easily overturned or subverted through apathy or lack of follow-up, so a continuous effort is required.*

Several citations of this quote on the Internet present a curious discrepancy, in that the word "impatience" is replaced with "patience." I find this variation to be intriguing, as it seems to me that both *courageous patience* and *courageous impatience* must be manifest for the greatest progress to occur. If one has a good idea, one must not only be both steadfast and patient to stay the course and see it through but also, at the same time, continue to push aggressively, even impatiently, lest the effort lose momentum and flag.

My idea for this book was based on the observation that computer-based clinical decision support (CDS) has the potential to be truly transformative in health care but that, despite considerable creativity and experimentation by enthusiasts over more than four decades, and convincing demonstration of effectiveness in particular settings, the adoption of CDS has proceeded at a snail's pace. This slow progress has not significantly accelerated even with major national and regional efforts in a number of nations to promote the use of the electronic health record (EHR), computer-based physician order entry (CPOE), electronic prescribing, and the personal health record (PHR), which are important substrates on which CDS can operate (and for which the prospect of CDS itself is a major driver). Some capabilities have made their way into commercial health information system products—examples include advice and warnings during CPOE, to ensure proper doses, avoid harmful interactions, or warn about allergies, the provision of alerts to providers when an abnormal laboratory result is found, and the use of order sets, or groupings of orders, for specific clinical problems and settings, such as coronary care unit admission or postoperative care after a hip replacement. Nonetheless, CDS usage remains spotty at best, most prevalent but by no means ubiquitous at academic medical centers, less so in community hospitals, and almost nonexistent in office practice. Although the public frequently now turns to the Internet for medical knowledge, automated decision support oriented to patients and consumers in terms of reminders, alerts, and patient-specific advice is similarly largely lacking.

What's wrong with this picture? Is CDS perhaps not really a good idea? Are the requirements for wide dissemination and use beyond reach? Or are there initiatives that can be undertaken that can change the dynamic, and significantly boost adoption? This book is an effort to address this conundrum. Our goal is to examine CDS in detail from many perspectives—its history, the motivations for it, the technologies, the psychology and human factors concerning deployment and use of CDS, the sociology and organizational and management considerations that can lead to successful deployment, the financial and economic drivers and constraints, the marketplace and business opportunities, and last, the role of communal and top-down initiatives such as standardization and creation of infrastructure and sharable resources.

The premise is that if CDS is truly a good idea, then the sluggish progress in adoption and use to date can only mean that we are in need of a new approach. But to develop a new approach to a complex, multifaceted problem, one that would have a better chance of success than current incremental, uncoordinated efforts, the effort will require the participation of many stakeholders representing a range of perspectives. It is much easier in such situations to preserve the status quo, or to introduce minor tweaks than to take concerted action—since tweakers can tinker with an existing system without needing to build a huge consensus, but concerted action requires a major effort to promote a shared vision. Only by finding a way for these participants to come together with a commonality of purpose can this good idea be driven forward with the duality of courageous patience and impatience that it requires.

This book is thus an effort to develop a common ground for addressing this challenge. It should be of most interest to health care organization managers, policy makers and other senior leadership, payers, government funding agencies and foundations focused on health care delivery, medical informatics researchers and students, information technology development managers, information systems and knowledge product and service vendors, and clinical investigators and health care providers more generally who have interest in the issues of health care quality, safety, and cost-effectiveness.

Let's look a bit more at the motivations for CDS and the challenge of aligning them. In this book, we adopt a view of CDS as decision support aimed at individual patient-specific health care. It is advice and guidance offered by computers (more properly, information and communication technology) to aid the problem solving and decision making of health care providers, patients, and the public (i.e., including those not currently patients). CDS is in most views not only a *good idea* but also an *essential* one. The most compelling reasons for CDS are to help practitioners avoid errors, optimize quality, and improve efficiency in health care. Many pressures fuel the need—the explosion of biomedical knowledge over the past several decades, the multiplicity of diagnostic and therapeutic choices available for patient care, the specialization and fragmentation of care, the time constraints on practitioners, regulatory and compliance demands, malpractice concerns, the increasing prevalence of multisystem diseases as the population ages, the spiraling costs of health care, the growing activism and involvement of individuals in their own health care, and the emerging capabilities for "personalized medicine" through genomics, biomarkers, and increasingly structured clinical phenotype data.

Although the descriptors used to characterize the aforementioned trends have a kind of desperate urgency to them, which collectively suggest hyperbole, the fact is that they do reflect the reality of health care today. Because the trends and their consequences impact on different stakeholders in various ways, however, the combined extent of the need perhaps has not been appreciated to its full degree by individuals, as a result of which there has been little impetus for a broad-based effort to address it.

It is important to understand the differences in perspective of the various stakeholders, and to recognize their motivations and needs. For practitioners, despite the benefits offered, there are a number of reasons why CDS is not unequivocally endorsed. Providers often become quite expert in their own particular subject domains, keep up with the literature, and don't feel a compelling need for computers to make recommendations. Despite this, they do generally accept the value of CDS to monitor their actions especially where the aim is to help avoid accidental errors; they welcome alerts for unexpected lab results—if false positives are kept to a minimum; they appreciate timely reminders for schedulable actions; and they take advantage of predefined order sets for frequent clinical situations they encounter. In general, and not surprisingly, the satisfaction with CDS by practitioners seems mostly related to the degree to which it is supportive, patient-specific and relevant, and provided in a way that doesn't interfere with care or require inordinate additional effort and time.

In some circumstances, however, the use of CDS not only requires extra time and effort by the practitioner, but the benefits of its use aren't seen as accruing to the practitioner or the patient. Examples are applications of CDS aimed at limiting orders for expensive tests and treatments. This purpose has been manifested in adoption of drug formularies, substitution of generic for brand name drugs, utilization review and utilization management, and requirements for preapproval/prior authorization from payers for imaging procedures, specialty referrals, or surgery as a function of clinical indication. These may be tolerated by practitioners as necessary evils, but they are regarded primarily as interference with medical judgment and are hardly embraced. Patients also exhibit disdain for and distrust of such applications. So although a societal net benefit may be at play, it is difficult to align support of institutions, payers, providers, and patients in cost containment circumstances.

Difficulties such as the above have been true of many information system innovations in health care. Given the frequently tenuous acceptance or tolerance of such technologies by those who must interact with them directly, instances in which their implementation has been poorly done have been quick to be seized on by unhappy users and critics as examples of why the adoption of such systems should be resisted. Before an information system innovation is introduced, careful thought and experience with testbeds and pilot implementations are needed. Planning, both locally and at an organizational or enterprise level, needs to be directed at issues of how to motivate the participants, and how to make the innovation function effectively, including considerations of process flow and work flow, responsibilities and prerogatives of the users, and ease of use and perception of benefit by them.

With respect to the deployment and support of CDS, it also appears that a major barrier to progress is a lack of recognition of how hard the problem is. On the surface, most CDS does not appear to be very difficult to do. Methods for provision of CDS have been the subject of study throughout its long history, and many useful approaches have been identified, explored, demonstrated, and evaluated. As I've noted, some of these have become highly successful at achieving the intended goals in operational settings. Although some forms of CDS are quite complex, by and large, the most effective approaches have indeed been relatively simple, such as the use of *if…then* rules for applications such as mentioned earlier for determining appropriateness of an action such as a medication order, for recognizing an abnormality of a laboratory result, or for generating a reminder for a test or procedure. Another straightforward form of CDS is the establishment of groups of orders into order sets for particular clinical situations. As also noted, these approaches have made their way into a number of commercial products.

The point that is often overlooked, however, is that robust, sustainable use of CDS is not at all simple, even for *if…then* rules or order sets, when one considers it not with respect to a single point in time but from a long-term maintenance and update perspective. This is difficult even in a single implementation at one site with a limited focus, but the complexity of managing CDS, and the knowledge embedded in it, increases dramatically when one considers its deployment and frequent update on an enterprise-wide basis, driven in large part by the continual efforts in most health care organizations to improve patient safety, quality, and cost-effectiveness. The knowledge assets underlying CDS are time-consuming and expensive to generate, voluminous, and subject to change, so sharing and reuse of them, once created, would be highly advantageous. The knowledge derived from multiple research studies and analyses must be collected, validated, and refined. The knowledge considered most useful for CDS must then be assembled, curated, and represented formally and unambiguously. Knowledge should be represented in standard form so that it can be disseminated and used widely, and it needs to be updated on a regular basis. As sites seek to use such knowledge in clinical applications, the tools for local adaptation of it, for integrating it into clinical applications, and for integrating instances of it, and a means for sharing the experiences of these activities, both positive and negative, should be made available.

These tasks can be considered to belong to three interacting but separate lifecycle processes: (i) knowledge generation and validation, (ii) knowledge management and dissemination, and (iii) CDS implementation and evaluation. Currently, much of the work involved in these tasks tends to occur haphazardly or without explicit delineation, and without formalization or infrastructure to support it. The provision of such capabilities on a sharable, communal basis is an area in which effort has barely begun, and where alignment of stakeholders is essential in order to make substantial progress. I believe that the lack of such capabilities is one of the primary impediments to driving widespread CDS adoption and use. A further key obstacle has been the slow progress in development of standards for representing knowledge for decision support, resulting in the perpetuation of multiple incompatible vendor system implementations, and the embedding of CDS capabilities in software in such a way that it is not easily and separately maintained and updated. The lack of such standards limits the ability to use knowledge, except after extensive adaptation and recoding for specific platforms and with the attendant maintenance problems thereof.

In the first two sections of this book, we introduce CDS in terms of its various purposes, its design, the motivations for using it, and the experiences over the years with implementation, both in academic settings and in commercial products. In subsequent sections, we consider the issues underlying knowledge generation, knowledge management, and CDS deployment, and approaches that have begun to be taken to formalize these processes. After examining efforts being undertaken in various organizations, we consider the prospect of mustering forces on a national or international scale in order to move ahead more rapidly. Is it feasible to do this and can stakeholders be aligned to support such an agenda? We conclude with suggestions about how such a communal process might be initiated on a modest scale in order to gain experience with it and to hopefully build support and momentum for larger-scale efforts.

In summary, then, with respect to the widespread adoption of CDS, that which may seem obvious, compelling, and straightforward to proponents is really a very complex, multifaceted challenge that must be attacked on many fronts. My hope is that this book will enable readers with a variety of backgrounds to gain an appreciation of both the nature of the challenge and the benefits of tackling it. Given that this odyssey will require the shared vision and cooperation of many stakeholders such as those I noted earlier, the intent is that parts of the book will address issues relevant to their various perspectives.

In writing this book, I am grateful for the participation of extraordinarily gifted and experienced colleagues who have contributed a number of excellent chapters. While I take responsibility for the overall vision of the road ahead espoused in this book, I know that all of the contributors have a passion for the prospect of widespread clinical decision support and subscribe to the general goal. I am proud and honored to have such wonderful colleagues, whom I count as friends and fellow travelers on this important journey.

Robert A. Greenes

SECTION I

Goals, methodologies, and challenges for clinical decision support and beyond

CHAPTER

1

Definition, purposes, and scope

Robert A. Greenes[a] *and Guilherme Del Fiol*[b]

[a]Biomedical Informatics, Arizona State University, Phoenix, AZ, United States [b]Department of Biomedical Informatics, University of Utah Health, University of Utah, Salt Lake City, UT, United States

1.1 Introduction: CDS And beyond

Computer-aided clinical decision support ("CDS") for health and healthcare has long been a dream – ever since computers first became available in the early 1960s, and visions of an all-knowing source of knowledge and wisdom that could intercede at appropriate times, to aid both patients and care providers. This vision has broadened over the years to include ongoing monitoring of health and wellness, and to encompass not just the care team but public health and environmental monitoring and advice. It has also broadened in terms of *use of knowledge to enhance health and healthcare delivery* by facilitating workflow, efficiency, and effectiveness through other processes involving timely and useful ways to obtain, organize, present, and apply relevant knowledge at the appropriate times in the care process.

We introduce the book with a conceptualization of health and healthcare as a **Learning Health System (LHS)** [1–3], in which: (a) health and health care practice yields **data**; (b) collecting and analyzing data yield **knowledge;** and **(c)** knowledge then can be applied in terms of guidance and decision support to improve **practice**. Together these form a virtuous cycle – the essence of an LHS. We can consider such a cycle as focusing on CDS/knowledge for **both** the health/well-being and the ongoing healthcare of **individual patients**, as well as for **populations**. The knowledge about populations can of course be applied to individual patients, and individual patients' data are combined to create population health data.

A project begun in 2018 under the auspices of the US Agency for Healthcare Research and Quality (AHRQ), known as AHRQ evidence-based Care Transformation Support (ACTS) initiative [4] has developed a future vision for how different kinds of resources can be brought together to achieve this goal. The conceptualization is based on the "5 rights" framework [5] for **CDS**. According to the CDS 5 rights framework[a]:

[a]See: https://sites.google.com/site/cdsforpiimperativespublic/cds.

https://doi.org/10.1016/B978-0-323-91200-6.00029-2

To improve targeted healthcare decisions/outcomes with well developed and deployed CDS interventions, the interventions must provide:

- *the **right information**,*
- *to the **right people**,*
- *in the **right intervention formats**,*
- *through the **right channels**,*
- *at the **right points in workflow**.*

We illustrate the vision from a care delivery perspective, adapted from the ACTS website[b] (reproduced by permission) by describing an example patient's journey through the health care system as the patient undergoes various health challenges. Fig. 1.1 depicts the future vision from both patient-centered and a care-team perspectives.

> Briefly, the ACTS example scenario considers Mae, a 63-year-old female who has been followed by the same primary care physician (PCP) for 10 years. She has been treated for hypertension and osteoarthritis. Recently, the PCP had to discontinue a nonsteroidal anti-inflammatory drug (NSAID) for her osteoarthritis pain due to progression of stage 3 chronic kidney disease (CKD). Acetaminophen has been ineffective, as has a recent trial of tramadol. The PCP elected to start her on low doses of opiates (oxycodone, 5 mg every 4 h). This was initially effective in that it reduced her pain, improved her function, and enhanced her overall quality of life. However, Mae told her PCP after several months that it wasn't working as well as it had at the start of treatment.

Note that the vision described by the ACTS project is based on projecting ahead from an evolving state of the art. Consider an initial state, about six decades ago, in which none of the processes described were aided by computer – as compared to the vision's goal state. The present state is between the two. Some processes have computer support, but interaction of a patient with the healthcare system still relies largely on contact by phone, and office plus follow-up visits. The episodes of healthcare are generally reactive to a current problem rather than planned and proactive.

The ACTS vision considers Mae's journey as she progresses through seven states of health/healthcare. Here we do not describe the seven states, but rather have abbreviated the description to include the key points in which decision support and knowledge-enhanced care processes are involved. (The interested reader may wish to review the full vision [4].) Following the description of the scenario, Table 1.1 summarizes the different steps in the scenario that can be supported by CDS tools, along with the "CDS 5 rights" for each step.

[b]https://digital.ahrq.gov/acts, last accessed 13 October 2022.

140
90
90
As part of daily life, the patient has convenient access to information and tools to manage health.
The patient can:
• Get answers to questions
• Track health data
• Get health guidance (diet, when and what screenings to get, exercise, etc.)
• Be proactive and participate with Care Team in developing a Comprehensive Shared Care Plan that is easily accessible, when needed

As normal routine, a Care Team can easily access resources and tools, informed by the best evidence, to aid decision-making and improve work satisfaction.
They can:
• Easily gather, integrate, prioritize, and review critical patient data (including patient gathered data) to enable proactive healthcare; including shared decision making with patients
• Easily make the right decisions, act and document care based on the latest guidance and evidence
• Optimize care for both individuals and populations via a population management perspective that helps identify and close care gaps.

FIG. 1.1 Future vision for patient care from patient-centered and care-team perspectives. *From the ACTS project report, AHRQ evidence-based Care Transformation Support (ACTS). Digital Healthcare Research; n.d. https://cmext.ahrq.gov/confluence//display/PUB/Individual+and+Population+Health+Journey [Accessed 13 October 2022], by permission.*

TABLE 1.1 Steps in the patient scenario that can be supported by CDS tools and the "CDS 5 rights" for each step.

Steps in patient scenario	The right information	To the right people	In the right intervention format	Through the right channel	At the right points in workflow
Self-assessment of pain and activity levels	Standardized assessment instrument	Patient	Documentation template	Phone app or Web-based patient portal	Daily assessment at home
Pain level increases	Recommendation to contact PCP to discuss alternative treatment	Patient	Alert	Phone app or Web-based patient portal	Upon entering pain self-assessment
Clinical staff communication and screening	Self-assessments trend and other relevant information (e.g., current medications, problems)	Clinical staff	Integrated visual display	EHR native visual display or 3rd party app launched within the EHR	On-demand when staff reviews patient messages as a part of clinic routine
Pre-visit preparation	Standardized pre-visit questionnaire	Patient	Documentation template	Phone app or Web-based patient portal	On-demand, any time prior to appointment
Pre-visit planning	Pain medication history, morphine equivalent dose, PDMP report	Clinical staff, PCP	Integrated visual display	EHR native visual display or 3rd party app launched within the EHR	App launched automatically upon chart opening as a part of pre-visit planning
Patient telehealth visit	Suggested alternative treatments, along with pros and cons for shared-decision-making	PCP, patient	Integrated visual display, shared-decision making tool (shared with patient's computer screen)	EHR native visual display or 3rd party app launched within the EHR	App launched automatically upon chart opening during telehealth visit, accessed on-demand
Preventive care	Patient-specific recommendations for preventive care	PCP, clinical staff	Provider reminders	EHR preventive care reminders	Always available on the EHR screen, addressed at any time during the patient visit
Preventive care	Orderable items	PCP, clinical staff	Order sets, standing orders	Actionable items within provider reminders	On-demand while reviewing provider reminders
Colorectal cancer screening	Notification that FIT test will be mailed along with instructions	Patient	Patient reminder	Phone app or Web-based patient portal	Automatic, every year prior to patient's due date for FIT test

Initial and ongoing. Mae periodically self-reports her pain and activity levels using a standardized instrument available through a phone app that was prescribed by her PCP as an order in the electronic health record (EHR) system. The app writes self-reported assessments to the EHR.

Change in status. At some point her pain levels start to increase and the app recommends Mae to contact her PCP to discuss an alternative treatment. Mae contacts her PCP's office through an app communication feature that auto-populates a care team message with information such as reason and pain level assessments. Care team members open the message and launch the provider-facing version of the app, which provide an integrated display including pain level assessments along with other relevant data such as current and previous pain medications, medication adherence measures, and relevant problems (e.g., CKD).

Preparation for a telehealth visit by patient. Based on Mae's status, the care team decides to schedule a telehealth visit sending a set of proposed dates and times that Mae can access and select through the patient portal. The telehealth appointment is followed with a "prescription" to use an app to provide some additional information prior to her visit (i.e., tools that screen for possible opioid use disorder (OUD) and depression and assess her level of pain and function). Results of the screening tools (saved into her record) reveal mild OUD, mild-to-moderate depression with no suicidality, and that Mae is having more trouble with housekeeping and meal preparation and has very few activities outside the home. All of this is communicated to the professional care team via the portal before her visit (Not using manual questionnaire at check-in).

Preparation for visit by care team. By the time Mae launches the patient portal app for a telehealth visit, the PCP and the care team have already reviewed her data using the pain assessment app during their daily morning huddle and created a plan for her visit. Besides Mae's self-reported assessments and relevant medication history, the app also computes the morphine equivalent dose for Mae's opioid prescriptions along with relevant information gathered in advance, shared via the portal, and integrated into Mae's EHR. This information helps populate a dashboard with the latest function/pain information and opioid use history, including a prescription drug monitoring program (PDMP) report.

Problem-based focus of the telehealth visit. The PCP and Mae begin the visit by reviewing the information provided and discussing any updates. The PCP tells Mae that the screening tools she completed confirm a significant loss of function and overall quality of life, as well as mild OUD and mild-to-moderate depression. The app suggests alternative treatments, such as the addition of duloxetine and physical therapy, along with a shared-decision-making display that helps the PCP and Mae discuss the pros and cons of each alternative. Because so much time has been saved by data gathering, reporting, and planning in advance, the PCP is able to spend more time discussing the diagnoses and treatment alternatives and jointly creating a care plan that Mae understands and agrees with.

Other considerations. While not the primary focus of Mae's visit, her care team also goes through a set of reminders for preventive care gaps and addresses those gaps through workflow-friendly order sets, protocols, and documentation templates. Through this process, the care plan also schedules future preventive care services that can be completed at home or care settings outside of office visits. The schedule includes recommended immunizations, an appointment for breast cancer screening, and a standing order to mail a fecal immunochemical test (FIT) to Mae's home. In a previous visit, Mae and her PCP had decided to use an annual FIT test instead of a colonoscopy every 10 years. The decision was supported by a shared decision-making tool that calculates personalized risks and benefits of FIT versus colonoscopy using a machine learning prediction model that uses Mae's habits, personal health history, and family health history as predictors.

The ACTS scenario in Ref. [4] goes on to consider episodes in which the patient's symptoms deteriorate at home, she has follow-up visits and additional procedures, hospitalization, specialized care, and extended care, for which coordination is required. Resources are provided to facilitate engagement by her and other family member caregivers.

The above scenario is introduced as a way to explain what this book is about. Essentially, we discuss ways to enable the various aspects of this vision – as well as care in other contexts that can portrayed by additional scenarios. Beginning with the historical social and technical underpinnings, we discuss both current status and future prospects for clinical decision support and the broader vision of knowledge-enhanced health and healthcare in a learning health system.

1.1.1 Definition and scope

Computer-based clinical decision support (CDS) can be defined as ***"the use of information and communication technologies to bring relevant knowledge to bear on the health care and well-being of a patient."***

There are several key words in the phrase "computer-based clinical decision support" and in the definition. Here, the term *computer* is really shorthand for *information and communication technologies (ICT)*, collectively. Clinical decision support can, of course, be provided by textbooks, teaching, manual feedback, and a variety of other methods; by *computer-based*, we mean that our focus is specifically on use of ICT as the *basis* for providing it. By *clinical decisions*, we mean those that bear on the management of health and health care of an *individual* recipient (the patient). Of course, the person may not be an active patient, for example, for consideration of environmental risks or wellness/prevention. (The phrase *healthcare consumer* is sometimes used). By *support*, we mean the *aiding* of, rather than the *making* of, decisions.

Note that although clinical decisions are the focus, it is often difficult to separate out aspects of the decision-making process that relate to business processes, workflow, local preferences, and other aspects of choice that enter into operationalizing the decision-making process and also into determining how support for the decision should be best provided.

Note further that aiding in the decision process rather than making decisions means that there is an intermediary – the recipient – in the loop, and that CDS is not a "closed-loop" process typically. There are some situations in which closed-loop decision support is possible, e.-g., in implantable devices such as pacemakers or drug infusion pumps, but the usual norm is for an open-loop process, with a human in the loop. Where this does not occur, this requires a considerably greater degree of testing, validation, and regulatory filing and approval, as a medical device (see Chapter 23).

CDS has a number of characteristics that apply to most, if not all, of the many ways it can occur. We will discuss these features in some detail in this and the following chapters in this section.

1. The general aim of CDS can be one or both of the following:
   (a) To make data about a patient easier to assess by, or more apparent to, a human. This includes approaches to accessing, organizing/grouping, summarizing, and visualizing data that facilitate decision making. This is a somewhat *passive* form of CDS.
   (b) To foster optimal problem-solving, decision-making, and action by the human. The exact nature of a particular form of CDS depends on its specific purpose. But the goal generally is to apply knowledge to aid the human user in selecting and carrying out recommended courses of action. This is a more *proactive* form of CDS.
2. The decision support is provided to a user—who may be a physician, a nurse, a laboratory technologist, a pharmacist, a patient (or family member or caregiver), or other individual with a need for it. (In this book we will also sometimes refer to a patient as a health care recipient to denote circumstances in which an individual is not actually undergoing patient care, but nonetheless must interact with the health care system.) In some instances, the user may be a computer program rather than a human user. Many possible settings can give rise to this need, such as a problem arising in clinical practice, a health maintenance/preventive care question of a patient, or a training/educational exercise.
3. A primary task of the computer is to select or group knowledge that is pertinent, and/or to process data to create the pertinent knowledge. To the extent that the computer can make the selection based on patient-specific data, the relevance of the CDS to the individual patient is enhanced.
4. The selection or grouping of knowledge and processing of data involve carrying out some sort of inferencing process, algorithm, rule, or association method.
5. The result of CDS is to perform some action, usually to make an assessment or a recommendation, although, as previous noted, this may be through a somewhat passive or a more proactive process. In some forms of CDS the action is implicit. A rule with an if and a then part is a form of CDS that has an explicit action, in the then part. An order set is a form of CDS that groups information that has an association, e.g., a set of orders that might be needed for a particular problem or indication. In this case, the action is that of assembling the individual items in the order set for presentation and consideration by the user. Another major form of CDS is a documentation template – either for input of data into structured forms or producing a report based on structured data elements. The action involved in using a documentation template for CDS is that of assembling the data elements needed for particular contexts in the document, either in an input form or for presentation –such as describing the characteristics of abdominal pain.

It is helpful to consider the realm of CDS in terms of a matrix, as depicted in Table 1.2: Rows indicate the general aim or purpose of a CDS instance (a CDS artifact or resource), such as making a diagnosis, deciding on treatment, or estimating prognosis. Columns indicate the various classes of methodologies that can be applied toward that purpose. Some examples of CDS artifacts for particular purposes are shown in the table, with an indication of which method(s) they rely on. We will discuss the classes of methodologies in Chapter 2 and, at a more technical level, in Section III of the book.

TABLE 1.2 Approaches to Clinical Decision Support may be considered in terms of (a) *Purpose* or Aim of the CDS, and (b) *Method* (i.e., decision model, inferencing process, execution engine) used to implement it.

Purpose		Methods						
		IR and search	Logical evaluation	Probability estimation	Artificial Intelligence	Algorithmic/ multi-step	Grouping	Visualization
	Examples							
Find info								
	Direct hyperlink	x						
	Infobuttons	x	x			x	x	
	Question-answering systems	x			x			
	Context-aware retrieval	x	x		x		x	
Make a decision								
	Diagnosis		x	x	x			
	Test selection		x	x		x	x	
	Choice of treatment		x	x	x	x	x	x
	Prognosis			x	x	x		x
Do calculations								
	Medication dosing	x	x		x	x	x	
	Radiation portal				x	x		x
	Surgical planning					x		x

	Index or score		x	x	x		x
Manage process							
	Checklist	x			x	x	x
	Flow sheet				x	x	x
	Guideline	x			x		x
	Protocol	x			x		x
Monitor							
	Alerts on abnormal events	x	x	x			
	Reminders	x					
	Error checking	x					
Organize info							
	Order sets	x				x	
	Structured forms	x				x	x
	Structured reports	x				x	x
	Dashboards	x			x	x	x
	Graphs and charts		x	x	x	x	x

1.1.2 The evolution of CDS approaches and experiences

The application of computers in health care is already quite an "old" pursuit, when assessed by the yardsticks of the electronics age where product lifecycles and the rise and fall of industries happens in months or years rather than decades. Efforts to automate aspects of health care began in earnest as far back as the early 1960s—more than 60 years ago. Until the past decade, however, the rate of adoption and degree of impact of computers and information technology in general, let alone for CDS in particular, remained low in comparison to the extent to which they became primary, even driving, forces in other fields such as engineering, physics, finance, and, over the last decade, personal communication, and daily life.

In the very earliest days of computer use in healthcare, the prospect that computers could play an active role in helping to solve problems and making decisions stimulated much interest and excitement, and it was among the initial motivators for pursuing the use of computers in healthcare. This was manifested both in terms of research and development as well as public imagination and attention – already raising concerns about automated diagnosis, cookbook medicine, robo-doctors, and loss of privacy.

It is interesting to look at the more than six-decade history in terms of evolution of a relationship between the human user and the computer, and the stages that the relationship has gone through. Table 1.3 shows these phases.

Over the many decades that have transpired, computers, of course, have become essential for record keeping, ordering, results review, charge capture, communication, and billing. Nonetheless, despite the growing need for it, clinical decision support, for a long time, did not have the impacts anticipated, either beneficially in terms of improved access to, or safety, quality, and cost-effectiveness of, health care, or negatively in terms of depersonalization. The changes in public policy, perception, environment, and technology landscape over the years, and how they have furthered or inhibited the advancement, adoption, and use of CDS over the years is explored further in Chapter 3, since this history provides a valuable perspective on what is needed for success beyond healthcare knowledge and technical methodologies themselves. Over the most recent decade, a number of factors (see last row in Table 1.3) have aligned to increase the momentum in development and adoption of CDS. These will also be discussed in Chapter 3, and, in terms of a vision of what the future holds for CDS and knowledge-enhanced health and healthcare, at the end of the book, in Chapter 30.

CDS has been recognized as essential for at least the following constituencies:

- **Health care practitioners**: Providers continually confront a wide range of challenges—seeking to make difficult diagnoses, avoid errors, ensure highest quality, maximize efficacy, and save money—all at the same time!
- **Patients and the public in general**: Individuals frequently have questions and needs in evaluating their health and in making decisions about it.
- **Healthcare delivery organizations, public health entities, and payers**: While these entities are not directly providing or receiving care, they are key participants in formulating, monitoring, and responding to health and healthcare trends, including quality, safety, efficacy, equity, access, and other concerns that require access to knowledge for decision support, at the population level as well as to inform individual care processes.

Many computer-based clinical decision aids have been developed and their usefulness evaluated over the past 60-plus years. Some CDS involves simple types of decision support

TABLE 1.3 Relationship between computers as source of clinical decision support and providers and recipients of health care.

Phase of relationship	Duration (approximate date ranges)	Hallmarks
A long infatuation	1960s–80s	Enthusiasm for clinical decision support, research, new ideas
A troubled courtship	1980s–90s	Successful implementations, evaluations showing benefit, but limited dissemination
Renewed passions	Late 1990s–early 2000s	Internet, the Web, knowledge explosion, safety, and quality agendas, evidence-based medicine (EBM)
Getting the support of the relatives (stakeholders)	2000s–present	National agendas, roll out of electronic health records (EHRs), computer-based provider order entry (CPOE), electronic prescribing (eRx); personal health records and patient portal access; international initiatives; health information exchange (HIE), and quality measurement and reporting
New parties to the relationship	Mid-2000s–present	Recognizing need for knowledge management (KM) and other necessary infrastructure; pushes for health information interoperability, standards, CDS marketplace, distributed knowledge resources, third-party innovators
Building the foundations for a lasting relationship: new drivers for adoption: older, wiser, and better equipped to enter into a mature union – and with a new perspective and vitality re-energizing it	Late 2000s–present	Health finance and health system transformation; top-down national initiatives, incentives, and regulations; genomics; focus on wellness, fitness, and prevention; apps and services; mobile health; telehealth; at-home diagnostic testing; biosensors and monitoring; "big data" and analytics; cognitive support and visualization and workflow enhancement initiatives. The Quintuple Aim and the Learning Health System movements.

like recognizing that a laboratory test result is out of normal range, or that a medication being ordered has a dangerous interaction with another one that a patient is taking, or determining that a patient is now due for a flu shot. In numerous studies, such checks, warnings, and reminders have been demonstrated to be beneficial.

In addition to methodologies for specific kinds of decision support, essential aspects of use of decision aids are how they are *integrated into the workflow of health and healthcare systems, with whom they interact, and whether the experience is regarded as beneficial*. This includes the effort, cost, and resources needed to perform the integration (and update/manage it over time), as well as experience of using the decision support. In Chapter 22, we discuss evaluation approaches to determining effectiveness of computer-based clinical decision support, that have been shown in published studies to reduce errors, encourage best practices, reduce costs, and provide a variety of other benefits. The chapter also reviews approaches to introducing decision support, evaluating it, and managing it over time.

Although most of the initial progress and success were in academic settings (see Chapter 2), increasing capability has found its way, over the years, into commercial products, as part of clinical information systems, or as components that can be added onto or integrated with clinical systems. Nonetheless, availability and, more significantly, use and fidelity to original CDS design with demonstrated benefits in clinical studies have been decidedly spotty and limited, not just for complex forms of decision support, but even for the simpler aids like alerts and reminders or drug interaction checks. Despite the benefits and promise of decision support, it can be a surprise to note that, over the past six decades, we have not seen broad adoption and ubiquitous use of *effective* CDS. Whereas flying an airplane would be unthinkable without decision support today, and even automobile driving depends increasingly on computer-based monitoring, warnings, alerts, reminders, and guidance, much of medical care is still practiced without it.

What has typically not been appreciated is that, even for the simplest forms of decision support, it takes a large effort to go from an initial implementation, aimed at showing that clinical decision support is effective in a particular application setting, to having the ability to provide ongoing management of decision support in the same setting. A further leap is required to move from that capability to wider deployment beyond a single application, even within a single institution or across a single enterprise. This becomes not just a big leap but a giant one, if the goal expands so as to address the possibilities of regional or national (or, do we dare contemplate, international) adoption of accepted clinical practices and guidelines—even for limited aspects of health care, such as appropriate utilization of imaging procedures or cost-effective prescribing of high-cost medications [6].

Challenges that are manageable with some effort in a single environment become much more difficult in a multi-institutional setting. These relate to maintenance and updating of the knowledge underlying decision support; managing the corpus of knowledge, in terms of conflicts, overlaps, and gaps; and determining the best ways to deploy various forms of decision support, in terms of their integration with practice and impact on efficiency and workflow. When it comes to broadly disseminating knowledge that has been well established, ideally in a form that is platform-independent, so that it can be adapted and used in multiple sites, the above issues are compounded, given the plethora or organizations and entities that could be users, their individual differences in organizational culture, preferences, structure, workflow, and the disparate non-standard electronic systems they have. However, addressing this last challenge, in particular, is essential to leveraging and making the effort involved in knowledge management economically feasible on a broad scale.

The story that has unfolded over the decades is one about a provocative, tantalizing, and yet frustrating relationship between:

(a) the source of clinical decision support –computing and information technology, on the one hand, and

(b) the potential recipients of such support – the practitioners of health care and their patients, and, more broadly, healthcare systems and public health.

Thus, as noted above, this book is, in a sense, a tale of that relationship, as depicted in Table 1.2 and described further in Chapter 3—the early allure (technical and medical benefits), the roots of attraction (the characteristics and needs of the parties), the realities of the long courtship (the cultural, social, and organizational milieu that have both encouraged and held back the relationship), the growth and adaptation that have occurred (improved understanding of requirements and limitations), the start of a mature union (the underpinnings and drivers that are now recognized as necessary for the relationship to thrive, and the beginning of convergence of all such drivers as an expected basis for achieving the long-sought benefits of this technology).

1.1.3 CDS and beyond

In the past decade of the 2010s, we saw – and continue to see in the 2020s – an explosion in the integration of computer and communication technology into all aspects of health and healthcare. This has been stimulated by the convergence of many factors, ranging from socio-political-economic to scientific and technical advances that are driving a need for more and better decision support and knowledge access. Table 1.4 identifies some of the key drivers for the interest and uptake we are now seeing. The array of choices, the tradeoffs, and the other mitigating factors that affect decisions are more complex and require more detailed knowledge than ever before. Also, as of this writing, the application and penetration into routine use of many of these advances have been accelerated by the COVID-19 pandemic of the early 2020s and the attendant increased reliance on such technologies in many sectors of the global economy, including health and healthcare.

One of the motivations for creating the third edition of this book is that much is now changing. The changes have included the advances in science and technology, as well as societal adoption of enabling technologies, such as those listed in Table 1.4, and the pace of advance is accelerating also, as a result of government and payer regulations and incentives, and investment and commercial opportunity. In addition, the ability to support clinicians and patients in their decision making and to drive health and healthcare improvement in quality, effectiveness, and value *writ large* now extend **beyond the approaches** that we have traditionally labeled as **"clinical decision support" (CDS)** that have, in prior editions, been our primary focus. Those have mainly been alerts, reminders, guidelines, order sets, documentation templates, and context-based retrieval of relevant literature. These remain the cornerstone of efforts to support clinicians in practice with timely, point-of-need, patient-specific advice, warnings, and information. The technical advances in recent decades are providing a greatly expanded range of data – e.g., from genomics, imaging, personal apps and sensors, and broader access to distributed health records, as input to these traditional approaches to decision support.

Beyond the above, there are new opportunities to include enhanced monitoring of state and context of both provider and patient, as well as of the environment, to anticipate needs more proactively, both in terms of thought processes and potential actions, at appropriate points in the workflow, and to facilitate the selection and performance of appropriate tasks. In addition, there is now increasing ability to leverage data analytics from population data sets and public health to tie results from quality monitoring, utilization rates, social-determinant-based patterns of health and disease, access to care, and outcomes, and to be able

TABLE 1.4 Drivers and opportunities for knowledge-enhanced health and healthcare.

Driver	Focus	Examples
Foundational technologies	• Architectures and platforms	• Distributed systems/IoT • Cloud-based architectures, resources, services
	• Methodologies	• Natural language processing • Image processing • Data analytics, machine learning/artificial intelligence
Data sources	• Sensors and app data • Genomics • Population databases • Narrative text • Images • Environment • Setting	• Person-specific measurements, trends, actions (e.g., meds, activity, blood pressure, heart rate, weight) • Demographic, climate, air quality, epidemiologic • Location, time/place, person(s), situational awareness
Interoperability and standards	• Knowledge representation • Vocabulary/terminology • Data exchange and integration	• Computable knowledge models • Terminologies (e.g., SNOMED, ICD, LOINC, RXNORM) • FHIR • APIs • SMART-on-FHIR • CDS Hooks
Conducive regulation, policy, social movements	• Health/health care priorities, payment/incentives	• Continuity of care and connected care – care across episodes and providers, and across systems • Wellness, prevention, and early proactive intervention • Meaningful Use – stimulating wide adoption of EHRs, accelerated standards adoption, need for CDS capabilities • 21st Century Cures Act – requiring patient access to data/patient apps, and measures against information blocking • Telehealth – advancing reimbursement, regulations, access, convenience, and responsiveness
	• Patient-centeredness	• Focus on patient priorities, preferences, and needs
	• Equity and access	• Responsiveness to social determinants of health • Deliberate design of CDS tools to maximize reach and effectiveness to ensure health equity • Investments to expand broadband internet access in rural/frontier areas
	• Quintuple Aim [7]	• Pursuit of better health, improved outcomes, lower costs, health equity, and clinician well-being (general theme encompassing much of the above)
	• The Learning Health System [3]	• A virtuous cycle in which Practice ➔ Data ➔ Knowledge. Data are evaluated to create knowledge, and the knowledge is then applied to practice, yielding continuous learning and improvement.

to use these results proactively to enhance care delivery. Two overarching themes driving health and healthcare improvement in the recent decades are the **Learning Health System**, and the **Quintuple Aim**, as indicated in Table 1.4. The Learning Health System [1,3] is a movement to enable a virtuous cycle of collecting data through practice, gaining knowledge through data analytics, applying the knowledge to practice, evaluating the results of that knowledge, and improving the process as result of this feedback, thus initiating a repeat of the cycle. It is an expansion and major step beyond the Evidence-Based Medicine movement of the 2000s (including practice-based evidence, for example). The Quintuple Aim [7] is another movement (expanded beyond an earlier Triple Aim, and then Quadruple Aim), to adopt practices that concurrently foster (a) better health, (b)improved outcomes, (c) lower costs, (d) health equity, and (e) clinician well-being. Together, these trends are the motivation for our adding the words "**and Beyond**" to the book title.

In this edition, we will examine the trajectory of CDS and what we refer to as "**knowledge-enhanced health and healthcare**" – which includes, but goes beyond, CDS. Understanding the complexity of getting CDS right [5] has grown over the years.

Our goal for prior editions of this book was to address the question of how to achieve broad impact of clinical decision support on patient safety, health care quality, and health care cost-effectiveness. This involved three primary foci:

1. Understanding of the issues involved in **identifying what kinds of decision** support and application of knowledge resources **are useful** for these purposes, as shown in implementations and evaluation studies.
2. Understanding the **problems and challenges** that must be addressed **in order to broadly disseminate** and replicate these successes, as well as to extend successful approaches to other settings so that the long-anticipated benefits can be realized.
3. Ultimately, for adoption and dissemination to occur, **identifying the key stakeholders and finding ways** to marshal the resources, commitments, and coordinated, sustained effort that are required.

Our expansion of coverage to the broader concept of knowledge enhancement, includes both healthcare delivery as well as maintenance of health and wellness. This encompasses not only CDS per se, but both **more actors, more spheres of activity, more sources of information, and more methodologies:**

1. Using context and setting to determine needs
2. Anticipating needs and intentions
3. Expanding the realm of information and knowledge available, from patient data, population analytics, public health, and other sources, for their own purposes as well as for enhancing the health and healthcare at the individual level
4. Integrating knowledge-enhanced views of data, trends, patterns, and potential needs with workflow

A white paper commissioned by the U.S. Office of the National Coordinator for Health Information Technology (ONC) on *A Roadmap for National Action on Clinical Decision Support* [8] preceded our first edition (in 2007), and was one of the first instances of a call for the creation of an organized national effort in the United States to develop and deploy needed infrastructure and alignment of the multiple stakeholders involved. This is significant in that subsequently in the United States (US), as well as elsewhere in the world, there have been growing national initiatives to install computer-based electronic health records (EHRs) and lay the foundation for technologies for quality measurement, interoperable exchange of information, patient engagement, and clinical decision support [9].

Over the decade following the publication of the First Edition, such national and international initiatives expanded, and, in addition, a number of new drivers emerged that stimulated and expanded efforts to develop, improve, and incorporate decision support and other means of application of knowledge to enhance health and health care. Our Second Edition (in 2014) examined these, which included the rise of national quality agendas in the United States; the growth of "big data", trends in health, wellness, and fitness, population health, and population management; the growth of genomics and individualized (personalized, precision) medicine; the recognition of the need for help in managing cognitive complexity and advances in methods to enhance understandability, usability, and workflow; and new architectures and models for deploying CDS – apps and services, mobile health, and advances in standards and interoperability.

Since the Second Edition, these trends and advances have continued, and have been further enhanced by drivers and opportunities identified in Table 1.4. Therefore, with this Third Edition, it appears timely to address both **CDS and Beyond** – i.e., expanding our focus to include other approaches to generating and incorporating knowledge in health and healthcare more broadly.

Our expanded focus is expressed in enhanced interaction among:

(a) the **individuals whose health and wellness are of primary focus** (variously referred to as the healthcare consumer, patient, or other terms that don't quite capture the fact that it is all of "us")
(b) **direct providers** of health and healthcare services
(c) **organizations** of such services
(d) **public health**
(e) **healthcare payers**
(f) other **regional, national, international, and commercial entities** involved in **providing** health and healthcare services and resources

In particular, we consider the ways in which:

- **data analytics** is producing measurements, insights, and knowledge that can be applied to quality measurement, population health, and the management of individual patients
- **data can be integrated and viewed** at different levels across settings and platforms to enable broader, more-informed decision-making
- **genomics, imaging, and personal sensors and apps** provide expanded features on which to base decisions
- expanded access to data and knowledge can help achieve the "**quintuple aim**" [7] and further the goals of a Learning Health System [3].

1.1.4 Scope and plan of this book

As a guide to readers, we provide the following overview of the various Sections of the book, and suggested foci for different audiences.

Section I. This Section introduces the purposes and goals for CDS&B and the circumstances in which they are most useful (this chapter), technical aspects involved in providing them (Chapter 2), and the practical aspects of implementing them and maintaining them (Chapter 3). In the latter, we focus on what the realistic goals for CDS&B should be, in what settings CDS&B would be most useful, and what forms of CDS&B are appropriate in each of the settings. As part of this examination, we consider national, organizational, and management priorities, incentives, obstacles, and feasibility of approaches that are needed to make the deployment of CDS&B successful, and how they have changed and are evolving.

We also elaborate (Chapter 4) on the emerging importance of a related goal – quality measurement and reporting. While not the same as CDS, since it is done retrospectively, periodic quality measurement and reporting can be a powerful driver for improvement even if indirectly, because it provides feedback and incentives for self-examination, increased awareness, and focus on care improvement to a user. Moreover, to the extent that quality measures are used as benchmarks for rating individual providers, practices, or hospitals that are available to the public, or as a metric for value-based reimbursement in emerging value-based health care finance reform, their use provides a very strong motivation for the adoption of CDS&B prospectively to maximize opportunity for meeting benchmark goals.

Lastly, although much of the book is somewhat United States (US)-centric, Chapter 5 examines international dimensions of this topic.

Section II. Here we address the various ways in which knowledge needed for CDS&B is acquired or generated. There are many possible ways to obtain knowledge, which we classify into three main categories of methods for deriving knowledge: (1) human-intensive techniques; that is, developed either by watching and analyzing human experts, or by debriefing and extracting knowledge from experts systematically (Chapter 6); (2) data-intensive techniques; that is, developed through data mining and analysis, and involving models that perform a task via methods that do not necessarily correspond to human approaches but which can be measured objectively against the data (Chapter 7); and (3) literature-derived; that is, developed by extraction, critical appraisal, and meta-analysis of information from the published literature (Chapter 8).

Section III. In this section we examine the technical approaches and challenges of CDS&B. In Chapters 9–18, we consider the methods of implementation of the various modes of CDS&B delivery, their history, and how they are evolving to become more standardized and interoperable, so that they can be used in a variety of proprietary and nonproprietary settings, facilitating wider dissemination and reuse of successful methods. Formalization and standardization are complicated processes, which involve both technical and political and economic considerations, and standardization initiatives tend to be quite slow, typically proceeding in fits and starts. Benefits of standardization of CDS&B (i.e., reuse and easier implementation across platforms and settings), depend on not only an agreed-upon method for representing the knowledge, but also an information model for the data elements referred to in the knowledge representation, a vocabulary or taxonomy of terms used to denote these data elements, and methods for communicating between a CDS&B tool and the clinical IT system for obtaining the necessary data elements and communicating the results of the evaluations by the CDS tool in terms of recommendations are actions that need to be performed. Thus in Section III, we review the state of the art of these various aspects of the field.

Section IV. This Section focuses on the challenges of integrating CDS&B into operational settings from clinical practice to healthcare delivery organizations to personal use to public health (Chapters 19–27). This Section deals with organizational and business issues involved in implementing decision support and other knowledge-enhancement methods. These issues involve consideration of change management and of the need for creating a culture that is supportive of CDS&B goals, the costs for development and maintenance of CDS&B capabilities, the business rationales and drivers for implementing CDS&B, and the liabilities and regulatory issues involved in providing CDS&B (or not providing it). We also consider other forces for widespread use of CDS&B, including the role of the empowered consumer/patient, and the CDS&B capabilities available for consumer use.

Section V. This final Section is about emerging models and infrastructure approaches to deal with the growing demand for CDS&B, the expanding complexity and range of possible uses, and the difficulty for any single organization to meet these challenges. Thus, we focus on new architectures and ways of thinking about CDS&B delivery and potential ways to enhance collaboration and sharing of tools and resources for CDS&B adoption and management. We consider two main technical factors needed beyond formal representation and standards for widespread adoption and use, (1) approaches for knowledge management (Chapter 28) and (2) architectures for integrating CDS&B into operational environments (Chapter 29). These issues need to be considered both on an enterprise scale, for larger health care organizations, as well as with respect to the challenges of national or even international approaches to capitalizing on research advances and accelerating the translation of effective methods into practice, and dissemination, maintenance, and update of knowledge resources in use operationally.

In the final chapter of Section V (Chapter 30), we consider the future, with projections based in large part on some of the trends identified in the last row of Table 1.3 in Table 1.4, and which are elaborated in prior chapters. Much of these fall into what we consider the "and Beyond" bucket, and is relatively new or has become increasingly evident as a driver for CDS&B adoption only since the publication of our prior two editions in 2007 and 2014. The rapidity which these factors have been seen to influence the direction of health care make it important to consider their role in accelerating CDS&B adoption. This includes both new policy directions and incentives by governments, health finance reforms, science advances particularly in pharmacogenomics, big data analytics, health and wellness trends, advances in mobile computing, the potential of the app culture for disruption of current health care IT architectures, and the standards and interoperability initiatives now taking shape based on the above drivers.

1.1.5 Goals of the book

By the end of the book, it should become apparent that the development, delivery, and support of CDS&B are substantial undertakings that require concerted efforts on many fronts in order to be successful. It is very easy to underestimate the problems of CDS&B, and then to become frustrated by the apparent lack of progress that one sees. Thus, one purpose of this book is to delineate these issues as a kind of call to action, so as to marshal the resources needed to systematically tackle the complexities involved. The challenges have long been dealt with on an ad hoc basis by various groups, many of which have been academic, who have been particularly motivated to work in this area. However, these efforts typically have been done without either the platform or the support for extending their approaches to the wider community. Until recently, there has been essentially no concerted effort to look at this problem systematically, to develop a roadmap and a plan for establishing a suitable infrastructure.

This book should be a useful resource for both those wishing to understand the current status and potential for CDS and those engaged in finding ways to incorporate it in current and future health information systems. We hope that it can play a role in enunciating requirements and stimulating a process of systematic design and development of needed infrastructure, both for local systems and regional, national, and international initiatives – because after all, ultimately success in managing and deploying CDS broadly will require articulation at all levels. Such goals as increasing patient safety, reducing unnecessary health care expenditures, reducing practice variation, and encouraging best practices make beautiful slogans, but they are difficult to achieve without the kind of coordinated efforts we discuss. Rather, the race to implement EHRs, CPOE, as well as various kinds of decision support on top of them, without the systematic attention to the problems and issues we have identified, may have unfortunate consequences if based on legacy architectures that are incapable of adaptation to new requirements of the health care system. On the other hand, of course, national-scale projects have a danger of over-promising, being too heavy-handed, and costing more than anticipated, and may have unfortunate unanticipated consequences.

Thus, our bottom-line message is that CDS&B is a complex topic – the rewards for doing it right are substantial, in terms of high-quality, safe, cost-effective health and healthcare, but achieving these outcomes requires attention on many fronts, technical, organizational, policy, regulatory, financial, and sociocultural. We envision this as a collective effort of many participants and stakeholders, hence we hope that this book has information that is useful for their various perspectives. We have a long road to travel from where we are now, although the field has made considerable progress in a short time as well as witnessed the development of several important new drivers (as judged in relation to our first edition in 2007 and second edition in 2014) so let the journey continue. Exciting destinations are ahead!

1.1.6 What we do not cover

There have been and continue to be inroads in many exciting areas. We only touch on some of them in this book. Our main focus is on capabilities needed for broad deployment of CDS&B to impact on health and healthcare safety, quality, and cost-effectiveness. There are a variety of other important kinds of decision support and knowledge use that we will not cover in depth, because they are outside of that scope or take us in directions that would be too diverse to cover adequately here. The two main categories that we omit are:

1. **Image and signal analysis** methods for interpreting clinical data in such diverse modalities as radiographs, cytology smears, immunoglobulin assays, electrocardiograms, and electroencephalograms. Such methods can recognize patterns, extract findings, and make diagnostic assessments or measurements.
2. **Treatment planning and image-guided intervention**, including radiation dosimetry, three-dimensional modeling, and virtual reality applications. The latter may include, for example, use of heads-up display for integrating images from MRI in real time with a surgeon's view of the operating field.

A key point with respect to these categories is that they tend to be *niche* applications; that is, those that interact only with a limited part of the health care system, such as an image processing method for a CT scanner or for use in a Picture Archiving and Communication System (PACS) workstation, or a signal processing method for ECG interpretation. Niche applications are more likely to be disseminated and adopted widely, when shown to be successful, than innovations, even simple ones, that impact on broader portions of the health care system. This suggests that the primary impediments to adoption are often not the technologies, but the sociotechnical difficulties of adaptation of the business, organizational, and cultural aspects of health care delivery in order to integrate the new technologies into them smoothly and effectively.

1.1.7 Audiences

This book is thus aimed at providing an appreciation of the whole landscape of clinical decision support and knowledge-enhanced health and healthcare, which we believe will be useful to:

(a) **strategic planners** seeking to develop consensus and infrastructure on a national scale
(b) **managers and implementers** of clinical decision support in medical centers and practices
(c) **developers** engaged in CDS, health services, implementation science, and informatics research
(d) **knowledge content providers** in identifying potential opportunities for making their content and services available
(e) **clinical information system vendors** in terms of the ways in which they can facilitate the development of the needed interfaces and application capabilities
(f) **informed users** of clinical information systems, to enable better appreciation of the potential and actual characteristics of CDS encountered in their own activities, and to prepare them to be advocates and participants in CDS and quality improvement initiatives
(g) **students and practitioners of biomedical informatics** with a focus on health and healthcare information systems, tools, and services.

In navigating this book, you may, of course, be interested in reading everything. Nonetheless, various sections and topics may be more relevant to different audiences. In particular, the following sections might be *skimmed or omitted* by audiences:

(a) **Strategic planners** may be *most int*erested in Sections I, IV, and V.
(b) **Managers and implementers** may be *less interested* in Section II.
(c) **Developers** may be *less interested* in Section IV.
(f) **Informed users** may be *most interested* in Sections I, II, and IV.

Nevertheless, we suggest this chapter, Chapters 2, 3, and 30 for all readers.

1.2 CDS and the human

A central characteristic of CDS is that it is intended to interact with and give advice to a human user or recipient. Our focus has been primarily on clinical decision support to health care providers, generally physicians and nurses, but also sometimes pharmacists, technologists, and other personnel. Many of the same principles apply to patient-centered decision support, with its added complexities of health care literacy, language, and mental models. Sometime CDS will involve processes for shared decision making between a health care provider and a patient, family member, or other caregiver. We discuss patient-centered decision support in Chapter 24.

The ability to provide advice to a physician has in many ways been a *disruptive innovation*, to borrow a phrase from the business world [10], in terms of traditional perceptions by physicians of their roles and responsibilities, and the practices and relationships that derive from those perceptions. Over the years, the co-author (RG) has collected cartoons clipped out of magazines and journals, portraying the use of computers in health care. Almost all of these relate to some sort of role of computers or information technology in making decisions. This focus no doubt exemplifies the way most people first think about computers if asked to consider their potential role in health care.

In one cartoon, a patient is consulting a computer and the computer is advising, "Take two interferon tablets and call me in the morning." In another, several surgeons around an operating table are discussing an operation they are about to perform. One is saying to the other, "Since it's been reported that 24% of surgery is unnecessary, let's only do 76% of the procedure." In yet another, a patient is entering a question into a computer and the response is "Not tonight—I have a headache." Another cartoon with an operating room venue shows a surgeon with two hearts, one in each hand, looking across the table at his colleague and saying, "Okay, the old one is in my left hand and the donor's is in my right, correct?" A final cartoon shows a doctor examining a patient. The patient has an arrow in his back, unseen by the doctor. The doctor says, "I'm pretty sure it's psychosomatic, but let's run some tests to be sure."

The humor in these cartoons has to do with the exaggeration of two opposing views of the relationship between the computer and the human, whereas a suitable relationship likely lies somewhere between these extremes.

1.2.1 Computer as omniscient sage

On the one hand, an extreme view is portrayed in which the computer's expertise is taken for granted, and its pronouncements are *accepted quite literally*; the computer is essentially *running the show*. The popular image of computers in medicine is that they are devices that are capable of storing vast amounts of information, performing lightning-fast computations, and making accurate decisions. The cartoons are funny because they start from that assumption and carry it to an absurd extreme. The computer may be seen as patronizing and even arrogant.

1.2.2 Computer as out-of-touch meddler

The opposite extreme is the traditional view of resistance by physicians to computers and to automated guidelines as being too simplistic and *representing "cookbook medicine"*—with the typical warning that computers are insensitive and incapable of recognizing the nuances of patient care, the role of physician judgment, and the prerogative of the physician as primary decision maker.

1.2.3 A more symbiotic view

In actuality, the use of computers in health care has taken a rather conservative, circumspect, and circuitous route to participating in clinical decision making. That route is clearly between the two extremes of the human ceding control to the computer versus the human not being willing to use the computer for decision support at all. Later in this chapter, we discuss a number of dimensions along which the nature of the human-computer interaction in CDS can be considered, including locus of control and degree of assertiveness. The extent to which the interaction is skewed in one direction or another along either of these or other dimensions will depend on the application and purpose, but it will rarely be entirely in one direction on all dimensions. As will be discussed in the historical perspective in Chapter 3, when CDS has been successfully deployed, the computers have been used primarily in an assistive (e.g., automating low-level information processing tasks), advisory or educational role, providing input to the practitioner or patient, who ultimately is responsible for making all decisions. This is why, when we discuss CDS in this book, it will largely be with that perspective—in other words, the emphasis is on decision support, *not* on decision making.

We mentioned earlier that CDS is a sort of disruptive technology. One manifestation of this is that there has been a change in patient attitudes toward their doctors, both in recognizing the limitations of knowledge and judgment of physicians as a group, and in increased tendency to question a physician's decisions and desire or demand to participate in the decision-making process. Further, whereas in the past, physicians were reluctant to consult an information source such as a textbook or a computer in front of a patient for fear that it be regarded by the patient as a sign of indecisiveness or lack of knowledge, now the ability to look up the latest information is regarded by both patients and doctors as necessary and desirable [11,12].

Only in limited circumstances in clinical medicine might one consider using the computer directly in a closed-loop fashion to collect data, analyze the data, make decisions, and take actions without human intervention. Probably the most notable exception is the implanted electrical cardiac pacemaker, which has algorithms for determining when to stimulate the heart automatically in response to heart rate or rhythm abnormalities. An implantable neurostimulator is a surgically placed device electric lead wires that deliver mild electrical signals to the epidural space near the spine to change pain signals as they travel from the spinal cord to the brain. Patient ventilators can do some automatic manipulations such as use of feedback control to adjust cycling thresholds to maintain a desired pressure level, adjustments to keep PEEP/CPAP pressure at specified levels to compensate for gas leaks, or modification of ventilator flow delivery rate to adapt to changes in patient inspiratory effort. However, any such usage of these closed-loop systems requires considerable caution and documentation of efficacy, and can be done only after an arduous process resulting in regulatory approval, from the Center for Devices and Radiological Health of the Food and Drug Administration (FDA) in the United States [13], or from comparable agencies in other nations [14].

But in most situations, the human remains in the middle of the decision-making loop. The guiding settings on a ventilator are still determined by a human, after viewing data obtained from the device and other information such as blood gas test results. Insulin infusion pump settings are still adjusted by humans, after reviewing laboratory and clinical data; even devices such as an "artificial pancreas" do semi-automated adjustment of insulin, but that is done within strict boundaries. For clinical decisions that are not integrated with embedded or connected devices, recommendations are not implemented without the express approval or action by the human.

1.2.4 Limitations of the technology

Until the recent decade, given limitations in the state of the art, computers could usually be expected to provide only relatively unsophisticated decision support, and computer models of human decision making remained limited. This relates primarily to two factors. First, many kinds of data and nuances regarding patient findings are either not captured by a computer or encoded in a form that the computer can interpret – such as content in narrative notes and reports and in diagnostic imaging – but which an experienced clinician not only has access to but can more effectively use. Second, some of the nuances of the decision process that could potentially be captured may not even have been encoded in the model used by the computer, but which a physician routinely considers. Because of the limitations of both the data available to the computer and the model used, it is thus important for the physician to check the reasonableness and appropriateness of a CDS recommendation before acting on it. Note, however, that with greater ranges of data from genomics, apps, sensors, images, and extracted from narrative notes, as well as population data relating to similar patients, this limitation is now changing, as we will cover in Chapters 7 (machine learning methods), 16 (population analytics), 17 (precision medicine), and 24 (patient-centered CDS).

A further difficulty is that some of the applications of CDS that have been pursued are quite complex, for example, determining a differential diagnosis, deciding on an optimal work up strategy, and doing treatment planning. Thus, it is not surprising that most of the success to date has been in the form of simpler kinds of CDS, where the modeling of the decision problem and capture of the nuances of data are much less challenging; and the problem is more deterministic, with tighter decision boundaries. Examples of such kinds of CDS include the use of single-decision rules in targeted settings, such as in CPOE for providing checks of medication doses against recommended ranges, or for verifying the absence of allergies or recognized drug interactions; or in checks of results of newly arrived laboratory results against normal ranges in order to alert physicians of abnormalities. Though these are considerably easier to implement, even they have subtleties and nuances that must be considered for successful implementation. For one thing, the simpler the rules, the less they take into account a variety of mitigating factors that affect the clinical significance of the potential recommendation. But increasing the patient-specificity and sophistication of the rules requires more data, which are often not readily available in computable format within the EHR. Also, to avoid redundant alerts and the well-known problem of "alert fatigue" [15] caused by too many, not sufficiently specific, or unhelpful alerts, the logic should take into account information such as past history of the condition, whether a similar alert was generated within a specified "alert-fatigue avoidance" time window, and the clinical workflow in which the alert would be most optimally displayed (e.g., who should see the alert, when, and in what format). Accounting for such factors makes the rules and the maintenance of them much more complicated, as a result of which the rules are no longer simple [15–17].

Last, replication of some of the approaches shown to be successful in early-adopter settings often has been problematic for several other reasons beyond the decision model and the availability of the data. Factors have been both technical, cultural, and organizational in nature, which we will discuss in greater detail next, and in subsequent chapters, particularly in Sections III and IV of this book. For example, as discussed briefly at the beginning of Section 1.2, when we consider the nature of computer interaction with human beings, an important issue that needs to be carefully addressed is how that interaction is regarded by the human user in terms of decision-making control, responsibility, and judgmental prerogatives. Other factors include the manner of interaction of the program with users in terms of impact on their ease of performance of operational tasks, time required, and effect on workflow procedures and processes, particularly as they relate to clinical IT services.

As we pointed out in Section 1.1, another reason for the gap in adoption has to do with underestimation of the complexity of the tasks involved in replicating an innovation such as CDS on a widespread basis, given the need for it to be well integrated with clinical information systems, actual workflow, and business and health care practice patterns in each site; and given the need for it to be readily updated and adapted to changing requirements. Challenges in deploying CDS at each site involve introducing it in a way that is acceptable to the individuals who will be required to use it, being sensitive to the culture, work style, time constraints, self-image, and other cultural and social factors of these individuals and the organizations in which they participate. This is also discussed further in Chapter 20 on implementation and governance, and Chapter 22 on evaluation.

1.2.5 Considerations regarding human-computer interaction

Computers have long been regarded ambivalently as both a boon and a possible threat with respect to their interaction with humans, particularly for decision making. The field of *artificial intelligence* was specifically created more than 60 years ago with the aim of exploring the nature of intelligence, including the processes of acquisition and representation of knowledge and the ability to do reasoning and problem solving [18]. This has involved both research studies and demonstrations centered on how to make intelligent computers as autonomous entities. We see a number of applications of this kind of pursuit in the form of advances in robotics, chess playing and other strategic game-playing programs, speech recognition, and automatic language translation. A notable example was the challenge by Watson, an IBM software system against three champion players on the popular television show *Jeopardy!*, in which the task is to state a correct question for which a trivia fact is provided as an answer. Needless to say, Watson defeated the humans, relying on vast storehouses of facts and relations, accumulated from text processing, and clever programming [19].

The problem of trying to build intelligent autonomous computers is a fascinating one, but this has less direct applicability in health care than the use of computers in partnership with humans. Many of the same methodologies are used as in the pursuit of autonomous intelligent computers, but an additional focus is on the nature of the interaction between the computer and the human [20]. Our concern here, as we have noted, is the role of the computer as a decision-*support* tool rather than as a decision-*making* entity.

What is the *best* way in which an "intelligent" computer should interact with a human to provide CDS? There are several possible modes of interaction, as we described at the beginning of Section 1.2: First, a computer can be *in charge* of the interaction, delivering recommendations or decisions that are expected to be carried out, and at the right point of time. This mode could be used, for example, in CPOE, when an attempt is made to order a medication with a dosage that is outside of therapeutic limits for safety. The computer should be able to either actively stop such orders from being processed or passively avoid them by not providing the means to enter (or select) such doses in the first place. In another sense, however, the choice of ordering the medication is still made by the physician, and it is typically only when the entered or attempted order needs to be overridden by computer because the dose is outside of therapeutic range that the computer exerts control of the interaction.

A slightly less assertive version of computer control is a mode in which a human decision maker can override the computer by providing a justification for an action to which the computer has raised a warning. An example of this mode of use in order entry systems is when a procedure is ordered by the user for an indication that is not recognized by the computer as being among those generally accepted, but which is nonetheless permissible [21].

Relaxing the constraints still further, consider the mode in which a computer presents a data entry form or dialog box to be completed by the user. The entries in the form are checked for validity, e.g., to ensure that they are within range of values expected for the requested items or that they match a controlled list of possible entries; or the computer may require that entries be chosen from a drop-down list. Thus, the kinds of allowed input are controlled by the computer, but within that scope, the user may make any desired choices. Following entry of data, subsequent displays or forms can be made available to the user, and the sequence and nature of the interaction guided by the computer in response to user entries. An example is a predefined order set for medications and procedures, e.g., for postoperative care after hip surgery, in which a physician might be interacting with the computer to order most of the procedures but also to customize some of the options. Standard dosage forms of medications might be offered, both to encourage those choices and to make it faster for the user, if he or she chooses them, but still allow entry of alternative dosing regimens.

If we consider shifting the focus of decision making further in the direction of user control, several modes of interaction are possible. In one mode, the user performs various actions, and the computer analyzes them in the background, displaying a warning when the action is considered dangerous or inappropriate. Among the earliest experiments aimed at refining this approach was a series of studies carried out by Miller and colleagues [22,23] on "critiquing systems". As discussed in Chapter 2, the primary applications investigated were in management of hypertension and anesthesia. In the critiquing mode, the computer made comments and recommendations for modifying notes and orders already created by the physician, which the physician could accept or reject. A more contemporary application is the typical circumstance in CPOE in which a physician is able to select choices for medications and doses directly, but the computer identifies some of the chosen orders or doses as contraindicated in this patient because of interactions or allergies or inappropriate dose [24], and advises the user of alternative actions that may be more appropriate.

A more passive mode of interaction is one in which the computer gathers statistics about the performance of users over some period of time and provides feedback to the users about how they compare with their peers. This can be regarded as more a motivational and social-pressure type of intervention than as CDS unless it is provided in a highly patient-specific context. It has been shown to have mixed success, and appears to be most effective when coupled with a concerted quality improvement initiative and buy-in by the physicians of this kind of decision support [25–28].

One of the primary targets for concerns about CDS invading a physician's autonomy is the use of clinical practice guidelines, especially those that are computer-based. Although clinical practice guidelines have abounded in magazines and journals, been distributed on compact disks (CDs), and are maintained on websites, they are rarely used directly in clinical practice by physicians, except as education or reference resources. In the care of a specific patient, the experience is often that the guideline does not capture the nuances of the patient, and/or does not embody what the physician believes to be best practices in his or her institution or in the current setting or to correspond with his or her own experience. There is some justification to these complaints. Clinical practice guidelines, however comprehensive they may be, usually cannot specify the details of every possible combination of circumstances that might arise in practice. If they could do so, nonetheless their rendering in a print or display medium for easy comprehension and use would be a significant challenge. Even for standardized guidelines such as those for hypertension management [29,30], a flow chart rendering of the various alternative pathways for management would take up many pages.

Another problem with clinical practice guidelines is that many of the characteristics of patients may be outside the scope of the guideline, for example, other concurrent diseases or medications that can alter the nature of the current condition; or the presence of findings that are not available to the computer-based guideline, for example, those that relate to nonverbal subjective assessments that can best be made face-to-face and are difficult to articulate in words. Finally, even the most well-researched, evidence-based guideline may not have achieved 100% consensus among experts, and alternative modes of care may exist that would be equally appropriate. Thus, guidelines can be best used in a mode in which they provide suggestions or advice when requested, but do not force compliance.

Due to limitations such as those just cited, the idea of clinical guidelines tends to conjure up some of the worst associations with the term "cookbook medicine" [31–33] that we mentioned at the outset of this chapter. At their basis, the objections relate to the view that medicine is too complex to be reduced to a set of algorithms or rules, and that it could never be codified to an extent similar, for example, to that which enables an autopilot to fly an airplane. But, in fact, no one is advocating the autopilot as a model for computer-based decision support in health care. Autopilot operation is successful in airplanes largely because the procedures and operations of normal flight are highly predictable, based on data that can be objectively gathered. As a result, rules for decision making can be fully specified and implemented. Autopilot systems also can be made "aware" of settings in which their use is not appropriate (e.g., takeoff and landing) and circumstances in which something happens that is outside of their realm of decision making, such as the occurrence of a combination of parameters for which there is no defined rule. In such situations, either automatically or through pilot initiative, they have a mode in which the pilot can take over control or override their operation. In the autopilot setting, the computer is more in control than the human most of the time, but the initiative can switch to the human. There are situations in health care that can approach this, for example, the aforementioned closed-loop systems of implanted cardiac pacemaker devices, or other applications that potentially are semi-automated, such as intravenous infusion systems for medication administration or patient ventilator management systems for adjusting O_2 levels or cycling thresholds of the ventilator. Also, guidelines and protocols can be used in a more *prescriptive* manner by physician assistants and nurse practitioners to deliver care in routine situations such as management of sore throat or routine prenatal care, where there is also a clear understanding of when to escalate the care to a more experienced provider.

In general, the cookbook response stems from misconceptions regarding the ways in which clinical practice guidelines can be useful in CDS. Computer-based guidelines can be made more patient-specific in several ways. First, they can be freed of the constraints of paper or screen display of static algorithms, by supporting identification of a variety of possible entry points and the eligibility/applicability criteria for these entry points, and by supporting more flexible means for browsing and navigation of pathways and access to explanatory materials [34]. Second, guidelines can be made arbitrarily more complex and nuanced with respect to patient findings (subject to limitations on available evidence, author expertise, and author fortitude in delineating all of these circumstances), since the need to render the guideline in static form is no longer an issue. Third, the guideline can be decomposed into parts that can be deployed in various application settings, such as those parts best used during CPOE, those that would be most appropriate as alerts or reminders, or those that should be considered during the patient assessment and progress-note-generation process [35,36]. Fourth, these parts can be specified precisely so that they can operate on entered or stored EHR data and produce their recommendations automatically [37]. The issues involved in automating clinical guidelines are discussed further in Chapter 10.

We will return to the general issue of interaction with the user when we discuss the process of integrating CDS into application environments later in this chapter.

1.3 Design and structure of CDS

Many opportunities exist for performing CDS. Two reviews in the early to mid-2000s developed taxonomies of features [38] and modes of use [39], respectively, for clinical decision support. The Kawamoto study [39] and a follow up to the Sim study [40], in particular, are noteworthy in identifying those forms of CDS that have been evaluated in clinical trials. Such schemas, as they are further refined, can be expected to be helpful in continually evaluating instances of CDS in terms of their focus and the settings and modes of their deployment to determine which are most effective.

In this section we propose our own schema for considering CDS, in terms of (1) its purpose, and (2) the architecture and component design elements required for providing it. Design elements include the decision model, knowledge content, data requirements, result specifications, and application environment factors affecting deployment and use. Since one of our main objectives is to understand what factors have held back widespread adoption of CDS and to identify ways of increasing adoption, we will analyze aspects of this schema to help us answer these questions. The questions to be addressed encompass the kinds of standards and infrastructure that may be needed, the kinds of business and organizational strategies that can be useful, and the kinds of approaches that can be used to encourage wider adoption.

1.3.1 Purpose

We now turn to a consideration of the many possible goals or purposes for which CDS might be intended. As noted in Section 1.1, purpose is somewhat orthogonal to the kinds of methodologies needed to carry it out, which are discussed in Chapter 2. The purposes in the rows of Table 1.1 are depicted again in Table 1.5, with a focus on the specific clinical activities for which a user might need CDS. Anticipating Chapter 2, Table 1.5 gives examples of the methodologies that can be used for these activities.

TABLE 1.5 Purposes of CDS with examples of approaches that may be used, and the methodologies involved.

Purpose	Examples of CDS approaches, using various methods or combinations of them
Find Information: Answer questions	Direct hyperlinks from context-specific settings, context-specific information retrieval, use of agents and information brokers, infobuttons as instance of the latter, or ultimately, a "personal guidance system"
Make a decision	Gathering data, analyzing the data, and providing recommendations for assessments or actions
• Diagnosis	Bayes theorem, algorithmic computation, heuristic reasoning, statistical data mining/pattern recognition methods
• Test selection	Decision analysis, logical rules/appropriateness criteria, and logistic models and belief networks for risk prediction (e.g., for screening decisions)
• Choice of treatment	For choosing among alternatives, decision analysis, and logical rules/ appropriateness criteria, including increasingly genotype considerations. For dose modifications for age or factors such as renal function, algorithmic computation. For dosimetry or dose distribution, algorithmic computation based on geometric and pharmacokinetic models, with use of heuristics and statistical methods for optimization
• Prognosis	Logistic regression, Markov modeling, survival analysis models, and quality of life assessment scoring methods
Manage Process: Optimizing process flow and workflow	Multistep algorithms, guidelines, and protocols, coordination of participants by workflow modeling, scheduling, and communication methods
Monitor: Track data and activity	Use of ECA rules, with background detection of events, in real-time or asynchronously, logical evaluation of conditions, and issuing of messages. Events can be a user activity such as choice selection or data entry, a result arrival, or the passage of time
Organize or summarize: Focus attention and enhance visualization	Organization and presentation of items in data entry, display, or reporting applications. May be done by use of sequences to encourage intended behaviors, by a process flow model such as an underlying guideline, and/or by visual groupings based on shared attributes such as purpose, medical subdomain, or application context. May also include dashboards, trend plots, or other summarization and visualization methods that make it easier to identify key elements needed for decision making

1.3.1.1 Find information: Answer questions

The simplest goal for CDS is to provide context-specific access to relevant information for a human user at the time of problem solving or decision making. Hyperlinks to specific resources at specific points in the interaction with a clinical IT system provide one such way to do this. An example would be a link to laboratory tests and their normal ranges, or to a list of medications in a hospital's formulary.

More sophisticated approaches involve using intermediate search tools, such as information agents or "bots" to go out to diverse sources and report back (like Web crawlers or spiders), and "information brokers," which can map queries to the formats required as input by external knowledge bases and then map the responses to a form recognized by the requesting source. The goal is to find resources relevant to a particular context, including patient-specific parameters. For example, in a lab test result review context, the display of a lab test result might be accompanied by an infobutton [41] (see Chapter 13) that, when selected, dynamically retrieves available resources about the test, such as normal ranges, textbook materials regarding the use and interpretation of the test, information about the diseases in which it is abnormal, and a list of MEDLINE references on the clinical use of the test. As discussed in Chapter 2, an HL7 standard for integrating EHRs with online knowledge resources via infobuttons [42] leveraging the EHR context in a consistent manner has been adopted, encouraging use of this approach. One can imagine increasingly sophisticated question answering systems [43–46] being automatically invoked in these contexts to provide highly specific information for decision making.

1.3.1.2 Make decisions—About diagnosis, test selection, choice of treatment, and prognosis

This purpose, in contrast to the more generic task of finding information just discussed, is specifically for help in analyzing information needed for a decision. This can be for a variety of types of decisions, including making diagnoses, selecting tests, planning therapy, and estimating prognosis.

We have noted that differential diagnosis was among the uses of CDS that most captivated interest in the earliest days of computer use in health care. An excellent book edited by Berner [47] focuses largely on diagnostic decision making, and reviews the many approaches that have been pursued. The basic goal of differential diagnosis is to deduce, from a set of findings, the diagnosis that best explains them. This is clearly an important task, not only to be able to select proper treatment but also to estimate prognosis and to give advice to the patient.

It should be recognized, however, that diagnosis is not usually a single event, but rather a *process* of continually refining knowledge about the patient by gathering data, performing tests, and reevaluating data, until sufficient confirmation is reached to take therapeutic action. Some approaches, such as decision analysis, appropriateness criteria, and clinical guidelines, have focused on structuring this process, rather than on the endpoint of diagnosis. Indeed, the decision table approach for representing a guideline [48], as illustrated in Table 2.3 in Chapter 2, doesn't even choose a diagnosis, but determines next actions based solely on combinations of findings. Issues that must be considered, in this view of diagnosis as a process, relate to the selection of appropriate tests based on cost, risk, inconvenience, and other factors versus the potential for information gain from the tests. When one considers the fact that the institution of treatment is also a diagnostic test, in terms of providing information about how the patient responds to it, it can be seen that the whole patient care process continually involves diagnosis in the form of ongoing reassessment. Prognosis estimation can also be regarded as a type of diagnostic assessment, in that it characterizes the patient's current state of health as one with a particular expected survival rate and quality of life.

For the purposes of exposition, we can divide the topic of making decisions into methods for hypothesis formation or refinement, both for diagnosis and prognosis estimation, and those aimed at performing an action (i.e., test selection, or choosing or detailing a treatment regimen), as depicted in Fig. 1.2.

1. **Diagnosis**. The process of diagnosis can be subdivided further into detection and classification. Although some screening recommendations remain controversial, the numbers of tests in use for screening purposes is sure to increase as a result of progress in understanding of the genetic basis of disease and development of biomarkers. Screening tests in common use include, among many others, testing for phenylketonuria (PKU) in newborns, mammography in older women or those with certain risk factors, colonoscopy in average-risk patients over age 50, and prostate-specific antigen (PSA) testing in older males. Generally, screening tests have been applied primarily to alert the user to the detection of the presence of disease, rather than to make detailed specific diagnoses. The typical approach in screening is to set a liberal operating point (decision threshold) on the Receiver Operating Characteristics (ROC) curve for considering the test to be positive, on the basis of the view that it is preferable to err in the direction of more false positives than to fail to detect cases with disease (false negatives). Although some assessment tools for supporting CDS in this realm (such as computer-aided detection and diagnosis (CAD) image processing methods in digital mammography [49,50]) try to classify the findings in terms of specific diagnosis, the ability in screening tests to make diagnoses is usually quite limited, and further testing typically is required. A particular caution is raised by Kohane et al. [51], with respect to whole-person genome mapping, because of the huge potential for

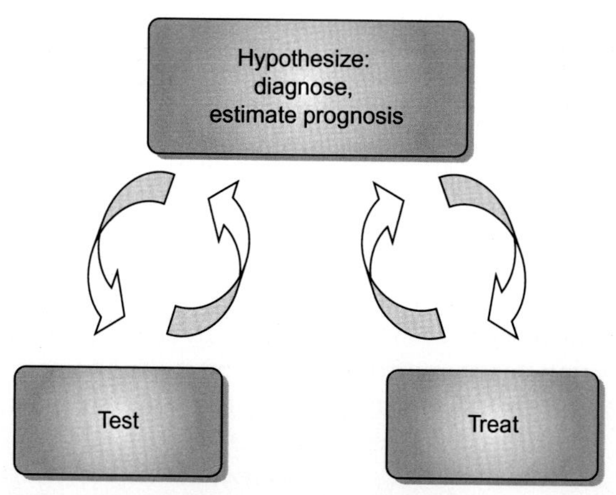

FIG. 1.2 Medical diagnosis as an iterative process of forming hypotheses and gathering data to confirm or refute the hypotheses. Prognosis estimation is a form of hypothesizing. Treatment is a form of testing, in that it also provides data that can help to confirm or refute hypotheses.

false positives and induced tests to further evaluate them, which the authors refer to as the "incidentalome".

Methods used for both detection and classification (diagnosis) include Bayesian probability revision, algorithmic approaches (e.g., for electrolyte imbalance), heuristic reasoning and weighting of findings (e.g., DXplain [52]), and statistical data mining/ pattern recognition methods such as logistic regression, classification and regression trees, and artificial neural networks.

2. **Test selection**. This set of clinical decision-making problems relates to whether and when to do screening, what test to use for screening or for diagnosis, and determination of need for and selection of follow-up testing, including referral to consultants/specialists. Test selection decision support needs to rely on prior work by researchers in technology assessment including ROC analysis, and comparative effectiveness research (CER) [53] studies to determine relative performance characteristics of competing alternatives, and work by health policy analysts and payers using clinical decision analysis and benefit-cost analysis to establish the optimal pathways that sequence the use of these in terms of maximizing expected utility, and formalizing these into a set of logic rules or appropriateness criteria.

   The decision to obtain screening tests or follow up tests may be modified on the basis of CDS tools, such as those for estimating a patient's risk of breast cancer based on risk prediction models [54,55], and policy recommendations/guidelines such as those of the American Cancer Society for early detection of cancer [56]. A major resource for such recommendations is the US Preventive Services Task Force which weighs evidence and publishes periodic recommendations on prevention and screening testing [57].

3. **Treatment decisions**. Needs for CDS include both picking the most appropriate therapy and determining dose or dosage administration regimen. Picking therapy involves tradeoffs of cost, risk, benefits, and patient preferences. Thus, as in the case of test selection, treatment recommendations can be developed by researchers, including economists and policy analysts, who rely on comparative effectiveness research, decision analysis, survival analysis, and benefit-cost analysis. Optimal strategies may be issued by professional societies and payers, and codified as logic rules and appropriateness criteria, or in the form of clinical practice guidelines.

   For dose determination, CDS can be used to constrain choices, make it easier to select preferred choices through displayed groupings such as order sets, or verify dosage requests as being within acceptable ranges. In some circumstances, CDS can be used to calculate dose modifications, for example, using pharmacogenomics-based rules that recognize the effect of particular gene makers on responsiveness of the patient to a dose, based on body surface area in pediatrics, or as a function of renal status or age; these rules tend to be algorithmic. Dose administration determination may involve more elaborate considerations, as in the case of an insulin sliding scale, or calculation of portals and beam configurations for radiation therapy. These tend to be algorithmic, but may use heuristics and pattern recognition and statistics for optimization.

4. **Prognosis estimation**. The CDS question here is to predict the likelihood of good outcomes, morbidity of various types, and mortality. Thus, prognostic evaluation of consequences should be an important consideration before treatment selection. Database prediction, as in very early systems like ARAMIS [58] and in modern-day databases aggregated across large patient populations, is an ideal approach when the data are

available and sufficiently structured. Methods of analysis include logistic regression, classification and regression trees, Bayesian network modeling, and artificial neural networks. Methods for modeling of the likelihood of future chance processes such as Markov modeling and for assessing quality of life, such as the calculation of quality-adjusted life years (QALYs) [59] or severity of illness, such as the Apache III score [60], are essential underpinnings.

1.3.1.3 Optimize process flow and workflow

We have mentioned multistep algorithms, guidelines, and protocols as a more complex form of CDS. They arise in and have use in a variety of settings, because medical care often requires sequences of tasks, with intermediate decision points and pathways. The intent is to help guide the user in the proper sequence of decisions and actions, to be sure all appropriate alternatives are considered, and to avoid proceeding along inappropriate pathways.

An example is the progression from less expensive and simpler tests to more expensive and invasive procedures in the evaluation of heart disease or breast cancer. Another is the initiation of single-drug therapy for hypertension, with adjustment, substitution, or addition/deletion of medications based on response, side effects, and complications. These are best conveyed to a user by clinical practice guidelines, flow charts, protocols, or flow sheets.

Sometimes multiple tasks, such as orders for tests and procedures, must be done concurrently, but the next step must await at least some of the results from those tests and procedures before proceeding; and the entire process may involve multiple participants (both human and information-system-based). In these settings, the coordination and communication among participants are important, and a goal is thus to improve workflow, and to maximize speed or efficiency. In this circumstance, augmenting guidelines with workflow management capabilities is desirable [61].

In other settings, data may be changing quite rapidly; for example, in an emergency or intensive care unit setting, the status of patients need to be able to be assessed at a glance in order to identify who needs near-term attention versus those that are less critical. CDS in the form of dashboards and alarms/alerts that portray these changing statuses can be helpful in such settings.

Lastly, in clinical trials and in some highly structured procedures, such as for routine management by a nurse or physician assistant of common conditions like sore throat or asthma, or complex processes such as renal dialysis, strict adherence to the steps of a protocol is desired, and a CDS tool can help to ensure that this occurs.

1.3.1.4 Monitor actions—guarding against errors, providing warnings, alerts, reminders, or feedback about performance

While decision support may provide information when sought, or by overtly asking for it (e.g., by invoking a differential diagnosis tool, a dose calculator, or a clinical guideline), another form of decision support works in the background without overt action of the user, and only interacts with the user when there is a reason to do so. This can occur either in the background of real-time interactive applications between a user and the computer, such as in CPOE, or can be asynchronous and decoupled from user actions (e.g., notification by paging or e-mail about arrival of an abnormal test result).

The computer essentially functions as a guardian or silent partner, monitoring the clinical context and what the user is doing, and either interrupting or contacting the user when situations arise that necessitate it. For example, if a potential order is determined to be hazardous (e.g., a medication interacts dangerously with one that the patient is already receiving, or to which he or she is allergic), a warning can be displayed. If a radiological procedure is being ordered for an indication that is not determined to be appropriate, e.g., ultrasound for question of appendicitis when a CT would be better, or if a medication is too expensive, is not on an approved formulary, or a generic medication would be an appropriate substitution, a recommendation for the alternative can be displayed. If a critical abnormal laboratory result is obtained, an alert can be triggered that notifies the physician so that appropriate action can be taken. The passage of time may cause reminders to be generated, as for periodic mammograms in a woman over 50, or for flu shots during the winter season for an elderly patient. If a patient has been in an ICU longer than an expected number of days for his or her diagnosis, the physician can be notified.

In all of these cases, the background alerts, reminders, or feedback may be considered useful or thought to be inappropriate or unjustified. A challenge in providing this kind of decision support is to minimize the situations in which they are inappropriate, lest the false alarms become annoying or in the worst case, are simply ignored because of their frequency. The conditions for generating messages must be properly defined to be as helpful and pertinent as possible, and the way in which alerts are provided must allow overriding of them when justified.

1.3.1.5 Focus attention and enhance visualization

Another form of CDS is quite indirect and subtle. This is the encouragement of best practices through use of techniques to organize and present information and options in such a way that they serve to enhance recognition of important aspects and remind or facilitate good choices. This may be done either by providing a framework for describing sequences of interactions between the computer and user, such as dialogs that are controlled by an underlying guideline or flowchart, by associative groupings of documentation elements on display screens or printed forms and documents, or by a variety of visualization methods.

With respect to **associative grouping** of elements, in particular, if it is done well, it can not only focus attention, but offer benefits of improved efficiency, because groups of items can be either selected as a unit when acceptable, or at least brought together for consideration and action decisions rather than each element needing to be sought for by a user and selected or entered individually. Order sets are an example, in that the grouping of orders for a particular indication, such as admission to the cardiac ICU, are pre-established for ease of use by a physician, and to be sure that the physician does not forget to consider including medications for control of pain and anxiety, and for anticoagulation, vital sign and ECG monitoring, diet, cardiac enzyme testing, and other typically considered tasks. If done well, such order sets have the virtue that they not only encourage desirable actions but also, by anticipating likely actions, facilitate workflow and save time.

Besides order sets, another application of this approach to CDS is the design of data entry forms for structured capture of information. Examples might be a form for recording a neonatal visit, an anesthesiology preoperative note, or a specialist referral for cardiac surgery evaluation. The form can include items that are automatically filled in, where possible, from stored data. It can include suggested items that are predetermined to be important, and thus serve as both a handy checklist for recording them and as a reminder to be sure to do so.

Yet another example is the generation of reports from structured elements, for example, the printout of a prescription order, or the production of a postoperative note or discharge summary. The design of the report is intended to be easy to automatically generate from stored elements, consistent in appearance, independent of the user who produced it, and well formatted and organized. Such predictability (as well as legibility) facilitates readability and usefulness.

Lastly, we include in this category all forms of **visualization** that tend to facilitate appreciation of data and their trends, and recognition of situations that may require attention or action. This includes graphics and trend plots, summarizations, dashboards, and flow sheets. The goal is to support cognitive tasks for managing complexity, usability of systems, and workflow to lead to optimal decision making.

1.3.2 Design of CDS: Components and interactions

We now consider the structural aspects of CDS. We do so by identifying a set of functional components of CDS and its invoking environment, and how these components interact. As we noted at the beginning of this chapter, this discussion is somewhat artificial, in that it presents an idealized model of how CDS should be created. What we mean by "idealized" is that, if the design of a CDS capability clearly identifies each of these components, separates them cleanly, and addresses the design of each component in a standardized way, the goals of widespread dissemination and use of CDS can be greatly facilitated. We focus on the idealized model while recognizing that much of CDS is not implemented that way. We maintain that all the components we discuss next that are needed for CDS are present in one form or another in any implementation, but they are not always separable, in terms of the actual software code that implements them. Nonetheless, we can consider these components and the functions they perform individually, at least from a conceptual point of view.

Another idealization we adopt is to refer to a unit of software that provides CDS as a CDS *module*. The heart of a CDS module is a method of transforming input parameters to a patient-specific output. To be modular, the CDS software should be cleanly separable from surrounding or invoking software code, communicating with it via a well-defined interface. Recognizing the many possible implementation methods for CDS that may not be modular at all, we nonetheless use this term to be able to direct our focus to the portion of software directly concerned with the provision of CDS functionality, and to the nature of the interactions by which it relates to the invoking environment.

To provide CDS, several tasks must be performed, as shown in Fig. 1.3:

- The CDS module is *initiated* or invoked by some process in the application environment.
- The module *obtains data* through an interface with the application environment, where the data are *entered* by a user, *retrieved* from the EHR or other source, or *provided directly* by the invoking entity. The latter might include context-specific information about the application, user, setting, and function being performed.

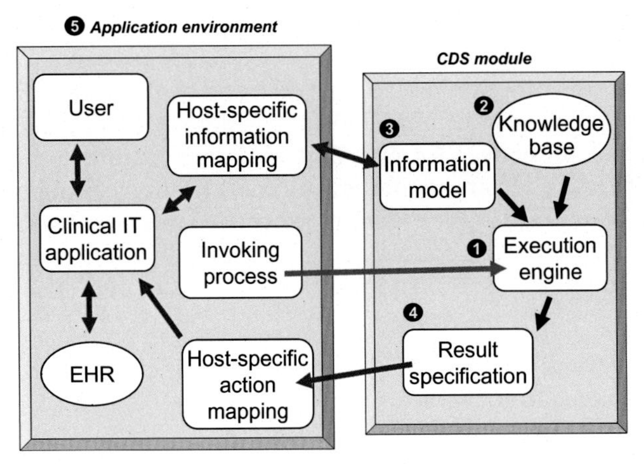

FIG. 1.3 A conceptual model of CDS design components and their interactions with the host application environment.

- The module *applies knowledge* (e.g., facts in the form of rules, algorithms, or semantic relations), either local to the module, or retrieved from a knowledge base.
- A process is then executed that *transforms* the input parameters and knowledge according to the specification of some sort of *decision method* to generate a patient-specific output. The decision method (or model) is usually embodied in an algorithm or computational procedure of some sort. It can carry out, for example, a retrieval, inference, classification, organization, or presentation/visualization process.
- The module then *produces a result* that must be communicated to the application environment. That result is usually a recommendation for action.

To carry out this process, the design of a CDS module conceptually has four design elements, or components, and operates in conjunction with an application environment, which is thus considered to be the fifth component (see Fig. 1.3). The application environment determines how and when the CDS module gets invoked, how it obtains data and communicates its results, how it interfaces with host software and hardware, and how it interacts with its users. The application environment can be so varied that the specification for this component is only defined with respect to CDS in terms of the nature of the CDS module's interactions with it.

1.3.2.1 Execution engine – Responsible for implementing the decision method/model

All kinds of CDS have some kind of execution paradigm, that is, a method of organizing or processing input information, to produce some kind of output, or result. The sequence in which data are requested and the algorithm or method for processing data depend on an underlying model of the decision problem. For example, an alert or reminder may be designed to be triggered by an event, such as a mouse click, the arrival of a lab result, or the passage of time, to obtain specific data items. It then evaluates a Boolean logical condition expression about the data, to determine the truth value for the expression. If the Boolean condition evaluates to "true," the alert or reminder may then cause an e-mail, page, or displayed message to be generated, in order to notify an appropriate physician. The *decision model* in this is *Boolean expression evaluation*.

A differential diagnosis program may collect data and evaluate the diagnostic possibilities using Bayes theorem: the Bayes theorem algorithm is the underlying decision model. A dose therapy calculation tool might use a formula that needs such parameters as body surface area, renal status, or age to make recommendations for modifications of medication dose for children, those with kidney failure or the elderly; the decision model is a computational formula.

The decision models that are used in CDS rely principally on classes of methods that are indicated in the columns of Fig. 1.2, and are described in more detail in Chapter 2. These include:

1. Information retrieval and search; that is, the model by which data and knowledge are used to select pertinent items to retrieve
2. Logical expression evaluation
3. Probabilistic and data-driven classification/prediction
4. Artificial intelligence: heuristics, expert systems, machine learning
5. Calculations, algorithms, and multistep processes
6. Associative grouping of elements; that is, the model determining what these associations are and under what conditions they are activated
7. Visualization: methods aimed at portraying data attributes and their trends, to facilitate recognition of situations that may require attention or action

Conceptually, we can consider that the decision model, to the extent that it involves computational processes, is embodied in an *execution engine*. The execution engine is the part of the CDS software that evaluates data to produce output. As we have noted, actual implementations may not cleanly separate this code from other parts of an application, or even from other parts of the CDS module itself, but advantages are to be realized if that can be done. Principally, this separation allows the execution engine to be refined and enhanced as improvements in the way it should operate become understood. Also, this provides flexibility and portability, in that the execution engine can be recoded and reimplemented in different platforms independently of other CDS parts, and can even be embodied in external services (see Chapter 29).

1.3.2.2 *Knowledge content resources*

Sometimes, as in an application for recommending electrolyte replacements in acid-base disorders, the calculations and sequences of actions are embedded in software code. However, as we noted earlier, it is often helpful to implement the general methodology of a particular decision model as an execution engine that can be used to apply the method to all knowledge of that type. For example, if the electrolyte replacement algorithm can be represented in a flow chart modeling formalism, then a guideline execution engine can run it. To be most flexible, ideally, the knowledge—the formulas or equations, the logic of production rules, the flow charts, and so on—should exist external to the "engine" that accepts inputs of that type, processes the knowledge, and produces a result. Separating the knowledge from the engine, when it is possible to do so, enables the engine to operate on a variety of similar kinds of knowledge.

- As a consequence, the knowledge resources can be managed independently. For example, they can be authored and edited through use of a knowledge editor tool.
- With appropriate editor functionality, an editor tool can display the knowledge in a form that is readable by a human subject expert rather than requiring the skills of a software engineer or other technical support person.
- If the knowledge is made transparent in this fashion, maintenance, review, and update are easier to do.
- If a standard format is used for encoding the knowledge, or for import and export of it from external repositories, the knowledge can be shared and disseminated.
- If the decision model evolves, e.g., in terms of the ability to use more refined or detailed knowledge, the knowledge base can be updated separately to incorporate those knowledge elements.

Knowledge content resources can be structured or unstructured, depending on the purpose and the computational requirements of the decision model.

We give somewhat technical examples here to illustrate the concept, and the following can be skipped by the less-interested reader. The approaches and methods referred to, the acronyms and names, and the standards-based models used are discussed elsewhere in the book. The *main point* is that the knowledge content resources needed for various methods, to carry out their function, must be structured in a way that the methods expect.

For example, if the purpose is simply to retrieve and display information in human-readable form, the only structure required might be the use of index terms or keywords, to facilitate retrieval by a search engine, although with no structure at all, a text-based search such as by Google® is still possible.

If the knowledge is a logical expression for a *production rule*, then the expression must obey the syntactic conventions needed to evaluate the expression in whatever language or formalism is used. In MYCIN, one of the earliest *rule-based expert systems* in medicine, production rules had the format IF *condition* THEN *action*, where *condition* was a *Boolean logical expression* with "certainty factors" associated with the terms. The execution engine could evaluate the conditional expression and had an algebra for combining the certainty factors to produce an updated certainty factor associated with the assertion in the *action* part of the rule [62], and could control the sequence of execution of rules through a goal-driven, backward chaining heuristic.

In *alerts and reminders* encoded in a standardized form using *Arden Syntax* [63], an *evoke* section defines triggering event(s), a *data* section specifies the data elements used, a *logic* section defines the procedure to evaluate the data elements in a programming language-style syntax, and an *action* section defines the task to be carried out if the logic section evaluates to *true*. An Arden Syntax interpreter or compiler could then serve as an execution engine to process a knowledge base of Arden Syntax rules, evaluating a rule when triggered by appropriate evoking conditions.

[Actually, languages and formats for if-then rules used in alerts, reminders, and for other purposes have gone through considerable evolution beginning with locally-developed formats, progressing to use of the Arden Syntax standard, to more database-independent approaches such as GELLO [64] and Health eDecisions [65]. More recently, standards based on HL7's Fast Healthcare Interoperability Resources (FHIR) [66] have enabled a variety of CDS architectures and capabilities, described in Ref. [67], such as integration of third-party apps within EHR systems (SMART on FHIR), triggering of CDS services upon user events in the EHR (CDS Hooks), and sharing and execution of knowledge artifacts (Clinical Reasoning Module and Clinical Quality Language [CQL])].

The knowledge base for a Bayesian diagnosis tool would be the prior probability distribution for the diseases to be considered, and the conditional probabilities of findings for each of the diseases [68–70].

A guideline interpretation engine that is designed to support traversal of a guideline and interactive acquisition of data and evaluation of conditions to determine next steps could operate on guidelines encoded in a knowledge base according to a formalism the interpretation engine understands. Examples of this are the guideline engines supporting representation formats discussed in Chapter 2 and reviewed further in Chapter 10, early approaches known as Pro*forma* [71], GLIF [72], EON [73], Asbru [74], and SAGE [75], among others, and the more recent FHIR-based Clinical Guidelines model [76].

1.3.2.3 *Information model*

CDS requires a precise specification of the kinds of information the computation model will utilize, which we refer to as its *information model*. The knowledge content resources typically contain statements, facts, conditional expressions, or other relations that refer to or operate on patient-specific data. If we formally specify the information model, this provides flexibility in that the same CDS resource can be used in more than one kind of setting; for example, interactively with a user as well as in background mode, retrieving data from the EHR, and in more than one platform and system environment. The specification must include not only the *format* of the data that the CDS module receives and uses but the *taxonomy* or coding scheme for its labels and also for any of its coded/categorical values, and the restrictions on value sets that are allowable. For example, if one is seeking to run a rule about medication interactions, it is important to know that they are encoded in RxNorm [77], or that a diagnosis is encoded in SNOMED-CT [78], or that a lab test is encoded in LOINC [79]. Value set restrictions may confine the medications of interest to angiotensin converting enzyme inhibitors, or diagnoses to cardiovascular diagnoses, or lab tests to particular enumerated tests. The specification should also involve *grounding* the data elements in terms of precise attributes like units, method of obtaining them, time frame, etc. This is discussed further in Chapter 11.

Note that beyond defining these requirements, the adaptations necessary for accessing or obtaining them are not the province of the CDS module but of the application environment. If the data are to be obtained by interaction with a user, the host may also need to include an external display name for a data entry field. If the input that is allowed needs to be validated (e.g., checked for limits of a numeric range, length of a text string, the presence of valid characters, or conformance with items on a predefined pick list or dictionary), then those criteria for validation (and the content of the pick list or dictionary) need to be adapted from the information model or added by the application environment.

If the data are to be retrieved from a stored repository, then either the information model used in the host application environment should be the same, or a process for mapping the data elements from it needs to be established. Some of the data elements may need to be transformed, if differences in the definitions of those elements in the host EHR and in the

CDS module's information model require it. For portability of CDS and integration with clinical systems, the use of a standardized information model such as the HL7 FHIR, is used to evaluate a CQL logic expression. In general, it is unlikely that real operational clinical information systems will use a standard reference information model (such as FHIR) directly in its native implementation. Thus, for interoperability between systems, or to use externally developed CDS, a mapping would need to be developed for each implemented vendor-specific system between that standard reference information model and the vendor system information model.

1.3.2.4 Result specification

Operation of the CDS execution engine is carried out with the goal of producing some output, whether in the form of retrieved resources, a calculated result, a recommended action, or a data entry screen or formatted report. Since the result is dynamically determined through execution, there needs to be an explicit process for determining how that output gets produced.

The result specification could be regarded as part of an expanded view of the information model, but we consider it separately because of its distinct role in the CDS process. Note that the result can of a CDS process can be a set of one or more values of clinical parameters, but it can alternatively be an action (possibly include the clinical values). Actions are verbs like notify, schedule, order, cancel, etc.

For example, the result of evaluating an Arden Syntax rule or CQL logic to *true* might be an action to be performed, such as to send a specific message to the attending physician. A calculation of dose of a medication based on adjustments for renal function or age might produce a result in terms of a modified recommended dose. A Bayesian differential diagnosis program's result would be the set of diagnoses with their posterior probabilities. Traversal of a clinical practice guideline algorithm based on evaluation of entered or retrieved patient-specific data values would produce a result at each step, indicating the optimal next step.

Thus, conveying the result to an application environment involves mapping of the result to the performance of actions or production of outputs that the application expects to carry out. This will largely be the responsibility of the application environment, as discussed in the next section. What is needed in the CDS module is a taxonomy of kinds of results that can be produced from decision support, to facilitate such host mappings. Some early work in the execution of clinical guidelines [35,80] produced such a taxonomy. This has also been pursued further in a CDS Taxonomy produced by the National Quality Forum Committee on CDS [81], and in the action ontology of the Health eDecision knowledge model [82]. More recently, the CDS Hooks standard provides an information model to represent the output of CDS services (referred to as CDS "cards"), which can include both conclusions in human-readable format, as well as assessments or actionable recommendations in computable format using FHIR resources [67].

Modularity of design is one of the reasons we consider the result specification in CDS separately from the mode of interaction with the user or applications in the application environment. Just as with the specification of the information model for data used in the decision model, separation of the result specification enables a decision support capability to be adapted to several possible modes of interaction in any of a variety of applications on different platforms. For example, the result could be provided in real-time in interactive applications, in the background in alert/reminder usages, or in batch mode in the production of reports or summaries. If the result is a set of one or more values, this might call for the EHR to be updated with that information. If the result is an action to be performed, this needs to be interpreted by the host environment in terms of its ways of performing that action. For example, an action may call for a physician to be notified about an abnormal value, which could be done by emailing to the physician, displaying an interruptive popup message on a screen, transmitting the recommendation to another application, or storing it in a pending task list to be seen when the physician next logs into the system.

1.3.2.5 Application environment

As we have seen, many of the features of CDS operation are not embodied in the CDS module itself but in the application environment that invokes it. The application environment determines how the CDS module communicates with a user, such as an interactive dialog, or obtains data from the EHR or other sources, and how it conveys its results. The application environment can also pass to the CDS module certain context data such as those describing the application setting, the user, and the purpose.

The degree of integration of CDS with applications is one of the most critical ones for determining success of CDS, yet too tight an integration limits the ability to achieve portability and reuse of CDS modules. We will begin the consideration of the nature of the interaction of a CDS module with the application environment by revisiting a simple example we used in Section 1.1 to illustrate the range of options to consider and the complexities involved—even for a simple form of CDS. The example relates to the set of rules regarding the handling of abnormal and critical laboratory test result values; in other words, results that exceed predefined limits requiring flagging or, for critical results, urgent attention. The knowledge regarding such abnormal values can be found in the literature, and the simplest form of decision support is the ability to retrieve references to such abnormal values. This could be in the form of bibliographic citations or Web sites displaying laboratory values and their accepted normal ranges and critical values. Ideally, those latter sites should also cite references to the literature about how to interpret them.

The least integrated way to make this information available would be to enable the user to access the Web and to do a search for it, using his or her own search terms. Slightly more integrated access would be a resources page, which would have a set of predefined links to useful reference information that could be accessed from the clinical IT system. To be of greater value, it would be useful to have access to this information at the point at which a physician is reviewing laboratory results for a patient. The difference between looking up lab result values to determine whether they are abnormal and having a direct link to a particular citation giving that information in the context in which it is needed—that is, when the lab result is being reviewed—is that in the latter case the information needed is preselected and automatically available to the user.

A more useful way to provide this information, which is done in most clinical systems, is to automatically flag the abnormal lab result on the report or display screen that is reviewed by the provider, by a symbol indicating that it is outside of normal ranges. This could then be combined with a link to the available citations to give further information. The flag indicating abnormality could be introduced by the clinical laboratory information subsystem or by the laboratory results report display application, using a formal logical condition expression that is evaluated by the computer to determine the presence of abnormality. Thus, it could occur at any of three points, either at the time the result is produced in the laboratory subsystem, when it is entered into the EHR, or when it is displayed.

Another way to deliver an abnormal result finding is by generating an alert message that is sent to the provider, perhaps by e-mail, text page, or interruptive alert. This requires integration into a clinical information system in such a way that information about the particular patient and the appropriate provider are able to be identified automatically. This would also allow more elaborate decision rules to determine whether the result is new or a repeat of an already abnormal value about which the physician has previously been notified, or if there are coexisting conditions that might explain the result (e.g., renal failure).

To be maximally useful, knowledge in the form of a decision rule such as that used for responding to abnormal laboratory test results would exist in a rules knowledge base and be triggered by a variety of different possible event scenarios—for example, the entry of an abnormal lab result into the patient's clinical record, a medication order interaction check with respect to existing lab results, or the flagging of abnormal laboratory results on review by a physician. With the knowledge in a knowledge base, different events such as result entry, order placement, or display of results could trigger evaluation of the rules indexed according to the various parameters of interest, to determine which, if any, might apply, and then to carry out actions that are appropriate based on the triggering application and depending on the result of the evaluation. For example, in an alerting application, evaluation of the rule to true would result in the generation of a warning message to the provider by e-mail, page, or other means of communication. In an interactive CPOE application, evaluation to true might generate a recommendation an immediate interactive popup or inline message to decrease the dose of a medication being prescribed. In the result display application, evaluation of truth would result in the flag symbol indicating abnormal result being appended to the value.

The first usage described, that of displaying a citation, simply requires that the knowledge about abnormal lab results be available in text form in some defined location (e.g., in a bibliographic database). It requires no computability, just the ability to retrieve it.

The second usage, access via a direct link from the result information display, can be done by manually identifying the appropriate citation to be displayed whenever abnormal results occur. To be more useful, however, retrieval based on the context, for example, could work as follows: By recognizing that the context is a laboratory results display application, the CDS tool could determine that information pertaining to abnormal laboratory results would be useful, and a specialized retrieval program could be designed to select the kind of information to be retrieved from a general retrieval search engine by passing context-specific parameters related to clinical laboratory abnormal results. This constitutes a kind of "information broker" function and is exemplified by the infobutton manager described in Chapter 2 and further elaborated in Chapter 13.

The third usage – that of flagging abnormal results – requires the presence of a formal computable expression that can be evaluated by the computer. This would be of a logical format such as "if lab test result y exceeds threshold a then return true." For this to work, the value of the lab test result of interest must be assigned to the parameter y and the upper normal range for the lab test result must be assigned to the parameter a. The software application must also know that if the result True is returned from the evaluation, then a flag value such as * or # should be appended to the display of the laboratory result. Thus, this application usage requires a formal expression, an evaluation engine, and a simple set of data parameters that can be passed to the evaluation engine and returned from it.

The fourth usage is the execution of a generalized rules interpreter in the context of an event-driven architecture, in which particular rules are evaluated as a result of triggering events, and the actions performed depending on the result of the evaluation are a function of the application that generated the trigger. This approach not only requires a knowledge base, an indexing scheme for accessing the rules in the knowledge base, and an execution engine, but it requires also a means of integration with various possible invoking applications, as well as, for example, in the case of notifications to physicians, the ability to invoke other applications. This usage has maximum flexibility and power, because the same piece of knowledge—in this case, any logical expression regarding what constitutes an abnormal laboratory test result value—can be used in a variety of different contexts. Thus, the knowledge rule itself only needs to be developed once, and can be maintained or updated if necessary in one place, and if properly set up, all the applications that utilize it can be identified, should there be a reason to change the rule in the knowledge base.

We can consider a variety of other kinds of clinical decision support usages that range from passive to active and from loose to tight integration with applications, and do a similar decomposition of the necessary elements. Work needs to be done to define the extent to which application-specific behavior can be further abstracted into a taxonomy of result types, as discussed earlier, so that more of the functionality can be moved into CDS modules rather than requiring custom interfacing and handling of results in the application environment.

1.3.3 Modes of interactions

One of the challenges in providing CDS is that of determining the most effective way to interact with humans, so that the advice is as patient-specific as possible, and delivered according to the five-rights paradigm [83], i.e., to the right person, at the right time, in the right format, and through the right channel. At the same time, it must be acceptable to the human user, by not requiring a lot of extra work, being disruptive to workflow, or being redundant. Also, given the role of decision *support* rather than decision *making*, advice must be given in a fashion that recognizes human decision-making prerogatives and avoids being inflexible or insistent when it is not necessary to do so.

Fig. 1.4 portrays a set of dimensions that can serve as a guide for thinking about the various aspects involved in providing CDS and interacting with users, which determine the extent and manner of integration of CDS with the clinical IT application environment. We consider each of these briefly.

1.3.3.1 Locus of control

A CDS instance can be *initiated* by a user when the need for help is recognized; for example, in seeking to find an answer to a question, or to obtain assistance in assessing a diagnostic or therapeutic decision. Or it can be initiated by the computer, usually in processes aimed at monitoring user actions to guard against errors or detecting suboptimal practices; for example, in detecting an inappropriate order, or a time interval at which a mammogram should be ordered or HbA1c level should be checked. The *locus of control* thus either resides with the

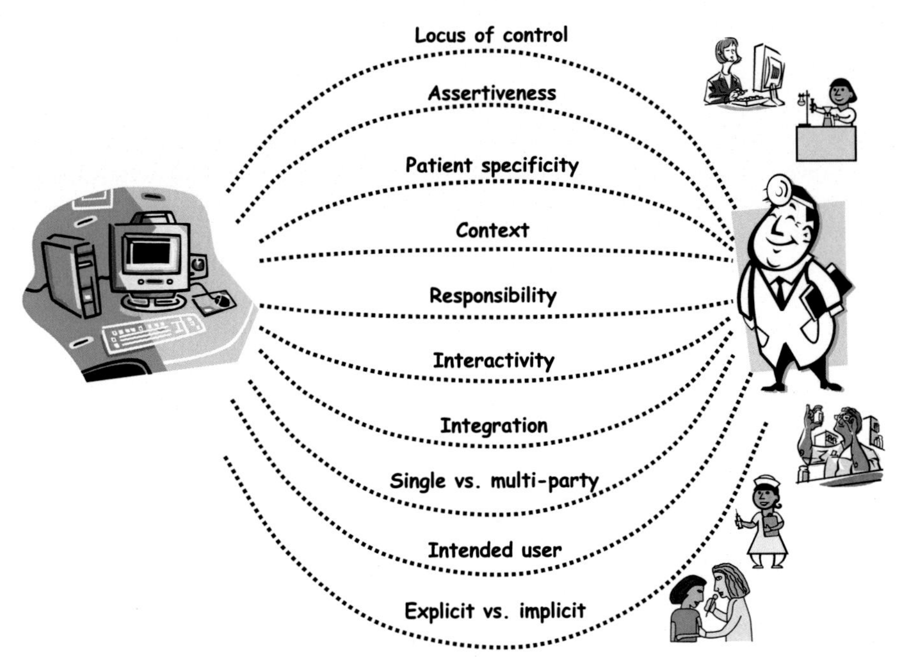

FIG. 1.4 Dimensions of computer-user interaction in CDS.

user or the computer in these examples. There are also intermediate situations, in which CDS resources are made available and have the means to automatically be context-aware, as in infobuttons or patient-specific guidelines, but it is up to the user to initiate their use.

1.3.3.2 Degree of assertiveness

Decision support capabilities can be provided to a user with varying degrees of insistence or "assertiveness." This only applies, of course, to settings in which the computer is the locus of control and initiates the CDS instance. The most passive way in which decision support can be offered would be to simply present a discussion of a topic that can be read by the user. One step up from this would be to present specific recommendations and advice, although again in a form that is simply to be read, such as a passive reminder for preventive care such as colorectal cancer screening. A slightly more insistent form of decision support would be an interruptive notification along with the requirement that the information provided be acknowledged. A further increase in assertiveness might involve a list of choices of action; for example, possible medication regimens, allowing "other" as an additional option or an override of the recommended dose of a medication if a justification is provided. Still more assertive would be forced choice among alternatives without the possibility of override. The most active form of decision support would be a closed loop process in which actions occur automatically in response to inputs, although it could be inspected or monitored by a user. Implanted cardiac pacemakers have this mode of operation, but for reasons we have noted previously, there are very few other instances of such closed-loop processes in routine use.

1.3.3.3 *Patient specificity*

Comprehensive knowledge base resources, such as the medical literature in PubMed, a guideline repository, or a collection of possible alerts and reminders, are important to the ability to provide robust decision support. But a challenge in delivering effective decision support is to be able to select resources that are relevant for a given patient and to determine when and where, or even if, they should be used in a particular setting.

To accomplish this, the CDS module needs to be able to obtain information about the clinical problems or findings, care setting (e.g., office visit, phone call, or hospitalization) and other patient-specific parameters, as specified by the logic of the rules under consideration. This involves the ability to access the EHR or obtain data from the user or from the patient's record.

1.3.3.4 *Context*

Beside patient characteristics and setting of the clinical encounter, context affects the nature of decision support in several ways. The principal contextual factor is the kind of application and the function being performed when CDS is invoked. For example, if the context is an interactive CPOE program, decision support would likely need to be very responsive so as not to perceptibly delay or impede the real-time interactivity. If the context is a background process aimed at collecting data on and evaluating conformity with care pathways in various units of a hospital (e.g., the coronary care unit or the postoperative orthopedic floor), it might be suitable to communicate deviations from expected targets to providers at the beginning of the workday, at rounds, or through care unit-level, multi-patient dashboards. Event-driven panic alerts indicating critical abnormal laboratory values would need to be communicated to providers by any available means as quickly as possible and require acknowledgment. Preventive screening or immunization reminders might have a much longer response window, such as within the context of a clinical encounter or directly to the patient via patient portal before the patient is overdue.

The *same* knowledge can be used in *different* contexts and practice settings. To return to an example discussed earlier, consider a rule that performs an action if a lab result exceeds a threshold: this might be used in an alert that is automatically triggered when a result is produced by the laboratory, or the rule might be used during the generation of a lab test result review display screen, for flagging abnormal results. The multiplicity of uses and contexts for decision support knowledge is one of the rationales for the need to construct knowledge bases containing collections of decision support content. The content in a knowledge base can be structured such that context, setting, medical problem, and CDS purpose, can enable specific knowledge to be retrieved and used when needed. Another consequence of this capability is that it would be necessary for only one instance of a particular item of core knowledge (e.g., a decision rule, to exist, making it easier to maintain the knowledge, review it, update it, and propagate changes to all the applications that use variations of it).

1.3.3.5 *Interactivity*

Somewhat related to the classification of decision support in terms of degree of assertiveness, patient specificity, and context is the degree of interactivity. Knowledge resources may be in the form of static human-readable information, such as text or a table, retrieved in response to specific search. Such a presentation could be viewed and examined but require no specific action by the user. An interactive mode of delivery of this same information would be one for which some entry or acknowledgment is required by the user.

A more interactive form of CDS might be a computational tool that produces a similar kind of text or tabular report, but which provides the ability to manipulate parameters that are entered into it, as, for example, a computation tool for drug dose calculation based on body surface area, or a tool for performing sensitivity analysis of a decision analysis model to determine how stable its decision is as a function of change in the estimate of prior probability of disease or intervention risk. Often, interactivity occurs in CDS in the form of requests for data to be entered or selection of options by the user.

Still other aspects relating to interactivity are illustrated by the various ways in which recommendations can be communicated to the user and the options available to the user for responding to or overriding recommendations. Some decision support messages might be provided in the form of noncritical alerts or reminders, generated in the background and only seen when a user next logs into the information system and requiring no further action on the user's part. Alternatively, for more important alerts or reminders, they might be sent by e-mail or text messaging. Various graphical renditions, dashboards, or summarizations might be available as options for viewing complex data and relationships, with interactive functions such as sorting, filtering, and details on-demand.

1.3.3.6 *Degree of integration with clinical IT applications*

A CDS capability can be integrated into the clinical information system to greater or lesser degree. CPOE is an example of an application with a need for a high degree of integration of CDS. A drug interaction checking tool would be most useful during CPOE if it were able to retrieve the patient's list of current medications. With access to EHR data, a CDS tool could also include automatic checks against allergies or other patient-specific contraindications.

1.3.3.7 *Single vs. multi-party focus*

Some applications of decision support, such as computer-based clinical practice guidelines, have the potential for optimizing the care process by suggesting appropriate next steps. In a busy practice environment or inpatient setting, automation of guidelines could also help to optimize workflow by coordinating scheduling and use of resources and activities of participants through communication and synchronization functions (e.g., don't do task B until task A is completed), and monitoring the times, delays, and statuses of expected events. Note that the parties involved may be human or computer-based (e.g., a scheduling system or messaging system).

Notification of alerts is another example of an application that may have a multi-party aspect. Typically, if a critical lab result needs to be acted on by someone, there is a set of processes defined for notifying a patient's primary physician or care team member about such an alert (e.g., some sequence of page, telephone message, e-mail, or text message, with requirement for acknowledgment), and a defined sequence for notification of other providers if that person does not respond within a specified period of time.

1.3.3.8 Intended user

Decision support for various purposes may be designed for different kinds of intended users; for example, direct support of physician decision making; aids to nurses, pharmacists, laboratory or radiology technologists, emergency medical technicians (EMTs), medical assistants or paramedics; reports of utilization of resources, errors or costs to managers; or information resources and decision aids for patients and the general public. The kind of knowledge involved, the decision-making approach used, and the mode of operation may vary considerably depending on user and purpose.

1.3.3.9 Explicit vs. indirect support

Calculation tools, guidelines, alerts, and reminders all are designed to give specific advice or recommendations. But another kind of decision support uses visualization methods to provide optimal organization, grouping, sequencing, or filtering of information presentation in order to support complex cognitive processes such as pattern matching, prioritization and associations. We mention again some examples of this mode of decision support, namely the use of structured data entry forms, order sets, templates for reports and summaries, dashboards, flow sheets, and protocols.

1.4 Other considerations

We have touched on a number of settings and contexts in which decision support could be used, but which will not be discussed in detail in this book. One of the primary other uses is in the realm of education and training. Not only can decision support knowledge bases be useful as educational reference tools, but the decision support can be used directly in a dynamic way in case-based problem-solving exercises—simulations of clinical problems requiring intervention by a user and feedback about the appropriateness of the actions taken. Methods analogous to CDS may be used to generate a range of variation of clinical parameters in a simulation, with the inclusion of a random component as well. CDS-like capabilities can be used in a critiquing mode, in which actions are performed by the user first and then evaluated by the CDS-like resource for conformance with the underlying decision model (e.g., a guideline). Or decision support may be used in what is known as an "intelligent tutoring system" mode of operation, to probe student responses or actions in terms of their similarity to prototypical problems in such situations, and to tease out the underlying misconceptions.

We will not delve further into image and signal processing, pattern recognition, and feature extraction, as these are largely embedded in niche applications, since our focus is on more generic CDS capabilities and the issues of deploying them in a health care enterprise.

Much of the development of CDS to date has been somewhat of an art form, with creative individuals identifying innovative and useful ways of providing it and showing effectiveness. Because of a lack of well-defined principles, the discovery process often has had to be replicated by others, sometimes with painful and disappointing results. The collective body of experience in the literature is nonetheless quite large.

Although there is new impetus to moving ahead, the lessons of the past need to be recognized if we are not to be destined to repeat the mistakes that have limited progress over the past 60 years or more. A goal of this book is to begin to move toward a formal understanding of the requirements for CDS, based on the lessons and experiences of the past, clarifying an understanding of the requirements for infrastructure, standards, and business/organizational strategies that will lead to success.

The task of providing and maintaining robust CDS capabilities is a long and complex undertaking. It is important not to oversimplify it, or to rush to deploy CDS without adequate preparation, lest unsatisfactory results occur, bad press be generated, and an era of discouragement take hold. We seek to increase awareness and understanding of what the effort requires and to begin a systematic approach to tackling the problems that have vexed the field and held it back over these many years.

References

[1] Friedman CP, Wong AK, Blumenthal D. Achieving a nationwide learning health system. Sci Transl Med 2010;2:57cm29. https://doi.org/10.1126/scitranslmed.3001456.

[2] Grossmann C, Powers B, McGinnis JM, editors. Digital infrastructure for the learning health system: The foundation for continuous improvement in health and health care: Workshop series summary. Washington, DC: National Academies Press (US). National Academy of Sciences; 2011.

[3] McGinnis JM, Fineberg HV, Dzau VJ. Advancing the learning health system. N Engl J Med 2021;385:1–5. https://doi.org/10.1056/NEJMp2103872.

[4] AHRQ evidence-based Care Transformation Support (ACTS). Digital Healthcare Research; n.d. https://digital.ahrq.gov/acts [Accessed 23 April 2022].

[5] Osheroff JA. CDS 5 rights. CDS/PI Collaborative., 2017, http://bit.ly/cds5rights. [Accessed 21 March 2022].

[6] Greenes RA. Why clinical decision support is hard to do. AMIA Ann Symp Proc 2006;1169–70.

[7] Coleman K, Wagner E, Schaefer J, Reid R, LeRoy L. Redefining primary care for the 21st century. Rockville, MD: Agency for Healthcare Research and Quality; 2016. p. 1–20. 16.

[8] Osheroff JA, Teich JM, Middleton B, Steen EB, Wright A, Detmer DE. A roadmap for national action on clinical decision support. J Am Med Inform Assoc 2007;14:141–5.

[9] Marcotte L, Seidman J, Trudel K, Berwick DM, Blumenthal D, Mostashari F, et al. Achieving meaningful use of health information technology: a guide for physicians to the EHR incentive programs. Arch Intern Med 2012;172:731–6.

[10] Christensen CM, Raynor ME. The innovator's solution. Cambridge, MA: Harvard Business School Press; 2003.

[11] Ogden J, Fuks K, Gardner M, Johnson S, McLean M, Martin P, et al. Doctors expressions of uncertainty and patient confidence. Patient Educ Couns 2002;48:171–6.

[12] Weaver RR. Informatics tools and medical communication: patient perspectives of "knowledge coupling" in primary care. Health Commun 2003;15:59–78.

[13] Hackett JL, Gutman SI. Introduction to the Food and Drug Administration (FDA) regulatory process. J Proteome Res 2005;4:1110–3.

[14] Altenstetter C. EU and member state medical devices regulation. Int J Technol Assess Health Care 2003;19:228–48.

[15] Embi PJ, Leonard AC. Evaluating alert fatigue over time to EHR-based clinical trial alerts: findings from a randomized controlled study. J Am Med Inform Assoc 2012;19:e145–8. https://doi.org/10.1136/amiajnl-2011-000743.

[16] Lee EK, Mejia AF, Senior T, Jose J. Improving patient safety through medical alert management: an automated decision tool to reduce alert fatigue. AMIA Ann Symp Proc 2010;2010:417–21.

[17] Kawamoto K, Hongsermeier T, Wright A, Lewis J, Bell DS, Middleton B. Key principles for a national clinical decision support knowledge sharing framework: synthesis of insights from leading subject matter experts. J Am Med Inform Assoc 2013;20:199–207. https://doi.org/10.1136/amiajnl-2012-000887.

[18] Feigenbaum EA, Feldman J, editors. Computers and thought. New York: McGrall Hill; 1963.

[19] IBMResearch. The DeepQA research team., 2013, http://researcher.ibm.com/researcher/view_project.php?id=2099. [Accessed 26 March 2022].

[20] Johnson H. Relationship between user models in HCI and AI. IEE Proc J Comput Digit Tech 1994;141:99–103.
[21] Harpole LH, Khorasani R, Fiskio J, Kuperman GJ, Bates DW. Automated evidence-based critiquing of orders for abdominal radiographs: impact on utilization and appropriateness. J Am Med Inform Assoc 1997;4:511–21.
[22] Miller PL. Critiquing anesthetic management: the "ATTENDING" computer system. Anesthesiology 1983;58:362–9.
[23] Miller PL, Black HR. Medical plan-analysis by computer: critiquing the pharmacologic management of essential hypertension. Comput Biomed Res 1984;17:38–54.
[24] Kuperman GJ, Teich JM, Gandhi TK, Bates DW. Patient safety and computerized medication ordering at Brigham and Women's Hospital. Jt Comm J Qual Improv 2001;27:509–21.
[25] Mugford M, Banfield P, O'Hanlon M. Effects of feedback of information on clinical practice: a review. BMJ 1991;303:398–402.
[26] Bindels R, Hasman A, Kester AD, Talmon JL, De Clercq PA, Winkens RA. The efficacy of an automated feedback system for general practitioners. Inform Prim Care 2003;11:69–74.
[27] Bodenheimer T. Interventions to improve chronic illness care: evaluating their effectiveness. Dis Manag 2003;6:63–71.
[28] Greenhalgh J, Long AF, Flynn R. The use of patient reported outcome measures in routine clinical practice: lack of impact or lack of theory? Soc Sci Med 2005;60:833–43.
[29] US Preventive Services Task Force. Screening for hypertension in adults: US preventive services task force reaffirmation recommendation statement. JAMA 2021;325:1650–6. https://doi.org/10.1001/jama.2021.4987.
[30] Unger T, Borghi C, Charchar F, Khan NA, Poulter NR, Prabhakaran D, et al. 2020 International Society of Hypertension global hypertension practice guidelines. J Hypertens 2020;38:982–1004. https://doi.org/10.1097/HJH.0000000000002453.
[31] Liang MH. From America: cookbook medicine or food for thought: practice guidelines development in the USA. Ann Rheum Dis 1992;51:1257–8.
[32] Harding J. Practice guidelines. Cookbook medicine. Physician Exec 1994;20:3–6.
[33] Costantini O, Papp KK, Como J, Aucott J, Carlson MD, Aron DC. Attitudes of faculty, housestaff, and medical students toward clinical practice guidelines. Acad Med 1999;74:1138–43.
[34] Abendroth TW, Greenes RA. Computer presentation of clinical algorithms. MD Comput 1989;6:295–9.
[35] Essaihi A, Michel G, Shiffman RN. Comprehensive categorization of guideline recommendations: creating an action palette for implementers. AMIA Annu Symp Proc 2003;220–4.
[36] Wang D, Peleg M, Bu D, Cantor M, Landesberg G, Lunenfeld E, et al. GESDOR—a generic execution model for sharing of computer-interpretable clinical practice guidelines. AMIA Annu Symp Proc 2003;694–8.
[37] Tu SW, Musen MA, Shankar R, Campbell J, Hrabak K, McClay J, et al. Modeling guidelines for integration into clinical workflow. Medinfo 2004;11:174–8.
[38] Sim I, Berlin A. A framework for classifying decision support systems. AMIA Annu Symp Proc 2003;599–603.
[39] Kawamoto K, Houlihan CA, Balas EA, Lobach DF. Improving clinical practice using clinical decision support systems: a systematic review of trials to identify features critical to success. BMJ 2005;330:765.
[40] Berlin A, Sorani M, Sim I. A taxonomic description of computer-based clinical decision support systems. J Biomed Inform 2006;39:656–67. https://doi.org/10.1016/j.jbi.2005.12.003.
[41] Cimino JJ, Li J, Bakken S, Patel VL. Theoretical, empirical and practical approaches to resolving the unmet information needs of clinical information system users. Proc AMIA Symp 2002;170–4.
[42] Del Fiol G, Huser V, Strasberg HR, Maviglia SM, Curtis C, Cimino JJ. Implementations of the HL7 Context-Aware Knowledge Retrieval ("Infobutton") Standard: challenges, strengths, limitations, and uptake. J Biomed Inform 2012;45:726–35. https://doi.org/10.1016/j.jbi.2011.12.006.
[43] Yu H, Cao YG. Automatically extracting information needs from ad hoc clinical questions. AMIA Ann Symp Proc 2008;96–100.
[44] Del Fiol G, Curtis C, Cimino JJ, Iskander A, Kalluri AS, Jing X, et al. Disseminating context-specific access to online knowledge resources within electronic health record systems. Stud Health Technol Inform 2013;192:672–6.
[45] IBMResearch. The DeepQA research team., 2013, http://researcher.ibm.com/researcher/view_project.php?id=2099.
[46] Jonnalagadda SR, Del Fiol G, Medlin R, Weir C, Fiszman M, Mostafa J, et al. Automatically extracting sentences from Medline citations to support clinicians' information needs. J Am Med Inform Assoc 2013;20:995–1000. https://doi.org/10.1136/amiajnl-2012-001347.

[47] Berner ES, editor. Clinical decision support systems: Theory and practice. 3rd ed. Springer; 2016.
[48] Shiffman RN, Greenes RA. Use of augmented decision tables to convert probabilistic data into clinical algorithms for the diagnosis of appendicitis. Proc Annu Symp Comput Appl Med Care 1991;686–90.
[49] Jiang Y, Nishikawa RM, Schmidt RA, Metz CE, Giger ML, Doi K. Improving breast cancer diagnosis with computer-aided diagnosis. Acad Radiol 1999;6:22–33.
[50] Giger ML. Computerized analysis of images in the detection and diagnosis of breast cancer. Semin Ultrasound CT MR 2004;25:411–8.
[51] Kohane IS, Masys DR, Altman RB. The Incidentalome: a threat to genomic medicine. JAMA 2006;296:212–5.
[52] Barnett GO, Cimino JJ, Hupp JA, Hoffer EP. DXplain. An evolving diagnostic decision-support system. JAMA 1987;258:67–74.
[53] Sullivan SD, Carlson JJ, Hansen RN. Comparative effectiveness research in the United States: a progress report. J Med Econ 2013;16:295–7. https://doi.org/10.3111/13696998.2012.754613.
[54] Rockhill B, Spiegelman D, Byrne C, Hunter DJ, Colditz GA, et al. Validation of the Gail model of breast cancer risk prediction and implications for chemoprevention. J Natl Cancer Inst 2001;93:358–66.
[55] Freedman AN, Seminara D, Gail MH, Hartge P, Colditz GA, Ballard-Barbash R, et al. Cancer risk prediction models: a workshop on development, evaluation, and application. J Natl Cancer Inst 2005;97:715–23.
[56] Smith RA, Cokkinides V, Eyre HJ. American Cancer Society guidelines for the early detection of cancer, 2004. CA Cancer J Clin 2004;54:41–52.
[57] USPSTF. U.S. Preventive Services Task Force. n.d. http://www.uspreventiveservicestaskforce.org/ [Accessed 25 March 2022].
[58] Bruce B, Fries JF. The Arthritis, Rheumatism and Aging Medical Information System (ARAMIS): still young at 30 years. Clin Exp Rheumatol 2005;23:S163–7.
[59] Richardson G, Manca A. Calculation of quality adjusted life years in the published literature: a review of methodology and transparency. Health Econ 2004;13:1203–10.
[60] Kim EK, Kwon YD, Hwang JH. Comparing the performance of three severity scoring systems for ICU patients: APACHE III, SAPS II, MPM II. J Prev Med Pub Health 2005;38:276–82.
[61] Ciccarese P, Caffi E, Quaglini S, Stefanelli M. Architectures and tools for innovative health information systems: the guide project. Int J Med Inform 2005;74:553–62.
[62] Shortliffe E. Computer-based medical consultations: MYCIN. New York: Elsevier; 1976.
[63] Hripcsak G. Arden syntax for medical logic modules. MD Comput 1991;8. 76, 78.
[64] Sordo M, Boxwala AA, Ogunyemi O, Greenes RA. Description and status update on GELLO: a proposed standardized object-oriented expression language for clinical decision support. Medinfo 2004;11:164–8.
[65] HL7. Health eDecisions., 2014, https://wiki.hl7.org/index.php?title=Health_eDecisions. [Accessed 3 August 2019].
[66] FHIR. FHIR release 3 (STU) welcome to FHIR (R)., 2014, http://www.hl7.org/implement/standards/fhir/. [Accessed 4 August 2018].
[67] Strasberg HR, Rhodes B, Del Fiol G, Jenders RA, Haug PJ, Kawamoto K. Contemporary clinical decision support standards using health level seven international fast healthcare interoperability resources. J Am Med Inform Assoc 2021;28:1796–806. https://doi.org/10.1093/jamia/ocab070.
[68] Warner HR, Toronto AF, Veasy LG. Experience with Baye's theorem for computer diagnosis of congenital heart disease. Ann N Y Acad Sci 1964;115:558–67.
[69] Lodwick GS. A probabilistic approach to the diagnosis of bone tumors. Radiol Clin North Am 1965;3:487–97.
[70] deDombal FT. Computer-aided diagnosis and decision-making in the acute abdomen. J R Coll Physicians Lond 1975;9:211–8.
[71] Fox J, Johns N, Lyons C, Rahmanzadeh A, Thomson R, Wilson P. PROforma: a general technology for clinical decision support systems. Comput Methods Programs Biomed 1997;54:59–67.
[72] Boxwala AA, Peleg M, Tu S, Ogunyemi O, Zeng QT, Wang D, et al. GLIF3: a representation format for sharable computer-interpretable clinical practice guidelines. J Biomed Inform 2004;37:147–61.
[73] Tu SW, Musen MA. Modeling data and knowledge in the EON guideline architecture. Medinfo 2001;10:280–4.
[74] Young O, Shahar Y, Liel Y, Lunenfeld E, Bar G, Shalom E, et al. Runtime application of Hybrid-Asbru clinical guidelines. J Biomed Inform 2007;40:507–26. https://doi.org/10.1016/j.jbi.2006.12.004.
[75] Tu SW, Campbell JR, Glasgow J, Nyman MA, McClure R, McClay J, et al. The SAGE Guideline Model: achievements and overview. J Am Med Inform Assoc 2007;14:589–98. https://doi.org/10.1197/jamia.M2399.
[76] CPG-IG. FHIR clinical guidelines. n.d. http://hl7.org/fhir/uv/cpg/ [Accessed 4 May 2022].

[77] NLM. RxNorm. Unified medical language system. n.d. http://www.nlm.nih.gov/research/umls/rxnorm/index.html [Accessed 26 March 2022].
[78] SNOMED. SNOMED CT. SNOMED International; n.d. https://www.snomed.org/snomed-ct/why-snomed-ct [Accessed 26 March 2022].
[79] Regenstrief. Logical Observation Identifiers Names and Codes (LOINC®). n.d. http://www.regenstrief.org/loinc/ [accessed 26 March 2022].
[80] Tu SW, Musen MA, Shankar R, Campbell J, Hrabak K, McClay J, et al. Modeling guidelines for integration into clinical workflow. Stud Health Technol Inform 2004;107:174–8.
[81] NQF. Driving quality and performance measurement—A Foundation for clinical decision support: A consensus report., 2013, http://www.qualityforum.org/Publications/2010/12/Driving_Quality_and_Performance_Measurement_-_A_Foundation_for_Clinical_Decision_Support.aspx. [Accessed 26 March 2022].
[82] HL7_HeD. HL7 implementation guide: clinical decision support knowledge artifact implementation guide, release 1; 2013.
[83] Osheroff JA, Teich JM, Levick D. Improving outcomes with clinical decision support: An implementer's guide. 2nd ed. Chicago, IL: Healthcare Information and Management Systems Society (HIMSS); 2012.

CHAPTER

2

Clinical decision support methods

Robert A. Greenes[a] *and Guilherme Del Fiol*[b]

[a]Biomedical Informatics, Arizona State University, Phoenix, AZ, United States [b]Department of Biomedical Informatics, University of Utah Health, University of Utah, Salt Lake City, UT, United States

2.1 Introduction

In Chapter 1 we reviewed the main purposes or aims for which clinical decision support (CDS) may be desired, as indicated by the columns of Table 1.2, which is reproduced below as Table 2.1 for convenience. In this chapter we examine the wide range of methods that have been explored, developed, and implemented over the years for addressing these aims. Classes of methods, corresponding to the columns of Table 2.1, will be described in a historical context.

The historical perspective for describing the methods is of interest because it corresponds to the development and evolution of the scientific and technical basis for the field – in terms of the primary research methodologies that have been proposed, tested, refined, extended, and in some cases deployed and evaluated in operational settings. Some of these themes represent key dimensions of activity in CDS currently, whereas others are related more to the underlying research to create the databases or knowledge used in CDS, and still others are of interest mainly in terms of their historical influence on current approaches. Note that, in the next chapter, we will focus on the major social, cultural, economic, management, and governmental influences on health care providers, health care delivery organizations, and the health care public that have contributed to the development of CDS implementation and adoption over the years, as well as the barriers to success.

2.2 Primary research methodologies that have been pursued and extended

Many different research approaches have been explored over the years to deliver clinical decision support. A number of these topics are explored in more depth in Sections III and IV

Clinical Decision Support and Beyond
https://doi.org/10.1016/B978-0-323-91200-6.00019-X

TABLE 2.1 CDS purposes, examples, and principal methodologies.

		Methods						
Purpose	**Examples**	**IR and search**	**Logical evaluation**	**Probability estimation**	**Artificial Intelligence**	**Algorithmic/ multi-step**	**Grouping**	**Visualization**
Find info								
	Direct hyperlink	x						
	Infobuttons	x	x			x	x	
	Question-answering systems	x			x			
	Context-aware retrieval	x	x		x		x	
Make a decision								
	Diagnosis		x	x	x			
	Test selection		x	x		x	x	
	Choice of treatment		x	x	x	x	x	x
	Prognosis			x	x	x		x
Do calculations								
	Medication dosing	x	x		x	x	x	
	Radiation portal				x	x		x
	Surgical planning					x		x
	Index or score			x	x	x		x
Manage process								
	Checklist		x			x	x	x
	Flow sheet					x	x	x
	Guideline		x			x		x
	Protocol		x			x		x
Monitor								
	Alerts on abnormal events		x	x	x			
	Reminders		x					
	Error checking		x					
Organize info								
	Order sets		x				x	
	Structured forms		x				x	x
	Structured reports		x				x	x
	Dashboards		x			x	x	x
	Graphs and charts			x	x	x	x	x

of this book, so our intent here is to give an overview of the various methodologies and how they relate to one another rather than attempt to be comprehensive.

2.2.1 Information retrieval

The ability to find information relevant to a problem is a basic form of CDS. MEDLINE is the classic example of an information retrieval resource. MEDLINE, the history of which was reviewed by Smith [1], became available in 1964. That first incarnation of MEDLINE was a bibliographic retrieval system for the biomedical literature intended for librarians, called MEDLARS, developed by the National Library of Medicine (NLM). This evolved to the MEDLINE online resource in the mid-to-late 1970s, which now is accessed in its most convenient and widely used form through the Web via PubMed, operated by the National Center for Biotechnology Information of the NLM. Many other online bibliographic resources, databases, and knowledge bases have been added to MEDLINE and to PubMed since the latter's beginning, and there are today, of course, many other online reference sources. Over the years, a growing array of textbooks, handbooks, other reference materials, and interactive medical knowledge resources has also become available online through subscriptions and other arrangements.

There are two main ways of doing basic information retrieval as illustrated in Fig. 2.1, taxonomy-based or ontology-based search, or text-based search.

2.2.1.1 *Taxonomy-based or ontology-based search*

The classic way of retrieving information resources relies on the use of a controlled vocabulary, consisting of terms (or *keywords*) by which the content has been pre-indexed.

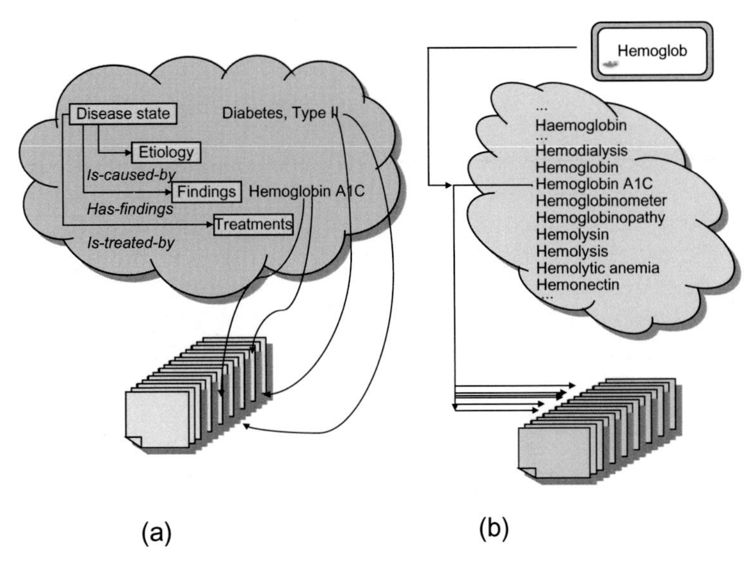

FIG. 2.1 (A) Ontology-based retrieval. A user can navigate terms and relations to select those wanted. Child terms can be included. (B) Text-based retrieval. Text entered by a user is looked up in text index to find occurrences.

The Medical Subject Headings (MeSH), the heading terms used to index articles stored in MEDLINE [2], have been in use for more than sixty years, since the early days of the precursor printed resource known as Index Medicus. The terms in MeSH are arranged hierarchically into trees covering a number of topic axes, and many terms actually occur in several MeSH subtrees. MeSH terms can be suggested by content authors, but the indexing for MEDLINE actually is done by specialists who choose the terms, in order to introduce consistency. This process is aided by automated tools for suggesting terms [3–5], development of which remains an area of research [6,7], or fully-automated indexing methods that have been studied more recently [8–10].

The user doing retrieval from a term-indexed resource must use the same resource to identify appropriate terms describing the topic of interest, and typically may create Boolean expressions comprised of logical combinations of terms, to do more complicated retrieval. For example, one might browse the MeSH tree to find terms for retrieving articles from PubMed, or browse therapeutic categories in a drug index, to find anti-hypertensive medications from a drug formulary, perhaps further indexed by mechanism of action such as diuretics, ACE inhibitors, and the like before selecting a particular medication for retrieval of its details. Since the MeSH tree is hierarchical, a term at a higher level can be used to connote the desire to "explode" the subtree at that point to be able to search for articles containing any lower-level terms as well. Use of MeSH terms was the primary way to retrieve information from MEDLINE in the past, and is one of the most precise ways to find relevant items, although free-text search of titles, abstracts or entire text using PubMed is now much more widely used. A disadvantage of term-based searching is that, since it relies on indexing by an expert (or, to some extent, automatic indexing), the choice of terms used to index may not correspond to the particular interest or focus of a user.

The term *taxonomy* refers to the use of a hierarchical classification of controlled terms to provide a conceptual framework for a domain of interest, as an aid to organization, analysis, or information retrieval. The hierarchical relations are various types of parent-child, such as *is-a*, *is-member-of*, or *is-part-of*. An *ontology* is similar to a taxonomy in that it is also a means of describing a knowledge domain. An ontology uses a controlled vocabulary or concept identifier to formally represent concepts that describe objects and the relations among them. Ontologies typically use richer semantic relationships than taxonomies, to describe concepts and their attributes, and have a set of formal rules and constraints about how terms are defined and relations are specified. Because of the potential richness of an ontology, it can be thought of as a knowledge representation, rather than just as a method to control the terms used for the domain's description. Thus, using an ontology can allow one to actually reason about a domain. For more detail on these topics, Chapter 11 discusses clinical vocabularies, taxonomies, and ontologies.

The MeSH tree is a taxonomic system. The terms used are controlled, but, as noted earlier, they do not represent unique concepts, in that the same term can occur in multiple subtrees of the MeSH hierarchy. Thus, tagging entities with MeSH terms themselves leaves some ambiguity. Using the controlled terms of an ontology would be a more precise way to do this, since each concept represented by the controlled term is unique. An alternative would be to specify the actual tree positions ("MeSH Tree Numbers") of the particular MeSH terms used, so that alternative uses of those terms are not considered.

If a taxonomy or ontology is present, we can traverse its classes and relations to find categories of interest. The Unified Medical Language System (UMLS) Metathesaurus [11,12], developed by the NLM, is the most comprehensive collection of terms from a variety of terminological systems. In the UMLS Metathesaurus, these various terms are mapped to unique UMLS concepts, with their own Concept Unique Identifiers (CUIs). The UMLS concepts are classified into semantic types using the UMLS Semantic Network [13], and semantic relations among types in the Semantic Network facilitate navigation of the UMLS. The UMLS, having incorporated and interrelated terms from many leading taxonomies, including ICD9/10, SNOMED, MeSH, and LOINC, provides a means for concept-based information retrieval rather than retrieval based simply on using terms. Such an approach also has the advantage that it can identify synonyms and alternative forms of the selected concepts to improve retrieval. An example of concept-based retrieval was the SAPHIRE system [14,15].

With structured markup formalisms such as XML, it is customary to include meta-level descriptors of many aspects of a document, such as author, date of creation or modification, source, domain of focus, and other attributes as fields, all of which can potentially be searched on, especially if the entries in those fields come from a controlled vocabulary.

2.2.1.2 Free text search

Direct text-based searching has become the most common way to search, since most Web-based search engines (such as Google® and Bing®), rely on this method. As we have noted, this is also now the most popular way to search PubMed, even though MeSH-heading-based retrieval also continues to be available. An obvious advantage of free text search is that it is the easiest to do, because it requires little effort by the user and almost no learning curve. If the search fails to provide appropriate results, then alternative terms can be tried. If the goal is to find *some* answers, and there is not concern about finding *all* answers, and the corpus is large, text search is very often sufficient, hence the rapid growth, popularity, and ubiquity of such search capabilities.

A disadvantage is that one does not know how many potential items were not retrieved due to the wrong choice of term for free text search, when there are multiple possible synonyms. *Recall* is defined as the fraction of the true relevant documents that were actually retrieved, analogous to sensitivity or true positive rate. Another disadvantage is that the retrieval may include false positives when the chosen search term has alternative meanings. *Precision* is defined as the fraction of true relevant documents in the set of all documents retrieved, analogous to a positive predictive value. Today's search engines, including PubMed [16], often combine elements of taxonomy/ontology-based search and free text search, typically with automated mapping of free-text search terms entered by a user to concepts and then using both free-text terms and concepts for searching.

2.2.1.3 Semi-automated or automated

Retrieval can be made more patient-specific and context/situation-specific, if additional conditions are used to constrain results (e.g., information about the application from which the query is invoked, the work setting, the user characteristics, and data about the patient). A user-initiated search in an application context could be modified automatically by adding terms related to the preceding, for example.

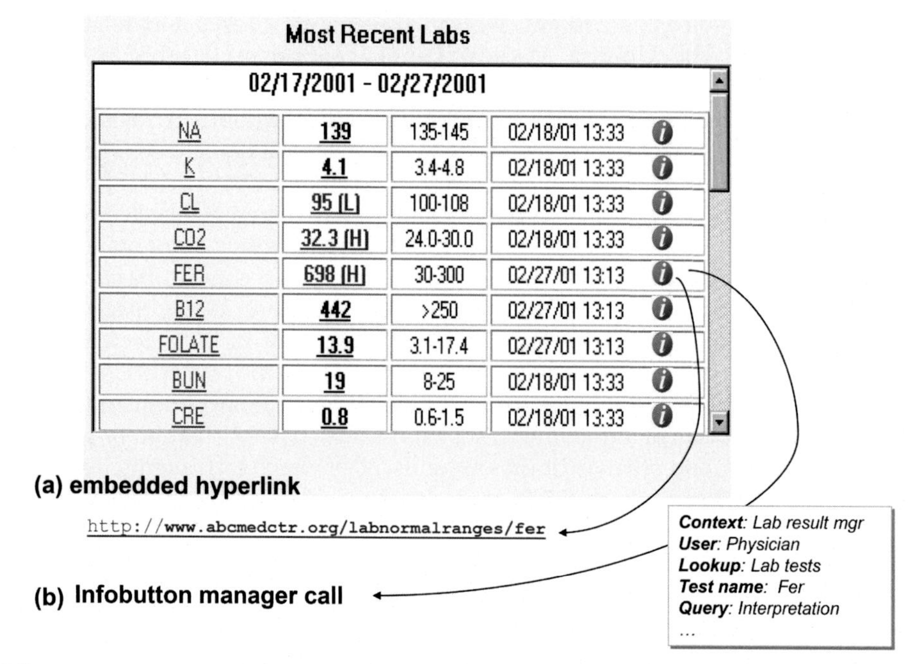

FIG. 2.2 Automated or semi-automated retrieval from within an application by (A) embedded predefined hyperlinks, or (B) infobutton manager invocation with context-specific parameters.

If the retrieval is initiated by an application, the most direct approach is a specific embedded link/hyperlink. To allow for more dynamic retrieval, the parameters for these conditions can be generated automatically by the application, thus enabling information retrieval to be more tightly integrated and even automated (see Fig. 2.2). The *infobutton* is an example of this—embedded within clinical systems, infobuttons can leverage information about the user's context (e.g., reviewing a laboratory test result, prescribing a medication) to automatically create search strategies relevant to common information needs in that particular context [17]. Infobuttons are discussed further in Chapter 13, which describes a standard protocol for implementing infobutton capabilities in clinical systems and knowledge resources [18].

2.2.1.4 *Question answering systems*

An area of considerable progress in recent years has been the ability to respond to user queries to retrieve articles of interest through improved natural language processing (NLP) understanding of the question and of the internal content of the information resources. Sophisticated question answering through NLP has been pursued in health care by Yu and others (See Chapter 13) [19,20]. IBM Corp. made a big splash in 2011 with the unveiling of its Watson system in a public competition against winners of the popular Jeopardy! television show, whom it defeated. Watson uses a collection of methods for text mining, association mining, and other methods to build up knowledge of a domain [21]. Subsequent efforts attempted to adapt this system to a variety of medical contexts [22–24], but with less success [25,26].

We can imagine a time in the near future where such question-answering systems are integrated into care, allowing a user to ask a query, with context of the user setting and patient parameters automatically supplied and with rapid retrieval of the desired information. Right now, most retrieval systems provide the documents that are most appropriate, but paragraph-level retrieval would be even better, and systems are beginning to approach that capability. Many health systems use popular commercially available information products that seek to answer clinical questions such as what to do in a patient with a particular combination of diseases and drugs who develops a specific complication, and which produce a short targeted answer. If this retrieval process could be even more automated and patient- and provider-specific, based on infobutton-like context information, such resources would be even more valuable [27].

2.2.2 Evaluation of logical conditions

Evaluation of logical conditions is among the most widely used forms of clinical decision support and occurs in a wide variety of settings. Many different ways to represent logical conditions have been explored.

2.2.2.1 *Decision tables*

One of the first themes to be pursued in clinical decision support was the idea of using logic as a way of refining and reducing the numbers of diagnostic possibilities. Let us consider n possible diseases D_i, where $I=1, \ldots, n$. We also consider m possible findings f_j, where $j=1, \ldots, m$.

For purposes of illustration here we make the simplifying assumption that a finding f_j can be only either positive, negative, or unspecified, denoted as $f_j=1$ or 2 or "–" (unspecified) for each j. For a finding to be unspecified in a definition, this means that its value is irrelevant to the disease.

If all the diseases under consideration can be characterized by their findings, then a vector F_i can be constructed consisting of all the values of individual findings f_j for disease D_i.

Now consider arranging the findings characterizing all the diseases in a decision table, as depicted in Table 2.2. Since each of the columns of the table represents a different disease, there are n such columns. Each of the rows represents a different finding, so there are m such rows.

The diseases can be sorted (i.e., the columns rearranged), based on the value of any finding. Table 2.3 is sorted by the first finding, f_1, so all the diseases for which f_1 has a value of *1* are in columns to the left of those for which its value is 2. All the diseases for which the finding doesn't matter have a "—" in the cell for that finding and appear to the right of those with either *1* or *2*.

Assume, for instance, that f_1 represents the sex of the patient, and *1* means *female*, and 2 means *male*; then this sorting has separated all diseases for which the sex of the patient must be female from those for which the patient must be male, and from those for which the disease can occur in either sex.

TABLE 2.2 Principal methodologies for clinical decision support and key developments.

Method class	Key methodologies
Information retrieval	Taxonomies, ontologies, text-based methods, context-specific retrieval, question understanding
Evaluation of logical conditions	Decision tables, Venn diagrams, logical expressions (event-condition-action rules)
Probabilistic and data-driven classification or prediction	Bayes theorem, decision theory, ROC analysis, belief networks, meta-analysis, and probabilistic inference from databases.
Artificial intelligence	Heuristic reasoning, expert systems, data mining/machine learning-derived models. (Note that the latter overlaps with probabilistic methods.)
Calculations, algorithms, and multistep processes	Process flow and workflow modeling, guideline formalisms and modeling languages
Associative groupings of elements	Report generators and document construction tools, document architectures, templates, markup languages, ontology tools, ontology languages. Also improved models of cognition and workflow, human factors, and usability. Information presented or requested based on an organizing schema or principle, such as indication for order sets, or clinical problem for note generation or reporting.
Visualization	Grouping and displaying information to facilitate appreciation of data and their associations, and recognition of situations that may require attention or action

TABLE 2.3 An example of a decision table for *n* diseases and *m* possible findings. For a disease, a finding can be positive, negative, or immaterial, symbolized by *1*, *2*, and –, respectively.

	Disease				
Finding	D_1	D_2	D_3	…	D_n
f_1	1	1	2		–
f_2	1	2	2		1
f_3	2	2	–		1
…					
f_m	–	1	1		2

One can do similar manipulations for combinations of findings by doing subsorts of the sorted columns based on the values of other findings. In this example, if some diseases can occur only in infant females, and finding f_2 corresponds to *1:* $age \leq 2$ and *2:* $age > 2$, then a subsort of the table that is already sorted by sex will subgroup columns by this age criterion, as shown in the table. This shows by inspection that only D_1 has this combination of findings values. Note that if a particular finding value (or combination of findings values) is pathognomonic (unique) for a disease, then the particular value(s) should be present in only one column for that row (or combination of rows).

In the classic 1959 *Science* article by Ledley and Lusted [28] and in Ledley's subsequent book [29], a similar discussion of the role of logic manipulation is presented as a prelude to other methodologies. Although Ledley and Lusted were particularly interested in probabilistic manipulation of findings to construct a differential diagnosis using Bayes theorem, they used logical manipulation to first reduce the range of possibilities that needed to be considered.

The representation of logic can be done in multiple ways. Decision tables as illustrated previously are one way to do that. An advantage of decision tables is the ability to sort and group columns with similar values in their rows. Another example of the way in which a decision table can be used is for representation of clinical practice guidelines, which are discussed further in Section 2.2.3. This use, pursued by Shiffman et al. in the 1990s [30–32], involves constructing a findings-action matrix in which the upper set of rows corresponds to individual findings, and the lower set of rows corresponds to possible therapeutic actions (see Table 2.4). Individual columns are used to represent each of the possible combinations of findings that could occur (actual diseases are not explicitly considered here, just findings). So if we have two possible findings that can be present or absent, four columns are needed to represent the possible combinations of the findings. If there are *n* findings, each of which can be present or absent, then there will need to be 2^n columns to represent the various combinations.

The cells in the action rows containing an X in a given column indicate that the follow-up test or treatment actions corresponding to those rows that are to be carried out given the particular findings in that column. (Note that more than one test or treatment may be appropriate.) Thus, each column represents a step in a clinical guideline: the conditions for being at that particular step (i.e., the eligibility or applicability criteria for a set of recommended actions) correspond to the combination of findings indicated in the findings

TABLE 2.4 A guideline represented as a decision table. There are three findings that can be present (indicated by 1) or absent (indicated by 2), so there are 3^2 possible combinations, represented by the 8 columns, C_i. The recommended actions for each C_i are indicated by an X in the rows corresponding to those actions, for that column.

Findings	C_1	C_3	C_3	C_4	C_5	C_6	C_7	C_8
f_1	1	1	1	1	2	2	2	2
f_2	1	1	2	2	1	1	2	2
f_3	1	2	1	2	1	2	1	2
Actions								
a_1	**X**		X	X			X	
a_2		X	X	X	X		X	
a_3	**X**			X	X			X

section of that column; the actions that are appropriate given those findings are indicated in the action part of that column. An *X* in an action cell means that the particular test or treatment should be carried out, and a blank means it should not. We can use this kind of decision table to look up any finding complex, and to determine what follow-up tests or treatments or other actions are applicable. Also, if we desire to determine under what conditions a particular test or treatment should be done, we can use this table simply to sort by actions, and to identify all the applicable findings complexes.

We can also use this approach to help design clinical guidelines and to check them for consistency and completeness. For example, we can simplify the table by eliminating redundant columns: if we inspect all the columns for which a particular test/treatment regimen is recommended versus those for which other test/treatment regimens are recommended, we can determine if particular findings are unique to that regimen. If the value of a finding does not distinguish between those columns for which the test/treatment regimen is recommended and those for which it is not, then that finding is irrelevant to the decision regarding that test/treatment regimen. Thus, the various columns representing alternative values for that finding can be collapsed into a single column with a "don't care" symbol (—) in the cell for that finding. For example, in Table 2.3, columns C_3 and C_7 have the same action specification, and their findings differ only with respect to f_1, so f_1 is irrelevant to the decision to carry out this test/treatment regimen, and the two columns can be replaced by a single column with "—" in the cell corresponding to f_1. Other manipulations can be done to identify inconsistencies, for example, columns with identical findings specifications with different action recommendations. Columns with combinations of conditions that do not have actions associated with them (e.g., C_6 in the table) can be considered as either omissions that need to be completed, or clinically impossible or irrelevant combinations for which the columns can be eliminated.

This discussion oversimplifies the process of using decision tables to represent clinical guidelines, because it does not consider sequences of tests, the performance characteristics of the tests, or the costs or risks of them and does not distinguish among the relative costs and utilities of various treatment options. Extensions of decision tables to create so-called "augmented decision tables" were pursued by Shiffman and Greenes [31] to deal with such issues.

2.2.2.2 Venn diagrams

Another well-known approach to representing clinical logic is by the use of Venn diagrams—shapes such as circles used to enclose groups of entities representing particular individual characteristics. The overlap of these shapes indicates the possible subgroups that can occur with combinations of these characteristics. If we consider a circle to represent those entities having attribute *A*, and another circle to represent those having attribute *B*, then the entities corresponding to the overlap of these two circles have both attribute *A* and attribute *B*. The area outside of either circle has neither attribute *A* nor attribute *B*. We can introduce a third circle representing attribute *C*, creating three more areas of overlap. Feinstein [33,34] pursued the use of Venn diagrams such as this for thinking about and teaching clinical judgment, as shown in Fig. 2.3A for three symptoms in acute rheumatic fever and (B) for four aspects of cancer, as adapted from his 1967 book. The process gets much more complex, however, when dealing with combinations of more than three or four logical entities, at which point visual display via Venn diagrams generally is considered to be impractical.

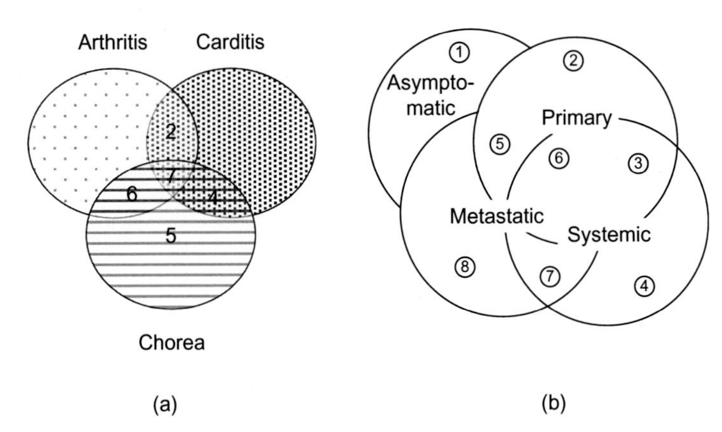

FIG. 2.3 The use of Venn diagrams to represent the spectrum of findings in disease. (A) The spectrum of rheumatic fever. (B) A simplified representation of the clinical spectrum of cancer. *Adapted from Feinstein,* Clinical Judgment, *1967.*

2.2.2.3 Logical expressions

The usual mode for representing logical conditions is by *logical expressions* composed of Boolean combinations of terms. Individual terms are of the form *parameter operator value*, where *parameter* denotes the entity being evaluated, *operator* denotes a comparison operation (e.g., =, >, *or* <) to be performed, and *value* is the object of the logical operation and may be another parameter or a literal. The evaluation of the term yields a result of *true* or *false*. Terms are combined into *expressions* of arbitrary complexity by Boolean logical operators (e.g., *and, or, not*; note that *not* is a unary operator on a term).

One common approach to CDS is to build *expert systems* based on the capture and representation of knowledge from human experts. This topic is discussed further in Chapter 6, but for our discussion here of logical approaches to CDS, we point out that a major form of representation of expert human-derived knowledge is as rule-based systems. A rule-based system consists of a set of statements called *production rules* of the form *If (condition) then (action)*, where *condition* is a Boolean logical expression, and an inferencing method (possibly as a separate inferencing engine) for processing sequences of rules in a knowledge base to arrive at a conclusion. For each production rule, if the logical expression evaluates to *true*, then the *action* is performed.

In the business community, rules of the *if…then* form, known as *event-condition-action (ECA)* rules, of the form:

On	*event*
If	*condition*
Then	*action*

have been used to trigger rule logic evaluation in response to events. ECA rules automatically perform actions when triggered if the stated conditions hold. The ECA formalism originated in the active database field (where an active database system is one that has mechanisms to be able to respond automatically to events either inside or outside the database system) [35].

Bailey et al. [36] review the history of ECA rules in the business community, noting that they also are used in workflow management, network management, personalization and publish/subscribe technology, and for specifying and implementing business processes. Similar methods to those of ECA rules have been used extensively in health care in CDS tools such as alerts and reminders.

Alerts and reminders

Although alerts and reminders can be implemented using methods such as decision tables and machine learning, most implementations in today's clinical systems use ECA rules. Alerts are most often used to notify the care team about critical situations, such as when entering a prescription for a medication to which the patient is allergic, or upon release of a laboratory test result that indicates a life threating condition. Typically an alert is triggered by an event that has been identified as a potential cause for a warning. Events can be data-driven (e.g., a new laboratory test result), time-driven (e.g., 24 h after a patient's hospital admission), or user-driven (e.g., ordering of a medication, opening a patient's record). Alerts can be triggered in real-time, with respect to a provider's interaction with the system, for example by the action of ordering a drug, or can be triggered in the background, e.g., upon availability of a laboratory test result or upon automatic collection of vital signs through sensors [37–39]. An example of a background alert might be:

> ON Potassium (K) test result, IF K $<$ 3.0 mg% AND Medications include Hydrochlorothiazide, THEN consider Potassium supplement AND Hydrochlorothiazide Dose Reduction.

Reminders are similar to alerts, but they are used more often as a check list of preventive care for which a patient is due. Reminders are often triggered upon user events (e.g., opening a patient's record) or time events, which causes a condition to be evaluated to determine whether a specific action is desirable. Examples are a reminder for an HbA1c test that is generated when the care team opens the record at the time of a clinic visit of a patient with diabetes who has not had the test in the past six months. Reminders can also be generated in "batch" mode at the population level, for example, by evaluating a target cohort to identify patients who are overdue for colorectal cancer screening and then notifying the care team or patients directly.

For either alerts or reminders, once evoked by an event or time trigger, a rule evaluator examines the condition part of the rule to determine whether it evaluates to *true*, in which case it carries out the specified action. The action may be, for example, to notify the provider of the result or some action to be taken. If this is a synchronous rule, a dialogue box or message may appear inline or as a popup. If asynchronous, notification would be typically by e-mail, text messaging, or a multi-patient dashboard.

Alert and reminder capabilities were first developed in the CARE system at Regenstrief Institute in Indianapolis. Arden Syntax is a representation scheme, explicitly based on the ECA formalism, for specifying alerts and reminders, in terms of their evoking conditions, the data elements referenced, the logic, and the action to be performed, and which has become a Health Level Severn (HL7) standard [40–42]. Arden Syntax and successor approaches for standardizing the representation of logic in conditional expressions are discussed in Chapter 9.

2.2.2.4 Other logic models

Other variations on logical manipulations address the fact that binary (true/false) logic is often insufficient to capture the nuances of medical knowledge. A data item may be present, or absent, but if it is absent, this may be because it is unknown, unavailable, or not obtained. Multivalued logic may be more appropriate to deal with such situations [43]; yet a review of the literature reveals that, although multivalued logic is widely used in engineering, particularly in circuit design, generally it has not been adopted in medicine. Another circumstance arises when the value of a data item may have an inherent uncertainty about it, reflected in a range of possible values. Because of the arbitrariness of asserting categories for some classifications, such as blood pressure characterized as *high*, *normal*, or *low*, fuzzy logic has been used to create qualitative degrees of membership to such categories [44]. The use of multivalued and fuzzy logic in rule-based formalisms such as the Arden Syntax is discussed further in Chapter 9. These notions also bear upon the need to incorporate probabilistic reasoning and uncertainty into CDS, as discussed next.

2.2.3 Probabilistic and data-driven classification or prediction

Most clinical judgments are not deterministic, and CDS needs to recognize the inherent variability of medical data, the imprecision of tests and measurements, and the fact that many principles of practice are based on limited evidence or just on expert opinion.

2.2.3.1 Updating probabilities based on evidence

As we have noted in Chapter 1, the *Science* paper, "Reasoning Foundations of Medical Diagnosis" [28], first introduced the idea of applying Bayes theorem to medical diagnosis. The core of Bayes theorem is the formula,

$$P'(D_i|F_j) = \frac{P(D_i)P(F_j|D_i)}{\sum_{k=1}^{n} P(D_k)P(F_j|D_k)}$$

where:

D_i is a particular disease of *n* mutually exclusive and exhaustive diseases
F_j is a set of findings
$P(D_i)$ is the prior probability of D_i
$P(F_j \mid D_i)$ is conditional probability of F_j given D_i
$P'(D_i \mid F_j)$ is the posterior probability of D_i, i.e., the probability of D_i given F_j

As we discussed in Section 2.2.1, the Ledley and Lusted method actually used a prior step before applying Bayes theorem, in that they first used logical manipulations to look for logical combinations of findings that would absolutely rule in (or rule out) certain diseases. In their formulation, the diseases D_i were considered to be mutually exclusive and exhaustive; that is, their probabilities sum to 1. Ledley and Lusted introduced a more general formulation in which combinations of diseases could occur, but in many cases the simplification is reasonable, or can be augmented by new "diseases" representing likely combinations of individual diseases. This makes it easier to develop the necessary set of estimates of prior probabilities of each disease.

Bayes theorem can be used to produce a posterior probability for each of the diseases under consideration, with the rank order of these thus corresponding to the relative magnitudes of the posterior probabilities. Since all possible diseases are considered, the sum of the posterior disease probabilities is equal to 1.0.

Bayes theorem is defined on findings complexes (the vector F_j for each possible combination of individual findings); this means that to be used operationally, conditional probabilities must be known for each such finding complex given each possible disease under consideration. Obtaining such joint probabilities is a monumental task, and databases containing such information are not usually available except in rare circumstances where only a few possible findings are under consideration. As a result of this, two other simplifying assumptions are typically made:

1. Each finding is considered to be conditionally independent of every other finding for a particular disease. This allows the probability of a combination of findings, given a disease, to be computed by multiplying the conditional probabilities of each individual finding given the disease.
2. The values for a finding are grouped into limited categories or bins, with the simplest categorization being binary (e.g., present, absent; male, female; age ≤ 20, age > 20).

These simplifying assumptions were used in the first applications of Bayes theorem to medical diagnosis, by Warner and colleagues in 1964 [45–48], for diagnosis of congenital heart disease. Assumption (1) is the most tenuous, in that it often does not hold in practice; as a result, some combinations of findings that are not actually conditionally independent tend to be over-supported when the independence assumption is used.

In differential diagnosis, it is often desirable to do some initial testing to narrow the range of diagnostic possibilities before proceeding to more definitive, and perhaps more invasive and expensive, tests. In the late 1960s, Gorry and Barnett (1968) explored the idea of a sequential Bayesian approach, in which certain tests were selected first, and based on the resulting diagnostic rankings, the method picked additional tests considered likely to contribute pertinent information based on various heuristics (e.g., reduction of entropy, or cost). The Iliad program, developed by H. Warner, Jr., and colleagues [49,50] in the mid to late 1980s, used Bayes theorem confined to smaller subdomains of findings in its diagnostic model, but was used more as an aid to teaching and quality assurance than for direct CDS.

Besides the early applications by Warner et al. in congenital heart disease diagnosis in 1964 and in bone tumor diagnosis by Lodwick in 1965, perhaps the most widely used application of Bayes theorem in health care was the acute abdominal pain diagnosis program by de Dombal et al. [47] in Leeds, United Kingdom. This program used a structured form to collect findings of patients presenting in an emergency room, and was shown to be moderately effective compared to unaided practitioners. However, subsequent studies of its use in other settings (including a nuclear submarine, where the challenge was to accurately diagnose surgical vs. non-surgical causes so as not to surface unnecessarily during long undersea missions) demonstrated vividly how dependent the performance of Bayes' theorem is on the base rates (prevalences) of the various diagnoses under consideration in the population on which it is being applied. For example, if the program was highly tuned to performing in a general emergency room in an inner city setting, base rates should reflect the expectation of some cases with pelvic inflammatory disease or ectopic pregnancy, but on a nuclear submarine, the base rates would be quite different.

2.2.3.2 *Decision analysis*

As early as Ledley and Lusted's 1959 article, the need to decide among treatment options by some sort of "value theory"-based rating of the alternatives was recognized. The formal methodology of statistical decision theory was just beginning to be developed at that time, as a successor to value theory. One of the pioneers of this was Howard Raiffa, a professor at Harvard Business School. His classic book on this topic, first published in 1970, has had a number of subsequent editions [51]. It wasn't until the mid-to-late 1970s, however, that the application of these methods for clinical decision making began to be done in earnest [52,53]. The essence of formal decision analysis is to develop a decision tree, in which a decision problem is structured in terms of branching sequences of decision nodes and chance nodes [54,55].

The initial node of the tree is often a decision, such as whether to treat a patient now, test further, or do nothing (see Fig. 2.4). Decision nodes are indicated by square boxes in the figure, and chance nodes by circles. An example would be a young boy with abdominal pain, where the question is whether he has appendicitis or nonspecific abdominal pain. If surgery is done immediately, there is a possibility that a normal appendix will be found. There is also some risk associated with surgery, along with cost and morbidity. If one decides to test further (e.g., perform a CT scan), there is still a possibility that the test will be wrong, but if one uses the test as a basis for deciding on surgery or doing nothing, different probabilities will be associated with the outcomes, and the desirability of the endpoints needs to account for the cost of the test. If one decides to do nothing, surgery may not be required, and expenditure, discomfort, and risk of surgery have been avoided. However, there is now a possibility that the appendix may rupture, requiring more urgent and risky surgery.

One of the main purposes of a decision tree is to lay out the sequences of decisions and possible outcomes at each step, so that the decision maker can focus on the critical variables (e.g., the probabilities of various branches, the costs or risks of the treatments or the performance characteristics of the tests). The tree is expanded until all branches reach points that

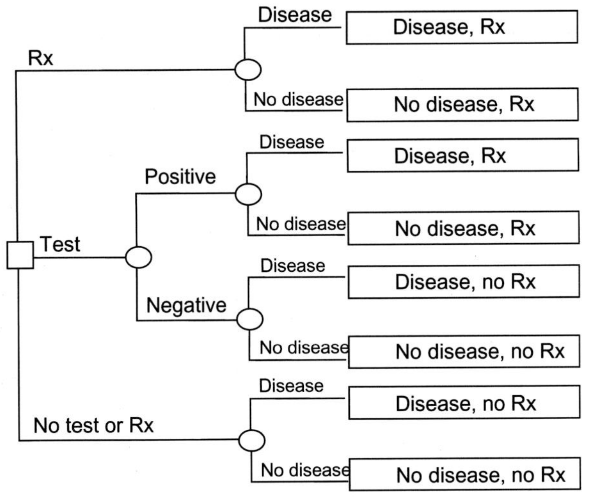

FIG. 2.4 Decision tree for a generic treat now vs. test further vs. do nothing decision problem. Square boxes indicate decision nodes; circles indicate chance nodes.

can be considered endpoints from the point of view of the analysis. The desirability of the endpoints is assessed, by assigning *utilities* to them [56]. All branches from a chance node are assigned conditional probabilities (the probability of that branch occurring given the state of the patient at the immediately preceding chance node). Since the decision tree is intended to determine what the initial decision should be, the solution proceeds by a method known as *fold-back analysis*. The process begins by looking at the distal branches of the tree. If the immediately preceding node of an endpoint is a chance node, then an *Expected Utility (EU)* is computed for the chance node as the sum, over all the branches arising from that node, of the product of the utility of each of the endpoints times the conditional probability associated with the branch leading to it. If the immediately preceding node of an endpoint is not a chance node but a decision node, the process is different. Since the optimal decision generally is considered to be one that maximizes EU, the EU of the decision node is determined to be the maximum EU of the branches extending from it, rather than the sum of the EUs of the branches. The chance node or decision node thus assigned an EU is now considered an endpoint, and the process continues backward in the tree to the next more proximal node, iteratively, until the initial decision node of the tree is reached. At that point, the optimal choice, the subtree with the maximum EU, is determined.

An important feature of the decision tree is that the conclusion about optimal choice will differ as a function of estimates of probability and utility. Thus it is important to do a *sensitivity analysis* on the various estimated parameters, to see how robust the decision is over reasonable ranges of key parameters, and to determine the *threshold* for a parameter [57], such as a probability estimate or a utility (depending on one's focus), which would cause the optimal choice to change. In the example of the young boy with abdominal pain, it is likely that over some range of probabilities of nonspecific causes, nonsurgical management would be optimal, whereas over some lower range of likelihood of nonspecific causes (with higher likelihood of appendicitis), the surgical option would be optimal, if only to avoid the infrequent complications of surgery for a ruptured appendix. There is thus some threshold probability at which the decision choice would change. Likewise, if the risk of surgery of either the simple type or that for ruptured appendix had a higher or lower frequency of undesirable outcomes (greater or lesser likelihood of complications or death), this would affect the point at which the choice to do surgery becomes optimal.

Multiple parameters can go into assessing a utility—for example, quality of life, length of life, and cost of treatment—in which case it may be necessary to rank the utility of each independently and then find a way to balance one against another or combine them into a single metric. Quality-adjusted life years (QALYs) [58–60] is a metric that has been used widely as a way to assess long-term consequences of various treatment options (controversial in that it can be associated with policy decisions about rationing of health care resources).

Other notable methods in decision analysis include the use of Markov modeling [61–63] for chance processes that may occur at unknown or multiple times in the future, and the declining exponential assessment of life expectancy (DEALE) [63], which enables the physician to collate various survival data with information on morbidity to determine a quality-adjusted expected survival for a potential management plan.

Although decision analysis has been occasionally used as a bedside decision support method, its primary use is for policy analysis [64,65] and the formulation of guidelines and rules. It is through these that the fruits of decision analysis are usually actually realized as CDS. However, another use of decision analysis is in the assessment of patient preferences [66–68], and in shared patient-doctor decision making (see Chapter 24) [68,69].

2.2.3.3 *Bayesian belief networks*

Recognizing the difficulties in developing conditional probabilities for Bayesian analysis, and seeking to model causality and the probabilistic dependencies of events on one another, Pearl, working at Stanford in the 1970s [70], developed the notion of Bayesian belief networks that depict explicitly the various dependencies in the form of an acyclic directed graph. The first application to medical problems was the work of Cooper in the 1980s [71]. This permits probabilities that are known to be explicitly entered into the network, others estimated where possible, and the remaining ones derived by inference from those that have been entered. Fig. 2.5 is an example of a Bayesian network being explored for breast cancer risk prediction.

2.2.3.4 *Technology assessment*

Probabilistic decision support has been further aided by advances in technology assessment including receiver operating curve (ROC) analysis of performance characteristics of diagnostic technologies. Although not decision support methods themselves, technology assessment techniques have been invaluable in improving Bayesian probability estimation, by providing systematic approaches to quantifying diagnostic procedures in terms of the conditional probabilities of test results given specific disease conditions [73–78].

2.2.3.5 *Database prediction*

As we introduced in Chapter 1, Table 1.4, "Data sources," and further discuss in Chapter 3, Section 3.1.6.3, health-related data are rapidly expanding from many sources, including genomics, imaging, biosensors, increased interoperability and exchange of EHR data, including unlocking through NLP of unstructured data embedded in narrative notes. This is giving rise to much attention to the potential of so-called "big data" for research, quality assessment, public health, and population based CDS, through analytics and predictive data modeling.

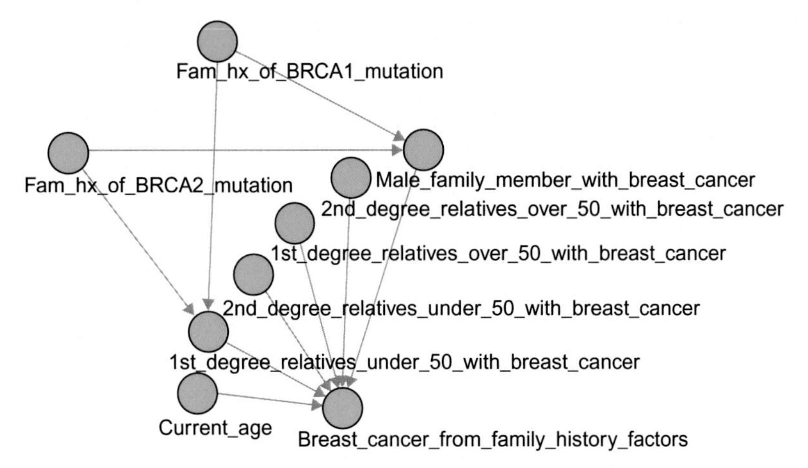

FIG. 2.5 Bayesian network for breast cancer risk prediction [72]. *Reproduced by permission of the author (Ogunyemi).*

Probabilistic decision making is critically dependent on the nature of the data on which it is based. The ideal situation is one in which large databases containing well-structured data are available that allow precise retrieval of patients similar to a current patient. Analysis of the responses of those patients to various treatments could then be used to decide upon the best treatment for the current patient. Two early approaches in the 1970s demonstrated the power of database prediction, that of Fries et al. [79–81] in rheumatological disease with a system known as ARAMIS, and that of Starmer, Rosati and colleagues [82,83] in evaluating patients with coronary disease. These systems used well-structured data on patient encounters to generate snapshots of outcomes of similar patients as a clinical tool when considering prognosis or possible treatments for a current patient.

More advanced methods of data analytics for predictive modeling often involve data mining and machine-learning, as discussed in Section 2.2.4.2.

2.2.3.6 *Evidence-based medicine and meta-analysis*

Any discussion of probabilistic methods would not be complete without recognizing the movement toward evidence-based medicine (EBM), which basically seeks to annotate any clinical action with appropriate justification in the literature. The term has been used since the early 1990s in a series of JAMA articles from the Evidence-Based Medicine Working Group, based at McMaster University [84,85]. EBM attempts to develop objective ways to ensure high quality and safety of medical practice, aims to reduce health care costs, and seeks to speed the transfer of clinical research into practice. Although the goals are worthwhile and have been broadly espoused, some concerns are that clinical trial settings may be dissimilar to situations encountered in practice, and that the role of clinical experience and judgment cannot be minimized.

A major activity in EBM is the Cochrane Collaboration, the formation of EBM centers for systematic review of clinical trials to make the results of the analyses widely available. The idea was proposed by A. Cochrane, a British epidemiologist, and the first Cochrane Center was established at Oxford University in 1992 [86]. A number of Cochrane Centers exist around the world, as well as evidence-based practice centers in the United States and Canada, funded by the U.S. Agency for Healthcare Quality (AHRQ).

Often there are no randomized clinical trials pertinent to a particular clinical question, or the studies that are available differ in some respects from one another and are insufficient in size to resolve the question. Thus, a major function of EBM is the use of meta-analysis of multiple studies to arrive at its conclusions. Methodologies for EBM and meta-analysis are described in further detail in Chapter 8.

2.2.4 Artificial intelligence

Artificial intelligence (AI) includes a variety of methods to develop decision models based on:

(a) extracting expert opinion from humans, e.g., in terms of a series of rules, or by collections of heuristics, or "rules of thumb"
(b) learning from databases through data mining and machine learning – although explored in the early days – have become the predominant focus of AI.

2.2.4.1 *Heuristic modeling and expert systems*

An alternative approach to modeling based on data is to develop models based on an attempt to emulate human expertise and reasoning processes. This is particularly needed in settings where there are insufficient data to be able to derive the needed estimates for probabilistic approaches, but where decisions nonetheless need to be made. An underlying motivation is the belief that it is useful to be able to understand cognitive processes of the human, and both to capture such knowledge and understand better what its limitations are. A secondary benefit is that, unlike probabilistic systems, heuristic systems can explain their conclusions, since their reasoning processes in fact are based on heuristics understandable to humans. These approaches are discussed in detail in Chapters 9 and 10.

Rule-based systems

Rule-based systems are an example of a heuristic approach, in which individual logical statements in the form of production rules (as described in Section 2.2.2) are obtained by observing human experts, or interviewing and debriefing them, and then combined in an attempt to emulate the reasoning processes of experts. A production rule may be of the form, Rule 7: *if x then diagnose y*, or Rule 17: *if m then assert n*. A rule-based system running in top-down (*backward chaining*) mode might start with a goal of seeking to diagnose y (see Fig. 2.6). That goal is the action part of a top-level rule. In order to determine if that rule can fire, the inference engine must establish that the antecedent condition x is true. Since x is typically a Boolean combination of terms, the engine must find data or other rules whose action parts, if asserted, would satisfy terms in condition x. To determine if those rules can fire, the inference engine then proceeds to try to evaluate their antecedents conditions in the same way. This process continues recursively until data are found that satisfy the conditions of a rule (or not found). If a rule invoked recursively cannot be satisfied, then that chain of reasoning is abandoned. If a rule's condition is established, higher level rules that depend on the rule's action part can then be evaluated, and the process proceeds up the chain seeking to establish the goal. If satisfying conditions are not found, the goal fails to be established, or data are requested that may be able to subsequently establish the goal. This was the approach used in the MYCIN system of Shortliffe et al. [87], which was the first medical application of a production rule system, with the goal of choosing appropriate antimicrobial therapy for a patient. Note that this system did not use purely logical rules, but also developed a model for dealing with uncertainty, which they called "certainty factors" [88]. The work on MYCIN led to a number of extensions, including a shell known as EMYCIN [89]; GUIDON, an intelligent tutoring application [90]; and TEIRESIAS, an explanation facility [91].

A rule-based system can also be executed in bottom-up (*forward chaining*) mode [92]. This is a useful approach in applications where data are arriving on a regular basis, as in patient monitoring [93]. A set of data is provided, and all production rules for which antecedent conditions refer to those data are evaluated. The results that are asserted from the rules that evaluate to true can serve as conditions for other rules that are then able to assert additional results, and the process continues until a rule is evaluated that establishes a goal, or until no more rules are able to be evaluated, and the process fails to reach a goal. Forward chaining is less efficient than backward chaining, in that many rules that are evaluated don't lead to actions.

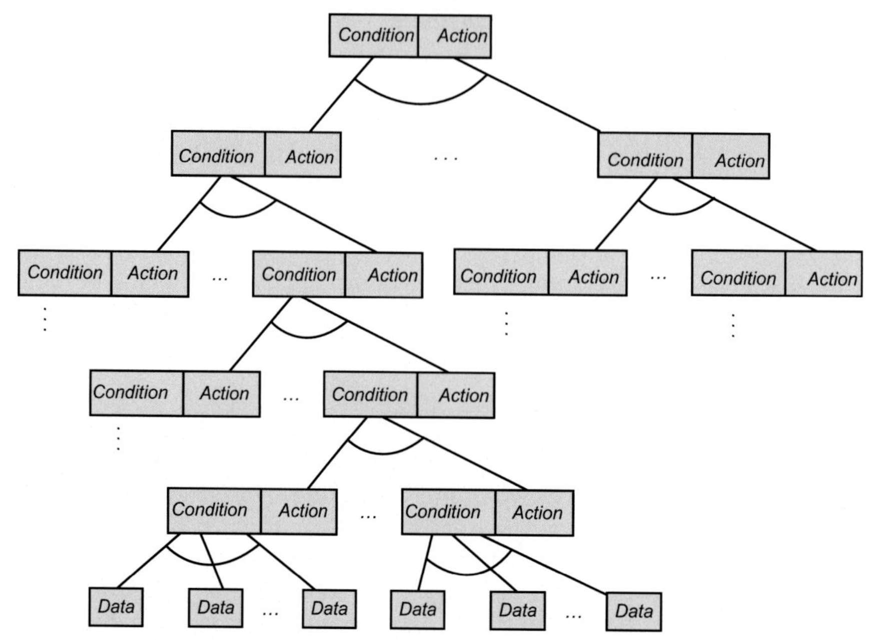

FIG. 2.6 Rule-based systems. In backward-chaining mode, a top-level goal is to perform an action indicated by the action part of the primary rule (e.g., make a diagnosis, select a treatment). This is possible if the condition part of the rule is satisfied. To do so, the inference engine looks for rules whose action parts in Boolean combination (indicated by the arc) satisfy the condition. Rules are invoked recursively in this manner until data are found or not found to satisfy a rule. If all the conditions in a chain are ultimately satisfied by data, the goal is established, otherwise not. In forward-chaining mode, rules are fired if data exist that satisfy their conditions. Higher level rules whose condition parts depend on the actions are evaluated, recursively, until a top-level goal is established. There may be more than one top-level goal, and none, one, or more can be satisfied by the data.

Other heuristic modeling approaches

Another modeling approach that generated considerable interest is the use of *frames*. As defined by Minsky [94], a frame is a data structure for representing a model of a stereotyped situation, represented as a network of nodes and relations, that pertain to various aspects of the situation, such as how to use the frame, expectations for the situation, and what to do about these expectations when they are satisfied. In medicine, we can think of frames as collecting attributes about diseases, patients, or other entities. We can then create relations among frames, such as ones portraying descriptive, causal, temporal, part/whole, or other relationships, and carry out reasoning processes by traversing the relations.

A frame-based representation was used in the 1970s in the Present Illness Program of Pauker et al. [95] and the Internist-1 program of Miller et al. [96], later refined and further developed by Miller et al. as QMR [97]. The general reasoning approach was to rank differential diagnoses by attempting to match a patient's findings with those of stored profiles of diseases. A heuristic scoring rule was used, which weighted the frequency of a particular finding in a given disease, the evoking strength of a finding (i.e., how strongly its presence should raise suspicion of the disease's presence), and the importance of the disease. The first measure is similar, in a qualitative way, to the conditional probability of a finding given a disease, and the second is analogous to a posterior probability of disease, given a finding, although developers of both systems explicitly avoided considering them as probabilities. The third factor is intended to reduce the likelihood of missing serious diagnoses that otherwise might be ranked low. Internist-1 did a better job when more than one disease coexists, by using a partitioning heuristic that allowed it to develop a set of competing hypotheses.

DXplain, the diagnostic system by Barnett et al. [98] also uses heuristic scoring rules to develop its differential diagnoses. CASNET/Glaucoma, developed beginning in the early 1970s by Kulikowski and colleagues [99] used a causal-association network as a basis for its reasoning. This involved a combination of statistical pattern recognition, inference networks with probabilistic scoring of hypotheses, a conceptual structure to represent disease processes, and a normative set of rules for inferring pathophysiological state from observed patterns of findings, and preferring a treatment based on observed findings. AI/Rheum, developed by Kingsland and Lindberg in the 1980s [100] used a knowledge representation system known as *criteria tables*. Findings or observations that could be entered by users were classified as Major or Minor decision elements for each disease in the knowledge base. The criteria for diagnosing any disease at particular levels of certainty are represented in a criteria table that indicates which observations could be present for the disease, are mandatory, or should cause the diagnosis to be excluded.

Other approaches attempted to develop causal models through deep reasoning; for example, for electrolyte and acid/base disorders [101], or qualitative simulations (e.g., for nephritic syndrome [102]). Ramoni et al. [103] have argued that with new deep knowledge from basic science, the possibilities for such approaches are now enhanced. A critiquing approach was pursued by Miller and colleagues, in hypertension [104] and in anesthesia management [105], in which clinical actions were proposed by users and then analyzed by computer.

A more recent approach to reasoning is the work of Fox et al. whose Pro*forma* system includes a set of modular components for modeling a decision process such as a guideline and uses a formalism for "argumentation"; that is, for combining arguments for and against a hypothesis, as a way of determining whether to accept the hypothesis [96,97]. This has some similarities to the criteria table approach used in AI/Rheum, described earlier.

A relatively new diagnostic decision support system, known as Isabel [106–108], employs pattern recognition methods to extract key concepts from textual descriptions to create a knowledge base that can be used to provide a differential diagnosis. The system began with a concentration on pediatrics, although it has expanded beyond that, and reportedly has over 25,000 users worldwide, accessed through a stand-alone Web interface (although it can be incorporated into an EHR). Isabel contains medical content on over 6500 diagnoses and heuristic rules such as applicability in particular age or gender groups. Clinical features entered by a user are matched with the knowledge base and filtered by the heuristic rules, and the disease entities thus matched are presented to the user for consideration. The system relies on a commercial concept matching software package known as Autonomy (Autonomy Corp, Cambridge, UK), which uses nonlinear adaptive digital signal processing derived from Bayesian inference and information theory to estimate the probability that a particular document is about a specific subject.

Semantic networks are a way of portraying relations among concepts. A frame-based approach, as described earlier in this subsection, can be equally well portrayed as a semantic net. Each attribute slot in a frame corresponds to a relation from a node in a semantic net to another node, and the slot filler is the value of that attribute, i.e., the concept of the related node. Ontologies are systems of concepts and their relations, and the semantic web is an extension of the World Wide Web to portray relationships among knowledge in the world. While applications in health care are not widespread, this technology is the basis for much work on medical ontologies at the heart of CDS, and semantic web-based knowledge bases are likely to become more prevalent, along with semantic web-based reasoning systems. Clinical terminologies, vocabularies, taxonomies, and ontologies are discussed in Chapter 11.

2.2.4.2 Data mining and machine learning

As discussed in Section 2.2.3.5, health-related data are rapidly expanding from many sources

Typically, databases are either not well structured, or contain so many parameters and dependencies that considerable effort must be made to extract meaningful information from them. Despite these limitations, interest in database prediction has exploded due to the massive growth in the sizes and numbers of variables in data repositories now available, stimulated in large part by the advances in molecular biology, and by genomics, in particular. Statistical techniques such as regression and nearest neighbor, as well as newer nonlinear techniques and use of fuzzy logic, have been refined and improved over many years, and are mainstays of database prediction. But statistical techniques are largely hypothesis-driven, with the aim of proving or disproving hypotheses, a tedious and time-consuming process. With the huge numbers of variables in some databases, the desirability of shifting the paradigm to one that focuses on examining a large number of features to *discover interesting hypotheses* has led to the pursuit of data mining approaches.

Data mining is a blend of statistical methods, artificial intelligence (AI) techniques, and database design/retrieval approaches [109]. Progress in *data mining* and *machine learning*, or "knowledge discovery from databases" (KDD) has been stimulated by development of AI methods such as *artificial neural networks* (ANNs). The principal tasks are to identify key features that are important for the classification or prediction problem, and to determine the way in which these features should be combined to create an output variable representing the classification or prediction.

We won't discuss the methods in detail here, since they are described in Chapter 7. Also, the methods are generally used to create models such as for classification (e.g., diagnosis), prediction (prognosis or outcome), or optimal choice. The models developed by these methods are then used in CDS.

The history of ANNs is a tortuous one. The notion of an ANN can be traced back to the work by McCulloch and Pitts [110] to model the human brain and cognitive processes by simulating neuronal pathways. As with their early work, subsequent research has continued to involve collaborations among computer scientists and neuroscientists, engineers, and psychologists. There was much excitement in 1958 when Rosenblatt reported on his work with *perceptrons* [111]. Using an input and output layer, and a middle "association layer", a perceptron could learn to associate specific inputs to particular output units. However, in a 1969 book by Minsky and Papert [112], the authors pointed out significant theoretical limitations in single layer perceptrons. Although the authors didn't generalize this limitation to multilayered systems, as discussed further in Chapter 11, algorithms for estimating weights associated with the nodes for such systems did not exist at the time. Their result had so much impact on the field that the enthusiasm for machine learning cooled substantially. Funding for neural network simulation dried up for many years, and work in this area was considered a waste of time. A two-decade period of disenchantment followed, with only minimal work in neuroscience-oriented research on pattern recognition occurring. Despite this, progress was slowly made. In 1974 Werbos developed a learning method based on a back-propagation method for his Harvard PhD thesis [113], and a different threshold function for the artificial neuron than that used in the single-layer perceptron. Progress in ANNs slowly continued over the next two decades. Stimulated by the need for such techniques by the tremendous growth in size of available databases, and demonstrations of success in recent years, the field once again has become robust and applications in a wide range of healthcare contexts are increasingly common [114], as described in Chapter 7.

Two general approaches are used in machine learning, *supervised* and *unsupervised learning*. In supervised learning, the model specifies that one set of features known as inputs will have an effect on another set of features known as outputs; that is, they are presumed to be connected by a causal chain, although other mediating variables may occur between them. In unsupervised learning, the features are all assumed to be at the end of the causal chain, and the task is to discover latent variables that predict them. It is possible, though, for input features and latent variables to be considered in combination as causes of the output features. In supervised learning, the goal is to find the connection between two sets of observations, whereas in unsupervised learning, it is possible to develop larger and more complex models, as well as to deal with the situation in which there is a large causal gap. As the model is built up, it may become easier, at higher levels of abstraction, to bridge the gap.

Besides ANNs, a variety of statistical and AI methods have been explored in various combinations for data mining and machine learning, including linear discriminant analysis, k-nearest neighbor, logistic regression, classification and regression trees, genetic algorithms, support vector machines, and deep learning. In Chapter 7 these methods are reviewed with particular reference to those that have proven most useful for generating knowledge for the purpose of CDS.

An approach that has lent itself to ready integration into clinical practice was the creation of a clinical prediction rule, derived from logistic regression techniques, but which may be reduced to a simple linear scoring rule [115]. A well-known example of this is the "Goldman rule" for evaluation of patients with chest pain in an emergency department to determine whether they should be admitted to the coronary intensive care unit [116]. This was particularly useful in the days when computing the score by hand was necessary; in these days of ubiquitous computers and smartphone, of course, the full logistic regression model can be used instead of the simple scoring rule derived from it.

A further extension of the use of population databases is the growth of population management methodologies, for identifying certain subgroups requiring particular intervention, for example, high risk patients, those who are high utilizers of services, or providers who are not achieving consistent or optimal outcomes. We discuss this growing area in Chapter 16.

2.2.5 Calculations, algorithms, and multistep processes

Many CDS approaches, such as those in Section 2.2.4, involve multistep processes that may include logical or computational processes, probability assessments, heuristic methods, or other methods, and often can be decomposed into those components. The added element that ties them together is an implicit or explicit "execution semantics" that governs the flow of control from one step to another.

In some cases, we can use the *flow chart* as a familiar and convenient representation paradigm for the execution semantics of multistep processes. In a flow chart, the execution is considered to be sequential from one step to the next, except at points at which decision steps indicate alternative choices, where the choice made is based on evaluation of a condition. To enable a flow chart representation to fulfill this role, we posit three features of the flow chart: (1) the processes that can occur at each step can be arbitrary; and (2) the model for choice at a decision step can be arbitrary as well, that is, it can involve any combination of logical, probabilistic and heuristic processes that are appropriate for the decision making task; and (3) any step can be decomposed into subprocesses, represented by their own flow charts. Other elements of flow charts are also needed to cover the semantics, such as iteration and stopping conditions, branching to two or more concurrent steps, and synchronization steps that wait for completion of concurrent steps.

As we consider various types of multistep processes in this section, it should be noted that an explicit representation of the execution semantics, whether by flow chart or other means, is often lacking. The flow of control is embedded in the application providing CDS, and the underlying flow can be inferred only through detailed inspection.

2.2.5.1 *Interactive dialogue and structured data entry control*

The simple flow chart with sequential execution except at decision steps is used much more frequently than may be initially recognized. This model of execution, in fact, underlies (either explicitly or implicitly) many, if not all, interactive dialogues between a user and a computer, in which questions are asked of the user, the responses are evaluated by the computer, and based on logical conditions associated with alternative answers, the computer's next output to the user is determined. This process continues, with further inputs by the user, and evaluations of next steps by the computer, based on logical conditions that determine them, until the computer is able to make a recommendation, or the user ends the session.

The earliest examples include the interactive history taking program developed in the 1960s by Slack and colleagues [117] for interviewing patients about their asthma history, and an editor/driver developed by Swedlow et al. for creating and administering branching question-answer dialogues to gather patient histories from patients [118,119]. Many developments have subsequently focused on the use of question/answer dialogue and form editors for managing human-computer interaction, validating responses, evaluating them to assess conditions for subsequent branching, and thus controlling the flow of execution from the current question/answer element or form to subsequent ones [120–123]. These approaches usually require specification, by a designer, of the underlying control structure, but typically this is done only on a per-step basis, in terms of conditions for branching at each step. Rarely is the linkage of the entire dialogue to an underlying flow chart made explicit, the exception being when a clinical practice guideline is used to drive a user-computer interaction, an example of which was Shiffman's use of a guideline for pediatric outpatient asthma care [124]. Fig. 2.7 depicts an explicit or implicit flow chart that represents the process flow underlying an interactive dialogue.

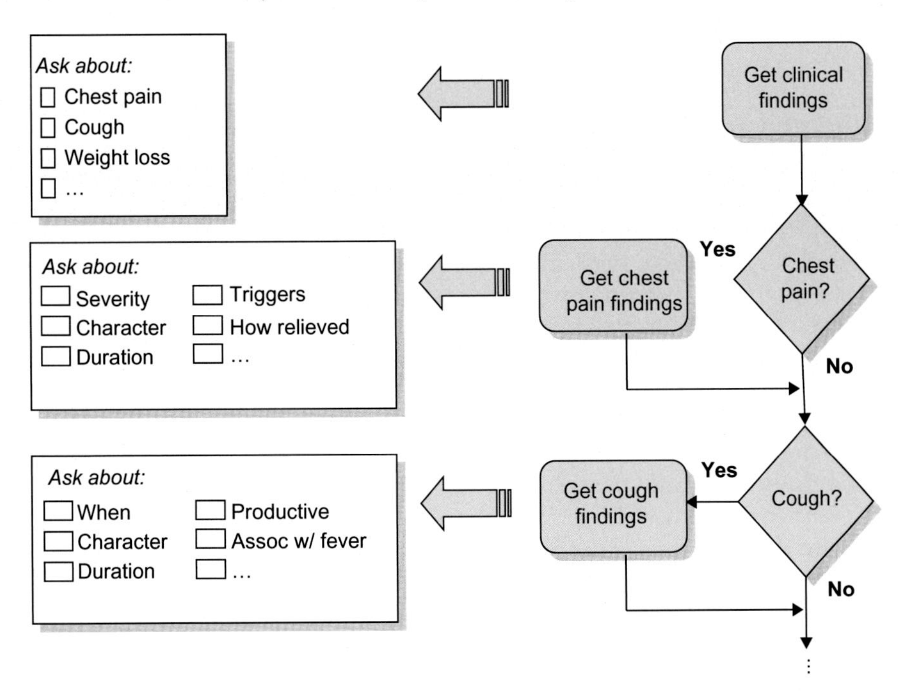

FIG. 2.7 Interactive dialogue guided by an implicit or explicit underlying flow chart or guideline.

The problem-oriented medical record developed by Weed [125], later implemented in computer form as a system called PROMIS [126], and the system for structured data entry for progress notes by Greenes et al. [127] and radiology reports by Bauman et al. [128] are also examples of structured environments for user data entry under implicit control of a system for managing process flow based on evaluation of inputs.

2.2.5.2 *Computer-based consultations*

Algorithms for performing complex tasks involving multiple steps also follow this model. An early acid-base/electrolyte disorder consultation program developed by Bleich [129,130] carried out an interactive dialogue with a physician, collecting data about a patient's electrolyte status, evaluating the data and performing calculations, and making recommendations regarding electrolyte replacement requirements. Although the software incorporated both the algorithm and the user interaction in a single program, the two components can be thought of as separate.

Formal algorithms and heuristic combinations of rules in multistep processes have also been used in many applications aimed at medication dose calculation and adjustment [131,132]. For example, Swartout developed a digitalis therapy advisor with explanation capability [133]. Other programs are used to calculate adjustment of medication dosage for children, the elderly, or those with renal failure [134], or to use pharamacokinetic models to determine optimal dosage regimens [135]. Another application of algorithms, heuristics, and multistep processes is in the control of devices, for example, for intravenous infusion of medications [136] or ventilator management [137,138].

2.2.5.3 Clinical practice guidelines

The use of clinical guidelines as an embodiment of clinical practice can be traced back to the work of Komaroff, Sherman, and Reiffen [139,140] at Beth Israel Hospital in Boston in the early 1970s, during which period a number of structured clinical protocols were developed to guide ambulatory practice, particularly for physician assistants. Another early setting in which clinical practice guidelines were deployed was for monitoring hospital inpatient utilization based on Diagnosis Related Groups (DRGs), in terms of specifying optimal patient care plans or clinical pathways, identifying deviations from these norms, and reviewing them [141,142].

The development of formal methods for specification of clinical practice guidelines was pursued by Margolis [143,144] and Abendroth and Greenes [145] in the 1980s, using a flow chart paradigm, and by Shiffman et al. in the early 1990s [30,31] in terms of the use of decision tables as described in Section 2.2.2. Formal guideline modeling languages with execution semantics were developed in the mid-1990s and subsequently, by many investigators, as reviewed in more detail in Chapter 16. These included work by Lobach et al. [146], Fox [147], Johnson, Purves, and colleagues [148], Shahar and colleagues [149,150]; Terenziani and colleagues [151]; and Tu et al. [152]. Tu and colleagues pursued the idea of modeling clinical guidelines by the construction of formal modular elements that represented the different kinds of tasks and decision processes that a guideline could carry out. The GUIDE system [153] has explored the use of guidelines as a way of specifying and managing workflow in a clinical environment. Other work has been focused on mark up of narrative guidelines to extract computable elements, notably the Guideline Elements Markup (GEM) approach developed by Shiffman et al. [154]. Still other work has focused on the specification of clinical trial protocols, which are essentially similar to clinical practice guidelines but are more prescriptive and include a randomization component that assigns patients to one or another arm of the flow chart [155,156].

The InterMed project in the late 1990s was a joint effort by biomedical informatics groups at Harvard, Columbia, and Stanford to develop a common formalism incorporating the best features of other available modeling languages for clinical guidelines, which could become a basis for interchange and sharing of guidelines. That effort resulted in the Guideline Interchange Format (GLIF), version 2 of which was published in 1998 [157], and version 3, in 2003–2004 [158,159].

Shortly after that time, activities aimed at seeking to create a common guideline modeling formalism were pursued by the Clinical Guidelines Special Interest Group of the HL7 standards development organization, now part of the HL7 Clinical Decision Support Working Group. The issues of clinical practice guideline modeling and standardization are described in Chapter 10. Peleg [160] characterized a number of common features of many guideline languages, in an effort to focus efforts at convergence on a common model. In addition to concerns about interchange and dissemination of guidelines, other foci have been on the issues of integrating guidelines into clinical environments and applications, such as the ATHENA system developed in collaboration between Stanford and the Palo Alto Veterans Administration Hospital [161] and the SAGE project, which was a collaboration between IDX Corporation and Stanford University, Mayo Clinic, University of Nebraska, and Intermountain Healthcare [162]. Currently, activity in HL7 is focused on a FHIR-based Clinical Guidelines model [163].

2.2.5.4 Biomedical signal and image processing

Another category of application of multistep processes is the analysis of biomedical signals and images. We will not explore this important but specialized category of decision support in this book. As we have noted in Section 1.1.6, subtopic 1, our reason for this relates to the highly specialized nature of these applications and their niche areas of focus—even though such specialized use has paradoxically enabled them to have had considerable impact (within those niches). Suffice it to say, however, that signal and image processing has had a very long history. One of the earliest applications was in the realm of ECG interpretation, which focused in particular on approaches to signal analysis of the ECG tracing. Another area of long interest that has become an increasingly important topic has been the pattern recognition and extraction of features from biomedical images (e.g., CT and MRI studies), volumetric modeling, dynamic imaging and computation of flow or metabolic activity or neuronal activation, and use of imaging interactively to guide surgery. Using computers for treatment planning, for example in radiation dosimetry, has also long been of interest and has involved considerable image analysis. These applications may involve a combination of algorithmic and logical operations, probabilistic methods, and heuristic reasoning. More recently, deep learning techniques have been investigated in a wide range of biomedical signal and medical imaging applications [164].

2.2.6 Associative groupings of elements

Associative grouping of elements include strategies for organizing data for workflow optimization. This can occur in a variety of applications, in data entry forms, screens for selecting or customizing an order set, structured reports, flow sheets, and other forms of input or output. We have already discussed in Section 2.2.5 the use of an explicit or implicit underlying guideline to control the sequencing of interaction in human-computer dialogues such as for structured data entry or consultation. Here we concentrate on the strategies for combining data elements into *groups* so that they are presented, viewed, and considered together by a user. In all of these uses, the grouping is guided by an underlying reason, such as similarity of topic, similarity of action required, or priority. For example, the elements in particular sections of a clinical history input form relate to family history of cancer, the cardiovascular findings, or the response to medications. An order set groups orders that relate to a particular situation (e.g., admission to the ICU after hip surgery), and includes orders that have particular intents or purposes, such as vital sign check, pain control, diet, activity restrictions, intravenous fluids, and oxygen, in addition to procedure-specific orders.

Although grouping is generally considered desirable in terms of the convenience it offers in entering data or making selections from a set of choices when arranged in a logical structure, or the clarity of presentation offered by reports organized in such a manner, it is not often thought of as a decision support method. Yet, we maintain that CDS is actually one of the more important benefits that can be achieved by the wise use of grouping. We refer to this as *associative grouping (AG)*, because of its purpose—to represent important associations and to elicit them in the mind of the user. AG can play a subtle but effective role in encouraging best practices simply by reminding the user to think about or include presented elements, or by making the right thing easy to do. The subtlety lies in the fact that the grouping of elements is not explicitly directive. The effectiveness derives from the consequences of using it: grouping elements together makes the desired behavior, that is, consideration of those elements (for either data entry or review purposes), the most convenient and efficient one to do. This is why we consider AG as a form of CDS.

The importance of AG as a CDS method has come to the fore partly as a result of the convergence of three recent trends:

1. Recognition of the large size and scope of effort required by an enterprise to manage its knowledge assets. This has been driven by new imperatives to deliver CDS, and has revealed how much of the knowledge in the enterprise is embedded in the way forms and reports are designed and used.
2. Recognition of the limitations on reuse of executable knowledge content without more precise definition of the data elements used. This has stimulated increased efforts in structured data capture, and thus also in methods for designing forms for that purpose.
3. Recognition of the importance of methods to facilitate cognition, usability, and workflow [165].

Groupings of data elements for presentation typically are described by *templates*. The way the specification of such groupings and associations is embodied in the structure of documents and the status of efforts to develop standards for document architectures and for templates are described in Chapter 12. The knowledge management associated with the organization and categorization of grouped content elements to facilitate best practices and encourage reuse is a relatively new activity, also described in that chapter, for which standardization efforts have not yet begun.

Conditional logic expressions are used in data entry applications for specifying constraints on allowable data types, formats, and ranges; and for specifying conditions for inclusion, omission or modification of certain elements in the presence of others (e.g., omission of pregnancy history items in a male patient).

The origin of document structuring approaches can be traced back to some of the earliest efforts to create forms for structured data entry and programs for report generation from data, and such capabilities are associated with most commercial database systems. In medicine, of course, the capture of structured data and its presentation in reports has faced the difficulty that much of the content does not lend itself to a high degree of structure. The effort to capture data in structured form is often unnatural and time-consuming to users, limiting the collection of data through structure data entry forms. Work in this area was briefly reviewed in Section 2.2.5, in the discussion on interactive dialogues and structured data entry control.

A compromise that often has been adopted is to divide the form into sections relating to various categories of information. Just that simple device provides a context for the data contained within each section that can be used as an aid to retrieval, analysis, and interpretation. The degree of structure can, of course, range from minimal top-level headings to much more detailed divisions into subheadings and sections within them, depending on the application. This is true for both input forms and output documents—there is often a tradeoff between the IT perspective seeking structure down to the level of specific data elements and the user perspective seeking usability and efficiency, and for reports, in terms of conciseness and clarity of presentation.

Reports and other documents produced from structured or semi-structured input typically retain the structure, although the content may be rearranged and presented differently. Sometimes form-based data are presented in reports in tabular or flow sheet format, but the data alternatively may be rendered as prose narrative. The decision regarding mode of presentation should ideally reflect user preferences; for example, the study by Bell and Greenes [166], which demonstrated physician preference to outline mode vs. narrative prose output of ultrasound procedure reports. The structure of the underlying data or sections of it facilitates some degree of manipulation for presentation purposes, and the ability to recast information in different views or for different media or form factors (e.g., printed report vs. display on a desktop monitor vs. smartphone output).

The most important capability for describing documents in a way that allows identification of its structure and its contained elements is the use of *markup languages*; that is, conventions for organizing and tagging sections or elements of a document. This is a complex topic beyond the scope of this book, but since some aspects of its history are relevant, they are briefly highlighted. *Standard Generalized Markup Language (SGML)*, which was developed and standardized by the International Organization for Standards (ISO) in 1986 [167] is widely used to facilitate management of large documents that are revised often and need to be printed in a variety of formats. A first version of *Hypertext Markup Language (HTML)* [168], used to describe pages in the World Wide Web, was specified in 1993 as a way of defining and interpreting tags consistent with SGML rules, although not a strict subset of SGML, and has gone through several updates.

Because the focus of HTML is just on the formatting and appearance of Web pages and not on their content, other capabilities were needed "to meet the needs of large-scale Web content providers for industry-specific markup, vendor-neutral data exchange, media-independent publishing, one-on-one marketing, workflow management in collaborative authoring environments, and the processing of Web documents by intelligent clients." The preceding statement was part of a 1997 press release [169] for the first specification of the *Extensible Markup Language (XML)* by the W3 Consortium. XML [170] is an universal format for structured documents and data on the Web, which is characterized as a subset of SGML. XML has been used extensively in healthcare data standards (e.g., HL7's Clinical Document Architecture) that support use cases such as structured data capturing, order sets, reports.

2.2.7 Visualization

The final category of methods involve the presentation of patient-specific information in intelligent information displays that are deliberately designed to support cognitive processes involved in decision making. Information displays can support a variety of cognitive processes, such as retrieval, sorting, clustering, and highlighting of relevant information; task prioritization; and recognition of patterns such as trends and associations. By automating simple, low-level cognitive tasks (e.g., clustering, sorting), visualization methods allow users to spare scarce cognitive resources such as the working memory and dedicate those resources to complex, high-level cognitive tasks, such as situation awareness and pattern recognition.

The EHR Meaningful Use incentive program led to a dramatic increase in the adoption of commercial EHR systems in the US [171]. Several benefits associated with the adoption of certain EHR functions such as CDS have been reported [172]. Yet, there have been substantial concerns regarding the lack of usability and cognitive support functions in those systems, which have been associated with poor outcomes such as significant increase in clinical documentation time and clinician burnout [173–175]. To address this issue, visualization methods that improve user efficiency and provide support for complex cognitive tasks have received increased attention. For example, among the 10 recommendations proposed by the American Medical Informatics Association (AMIA) EHR-2020 Task Force is to "Improve the designs of interfaces so that they support and build upon how people think (i.e., cognitive-support design)" [176]. A comprehensive description of visualization is provided in a new Chapter (Chapter 14) that has been added to the 3rd edition of this book. Visualization methods are also a critical component in the design and adoption of CDS tools that leverage artificial intelligence (AI) methods, addressing challenges such as the "black box" nature of machine learning methods, which is unacceptable in many healthcare applications, and supporting augmented intelligence in non-closed-looped systems [114]. Visualization methods in AI are covered in Chapters 7 (section on explainable AI) and 14.

2.3 Conclusion

In this chapter, we have reviewed a wide variety of technical approaches that have been explored for developing CDS and related technologies over more than a half century. Some of the approaches have been focused on generating and representing knowledge, others more on how to deliver and use it. In Chapter 1, we explored the many uses of CDS and the components that comprise it. Understanding of these features, together with the methodologies available, will prepare us to more effectively design both CDS capabilities and health care IT environments so that the two can be harmoniously integrated.

Note also, in reference to the diagram of the idealized components of CDS in Fig. 1.3 in Chapter 1, the Execution Engine is the component responsible for carrying out the Method. This is typically done in conjunction with the Knowledge Base, which contains structured data or knowledge needed by the Method, e.g., prior and conditional probabilities for Bayes theorem, the trigger, logical expression, and actions needed for an alert or reminder, the particular order parameters needed for an order set, such as indication and schema for the orders involved, or the retrieval resources needed for an infobutton manager. The form of the Result Specification in the diagram is also dependent on the Method used by the Execution Engine, e.g., a true/false value, a calculated quantity such as a probability or dose, or an order set.

Given a sociocultural imperative to implement robust, widespread CDS to foster health care safety, quality, and efficacy, yet recognizing the slow progress to date, we next set about to describe what will be necessary to accomplish this. The ability to integrate CDS and other knowledge resources into workflow and to support the process flow and cognitive needs of the user depends on an understanding of the organization, its policies and governance, individual responsibilities, coordination among individuals, and the underlying technology, including the EHR platform as well as other resources. These have evolved substantially over many years. Chapter 3 describes where we have been, how we got to the present, and what the current opportunities and challenges are. Chapter 30 describes what we anticipate for the future, and how to prepare for it.

References

[1] Smith C. An evolution of experts: MEDLINE in the library school. J Med Libr Assoc 2005;93:53–60.

[2] Coletti MH, Bleich HL. Medical subject headings used to search the biomedical literature. J Am Med Inf Assoc 2001;8:317–23.

[3] Joubert M, Peretti AL, Gouvernet J, Fieschi M. Refinement of an automatic method for indexing medical literature—a preliminary study. Stud Health Technol Inf 2005;116:683–8.

[4] Yang Y. An evaluation of statistical approaches to MEDLINE indexing. Proc AMIA Annu Fall Symp 1996;358–62.

[5] Aronson AR, Mork JG, Gay CW, Humphrey SM, Rogers WJ. The NLM indexing initiative's medical text indexer. Medinfo 2004;11:268–72.

[6] Wahle M, Widdows D, Herskovic JR, Bernstam EV, Cohen T. Deterministic binary vectors for efficient automated indexing of MEDLINE/PubMed abstracts. AMIA Annu Symp Proc 2012;2012:940–9.

[7] Huang M, Neveol A, Lu Z. Recommending MeSH terms for annotating biomedical articles. J Am Med Inf Assoc 2011;18:660–7. https://doi.org/10.1136/amiajnl-2010-000055.

[8] Dai S, You R, Lu Z, Huang X, Mamitsuka H, Zhu S. FullMeSH: improving large-scale MeSH indexing with full text. Bioinformatics 2020;36:1533–41. https://doi.org/10.1093/bioinformatics/btz756.

[9] You R, Liu Y, Mamitsuka H, Zhu S. BERTMeSH: deep contextual representation learning for large-scale high-performance MeSH indexing with full text. Bioinformatics 2021;37:684–92. https://doi.org/10.1093/bioinformatics/btaa837.

[10] Peng S, Mamitsuka H, Zhu S. MeSHLabeler and DeepMeSH: recent Progress in Large-Scale MeSH Indexing. In: Mamitsuka H, editor. Data mining for systems biology: Methods and protocols. New York, NY: Springer New York; 2018. p. 203–9. https://doi.org/10.1007/978-1-4939-8561-6_15.

[11] Lindberg DA, Humphreys BL, McCray AT. The unified medical language system. Methods Inf Med 1993;32:281–91.

[12] Humphreys BL, Del Fiol G, Xu H. The UMLS knowledge sources at 30: indispensable to current research and applications in biomedical informatics. J Am Med Inform Assoc 2020;27:1499–501. https://doi.org/10.1093/jamia/ocaa208.

[13] McCray AT, Nelson SJ. The representation of meaning in the UMLS. Methods Inf Med 1995;34:193–201.

[14] Hersh WR, Greenes RA. SAPHIRE—an information retrieval system featuring concept matching, automatic indexing, probabilistic retrieval, and hierarchical relationships. Comput Biomed Res 1990;23:410–25.

[15] Hersh WR, Hickam DH. A comparison of two methods for indexing and retrieval from a full-text medical database. Med Decis Making 1993;13:220–6.

[16] Ebbert JO, Dupras DM, Erwin PJ. Searching the medical literature using PubMed: a tutorial. Mayo Clin Proc 2003;78:87–91. https://doi.org/10.4065/78.1.87.

[17] Del Fiol G, Haug PJ, Cimino JJ, Narus SP, Norlin C, Mitchell JA. Effectiveness of topic-specific infobuttons: a randomized controlled trial. J Am Med Inf Assoc 2008;15:752–9. https://doi.org/10.1197/jamia.M2725.

[18] Del Fiol G, Huser V, Strasberg HR, Maviglia SM, Curtis C, Cimino JJ. Implementations of the HL7 context-aware knowledge retrieval ("infobutton") standard: challenges, strengths, limitations, and uptake. J Biomed Inf 2012;45:726–35. https://doi.org/10.1016/j.jbi.2011.12.006.

[19] Yu H, Cao YG. Automatically extracting information needs from ad hoc clinical questions. AMIA Annu Symp Proc 2008;96–100.

[20] Liu F, Antieau LD, Yu H. Toward automated consumer question answering: automatically separating consumer questions from professional questions in the healthcare domain. J Biomed Inf 2011;44:1032–8. https://doi.org/10.1016/j.jbi.2011.08.008.

[21] High R. The era of cognitive systems: An inside look at IBM Watson and how it works. n.d.
[22] Hoyt RE, Snider D, Thompson C, Mantravadi S. IBM Watson analytics: automating visualization, descriptive, and predictive statistics. JMIR Public Health Surveill 2016;2:e157. https://doi.org/10.2196/publichealth.5810.
[23] Somashekhar S, Sepúlveda M-J, Puglielli S, Norden A, Shortliffe E, Kumar CR, et al. Watson for oncology and breast cancer treatment recommendations: agreement with an expert multidisciplinary tumor board. Ann Oncol 2018;29:418–23.
[24] Piotrkowicz A, Johnson O, Hall G. Finding relevant free-text radiology reports at scale with IBM Watson Content Analytics: a feasibility study in the UK NHS. J Biomed Semant 2019;10:1–9.
[25] Strickland E. IBM Watson, heal thyself: how IBM overpromised and underdelivered on AI health care. IEEE Spectr 2019;56:24–31. https://doi.org/10.1109/MSPEC.2019.8678513.
[26] Schmidt C. MD Anderson breaks with IBM Watson, raising questions about artificial intelligence in oncology. J Natl Cancer Inst 2017;109.
[27] Del Fiol G, Mostafa J, Pu D, Medlin R, Slager S, Jonnalagadda SR, et al. Formative evaluation of a patient-specific clinical knowledge summarization tool. Int J Med Inf 2016;86:126–34. https://doi.org/10.1016/j.ijmedinf.2015.11.006.
[28] Ledley RS, Lusted LB. Reasoning foundations of medical diagnosis; symbolic logic, probability, and value theory aid our understanding of how physicians reason. Science 1959;130:9–21.
[29] Ledley R. Use of computers in biology and medicine. New York: McGraw-Hill; 1965.
[30] Shiffman RN, Greenes RA. Use of augmented decision tables to convert probabilistic data into clinical algorithms for the diagnosis of appendicitis. Proc Annu Symp Comput Appl Med Care 1991;686–90.
[31] Shiffman RN, Greenes RA. Rule set reduction using augmented decision table and semantic subsumption techniques: application to cholesterol guidelines. Proc Annu Symp Comput Appl Med Care 1992;339–43.
[32] Shiffman RN, Leape LL, Greenes RA. Translation of appropriateness criteria into practice guidelines: application of decision table techniques to the RAND criteria for coronary artery bypass graft. Proc Annu Symp Comput Appl Med Care 1993;248–52.
[33] Feinstein AR. "Clinical judgment" revisited: the distraction of quantitative models. Ann Intern Med 1994;120:799–805.
[34] Feinstein A. Clinical judgment. Baltimore, MD: Williams & Wilkins Co.; 1967.
[35] Chakravarthy S. Early active database efforts: a capsule summary. IEEE Trans Knowl Data Eng 1995;7:1008–10.
[36] Bailey J, Poulovassilis A, Wood P. An event-condition-action language for XML. New York, NY: Association for Computing Machinery (ACM); 2002. p. 486–95.
[37] McDonald CJ. Protocol-based computer reminders, the quality of care and the non-perfectability of man. N Engl J Med 1976;295:1351–5.
[38] Haug PJ, Gardner RM, Tate KE, Evans RS, East TD, Kuperman G, et al. Decision support in medicine: examples from the HELP system. Comput Biomed Res 1994;27:396–418.
[39] Kuperman GJ, Teich JM, Bates DW, Hiltz FL, Hurley JM, Lee RY, et al. Detecting alerts, notifying the physician, and offering action items: a comprehensive alerting system. Proc AMIA Annu Fall Symp 1996;704–8.
[40] Hripcsak G. Arden syntax for medical logic modules. MD Comput 1991;8. 76, 78.
[41] Hripcsak G, Ludemann P, Pryor TA, Wigertz OB, Clayton PD. Rationale for the Arden syntax. Comput Biomed Res 1994;27:291–324.
[42] Jenders RA, Huang H, Hripcsak G, Clayton PD. Evolution of a knowledge base for a clinical decision support system encoded in the Arden syntax. Proc AMIA Symp 1998;558–62.
[43] Gensler H. Introduction to logic. New York: Routledge; 2002.
[44] Zadeh L. Fuzzy sets. Inf Control 1965;8:338–53.
[45] Warner HR, Toronto AF, Veasy LG. Experience with Baye's theorem for computer diagnosis of congenital heart disease. Ann N Y Acad Sci 1964;115:558–67.
[46] Lodwick GS. A probabilistic approach to the diagnosis of bone tumors. Radiol Clin North Am 1965;3:487–97.
[47] deDombal FT. Computer-aided diagnosis and decision-making in the acute abdomen. J R Coll Physicians Lond 1975;9:211–8.
[48] Gorry GA, Barnett GO. Experience with a model of sequential diagnosis. Comput Biomed Res 1968;1:490–507.
[49] Guo D, Lincoln MJ, Haug PJ, Turner CW, Warner HR. Exploring a new best information algorithm for Iliad. Proc Annu Symp Comput Appl Med Care 1991;624–8.
[50] Warner Jr HR. Iliad: moving medical decision-making into new frontiers. Methods Inf Med 1989;28:370–2.
[51] Raiffa H. Decision analysis: Introductory readings on choices under uncertainty. New York: McGraw Hill; 1997.

[52] Schwartz WB, Gorry GA, Kassirer JP, Essig A. Decision analysis and clinical judgment. Am J Med 1973;55:459–72.
[53] Pauker SG. Coronary artery surgery: the use of decision analysis. Ann Intern Med 1976;85:8–18.
[54] Kassirer JP. The principles of clinical decision making: an introduction to decision analysis. Yale J Biol Med 1976;49:149–64.
[55] Pauker SG, Kassirer JP. Clinical application of decision analysis: a detailed illustration. Semin Nucl Med 1978;8:324–35.
[56] Plante DA, Kassirer JP, Zarin DA, Pauker SG. Clinical decision consultation service. Am J Med 1986;80:1169–76.
[57] Pauker SG, Kassirer JP. The threshold approach to clinical decision making. N Engl J Med 1980;302:1109–17.
[58] Miyamoto JM, Eraker SA. Parameter estimates for a QALY utility model. Med Decis Mak Int J Soc Med Decis Mak 1985;5:191–213. https://doi.org/10.1177/0272989X8500500208.
[59] Smith A. Qualms about QALYs. Lancet Lond Engl 1987;1:1134–6. https://doi.org/10.1016/s0140-6736(87)91685-0.
[60] Weinstein MC. A QALY is a QALY—or is it? J Health Econ 1988;7:289–90. https://doi.org/10.1016/0167-6296(88)90030-6.
[61] Beck JR, Pauker SG. The Markov process in medical prognosis. Med Decis Making 1983;3:419–58.
[62] Pauker SG, Kassirer JP. Decision analysis. N Engl J Med 1987;316:250–8.
[63] Beck JR, Pauker SG, Gottlieb JE, Klein K, Kassirer JP. A convenient approximation of life expectancy (the "DEALE"). II. Use in medical decision-making. Am J Med 1982;73:889–97.
[64] Weinstein MC. Cost-effectiveness analysis for clinical procedures in oncology. Bull Cancer 1980;67:491–500.
[65] Weinstein MC. Methodologic issues in policy modeling for cardiovascular disease. J Am Coll Cardiol 1989;14:38A–43A.
[66] Pauker SG, McNeil BJ. Impact of patient preferences on the selection of therapy. J Chronic Dis 1981;34:77–86.
[67] Eraker SA, Politser P. How decisions are reached: physician and patient. Ann Intern Med 1982;97:262–8.
[68] Fortin JM, Hirota LK, Bond BE, O'Connor AM, Col NF. Identifying patient preferences for communicating risk estimates: a descriptive pilot study. BMC Med Inf Decis Mak 2001;1:2.
[69] Col NF, Eckman MH, Karas RH, Pauker SG, Goldberg RJ, Ross EM, et al. Patient-specific decisions about hormone replacement therapy in postmenopausal women. JAMA 1997;277:1140–7.
[70] Pearl J. Probabilistic reasoning in intelligent systems: Networks of plausible inference. San Francisco, CA: Morgan Kaufmann; 1988.
[71] Cooper GF. A diagnostic method that uses causal knowledge and linear programming in the application of Bayes' formula. Comput Methods Programs Biomed 1986;22:223–37.
[72] Ogunyemi O, Chlebowski R, Matloff E, Schnabel F, Orr R, Col N. Creating Bayesian network models for breast cancer risk prediction. Cancer risk prediction models. A workshop on development, evaluation, and application. National Cancer Institute; 2004.
[73] Metz CE. Basic principles of ROC analysis. Semin Nucl Med 1978;8:283–98.
[74] Swets JA. ROC analysis applied to the evaluation of medical imaging techniques. Invest Radiol 1979;14:109–21.
[75] Begg CB, Greenes RA. Assessment of diagnostic tests when disease verification is subject to selection bias. Biometrics 1983;39:207–15.
[76] McNeil BJ, Hanley JA. Statistical approaches to the analysis of receiver operating characteristic (ROC) curves. Med Decis Making 1984;4:137–50.
[77] Greenes RA, Begg CB. Assessment of diagnostic technologies. Methodology for unbiased estimation from samples of selectively verified patients. Invest Radiol 1985;20:751–6.
[78] Hanley JA. Receiver operating characteristic (ROC) methodology: the state of the art. Crit Rev Diagn Imaging 1989;29:307–35.
[79] Fries JF. The chronic disease data bank: first principles to future directions. J Med Philos 1984;9:161–80.
[80] Dannenberg AL, Shapiro AR, Fries JF. Enhancement of clinical predictive ability by computer consultation. Methods Inf Med 1979;18:10–4.
[81] Bruce B, Fries JF. The Arthritis, Rheumatism and Aging Medical Information System (ARAMIS): still young at 30 years. Clin Exp Rheumatol 2005;23:S163–7.
[82] Rosati RA, McNeer JF, Starmer CF, Mittler BS, Morris Jr JJ, Wallace AG. A new information system for medical practice. Arch Intern Med 1975;135:1017–24.
[83] Starmer CF, Rosati RA, McNeer JF. A comparison of frequency distributions for use in a model for selecting treatment in coronary artery disease. Comput Biomed Res 1974;7:278–93.

[84] Guyatt GH, Sackett DL, Cook DJ. Users' guides to the medical literature. II. How to use an article about therapy or prevention. A. Are the results of the study valid? Evidence-Based Medicine Working Group. JAMA 1993;270:2598–601.
[85] Oxman AD, Sackett DL, Guyatt GH. Users' guides to the medical literature. I. How to get started. The Evidence-Based Medicine Working Group. JAMA 1993;270:2093–5.
[86] Herxheimer A. The Cochrane collaboration: making the results of controlled trials properly accessible. Postgrad Med J 1993;69:867–8.
[87] Shortliffe EH, Davis R, Axline SG, Buchanan BG, Green CC, Cohen SN. Computer-based consultations in clinical therapeutics: explanation and rule acquisition capabilities of the MYCIN system. Comput Biomed Res 1975;8:303–20.
[88] Shortliffe E. Computer-based medical consultations: MYCIN. New York: Elsevier; 1976.
[89] van Melle W, Shortliffe EH, Buchanan BG. EMYCIN: a knowledge engineer's tool for constructing rule-based expert systems. In: Buchanan BG, Shortliffe EH, editors. Rule based expert systems: The Mycin experiments of the Stanford heuristic programming project. Reading, MA: Addison Wesley; 1984. p. 302–13.
[90] Clancey W. Knowledge-based tutoring: The GUIDON program. Cambridge, MA: MIT Press; 1987.
[91] Davis R, Lenat D. Knowledge-based Systems in artificial intelligence: AM and TEIRESIAS. New York: McGraw-Hill; 1982.
[92] Bartels PH, Hiessl H. Expert systems in histopathology. II. Knowledge representation and rule-based systems. Anal Quant Cytol Histol 1989;11:147–53.
[93] Rudowski R, Frostell C, Gill H. A knowledge-based support system for mechanical ventilation of the lungs. The KUSIVAR concept and prototype. Comput Methods Programs Biomed 1989;30:59–70.
[94] Minsky M. A framework for representing knowledge. In: Winston P, editor. The psychology of computer vision. New York: McGraw-Hill; 1975. p. 211–77.
[95] Pauker SG, Gorry GA, Kassirer JP, Schwartz WB. Towards the simulation of clinical cognition. Taking a present illness by computer. Am J Med 1976;60:981–96.
[96] Miller RA, Pople Jr HE, Myers JD. Internist-1, an experimental computer-based diagnostic consultant for general internal medicine. N Engl J Med 1982;307:468–76.
[97] Miller R, Masarie FE, Myers JD. Quick medical reference (QMR) for diagnostic assistance. MD Comput 1986;3:34–48.
[98] Barnett GO, Cimino JJ, Hupp JA, Hoffer EP. DXplain. An evolving diagnostic decision-support system. JAMA 1987;258:67–74.
[99] Kulikowski CA, Weiss SM. Representation of expert knowledge for consultation: the CASNET and EXPERT projects. In: Szolovits P, editor. Artificial intelligence in medicine. Boulder, CO: Westview Press; 1982. p. 21–56.
[100] Kingsland 3rd LC, Lindberg DA, Sharp GC. AI/RHEUM. A consultant system for rheumatology. J Med Syst 1983;7:221–7.
[101] Patil RS. Causal reasoning in computer programs for medical diagnosis. Comput Methods Programs Biomed 1987;25:117–23.
[102] Kuipers B, Kassirer J. Causal reasoning in medicine: analysis of a protocol. Cognit Sci 1984;8:363–85.
[103] Ramoni M, Riva A. Basic science in medical reasoning: An artificial intelligence approach. Adv Health Sci Educ Theory Pract 1997;2:131–40.
[104] Miller PL, Black HR. Medical plan-analysis by computer: critiquing the pharmacologic management of essential hypertension. Comput Biomed Res 1984;17:38–54.
[105] Miller PL. Critiquing anesthetic management: the "ATTENDING" computer system. Anesthesiology 1983;58:362–9.
[106] Ramnarayan P, Tomlinson A, Kulkarni G, Rao A, Britto J. A novel diagnostic aid (ISABEL): development and preliminary evaluation of clinical performance. Medinfo 2004;11:1091–5.
[107] Vardell E, Moore M. Isabel, a clinical decision support system. Med Ref Serv Q 2011;30:158–66. https://doi.org/10.1080/02763869.2011.562800.
[108] Sibbald M, Monteiro S, Sherbino J, LoGiudice A, Friedman C, Norman G. Should electronic differential diagnosis support be used early or late in the diagnostic process? A multicentre experimental study of Isabel. BMJ Qual Amp Saf 2021. https://doi.org/10.1136/bmjqs-2021-013493. bmjqs-2021-013493.
[109] Hill T, Lewicki P. Statistics: Methods and applications. Tulsa, OK: Statsoft, Inc.; 2006.
[110] McCulloch W, Pitts W. A logical calculus of the ideas immanent in nervous activity. Bull Math Biophys 1943;5:115–33.

[111] Rosenblatt F. The perceptron: a probabilistic model for information storage and organization in the brain. Psychol Rev 1958;65:386–408.
[112] Minsky M, Papert S. Perceptrons. MIT Press; 1969.
[113] Werbos P. Beyond regression: New tools for prediction and analysis in the behavioural science. Harvard University; 1974.
[114] Matheny ME, Whicher D, Thadaney IS. Artificial intelligence in health care: a report from the National Academy of Medicine. JAMA 2020;323:509–10. https://doi.org/10.1001/jama.2019.21579.
[115] Wasson JH, Sox HC, Neff RK, Goldman L. Clinical prediction rules. Applications and methodological standards. N Engl J Med 1985;313:793–9.
[116] Lee TH, Juarez G, Cook EF, Weisberg MC, Rouan GW, Brand DA, et al. Ruling out acute myocardial infarction. A prospective multicenter validation of a 12-hour strategy for patients at low risk. N Engl J Med 1991;324:1239–46.
[117] Slack WV, Hicks GP, Reed CE, Van Cura LJ. A computer-based medical-history system. N Engl J Med 1966;274:194–8.
[118] Swedlow DB, Barnett GO, Grossman JH, Souder DE. A simple programming system ("driver") for the creation and execution of an automated medical history. Comput Biomed Res 1972;5:90–8.
[119] Grossman JH, Barnett GO, McGuire MT, Swedlow DB. Evaluation of computer-acquired patient histories. JAMA 1971;215:1286–91.
[120] van Mulligen EM, Stam H, van Ginneken AM. Clinical data entry. Proc AMIA Symp 1998;81–5.
[121] Poon AD, Fagan LM. PEN-Ivory: the design and evaluation of a pen-based computer system for structured data entry. Proc Annu Symp Comput Appl Med Care 1994;447–51.
[122] Kahn Jr CE. A generalized language for platform-independent structured reporting. Methods Inf Med 1997;36:163–71.
[123] Bell DS, Greenes RA, Doubilet P. Form-based clinical input from a structured vocabulary: initial application in ultrasound reporting. Proc Annu Symp Comput Appl Med Care 1992;789–90.
[124] Shiffman RN. Towards effective implementation of a pediatric asthma guideline: integration of decision support and clinical workflow support. Proc Annu Symp Comput Appl Med Care 1994;797–801.
[125] Weed LL. Medical records that guide and teach. N Engl J Med 1968;278:593–600.
[126] Schultz J. A history of the Promis technology: An effective human interface. Palo Alto, CA: ACM Press; 1986. p. 159–82.
[127] Greenes RA, Barnett GO, Klein SW, Robbins A, Prior RE. Recording, retrieval and review of medical data by physician-computer interaction. N Engl J Med 1970;282:307–15.
[128] Bauman R, Pendergrass H, Greenes R, Kalayan R. Further development of an on-line computer system for radiology reporting. DHEW Publication no. (FDA)73-8018; 1972. p. 409–22.
[129] Bleich HL. The computer as a consultant. N Engl J Med 1971;284:141–7.
[130] Bleich HL. Computer evaluation of acid-base disorders. J Clin Invest 1969;48:1689–96.
[131] Walton RT, Harvey E, Dovey S, Freemantle N. Computerised advice on drug dosage to improve prescribing practice; 2001.
[132] Walton R, Dovey S, Harvey E, Freemantle N. Computer support for determining drug dose: systematic review and meta-analysis. BMJ 1999;318:984–90.
[133] Swartout WR. A Digitalis therapy advisor with explanations; 1977. Technical Report TR-176.
[134] Chertow GM, Lee J, Kuperman GJ, Burdick E, Horsky J, Seger DL, et al. Guided medication dosing for inpatients with renal insufficiency. JAMA 2001;286:2839–44.
[135] Jelliffe RW, Schumitzky A, Bayard D, Milman M, Van Guilder M, Wang X, et al. Model-based, goal-oriented, individualised drug therapy. Linkage of population modelling, new "multiple model" dosage design, Bayesian feedback and individualised target goals. Clin Pharmacokinet 1998;34:57–77.
[136] Larsen GY, Parker HB, Cash J, O'Connell M, Grant MC. Standard drug concentrations and smart-pump technology reduce continuous-medication-infusion errors in pediatric patients. Pediatrics 2005;116:e21–5.
[137] Uckun S. Intelligent systems in patient monitoring and therapy management. A survey of research projects. Int J Clin Monit Comput 1994;11:241–53.
[138] Rutledge G, Thomsen G, Farr B, Tovar M, Sheiner L, Fagan L. VentPlan: a ventilator-management advisor. Proc Annu Symp Comput Appl Med Care 1991;869–71.
[139] Komaroff AL, Black WL, Flatley M, Knopp RH, Reiffen B, Sherman H. Protocols for physician assistants. Management of diabetes and hypertension. N Engl J Med 1974;290:307–12.

[140] Sherman H, Komaroff A. Ambulatory care protocols as management tools. Health Care Manage Rev 1976;1:47–52.

[141] Huertas-Portocarrero D, Ruiz PP, Marmol JP. Concurrent clinical review: using microcomputer-based DRG-software. Health Policy 1988;9:211–7.

[142] Tan JK, McCormick E, Sheps SB. Utilization care plans and effective patient data management. Hosp Health Serv Adm 1993;38:81–99.

[143] Gottlieb LK, Margolis CZ, Schoenbaum SC. Clinical practice guidelines at an HMO: development and implementation in a quality improvement model. QRB Qual Rev Bull 1990;16:80–6.

[144] Margolis CZ. Uses of clinical algorithms. JAMA 1983;249:627–32.

[145] Abendroth TW, Greenes RA. Computer presentation of clinical algorithms. MD Comput 1989;6:295–9.

[146] Lobach DF, Gadd CS, Hales JW. Structuring clinical practice guidelines in a relational database model for decision support on the Internet. Proc AMIA Annu Fall Symp 1997;158–62.

[147] Fox J, Johns N, Lyons C, Rahmanzadeh A, Thomson R, Wilson P. PROforma: a general technology for clinical decision support systems. Comput Methods Programs Biomed 1997;54:59–67.

[148] Johnson PD, Tu S, Booth N, Sugden B, Purves IN. Using scenarios in chronic disease management guidelines for primary care. Proc AMIA Symp 2000;389–93.

[149] Shahar Y, Miksch S, Johnson P. An intention-based language for representing clinical guidelines. Proc AMIA Annu Fall Symp 1996;592–6.

[150] Hatsek A, Shahar Y, Taieb-Maimon M, Shalom E, Klimov D, Lunenfeld E. A scalable architecture for incremental specification and maintenance of procedural and declarative clinical decision-support knowledge. Open Med Inf J 2010;4:255–77. https://doi.org/10.2174/1874431101004010255.

[151] Terenziani P, Montani S, Bottrighi A, Molino G, Torchio M. Applying artificial intelligence to clinical guidelines: the GLARE approach. Stud Health Technol Inf 2008;139:273–82.

[152] Tu SW, Musen MA. Modeling data and knowledge in the EON guideline architecture. Medinfo 2001;10:280–4.

[153] Ciccarese P, Caffi E, Quaglini S, Stefanelli M. Architectures and tools for innovative Health Information Systems: the Guide Project. Int J Med Inf 2005;74:553–62.

[154] Shiffman RN, Karras BT, Agrawal A, Chen R, Marenco L, Nath S. GEM: a proposal for a more comprehensive guideline document model using XML. J Am Med Inf Assoc 2000;7:488–98.

[155] Greenes RA, Tu S, Boxwala A, Peleg M, Shortliffe EH. Toward a shared representation of clinical trial protocols: application of the GLIF guideline modeling framework. In: Silva J, Ball M, Chute CG, Douglas J, Langlotz C, Niland J, et al., editors. Cancer informatics: Essential technologies for clinical trials. New York: Springer-Verlag New York; 2002.

[156] Hickam DH, Shortliffe EH, Bischoff MB, Scott AC, Jacobs CD. The treatment advice of a computer-based cancer chemotherapy protocol advisor. Ann Intern Med 1985;103:928–36.

[157] Ohno-Machado L, Gennari JH, Murphy SN, Jain NL, Tu SW, Oliver DE, et al. The guideline interchange format: a model for representing guidelines. J Am Med Inf Assoc 1998;5:357–72.

[158] Boxwala AA, Peleg M, Tu S, Ogunyemi O, Zeng QT, Wang D, et al. GLIF3: a representation format for sharable computer-interpretable clinical practice guidelines. J Biomed Inf 2004;37:147–61. https://doi.org/10.1016/j.jbi.2004.04.002.

[159] Peleg M, Tu S, Bury J, Ciccarese P, Fox J, Greenes RA, et al. Comparing computer-interpretable guideline models: a case-study approach. J Am Med Inf Assoc 2003;10:52–68.

[160] Peleg M. Computer-interpretable clinical guidelines: a methodological review. J Biomed Inf 2013;46:744–63. https://doi.org/10.1016/j.jbi.2013.06.009.

[161] Goldstein MK, Coleman RW, Tu SW, Shankar RD, O'Connor MJ, Musen MA, et al. Translating research into practice: organizational issues in implementing automated decision support for hypertension in three medical centers. J Am Med Inf Assoc 2004;11:368–76.

[162] Tu SW, Musen MA, Shankar R, Campbell J, Hrabak K, McClay J, et al. Modeling guidelines for integration into clinical workflow. Stud Health Technol Inf 2004;107:174–8.

[163] CPG-IG. FHIR clinical guidelines. n.d. http://hl7.org/fhir/uv/cpg/ [Accessed 4 May 2022].

[164] Chan H-P, Samala RK, Hadjiiski LM, Zhou C. Deep learning in medical image analysis. Adv Exp Med Biol 2020;1213:3–21. https://doi.org/10.1007/978-3-030-33128-3_1.

[165] Stead WS, Lin H, editors. Computational technology for effective health care: Immediate steps and strategic directions. Washington, DC: National Research Council; 2009.

[166] Bell DS, Greenes RA. Evaluation of UltraSTAR: performance of a collaborative structured data entry system. Proc Annu Symp Comput Appl Med Care 1994;216–22.
[167] ISO. Information processing—Text and office systems—Standard Generalized Markup Language (SGML). International Organization for Standardization; 1986. http://www.iso.org/iso/en/CatalogueDetailPage.CatalogueDetail?CSNUMBER=16387&ICS1=35&ICS2=240&ICS3=30.
[168] Berners-Lee T, Connolly D. Hypertext markup language (HTML): A representation of textual information and metainformation for retrieval and interchange., 1993, http://www.w3.org/MarkUp/draft-ietf-iiir-html-01.txt. [Accessed 31 October 2017].
[169] W3C. W3C issues XML1.0 as a proposed recommendation., 1997, http://www.w3.org/Press/XML-PR.
[170] W3C. Extensible markup language (XML)., 2006, http://www.w3.org/XML/.
[171] Adler-Milstein J, Jha AK. HITECH Act drove large gains in hospital electronic health record adoption. Health Aff (Millwood) 2017;36:1416–22.
[172] Jones SS, Rudin RS, Perry T, Shekelle PG. Health information technology: an updated systematic review with a focus on meaningful use. Ann Intern Med 2014;160:48–54.
[173] Melnick ER, Dyrbye LN, Sinsky CA, Trockel M, West CP, Nedelec L, et al. The association between perceived electronic health record usability and professional burnout among US physicians. Mayo Clin Proc 2020;95:476–87. https://doi.org/10.1016/j.mayocp.2019.09.024.
[174] Yan Q, Jiang Z, Harbin Z, Tolbert PH, Davies MG. Exploring the relationship between electronic health records and provider burnout: a systematic review. J Am Med Inform Assoc 2021;28:1009–21. https://doi.org/10.1093/jamia/ocab009.
[175] Kroth PJ, Morioka-Douglas N, Veres S, Babbott S, Poplau S, Qeadan F, et al. Association of electronic health record design and use factors with clinician stress and burnout. JAMA Netw Open 2019;2:e199609. https://doi.org/10.1001/jamanetworkopen.2019.9609.
[176] Payne TH, Corley S, Cullen TA, Gandhi TK, Harrington L, Kuperman GJ, et al. Report of the AMIA EHR-2020 Task Force on the status and future direction of EHRs. J Am Med Inform Assoc 2015;22:1102–10. https://doi.org/10.1093/jamia/ocv066.

CHAPTER

3

The journey to broad adoption

Robert A. Greenes[a] *and Guilherme Del Fiol*[b]

[a]**Biomedical Informatics, Arizona State University, Phoenix, AZ, United States** [b]**Department of Biomedical Informatics, University of Utah Health, University of Utah, Salt Lake City, UT, United States**

3.1 The tale of a relationship

Our purpose in this chapter is to address the *third aspect* of Clinical Decision Support (CDS) and Beyond. We addressed *(1) purposes and goals* for CDS (and other knowledge-enhancing capabilities – which we will refer to simply as CDS here) in Chapter 1; and *(2) classes of methods for implementing* various kinds of CDS in Chapter 2. In this chapter, we describe ***(3) the challenges of incorporating CDS into operational environments*** – how to interface with the user, and when and where to do so, to be maximally effective.

This relates to the arrows in our diagram of the components of CDS (see Fig. 1.3), ***between the CDS Module and the host Application Environment (AE)***. The AE is typically the Electronic Health Record (EHR) system, but can be a patient portal, a personal device app, a population health platform, or other host. The arrows relate to how the CDS or knowledge tool is invoked, how information about the AE, context, user, and patient are communicated between the AE and the CDS module, and how the result of the CDS module is acted upon by the AE.

Much of the success and failure of CDS is related to the sociotechnical interplay between the AE, the CDS module, and the AE's users in the context of their work environment. The discussion that follows relates to the nature of this interaction, whether the CDS is truly a separable module or not. Poorly integrated, CDS may occur at the wrong time, be presented to a user who is not able or willing to carry out the decisions supported by the CDS, be presented in a suboptimal format, be insufficiently patient-specific, or interface poorly with the workflow of users of the AE. Much of this is dependent also on the technology available, which has evolved over the decades. Many of the good ideas for CDS came in the earliest years but could not be readily accomplished until many years later – i.e., they were ahead of their time. Newer technologies, such as devices, platforms, and methods of communication (invocation and passing of data) have, in turn, opened up new opportunities.

Clinical Decision Support and Beyond
https://doi.org/10.1016/B978-0-323-91200-6.00025-5

This chapter discusses the current state and how we got here, as well as the current opportunities and challenges. At the end of the book, in Chapter 30, we look ahead at where we are headed, in terms of a vision of the future, relating in part to the ACTS stakeholders vision [1] introduced in Chapter 1.

As we wrote in Chapter 1, we use a somewhat contrived metaphor of an evolving relationship between computing and humans to characterize the many phases of exploration, implementation, evaluation, and refinement, with repeats, that have taken place over time, as depicted in Table 3.1 (replicating, for convenience, Table 1.3).

TABLE 3.1 Relationship between computers as source of clinical decision support and providers and recipients of health care.

Phase of relationship	Duration (approximate date ranges)	Hallmarks
A long infatuation	1960s–80s	Enthusiasm for clinical decision support, research, new ideas
A troubled courtship	1980s–90s	Successful implementations, evaluations showing benefit, but limited dissemination
Renewed passions	Late 1990s–early 2000s	Internet, the Web, knowledge explosion, safety, and quality agendas, evidence-based medicine (EBM)
Getting the support of the relatives (stakeholders)	2000s–present	National agendas, roll out of electronic health records (EHRs), computer-based provider order entry (CPOE), electronic prescribing (eRx); personal health records and patient portal access; international initiatives; health information exchange (HIE), and quality measurement and reporting
New parties to the relationship	Mid-2000s–present	Recognizing need for knowledge management (KM) and other necessary infrastructure; pushes for health information interoperability, standards, CDS marketplace, distributed knowledge resources, third-party innovators
Building the foundations for a lasting relationship: new drivers for adoption: older, wiser, and better equipped to enter into a mature union – and with a new perspective and vitality re-energizing it	Late 2000s–present	Health finance and health system transformation; top-down national initiatives, incentives, and regulations; genomics; focus on wellness, fitness, and prevention; apps and services; mobile health; telehealth; at-home diagnostic testing; biosensors and monitoring; "big data" and analytics; cognitive support and visualization and workflow enhancement initiatives. The Quintuple Aim and the Learning Health System movements.

3.1.1 A long infatuation – 1960s–1980s

Many attractive scenarios have been posited for use of CDS over its more than 60-year history. Consider a doctor seeing a patient with a skin rash. The physician could benefit from a variety of forms of CDS in this setting; for example, he or she could be presented with a set of information about similar eruptions, their differential diagnosis, and suggestions about the next steps for evaluating them. This could be in the form of an atlas of images from a Web site, a textbook reference, an interactive differential diagnosis software, or a clinical practice guideline for skin rashes. The interactive software and practice guideline, if used together such that guideline steps are selected based on patient-specific information entered by the patient or provider, or even by image processing software based on a photo of the patient's lesion, would allow the decision support to be customized for that patient. It could be used to directly make recommendations for treatment or to trigger the display of a set of potential orders for an oral prescription medication, topical remedies, dietary recommendations, and other associated activities; it could perhaps also identify educational materials to be made available to the patient as a paper handout or through the patient portal.

For decision support resources such as those referred to in the preceding example to work optimally, they should be based on high-quality, evidence-based medical knowledge. Further, items of knowledge should ideally be automatically selected or derived based on the clinical context and the particular findings of the patient so that the CDS is as relevant to the user and patient-specific as possible.

3.1.1.1 Simple and complex

The roles and proper uses of CDS have intrigued investigators from the early days [2–4]. The preponderant applications of CDS have been of the more *straightforward* variety we have alluded to already. Computers can be used for information retrieval, by providing search capabilities to find answers to specific clinical questions. They can do very basic error checks, enabling them to be guardians of safety—to detect problems when they occur or to prevent them altogether. A particularly valuable yet simple task is to perform data entry validation, as in the checking of a requested dose in a physician-entered medication order against predefined limits. Another practical and uncomplicated function is to continuously monitor new test results in a clinical laboratory, to detect conditions such as a critically low potassium level that require urgent notification of the patient's care team. Yet another is to identify conditions that should trigger reminders such as for scheduling an annual mammogram in a woman over 50 or for giving a flu shot to an elderly patient in the winter flu season.

More complex uses are also of value. We only sample these, either giving early examples or representative ones. The literature is replete with papers describing many ingenious and effective explorations of the potential for CDS. Among these, the idea of putting the computer to work to help make difficult diagnoses has been especially intriguing from the earliest days of computer use. In fact, from those earliest days up to the present, if one were to ask a layperson how a computer could be most useful for decision support in medicine, chances are that the person would say that it would be for making diagnoses. One of my (RG) own first exposures to CDS in clinical medicine was the seminal paper in *Science* by Ledley and Lusted, published in 1959, entitled "Reasoning Foundations of Medical Diagnosis" [5]. This manuscript explored a combination of logical manipulation and probability, in particular, Bayes theorem, to identify most likely diagnoses given a particular set of findings. Over the ensuing four and a half decades, multiple applications and extensions of the approach have occurred, as well as development and evaluation of a number of alternative models for differential diagnosis [2,3,6,7]. It is interesting to note that in addition to these activities, some of the earliest developments were in the specialized area of electrocardiographic (ECG) diagnosis [4,8], which involved signal processing and analysis of the ECG tracing, whereas the other efforts all dealt with clinical diagnosis requiring entry of findings by a user. As we will discuss further in Section 3.1.2.2, the singularity of focus of ECG analysis and lack of need for human data entry likely contributed to the wider adoption and use of computer-based ECG interpretation from its early days and continuing to today than other clinical diagnostic applications.

Beyond diagnosis, the computer can support a variety of other complex decision-making tasks. Ledley and Lusted's 1959 paper also introduced "value theory", which was one of the precursors of statistical decision analysis [9,10], in which a tree of possible choices and outcomes is constructed, where outcomes are rated according to values (later called "utilities") that summarize relative benefit along some metric like dollars, mortality, or "quality adjusted life years" (QALYs) [11]. The utilities of distal branches are weighted by the probabilities of the various outcomes using Bayes theorem, and the process "folded back" toward the root of the tree, to identify optimal decision choices. Various statistical and artificial intelligence and guideline development projects were carried out in the 1970s and 1980s, to determine optimal workup strategy [12,13] for evaluating a clinical problem (e.g., staging of colon cancer in an elderly man or evaluation of a breast lump in a young woman). Others developed tools to assist in selecting treatment [14], or in evaluating alternative treatment strategies [15]. For treatment itself, the computer was shown to be useful to perform detailed treatment plans, in terms of dose calculations for chemotherapy [16] or detailed 3D modeling and dosimetry calculations for radiation therapy [17]. It could provide estimates of prognosis and risk of complications for alternative treatments [18,19].

In complex decision-making problem areas such as workup, diagnosis, treatment, and long-term management, just the ability to organize and coordinate the sequence of steps for performing various actions, evaluating results, and making choices of next steps is valuable. Thus, interest in decision support in the form of clinical practice guidelines grew particularly in the 1990s–2000s [20–26]. Guidelines could be used to embody best practices, with the hope that their use would improve health care quality, reduce variation, and improve efficiency and workflow.

3.1.1.2 *Evidence of usefulness*

The usefulness of CDS for the kinds of applications just mentioned, as well as many others, has been demonstrated in a number of evaluation studies. Among early developers, three US academic institutions pioneered the practical use of CDS in clinical systems and systematically evaluated them: Brigham and Women's Hospital in Boston, MA; Intermountain Healthcare, UT, and Regenstrief Institute, Indianapolis, IN, primarily with respect to the use of alerts and reminders, drug interaction and dosing checks in computer-based provider order entry (CPOE), and use of order sets to enhance quality and safety. The experiences at the three sites at Regenstrief, Brigham and Women's, and Intermountain were reviewed in Chapters 5–7 respectively, in the Second Edition of this book [27]. Chapter 22 of the current edition highlights the evaluation methods and results from those studies as well as more recent efforts.

Comparative evaluation studies for various types of systems also have been carried out, for example, for differential diagnosis [3,28,29], information retrieval for clinical questions [30–32], clinical guidelines [21,33–35], and CDS in general [36–41]. Success factors for CDS appear to relate to the degree of patient-specificity (with appropriate true positive vs. false positive rate), degree of integration with clinical practice workflow (without excessive demand for additional or extraneous effort by practitioners), and delivery at the point of need (in the appropriate context and format, and at the time it is maximally useful or able to be acted upon).

Typical of studies of CDS effectiveness was a US government-sponsored review of clinical trials of CDS [42], which concluded that it was not possible to show significant impact on health outcomes such as morbidity and mortality. It is not surprising that demonstrating such an effect is an elusive goal, since there is usually significant temporal separation between the time at which a CDS intervention occurs and the realization of a health outcome – and many events and decisions have likely intervened. Nonetheless, a positive effect on process and short-term outcomes could often be shown, like reduction in errors, decreased costs, reduction of readmission rates, increased efficiency, increased compliance with evidence-based practice and the like. Thus, it could be argued that such measures should be considered the best way to evaluate effectiveness of CDS.

3.1.2 A troubled courtship – 1980s–90s

Demonstration of success, by any measure, does not always translate into widespread dissemination and adoption. As found in the early experiences of pioneering academic institutions that first developed and introduced CDS into practice, effective integration of CDS capabilities in clinical information systems is a tortuous process of experimentation and refinement. These institutions have studied CDS extensively to determine what works and what doesn't and have evolved their approaches over the years.

Some of these approaches have made their way into commercial offerings, but many haven't. By and large, the predominant capabilities for CDS that are available in commercial systems are simple alerts and reminders, order sets, infobuttons, and documentation templates – methodologies for which we will discuss briefly in Chapter 2 and in more detail in Section III. However, the consistency and uniformity of adoption are highly variable. Furthermore, usage even in the most advanced settings barely scratches the surface in terms of what is possible to achieve with CDS.

Without concerted policy and financial and regulatory drivers, the adoption of interoperability standards, and deployment of common or shared infrastructure, as have more recently occurred (see Section 3.1.5 of this chapter and Section III of this book), the natural tendency appears to be a slow and piecemeal process of dissemination and adoption of CDS. Reasons are both technical and nontechnical and appear to relate to the complexity of providing CDS. This is true not only for *inherently* complex types of CDS such as differential diagnosis and treatment selection, but even for more simple forms such as alerts and reminders.

In fact, a central thesis of this book is that the difficulty in deploying and disseminating CDS is in large part due to the lack of recognition of how hard the job is and lack of availability of widely adopted interoperability standards, tools and resources to make this job easier. The many reasons for lack of penetration and the requirements for breaking down barriers, in order to move ahead and realize the potential for CDS, are examined in later chapters.

3.1.2.1 Recognizing that even simple CDS is a hard problem

As we have noted, the prospect of providing simple forms of CDS appears deceptively easy; for example, the incorporation of an *if…then* rule for checking whether a laboratory test result exceeds a threshold for abnormality (e.g., IF serum K+ > 5.2 mEq/L THEN notify physician). And, in fact, deploying such a rule in a *single* computer system is relatively straightforward to do.

The problems relate to everything *else* surrounding this act—the necessity of considering not just the single use of the logic in an application at a point in time, but also:

- how the rule is intended to interact with users;
- what kind of user should respond to the output of the rule;
- whether the use of the rule is cost-effective (e.g., is this a critical enough condition to require an interruptive alert);
- how to tune it so that there are not too many false positives yet the appropriate numbers of true positives (e.g., by adjusting the threshold for alerting on abnormal K+ to 5.4 mEq/L, or not firing if a prior K+ was also high and this one is trending downward);
- how the knowledge underlying the rule will be maintained and updated (e.g., if the analytic method or units change, or data about stratifying normal range allow the threshold to be adjusted for different patient characteristics);
- how the rule will be encoded or interfaced with the application that will use it;
- how it relates to other rules and other patient data (e.g., medications the patient is taking);
- how it can be deployed in other applications or in other system platforms (with different programming languages and architectures and developer conventions);
- how knowledge and CDS approaches that are found to be effective can be disseminated and used more broadly, in terms of whether there are sustainable, viable mechanisms for doing this that are supportable by commercial or other means.

In addition, interoperability issues need to be resolved so that data required for the execution of the rule (e.g., all relevant measures of serum potassium) can be retrieved from data sources in the settings in which the rule is deployed.

More broadly, we need to view CDS as a complex sociotechnical problem that entails many detailed questions related to the interaction between humans and the computer. This interaction is in turn reflected in the 5 Rights [43] for CDS – how, when, where, and in what form the AE invokes the CDS, and the extent to which the CDS is specific to the patient. For any rule, for example, who should see the output of the rule, how (e.g., computer screen, text message), where (e.g., lab results, patient overview, multi-patient dashboard at a hospital unit), and when (e.g., upon opening the patient's chart, when result is released from the lab). Answers to these questions may be different based on the clinical context, care setting, etc.

These issues pertain to all forms of CDS that we shall examine, ranging from simple *if…then* rules to the more complex forms of CDS. Going from a single instance of use with demonstration of effectiveness to continued use over time, with the need for maintaining and updating of it, to deployment in more than one setting, and to possible adaptation for other uses are all challenging tasks that must be addressed if the desire is to make best-practice CDS broadly available.

One of the challenges for CDS cited earlier is its dependency on the computer application environment in which it is to be deployed. This relates to more than just hardware and software platform, programming language, and architecture. As noted in Section 3.1.1.2, to be optimally effective, CDS needs to be highly patient-specific and delivered at the point in patient care where it is most appropriate or most likely to be needed. Thus, provider-oriented CDS depends on the existence of EHRs, and on the suite of applications that are available in the EHR system in which CDS can be incorporated. For example, use of CDS for identifying potential drug–drug interactions is most effective if interfaced with CPOE, but if that capability is not available, an alternative way to introduce the CDS may need to be sought, such as incorporating it into an application based in the pharmacy. In this latter case, dangerous interactions are identified before the order is filled rather than at the time of entry of the order; although not optimal, the overall effect may still be positive.

3.1.2.2 *Implications for the technology of CDS*

Given the importance of the interaction between the CDS module (the computational part of CDS) and the AE (the invoker and recipient of the CDS result), we now turn our attention to the major ways in which invoking and receiving CDS and inserting the results into the process flow and workflow of the user are accomplished.

To understand in more detail why CDS deployment on a broad scale is difficult, we will begin by looking at some examples of widely implemented and useful CDS artifacts:

- The **performance of calculations** when needed (e.g., computing the creatinine clearance or adjusting a drug dose in an infant, elderly patient, or patient with renal failure)
- **Evaluation of simple conditional expressions** in order to provide immediate feedback (e.g., in CPOE, a message to the physician when exceeding a recommended drug dose limit or attempting to order a medication in the presence of a recognized interaction with another medication, or an allergy)
- **Triggering the evaluation** of a conditional expression to generate an alert or reminder that is communicated to a provider or patient (e.g., the presence of a critical abnormal lab result)

These forms of CDS all have specific purposes, whether it is to provide alerts of critical values, remind physicians about timely actions, or perform useful calculations. They also all have the five components required as delineated in Fig. 1.3: an inferencing method (e.g., evaluation of a formula or a Boolean conditional expression), a knowledge base (e.g., the formulas or conditional expressions themselves), an information model (the data elements needed), a result specification (e.g., the format of the calculated result of a formula or the possible truth values for conditional expressions), and an application environment, in terms of how the CDS will be interfaced with the users, how it will be triggered or invoked, a way of obtaining data from a host information system or from a user, and a means for providing results to a user.

A common theme for both simple and complex types of CDS has been the set of challenges of providing formal specification, not only for each of these components, but also for creating common approaches and tools for incorporating CDS into clinical IT environments and for adapting the CDS to diverse environments. Desirability of doing so is that it enhances the potential for interoperability and sharing.

3.1.3 Rekindled passions – Late 1990s–early 2000s

Some of the major advances in computer use in health care during the past 6 or more decades have dealt with relatively mundane matters such as approaches to capturing and storing information, communicating it, retrieving it, and producing and distributing reports. These capabilities have greatly reduced transcription errors, improved legibility of reports, eliminated redundancy, facilitated billing and financial functions, and provided a wide variety of other benefits, which indirectly do, of course, affect patient safety, and health care quality, and cost-effectiveness.

Up until the beginning of the 2000s, as mentioned earlier, adoption of CDS was largely piecemeal and often research-driven or based on opportunity or particular local need. Estimates of adoption vary, depending on what one considers to represent an EHR (ranging from the ability to review laboratory and radiology results or other limited functionality to that of systems that include CPOE), and with adoption in the US in the early 2000s inexactly estimated to be from below 10% to close to 40% [44–47], and more recently significantly higher [48]. Rates were higher in some other countries, particularly those with national health care systems, and adoption was already beginning to accelerate in the US due to several initiatives led by the federal government, industry consortia, payer groups, and other stakeholders.

3.1.3.1 *Socioeconomic drivers*

Beginning in the late 1990s, one of the most important drivers for increased adoption and use of the EHR, and of CDS, was a series of studies and reports on medication errors, patient safety, and care quality. This began with a landmark report from the Institute of Medicine (IOM, now known as the National Academy of Medicine (NAM)), *To Err is Human* [49] – and in the lay press, regarding the frequency of preventable medical errors. When this report was released in 1999, containing an estimate that medical errors of all sorts led to close to 98,000 US deaths per year – which was more than the combined deaths from breast cancer and highway accidents – it resulted in sensational headlines in leading newspapers and television.

This finding and the subsequent discussions contributed to a growing sense of desperation and urgency that something needed to be done. The intention of the report, however, was to call attention, not to the errors themselves as much as to patient safety, and how the system needs to provide the ability to ensure safety as part of healthcare quality. Added to this was the recognition of variations in the quality of medical practice—again brought to the forefront by an IOM report, *Crossing the Quality Chasm* [50] and by a continuing series of subsequent IOM/NAM reports; the 12 books are known collectively as the "Quality Chasm" series, further documenting the lack of availability of even basic care to large segments of our population.

Another notable initiative was the "100K Lives Campaign" begun in 2005 by the Institute for Healthcare Improvement,[a] which called for concerted effort by hospital and other health care organizations to adopt six specific approaches to improving health care safety and quality aimed at saving 100,000 lives over an 18-month period (and every 18 months thereafter). Subsequent approaches which targeted specific problems including acute myocardial

[a] http://www.ihi.org, last accessed 5/17/2022.

infarction, adverse drug events, central line infections, surgical site infections, and ventilator-associated pneumonias, and a more generic initiative for rapid response teams at the first sign of patient decline, all relied on the ability to apply best practices relying on specific knowledge at the point of care.

In this context, many factors contributed to the growing recognition of importance of CDS, but root causes appeared to stem from the relentlessly increasing complexity, costs, and constraints on the delivery of health care that were occurring. The stresses and strains included steadily growing demand for medical services generally, especially as a result of aging of the population, with attendant increased frequency and multiplicity of chronic diseases. With the growing complexity of care of patients, more specialists were involved, resulting in an ever-increasing fragmentation of the health care process. The range of diagnostic and therapeutic technologies and medications available were increasing, and becoming more expensive, contributing additional financial strain to a system already under pressure. Doctors were seeing more patients with decreased time available for each patient, and with more paperwork, including that associated with increasing regulatory compliance. Added to this was increase in health care malpractice awards and costs of insurance These trends were accelerating toward the end of the last century, and compounded, of course, by the explosion of biomedical knowledge that occurred in the early genomic (and continuing into the post-genomic) era, the growth of the Internet and ubiquitous computing, and a better informed and demanding public.

3.1.3.2 Evolution of foundational technologies on which to develop CDS

The key foundation on which to enable CDS and other knowledge enhancements is, of course, the **EHR** itself. But particular components of EHR systems are especially important as the means to deliver CDS. During this period of the evolution, these included **CPOE**, electronic prescription writing and communication (electronic prescribing, e-prescribing, or **eRx**), and approaches to enabling patient-centered records, with individuals having access to and contribution of data to their records. Approaches to the latter during this period included the idea of a Personal Health Record (**PHR**), and access to **patient portals** providing access to EHRs.

- **CPOE** is the use of the computer to enable a physician or other provider to enter orders (for medications, procedures, or other actions), thus ensuring that they are legible, so they can be unambiguously stored in the EHR, communicated to pharmacies or other entities responsible for carrying them out, monitored for completion, and billed. CPOE is a valuable foundational platform on which to incorporate a wide variety of CDS capabilities, as we shall explore, but the basic CPOE functionality does not itself include it. The rate of adoption of CPOE in the early 2000s is reviewed elsewhere [51,52] but lagged that of EHR by a considerable degree, especially in ambulatory and community settings where it was almost non-existent. (As we shall discuss in Section 3.1.4, this situation changed dramatically in the US and in some other nations with national policy agendas and incentive programs in the mid-to-late 2000s, perhaps most dramatically by requiring that ambulatory practices include CPOE in their EHR functionality.)
- **eRx** has some features shared with CPOE, but focuses on the entry of prescriptions into a computer, and may include the ability to communicate prescription information to

pharmacy benefit management (PBM) systems for approval of insurance coverage, and to pharmacies for filling of the prescriptions, as well as storing the information in an EHR if available, and printing of hard copy [53–55].

- **Person-controlled records** as a concept encompasses a variety of approaches. Early enthusiasm for **PHRs** as a potential capability for enabling access by patients and the health care public to maintain their own health care records stemmed from a paper by Szolovits and colleagues in 1994 [56] of a "Guardian Angel" that would maintain a **lifetime health record**. This would be as an alternative to, or adjunct that synchronized with, a patient's (possibly multiple) EHRs that typically have narrow, single-enterprise views of patients. The person-controlled longitudinal record was seen as the primary means for hosting the lifetime record of a person [57–59]. A Guardian Angel prototype was implemented in 2001 [60] and implemented as Indivo at Children's Hospital in Boston [61,62]. Yasnoff advocated this model, which he referred to as a "health record bank" [59,63], and it was also supported by the MyDataCan project of Sweeney which leveraged Indivo [64].

  Although PHRs could be valuable in that they potentially provide a single place for viewing all the various aspects of a patient's health and health care, and, importantly, serve as comprehensive substrate for CDS functionality, the PHR has not had traction in the real world. There was considerable hope for this technology in the mid-to-late 2000s, but adoption has been slow and non-standard, because of the lack of a compelling business model or way of sustaining an untethered patient-controlled lifetime PHR [65–68]. Questions of who would pay for it and how to make a viable business out of it caused some ambitious commercial entrants into the field to subsequently pull out.
- **Patient portals** began, in this period, to supplant the notion of a stand-alone PHR as a primary means to allow individuals to access their own data. These are, to an extent, PHRs, but they are "tethered" to a provider-based or medical center-based EHR system, providing browser- or app-based access to an EHR. Portals typically include the ability to view clinical data and to communicate with their care team for such tasks as scheduling, prescription refills, obtaining educational materials, and sending and receiving secure emails. But they only manage the data of the patient associated with that particular medical center/health system or provider, and thus do not fulfill the role of a lifetime health record for a patient. They also typically do not provide a means for tracking of an individual's own health and wellness data, directly, although various tools have been developed to allow patient data from apps and sensors (such as weight, heart rate, blood pressure, and blood glucose) to be incorporated into EHRs, as later development discussed in Section 3.1.6 enables.

3.1.4 Getting the support of the relatives (stakeholders) – 2000s–present

The primary initiatives in the early-mid 2000s were stimulus programs aimed at **broad adoption** of EHRs. Until the 2010s, the smaller community hospitals and office practices relied on computers primarily for financial and billing purposes. As studies such as those cited in Section 3.1.3 have shown, acceptance of the EHR and of clinical IT systems in smaller hospitals and office practices was limited by concerns about cost of implementation and of ongoing support, the technical experience needed, the extra time and effort required to use it, the confusion in the marketplace due to a plethora of nonstandard systems offerings by many vendors, and lack of positive incentives for introducing these systems.

The emphasis by stimulus programs on broad adoption of EHRs was significant in that the **majority of patient encounters** occur in smaller office and community settings. Although estimates of the magnitude of the opportunity are difficult to obtain, it stands to reason that, in the community setting, though acuteness of the problems is typically less, the potential for optimizing care, preventing disease, and avoiding acute problems requiring medical center admission is potentially quite large. A study in the early 2000s [69] projected that ambulatory CPOE in the US, for example, could result in the avoidance of 2 million adverse drug events and 190,000 hospitalizations, and savings of $44 billion per year. In the US, during that period, various legislative proposals, and national and regional campaigns and programs sponsored by professional specialty organizations, health insurers, and employer-based consortia such as the Leapfrog group[b] were promoting adoption of the EHR, CPOE, and related initiatives.

The ONC was established by the US government in April 2004, with the aims of aligning the various stakeholders to provide incentives and support for widespread adoption of the EHR, for establishing a National Health Information Network (NHIN), and for the systems that depend on it, such as CPOE, eRx, and the PHR, and for building national consensus on standards for interoperability of clinical systems, and functionality such as **Health Information Exchange (HIE)** between EHRs, **clinical data repositories**, and **federated architectures for virtually integrating** databases and functionality across systems.

- **HIE** is both an informal term describing what needs to be done to coordinate care among venues, and a designation of a set of initiatives sponsored by the US ONC beginning in the late 2000s to foster the development in each state and territory of the US of a sustainable method of HIE. It is beyond our scope to discuss this in detail, but as of this writing it has had mixed success [70–73]. Although there are hub-based HIE implementations, in which a repository of patient summary data from different sites is able to be accessed by other sites, the most prevalent approach is to provide direct point-to-point communication through a secure email protocol (known as DIRECT) [74]. This can be used, for example, by a primary care provider to send an electronic referral to a specialist, or for exchange of summary data on patients between sites, in the form of a Continuity of Care Document (CCD) based on the HL7 CCDA (Consolidated Clinical Data Architecture) model [75,76].
- **Clinical data repositories** are another approach that harvests data from multiple systems, integrates them, resolving patient identification issues, aggregates them, and maintains them. Some HIEs have this hub-type model. This is also done within health care enterprises that have multiple disparate EHR systems, and in regional and payer sites to provide the data for analytics and predictive modeling.

As noted, a major thrust of HIT efforts since the early 2000s has been to encourage broad adoption of EHRs, and clinical systems and applications based on the EHR that provide various kinds of functionality. This effort was greatly stimulated by various government-led initiatives and incentive plans in several nations. In the UK this primarily took the form of

[b]http://www.leapfroggroup.org, last accessed 6/4/2022.

roll-out of EHRs for general practice and office use, with only secondary attention to adoption in acute care hospitals (see Chapter 5). In the US, adoption was already most prevalent in the large academic medical centers, so attention was focused on universal adoption, including smaller hospitals and office practices.

The most notable initiative in the US was the HITECH Act of 2009 [77,78], which established, among other things, an incentive structure and timetable for hospital and practice adoption of EHRs and **"Meaningful Use" (MU)** of them. Basically, the model was that hospitals and providers receive incentive payments for adoption of EHRs, provided that the systems are certified to provide capabilities for MU and that the users attest to their actual meaningful use of these capabilities. Subsequent benchmarks would convert incentives to reductions in reimbursement for failure to meet them. This began as an iterative process, with criteria for what constitutes MU going through at least 3 stages, beginning in 2012 (stage 1), and progressing in 2014 (stage 2) and 2016 (stage 3), with more expected subsequently. MU criteria included many components, involving patient access to data, health information exchange, use of CPOE, and public health reporting, but for our discussion here, most relevant were the requirements for demonstration of use of CDS and quality measurement and reporting as essential components, with the criteria for compliance on each increasing through the three stages.

As a result of incentives such as the EHR Meaningful Use Program, the adoption of EHR systems increased dramatically. By 2017, over 86% of US physician practices had an EHR and by 2013, 94% of hospitals had a certified EHR [79,80]. Subsequently, there was pushback on the MU criteria and timeline for achieving the benchmarks, resulting in change of regulatory and incentive strategies, discussed in Section 3.1.6.

We also note that the widespread use of EHRs that resulted from these initiatives ushered in a different transition, in which most of the innovation that had been occurring in academic medical centers on largely custom-built EHRs and health IT environments, essentially ended, as the medical centers transitioned to EHRs from commercial providers over the most recent decade or so. A significant consequence of this has been that the opportunities for innovation have shifted from primary development of new functionality in an EHR to the ability to layer new functionality on top of, or interface with, a vendor-provided EHR. We will discuss further the implications of that in Section 3.1.6.

3.1.5 New parties to the relationship – Mid-2000s–present

As we have noted earlier, it is relatively easy to implement a CDS capability directly into a single host IT environment. Nonetheless, if the organization has multiple host environments (commercial or home-grown), the CDS implementation must be redone for every system in which it is to be deployed. Such a situation was rather common in large integrated health care networks that sprung up in the US, beginning in the 1990s, for example, typically involving an academic medical center, several community hospitals, and participating practices, imaging centers, and other facilities, that were acquired or agreed to partner over time, and which often had their own systems prior to integration.

That situation has gradually disappeared with the ascendancy of vendor-provided EHRs in the recent decade, but large enterprises continue to have multiple clinics, hospitals, and connected offices with their own instances of the EHR, even though provided by the same vendor. Because of local workflows, staffing, priorities, and other factors, the individual entities may need to customize or localize the way that CDS modules are inserted into the workflow and interact with the users.

If external knowledge bases are to be used or adapted locally, they must be rendered into a form that is compatible with the local host IT environment(s). If the same knowledge can be used to provide different kinds of CDS in various applications (e.g., a medication dose that should be adjusted based on a lab result, which adjustment might be of importance to order entry as well as to a lab result alerting application), it needs to be recoded or invoked in each of those contexts. If the knowledge is primarily embedded in the applications or in a programming language or other technical specification (e.g., an XML rendition), it is of course not readily transparent or viewable by subject experts, and is difficult to review, maintain, or update when necessary.

Looking at early pioneering health care systems in the US such as at Regenstrief Institute, Brigham and Women's Hospital, and Intermountain Healthcare, mentioned in Section 3.1.1.2, despite successful implementation and positive evaluation of CDS in these local settings, their positive experiences did not result in widespread penetration and usage of the approaches elsewhere. It became clear soon thereafter that a number of infrastructural capabilities must be present for such widespread use, but which largely did not exist.

To a large extent, this situation is still true. Thus, a major theme we explore here is the infrastructure required for successful CDS deployment both locally and more broadly, such as for accepted standards of care – as well as for long-term management and update. The latter has been a relatively new area of focus for clinical informatics (the most recent two decades), in which a few institutions had been taking the lead, but which was deemed less critical by most medical centers and practices, in which only a modest number of CDS alerts, reminders, and order sets were implemented.

3.1.5.1 *Addressing knowledge management and infrastructure*

As knowledge resources and their use continue to grow, we believe that properly addressing "**knowledge management**" (KM) is an essential element necessary to move CDS and Beyond ("CDS&B" – including CDS and other resources for knowledge-enhanced health and healthcare) to a higher level of activity and focus. KM has been of interest in the business community for many years [81], yet until the two decades, it had rarely been addressed with respect to clinical knowledge except in a handful of medical centers pioneering such work. Earliest references are in the late 1980s and early 1990s [82,83].

A comprehensive exposition on the topic of KM in health care is a book by Bali [84]. Citing Gupta [81], Bali points out that "the cornerstone of any KM project is to transform tacit knowledge to explicit knowledge so as to allow its effective dissemination." As we have mentioned, a few US medical centers made early commitments to KM, notably Partners Healthcare In Boston, MA; Mayo Clinic, in Rochester MN; and Intermountain Healthcare, in Salt Lake City, UT, as discussed in detail in our Second Edition [27]. Efforts were also undertaken to create multi-institutional collaborative projects for knowledge management, including development of tools, knowledge bases, and promulgation of standards for knowledge representation [85,86]. Currently, however, many of those efforts are scaled back or no longer active, given that EHR implementation has generally moved from homegrown systems such as those of the pioneers, toward reliance on commercial offerings. Such efforts, also initiated largely in centers with homegrown systems, have also needed to shift to a situation in which most EHRs are vendor-supplied, and the knowledge management needs to be done either using the tools provided by the vendor or using external knowledge management resources that can interface with those EHRs.

Chapter 28 reviews the current approaches to clinical knowledge management, and Chapter 29 discusses ways in which such resources are integrated into EHRs to deliver them in practice. Overall, in this book, we discuss the infrastructure needed to support the whole knowledge resource life cycle, from knowledge generation to validation, obtaining consensus on it, documentation, organization, representation, dissemination, and update. How these functions are done, both on a local scale as well as a larger scale involving professional specialty organizations, and national or even international entities, is critical to the ability to create robust repositories of knowledge. Business models and commercial offerings for sustaining such activities are now becoming available.

The separation of a knowledge base from the engine that delivers it is also highly desirable, because it allows the two to be separately maintained and refined. This is indicated in our idealized diagram of the components of a CDS module (see Fig. 1.3). As we discussed in the opening section of this chapter, a separation of the CDS execution engine from the clinical application environment is also desirable in some circumstances, in that such separation can facilitate dissemination and installation of generic execution engines that are able to communicate through message protocols or Web service interfaces to the clinical information systems with which they interact [87–89]. The benefits of doing this are explored in Chapter 29.

3.1.5.2 Three intersecting and interacting lifecycles

A primary goal of this book is to consider the range of issues and challenges involved in moving from the general concept of CDS&B to implementation, maintenance, dissemination, and update. This involves understanding of the "lifecycles" of CDS and other knowledge-enhancing efforts. Notice that we use the plural here. Three aspects of CDS&B that we have touched on earlier appear to progress as if they have their own lifecycles:

- Knowledge generation, refinement, and update
- Clinical decision support and knowledge enhancement method development and refinement
- Knowledge management and dissemination

These three lifecycles are somewhat interdependent (see Fig. 3.1), but they evolve at different paces, have their own constituencies, and involve separate processes.

Knowledge generation, refinement, and update. The knowledge underlying CDS&B can be generated in a variety of possible ways. These are examined in Section II of this book. In general, the knowledge is initially unstructured and unassembled, or even only implicit, and must be extracted (from experts, from databases, or from the literature), organized and synthesized, analyzed for consistency and accuracy, and represented in an unambiguous form that can be computer-interpretable and acted upon. Any synthesis of knowledge about a topic should have an appropriate expiration date, at which time the sources should be re-reviewed and the knowledge updated if necessary. Thus, each item of knowledge must go through a continuous lifecycle process (see Fig. 3.2). In addition, the introduction of machine learning

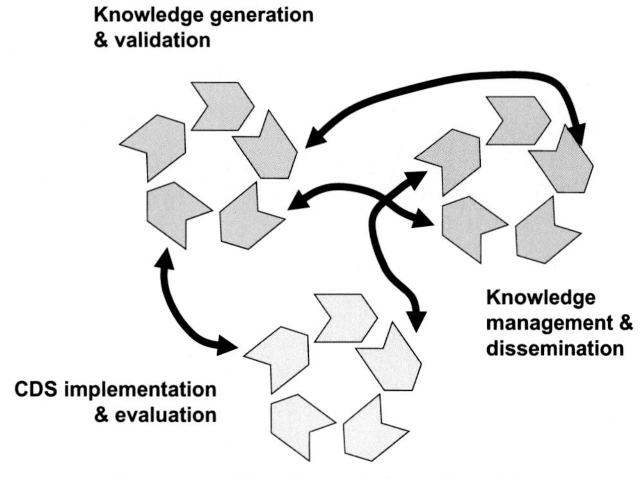

FIG. 3.1 Three intersecting and interacting lifecycles underlying clinical decision support systems (and other knowledge enhanced health/healthcare) technology.

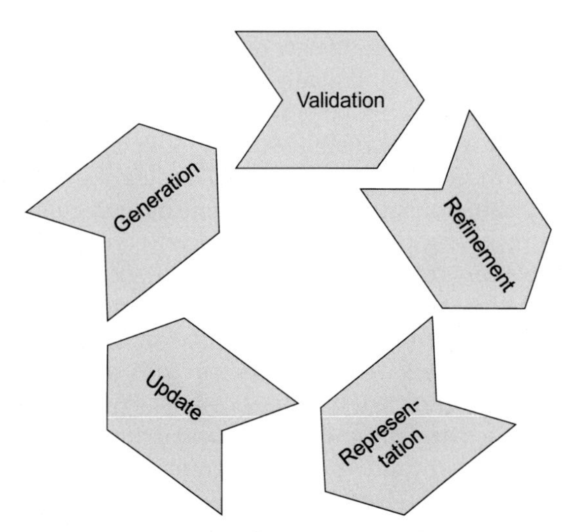

FIG. 3.2 The lifecycle of knowledge generation, refinement, and update.

methods creates further challenges in knowledge refinement and update that can be very difficult to predict, e.g., model accuracy deterioration due to issues such as lack of external validation or prospective evaluation, as well as data drift.

CDS&B method development, implementation, and refinement. This process is a very complicated one. A lifecycle is involved in developing a model for providing CDS or other knowledge enhancement capability (see Fig. 3.3), in terms of the inferencing method and intended decision support/knowledge content delivery approach. The inferencing model and its detailed methodology may evolve over time as nuances of the decision problem or knowledge-usage setting are recognized that require more or different parameters, or more

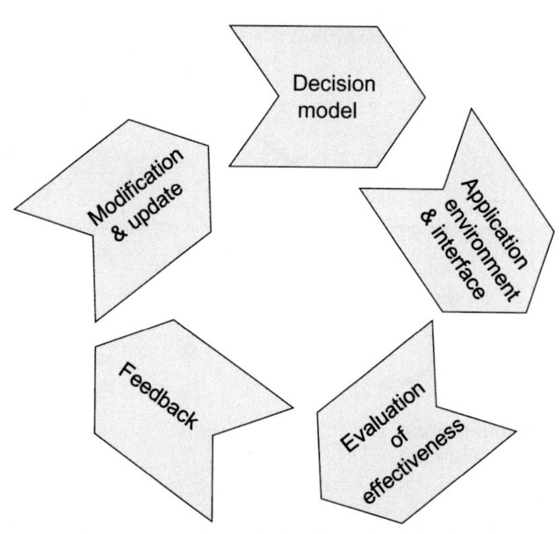

FIG. 3.3 The lifecycle of clinical decision support method development and refinement.

complex computational or logical manipulations, or that require it to reformulate its advice for delivery in different ways to accommodate new needs. The optimal application environments in which the CDS or other knowledge enhancement capability is to be used must be determined, often by experimentation and pilot use, feedback, and refinement. The goals are to learn how best to integrate the knowledge capability into specific clinical settings, and to evaluate effectiveness. This process must continuously iterate. The data sources may evolve, e.g., as data are collected from patients or biosensors, as these become more prevalent for home monitoring, or as natural language processing (NLP) methods become more sophisticated in extracting needed data from narrative text reports and notes in the EHR [90–92].

The representation scheme, mappings, and modes of integration of the knowledge resource usage into host IT systems may evolve over time and must be updated. The methods for delivering CDS or other knowledge enhancement will depend on the availability and adoption of capabilities for integrating it into applications. For example, CPOE is most desirable as a locus for performing drug–drug, drug–lab, and drug–allergy checks to provide real-time feedback to physicians. Alerts and reminders require some kind of event or time trigger to cause the logic to be evaluated. Use of interactive data checks in structured data entry, groupings of knowledge in structured data entry forms and reports (documentation templates) and in order sets, and methods for process and workflow optimization each require appropriate application environments in which to provide their capabilities. All these settings for incorporating knowledge use need to be continually revised and updated as experience is gained with approaches that are successful versus those that are not.

There is considerable interest and activity in delivering CDS as an external service, e.g., through web service calls, as noted above and discussed further in Chapter 29. This has the potential to make it easier to separate this capability from underlying systems, and thus evolve more quickly through competition. Yet to be successful, such services must still require that issues of data source, workflow adaptation, and integration with the user experience be accommodated.

Knowledge content management and dissemination. Although we have considered the lifecycle of individual knowledge content resources earlier, there is another task related to the corpus of knowledge in use in an enterprise and other knowledge that is being prepared for use. Imagine all the knowledge resources incorporated in or invoked by various applications throughout an enterprise, as indicated in Fig. 3.4. A subject expert in diabetes now wants to have the institution provide a set of checks and reminders for compliance with quality-oriented guidelines, such as periodic testing of a patient's HbA1c, eye examination, and foot examination. It is important not only to decide what the guiding knowledge should be, in terms of rules logic, order sets, and structured documentation templates, but how this relates to similar knowledge resources that may already be implemented. What is needed is a means of curation of knowledge resources, to identify those existing items of knowledge pertaining to a topic of interest that are already in use, as an aid to the subject expert in creating new knowledge or refining existing knowledge, to avoid redundancy, to ensure consistency and avoid contradictions, and to recognize gaps where the opportunity for additional use of knowledge, e.g., as CDS, may be needed. Another example may be more refined guidelines for drug prescribing based on genomic testing for responsiveness of various drugs to particular genotypes. This would require that all existing prescribing rules, order sets, and other knowledge resources touching on the involved class of drugs be retrieved and re-examined. It may be useful to have a formal editorial process, with panels of experts, peer review and approval mechanisms in place, in order to accept new knowledge into a system. In short, a resource is needed to facilitate content management and collaborative authoring and review. Once knowledge is implemented in applications, it is necessary to keep track of where it is used, to be able to identify those instances when updates are required. This is part of function of the KM environment described in Chapter 28 in Section V of this book.

Interactions among the lifecycles. As depicted in Fig. 3.1, multiple interactions among the three lifecycles occur, because new knowledge needs call for new CDS or other knowledge usage delivery methods, and the opportunity to deliver new kinds of CDS or knowledge usage call for the availability of new kinds of knowledge; these in turn create the demand for new knowledge management, update, and sharing capabilities; and such capabilities in turn

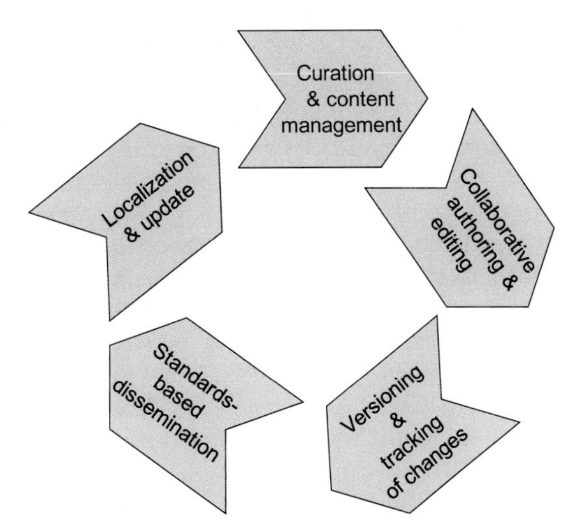

FIG. 3.4 The lifecycle of knowledge management and dissemination.

provide new sources for knowledge to be incorporated in local environments and utilized in practice.

Support for the knowledge generation, knowledge usage implementation and refinement, and knowledge management capabilities and their evolution through lifecycle processes can be expected to be a major part of the organizational, financial, and societal commitment needed to make broad use of high-quality CDS and other knowledge-enhancing capability a reality (see Fig. 3.4). These other aspects are also discussed om Section V of this book.

3.1.5.3 Broader infrastructure initiatives

The above efforts are sufficiently large that even large enterprises won't generally be able to tackle them on their own. One can predict that this will be a major issue going forward, as the needs for CDS&B continue to grow, as the complexity increases, and as the common infrastructure and resources needed for supporting it are increasingly recognized and understood.

On a broad scale, it is desirable to maintain common repositories of computer-interpretable, unambiguous knowledge content (e.g., guidelines, decision rules, order sets, and documentation templates) for use across an enterprise that has multiple information system platforms, or variations of the ways it is implemented within an enterprise among different instances of the same EHR platform – which is now the more common situation. Ideally these should also be linked to original source documents with human-readable recommendations citing evidence-based studies as their basis. A still more ambitious goal would be to have regional, or national (or even international) repositories of knowledge that are maintained and supported by government agencies, insurers, or professional specialty or disease-focused organizations.

For this to be feasible, such knowledge resources should be made available in a common format that is capable of being adapted to different platforms. This requires development and refinement of standards for representation of the knowledge. The notion of reuse would also benefit greatly from tools or standard approaches for adapting the content for different platforms, adapting to local customs or work processes, interfacing it to host patient databases, and perhaps invoking CDS through external services interfaces.

The Clinical Decision Support Consortium [85] is an example of a project that had this goal, during its years of support from 2007 to 2013. Other consortium efforts have arisen in the intervening decade or more. Consortia are well-intentioned but are hard to sustain. At the time of this writing, the way in which communal and national/international efforts will coalesce to provide such support on an ongoing basis is unclear. There is also considerable work in the standards community on methods of knowledge representation and for interchange and sharing of knowledge artifacts, which we explore in Section IV. A noteworthy ongoing activity led by HL7, with support from the US Agency for Healthcare Research and Quality (AHRQ) and the ONC, seeks to develop a common format for sharing of best-practice computable knowledge artifacts. Standards that have resulted from this initiative [93] include the Clinical Quality Language (CQL) and PlanDefinition resources based on the HL7 Fast Health Interoperability Resources (FHIR) standard. In addition, to promote the authoring and sharing of knowledge artifacts, the AHRQ created CDS Connect [94], a public repository of CDS knowledge artifacts represented in these standard formats.

3.1.6 Building the foundations for a lasting relationship: New drivers for adoption – Late 2000s–present

Several unrelated but converging factors have arisen or become more significant beginning in the late 2000s, largely since the publication of our first edition, that collectively greatly expand the need and the opportunities for CDS&B. A number of these factors relate to the transformation of health care systems themselves, which are slowly transitioning under the influence of drivers described in Table 1.4. This trend is most pronounced in the US, perhaps, because of the unsustainability of its ever-increasing costs and inefficiencies, and its uneven and overall suboptimal quality. But the forces leading to it that we will identify below are relentless and ubiquitous, and will alter the landscape throughout the world, albeit at differing rates.

3.1.6.1 A more holistic view of health and healthcare

Returning to our metaphor of a relationship between computer and health care user, the relationship itself is becoming ever more complicated under the trends and forces that are converging, requiring an expanding foundation and ever more alignment and coordination among stakeholders. A transition in health and healthcare, albeit slow and episodic, is underway in the most recent decade to a more proactive, patient-centered, holistic model, in contrast to the prevalent older model of reactive episodic care. Increased emphasis is now being focused on, for example:

- Health and wellness and more proactive management of disease
- Continuity and connectedness of care
- Improved access to healthcare through automated advice and apps as well as telehealth services
- Patient preferences and patient satisfaction
- Social determinants of health.

The rise of a more holistic view of health and healthcare has also gained recognition and acceptance as a result of a variety of trends related to the learning health system [95–97]—and the quintuple aim [98] introduced in Section 1.1.3 of Chapter 1. The prospect of a more holistic approach appears more attainable because of the feasibility of using large databases to track disease trends, patterns of care, outcomes, quality metrics, and disparities in access and use of health resources (see Chapters 4, 7, 16, 25–27). The need for a holistic approach has also been greatly stimulated by the COVID-19 epidemic of the early 2020s, which increased the importance of public health information, personal tracking, and case surveillance [99–101].

The slow process of adoption and relative lack of impact of CDS over the decades began to make it clear, by the early to mid-2000s, that, despite new urgency and enthusiasm, we must not then rush into CDS without providing a suitable foundation. As we have noted, providing even "simple" CDS is a hard problem, if it is to be done well, accepted by users, integrated into practice, and capable of being maintained over time. Being done poorly, creating extra work for providers, being regarded as interrupting work or supplying unneeded or inappropriate warnings (by being insufficiently patient-specific), and causing "alert fatigue" all will serve to create negative responses and make it even harder to achieve broad adoption.

As we noted in the introduction to this chapter, among the five structural components of CDS (see Fig. 1.3), the interfaces between the application environment and CDS modules are most critical yet hardest to implement well. Successful deployment of CDS for provider use requires effective coupling and interaction of all the components with the functions and operations of clinical practice. The clinical data needed must be either entered by a user or obtained through access to an EHR or other data source. The actions to be carried out as a result of CDS need to be communicated to the appropriate entities: if in the form of recommendations, the users must be notified; if in the form of tasks to be performed by the IT system, the target applications must be notified.

For these kinds of interactions to work, the specification of the data elements needed by CDS must be compatible with those in the IT system, and the actions that CDS determines should be performed must be capable of being carried out by the IT system. This means that either the CDS specification must be highly specific to the host IT environment, in order to ensure compatibility, or that an agreed upon means of mapping of clinical IT data to those data parameters needed by CDS must exist, including an agreed-upon mapping between a recommendation or other action determined necessary by CDS and the particular functions the IT system must perform to carry it out, and an unambiguous, functional method of communicating that action.

For acceptance and adoption to be successful, the coupling and interaction of the CDS method's operation and clinical IT systems ideally should be done in a way that does not disrupt the workflow and practice patterns of the intended users, and the CDS capability should be perceived as helpful or actually enhancing workflow and perhaps facilitating the launch of actions that are likely to be needed, i.e., "making the right thing the easy thing to do". Three significant publications in the early to mid-2000s, demonstrating negative consequences resulting from the implementation of CPOE in leading medical centers, highlighted the difficulties in implementing any intervention that requires that providers devote effort to it, the need to understand the nature of their activities, processes and workflow patterns, and the importance of ensuring that the interventions are perceived by them as having a net benefit. The implementation of CPOE at Cedars-Sinai Medical Center, in Los Angeles in 2002, was perceived as too fast and triggered a physician revolt causing the system to be taken down [102]; an implementation at the Hospital of the University of Pennsylvania, in Philadelphia, was found, paradoxically, to increase medication errors [103]; and an implementation at the Children's Hospital at the University of Pittsburgh was found to coincide with an increased incidence of mortality in a neonatal critical care unit [104]. These reports, and the considerable discussions that ensued within the clinical informatics community, both as blogs and published articles, illuminated many reasons why undesirable and unintended consequences can occur, and suggestions about how they might have been avoided (see, for example, published comments on the Koppel et al. report and the authors' response to them [45,105–107]). However, the unfortunate fact is that very little had been published in terms of scientific and rigorous approaches to identifying the right versus wrong ways to carry out such implementations. Although the focus of these three experiences was CPOE, the same issues pertain to CDS, as discussed in Section IV of this book. The understanding of cognitive behavior; the experience of inserting into workflow and evaluation of process as well as impact; and the organizational, management, and governance issues that need to be considered in order to create a positive and constructive approach to implementation are discussed in depth in Section IV. We basically need an "implementation science" to understand how to do this kind of intervention well. This needs to focus on a lot more than the technology – i.e., the psychological, social, cultural, organizational, and business drivers that can either be aligned for success or spell doom for the project.

3.1.6.2 *New models of health care finance*

Perhaps the most significant health/health care trend in the US related to our topic is the rise of new models for value-based health care reimbursement that focus on wellness, disease prevention, and early proactive intervention for disease, as well as reduction of costs of health care by reducing inappropriate or excess use or services. Reimbursement for care in value-based models is basically considered as a capitated (lump-sum) payment for a specific condition, and participants are at risk if cost of care exceeds that amount but share in the revenue if costs are reduced. Capitated payments are not new, since Health Maintenance Organizations widely used this approach in the 1980s and 1990s, but fell into disfavor [108] when it began to be recognized that imposition of policies and procedures were too often dictated more by costs than benefits and were seen as being imposed by the payers rather than by engaging participation of the providers and patients in policy formulation. However, the traditional fee-for-service reimbursement model that has been prevalent in the US and in many countries does not have built-in incentives to optimize health and wellness, or to reduce costs. Quite the opposite occurs, of course, when the quest for information such as by doing another procedure, or the desire to treat have no brakes on them, to force consideration of alternative more cost-effective pathways. Thus, the pay-for-value model, if successful, is revolutionary in its effect on how efforts to promote health and deliver health care are organized and carried out.

Optimization of care under such a model should not be based on local optimization of a practice, hospital, or other care delivery entity, as is the traditional model, because optimal management of care requires longitudinal integrated data across venues, including the home, workplace, medical office, hospital, emergency department, imaging facility, rehabilitation center, and extended care facilities. Safety, quality, and cost-effectiveness of care are of major import, and there should be positive incentives to avoid health care in the first place, by putting primary emphasis on wellness, disease prevention, and active patient self-management of disease and early detection of problems. The Patient-Centered Medical Home (PCMH) [109–111] is an organizational model that focuses on general practice-based coordination of all aspects of a patient's health care. Accountable Care Organizations (ACOs) [109,112–115] are reimbursement models that pay capitated amounts to a practice for management of patients with specific diseases, and it is up to the ACO to determine how best to manage the care in coordination with the patient and among specialists, hospitals, and other participants, so as to achieve best possible outcomes. The practices are at risk for exceeding the budget, but if they are cost-effective, they get to retain unspent funds.

There are many ways to organize a PCMH or to structure the business organization of an ACO, but at bottom, what they require is a change in the basic thinking, with a focus on keeping patients well, early detection of problems, and consideration of the impact on value (defined as quality/cost) of every potential action. They also require an unprecedented degree of knowledge about risks, benefits, and costs, of every potential action, the ability to have as much knowledge about the patient as possible across his/her lifetime and all encounters with the health care system, especially as they pertain to the diseases under management, and coordination of care across the entire spectrum from home care to office encounters to emergency and hospital use, and to extended care facilities [116,117]. Because of the incentives, communication is essential, and traditionally unreimbursed activities such as phone calls or secure email consultation may be highly cost-effective in this model. Because of the need for continuous tracking of patients and the need to proactively assess value of potential interventions, advanced IT infrastructures and integrated databases are called for that are not easily met by existing siloed, enterprise- or practice-focused EHRs, and CDS is essential – focused not only on providers, but on workflow optimization, team coordination, and patient self-care and health promotion.

How quickly this transformation will occur is difficult to estimate, because of its highly disruptive nature and the intense politics and huge financial resources that are involved. But it is already beginning in the US, and in many respects, the changes called for are inevitable. Aspects of this model are already in place in more nationally-driven health systems elsewhere (see Chapter 5).

3.1.6.3 *Science and technology drivers and enablers*

In the context of the socioeconomic changes in the perspective of health and healthcare, a number of capabilities, some within the healthcare system and some external to it, have arisen and matured to the point that they are bringing about new thinking about what is possible to do in the realm of knowledge-enhanced health and healthcare.

Clearly, some science and technology advances are the result of focused initiatives and priorities, whereas others arise, possibly for other purposes, but dramatically change the ways of thinking about current priorities. Some even bring about new possibilities themselves that can then become priorities. Consider the evolution of the smartphone, and the multiple functions such as GPS, camera, flashlight, and information access that we can no longer live without. Some of the advances we describe here are changing the very landscape of possibilities.

- **Precision medicine**

Genomics, proteomics, and the potential for highly refined biomarkers for diagnosis and much more specific characterization of disease through advances in understanding of disease expression and regulation are also leading to advanced diagnostics, prevention and treatment tailored to the individual, especially with the emerging growth of whole genome sequencing as part of routine primary care. These trends lead to the potential for highly individualized care – approaching the N of 1 [118–124], where each person is unique and care can be tailored specifically to that individual. Previously referred to as *personalized medicine*, ***precision medicine*** is now considered the preferred term.

- **Personal data sources**

Biosensors are becoming much more intelligent and capable, and embedded devices for auto-administration of electrical impulses (pacemakers, pain inhibitors), laboratory analysis, or drug infusion are becoming available. There is a huge growth in fitness trackers which measure vital signs as well as other parameters such as activity. The "Quantified Self" movement and so-called "extreme lifelogging" carry this to the ultimate level of tracking everything imaginable [125,126]. This is a group of individuals who track every aspect of their lives, in the belief that this is not only of intrinsic or social value but can perhaps shed light on risks for diseases or status in developing them.

More mainstream is the growth in personal health apps and biometric sensors as data sources. Personal health apps, as just one aspect of a growing mobile health (mHealth) movement, are a relatively new capability, which have evolved from personal fitness/activity tracking to monitoring and capturing inputs from an increasing variety of sensors, including motion, position, environmental conditions, as well as clinical measurements such as blood oxygen level, heart rate, and electrocardiogram. Other sensors can be interfaced, such as for blood pressure, weight, and blood glucose. In addition, the ability to incorporate approved measurements in the patient's EHR is available in some EHR systems, as well as the ability to pull selected information from the EHRs into the personal app [127].

As a result, we can expect enormous growth in the quantity and variety of personal data, not only about health, but all aspects of one's life. These initiatives in combination demand repositories for their storage, and, as a result, have begun to provide a new perspective on ways to achieve the longitudinal personal health record (see Section 3.1.3.2), in terms of data banks, in cloud storage, managed by third parties.

As an aside, which we won't discuss further here, it should be noted that mHealth encompasses a variety of other apps and services available to users of smartphones and tablets [128–131] *– for health and medical information and advice, decision support, health/wellness challenges, gaming, social networking, and other purposes. Many apps are also being developed for providers to be able to access data about their patients and provide decision support and communication/scheduling capabilities while mobile. Because of the low investment and ease of access, this has also become a very promising technology for deployment of health IT capabilities in low and middle-income countries* [132,133] *(see also Chapter 5).*

- **Other health data sources**

We include here, for completeness, a variety of other data sources – all major topics in themselves, which we will also not discuss in further detail. These sources encompass major advances in feature extraction, pattern recognition, and deep learning from clinical images; public health and environmental data such as air quality, exposures, and risks; demographic setting; and social determinants of health.

- **Population databases**

Because of the growth of the above sources of data and others, we have entered the era of "big data" – a term that briefly became a mantra for opportunities to do advanced analytics, predictive modeling, and application to decision making not previously possible, as well as caveats about need for standards and infrastructure and concerns for privacy [134,135]. Data streams are coming from increased EHR adoption, enhanced structure of EHR content, advances in NLP to extract structure from narrative note content in EHRs as discussed in Section 3.1.5.1, adoption of standards for data exchange and interoperability, and new sources such as parameters extracted from imaging, from genomics and other "-omics" analyses, from environmental monitoring, and from growth of biosensors and home monitoring devices.

- **Data analytics-based decision support**

Population databases are essential for individualized care to be achieved, because population analyses or even well-designed clinical trials on subpopulations are not sufficiently granular to optimize care for an individual. We need to be able to rapidly identify the much smaller number of individuals most like our patient to be able to have maximal ability to optimize care. As we discuss in Chapter 16, population data are also needed for identifying categories of patients that are at particular risk, or are high utilizers of care, or providers that are outliers in terms of the costs or outcomes of their care delivery. Further, tracking of health and healthcare disparities and social determinants of disease are needed to adapt interventions to enhance equity and access. Also, as examined in Chapter 27, public health data are useful as guidance for health and healthcare on an individual basis, demonstrated vividly by the COVID-19 pandemic of the early 2020s.

- **Rise of an "app culture"**

In most activities in our lives, it is clear that we are immersed in an "app culture", in which a plethora of functionalities are available through mobile devices, plug-ins to desktop apps, web apps, etc. Mobile health apps, sensors, and other devices have already been mentioned in the context of personal health records and so-called personal mHealth applications.

The growth of an app culture in health/healthcare is now going considerably beyond this, to address the delivery of care as well, by being incorporated into or interacting with the IT infrastructure and participating as a mode of delivery of health care. Most health care delivery organizations and practices have, as we described earlier, installed or are relying on EHRs, and are connecting them to each other through various forms of health information exchange (see Section 3.1.4). Yet the various EHRs, commercial and otherwise, have been largely incompatible, and there is no truly integrated EHR/longitudinal PHR for covering the lifespan of a patient. Also, many EHR systems are products that have evolved over the past 30 or more years and are highly integrated monolithic systems. The EHR vendors, of course, also have incentives to keep the functionality of the system bundled, and to resist the ability to provide interfaces to add-on components or services by third parties.

Yet the need for integrative data views, advanced decision support, care coordination across care venues, and management of complexity of care are stimulating a growing add-on app industry, as well as app development in some of the major health care delivery organizations, to extract data from underlying EHRs, integrate across several EHRs in their enterprises, and provide additional services. As noted above, an "app" in this context is not just a mobile app, but anything that can be implemented and obtained independently of an integrated system.

One of the ONC-funded initiatives coming out of the 2009 HITECH Act was one of the Strategic Health Advanced Research Project (SHARP) grants to a group at Harvard, called for a project called SMART, aimed at stimulating app store-like marketplace [136]. Coupled with FHIR, a standard called SMART on FHIR [137] provides a complete solution for the interoperability of 3rd party apps with clinical systems. While SMART supports app-level interoperability capabilities such as single-sign and app launching, FHIR provides standard specifications for data access and data representation. Another relevant effort in the standards space is the HL7 CDS Hooks standard. CDS Hooks specifies a set of standard user triggers known as "hooks" (e.g., opening a patient's record, entering an order, scheduling an appointment) that EHR systems can implement to send requests to external CDS Web services. EHR support for SMART on FHIR and CDS Hooks is rapidly increasing as well as CDS applications that leverage these standards [138]. Chapter 15 discusses SMART on FHIR, CDS Hooks, and the role of other standards for CDS.

- **Federated architectures**

Returning to the issue of personal health records and the prospect of a longitudinal, integrated lifetime record specifically, a **federated architecture is** an approach, which does not create a physical data repository or rely on an intermediary such as an HIE service, but rather assembles data on demand by query of multiple sites. When this had been done in the past, it was on an ad hoc (non-standard) basis within enterprises that had disparate EHR systems.

In the early 2010s, a US ONC-sponsored Standards and Interoperability Framework initiative known as Query Health [139] developed models and reference implementations for a more formal standards-based approach. More recently, as part of the shift from the usage-focused Meaningful Use incentive program in the US in the 2010s, usage incentives currently are more focused on technological enablement, in the form of the HL7 US Core FHIR Implementation Guide[c] and the 21st Century Cures Act.[d]

The US Core specifies profiles that define the minimum set of requirements for FHIR data elements, along with RESTful interactions for data retrieval to be supported by EHR systems in the US. The 21st Century Cures Act includes legislation that governs relevant topics for CDS such as prohibiting information blocking by EHR systems; and requiring health care systems to provide patients with access to their data through mechanisms such as 3rd party patient apps. Given growing support of this approach by vendors and users of EHR systems, we can expect the health/healthcare IT environment to be an increasingly federated architecture, further facilitating not only the widespread use of apps, but also enabling a "virtual" integrated record through federated access to multiple data sources.

- **Cognitive modeling**

Another aspect of CDS&B that has gained recent prominence is the recognition of the need for enhanced tools for managing the complexity of care in terms of cognitive models, usability, and workflow optimization needs. Although recognized for many years, need for this was strongly articulated by a US National Research Council 2009 report [140]. Another ONC-funded SHARP project at University of Texas Houston, with multiple external collaborators, focused on this work, and there is a growing cadre of individuals working in the area of human factors, usability, data visualization, and workflow enhancement. Chapter 19 summarizes this realm of activity.

[c]https://www.hl7.org/fhir/us/core, last accessed 6/4/2022.

[d]https://www.federalregister.gov/documents/2020/05/01/2020-07419/21st-century-cures-act-interoperability-information-blocking-and-the-onc-health-it-c, last accessed 6/4/2022.

3.2 Where we go from here

The six decades of on-again, off-again progress through the various phases of the relationship between user and computer/communications environment that we have described in this chapter have, in perspective, resulted in considerable achievement. The seeds of many important ideas began early on, often before technology was able to fully exploit them. Other ideas met resistance for socioeconomic reasons, e.g., competition, regulatory hurdles, lack of financial incentives or barriers.

The most recent two decades, described in Section 3.1.6, have seen many capabilities develop that were not on the horizon or in some cases, even anticipated, in the earlier period. They are significant, in that health/healthcare incorporation and adoption of these capabilities has been steadily gaining momentum.

In Chapter 30, we will describe our vision of how knowledge-enhanced health and healthcare will evolve over the next decade. Much of this will be approaches that address the complexity to truly enable the CDS 5 rights [43]. Setting, context, and activity-awareness will aid participants in health and healthcare with timely access to relevant organization and presentation of data, interpretations, and suggested next activities, by harnessing the multiple resources and developments described in this book.

References

[1] AHRQ evidence-based Care Transformation Support (ACTS). Digital Healthcare Research. n.d. https://digital.ahrq.gov/acts. [Accessed 23 April 2022].
[2] Gorry GA, Barnett GO. Experience with a model of sequential diagnosis. Comput Biomed Res 1968;1:490–507.
[3] deDombal FT. Computer-aided diagnosis and decision-making in the acute abdomen. J R Coll Physicians Lond 1975;9:211–8.
[4] Caceres CA. Electrocardiographic analysis by a computer system. Arch Intern Med 1963;111:196–202.
[5] Ledley RS, Lusted LB. Reasoning foundations of medical diagnosis; symbolic logic, probability, and value theory aid our understanding of how physicians reason. Science 1959;130:9–21.
[6] Warner HR, Toronto AF, Veasy LG. Experience with Baye's theorem for computer diagnosis of congenital heart disease. Ann N Y Acad Sci 1964;115:558–67.
[7] Lodwick GS. A probabilistic approach to the diagnosis of bone tumors. Radiol Clin North Am 1965;3:487–97.
[8] Pipberger HV, Stallmann FW, Yano K, Draper HW. Digital computer analysis of the normal and abnormal electrocardiogram. Prog Cardiovasc Dis 1963;5:378–92.
[9] Pauker SG, Kassirer JP. Decision analysis. N Engl J Med 1987;316:250–8.
[10] Raiffa H. Decision analysis: Introductory readings on choices under uncertainty. New York: McGraw Hill; 1997.
[11] Torrance GW, Feeny D. Utilities and quality-adjusted life years. Int J Technol Assess Health Care 1989;5:559–75.
[12] Greenes RA. Computer-aided diagnostic strategy selection. Radiol Clin North Am 1986;24:105–20.
[13] Richards RJ, Hammitt JK, Tsevat J. Finding the optimal multiple-test strategy using a method analogous to logistic regression: the diagnosis of hepatolenticular degeneration (Wilson's disease). Med Decis Mak: An International Journal of the Society for Medical Decision Making 1996;16:367–75.
[14] Shortliffe EH, Davis R, Axline SG, Buchanan BG, Green CC, Cohen SN. Computer-based consultations in clinical therapeutics: explanation and rule acquisition capabilities of the MYCIN system. Comput Biomed Res 1975;8:303–20.
[15] Kassirer JP, Moskowitz AJ, Lau J, Pauker SG. Decision analysis: a progress report. Ann Intern Med 1987;106:275–91.
[16] Knaup P, Wiedemann T, Bachert A, Creutzig U, Haux R, Schilling F. Efficiency and safety of chemotherapy plans for children: CATIPO—a nationwide approach. Artif Intell Med 2002;24:229–42.

[17] Ten Haken RK, Fraass BA, Kessler ML, McShan DL. Aspects of enhanced three-dimensional radiotherapy treatment planning. Bull Cancer 1995;82(Suppl. 5):592s–600s.
[18] Resnic FS, Popma JJ, Ohno-Machado L. Development and evaluation of models to predict death and myocardial infarction following coronary angioplasty and stenting. Proc AMIA Symp 2000;690–3.
[19] Inza I, Merino M, Larranaga P, Quiroga J, Sierra B, Girala M. Feature subset selection by genetic algorithms and estimation of distribution algorithms. A case study in the survival of cirrhotic patients treated with TIPS. Artif Intell Med 2001;23:187–205.
[20] Ohno-Machado L, Gennari JH, Murphy SN, Jain NL, Tu SW, Oliver DE, et al. The guideline interchange format: a model for representing guidelines. J Am Med Inform Assoc 1998;5:357–72.
[21] Shiffman RN, Liaw Y, Brandt CA, Corb GJ. Computer-based guideline implementation systems: a systematic review of functionality and effectiveness. J Am Med Inform Assoc 1999;6:104–14.
[22] Miller PL, Frawley SJ, Sayward FG. Informatics issues in the national dissemination of a computer-based clinical guideline: a case study in childhood immunization. Proc AMIA Symp 2000;580–4.
[23] Bernstam E, Ash N, Peleg M, Tu S, Boxwala AA, Mork P, et al. Guideline classification to assist modeling, authoring, implementation and retrieval. Proc AMIA Symp 2000;66–70.
[24] Greenes RA, Peleg M, Boxwala A, Tu S, Patel V, Shortliffe EH. Sharable computer-based clinical practice guidelines: rationale, obstacles, approaches, and prospects. Medinfo 2001;10:201–5.
[25] Hasman A. Computer-interpretable guidelines. Stud Health Technol Inform 2013;190:3–7.
[26] Peleg M, Shahar Y, Quaglini S. Making healthcare more accessible, better, faster, and cheaper: the MobiGuide project. Eur J EPractic 2013;e20:5–20.
[27] Greenes RA, editor. Clinical decision support: The road to broad adoption. 2nd ed. New York: Elsevier; 2014.
[28] Berner ES, Webster GD, Shugerman AA, Jackson JR, Algina J, Baker AL, et al. Performance of four computer-based diagnostic systems. N Engl J Med 1994;330:1792–6.
[29] Friedman C, Elstein A, Wolf F, Murphy G, Franz T, Fine P, et al. Measuring the quality of diagnostic hypothesis sets for studies of decision support. Medinfo 1998;9(Pt 2):864–8.
[30] Haynes R, McKibbon K, Wilczynski N, Walter S, Werre S. Optimal search strategies for retrieving scientifically strong studies of treatment from MEDLIN. BMJ 2005;330:1179–82.
[31] Hung PW, Johnson SB, Kaufman DR, Mendonca EA. A multi-level model of information seeking in the clinical domain. J Biomed Inform 2008;41:357–70. https://doi.org/10.1016/j.jbi.2007.09.005.
[32] Hunt S, Cimino JJ, Koziol DE. A comparison of clinicians' access to online knowledge resources using two types of information retrieval applications in an academic hospital setting. J Med Libr Assoc 2013;101:26–31. https://doi.org/10.3163/1536-5050.101.1.005.
[33] Sintchenko V, Coiera E, Iredell JR, Gilbert GL. Comparative impact of guidelines, clinical data, and decision support on prescribing decisions: an interactive web experiment with simulated cases. J Am Med Inform Assoc 2004;11:71–7.
[34] Heselmans A, Van de Velde S, Donceel P, Aertgeerts B, Ramaekers D. Effectiveness of electronic guideline-based implementation systems in ambulatory care settings—a systematic review. Implement Sci 2009;4:82. https://doi.org/10.1186/1748-5908-4-82.
[35] Damiani G, Pinnarelli L, Colosimo SC, Almiento R, Sicuro L, Galasso R, et al. The effectiveness of computerized clinical guidelines in the process of care: a systematic review. BMC Health Serv Res 2010;10:2. https://doi.org/10.1186/1472-6963-10-2.
[36] Garg AX, Adhikari NK, McDonald H, Rosas-Arellano MP, Devereaux PJ, Beyene J, et al. Effects of computerized clinical decision support systems on practitioner performance and patient outcomes: a systematic review. JAMA 2005;293:1223–38.
[37] Kawamoto K, Houlihan CA, Balas EA, Lobach DF. Improving clinical practice using clinical decision support systems: a systematic review of trials to identify features critical to success. BMJ 2005;330:765.
[38] Sittig DF, Krall MA, Dykstra RH, Russell A, Chin HL. A survey of factors affecting clinician acceptance of clinical decision support. BMC Med Inform Decis Mak 2006;6:6.
[39] Jaspers MW, Smeulers M, Vermeulen H, Peute LW. Effects of clinical decision-support systems on practitioner performance and patient outcomes: a synthesis of high-quality systematic review findings. J Am Med Inform Assoc 2011;18:327–34. https://doi.org/10.1136/amiajnl-2011-000094.
[40] Cresswell K, Majeed A, Bates DW, Sheikh A. Computerised decision support systems for healthcare professionals: an interpretative review. Inform Prim Care 2012;20:115–28.

[41] Roshanov PS, Fernandes N, Wilczynski JM, Hemens BJ, You JJ, Handler SM, et al. Features of effective computerised clinical decision support systems: meta-regression of 162 randomised trials. BMJ (Clin Res Ed) 2013;346:f657. https://doi.org/10.1136/bmj.f657.
[42] Bright TJ, Wong A, Dhurjati R, Bristow E, Bastian L, Coeytaux RR, et al. Effect of clinical decision-support systems: a systematic review. Ann Intern Med 2012;157:29–43. https://doi.org/10.7326/0003-4819-157-1-201207030-00450.
[43] Osheroff JA. CDS 5 rights. CDS/PI collaborative., 2017, http://bit.ly/cds5rights. [Accessed 21 March 2022].
[44] Ash JS, Bates DW. Factors and forces affecting EHR system adoption: report of a 2004 ACMI discussion. J Am Med Inform Assoc 2005;12:8–12.
[45] Bates DW. Computerized physician order entry and medication errors: finding a balance. J Biomed Inform 2005;38:259–61.
[46] Berner ES, Detmer DE, Simborg D. Will the wave finally break? A brief view of the adoption of electronic medical records in the United States. J Am Med Inform Assoc 2005;12:3–7.
[47] Middleton B, Hammond WE, Brennan PF, Cooper GF. Accelerating U.S. EHR adoption: how to get there from here. recommendations based on the 2004 ACMI retreat. J Am Med Inform Assoc 2005;12:13–9.
[48] Shi Y, Amill-Rosario A, Rudin RS, Fischer SH, Shekelle P, Scanlon D, et al. Health information technology for ambulatory care in health systems. Am J Manag Care 2020;26:32–8. https://doi.org/10.37765/ajmc.2020.42143.
[49] Kohn L, Corrigan J, Donaldson M, Institute of Medicine (US) Committee on Quality of Health Care in America, editors. To err is human: Building a safer health system. Washington, DC: National Academies Press; 1999.
[50] IOM. Crossing the quality chasm: A new health system for the 21st century. Washington, DC: National Academy Press; 2001.
[51] Kaushal R, Shojania KG, Bates DW. Effects of computerized physician order entry and clinical decision support systems on medication safety: a systematic review. Arch Intern Med 2003;163:1409–16.
[52] Ash JS, Gorman PN, Seshadri V, Hersh WR. Computerized physician order entry in U.S. hospitals: results of a 2002 survey. J Am Med Inform Assoc 2004;11:95–9.
[53] Schectman JM, Schorling JB, Nadkarni MM, Voss JD. Determinants of physician use of an ambulatory prescription expert system. Int J Med Inform 2005;74:711–7.
[54] Teich JM, Osheroff JA, Pifer EA, Sittig DF, Jenders RA. Clinical decision support in electronic prescribing: recommendations and an action plan: report of the joint clinical decision support workgroup. J Am Med Inform Assoc 2005;12:365–76.
[55] Wang CJ, Marken RS, Meili RC, Straus JB, Landman AB, Bell DS. Functional characteristics of commercial ambulatory electronic prescribing systems: a field study. J Am Med Inform Assoc 2005;12:346–56.
[56] Szolovits P. Guardian angel: Patient-centered health information systems. Massachusetts Institute of Technology, Laboratory for Computer Science; 1994.
[57] Kimmel Z, Greenes RA, Liederman E. Personal health records. J Med Pract Manage 2005;21:147–52.
[58] Tang PC, Ash JS, Bates DW, Overhage JM, Sands DZ. Personal health records: definitions, benefits, and strategies for overcoming barriers to adoption. J Am Med Inform Assoc 2006;13:121–6.
[59] Yasnoff W. Are health records banks the answer? Health Data Manag 2008;16:23.
[60] Riva A, Mandl KD, Oh DH, Nigrin DJ, Butte A, Szolovits P, et al. The personal internetworked notary and guardian. Int J Med Inform 2001;62:27–40.
[61] Mandl KD, Simons WW, Crawford WC, Abbett JM. Indivo: a personally controlled health record for health information exchange and communication. BMC Med Inform Decis Mak 2007;7:25. https://doi.org/10.1186/1472-6947-7-25.
[62] Weitzman ER, Kaci L, Mandl KD. Acceptability of a personally controlled health record in a community-based setting: implications for policy and design. J Med Internet Res 2009;11:e14. https://doi.org/10.2196/jmir.1187.
[63] Yasnoff WA, Sweeney L, Shortliffe EH. Putting health IT on the path to success. J Am Med Assoc 2013;309:989–90. https://doi.org/10.1001/jama.2013.1474.
[64] Sweeney LA. MyDataCan., 2013, http://mydatacan.org/.
[65] Haggstrom DA, Saleem JJ, Russ AL, Jones J, Russell SA, Chumbler NR. Lessons learned from usability testing of the VA's personal health record. J Am Med Inform Assoc 2011;18(Suppl. 1):i13–7. https://doi.org/10.1136/amiajnl-2010-000082.
[66] Kim J, Jung H, Bates DW. History and trends of "personal health record" research in PubMed. Healthc Inform Res 2011;17:3–17. https://doi.org/10.4258/hir.2011.17.1.3.

[67] Krist AH, Peele E, Woolf SH, Rothemich SF, Loomis JF, Longo DR, et al. Designing a patient-centered personal health record to promote preventive care. BMC Med Inform Decis Mak 2011;11:73. https://doi.org/10.1186/1472-6947-11-73.

[68] Day K, Gu Y. Influencing factors for adopting personal health record (PHR). Stud Health Technol Inform 2012;178:39–44.

[69] Johnston D, Pan E, Middleton B, Walker J, Bates D. The value of computerized provider order entry in ambulatory settings. Boston, MA: Center for Information Technology Leadership; 2003.

[70] Dixon BE, Zafar A, Overhage JM. A framework for evaluating the costs, effort, and value of nationwide health information exchange. J Am Med Inform Assoc 2010;17:295–301. https://doi.org/10.1136/jamia.2009.000570.

[71] Dullabh P, Hovey L. Large scale health information exchange: implementation experiences from five states. Stud Health Technol Inform 2013;192:613–7.

[72] Payne TH, Lovis C, Gutteridge C, Pagliari C, Natarajan S, Yong C, et al. Status of health information exchange: a comparison of six countries. J Glob Health 2019;9:0204279. https://doi.org/10.7189/jogh.09.020427.

[73] Everson J, Butler E. Hospital adoption of multiple health information exchange approaches and information accessibility. J Am Med Inform Assoc 2020;27:577–83. https://doi.org/10.1093/jamia/ocaa003.

[74] National Rural Health Resource Center. Direct guide. n.d.

[75] Richel W. A new approach to clinical interop in stage 2 meaningful use., 2013, http://blogs.gartner.com/wes_rishel/2012/03/19/a-new-approach-to-clinical-interop-in-stage-2-meaningful-use/.

[76] D'Amore JD, Mandel JC, Kreda DA, Swain A, Koromia GA, Sundareswaran S, et al. Are meaningful use stage 2 certified EHRs ready for interoperability? Findings from the SMART C-CDA Collaborative. J Am Med Inform Assoc 2014;21:1060–8. https://doi.org/10.1136/amiajnl-2014-002883.

[77] Pipersburgh J. The push to increase the use of EHR technology by hospitals and physicians in the United States through the HITECH Act and the Medicare incentive program. J Health Care Finance 2011;38:54–78.

[78] Burde H. Health law the hitech act—an overview. Virtual Mentor 2011;13:172–5. https://doi.org/10.1001/virtualmentor.2011.13.3.hlaw1-1103.

[79] Office-based Physician Electronic Health Record Adoption., 2021. [Accessed 13 November 2021], 2021, at https://www.healthit.gov/data/quickstats/office-based-physician-electronic-health-record-adoption. [Accessed 17 May 2022].

[80] Burwell SM. Setting value-based payment goals—HHS efforts to improve U.S. health care. N Engl J Med 2015;372:897–9. https://doi.org/10.1056/NEJMp1500445.

[81] Gupta B, Iyer LS, Aronson J. Knowledge management: practices and challenges. Ind Manag Data Syst 2000;100:17–21.

[82] Greenes RA, Tarabar DB, Krauss M, Anderson G, Wolnik WJ, Cope L, et al. Knowledge management as a decision support method: a diagnostic workup strategy application. Comput Biomed Res 1989;22:113–35.

[83] Chute CG, Cesnik B, van Bemmel JH. Medical data and knowledge management by integrated medical workstations: summary and recommendations. Int J Biomed Comput 1994;34:175–83.

[84] Bali R. Clinical knowledge management: Opportunities and challenges. Hershey, PA: Idea Group Publishing; 2005.

[85] Middleton B. The clinical decision support consortium. Stud Health Technol Inform 2009;150:26–30.

[86] Greenes R, Bloomrosen M, Brown-Connolly NE, Curtis C, Detmer DE, Enberg R, et al. The morningside initiative: collaborative development of a knowledge repository to accelerate adoption of clinical decision support. Open Med Inform J 2010;4:278–90. https://doi.org/10.2174/1874431101004010278.

[87] Kawamoto K, Honey A, Rubin K. The HL7-OMG Healthcare Services Specification Project: motivation, methodology, and deliverables for enabling a semantically interoperable service-oriented architecture for healthcare. J Am Med Inform Assoc 2009;16:874–81. https://doi.org/10.1197/jamia.M3123.

[88] Kawamoto K, Jacobs J, Welch BM, Huser V, Paterno MD, Del Fiol G, et al. Clinical information system services and capabilities desired for scalable, standards-based, service-oriented decision support: consensus assessment of the Health Level 7 clinical decision support Work Group. AMIA Ann Symp Proc 2012;2012:446–55.

[89] Paterno MD, Goldberg HS, Simonaitis L, Dixon BE, Wright A, Rocha BH, et al. Using a service oriented architecture approach to clinical decision support: performance results from two CDS Consortium demonstrations. AMIA Ann Symp Proc 2012;2012:690–8.

[90] Demner-Fushman D, Chapman WW, McDonald CJ. What can natural language processing do for clinical decision support? J Biomed Inform 2009;42:760–72. https://doi.org/10.1016/j.jbi.2009.08.007.

[91] Smith JC, Spann A, McCoy AB, Johnson JA, Arnold DH, Williams DJ, et al. Natural language processing and machine learning to enable clinical decision support for treatment of pediatric pneumonia. AMIA Annu Symp Proc 2020;2020:1130–9.
[92] Doan S, Conway M, Phuong TM, Ohno-Machado L. Natural language processing in biomedicine: a unified system architecture overview. Methods Mol Biol 2014;1168:275–94. https://doi.org/10.1007/978-1-4939-0847-9_16.
[93] Strasberg HR, Rhodes B, Del Fiol G, Jenders RA, Haug PJ, Kawamoto K. Contemporary clinical decision support standards using health level seven international fast healthcare interoperability resources. J Am Med Inform Assoc 2021;28:1796–806. https://doi.org/10.1093/jamia/ocab070.
[94] Lomotan EA, Meadows G, Michaels M, Michel JJ, Miller K. To share is human! Advancing evidence into practice through a national repository of interoperable clinical decision support. Appl Clin Inform 2020;11:112–21. https://doi.org/10.1055/s-0040-1701253.
[95] McGinnis JM, Fineberg HV, Dzau VJ. Advancing the learning health system. N Engl J Med 2021;385:1–5. https://doi.org/10.1056/NEJMp2103872.
[96] Friedman CP, Wong AK, Blumenthal D. Achieving a nationwide learning health system. Sci Transl Med 2010;2:57cm29. https://doi.org/10.1126/scitranslmed.3001456.
[97] IOM. Health IT and patient safety: Building safer systems for better care., 2011, http://www.nationalacademies.org/hmd/Reports/2011/Health-IT-and-Patient-Safety-Building-Safer-Systems-for-Better-Care.aspx. [Accessed 18 April 2017].
[98] Coleman K, Wagner E, Schaefer J, Reid R, LeRoy L. Redefining primary care for the 21st century. Rockville, MD: Agency for Healthcare Research and Quality; 2016. p. 1–20. 16.
[99] Doraiswamy S, Abraham A, Mamtani R, Cheema S. Use of telehealth during the COVID-19 pandemic: scoping review. J Med Internet Res 2020;22:e24087. https://doi.org/10.2196/24087.
[100] Horn DM, Haas JS. Covid-19 and the mandate to redefine preventive care. N Engl J Med 2020;383:1505–7. https://doi.org/10.1056/NEJMp2018749.
[101] Inkster B, O'Brien R, Selby E, Joshi S, Subramanian V, Kadaba M, et al. Digital health management during and beyond the COVID-19 pandemic: opportunities, barriers, and recommendations. JMIR Mental Health 2020;7:e19246. https://doi.org/10.2196/19246.
[102] Shabot MM. Ten commandments for implementing clinical information systems. Proc (Bayl Univ Med Cent) 2004;17:265–9.
[103] Koppel R, Metlay JP, Cohen A, Abaluck B, Localio AR, Kimmel SE, et al. Role of computerized physician order entry systems in facilitating medication errors. JAMA 2005;293:1197–203.
[104] Han YY, Carcillo JA, Venkataraman ST, Clark RS, Watson RS, Nguyen TC, et al. Unexpected increased mortality after implementation of a commercially sold computerized physician order entry system. Pediatrics 2005;116:1506–12.
[105] Horsky J, Zhang J, Patel VL. To err is not entirely human: complex technology and user cognition. J Biomed Inform 2005;38:264–6.
[106] Koppel R, Localio AR, Cohen A, Strom BL. Neither panacea nor black box: responding to three Journal of Biomedical Informatics papers on computerized physician order entry systems. J Biomed Inform 2005;38:267–9.
[107] Nemeth C, Cook R. Hiding in plain sight: what Koppel et al. tell us about healthcare IT. J Biomed Inform 2005;38:262–3.
[108] Marquis MS, Rogowski JA, Escarce JJ. The managed care backlash: did consumers vote with their feet? Inquiry: A Journal of Medical Care Organization, Provision and Financing 2004;41:376–90.
[109] Helfgott AW. The patient-centered medical home and accountable care organizations: an overview. Curr Opin Obstet Gynecol 2012;24:458–64. https://doi.org/10.1097/GCO.0b013e32835998ae.
[110] Wagner EH, Coleman K, Reid RJ, Phillips K, Abrams MK, Sugarman JR. The changes involved in patient-centered medical home transformation. Prim Care 2012;39:241–59. https://doi.org/10.1016/j.pop.2012.03.002.
[111] Pourat N, Lavarreda SA, Snyder S. Patient-centered medical homes improve care for adults with chronic conditions. Policy Brief (UCLA Center for Health Policy Research) 2013;1–8.
[112] Lowell KH, Bertko J. The Accountable Care Organization (ACO) model: building blocks for success. J Ambul Care Manage 2010;33:81–8. https://doi.org/10.1097/JAC.0b013e3181c9fb12.
[113] Reddy J, Kennedy K. From medical home to ACO: a physician group's journey. Healthc Financ Manag: Journal of the Healthcare Financial Management Association 2013;67:38–42.

[114] Colla CH, Fisher ES. Moving forward with accountable care organizations: some answers, more questions. JAMA Intern Med 2017;177:527–8. https://doi.org/10.1001/jamainternmed.2016.9122.
[115] Kaufman BG, Spivack BS, Stearns SC, Song PH, O'Brien EC. Impact of accountable care organizations on utilization, care, and outcomes: a systematic review. Med Care Res Rev 2019;76:255–90. https://doi.org/10.1177/1077558717745916.
[116] Parton R, Ravi S. ACO rule has big implications for IT. New population health tools needed to effectively manage ACOs. Health Manag Technol 2012;33:12–4.
[117] Kraschnewski JL, Gabbay RA. Role of health information technologies in the patient-centered medical home. J Diabetes Sci Technol 2013;7:1376–85.
[118] Willke RJ, Zheng Z, Subedi P, Althin R, Mullins CD. From concepts, theory, and evidence of heterogeneity of treatment effects to methodological approaches: a primer. BMC Med Res Methodol 2012;12:185. https://doi.org/10.1186/1471-2288-12-185.
[119] Brannon AR, Sawyers CL. "N of 1" case reports in the era of whole-genome sequencing. J Clin Invest 2013. https://doi.org/10.1172/jci70935.
[120] Duan N, Kravitz RL, Schmid CH. Single-patient (n-of-1) trials: a pragmatic clinical decision methodology for patient-centered comparative effectiveness research. J Clin Epidemiol 2013;66:S21–8. https://doi.org/10.1016/j.jclinepi.2013.04.006.
[121] Williams MS, Buchanan AH, Davis FD, Faucett WA, Hallquist MLG, Leader JB, et al. Patient-centered precision health in a learning health care system: Geisinger's genomic medicine experience. Health Aff 2018;37:757–64. https://doi.org/10.1377/hlthaff.2017.1557.
[122] Murray MF. The path to routine genomic screening in health care. Ann Intern Med 2018;169:407–8. https://doi.org/10.7326/M18-1722.
[123] Manolio TA, Rowley R, Williams MS, Roden D, Ginsburg GS, Bult C, et al. Opportunities, resources, and techniques for implementing genomics in clinical care. Lancet 2019;394:511–20. https://doi.org/10.1016/S0140-6736(19)31140-7.
[124] Abul-Husn NS, Kenny EE. Personalized medicine and the power of electronic health records. Cell 2019;177:58–69. https://doi.org/10.1016/j.cell.2019.02.039.
[125] Bell G. Your life, uploaded: The digital way to better memory, health, and productivity. New York: Plume (Penguin Group); 2010.
[126] QS. Quantified Self: Self-knowledge through numbers., 2012, http://quantifiedself.com/.
[127] Tiase VL, Hull W, McFarland MM, Sward KA, Del Fiol G, Staes C, et al. Patient-generated health data and electronic health record integration: a scoping review. JAMIA Open 2020;3:619–27. https://doi.org/10.1093/jamiaopen/ooaa052.
[128] Martinez F. Developing a full-cycle mHealth strategy. Front Health Serv Manage 2012;29:11–20.
[129] Pierce N. Keeping up with a fast-moving target—mHealth. Front Health Serv Manage 2012;29:28–32.
[130] Marcolino MS, Oliveira JAQ, D'Agostino M, Ribeiro AL, Alkmim MBM, Novillo-Ortiz D. The impact of mHealth interventions: systematic review of systematic reviews. JMIR Mhealth Uhealth 2018;6:e8873. https://doi.org/10.2196/mhealth.8873.
[131] Rowland SP, Fitzgerald JE, Holme T, Powell J, McGregor A. What is the clinical value of mHealth for patients? Npj Digit Med 2020;3:1–6. https://doi.org/10.1038/s41746-019-0206-x.
[132] Littman-Quinn R, Luberti AA, Kovarik C. mHealth to revolutionize information retrieval in low and middle income countries: introduction and proposed solutions using Botswana as reference point. Stud Health Technol Inform 2013;192:894–8.
[133] Alam MZ, Hoque MR, Hu W, Barua Z. Factors influencing the adoption of mHealth services in a developing country: a patient-centric study. Int J Inf Manag 2020;50:128–43. https://doi.org/10.1016/j.ijinfomgt.2019.04.016.
[134] Ganapathiraju MK, Orii N. Research prioritization through prediction of future impact on biomedical science: a position paper on Inference-Analytics. GigaScience 2013;2:11. https://doi.org/10.1186/2047-217x-2-11.
[135] Hoffman S, Podgurski A. Big bad data: law, public health, and biomedical databases. J Law Med Ethics: A Journal of the American Society of Law, Medicine & Ethics 2013;41(Suppl. 1):56–60. https://doi.org/10.1111/jlme.12040.

[136] Mandl KD, Mandel JC, Murphy SN, Bernstam EV, Ramoni RL, Kreda DA, et al. The SMART Platform: early experience enabling substitutable applications for electronic health records. J Am Med Inform Assoc 2012;19:597–603. https://doi.org/10.1136/amiajnl-2011-000622.
[137] Mandel JC, Kreda DA, Mandl KD, Kohane IS, Ramoni RB. SMART on FHIR: a standards-based, interoperable apps platform for electronic health records. J Am Med Inform Assoc 2016;23:899–908. https://doi.org/10.1093/jamia/ocv189.
[138] Taber P, Radloff C, Del Fiol G, Staes C, Kawamoto K. New standards for clinical decision support: a survey of the state of implementation. Yearb Med Inform 2021;30:159–71. https://doi.org/10.1055/s-0041-1726502.
[139] Query Health. S&I Framework; 2013. Retrieved 20 October 2013 from http://wiki.siframework.org/Query+Health.
[140] Stead WS, Lin H, editors. Computational technology for effective health care: Immediate steps and strategic directions. Washington, DC: National Research Council; 2009.

CHAPTER

4

The role of quality measurement and reporting feedback as a driver for care improvement

Floyd Eisenberg
iParsimony, LLC, Washington, DC, United States

4.1 Introduction

The purposes of clinical decision support (CDS) and knowledge-enhanced healthcare include improvement of quality, safety, and cost-effectiveness of care processes. We will in this chapter include under the term Quality Measurement (QM) measures of appropriateness of care from all three perspectives of quality per se, safety, and cost-effectiveness.

The Quality Improvement Ecosystem (Fig. 4.1) shows the relationship of evidence, guidelines, CDS and QM. It initiates with information and evidence from research, public health surveillance, data mining, and other analyses performed by public health, academic institutions, payers, and others [1]. Professional societies, public health agencies, and governmental bodies publish such information to assure awareness among consumers, healthcare practitioners, and healthcare organizations about management recommendations to address the clinical topic. One method of such publication is the issuance of clinical guidelines, typically based on collaboration among clinical subject matter experts, terminologists, informaticists, clinicians and consumers. To assist with implementation and adherence to guideline recommendations, professional societies, clinical system vendors, healthcare organizations, and others translate clinical guideline components into CDS artifacts, in order to incorporate relevant, evidence-based, and patient-specific clinical recommendations and actions as part of clinical workflow. To support local populations, clinical processes, and jurisdictional requirements local sites coordinate the recommendations to impact clinical care more effectively, by aiming that CDS not replace clinician judgment, but rather provides information to assist care team members during appropriate times in the care process. The aim is to help manage the

Clinical Decision Support and Beyond
https://doi.org/10.1016/B978-0-323-91200-6.00002-4

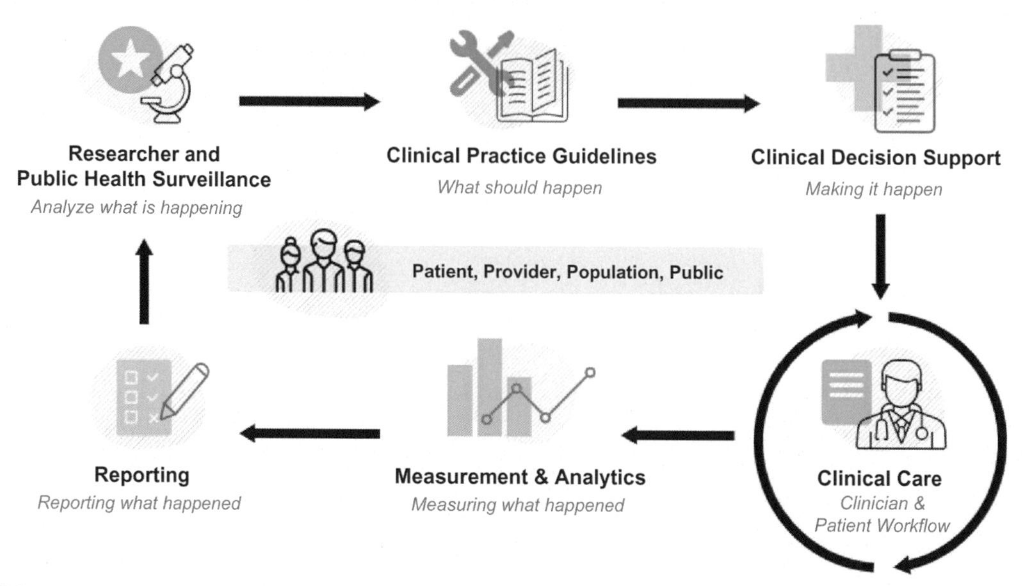

FIG. 4.1 Quality Improvement Ecosystem.

complex and expanding volume of biomedical and person-specific data needed to make timely, informed, and higher quality decisions based on current clinical science. Healthcare organizations, payers, public health, and others also determine methods for evaluating what successful implementation means (i.e., whether the clinical care ultimately provided included processes that addressed the intent of the guideline and if it achieved the desired outcomes). Reporting aggregate measurement analytic results helps to close the loop and enable continuous improvement. Ultimately, this information may then serve as part of the evidence base that updates the guidelines.

Fig. 4.1 may also be interpreted as an interpretation of the Learning Health System, a health system in which internal data and experience are systematically integrated with external evidence, putting knowledge into practice [2]. Other chapters in the book will address approaches to CDS and knowledge-enhanced care. This chapter will focus on the QM and reporting considerations.

Measurement may be patient-based, i.e., focused on what happens to specific patients over time, or evaluating episodes of care such as all hospitalizations for a specific diagnosis. They may also be used for surveillance or to evaluate resources such as healthcare organizations, communities or even devices. Although the basic principles discussed can apply to any type of measure, this chapter will focus primarily on patient-based measures since those are more amenable to direct CDS intervention.

Metrics often enable a comparison to other entities and individuals, and thus inform the recipient of the degree to which a practice may be an outlier. Recognizing that one has outlier status raises a flag that often induces some behavior or procedure modification to address it. Sometimes, of course, the comparisons are not apt, for example if the comparands are different in terms of patient demographics, case mix, severity mix, or due to other reasons, but if they are sufficiently similar, then the recipient will typically pay attention to the feedback.

Thus, QM and reporting are by themselves a passive form of CDS. As a first line of using electronic health record systems to improve care, therefore, there is widespread adoption of efforts to use collected patient data to create quality measures and provide feedback and reporting of them to recipients. Moreover, as we noted in Chapter 1, ratings on quality measures are increasingly used as benchmarks for rating individual providers, practices, or hospitals that are available to the public, or as a metric for value-based reimbursement in emerging ACO-type models of health care finance reform. Thus, they provide a very strong motivation for the adoption of CDS prospectively to maximize opportunity for meeting benchmark goals.

In this chapter, we discuss QM (and reporting) both with respect to it being a form of CDS and how it differs from CDS in the traditional sense.

4.2 Quality measures and clinical decision support: Similarities and differences

QM and CDS rules and interventions are created based on the same goal – to assure that the best care is provided to patients and that they recover as quickly as possible and/or that they can live their lives as fully as possible. Each set of criteria for QM and CDS requires effort to define and to process it, and each has the potential to affect what kinds of data clinical systems need to capture in structured form and how physicians, nurses, other clinicians and patients record information in electronic health records (EHRs) or personal health records (PHRs).

Therefore, it is important to take care that anything measured and any CDS intervention recommendations truly add value to the patient and those providing his or her care. Most QM and CDS work is grounded on evidence-based studies published in reputable medical journals, from which recommendations for optimal care are developed, often in the form of clinical guidelines, by specialty societies. These guidelines recommend how to care for patients with specific diseases or conditions. Generally, QM evaluates whether the care process (screening, testing, or treatment) given by the measured entity over time is consistent with the guideline recommendations. QM is therefore retrospective, evaluating care that has already been delivered. Conversely, traditional CDS evaluates steps in the care process prospectively and recommends what might be done next to assure that care will be consistent with the guideline recommendations, concurrently as the care is being delivered. Both quality measures and CDS rules are generally based on the same evidence and guidelines. An example will help paint the picture so the reader can understand what is similar and what is different between QM and CDS.

4.3 Creating a quality measure

To show the relationship between QM and CDS, let's take an example that shows how a physician's practice might use both to improve the care of teenagers with asthma. The United States (US) National Heart Lung and Blood Institute (NHLBI) defines asthma as "a chronic (long-term) lung disease that inflames and narrows the airways...caus[ing] recurring periods of wheezing (a whistling sound when you breathe), chest tightness, shortness of breath, and coughing.... affecting more than 25 million people in the US, of which about 7 million are children" [3]. Treatment recommendations were updated in 2021 [4]. Asthma attacks can be severe and require hospitalization and even cause the patient to die. Even with treatment, people with asthma often have difficulty with normal daily activities and especially with physical activity. So, let's break down the kind of information that an EHR might contain, to help decide if a patient's asthma treatment is working.

The first step to evaluate treatment for asthma is to define the terms. Some patients with asthma are fairly stable, requiring treatment only occasionally and only when exposed to certain substances to which they are allergic, such as tree pollen. Other patients have symptoms frequently and require treatment all the time. While asthma can be intermittent or severe and persistent, to keep this example simple, we will consider here any current, or active, diagnosis of asthma. We will also assume that the doctor records the diagnosis of asthma in the master problem list of the patient's EHR. Let's further narrow the evaluation to teenagers because they have specific challenges managing chronic illness. We will assume the definition of teenager as being between ages 12 and 17. Since age changes as time advances, age is a *derived* element; it is calculated by subtracting the patient's birth date from the current date. Now we know which patients we need to evaluate. The quality measure defines these patients as the ***denominator***, or the specific population or cohort that will be evaluated. CDS rules require the same definition to define the population, or cohort, for whom a specific type of care is required. In applying a CDS rule to an individual patient, the rule would need this definition to determine eligibility of a given patient for the CDS recommendation.

The next step is to decide how to define what we want to measure about teenagers with asthma. Treatment generally includes medications, typically oral medications, and inhalers (devices that provide medication directly to the breathing tubes as patients inhale or breathe in). Some inhalers include medications to treat at specific times when the patient has difficulty breathing; these are called rescue inhalers. Treatment often involves a combination of medications; ideally, the combinations are adjusted for each patient, based on his or her improvement in symptoms. Additional treatment can include changes in the home setting to remove things that can cause asthma attacks, and changes in activities inside and outside the home and school to decrease exposure to situations that cause attacks. One option is to measure whether the treatment has been acceptable based on the treatment guidelines published for asthma. Evaluating the treatment given is a ***process*** measure, i.e., determination of the percentage of the eligible care episodes in which the doctor prescribed the right medication and the right evaluation of the patient's home and school. The challenge is that there are many details that must be evaluated to decide if the doctor followed the guideline or whether there were acceptable reasons when he or she varied from the guideline. A more significant challenge is that a process measure only indicates that the doctor provided the care that should be expected; it does not show that the patient has responded to the treatment. A measure that determines the response to treatment evaluates the ***outcome***. Outcome measures help determine if patients are improving or at least doing as well as can be expected based on the amount of disease that is present.

To construct an outcome measure for asthma, we need to know what might be used to evaluate the success of treatment. Success is generally handled by asking the patient how often, how long and how severe their symptoms are, to enable comparison over time. Many doctors use some standardized tools to determine the success or failure of treatment. One tool is the peak flow meter. This small device shows how well air moves out of the lungs. The patient blows into the device, which scores how well the lungs are working at the time of the test, the peak flow number. Thus, it would be useful to be able to monitor the disease activity by following the trend of peak flow number results over time. However, the peak flow results show what is happening at a particular instant of time and, unless there are symptoms at that instant, the results are typically normal. So peak flow results may not help to create a trend to evaluate outcomes. Other tools have been developed to evaluate and validate how frequently and how long patients experience symptoms. Some of these tools are based on information patients tell their doctors; others are simple questions the patients can answer themselves. These tools are basically questionnaires that ask about the number of times the patient wakes up at night short of breath, the number of days he or she has missed school because of breathing troubles, the types of activities (such as sports) in which the patient has been participating, and similar questions about how the disease is affecting daily activities [5–7]. Following results of the patient's experience over time is another way to decide if the patient's outcome has changed. Another item that may help decide if the patient is doing well is how often he or she needs to use a rescue inhaler. Since patients only use rescue inhalers when they have difficulty breathing, an increase in use suggests worsening of symptoms, and a decrease in use suggests improvement in symptoms overall. Another outcome that suggests worsening of symptoms is the use of hospital emergency department visits or admissions to the hospital for difficulty breathing.

As we can see, there are several ways to evaluate the care given to teenagers with asthma. Measuring the process of care is complicated, since care is often customized specifically, based on each patient's need. There is no simple "one size fits all" process to evaluate a patient's care, and even if we were successful at evaluating all processes correctly, the result would not necessarily show that patients are healthier, but only that the guideline was followed. Therefore, we decide to measure what is important to the patient, i.e., the outcomes, in this case that he or she has fewer symptoms due to asthma. We saw that there are two ways to know that the symptoms are less, or at least stable (i.e., no worse) when comparing one period with another. One method is to ask how often the patient uses the rescue inhaler. Since each inhaler only contains a small number of doses, the number of times the patient refills the prescription for the rescue inhaler may be a way to know how often he or she is using it. That means there isn't any work the patient (or the patient's parents) must do to be sure we have the information we need. We only need to look at the number of rescue inhaler refills that have been requested from the pharmacy over the past month. The other method is to ask the patient (sometimes with his or her parents' help) every month about daily activities as discussed above.

4.4 Constructing the quality measure equation

We have discussed what we want to measure for all teenaged patients with asthma. Suppose we were a large business, and we were concerned that many employees were taking time off from work because their children were sick and home from school with asthma complications. We might want to work with our health plan to find the doctors whose practices managed their patients so that they had the least complications. Working with the health plan we determine that there are several ways to look at performance of physician practices. First might be the proportion of practices that monitor use of rescue medications and patients' compliance with completing the functional status assessments. This *process* measure evaluates if the practice is doing what is expected. Let's look at the components of a proportion measure [8].

Initial patient population: Some measures are part of *sets*, or groups of measures that, taken together, provide a comprehensive picture of care provided for a specific condition. The *initial population* describes the characteristics of all patients evaluated by the set. We defined that population as all patients with asthma who have had at least two visits to the practice during the year. The two-visit requirement indicates that the practice is involved in the patients' care.

$$\text{Initial Population}: \quad \text{Two Outpatient Encounters} + \text{Asthma}$$

Denominator: Next is the *denominator*, the population evaluated by the individual measure. In many measures, the denominator is the same as the initial population. In this example, the set evaluates all patients with asthma, the denominator, however, narrows the population to include only teenagers. Denominators define age ranges, diagnoses, procedures, time windows, and other information to delineate the people to include in the measure.

$$\text{Denominator}: \quad \text{Teenagers} + \text{Asthma}$$

Denominator exclusion: We might want to exclude some patients because they may incorrectly be identified as asthmatics in the EHR even though they have other conditions that cause their problems. In this example, a *denominator exclusion* includes all patients with cystic fibrosis. Denominator exclusions are those patients identified in the denominator definition but who should not receive the process or are not eligible for the outcome for some other reason.

$$\text{Denominator exclusion}: \quad \text{Cystic Fibrosis}$$

Numerator: Next, we describe the *numerator*, the process, condition, event, or outcome that satisfies the measure focus or intent. In our example, we will define the numerator as all patients for whom rescue inhaler medication is used and for whom functional status evaluation is available in the record, at least twice in the previous six months. To evaluate each element, we will use a measure set, one measure each for rescue inhaler use and for functional status evaluation completion.

$$\begin{aligned}\text{Numerator(Measure 1 of set)}: \quad & \text{Documented rescue inhaler use} \\ & \text{at least twice in the previous six months}\end{aligned}$$

$$\begin{aligned}\text{Numerator(Measure 2 of set)}: \quad & \text{Documented functional status evaluation} \\ & \text{at least twice in the previous six months}\end{aligned}$$

Denominator Exception: *Denominator exceptions* are those conditions that should remove a patient, procedure, or unit of measurement from the denominator only if the numerator criteria are not met. In this example, if there are reasons why a teenager with asthma might have extenuating circumstances for not using the rescue inhaler or filling out the functional status evaluations, this individual can be removed from the denominator, so the practice's performance is not adversely affected due to such issues. In our example, we might list as exceptions patients who were hospitalized during the measurement period since the information may not be available while the patient was in the hospital. However, if the information required by the numerator is present, it may be important to allow those patients to be included in the denominator. Denominator exceptions are not used in all measures and some experts disagree with allowing exceptions, i.e., conditions are either completely excluded in the denominator exclusions or they are not excluded at all.

$$\text{Denominator exception}: \quad \text{Hospitalization}$$

In summary, we can look at our measure as an equation:

Proportion Measure:

$$\frac{\text{Numerator}}{[\text{Denominator} - \text{Denominator Exclusions}] - \text{Denominator Exceptions}}$$

Proportion Measure Example 1 (numerator 1):

$$\frac{\text{Rescue Inhaler Use Documented at least twice in the prior six months}}{[(\text{Teenagers} + 2\ \text{Outpatient Visits}) - \text{Cystic Fibrosis}] - \text{Hospitalized}}$$

Proportion Measure Example 2 (numerator 2):

$$\frac{\text{Functional Status Evaluation Documented at least twice in the prior six months}}{[(\text{Teenagers} + 2\ \text{Outpatient Visits}) - \text{Cystic Fibrosis}] - \text{Hospitalized}}$$

We have discussed a process measure we can evaluate. Let's look at another way to evaluate treatment of teenage patients with asthma, using *outcome* measures. Outcomes are the result of care provided, such as ability to function or survival rates. Some *intermediate outcomes* can also be evaluated, such as the frequency of rescue inhaler use. In this example, less rescue inhaler use has been identified as an indicator that asthma treatment is successful; hence, rescue inhaler use is an intermediate outcome. We can use a proportion measure to evaluate the percentage of patients in a physician's practice who use rescue inhalers less than 3 times per month, or who have functional status evaluation results greater than a specific value. As with the process measure example, such a proportion measure has a denominator, denominator exclusions, a numerator, and denominator exceptions.

Proportion Measure Example 3:

$$\frac{\text{Rescue Inhaler} \leq 3 \text{ times per month } (\geq 6 \text{ months of treatment})}{[(\text{Teenagers} + 2\ \text{Outpatient Visits} + \text{Asthma}) - \text{Cystic Fibrosis}] - \text{Hospitalized}}$$

Proportion Measure Example 4:

$$\frac{\text{Function Status Evaluation} \geq \mathit{SpecificValue}\,(\geq 6 \text{ months of treatment})}{[(\text{Teenagers} + 2\ \text{Outpatient Visits} + \text{Asthma}) - \text{Cystic Fibrosis}] - \text{Hospitalized}}$$

These proportion measures may be limiting, since each patient is either a success or failure given that definition. We might want to evaluate the median number of rescue inhaler uses and the median functional status result for all patients under treatment for at least 6 months. That way, no patient is a success or failure, but rather, a point on a continuum. Such a measure is a *continuous variable* measure. Continuous variable measures have slightly different components:

- **Measure population**: *Measure population* is used only in continuous variable eMeasures. Like an initial patient population and a denominator, it is a description of the patients who are being measured, e.g., all teenagers with asthma seen in the practice for at least 6 months. Note, 6 months was chosen to give the practice some time to adjust treatment for new patients. All exclusions for a continuous variable measure are incorporated in the measure population description.
- **Measure Observations:** *Measure observations* are used only in continuous variable eCQMs. They are analogous to a numerator in that they describe the process or outcome that is being evaluated. In this case the measure observations are the median functional status value for each patient (using the most recent value for each), and the median number of rescue inhaler uses for each patient (using the most recent month's results).

In summary, we can look at our continuous variable measures as equations:
Continuous Variable Measure:

$$\frac{\text{Measure Observations}}{[\text{Measure Population} - \text{Exclusions}]}$$

Continuous Variable Measure Example 1:

$$\frac{\text{Median Rescue Inhaler Use per Month } (\geq 6 \text{ months of treatment})}{[(\text{Teenagers} + 2 \text{ Outpatient Visits} + \text{Asthma}) - \text{Cystic Fibrosis}]}$$

Continuous Variable Measure Example 2:

$$\frac{\text{Median Function Status Evaluation Score } (\geq 6 \text{ months of treatment})}{[(\text{Teenagers} + 2 \text{ Outpatient Visits} + \text{Asthma}) - \text{Cystic Fibrosis}]}$$

Now that we have defined a few measures, let's think through several decision points we can use to impact the results using CDS proactively. Each use of a rescue inhaler raises some concern, but more than two uses in a short period might be an indicator that a case manager should speak with the patient about other potential interventions. For the example, let's use two uses in one day as the trigger to educate the patient or to prompt a call by the case manager. Similarly, worsening results on the functional status evaluation tool may be another trigger for asking additional questions, providing education, or prompting a discussion with the case manager. The input data are those that calculate the denominator and the results that help calculate the numerator compliance. The interventions and action steps are those noted – asking additional questions, providing education specific to the patient's needs, or prompting a call from the case manager. All these CDS interventions help ensure improved overall performance in quality measure results.

4.5 Translating measure concepts into interoperable structures and definitions

Beyond defining the measures and calculations, there are some basic prerequisites to effect implementation of the ecosystem presented in Fig. 4.1 through an interoperable, standards-based architecture. These include structural components to contain and share content, metadata to describe the intent and justification for the artifact, general description of the content, data criteria, terminology, and methods for sharing, displaying, and interacting with the content. Initial attempts developed at Health Level Seven (HL7) International included various standards for measures – Health Quality Measure Framework (HQMF) [9] for measures and Quality Reporting Document Architecture for single reports (Category I) [10] and aggregate reports (Category III) [11]. Further, measures used a data model for measures and expression language called Quality Data Model (QDM) [12], CDS standards included GELLO [13], Arden Syntax [14], Virtual Medical Record (vMR) Logical Model [15], and Virtual Medical Record (vMR) for Clinical Decision Support (vMR-CDS) XML Specification [16]. Fig. 4.2 displays the standards environment as of approximately 2017.

The Clinical Quality Framework (CQF) initiative started by the Office of the National Coordinator for Health Information Technology (ONC) and incorporated into the Health Level Seven (HL7) environment re-evaluated the structure, leading to a common model for managing data and expressions for both CDS and QM and reporting [17]. Fig. 4.3 shows the CQF direction for harmonization to enable these common components.

The first activity created a Quality Common Metadata Conceptual Model [18] to address both CDS and measurement. Next was to create a common expression language, Clinical Quality Language to address simple and complex logic representations [19]. The CQF community also developed Quality Improvement Core (QI-Core) from a combination of QDM and vMR to provide a common data model [20] for use in HL7 Fast Health Interoperability Resources (HL7®FHIR®) [21].

Fig. 4.4 shows the transition in progress as measure activity moves to FHIR. In addition to using QI-Core instead of QDM, the measures will replace HQMF with the FHIR Quality Measure [22] and the FHIR report (FHIR Data Exchange for Quality Measures) [23].

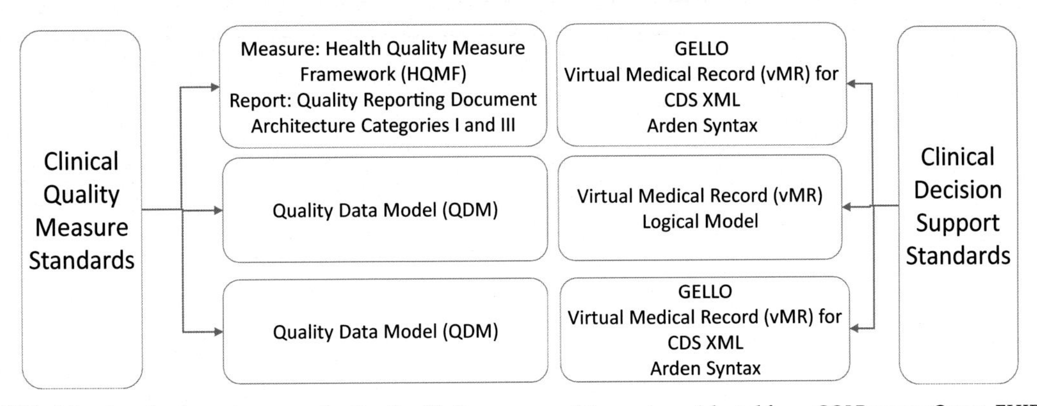

FIG. 4.2 Standards environment for the Quality Improvement Ecosystem. *Adapted from eCQI Resource Center, FHIR (https://ecqi.healthit.gov/fhir).*

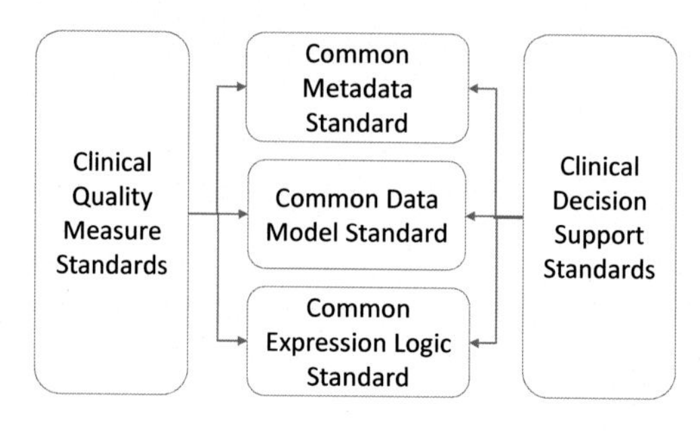

FIG. 4.3 Clinical Quality Framework harmonization model. *Adapted from eCQI Resource Center, FHIR (https://ecqi.healthit.gov/fhir).*

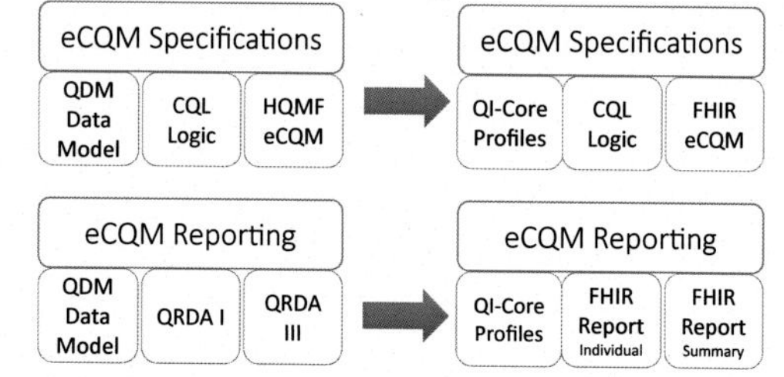

FIG. 4.4 Electronic Clinical Quality Measures (eCQMs) transition to FHIR. *Adapted from eCQI Resource Center, FHIR (https://ecqi.healthit.gov/fhir).*

Since CDS artifacts can use the same expressions based on QI-Core profiles and CQL logic expressions, implementers can more easily retrieve the required data for concurrent care decision making (CDS) and for retrospective analysis (measurement). HL7 FHIR Clinical Guidelines provides further information about how these structures enable development and implementation guidelines [24].

Understanding the structure and data model to express the content is important, but not sufficient. The terminology used to define data needed for evaluation of measure components is critical. For example, an observable entity (a clinical intervention, or a question or assessment requiring an answer or result), should generally use the Logical Observation Identifiers Names and Codes (LOINC®) code system [25]. Findings and observations that represent answers to such observable entities should use SNOMED-CT [26] and the hierarchy that best describes the concept findings [27]. The CMS Measures Management Blueprint table 17 provides details about which terminology and respective hierarchy to consider for different types of data elements and attributes [4].

Many data collection tools exist to capture health-related measures, including those to determine asthma activity status, including EHR templates, smart phone applications, online data collection tools, paper forms, and some are available via APIs to incorporate the tools into other data collection systems [28]. Further, several standard, validated tools exist for use in measures to evaluate outcomes, or in CDS to suggest interventions based on existing results, or to suggest patient evaluation in the absence of such information [5–7,29–32]. While appropriate tools may exist to capture useful patient status and outcomes, access to such information in existing clinical data stores is not necessarily consistent within and across settings. EHR products and local implementations vary with respect to the extent to which they enable capture of such content; some may use only paper-based tools, others may capture only the calculated result of each tool as a single observation value, and others may allow access to all data including the granular components of such tools. At the time of publication, the existing requirements for interoperability, the United States Core Data for Interoperability (USCDI) version 2, do not include a requirement for sharing a questionnaire or evaluation tool [33]. If the data required for the measure or CDS is not specifically identified in USCDI, software may not share such data in a common format.

Readers are encouraged to investigate and participate in current collaboration efforts among the QM and research communities to harmonize efforts to retrieve data from clinical data stores for analysis and research. Clinical organizations evaluating quality performance use measures such as the Healthcare Effectiveness Data and Information Set (HEDIS®) [34]. Organizations reporting HEDIS data retrospectively review existing records which may include high-cost and time-consuming manual data abstraction from clinician records. Natural language processing and data extraction to develop a reproducible data set has value and managing these data sets for further analysis using the research-based Observational Medical Outcomes Partnership (OMOP) common data model [35] may help ease the burden of data collection as well as identify areas for performance improvement using CDS [36]. Researchers use OMOP to define the information needed to address clinical research and perform significant curation of real-world data to assure they create data sets consistent with the data element specifications required. Such curation is time consuming and somewhat analogous to the data abstraction efforts required for HEDIS reporting. The HL7® and OHDSI® collaborative effort to develop bi-directional mapping from the OMOP data model to FHIR® is intended to harmonize approaches to reduce data curation and retrieval efforts [37]. The effort includes a detailed use case to evaluate the feasibility of export existing OMOP research cohort definitions (called phenotypes) into FHIR with CQL to retrieve data from existing clinical data stores and reporting the data as an OMOP data set for analysis [38]. The same OMOP phenotypes could provide valuable expressions for CDS artifacts as well. This work started in late 2021.

4.6 Identifying CDS interventions based on the quality measure

Measures are good indicators of how well (or poorly) we are doing to improve our patients' health. If we really want to enable the EHR system to encourage the right things to happen to help improve the outcomes we measure, we need to use CDS, the focus of this book. So, given our example of the adolescent patient with asthma, we need to identify those points in the disease and the process of providing care in which we can use the EHR system to help the right thing happen. We also need to decide what is best done by whom to know the kind of information we can have the EHR system provide, to whom, when, and in the right way, so it has the greatest chance of changing behavior. We will continue to use our example to help explain these concepts.

Let's start with remembering what we are measuring. Ultimately QM is highly tied to the health care financing/reimbursement model. Fee-for-service reimbursement, which has been prevalent in the US for several decades, reimburses providers for what they do, independent of its appropriateness or outcome (although there may be a need for prospective approval before undertaking certain care processes). In the US in the 1970s through 1990s a "capitated" model of care reimbursement replaced fee-for-service payments to providers. This capitation model provided a fixed monetary payment for all care provided to each assigned patient for a specified period in many health maintenance organizations (HMOs) [39]. Some models provided modified payments for patients with more complex diseases or problems [40]. It was up to the practice to determine how best to use capitated funds for a given patient. Fee-for-service tends to stimulate over-use, whereas capitation tends to stimulate under-use.

What we want ideally is to incentivize "appropriate use" of health care, and one way to achieve that is to develop reimbursement models that reward providers (and patients) based on adherence to quality measures and achievement of desired outcomes. In emerging models of health care financing, such as Accountable Care Organizations (ACOs), groups of hospitals, doctors, and other healthcare providers organize to provide coordinated care to patient members. The ACO generally receives a payment for all services provided to patients and assigns payments to participating providers and hospitals [41]. The idea of pay for value is primary, and quality measures are thus essential to determining the amount of money a health plan will pay a practice. New models of payment for primary care will continue to assess quality care performance. An example is a recent Primary Care First model [42]. There is also a strong incentive, beyond the use of quality measures, to put into place methods of CDS that will foster the desired adherence to quality. For example, these models tend to focus on and stress preventive and early intervention approaches to health and healthcare that are more proactive rather than primarily reactive to episodes of worsening health status.

In our asthma example, where available measures evaluate how well the patients can function during the day and how often they have symptoms bad enough to use their rescue inhalers, there are lots of additional things we can do to educate the patients and their families and to assure they are on the right medications. But we can start with the closest thing to what is being measured. For example, a patient could be asked to provide a weekly update about symptoms using a form-based tool that asks about daily activities. The form used for the update can calculate a score. Any week during which the results of that score decreases is an opportunity to consider doing something differently. So, the patient's completion of the survey tool is a *trigger* to the EHR system to check results and determine what might be done. The patient's failure to complete the survey might also be a *trigger*, but the recommendations, or intervention might be different.

So, now we have instructed an EHR-based CDS system to look for a *trigger*, entry of weekly symptoms. That trigger can be set to actively look for an entry every 7 days or to passively wait for the survey results to be entered. Which to use depends on how closely one plans to monitor the patient's care. Here let us assume that we select the more proactive approach, looking for an event to happen every 7 days, and giving the patient a day leeway to be late on entering the information. That way the system can make recommendations even if the information is missing.

Now that we have defined the triggers, let's think about what information the CDS rule will need to evaluate, i.e., the *input data*. In this case we want to know the pattern of survey results over the last month. That will help determine if there really is a change and if that change suggests a downward trend or just a one-time issue. If there is a trend, we probably want to know what medications the patient is expected to be taking. Thus, the active medication list in the EHR is another set of needed input data. If there is a personal health record (PHR) the patient is using to record his or her own health, we would want to know if there are other medications or new exposures the patient has recorded – more input data. We also want to look at other information that we can define from the record, including the number of refills for the rescue inhaler, and the patient's peak flow values over the past month. We can also look to see if the patient has gone to a hospital emergency department or an urgent care clinic for asthma symptoms over the past month. That is a large amount of input data. It is important to note that not all this information is available within a typical EHR – such as the data about refills at a pharmacy or that entered into a PHR. So, depending on the degree of integration and access to information from multiple sources, some of the proposed interventions would need to be modified.

If the input data described above are all potentially available about our patients with asthma, we need to decide what we need to do for the patients – what interventions should be recommended, to whom, and when. There are a lot of things we might be able to do, but CDS developers can look at prior studies as evidence about what might work and what might be interesting but not have much success. In many cases, there is only limited evidence about useful *interventions* when we start out. So, some of the efforts may be needed even to develop the evidence. That means we try them for a while and see if results of the measures change for the better or worse. From those results we can decide if we should continue each *intervention*, modify it, or abandon it entirely (keeping a list so we don't repeat ineffective *interventions* in the future). Here are some *interventions* that might be considered along with the *action steps* that people can take to improve the management of the condition:

1. Notify (the notify intervention generally requires identifying a person to make aware that something has occurred)
   a. The case manager (a nurse who coordinates the care a patient receives and checks in with the patient regularly to see how the patient is doing):
      i. Action Step: If the survey tool isn't completed every 7–10 days, remind the patient through an automated reminder. If there is still no response, notify a case manager. The case manager can then intervene to get the answers needed.
   b. The patient:
      i. Action Step: Provide a reminder to complete the survey and ask the patient to contact the case manager and/or the physician if there is some change in status that may cause trouble in completing the survey, or in managing the asthma symptoms.
2. Inform (the inform intervention generally requires identifying the person to inform, often to educate):
   a. The patient:
      i. Action Step: Provide education materials, e.g., about how to manage sports activities with asthma. The method to provide these materials may include use of a standard called the *infobutton* (see Chapter 13).
   b. The case manager:
      i. Action Step: Provide education materials to the case manager about how to advise the patient, e.g., about sports activities. The method may also use the *infobutton* standard.

3. Record (the record intervention also requires identification of the person, or the device to record the information – in this case, the EHR may be recording that the information available requires no intervention on the part of any person; the information is stored by the EHR as a successfully completed step in a guideline process):
   a. The EHR:
      i. Action Step: Record the successful completion of the survey tool by the patient in that the result shows a stable, or improved status.

4.7 A CDS rule component taxonomy

QM can influence the construction of many kinds of CDS and knowledge-enhanced care, ranging from context-aware recommendations to guidelines, order sets, and documentation templates, for example. But the primary relationship is with the rules-based formalism for CDS, because of similarities in the concepts, measures, and calculations we have discussed. Thus, in this section we seek to clarify the relationship between CDS rules and QM by examining a taxonomy of components of both and comparing them. Several taxonomies, in fact, ontologies, of CDS rules components have been developed for various purposes to facilitate authoring of rules, sharing of components, or enhanced indexing of rules repository. We will discuss other approaches in Chapter 28, for enterprise knowledge management, and rule-based CDS in Chapter 9. Among recent approaches is CDS Hooks, an HL7 specification based on Web services and FHIR used to make CDS available to providers based on specific triggers (i.e., hooks) [43]. A continuous integration build of the CDS Hooks specification is available for evaluating new hooks, or triggers [44]. Chapter 15 discusses CDS Hooks and related standards in more detail.

For our purposes here, we consider the following components of a CDS rule: the *trigger* (the activity or event that initiates the CDS process), the *input data* (the information the CDS process needs to evaluate), the *intervention* (the activity the EHR system performs directly) and the *action steps* (those items the EHR system offers to people involved in the patient's care so they can decide the appropriate action and perform it). Each *intervention* is accompanied by an *action step* that can be taken by a person with a specific role in caring for the patient, or the EHR system or other device used as a tool to support the patient's care.

We can see that the *triggers* and the *input data* for CDS can be the same information used by the quality measures. The *interventions* and the *action steps* are different. These CDS components are basically functions that the EHR system can perform directly or that the EHR system can process once instructed by the person (or *actor*) involved in caring for the patient. *Actors* can include anyone involved in the care including the physician, a nurse, a specialized nurse with a broader case manager role, a physical therapist, a respiratory therapist, a dietician, the patient, the patient's parent, or, if programmed to do so, the EHR system itself. In this sense, CDS can be considered the *effector* arm of quality, providing the actionable items that can directly enhance the care process and the outcomes. Let's look at one more example of a set of CDS interventions for our adolescents with asthma to be sure this framework for understanding the relationship between CDS and QMs is clear.

The previous CDS activities in our example were focused on the survey tool used to determine if the adolescent patient's disease limited his or her routine activities. This next set of interventions will help evaluate and modify medical treatment as determined to be appropriate by the patient's physicians. In this case, there are two *triggers* that can initiate the process. The first is a poor result identified by the weekly survey tools, indicating that the patient has had significantly increased difficulty with daily activities for 3 weeks in a row. The second *trigger* is a refill for the prescription for the patient's asthma rescue inhaler occurring more frequently than expected. Perhaps the refill is merely because the patient is planning a two-month summer vacation trip cross-country, and he or she wants to have the extra inhaler available if needed. But the refill information comes from the pharmacy directly and it most often will not include a statement of the reason. The input data are basically the same: the results of the survey tool over the past month, the information the patient included in the personal health record (PHR) about new exposures, and information regarding whether there were any visits to a hospital emergency room or urgent care clinic over the past month.

The *interventions* and *action steps*, then, are more specific to the medications and methods to avoid exposure to things that could exacerbate the patient's asthma symptoms.

1. Notify (the case manager or the physician, or both)
   a. Action Step: Provide the current medication list for the physician or case manager to determine if a change to the routine medications is appropriate currently.
   b. Action Step: Provide the case manager with the patient's contact information, to explore with the patient (and/or his or her parents) what is happening that might be causing the increase in use of the rescue inhaler.
2. Educate (the patient)
   a. Action Step: Provide the patient with information about what might be causes for increased used of rescue inhalers so the patient can participate in his or her own care and modify behaviors that might be causing increased symptoms.
   b. Action Step: Provide the patient with access to an online support group with other adolescents to help him or her work with others in the same situation to find a way to live a more 'normal' life with fewer symptoms and without as much need for rescue medication.

4.8 The details about the CDS rules component taxonomy

We have discussed the relationships between QM and CDS rules, as well as the components of CDS rules, in terms of their relationship to QM, in terms of a *taxonomy*. In this framework, *triggers* initiate a CDS rule, *input data* are what the rule uses to evaluate what needs to happen, and *interventions* are what the rule tells the computer system to perform to give the provider the *action steps* he or she can take to help the patient improve (Fig. 4.5) [45].

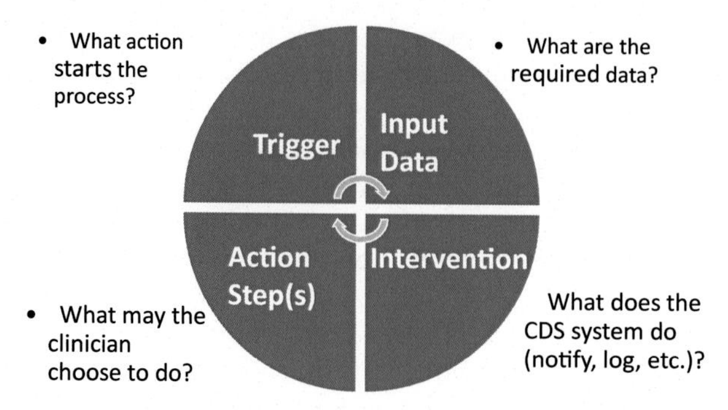

FIG. 4.5 CDS rules require four components: The *trigger* is the activity or event that initiates a decision support rule. Trigger events can be the entry of new information into the EHR including observations, orders or new results automatically entered from other software (e.g., laboratory results), or other actions such as logging into the computer. Triggers can also be set based on the time elapsed between two events so that a missing event at the expected time can trigger CDS. A CDS rule then evaluates *input data* to evaluate what actions the EHR system should take. Based on the input data, the rule is programmed to perform *interventions* to deliver information to a physician, a nurse, other clinicians, or the patient. Interventions include displaying information in a specific way, notifying a clinician, or merely logging the results. The interventions give the recipient *action steps* that he or she can perform. In any given decision support rule, the action steps chosen can be triggers for other rules. Actions can include collecting additional information, requesting services, acknowledging information that is present, documenting information or communicating with the patient or other clinicians.

4.9 The CDS rules component taxonomy as a driver for quality measurement and care improvement

The QM lifecycle incorporates these same four components on a more macro level, i.e., from the perspective of the organizational, public health system, government, or accrediting body. The *trigger* for a measure is traditionally based on published information about gaps in care in the country or the region. The US Department of Health and Human Services (DHHS) addresses measurement based on the National Quality Strategy, driven by three aims: better care, healthy people/health communities, and affordable care [46]. The Measure Application Partnership (MAP) [47], established by the Affordable Care Act [48]. is a public-private partnership convened by the National Quality Forum (NQF), a consensus development organization that reviews, endorses, and recommends health care performance measures [49]. MAP provides input to the DHHS about what performance measures it should select for public reporting and performance-based payment programs as required in the Affordable Care Act. It also addresses areas of care important to the National Quality Strategy for which there are no measures. Once identified, these gaps drive new measure development by private organizations and by DHHS programs. Measures that address these gaps are developed based on clinical knowledge and research.

Measures evaluate performance on an annual basis for national governmental programs and accrediting agencies (such as the Joint Commission, a not-for-profit organization that accredits healthcare organizations and programs) [50] that evaluates the quality and safety of care delivered in hospitals and provides certification that established standards are met. Thus, the trigger is the beginning of the measurement year, or cycle. A hospital might want to trigger the measure at different times to determine how the organization is performing at various points along the way. The same *input data* used by CDS is used in measure specification to decide which patients to include, which to exclude, and whether each included patient has received the care expected or has shown improvement. The final quality report is then used by the organization as a basis for implementing system-wide changes, or *interventions*, so that results will improve during the next cycle of measurement. The interventions generate *actions* by physicians and nurses to improve the rate at which expected orders, documentation, or communications are performed. While CDS addresses what should be done for an individual patient, the quality improvement process addresses trends in performance across many patients and identifies processes that can be improved.

In summary, Fig. 4.6 shows the close linkage between QM and CDS. Both are driven by the same clinical knowledge and clinical data. Each requires similar data, and each plays a role in evaluating clinical performance. CDS and QM enable reuse of captured data to enhance clinical performance and report on quality processes and outcomes. Both are based on the same clinical practice guidelines or recommendations to evaluate the next step in care (CDS) and assess how care was performed or the outcome achieved (QM). Clinical organizations often use CDS to improve provider performance scores for measure results submitted to outside agencies.

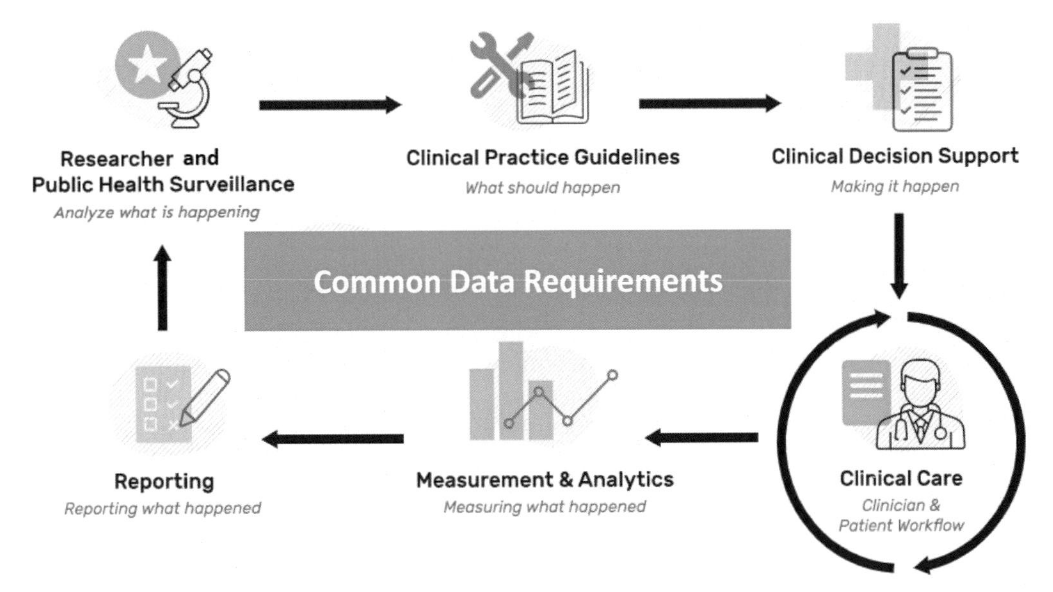

FIG. 4.6 Relationship of clinical knowledge, CDS and quality measures to encourage data capture and reuse to enhance clinical performance and report on quality processes and outcomes. The figure shows a close relationship between QM and CDS, as both are often based on the same clinical practice guidelines or recommendations. Both use similar information to evaluate the next step in care (CDS) or assess how care was performed (QM). Often, CDS processes are created to improve the provider's performance scores in the quality measure scores reported to outside agencies.

4.10 Assuring the quality of a measure

To ensure that measures meet the needs of those who use them, it is important to manage the quality, the impact, and the value of the information they produce. Most US government programs exclusively use measures that have been rigorously evaluated for their value by the NQF. NQF employs a formal Consensus Development Process (CDP) to endorse measures for quality performance and public reporting [51]. Criteria are summarized below:

1. **Importance to measure and report:** the extent to which the measure focus is important to make significant gains in health care quality (safety, timeliness, effectiveness, efficiency, equity, patient centeredness) and to improve health outcomes for a specific high-impact aspect of health care, where there is variation in overall poor performance.
2. **Scientific acceptability of the measure properties:** the extent to which the results of the measure are consistent (reliable) and credible (valid) if it is implemented as specified.
3. **Usability:** the extent to which those who will use the measure (e.g., consumers, purchasers, providers, policy makers) can understand the results and find them useful, to make meaningful decisions.
4. **Feasibility:** the extent to which the data required to compute the measure are readily available, retrievable without undue burden, and can be implemented for performance measurement. For clinical data, the elements should be captured as a byproduct of routine care processes. NQF has specifically defined feasibility requirements for electronic measures designed specifically for EHRs. Those requirements basically indicate that data elements should be available, accurate, collected as part of the clinical process, and use nationally accepted vocabularies [52]. NQF further provided guidance on more effective methods to evaluate feasibility of eCQMs [53].

4.11 Driving care improvement

Several teaching hospitals have used CDS and QM to improve care of their patients. We have seen how some of the national initiatives have been addressing QM to drive better care and lower costs. The Academy of Medicine publishes proven techniques for improving quality among collaborating organizations including improving diagnosis accuracy and reducing error in clinical practice [36]. The Centers for Medicare and Medicaid Services (CMS) established value-based programs for providers that incorporate performance measurement [54,55].

References

[1] Health Level Seven. Quality measure implementation guide (STU4), 2022, http://hl7.org/fhir/us/cqfmeasures/.

[2] Agency for Healthcare Research and Quality. About learning health systems, 2019. [Internet]. [cited 04 January 2022]. Available from: https://www.ahrq.gov/learning-health-systems/about.html.

[3] National Heart Lung and Blood Institute. Focused updates to the asthma management guidelines—A report from the National Asthma Education and Prevention Program Coordinating Committee Expert Panel Working Group 2021, 2020. [Internet]. [cited 30 November2021]. Available from: https://www.nhlbi.nih.gov/health-topics/all-publications-and-resources/clinician-guide-2020-focused-updates-asthma-management-guidelines.

[4] National Heart Lung and Blood Institute. Asthma, 2020. [Internet]. [cited 14 December 2021]. Available from: http://www.nhlbi.nih.gov/health/health-topics/topics/asthma/.
[5] Allergy and Asthma Foundation of America. Asthma personalized assessment and control tool (PACT), 2021. [Internet]. [cited 27 December 2021]. Available from: http://www.asthmapact.org/.
[6] Wildfire JJ, et al. Development and validation of the Composite Asthma Severity Index—an outcome measure for use in children and adolescents. J Allergy Clin Immunol 2012;129(3):694–701.
[7] Yeats KB, et al. Construction of the Pediatric Asthma Impact Scale (PAIS) for the Patient-Reported Outcomes Measurement Information System (PROMIS). J Asthma 2010;47(3):295–302.
[8] Centers for Medicare and Medicaid Services. CMS measures management system blueprint, version 17.0, 2021. [Internet]. [cited 30 November 2021]. Available from: https://www.cms.gov/Medicare/Quality-Initiatives-Patient-Assessment-Instruments/MMS/MMS-Blueprint.
[9] Health Level Seven. HL7 version 3 standard: Representation of the health quality measure format (eMeasure) release 1, 2017. [Internet]. [cited 12 December 2021]. Available from: https://www.hl7.org/implement/standards/product_brief.cfm?product_id=97.
[10] Health Level Seven. HL7 CDA® R2 implementation guide: Quality reporting document architecture—Category I (QRDA I)—US realm, 2021. [Internet]. [cited 12 December 2021]. Available from: https://www.hl7.org/implement/standards/product_brief.cfm?product_id=35.
[11] Health Level Seven. HL7 CDA®R2 implementation guide: Quality reporting document architecture—Category III (QRDA-III) release 1—US realm, 2021. [Internet]. [cited 12 December 2021]. Available from: https://www.hl7.org/implement/standards/product_brief.cfm?product_id=286.
[12] eCQI Resource Center. Quality data model, 2021. [Internet]. [cited 12 December 2021]. Available from: https://ecqi.healthit.gov/qdm.
[13] Health Level Seven. HL7 version 3 standard: GELLO, a common expression language, release 2, 2010. [Internet]. [cited 12 December 2021]. Available from: https://www.hl7.org/implement/standards/product_brief.cfm?product_id=, 338.
[14] Health Level Seven. Arden Syntax v2.10 (Health Level Seven Arden Syntax for medical logic systems), 2019. [Internet]. [cited 12 December 2021]. Available from: https://www.hl7.org/implement/standards/product_brief.cfm?product_id=372.
[15] Health Level Seven. HL7 version 3 standard: Clinical decision support: Virtual medical record (vMR) logical model, release 2, 2018. [Internet]. [cited 12 December 2021]. Available from: https://www.hl7.org/implement/standards/product_brief.cfm?product_id=, 338.
[16] Health Level Seven. HL7 version 3 standard: Virtual medical record (vMR) for clinical decision support (vMR-CDS) XML specification, release 1, 2017. [Internet]. [cited 12 December 2021]. Available from: https://www.hl7.org/implement/standards/product_brief.cfm?product_id=342.
[17] Health Level Seven. Clinical quality framework, 2021. [Internet]. [cited 12 December 2021]. Available from: https://confluence.hl7.org/display/CQIWC/Clinical+Quality+, Framework.
[18] Health Level Seven. HL7 specification: Clinical quality common metadata conceptual model, release 1, 2015. [Internet]. [cited 12 December 2021]. Available from: http://www.hl7.org/implement/standards/product_brief.cfm?product_id=391.
[19] Health Level Seven. HL7 Specification: Clinical Quality Language, Release 1, 2019. [Internet]. [cited 12 December 2021]. Available from: http://www.hl7.org/implement/standards/product_brief.cfm?product_id=400.
[20] Health Level Seven. HL7 FHIR profile: quality, release 1—US realm, 2015. [Internet]. [cited 12 December 2021]. Available from: http://www.hl7.org/implement/standards/product_brief.cfm?product_id=415.
[21] Health Level Seven. Fast health interoperability resources (FHIR® R4), 2019. [Internet]. [cited 12 December 2021]. Available from: http://www.hl7.org/implement/standards/product_brief.cfm?product_id=491.
[22] Health Level Seven. HL7 FHIR implementation guide: Quality measures, release 1, 2019. [Internet]. [cited 12 December 2021]. Available from: http://www.hl7.org/implement/standards/product_brief.cfm?product_id=511.
[23] Health Level Seven. HL7 FHIR implementation guide: Data exchange for quality measures, 2021. [Internet]. [cited 12 December 2021]. Available from: http://www.hl7.org/implement/standards/product_brief.cfm?product_id=511.
[24] Health Level Seven. HL7 FHIR clinical guidelines, 2021. [Internet]. [cited 14 December 2021]. Available from: http://hl7.org/fhir/uv/cpg/.
[25] LOINC. Logical observation identifiers names and codes (LOINC®), 2021. [Internet]. [cited 14 December 2021]. Available from: https://loinc.org.

[26] SNOMED. Systematized nomenclature of medicine, 2021. [Internet]. [cited 14 December 2021]. Available from: https://www.snomed.org.

[27] Willett D, Kannan V, Chu L, et al. SNOMED CT concept hierarchies for sharing definitions of clinical conditions using electronic health record data. Appl Clin Inform 2018;9(3):667–82. [Internet]. [cited 14 December 2021]. Available from https://www.ncbi.nlm.nih.gov/pmc/articles/PMC6115233/.

[28] Health Measures. Administration platforms, 2021. [Internet]. [cited 13 December 2021]. Available from: https://www.healthmeasures.net/implement-healthmeasures/administration-platforms.

[29] PROMIS. Pediatric item bank—Asthma—Version 2.0 Tscore, 2010. [Internet]. [cited 13 December 2021]. Available from: https://loinc.org/91538-9/.

[30] PROMIS. PROMIS parent proxy asthma impact—Version 2.0 T-score, 2021. [Internet]. [cited 13 December 2021]. Available from: https://loinc.org/91426-7/.

[31] PROMIS. PROMIS pediatric short form—Asthma 8a—Version 1.0, 2010. [Internet]. [cited 13 December 2021]. Available from: https://loinc.org/62208-4/.

[32] Optum. Asthma control test, 2021. [Internet]. [cited 13 December 2021]. Available from: https://loinc.org/82674-3/.

[33] Office of the National Coordinator for Health Information Technology. United States core data for interoperability (USCDI) version 2, 2021. [Internet]. [cited 14 December 2021]. Available from: https://www.healthit.gov/isa/sites/isa/files/2021-07/USCDI-Version-2-July-2021-Final.pdf.

[34] National Committee for Quality Assurance. HEDIS measures and technical resources, 2020. [Internet]. [cited 14 December 2021]. Available from: https://www.ncqa.org/hedis/measures/.

[35] Observational Health Data Sciences and Informatics. Observational medical outcomes partnership (OMOP), 2021. [Internet]. [cited 14 December 2021]. Available from: https://www.ohdsi.org/data-standardization/the-common-data-model/.

[36] Kostka K, Beaulah S, Marshall E, Hamlin B, [Internet]. Reimagining quality reporting using the OMOP CQM NOTE_NLP table: a pilot. In: Poster session presented at OHDSI 2019 US symposium showcase; 2019. [cited 27December2029]. Available from: https://www.ohdsi.org/2019-us-symposium-showcase-15/.

[37] Health Level Seven and OHDSI. FHIR-OMOP Working Group—Home, 2021. [Internet]. [cited 14 December 2021]. Available from: https://confluence.hl7.org/display/OOF/FHIR-OMOP+Working+Group+-+Homepage.

[38] Health Level Seven and OHDSI. FHIR-OMOP Digital Quality Measurement, 2021. [Internet]. [cited 14 December 2021]. Available from: https://confluence.hl7.org/display/OOF/FHIR-OMOP+Digital+Quality+Measurement.

[39] American College of Physicians. Understanding capitation, 2021. [Internet]. [cited 27 December 2021]. Available from: https://www.acponline.org/about-acp/about-internal-medicine/career-paths/residency-career-counseling/resident-career-counseling-guidance-and-tips/understanding-capitation.

[40] Academy of Medicine. Driving quality improvement efforts, 2021. [Internet]. [cited 27 December 2021]. Available from: https://www.improvediagnosis.org/driving-quality-improvement-efforts/.

[41] Centers for Medicare and Medicaid Services. Accountable care organizations, 2021. [Internet]. [cited 27 December 2021]. Available from: https://www.cms.gov/Medicare/Medicare-Fee-for-Service-Payment/ACO.

[42] Centers for Medicare and Medicaid Services. CMS primary care first model options, 2021. [Internet]. [cited 14 December 2021]. Available from: https://innovation.cms.gov/innovation-models/primary-care-first-model-options.

[43] Health Level Seven. CDS hooks 1.0 release, 2019. [Internet]. [cited 10 December 2019]. Available from: https://cds-hooks.hl7.org/1.0/.

[44] Health Level Seven and Boston Children's Hospital. CDS hooks continuous integration build, 2018. [Internet]. [cited 10 December 2021]. Available from: https://cds-hooks.org.

[45] National Quality Forum. Driving quality and performance measurement—A foundation for clinical decision support: A consensus report, 2010. [Internet]. [cited 14 December 2021]. Available from: http://www.qualityforum.org/Publications/2010/12/Driving_Quality_and_Performance_Measurement_-_A_Foundation_for_Clinical_Decision_Support.aspx.

[46] Agency for Healthcare Research and Quality. National quality strategy overview, 2017. [Internet]. [cited 14 December 2021]. Available from: https://www.ahrq.gov/sites/default/files/wysiwyg/NQS_overview_slides-2017.pdf.

[47] National Quality Forum. Measure application partnership, 2021. [Internet]. [cited 14 December 2021]. Available from: https://www.qualityforum.org/map/.

[48] Healthcare.gov. Affordable care act, 2021. [Internet]. [cited 27 December 2021]. Available from: https://www.healthcare.gov/glossary/affordable-care-act/.
[49] National Quality Forum. National Quality Forum, 2013. [Internet]. [cited 14 December 2014]. Available from: http://www.qualityforum.org/Home.aspx.
[50] The Joint Commission. The Joint Commission, 2021. [Internet]. [cited 27 December 2021]. Available from: http://www.jointcommission.org/.
[51] National Quality Forum. Consensus development process, 2013. [Internet]. [cited 14 December 2021]. Available from: http://www.qualityforum.org/Measuring_Performance/Consensus_Development_Process.aspx.
[52] National Quality Forum. Value set harmonization, 2016. [Internet]. [cited 14 December 2021]. Available from: https://www.qualityforum.org/Value_Set_Harmonization.aspx.
[53] National Quality Forum. eMeasure feasibility assessment: Technical report, 2013. [Internet]. [cited 14 December 2021]. Available from: http://www.qualityforum.org/Publications/2013/04/eMeasure_Feasibility_Assessment.aspx.
[54] Centers for Medicare and Medicaid Services. Value-based programs 2021, 2021. [Internet]. [cited 14 December 2021]. Available from: https://www.cms.gov/Medicare/Quality-Initiatives-Patient-Assessment-Instruments/Value-Based-Programs/Value-Based-Programs.
[55] VanLare JM, Conway PH. Value-based purchasing—national programs move from volume to value. N Engl J Med 2012;367(4):292–5.

CHAPTER

5

International dimensions of clinical decision support systems

Farah Magrabi[a], *Kathrin Cresswell*[b], *and Hamish S.F. Fraser*[c]

[a]Australian Institute of Health Innovation, Macquarie University, Sydney, NSW, Australia [b]Usher Institute, The University of Edinburgh, Edinburgh, United Kingdom [c]Brown Center for Biomedical Informatics, Brown University, Providence, RI, United States

5.1 Introduction

This book has a somewhat United States (US)-centric orientation, because of the large amount of activity in clinical decision support and knowledge-based systems that has been ongoing in the US over several decades. This chapter is intended to provide a broader context by describing the kinds of activities that are occurring elsewhere in the world as well.

The recent interest in artificial intelligence (AI) and machine learning (ML), resurgent after many years of dormancy, has brought increased focus on the importance of clinical decision support systems (CDSSs). Regardless of whether they are based on machine-learned algorithms or human-engineered knowledge, CDSSs are *computer programs which aid human information processing* – ranging from systems that acquire information, analyze information, provide options for decisions, through to systems with the capability of making decisions entirely on their own [1]. Indeed, with safer and more effective integration of CDSSs into clinical workflows, there is potential to realize a return on the decades of investments to transform health services across the world.

International experience with the digitalization of health services and implementation of CDSSs can offer useful insights for contemporary ML-enabled applications as well as future systems that are expected to dynamically adapt to their working context. Outside the US, large-scale programs to digitalize health services in other high-income nations have been underway for many decades. These large-scale implementations offer a wide variety of settings to examine how the local context including social and technical aspects of health systems or *sociotechnical factors* influence use and impact of CDSSs, and can be studied to identify principles in implementation and use which can be transferred across contexts. The US

Clinical Decision Support and Beyond
https://doi.org/10.1016/B978-0-323-91200-6.00013-9

experience, in contrast, largely comes from centers of clinical excellence where the sociotechnical context appears to be fairly uniform in terms of mature customizable health information infrastructures as well as implementation expertise, and therefore more conducive to adoption and use of CDSSs. The experience from low- and middle-income countries (LMICs) is as valuable, given that many high-income nations are tackling social disparities in wealth and access to health services in their own systems.

This chapter gives a snapshot of the international experience with CDSSs by examining implementation and use in Australia, New Zealand, Europe and LMICs focusing the latter on East Africa and Haiti. We examine a variety of nations to highlight current and emerging applications of CDSSs as well as the principles in implementation and use that are transferable across international contexts. Australia and New Zealand are covered in Sections 5.2–5.4. Sections 5.5–5.7 cover the European experience. Emerging uses in high income nations are described in Section 5.8. LMICs are presented in Sections 5.9–5.11. We provide an overview of health systems and the digital health context followed by the applications of CDSSs. Section 5.12 summarizes the chapter and gives an overview of the principles of CDSSs implementation and use.

5.2 Healthcare system organization in Australia and New Zealand

As with any other intervention, the adoption and use of CDSSs to improve care delivery and patient outcomes is influenced by the sociotechnical context of implementation as well as wider health system considerations [1]. The health systems in Australia and New Zealand have a similar structure. Universal health care schemes provide free public hospital care as well as substantial coverage for a variety of other services [2,3]. Private health insurance is voluntary and is used for co-payments and ancillary services such as dental, optical and physiotherapy by a considerable proportion of the population (Australia: 44%; New Zealand: 33%).

Australia's population of 25.7 million is concentrated in the major cities with almost one-fourth spread across vast regional and remote areas about the same size as continental USA. While New Zealand's population of 5 million is also concentrated in cities, it is mostly contained to the two main islands about the size of the US state of Colorado. In both nations the most prominent disparities in health outcomes are between indigenous peoples[a] and the rest of the population, and these are widely acknowledged as being unacceptable.

Public hospitals are managed by state, territory and local governments. Alongside hospitals there are specialized health services at a state or local level e.g. for indigenous health; cancer; child, youth and family health; public health; and telehealth. A standout example of these specialized health services is the dedicated health protection units which were highly effective in controlling local community spread of the virus during the first wave of COVID-19 pandemic in the Australian state of New South Wales (NSW) in 2020 [4]. Primary care is coordinated at a local level, there are 31 primary health networks across Australia, and 20

[a] In Australia, Aboriginal and Torres Strait Islander peoples; in New Zealand, Maori and Pacific Island people.

district health boards in New Zealand. Australian patients are free to go to any available general practitioner (GP) whereas in New Zealand, patients typically enroll with a general practice on an ongoing basis.

5.3 Digital health context in Australia and New Zealand

A pre-existing health information infrastructure including standards-based electronic health records (EHR) is foundational to creating effective CDSSs. One way to integrate CDSS into clinician workflow is by embedding or providing access to tools via an EHR or another clinical information system (CIS). Indeed, CDSSs that provide patient-specific recommendations often rely on CISs to supply their data. This section examines the nature and extent health information infrastructures as well the way they represent their data (*clinical terminologies*) and their ability to communicate, exchange data, and use the information that has been exchanged (*interoperability*).

5.3.1 Existing health information infrastructures

In both nations, the digitalization of health services is at an advanced stage. General practice is highly computerized, with 87% of Australian doctors operating a paperless practice [5]. The use of EHRs is incentivized by government payments including for data records and clinical coding. Most general practice CISs are provided by local developers via on premise or cloud-based implementations.

In contrast, the provision of hospital CISs is dominated by large, mainly North-American suppliers, often requiring systems to be re-configured to local needs (e.g. Cerner Corporation, Epic Systems Corporation, InterSystems and All Scripts). CIS set up varies by jurisdiction, where commercial software systems are procured from one or more vendors and then adapted to local requirements. The hub and spoke model where a variety of commercial CISs are integrated with home-grown interfaces is a commonly employed configuration. Governments continue with large projects to implement or replace CISs [6,7]. There are some efforts to standardize CISs at a jurisdictional level. For example, the state of NSW has recently embarked on an ambitious program to bring together the different instances of its EHR, patient administration and multiple laboratory information systems onto a single platform.

While existing CISs do not support direct sharing of records between hospitals and general practice, locally developed CISs that cover both primary care and hospitals tend to provide better support for the continuity of care (e.g. Telstra Health). Other features in locally developed systems include specific support for complex care needs in indigenous and remote communities. For instance, these CISs are specifically designed to be used by multi-disciplinary teams to manage chronic conditions such as renal disease; are configured to record data required to provide culturally specific and culturally sensitive care; and can be integrated with other systems such as transport management systems to facilitate referrals and follow-up in rural and remote communities.

While telehealth has been an important modality for providing services to rural and remote communities, the COVID-19 pandemic has resulted in a rapid uptake of virtual care across both countries. GPs and hospitals are using a wide variety of configurations including real-time via telephone and video conference; remote patient monitoring; as well as store-and-forward where patient information is gathered and communicated asynchronously.

For patients, Australia has an opt-out national *shared record* for prescription information, medical notes, referrals, and diagnostic imaging reports. At the time of writing, 90% of the population had a shared record, but its suitability for supporting CDSSs is low as most clinical information is not stored in computable form [8]. In contrast New Zealand's approach is decentralized, the majority of general practices have implemented a patient portal which provides patients with access to medical records, test results and facilitates appointment booking, prescription refills and email [9].

5.3.2 Clinical terminologies

In both countries SNOMED CT (Systematized Nomenclature of Medicine – Clinical Terms) is the endorsed and recommended standard for representing data in CISs. Australia has a digital health agency that oversees standards as part of its remit to ensure nationally consistent and interoperable systems [8]. It has its own extension, SNOMED CT AU and its own medicines terminology extension, the Australian Medicines Terminology. However, local adoption remains an ongoing challenge. Only a small number of implemented systems are mature enough to allow full integration of SNOMED CT [10]. Hospital systems, for the most part, still use the International Classification of Diseases rather than a clinical terminology. New Zealand also has its own version, SNOMED CT NZ and a medicines terminology that is overseen by the national health information standards organization [11]. Adoption is not widespread, primary and community care systems are currently being migrated to SNOMED CT from Read Codes which have been used since the 1990s.

5.3.3 Data exchange

Despite high use of digital health technology there is only fragmented interoperability between CISs in different parts of the health system and efforts are underway to improve interoperability. Current CISs do not effectively support the journey through the healthcare system for patients as well as continuity of care when patients move from one provider to another. Consequently, the information presented to clinicians may often be incomplete because it is stored on multiple systems that do not communicate making CISs and the CDSSs they support potentially unsafe [12]. By and large, current CDSSs are limited to structured, clinical data in EHRs. The need to also access data across the care continuum is widely acknowledged as is the need to incorporate non-clinical characteristics and situations (including homelessness, cultural background, and socioeconomic status) for alerts to clinicians e.g. to recall patients for further screening or management.

Even though the Fast Healthcare Interoperability Resources (FHIR) standard is an Australian invention and there is some local adoption of HL7 FHIR as the standard for information exchange, its uptake is not yet widespread. Both countries have programs underway based on modern web services and FHIR data exchange standards and clinical terminologies [8,11]. In Australia, FHIR data exchange standards are being developed for child health, medications, patient administration as well as general practice, and there are now calls for its use in the aged care sector [13].

At an organizational level there is some use of FHIR in hospitals for sharing information between CISs that do not communicate e.g. for transfer of medications information between the intensive care record and EHR to automate the workflow of checking medication orders as patients move from intensive care units to wards [14]. Programs are also underway to improve connectivity between hospitals, community and private healthcare settings. For example, the NSW HealtheNet clinical portal enables exchange information across local health districts, secure messaging with GPs and via the national shared record. Organizational and jurisdictional boundaries are also being bridged by mapping diverse hospital datasets to a common data model (e.g. OHDSI CDM) enabling real-time analytics at scale [15].

5.3.4 Implications for patient safety governance of CDSSs

Patient safety governance refers to the approaches taken to minimize the risk for patient harm across an organization or health system [16]. This section examines approaches to standardization of CISs and operational oversight which are the two main safety governance approaches that are relevant to CDSSs.

5.3.4.1 Regulation of software as a medical device

Regulation of clinical software systems as medical devices provides formal oversight over the safety and effectiveness of CDSSs. Australian regulation of medical devices is risk-based and moving toward international harmonization. While the vast majority of contemporary CISs and CDSSs are not considered as medical devices or excluded in law, emerging CDSS incorporating machine learning algorithms are clearly identified as medical devices subject to regulation [17]. Similar efforts in New Zealand to regulate software as a medical device are in their infancy.

The rules setting out regulatory requirements for CDSSs in Australia are rather complex [17]. Despite recent changes to specifically address software-based medical devices, most CISs and current CDSSs that are based on published clinical rules or static data are not regarded as medical devices and therefore not regulated. For example, a CDSS that digitizes a clinical guideline decision tree for stroke management and recommends a management plan based on decision selections by a clinician is not considered a medical device. While CDSSs specifically intended to be used for medical purposes (e.g. diagnosis or treatment) are regarded as a medical device, they may be considered low-risk and *excluded* by declaring them not to be medical devices in law. Examples of such excluded CDSSs are prescribing software used by doctors as well as modules that provide alerts for adverse drug reactions based on established rules and guidelines.

Other CDSSs regarded as medical devices may be *exempt* i.e. not subject to all the regulatory requirements [17]. For instance, a CDSS does not need to be registered as a medical device if does not directly process or analyze a medical image or a signal from another medical device; and is solely used to provide or support a recommendation to a clinician; and does not replace the clinical judgment of the user in relation to making a diagnosis or decision about the treatment of patients. An example of an exempt CDSS is a surgical workflow system that is based on an accepted clinical guideline, and analyses an individual patient's test results to determine which steps of the surgery are required and presents recommendations that the surgeon can choose to override. However, developers must ensure that such CDSSs meet medical device safety and performance requirements and notify the regulator about the system and report adverse events.

5.3.4.2 *Clinical software standards*

The lack of standards for the design and development of CISs and CDSSs means both functionality and usability remain unaddressed. There is growing recognition of the impact of widely varying functionality and the poor usability of current systems on clinical workflow and patient safety [18]. A noteworthy response are the requirements for usability of CISs developed by the Australian College of GPs [19]. However, there are no similar guidelines addressing functionality of the wide variety of CISs used in hospitals or other parts of the health system.

5.3.4.3 *Safety governance over CDSS deployment and use*

Beyond regulation and standards which can provide safety governance over CDSS development, there are few formal mechanisms to specifically govern deployment and use of CDSSs in clinical settings. A notable example that specifically addresses the configuration of CDSSs are the national guidelines for safe implementation of medications management systems [19]. Another example are practice standards developed by the Australian and New Zealand college of radiologists to assist organizations integrate the use of AI-based CDSSs in real-world settings [20]. Notably, these standards are guided by a set of ethical principles for the use of AI in medicine and represent a useful approach for managing risks and for operational oversight.

AI-based CDSSs have also brought attention to ethical concerns, though there is a rich tradition of studying ethical issues in medical informatics [21]. Indeed, the expectation for CDSS developers, implementers and users to make explicit their commitment to safe, effective, secure, fair, transparent, and sustainable systems will only grow in coming years [22]. To this end, Australia has developed a voluntary ethics framework for AI that is applicable to all industries [23]. New Zealand's approach is perhaps more practicable, whereby it has established a standard to explicitly govern the use of algorithms by public agencies called the Algorithm charter for Aotearoa New Zealand [24]. The Charter commits agencies to a range of measures, including explaining how decisions are informed by algorithms and embeds a unique Te Ao Māori (indigenous) perspective in programs to develop and use algorithms.

5.4 CDSSs in Australia and New Zealand

Current CDSSs can be broadly distinguished into two groups: *task-specific CDSSs* which support individual clinical tasks such as prescribing and information retrieval; and *process support CDSSs* which facilitate multiple tasks or clinical management of a specific condition, representing a problem-driven approach to CDSS use. For instance, a CDSS that supports primary and secondary prevention for individuals at high risk of cardiovascular disease may incorporate facilities to support risk assessment as well as medications management and monitoring tasks. Process support CDSSs are geared toward improving the quality and safety of care by making sure that care is delivered, at the right time and in the right way [1]. This section gives an overview of task-specific as well as process support CDSSs in general practice and hospitals.

5.4.1 Task-specific CDSSs in general practice

Current CDSSs in general practice are mostly task-specific. Here decision support facilities within locally developed, commercial CISs provide doctors with a wide variety of patient-specific alerts, reminders, and recommendations for investigations and interventions on their desktop. *Medication prescribing* is an important focus area for task-specific CDSSs. Based on human-engineered rules, CISs have the ability to generate drug interaction alerts based on comprehensive drug databases, although users are able to set the level of drug interaction alerts, and in several systems turn them off altogether. The use and availability of drug calculators within CISs (weight/dose calculators or warfarin calculators) is variable [10]. Due to a lack of standards, there is no consistency in functionality, and there are concerns with the overall quality, currency as well as consistency of advice presented by different CISs [25]. GPs also use mobile devices to access drug reference information [26], though access to resources outside CISs is seen to be time consuming by some, creating a barrier to use.

Another task commonly supported by CDSSs is *evidence-retrieval*. Here CISs provide general immunization advice, travel information, and text-based health resources as well as links to a vast array of online resources including textbooks and clinical guidelines via the College of GPs [27].

Decision support for *risk assessment* is provided via standalone calculators that are available online or as apps. One such example is a nationally endorsed calculator for computing an absolute cardiovascular risk score, while this system supports preventative health assessments it requires clinicians to enter patient information [28]. Currently most risk assessment CDSSs are based on conventional algorithms, although pooled cohorts of local data are enabling the development of ML-based risk prediction models that are better calibrated to local populations [29]. New Zealand has an algorithm hub that usefully collects a wide variety of clinical models and algorithms to make them more accessible to clinicians [30].

5.4.2 Task-specific CDSSs in hospitals

Like general practice, most CDSSs in hospitals are task-specific and largely aimed at doctors. Here facilities within commercially available CISs support basic automation such as to create tasks, referrals and orders from observations entered and to trigger early warning alerts about patient deterioration based on vital signs [6,31]. *Risk assessment* is an important area, CDSSs generate assessment scores from risk screening for pressure injuries, falls, substance use and delirium. However, the implementation and use of such systems is variable. One possible reason is the non-prescriptive nature of current clinical standards around the use of available algorithms. For example, clinical standards set out by the Australian safety commission to reduce unwarranted variation in care for problems like hip fracture, venous thromboembolism (VTE) prevention, and delirium, do not prescribe specific algorithms leading to disparity in clinical workflows between hospitals [32]. Therefore, decision support requirements for risk assessment need to be specifically determined and managed at a local hospital level, despite state-wide EHRs.

Another challenge for hospital CDSSs is the poor quality of clinical data. In many CISs, patient notes are maintained using templates, therefore key clinical information (e.g. allergies, weight) may be documented in multiple locations within the EHR and there is no universally acknowledged source of truth. This fragmentation of data within EHRs is exacerbated by inconsistencies in user interfaces with respect to data entry as well as the display of patient information.

In contrast, CDSSs for *medication management* rest on more solid foundations. Australia has national standards for in-patient medication charts which are well-supported by some CISs providing a comprehensive view of patient medications [33]. Standardized charts are complemented by national guidelines for on-screen display of medicines information, a tall man lettering list as well as recommendations for terminology, abbreviations and symbols used in medicines documentation [34]. Some hospitals now have more than a decade of experience using electronic medication management systems which have been shown to reduce prescribing errors but the effectiveness of decision support in reducing medication-related harm is yet to be realized [35]. Recent implementations have demonstrated a reduction in mortality following implementation, but the role of CDDSs is unclear [36]. The most frequently implemented category of decision support are drug-drug interaction alerts with some systems incorporating >15,000 alerts to warn prescribers of potential DDIs. Responding to concerns about alert fatigue, new implementations have taken a more thoughtful approach and limited prescribing alerts to allergy checking, pregnancy warnings, therapeutic duplication, some dose-range checking, and a limited number of local decision-support rules (e.g. antibiotic stewardship guidelines). Multipronged programs including data driven approaches to optimize CDSSs for medications are also underway [37,38]. In pathology and imaging, task-specific CDSSs are confined to order sets and basic alerting e.g. for imaging orders in women of child-bearing age.

Evaluations of electronic medication management systems have shown that CDSS use is highly context-specific. In one study which examined use during ward rounds, more than 50% of medication orders generated CDSS alerts but fewer than 20% were read and no orders were changed [39]. This was because the CDSS was largely rendered obsolete in the context of a ward round - medication decisions were made by senior clinicians, while the CDSS alerts were received by junior doctors entering the orders and were not communicated to the senior clinicians. A follow-up study of junior doctors prescribing alone at night, at the same hospital, found that 80% of alerts were read and 5% of medication orders were changed in response, demonstrating that attention to both the content and context of CDSS use is required [40].

Evidence retrieval in hospitals is supported via provider portals that are accessible from within CISs. Portals provide information and a wide variety of resources to support evidence-based practice at the point of care (treatment guidelines, drug reference, patient education, mobile apps etc.). For example, the NSW Clinical Information Access Portal which is available to all health professionals in the public system has been operational for more than 20 years and has provided an opportunity to examine decision support use on a large scale [41]. Early post-implementation studies showed that use was influenced by organizational, professional and cultural factors rather than technical factors [42,43]. These included the presence of champions, organizational cultures which supported evidence-based practice, and the database searching skills of individual clinicians. A follow-up study showed that team functioning had the greatest impact on the effective use of online evidence in terms of reported experiences of improved patient care following system use [44].

5.4.3 Process support CDSSs in general practice

In general practice, process support CDSSs typically *harness EHR data* to facilitate quality improvement activities targeting chronic conditions such as cardiovascular disease, chronic kidney disease and type 2 diabetes. These activities may be incentivized, for instance, in 2019 the Australian government introduced incentives bringing increased focus on such activities.

An example process support CDSS is PREDICT which gives a personalized 5-year cardiovascular disease (CVD) event risk estimate from a regression model and then applies rules to provide individualized recommendations about treatment [45]. Another comprehensive system is the Severe Asthma Global Evaluation (SAGE) which incorporates: a questionnaire battery for patient completion before clinical consultation; asthma and comorbidity modules; a clinical summary page in an asthma management module; a nurse educator module; a structured panel discussion record; and an automatically generated report incorporating all key data [46].

Process-support CDSSs may also *incorporate clinical measurements*. In one study, GPs and practice nurses offered screening for atrial fibrillation with a smartphone-based single-lead ECG to patients ≥65 years of age in rural practices. The program successfully screened 34% of eligible patients and was shown to be cost-effective [47]. Another feature is *audit and feedback*, where CDSS provide feedback to clinicians about their prescribing rates compared with the normative data of their peers, to influence behavior alongside guideline concordant recommendations for providing care during consultations [48]. Such tools can be further enhanced with *clinical dashboards* that enable seamless integration of CDSSs in the EHR environment. In one study, use of a co-design approach ensured that CDSS tools were integrated with the EHR and provided access to guidelines within the clinical workflow [49].

Process support systems enable a problem driven approach to CDSS implementation and use, providing clinicians with patient-specific information that can change decision-making, thereby enhancing care delivery care delivery and improving patient outcomes [50]. However, the long-term use and sustainability of even seemingly simple CDSSs remains elusive in the current payment system which rewards GPs on a fee-for-services basis (Box 5.1) [51].

Recently a data-driven approach to process support CDSSs has emerged. This is based on longitudinal, de-identified and structured EHR data provided by participating practices. Examples include the MedicineInsight program involving 700 practices and 3300 clinicians [52]; and the POLAR program via primary health networks in the states of Victoria and New South Wales [53]. Here individual clinician and practice level quality improvement activities are supported by real-time reports comparing practice activity with best practice clinical guidelines in areas such as asthma, diabetes, anxiety disorder and opioid use. As the POLAR approach is EHR agnostic it has potential to be scaled up, and has been used to examine the impact of the Australian bushfires, the COVID-19 pandemic, and Prostate-specific antigen (PSA) testing at a population level.

BOX 5.1

Case study to illustrate the complexity of integrating a simple process support CDSS in the Australian context [51].

HealthTracker is a CDSS that calculates the patient's absolute risk of heart attack or stroke and then automatically synthesizes recommendations from all relevant clinical guidelines to provide tailored management advice specific to the patient's circumstances. It uses a simple traffic light system to provide alerts to GPs about management recommendations.

Studies undertaken over a 10-year period demonstrate the complexity of implementing and sustaining use of Healthtracker for cardiovascular disease prevention in Australian general practice. Following initial clinical validation studies as well as a pilot implementation, a cluster randomized controlled trial of HealthTracker vs usual care was undertaken in 60 practices. In the intervention group, use of the CDSS resulted in a 10% increase in the number of patients receiving appropriate and timely measurement of cardiovascular risk factors. While there was also a small increase in the percentage of people at high risk of cardiovascular disease receiving recommended medication prescriptions, the overall uptake and sustainability of the system in real-world settings was patchy. Follow-up studies of 41 sites found that use of the CDSS under real-world conditions varied widely between clinicians and practices, and was influenced by multiple factors including the organizational mission and history, leadership, team environment, and the technology.

While HealthTracker had a highly usable interface and its recommendations were actionable, it was *not easily integrated into existing clinical workflows* as well as local work practices for quantifying risk, advising patients, and prescribing medication. During the trial, clinicians experienced many *technical glitches* which were reported to be frustrating and interfered with their use of the CDSS during consultations. Moreover, the inbuilt *algorithms did not facilitate an individualized approach to patient care*, as they prioritized biometrics and family history over personal and cultural context. The *lack of interoperability* with the CIS was also seen to be a barrier to seamless use. While the decision support was specifically designed to read data from two main CISs covering 80% of the Australian market, the risk score and management plan could not be saved back into the patient's electronic record. Use of the CDSS was found to increase the length of consultations and could have a negative financial value for GPs in the current payment system which operates on a fee-for-service basis.

Large-scale curations of EHR data are also enabling the development of sophisticated ML-based CDSSs. Take for instance, a tool to predict the risk of ED presentation which provides an explanation by listing features used to make decisions and incorporates feedback on data missing from the record, prompting GPs to complete missing information [54]. Another data driven approach to clinical audit is based on administrative data about medicines, pathology and imaging from the national shared health record (My Health Record) combined with educational interventions to promote their rational use [55].

EHR data extractions are also being extended to track patient journeys across the care continuum to overcome the lack of interoperability between current electronic record systems. An example is the Lumos linkage program to support integration of services across care settings [56]. Here general practice EHR data is being linked to other health system records including hospitals, emergency departments, mental health services, ambulance records, non-admitted patient services, integrated care enrolments, cancer registrations and mortality records. Although the technology is vendor independent it is not real-time, with linkages being undertaken twice per year.

5.4.4 Process support CDSS in hospitals

A simple way to provide process support in hospitals is via clinical pathways for specific areas such as respiratory disease, renal failure, stroke, cardiology and hip fracture [6]. The use of clinical pathways is often led by nurses and may involve allied health professionals like physiotherapists. Recently these CDSSs incorporate patient reported outcome measures and may also be shared with patients. A major issue with current clinical pathways is that they are not specifically designed for the EHR environment. Essentially when paper-based pathways are transported online they are not seamlessly integrated with the EHR, requiring patient information to be re-entered by clinicians and creating documentation burden.

Another example is in antimicrobial stewardship programs to improve the appropriate use of antibiotics and reduce antibiotic resistance. Here the CDSS—which can be implemented as a standalone system or integrated with CISs—guides prescribers on the appropriate use of antimicrobials, generates approvals; and also provides facilities to restrict use based on the spectrum of action, potential toxicity, and cost. A 4-year trial of one such CDSS as part of an antimicrobial stewardship program across five hospitals along with the concurrent distribution of clinical guidelines improved antimicrobial use and decreased costs, with no observable increase in the length of hospital stay or mortality [57].

Process support systems have also been applied to monitor the safety of medical treatments. An example is a CDSS to monitor the autoimmune adverse effects of alemtuzumab treatment in multiple sclerosis patients which requires regular pathology monitoring [58]. This system combined an automated pathology-monitoring system to prompt and track pathology collection and provided both prescribers and patients with customized alerts about abnormal results for identified risks. An Australia wide-trial showed that automated analysis of pathology results was significantly faster than standard care involving neurologist review, with the system correctly identifying and alerting clinicians and patients about abnormalities.

Another patient safety monitoring example is a national real time prescription monitoring system to monitor the prescribing and dispensing of high-risk medicines such as opioids and benzodiazepines [59]. Here state-based CDSSs link to a national data exchange which captures information from state and territory regulatory systems as well as prescribing and dispensing software. Clinicians are alerted if there is a concern about patients relating to a prescribed medication.

5.4.5 Section summary

Despite systemic differences in the organization and delivery of health services, Australia and New Zealand have had a similar experience with CDSSs. In general practice and hospitals, CDSSs are largely aimed at doctors, they are task-specific using facilities within commercially available CISs to support individual clinical tasks such as prescribing medications, information retrieval and risk assessment. Representing a technology-driven approach, commercial CISs provide general decision support functionality (e.g. alerts and reminders) that needs to be carefully configured to minimize issues with over-alerting and to ensure that alerts are actionable. While there is an increased awareness about the importance of system usability and the need to integrate CDSSs with clinical workflows among developers and implementers, these factors remain as the main barriers to routine use, as do issues with the quality, currency and consistency of advice.

As EHR implementations have matured, implementation and use of CDSSs appears to be shifting to a problem-driven approach with *process support CDSSs* that facilitate a bundle of clinical tasks or clinical management of specific conditions. Here clinical management is built into CDSSs functionality. These systems tend to be better integrated with the EHR, they are deliberately and thoughtfully being designed in partnership with clinicians and patients incorporating behavior change strategies and multiple human-computer interaction methods—including think aloud interviews, prototyping and usability testing—resulting in systems that are better integrated with clinical workflow and more usable. Process support CDSS are also being used for clinical management and safety monitoring at the jurisdictional (e.g. antimicrobial stewardship) and national levels (e.g. high risk-medications).

Despite high use of digital health technology, current CDSSs are limited to structured clinical data in EHRs. Local adoption of clinical terminologies remains an ongoing challenge. Only a small number of implemented systems are mature enough to allow full integration of SNOMED CT. There is only fragmented interoperability between CISs in different parts of the health system and efforts are underway to enable data exchange. CDSSs will also need to be adapted for virtual care that may run in parallel with in person care. For instance, virtual care is likely to affect data quality as data collection may shift to a pre-consultation activity involving nurses with an increased role for patients in documenting data. Clinical workflow is another aspect that will need consideration as CDSSs will need to be seamlessly integrated with in-person and virtual care [60].

5.5 Healthcare system organization in Europe

Europe is characterized by large variability across countries with a range of political contexts and health systems, making it hard to extrapolate general trends in relation to CDSS. In addition, most existing CDSS literature is in English resulting in a lack of accessibility. Europe as a region consists of 44 countries (not including dependencies and other territories) [61], with 24 official languages [62]. Each country has municipalities and there are around 160,000 local governments across Europe. Some of these have a unitary governmental system (with varying degrees of power across local governments) and others have a federal model (where each entity has its own government) [63]. 27 European countries are part of the European Union (EU), characterized by common policies and a single digital market strategy.

Each European country has a different health system organization, which is not governed by the EU. Most countries have universal health coverage, but there are differences in the role

BOX 5.2

Examples of health system organization in European countries.

Eastern Europe: government finances health services, but often struggle with high reliance on the private sector and low quality of care.

France: healthcare provided through salary deductions, central funds administered by government, citizens can take additional private insurance.

Germany: obligatory social insurance, regional decentralized decision-making, private insurance companies [68].

Netherlands: health plans regulated by the government, citizens required to have basic cheap but mandatory health insurance, can purchase additional services on top of this.

Nordic region (Denmark, Finland, Iceland, Norway, Sweden): democratized, focus on public involvement, decentralized decision making to regional authorities [69].

United Kingdom (UK): publicly funded health system, centrally regulated, free to all at the point of care.

of government and private providers (Box 5.2) [64]. There are also differences in the delivery of care and in the way medicine is practiced, especially between Central and Eastern Europe [65–67].

5.6 Digital health context in Europe

5.6.1 Existing health information infrastructures in Europe

There is large variability of EHR use across European countries. For example, high use rates can be seen in Denmark, Estonia, Finland, Greece, Spain, Sweden, and the UK; while Croatia and Poland have low use rates [70]. This reflects a general trend surrounding limited health information infrastructures in Eastern Europe [71]. Use of primary care EHRs in the EU is high, with around 80% of practices using computerized systems [70].

Most European countries have national eHealth strategies and are working toward implementing EHRs and CDSSs a central part of these. However, strategies vary across countries and over time within countries. They range from national EHR implementations to open platform implementations [70,72–78]. In addition, the European Commission increasingly promotes sharing of health information across EU member states.

These different approaches to building health information infrastructures across Europe reflect different strategies to implementing CDSSs, as existing health information infrastructures determine how CDSSs integrate with existing technological and social systems. This is illustrated by work from Aanestad and colleagues who describe how different countries in Europe have implemented ePrescription services and CDSSs [79], in some cases digitizing paper prescriptions and then gradually implementing additional functionalities such as CDSSs on top of these.

BOX 5.3

Types of CDSS in Europe.

CDSSs integrated in hospital EHRs: e.g. Allscripts Healthcare Solutions, Cerner Corporation, Epic Systems Corporation, Meditech, McKesson.

CDSSs integrated in primary care EHRs: e.g. in the UK there is inbuilt decision support in primary care systems including Egton Medical Information Systems (EMIS), SystmOne, and Vision.

Commercially available CDSSs as standalone and/or as EHR-integrated solution (often focusing on specialties e.g. radiology/pathology): Koninklijke Philips N.V., Elsevier B.V., Wolters Kluwer Health,Siemens Healthineers (Germany), IBM, Zynx Health.

Home-grown CDSSs in specific settings.

CDSSs in Europe consist of mainly localized initiatives within a healthcare setting/system or within a specific software that holds clinical data. The main types of CDSSs used include electronic implementation of guidelines to support clinicians and CDSSs to support prescribing and medical tests. These are mostly rule-based and linked to EHRs. Box 5.3 outlines the types of CDSS commonly used in Europe. One or two major vendors in primary and secondary care dominate markets in most European countries. These have CDSS scripts within their software that are vendor central. There are increasing efforts toward open source vendor neutral CDSSs [80,81].

5.6.2 Clinical software regulatory standards in Europe

Since 2021, the EU Medical Device Regulation (MDR) applies to medical device manufacturers, importers and distributors. It applies the notion of software, including CDSS, as a medical device and requires clinical evaluation of systems, including by institutions developing in-house CDSS and vendors who want to enter the European market. These requirements include a need to demonstrate that systems fulfill certain safety and performance standards. Many institutions currently struggle to navigate and conform with these new laws [82]. The EU takes a precautionary risk-based approach to developing CDSSs, where different types of CDSSs fall under different legislations and regulations (e.g. data driven and knowledge based) [83]. AI has only recently been defined by the European Commission and is currently treated like any other software, although no approaches to regulation are evolving.

5.6.3 Clinical terminologies in Europe

There is a diverse and variable use of structures in Europe. While some countries (such as France, Estonia, Slovakia, Denmark) have most of their EHRs structured [84], others, including some eastern European countries are experiencing inconsistencies in terminology standards used, and have different information models and datasets across settings [84]. In some cases there are no national drug catalogues, and there is also no EU-wide standardized drug/ingredient list. Knowledge bases and guidelines are often developed as part of national projects [85], but there are some efforts of joining different data repositories across Europe [86].

ICD-10 is used for diagnosis in many EU member states, and there is an increasing use of SNOMED-CT for medications (e.g. in the UK and Spain) [87]. The European Commission is now planning to promote semantic interoperability for care across borders, and there is consideration of a EU-wide deployment of SNOMED CT as a reference terminology [88].

Although some countries, such as providers in England, are mandated to use SNOMED CT in EHRs [89], there are some concerns around using it for clinician facing applications such as CDSSs due to its granularity and fears that it may burden users. Suggested mitigation strategies have included mapping local clinical vocabularies to SNOMED CT [90].

5.6.4 Data exchange in Europe

Overall there is fragmentation and lack of interoperability as European states are responsible for their own strategy. However, there are some examples surrounding professionally led cross-border initiatives. For example, the European Society for Radiology has developed guidelines and a CDSS tool that hospitals can integrate [91].

Within countries, there is limited inter and intra-organizational interoperability, with some exceptions of small centrally-led countries such as Estonia [92]. Large scale efforts in other countries to promote interoperability have not achieved their objectives (e.g. the National Programme for Information Technology in England) [93].

Health information in most countries is held within organizational silos, but there is some sharing of limited information across settings emerging within countries (e.g. the National Summary Care Records in Norway and England) [94,95]. Initiatives to promote sharing of more detailed information is ongoing in some countries, but is limited to regions in most instances.

FHIR implementations are ongoing but specifications are developed by each country, resulting in a mixture of local terminologies and slow implementations due to a lack of national integrated drive [96]. An exception is the FHIR based national Personal Health Record Platform in Finland, and the EU FHIR based Digital COVID Certificate. Barriers to interoperability do not only include the range of systems and contexts, but also challenges associated with language and localization of systems [97,98].

5.7 CDSSs in Europe

5.7.1 Task-specific CDSS implementation and use in Europe

Task-specific CDSSs in Europe are used in primary care and in hospitals. These are currently mainly medicine-related and uses are highly variable. High use rates can be observed in countries with national strategies and large-scale uses of commercially available EHR systems [99,100].

In primary care, there are systems that integrate with primary care systems and standalone CDSSs. Examples of integrated systems include Cambio Healthcare Systems in Sweden,

BOX 5.4

Examples of widely used web-or app-based CDSSs in European primary care.

Country	Application
France	Web-based CDSS app for antibiotic prescribing in primary care that can be downloaded [103]
Poland	Web-based CDSS for assessing mental-health risks [104]
Iceland	Osteoporosis Risk Advisor App [105]
Norway	Web-based chronic obstructive pulmonary disease (COPD) management CDSS giving diagnostic and treatment advice [106]
UK	Web-based orthodontic treatment plans for dentists [107]

which is the supplier of the EHRs in the county of Östergötland, and provides guideline-based CDSSs for 43 primary care clinics [101]. Big commercial suppliers with integrated CDSSs holding the majority of the market share in the UK are Egton Medical Information Systems (EMIS), SystmOne, and Vision.

Standalone CDSSs can be either web- or app-based. Some suppliers offer a mixture of products interfacing with primary care systems and web-based systems [102]. Many primary care web- or app-based CDSSs are developed and tested in clinical settings, but most fail to scale. Some exceptions are summarized in Box 5.4.

European hospitals also have mainly medicines-related CDSSs, but increasingly implement radiology and oncology CDSSs with AI/ML functionality to support diagnosis. Systems used can be divided into CDSSs integrated with commercially available hospital EHRs, home-grown CDSSs interfacing with EHRs, commercially available systems interfacing with EHRs, and stand-alone web- or app-based CDSSs. Examples of each of these used in European hospitals are given in Table 5.1.

5.7.2 CDSSs across settings used in Europe

In Europe, there are many local efforts integrating data from across different care settings and conditions [120]. These mainly include the use of dashboards to promote data-driven decision making within specific care settings, but depend on the sophistication of existing health information infrastructures. Examples include population health management in UK primary care, where messages generated by primary care system CDSSs can be viewed at aggregated levels through the Clinical Commissioning Group (CCG) medicines management teams who can then tailor alerts to local needs [121]. Other examples include systems that draw on specific specialties and disease areas. For instance, the Wise Antimicrobial Stewardship Program Support System (WASPSS) in Spain, collects aggregate data from several hospitals to develop CDS guidelines and prediction models [122].

TABLE 5.1 Examples of different types of CDSSs used in European hospitals.

CDSSs integrated with commercially available hospital EHRs	Home-grown systems interfacing with EHRs (either commercially available EHR or home-grown EHR)	Commercially available systems interfacing with EHRs	Stand-alone web- or app-based CDSSs
Cerner Corporation and Epic Systems Corporation electronic prescribing modules are implemented across Europe Slovenia: Ljubljana Children's Hospital use Marand's Think!Clinical™ and Think! Meds™, including electronic prescribing functionality, based on OpenEHR	Switzerland: CDSSs for antimicrobial prescriptions, integrated into EHRs in Geneva University Hospitals and Ticino Regional Hospitals [108] Austria: CDSS in Vienna General Hospital, implemented three types of CDSS around monitoring healthcare-associated infections in intensive care, risk prediction in cutaneous melanoma metastases, and a dosage CDSS for medications used after kidney transplantation, system interfaces with Systems Applications and Products in Data Processing (SAP) system developed by Siemens Health Services [109] Belgium: home-grown electronic prescribing drug-drug-interaction CDSS module, one hospital with a home grown EHR system (Primuz) [110] France: venous thromboembolism prophylaxis guidelines using a CDSS [111] Italy: laboratory test ordering system (PROMETEO) [112] Various other examples of home-grown CDSS systems (including post-operative care, neonatal nutrition planning, ventilator management, laboratory CDSS and diagnosis CDSS) [107]	Netherlands: Gaston prescribing CDSS [113] UK: Philips warning systems for clinical deterioration [114] Germany: Pharmacogenomics CDSS [115] – can be either integrated or separate (web or mobile app) UK: Wellsky (formerly JAC) electronic Prescribing functionality	Spain: web-based CDSS for medication dispensing [116] Scotland: web-based CDSS mobile app maternity risk assessment [117] Portugal: dosing CDSS in a network of three hospitals in Lisbon [118] Across Europe: web-based radiology referral guidelines [118,119] France: CDSS for treatment and monitoring of drug poisoning [107]

Use of dashboards has significantly increased with COVID-19, but the degree to which these are used routinely in clinical settings for decision support is unclear [123]. There are also various efforts at integrating genomics data with EHRs to facilitate clinical decision making, e.g. around diagnosis of rare inherited diseases and cancer [124,125].

5.7.3 National and cross-country CDSSs in Europe

There are several national guidelines for decision support and knowledge bases used across Europe with different levels of integration, including interfaces with existing systems and standalone systems. For example, Estonia has an integrated national CDSS architecture [126]. Sweden has a national ePrescription infrastructure with interaction control at the point of prescribing, and a national CDSS for pharmacy chains, which analyses prescriptions from a national repository and flags drug-related problems [127,128].

The national CDSS platform EBMeDS, owned by the Finish medical Society Duodecim, is used in Finland, Belgium and Estonia [129–131]. It integrates with primary care EHRs and includes alerts at the point of prescribing based on drug information, some guideline links and general reminders. Another example of a national drug database is the Dutch G-Standaard [132], which can be integrated within primary and secondary care systems.

An example of a standalone knowledge base can be found in Norway. The electronic health library UpToDate provides access to guidelines via web and app interfaces. It is free to access and operated by Norwegian Institute of Public Health [133,134]. The country is currently working on developing a more integrated national CDSS architecture [135,136].

Similar plans around national CDSSs are taking place in Denmark, where the National eHealth Authority is planning centralized CDSS for primary care and hospital prescribing, which will interface with various existing electronic prescribing systems [137].

There are also some examples of failures to implement national CDSSs. Common underlying factors relate to sociotechnical considerations such as lack of usability, lack of national political drive, and issues with integration with existing systems. One example here is the Prescribing Rationally with Decision Support in General Practice Study (PRODIGY) in the UK. This was a national guideline based primary care CDSS commissioned by government, but studies found no impact of the CDSS on patient outcomes and low use rates among primary care physicians [138]. Underlying reasons were lack of system usability, timing of alerts, and perceived lack of helpfulness of alerts [139]. Similarly, Scotland had considered a national CDSS initiative in the Scottish National Decision Support Programme. The program was halted due to a lack of political drive and integration issues with existing suppliers, even though the proposed system had good traction with clinicians [140].

There are also some examples of cross-country CDSSs in Europe. These tend to relate to specific specialties and include an mHealth CDSS for Parkinson's disease, a cancer database combining bioinformatics to help decision making of oncologists, pharmacogenomics, and radiology CDSSs [91,141–144].

5.7.4 Section summary

Europe is a diverse area consisting of many different countries. These vary not only in cultures and languages but also in health system organization and existing health information infrastructures. Consequently, approaches to implementing and optimizing CDSSs are highly variable, although central to most IT strategies. Challenges include integration with existing systems, localization efforts tailoring systems to local needs and existing practices, and political drive to implement systems. Notable developments include national efforts of implementing CDSSs (particularly in Scandinavian countries and those with national strategies and established infrastructures) and also some CDSSs operating across international boundaries driven by professional agendas.

5.8 Emerging CDSS uses in high-income nations

Although many AI/ML-based CDSSs are currently in development [145–147], only some are currently implemented and routinely used in clinical settings, and there is a lack of evidence in relation to effectiveness and impacts [148]. AI/ML-based diagnosis CDSS relating to images and integrating with specialty systems in hospitals are most common (e.g. radiology and oncology). An example is Aidence, a system that helps radiologists decide if a patient is at risk or needs further investigation and is currently implemented in Dutch and UK hospitals [149]. Another example is Annalise CXR, an AI-based system that is intended to assist clinicians with the interpretation of chest radiographs and provide notification of suspected findings. Unlike traditional computer aided detection systems, which are restricted to a small number of findings for specific conditions, Annalise CXR detects 124 radiological findings and provides localization information for relevant findings [150]. Over the next few years we are likely to see increased integration of such CDSSs into health information infrastructures [151], but there are important issues that need to be navigated, particularly surrounding trust, explainability, regulation and bias.

There have also been some examples of failures of AI-based decision support. For example the Google DeepMind Streams app designed in the UK to alert clinicians to acute kidney injury was abandoned after it became apparent that it breached data protection regulations [152]. The developer had been given access to sensitive patient data during the app's development.

CDSSs are also being considered outside clinical settings. In Australia, for instance, the need for CDSSs in aged care is receiving considerable attention following a public enquiry into its quality and safety [153]. The enquiry specifically identified deficiencies in data practices within the sector as a target for improvement [13]. Development and use of CDSSs to care for this vulnerable population is likely to grow alongside indigenous as well as culturally and linguistically diverse groups.

5.9 Healthcare system organization in low- and middle-income countries: Focus on East Africa and Haiti

LMICs include a wide range of environments and levels of resources. These range from private hospitals in more prosperous cities which may resemble institutions in the US or Europe, to extremely disadvantaged communities seen in sub-Saharan Africa, South Asia or Haiti for example (Box 5.5). Likewise, information systems in the better-off sites tend to

BOX 5.5

Examples of health system organization in low- and middle-income countries.

Kenya: a mixed health care economy with government funded, privately funded and Non-Government Organization (NGO)/donor supported health care.

Rwanda: obligatory social insurance called Mutuelle de Santé that all citizens must contribute a modest annual payment to unless they are severely impoverished in which case the Ministry of Health (MOH) covers them. Also NGO and international donor support is mostly routed through or under management of the MOH.

Haiti: There is a very fragmented health system with many small NGOs and some major health care providers with international support. The latter include Partners In Health/Zanmi Lasante providing care in the Central Plateau area, Centres Gheskio in Port Au Prince area, and the St Boniface hospital foundation in the Southern Peninsula. HIV care is funded by US PEPFAR program and the Global Fund (GFATM) with over 120 clinics providing care managed by CHARESS and iTech/University of Washington, in addition to the PIH/ZL sites.

resemble those in richer environments and may actually use the same products. Some large commercial vendors have subsidiaries and installations in countries like South Africa, and there is an active commercial EHR sector in India and Kenya [154]. The VistA system developed by the US Veterans Administration is also used in Mexico, Jordan [155], and some other LMICs. Major differences in health care requirements, systems and experience are however seen in low-income environments which are the main focus here.

Health care in low-income countries is undergoing major changes with dramatic improvements in outcomes for a number of diseases, and development of new paradigms for care [156,157]. These changes are driven by a number of factors, including increased recognition of the critical need for basic health care to allow countries to develop economically and socially [158]. Other drivers are the response to the HIV pandemic, large-scale initiatives to tackle malaria and tuberculosis including drug-resistant strains, and the growing awareness of the impact of non-communicable diseases worldwide [158]. The vast majority of health care needs in LMICs are similar to those of developed countries, hence treatments exist for them even if they are used infrequently, such as for multidrug-resistant tuberculosis (MDR-TB), hypertension or leukemia. The recognition that the major barriers relate to *getting the available treatments to patients* has led to the field of "Global Health Delivery" [159]. This takes into account not only the staff shortages, logistical challenges, and gaps in available resources for health care, but also the unique advantages that can be presented by trained community health care workers, low costs for generic medications, and the almost ubiquitous access to mobile phones.

Shortages of trained staff are especially critical to the delivery of effective health care in sub-Saharan Africa. Whereas developed countries typically have 240 or more doctors per

100,000 citizens, many countries in sub-Saharan Africa such as Malawi have less than 10 [160]. Shortages are similar for nurses, pharmacists, laboratory technicians, radiographers, IT staff and managers. Although lack of resources to train staff is a key factor, low wages lead to many staff leaving to work in richer countries, and the distribution of staff remaining is heavily skewed to larger cities. At the same time, many of the most vulnerable patients are in the poorest and most rural areas or are members of disadvantaged groups including ethnic minorities, migrant populations, and sex workers [161]. Ensuring that all patients receive good quality care also requires documenting health care needs and delivery of care, and addressing the gaps detected. This has driven many organizations to implement health information systems even in the remotest sites, which, along with the shortage of trained staff, has created a particular opportunity for effective decision support tools. Challenges with severe poverty, lack of infrastructure and high levels of HIV and TB are similar in Haiti to many countries in Sub-Saharan Africa, and the funding sources (such as the US PEPFAR program, the Global Fund) and informatics tools to address these problems are also similar in many cases and some were developed in Haitian health facilities.

5.10 Health information systems in LMICs

Over the past two decades health information systems (HIS) have been deployed in virtually every country worldwide although the nature of the systems and the distribution and level of use varies greatly [162]. Critical enablers were minimum levels of infrastructure particularly electrical power, the availability of lower power consumption IT hardware, and access to basic internet access at least in larger facilities. Initial use of HIS was strongly driven by the needs of disease specific programs being scaled up through international funding mechanisms supporting care for HIV/AIDS, tuberculosis, and malaria (US PEPFAR Program, the Global Fund and UK DFID for example). In the first decade this supported early success stories and rapid innovation, but also left a legacy of vertical disease specific programs in many countries.

Health information systems in LMICs typically fall into 4 main categories:

- District and national reporting systems typically collecting and analyzing aggregate data.
- EHR systems ranging from locally developed disease specific systems to national deployments of scalable standards-based systems.
- Mobile health tools ranging from simple text-message based systems, to customizable platforms, and sophisticated CDSSs, some of which are AI based.
- Supporting systems including laboratory information systems, pharmacy systems for stock management and dispensing, radiology information systems, and systems for disease surveillance.

There are typically multiple examples of all 4 classes of systems but a higher degree of standardization of national reporting systems and EHR systems (1 and 2). Dividing lines are not strongly demarcated, with pharmacy and lab systems often incorporated into EHRs, and many PC based systems have components accessible on mobile devices. In this section we will focus on several low or lower middle-income countries that have important health informatics projects and their evolving use of CDSSs.

A number of organizations have pioneered the development and implementation of health information systems in resource-poor environments. A major factor in this work was the expansion of care for HIV and MDR-TB in many countries over the last decade. Poor quality paper records made management, follow-up of patients and reporting on patient outcomes especially difficult. These issues are particularly important for chronic diseases like MDR-TB which require two years or more of care with complex drug regimens, and HIV which requires lifelong care. As treatment programs grow to hundreds or thousands of patients, effective health information systems become essential to ensure tracking of patient enrolment, detection of patients who are lost to follow-up, access to laboratory data, and the correct prescribing of drug regimens. Monitoring adverse drug events, treatment failure and complications are also important.

Many of the first generation of HIS in low-income settings were designed to address these issues. Customized EHRs, mobile health systems based on text messaging, and national reporting systems were all developed for HIV care, with similar systems for MDR-TB and sometimes for malaria. One of the earliest examples of a health information system used to support chronic disease management in a LMIC was the PIH-EHR. This was developed by the health care NGO Partners In Health (PIH) and first implemented in 2001 in Peru [163]. This web-based EHR system was designed to support clinical care of MDR-TB as it was scaled up from 71 patients [164] to many thousands, supported by funding from the Gates Foundation and subsequently the Global Fund for AIDS, TB, and Malaria. The system recorded patient demographics, basic clinical assessment, TB laboratory data, drug regimens, and treatment status. A second version of this system with similar functionality called the HIV-EHR was created in 2003 to support HIV care in Haiti [165]. Both versions of the system were extended to include reporting to clinicians, managers and funders, and to support the management of medication supplies.

Several other EHR systems have had a significant impact in low-income countries. Baobab Health Systems pioneered a touch screen EHR system in Malawi which was used directly by registration staff, nurses, clinicians and pharmacists for HIV care [166]. It was also adapted to support diabetes and TB care. The DREAM system was an EHR designed to support HIV care and used in several African countries [167]. In Zambia, a system called SMARTcare was implemented in 2006 to support HIV care and was eventually scaled nationwide [168]. It was subsequently implemented for more general care in hospitals in Ethiopia, but found in a recent evaluation study to have low usage and poor usability rating [169]. In Haiti, an EHR system called iSante was developed by iTech/University of Washington, the US CDC, and the Haitian Ministry of Health to support HIV care. It was successfully rolled out to more than 120 clinics [170].

The next generation of EHR systems were built with the recognition that a whole range of diseases and clinical care programs needed to be supported. This was driven by the recognition of the rapid growth of non-communicable diseases even in low-income settings, the need to improve poor outcomes in pregnancy, childbirth and infancy, and the growth of surgery and cancer care. There was frequent criticism that the systems, infrastructure and training from disease specific programs such as HIV care were not helping the majority of patients. At the same time it was recognized that well designed general purpose HIS including EHR systems could leverage existing infrastructure and skills to support a broad range of clinical care needs [171]. Subsequent to the work on MDR-TB and HIV, PIH along with the Rwandan Biomedical Centre partnered with the Regenstrief Institute in Indiana, the AMPATH project in Kenya, and the South African Medical Research Council to create the OpenMRS open source EHR system [172,173]. OpenMRS is a modular EHR system built around a concept dictionary to allow very flexible representation of clinical data, and provides support for a range of medical data coding standards. OpenMRS has now been implemented clinically in more than 45 LMICs [174,175] and in several sites in the US [176]. Most of the initial sites used OpenMRS to support HIV treatment, but the system has now been customized to support treatment of TB, MDR-TB, heart disease, diabetes and cancer. The iSante EHR has been expanded to support primary care and women's health in Haiti; it is currently being replaced with an OpenMRS based system called iSante Plus. Another well-established open source EHR is OpenClinic GA which has been deployed in several larger hospitals in West African countries as well as Rwanda and Burundi and has been used to track health insurance coverage [177]. A survey in Kenya documented a range of EHR systems in use in 2020 [154].

National reporting systems have been focused in many cases on HIV care, for example TRACnet in Rwanda [178] and MESI in Haiti [179]. Subsequently these systems have largely been replaced by a general purpose analysis and reporting system DHIS2 [180]. This system was developed by HISP in South Africa and Norway to collect aggregate data from health facilities and communities with tools for data analysis and mapping. It is now deployed in 80 countries and has been augmented with tools to track individual patient level data for disease surveillance purposes such as for Ebola. In 2020–2021 it was adapted to track testing, quarantining and vaccination for COVID-19 in Sri Lanka [181], Rwanda and other countries. In tandem with the growth of EHR systems has been the development of systems to collect laboratory data, pharmacy dispensing and warehousing systems, systems for tracking insurance, and for disease surveillance. These systems are used widely in health facilities and broader health systems. There are a diversity of types of system including proprietary and some open source software.

All these systems have had to overcome many challenges due to poor infrastructure, limited resources, and lack of trained staff. In most implementations to date, the data have been collected by clinical staff on paper forms and entered by dedicated data entry staff. System outputs are usually in the form of printed reports and patient summaries. Baobab Health was an early pioneer of direct data entry by users, including clinicians, pharmacists, and patient registration staff. Some health facilities run by PIH in Rwanda and Haiti, and all facilities at AMPATH in Kenya also support direct clinician use of OpenMRS. There are established commercial EHR systems used at the point of care in Nigeria and Kenya.

5.10.1 Mobile health (mHealth) systems

A newer strategy for dealing with the challenges of poor infrastructure and communications is the use of mobile phones or other hand-held devices to collect, manage, and transmit data. Over a decade ago, pioneering projects used Personal Digital Assistants in clinical care and surveillance projects in Peru, South Africa, Kenya and Uganda. More recently there has been an explosion in the availability of mobile phones in LMICs, with countries like Haiti and Rwanda going from almost no mobile phone access outside big cities to near universal coverage in half a decade. This has profoundly altered not only the tools available to manage information but also the perception of what is possible in very resource-poor environments. It has also opened up a wide range of new approaches to delivering decision support to health care workers and to patients themselves. MomConnect is an example of a national scale mHealth application in South Africa. It provides text messages to pregnant women appropriate to the stage of pregnancy. It was built using open standards and an open source health information exchange that can connect any standards based front end application to a standards based database [182].

5.10.2 Clinical software standards in LMICs

Software standards for healthcare are currently limited especially in low income settings. Some countries have standards on data confidentiality, security and sharing. For example, Rwanda prevents clinical data from being stored on servers outside the country; data must be stored in a government run data center. WHO has a database of medical device regulation in different countries (the example from Rwanda is here [183]), but medical software such as EHRs and CDSS are not clearly covered at present. The Kenya Bureau of Standards does cover medical software [184] including Primary Health Care Electronic Medical Records Systems, and Health Informatics Information Security Management in Health Using ISO IEC 27002.

5.10.3 Clinical terminologies in LMICs

Initial work on health information systems in LMICs focused on the effective capture of clinical data relevant to the local context. Due to the limited staff experience in typing, the short time available, and the need for generating reports, the data forms were typically structured with limited free text. While potentially suitable for use in decision support systems, there was little or no standardization. The development of OpenMRS started this way with clinical data concepts developed and added to the concept dictionary in each different project. This changed with the development by Andrew Kanter and colleagues at Columbia University of the Columbia International eHealth Laboratory (CIEL) concept dictionary. This brought two key improvements, (1) standardization of the design and wording of concepts to improve clarity and data re-use (2) the mapping of concepts to clinical vocabularies. This mapping was greatly aided by intellectual property donated by Intelligent Medical Objects inc. [185] This approach has been further strengthened by creation of the Open Concept Lab (OCL) [186] which has created tools to allow users of OpenMRS (and potentially other HIS) to receive updates of concepts and add groups of concepts for specific disease areas such as oncology or diabetes.

5.10.4 Standard vocabularies

Use of International Classification of Disease [187] codes is supported by the WHO and has a long history in global health. EHR systems typically use ICD-10 currently although there is initial work to support ICD-11 [188]. More recently a range of international standards have been used in HIS systems in LMICs. These include SNOMED-CT which has been shown to be effective in coding diagnoses and clinical findings [189], and LOINC [190] particularly for coding laboratory data. A variety of coding systems exist for medications. For example, the US national Library of Medicine supported RxNorm [191] is used in a range of projects. The OpenMRS EHR supports the use of ICD-10 and -11, LOINC, RxNorm and SNOMED-CT through the mappings in the CIEL concept dictionary. DHIS2 has a comprehensive list of standards it supports and interoperability with a wide range of other health information systems used in LMICs [192].

5.10.5 Data exchange in LMICs

Some EHR systems in LMICs have supported open standards for data exchange for over a decade, primarily HL7 version 2 for laboratory data exchange. There has also been interoperability between the OpenMRS EHR and DHIS2 since 2009. The development of HL7 FHIR has created new opportunities for effective interoperability. These include upgrading interoperability between EHRs and laboratory information systems [193], linking the Bahmni version of OpenMRS to an open source insurance management system in Nepal (see Digital Square below), and supporting data exchange between the OpenMRS "backend" server and the MicroFrontEnds UI. There is strong interest in replacing the links between OpenMRS and mHealth applications like OpenSRP or CommCare with FHIR based links, and using FHIR based data exports to create OMOP format [194] data for clinical research.

5.10.6 Open source software and global goods in LMICs

The increasing importance of HIS developed with open source software and open design principles in the global health context has led to initiatives to recognize those benefits and optimize the ability of these systems to work together. For a number of systems these benefits include the ability to interoperate and share data. Fig. 5.1 shows the currently approved list of Global Goods recognized by Digital Square [195]. At the health facility level they include EHRs, as well as laboratory, pharmacy and logistics management systems (largely for medications and medical supplies). Also included are mHealth systems for community health management, disease surveillance systems, and tools for human resources and finance. Health data exchange is well represented including OpenHIE and OpenHIM which support data exchange between systems shown by (*). mHero and MomConnect are examples of applications created from several component systems. The Bahmni EHR includes OpenMRS, the OpenELIS lab system, and two additional open source systems, Odoo for logistics and DCM4CHE for radiology image management. This type of digital ecosystem, especially when built using a shared clinical terminology manager like the OpenConceptLab, makes sharing of CDSS components much more feasible than closed, proprietary systems. An additional resource for the creation and use of software for global health is the "Guideline for Digital Health Development". Two reviews of open source EHRs in use in LMICs have described the range of available functions and breath of use of leading systems [196,197].

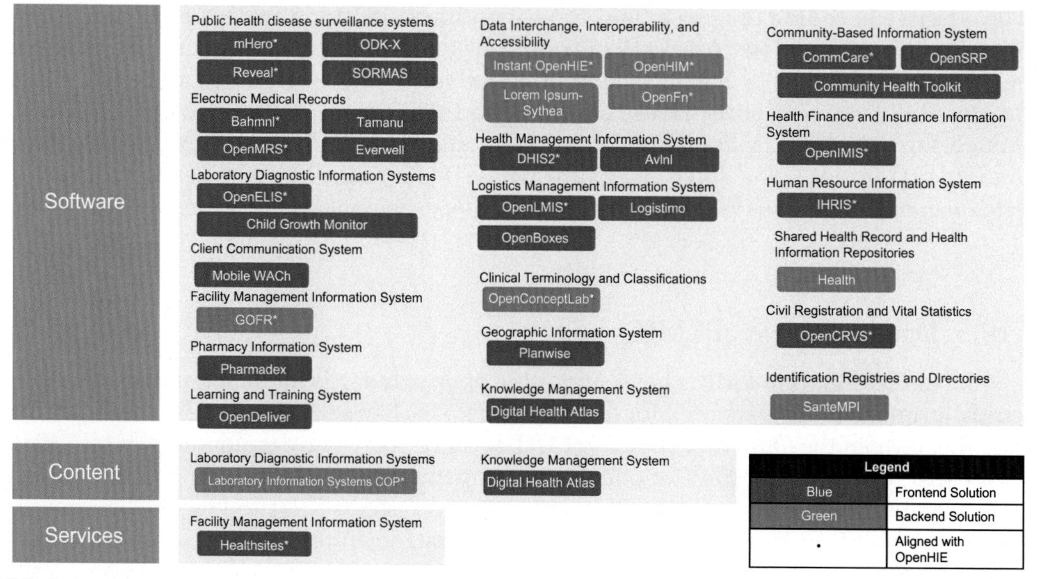

FIG. 5.1 Digital square list of approved Global Goods [195].

5.11 CDSSs in low- and middle-income countries

The use of CDSS in LMICs can also be classified within the four levels introduced earlier: (1) gathering information from different sources needed for making a decision, (2) presenting information with trends and alerts, (3) using a knowledge base to automatically generate suggestions based on a patient's data, (4) using inference models and prognostic predictions to make recommendations. Initial work on decision support for health care workers focused on collecting and displaying critical data for decision making. Perhaps the most fundamental benefit of an EHR system is providing the clinician with complete and timely data to support patient care whenever, and increasingly wherever it is needed. A common cause of delayed or poor-quality health care is the lack of information on previous diagnoses, laboratory results, current or previous medications, or previous surgical procedures. Important examples in East Africa and Haiti are out-of-date CD4 counts or (more recently) viral load results causing a clinician to miss deterioration in a patient with HIV/AIDS. Studies in Kenya and Rwanda have shown benefits of EHRs that generate patient summaries showing up-to-date CD4 count results [198,199].

Decision support can include a wide variety of information and delivery methods. In resource-poor environments, communications and transport are typically very poor, and clinical data such as laboratory results can easily go missing. Patients may default from care for a variety of reasons, typically due to barriers to accessing care; determining who is "lost to follow-up" can be difficult. Increasingly, clinical guidelines are being developed and implemented with the intent to improve quality of care for complex diseases by staff who are often inexperienced and under-trained. However, adherence to paper guidelines is often poor, and there is evidence in developed countries that automated reminders can improve compliance [200]. Another key function of information systems, especially EHR systems, is the improved ability to monitor access to and quality of care. This can reveal gaps in care as well as provide tools to measure the impact of improvements including decision support.

To address the challenges of poor infrastructure and lack of IT experience in low-income settings, decision support has been provided to clinicians in a variety of ways. These include printed patient summaries or consult sheets that incorporate recent clinical and laboratory data, and may include alerts for patients requiring a change in management, such as starting antiretroviral (ARV) therapy. More recently summaries are usually viewed online with point of care EHRs or tablet devices, such as the AMPATH EHR. Clinicians can also receive e-mail alerts, for example, warning of low CD4 counts or new Tuberculosis lab results [199] and text message alerts [201]. Many clinics create regular reports on all patients in a health facility with a specific clinical problem, including data on abnormal laboratory results or loss of follow-up. These have been used for quality improvement initiatives by clinical teams such as PIH/IMB in facilities caring for HIV patients in Rwanda [156].

Blaya et al. evaluated eChasqui, a laboratory reporting system for TB culture and Drug Sensitivity Results (DST) for clinics in Peru [199]. The system includes tools to track specimen samples, reports for quality control in the laboratories, and e-mail alerts to clinicians for new test results. A cluster randomized controlled trial was performed with 12 intervention and 20 control clinics, and a total of 1671 patients were enrolled. The intervention health centers took significantly less time to receive the DST results (median 11 vs. 17 days, $p < 0.001$) and the culture results (5 vs. 8 days, $p < 0.001$). There were 47% fewer DSTs that took over 60 days to arrive ($p = 0.12$). In addition, the patients in intervention health centers had a 20% reduction in time to culture conversion ($p = 0.047$) and there was a greater than 80% reduction in errors in lab results. The system was subsequently expanded from 12 to 259 clinics. A study of an order entry system for MDR-TB drug regimens in the same EHR showed a significant drop in errors compared to a paper-based system (17.4% vs. 3.1%, $p = 0.0074$). A follow up study of log files of alerts and drug prescriptions was carried out 3 years later. It showed that 78% of warnings were overridden by staff and 7% were definitely responded to, a similar behavior to other studies of alerts [202].

Were et al. studied the impact of reminders on the care of pediatric HIV patients in Kenya [203]. All patients had records in the OpenMRS EHR system and had paper summaries generated and printed. Reminders were generated for all patients but were only shown on the intervention group's summaries. They addressed the following issues:

- overdue 6-week HIV DNA polymerase chain reaction tests
- 18-month enzyme-linked immunosorbent assay (ELISA) antibody tests
- CD4 tests
- routine laboratory studies
- chest radiographs
- initiating antiretroviral therapy
- referring malnourished children for nutritional evaluation and assistance

They randomized 1611 patients that were seen by a total of 30 different providers. A fourfold increase was observed in the completion of overdue clinical tasks over the five months of the study when the providers were exposed to reminders (68% intervention vs. 18% control, $p = 0.001$). In addition, orders were placed at an earlier stage for the intervention group (77 days, SD 2.4 days) versus the control group (104 days, SD 1.2 days) ($p=0.001$). Oluoch and colleagues evaluated the impact of CDSS in an EHR system for clinicians managing HIV patients on anti-retroviral (ARV) treatment in 13 clinics in Kenya with 41,062 patients. Clinics were randomly assigned to the control group ($n=6$) or the intervention group ($n=7$). Results showed 1125 patients (11%) in the control group and 1342 (12%) in the intervention group had immunological treatment failure with 332 (30%) versus 727 (54%), respectively, receiving appropriate clinical actions. As treatment failure is associated with high mortality and also risk of passing on resistant strains of HIV, these results have important potential impact.

CDSS in OpenMRS: Decision support has been part of OpenMRS deployments from first applications in Kenya and Rwanda. Initially CDSS rules were hard coded into patient summaries and reports. These included flags for missing CD4 counts for HIV patients, and other abnormal laboratory results. A specific module called "Patient Flags module" has been used to post notices in the chart about patients meeting certain criteria for action. "Bahmni", a version of OpenMRS used in India, Nepal, Jordan and other countries, has developed and used order sets. These typically are part of medication order entry, and provide additional orders such as for extra required medications or laboratory tests to ensure patient safety. A version of OpenMRS was developed to support treatment of MDR-TB for the EndTB project. This system includes a guided long-term protocol, i.e., a clinical pathway or care plan similar to those described earlier in the chapter. Examples of the use of CDSS in OpenMRS include the following:

- Screening applications for HIV, TB and non-communicable diseases in Malawi, with recommendations for next steps in patient management.
- PIH emergency department triage module used in the University Hospital Mirebalais, Haiti based on the South African Triage Scale, which generates a stoplight icon for more serious patients [204].
- An order entry system for chemo-therapy agent prescribing developed for use in Rwanda in 2011 and also used in Haiti. Recent research in Uganda has focused on user requirements for improved EHRs and CDSS in an oncology unit [205].
- The iDeliver project supporting skilled birth attendants in Kenya in maternal care and delivery using clinical pathways in OpenMRS [206].

The new MicroFrontEnds user interface in OpenMRS includes color coded alerts for abnormal laboratory results. In the next stage of development it will include a decision support server incorporating a range of standard decision support tools. It will also support the use of clinical prediction models based on machine learning as part of a new project to improve HIV care [207].

5.11.1 CDSSs delivered by mobile devices: mHealth

The rapid expansion of the use of mHealth in LMICs has provided an effective platform for a very wide range of clinical decision support tools for a diverse range of users including physicians, nurses, clinical officers, and laboratory technicians. Two of the most important groups using mHealth in LMICs are patients and community health care workers (CHWs). Labrique and colleagues developed a framework of 12 types of mHealth applications shown in Fig. 5.2 [208]. These include systems used by the full range of stakeholders and address a wide range

FIG. 5.2 Twelve Common mHealth and ICT applications [208].

of needs from clinical care, telehealth, training, pharmacy and supply chain management, and personnel management. A key goal for mHealth is interoperability with other HIS within the same health system such as providing an mHealth portal to an EHR, pharmacy dispensing system, or laboratory information system.

Early studies by DeRenzi et al. showed evidence of impact of text messages on CHWs practice [209]. A randomized controlled trial was conducted to assess the effect of SMS reminders on CHW follow-up of clients' health issues in their homes. The trial showed a significant reduction in the delay in patient follow-up. The reminders resulted in an 86% reduction in the average number of days a CHW's clients were overdue (9.7 to 1.4 days), with only a small number of cases ever escalating to the supervisor. Benefit was greatest if escalation to the CHW's supervisor occurred when there was no response to the SMSs. Cost was low, $0.84 per patient/year, and could be reduced further with better organization of messages and improved contracts with phone companies. A qualitative study showed that there were often valid reasons for non-response to the reminders, such as the client or CHW traveling, the CHW being busy or sick, or technical problems with the phone. Zurovac et al. [201] studied the effect of sending twice-daily text messages for six months with advice on malaria drug management to health care workers. In a randomized controlled trial of 2269 staff, they showed that adherence to the guidelines (including treatment, dispensing, and counseling tasks) increased by 23.7% immediately after the intervention and 24.5% at six months. The messages were general in nature rather than specific to a particular patient. A study by Seidenberg and colleagues [210] evaluated the time it took to get results of HIV testing of infants back to rural clinics in Zambia. After implementing a system to send SMS messages with results directly to clinics they showed a drop from 44.2 days pre-implementation to 26.7 days with the mHealth application.

Poor adherence to antiretroviral therapy is a major concern worldwide and can lead to poor patient outcomes and also drug resistance. Lester et al. performed an RCT of weekly SMS reminders for ARV adherence to patients in Kenya with follow-up by phone if no response [211]. The outcomes were self-reported ARV adherence (>95% of prescribed doses in the past 30 days at both six and twelve month follow-up visits) and plasma HIV-1 viral RNA load suppression. Adherence to ARV was reported in 168 of 273 (61.5%) patients receiving the SMS intervention compared with 132 of 265 (49.8%) in the control group (RR 0.81, 95% CI 0.69–0.94; $p=0.006$). Suppressed viral loads were found in 156 of 273 patients in the SMS group and 128 of 265 in the control group, with an RR for virologic failure of 0.84 (95% CI 0.71–0.99; $p=0.04$). Similar results were shown by Pop-Eleches and colleagues in an a RCT of four SMS reminders to patients for ARV adherence in Uganda [212]. Further studies have shown the importance of timing, length and wording of text messages, that they be patient specific [213] and the need for human backup if messages are not responded to.

Increasingly this type of application uses an android smart phone and a standard mHealth platform such as CommCare [214], ODK, OpenSRP [215] or others. The impact of CDSS delivered by mHealth has been studied for a wide range of health conditions. A Cochrane review of the evidence for impact of this type of CDSS found 8 RCTs including for cardiovascular diseases, gastrointestinal risk assessment, and maternal and child health. Quality of studies was generally poor and there was limited evidence of clinical impacts. Clearly a wider range of more rigorous studies are required to determine which applications should be adopted and scaled up. There is recent guidance on the best approaches to scaling up mHealth and other HIS application [216]. It is possible that the more positive results with CDSSs delivered through EHRs reflects the need for tight integration into health facility workflow and coordination of clinical data collection and management.

5.11.2 Future trends

The WHO is supporting work on standardized computable guidelines to allow reuse of decision support components. This will build on the standardization of core data sets for specific diseases as well as the use of standard coding systems as described here and be supported by a number of Global Goods systems in Fig. 5.2. It will build on the CDS-Hooks [217] standard to trigger or invoke CDS from within the clinical workflow. A similar approach is seen in a shared repository of reusable computable guidelines created by the US Agency For Healthcare Research and Quality (AHRQ) [218].

Another key area of research and early implementation is in the use of machine learning techniques based on large clinical data sets. These tools have to date mostly been developed in the US, China and some European countries, therefore tailoring of CDSS will likely be necessary for specific populations and regions [219] and to ensure all patient groups are appropriately handled. This will be particularly important for imaging orientated CDSS such as deep learning models for radiology [220] or pathology. An important example of the use of machine learning techniques in LMICs is the use of large EHR derived data sets to create prediction models e.g. for lost to follow up for HIV patients in Haiti [221,222] and Kenya [207]. Models have also been successfully created to predict severe dehydration in children with diarrheal disease in countries like Bangladesh and Rwanda [223]. While this is a rapidly developing field there is limited evidence of clinical benefits to date.

There is a high level of innovations in the mHealth space worldwide with applications for many disease types and clinical settings. Increasingly innovation is occurring in LMICs such as Kenya, Uganda [224] and many areas of South Asia. An area of strong interest from practitioners and funding agencies is the development of CDSS based on theories of behavior change [225], particularly in systems directly used by patients. These CDSS applications should be more effective at influencing positive changes in health care delivery and compliance with care. Recently many mHealth projects for global health funded by the US National Institutes of Health (NIH) and other funders have followed a two-phase pattern. Phase one involves the development, user testing and initial clinical validation of a CDSS application (process evaluation), and in phase two there is a large, typically cluster randomized controlled trial to assess clinical impact and the potential for scaleup.

Another emerging area is diagnostic decision support systems designed for health care providers or patients. Missed diagnosis has been shown to be a major cause of medical errors in the US [226] and many other countries. There has been a rapid scale up of diagnostic apps for patient use (termed Symptom Checkers) in high income countries particularly the UK, the US and Germany. These systems build on many decades of development of diagnostic decision support for physicians [227] as well as newer algorithms, access to large clinical data sets, and automated literature searching and data abstraction. Initial evaluation studies have suggested that some symptom checker apps may approach (but not yet equal) primary care doctors diagnosis and triage decisions, when relying only on symptom data [228,229]. With the acute shortage of doctors and other trained diagnosticians in LMICs, symptom checkers may offer helpful advice to patients on seeking care. Babylon Health in the UK has run a telephone helpline called Babyl in Rwanda for more than 3 years with some data showing good acceptance from patients. There is however no published evidence on its performance, and discussion with health informaticians in Rwanda suggest that there is limited algorithmic support for call staff. Ada Health based in Berlin has started pilot studies of their symptom checker app in Dar es Salam, Tanzania. They initially customized the Ada knowledge base with more than 100 additional diagnoses that are important in East Africa. Currently Ada is running two observational studies in the main teaching hospital in Dar es Salam supported by the Fondation Botnar, one on patient use and one on use by clinical officers (similar to physician assistants). It is essential that symptom checkers are rigorously evaluated before use by vulnerable patients [230], especially in low income settings.

The fundamentals are in place for successful deployment, usage and impact of CDSS in LMICs, but significant barriers remain for successful scale up especially in low-income countries and underserved communities. Adopting common data standards allows use of CDSS on different versions of HIS platforms such as EHRs. Data quality and completeness, while recognized in many projects, requires effective system design and constant attention to training and support of staff. Finally close attention is required to UI design and usability testing, along with clinical evaluation studies. The "mHealth Evidence Reporting and Assessment (mERA) checklist" provides a framework for effective communication of evidence on the effectiveness and generalizability of mHealth and HIS applications [231].

5.12 Conclusions

The majority of current CDSSs aid clinicians by analyzing information and recommending possible actions for clinical management. While the acquisition of some clinical information may be automated (e.g. vital signs from medical devices) CDSSs are highly reliant on the data entered by clinicians into EHRs and other CISs, and require clinicians to confirm the information they provide and be responsible for decisions. The capability to make decisions entirely on their own is confined to very basic tasks such as referrals and orders. Clinical knowledge is mostly human-engineered and dominated by simple rule-based approaches, but there is some use of statistical models. At the same time significant investments and efforts to implement CDSSs are continuing with an increasing focus on AI and ML algorithms but few are in routine use and there is limited evidence of clinical benefits to date.

The international experience in both high-income and LMICs shows that approaches to implementing CDSSs are influenced by health system organization, existing health information infrastructure and organizational culture. In high income nations, top-down approaches where health care systems are centrally administered, particularly at national or regional level seem to be more successful as they are usually underpinned by national health information infrastructure. In LMICs, mobile devices have provided an effective platform to deliver CDSSs in the hands of a diverse range of healthcare workers. Countries with established health information infrastructures have more sophisticated approaches and plans to implement CDSSs, often working closely with vendors to ensure systems are optimized to local contexts. Even so, open source EHRs are facilitating CDSS integration in many nations. And most nations have programs underway to improve interoperability based on FHIR data exchange standards and clinical terminologies.

The approach to CDSSs in high-income nations has historically been tied to EHR implementation and has largely been technology-driven with task-specific systems that continue to struggle with similar issues that have not changed much over the last decade or so. These include poor usability as well as the tension between too many and not enough alerts (alert fatigue), and poor integration of CDSSs within existing workflows and the resulting risk of workarounds. Some of these issues emerge from a lack of integration/interfacing of systems and some from a lack of adaptability/customizability of commercially available systems. While home-grown systems can be tailored to local as well as clinician specific needs, this poses challenges for interoperability of systems across settings and scale-up.

In contrast, the approach to CDSS use in LMICs has been more problem-driven with systems typically deployed to address specific areas of clinical need. As EHR implementations have been optimized, development and use of CDSSs in high income nations appears to be shifting to a problem-driven approach with increasing use of process support CDSSs that facilitate a bundle of clinical tasks or clinical management of specific conditions. Here clinical management is built into CDSSs functionality resulting in systems that are better integrated with the EHR. Contemporary systems are also increasingly being designed in partnership with clinicians and patients and the need to ensure system usability and integration with clinical workflow is widely understood and being addressed with formative evaluations. While significant challenges remain for large-scale deployment, use and impact of CDSS, there is better understanding of the principles of CDSSs implementation and use (Box 5.6). And the sociotechnical context in high and LMICs is increasingly becoming more favorable to harness the benefits of AI and ML-based systems, although issues with trust, explainability, regulation and trust need to be negotiated.

BOX 5.6

Eight principles for effective implementation and use of CDSSs.

1. Ensure health information infrastructure can support intended CDSS.
2. Take a problem-driven approach, a CDSS should address specific clinical needs.
3. Support broader clinical management rather than isolated tasks.
4. Design and implement systems in partnership with clinicians and patients.
5. Integrate CDSSs with clinical workflow.
6. Identify and address issues with system usability.
7. Evaluate early to ensure system fits local requirements and address any issues.
8. Ensure CDSS is suitably embedded i.e. its use and utility in a particular context is established before assessing impact on care delivery and patient outcomes.

Acknowledgments

We wish to thank the following individuals for their invaluable advice and input to assemble this chapter: Jos Aarts, Gabriel Antoja, Ian Bacher, Melissa Baysari, Rong Chen, Enrico Coiera, Andrew Georgiou, Thomas Gitter, Louise Grannell, Chris Pearce, Guy Tsafnat, Julien Venne and Jim Warren.

References

[1] Coiera E. Guide to health informatics. 3rd ed. Boca Raton, FL: CRC Press, Taylor & Francis Group; 2015.
[2] The Commonwealth Fund, International health care system profiles: New Zealand. Available from: https://www.commonwealthfund.org/international-health-policy-center/countries/new-zealand [last accessed 1 November 2021].
[3] The Commonwealth Fund, International health care system profiles: Australia. Available from: https://www.commonwealthfund.org/international-health-policy-center/countries/australia [last accessed 1 November 2021].
[4] Stobart A, Duckett S. Australia's response to COVID-19. Health Econ Policy Law 2021;1–12.
[5] The Royal Australian College of General Practitioners. General practice: Health of the nation 2019. East Melbourne, VIC: RACGP; 2019.
[6] Electronic Medical Record—eHealth NSW. Available from: https://www.ehealth.nsw.gov.au/programs/clinical/emr-connect/emr [last accessed 1 November 2021].
[7] Bonello K, Riley M, McBain D, Lee J, Prasad N, Campbell S, Barker E, Robinson K. Implementation status of hospital EMRs: findings from a survey of public hospitals in Victoria, Australia. In: 38th national conference 2021; 2021. p. 23.
[8] The Australian Digital Health Agency. Available from: https://www.digitalhealth.gov.au/ [last accessed 1 November 2021].
[9] Bowden T, Coiera E. Comparing New Zealand's 'Middle Out' health information technology strategy with other OECD nations. Int J Med Inform 2013;82(5):e87–95.
[10] Pearce CM, de Lusignan S, Phillips C, Hall S, Travaglia J. The computerized medical record as a tool for clinical governance in Australian primary care. Interact J Med Res 2013;2(2):e26.
[11] Health information standards, New Zealand Ministry of Health. Available from: https://www.health.govt.nz/our-work/digital-health/digital-health-sector-architecture-standards-and-governance/health-information-standards-0 [last accessed 1 November 2021].

[12] Magrabi F, Ford D, Arachi D, Williams H. Death despite known drug allergy. In: Johnson JK, Haskell HW, Barach PR, editors. Case studies in patient safety: Foundations for core competencies. Burlington, MA: Jones & Bartlett Learning; 2015. p. 247–60.
[13] Dendere R, Hargrave M, Ferris J, Ebrill K, Frean I, Gray L. A compelling case for the development and adoption of data standards and interoperability in the Australian aged care sector—White Paper. Brisbane: The University of Queensland; 2021.
[14] McDonald K. eHealth NSW rolling out eTOC interoperability between ICU and wards. Pulse+IT; 3 June 2021.
[15] The Australian Health Research Alliance (AHRA)—Transformational Data Collaboration. Available from: https://machaustralia.org/projects/transformational-data-collaboration/ [last accessed 1 November 2021].
[16] Coiera E, Magrabi F. Information system safety. Guide to Health Informatics. Boca Raton, FL: CRC Press, Taylor & Francis Group; 2015. p. 195–220.
[17] Clinical decision support software: Scope and examples. Therapeutic Goods Administration, Department of Health, Australian Government; October 2021. Version 1.1, Available from: https://www.tga.gov.au/resource/clinical-decision-support-software. [last accessed 1 November 2021].
[18] Kim MO, Coiera E, Magrabi F. Problems with health information technology and their effects on care delivery and patient outcomes: a systematic review. J Am Med Inform Assoc 2017;24(2):246–50.
[19] The Royal Australian College of General Practitioners: Practice Technology and Management—Minimum requirements for general practice clinical information systems to improve usability. Available from: https://www.racgp.org.au/running-a-practice/technology/business-technology/minimum-requirements-for-cis [last accessed 1 November 2021].
[20] Royal Australian and New Zealand College of Radiologists. Standards of practice for artificial intelligence, version 1, 2020. Available from: https://www.ranzcr.com/college/document-library/standards-of-practice-for-artificial-intelligence. [last accessed 1 November 2021].
[21] Goodman KW, Miller RA. Ethics and health informatics: users, standards, and outcomes. In: Medical informatics. Springer; 2001. p. 257–81.
[22] Ethics and governance of artificial intelligence for health: WHO guidance. Geneva: World Health Organization; 2021. Licence: CC BY-NC-SA 3.0 IGO.
[23] Australia's Artificial Intelligence Ethics Framework. Available from: https://www.industry.gov.au/data-and-publications/australias-artificial-intelligence-ethics-framework [last accessed 1 November 2021].
[24] Algorithm charter for Aotearoa New Zealand. Available from: https://data.govt.nz/toolkit/data-ethics/government-algorithm-transparency-and-accountability/algorithm-charter/ [last accessed 1 November 2021].
[25] The Royal Australian College of General Practitioners: Position Statement on Electronic clinical decision support in general practice. Available from: https://www.racgp.org.au/advocacy/position-statements/view-all-position-statements/clinical-and-practice-management/electronic-clinical-decision-support [last accessed 1 November 2021].
[26] The Royal Australian College of General Practitioners. Views and attitudes towards technological advances in general practice: Survey report 2018. East Melbourne, VIC: RACGP; 2019.
[27] The Royal Australian College of General Practitioners: Clinical Resources. Available from: https://www.racgp.org.au/ [last accessed 1 November 2021].
[28] Australian absolute cardiovascular disease risk calculator. Available from: https://www.cvdcheck.org.au/ [last accessed 30 October 2021].
[29] Sajeev S, Champion S, Beleigoli A, Chew D, Reed RL, Magliano DJ, Shaw JE, Milne RL, Appleton S, Gill TK, Maeder A. Predicting Australian adults at high risk of cardiovascular disease mortality using standard risk factors and machine learning. Int J Environ Res Public Health 2021;18(6).
[30] New Zealand Algorithm Hub. Available from: https://algorithmhub.co.nz/ [last accessed 5 November 2021].
[31] Cheng DR, South M. Electronic task management system: a pediatric institution's experience. Appl Clin Inform 2020;11(5):839–45.
[32] Clinical Care Standards, Australian Commission on Safety and Quality in Health Care. Available from: https://www.safetyandquality.gov.au/standards/clinical-care-standards [last accessed 10 November 2021].
[33] National standard medication charts, Australian Commission on Safety and Quality in Health Care. Available from: https://www.safetyandquality.gov.au/our-work/medication-safety/medication-charts/national-standard-medication-charts [last accessed 10 November 2021].
[34] National Guidelines for On-Screen Display of Medicines Information, Australian Commission on Safety and Quality in Health Care. Available from: https://www.safetyandquality.gov.au/our-work/e-health-safety/national-guidelines-screen-display-medicines-information [last accessed 10 November 2021].

[35] Westbrook JI, Baysari MT. Nudging hospitals towards evidence-based decision support for medication management. Med J Aust 2019;210(Suppl. 6):S22–s4.
[36] South M, Cheng D, Andrew L, Egan N, Carlin J. Decreased in-hospital mortality rate following implementation of a comprehensive electronic medical record system. J Paediatr Child Health 2022;58(2):332–6.
[37] Baysari MT, Zheng WY, Li L, Westbrook J, Day RO, Hilmer S, Van Dort BA, Hargreaves A, Kennedy P, Monaghan C, Doherty P, Draheim M, Nair L, Samson R. Optimising computerised decision support to transform medication safety and reduce prescriber burden: study protocol for a mixed-methods evaluation of drug-drug interaction alerts. BMJ Open 2019;9(8):e026034.
[38] Chin PKL, Chuah Q, Crawford AM, Clendon OR, Drennan PG, Dalrymple JM, Barclay ML, Doogue MP. Clinical decision support in a hospital electronic prescribing system informed by local data: experience at a tertiary New Zealand centre. Intern Med J 2020;50(10):1225–31.
[39] Baysari MT, Westbrook JI, Richardson KL, Day RO. The influence of computerized decision support on prescribing during ward-rounds: are the decision-makers targeted? J Am Med Inform Assoc 2011;18(6):754–9.
[40] Jaensch SL, Baysari MT, Day RO, Westbrook JI. Junior doctors' prescribing work after-hours and the impact of computerized decision support. Int J Med Inform 2013;82(10):980–6.
[41] Clinical Information Access Portal, NSW, Australia. Available from: https://www.ciap.health.nsw.gov.au/about/usage.html [last accessed 2 November 2021].
[42] Gosling AS, Westbrook JI, Coiera EW. Variation in the use of online clinical evidence: a qualitative analysis. Int J Med Inform 2003;69(1):1–16.
[43] Westbrook JI, Gosling AS, Coiera E. Do clinicians use online evidence to support patient care? A study of 55,000 clinicians. J Am Med Inform Assoc 2004;11(2):113–20.
[44] Gosling AS, Westbrook JI, Braithwaite J. Clinical team functioning and IT innovation: a study of the diffusion of a point-of-care online evidence system. J Am Med Inform Assoc 2003;10(3):244–51.
[45] Pylypchuk R, Wells S, Kerr A, Poppe K, Riddell T, Harwood M, Exeter D, Mehta S, Grey C, Wu BP, Metcalf P, Warren J, Harrison J, Marshall R, Jackson R. Cardiovascular disease risk prediction equations in 400 000 primary care patients in New Zealand: a derivation and validation study. Lancet 2018;391(10133):1897–907.
[46] Denton E, Hore-Lacy F, Radhakrishna N, Gilbert A, Tay T, Lee J, Dabscheck E, Harvey ES, Bulathsinhala L, Fingleton J, Price D, Gibson PG, O'Hehir R, Hew M. Severe asthma global evaluation (SAGE): an electronic platform for severe asthma. J Allergy Clin Immunol Pract 2019;7(5):1440–9.
[47] Orchard J, Li J, Freedman B, Webster R, Salkeld G, Hespe C, Gallagher R, Patel A, Kamel B, Neubeck L, Lowres N. Atrial fibrillation screen, management, and guideline-recommended therapy in the rural primary care setting: a cross-sectional study and cost-effectiveness analysis of eHealth tools to support all stages of screening. J Am Heart Assoc 2020;9(18):e017080.
[48] Hunter B, Biezen R, Alexander K, Lumsden N, Hallinan C, Wood A, McMorrow R, Jones J, Nelson C, Manski-Nankervis JA. Future Health Today: codesign of an electronic chronic disease quality improvement tool for use in general practice using a service design approach. BMJ Open 2020;10(12):e040228.
[49] Bonner C, Fajardo MA, Doust J, McCaffery K, Trevena L. Implementing cardiovascular disease prevention guidelines to translate evidence-based medicine and shared decision making into general practice: theory-based intervention development, qualitative piloting and quantitative feasibility. Implement Sci 2019;14(1):86.
[50] Coiera E. Assessing technology success and failure using information value chain theory. Stud Health Technol Inform 2019;263:35–48.
[51] Abimbola S, Patel B, Peiris D, Patel A, Harris M, Usherwood T, Greenhalgh T. The NASSS framework for ex post theorisation of technology-supported change in healthcare: worked example of the TORPEDO programme. BMC Med 2019;17(1):233.
[52] Busingye D, Gianacas C, Pollack A, Chidwick K, Merrifield A, Norman S, Mullin B, Hayhurst R, Blogg S, Havard A, Stocks N. Data Resource Profile: MedicineInsight, an Australian national primary health care database. Int J Epidemiol 2019;48(6). 1741–h.
[53] Pearce C, McLeod A, Rinehart N, Ferrigi J, Shearer M. What a comprehensive, integrated data strategy looks like: the population level analysis and reporting (POLAR) program. Stud Health Technol Inform 2019;264:303–7.
[54] Pearce C, McLeod A, Rinehart N, Patrick J, Fragkoudi A, Ferrigi J, Deveny E, Whyte R, Shearer M. POLAR diversion: using general practice data to calculate risk of emergency department presentation at the time of consultation. Appl Clin Inform 2019;10(1):151–7.

[55] Bonney A, Metusela C, Mullan J, Barnett S, Rhee J, Kobel C, Batterham M. Clinical and healthcare improvement through My Health Record usage and education in general practice (CHIME-GP): a study protocol for a cluster-randomised controlled trial. Trials 2021;22(1):569.
[56] Correll P, Feyer A-M, Phan P-T, Drake B, Jammal W, Irvine K, Power A, Muir S, Ferdousi S, Moubarak S, Oytam Y, Linden J, Fisher L. Lumos: a statewide linkage programme in Australia integrating general practice data to guide system redesign. Integr Healthc J 2021;3(1):e000074.
[57] Bond SE, Chubaty AJ, Adhikari S, Miyakis S, Boutlis CS, Yeo WW, Batterham MJ, Dickson C, McMullan BJ, Mostaghim M, Li-Yan Hui S, Clezy KR, Konecny P. Outcomes of multisite antimicrobial stewardship programme implementation with a shared clinical decision support system. J Antimicrob Chemother 2017;72(7):2110–8.
[58] Reddel SW, Barnett MH, Riminton S, Dugal T, Buzzard K, Wang CT, Fitzgerald F, Beadnall HN, Erickson D, Gahan D, Wang D, Ackland T, Thompson R. Successful implementation of an automated electronic support system for patient safety monitoring: the alemtuzumab in multiple sclerosis safety systems (AMS3) study. Mult Scler 2019;25(8):1124–31.
[59] National Real Time Prescription Monitoring, Australia. Available from: https://www.health.gov.au/initiatives-and-programs/national-real-time-prescription-monitoring-rtpm [last accessed 10 November 2021].
[60] Aabel B, Abeywarna D. Digital cross-channel usability heuristics: improving the digital health experience. J Usability Stud 2018;13(2):52–72.
[61] Countries in Europe. Available from: https://www.worldometers.info/geography/how-many-countries-in-europe/ [last accessed 1 November 2021].
[62] European Languages: Exploring the Languages of Europe. Available from: https://www.tomedes.com/translator-hub/european-languages [last accessed 1 November 2021].
[63] Local governments in Europe. Available from: https://barometre-reformes.eu/en/local-authorities/ [last accessed 1 November 2021].
[64] Around the World in Healthcare Systems: Europe. Available from: https://www.marshmclennan.com/insights/publications/2021/february/around-the-world-in-healthcare-systems-europe.html [last accessed 1 November 2021].
[65] Profiles of General Practice in Europe. Available from: https://www.nivel.nl/sites/default/files/bestanden/profiles-of-general-practice-in-europe.pdf [last accessed 1 November 2021].
[66] Aarts J, Koppel R. Implementation of computerized physician order entry in seven countries. Health Aff (Millwood) 2009;28(2):404–14.
[67] Bautista MC, Lopez-Valcarcel BG. Review of medical professional organizations in developed countries: problems of decentralized membership registers. AIMS Public Health 2019;6(4):437–46.
[68] Available from: https://www.europarl.europa.eu/workingpapers/saco/pdf/101_en.pdf [last accessed 1 November 2021].
[69] Branding Nordic Healthcare Strongholds. Available from: https://norden.diva-portal.org/smash/get/diva2:1297054/FULLTEXT01.pdf [last accessed 1 November 2021].
[70] Health at a Glance: Europe. Available from: https://www.oecd-ilibrary.org/social-issues-migration-health/health-at-a-glance-europe-2018_health_glance_eur-2018-en [last accessed 1 November 2021].
[71] Healthcare IT System in Eastern Europe. Available from: https://healthmanagement.org/c/it/issuearticle/healthcare-it-system-in-eastern-europe [last accessed 1 November 2021].
[72] Future Digital Health in the EU. Available from: https://www.espon.eu/sites/default/files/attachments/Scientific%20annexes.%20TG%202019%2003%2025_final%20version_0.pdf [last accessed 1 November 2021].
[73] e-estonia. Available from: https://e-estonia.com/solutions/healthcare/e-health-record/ [last accessed 1 November 2021].
[74] e-Health in France: Spotlight on the National Healthcare Digitalization Strategy. Available from: https://healthadvancesblog.com/2020/03/24/e-health-in-france/ [last accessed 1 November 2021].
[75] Overview of the national laws on electronic health records in the EU Member States National Report for Austria. Available from: https://ec.europa.eu/health/sites/default/files/ehealth/docs/laws_austria_en.pdf [last accessed 1 November 2021].
[76] Defining an Open Platform. Available from: https://apperta.org/openplatforms/ [last accessed 1 November 2021].

[77] Catalonia's digital health strategy and new information system model based on #openEHR standard. Available from: https://echalliance.com/catalonias-digital-health-strategy-and-new-information-system-model-based-on-openehr-standard/ [last accessed 1 November 2021].
[78] Who has adopted Open Platforms? Available from: https://inidus.com/who-has-adopted-open-platforms/ [last accessed 1 November 2021].
[79] Aanestad M, Grisot M, Hanseth O, Vassilakopoulou P. Information infrastructures and the challenge of the installed base. In: Aanestad M, Grisot M, Hanseth O, Vassilakopoulou P, editors. Information infrastructures within European health care: Working with the installed base. Cham (CH): Springer; 2017. p. 25–33. Copyright 2017, The Author(s).
[80] Open Clinical Decision Support (OpenCDS). Available from: https://www.opencds.org/ [last accessed 1 November 2021].
[81] Release of open-source software artefacts for decision support solutions. Available from: https://upgx.eu/wp-content/uploads/2018/08/Deliverable-D7.6_P9-MUW.pdf [last accessed 1 November 2021].
[82] Clinical Decision Support Software Regulatory landscape in Europe from May 26th 2020. Available from: https://bigmed.no/assets/Reports/clinical_decision_support_software.pdf [last accessed 1 November 2021].
[83] The use of AI in healthcare: A focus on clinical decision support systems. Available from: https://recipes-project.eu/sites/default/files/2020-11/D2_3_AI_In_Healthcare%28CDSS%29_HarvardStyle.pdf [last accessed 1 November 2021].
[84] Strengthening Health Information Infrastructure for Health Care Quality Governance. Available from: https://www.oecd.org/publications/strengthening-health-information-infrastructure-for-health-care-quality-governance-9789264193505-en.htm [last accessed 1 November 2021].
[85] Hoffmann M, Vander Stichele R, Bates DW, Björklund J, Alexander S, Andersson ML, Auraaen A, Bennie M, Dahl ML, Eiermann B, Hackl W. Guiding principles for the use of knowledge bases and real-world data in clinical decision support systems: report by an international expert workshop at Karolinska Institutet. Expert Rev Clin Pharmacol 2020;13(9):925–34.
[86] EOSC Portal—A gateway to information and resources in EOSC. Available from: https://www.eosc-portal.eu/ [last accessed 1 November 2021].
[87] Semantic strategy in EU and SNOMED CT. Available from: https://confluence.ihtsdotools.org/download/attachments/73368385/SnomedCtShowcase2014_Present_14100.pdf?version=1&modificationDate=1535032086000&api=v2 [last accessed 1 November 2021].
[88] Kalra D, Schulz S, Karlsson D, Vander SR, Cornet R, Rosenbeck GK, Cangioli G, Chronaki C, Thiel R, Thun S, Stroetmann V. Assessing SNOMED CT for large scale eHealth deployments in the EU. Brussels: European Union; 2016.
[89] SNOMED CT. Available from: https://digital.nhs.uk/services/terminology-and-classifications/snomed-ct [last accessed 1 November 2021].
[90] Delvaux N, Vaes B, Aertgeerts B, Van de Velde S, Vander Stichele R, Nyberg P, Vermandere M. Coding systems for clinical decision support: theoretical and real-world comparative analysis. JMIR Form Res 2020;4(10):e16094.
[91] Clinical Decision Support using European Imaging Referral Guidelines. Available from: https://www.myesr.org/esriguide [last accessed 1 November 2021].
[92] X-Road—A Complete Solution for Inter-organizational Information Exchange. Available from: https://cyber.ee/research/reports/T-4-1_X-Road_complete_solution_for_inter-organizational_information_exchange.pdf [last accessed 1 November 2021].
[93] Sheikh A, Cornford T, Barber N, Avery A, Takian A, Lichtner V, Petrakaki D, Crowe S, Marsden K, Robertson A, Morrison Z, Klecun E, Prescott R, Quinn C, Jani Y, Ficociello M, Voutsina K, Paton J, Fernando B, Jacklin A, Cresswell K. Implementation and adoption of nationwide electronic health records in secondary care in England: final qualitative results from prospective national evaluation in "early adopter" hospitals. BMJ 2011;343:d6054.
[94] Dyb K, Warth LL. The Norwegian National Summary Care Record: a qualitative analysis of doctors' use of and trust in shared patient information. BMC Health Serv Res 2018;18(1):252.
[95] Greenhalgh T, Stramer K, Bratan T, Byrne E, Russell J, Potts HW. Adoption and non-adoption of a shared electronic summary record in England: a mixed-method case study. BMJ 2010;340:c3111.

[96] Interoperability out-of-the-box? Assessing FHIR. Available from: https://www.cocir.org/fileadmin/Publications_2021/21035_COC_FHIR_28JUNE_PAPER_FINAL.pdf [last accessed 1 November 2021].
[97] Mozaffar H, Williams R, Cresswell K, Morrison Z, Bates DW, Sheikh A. The evolution of the market for commercial computerized physician order entry and computerized decision support systems for prescribing. J Am Med Inform Assoc 2016;23(2):349–55.
[98] Lost in translation: Epic goes to Denmark. Available from: https://www.politico.com/story/2019/06/06/epic-denmark-health-1510223 [last accessed 1 November 2021].
[99] New funding to help hospitals introduce digital prescribing: Available from: https://www.gov.uk/government/news/new-funding-to-help-hospitals-introduce-digital-prescribing [last accessed 1 November 2021].
[100] Chapter 5: Digitally-enabled care will go mainstream across the NHS. Available from: https://www.longtermplan.nhs.uk/online-version/chapter-5-digitally-enabled-care-will-go-mainstream-across-the-nhs/ [last accessed 1 November 2021].
[101] Karlsson LO, Nilsson S, Bång M, Nilsson L, Charitakis E, Janzon M. A clinical decision support tool for improving adherence to guidelines on anticoagulant therapy in patients with atrial fibrillation at risk of stroke: a cluster-randomized trial in a Swedish primary care setting (the CDS-AF study). PLoS Med 2018;15(3):e1002528.
[102] DXS. Available from: https://www.dxs-systems.co.uk/ [last accessed 1 November 2021].
[103] Delory T, Jeanmougin P, Lariven S, Aubert JP, Peiffer-Smadja N, Boëlle PY, Bouvet E, Lescure FX, Le Bel J. A computerized decision support system (CDSS) for antibiotic prescription in primary care-Antibioclic: implementation, adoption and sustainable use in the era of extended antimicrobial resistance. J Antimicrob Chemother 2020;75(8):2353–62.
[104] Mental Health Decision Support for Everyone. Available from: https://www.egrist.org/ [last accessed 1 November 2021].
[105] Rajput VK, Dowie J, Kaltoft MK. Are clinical decision support systems compatible with patient-centred care? Stud Health Technol Inform 2020;270:532–6.
[106] Vijayakumar VK, Mustafa T, Nore BK, Garatun-Tjeldstø KY, Næss Ø, Johansen OE, Aarli BB. Role of a digital clinical decision-support system in general practitioners' management of COPD in Norway. Int J Chron Obstruct Pulmon Dis 2021;16:2327–36.
[107] Acute Care Systems. Available from: https://www.is.umk.pl/~duch/ref/PL/_9ai-med/list-main.html [last accessed 1 November 2021].
[108] Catho G, Centemero NS, Waldispühl Suter B, Vernaz N, Portela J, Da Silva S, Valotti R, Coray V, Pagnamenta F, Ranzani A, Piuz MF, Elzi L, Meyer R, Bernasconi E, Huttner BD. How to develop and implement a computerized decision support system integrated for antimicrobial stewardship? Experiences from two Swiss hospital systems. Front Digit Health 2020;2:583390.
[109] Schuh C, de Bruin JS, Seeling W. Clinical decision support systems at the Vienna General Hospital using Arden Syntax: design, implementation, and integration. Artif Intell Med 2018;92:24–33.
[110] Muylle KM, Gentens K, Dupont AG, Cornu P. Evaluation of an optimized context-aware clinical decision support system for drug-drug interaction screening. Int J Med Inform 2021;148:104393.
[111] Durieux P, Nizard R, Ravaud P, Mounier N, Lepage E. A clinical decision support system for prevention of venous thromboembolism: effect on physician behavior. JAMA 2000;283(21):2816–21.
[112] Bellodi E, Vagnoni E, Bonvento B, Lamma E. Economic and organizational impact of a clinical decision support system on laboratory test ordering. BMC Med Inform Decis Mak 2017;17(1):179.
[113] Clinical decision support systems for 'making it easy to do it right'. Available from: https://pure.tue.nl/ws/portalfiles/portal/3995882/781514.pdf#page=44 [last accessed 1 November 2021].
[114] Rapid response to clinical deterioration with early warning scoring. Available from: https://www.philips.co.uk/healthcare/clinical-solutions/early-warning-scoring?origin=7_700000002227746_71700000082651121_58700007025690103_43700063209361797&dmcm=Cj0KCQjw- [last accessed 1 November 2021].
[115] Hinderer M, Boerries M, Boeker M, Neumaier M, Loubal FP, Acker T, Brunner M, Prokosch HU, Christoph J. Implementing pharmacogenomic clinical decision support into German hospitals. In: Building continents of knowledge in oceans of data: The future of co-created eHealth. IOS Press; 2018. p. 870–4.
[116] Rogero-Blanco E, Lopez-Rodriguez JA, Sanz-Cuesta T, Aza-Pascual-Salcedo M, Bujalance-Zafra MJ, Cura-Gonzalez I. Use of an electronic clinical decision support system in primary care to assess inappropriate

polypharmacy in young seniors with multimorbidity: observational, descriptive, cross-sectional study. JMIR Med Inform 2020;8(3):e14130.

[117] Yaëlle C, Thomas M, Brian M. A clinical decision support system for maternity risk assessment developed for NHS Scotland. Prog Asp Pediatr Neonatol 2018;1(5):29–43. https://doi.org/10.32474/PAPN.2018.01.000124. PAPN. MS. ID. 000124. 92. UPINE PUBLISHERS Open Access L Progressing Aspects in Pediatrics and Neonatology.

[118] DoseMe Signs 3-Year Agreement with Portugal's National Health Service. Available from: https://doseme-rx.com/news-media/articles/portugals-national-health-service [last accessed 1 November 2021].

[119] Donoso L. ESRiGuide—clinical decision support for imaging referral guidelines in Europe. Roentegenol Radiol 2014;2:93–4.

[120] Schaaf J, Sedlmayr M, Sedlmayr B, Prokosch HU, Storf H. Evaluation of a clinical decision support system for rare diseases: a qualitative study. BMC Med Inform Decis Mak 2021;21(1):65.

[121] Jeffries M, Salema NE, Laing L, Shamsuddin A, Sheikh A, Avery A, Chuter A, Waring J, Keers RN. The implementation, use and sustainability of a clinical decision support system for medication optimisation in primary care: a qualitative evaluation. PLoS One 2021;16(5):e0250946.

[122] Segura BC, Morales A, Juarez JM, Campos M, Palacios F. WASPSS: a clinical decision support system for antimicrobial stewardship. In: Recent advances in digital system diagnosis and management of healthcare. IntechOpen; 2020.

[123] Ivanković D, Barbazza E, Bos V, Brito Fernandes Ó, Jamieson Gilmore K, Jansen T, Kara P, Larrain N, Lu S, Meza-Torres B, Mulyanto J, Poldrugovac M, Rotar A, Wang S, Willmington C, Yang Y, Yelgezekova Z, Allin S, Klazinga N, Kringos D. Features constituting actionable COVID-19 dashboards: descriptive assessment and expert appraisal of 158 public web-based COVID-19 dashboards. J Med Internet Res 2021;23(2):e25682.

[124] Congenica. Available from: https://www.congenica.com/solutions/ [last accessed 4 November 2021].

[125] IBM's Watson for Genomics Launches in First European Hospital. Available from: https://www.labiotech.eu/trends-news/ibm-watson-genomics-european-hospital/ [last accessed 4 November 2021].

[126] The best public sector digital service in Estonia is supporting doctors. Available from: https://e-estonia.com/the-best-public-sector-digital-service-in-estonia-is-supporting-doctors/ [last accessed 1 November 2021].

[127] European countries on their journey towards national eHealth infrastructures. Available from: https://www.ehealth-strategies.eu/report/eHealth_Strategies_Final_Report_Web.pdf [last accessed 1 November 2021].

[128] Ohlund SE, Astrand B, Petersson G. Improving interoperability in ePrescribing. Interact J Med Res 2012;1(2):e17.

[129] Heselmans A, Delvaux N, Laenen A, Van de Velde S, Ramaekers D, Kunnamo I, Aertgeerts B. Computerized clinical decision support system for diabetes in primary care does not improve quality of care: a cluster-randomized controlled trial. Implement Sci 2020;15(1):5.

[130] Duodecim Clinical Decision Support EBMEDS launched in Estonia—a step towards more patient-centered healthcare. Available from: https://www.ebmeds.org/en/2020/05/13/duodecim-clinical-decision-support-ebmeds-launched-in-estonia-a-step-towards-more-patient-centered-healthcare/ [last accessed 1 November 2021].

[131] Duodecim Medical Publications Ltd. Available from: https://www.duodecim.fi/ [last accessed 1 November 2021].

[132] The G-Standaard: the medicines standard in healthcare. Available from: https://www.knmp.nl/producten/gebruiksrecht-g-standaard/informatie-over-de-g-standaard/the-g-standaard-the-medicines-standard-in-healthcare [last accessed 1 November 2021].

[133] Clinical decision support goes mobile for clinicians across Norway. Available from: https://www.wolterskluwer.com/en/news/clinical-decision-support-goes-mobile-for-clinicians-across-norway [last accessed 1 November 2021].

[134] Clinical Decision Support Goes Mobile for Clinicians Across Norway with Free Access to UpToDate Anywhere. Available from: https://www.bloomberg.com/press-releases/2020-02-19/clinical-decision-support-goes-mobile-for-clinicians-across-norway-with-free-access-to-uptodate-anywhere [last accessed 1 November 2021].

[135] Marco-Ruiz L, Malm-Nicolaisen K, Makhlysheva A, Pedersen R. Towards a national clinical decision support framework for Norway: expert assessment and proposed architecture. In: eTELEMED; 2020.

[136] National Clinical Decision Support Framework for Norway: Expert Assessment and Proposed Architecture. Available from: https://www.iaria.org/conferences2020/fileseTELEMED20/eTELEMED_40079.pdf [last accessed 1 November 2021].

[137] National Medication Decision Support System (Denmark). Available from: https://confluence.ihtsdotools.org/pages/viewpage.action?pageId=123898364 [last accessed 1 November 2021].

[138] Eccles M, McColl E, Steen N, Rousseau N, Grimshaw J, Parkin D, Purves I. Effect of computerised evidence based guidelines on management of asthma and angina in adults in primary care: cluster randomised controlled trial. BMJ 2002;325(7370):941.

[139] Rousseau N, McColl E, Newton J, Grimshaw J, Eccles M. Practice based, longitudinal, qualitative interview study of computerised evidence based guidelines in primary care. BMJ 2003;326(7384):314.

[140] Cresswell K, Callaghan M, Mozaffar H, Sheikh A. NHS Scotland's Decision Support Platform: a formative qualitative evaluation. BMJ Health Care Inform 2019;26(1).

[141] Timotijevic L, Hodgkins CE, Banks A, Rusconi P, Egan B, Peacock M, Seiss E, Touray MML, Gage H, Pellicano C, Spalletta G, Assogna F, Giglio M, Marcante A, Gentile G, Cikajlo I, Gatsios D, Konitsiotis S, Fotiadis D. Designing a mHealth clinical decision support system for Parkinson's disease: a theoretically grounded user needs approach. BMC Med Inform Decis Mak 2020;20(1):34.

[142] Seven European centres develop a web-based decision-making support tool for oncologist to improve treatment selection. Available from: https://cancercoreeurope.eu/seven-european-centres-develop-a-web-based-decision-making-support-tool-for-oncologist-to-improve-treatment-selection/ [last accessed 1 November 2021].

[143] Blagec K, Koopmann R, Crommentuijn-van Rhenen M, Holsappel I, van der Wouden CH, Konta L, Xu H, Steinberger D, Just E, Swen JJ, Guchelaar HJ, Samwald M. Implementing pharmacogenomics decision support across seven European countries: the Ubiquitous Pharmacogenomics (U-PGx) project. J Am Med Inform Assoc 2018;25(7):893–8.

[144] Tamborero D, Dienstmann R, Rachid MH, Boekel J, Baird R, Braña I, De Petris L, Yachnin J, Massard C, Opdam FL, Schlenk R, Vernieri C, Garralda E, Masucci M, Villalobos X, Chavarria E, Calvo F, Fröhling S, Eggermont A, Apolone G, Voest EE, Caldas C, Tabernero J, Ernberg I, Rodon J, Lehtiö J. Support systems to guide clinical decision-making in precision oncology: the Cancer Core Europe Molecular Tumor Board Portal. Nat Med 2020;26(7):992–4.

[145] Integration of clinical decision support systems in Dutch radiology departments. Available from: https://essay.utwente.nl/77205/1/Berkel_MA_EEMCS.pdf [last accessed 1 November 2021].

[146] Wellbeing Software Partners with contextflow to Deliver Dynamic AI Clinical Decision Support to Radiologists. Available from: https://www.einnews.com/pr_news/551362056/wellbeing-software-partners-with-contextflow-to-deliver-dynamic-ai-clinical-decision-support-to-radiologists [last accessed 1 November 2021].

[147] AI-Pathway Companion Lung Cancer. Available from: https://www.siemens-healthineers.com/digital-health-solutions/digital-solutions-overview/clinical-decision-support/ai-pathway-companion-lung-cancer [last accessed 1 November 2021].

[148] French government gets ready for AI in healthcare. Available from: https://healthcare-in-europe.com/en/news/french-government-gets-ready-for-ai-in-healthcare.html [last accessed 1 November 2021].

[149] AI-powered clinical applications for the oncology pathway. Available from: https://www.aidence.com/ [last accessed 1 November 2021].

[150] Seah JCY, Tang CHM, Buchlak QD, Holt XG, Wardman JB, Aimoldin A, Esmaili N, Ahmad H, Pham H, Lambert JF, Hachey B, Hogg SJF, Johnston BP, Bennett C, Oakden-Rayner L, Brotchie P, Jones CM. Effect of a comprehensive deep-learning model on the accuracy of chest x-ray interpretation by radiologists: a retrospective, multireader multicase study. Lancet Digit Health 2021;3(8):e496–506.

[151] Lyell D, Coiera E, Chen J, Shah P, Magrabi F. How machine learning is embedded to support clinician decision making: an analysis of FDA-approved medical devices. BMJ Health Care Inform 2021;28.

[152] UK class action-style suit filed over DeepMind NHS health data scandal. Available from: https://techcrunch.com/2021/09/30/uk-class-action-style-suit-filed-over-deepmind-nhs-health-data-scandal/?guccounter=1&guce_referrer=aHR0cHM6Ly93d3cuZ29vZ2xlLmNvbS8&guce_referrer_sig=AQAAAIRPQLfZcfBASjS5G8r-tgBd-R98tgIise6dOEA7QGYSzFsnKPCro4U2up-KsmHHGYz3 [last accessed 9 November 2021].

[153] Final Report of the Royal Commission into Aged Care Quality and Safety. Available from: https://agedcare.royalcommission.gov.au/ [last accessed 12 November 2021].

[154] Muinga N, Magare S, Monda J, English M, Fraser H, Powell J, Paton C. Digital health Systems in Kenyan Public Hospitals: a mixed-methods survey. BMC Med Inform Decis Mak 2020;20(1):1–14.
[155] Hakeem Program. Available from: https://ehs.com.jo/hakeem-program [last accessed 11 November 2021].
[156] Rich ML, Miller AC, Niyigena P, Franke MF, Niyonzima JB, Socci A, Drobac PC, Hakizamungu M, Mayfield A, Ruhayisha R. Excellent clinical outcomes and high retention in care among adults in a community-based HIV treatment program in rural Rwanda. J Acquir Immune Defic Syndr 2012;59(3):e35–42.
[157] Farmer PE, Nutt CT, Wagner CM, Sekabaraga C, Nuthulaganti T, Weigel JL, Farmer DB, Habinshuti A, Mugeni SD, Karasi J-C. Reduced premature mortality in Rwanda: lessons from success. BMJ 2013;346.
[158] Maher D, Ford N, Unwin N. Priorities for developing countries in the global response to non-communicable diseases. Glob Health 2012;8(1):1–8.
[159] Kim JY, Farmer P, Porter ME. Redefining global health-care delivery. Lancet 2013;382(9897):1060–9.
[160] Physicians per 1000 people. Available from: https://data.worldbank.org/indicator/SH.MED.PHYS.ZS [last accessed 11 November 2021].
[161] Strasser R, Kam SM, Regalado SM. Rural health care access and policy in developing countries. Annu Rev Public Health 2016;37:395–412.
[162] Global diffusion of eHealth: Making universal health coverage achievable. Report of the third global survey on eHealth. Geneva: World Health Organization; 2016. Licence: CC BY-NC-SA 3.0 IGO.
[163] Fraser HS, Jazayeri D, Mitnick CD, Mukherjee JS, Bayona J. Informatics tools to monitor progress and outcomes of patients with drug resistant tuberculosis in Peru. Proc AMIA Symp 2002;270–4.
[164] Mitnick C, Bayona J, Palacios E, Shin S, Furin J, Alcántara F, Sánchez E, Sarria M, Becerra M, Fawzi MC, Kapiga S, Neuberg D, Maguire JH, Kim JY, Farmer P. Community-based therapy for multidrug-resistant tuberculosis in Lima, Peru. N Engl J Med 2003;348(2):119–28.
[165] Fraser HS, Jazayeri D, Nevil P, Karacaoglu Y, Farmer PE, Lyon E, Fawzi MK, Leandre F, Choi SS, Mukherjee JS. An information system and medical record to support HIV treatment in rural Haiti. BMJ 2004;329(7475):1142–6.
[166] Douglas GP, Gadabu OJ, Joukes S, Mumba S, McKay MV, Ben-Smith A, Jahn A, Schouten EJ, Landis Lewis Z, van Oosterhout JJ. Using touchscreen electronic medical record systems to support and monitor national scale-up of antiretroviral therapy in Malawi. PLoS Med 2010;7(8):e1000319.
[167] Nucita A, Bernava GM, Bartolo M, Masi FDP, Giglio P, Peroni M, Pizzimenti G, Palombi L. A global approach to the management of EMR (electronic medical records) of patients with HIV/AIDS in sub-Saharan Africa: the experience of DREAM software. BMC Med Inform Decis Mak 2009;9(1):1–13.
[168] The SmartCare electronic health record system. Available from: https://www.moh.gov.zm/?page_id=5265 [Last accessed 27 November 2021].
[169] Tilahun B, Fritz F. Comprehensive evaluation of electronic medical record system use and user satisfaction at five low-resource setting hospitals in Ethiopia. JMIR Med Inform 2015;3(2):e4106.
[170] Matheson AI, Baseman JG, Wagner SH, O'Malley GE, Puttkammer NH, Emmanuel E, Zamor G, Frédéric R, Coq NR, Lober WB. Implementation and expansion of an electronic medical record for HIV care and treatment in Haiti: an assessment of system use and the impact of large-scale disruptions. Int J Med Inform 2012;81 (4):244–56.
[171] Fraser H, Biondich P, Moodley D, Choi S, Mamlin B, Szolovits P. Implementing electronic medical record systems in developing countries. J Innov Health Inform 2005;13(2):83–95.
[172] Allen C, Jazayeri D, Miranda J, Biondich PG, Mamlin BW, Wolfe BA, Seebregts C, Lesh N, Tierney WM, Fraser HS. Experience in implementing the OpenMRS medical record system to support HIV treatment in Rwanda. Stud Health Technol Inform 2007;129(1):382.
[173] Mamlin BW, Biondich PG, Wolfe BA, Fraser H, Jazayeri D, Allen C, Miranda J, Tierney WM. Cooking up an open source EMR for developing countries: OpenMRS—a recipe for successful collaboration. AMIA Annu Symp Proc 2006;2006:529–33.
[174] Verma N, Mamlin B, Flowers J, Acharya S, Labrique A, Cullen T. OpenMRS as a global good: impact, opportunities, challenges, and lessons learned from fifteen years of implementation. Int J Med Inform 2021;149:104405.
[175] Seebregts CJ, Mamlin BW, Biondich PG, Fraser HS, Wolfe BA, Jazayeri D, Allen C, Miranda J, Baker E, Musinguzi N. The OpenMRS implementers network. Int J Med Inform 2009;78(11):711–20.
[176] Bauer NS, Ofner S, Pottenger A, Carroll AE, Downs SM. Follow-up of mothers with suspected postpartum depression from pediatrics clinics. Front Pediatr 2017;5:212.

[177] Karara G, Verbeke F, Nyssen M. ICT-enabled universal health coverage monitoring and evaluation in sub-Saharan health facilities: study in 8 reference hospitals of Rwanda, Burundi, the Democratic Republic of Congo and Mali. J Health Sci 2017;2017(5):215–26.
[178] Nsanzimana S. Linkage to and retention in HIV care and treatment in the Rwanda national HIV program. University of Basel; 2018.
[179] Wang Y, Barnhart S, Francois K, Robin E, Kalou M, Perrin G, Hall L, Koama JB, Marinho E, Balan JG. Expanded access to viral load testing and use of second line regimens in Haiti: time trends from 2010–2017. BMC Infect Dis 2020;20(1):1–13.
[180] Braa J, Sahay S. The DHIS2 open source software platform: Evolution over time and space. LF Celi, Global Health Informatics; 2017. p. 451.
[181] Kobayashi S, Falcón L, Fraser H, Braa J, Amarakoon P, Marcelo A, Paton C. Using open source, open data, and civic technology to address the COVID-19 pandemic and infodemic. Yearb Med Inform 2021.
[182] Seebregts C, Dane P, Parsons AN, Fogwill T, Rogers D, Bekker M, Shaw V, Barron P. Designing for scale: optimising the health information system architecture for mobile maternal health messaging in South Africa (MomConnect). BMJ Glob Health 2018;3(Suppl. 2):e000563.
[183] Medical device country profile Rwanda, WHO. Available from: https://www.who.int/medical_devices/countries/regulations/rwa.pdf [last accessed 15 November 2021].
[184] MOH K. Kenya Bureau of standards—Services. Government of Kenya; 2021. Available from: https://www.kebs.org/index.php?option=com_content&view=article&id=178&Itemid=243.
[185] Intelligent Medical Objects. Available from: https://www.imohealth.com/ [last accessed 11 November 2021].
[186] Open Concept Lab. Available from: https://digitalhealthatlas.org/en/-/projects/661/published [last accessed 15 November 2021].
[187] International Statistical Classification of Diseases and Related Health Problems (ICD), WHO. Available from: https://www.who.int/standards/classifications/classification-of-diseases [last accessed 15 November 2021].
[188] Mugisha M, Byiringiro JB, Uwase M, Abizeyimana T, Ndikubwimana B, Karema N, Kayiganwa M, Tran Ngoc C, Kostanjsek N, Celik C. Integration of international classification of diseases version 11 application program interface (API) in the Rwandan electronic medical records (openMRS): findings from two district hospitals in Rwanda. In: The importance of health informatics in public health during a pandemic. IOS Press; 2020. p. 280–3.
[189] Elkin PL, Brown SH, Husser CS, Bauer BA, Wahner-Roedler D, Rosenbloom ST, Speroff T. Evaluation of the content coverage of SNOMED CT: ability of SNOMED clinical terms to represent clinical problem lists. Mayo Clinic Proc 2006;81(6):741–8.
[190] LOINC: The international standard for identifying health measurements, observations, and documents. Available from: www.loinc.org [last accessed 15 November 2021].
[191] RxNorm, National Library of Medicine, Bethesda, Maryland. Available from: https://www.nlm.nih.gov/research/umls/rxnorm/index.html [last accessed 11 November 2021].
[192] Integration & Interoperability with DHIS2. Available from: https://dhis2.org/integration/ [last accessed 11 November 2021].
[193] Bacher I, Mankowski P, White C, Flowers J, Fraser HS. A new FHIR-based API for OpenMRS. In: AMIA clinical informatics conference; May 2021; 2021.
[194] FitzHenry F, Resnic F, Robbins S, Denton J, Nookala L, Meeker D, Ohno-Machado L, Matheny M. Creating a common data model for comparative effectiveness with the observational medical outcomes partnership. Appl Clin Inform 2015;6(03):536–47.
[195] Digital Square Global Goods Guidebook 2019, PATH, Seattle. Available from: https://digitalsquare.org/global-goods-guidebook [last accessed 15 November 2021].
[196] Syzdykova A, Malta A, Zolfo M, Diro E, Oliveira JL. Open-source electronic health record systems for low-resource settings: systematic review. JMIR Med Inform 2017;5(4):e8131.
[197] Purkayastha S, Allam R, Maity P, Gichoya JW. Comparison of open-source electronic health record systems based on functional and user performance criteria. Healthc Inform Res 2019;25(2):89–98.
[198] Were MC, Shen C, Tierney WM, Mamlin JJ, Biondich PG, Li X, Kimaiyo S, Mamlin BW. Evaluation of computer-generated reminders to improve CD4 laboratory monitoring in sub-Saharan Africa: a prospective comparative study. J Am Med Inform Assoc 2011;18(2):150–5.
[199] Blaya JA, Shin SS, Yagui M, Contreras C, Cegielski P, Yale G, Suarez C, Asencios L, Bayona J, Kim J, Fraser HS. Reducing communication delays and improving quality of care with a tuberculosis laboratory information system in resource poor environments: a cluster randomized controlled trial. PLoS One 2014;9(4):e90110.

[200] Nair BG, Newman SF, Peterson GN, Wu WY, Schwid HA. Feedback mechanisms including real-time electronic alerts to achieve near 100% timely prophylactic antibiotic administration in surgical cases. Anesth Analg 2010;111(5):1293–300.

[201] Zurovac D, Sudoi RK, Akhwale WS, Ndiritu M, Hamer DH, Rowe AK, Snow RW. The effect of mobile phone text-message reminders on Kenyan health workers' adherence to malaria treatment guidelines: a cluster randomised trial. Lancet 2011;378(9793):795–803.

[202] Kerrison F, Fraser HS. The impact of uncertain diagnostic results on responses to a decision support system for TB drug prescribing. AMIA Annu Symp Proc 2008;949.

[203] Were MC, Nyandiko WM, Huang KT, Slaven JE, Shen C, Tierney WM, Vreeman RC. Computer-generated reminders and quality of pediatric HIV care in a resource-limited setting. Pediatrics 2013;131(3):e789–96.

[204] Rouhani SA, Aaronson E, Jacques A, Brice S, Marsh RH. Evaluation of the implementation of the South African Triage System at an academic hospital in central Haiti. Int Emerg Nurs 2017;33:26–31.

[205] Kabukye JK, Koch S, Cornet R, Orem J, Hagglund M. User requirements for an electronic medical records system for oncology in developing countries: a case study of Uganda. AMIA Annu Symp Proc 2018; 2017:1004–13.

[206] Bartlett L, Avery L, Ponnappan P, Chelangat J, Cheruiyot J, Matthews R, Rocheleau M, Tikkanen M, Allen M, Amendola P. Insights into the design, development and implementation of a novel digital health tool for skilled birth attendants to support quality maternity care in Kenya. Family Med Community Health 2021;9(3).

[207] Kimaina A, Dick J, DeLong A, Chrysanthopoulou SA, Kantor R, Hogan JW. Comparison of machine learning methods for predicting viral failure: a case study using electronic health record data. Stat Commun Infect Dis 2020;12(s1).

[208] Labrique AB, Vasudevan L, Kochi E, Fabricant R, Mehl G. mHealth innovations as health system strengthening tools: 12 common applications and a visual framework. Glob Health Sci Pract 2013;1(2):160–71.

[209] DeRenzi B, Findlater L, Payne J, Birnbaum B, Mangilima J, Parikh T, Borriello G, Lesh N. Improving community health worker performance through automated SMS. In: Proceedings of the fifth international conference on information and communication technologies and development. Atlanta, GA: Association for Computing Machinery; 2012. p. 25–34.

[210] Seidenberg P, Nicholson S, Schaefer M, Semrau K, Bweupe M, Masese N, Bonawitz R, Chitembo L, Goggin C, Thea DM. Early infant diagnosis of HIV infection in Zambia through mobile phone texting of blood test results. Bull World Health Organ 2012;90:348–56.

[211] Lester RT, Ritvo P, Mills EJ, Kariri A, Karanja S, Chung MH, Jack W, Habyarimana J, Sadatsafavi M, Najafzadeh M. Effects of a mobile phone short message service on antiretroviral treatment adherence in Kenya (WelTel Kenya1): a randomised trial. Lancet 2010;376(9755):1838–45.

[212] Pop-Eleches C, Thirumurthy H, Habyarimana JP, Zivin JG, Goldstein MP, De Walque D, Mackeen L, Haberer J, Kimaiyo S, Sidle J. Mobile phone technologies improve adherence to antiretroviral treatment in a resource-limited setting: a randomized controlled trial of text message reminders. AIDS (London, England) 2011;25 (6):825.

[213] Mbuagbaw L, Thabane L, Ongolo-Zogo P, Lester RT, Mills EJ, Smieja M, Dolovich L, Kouanfack C. The Cameroon Mobile Phone SMS (CAMPS) trial: a randomized trial of text messaging versus usual care for adherence to antiretroviral therapy. PLoS One 2012;7(12):e46909.

[214] CommCare. Available from:https://www.dimagi.com [last accessed 11 November 2021].

[215] OpenSRP—Open-source smart register platform (SRP). Available from: www.smartregister.org [last accessed 11 November 2021].

[216] Labrique AB, Wadhwani C, Williams KA, Lamptey P, Hesp C, Luk R, Aerts A. Best practices in scaling digital health in low and middle income countries. Glob Health 2018;14(1):1–8.

[217] CDS-Hooks. Available from: https://cds-hooks.org/ [last accessed 15 November 2021].

[218] Lomotan EA, Meadows G, Michaels M, Michel JJ, Miller K. To share is human! Advancing evidence into practice through a national repository of interoperable clinical decision support. Appl Clin Inform 2020;11 (01):112–21.

[219] Futoma J, Simons M, Panch T, Doshi-Velez F, Celi LA. The myth of generalisability in clinical research and machine learning in health care. Lancet Digit Health 2020;2(9):e489–92.

[220] Tariq A, Purkayastha S, Padmanaban GP, Krupinski E, Trivedi H, Banerjee I, Gichoya JW. Current clinical applications of artificial intelligence in radiology and their best supporting evidence. J Am Coll Radiol 2020;17 (11):1371–81.

[221] Puttkammer N, Simoni JM, Sandifer T, Chéry JM, Dervis W, Balan JG, Dubé JG, Calixte G, Robin E, François K. An EMR-based alert with brief provider-led ART adherence counseling: promising results of the InfoPlus adherence pilot study among Haitian adults with HIV initiating ART. AIDS Behav 2020;24(12):3320–36.
[222] Ridgway JP, Lee A, Devlin S, Kerman J, Mayampurath A. Machine learning and clinical informatics for improving HIV care continuum outcomes. Curr HIV/AIDS Rep 2021;1–8.
[223] Levine AC, Barry MA, Gainey M, Nasrin S, Qu K, Schmid CH, Nelson EJ, Garbern SC, Monjory M, Rosen R. Derivation of the first clinical diagnostic models for dehydration severity in patients over five years with acute diarrhea. PLoS Negl Trop Dis 2021;15(3):e0009266.
[224] Mugabirwe B, Flickinger T, Cox L, Ariho P, Dillingham R, Okello S. Acceptability and feasibility of a mobile health application for blood pressure monitoring in rural Uganda. JAMIA Open 2021;4(3):ooaa068.
[225] Mbuthia F, Reid M, Fichardt A. Development and validation of a mobile health communication framework for postnatal care in rural Kenya. Int J Africa Nurs Sci 2021;100304.
[226] Singh H, Meyer AN, Thomas EJ. The frequency of diagnostic errors in outpatient care: estimations from three large observational studies involving US adult populations. BMJ Qual Saf 2014;23(9):727–31.
[227] Berner ES, Webster GD, Shugerman AA, Jackson JR, Algina J, Baker AL, Ball EV, Cobbs CG, Dennis VW, Frenkel EP. Performance of four computer-based diagnostic systems. N Engl J Med 1994;330(25):1792–6.
[228] Gilbert S, Mehl A, Baluch A, Cawley C, Challiner J, Fraser H, Millen E, Montazeri M, Multmeier J, Pick F, Richter C, Turk E, Upadhyay S, Virani V, Vona N, Wicks P, Novorol C. How accurate are digital symptom assessment apps for suggesting conditions and urgency advice? A clinical vignettes comparison to GPs. BMJ Open 2020;10 (12):e040269.
[229] Ceney A, Tolond S, Glowinski A, Marks B, Swift S, Palser T. Accuracy of online symptom checkers and the potential impact on service utilisation. PLoS One 2021;16(7):e0254088.
[230] Fraser H, Coiera E, Wong D. Safety of patient-facing digital symptom checkers. Lancet 2018;392(10161):2263–4.
[231] Agarwal S, Lefevre AE, Labrique AB. A call to digital health practitioners: new guidelines can help improve the quality of digital health evidence. JMIR Mhealth Uhealth 2017;5(10):e6640.

SECTION II

Sources of knowledge for clinical decision support and beyond

CHAPTER

6

Human-intensive techniques

Vimla L. Patel[a,b,c], *Jane Shellum*[d], *Timothy Miksch*[c,d], *and Edward H. Shortliffe*[b,c,e]

[a]Cognitive Studies in Medicine and Public Health, The New York Academy of Medicine, New York, NY, United States [b]Department of Biomedical Informatics, Columbia University, New York, NY, United States [c]Biomedical Informatics, Arizona State University, Phoenix, AZ, United States [d]Information Technology, Mayo Clinic, Rochester, MN, United States [e]Department of Population Health Sciences, Weill Cornell Medical College, New York, NY, United States

6.1 Introduction

This section presents chapters that discuss the various ways that knowledge is derived to serve as a basis for clinical decision support. In this first chapter, we focus on human expertise as a source of knowledge. Subsequent chapters explore data-intensive methodologies, and techniques for synthesizing the collected knowledge of the medical literature. We then further explore how advances in genomics can offer new types of knowledge for clinical decision making and, in turn, how large databases and data-intensive methodologies can be applied to personal and population-based decision making.

Two key determinants are involved in enhancing human effectiveness in clinical decision making: how much the experts know, and how well they apply what they know when devising solutions to problems that may arise. Thus, as we consider the creation of optimal decision support systems, we must similarly consider both the knowledge that those systems embody and the processes they adopt when applying that knowledge. A system can be "dumb" if the knowledge it needs is lacking or faulty, and it can demonstrate "poor judgment" if it reaches inappropriate conclusions despite a wealth of necessary factual knowledge. It means little if we cram huge amounts of knowledge into a system but the program subsequently cannot use it wisely or appropriately.

https://doi.org/10.1016/B978-0-323-91200-6.00030-9

This chapter focuses on human-intensive techniques for acquiring knowledge from a variety of knowledge sources. Much of such work deals with the acquisition of knowledge for the purpose of encoding it to be used in decision support systems. As is described later in this chapter, there are also approaches that deal with the role of experts in creating evidence-based guidelines and institutional protocols. Sometimes such knowledge is created and formalized by expert panels. The emphasis is on what we can learn through interaction with human beings who are excellent at the same task for which the knowledge product is intended. As noted, that means that we need to understand both the factual knowledge that is required to solve the relevant problems and the judgmental knowledge that characterizes an excellent decision maker. Such decision makers are expected to get to the heart of a problem effectively, to discard irrelevant information, and to demonstrate an ability to be creative rather than to solve problems by rote formula every time they arise.

Whereas the next two chapters discuss analytical methods for identifying new or relevant knowledge from databases or the literature, such as data mining techniques, machine learning, and meta-analysis, we focus here on the elicitation of knowledge by interacting with expert human beings. This involves analyzing their behaviors, inferring their beliefs and knowledge, asking them to explain their thought processes and actions, shadowing them while they perform expert tasks, and applying formal or informal methods for extracting from those behaviors and explanations the factual and judgmental knowledge that they appear to be applying. Such interactions can be undertaken by human beings interacting with experts (often called *knowledge elicitation* by cognitive scientists, or *knowledge engineering* by computer scientists) or, less commonly, by computer programs that experts can use to convey what they know for capture in a computer-based representation (often called *interactive transfer of expertise*). This chapter focuses on the former processes, with brief discussion of the history of computer-based transfer of expertise from human being to computer in Section 6.4. Chapters 7 and 8 discuss current approaches to the inference of new knowledge by analyzing large data sets.

With the exploding current interest in machine learning that develops decision models using numerical or statistical methods, leveraging large training datasets, the emphasis on explicit acquisition and engineering of computer-encoded knowledge has decreased. However, even leading deep learning scientists recognize that knowledge and models of human reasoning are too often overlooked [1]. The resurgence of interest in knowledge acquisition is reflected in recent volumes [2,3] and in scientific journals[a] that continue to publish papers in this area. Yet much of the seminal and definitional work was undertaken in the 1990s or earlier, as is reflected in many of the references provided in the sections that follow and in Section 6.4.

There are a number of reasons why one may want to capture expert knowledge [4]. These include:

- **Knowledge preservation**. We want to capture "wisdom," which develops with expertise. Such knowledge is typically experiential, too often undocumented, and we lose it once the expert retires or otherwise leaves the job.

[a]See, for example, *Data & Knowledge Engineering*, https://www.sciencedirect.com/journal/data-and-knowledge-engineering.

- **Knowledge sharing**. Captured expert knowledge, meaningfully represented, can be reused in training programs, where trainees can be taught to develop expert strategies and functional efficiency. Such knowledge also can be shared among those who need to use it for a wide variety of decision-making tasks.
- **Knowledge to form the basis for decision aids**. New technology can be created based on the expert knowledge to help practitioners to make better decisions. The technology, properly implemented, must embody the concepts, principles, and procedures of the work domain.
- **Knowledge that reveals underlying skills**. As the use of expert knowledge is explicated, it also reveals underlying strategies and skills, and how heuristics and intuition are applied in practice.
- **Understandability of human-derived knowledge**. As opposed to the "logic" underlying data-intensive or probabilistic techniques, the rationale for a judgment by a human expert can be explained in a form that makes sense and can thus be more readily accepted by a user.

Although the computer-based representation of knowledge is covered in Section IV of this volume, it is difficult to discuss the acquisition of knowledge without considering the representational issues that motivate and guide the acquisition process. Furthermore, the entire effort to capture and utilize knowledge in computer programs is predicated on the recognition that knowledge has a central role to play in providing tailored guidance through decision support systems.[b] For example, cognitive psychologists have recognized the centrality of domain-specific knowledge in the skilled solving of complex problems [5,6]. Researchers in artificial intelligence have been noting for decades that "knowledge is power" [7] and that general representations and search strategies, once a primary focus in that field, are limited in their ability to create intelligent behavior in machines [8]. Knowledge-dependent computer applications, such as *expert systems* that use expert knowledge to perform complex problem solving and decision making tasks [9], were first introduced in the 1970s and 80s for use when the real experts are scarce, expensive, inconsistent, or simply unavailable on a routine basis. This characterization begs the questions "What is an expert?" and "How do we distinguish the knowledge and abilities of experts from those who are novices, or less expert, in a field?" Although we can easily agree that experts are those who have special skills or knowledge derived from extensive experience in their domain of expertise, their ability to achieve accurate and reliable performance also shows flexibility and adaptivity in their environment that is difficult to explain by factual knowledge alone. We recognize that experts know "how" and "when", not just "what," and any attempt to capture knowledge for computer representation and use must recognize that these two general classes of knowledge are equally important.

Knowledge acquisition is a very general term that may be defined as the process of identifying and eliciting knowledge from existing sources—from domain experts, from documents, or inferred from large datasets—and subsequently recording or encoding that knowledge so that it can be verified, validated, and utilized. This volume discusses the design

[b]Summarizing and sharing both knowledge and expert guidance can of course also be offered in documents such as written guidelines or decision diagrams. The elicitation of pertinent knowledge and the presentation of such approaches are summarized in Sections 6.3.4 and 6.5.

and implementation of such knowledge-based approaches and the evaluation of their performance. However, reproducible methods to acquire such knowledge, and to assure its accuracy, are typically discussed separately, even though they are intimately related to the design and production of decision support methodologies. A knowledge base used in a clinical decision-support system might contain knowledge structures that represent potential findings and diagnoses and the relationships among them (*conceptual* or *factual* knowledge). They may also utilize guidelines or algorithms that operate on such knowledge structures (*procedural* knowledge), and possibly offer an application logic that is used to apply guidelines and algorithms to the underlying conceptual structure. All these types of knowledge must be combined to achieve a functioning decision-support facility, and the elicitation of knowledge needed for expert performance must address each type of knowledge, not just "facts".

In addition to the knowledge used within a clinical decision support system, another type of knowledge is essential when used in practice: knowledge about the socio-cultural environment in which the system will operate (*contextual* knowledge). This knowledge is required to ensure that the system delivers the five "rights" of clinical decision support: the right information, to the right person, in the right intervention format, through the right channel, at the right time within the workflow [10]. All these types of knowledge must be combined to achieve a functioning decision support facility.

Note that some knowledge is local to an institution or organization and needs to be incorporated for effective and relevant decision guidance. This includes both (a) factual knowledge, such as medications available or preferred in an organization's formulary (including resultant potential drug-drug interactions) and approved order sets for particular indications; and (b) procedural knowledge, such as how guidance is triggered or is incorporated into workflow, and with which entities it interacts. Local decision support often involves knowledge resources derived or obtained from other sources, human or otherwise, and modified to conform to institutional preferences and organizational/business logic. This adaptation itself is a human-centered process, carried out by individuals, teams, or committees in an organization charged with that responsibility. The principles of this chapter accordingly apply to this localization and adaption process as well as to the creation of the more generic (non-localized) knowledge from which it is derived.

The techniques and theories that enable knowledge elicitation can be viewed within the context of the process illustrated in Fig. 6.1. The process begins with methods to "extract" knowledge from human experts (knowledge acquisition [*KA*] or knowledge elicitation [*KE*]), followed by the representation of that knowledge (*KR*) in a computationally tractable form that supports knowledge-based agents or applications [11]. Many people would then include the verification and validation of the output of those knowledge-based agents or applications as part of the complete process, since they provide feedback regarding the quality of the contents of the underlying knowledge structures.

There is a variant of Fig. 6.1 in which the knowledge is acquired not from a single expert collaborator but rather from a group of experts, perhaps through a consensus development process or by studying several experts and merging what one has learned into a single knowledge base. The field of cognitive science offers several methods for understanding the reasoning processes, mental models, and knowledge used by experts when they solve problems, as well as for dealing with team decision making and consensus development. The following

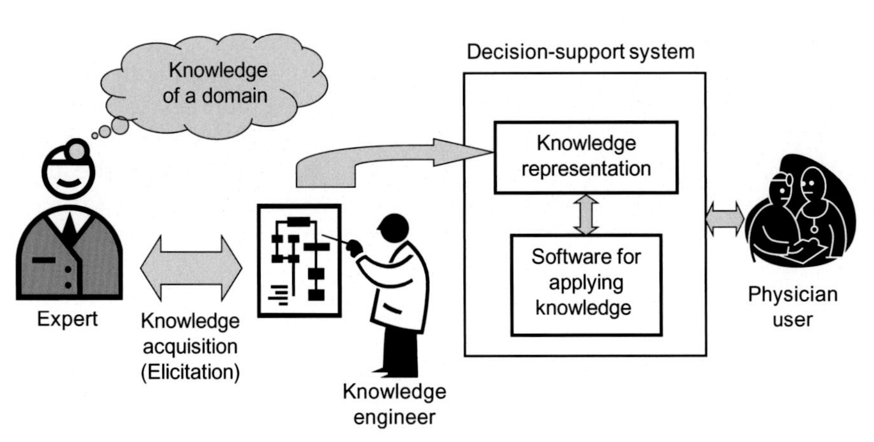

FIG. 6.1 The classical view of knowledge engineering, in which an individual who knows the technical details of a system's representational conventions also has the skills of interviewing and observation necessary to work closely with an expert (or a group of experts) in order to obtain the needed knowledge and to convert it to a computationally useful form.

subsections present some of those notions. There are also formal methods by which experts work together, supported by the literature and formal research studies, to reach consensus in formulating knowledge (e.g., the process of evidence-based guideline development [12]). This applies, for example, to the processes by which organizations and practices both create and adapt knowledge for use, through assigned personnel, teams, or committees, as noted above. The acquisition and representation of consensus guidelines is further discussed in Chapter 10.

Finally, there has been substantial work to develop computer programs that acquire knowledge directly from experts (see Fig. 6.2). Termed *knowledge acquisition systems* or *knowledge authoring systems* (see Section 6.4), these programs are intended to fill the role of knowledge engineer, providing human beings with a computational environment for assessing

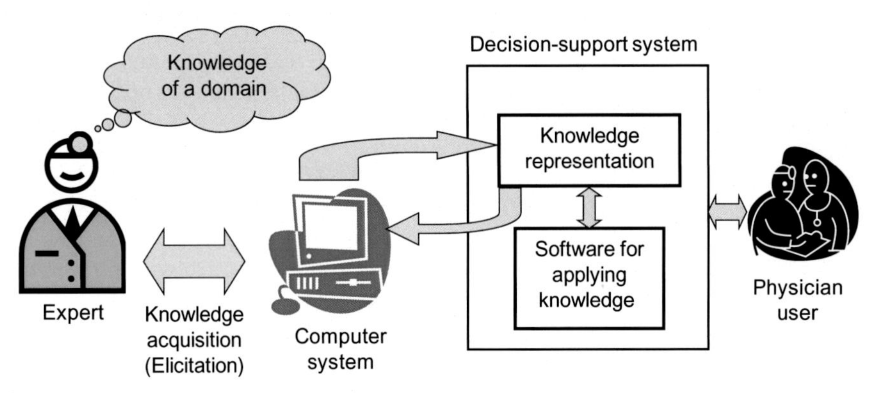

FIG. 6.2 The interactive transfer of expertise using a computer program for knowledge acquisition. Note that such programs will generally both create new knowledge *and* use pre-existing knowledge to guide the knowledge acquisition process. See also Section 6.4.

what knowledge is missing from a system and transferring their knowledge so that it can be encoded for that system's use. Such programs may be tightly coupled with the decision support system itself, allowing the system's decision-making abilities to be assessed and debugged as part of the knowledge acquisition/enhancement process. They always rely on access to the preexisting knowledge in the system, as is indicated by the arrows going in both directions between the computer and the knowledge base in the figure.

6.2 Theoretical basis for knowledge acquisition

We now focus on the frequently cited theoretical basis that underlies the numerous methods and techniques that exist to elicit domain knowledge from sources such as relevant experts. The currently accepted psychological basis for KA depends on defining and acknowledging the concept of expertise. Two major goals of expertise research have been (a) to understand what distinguishes outstanding individuals in a domain from those who may be less outstanding and (b) to characterize the development of expertise. This approach originated with the pioneering research of deGroot [13] in the domain of chess, from which it extended to investigations of expertise in a range of content domains, including physics [14,15], music [16], sports [17], and medicine [18,19]. This research has shown that, on average, the achievement of expert levels of performance in any domain requires about ten years of full-time experience. An "expert" is someone who has achieved a high level of proficiency, as indicated by various measures, such as international "Elo" ratings in chess,[c] world rankings in various athletic endeavors, and certification by a sanctioned licensing body, as in medical subspecialties.

6.2.1 The nature of expertise

In medicine, the expert–novice paradigm has contributed to our understanding of the nature of medical expertise and skilled clinical performance. Expert physicians have extensive general knowledge of medicine (acquired through medical school and residency training) and deep, detailed knowledge of their relatively narrow areas of specialization (acquired from both training and clinical experience). Every experienced physician has acquired common wisdom and medical knowledge as well as certain mastery in the application of medical skills; this constitutes generic expertise. Investigators have suggested the following classification of levels of expertise [20]:

- A *beginner* is a person who has only routine, lay knowledge of a domain; an example is a typical patient.
- A *novice* is someone who has begun to acquire the prerequisite knowledge assumed in the domain, such as a medical student; novices have a basic familiarity with the core concepts, the language, and to a lesser extent, the culture of medicine.
- An *intermediate* is above the beginner level but below the sub-expert level and is typically a senior medical student or a junior resident.

[c]Named for the system's creator, Árpád Élő, a Hungarian-born American physics professor.

- A *sub-expert* (e.g., a specialist solving a clinical problem outside his or her domain of expertise) possesses generic knowledge and experience that exceeds that of an intermediate but lacks specialized knowledge of the medical subdomain in question.
- An *expert* (e.g., a cardiologist or an experienced intensive care nurse) has specialized knowledge of the subdomain in addition to broad generic knowledge.

The development of expertise has been shown to follow a somewhat counterintuitive trajectory. It is often assumed that the novice becomes an expert by a steady, gradual accumulation of knowledge and fine-tuning of skills. That is, as a person becomes more familiar with a domain, his or her level of performance (e.g., accuracy and quality) gradually increases. It turns out, however, that one generally can document a degradation in performance as a subject moves from novice to expert. This has been referred to as the *intermediate effect* [20,21]. It has been repeatedly demonstrated that superior expert performance is mediated by highly structured and richly interconnected domain-specific knowledge. Experts' knowledge is hierarchical and densely interconnected, which allows new pieces of information to become well integrated. Given that a novice's knowledge base is sparse and an expert's knowledge base is intricately interconnected, an intermediate may have many of the pieces of knowledge in place but lack the extensive connectedness of an expert, leading to the intermediate effect just mentioned. For example, expert cardiologists are routinely called upon to integrate clinical findings at various levels of aggregation, from biochemical abnormalities evidenced in blood tests to perturbations at the system level to clinical manifestations as expressed in the patient's complaints. After the performance degradation phase due to the intermediate effect, practitioners develop the missing connections among concepts in their knowledge base and, as they gain experience in the execution of a task, their performance becomes increasingly smooth, efficient, and automatic.

A great deal of experts' knowledge is finely tuned and highly automated, enabling them to execute a set of procedures in an efficient, yet highly adaptive manner, which is sensitive to shifting contexts. They can readily filter out irrelevant information. Novices, as opposed to intermediates, do not conduct irrelevant searches, simply because they lack knowledge rich enough to generate such searches. Studies demonstrate that expert performance is not a result of generally superior memory skills, but it is a function of a well-organized knowledge base adapted to recognizing familiar configurations of stimuli. The nature of experts' organized knowledge can also account for their superior perceptions of patterns. This is demonstrated compellingly in studies of expert radiologists, where they can be shown to look at the x-ray image at a glance, to develop an immediate impression, and then to search the image for findings that fail to fit or that otherwise modify the initial impression. For more details on the nature of expertise, refer to several of the key books in the field [22–26].

One of the things that domain experts know about is the procedures they use in their practices. They learn many "heuristics" or rules of thumb [27]. These compiled, top-level procedures can lead experts to skip steps when they describe the processes by which they carry out their task. Some such heuristics are shared with other experts, but others are ones they have created on their own [18]. In addition, experts have meta-cognitive awareness of their own strategies and how they manage their resources [28]. Meta-cognition refers to the collection of cognitive process and functions that individuals use when thinking about their own cognition (about the way that they think).

Thus, when such experts work with knowledge engineers or KA programs, their goal is to articulate their existing knowledge so that a computer or another decision support mechanism is able to replicate or support human expert performance in the task for which they have specialized expertise. Given the complexity of the types of knowledge and perceptual issues that characterize human expertise, it is challenging to capture such knowledge and to encode it for computer use (or to articulate for other guidance methods) so that expert performance by a third party can be achieved.

6.2.2 Role of mental models

One of the significant challenges of knowledge representation, especially from a cognitive perspective, is to devise mechanisms for capturing and representing products of human clinical comprehension. For example, during a clinical diagnosis task, clinicians perceive, focus on, comprehend, and create solutions using available patient information. The summary results of this process (e.g., a diagnosis or an assessment and plan) are often documented for future users (e.g., another physician during a later shift). However, the intermediate thought processes (i.e., how the diagnosis was reached, or what relationships among the available data were considered) are quite difficult to capture and use for the future. Such a representation would be useful not only for characterizing the nature of knowledge that is used for clinical diagnosis, but also for developing intelligent applications that can support decision making. While much has been written about rule-based, probabilistic and knowledge-based solutions, there has been very little research on how to capture and distinguish among the corresponding mental models of clinicians. Insights on this topic are especially important given the variability in clinicians' expertise in a given disease. For example, there will be differing mental models of a disease between a specialist in the relevant discipline and a less experienced individual such as a house officer or even a primary-care physician. The patient's mental model of his or her own disease is of course even more rudimentary.

Such mental models are especially useful for documenting the basis for the diagnosis and for sharing information regarding patients and their care transitions. These models are designed to answer questions such as "how does this work?" or "what will happen if I take the following action?" or "why is the patient's blood glucose level so high given the medications he is currently receiving?" Running a computer-based model corresponds to a process of mental simulation for generating possible future states of a system from an observed or hypothetical state. An individual's mental models provide predictive and explanatory capabilities of the function of a given system [21].

6.2.3 Team-based decisions and shared knowledge

The development of clinical guidelines and other decision support tools involves multiple team players such as attending physicians, consultants, clinical trainees, computer scientists, and psychologists with a range of expertise, unique vocabularies, and specific mental models. Shared mental models are an extension of the mental model concept and reflect the shared and collective knowledge of a team. They provide reciprocal expectations, which enable teams to coordinate and to make predictions about the behavior and needs of their colleagues

[29]. Individual mental models can be studied through a wide range of experimental tasks that involve prediction and explanation. Shared or team mental models are best captured using naturalistic or quasi-naturalistic methods[d] that characterize communication and collective expertise. The study of teams necessitates a convergence of methods that focus on both individual and collective performance.

6.3 Cognitive task analysis

What, then, are the approaches that have allowed non-experts to analyze, understand, and capture (e.g., by encoding those approaches in a computer system or by summarizing them in a decision diagram) the ways that individual experts make decisions? The general approach that cognitive scientists use in analyzing the basis for human performance is known as *cognitive task analysis* (CTA). Its purpose is to capture the way the mind works—to capture *cognition*. CTA should describe the basis for skilled performance that is being studied. The methods in this field are varied, and a detailed exposition is beyond the scope of this book. In using CTA, cognitive scientists try to capture what people are thinking about, what they are paying attention to, the strategies they are using in making decisions, what they are trying to accomplish, what information they discard, and what they know about the way a process works [4]. The three key aspects of CTA are (1) knowledge elicitation, (2) data analysis, and (3) knowledge representation, where in the generic case the representation of knowledge conforms to formal criteria and methods that may not be inherently computational, even though they might provide insight when one is constructing a computer system's knowledge base in the same domain. Cognitive scientists will utilize one of a variety of knowledge representation schemes to describe and capture what they have learned and to compare the expertise and reasoning processes of individuals (for example, novices versus experts when presented with identical problems). In the following sections, we briefly describe each of these three key aspects of CTA.

6.3.1 Knowledge elicitation (KE) methods

Conducting KE studies is often complex and resource intensive. As a result, it is important to select the appropriate KE methods and tools at the outset of such projects to ensure that the end product is amenable to the planned application domain. One of the key issues to consider when planning a KE study is the source of the knowledge to be elicited. The use of domain experts is probably the most common and simultaneously problematic source of knowledge [30]. This approach presupposes the selection of individuals with sufficient domain knowledge, interest in participating in the KE process, and minimal bias—a combination of attributes not always easily attained.

Further complicating the use of domain experts is the frequent need to collect knowledge from multiple experts. Groups of experts often are needed to mitigate the problems associated

[d] *Naturalistic methods* involve observing subjects in their natural environment. Sometimes compromises are required because of the nature of that environment, in which case one attempts to carry out experiments in settings designed to be as close as possible to the natural one—i.e., *quasi-naturalistic*.

with using single experts, as described later in this chapter [31], which may lead to knowledge elicitation with incomplete or potentially ineffective contents. However, though the use of multiple experts has the potential benefit of utilizing group synergies to generate consensus [32,33], it is also not without potential pitfalls. Most notable are the difficulties surrounding the merging of multiple experts' knowledge [33] and the potential for the resulting knowledge to represent a single expert's opinion or input, rather than a true group consensus [31]. Despite these potential concerns, the benefits of using multiple experts in the knowledge elicitation process generally outweigh the disadvantages.

Straightforward interview techniques often are used because they require a minimum level of resources, can be performed in a relatively short period, and can yield a significant amount of qualitative knowledge. The disadvantages of interview techniques include a frequent lack of quantitative data, which are needed for the input into the next step in the process. Furthermore, the results often can be biased due to the framing or presentation of questions or the selection of topics that are of interest only to researchers [32–35]. However, perhaps most importantly, interviews simply lead to introspective opinions of the collaborating experts, and the knowledge elicited may *not* correspond to what they actually do when solving problems in the domain. For this reason, most knowledge engineers and psychologists who perform knowledge elicitation would prefer to observe the experts as they carry out tasks, either in simulated or "real world" environments. In order to gain insight into their mental processes, the experts may be asked to talk aloud about what they are doing and thinking *while they are performing the task*. In the world of cognitive science, the capture of such responses generated during problem solving is known as applying a *think-aloud protocol* [36].

In contrast, ethnographic evaluations of expert performance are naturalistic observational studies conducted in context, with a minimum of knowledge engineer or psychologist involvement in the workflow or situation under consideration ("fly on the wall" observations). Such studies also implicitly evaluate the knowledge used by those experts and have been applicable in a variety of domains, ranging from air traffic control systems to complex healthcare delivery applications [37–40]. One of the primary benefits of contemporary ethnographic research methods is that they are specifically tailored to minimize potential observational or researcher-induced biases (e.g., the Hawthorne effect[e]), while maximizing the role of collecting information in context, providing situation-specific knowledge. The resulting qualitative data generated by observational studies are often characterized as being "rich" or "concrete" [41]. The advantages of observational techniques are similar to those of interviews in that they require a minimum of resources, and further, provide for the capture of generally unbiased and contextual information. Yet the disadvantages are again similar to what occurs in interviews, in that they are time-intensive and do not easily yield large amounts of quantitative data. When quantitative data are generated from the observational studies, it is often a time- and resource-intensive task to code generated transcripts in order to extract data. Furthermore, in the absence of think-aloud protocols, it is left to the researchers to infer thought processes and knowledge structures from the behaviors that they have observed. However, one could debrief the subjects after the observations, using specific probes (questions) to get

[e]The phenomenon in which in which individuals modify an aspect of their behavior in response to their awareness of being observed.

their interpretations and thereby to check for accuracy. Note that this aspect of data collection does not capture the thought process, but captures their explanations of the process.

6.3.1.1 Group techniques

A number of group techniques for expert KE have been reported, including brainstorming [42], nominal group studies [43,44], presentation discovery [45], Delphi studies [46], consensus decision making [47], and computer-aided group sessions [48]. All of these techniques focus on the elicitation of consensus-based knowledge. It has been argued that such consensus-based knowledge is superior to the knowledge that can be gained from a single expert, since the group techniques used to generate such knowledge can reduce individual biases, increase the potential for the incorporation of multiple lines of reasoning, and account for potentially incomplete domain knowledge on the part of individuals [47]. Besides gaining consensus-based knowledge from a team of experts, such as expert physicians, there is also the potential to gain insight about shared interactions from teams who represent multiple areas of expertise (see the earlier discussion on Mental Models). However, conducting such group-technique KE studies can be difficult; it may be challenging to recruit appropriate experts to participate or to schedule mutually agreeable times and locations for such groups to meet. Furthermore, a forceful or coercive minority of experts or single experts might exert disproportionate influence over the contents of the resulting knowledge collection [31].

6.3.1.2 Biases in logical and probabilistic reasoning

In clinical medicine, much of what experts report during knowledge elicitation is inherently uncertain. Although physicians, including experts in specific clinical subdomains, have been shown to be poor at the formal estimation of probabilities associated with relationships [49,50], they will frequently use terms that show that they are managing uncertainty in their approach to problems (e.g., "suggests," "supports," "goes against," "often," "evokes the possibility"). Despite the challenges, many knowledge engineers and psychologists have sought to obtain true probabilities from experts as part of their knowledge elicitation activities. In addition to poor estimation of probabilities by human beings, bias in their probabilistic reasoning has also been well documented [51,52], and types of bias have been categorized [53,54]. These bias types include tendencies (a) to allow undue influence of cognitive availability (recency) of information, mistaking this characteristic for frequency; (b) to anchor judgments on initial estimates; (c) to assess the likelihood of an event based on familiarity or stereotypic representativeness rather than objective frequency; and (d) to overestimate the frequency of rare events.

Following the demonstrations of Tversky and Kahneman, some researchers speculated that various biases might also be manifest in experts [55], and they suggested that knowledge engineers should avoid the use of probabilistic or statistical judgments in knowledge elicitation altogether [56]. The work on probabilistic reasoning bias became a red flag, because the notion of uncertainty is crucial in many expert systems [57–59]. For example, in diagnostic problems one may need to formulate such rules as: "If the patient has spots, then the patient has measles with certainty X" (see, for example, the *certainty factor* uncertainty model used in the MYCIN expert system [60] and subsequently applied in many other domains, both within and outside medicine). If experts provide biased probability estimates, there could be substantial problems for those building systems containing rules that are triggered when particular probability values are in effect for specific variables.

In many applications, statistical judgment and the sorts of judgments involved in decision analysis are contrived in that they can take experts away from their usual way of thinking about problems. However, some investigators have argued that people have little trouble in giving probabilities, and that decision analysis can be used in knowledge elicitation [55], where the focus is on improving judgment by making decision processes and judgment criteria explicit. Some researchers have expressed doubt that the biases in probabilistic reasoning that have been observed in laboratory research occur with the same frequency and magnitude in any real-world problem solving situations [61,62].

Bias in logical reasoning also has been observed in the laboratory, where many problems have been observed [53,55,63,64]:

- A tendency to assign undue weight to the first evidence obtained
- Over-reliance on variables that have taken on extreme values
- The tendency to seek evidence that confirms the current hypothesis
- The tendency to reason about only one or two hypotheses at a time
- The tendency to be overconfident
- The desire to maintain consistency with prior hypotheses even if that means devaluing, distorting, or ignoring important information
- Belief in illusory correlations
- The tendency to be overly conservative
- Basing conclusions on hindsight

In their studies of medical decision making, Schwartz and Griffin [65] cited over 20 relevant papers supposedly demonstrating that experts rely on heuristics. In fact, they argued that experts do not seem to be prone to biases to such an extent that the concern should have practical import in knowledge elicitation work. However, most workers believe that such biasing tendencies are sufficiently common that they must be considered as confounders during the knowledge elicitation process.

6.3.2 Data analysis methods

6.3.2.1 Protocol and discourse analysis

The techniques of protocol and discourse analysis are very closely related. They are both concerned with the elicitation of knowledge from individuals while they are engaged in problem solving or reasoning tasks (i.e., *think-aloud* studies, as mentioned earlier). Such analyses may be performed in order to determine the conceptual entities and relationships between those entities used by individuals while they reason about a problem domain. The basic premises of these techniques are derived from the domains of psychology and cognitive science [66–68]. In this approach, not only are a job's task activities charted, but also problem solvers are instructed to explain what they are doing and thinking while they are performing the task. The think-aloud procedure generates a response protocol, which is a recording of the deliberations that is subsequently transcribed and analyzed for propositional content and semantic content. The process of verbalization typically does not significantly affect the normal course of cognitive processes [36], and it can yield information about the reasoning sequences and goal structures in experts' problem solving [20,69].

The think-aloud problem solving/protocol analysis technique has been used extensively in cognitive research on medical expertise [58,70–72]. For example, Kuipers and Kassirer [71] found that, in a routine case, experts tended to produce very sparse protocols that did not provide much basis for characterizing reasoning patterns. The authors suggested that expert knowledge is so compiled (as though associated with a single process) that it is difficult to articulate intermediate steps. This led to using clinical probes to elicit constrained information within the think-aloud paradigm [66]. Patel, Arocha, and Kaufman [18] showed that experts interpret clinical data from the first few segments of the patient problem evaluation process in terms of high-level hypotheses, which they later evaluate. This serves to partition the problem into manageable units, thus reducing the load on working memory. In contrast, experts out of their domain of expertise (sub-experts) generate hypotheses mostly at lower levels, and they keep generating new hypotheses instead of evaluating and discarding some of them.

During such protocol analysis studies, the recorded explanations by subjects are codified for analysis at varying levels of granularity [20,73,74]. Discourse analysis is the process by which an individual's intended meaning within a body of text or some other form of narrative discourse is analyzed into discrete units of thought (propositions). These units are then analyzed according to the context in which those units appear (propositional relations in semantic structures) as well as the quantification and description of the relationships existing among those same units [75,76]. The advantage of this approach to conceptual knowledge elicitation is that it situates the overall elicitation process within the broader distributed socio-cognitive context in which individuals perform real-world reasoning and problem solving [68,77].

6.3.2.2 Concept analysis

In recent years, some CTA researchers have adopted a technique called concept mapping as a method of both eliciting and representing knowledge [4,78]. The modern idea of a concept map can be interpreted as a "user-friendly" expression of meaning in a text. Concept maps have been used in many studies of the psychology of expertise, and this work has shown that these maps can support the formation of consensus among experts [79]. Concept maps constructed by domain experts clarify what they wish to express, and they eventually show high levels of agreement [80]. In concept mapping knowledge elicitation, the researchers help the domain practitioners to build a representation based on their domain knowledge, merging the activities of knowledge elicitation and representation. This technique also is proven to be useful as a tool for creating knowledge-based performance-support systems [81,82]. Concept maps are labeled node-link structures, like semantic networks described later in the chapter, but are less formal than the networks based on formal propositional representations.

6.3.2.3 Verification and validation of knowledge acquisition

As mentioned earlier, the process of verification and validation of knowledge is ideally, and most effectively, applied throughout the entire knowledge engineering spectrum. Therefore, it is important to understand the types of verification and validation metrics and techniques available for use within the specific context of KA. *Verification* is the evaluation of a knowledge-based system to ensure that it satisfies the end-user or domain-specific requirements used to define the design of that system (logical consistency, general notions of completeness, avoidance of redundancy, and the like). *Validation* is the evaluation of a knowledge-based system to ensure that it satisfies an external criterion of correctness, e.g., the end-user or domain-specific requirements that are intended to be realized upon implementation and refinement of that system. An example of a validation measurement for a knowledge-based system would be the concordance between the system's reasoning concerning a given set of "real world" input data in comparison to the reasoning that would be used by a domain expert assessing the same input data within the same real-world context. These notions were pioneered in the early days of expert systems when MYCIN [83] pursued these kinds of knowledge base validation experiments [84,85].

To summarize the distinction, verification is the evaluation of whether a knowledge-based system meets the perceived requirements of the end users or application domain, and validation is the evaluation of whether that system meets the realized (e.g., real-world) requirements of the end users or application domain. However, in both instances, similar evaluation metrics may be used. A number of critical verification and validation criteria exist, such as multiple-source or expert agreement, degree of interrelatedness of the knowledge, and consistency of the generated knowledge.

6.3.2.4 Heuristic methods

The most commonly used approach to evaluating knowledge is the use of heuristic evaluation criteria [86]. The advantage of this approach is the obvious simplicity of the evaluation method (e.g., knowledge engineers or experts may manually review the knowledge generated and determine if its contents are consistent with the heuristics actually used during expert performance of the related tasks). However, methods for doing this are informal and limited in their tractability when applied to large knowledge sets, since they are difficult, if not impossible, to automate. Furthermore, they make comparison of knowledge "quality" across multiple sets infeasible because of the qualitative nature of the evaluation results being generated.

6.3.3 Representational methods

Cognitive task analysis also speaks to representation of interpreted data, rather than just the collection of primary data. CTA techniques generally provide abstract frameworks that assume particular types of knowledge structures as well as underlying reasoning processes.

In the representation of verbal data (either declarative or procedural), investigators have made use of two principal kinds of formalisms: propositional representations and semantic networks. The utility of both approaches arises from the recognition that a given piece of discourse can have many related ideas embedded within it. A propositional representation provides a means for representing these ideas, and the relationships among them, in an explicit fashion. In addition, it provides a way of classifying and labeling these ideas. Systems of propositional analysis [87,88] are essentially languages that provide a uniform notation and classification for propositional representations. In all these approaches, as in case grammars, a proposition is the smallest idea unit, denoted as a relation (predicate) over a set of arguments (concepts).

Sowa's system of conceptual graphs provides another example of a language of this type [5]. Although there are notational differences in the formalisms, the underlying assumption is that propositions correspond to the basic units of the representation of discourse and form manageable units of knowledge representation.

The primary challenge is to represent the structure of verbal or written data arising from observations and interviews as well as from think-aloud protocols. The first stage of analysis involves generating a propositional representation of the acquired text. This is then transformed into a semantic network representation. The network consists of propositions that describe attribute characteristics, which form the nodes of the network, and propositions that describe relational information, which form the links.

The primary relations of interest in these networks are binary dependency relations, specifically, *causal*, *conditional*, and Boolean connectives (*and*, *inclusive or*, and *exclusive or* relations). In addition, algebraic relations (e.g., *greater than*), identifying relations, and categorical relations (i.e., category membership, part-whole relations) can be expressed. One can also distinguish between the *source* of a process and the *result* of a process. Uncertainty in relations can be represented by modal qualifiers (e.g., *can*), and truth values can be indicated when they deviate from the default value (truth with certainty).

A semantic network is a directed graph formed by nodes and by labeled connecting paths. Nodes may represent either clinical findings or hypotheses, whereas the paths represent directed connections between such nodes. These networks also provide a relatively precise means for characterizing the directionality of reasoning [66,72].

Fig. 6.3 shows a semantic structure generated using discourse analysis to understand the implied and explicit knowledge contained in a specific text taken from a think-aloud protocol. The example is based on a diagnostic explanation offered by a psychiatrist when presented

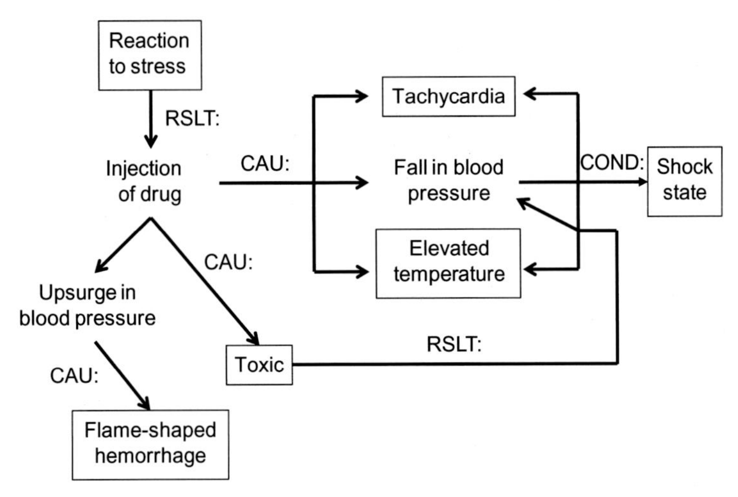

FIG. 6.3 Semantic analysis of a clinical text. In the diagram, solid rectangles indicate cues from the text, broken lines indicate diagnostic hypotheses, and arrows indicate directionality of relations. **COND:** = *conditional* relation, **CAU:** = *causal* relation, **RSLT:** = *resultive* relation. In this case, the text is taken from an explanation protocol provided by a psychiatrist who had been challenged by a case from the field of cardiology: *"The patient has been reacting to stress, likely by his injecting a drug (or drugs), which has resulted in tachycardia, a fall in blood pressure, and elevated temperature. These findings are due to the toxic reaction caused by the injected drugs. He is in or near shock. The flame-shaped hemorrhage may represent a sequel of an upsurge in blood pressure possibly as a result of his injection of drugs."*

with a case from cardiology [89]. The case is not within the subject's domain of specialization, and the diagnosis of a *shock state* is inaccurate. Accuracy is often associated with coherent network structures and inaccuracy is shown to lack this coherency [72]. Because of the inaccurate diagnosis, the representation is lacking in coherence and contains one possible inconsistency; that is, a patient cannot have both high and low blood pressure at the same time. Furthermore, the underlying mechanism that explains the signs and symptoms in this patient is attributed to toxicity of drugs that the patient has injected in an effort to respond to external psychological stress. This is not an accurate description of the patient's problem. Such representational networks generated from explanation protocols can help us to identify the behavior of experts out of their domain of expertise, the associated reasoning, and the nature errors generated.

The diagram consists of nodes linked by arrows. The arrows have labels indicating the relationship between nodes. The two most important are **CAU:** and **COND:**. The arrows labeled **CAU:** represent causal relations, and those labeled **COND:** represent conditional relations. **CAU:** means that the source node causes the target (e.g., *upsurge in blood pressure* causes *flame-shaped hemorrhage*), and **COND:** means that the source node is an indicator of the target (e.g., *tachycardia* indicates *shock state*). A difference between the two relates to the strength of implication: **COND:** expresses a directional conditionality, $P1 \rightarrow P2$, which implies if proposition $P1$ is true then $P2$ is true. **CAU:** $P1 \rightarrow P2$, is a stronger relation indicating that one variable, $P2$, is a functional result of another, $P1$.

6.3.4 Practical applications of representational approaches

As a practical matter, the goal of knowledge representation is to capture and convey knowledge so that the user's interpretation is consistent with the intended meaning. In the context of clinical decision support, knowledge representation is intended to support an expert reasoning process that leads to a suggested decision. In practical situations, the selected approach will depend on the intended user and the type of knowledge (conceptual, procedural, or strategic). In capturing clinical knowledge for the purposes of developing clinical decision support, the first step is to capture the concepts used within the domain. Intuitively, concepts are meaningful abstractions that refer to an entity (the referent) and are represented by a symbol (representation). It is useful to represent conceptual knowledge as a graph in which the concepts form the nodes and the relationships between the concepts, which form the links. These representations can provide a mechanism for validation by the Subject Matter Expert (SME) and can be used as the basis for developing a domain ontology that could serve to define the concepts used in the procedural knowledge, or could be directly reasoned over.

As described earlier, procedural knowledge plays an important role in clinical decision support. There are multiple ways of expressing business rules. In health care, Clinical Quality Language[f] offers a combination of precision and human-readability that allows for faithful representation of clinical rules (see Chapter 9 for details). However, when the rules are very complex, alternative representations may allow for easier validation by a subject matter expert (SME) while still being directly executable. Decision tables allow the display of multiple

[f]See http://cql.hl7.org/.

rules in a way that makes all the inputs and values clear. Decision Model and Notation is a standard for visually representing decision logic and the dependencies between decisions.

Procedural knowledge, the sequencing and orchestration of tasks, either physical or cognitive, is best represented visually [90]. (Several approaches for formally modeling processes have been explored and are discussed in detail in Chapter 10.) Two standards -based approaches are the Health Level 7 Fast Healthcare Interoperability Resources (HL7 FHIR) plan definition resource and the BPM+ family of process modeling languages (Business Process Model and Notation, Decision Model and Notation, and Case Management Model and Notation). Previous attempts have been made to model care processes using business process modeling, but the complexity of task orchestration in health care poses challenges. Tasks may be partially, or conditionally performed, and multiple tasks may be performed in parallel [6]. However, the more recent Case Management Model and Notation (CMMN) standard complements Business Process Model and Notation and addresses the constraints imposed using BPM alone. These two standards paired with the Decision Model and Notation standard for the representation of decision rules have been successfully used in creating models of clinical processes [8,91].

Because of its complexity, the contextual knowledge captured through observational studies and ethnographic approaches is best represented in nonlinear activity diagrams which include details of the people involved, the care settings, the information flows, the artifacts used, handoffs, and other knowledge assets critical to ensuring that CDS interventions are effective.

6.4 History and evolution of computer-based knowledge acquisition

The knowledge contained in any large-scale decision support system is so extensive and complex that it has become unreasonable to consider managing such knowledge bases manually. As a result, specialized environments have been constructed that allow trained individuals to enter new knowledge, and maintain or "curate" what is already there. Such systems often require structural knowledge of a domain over which the inferential knowledge is overlaid. Today, that structural knowledge, which defines the concepts in a domain and some aspects of the hierarchical relationships among them, is known as an *ontology* of that domain. Knowledge base developers and maintainers typically begin with the creation of a basic ontology for a field and then build inferential structures and relationships that allow a knowledge system to draw conclusions and generate advice. These knowledge representation issues are discussed in several chapters in Section IV.

We mention this topic here because there is a continuum in the development of computer systems for knowledge acquisition between those that are used for entering knowledge acquired through another means and those that actually interact with experts to extract, encode, and maintain that knowledge. Today, systems in the former category dominate, among which the well-known Protégé system is an important example. Protégé supports the creation of ontologies and the encoding of related complex knowledge in a domain [92,93]. But it would be rare to identify clinical experts who would be able to sit down with Protégé and "teach" it what they know about their domains of expertise. Protégé is for programmers and knowledge engineers to use after they have identified the knowledge that needs to be encoded.

The notion of obtaining knowledge directly from experts using an interactive dialog had its roots in the field of artificial intelligence in the early 1970s. For example, Carbonell pioneered the notion of computer-based mixed-initiative dialogs, focusing on educational uses but recognizing that an ability to interview and interact with a knowledgeable user had broad implications for computational extraction of knowledge, as well as its conveyance in the educational setting [94]. Heavily influenced by Carbonell's work, Shortliffe later experimented with a mixed-initiative dialog system that would allow physicians to teach the MYCIN program new knowledge (rules) in the domain of infectious disease therapy [83]. The basic notion was that an expert ought to be able to challenge MYCIN with a new patient case, use its explanation facilities to determine what faulty or missing knowledge explained any errors in performance (see Chapter 18 in Ref. [95]), and then to enter new or corrected rules for MYCIN to incorporate into its knowledge base.

MYCIN's early foray into knowledge acquisition was later extensively expanded and enhanced by Davis [96]. It was he who coined the phrase "interactive transfer of expertise" to describe the notion of an expert interacting with an intuitive, natural interface implemented in a computer program to "teach" the machine about his or her knowledge of a domain. Several examples can be found in Chapter 9 of Ref. [95] and in Davis' summary article, which shows how the approach could be used in a totally different domain [96]. At the peak of interest in these interactive approaches, there was also a journal entitled *Knowledge Acquisition* that published six volumes between 1989 and 1994.[g]

Although the performance of these early programs was promising, the complexity of their creation, maintenance, and use made it difficult to get experts to work with them directly. They much preferred to work with knowledge engineers and psychologists who used the knowledge elicitation techniques we have previously described. Thus, in the 1980s, there was a gradual move toward creating powerful knowledge authoring and editing tools that could be used by knowledge engineers *after* they had elicited the pertinent knowledge from human experts. The introduction of graphical user interfaces encouraged the adaptation of visual programming concepts for use in knowledge base construction and maintenance. One of the earliest efforts was Musen's creation of OPAL, a graphical authoring environment for entering and maintaining cancer chemotherapy research protocols [97], which was later generalized to be used for knowledge entry and editing in any domain. That work evolved to become Protégé, which is today heavily used for ontology construction and maintenance [98].

Today, although experiments continue, it is a rare knowledge-elicitation tool that is designed and successfully implemented for use directly by physicians or other clinical experts. Early work has begun to explore computational ethnography data mining to be able to identify patterns in event logs and other digital forms of capture [99]. Combining this with observational studies can provide scaled insights into patterns of how clinicians work [100], thereby providing insights and documentation of how knowledge is applied. We see continued emphasis on the specialized skills of individuals who know the computational systems but who also have the interpersonal skills, and ability to learn about what is often a new domain to them, in order to work closely with experts, and groups of individuals, in order to elicit the knowledge that is needed for medical decision support. In addition, there is today

[g]See https://www.sciencedirect.com/journal/knowledge-acquisition.

a great deal of work that seeks to derive new knowledge from large datasets, especially in the modern era of "big data". Many of these approaches are discussed in the remaining chapters in Section III.

6.5 Example

To this point in the chapter, the focus has been on techniques and methods for acquiring and representing knowledge. The following is a brief example of how this knowledge management can be applied to clinical practice. Atrial fibrillation (Afib) is an irregular and often rapid heartbeat that can increase risk of strokes, heart failure, and other cardiac complications [101]. The treatment knowledge begins at the conceptual level with the various entities involved. For example, Afib can cause a clot, which can lead to a stroke. Warfarin is an anticoagulant that can prevent a clot.

From a procedural perspective, an organization may engage subject matter experts to create rules, policies, care process models, and order sets to guide clinicians for treatment options and recommendations. Rules can be simple or complex if-then statements:

IF the patient is male,
AND the patient has a stroke risk score (CHA_2DS_2VASc)[h] of 1
THEN anticoagulation treatment should be considered.

Care process models can take the form of a flowchart, guiding the clinician through a series of steps with decision options along the way: An ECG positive result leads to validating that the heart rate is within an acceptable range, leading to a physical evaluation, and so on. From a strategic and environmental perspective, the organization needs to publish the knowledge in ways to facilitate the application of knowledge in the course of care. Rules and order sets can be embedded into the EHR to provide real-time notification to clinicians.

Elements of knowledge are elicited in multiple ways. Review of text in medical literature allows the knowledge engineer to document concepts to be validated by subject matter experts. For example, if a text refers to sleep apnea, the knowledge engineer might interview a subject matter expert to determine whether the type of sleep apnea (central vs. obstructive) is an important distinction. The validation process can be through semi-structured interviews and cognitive task analysis. In the same manner, rules can be documented and validated. For example, this could involve a rule relating to the decision on whether to send a patient to the ER based on criteria that include the presence of hemodynamic instability, chest pain, or syncope.

To understand the processes and procedures, additional steps are needed to directly observe using rapid ethnography, to ask questions, and to ask the subject to "think-aloud" while performing their work. A knowledge engineer may ask a cardiologist what data elements are necessary to generate a CHA_2DS_2VASc score,[h] and where each of those may be found. It is also important to know how those elements are entered into the risk calculator. This activity is also

[h]The CHA_2DS_2VASc score is used to predict the risk of stroke in patients with atrial fibrillation based on congestive heart failure, hypertension, age, diabetes mellitus, stroke/transient ischemic attack/thromboembolism, sex category, and vascular disease [102].

necessary for the strategic and environmental aspects to best understand how the knowledge should be introduced into the workflow. The observer should understand what artifacts and tools are used by the clinical team so they can ensure that the knowledge is delivered to the right person and place, in the right format and medium. This knowledge can be documented in multiple ways such as transcribed interviews, thematic analysis, activity diagrams, and video and digital images to communicate to other clinicians, leadership, IT staff, and future knowledge engineers.

6.5.1 Case study: The Mayo Clinic knowledge management program

One challenge in knowledge elicitation for clinical decision support is to develop a process that will assure trust and acceptance. As an example, the Mayo Clinic has developed a knowledge management program that creates and maintains a body of evidence, and expert opinion-based best-practice clinical knowledge, referred to as "core clinical knowledge". This content is delivered through a point-of-care resource called AskMayoExpert (AME), and it also serves as the basis for text-based derivatives, such as patient-education materials and consumer health information, and for other forms of knowledge delivery such as order sets and clinical decision support rules.

The first step in knowledge elicitation is to define the scope of the knowledge to be elicited. The goal is not to create an encyclopedic knowledge base, but rather to focus on the knowledge that answers clinical questions that arise in practice. Multiple approaches have been used to scope the knowledge base, including reviews of the reasons for consultation requests, questions subspecialists are most often asked, teaching materials, and a taxonomy of clinical questions [103].

Knowledge elicitation is done by advanced practice providers (APPs), partnered with editors who also work with subject matter experts. The inclusion of the APPs in the elicitation process helps to ensure that the knowledge is not only clear and concise, but actionable at the point of care. Their clinical knowledge enables them to ask targeted, insightful questions and quickly to grasp the answers, reducing the amount of subject matter expert time required.

The content is then vetted by knowledge content boards (KCBs) whose members are select groups of experienced and highly recognized faculty from each specialty and subspecialty. At the time of this writing, there are over 50 KCBs. In addition to the vetting knowledge, these groups are responsible for processing user feedback and rapidly incorporating new knowledge.

While every effort is made to streamline the process, it still requires a significant time commitment from the clinical staff. The institutional leadership has provided the members of the KCBs with dedicated time to review and update the knowledge on an ongoing basis, indicative of the value the institution places on the knowledge management. KCB participation is considered an academic contribution by the Mayo Clinic Academic Appointments and Promotions Committee.

6.6 Conclusions

In the modern world, knowledge management has become a major focus of activity in diverse businesses, including health care. Because of the effort required to develop and validate such knowledge, there is growing recognition of the need to share knowledge components when they are developed and optimally to involve experts in providing, assessing, and maintaining knowledge that is needed. Although we are creating large institutional, local, regional, and national databases, only some of the knowledge that we require to inform practice and policy can be derived solely by analyzing those data or the literature (see Chapters 7 and 8). Many areas of clinical endeavor still depend heavily on the kind of judgmental knowledge and experience that is difficult to acquire from anyone other than those who have the wisdom and efficiency that comes with experience and lifelong learning. Thus, despite the formal analytical methods that are appropriately being used to make sure that we learn as much as we can from our accumulated experience stored in pooled databases and in the literature, knowledge elicitation from experts, and groups of experts, will continue to be a crucial component of knowledge creation and management for clinical decision support. The early promise of computer-based transfer of expertise to knowledge systems has not been borne out, although significant research opportunities and potential continue to exist. The re-emergence of such systems may be facilitated by our increasing knowledge of human problem-solving methods and by enabling improvements in technology. For now, however, it is the direct interaction among experts, and between experts and knowledge engineers, that will serve a crucial role in assuring the development of high quality and accepted knowledge bases. These, in turn, will enable the development and effective use of decision support systems.

References

[1] Bengio Y, Lecun Y, Hinton G. Deep learning for AI. Commun ACM 2021;64(7):58–65.

[2] Schmalhofer F. Constructive knowledge acquisition: A computational model and experimental evaluation. 1st ed. Mahwah, NJ: Psychology Press; 1997. 328 p.

[3] Tecuci G, Boicu M, Schum D. Knowledge engineering: Building cognitive assistants for evidence-based reasoning. New York: Cambridge University Press; 2016.

[4] Crandall B, Klein G, Hoffman RR. Working minds: A Practitioner's guide to cognitive task analysis. 1st ed. Cambridge, MA: A Bradford Book; 2006. 332 p.

[5] Sowa J. Conceptual structures: Information processing in mind and machine. Reading, MA: Addison-Wesley; 1984.

[6] Gooch P, Roudsari A. Computerization of workflows, guidelines, and care pathways: a review of implementation challenges for process-oriented health information systems. J Am Med Inf Assoc 2011;18(6):738–48.

[7] Feigenbaum EA. The art of artificial intelligence: themes and case studies of knowledge engineering. In: Proceedings of the fifth international joint conference on artificial intelligence | IJCAI [Internet]. Cambridge, MA: Massachusetts Institute of Technology; 1977. p. 1014–30. Available from: https://www.ijcai.org/Proceedings/77-2/Papers/092.pdf.

[8] Sooter L, Hasley S, Lario R, Rubin K, Hasić F. Modeling a clinical pathway for contraception. Appl Clin Inform 2019;10(5):935–43.

[9] Duda RO, Shortliffe EH. Expert systems research. Science 1983;220(4594):261–8.

[10] Osheroff J, Levick D, Saldana L, Velasco F, Sittig D, Rogers K, et al. Improving outcomes with clinical decision support. 2nd ed. Boca Raton, FL: HIMSS Publishing; 2012.

[11] Hoffman RR, Shadbolt NR, Burton AM, Klein G. Eliciting knowledge from experts: a methodological analysis. Organ Behav Hum Decis Process 1995;62(2):129–58.

[12] Peleg M, Gutnik LA, Snow V, Patel VL. Interpreting procedures from descriptive guidelines. J Biomed Inform 2006;39(2):184–95.

[13] deGroot AD. Thought and choice in chess. The Hague: Mouton; 1965.

[14] Chi MTH, Feltovich PJ, Glaser R. Categorization and representation of physics problems by experts and novices*. Cognit Sci 1981;5(2):121–52.
[15] Larkin J, McDermott J, Simon DP, Simon HA. Expert and novice performance in solving physics problems. Science 1980;208(4450):1335–42.
[16] Mishra J. Musical expertise. In: The Oxford handbook of expertise [Internet]. New York, NY: Oxford University Press; 2019. Available from: https://www.oxfordhandbooks.com/view/10.1093/oxfordhb/9780198795872.001.0001/oxfordhb-9780198795872-e-25.
[17] Farrow D, Reid M, Buszard T, Kovalchik S. Charting the development of sport expertise: challenges and opportunities. Int Rev Sport Exerc Psychol 2018;11(1):238–57.
[18] Patel V, Arocha JF, Kaufman D. Diagnostic reasoning and medical expertise. In: Medin DL, editor. Psychology of learning and motivation—Advances in research and theory, vol. 31(C); 1994. p. 187–252.
[19] Patel VL, Kaufman DR, Kannampallil TG. Diagnostic reasoning and expertise in health care. In: The Oxford handbook of expertise [Internet]. New York, NY: Oxford University Press; 2019. Available from: https://www.oxfordhandbooks.com/view/10.1093/oxfordhb/9780198795872.001.0001/oxfordhb-9780198795872-e-27.
[20] Patel VL, Groen GJ. The general and specific nature of medical expertise: a critical look. In: Ericsson KA, Smith J, editors. Toward a general theory of expertise: Prospects and limits. New York, NY: Cambridge University Press; 1991. p. 93–125.
[21] Patel VL, Kaufman DR. Cognitive informatics. In: Shortliffe EH, Cimino JJ, editors. Biomedical informatics: computer applications in health care and biomedicine [Internet]. Cham: Springer International Publishing; 2021. p. 121–52. Available from: https://doi.org/10.1007/978-3-030-58721-5_4.
[22] MTH C, Glaser R, Farr MJ, editors. The nature of expertise. Hillsdale, NJ: Lawrence Erlbaum Associates, Inc; 1988. xxxvi, 434 p.
[23] Ericsson KA, editor. The road to excellence: The acquisition of expert performance in the arts and sciences, sports, and games. 1st ed. Mahwah, NJ: Lawrence Erlbaum Associates; 1996. 369 p.
[24] Ericsson KA, Smith J. Toward a general theory of expertise : Prospects and limits [Internet]. New York, NY: Cambridge University Press; 1991. [cited 2021 Dec 16]. Available from: https://www.semanticscholar.org/paper/Toward-a-general-theory-of-expertise-%3A-prospects-Ericsson-Smith/1fa47a1c37bf427edb2e11218f0f0f3d0fa433f6.
[25] Feltovich PJ, Ford KM, Hoffman RR, editors. Expertise in context. Cambridge, MA: AAAI Press; 1997. 608 p. (American Association for Artificial Intelligence).
[26] Ericsson KA, Hoffman RR, Kozbelt A, Williams AM, editors. The Cambridge handbook of expertise and expert performance. 2nd ed. Cambridge, United Kingdom ; New York, NY, USA: Cambridge University Press; 2018. 984 p.
[27] Chapman GB, Elstein AS. Cognitive processes and biases in medical decision making. In: Decision making in health care: Theory, psychology, and applications. Cambridge series on judgment and decision making. New York, NY: Cambridge University Press; 2000. p. 183–210.
[28] Glaser R. Changing the agency for learning: acquiring expert performance. In: Ericsson KA, editor. The road to excellence: The acquisition of expert performance in the arts and sciences, sports, and games. Hillsdale, NJ: Lawrence Erlbaum Associates, Inc; 1996. p. 303–11.
[29] Cannon-Bowers JA, Salas E, Converse S. Shared mental models in expert team decision making. In: Castellan NJ, editor. Individual and group decision making. Psychology Press; 1993.
[30] Scott AC, Clayton JE, Gibson EL. A practical guide to knowledge acquisition. First Printing edition, Reading, MA: Addison-Wesley; 1991. 528 p.
[31] Liou YI. Knowledge acquisition: issues, techniques and methodology. SIGMIS Database 1992;23(1):59–64.
[32] Boy GA. The group elicitation method for participatory design and usability testing. Interactions 1997;4(2):27–33.
[33] Shepherd MM, WmB M. Group consensus: do we know it when we see it? In: Proceedings of the 37th annual Hawaii international conference on system sciences, 2004; 2004. p. 7.
[34] Hawkins D. An analysis of expert thinking. Int J Man Mach Stud 1983;18(1):1–47.
[35] Wood WC, Roth RM. A workshop approach to acquiring knowledge from single and multiple experts. In: Proceedings of the 1990 ACM SIGBDP conference on trends and directions in expert systems [Internet]. (SIGBDP

'90). New York, NY: Association for Computing Machinery; 1990. p. 275–300. Available from: https://doi.org/10.1145/97709.97730.
[36] Ericsson KA, Simon HA. Protocol analysis: Verbal reports as data. Revised Edition, Cambridge, MA: A Bradford Book; 1993. 500 p.
[37] Liszka AH, Stubblefield WA, Kleban SD. GMS: preserving multiple expert voices in scientific knowledge management. In: Proceedings of the 2003 conference on designing for user experiences [Internet]. (DUX '03). New York, NY: Association for Computing Machinery; 2003. p. 1–4. Available from: https://doi.org/10.1145/997078.997098.
[38] Cohen T, Blatter B, Almeida C, Shortliffe E, Patel V. A cognitive blueprint of collaboration in context: distributed cognition in the psychiatric emergency department. Artif Intell Med 2006;37(2):73–83.
[39] Hughes J, King V, Rodden T, Andersen H. The role of ethnography in interactive systems design. Interactions 1995;2(2):56–65.
[40] Laxmisan A, Malhotra S, Keselman A, Johnson TR, Patel VL. Decisions about critical events in device-related scenarios as a function of expertise. J Biomed Inform 2005;38(3):200–12.
[41] Iqbal R, Gatward R, James A. A general approach to ethnographic analysis for systems design. In: Proceedings of the 23rd annual international conference on design of communication: documenting & designing for pervasive information [Internet]. (SIGDOC '05). New York, NY: Association for Computing Machinery; 2005. p. 34–40. Available from: https://doi.org/10.1145/1085313.1085324.
[42] Osborn AF. Applied imagination: Principles and procedures of creative thinking. 1st ed. Charles Scribner's Sons; 1953. 317 p.
[43] Delbecq AL, Ven AHVD, Gustafson DH. Group techniques for program planning: A guide to nominal group and Delphi processes. 1st ed. Middleton, WI: Green Briar Press; 1986. 174 p.
[44] Jones J, Hunter D. Using the Delphi and nominal group technique in health services research. In: Qualitative research in health care. London: BMJ Books; 1999. p. x, 156.
[45] Payne PRO, Starren JB. Quantifying visual similarity in clinical iconic graphics. J Am Med Inform Assoc 2005;12(3):338–45.
[46] Adelman L. Measurement issues in knowledge engineering. IEEE Trans Syst Man Cybern 1989;19(3):483–8.
[47] McGraw KL, Seale MR. Knowledge elicitation with multiple experts: considerations and techniques. Artif Intell Rev 1988;2(1):31–44.
[48] Adams L, Toomey L, Churchill E. Distributed research teams: meeting asynchronously in virtual space. J Comput Mediat Commun 1999;4(4).
[49] Berwick DM, Fineberg HV, Weinstein MC. When doctors meet numbers. Am J Med 1981;71(6):991–8.
[50] Leaper DJ, De Dombal FT, Horrocks JC, Staniland JR. Computer-assisted diagnosis of abdominal pain using estimates provided by clinicians. Br J Surg 1972;59(11):897–8.
[51] Kahneman D, Tversky A. On the study of statistical intuitions. Cognition 1982;11(2):123–41.
[52] Lichtenstein S, Fischhoff B. Training for calibration. Organ Behav Hum Perform 1980;26(2):149–71.
[53] Fraser JM, Smith PJ, Smith JW. A catalog of errors. Int J Man Mach Stud 1992;37(3):265–307.
[54] Payne VL, Patel VL. Enhancing medical decision making when caring for the critically ill: the role of cognitive heuristics and biases. In: Patel VL, Kaufman DR, Cohen T, editors. Cognitive informatics in health and biomedicine: case studies on critical care, complexity and errors [Internet]. (Health informatics). London: Springer; 2014. p. 203–31. Available from: https://doi.org/10.1007/978-1-4471-5490-7_10.
[55] Fischhoff B. Eliciting knowledge for analytical representation. IEEE Trans Syst Man Cybern 1989;19(3):448–61.
[56] Hink RF, Woods DL. How humans process uncertain knowledge: an introduction. AI Mag 1987;8(3):41.
[57] Fox J. Knowledge, decision making, and uncertainty. In: Gale WA, editor. Artificial intelligence and statistics. United States: Addison-Wesley Pub Co Inc; 1986. p. 57–76.
[58] Kuipers B, Moskowitz AJ, Kassirer JP. Critical decisions under uncertainty: representation and structure. Cognit Sci 1988;12(2):177–210.
[59] Zadeh LA, Kacprzuk J. Fuzzy logic for the management of uncertainty Wiley professional computing series. 1st ed. John Wiley & Sons; 1992.
[60] Shortliffe EH, Buchanan BG. A model of inexact reasoning in medicine. Math Biosci 1975;23(3):351–79.
[61] Beyth-Marom R, Arkes HR. Being accurate but not necessarily Bayesian: comments on Christensen-Szalanski and Beach. Organ Behav Hum Perform 1983;31(2):255–7.

[62] Christensen-Szalanski JJJ, Beach LR. The citation bias: fad and fashion in the judgment and decision literature. Am Psychol 1984;39(1):75–8.
[63] Evans JSBT. Bias in human reasoning: Causes and consequences. Psychology Press; 1989.
[64] Johnson-Laird PN. Mental models: Towards a cognitive science of language, inference, and consciousness. Cambridge, MA: Harvard University Press; 1986. 513 p.
[65] Schwartz S, Griffin T. Medical thinking: The psychology of medical judgment and decision making. Softcover reprint of the original 1st ed. 1986 edition, New York, NY: Springer; 2011. 277 p.
[66] Groen GJ, Patel VL. The relationship between comprehension and reasoning in medical expertise. In: Chi MTH, Glaser R, Farr MJ, editors. The nature of expertise. Hillsdale, NJ: Lawrence Erlbaum Associates, Inc; 1988. p. 287–310.
[67] Kintsch W, Greeno JG. Understanding and solving word arithmetic problems. Psychol Rev 1985;92(1):109–29.
[68] Patel VL, Arocha JF, Kaufman DR. A primer on aspects of cognition for medical informatics. J Am Med Inform Assoc 2001;8(4):324–43.
[69] Patel RM. Cognitive models of directional inference in expert medical reasoning. In: Ford K, Feltovich P, Hoffman R, editors. Human & machine cognition. Hillsdale, NJ: Lawrence Erlbaum Associates; 1997. p. 67–99.
[70] Johnson PE, Duran AS, Hassebrock F, Moller J, Prietula M, Feltovich PJ, et al. Expertise and error in diagnostic reasoning. Cognit Sci 1981;5(3):235–83.
[71] Kuipers B, Kassirer JP. Causal reasoning in medicine: analysis of a protocol. Cogn Sci: A Multidisciplinary Journal of Artificial Intelligence, Psychology and Language 1984;8(4):363–85.
[72] Patel VL, Groen GJ. Knowledge based solution strategies in medical reasoning. Cognit Sci 1986;10(1):91–116.
[73] Feltovich PJ, Spiro RJ, Coulson RL. The nature of conceptual understanding in biomedicine: the deep structure of complex ideas and the development of misconceptions. In: Evans DA, Patel VL, editors. Cognitive science in medicine: Biomedical modeling [Internet]. Cambridge, MA: MIT Press; 1989. https://doi.org/10.7551/mitpress/1878.003.0007. Available from:.
[74] Polson P, Lewis C, Rieman J, Wharton C. Cognitive walkthroughs: a method for theory-based evaluation of user interfaces. Int J Man Mach Stud 1992;36(5):741–73.
[75] Alvarez R. Discourse analysis of requirements and knowledge elicitation interviews. In: Proceedings of the 35th annual Hawaii international conference on system sciences. IEE Computer Society; 2002. p. 10. Available from: 2002, https://ieeexplore.ieee.org/document/994388.
[76] Davidson JE. Topics in discourse analysis. Canada: University of British Columbia; 1977.
[77] Patel VL, Kaufman DR, Arocha JF. Emerging paradigms of cognition in medical decision-making. J Biomed Inform 2002;35(1):52–75.
[78] Novak JD. Concept maps and Vee diagrams: two metacognitive tools to facilitate meaningful learning. Instr Sci 1990;19(1):29–52.
[79] Gordon SE, Schmierer KA, Gill RT. Conceptual graph analysis: knowledge acquisition for instructional system design. Hum Factors 1993;35(3):459–81.
[80] Gordon SE. Implications of cognitive theory for knowledge acquisition. In: Hoffman RR, editor. The psychology of expertise: cognitive research and empirical AI [Internet]. New York, NY: Springer; 1992. p. 99–120. Available from: https://doi.org/10.1007/978-1-4613-9733-5_6.
[81] Cañas A, Coffey J, Reichherzer T, Hill G, Suri N, Carff R, et al. El-Tech: a performance support system with embedded training for electronics technicians. In: Proceedings of the eleventh international Florida Artificial Intelligence Research Society conference, May 18–20, 1998, Sanibel Island, Florida, USA [Internet]. Menlo Park, CA: AAAI; 1988. Available from: https://www.aaai.org/Library/FLAIRS/1998/flairs98-015.php.
[82] Dorsey DW, Campbell GE, Foster LL, Miles DE. Assessing knowledge structures: relations with experience and posttraining performance. Hum Perform 1999;12(1):31–57.
[83] Shortliffe EH. Computer-based medical consultations: MYCIN. New York, NY: American Elsevier; 1976. 286 p.
[84] Yu VL, Buchanan BG, Shortliffe EH, Wraith SM, Davis R, Scott AC, et al. Evaluating the performance of a computer-based consultant. Comput Programs Biomed 1979;9(1):95–102.
[85] Yu VL, Fagan LM, Wraith SM, Clancey WJ, Scott AC, Hannigan J, et al. Antimicrobial selection by a computer. A blinded evaluation by infectious diseases experts. JAMA 1979;242(12):1279–82.
[86] Nielsen J. Usability engineering. 1st ed. Boston, MA: Morgan Kaufmann; 1993. 376 p.
[87] Frederiksen CH. Representing logical and semantic structure of knowledge acquired from discourse. Cogn Psychol 1975;7(3):371–458.

[88] Kintsch W. The representation of meaning in memory. Hillsdale, NJ: Lawrence Erlbaum Associates; distributed by Halsted Press Division, Wiley, New York; 1974.
[89] Patel VL, Groen GJ, Arocha JF. Medical expertise as a function of task difficulty. Mem Cognit 1990;18 (4):394–406.
[90] Kremer R. KAW98: visual languages. In: Proc of 11th workshop on knowledge acquisition, modeling and management (KAW'98) [Internet]. Banff, AB: Morgan Kaufmann; 1998. [cited 2022 Feb 7]. Available from: http://ksi.cpsc.ucalgary.ca/KAW/KAW98/kremer/.
[91] McClay J, Goyal P. Piloting implementation and dissemination of best practice guidelines using BPM+Health. J Clin Transl Sci 2020;4(s1):141–2.
[92] Musen MA. Dimensions of knowledge sharing and reuse. Comput Biomed Res 1992;25(5):435–67.
[93] Tudorache T, Nyulas C, Noy NF, Musen MA. WebProtégé: a collaborative ontology editor and knowledge acquisition tool for the web. Semant Web 2013;4(1):89–99.
[94] Carbonell JR. AI in CAI: an artificial-intelligence approach to computer-assisted instruction. IEEE Trans Man Mach Syst 1970;11(4):190–202.
[95] Buchanan BG, Shortliffe EH, editors. Rule based expert systems: The Mycin experiments of the Stanford heuristic programming project. 1st ed. Reading, MA: Addison-Wesley; 1984. 748 p.
[96] Davis R. Interactive transfer of expertise: acquisition of new inference rules. Artif Intell 1979;12(2):121–57.
[97] Musen MA, Combs DM, Walton JD, Shortliffe EH, Fagan LM. OPAL: toward the computer-aided design of oncology advice systems. Proc Annu Symp Comput Appl Med Care 1986;26:43–52.
[98] Musen MA, Protégé Team. The Protégé project: a look back and a look forward. AI Matters 2015;1(4):4–12.
[99] Zheng K, Hanauer D, Weibel N. Computational ethnography: Automated and unobtrusive means for collecting data in situ. In: Patel VL, Kannampallil T, Kaufman K, editors. Human–computer interaction in biomedicine and healthcare. Switzerland: Springer Nature; 2015.
[100] Grando M, Vellore V, Duncan B, Kaufman D, Furniss S, Doebbeling B. Study of EHR-mediated workflows using ethnography and process mining methods. Health Informatics J 2021;27(2). 146045822110082.
[101] Atrial fibrillation: Symptoms & causes [Internet]. Mayo Clinic; 2021. Available from: https://www.mayoclinic.org/diseases-conditions/atrial-fibrillation/symptoms-causes/syc-20350624.
[102] Olesen JB, Torp-Pedersen C, Hansen ML, Lip GYH. The value of the CHA2DS2-VASc score for refining stroke risk stratification in patients with atrial fibrillation with a CHADS2 score 0-1: a nationwide cohort study. Thromb Haemost 2012;107(6):1172–9.
[103] Ely JW, Osheroff JA, Chambliss ML, Ebell MH, Rosenbaum ME. Answering physicians' clinical questions: obstacles and potential solutions. J Am Med Inform Assoc 2005;12(2):217–24.

CHAPTER

7

Data-driven approaches to generating knowledge: Machine learning, artificial intelligence, and predictive modeling

Michael E. Matheny[a,b], *Lucila Ohno-Machado*[c], *Sharon E. Davis*[a], *and Shamim Nemati*[d]

[a]Biomedical Informatics, Vanderbilt University Medical Center, Nashville, TN, United States [b]Geriatrics, Research, Education, and Clinical Care Service, Tennessee Valley Healthcare System VA, Nashville, TN, United States [c]Section on Biomedical Informatics and Data Science, Yale School of Medicine, New Haven, CT, United States [d]Department of Medicine, University of California San Diego, San Diego, CA, United States

7.1 Introduction

Clinical decision support (CDS) systems must rely on knowledge that originates from a variety of sources, including domain literature, expert knowledge, and statistical analysis of data. However, selecting the sources and integrating this knowledge into a functional system is not a trivial task. The earliest systems for medical diagnosis, in the 1960s, used Bayesian probability models (as discussed in Chapter 2), relying on either databases of patient data or subjective estimates of prior and conditional probabilities from human experts. In most early CDS systems developed in the 1970s and 1980s, because of the dearth of available data and the interest at the time in the burgeoning field of artificial intelligence, knowledge was acquired directly from medical experts (as discussed further in Chapter 6). In some systems, pioneered by an antibiotic treatment advisor program known as MYCIN [1], knowledge was encoded in the form of rules that were triggered and chained according to an embedded or an external

Clinical Decision Support and Beyond
https://doi.org/10.1016/B978-0-323-91200-6.00031-0

inference engine [2]. In MYCIN, rules were expert-derived and had associated certainty factors, a mathematical formulation of a quasi-statistical representation of degree of belief, developed by Shortliffe et al. [1]. Various approaches to medical diagnostic reasoning from this period such as the Present Illness Program [3], Internist/QMR [4,5], DXplain [6], and others that were developed subsequently were constructed from a set of physician-based assessments of (a) the strength with which clinical findings evoke a certain diagnosis, (b) prevalence of diseases, and (c) related indices. As noted by Heckerman and colleagues, the formal mathematical definitions of these indices in terms of probabilities have not been fully elucidated [7].

Even when newer knowledge representation strategies incorporating statistical data, such as Bayesian networks [8], were proposed by some researchers in the late 1980s [9,10], definition of the graph structure and probabilities involved in the model were usually still assessed by experts. For example, Shwe and colleagues [11] "translated" the QMR representation into Bayesian networks, and showed that the same diagnostic quality could be achieved with a representation that made explicit important modeling assumptions. However, the popularity of Bayesian networks in the medical community did not grow as expected, and this type of model is not used in practice. Algorithms for learning Bayesian networks from data have evolved in the past few decades but are primarily used in research applications [12–14].

While newer clinical decision support systems learn from data and no longer rely on the rule-based paradigm, CDS tools with knowledge represented in the form of single IF/THEN rules (see Chapter 9) are still the most commonly used. However, more elaborated rules used to represent computer-interpretable guidelines (see Chapter 10) which chain together steps in a care process using branching decision rules, are not yet common in practice, primarily because of scalability challenges in maintaining the rule content across various settings and workflow integration challenges, which some initiatives like FHIR are trying to overcome. With rule-based CDS, although probabilistic considerations may have been considered when constructing the rules or guidelines, using evidence-based medicine techniques (see also Chapter 8), the algorithms tend to be stated in deterministic fashion, without associated probabilities. A summary of the historical timeline of artificial intelligence and machine learning in Fig. 7.1.

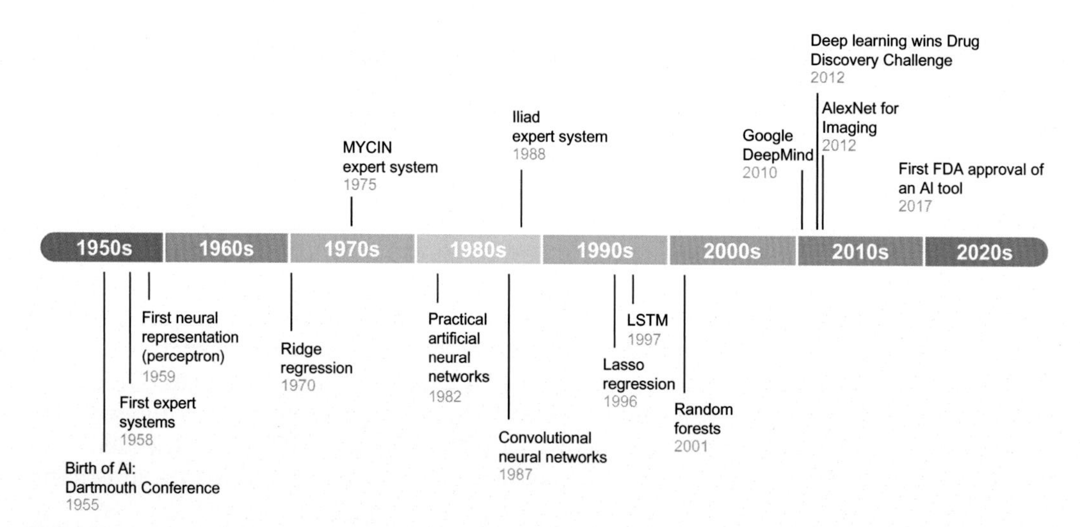

FIG. 7.1 Historical timeline of artificial intelligence and machine learning innovations and important events.

There are at least four factors that contribute to the continued reliance on expert assessments for the construction of tools such as CDS alerts and guidelines:

- Many clinical recommendations, such as for preventive maintenance or processes of care, like regularly checking a HgbA1c on a diabetic patient, or a urine drug screen in someone taking narcotics, do not require complex, probabilistic algorithms. A systematic review of such alerts and reminders directed at patients concludes that these technologies are useful [15]. Also, as noted previously, localization of rules for organizational or business process preferences and for integration into workflow are based on human experts.
- Data used to build data-driven CDS models may not represent the population where the model will be used [16]. Evidently, this can also be a limitation of models based on human experts, who may translate data from specific studies into rules.
- Lack of model explainability raises concerns that a "black box" approach may not be acceptable for billing or legal purposes. For example, the General Data Protection Rule (GDPR) in the European Union requires models to provide explanations when used to make decisions about individuals [17].
- Lack of support systems and tools in the electronic health record systems to support integration of complex predictive analytics into CDS, and the ability to scale and maintain algorithm performance over time. Simple regression models are already available, but AI-based ones are not yet featured in the most popular EHR systems [18]. Utilization of external resources to run the models, e.g., via a FHIR interface, could solve this problem.

Systems derived from human knowledge in which non-probabilistic rules are defined by experts may be more clearly understandable by clinicians and less prone to "unfairness". For example, if an expert can articulate all the rules that were used to make a diagnosis and how they were chained, then a system based on these rules can potentially explain its reasoning in a way that clinicians would be able to understand and accept [19]. By articulating the rules, it becomes clear whether differences in race, ethnicity, or other factors change the predictions or decision, which is not always true when these factors are considered in complex networks in which the influence of particular variables cannot always be reliably assessed for a particular case [20]. Whether understanding and agreement by clinicians is necessary for the underlying logic in CDS systems to be useful remains a controversial issue [21].

While CDS applications currently in use frequently rely on deterministic rules for their "logic", these are being increasingly combined or even replaced with data-driven, AI models. For example, AI models that analyze imaging data are increasingly used in practice, and many have binary outputs such as "pneumonia." Many of these neural network-based models produce continuous outcomes that are transformed into binary outcomes by a simple rule. This rule could change depending on the use case. For screening purposes, the threshold may be different than for diagnostic purposes.

There have been extensive new developments in statistical and machine learning research in the past few decades [22]. These advances have coincided with improvements in data quality and quantity from the implementation of large repositories of structured electronic data, some of which are based on domain-specific data element standards [23–25]. Increased availability of data has allowed further development of several models that can detect patterns in biomedical data and generalize well to previously unseen cases. Clinical decision support systems that rely on patterns that are recognized in these data are now available in virtually every medical specialty [26–31]. Just as in the foregoing discussion relating rule-based systems and more sophisticated knowledge representation paradigms, simple understandable models (e.g., linear and logistic regression and linear score systems) are still used in practice, while more sophisticated machine learning models (e.g., deep neural networks), are typically used for imaging, instrument signals (e.g., ICU monitors) and other less tractable problems [32–34]. Reviews of AI models that have been integrated into clinical workflows still do not differentiate between those used during studies and those used on a routine basis [35].

The more complex the setting, the more data are needed. The most popular data-driven models have achieved success in domains in which structured data are abundant, and the decisions are made at times in which a snapshot of these data can help identify specific patterns. This is, of course, becoming especially true in the era of genomics and precision medicine, as we discuss in Chapter 17, in which it is both important and increasingly feasible to base decisions about a patient on comparison to the experiences of a subset of maximally similar patients. The number of factors to be considered and their combinatorics make the task of developing a rule set to cover all the variations in genomic risk factors, presence of disease, stage of disease, comorbidities, treatments, and complications of treatments more and more intractable. Thus, there is a trade-off between explainability and the ability to solve complex problems. Chapter 16 discusses the growing opportunities for harnessing population health data for decision support.

In this chapter, we will review the methodologies of the most commonly used diagnostic and prognostic models in the medical domain and discuss specific strengths and weaknesses of alternative modeling methods. Popular examples of some modeling methods will be discussed, with a focus on models and modeling methods that have been utilized in practice. We conclude with a discussion on current directions for the field.

Note the absence of sections dedicated to other topics that have received wide coverage in the computer science literature, but that in fact have limited representation in clinical informatics applications and are beyond the scope of an introductory chapter. Some example omissions include discussion of rule-induction algorithms, optimization techniques such as genetic algorithms and evolutionary computing [36], and formalism extensions such as fuzzy logic [37] and rough sets [38]. Refer to statistical and machine learning textbooks for a review on these topics.

7.2 Types of machine learning

A myriad of publications in the scientific and lay literature can now be found under the rubric of "data mining". Data mining techniques are pattern recognition techniques intended to find correlations and relationships in the plethora of data. The term is intriguing, but also somewhat misleading. Most pattern recognition or predictive models used in clinical domains are confirmatory rather than exploratory in nature. The distinction between unsupervised and supervised learning models is directly related to this issue.

Unsupervised learning models are not based on pre-defined classifications and are used frequently for exploratory data analyses in domains in which knowledge is sparse. For example, high-throughput data are often subject to unsupervised learning modeling so that "clusters" of variables or features can be revealed without guidance from the users or the existing literature. The objective is to unveil hidden patterns in the data that were not previously anticipated and label these patterns "a posteriori".

This is in sharp contrast with *supervised learning* models, in which the objective is to determine how to best classify objects with pre-defined labels representing classes of interest (e.g., malignant versus benign cases) using the data at hand. Unsupervised learning models, particularly those in the deep learning family, have been recently applied within CDS, but these are generally used for dimensionality reduction, feature generation, or image segmentation (area identification) in sequence with a downstream supervised learning algorithm to support clinical decision support systems.

Reinforcement learning (RL) is the third major area of machine learning, in which a computer construct named an agent (or model), acts as a decision maker, and operates within a specified environment with a series of decision points in which the available choices have to be optimized to maximize a long-term reward [39]. The reward function is mathematically defined as a set of positive and negative feedbacks that are obtained from the environment in response to an action by the agent. In the setting of conditional independence among the states, RL can be framed as a Markov Decision Process [40], and the agent is trained through iteration over and over through the environmental state, decisions, transition states, and rewards [41]. There have been some promising applications of RL in prescriptive analytics, and we anticipate this to grow rapidly in the coming years, but it is still relatively nascent, and many challenges remain [42,43]. The main reasons are limitations in (1) how learning from observational data generated under fixed policy (termed "Off-Policy Evaluation") rather than repeated experiments with exploration and exploitation trade-offs, (2) limited observability of the environment (hidden knowledge), (3) and challenges specifying an appropriate reward function that balances short-term resolution of illness and long-term overall health and quality of life. We direct the readers to several excellent reviews of the exploration of reinforcement learning in healthcare [44,45].

Most models in current use for clinical decision support have been based either on expert knowledge or supervised learning models. Thus, we focus on the latter for the remainder of this chapter.

7.3 Frameworks for developing machine learning models in healthcare

In recent years, the severe mismatch between the volume of machine learning models developed and published and the few that achieve successful use and sustainment in clinical practice has been highlighted through numerous publications. For these reasons, there has been a growing emphasis on considering AI/ML model conceptualization, development, implementation, and sustainment as a lifecycle, incorporating best practices from many disciplines including healthcare organization and delivery, human factors, human computer interaction, implementation science, continuous quality improvement, computer science, heath information technology, biomedical informatics, and biostatistics.

The National Academy of Medicine (NAM) published an environmental scan, challenges, and opportunities for artificial intelligence in healthcare in 2019, and the need to follow a

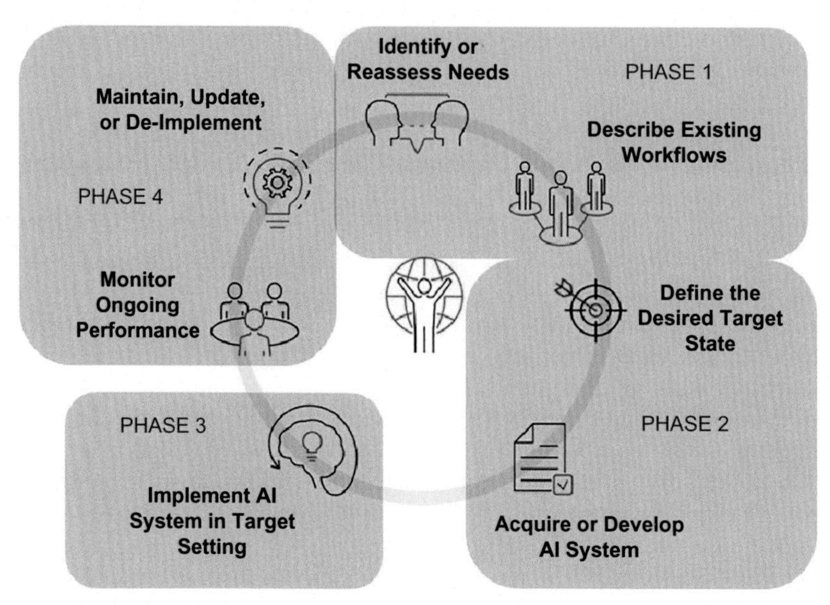

FIG. 7.2 National academy of medicine healthcare AI modeling lifecycle. *Adapted and reproduced with permission Matheny ME, Thadaney Israni S, Ahmed M, Whicher D. Artificial intelligence in health care: the hope, the hype, the promise, the peril. Washington, DC: National Academy of Medicine; 2019.*

careful reasoned course as part of the modeling lifecycle was a key component of that [46]. Fig. 7.2 is an adaptation of the NAM AI Modeling Lifecycle organized into phases, illustrates the proposed AI implementation lifecycle.

There are four key conceptual phases in the implementation lifecycle of healthcare AI, reviewed here along with a table of key recent references mapped to this framework. These include: (1) assessing organizational needs, the current state, and the desired future state, (2) AI model development, (3) model implementation, and (4) modeling performance surveillance and maintenance. *Phase 1* ensures that the models will address a clinical need, be appropriate in the context of the population in which they are used, and that the organization has adequate resources to deploy and sustain the tools and workflow changes required. *Phase 2* focuses on developing the target state for the tool to achieve, the data to use, the outcome of interest (the labels for supervised learning), and the design choices around the model to be used, execution of the model training, and required performance for success. *Phase 3* focuses on organizational implementation, and requires heavy integration with information technology, electronic health record, and clinical and patient user needs and perspectives. *Phase 4* focuses on the maintenance and sustainment of implemented models. There is a growing literature on systematic drift of data over time, and the COVID-19 pandemic certainly highlighted challenges when shifts occur rapidly. It is important to conduct surveillance and maintenance of the data drifts and model performance [16,47–49].

In the remainder of this chapter, we will focus on Phase 2 of the lifecycle, with a small section on Phase 4, and other chapters of the book will focus more on other Phases in the context of clinical decision support. Please see the provided references for further reading outside of this focus.

7.4 Learning from data

Artificial intelligence, statistical, and machine learning pattern recognition algorithms recognize regularities, patterns, and associations in data and construct a model that selects (classifies) or estimates the likelihood (predicts) in new cases. In this section, we describe some of the most commonly encountered and impactful issues and considerations for the selection and use of data for AI/ML model development.

To understand how a model can be derived from data, it is useful to construct an artificial example. Suppose a researcher does not know the range of normal values for a new diagnostic test, but she does have a large data set indicating, for a set of patients, the value of the test and the actual diagnosis for each patient. Also suppose that there are missing and noisy data in the data set. The task is to determine the range of normal values for the test, so that when anyone examines the value for a new patient, it would be possible to declare, with a certain level of confidence, whether the result pointed to an abnormality or not. While one might not need a sophisticated model to answer this simple question, it would be necessary to review all labeled data to determine optimal thresholds to label a result as "normal" or "abnormal."

This analysis can extend to several tests and clinical findings, and multiple possible diagnoses, in which case the task is to find optimal combinations of values that are most frequently associated with particular diagnoses, since a single test or clinical finding in isolation may not suffice. Researchers would have to examine several thousands of records containing dozens of attributes for each patient to determine which combinations of variable values seem to be most likely associated with each diagnosis. Given time and memory limitations, it might be difficult to build this type of classifier. For this type of problems, utilizing multivariate techniques that "learn" from data can be very helpful.

7.4.1 Representative data selection

It is important to first be clear about the purpose and intent of the clinical decision support to be implemented, and how the machine learning algorithm is to be used, what is the cohort of patients to be scored using the algorithm, and the definition of the outcome (dependent variable) that is to be predicted [50]. Although it is intuitive to state that the performance and generalizability of AI models is directly related to the characterization and composition from the data in which they were built, many instances of application of these models go beyond or outside of those boundaries. In some cases, model development and reporting transparency is insufficient to even be able to assess which populations the model could or could not be used on [51]. This brings into focus the need to have detailed reporting of data cohort characteristics, composition, and an understanding of how and why the data were collected, in order to assess how and where the algorithms trained on that data can be used.

Other factors affecting data representation and model generalizability include differences in local populations, EHR systems, coding definitions, laboratory equipment and assays, as well as variations in clinical and administrative practices. For instance, in a recent study aimed at detecting abnormal chest radiographs, the specificity of an ML model at a fixed operating point varied widely, from 0.566 to 1.000, across five independent datasets [52,53]. Although less common, clinical ML models that have been validated across different healthcare systems are either re-trained from scratch or are fine-tuned (via transfer learning) on every new patient cohort, or when applied out-of-the-box often exhibit significant degradation in performance [54,55]. While transfer learning (a machine learning technique that allows for optimizing/fine-tuning parameters of a neural network model on data from a new institution) can improve model performance, fine-tuning of an algorithm requires obtaining gold-standard labels for supervised learning which are often expensive to obtain [56].

7.4.2 Data bias

One of the most concerning issues in developing AI and ML algorithms for use are biases that are embedded into the data through why and how the data are collected [57]. Such embedded biases can then be reproduced unintentionally through algorithms training on that data [58,59]. This can result in unintended consequences, for example in some epidemiological studies, patients with prevalent depression are more commonly women, however this could result in skewed risk prediction because of symptoms that occur more commonly among women, which can over-predict risk [60].

Another issue is that AI/ML systems seek to predict based on complex systems and in which the data are only partially observable. Without direct ascertainment, this can result on the AI/ML algorithm learning complex correlations from co-linear or other variables associated with the direct data needed, and these proxies, while measurable, sometimes have applications that have unintended consequences. For example, one algorithm that was used to assign extra health care resources to patients in need used healthcare utilization as the outcome of interest, but this generated a racially biased algorithm against Black patients, who were not receiving as many resources to begin with. After this algorithm was changed to target clinical condition severity to direct healthcare resources, the bias was largely mitigated [61].

7.4.3 Ethical and equitable use of data

The highly appropriate focus on the ethical, transparent, and equitable use of AI/ML algorithms to support healthcare has drastically increased in the last few years [62,63]. It is important to have an ethical framework in place and requirements regarding transparency for healthcare system implementers, as well as end-users and patients.

While no chapter on AI/ML CDS implementation would be complete without highlighting the need for consideration of ethics from conception to implementation to maintenance, detailed discussion of this is beyond the scope of this Chapter. We refer the reader to important initiatives and guidelines for the use of ethical AI, namely the Partnership on AI, Open AI, the Ethics and Governance of Artificial Intelligence Initiative, and the Principles for Accountable Algorithms, just to name a few well-known ones [64].

7.5 Overview of machine learning modeling methods

Artificial intelligence techniques such as those commonly referred to as *machine learning* techniques have been explored to address some potential limitations of standard modeling techniques. Although logistic regression is sometimes thought of outside of "machine learning" models, it is in fact one of the simplest of the family of supervised ML algorithms. Because it is intuitive and relatively simple to describe its internal operation, we will first focus on the family of LR algorithms below.

Beyond LR, there are several machine learning techniques of increasing complexity and decreasing transparency that represent, transform, and analyze input data to classify or predict. Foundationally, many of these methods rest on the fundamental mathematical constructs of a classification tree and the basic unit of an artificial neural network, the perceptron. We will highlight the key functions and considerations of these fundamental units, and provide modern examples of how these have been extended to achieve best-of-breed performance in the current healthcare artificial intelligence community. For classification trees, modern extensions that we will highlight will be Random Forests and Gradient Boosting, and for the artificial neural network/perceptron, we will describe both historical forward and backward feed artificial neural networks as well as some of the more modern deep learning architectures.

7.5.1 Logistic regression

Logistic regression is historically the most popular method for constructing predictive models in medicine [65]. This type of classification model usually deals with binary outcomes such as diagnosis of a certain disease or condition (e.g., myocardial infarction), or prognosis within a certain period of time (e.g., death while in hospital). Using a large number of training cases, it is possible to estimate the parameters of a logistic regression model with a certain level of confidence and estimate the future performance of the model in previously unseen cases. The level of confidence will depend on the number and quality of cases (e.g., presence of outliers and noise), as well as how well the model fits the training data.

The logistic function links i predictors, or independent variables, each denoted by x_i and collectively represented by the vector $\mathbf{x}$, to the dependent variable being predicted, represented by Y using the logistic function as in the equation below:

$$Y = \frac{1}{1 + e^{-(\beta\mathbf{x}+c)}}$$

This function tries to model a step function with two possible values for Y, and it is therefore used to classify binary outcomes. The resulting function is a continuous value from 0 to 1 along a sigmoid curve. Fig. 7.3 illustrates a logistic regression model (a sigmoidal function), and this function is also one of the possible functions used within the nodes of artificial neural networks, which we will describe in Section 4. In most models, Y is a binary variable representing patient status as having a certain disease or condition ($Y=1$) or not ($Y=0$), or prognostic class, and the vector $\mathbf{x}$ represents the clinical, laboratory, and demographic predictors (e.g., x_1 may represent *age*, x_2 may represent *TSH*, and so on). The vector $\boldsymbol{\beta}$ represents the coefficients that are estimated for each predictor and c is a constant. The parameters of the logistic function are usually obtained by maximum likelihood estimation using iterative algorithms [66]. The coefficients correspond directly to the log of the odds ratio associated with each variable. The parameter c calibrates the model for the baseline rate of the outcome of interest. These features make the model somewhat easy to interpret, since the sign and

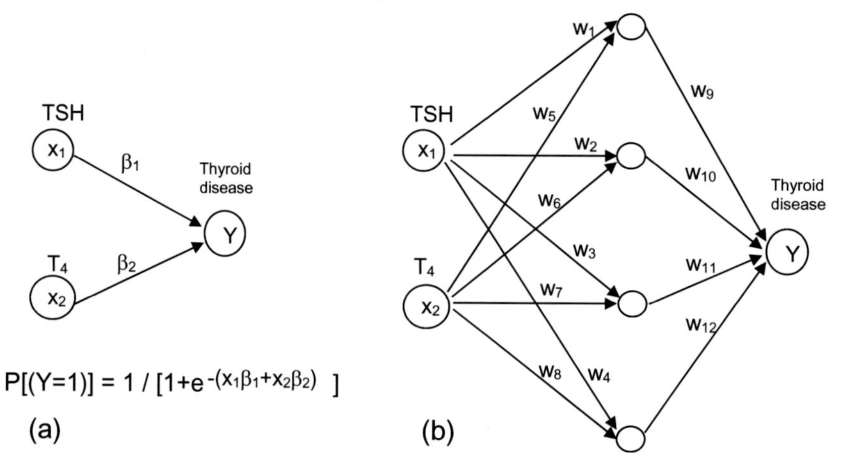

FIG. 7.3 (A) Example of a simple bivariate logistic regression model (no intercept is included for simplicity). (B) Example of an artificial neural network constructed for the same purpose.

magnitude of the coefficients (when standardized) can provide direct indication of how much each particular predictor is associated with an increased risk of a certain outcome (e.g., large positive coefficients will usually increase the probability of $Y=1$ for variables such as those representing most laboratory assays).

In general, when selecting input variables for use in the model, the number of cases needs to exceed the number of variables; a well-known heuristic is that the number of variables utilized in a model should not exceed one tenth of the number of cases. Of note, however, we refer the reader to simulation work that called this into question, and provides a robust discussion of some of the underlying interactions that make a one size fits all heuristic unrealistic [67].

For certain data, predictors may need to be combined in interaction terms or transformed so that a good fit to the data can be obtained. Consider the example in Fig. 7.4: a laboratory test value that is considered normal if within a certain range (e.g., *TSH* within 0.4–6 μU/mL), and abnormal otherwise. Even in this simple univariate problem of classifying the values into normal and abnormal, a logistic regression model in which variables are not transformed will not be able to correctly classify all cases, even in the absence of noise. The reason is simple: the logistic regression function is monotonic and would necessarily classify a portion of the abnormal cases (either the low or high values) as being normal. However, a simple transformation of the variable, in this case, a quadratic, might allow the logistic regression model to correctly classify all cases. Without variable transformation, the logistic regression function will always miss one of the extremes of values, misclassifying values within that range as not "normal". A simple quadratic transformation can make logistic regression work for this example.

Fig. 7.5 illustrates a bivariate problem in which values for two laboratory tests must be within a certain range for the patient to be considered healthy. In this example, both free T_4 and *TSH* need to be within normal limits for the classification "euthyroid" to be made.

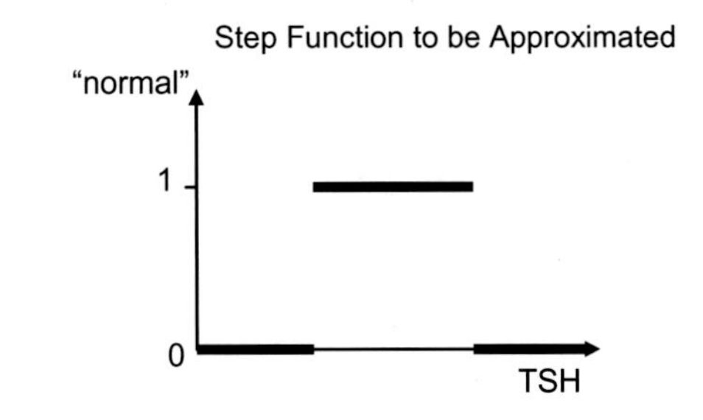

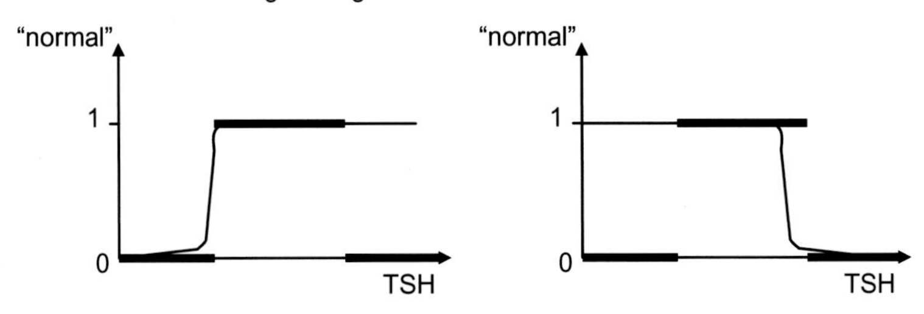

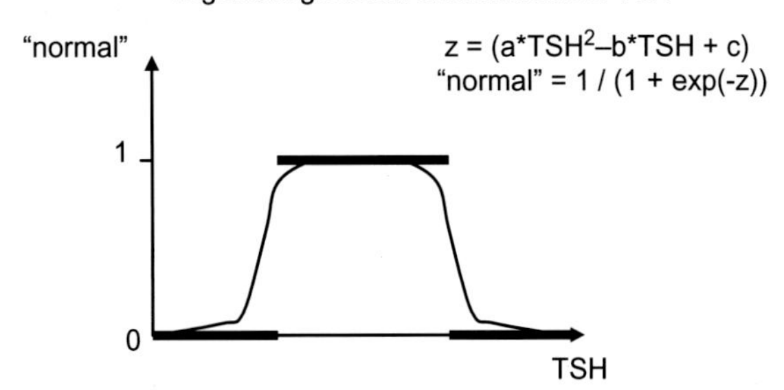

FIG. 7.4 A step function (bold) indicating "normal" laboratory values within a certain range. The step function is overlaid with logistic functions for illustration purposes.

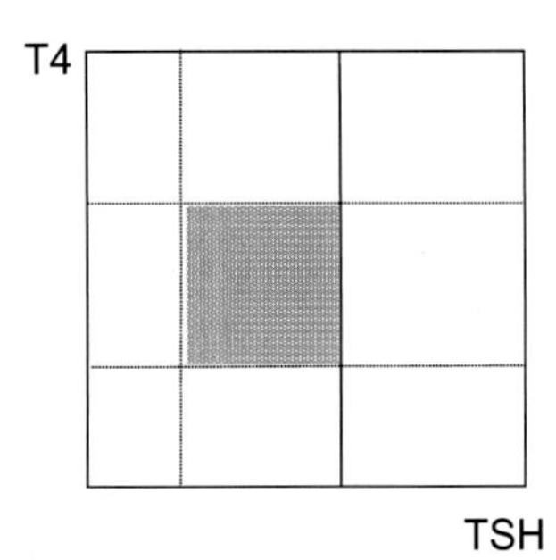

FIG. 7.5 Simplified bivariate example. For a case to be considered "euthyroid" (shaded area), values for both tests must be within a certain range. Without variable transformation, logistic regression will not work for all cases because the problem is not linearly separable.

It is easy to see that no single line would separate the shaded area from the rest, which means that no linear model can produce correct classifications for all cases. Variable transformations or interaction terms are necessary.

Problems that are not solvable by linear or semi-linear models such as logistic regression without variable transformations or addition of interaction terms are known as non-linearly separable problems. Although logistic regression models can be used in linearly non-separable problems, pre-determining which transformations or interactions are necessary is a laborious and computationally expensive process. Furthermore, the interpretation of a model that uses transformed variables or interaction terms is difficult. For these reasons, many models used in practice do not make use of interaction or transformed terms. However, strong performance increases occur in modeling methods that can explicitly transform variables into non-linear representations, and recent ensemble methods that leverage feature building into non-linear space for use in LR have shown that this family of methods can still retain strong performance by addressing this issue [68].

Finally, no discussion of the family of logistic regression methods would be complete with the modern use of L1 and L2 penalization methods, also called LASSO, Ridge Regression, and Elastic Nets (both together) [69–71]. Ridge regression shrinks the sum of the squares of the regression coefficients to be less than a fixed value to try to account for over-fitting but does not remove any variables. LASSO both shrinks variable coefficients and removes variables (by shrinking the coefficient to 0) by forcing the sum of the absolute value of the regression coefficients to be less than a fixed value. These methods supplanted the prior practice of step-wise selection, which has been shown to have significant biases in operation [72,73].

7.5.2 Classification trees

Classification trees recursively and univariately partition cases into two subgroups [74]. At each branch in an upside-down tree, as illustrated in Fig. 7.6, the attribute-value pair that best partitions the cases into the categories of interest (e.g., "euthyroid" or not) is chosen. A simple step function assigns "yes" or "no" to the criterion in question (e.g., TSH > 6 = yes). This is repeated until the partitions that represent the "leaves" of the tree have only cases from a single category. Fig. 7.5 illustrates the simplified bivariate example from Fig. 7.4 above. The first attribute-value pair to be chosen is (TSH, 6μU/ml). Cases in the right branch/leaf (TSH > 6) are not euthyroid. Cases in the left branch (TSH ≤ 6) may be euthyroid or not. The next attribute is T_4 at 12.5. Cases in the right branch/leaf ($T_4 > 12.5$) are not euthyroid. Cases in the left branch may be euthyroid. The pair (TSH, 0.4 μU/mL) is then chosen, and cases are classified into "Not euthyroid" if TSH ≤ 0.4. Otherwise, (T_4, 6.5 is chosen) and those cases with $T_4 > 6.5$ are classified as "euthyroid".

Note that classification trees can solve non-linearly separable problems, since the number of branches is not limited. However, given their limitation of using only univariate cuts at each branching point, there may be too many branches for the tree to be easy to interpret. Pruning algorithms have been developed to address this issue [75]. The Goldman tree (shown in Fig. 7.7) for deciding whether a patient with chest pain should be admitted to the emergency department is the prime example of an application of a classification tree [29]. This study identified nine important clinical factors that enabled the system to correctly categorize

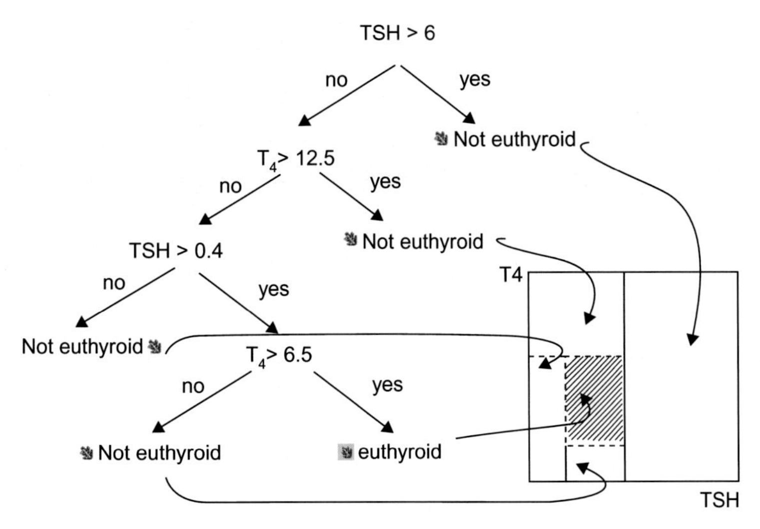

FIG. 7.6 Classification tree for the bivariate outcome problem illustrated in Fig. 7.5. Cases are recursively partitioned according to the attribute-value pair that best divides the cases into "euthyroid" or not. The resulting partitions can be easily visualized in this simplified 2-dimensional problem.

all (60) patients with myocardial infarction (MI) in the sample (482). Sensitivity was an absolute priority in this model, and a portion (71) of patients without MI was categorized as false positives. The clinical factors were: age, duration of pain, chest pain +/− radiation, presence of diaphoresis, history of angina (and severity of pain) or prior MI, local pressure causes reproduction of pain, EKG ST-segment changes, Q waves, or T-wave changes not known to be old.

While classification trees can fit complex problem spaces and provide a human-interpretable graphical model, any single tree often underperforms when applied to new data. Classification trees are also highly sensitive to small changes in the training data. For example, if we rebuilt the tree in Fig. 7.7 with a random sample of observations drawn from the original training data, the tree may split on a variable that wasn't selected for a split in the original tree or may select different age cutoffs at certain nodes. These limitations of single classification trees motivate several widely implemented extensions of tree-based models that rely on bagging and boosting to improve generalizability—random forests and gradient boosting. The similarities and differences between these tree-based approaches are highlighted in Fig. 7.8.

In machine learning, bagging models refers to the process of drawing wisdom from the crowd. In this case, the "crowd" being many submodels that individually may not provide acceptable performance, but together provide accurate classifications when predictions are combined across submodels [76]. Random forests (RF), introduced by Breiman [77], use bagging to improve the generalizability of classification trees by leveraging the sensitivity of trees to perturbations in training data. A forest (i.e. crowd) of individual classification trees are each constructed with a random subsample of candidate predictors and a bootstrap sample of the training data. Predictions for new data are then based on the majority vote across the forest of trees, with improved classification accuracy and generalizability over any particular

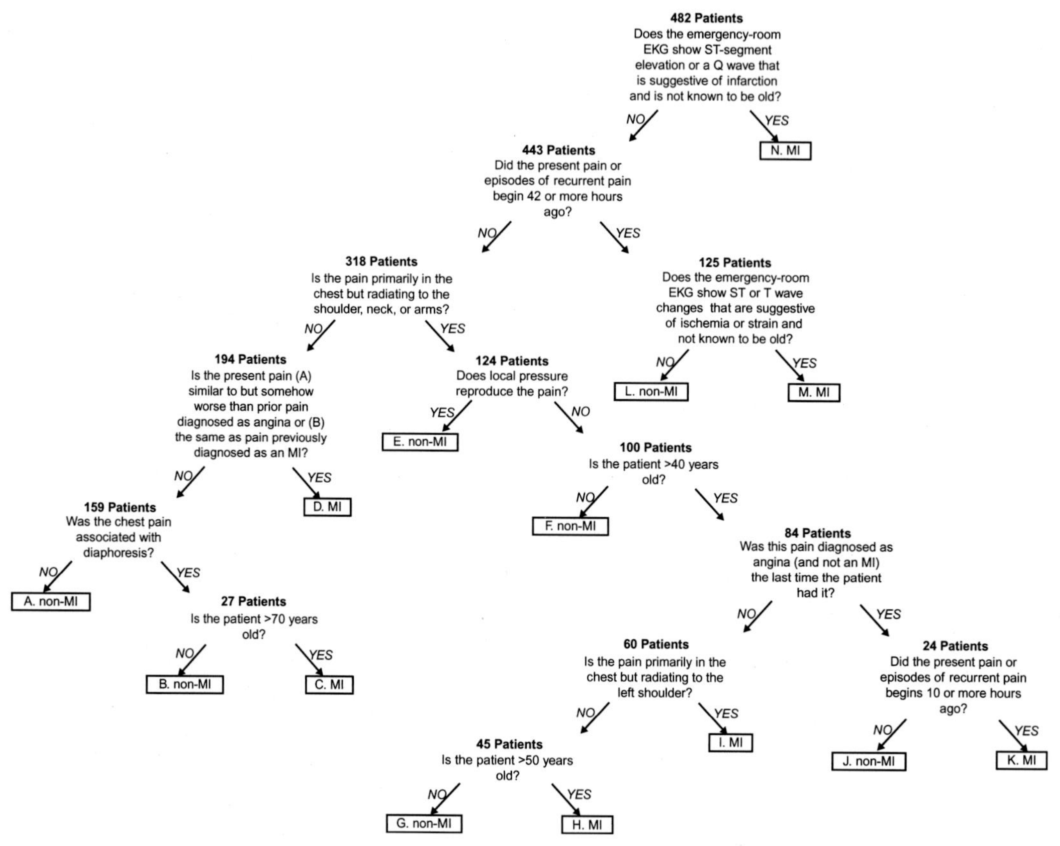

FIG. 7.7 Computer-derived decision tree for the classification of patients with acute chest pain. "Each of the 14 letters (A through N) identifies a terminal branch of the tree." In the Goldman study, seven terminal branches (C, D, H, I, K, M, and N) contained all the patients with acute myocardial infarction, along with a portion of the patients with other diagnoses. *Reproduced (with permission) from Goldman L, Weinberg M, Weisberg M, et al. A computer-derived protocol to aid in the diagnosis of emergency room patients with acute chest pain. NEJM 1982;307:588–596.*

submodel. Cross-validation is required to determine the number if trees in the RF and the number of candidate predictors to be considered by each tree. Balancing the number of trees and predictors is key to achieving generalizable, accurate models without computation time becoming overly costly [78,79]. Tuned properly, RFs support complex predictor-outcome associations, high-order interactions among predictors, and the consideration of any number of predictors, even when predictors outnumber training observations [77,80–82]. As a result, RF models have increasingly been applied across biomedical and clinical domains [83–86].

Introduced by Friedman in 2001 [87], gradient boosting takes a different approach to improving upon the single weak prediction model such as a single classification tree. Rather than building a crowd of similar models, boosting approaches build a series of models in which those observations that are misclassified by prior models receive more attention or weight when training the next model in the series [87]. In this way, those observations that were hardest to classify and not well characterized by one model get more attention in

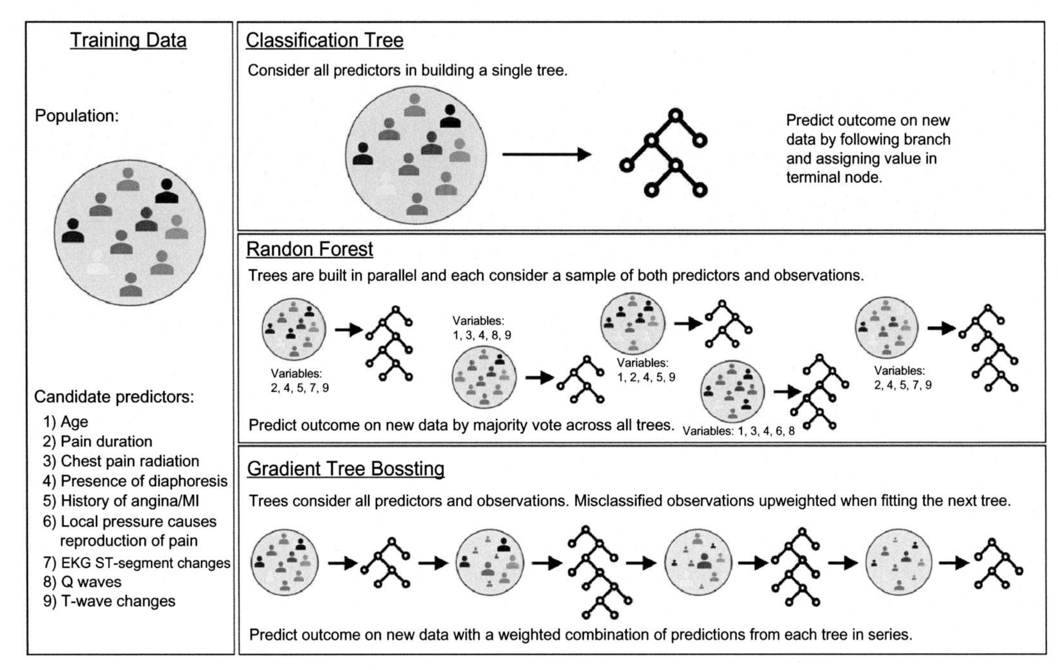

FIG. 7.8 Given a set of training data and predictors (left pane), this figure shows how the data are utilized in various machine learning methods that use classification trees.

how the next model learns classification rules in the hope of better characterizing the challenging prediction region. Predictions for new data are developed by combining predictions across the series of submodels, giving more weight to high performing submodels. In contrast to random forests, which builds a random set of trees, gradient tree boosting approaches build an informed series of trees and stops when no further performance gains are realized. Thus the boosting approach tends to be faster than bagging. The popular Xgboost implementation described by Chen and Guestrin in 2016 further enhances the speed of the boosting approach with computational efficiencies and parallelization [88]. Teams implementing Xgboost have dominated data science competitions in recent years and the method is quickly gaining attention in clinical decision support applications [89–92].

7.5.3 Support vector machines

Since Vapnik and colleagues described support vector machines (SVMs) and formalized optimization techniques in the early 1990s [93], SVMs they have proved popular and successful across a number of machine learning domains. SVMs take a unique, but intuitive geometric approach to learning how to classify data. For example, if all training data are viewed on a plane with coordinates defined by predictor variables, there exist many potential lines that could separate the outcome categories (i.e., cases and non-cases). If training data are viewed on a three-dimensional space, then planes separate the outcome categories, and so on for higher dimensions. To maximize generalizability on new data, in the example of data represented on a plane, SVMs seek the line that separates the outcomes with the widest margin, or distance, between the line and the nearest training datapoints on either side. Those training points nearest the separating line are called the support vectors, and are the only data required to define the final model. While visually less intuitive, this design easily extends to datasets with many predictors using high dimensional spaces partitioned by complex hyperplanes to divide the training points.

SVMs have several advantages that motivated their uptake, extension, and widespread use in machine learning research. First, SVMs can incorporate nonlinearity and complex associations through kernel functions that map the training data to additional dimensions prior to defining the separating hyperplane. Second, to accommodate the common issue of categories that may not be perfectly separable, SVM constraints can be relaxed by defining a soft margin that allows some training points to fall within the margin around the separating hyperplane. The flexibility of this soft margin is controlled by users through a hyperparameter that balances generalizability and error tolerance. Finally, SVMs can be solved through quadratic programming techniques that are simple and widely implemented.

Other machine learning techniques increased in popularity as the limitations of SVMs become more apparent. On modestly sized datasets (e.g. thousands or tens of thousands observations), SVMs are easily and quickly trained through quadratic programming. However, as training sample sizes grow into the hundreds and millions of observations, SVMs become impractical or intractable to train. Methods focused on linear kernel SVMs have been developed to enable the application of SVMs in large datasets [94,95]. Another challenge to developing SVM models is selecting and parameterizing the kernel, which can have important impacts on model performance [96–99]. Numerous potential kernel functions are available, with limited recommendations for selecting a kernel form [99]. Even having selected a kernel, for large samples sizes, the traditional resampling methods for tuning kernel parameters are computationally expensive. Finally, SVM outputs need transformation to serve as proxies for probabilities. Even when they are transformed, these "probabilities" are often not well calibrated, thus requiring further transformation for use in practice [100].

Research applications have utilized SVMs across clinical domains, including for cancer diagnosis, screening, and treatment [96,101–103]; drug discovery [97]; mental health diagnosis and treatment [104]; and diabetes diagnosis and treatment [105,106]. While SVMs have been popular in biomedical research for decades [107], there has been limited translation of SVM-based models into clinical decision support applications.

7.5.4 Artificial neural networks and deep learning

The use of artificial neural networks (ANNs) has been reported in many medical domains [108–112]. ANNs are highly flexible models composed of several processing units. Each of these units processes incoming information and may propagate information forward if warranted by their activation function. The most common activation function is the logistic, which has been presented in Section 11.3. The logistic function tries to model a step function that has been widely used to represent the electrical conduction in real neurons, which only propagate electric impulses if a certain threshold value is achieved. Although it is possible to build ANNs without utilizing an intermediate "hidden" layer of neurons, the flexibility of ANNs comes from the inclusion of more than one non-linear "hidden" node in this layer.

In fact, the limitation of perceptrons, which were precursors to ANNs and were subject of much interest in the mid-1950s, was noted by several authors [113]. The same authors noted that multi-layered perceptrons did not suffer from this limitation, but at that time there were no algorithms for estimating weights of multilayered perceptrons. The field was stagnant until Rumelhart [114] published the back-propagation algorithm in the mid-1980s. In the following two decades, a plethora of successful applications were reported in and out of the medical literature, but many of these research models did not translate into real clinical applications. Some, however, have been evaluated in real applications, such as automated analysis of pap smears [30,31]. For these types of ANNs, there is no theoretical advantage of using ANNs over logistic regression in binary classification problems unless the ANNs have a hidden layer of non-linear neurons.

Fig. 7.3 illustrated the similarities and differences between binary logistic regression and these types of ANNs with a single output unit. ANNs and logistic regression models have several differences: (1) the activation function of the output unit needs not be the logistic in ANNs; (2) ANNs have intermediate processing units, often called hidden units or hidden nodes; and (3) ANNs can have multiple output units, so different classification problems can be modeled with a single network (although one could argue that polytomous logistic regression also allows for multiple outcomes to be modeled).

The hidden units in ANNs operate between the inputs and the outputs to process information to be sent to the output unit. Fig. 7.9 illustrates how an ANN might solve the linearly nonseparable problem of classifying cases into "euthyroid" or not based on values of two laboratory tests, as illustrated in Fig. 7.5. In this example, the activation functions of the intermediate layer of neurons correspond to the branching points that define the partitions of the classification tree presented in Section 7.5.2, but this will often not be the case. Furthermore, in the example we used step functions in the hidden layer. Step functions are not linear, and the use of non-linear step function combination offers a potential advantage over logistic regression, which does not have a hidden layer.

The outputs of these hidden layer functions are multiplied by the weights that lead into the output node and summed to serve as input to the output node, which classifies cases into "euthyroid" or not. We do not represent every possible weight between the input layer and the hidden layer, to allow better visualization in the picture, but the reader can, for purposes of simplicity, assume here that the non-displayed connections are associated with null weights.

For many complex and temporal problems, a single hidden layer in the ANN may not be sufficient to capture the nuance of input and output relationships. For example, image analysis, text evaluation, and translation problems require hierarchies of features and construction of features that are unknown or intractable to specify [46]. Given a large training set, multilayer ANNs are able to abstract such information and create deeply complex models that are proving to provide new insights and opportunities for decision support. In an ANN with a single hidden layer, each node of the hidden layer receives an activation level from the weighted combination of inputs. When multiple hidden layers are included in a network structure, the activation levels in the nodes of each layer serve as inputs that are weighted and combined (possibly with additional external inputs) to set the activation levels of the nodes in the next connected layer. As information propagates through the network, high order interactions and nonlinearity can be accounted for without any user specification. Highly complex network structures are possible with fully connected layers, recursively connected layers, and connections that skip select layers.

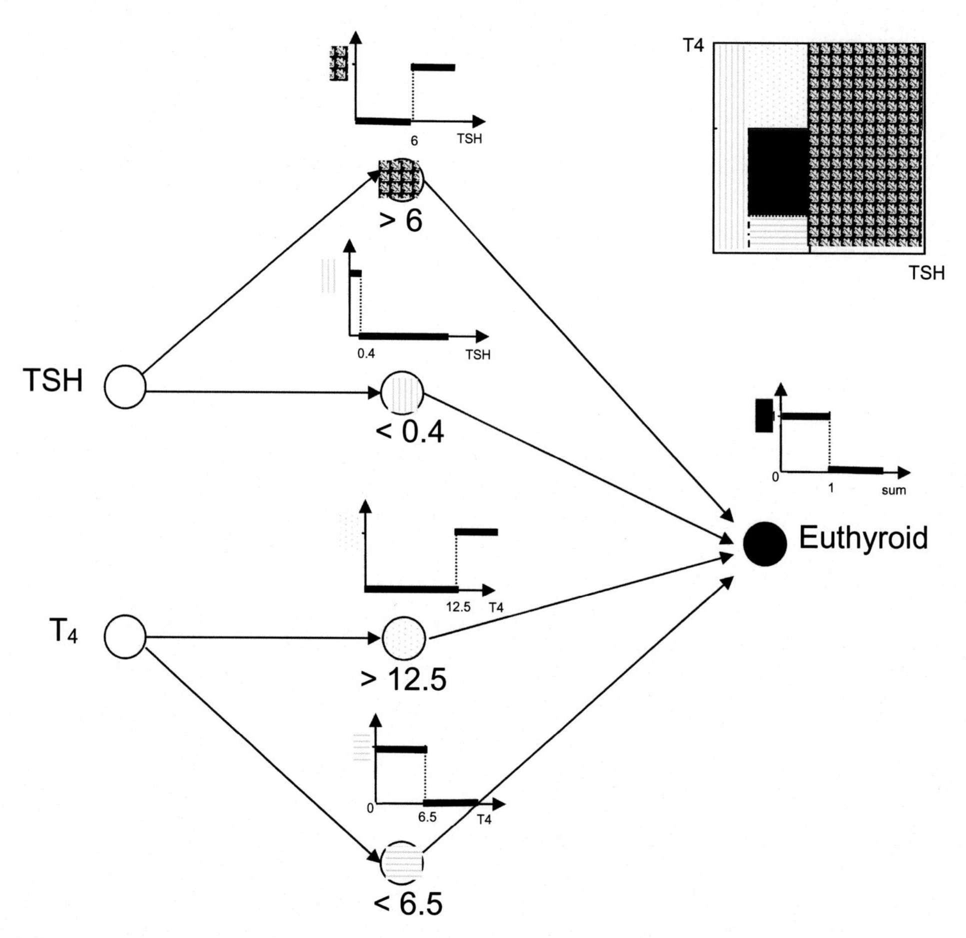

FIG. 7.9 Artificial neural network with a hidden layer of nodes. For didactic purposes, activation functions in this example correspond to step functions that define partitions similar to the ones in the classification tree. Corresponding sigmoid (logistic) functions would be used in practice. As opposed to the classification tree example, the partitions here are overlapping. The outputs of the step functions are multiplied by their respective weights and combined as inputs to the output unit. The output unit has a step function that determines whether a case is "euthyroid" or not.

Termed "deep learning", such multilayer ANNs are gaining interest across the data science and biomedical communities. While the added complexity of multiple hidden layers and multiple connection layouts between layers is seen as a "black box" by some concerned users [46], deep learning models have proven useful across clinical domains. Deep learning models have accurately assessed suicide risk from social media posts [115], automated diagnostic coding of clinical notes [116], and evaluated radiological images for cancer screening [117,118]. Deep learning models are even used to generate synthetic EHR data to support development of prediction models for clinical decision support without accessing private patient information [119].

Recurrent neural networks (RNNs) are one type of deep learning model that has become popular in biomedical research involving temporal data. RNNs are composed of a string of smaller networks (sub-units), one per timepoint, that are connected into a sequential series. Each sub-unit within an RNN handles information at a specific timepoint and may produce a predicted outcome for the current time, although only predictions at certain timepoints may be of interest to the user. Each sub-unit of the RNN takes in observed features (inputs) for the specified time and the activation level in the previous sub-unit (i.e., from the prior time point), creating the current sub-unit's activation level through a weighted combination of these data. In this way, RNNs process temporal data by combining inputs from each time point with information "remembered" from the prior time points to predict outcomes over time.

For long time series, a key challenge to fitting RNNs involves a phenomenon called "vanishing gradients." The weights in any neural network are estimated through backpropagation using gradient descent. Gradient descent estimates model parameters by incrementally adjusting parameters toward those values that minimize error [120]. As data is repeatedly processed forward through an ANN or RNN, estimated outcomes for each observation are constructed using current parameter values and the gradient of a loss function is evaluated with these estimates. These gradients are fed backward through the network to adjusted parameters based on the gradient values proportional to some learning rate. By repeating this process multiple times, parameter estimates step toward optimal values that minimize loss [120]. In the case of RNNs, these gradient values exponentially decay and can become incredibly small as they are propagated back many timepoints (>10 temporal units) [121], making the steps toward ideal parameter values very slow and reaching convergence impractical or impossible [122,123].

Several extensions of the basic RNN approach have been developed to address this limitation, with long short-term memory networks (LSTMs) proving particularly successful. Developed by Hochreiter and Schmidhuber, they avoid the vanishing gradient problem by making use of a specialized memory unit (or cell), an input gate, an output gate, and a forget gate, which together enable the state of a cell to stay locked over arbitrary long time-steps and thus the gradients to also flow unchanged [124]. For a data point at time *t*, the units of the LSTM calculate a new activation level by adding the activation level at time *t-1* with time *t*'s weighted input information. LSTMs units may include a "forgetting" feature which is constructed with separate weights than the activation level and can determine when to clear a unit's prior activation information. LSTMs can support temporally distant dependencies (e.g., over 1000 time steps), such as those in natural language models.

Clinical decision support models based on LSTMs have been developed for detecting sepsis [125,126], evaluating pain from speech patterns [127], diagnosing congestive heart failure and cardiac arrhythmias from ECG signals [128,129], and identifying sleep stages and patterns from polysomnography signals [130]. For an in-depth discussion of LSTMs, we refer readers to a tutorial by Sherstinsky [131].

7.6 Prediction models in medicine

In this section, we will discuss some applications of the modeling techniques described above, although each clinical example will not include all methods. First, as a foundation, the most common indices of model performance are discrimination and calibration. Discrimination assesses how well the models can potentially discriminate positive and negative cases in general. Models that estimate higher probabilities of outcome "1" for cases that had that outcome have high discrimination, which is usually measured as the area under the ROC curve [132]. Calibration assesses how close the model's estimated probability is to the "true" underlying probability of the outcome of interest. Since the true probability is not observable, several approaches aim to estimate true probabilities from observed data with varying levels of rigor [133]. Common calibrations measures such as the observed to expected outcome ratio and the Hosmer-Lemeshow test are limited in their utility, particularly for clinical decision-making applications [133–135]. Graphical measures of calibration provide a more nuanced understanding of the alignment of predicted and observed probabilities across the range of risk [133,136,137]. With graphical measures difficult to compare across studies, summary measures such as the estimated calibration index (ECI) have been developed to provide a measure of miscalibration that is weighted to emphasize model performance in the ranges of risk most relevant to patient populations. ECI is calculated as the mean squared difference between predicted probabilities and estimated observed probabilities from the nonlinear calibration curves. Fig. 7.10 displays a calibration plot along with measures of both discrimination and calibration for a model predicting acute kidney injury in a national inpatient cohort of US veterans. Huang et al. provide a brief review of calibration measures [139].

7.6.1 ICU severity of illness prediction

Historically, the Acute Physiology and Chronic Health Evaluation series of models (APACHE-III [140] and APACHE IV [141]) were some of the most widely used logistic regression-based predictive models. This makes sense in the clinical context, where the ICU has been an environment with earlier adoption of granular structured data with many levels of monitoring of patient condition and state changes. These environments naturally led

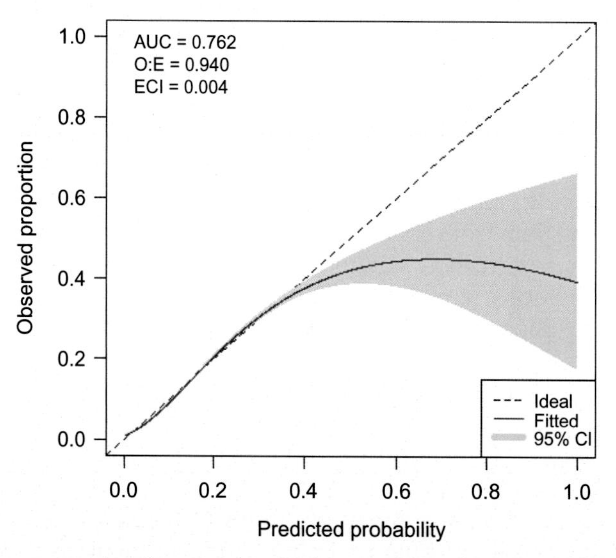

FIG. 7.10 Calibration curves and performance metrics for a model predicting acute kidney injury among an inpatient national cohort of US veterans for the year after model development [138].

themselves to early work in AI/ML and the short time windows experiencing adverse events and death also lent themselves to prediction.

Other prognostic systems for the adult ICU, more common in Europe, are the Simplified Acute Physiologic Score SAPS-3, and the Mortality Prediction Model MPM-III. The Sequential Organ Failure Assessment SOFA model has also been used to assess organ function over time. These models or their earlier versions have been extensively compared all over the world in disparate patient populations. Numerous reviews and comparisons among these models have been published to date [142–163].

Both APACHE and SOFA scoring systems, and many other AI/ML algorithms remain in use today for research, quality control, and clinical applications, and undergo regular re-calibration and assessment in different environments. We direct the reader to some excellent summaries of the literature for the use of ML in the ICU, namely a systematic review that found 285 publications leveraging ML in ICU prediction, including 77 for predicting complications and 70 for mortality [164]. In this study, all of the methods we have described in this chapter were commonly used, from neural networks (72), to random forests (29), to decision tree based ensemble methods (34). However, another systematic review of 494 studies of the use of ML in the ICU found that very few were in actual clinical practice beyond the longstanding methods, and evaluations of clinical implementation were scant [165].

As a further example of the traditional scoring models' ubiquity, recently, qSOFA, SOFA, APACHE-II, and SAPS-II were evaluated for accuracy among COVID-19 admissions in the early months of the pandemic in 2020, and found good discrimination with APACHE-II showing the best with an AUC of 0.851 for the need for ICU admission [166]. Many algorithms have shown severe mis-classification and mis-calibration as the COVID pandemic induced heretofore unheard of magnitudes of data shifts, but models that used physiologic variables as a general marker of severity of illness and short time windows of prediction remained more robust to these changes.

Some of the other examples of CDS systems based on advanced machine learning techniques include early prediction of sepsis [167], respiratory failure [168], and AKI [169,170], as well as algorithms for ED triage [171], prediction of cardiac arrest [172], mortality and length of stay [173], 30 days readmission prediction [174,175], and inpatient fall risk assessment [176], among others. However, high-quality evidence via randomized clinical trials and regulatory approvals are scant—a significant barrier to the wider adoption of such systems in clinical practice.

7.6.2 Cardiovascular disease risk

Another category of extremely well-known prediction tools in medicine provides estimates to patients of the risk of developing future heart disease. Although over 100 risk prediction models have been developed for this purpose, US medical practice has almost exclusively used the family of 10-year heart disease risk models developed from patients in one of the most famous patient cohorts who have been followed in the community of Framingham, Massachusetts, since the early 1950s. From oldest to most recent, these include models developed from the initial examination of the Framingham Offspring Study (FOS) [177], the 11th examination of the original Framingham cohort (FC) [178], or 1st examination of the FOS and 11th examination of the FC [179], or the 1st or 3rd examination of the FOS and the 11th examination of the FC [180]. Worldwide, other well-known models have been developed from the PROCAM [181] (Germany), UKPDS [182] (United Kingdom), and QRESEARCH [183,184] (United Kingdom) cohorts. All of the well-known models developed in this domain are based on Cox proportional hazards and logistic regression methods [185].

The widespread use of these models is related to a number of key factors that influence the utility and generalizability of the prediction model. First, the modeled outcome is of paramount importance, since heart disease is the number one cause of mortality in the US, accounting for 161 deaths per 100,000 people. Effective treatments exist for many of the outcome predictors, such as hypertension, hyperlipidemia, and smoking. Second, the patient population that was used in model development is in many ways representative of the American population that received regular medical care. The Framingham cohort was an excellent source of data because the longitudinal nature of the cohort allowed reliable discrimination of patients at higher risk but who had not yet presented any sign or symptom of heart disease. One of the primary limitations of the cohort was the lack of racial diversity.

External validation of these models showed good discrimination and moderate calibration, with some limitations when applied to populations with significantly different demographics and specific co-morbidities (such as diabetes) [186–189]. Recalibration strategies were used to remediate this problem [190,191]. Perhaps more concerning in this domain was the tendency to externally validate the models with different outcome definitions, most commonly seen between more stringent primary modeling outcomes and more relaxed external validation outcomes [190,192,193]. For example, in some cases the models were developed on Hard CHD, which included sudden CHD death or myocardial infarction only, but then externally validated on Total CHD, which included Hard CHD outcomes as well as unstable angina and angina pectoris.

These models have been used by a number of medical associates to establish guidelines of care [194]. In addition, the models were distributed as simple equations that could be quickly scanned by clinicians and patients, or embedded in calculators or computer-based software [195]. However, these models did not include variables that are increasingly recognized as important toward the development of a personalized approach to medicine: social determinants of health and genetic information were left out. A review of cardiovascular models that include social determinants of health indicates that there is added benefit of include these variables [196]. Increasing interest in polygenic risk scores for cardiovascular outcomes will likely result in publication of several models [197]. The extent to which these models will be used in practice remains unclear, as issued related to data harmonization and availability of genetic information at large scale in different populations still prevent their widespread dissemination [198–200].

7.6.3 Pneumonia severity-of-illness index

Another logistic regression risk model example that has had a significant impact in the emergency department for both work flow (documentation requirements) and treatment is the Pneumonia Severity Index (PSI) developed from the Pneumonia Patient Outcomes Research Team (PORT) [201].

The team developed a prediction rule for the risk of death within 30 days for adult patients with community-acquired pneumonia. This disease is diagnosed in approximately 4 million adults each year in the US, and over 600,000 of the diagnosed patients are hospitalized [202]. The aggregate cost of hospitalization for this disease was estimated at 4 billion dollars per year [203,204]. The results of the PORT study suggested that, if the risk model had been used to treat patients based on the risk categories suggested, 26–31% of patients who were hospitalized for care could have been treated safely as outpatients, and an additional 13–19% could have been hospitalized only for brief observation [201].

The key factors that led to the widespread use of this risk prediction tool were a combination of coinciding interest in evidence-based medical practice and in cost containment, as well as the high quality of the risk prediction tool. The model was validated on over 50,000 patients in 275 US and Canadian hospitals in the PORT study. Prior pneumonia risk prediction tools had suffered from small development population sizes [205–208] and limited external validation [207–209].

The model has been widely used, and incorporated in both paper [210] and electronic [211] decision support tools for use in determining hospital admission from an emergency department. A number of subsequent multi-center randomized prospective studies have supported the use of the PSI as an appropriate admission tool [212,213]. It was incorporated into the American Thoracic Society's (ATS) Community-Acquired Pneumonia guidelines [214], although the society emphasized limitations of the model in populations that were not well represented in the development data set (such as outpatient clinic patients), echoing findings from a few studies [215]. PSI was incorporated into the Infectious Diseases Society of America/ATS consensus guidelines [216] in 2007. Subsequent meta-analyses showed that the PSI has similar performance to CURB65 and CRB65, which are alternative tools [217–219]. In addition to these tools, there are a number of factors that physicians must take into account. Such as the presence of coexisting conditions, patients' preferences, and inadequate home support [220]. Cooper and colleagues [221] reported that several types of classifiers can achieve similar performance in this domain.

7.6.4 Machine learning for diagnostic prediction in medical imaging

Image recognition arguably is one of the most successful and mature application areas of modern machine learning and deep learning [222,223]. The rapid rise of the use of convolutional neural networks began with LeNet in 1998 [224], and saw widespread evaluation and development after AlexNet trained on ImageNet almost halved the error rate for object recognition [225].

Thus, it is not surprising that such techniques have found many applications in medical imaging, including diabetic retinopathy detection [226], skin cancer classification [227], histology-based breast cancer detection [228], automated tissue characterization [229], lesion segmentation [230], detection of pulmonary nodules [231], and histology-based outcome prediction in patients with cancer [232]. A very recent systematic review of the diagnostic accuracy of deep learning in medical imaging found that a total of 503 studies met criteria, among them 115 were in respiratory medicine, 82 in ophthalmology, 82 in breast cancer, and 78 in neurology/neurosurgery [233]. For these domains, AUCs ranged between 0.933 and 1.0

for ophthalmology imaging, 0.864–0.937 for lung nodules and lung cancer, and 0.868–0.909 for breast cancer, but noted extreme heterogeneity in methodology, terminology, and outcome measures.

However, a major bottleneck for successful application of AI techniques to medical imaging is the scale and quality of labeled images [234–236]. Crowdsourcing has been proposed as a solution to scalable data labeling, and advanced machine learning techniques have been proposed to tackle the tradeoffs between label volume and quality [237].

While many existing studies have focused on imaging data, recent surveys of radiologists, as well as other imaging-based medical specialties such as pathology, ophthalmology, and dermatology, have concluded that clinical information are essential for accurate interpretation of imaging data [238]. As such, contextualization of medical imaging findings by integrating clinical history and laboratory data is another area of active research, where deep learning-based multimodal data fusion techniques play an important role [239].

7.7 Machine learning CDS implementation

The path from conception of an ML/AI solution to implementation at the bedside and clinical adoption is long and fraught with a dynamic set of sociotechnical challenges [165,240]. This is partially due to the culture of medicine (which is evidence-based and slow), but also due to the changing landscape of the underlying ML and software as a medical device (SaMD) technologies and requirements of regulatory oversight. This implementation, delivery and dissemination process provides another opportunity for knowledge generation, not only regarding efficacy and effectiveness of such digital solutions but also to ensure a safe, symbiotic and equitable CDS delivery to the end-users. Recent studies have suggested that research-based evaluations of ML/AI tools may not be representative of algorithm acceptance or performance in a real-world clinical settings when patient care is at stake [241]. Furthermore, it has been argued that more realistic model performance criteria (e.g., utility metrics) are needed to effectively evaluate the performance of prediction models in real-life scenarios [242]. In response, real-time clinical implementation studies are gaining increasing attention, with an emphasis on blending quality improvement (QI) and implementation studies [243]. Such studies, however, have yet to become standard practice.

As noted above, *Phase 3* of the AI/ML Lifecycle includes both software implementation aspects of ML/AI algorithms and clinical workflow integration. For simplicity, here we focus on the software implementation of such systems with an emphasis on interoperability, portability, surveillance, and change management.

Over the past decade, there has been an increasing emphasis on *portability* and *interoperability* of CDS tools with the goal of facilitating clinical evidence-generation via multicenter clinical trials. However, across sites the data elements serving as input to ML/AI algorithms may be generated by different types of bedside devices, at different sampling frequencies, and stored in different formats. Fig. 7.11 illustrates how modern clinical informatics fulfills the requirements of interoperability and portability by relying on data standards and exchange protocols such as Health Level (HL7) Fast Healthcare Interoperability Resources (FHIR) standard framework and software containerization for portability [244,245]. FHIR

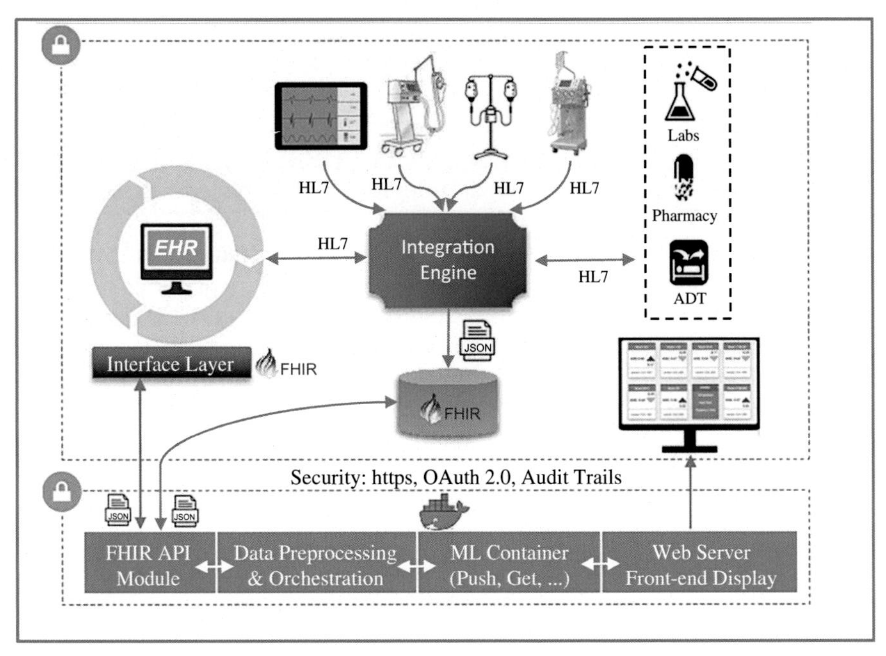

FIG. 7.11 Data integration for real-time analytics.

defines a collection of standards for both storing clinical data (i.e., FHIR resources), as well as a set of REpresentational State Transfer (REST) Application Programming Interface (API) protocols for the exchange of these resources (see Chapter 11 for a more detailed description of interoperability standards and Chapter 15 for a discussion about the role of these standards in CDS implementation and sharing). The use of containerized microservices removes the need to install distinct applications and their associated dependencies on a host machine at various deployment sites. It also allows the system to leverage the inherent scalability and fault tolerance of containerized applications. Containerization and APIs allow for defining single points of access (via predetermined ports) and restricting access to specific microservices. These access policies are defined in advance, and access/audit logs are generated to ensure adherence to predefined policies. Furthermore, this approach minimizes the potential for denial of service attacks by ignoring the service invocations from unknown sources. Lastly, a front-end user interface (UI) enables the clinical team to view and interaction with the clinical risk scores and algorithm recommendations (see Chapter 14 for a description of visualization and integrated information displays and their potential roles in presenting AI outputs for clinical decision support).

A typical real-time analytic system may follow the following steps to produce a risk score: (1) A workflow event (or a "hook" [246]) such as "patient-view" (opening the chart) triggers one or more API calls via the FHIR protocol (or other web Services) to extract data on a patient over a temporally defined observation window; (2) data preprocessing is used to remove outlier observations and generate features for model prediction; (3) the feature vector is then passed to the ML/AI algorithm to produce risk scores in addition to other useful information such as model confidence and the top contributing factors to risk (i.e., model explanations); (4) a user interface (UI) organizes these information (e.g., by sorting patients according to the risk) and displays it in a user friendly format. Under the CDS Hooks framework, the CDS service may also receive the ML/AI output and pass it on back to the EHR in a CDS response. Alternatively, model outputs can be sent back to the EHR system via the "device interface" HL7 protocol. For more detail see Chapter 29 (Integration of Knowledge Resources: Architectures) and Chapter 15 (The Role of Standards).

User-centered UI/UX design and usability studies provide an opportunity for knowledge generation to elucidate factors contributing to communicating algorithmic findings with the clinical end-users, and to explore workflow design choices that may maximize actionability and minimize information overload (see Chapter 14 for a description of ways in which AI/ML outputs can be presented to clinical users). Within the user-centered framework it is widely acknowledged that people are active participants who form their own mental models of how AI/ML systems work. For instance, users may prefer systems that provide transparent and complete explanations of a model's recommendations, where they are given the opportunity to review and potentially modify the system (even when the modification have no effects [247]. A recent study demonstrated that detecting outlier cases and showing users an outlier focused message better enabled them to detect and correct for potential spurious predictions by an AI model. Determining the optimal level of information exchange and interaction between AI/ML agents and clinical users and at the right conceptual level (or "intermediate constructs") remains an open area of research [248].

It is unsafe and/or ineffectual to implement ML/AI algorithms into practice without a coordinated surveillance and updating plan. Such concerns motivate the inclusion of Phase 4 of the AI Modeling Lifecycle to formalize the critical importance of ongoing monitoring of model performance and utility; model updating to restore performance and utility in evolving clinical environments; and de-implementation planning for models no longer providing clinical benefit for users and patients.

The performance of complex machine learning algorithms has been shown to degrade over time [48,138,249–251]. Temporal changes in clinical practice patterns, data documentation and measurement practices, and in the distribution of patient features in the population lead to dataset shift and impact the accuracy of the relationships ML/AI models learned under prior data conditions [49,252,253]. Moreover, similar differences between training and application populations may harm the generalizability of models applied across different geographic and demographic contexts [254,255]. A successful implementation of model-based CDS systems thus requires both initial assessment of model performance on the target population and continuous prospective evaluations of data fidelity and model behavior.

One approach involves restricting the application of models to patients "similar enough" to those included in the training data. Fig. 7.12 shows an example of a deep learning prediction model which utilizes a statistical outlier detection framework to limit its predictions to instances similar to those previously seen during model training [256]. The *conformal prediction* module utilizes a set of data representations (aka, conformal set) to quantify explicitly the conformity of new patient-level feature vectors to the previously seen examples within the training cohort. The model assigns a risk score to those input features that are deemed conformant and rejects the remaining samples as *indeterminates*. Moreover, the rejection statistics (over predefined weekly or monthly intervals) are used to trigger an algorithm change

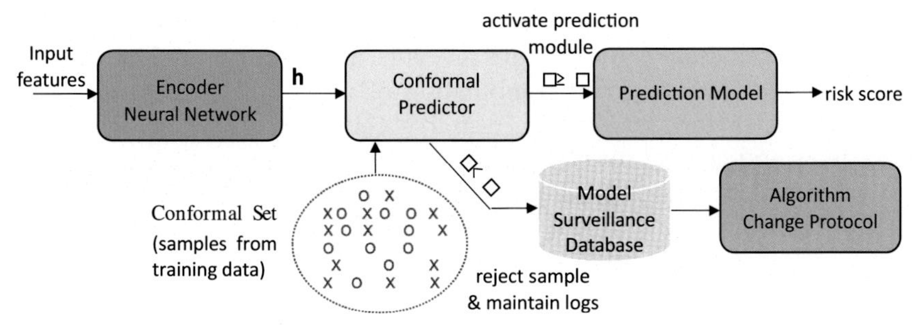

FIG. 7.12 Model surveillance and update.

protocol (ACP), which follows the developer's detailed plan to update the model in a safe and effective manner.

An alternative design to ongoing model assessment takes a less technical, more human-centered approach [49]. Governance panels and knowledge management teams can identify planned changes to clinical guidelines and revisions to coding or data collection practices that directly impacted implemented CDS models (see Chapters 20 and 28 for detailed discussions of CDS governance and knowledge management respectively). When changes are identified, models can be evaluated or updated to ensure relevant patient features are appropriately included in the models. Reports from end-users can highlight degrading model performance when predictions are noted as increasingly misaligned with clinical experience. Such reporting can trigger investigations into model performance and relevant data pipelines, revealing options for model updating, technical corrections to input datastreams, or model de-implementation.

A final approach to model maintenance involves updating models in response to deteriorating accuracy. Several such designs have been proposed. Pre-defined scheduled updating, in which models are retrained at specified timepoints (e.g., annual), is the simplest and most common implementation [257–259]. However, several studies highlight such a design may not align updates with model needs (e.g., updating during periods of performance stability or not updating soon enough when performance deteriorates) and may be prone to overfitting [48,260–263]. As an alternative, a surveillance-based design can monitor model performance and trigger updating in response to deterioration as it occurs. Methods to support a cycle of identifying performance drift, select a relevant dataset on which to train updates, and recommend updating methods that promote prospective model performance have been developed and continue to evolve [260,261,264,265].

Each of the three approaches to sustaining the utility of ML/AI CDS tools has benefits and challenges. A key limitation of the outlier detection approach is the restriction of CDS recommendations to a subset of the population and difficulty recognizing changes in feature-outcome relationships which may harm model performance even among those patients with similar feature distributions. Relying on governance panels, knowledge management teams, and end-users is resource intensive and possibly error prone, especially as the complexity and number of model-based CDS applications grow. Informal or missed changes in clinical practice may also be difficult to identify before model accuracy is seriously impacted, which may harm user trust and confidence in CDS tools. The data-driven surveillance approach may not respond as quickly or successfully as a proactive governance and knowledge management team to abrupt changes that require model revision. Successful model-based CDS applications will most likely involve some combination of these approaches depending on the risk/safety tolerance in the application, resource constraints, and culture/connection of healthcare systems and clinical information system teams.

It is also important to recognize that implementation of AI/ML algorithms can induce changes in clinical workflow and practice patterns that can alter the distribution of data and applicability of predictions. Moreover, successful CDS tools using model-based predictions will alter the observed outcomes of patients, challenging our understanding of model performance. If algorithms are allowed to periodically or continuously learn from the data (a key advantage of machine learning algorithms) unintended feedback loops may emerge. How best to incorporate new data into models, through either continuous learning or intermitted updating, in light of confounding by intervention is an open area of research [266].

7.8 Conclusion

The utilization of statistical and machine learning techniques to discover knowledge from existing clinical data has become an integral component of biomedical informatics. The techniques for constructing and evaluating classification and prediction models are constantly evolving, and there are few theoretical justifications to prefer one learning technique over another. In the last 5 years, machine learning models have been widely developed and are touted to be a panacea for diagnosis, treatment, and a variety of clinical applications. However, successful, wide implementation and sustainment of these technologies has yet to be achieved, owing to numerous issues that are largely outside of the domain of the core technical algorithms and more to do with the sociotechnical environment, data choices, and intended versus unintended consequences. To that end, the grand challenges in the use of AI/ML in clinical decision support at this time are:

- How can large scale data resources be leveraged for training AI/ML algorithms in ways that are ethical, representative of the populations they seek to be used in, and do not carry a host of unintended biases with them into implementation?
- It is important to understand the key algorithmic assumptions and limitations when choosing to deploy AI/ML in clinical decision support applications, and how these are effectively implemented into clinical workflows and systems is a field still in its infancy
- Developing scalable systems for surveillance and maintenance of growing concurrently applied AI/ML algorithms will be computationally complex, and how will the interactions BETWEEN these algorithms be managed as the system complexity quickly grows.

In conclusion, it is more important than ever before that, clinicians utilize classification and prediction models in clinical practice. However, the integration of these tools in a user-centric and integrated way to achieve the ends desired during design remains a challenge. The effective utilization of CDS systems depends on their seamless integration in a computer environment that is effectively used by practicing clinicians; in order to provide counseling at the individual or population management level, predictive models have to achieve increases in transparency, equity, and trust so that the estimates are acceptable from a clinical perspective. However, given the rapid pace of technological advances in biomedicine and the increasing utilization of computers by health care providers, it is our hope that smarter and more effective implementation strategies become standard that promotes the successful, wide use of these technologies to support clinical care and reduce healthcare system costs.

References

[1] Shortliffe EH, Davis R, Axline SG, Buchanan BG, Green CC, Cohen SN. Computer-based consultations in clinical therapeutics: explanation and rule acquisition capabilities of the MYCIN system. Comput Biomed Res 1975;8(4):303–20.
[2] Shortliffe EH. Computer-based medical consultations, MYCIN. New York, NY: Elsevier; 1976.
[3] Pauker SG, Gorry GA, Kassirer JP, Schwartz WB. Towards the simulation of clinical cognition. Taking a present illness by computer. Am J Med 1976;60(7):981–96.
[4] Miller RA, Pople Jr HE, Myers JD. Internist-1, an experimental computer-based diagnostic consultant for general internal medicine. N Engl J Med 1982;307(8):468–76.
[5] Miller R, Masarie FE, Myers JD. Quick medical reference (QMR) for diagnostic assistance. MD Comput 1986;3 (5):34–48.
[6] Barnett GO, Cimino JJ, Hupp JA, Hoffer EP. DXplain. An evolving diagnostic decision-support system. JAMA 1987;258(1):67–74.
[7] Heckerman D, Miller RA. Towards a better understanding of the INTERNIST-1 Knowledge Base. Medinfo 1986;86.
[8] Pearl J. Probabilistic reasoning in intelligent systems. San Mateo, CA: Morgan-Kaufmann; 1988.
[9] Heckerman D. A tractable inference algorithm for diagnosing multiple diseases. In: Proceedings of fifth conference on uncertainty in artificial intelligence; 1990. p. 163–71.
[10] Beinlich IA, Suermondt HJ, Chavez RM, Cooper GF. The ALARM monitoring system: a case study with two probabilistic inference techniques for belief networks. In: Proceedings of the second European conference on artificial intelligence in medicine; 1989. p. 247–56.
[11] Swhe M, Middleton B, Heckerman D, Henrion M, Horvitz E, Lehmann H. Probabilistic diagnosis using a reformulation of the INTERNIST-1/QMR knowledge base I: the probabilistic model and inference algorithms. Methods Inf Med 1991;30:241–55.
[12] Cooper G, Herskovits E. A Bayesian method for the induction of probabilistic networks from data. Machine Learn 1992;9:309–47.
[13] Buntine WL. A guide to the literature on learning probabilistic networks from data. IEEE Trans Knowl Data Eng 1996;8:195–210.
[14] Moore A, Lee MS. Cached sufficient statistics for efficient machine learning with large datasets. JAIR 1998;8:67–91.
[15] Perri-Moore S, Kapsandoy S, Doyon K, et al. Automated alerts and reminders targeting patients: a review of the literature. Patient Educ Couns 2016;99(6):953–9.
[16] Kelly CJ, Karthikesalingam A, Suleyman M, Corrado G, King D. Key challenges for delivering clinical impact with artificial intelligence. BMC Med 2019;17(1):195.
[17] Casey B, Farhangi A, Vogl R. Rethinking explainable machines: the GDPR's 'Right to Explanation' debate and the rise of algorithmic audits in Enterprise. Berkeley Technol Law J 2019;34(1):143–88.
[18] Wong A, Otles E, Donnelly JP, et al. External validation of a widely implemented proprietary sepsis prediction model in hospitalized patients. JAMA Intern Med 2021;181(8):1065–70.
[19] Clancey WJ. The epistemology of a rule-based expert system: a framework for explanation. Artif Intell 1983;20:215–51.
[20] Amann J, Blasimme A, Vayena E, Frey D, Madai VI, the Precise Qc. Explainability for artificial intelligence in healthcare: a multidisciplinary perspective. BMC Med Inform Decis Mak 2020;20(1):310.
[21] Singh A, Sengupta S, Lakshminarayanan V. Explainable deep learning models in medical image analysis. J Imaging 2020;6(6):52.
[22] Vapnik VN. The nature of statistical learning theory. New York, NY: Springer-Verlag; 1995.

[23] Cannon CP, Battler A, Brindis RG, et al. American College of Cardiology key data elements and definitions for measuring the clinical management and outcomes of patients with acute coronary syndromes. A report of the American College of Cardiology Task Force on clinical data standards (acute coronary syndromes writing committee). J Am Coll Cardiol 2001;38(7):2114–30.

[24] Wattigney WA, Croft JB, Mensah GA, et al. Establishing data elements for the Paul Coverdell National Acute Stroke Registry: part 1: proceedings of an expert panel. Stroke 2003;34(1):151–6.

[25] Pollock DA, Adams DL, Bernardo LM, et al. Data elements for emergency department systems, release 1.0 (DEEDS): a summary report. DEEDS writing committee. J Emerg Nurs 1998;24(1):35–44.

[26] Knaus WA, Draper EA, Wagner DP, Zimmerman JE. APACHE II: a severity of disease classification system. Crit Care Med 1985;13(10):818–29.

[27] Grundy SM, Pasternak R, Greenland P, Smith Jr S, Fuster V. Assessment of cardiovascular risk by use of multiple-risk-factor assessment equations: a statement for healthcare professionals from the American Heart Association and the American College of Cardiology. Circulation 1999;100(13):1481–92.

[28] Shaw RE, Anderson HV, Brindis RG, et al. Development of a risk adjustment mortality model using the American College of Cardiology-National Cardiovascular Data Registry (ACC-NCDR) experience: 1998-2000. J Am Coll Cardiol 2002;39(7):1104–12.

[29] Goldman L, Weinberg M, Weisberg M, et al. A computer-derived protocol to aid in the diagnosis of emergency room patients with acute chest pain. NEJM 1982;307:588–96.

[30] Baxt WG. Use of an artificial neural network for the diagnosis of myocardial infarction. Ann Intern Med 1991;115:845–8.

[31] O'Leary TJ, Tellado M, Buckner SB, Ali IS, Stevens A, Ollayos CW. PAPNET-assisted rescreening of cervical smears: cost and accuracy compared with a 100% manual rescreening strategy. JAMA 1998;279:235–7.

[32] Esteva A, Chou K, Yeung S, et al. Deep learning-enabled medical computer vision. NPJ Digital Med 2021;4(1):5.

[33] Liu X, Gao K, Liu B, et al. Advances in deep learning-based medical image analysis. Health Data Sci 2021;2021:8786793.

[34] Rim B, Sung N-J, Min S, Hong M. Deep learning in physiological signal data: a survey. Sensors 2020;20(4):969.

[35] Sendak MP, D'Arcy J, Kashyap S, et al. A path for translation of machine learning products into healthcare delivery. EMJ Innov 2020;19. https://doi.org/10.33590/emjinnov/19-00172.

[36] Koza JR. Genetic programming: On the programming of computers by means of natural selection. Cambridge, MA: MIT Press; 1992.

[37] Zadeh LA. Fuzzy logic, neural networks, and soft computing. Commun ACM 1994;37:77–84.

[38] Pawlak Z. Rough sets. Int J Inf Comput Sci 1982;11:341–56.

[39] Montague PR. Reinforcement learning: an introduction, by Sutton, R.S. and Barto, a.G. Trends Cogn Sci 1999;3 (9):360.

[40] Howard R. Dynamic programming and Markov process. New York, USA: MIT Press and Wiley; 1960.

[41] Kiumarsi B, Vamvoudakis KG, Modares H, Lewis FL. Optimal and autonomous control using reinforcement learning: a survey. IEEE Trans Neural Netw Learn Syst 2018;29(6):2042–62.

[42] Nemati S, Ghassemi MM, Clifford GD. Optimal medication dosing from suboptimal clinical examples: a deep reinforcement learning approach. In: Annual international conference of the IEEE engineering in medicine and biology society IEEE engineering in medicine and biology society annual international conference, 2016; 2016. p. 2978–81.

[43] Gottesman O, Johansson F, Komorowski M, et al. Guidelines for reinforcement learning in healthcare. Nat Med 2019;25(1):16–8.

[44] Liu S, See KC, Ngiam KY, Celi LA, Sun X, Feng M. Reinforcement learning for clinical decision support in critical care: comprehensive review. J Med Internet Res 2020;22(7), e18477.

[45] Zhou SK, Ngan Le H, Luu K, Nguyen HV, Ayache N. Deep reinforcement learning in medical imaging: A literature review, 2021. 2021:arXiv:2103.05115 https://ui.adsabs.harvard.edu/abs/2021arXiv210305115Z. [Accessed 1 March 2021].

[46] Matheny ME, Thadaney Israni S, Ahmed M, Whicher D. Artificial intelligence in health care: The Hope, the hype, the promise, the peril. Washington, DC: National Academy of Medicine; 2019.

[47] Subbaswamy A, Saria S. From development to deployment: dataset shift, causality, and shift-stable models in health AI. Biostatistics 2019;21(2):345–52.

[48] Davis SE, Greevy RA, Lasko TA, Walsh CG, Matheny ME. Comparison of prediction model performance updating protocols: using a data-driven testing procedure to guide updating. In: Proceedings of the AMIA annual symposium; 2019. p. 1002–10.

[49] Finlayson SG, Subbaswamy A, Singh K, et al. The clinician and dataset shift in artificial intelligence. N Engl J Med 2021;385(3):283–6.
[50] Hernandez-Boussard T, Bozkurt S, Ioannidis JPA, Shah NH. MINIMAR (MINimum information for medical AI reporting): developing reporting standards for artificial intelligence in health care. J Am Med Inform Assoc: JAMIA 2020;27(12):2011–5.
[51] Bozkurt S, Cahan EM, Seneviratne MG, et al. Reporting of demographic data and representativeness in machine learning models using electronic health records. J Am Med Inform Assoc: JAMIA 2020;27(12):1878–84.
[52] Hwang EJ, Park S, Jin KN, et al. Development and validation of a deep learning-based automated detection algorithm for Major thoracic diseases on chest radiographs. JAMA Netw Open 2019;2(3), e191095.
[53] Zech JR, Badgeley MA, Liu M, Costa AB, Titano JJ, Oermann EK. Variable generalization performance of a deep learning model to detect pneumonia in chest radiographs: a cross-sectional study. PLoS Med 2018;15(11), e1002683.
[54] Oh J, Makar M, Fusco C, et al. A generalizable, data-driven approach to predict daily risk of *Clostridium difficile* infection at two large academic health centers. Infect Control Hosp Epidemiol 2018;39(4):425–33.
[55] Wardi G, Carlile M, Holder A, Shashikumar S, Hayden SR, Nemati S. Predicting progression to septic shock in the emergency department using an externally generalizable machine-learning algorithm. Ann Emerg Med 2021;77(4):395–406.
[56] Holder AL, Shashikumar SP, Wardi G, Buchman TG, Nemati S. A locally optimized data-driven tool to predict sepsis-associated vasopressor use in the ICU. Crit Care Med 2021.
[57] Agniel D, Kohane IS, Weber GM. Biases in electronic health record data due to processes within the healthcare system: retrospective observational study. BMJ 2018;361, k1479.
[58] Gianfrancesco MA, Tamang S, Yazdany J, Schmajuk G. Potential biases in machine learning algorithms using electronic health record data. JAMA Intern Med 2018;178(11):1544–7.
[59] Cirillo D, Catuara-Solarz S, Morey C, et al. Sex and gender differences and biases in artificial intelligence for biomedicine and healthcare. npj Digital Med 2020;3(1):81.
[60] Martin LA, Neighbors HW, Griffith DM. The experience of symptoms of depression in men vs women: analysis of the National Comorbidity Survey Replication. JAMA Psychiat 2013;70(10):1100–6.
[61] Obermeyer Z, Powers B, Vogeli C, Mullainathan S. Dissecting racial bias in an algorithm used to manage the health of populations. Science (New York, NY) 2019;366(6464):447–53.
[62] Gibney E. The battle for ethical AI at the world's biggest machine-learning conference. Nature 2020; 577(7792):609.
[63] Ouchchy L, Coin A, Dubljević V. AI in the headlines: the portrayal of the ethical issues of artificial intelligence in the media. AI Soc 2020;35(4):927–36.
[64] Jobin A, Ienca M, Vayena E. The global landscape of AI ethics guidelines. Nat Mach Intell 2019;1(9):389–99.
[65] Lemeshow S, Le Gall JR. Modeling the severity of illness of ICU patients. A systems update. JAMA 1994; 272(13):1049–55.
[66] Hosmer DW, Lemeshow S. Applied logistic regression. New York: Wiley; 1989.
[67] van Smeden M, de Groot JAH, Moons KGM, et al. No rationale for 1 variable per 10 events criterion for binary logistic regression analysis. BMC Med Res Methodol 2016;16(1):163.
[68] Levy JJ, O'Malley AJ. Don't dismiss logistic regression: the case for sensible extraction of interactions in the era of machine learning. BMC Med Res Methodol 2020;20(1):171.
[69] Tibshirani R. Regression shrinkage and selection via the Lasso. J R Stat Soc B Methodol 1996;58(1):267–88.
[70] Hoerl AE, Kennard RW. Ridge regression: applications to nonorthogonal problems. Dent Tech 1970;12 (1):69–82.
[71] Zou H, Hastie T. Regularization and variable selection via the elastic net. J R Stat Soc Series B Stat Methodology 2005;67(2):301–20.
[72] Flom P, Cassell D. Stopping stepwise: Why stepwise and similar selection methods are bad, and what you should use. NESUG 2007 Proceedings; 2007.
[73] Dean PF, Edward IG. The risk inflation criterion for multiple regression. Ann Stat 1994;22(4):1947–75.
[74] Breiman L, Friedman J, Olshen R, Stone C. Classification and regression trees. Wadsworth and Brooks; 1984.
[75] Gelfand SB, Ravishankar CS, Delp EJ. An iterative growing and pruning algorithm for classification tree design. IEEE Trans Pattern Anal Mach Intell 1991;13:163–74.
[76] Breiman L. Bagging predictors. Mach Learn 1996;24(2):123–40.
[77] Breiman L. Random forests. Mach Learn 2001;45(1):5–32.

[78] Bernard S, Heutte L, Adam S. Influence of hyperparameters on random forest accuracy. In: International workshop on multiple classifier systems; 2009.

[79] Probst P, Boulesteix AL. To tune or not to tune the number of trees in a random forest. J Mach Learn Res 2017;18 (1):6673–90.

[80] Sajda P. Machine learning for detection and diagnosis of disease. Annu Rev Biomed Eng 2006;8:537–65.

[81] Malley JD, Kruppa J, Dasgupta A, Malley KG, Ziegler A. Probability machines: consistent probability estimation using nonparametric learning machines. Methods Inf Med 2012;51(1):74–81.

[82] Breiman L. Statistical modeling: the two cultures. Stat Sci 2001;16(3):199–231.

[83] Goldstein BA, Hubbard AE, Cutler A, Barcellos LF. An application of random forests to a genome-wide association dataset: methodological considerations & new findings. BMC Genet 2010;11:49.

[84] Khalilia M, Chakraborty S, Popescu M. Predicting disease risks from highly imbalanced data using random forest. BMC Med Inform Decis Mak 2011;11:51.

[85] Qi Y. Random forest for biomedical informatics. In: Zhang C, Ma Y, editors. Ensemble machine learning. Boston, MA: Springer; 2012.

[86] Sarica A, Cerasa A, Quattrone A. Random Forest algorithm for the classification of neuroimaging data in Alzheimer's disease: a systematic review. Front Aging Neurosci 2017;9:329.

[87] Friedman JH. Greedy function approximation: a gradient boosting machine. Ann Stat 2001;1189–1232.

[88] Chen T, Guestrin C. Xgboost: a scalable tree boosting system. In: 22nd ACM SIGKDD international conference on knowledge discovery and data mining; 2016.

[89] Fernandes M, Vieira SM, Leite F, Palos C, Finkelstein S, Sousa JMC. Clinical decision support Systems for triage in the emergency department using intelligent systems: a review. Artif Intell Med 2020;102, 101762.

[90] Fitriyani NL, Syafrudin M, Alfian G, Rhee J. HDPM: an effective heart disease prediction model for a clinical decision support system. IEEE Access 2020;8:133034–50.

[91] Hou N, Li M, He L, et al. Predicting 30-days mortality for MIMIC-III patients with sepsis-3: a machine learning approach using XGboost. J Transl Med 2020;18(1):462.

[92] Sharma A, Verbeke W. Improving diagnosis of depression with XGBOOST machine learning model and a large biomarkers Dutch dataset (n = 11,081). Front Big Data 2020;3:15.

[93] Boser BE, Guyon IM, Vapnik VN. A training algorithm for optimal margin classifiers. In: Paper presented at: fifth annual workshop on computational learning theory; 1992.

[94] Menon AK. Large-scale support vector machines: algorithms and theory. Research Exam University of Califormia San Diego; 2009. p. 117.

[95] Hsieh CJ, Chang KW, Lin CJ, Keerthi SS, Sundararajan S. A dual coordinate descent method for large-scale linear SVM. In: Paper presented at: 25th international conference on machine learning; 2008.

[96] Huang S, Cai N, Pacheco PP, Narrandes S, Wang Y, Xu W. Applications of support vector machine (SVM) learning in cancer genomics. Cancer Genomics Proteomics 2018;15(1):41–51.

[97] Maltarollo VG, Kronenberger T, Espinoza GZ, Oliveira PR, Honorio KM. Advances with support vector machines for novel drug discovery. Expert Opin Drug Discov 2019;14(1):23–33.

[98] Hussain M, Wajid SK, Elzaart A, Berbar M. A comparison of SVM kernel functions for breast cancer detection. In: Paper presented at: eighth international conference computer graphics, imaging and visualization; 2011.

[99] Ali S, Smith-Miles K. A meta-learning approach to automatic kernel selection for support vector machines. Neurocomputing 2006;70:173–86.

[100] Jiang X, Menon A, Wang S, Kim J, Ohno-Machado L. Doubly optimized calibrated support vector machine (DOC-SVM): an algorithm for joint optimization of discrimination and calibration. PLoS ONE 2012;7(11), e48823.

[101] Magalhaes C, Mendes J, Vardasca R. The role of AI classifiers in skin cancer images. Skin Res Technol 2019;25 (5):750–7.

[102] William W, Ware A, Basaza-Ejiri AH, Obungoloch J. A review of image analysis and machine learning techniques for automated cervical cancer screening from pap-smear images. Comput Methods Programs Biomed 2018;164:15–22.

[103] Ozer ME, Sarica PO, Arga KY. New machine learning applications to accelerate personalized medicine in breast cancer: rise of the support vector machines. OMICS 2020;24(5):241–6.

[104] Shatte ABR, Hutchinson DM, Teague SJ. Machine learning in mental health: a scoping review of methods and applications. Psychol Med 2019;49(9):1426–48.

[105] Kavakiotis I, Tsave O, Salifoglou A, Maglaveras N, Vlahavas I, Chouvarda I. Machine learning and data mining methods in diabetes research. Comput Struct Biotechnol J 2017;15:104–16.
[106] Woldaregay AZ, Arsand E, Walderhaug S, et al. Data-driven modeling and prediction of blood glucose dynamics: machine learning applications in type 1 diabetes. Artif Intell Med 2019;98:109–34.
[107] Jiang F, Jiang Y, Zhi H, et al. Artificial intelligence in healthcare: past, present and future. Stroke Vasc Neurol 2017;2(4):230–43.
[108] Frize M, Ennett CM, Stevenson M, Trigg HC. Clinical decision support systems for intensive care units: using artificial neural networks. Med Eng Phys 2001;23(3):217–25.
[109] Dybowski R, Weller P, Chang R, Gant V. Prediction of outcome in critically ill patients using artificial neural network synthesised by genetic algorithm. Lancet 1996;347(9009):1146–50.
[110] Tu JV, Guerriere MR. Use of a neural network as a predictive instrument for length of stay in the intensive care unit following cardiac surgery. Comput Biomed Res 1993;26(3):220–9.
[111] Kayaalp M, Cooper GF, Clermont G. Predicting ICU mortality: a comparison of stationary and nonstationary temporal models. In: Proceedings/AMIA, annual symposium; 2000. p. 418–22.
[112] Fraser RB, Turney SZ. An expert system for the nutritional management of the critically ill. Comput Meth Progr Biomed 1990;33(3):175–80.
[113] Minsky ML, Papert S. Perceptrons. Cambridge, MA: MIT Press; 1969.
[114] Rumelhart DE, Hinton GE, Williams RJ. Learning representations by back-propagating errors. Nature 1986;323:533–6.
[115] Du J, Zhang Y, Luo J, et al. Extracting psychiatric stressors for suicide from social media using deep learning. BMC Med Inform Decis Mak 2018;18(Suppl 2):43.
[116] Huang J, Osorio C, Sy LW. An empirical evaluation of deep learning for ICD-9 code assignment using MIMIC-III clinical notes. Comput Methods Programs Biomed 2019;177:141–53.
[117] Kadir T, Gleeson F. Lung cancer prediction using machine learning and advanced imaging techniques. Transl Lung Cancer Res 2018;7(3):304–12.
[118] Hosny A, Parmar C, Coroller TP, et al. Deep learning for lung cancer prognostication: a retrospective multi-cohort radiomics study. PLoS Med 2018;15(11), e1002711.
[119] Miotto R, Li L, Kidd BA, Dudley JT. Deep patient: an unsupervised representation to predict the future of patients from the electronic health records. Sci Rep 2016;6:26094.
[120] Ruder S. An overview of gradient descent optimization algorithms. arXiv preprint arXiv: 160904747; 2016.
[121] Gers FA, Schraudolph NN, Schmidhuber J. Learning precise timing with LSTM recurrent networks. J Mach Learn Res 2002;3(1):115–43.
[122] Hochreiter S. The vanishing gradient problem during learning recurrent neural nets and problem solutions. Int J Uncertain Fuzziness Knowl-Based Syst 1998;6(02):107–16.
[123] Pascanu R, Mikolov T, Bengio Y. On the difficulty of training recurrent neural networks. In: International conference on machine learning; 2013. p. 1310–8.
[124] Hochreiter S, Schmidhuber J. Long short-term memory. Neural Comput 1997;9(8):1735–80.
[125] He Z, Du L, Zhang P, Zhao R, Chen X, Fang Z. Early sepsis prediction using ensemble learning with deep features and artificial features extracted from clinical electronic health records. Crit Care Med 2020;48(12): e1337–42.
[126] Shashikumar SP, Josef CS, Sharma A, Nemati S. DeepAISE—an interpretable and recurrent neural survival model for early prediction of sepsis. Artif Intell Med 2021;113, 102036.
[127] Tsai FS, Weng YM, Ng CJ, Lee CC. Embedding stacked bottleneck vocal features in a LSTM architecture for automatic pain level classification during emergency triage. In: Paper presented at: seventh international conference on affective computing and intelligent interaction (ACII); 2017.
[128] He R, Liu Y, Wang K, et al. Automatic cardiac arrhythmia classification using combination of deep residual network and bidirectional LSTM. IEEE Access 2019;7:102119–35.
[129] Wang L, Zhou X. Detection of congestive heart failure based on LSTM-based deep network via short-term RR intervals. Sensors (Basel) 2019;19(7).
[130] Zhang L, Fabbri D, Upender R, Kent D. Automated sleep stage scoring of the sleep heart health study using deep neural networks. Sleep 2019;42(11).
[131] Sherstinsky A. Fundamentals of recurrent neural network (RNN) and long Short-term memory (LSTM) network. Phys D: Nonlin Phenom 2020;404.

[132] Hanley JA, McNeil BJ. The meaning and use of the area under a receiver operating characteristic (ROC) curve. Radiology 1982;143(1):29–36.

[133] Van Calster B, Nieboer D, Vergouwe Y, De Cock B, Pencina MJ, Steyerberg EW. A calibration hierarchy for risk models was defined: from utopia to empirical data. J Clin Epidemiol 2016;74:167–76.

[134] Paul P, Pennell ML, Lemeshow S. Standardizing the power of the Hosmer-Lemeshow goodness of fit test in large data sets. Stat Med 2013;32(1):67–80.

[135] Van Calster B, Vickers AJ. Calibration of risk prediction models: impact on decision-analytic performance. Medical Decision Making: Int J Soc Med Decision Making 2015;35(2):162–9.

[136] Steyerberg EW, Vickers AJ, Cook NR, et al. Assessing the performance of prediction models: a framework for traditional and novel measures. Epidemiology 2010;21(1):128–38.

[137] Austin PC, Steyerberg EW. Graphical assessment of internal and external calibration of logistic regression models by using loess smoothers. Stat Med 2014;33(3):517–35.

[138] Davis SE, Lasko TA, Chen G, Siew ED, Matheny ME. Calibration drift in regression and machine learning models for acute kidney injury. J Am Med Inform Assoc 2017;24(6):1052–61.

[139] Huang Y, Li W, Macheret F, Gabriel RA, Ohno-Machado L. A tutorial on calibration measurements and calibration models for clinical prediction models. J Am Med Inform Assoc 2020;27(4):621–33.

[140] Knaus WA, Wagner DP, Draper EA, et al. The APACHE III prognostic system. Risk prediction of hospital mortality for critically ill hospitalized adults [see comment]. Chest 1991;100(6):1619–36.

[141] Zimmerman JE, Kramer AA, McNair DS, Malila FM. Acute physiology and chronic health evaluation (APACHE) IV: hospital mortality assessment for today's critically ill patients. Crit Care Med 2006;34(5): 1297–310.

[142] Vincent JL, Moreno R, Takala J, et al. The SOFA (sepsis-related organ failure assessment) score to describe organ dysfunction/failure. On behalf of the working group on sepsis-related problems of the European Society of Intensive Care Medicine. Intensive Care Med 1996;22(7):707–10.

[143] Ohno-Machado L, Resnic FS, Matheny ME. Prognosis in critical care. In: Yarmush ML, DIller KR, editors. Annual review of biomedical engineering, vol. 8. Palo Alto, CA: Nonprofit Publisher of the Annual Review of TM Series; 2006.

[144] Castella X, Gilabert J, Torner F, Torres C. Mortality prediction models in intensive care: acute physiology and chronic health evaluation II and mortality prediction model compared. Crit Care Med 1991;19(2):191–7.

[145] Rowan KM, Kerr JH, Major E, McPherson K, Short A, Vessey MP. Intensive Care Society's acute physiology and chronic health evaluation (APACHE II) study in Britain and Ireland: a prospective, multicenter, cohort study comparing two methods for predicting outcome for adult intensive care patients. Crit Care Med 1994;22(9): 1392–401.

[146] Wilairatana P, Noan NS, Chinprasatsak S, Prodeengam K, Kityaporn D, Looareesuwan S. Scoring systems for predicting outcomes of critically ill patients in northeastern Thailand. Southeast Asian J Trop Med Public Health 1995;26(1):66–72.

[147] Del Bufalo C, Morelli A, Bassein L, et al. Severity scores in respiratory intensive care: APACHE II predicted mortality better than SAPS II. Respir Care 1995;40(10):1042–7.

[148] Castella X, Artigas A, Bion J, Kari A. A comparison of severity of illness scoring systems for intensive care unit patients: results of a multicenter, multinational study. The European/north American severity study group. Crit Care Med 1995;23(8):1327–35.

[149] Moreno R, Apolone G, Miranda DR. Evaluation of the uniformity of fit of general outcome prediction models. Intensive Care Med 1998;24(1):40–7.

[150] Nouira S, Belghith M, Elatrous S, et al. Predictive value of severity scoring systems: comparison of four models in Tunisian adult intensive care units. Crit Care Med 1998;26(5):852–9.

[151] Tan IK. APACHE II and SAPS II are poorly calibrated in a Hong Kong intensive care unit. Ann Acad Med Singapore 1998;27(3):318–22.

[152] Patel PA, Grant BJ. Application of mortality prediction systems to individual intensive care units. Intensive Care Med 1999;25(9):977–82.

[153] Vassar MJ, Lewis Jr FR, Chambers JA, et al. Prediction of outcome in intensive care unit trauma patients: a multicenter study of acute physiology and chronic health evaluation (APACHE), trauma and injury severity score (TRISS), and a 24-hour intensive care unit (ICU) point system. J Trauma-Injury Infect Crit Care 1999;47(2):324–9.

[154] Katsaragakis S, Papadimitropoulos K, Antonakis P, Strergiopoulos S, Konstadoulakis MM, Androulakis G. Comparison of acute physiology and chronic health evaluation II (APACHE II) and simplified acute physiology score II (SAPS II) scoring systems in a single Greek intensive care unit. Crit Care Med 2000;28(2):426–32.
[155] Livingston BM, MacKirdy FN, Howie JC, Jones R, Norrie JD. Assessment of the performance of five intensive care scoring models within a large Scottish database. Crit Care Med 2000;28(6):1820–7.
[156] Capuzzo M, Valpondi V, Sgarbi A, et al. Validation of severity scoring systems SAPS II and APACHE II in a single-center population. Intensive Care Med 2000;26(12):1779–85.
[157] Markgraf R, Deutschinoff G, Pientka L, Scholten T. Comparison of acute physiology and chronic health evaluations II and III and simplified acute physiology score II: a prospective cohort study evaluating these methods to predict outcome in a German interdisciplinary intensive care unit [see comment]. Crit Care Med 2000;28(1):26–33.
[158] Beck DH, Smith GB, Pappachan JV, Millar B. External validation of the SAPS II, APACHE II and APACHE III prognostic models in South England: a multicentre study. Intensive Care Med 2003;29(2):249–56.
[159] Keegan MT, Gajic O, Afessa B. Comparison of Apache iii and iv, Saps 3 and Mpm0iii, and influence of resuscitation status on model performance. Chest 2012.
[160] Vasilevskis EE, Kuzniewicz MW, Cason BA, et al. Mortality probability model III and simplified acute physiology score II: assessing their value in predicting length of stay and comparison to APACHE IV. Chest 2009;136(1):89–101.
[161] Hwang SY, Lee JH, Lee YH, Hong CK, Sung AJ, Choi YC. Comparison of the sequential organ failure assessment, acute physiology and chronic health evaluation II scoring system, and trauma and injury severity score method for predicting the outcomes of intensive care unit trauma patients. Am J Emerg Med 2012;30(5):749–53.
[162] Costa e Silva VT, de Castro I, Liano F, Muriel A, Rodriguez-Palomares JR, Yu L. Performance of the third-generation models of severity scoring systems (APACHE IV, SAPS 3 and MPM-III) in acute kidney injury critically ill patients. Nephrol Dial Transplant 2011;26(12):3894–901.
[163] Shrope-Mok SR, Propst KA, Iyengar R. APACHE IV versus PPI for predicting community hospital ICU mortality. Am J Hosp Palliat Care 2010;27(4):243–7.
[164] Shillan D, Sterne JAC, Champneys A, Gibbison B. Use of machine learning to analyse routinely collected intensive care unit data: a systematic review. Crit Care 2019;23(1):284.
[165] van de Sande D, van Genderen ME, Huiskens J, Gommers D, van Bommel J. Moving from bytes to bedside: a systematic review on the use of artificial intelligence in the intensive care unit. Intensive Care Med 2021;47(7):750–60.
[166] Wilfong EM, Lovly CM, Gillaspie EA, et al. Severity of illness scores at presentation predict ICU admission and mortality in COVID-19. J Emerg Crit Care Med 2020;5.
[167] Goh KH, Wang L, Yeow AYK, et al. Artificial intelligence in sepsis early prediction and diagnosis using unstructured data in healthcare. Nat Commun 2021;12(1):711.
[168] Shashikumar SP, Wardi G, Paul P, et al. Development and prospective validation of a deep learning algorithm for predicting need for mechanical ventilation. Chest 2021;159(6):2264–73.
[169] Brown JR, MacKenzie TA, Maddox TM, et al. Acute kidney injury risk prediction in patients undergoing coronary angiography in a National Veterans Health Administration Cohort with External Validation. J Am Heart Assoc 2015;4(12).
[170] Tomašev N, Glorot X, Rae JW, et al. A clinically applicable approach to continuous prediction of future acute kidney injury. Nature 2019;572(7767):116–9.
[171] Levin S, Toerper M, Hamrock E, et al. Machine-learning-based electronic triage more accurately differentiates patients with respect to clinical outcomes compared with the emergency severity index. Ann Emerg Med 2018;71(5):565–574.e562.
[172] Churpek MM, Yuen TC, Winslow C, Meltzer DO, Kattan MW, Edelson DP. Multicenter comparison of machine learning methods and conventional regression for predicting clinical deterioration on the wards. Crit Care Med 2016;44(2):368–74.
[173] Rajkomar A, Oren E, Chen K, et al. Scalable and accurate deep learning with electronic health records. NPJ Digit Med 2018;1:18.
[174] Hao S, Wang Y, Jin B, et al. Development, validation and deployment of a real time 30 day hospital readmission risk assessment tool in the Maine healthcare information exchange. PLOS ONE 2015;10(10), e0140271.

[175] Matheny ME, Ricket I, Goodrich CA, et al. Development of electronic health record-based prediction models for 30-day readmission risk among patients hospitalized for acute myocardial infarction. JAMA Netw Open 2021;4 (1):e2035782.
[176] Cho I, Boo EH, Chung E, Bates DW, Dykes P. Novel approach to inpatient fall risk prediction and its cross-site validation using time-variant data. J Med Internet Res 2019;21(2), e11505.
[177] Kannel WB, Feinleib M, McNamara PM, Garrison RJ, Castelli WP. An investigation of coronary heart disease in families. The Framingham offspring study. Am J Epidemiol 1979;110(3):281–90.
[178] Anderson KM, Wilson PW, Odell PM, Kannel WB. An updated coronary risk profile. A statement for health professionals. Circulation 1991;83(1):356–62.
[179] Wilson PW, D'Agostino RB, Levy D, Belanger AM, Silbershatz H, Kannel WB. Prediction of coronary heart disease using risk factor categories. Circulation 1998;97(18):1837–47.
[180] D'Agostino Sr RB, Vasan RS, Pencina MJ, et al. General cardiovascular risk profile for use in primary care: the Framingham heart study. Circulation 2008;117(6):743–53.
[181] Assmann G, Cullen P, Schulte H. Simple scoring scheme for calculating the risk of acute coronary events based on the 10-year follow-up of the prospective cardiovascular Munster (PROCAM) study. Circulation 2002;105 (3):310–5.
[182] Stevens RJ, Kothari V, Adler AI, Stratton IM. The UKPDS risk engine: a model for the risk of coronary heart disease in type II diabetes (UKPDS 56). Clin Sci (Lond) 2001;101(6):671–9.
[183] Simmons RK, Sharp S, Boekholdt SM, et al. Evaluation of the Framingham risk score in the European prospective investigation of cancer-Norfolk cohort: does adding glycated hemoglobin improve the prediction of coronary heart disease events? Arch Intern Med 2008;168(11):1209–16.
[184] Hippisley-Cox J, Coupland C, Vinogradova Y, Robson J, May M, Brindle P. Derivation and validation of QRISK, a new cardiovascular disease risk score for the United Kingdom: prospective open cohort study. BMJ 2007;335 (7611):136.
[185] Cox DR, D O. Analysis of survival data. New York, NY: Chapman & Hall; 1984.
[186] Guzder RN, Gatling W, Mullee MA, Mehta RL, Byrne CD. Prognostic value of the Framingham cardiovascular risk equation and the UKPDS risk engine for coronary heart disease in newly diagnosed type 2 diabetes: results from a United Kingdom study. Diabet Med 2005;22(5):554–62.
[187] Stephens JW, Ambler G, Vallance P, Betteridge DJ, Humphries SE, Hurel SJ. Cardiovascular risk and diabetes. Are the methods of risk prediction satisfactory? Eur J Cardiovasc Prevent Rehabil 2004;11(6):521–8.
[188] Lenz M, Muhlhauser I. Cardiovascular risk assessment for informed decision making. Validity of prediction tools. Med Klin 2004;99(11):651–61.
[189] Song SH, Brown PM. Coronary heart disease risk assessment in diabetes mellitus: comparison of UKPDS risk engine with Framingham risk assessment function and its clinical implications. Diabet Med 2004;21(3):238–45.
[190] Ridker PM, Buring JE, Rifai N, Cook NR. Development and validation of improved algorithms for the assessment of global cardiovascular risk in women: the Reynolds risk score. JAMA 2007;297(6):611–9.
[191] Paynter NP, Chasman DI, Buring JE, Shiffman D, Cook NR, Ridker PM. Cardiovascular disease risk prediction with and without knowledge of genetic variation at chromosome 9p21.3. Ann Intern Med 2009;150(2):65–72.
[192] Berry JD, Lloyd-Jones DM, Garside DB, Greenland P. Framingham risk score and prediction of coronary heart disease death in young men. Am Heart J 2007;154(1):80–6.
[193] Denes P, Larson JC, Lloyd-Jones DM, Prineas RJ, Greenland P. Major and minor ECG abnormalities in asymptomatic women and risk of cardiovascular events and mortality. JAMA 2007;297(9):978–85.
[194] Grundy SM, Balady GJ, Criqui MH, et al. Primary prevention of coronary heart disease: guidance from Framingham: a statement for healthcare professionals from the AHA Task Force on Risk Reduction. American Heart Association. Circulation 1998;97(18):1876–87.
[195] Hingorani AD, Vallance P. A simple computer program for guiding management of cardiovascular risk factors and prescribing. BMJ 1999;318(7176):101–5.
[196] Zhao Y, Wood EP, Mirin N, Cook SH, Chunara R. Social determinants in machine learning cardiovascular disease prediction models: a systematic review. Am J Prev Med 2021;61:596–605.
[197] Levin MG, Rader DJ. Polygenic risk scores and coronary artery disease. Circulation 2020;141(8):637–40.
[198] Wand H, Knowles JW, Clarke SL. The need for polygenic score reporting standards in evidence-based practice: lipid genetics use case. Curr Opin Lipidol 2021;32(2).

[199] Knowles JW, Ashley EA. Cardiovascular disease: the rise of the genetic risk score. PLoS Med 2018;15(3): e1002546.
[200] Aragam KG, Natarajan P. Polygenic scores to assess atherosclerotic cardiovascular disease risk: clinical perspectives and basic implications. Circ Res 2020;126(9):1159–77.
[201] Fine MJ, Auble TE, Yealy DM, et al. A prediction rule to identify low-risk patients with community-acquired pneumonia. N Engl J Med 1997;336(4):243–50.
[202] Garibaldi RA. Epidemiology of community-acquired respiratory tract infections in adults. Incidence, etiology, and impact. Am J Med 1985;78(6B):32–7.
[203] Dans PE, Charache P, Fahey M, Otter SE. Management of pneumonia in the prospective payment era. A need for more clinician and support service interaction. Arch Intern Med 1984;144(7):1392–7.
[204] La Force FM. Community-acquired lower respiratory tract infections. Prevention and cost-control strategies. Am J Med 1985;78(6B):52–7.
[205] Daley J, Jencks S, Draper D, Lenhart G, Thomas N, Walker J. Predicting hospital-associated mortality for Medicare patients. A method for patients with stroke, pneumonia, acute myocardial infarction, and congestive heart failure. JAMA 1988;260(24):3617–24.
[206] Keeler EB, Kahn KL, Draper D, et al. Changes in sickness at admission following the introduction of the prospective payment system. JAMA 1990;264(15):1962–8.
[207] Kurashi NY, Al-Hamdan A, Ibrahim EM, Al-Idrissi HY, Al-Bayari TH. Community acquired acute bacterial and atypical pneumonia in Saudi Arabia. Thorax 1992;47(2):115–8.
[208] Fine MJ, Hanusa BH, Lave JR, et al. Comparison of a disease-specific and a generic severity of illness measure for patients with community-acquired pneumonia. J Gen Intern Med 1995;10(7):359–68.
[209] Marrie TJ, Durant H, Yates L. Community-acquired pneumonia requiring hospitalization: 5-year prospective study. Rev Infect Dis 1989;11(4):586–99.
[210] Dean NC, Suchyta MR, Bateman KA, Aronsky D, Hadlock CJ. Implementation of admission decision support for community-acquired pneumonia. Chest 2000;117(5):1368–77.
[211] Aronsky D, Chan KJ, Haug PJ. Evaluation of a computerized diagnostic decision support system for patients with pneumonia: study design considerations. J Am Med Inform Assoc 2001;8(5):473–85.
[212] Marrie TJ, Lau CY, Wheeler SL, Wong CJ, Vandervoort MK, Feagan BG. A controlled trial of a critical pathway for treatment of community-acquired pneumonia. CAPITAL study investigators. Community-acquired pneumonia intervention trial assessing levofloxacin.[see comment]. JAMA 2000;283(6):749–55.
[213] Atlas SJ, Benzer TI, Borowsky LH, et al. Safely increasing the proportion of patients with community-acquired pneumonia treated as outpatients: an interventional trial. Arch Intern Med 1998;158(12):1350–6.
[214] Niederman MS, Mandell LA, Anzueto A, et al. Guidelines for the management of adults with community-acquired pneumonia. Diagnosis, assessment of severity, antimicrobial therapy, and prevention. Am J Respir Crit Care Med 2001;163(7):1730–54.
[215] Marras TK, Gutierrez C, Chan CK. Applying a prediction rule to identify low-risk patients with community-acquired pneumonia. Chest 2000;118(5):1339–43.
[216] Mandell LA, Wunderink RG, Anzueto A, et al. Infectious Diseases Society of America/American Thoracic Society consensus guidelines on the management of community-acquired pneumonia in adults. Clin Infect Dis 2007;44(Suppl 2):S27–72.
[217] Chalmers JD, Singanayagam A, Akram AR, et al. Severity assessment tools for predicting mortality in hospitalised patients with community-acquired pneumonia. Systematic review and meta-analysis. Thorax 2010;65(10):878–83.
[218] Loke YK, Kwok CS, Niruban A, Myint PK. Value of severity scales in predicting mortality from community-acquired pneumonia: systematic review and meta-analysis. Thorax 2010;65(10):884–90.
[219] Chalmers JD, Mandal P, Singanayagam A, et al. Severity assessment tools to guide ICU admission in community-acquired pneumonia: systematic review and meta-analysis. Intensive Care Med 2011;37(9): 1409–20.
[220] Halm EA, Atlas SJ, Borowsky LH, et al. Understanding physician adherence with a pneumonia practice guideline: effects of patient, system, and physician factors. Arch Intern Med 2000;160(1):98–104.
[221] Cooper GF, Abraham V, Aliferis CF, et al. Predicting dire outcomes of patients with community acquired pneumonia. J Biomed Inform 2005;38:347–66.
[222] LeCun Y, Bengio Y, Hinton G. Deep learning. Nature 2015;521(7553):436–44.

[223] Bluemke DA, Moy L, Bredella MA, et al. Assessing radiology research on artificial intelligence: a brief guide for authors, reviewers, and readers-from the radiology editorial board. Radiology 2020;294(3):487–9.
[224] LeCun Y, Boser B, Denker JS, et al. Backpropagation applied to handwritten zip code recognition. Neural Comput 1989;1(4):541–51.
[225] Krizhevsky A, Sutskever I, Hinton GE. ImageNet classification with deep convolutional neural networks. Commun ACM 2017;60(6):84–90.
[226] Gulshan V, Peng L, Coram M, et al. Development and validation of a deep learning algorithm for detection of diabetic retinopathy in retinal fundus photographs. JAMA 2016;316(22):2402–10.
[227] Esteva A, Kuprel B, Novoa RA, et al. Dermatologist-level classification of skin cancer with deep neural networks. Nature 2017;542(7639):115–8.
[228] Albarqouni S, Baur C, Achilles F, Belagiannis V, Demirci S, Navab N. AggNet: deep learning from crowds for mitosis detection in breast cancer histology images. IEEE Trans Med Imaging 2016;35(5):1313–21.
[229] Anthimopoulos M, Christodoulidis S, Ebner L, Christe A, Mougiakakou S. Lung pattern classification for interstitial lung diseases using a deep convolutional neural network. IEEE Trans Med Imaging 2016;35 (5):1207–16.
[230] Brosch T, Tang LY, Youngjin Y, Li DK, Traboulsee A, Tam R. Deep 3D convolutional encoder networks with shortcuts for multiscale feature integration applied to multiple sclerosis lesion segmentation. IEEE Trans Med Imaging 2016;35(5):1229–39.
[231] Sihong C, Jing Q, Xing J, et al. Automatic scoring of multiple semantic attributes with multi-task feature leverage: a study on pulmonary nodules in CT images. IEEE Trans Med Imaging 2017;36(3):802–14.
[232] Mobadersany P, Yousefi S, Amgad M, et al. Predicting cancer outcomes from histology and genomics using convolutional networks. Proc Natl Acad Sci U S A 2018;115(13):E2970–e2979.
[233] Aggarwal R, Sounderajah V, Martin G, et al. Diagnostic accuracy of deep learning in medical imaging: a systematic review and meta-analysis. npj Digital Med 2021;4(1):65.
[234] Greenspan H, Ginneken BV, Summers RM. Guest editorial deep learning in medical imaging: overview and future promise of an exciting new technique. IEEE Trans Med Imaging 2016;35(5):1153–9.
[235] Stead WW. Clinical implications and challenges of artificial intelligence and deep learning. JAMA 2018;320 (11):1107–8.
[236] Tizhoosh H, Pantanowitz L. Artificial intelligence and digital pathology: challenges and opportunities. J Pathol Inform 2018;9(1):38.
[237] López-Pérez M, Amgad M, Morales-Álvarez P, et al. Learning from crowds in digital pathology using scalable variational gaussian processes. Sci Rep 2021;11(1):11612.
[238] Comfere NI, Peters MS, Jenkins S, Lackore K, Yost K, Tilburt J. Dermatopathologists' concerns and challenges with clinical information in the skin biopsy requisition form: a mixed-methods study. J Cutan Pathol 2015;42 (5):333–45.
[239] Huang SC, Pareek A, Seyyedi S, Banerjee I, Lungren MP. Fusion of medical imaging and electronic health records using deep learning: a systematic review and implementation guidelines. NPJ Digit Med 2020;3:136.
[240] Fleuren LM, Thoral P, Shillan D, Ercole A, Elbers PWG. Machine learning in intensive care medicine: ready for take-off? Intensive Care Med 2020;46(7):1486–8.
[241] McIntosh C, Conroy L, Tjong MC, et al. Clinical integration of machine learning for curative-intent radiation treatment of patients with prostate cancer. Nat Med 2021;27(6):999–1005.
[242] Ko M, Chen E, Agrawal A, et al. Improving hospital readmission prediction using individualized utility analysis. J Biomed Inform 2021;119, 103826.
[243] Tarabichi Y, Cheng A, Bar-Shain D, et al. Improving timeliness of antibiotic administration using a provider and pharmacist facing sepsis early warning system in the emergency department setting: a randomized controlled quality improvement initiative. Crit Care Med 2021.
[244] Braunstein ML. Healthcare in the age of interoperability: the promise of fast healthcare interoperability resources. IEEE Pulse 2018;9(6):24–7.
[245] Saltz J, Sharma A, Iyer G, et al. A containerized software system for generation, management, and exploration of features from whole slide tissue images. Cancer Res 2017;77(21):e79–82.
[246] Strasberg HR, Rhodes B, Del Fiol G, Jenders RA, Haug PJ, Kawamoto K. Contemporary clinical decision support standards using health level seven international fast healthcare interoperability resources. J Am Med Inform Assoc: JAMIA 2021;28(8):1796–806.

[247] Vaccaro K, Huang D, Eslami M, Sandvig C, Hamilton K, Karahalios K. The illusion of control: placebo effects of control settings. In: Proceedings of the 2018 CHI conference on human factors in computing systems, Montreal QC, Canada; 2018.

[248] Patel VL, Kannampallil TG. Cognitive approaches to clinical data Management for Decision Support. Information quality in e-health: 7th conference of the workgroup human-computer interaction and usability engineering of the Austrian Computer Society. Graz, Austria: USAB; 2011. p. 1–13.

[249] Hickey GL, Grant SW, Murphy GJ, et al. Dynamic trends in cardiac surgery: why the logistic euroscore is no longer suitable for contemporary cardiac surgery and implications for future risk models. Eur J Cardiothorac Surg 2013;43(6):1146–52.

[250] Booth S, Riley RD, Ensor J, Lambert PC, Rutherford MJ. Temporal recalibration for improving prognostic model development and risk predictions in settings where survival is improving over time. Int J Epidemiol 2020;49 (4):1316–25.

[251] Minne L, Eslami S, De Keizer N, De Jonge E, De Rooij SE, Abu-Hanna A. Effect of changes over time in the performance of a customized SAPS-II model on the quality of care assessment. Intensive Care Med 2012;38 (1):40–6.

[252] Quinonero-Candela J, Sugiyama M, Schwaighofer A, Lawrence N. Dataset shift in machine learning. Cambridge, MA: The MIT Press; 2009.

[253] Song H, Thiagarajan JJ, Kailkhura B. Preventing failures by dataset shift detection in safety-critical graph applications. Front Artif Intell 2021;4, 589632.

[254] Kappen TH, Vergouwe Y, van Klei WA, van Wolfswinkel L, Kalkman CJ, Moons KG. Adaptation of clinical prediction models for application in local settings. Med Decision Making: Int J Soc Med Decision Making 2012;32(3):E1–10.

[255] Janssen KJ, Vergouwe Y, Kalkman CJ, Grobbee DE, Moons KG. A simple method to adjust clinical prediction models to local circumstances. Can J Anesth 2009;56(3):194–201.

[256] Shashikumar SP, Wardi G, Malhotra A, Nemati S. Artificial intelligence sepsis prediction algorithm learns to say "I don't know". medRxiv 2021. 2021.2005.2006.21256764.

[257] Hannan EL, Cozzens K, King 3rd SB, Walford G, Shah NR. The New York state cardiac registries: history, contributions, limitations, and lessons for future efforts to assess and publicly report healthcare outcomes. J Am Coll Cardiol 2012;59(25):2309–16.

[258] Siregar S, Nieboer D, Vergouwe Y, et al. Improved prediction by dynamic modelling: an exploratory study in the adult cardiac surgery database of the Netherlands association for cardio-thoracic surgery. Circ Cardiovasc Qual Outcomes 2016;9(2):171–81.

[259] Jin R, Furnary AP, Fine SC, Blackstone EH, Grunkemeier GL. Using Society of Thoracic Surgeons risk models for risk-adjusting cardiac surgery results. Ann Thoracic Surg 2010;89(3):677–82.

[260] Vergouwe Y, Nieboer D, Oostenbrink R, et al. A closed testing procedure to select an appropriate method for updating prediction models. Stat Med 2017;36(28):4529–39.

[261] Davis SE, Greevy RA, Fonnesbeck C, Lasko TA, Walsh CG, Matheny ME. A nonparametric updating method to correct clinical prediction model drift. J Am Med Inform Assoc 2019;26(12):1448–57.

[262] Toll DB, Janssen KJ, Vergouwe Y, Moons KG. Validation, updating and impact of clinical prediction rules: a review. J Clin Epidemiol 2008;61(11):1085–94.

[263] Van Calster B, Van Hoorde K, Vergouwe Y, et al. Validation and updating of risk models based on multinomial logistic regression. Diagn Progn Res 2017;1(2).

[264] Jenkins DA, Martin GP, Sperrin M, et al. Continual updating and monitoring of clinical prediction models: time for dynamic prediction systems? Diagn Progn Res 2021;5(1):1.

[265] Davis SE, Greevy Jr RA, Lasko TA, Walsh CG, Matheny ME. Detection of calibration drift in clinical prediction models to inform model updating. J Biomed Inform 2020;112, 103611.

[266] Lenert MC, Matheny ME, Walsh CG. Prognostic models will be victims of their own success, unless. J Am Med Inform Assoc: JAMIA 2019;26(12):1645–50. https://doi.org/10.1093/jamia/ocz145.

CHAPTER

8

Modernizing evidence synthesis for evidence-based medicine

Ian Jude Saldanha[a,b], *Gaelen P. Adam*[a], *Christopher H. Schmid*[c], *Thomas A. Trikalinos*[a,c], *and Kristin J. Konnyu*[a]

[a]Health Services, Policy, and Practice, Brown University School of Public Health, Providence, RI, United States [b]Epidemiology, Brown University School of Public Health, Providence, RI, United States [c]Biostatistics, Brown University School of Public Health, Providence, RI, United States

8.1 Introduction

The previous two chapters discussed the use of human expertise to establish the knowledge base for clinical decision support and the use of data mining and machine learning to derive such knowledge from databases. The other major, and generally more dependable, source of knowledge is the accumulated evidence contained in the medical literature. The aim of evidence-based medicine (EBM) is to inform clinical practice in light of all available evidence [1]. According to the late Dr. David Sackett, widely considered to be the father of EBM, EBM refers to the *integration* of the best research evidence with clinical expertise and patient values [2].

Systematic reviews are research efforts whose goal is to identify and synthesize all relevant studies that fulfill pre-specified eligibility criteria to answer specific research questions [1,3]. Systematic reviews are the foundation of EBM. Creating systematic reviews involves identifying, selecting, appraising, and summarizing the evidence from similar, but separate, studies that address the same research question. When appropriate, systematic reviews include meta-analyses (i.e., statistical synthesis of results from multiple studies). Meta-analyses provide concise summaries of treatment effects, typically with greater precision than those of individual studies. Systematic reviews are used to inform decisions at all levels of health care, from bedside individualized care to policymaking. They have gained wide acceptance as a practical way to provide reliable and comprehensive syntheses of the medical evidence base.

Clinical Decision Support and Beyond
https://doi.org/10.1016/B978-0-323-91200-6.00006-1

Although systematic reviews are crucial for integrating all relevant research findings into practice, the exponential expansion of the biomedical literature has made producing and maintaining such reviews increasingly onerous. As the amount of published evidence increases, so does the labor required to synthesize it. Increasingly, systematic reviewers look beyond just journal articles and incorporate additional information from other sources, such as clinical trial and other study registries and databases, protocols, conference abstracts, and contact with study authors [4,5]. Higher standards and improved methods for evaluating and critiquing study conduct, analysis, and reporting [1,6] have further increased the labor involved in conducting systematic reviews, exacerbating the problems caused by information overload.

This chapter provides a basic introduction to systematic review and meta-analysis methods but does not attempt a comprehensive exposition. Interested readers should consult existing texts for more thorough discussions of the subject [6–10]. Instead, we focus on methods for providing the evidence base for decision support, and the potential for medical informatics to facilitate and optimize the conduct and maintenance of evidence syntheses.

This need could not be more pressing. For example, Medline – a database of biomedical literature – added more than 16,000 new randomized trials in 2021 alone, and the increasing trajectory of publication rates shows no signs of slowing. On average, a decade ago it was estimated that 75 new trials and 11 systematic reviews were published every day [11,12]; current estimates are considerably higher, with a million new records added to Medline alone annually since 2017 [13]. If we are to keep up with the literature, we must make our tools and approaches to identifying, synthesizing, and disseminating relevant evidence more efficient. Some of these tools and approaches are described in this chapter.

8.2 Systematic reviews and meta-analysis: The premise and promise

Systematic reviews and meta-analyses grew out of the need to synthesize the large volume of biomedical literature as it was expanding rapidly during the second half of the 20th century [14–16]. Clinicians are faced with numerous studies on medical interventions and diagnostic tests, often with imprecise or contradictory results. The basic premise of the systematic review is that it is a comprehensive, rigorous, and unbiased review and synthesis of up-to-date evidence, thus providing the most reliable information to inform medical practice.

Meta-analysis is a set of statistical techniques for combining quantitative results from several studies to address specific questions. Meta-analysis represents a natural way to combine the results of studies included in a systematic review. Typically, one is interested in synthesizing estimates of treatment effects. For example, the relative risk is the ratio of the risk of an outcome event (e.g., death) among individuals in a treated group compared with the risk of the same outcome among individuals in a control group. Meta-analysis can be used to provide more precise overall estimates of effect than individual studies due to the combination of study effect estimates into one overall meta-analytic estimate. This is particularly important when treatment effects are not large [17]. The most basic meta-analysis computes a weighted average of the effect estimates from each study to produce an overall estimate of treatment efficacy. Although a variety of approaches to weighting the estimates from individual studies exist, the most popular approaches incorporate weighting by the inverse of the variance in each study, which roughly corresponds to weighting by their sample sizes (i.e., studies with larger samples contribute greater weight to the overall meta-analytic estimate) [18].

Conducting a meta-analysis is not always appropriate in a systematic review. For example, if the study designs, study populations, treatment doses/durations/frequencies, outcome measurement methods, or study results indicate considerable heterogeneity (or variability) across studies, meta-analysis may be unable to properly capture this variability (see Section 8.4.5.1). When the number of studies is very small, a meta-analysis may also be inappropriate because the assumptions behind the method may not hold. Finally, it may become apparent that results of key outcomes or whole studies are unavailable, so any summary would represent a biased estimate of the true effect.

Meta-analyses, when appropriate, can be conducted using individual participant data (IPD), person-level information on study participants (e.g., treatments, outcomes, covariates) contributed by study investigators, or with aggregate data, study-level summaries that can often be extracted from published study reports [18–21]. Both approaches rely on systematic reviews to identify and select relevant research studies and extract data. However, aggregate data meta-analyses can be conducted faster and at a fraction of the cost of IPD analyses; for that reason, they represent the vast majority of published meta-analyses [8,22–24]. IPD meta-analyses have several advantages, such as their ability to standardize outcome and covariate definitions across studies, more refined modeling of the effect of treatments and covariates on outcomes, and the ability to explore treatment effect modification by participant-level covariates [25]. Nonetheless, for the remainder of this chapter we will focus on meta-analysis as it has traditionally been conceived and practiced (i.e., using aggregate, study-level data). The results of meta-analyses are usually presented visually in forest plots, an example of which [26] is shown in Fig. 8.1. Forest plots display individual effect estimates and

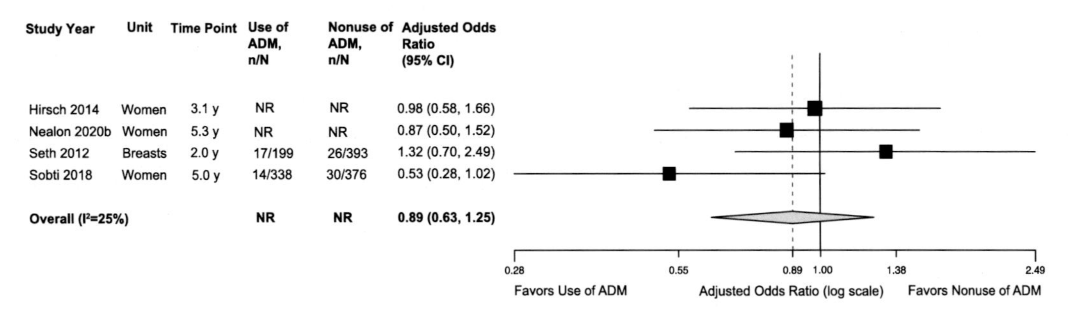

FIG. 8.1 An example of a Forest plot from a systematic review addressing the use versus nonuse of an acellular dermal matrix (ADM) during breast reconstruction surgery [27]. The outcome of interest is development of necrosis. The corresponding point estimates (of the adjusted odds ratio [OR] for the outcome of interest) are demarcated by squares sized proportionally to the weight assigned to the study. The horizontal lines passing through the squares represent the 95% confidence intervals of the estimates. The vertical solid line represents the null value suggesting no treatment effect (adjusted OR of 1). The vertical dotted line provides the overall meta-analytic (pooled) estimate of the adjusted OR, which is denoted by the center of the diamond. The width of the diamond represents the 95% confidence interval of the overall meta-analytic estimate. The meta-analysis is conducted using a common-effect model. Abbreviations: ADM = acellular dermal matrix, CI = confidence interval, I^2 = measure of statistical heterogeneity (% of total variability that is due to between-study variability), NR = not reported, OR = odds ratio, y = years.

corresponding confidence intervals from each study. They also display the overall (meta-analytic) estimate, which provides a statistical summary of the studies.

Systematic reviews and meta-analyses have contributed important insights to interpreting clinical trial results and have made important impacts on clinical practice, health policy, and biomedical research. A classic but highly instructive example of this is illustrated by a cumulative meta-analysis, in which studies are added to a meta-analysis sequentially, revealing trends in the pooled effect estimate over time (as additional trials are conducted). An example is shown in Fig. 8.2. A cumulative meta-analysis of thrombolytic therapy for acute myocardial infarction demonstrated that the efficacy of a particular treatment (streptokinase therapy) could have been confidently established using meta-analysis well before two large clinical trials confirmed it [28]. That is, a meta-analysis combining 33 studies (most of which reported nonsignificant findings) found a highly statistically significant result of approximately 20% reduction in overall mortality, and thus could have resulted in earlier adoption of the therapy and the saving of thousands of patient lives had this analysis been conducted prior to the two later definitive studies. An extension of this work also showed that meta-analysis could detect signals of treatment effectiveness earlier than clinical "experts" [29].

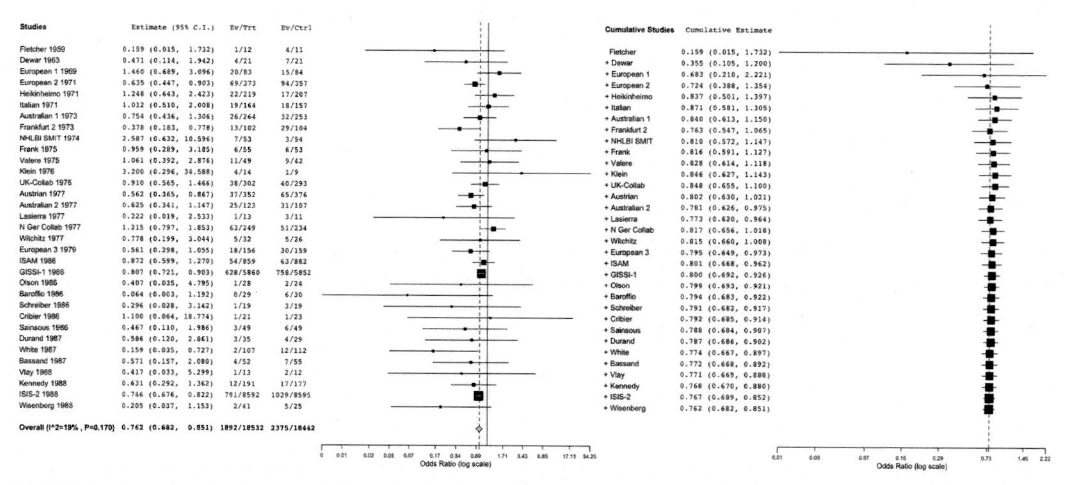

FIG. 8.2 An example of a cumulative meta-analysis. In cumulative meta-analysis, one updates and displays meta-analysis results after the inclusion of each new study. In this example, the method readily demonstrates statistically significant evidence of the efficacy of intravenous streptokinase in reducing overall mortality, and that such a conclusion might have been reached as early as 1973, after combining the first eight trials involving about 2400 patients. Additional studies, including the two largest trials, did not substantially alter the estimate of the magnitude of the treatment effect, instead mostly narrowing the confidence interval of the estimate. Thus, routinely updated meta-analyses could provide the earliest indication of the benefit or harm of an intervention, thereby minimizing the ethical issue of conducting additional trials in areas where there is already sufficient evidence (for or against a procedure), saving lives, and avoiding the expenditure of unnecessary resources.

8.3 Uses of systematic reviews and meta-analyses

EBM methods have been widely disseminated throughout health care and beyond. Initially used to assess evidence in clinical medicine, they have now been applied to virtually every field of medicine and healthcare. Systematic reviews have been used to synthesize quantitative as well as qualitative evidence regarding interventions, diagnostic tests, prognosis, risk factors (e.g., secondhand cigarette smoke and cancers), incidence/prevalence, and other areas. They have covered virtually every clinical discipline. These reviews have synthesized various types of studies, such as randomized controlled trials, cohort studies, case-control studies, single-group studies, case reports, and qualitative studies.

In addition to adopting EBM, whose goal is that the best research informs clinical practice, we also need to conduct 'evidence-based research.' Evidence-based research refers to the use of prior research in a systematic and transparent way to inform new research so that the new research answers the questions of greatest importance [30,31]. For randomized controlled trials to be truly evidence-based, they should be informed by systematic reviews of prior relevant evidence. This avoids unnecessary duplication and provides the strongest and most ethical justification for the new trial, avoiding randomizing patients to control treatments that have been proven to be suboptimal. Indeed, as of 2021, the National Institute of Health Research (NIHR), the United Kingdom's major public funder of randomized controlled trials, explicitly requires justification for new research, emphasizing that it "*will only fund primary research where the proposed research is informed by a review of the existing evidence*" [32].

From a methodological standpoint, systematic review methods and meta-analysis techniques can be used to evaluate the impact of study design characteristics on estimates of effectiveness. For example, "meta-epidemiological" studies use systematic review methods to systematically identify primary studies and empirically evaluate the impact of methodological factors, such as masking (e.g., of outcome assessors) and allocation concealment on the results of randomized controlled trials [33,34].

Clearly, then, systematic reviews can provide valuable insights into what works in health care and how research is done. However, as discussed, the main challenge is the large amounts of resources required to produce quality systematic reviews.

8.4 Steps of a systematic review

We now describe each step involved in producing systematic reviews and highlight where informatics has been applied and opportunities for further advances. A systematic review follows a set of highly structured and reproducible steps (to minimize bias) that are described in a well-defined protocol (ideally registered and/or published in a public domain) to identify, select, appraise, and synthesize the available evidence (Fig. 8.3 depicts this process schematically). Each step presents a unique set of challenges. Systematic reviews begin with the formulation of a precise research question and end with a report of the synthesis of the available relevant evidence, which as discussed may include a quantitative synthesis (meta-analysis).

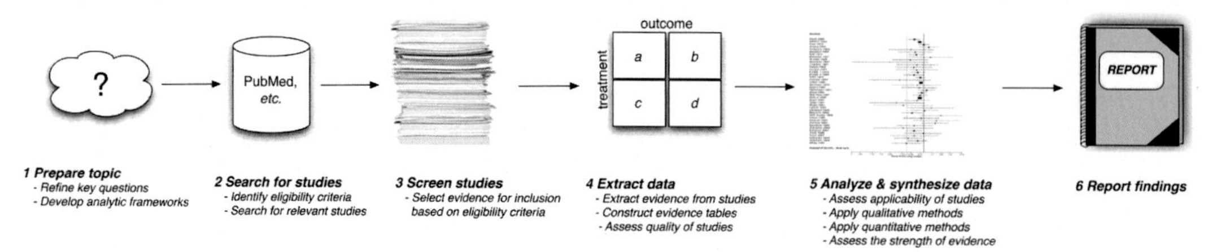

FIG. 8.3 A schematic of the steps of a systematic review. Researchers first formulate the precise clinical question to be answered and from this derive a search strategy. This search strategy is then implemented to retrieve potentially eligible studies from literature databases, such as Medline, and sources of unpublished studies, such as preprint servers and the gray literature. Only a fraction of the studies retrieved via the search strategy will be eligible for inclusion in the review; researchers must read through all the retrieved records (and then the full texts of potentially relevant records) to determine which studies to include in the systematic review. Next, researchers extract the data elements of interest from the text of the included studies. The next step is to narratively and, if appropriate, statistically synthesize these extracted elements. Finally, researchers report the methods and findings of the systematic reviews in a report. Not shown here is "downstream" work, i.e., keeping the systematic review current.

Systematic reviews differ from traditional narrative reviews. The latter typically cover a broad range of issues such as etiology, pathology, methods of diagnosis, range of treatments available, and prognosis. Rather than attempting to provide a broad overview of a topic, systematic reviews seek to answer one or a few focused research question(s) [35,36]. Although unique methodological issues arise in reviewing studies addressing different questions and data types, the fundamentals of the systematic review process remain the same. Here, we focus on the basic principles and methods of systematic review and meta-analysis of intervention studies.

8.4.1 Formulating the research question

Formulating the research question is the most critical step in any systematic review or meta-analysis. The question should be clinically important, address the needs of various relevant stakeholders, and be framed in a way that it could, at least potentially, be directly answered with available studies. For systematic reviews of intervention effectiveness, the PICO (Population, Intervention, Comparator, Outcome) formalism is widely used to define the elements of key research question(s). (See Box 8.1.) Various researchers add mnemonic initials for capturing information on study designs (D; e.g., randomized controlled trials), timing (T; e.g., duration of follow-up), and study setting (S; e.g., outpatient), but the essence is unchanged [37,38].

The research question thus formulated will guide every subsequent phase of the review process, from searching the literature to interpreting and reporting the results. A typical literature search may start off with hundreds or thousands of citations retrieved from Medline or other biomedical literature databases, via a purposefully broad query designed to ensure the capture of most, if not all, studies relevant to the research question. Most of these studies will not be relevant. The PICO(DTS) review criteria serve as an eligibility sieve through which only relevant studies pass. For example, the question "what drugs should be used to treat hypertension?" is not directly answerable from an EBM perspective. Instead, a

BOX 8.1

Medical informatics potential: Automatically inferring PICO elements.

The PICO formulism seeks to formalize the eligibility criteria for systematic reviews. PICO (and its variants, such as PICO(DTS)) elements are thus a particular type of structure latent in the free text of biomedical articles. It would be ideal if clinical studies' PICO elements were known and stored explicitly in a computable format, since this would greatly facilitate informatics applications for EBM. For example, it would simplify searching for relevant literature. However, this information is not readily available, and efforts are therefore underway to design natural language processing (NLP) systems to automatically infer PICO elements from free text. Numerous approaches have been pursued, ranging from inferring PICO elements for examining only the abstract versus the whole document and assigning PICO to the document level versus to specific sentences in the text. The more successful approaches treat the task as a "supervised machine learning" problem [39]. Several works have investigated the PICO task from this perspective with success [15,40–43].

well-formulated systematic review question would ask, "How does nifedipine compare with hydrochlorothiazide in patients with moderately elevated diastolic blood pressure in clinical trials evaluating long-term (1-year or more) mortality and morbidity?" The first question is broad and unfocused, while the second is precise and explicitly articulates the eligibility criteria.

In practice, we do want an answer to the first, broad question; but informing this answer with evidence requires asking the more specific questions. Moreover, the broad question often has many different answers depending on the context in which it is asked. Sometimes evidence regarding such questions is not available, thus suggesting areas in need of future research. Formulating a research question is an iterative process that requires engagement with relevant stakeholders (e.g., clinicians, patients, researchers, guideline developers, payors). Often, formulating a question broad enough to both meet the needs of stakeholders and merit spending precious resources but specific enough to be meaningfully answered by the available evidence requires compromise. A question that is too narrowly focused may directly apply only to a small subset of patients and may yield little relevant evidence. By contrast, many studies may address a broadly formulated question but may not apply clearly to specific populations. The complexity and heterogeneity of a large set of studies may limit the organization and clarity of the synthesis and its actionable messages.

8.4.2 Searching the literature

A key principle of a quality systematic review is that it should consider *all* existing literature that is relevant to the research question. However, in practice, the comprehensiveness of

BOX 8.2

Medical informatics potential: Optimizing literature searches.

Due to the torrential rate of expansion of published biomedical literature, it is difficult to keep up with new information, both in terms of conducting new reviews and updating existing reviews [11]. Much of this workload is due to the requirement that systematic reviews be as comprehensive as possible, encouraging broad literature searches. Reviewers must then screen all the articles retrieved from such searches to assess their relevance. This translates into a huge amount of human effort. Making the literature search process more efficient via computational methods is therefore an imperative, and one in which informatics researchers can play a key role. Indeed, there has been considerable research in this direction [44]. Several freely available tools now exist to streamline the task of identifying relevant studies from a pool of query results. These tools address the problem as a classification task, in which the aim is to train a model to automatically discriminate between "relevant" and "irrelevant" studies [45–48]. The workload involved in the literature search phase of updating existing systematic reviews can be cut in half by such methods, without missing relevant literature [40,49]. Increasingly, tools are also being developed to facilitate the initial step of developing and refining a search query. At present, most researchers iteratively develop a search query by hand, balancing the desire to be broad against practical time constraints. Increasingly, tools that leverage text mining and seed sets of citations are being incorporated into search design workflows to decrease the time needed to design the search query and reduce the number of citations to screen [50–52].

a literature review is constrained by the time and resources available to the systematic review team. Exhaustively searching and screening through the literature can be a daunting task due to the large volume of published literature (see Box 8.2) – particularly for complex or broad questions (e.g., policy questions or comparative treatments for clinical conditions) or questions relating to rapidly growing clinical topics (e.g., COVID-19). The literature search strategy and the choice of eligibility criteria for the review should therefore be guided by careful forethought and understanding of the nature of the evidence. Literature searches generally begin with a search of the Medline database of biomedical literature, often via the PubMed or OVID web interfaces. Medline indexes thousands of biomedical journals that are most likely to be useful in systematic reviews of clinical and health policy topics. As of February 2022, Medline has indexed over 33 million articles. Systematic reviews that include clinical trials also search the Cochrane Central Register of Controlled Trials (CENTRAL), part of the Cochrane Library, which, as of February 2022, has indexed almost two million trials (see Section 8.5). The Cochrane Library is a product of volunteer contributors, hand-searching journals to identify clinical trials not indexed originally by Medline. This database can be searched using standard search terms and Boolean operators and is updated quarterly. The Excerpta Medica Database (EMBASE), based in Europe, is another frequently searched database. In addition to these general databases, it is recommended to search databases relevant to the topic (e.g., PsycINFO for psychological topics).

The importance of including non-English language articles in a systematic review is also likely to be topic dependent. In general, the exclusion of non-English language articles in a meta-analysis has not been found to change summary results substantially [53–55]. For certain topics, unrestricted inclusion of articles in all languages may actually lead to a biased assessment of the overall effect, if published research in certain countries tends to include only positive results [56]. On the other hand, topics of particular importance in certain regions will need to consider publications in the regional language(s) (e.g., studies of acupuncture in China). Computational tools, such as intelligent machine translation, may in some cases be sufficiently accurate to facilitate incorporation of diverse foreign-language literature into systematic reviews [57].

Keeping systematic reviews up to date with newly published evidence is an important practical problem. If a resource is to be kept "live," researchers must periodically query the relevant databases to identify newly published eligible citations and then extract relevant information – an onerous requirement [58].

8.4.3 Screening

As mentioned, screening titles and abstracts and then full-text articles to obtain the set of studies that are eligible and will be included in the systematic review is a daunting task. Reviewers typically need to screen many thousands of abstracts. Various tools have been developed to facilitate the screening process. These tools typically present the next unscreened abstract to screeners such that two individuals screen each abstract independently and subsequently resolve any conflicts in their determinations of eligibility. Examples of tools include Abstrackr, Covidence®, and PICO Portal. For example, Abstrackr, developed in 2012, is an open-source software that manages citations screening tasks online, providing functions for single or redundant screening, term highlighting, and resolution of conflicting annotations, and progress monitoring. While reviewers screen abstracts, the software uses their annotations to learn an ensemble of machine-learning classifiers in a dynamic fashion. Once a new batch of abstracts has been screened, the machine learning classifier is re-trained, updates its predictions about the relevance of the yet unscreened abstracts, and prioritizes for screening those that are more likely to be relevant [48]. This allows for front-loading of likely-eligible studies, which greatly decreases the amount of time needed to conduct abstract screening [59]. This iterative training of the machine learning algorithms continues until an empirical stopping rule is reached, e.g., once the algorithm predicts no more relevant papers among the unscreened ones and, thereafter, no relevant citations are found among the next 400 consecutive citations. Empirically, this threshold is typically reached when about half the abstracts have been screened.

8.4.4 Extracting data

Data needed for description of the populations and interventions studied as well as the statistics needed to carry out meta-analyses must be extracted from all studies included in the systematic review. For example, the efficacy of a treatment to reduce mortality might be measured by the relative risk, the ratio of the risks of mortality in the treatment and control groups. For meta-analysis, one would need to extract the relative risk and corresponding confidence interval from each study being synthesized. These statistics are sometimes reported explicitly but at other times must be calculated from reported group data (e.g., the proportion of events in treatment and control groups for dichotomous outcomes). Data extraction is a tedious and time-consuming process, and prone to human error. Considerable skill and experience are required to ensure the accuracy of the extracted data. It has been recommended that the data extracted from each study should be independently verified, with reconciliation of any discrepancies [1,60–63]. However, this process increases costs and time. In addition to the metric(s) of interest, the data elements extracted for systematic reviews and meta-analyses include information pertaining to the PICO parameters discussed earlier as well as to methodological issues regarding the design and conduct of studies. This information should be made available to readers so that they can draw conclusions about the methodological quality and applicability of a study and the reliability of the results. To collect this information, a data collection form is typically developed for each systematic review.

Data extraction is often a challenging exercise (see Box 8.3). Often, data reported across studies are not standardized, and important information is missing. The same information may be reported inconsistently within or across report of a study, leading to uncertainties about the correct answer. The need for subjective judgment when extracting data can lead to inconsistencies and errors in extraction.

Assessment of the risk of bias in included studies is a key aspect of conducting systematic reviews because it helps evaluate the strength of the evidence identified [6]. Various tools and software facilitate data extraction and risk of bias assessment. These vary in functionality and sophistication, including word processors (e.g., Microsoft Word®, Google Documents®), spreadsheets (e.g., Microsoft Excel®, Google Sheets®), stand-alone relational databases (e.g., Microsoft Access®), and online, server-based relational databases (e.g., Systematic Review Data Repository Plus [SRDR+], DistillerSR®, Covidence®, EPPI-Reviewer). It is generally recommended that systematic reviewers use online, server-based relational databases because of their specific development for systematic reviews, flexibility, online collaboration functionality, and built-in study-design specific risk of bias assessment tools. To our knowledge, SRDR+, funded by the United States Agency for Healthcare Research and Quality (AHRQ), is the only free, open-source systematic review platform for extracting data and sharing extracted data (open-access) [64,65].

8.4.5 Conducting a meta-analysis

Having reviewed the systematic review steps that collect the available evidence, we now turn our attention to statistical methods for performing meta-analysis, i.e., the quantitative synthesis of results from the included studies [70]. The main goal of a meta-analysis is to summarize knowledge about the treatment effect provided by the studies. The simplest summary averages the estimates of the effects extracted from each study. This summary provides the most appropriate single estimate of the treatment effect in a set of studies and can be formulated as either a weighted or unweighted average. Weighting implies that some studies provide more information about the average than others, usually because they are larger and

BOX 8.3

Medical informatics potential: Modernizing data extraction.

Data extraction requires a huge amount of time and resources. Compounding the problem, data elements from the same studies are often extracted multiple times by different groups, duplicating efforts. The Systematic Review Data Repository (SRDR+; available at https://srdrplus.ahrq.gov) is an online extraction tool and data repository that facilitates the extraction process and makes extracted data accessible to researchers worldwide, free of charge [64,65]. Such an open access repository of all the data extracted from studies is intended to avoid unnecessary redundancies across the systematic review enterprise. In this way, extracted data can be searchable and re-usable. Other offerings for data extraction exist; Distiller SR® (https://www.evidencepartners.com), for example, facilitates data extraction (although it does not provide a means of making the extracted data publicly available) and is only available through license.

Another related aspect of data extraction is automatic extraction of aspects of study information, such as identifying the study design or the risk of bias. The idea is like that of the PICO element extraction described above: map free text to clinically relevant variables [66]. Some examples include:

1. An "RCT classifier" has been developed to distinguish between randomized controlled trials, quasi-randomized controlled trials, and non-randomized studies [67].
2. RobotReviewer is a machine learning system that automatically assesses the risk of bias in clinical trials. Evidence suggests that RobotReviewer, when used in conjunction with validation by humans, can improve the efficiency of the data extraction process with reasonable accuracy [68,69].

Each step in the systematic review process is labor-intensive but necessary to ensure the production of high-quality reviews. Use of medical informatics tools may increase the efficiency of the systematic review process without sacrificing its accuracy.

thus provide more precise effect estimates. The two most common weighting schemes lead to the common-effect estimate (formerly called a fixed-effect estimate) and the random-effects estimate.

The common-effect estimate is obtained by weighting each study estimate by its precision, the inverse of its variance. Because the precision is proportional to the sample size, weighting by precision is akin to weighting by sample size. The bigger studies get the most weight. Use of these weights implies that each study estimates the same quantity, in other words the treatment effects share a common value. Often, the effects are assumed to follow a normal distribution because they are estimated by study sample means which are approximately normally distributed by the central limit theorem. Therefore, each observed study effect $Y_i \sim N(\theta, v_i)$, where θ is the common effect and v_i is the variance of study I (Fig. 8.4A).

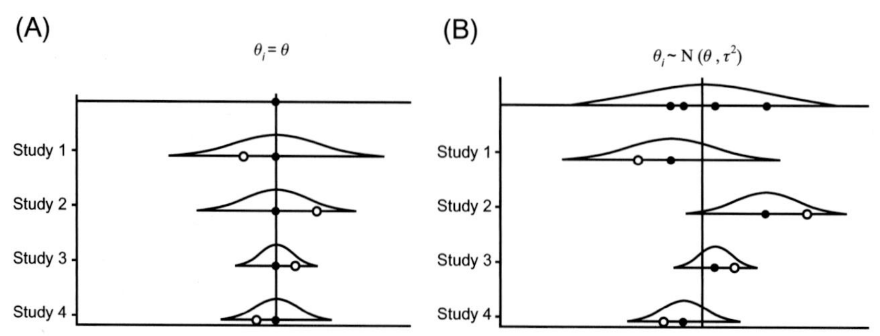

FIG. 8.4 Schematic representation of meta-analysis models. The figure shows a hypothetical meta-analysis of four studies, represented by four normal distributions. **(A) Common-effect meta-analysis.** The common-effect model assumes that the true (parameter) effect in all four studies (black circles) is the same (θ). The observed effect in each study (white circles) deviates from the common effect θ because of sampling variability. **(B) Random-effects meta-analysis.** The random-effects model assumes that the study-specific true effects (black circles) are drawn from a common distribution (typically assumed to be normal) as shown in the top curve. The true effects in each study differ (i.e., are heterogeneous), perhaps reflecting differences in the populations, interventions, or other factors. The observed effect in each study (white circle) deviates from its own underlying true effect (black circle) and from the grand mean θ, as shown in the individual study curves in the figure. The goal is to learn about both the mean and the variance (τ^2) of the distribution of true effects θ_i as well as sometimes about the individual true effects themselves.

The common-effect model assumes that there is a single "true" effect latent in all the studies. Usually, though, this assumption is unreasonable; studies may differ in the populations they enroll, the ways they carry out interventions, and/or the medical settings in which they are conducted. Instead, it is more likely that each study captures a different latent true effect. To determine an average effect, one may then assume that these individual effects are drawn from a distribution of random study effects, whose mean is called the random-effects estimate. Under these assumptions, the observed effect for study i is $Y_i \sim N(\theta, \tau^2 + v_i)$, where θ is the overall mean effect, τ^2 is the variance of the distribution of true effects about θ, and v_i is the variance of study i. This is referred to as the random-effects model (Fig. 8.4B). In a random-effects meta-analysis, one must estimate both θ and τ^2. This can be done in many ways, including restricted maximum likelihood, empirical Bayes, and other methods (e.g., see Ref. [71]). If an assumption of a common distribution is believed to be unreasonable, then one could still estimate an average effect as an unweighted average of the individual study effects under the assumption that the effects are unrelated and therefore each contributes independent information about the average [8].

8.4.5.1 Exploring heterogeneity

The common-effect assumption is unlikely to be tenable; no two studies are identical in design. Studies are manifestations of different types of patients, different treatment deliveries, different study settings, and differences due to many other possible factors. Thus, differences in results across studies in a meta-analysis are to be expected, and one should try to understand the reasons for these differences. When significant heterogeneity of treatment effects among trials is present, a single estimate of treatment effect, using either a common-effect model or a random-effects model, fails to distinguish the factors involved. As an alternative to either ignoring differences, as in the common-effect model, or averaging over potentially important data patterns, as in the random-effects model, analysis of subgroups by meta-regression may be considered to explore heterogeneity across studies [72,73].

Subgroups within individual studies may be too small to yield significant results. By combining similarly defined subgroups across several studies, a meta-analysis can reveal consistent trends and statistical significance when combined. Age and sex provide natural divisions into informative subgroups for which data are frequently available. However, subgroup results may not be consistently reported across studies. Therefore, meta-analyses of subgroup data should be viewed with caution because their summaries may be based on selectively reported significant subgroup results.

Meta-regression is a technique for performing a regression analysis to assess the relationship between the treatment effects and relevant study characteristics (e.g., study design, proper blinding), participant characteristics (e.g., age, severity of illness), and intervention characteristics (e.g., dosage, duration of treatment) [72,74–76]. The method provides a means to explore sources of heterogeneity and provides a test for their statistical significance and can therefore "explain" discrepancies that may be found across studies. A significant meta-regression effect for a particular characteristic (i.e., a significant regression coefficient) suggests an interaction between the treatment and that characteristic. Because meta-regression models rely on the summary results of published studies, they can describe only study-level, not patient-level, variation in risk factors. They are therefore most useful for studying characteristics that differ across studies (e.g., drug dosages or study protocol items like blinding) [75]. As an analysis at the study-level, meta-regression has limitations. These include its small sample size determined by the number of available studies, the lack of consistently reported study factors, and/or reporting of a biased set found to interact with treatment in a given study. Key risk factors that vary across patients within a study can only be measured as aggregate values, such as mean age and proportion of participants of each sex. Associations with aggregate variables across studies do not necessarily imply associations with the individual variables within studies. Assuming they do can lead to aggregation bias [20]. For instance, two studies with the same average age of patients might have very different age distributions. If risk is concentrated in older individuals and if one of the studies had a higher proportion of older individuals (but the same mean age), then the aggregate risk factor would not capture the difference in risk. Even if studies had similar age distributions so that aggregation bias was not present, the variation of average age across studies might be small (e.g., the mean patient age in many cardiovascular studies is close to 50 years), so aggregate factors might be unable to differentiate among heterogeneous study results.

The common practice of stratifying patients by subgroups of risk factors, conducting separate analyses within each stratum, and then comparing the stratified subgroup results only provides a valid statistical comparison if an appropriate statistical test of the differences between the subgroup results is carried out. Meta-regression provides such an appropriate test. The practice of testing each subgroup against a common reference (such as no difference) separately and then comparing the results of the two statistical tests is not valid, however. For instance, determining that one subgroup estimate is significantly different from zero and the other does not necessarily imply that the two estimates are different from each other. The second one may just be very imprecisely estimated.

8.4.5.2 Meta-analysis of complex datasets

In the preceding sections, we reviewed basic meta-analysis methods for analyses with a single outcome of interest comparing only two treatments. However, data and the questions to be asked of them are often more complex. There may be interest in multiple (possibly correlated) outcomes and multiple competing interventions. This requires the use of multivariate meta-analysis methods [77,78]. Similar methods are also required for meta-analysis of diagnostic test performance because classification accuracy can be measured with multiple metrics (e.g., sensitivity and specificity) that may be correlated [79–81]. For example, consider that for a given condition, multiple treatments may be available, and all might potentially be the best. But how can we know which is? One could attempt a series of meta-analyses, each examining a pair of treatments to form a coherent picture. However, the picture is usually complicated and obscured by demographic differences between the patients involved in each comparison, variation in the eras in which the comparisons were made, the handling of multi-arm trials that encompass different formulations and/or different dosages, as well as statistical issues, such as multiple testing. Perhaps most importantly, some pairs of treatments may never have been directly compared to each other.

Network meta-analysis provides a framework for elegantly addressing these challenges, providing valid estimation in a single model under explicitly formulated assumptions. Such a model allows comparison of multiple interventions, incorporation of indirect evidence, increased precision of estimation, and principled ranking of treatments [82–87]. Network meta-analysis is predicated on the idea of indirect effect estimation. If treatment A is compared to treatment B and B is compared to some other treatment C, then we can estimate the relative efficacy of A to C indirectly based on their comparisons with the common reference B. This is illustrated by Fig. 8.5, which shows an example of a network meta-analysis of 15 alternative interventions for patients with glaucoma.

With network meta-analysis one can obtain estimates of the treatment effect between interventions that have never been directly compared by considering relative comparisons to other treatments in the network. In the example illustrated by Fig. 8.5, latanoprost (a relatively thoroughly studied drug) was not compared in head-to-head trials with apraclonidine (a newer drug) [88]. Yet, by analyzing all the data jointly, one can obtain an indirect treatment effect between these two treatments. For more details on network meta-analysis, we refer the interested reader to existing resources [8,10,83,86,89] (see Box 8.4).

8.5 Accessing systematic reviews, meta-analyses, and field synopses

Journal articles describing systematic reviews and meta-analyses, like other journal articles, are increasingly available online soon after (and sometimes before) they are published. However, identifying them in databases may be challenging; although several search filters have been developed, no singular accepted filter is specific enough to only identify truly 'systematic' reviews. One can simply add 'AND systematic[sb]' to Medline searches, but not all records that are tagged as systematic reviews truly are systematic reviews and vice-versa.

The Cochrane Library is the product of Cochrane, which is an international voluntary organization with the aim of identifying, synthesizing, and disseminating information about

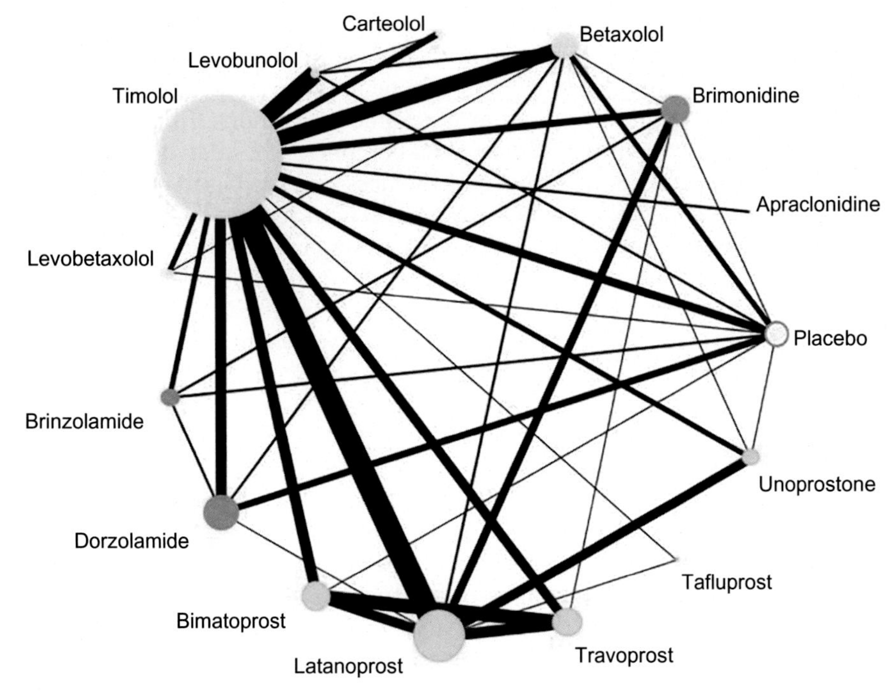

FIG. 8.5 Example of a network of alternative interventions for treating patients with glaucoma, an eye disease characterized by damage to the optic nerve [88]. The outcome is intraocular pressure. In the graph, the nodes (circles) represent 15 interventions (placebo and the 14 active treatments). The treatments are color coded by drug class. Treatments belonging to the same drug class are colored the same and are positioned close to each other. Edges connect treatments that have been compared head-to-head in randomized controlled trials. The thickness of each edge is proportional to the number of trials that have compared the two treatments that the edge connects. The size of each node is proportion to the total number of trial participants randomized to that treatment (across all trials).

BOX 8.4

Medical informatics potential: Treatment networks.

The study of treatment networks in the context of clinical meta-analysis represents an area naturally amenable to collaboration with computer science and informatics. Graph theory, for example, has importance in this context (see, e.g., Ref. [90]). Practical tasks that will require informatics contributions include methods for facilitating treatment network storage, navigation, representation, and automatic generation of complex statistical models from complex datasets [91].

the effects of health-care interventions. The Cochrane Library (https://www.cochranelibrary.com) currently represents the single most comprehensive, routinely updated source of high-quality systematic reviews. It contains over 7500 systematic reviews [92]. Cochrane systematic reviews are indexed in Medline.

A major EBM initiative in the United States was undertaken by the Agency for Healthcare Research and Quality (AHRQ) in 1997, when it created the Evidence-based Practice Center (EPC) Program to produce evidence reports and technology assessments. The EPCs (currently numbering nine), primarily based at academic institutions, function as contractors to AHRQ. Through engagement with diverse stakeholders, EPCs develop evidence reports, technology assessments, and comparative effectiveness reviews based on rigorous, comprehensive reviews of relevant scientific literature [93]. These reports are intended to facilitate the development of clinical practice guidelines, inform payment coverage decisions, provide educational materials and tools, and develop research agendas. Hundreds of these reports covering a wide array of topics have been completed and indexed in Medline. These reports are freely available at the AHRQ web site (https://www.ahrq.gov/research/findings/evidence-based-reports/index.html). Because these reports tend to be long and dense to read, AHRQ is currently pilot testing interactive web-based report presentations to supplement the traditional reports.

In addition to Medline, the Cochrane Library, and AHRQ, other searchable databases for identifying systematic reviews include Epistemonikos (https://www.epistemonikos.org/), the Joanna Briggs Institute Evidence-based Practice (EBP) Database (http://know.lww.com/JBI-resources.html), the Campbell Collaboration Library (https://campbellcollaboration.org/library.html), and the Open Science Framework (https://osf.io/).

Until July 2018, AHRQ managed the National Guideline Clearinghouse (NGC), a web site that provided a database of evidence-based clinical practice guidelines. This database contained over 3000 guidelines from various organizations around the world. Unfortunately, funding for NGC lapsed, and the website has been taken down [94]. The Emergency Care Research Institute (ECRI), an independent nonprofit organization, now maintains a centralized repository of evidence-based guidelines. This repository, named the ECRI Guidelines Trust, contains nearly 2000 guidelines from approximately 200 guideline development organizations covering approximately 50 medical specialty areas [95].

8.6 Systematic reviews, meta-analyses, and the evidence ecosystem

It is important that the data aggregated in systematic reviews are computable and interoperable in a way that facilitates data exchange within the rest of the broader evidence ecosystem. Such an ecosystem includes a cycle of endeavors: producing primary evidence (e.g., studies such as clinical trials), synthesizing the evidence (e.g., systematic reviews), creating trustworthy guidance (e.g., clinical practice guidelines), disseminating and implementing trustworthy guidelines (e.g., through clinical decision support [CDS] tools), evaluating practice (e.g., registry data, quality improvement research), and returning to the generation of additional primary evidence (Fig. 8.6) [96].

Fast Healthcare Interoperability Resources (FHIR) is a standard for electronic exchange of healthcare information that has been developed by Health Level 7 International (HL7), a standards development organization. FHIR is quickly becoming the globally accepted standard for interoperability of healthcare data. To help realize the vision of interoperable data in the

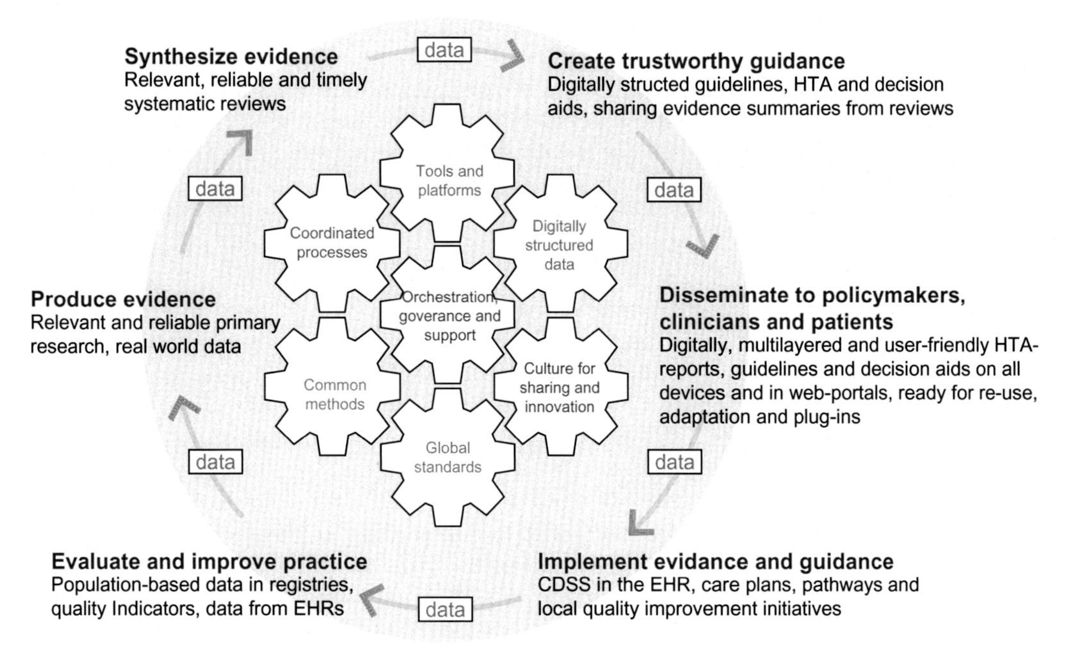

FIG. 8.6 Illustration of the evidence ecosystem [96].

evidence ecosystem, the Evidence-Based Medicine on FHIR (EBMonFHIR) project, approved by HL7 in 2018, includes a broad set of stakeholders, such as health informatics experts, computer scientists, clinicians, researchers, systematic reviewers, guideline developers, clinical decision support developers, terminology developers, patients, and others [97,98]. The EBMonFHIR project workgroup is developing standards for most types of data relevant to evidence synthesis, such as citation information, study design, evidence variables (definitions of populations, interventions/exposures, and outcomes), results, statistics, risk of bias, and strength (or certainty) of evidence.

As discussed in Section 8.4.3, SRDR+ serves as an open-source platform for extracting data during systematic reviews as well as a repository for sharing data. Plans are currently in place to enhance the SRDR+ platform such that the data in it, and exported from it, are fully formatted according to the FHIR standard. It is envisioned that such formatting will greatly facilitate the incorporation of data and knowledge from evidence syntheses into clinical practice guidelines, CDS tools, and other elements of the evidence ecosystem.

8.7 Conclusion

EBM is now part of the healthcare fabric as an invaluable approach to informing practices and policies. Although methodological and technological issues remain, there are no longer debates on whether systematic reviews are useful or whether meta-analysis is a valid statistical method to combine evidence. However, their limitations must be recognized. EBM requires data. It is sometimes discouraging to carry out a systematic review and find that there are few or no studies of sufficient quality from which to draw conclusions that could be useful for clinical practice. From the perspective of a user looking for systematic reviews to guide patient management, many clinical questions have yet to be adequately addressed, and this lack of evidence may also be disappointing.

The practice of EBM requires a specific skillset. Systematic reviews and meta-analyses of randomized controlled trials often are based on studies with strict criteria for participant eligibility. The applicability of their results to the general population may be uncertain. Interpreting results from systematic reviews for care of individual patients, as well as understanding the jargon and methods of systematic review and meta-analysis, requires training. The quality of systematic reviews also varies. Guidelines have been developed and widely adopted to improve the reporting of systematic reviews of randomized controlled trials (e.g., Preferred Reporting Items for Systematic Reviews and Meta-analysis [PRISMA-2]) [99,100], systematic reviews of diagnostic test accuracy studies (e.g., PRISMA-DTA) [101], systematic reviews that include network meta-analysis [102], and systematic reviews of observational studies (Meta-Analysis of Observational Studies in Epidemiology [MOOSE]) [103], among others (see full list on the website of the Enhancing the Quality and Transparency of Health Research [EQUATOR] Network [104]).

The practice of EBM should itself be evidence-based. Methodologies used in meta-analyses often are adapted from other areas without further evaluation. Methodological decisions are often based on assumptions that are not supported by evidence (e.g., including all languages in a systematic review may not necessarily be desirable, as noted earlier). The large number of systematic reviews and meta-analyses published over the last 30 years has fostered the development of better synthesis methods and appreciation of methodological issues; methodologists have performed many empirical studies to elucidate how best to synthesize evidence.

With respect to dissemination of syntheses, the availability of online EBM products such as Cochrane reviews and AHRQ evidence reports and the more general push to make science more open access have greatly facilitated immediate access to summaries of the evidence. However, users may still need to spend hours sifting through these publications to digest the information. System-wide implementation of resources to assist users to identify relevant evidence is needed to increase the impact of EBM in real-world settings.

The recent decade has witnessed the onset and advancement of methodologies and tools for conducting 'living systematic reviews' [58]. These types of reviews, which are typically updated every few weeks incorporate findings from studies soon after they become available. Technological tools to automate some systematic review steps are being rapidly developed and integrated into production of systematic reviews.

Evaluating and summarizing clinical evidence and using the analyses for patient care or health policy decisions are complex activities. The explosion of published biomedical literature and the increasingly complex nature of the evidence contained therein have made performing syntheses more complex still. Work remains to be done to fully realize the vision of EBM as articulated by the father of EBM, Dr. David Sackett: the true *integration* of the best research evidence with clinical expertise and patient values.

References

[1] Institute of Medicine. In: Eden J, Levit L, Berg A, Morton S, editors. Finding what works in health care: Standards for systematic reviews. Washington, DC: The National Academies Press; 2011. 340 p.
[2] Sackett D, Strauss S, Richardson WS. Evidence-based medicine: How to practice and teach EBM. 2nd ed. Edinburgh: Churchill Livingstone; 2000.
[3] Lasserson TJ, Thomas J, Higgins JPT. Starting a review. In: Higgins JPT, Thomas J, Chandler J, Cumpston M, Li T, Page MJ, et al., editors. Cochrane handbook for systematic reviews of interventions version 60. 2nd ed. John Wiley & Sons; 2019 [chapter 1].
[4] Mayo-Wilson E, Fusco N, Li T, Hong H, Canner JK, Dickersin K. Multiple outcomes and analyses in clinical trials create challenges for interpretation and research synthesis. J Clin Epidemiol 2017;86:39–50.
[5] Mayo-Wilson E, Li T, Fusco N, Dickersin K. Practical guidance for using multiple data sources in systematic reviews and meta-analyses (with examples from the MUDS study). Res Synth Methods 2018;9(1):2–12.
[6] Higgins JPT, Thomas J, Chandler J, Cumpston M, Li T, Page MJ, et al. Cochrane handbook for systematic reviews of interventions version 6.2. 2nd ed. John Wiley & Sons; 2019.
[7] Egger M, Schneider M, Davey SG. Spurious precision? Meta-analysis of observational studies. BMJ 1998;316 (7125):140–4.
[8] Schmid CH, Stijnen T, White IR. Handbook of meta-analysis. 1st ed. CRC handbooks of modern statistical methods, Chapman & Hall; 2019.
[9] Wilson DB, Lipsey MW. Practical meta-analysis. Thousand Oaks, CA: Sage Publications; 2000.
[10] Welton NJ, Sutton AJ, Cooper NJ, Abrams KR, Ades AE. Evidence synthesis for decision making in healthcare. Wiley; 2012.
[11] Bastian H, Glasziou P, Chalmers I. Seventy-five trials and eleven systematic reviews a day: how will we ever keep up? PLoS Med 2010;7(9):e1000326.
[12] Boudin F, Nie J-Y, Dawes M, editors. Positional language models for clinical information retrieval. Proceedings of the 2010 conference on empirical methods in natural language processing; 2010.
[13] Ossom Williamson P, Minter CIJ. Exploring PubMed as a reliable resource for scholarly communications services. J Med Libr Assoc 2019;107(1):16–29.
[14] Cook DJ, Mulrow CD, Haynes RB. Systematic reviews: synthesis of best evidence for clinical decisions. Ann Intern Med 1997;126(5):376–80.
[15] Demner-Fushman D, Lin J. Answering clinical questions with knowledge-based and statistical techniques. Comput Linguist 2007;33(1):63–103.
[16] Kim SN, Martinez D, Cavedon L, editors. Automatic classification of sentences for evidence based medicine. Proceedings of the ACM fourth international workshop on data and text mining in biomedical informatics; 2010.
[17] Collins R, Gray R, Godwin J, Peto R. Avoidance of large biases and large random errors in the assessment of moderate treatment effects: the need for systematic overviews. Stat Med 1987;6(3):245–54.
[18] Deeks JJ, Higgins JPT, Altman DG. Analysing data and undertaking meta-analyses. In: Higgins JPT, Thomas J, Chandler J, Cumpston M, Li T, Page MJ, et al., editors. Cochrane handbook for systematic reviews of interventions. 2nd ed. John Wiley & Sons; 2019 [chapter 10].
[19] Peto R, Collins R, Gray R. Large-scale randomized evidence: large, simple trials and overviews of trials. Ann N Y Acad Sci 1993;703:314–40.
[20] Schmid CH, Stark PC, Berlin JA, Landais P, Lau J. Meta-regression detected associations between heterogeneous treatment effects and study-level, but not patient-level, factors. J Clin Epidemiol 2004;57(7):683–97.
[21] Stewart LA, Tierney JF. To IPD or not to IPD? Advantages and disadvantages of systematic reviews using individual patient data. Eval Health Prof 2002;25(1):76–97.
[22] Kovalchik SA. Survey finds that most meta-analysts do not attempt to collect individual patient data. J Clin Epidemiol 2012;65(12):1296–9.
[23] Riley RD, Tierney JF, Stewart LA. Individual participant data meta-analysis: A handbook for healthcare research. 1st ed. Wiley; 2019.
[24] Simmonds M, Stewart G, Stewart L. A decade of individual participant data meta-analyses: a review of current practice. Contemp Clin Trials 2015;45(Pt A):76–83.
[25] van Walraven C. Individual patient meta-analysis—rewards and challenges. J Clin Epidemiol 2010;63(3):235–7.
[26] Lewis S, Clarke M. Forest plots: trying to see the wood and the trees. BMJ 2001;322(7300):1479–80.

[27] Saldanha IJ, Cao W, Broyles JM, Adam GP, Bhuma MR, Mehta S, et al. AHRQ comparative effectiveness reviews. In: Breast reconstruction after mastectomy: A systematic review and Meta-analysis. Rockville, MD: Agency for Healthcare Research and Quality (US); 2021.
[28] Lau J, Antman EM, Jimenez-Silva J, Kupelnick B, Mosteller F, Chalmers TC. Cumulative meta-analysis of therapeutic trials for myocardial infarction. N Engl J Med 1992;327(4):248–54.
[29] Antman EM, Lau J, Kupelnick B, Mosteller F, Chalmers TC. A comparison of results of meta-analyses of randomized control trials and recommendations of clinical experts. Treatments for myocardial infarction. JAMA 1992;268(2):240–8.
[30] Lund H, Brunnhuber K, Juhl C, Robinson K, Leenaars M, Dorch BF, et al. Towards evidence based research. BMJ 2016;355:i5440.
[31] Robinson KA, Goodman SN. A systematic examination of the citation of prior research in reports of randomized, controlled trials. Ann Intern Med 2011;154(1):50–5.
[32] National Institute for Health Research. HTA Programme stage 1 guidance notes (REALMS) [updated April 27, 2021]. Available from: https://www.nihr.ac.uk/documents/hta-programme-stage-1-guidance-notes-realms/27147.
[33] Savović J, Jones HE, Altman DG, Harris RJ, Jüni P, Pildal J, et al. Influence of reported study design characteristics on intervention effect estimates from randomized, controlled trials. Ann Intern Med 2012;157(6):429–38.
[34] Sterne JA, Jüni P, Schulz KF, Altman DG, Bartlett C, Egger M. Statistical methods for assessing the influence of study characteristics on treatment effects in 'meta-epidemiological' research. Stat Med 2002;21(11):1513–24.
[35] Breslow RA, Ross SA, Weed DL. Quality of reviews in epidemiology. Am J Public Health 1998;88(3):475–7.
[36] Mulrow CD. The medical review article: state of the science. Ann Intern Med 1987;106(3):485–8.
[37] Krupski TL, Dahm P, Fesperman SF, Schardt CM. How to perform a literature search. J Urol 2008;179(4):1264–70.
[38] Matchar DB. Introduction to the methods guide for medical test reviews. J Gen Intern Med 2012;27(Suppl. 1):S4–10 [chapter 1].
[39] Wallace BC, Kuiper J, Sharma A, Zhu MB, Marshall IJ. Extracting PICO sentences from clinical trial reports using supervised distant supervision. J Mach Learn Res 2016;17.
[40] Kim SN, Martinez D, Cavedon L, Yencken L, editors. Automatic classification of sentences to support evidence based medicine. BMC Bioinformatics 2011;1:13–21. BioMed Central.
[41] Marshall IJ, Nye B, Kuiper J, Noel-Storr A, Marshall R, Maclean R, et al. Trialstreamer: a living, automatically updated database of clinical trial reports. J Am Med Inform Assoc 2020;27(12):1903–12.
[42] Nye B, Jessy Li J, Patel R, Yang Y, Marshall IJ, Nenkova A, et al. A corpus with multi-level annotations of patients, interventions and outcomes to support language processing for medical literature. Proc Conf Assoc Comput Linguist Meet 2018;2018:197–207.
[43] Singh G, Marshall IJ, Thomas J, Shawe-Taylor J, Wallace BC. A neural candidate-selector architecture for automatic structured clinical text annotation. Proc ACM Int Conf Inf Knowl Manag 2017;2017:1519–28.
[44] Adam GP, Wallace BC, Trikalinos TA. Semi-automated tools for systematic searches. Methods Mol Biol 2022;2345:17–40.
[45] Bekhuis T, Demner-Fushman D. Towards automating the initial screening phase of a systematic review. MEDINFO 2010;2010:146–50.
[46] Cohen AM, Hersh WR, Peterson K, Yen P-Y. Reducing workload in systematic review preparation using automated citation classification. J Am Med Inform Assoc 2006;13(2):206–19.
[47] Wallace BC, Small K, Brodley CE, Trikalinos TA, editors. Active learning for biomedical citation screening. Proceedings of the 16th ACM SIGKDD international conference on Knowledge discovery and data mining; 2010.
[48] Wallace BC, Trikalinos TA, Lau J, Brodley C, Schmid CH. Semi-automated screening of biomedical citations for systematic reviews. BMC Bioinformatics 2010;11(1):1–11.
[49] Wallace BC, Small K, Brodley CE, Lau J, Schmid CH, Bertram L, et al. Toward modernizing the systematic review pipeline in genetics: efficient updating via data mining. Genet Med 2012;14(7):663–9.
[50] Hausner E, Waffenschmidt S, Kaiser T, Simon M. Routine development of objectively derived search strategies. Syst Rev 2012;1:19.
[51] Paynter R, Bañez LL, Erinoff E, Lege-Matsuura J, Potter S. Commentary on EPC methods: an exploration of the use of text-mining software in systematic reviews. J Clin Epidemiol 2017;84:33–6.
[52] Paynter RA, Featherstone R, Stoeger E, Fiordalisi C, Voisin C, Adam GP. A prospective comparison of evidence synthesis search strategies developed with and without text-mining tools. J Clin Epidemiol 2021;139:350–60.

[53] Jüni P, Holenstein F, Sterne J, Bartlett C, Egger M. Direction and impact of language bias in meta-analyses of controlled trials: empirical study. Int J Epidemiol 2002;31(1):115–23.
[54] Moher D, Jadad AR, Nichol G, Penman M, Tugwell P, Walsh S. Assessing the quality of randomized controlled trials: an annotated bibliography of scales and checklists. Control Clin Trials 1995;16(1):62–73.
[55] Morrison A, Polisena J, Husereau D, Moulton K, Clark M, Fiander M, et al. The effect of English-language restriction on systematic review-based meta-analyses: a systematic review of empirical studies. Int J Technol Assess Health Care 2012;28(2):138–44.
[56] Vickers A, Goyal N, Harland R, Rees R. Do certain countries produce only positive results? A systematic review of controlled trials. Control Clin Trials 1998;19(2):159–66.
[57] Balk EM, Chung M, Hadar N, Patel K, Yu WW, Trikalinos TA, et al. AHRQ methods for effective health care. Accuracy of data extraction of non-English language trials with Google translate. Rockville, MD: Agency for Healthcare Research and Quality (US); 2012.
[58] Elliott JH, Synnot A, Turner T, Simmonds M, Akl EA, McDonald S, et al. Living systematic review: 1. Introduction—the why, what, when, and how. J Clin Epidemiol 2017;91:23–30.
[59] Wallace BC, Small K, Brodley CE, Lau J, Trikalinos TA, editors. Deploying an interactive machine learning system in an evidence-based practice center: abstrackr. 2nd ACM SIGHIT international health informatics symposium; 2012.
[60] E JY, Saldanha IJ, Canner J, Schmid CH, Le JT, Li T. Adjudication rather than experience of data abstraction matters more in reducing errors in abstracting data in systematic reviews. Res Synth Methods 2020;11 (3):354–62.
[61] Li T, Saldanha IJ, Jap J, Smith BT, Canner J, Hutfless SM, et al. A randomized trial provided new evidence on the accuracy and efficiency of traditional vs. electronically annotated abstraction approaches in systematic reviews. J Clin Epidemiol 2019;115:77–89.
[62] Li T, Vedula SS, Hadar N, Parkin C, Lau J, Dickersin K. Innovations in data collection, management, and archiving for systematic reviews. Ann Intern Med 2015;162(4):287–94.
[63] Saldanha IJ, Schmid CH, Lau J, Dickersin K, Berlin JA, Jap J, et al. Evaluating Data Abstraction Assistant, a novel software application for data abstraction during systematic reviews: protocol for a randomized controlled trial. Syst Rev 2016;5(1):196.
[64] Ip S, Hadar N, Keefe S, Parkin C, Iovin R, Balk EM, et al. A web-based archive of systematic review data. Syst Rev 2012;1:15.
[65] Saldanha IJ, Smith BT, Ntzani E, Jap J, Balk EM, Lau J. The Systematic Review Data Repository (SRDR): descriptive characteristics of publicly available data and opportunities for research. Syst Rev 2019;8(1):334.
[66] Jonnalagadda SR, Goyal P, Huffman MD. Automating data extraction in systematic reviews: a systematic review. Syst Rev 2015;4:78.
[67] Marshall IJ, Noel-Storr A, Kuiper J, Thomas J, Wallace BC. Machine learning for identifying randomized controlled trials: an evaluation and practitioner's guide. Res Synth Methods 2018;9(4):602–14.
[68] Marshall IJ, Kuiper J, Wallace BC. RobotReviewer: evaluation of a system for automatically assessing bias in clinical trials. J Am Med Inform Assoc 2016;23(1):193–201.
[69] Soboczenski F, Trikalinos TA, Kuiper J, Bias RG, Wallace BC, Marshall IJ. Machine learning to help researchers evaluate biases in clinical trials: a prospective, randomized user study. BMC Med Inform Decis Mak 2019;19 (1):96.
[70] Laird NM, Mosteller F. Some statistical methods for combining experimental results. Int J Technol Assess Health Care 1990;6(1):5–30.
[71] Sidik K, Jonkman JN. A comparison of heterogeneity variance estimators in combining results of studies. Stat Med 2007;26(9):1964–81.
[72] Greenland S, O'Rourke K. On the bias produced by quality scores in meta-analysis, and a hierarchical view of proposed solutions. Biostatistics 2001;2(4):463–71.
[73] Lau J, Ioannidis JP, Schmid CH. Summing up evidence: one answer is not always enough. Lancet 1998;351 (9096):123–7.
[74] Schmid CH. Exploring heterogeneity in randomized trials via meta-analysis. Drug Inf J 1999;33(1):211–24.
[75] Thompson SG, Higgins JP. How should meta-regression analyses be undertaken and interpreted? Stat Med 2002;21(11):1559–73.
[76] van Houwelingen HC, Arends LR, Stijnen T. Advanced methods in meta-analysis: multivariate approach and meta-regression. Stat Med 2002;21(4):589–624.
[77] Jackson D, Riley R, White IR. Multivariate meta-analysis: potential and promise. Stat Med 2011;30(20):2481–98.

[78] Mavridis D, Salanti G. A practical introduction to multivariate meta-analysis. Stat Methods Med Res 2013;22(2):133–58.
[79] Rutter CM, Gatsonis CA. A hierarchical regression approach to meta-analysis of diagnostic test accuracy evaluations. Stat Med 2001;20(19):2865–84.
[80] Chu H, Nie L, Cole SR, Poole C. Meta-analysis of diagnostic accuracy studies accounting for disease prevalence: alternative parameterizations and model selection. Stat Med 2009;28(18):2384–99.
[81] Reitsma JB, Glas AS, Rutjes AW, Scholten RJ, Bossuyt PM, Zwinderman AH. Bivariate analysis of sensitivity and specificity produces informative summary measures in diagnostic reviews. J Clin Epidemiol 2005;58(10):982–90.
[82] Bagg MK, Salanti G, McAuley JH. Comparing interventions with network meta-analysis. J Physiother 2018;64(2):128–32.
[83] Chaimani A, Caldwell DM, Li T, Higgins JPT, Salanti G. Undertaking network meta-analyses. In: Higgins JPT, Thomas J, Chandler J, Cumpston M, Li T, Page MJ, et al., editors. Cochrane handbook for systematic reviews of interventions version 60. 2nd ed. John Wiley & Sons; 2019 [chapter 11].
[84] Higgins JP, Whitehead A. Borrowing strength from external trials in a meta-analysis. Stat Med 1996;15(24):2733–49.
[85] Lu G, Ades AE. Combination of direct and indirect evidence in mixed treatment comparisons. Stat Med 2004;23(20):3105–24.
[86] Lumley T. Network meta-analysis for indirect treatment comparisons. Stat Med 2002;21(16):2313–24.
[87] White IR, Turner RM, Karahalios A, Salanti G. A comparison of arm-based and contrast-based models for network meta-analysis. Stat Med 2019;38(27):5197–213.
[88] Li T, Lindsley K, Rouse B, Hong H, Shi Q, Friedman DS, et al. Comparative effectiveness of first-line medications for primary open-angle glaucoma: a systematic review and network meta-analysis. Ophthalmology 2016;123(1):129–40.
[89] Salanti G. Indirect and mixed-treatment comparison, network, or multiple-treatments meta-analysis: many names, many benefits, many concerns for the next generation evidence synthesis tool. Res Synth Methods 2012;3(2):80–97.
[90] Rücker G. Network meta-analysis, electrical networks and graph theory. Res Synth Methods 2012;3(4):312–24.
[91] van Valkenhoef G, Lu G, de Brock B, Hillege H, Ades AE, Welton NJ. Automating network meta-analysis. Res Synth Methods 2012;3(4):285–99.
[92] Cochrane. About us. Available from: https://www.cochrane.org/about-us.
[93] Agency for Healthcare Research and Quality Effective Health Care (EHC) Program. Evidence-based Practice Centers [updated December 2021]. Available from: https://effectivehealthcare.ahrq.gov/about/epc.
[94] Agency for Healthcare Research and Quality Guidelines and Measures Updates. Information about the National Guideline Clearinghouse (NGC) [updated September 2018]. Available from: https://www.ahrq.gov/gam/updates/index.html.
[95] ECRI. About ECRI Guidelines Trust. Available from: https://guidelines.ecri.org/about.
[96] Vandvik PO, Brandt L. Future of Evidence Ecosystem Series: evidence ecosystems and learning health systems: why bother? J Clin Epidemiol 2020;123:166–70.
[97] Alper BS, Dehnbostel J, Afzal M, Subbian V, Soares A, Kunnamo I, et al. Making science computable: developing code systems for statistics, study design, and risk of bias. J Biomed Inform 2021;115:103685.
[98] Alper BS, Richardson JE, Lehmann HP, Subbian V. It is time for computable evidence synthesis: the COVID-19 Knowledge Accelerator initiative. J Am Med Inform Assoc 2020;27(8):1338–9.
[99] Page MJ, McKenzie JE, Bossuyt PM, Boutron I, Hoffmann TC, Mulrow CD, et al. The PRISMA 2020 statement: an updated guideline for reporting systematic reviews. Rev Esp Cardiol (Engl Ed) 2021;74(9):790–9.
[100] Page MJ, Moher D, McKenzie JE. Introduction to PRISMA 2020 and implications for research synthesis methodologists. Res Synth Methods 2021.
[101] McInnes MDF, Moher D, Thombs BD, McGrath TA, Bossuyt PM, Clifford T, et al. Preferred reporting items for a systematic review and meta-analysis of diagnostic test accuracy studies: the PRISMA-DTA statement. JAMA 2018;319(4):388–96.
[102] Hutton B, Salanti G, Caldwell DM, Chaimani A, Schmid CH, Cameron C, et al. The PRISMA extension statement for reporting of systematic reviews incorporating network meta-analyses of health care interventions: checklist and explanations. Ann Intern Med 2015;162(11):777–84.
[103] Stroup DF, Berlin JA, Morton SC, Olkin I, Williamson GD, Rennie D, et al. Meta-analysis of observational studies in epidemiology: a proposal for reporting. Meta-analysis Of Observational Studies in Epidemiology (MOOSE) group. JAMA 2000;283(15):2008–12.
[104] EQUATOR Network. Reporting guidelines for main study types. Available from: https://www.equator-network.org.

SECTION III

The technology of clinical decision support and beyond

CHAPTER

9

Decision rules and expressions

Robert A. Jenders[a,b,c,d] and Bryn Rhodes[e]

[a]Center for Biomedical Informatics, Charles Drew University, Los Angeles, CA, United States [b]Medicine, Charles Drew University, Los Angeles, CA, United States [c]Clinical and Translational Science Institute, University of California, Los Angeles, CA, United States [d]Department of Medicine, University of California, Los Angeles, CA, United States [e]Alphora, Orem, UT, United States

9.1 Introduction

Deterministic reasoning is a key type of decision-making process in which a decision-maker applies branching logic and deduction against the information of a particular situation in order to arrive at a plan of action. A decision rule is a representation of knowledge in a particular domain that encapsulates the flow of logic employed in deterministic reasoning to make a decision. Decision rules, then, represent a form of algorithm, typically represented as discriminating questions or logical IF-THEN statements that may be followed to reach some conclusion. They map the circumstances of a particular situation, such as the case of an ill patient for whom a diagnosis must be chosen, to a particular choice, whether that be a diagnosis, a treatment plan or an inferred observation that, in turn, may lead to another decision.

In a computer-based clinical decision support (CDS) system, decision rules often are represented in one of two formats: procedures and production rules. Like a subroutine in a programming language, a procedure is a collection of references to data together with logical statements that manipulate them and execute, largely serially, using control structures to direct the flow of decision making through the procedure. In a system based on production rules, each unit of knowledge is a single IF-THEN logical statement, and an inference engine, evaluating the available data and statements, chooses which statement to execute next.

Although these formalisms have been applied to address a wide range of problems, lack of specificity for the medical domain and lack of standardization have impaired both use and sharing of knowledge bases encoded using them. Recognizing these impediments led in

Clinical Decision Support and Beyond
https://doi.org/10.1016/B978-0-323-91200-6.00011-5

the 1990s to development of a standard approach that combines these formalisms, represented by the Arden Syntax. Perceived limitations with this standard and the need to encode a growing body of computable clinical practice guidelines has led to the examination of other approaches, including the use of a standard expression language in the context of a guideline formalism.

This chapter examines the use of decision rules as a knowledge representation formalism for CDS. The details of such a formalism are explored, including inference mechanisms that are employed in order to make decisions using knowledge encoded in this fashion. Facilitating the use of this approach through standardization, with an emphasis on the Arden Syntax and expression languages such as the Clinical Quality Language (CQL), is explored. Additional efforts to facilitate transfer of knowledge so encoded through the use of standard data models are reviewed. Advantages and disadvantages of these approaches are explored.

9.2 Procedural knowledge

As noted by Miller [1], some of the earliest work in implementing decision rules for CDS used procedures written in conventional programming languages. Two key features characterize this representation. First, clinical knowledge and inferencing or control knowledge are mixed in the same representation. This means that instructions to the computer about how to use the clinical knowledge, such as which statement to execute next, is mixed with logical statements about the clinical domain, such as a laboratory test threshold that must be exceeded in order for the diagnosis of a particular disease state to be made.

Second, the flow of control is made explicit. A procedure typically is a series of statements that are executed serially—in the order that they appear in the unit of knowledge. Control statements, such as GO TO and iterations, interrupt the serial execution but still specify explicitly the next statement to be executed, although that may be dependent on data available only at the time of execution. Control knowledge includes not only specification of the flow of execution but also how communication with users occurs (e.g., synchronously via a computer terminal), conditions under which the procedure will execute (e.g., when called from an electronic medical record) and methods for displaying output (e.g., sending a message to a clinician).

Decision rules characterized by an explicit flow of control in accord with a series of branching questions or logical statements sometimes are represented graphically as decision trees (Fig. 9.1) or flow charts. In a typical decision tree, each node in the tree may ask a different yes/no question, and the appropriate branch of the tree is followed depending on the response to the question. Ultimately, a conclusion of the decision rule is reached when the traversal encounters a terminal or leaf node of the tree that offers no further refining questions.

This approach offers many advantages. Nearly any programming language that supports subroutines, functions or procedures can be used to encode the clinical knowledge in executable format. In turn, this means that commonly available programming tools for these languages, such as compilers or debuggers, can be used. If a programming language used is one that is supported on many different types of computers, development and maintenance of the knowledge can be done on multiple platforms without the need to acquire specialized software. Further, because flow control is explicit, the knowledge engineer can tightly control

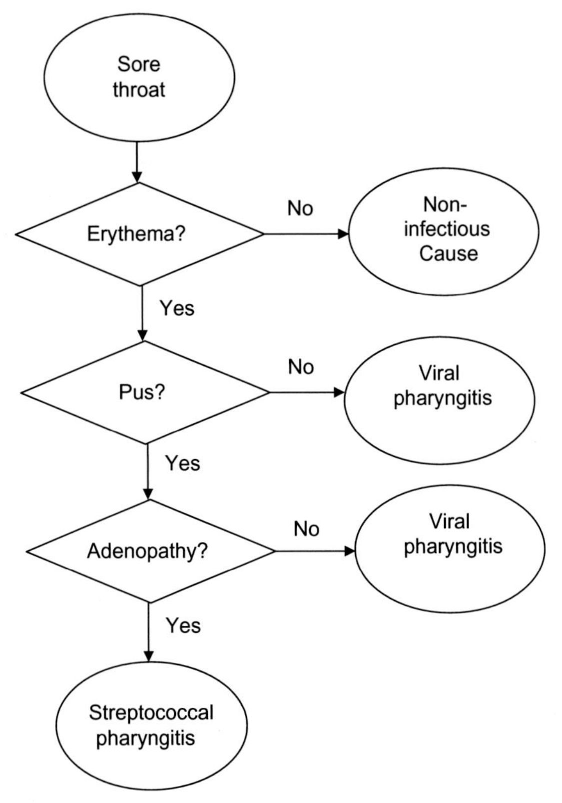

FIG. 9.1 Decision rule represented as a decision tree. The decision rule helps determine the diagnosis in a case of a patient with a sore throat based on physical examination findings.

the order of execution of statements, improving the predictability of the results of executing the software and thus improving its accuracy. Moreover, conventional programming languages, such as C++ or Java, typically offer libraries of pre-programmed functions to perform common tasks, such as retrieving data from databases, thus facilitating the interface between the decision rule and data repositories.

However, while advantageous in many respects, the procedural approach to knowledge representation also has significant flaws. Key among these is the aforementioned mixture of control and clinical knowledge. This makes it difficult to acquire and to maintain the knowledge, because the author must be familiar not only with the clinical domain but also with the syntax and control features of the programming language. Moreover, subsequent edits of the clinical knowledge may inadvertently alter the control structures embedded in the same statements, thus adversely affecting the execution and possibly the accuracy of the CDS. Also, any changes to the clinical knowledge may require recompilation of the software for the decision rule—an expensive and time-consuming process, magnified if the updated decision rule then must be distributed to many different places.

To avoid these challenges, many CDS systems employ an architecture that separates control knowledge from clinical knowledge. This allows the clinical knowledge to be maintained separately from the control knowledge, thus allowing the clinical domain expert or knowledge engineer to focus on just the expert decision rules without having to be concerned about control structures or the need to recompile the entire CDS system each time a new decision rule is introduced or an old one updated. Moreover, the clinical knowledge can be represented in a format more understandable to clinical experts than a typical programming language, thus facilitating validation of executable clinical knowledge. An early form of knowledge representation that fulfills these advantages is the production rule system.

9.3 Knowledge as production rules

Production rules were first studied in the 1940s, when they were developed as axioms that could be used to rewrite strings as part of the specification of a formal grammar. Because each such rule specified a new string that could be produced based on an extant string compliant with the grammar, these axioms were known as production rules [2]. Applied to solving problems, a production rule maps from the characteristics of a situation to the behavior that should be performed or the conclusion that should be reached in that situation. Consequently, they are sometimes called condition-action rules.

The conventional format for a production rule is the IF-THEN statement:

IF <condition> THEN <action>,

where <condition> represents a logical statement that, if true, leads to the <action> being undertaken. The condition part is sometimes known as the left-hand side (LHS) of the statement, while the action is known as the right-hand side (RHS). The condition may be a simple, single comparison involving data available to the CDS system, or it may be an arbitrarily complex statement in Boolean logic, using association, conjunction, disjunction and negation related to data (Fig. 9.2). Typical clinical conditions might be:

potassium > 5.5
(potassium > 5.5) and (creatinine < 2.0)
(diagnosis = 'acute renal failure') or ((potassium > 5.5) and (creatinine < 2.0))

IF NOT (erythema) AND NOT (pus) AND NOT (adenopathy) THEN
CONCLUDE "non-infectious cause"

IF erythema AND NOT (pus) AND NOT (adenopathy) THEN
CONCLUDE "viral pharyngitis"

IF erythema AND pus AND NOT adenopathy THEN
CONCLUDE "viral pharyngitis"

IF erythema AND pus AND adenopathy THEN
CONCLUDE "streptococcal pharyngitis"

FIG. 9.2 Decision rule represented as production rules. This collection of production rules represents the same knowledge as the decision tree in Fig. 9.1. Each rule associates a Boolean condition that evaluates to true or false with an action (in this case, a diagnosis), assuming a patient with a complaint of sore throat. The terms "erythema," "pus" and "adenopathy" are Boolean variables that evaluate to true or false based on data available to the CDSS.

The action or RHS of a production rule may be an instruction to generate a message, usually a recommendation that some action be performed by a person, or a conclusion, typically represented by an assignment statement, that contributes another fact or data element available to the CDS system. Typical actions include

write 'Consider reducing the dose of the drug'
diagnosis: = 'acute renal failure'
creatinine_clearance: = 54

In effect, the RHS is a modification to be performed to the data available to the CDS system or an output to be produced if the rule were to execute or "fire." The effect of such an action may be to negate a previously established data element or conclusion. In addition, like the LHS of a rule, the RHS may be arbitrarily complex and consist of several actions. A knowledge base represented using production rules would consist of a collection of these condition-action statements. The data elements against which the knowledge base would apply may consist of data about a patient, possibly retrieved by the CDS system from a clinical data repository.

The operation of a production rule CDS system consists of repeated cycles of match, select and execute, applying the knowledge base against data available to the CDS system in order to reach a desired conclusion, such as establishing a diagnosis or recommending a treatment. In the first step, matching, the LHS of the rules is compared to data available to the CDS system to see which ones could be executed. Because often more than one rule may be eligible for execution, the result of the match may be a conflict set: a collection of rules that are all true and eligible for execution at the same time. Because a production rule system, like the central processing unit of a computer, typically can execute only one instruction at a time, the second step in the process—selection—then occurs. Sometimes called conflict resolution when applied to a conflict set, the selection process identifies which rule will be executed next. Finally, one or more rules are executed, with the result specified by the RHS being carried out. In the case where the RHS specifies a conclusion or an assignment, this new fact or data element, in addition to whatever other new data may have been acquired by the CDS system from sources external to it, may render the LHS of additional rules true (or change those presently true to false), and the cycle begins anew.

Rules may be applied against data in one of two basic ways or inferencing mechanisms. In forward chaining, the inference engine of the CDS system attempts to match data elements against the LHS of rules, executing the actions of those rules that match until some goal state—for example, establishment of a diagnosis—is reached. In backward chaining, the inference engine initially finds rules that conclude whatever goal state the system is attempting to satisfy, and then it tries to ascertain which LHS of these, if any, can be satisfied by data. Forward chaining typically is employed when there is a large amount of data relative to the possible conclusions to be drawn from those data or if the CDS system is triggered or driven by the arrival of new data. By contrast, if the CDS system is used to critique a selection such as a treatment or a diagnosis made by a clinician, then backward chaining might be used.

The hallmark feature of a production-rule CDS system that distinguishes it from one that uses procedural knowledge is that each IF-THEN rule is independent of every other one and can be executed without regard to the execution state of any other rule. Thus, the order of execution of rules cannot be guaranteed. Some systems do include features, such as meta-rules or priority scores, to try to force a certain order of execution, particularly during conflict resolution, but even in these situations the order of execution cannot be predetermined completely.

One variation on the condition-action rule formalism is the event-condition-action (ECA) rule, used to represent expert knowledge in databases and in World Wide Web programming [3]. In this variation, an event is defined that specifies when the conditions should be evaluated, and if the conditions are true at that point the action is undertaken.

Just as a decision tree is a graphical representation of procedural knowledge, a decision table may be used to summarize the knowledge in a production-rule knowledge base. A decision table is a graphical structure in which each column is headed by a data element deemed important in making decisions in a particular domain, along with a column for the action (Table 9.1). Each row in the table is equivalent to a single IF-THEN rule. It associates the values of one or more of the variables (LHS), not all of which need be represented in any given tuple, with an action (RHS). (Note that sometimes decision tables are represented in a flipped configuration in which rows are condition and action alternatives, and the columns represent the tuples.) Indeed, this technique has sometimes been used to identify duplicate rules in a production rule database by allowing easy detection of those that have the same conditions and actions. This allows compression of the resulting knowledge base, which facilitates maintenance of the knowledge base. It also can be used to identify missing values for some of the variables in the LHS, as an aid to ensuring completeness of the production rule set, as well as conflicts, in terms of an identical LHS but different actions.

The key advantage of a production-rule CDS system over a procedural representation is that the representation of knowledge is independent of the control knowledge needed to operate the CDS system and manage the inferencing process. Because of this, production rule knowledge bases can be acquired, maintained and shared without having to alter or recompile the inference engine or the CDS system itself. Also, the rules are represented in a way that resembles natural language (using only IF-THEN logical statements). While IF-THEN statements may be similar to those in a programming language, other statement types used in a programming language are not included in the rules. This makes it easier for the clinical domain expert to manipulate and understand the knowledge more directly than would be the case with procedural knowledge, in which logic and control statements are intertwined. This feature supports relatively easy acquisition of expert knowledge as production rules, because they are encoded in a format familiar to most people. Indeed, this

TABLE 9.1 Decision rule represented as a decision table. This decision table represents the same knowledge as the decision tree in Fig. 9.1. Each tuple of the table represents an association between specific values for clinically important variables (conditions) and the diagnosis (action) that can be inferred from those findings.

Erythema?	Pus?	Adenopathy?	Diagnosis
No	No	No	Non-infectious cause
Yes	No	No	Viral pharyngitis
Yes	Yes	No	Viral pharyngitis
Yes	Yes	Yes	Streptococcal pharyngitis

resemblance to natural language provides another advantage of production-rule CDS systems: easy provision of explanation of reasoning to the user. The CDS system can collect all the rules that fire and display them in their order of execution, which allows the recipient of system advice to see the chain of reasoning and how the system's conclusion was reached. Finally, the modularity of production rules allows them to be manipulated individually, without needing to edit a large amount of procedural code in the process.

However, the independence of production rules also is a disadvantage. Because of this and the sometimes-unpredictable way that the rules may interact under various combinations of input data, the output of a production-rule CDS system may be difficult to predict. Indeed, changes to a single rule may have difficult-to-predict interactions with other rules, leading to unexpected changes in CDS system behavior. This challenge is magnified when the knowledge base grows beyond a hundred or so rules, as would be required for a CDS system addressing any meaningful set of clinical problems. A large number of rules makes it difficult for a knowledge engineer to locate related rules, so that the effect of any changes in the knowledge base can be understood. To cope with this issue, many rule-based CDS systems offer special tools for managing the knowledge base, which allow searching for related rules, or provide simulations to predict the response to knowledge base changes under various conditions. A further disadvantage of the production-rule approach is that it uses declarative logic. In its conventional form as described here, production rules do not incorporate probabilities. On the other hand, much of medical decision-making involves probabilistic reasoning to a certain extent. To compensate for this defect, some production-rule systems have incorporated measures of probability or belief, such as certainty factors, as part of the rule format. In this way, not only does each rule identify some consequent, but also it assigns a degree of certainty to that consequent. The inference engine then must take into account these factors and their propagation, as rules are chained together in order to make a recommendation to the clinician with a corresponding estimate of certainty.

A seminal system that demonstrated the use of decision rules implemented as production rules was MYCIN [4]. MYCIN was a computer-based consultation system developed in the mid-1970s that gave advice about diagnosis and treatment of infectious diseases. MYCIN used primarily backward chaining to reach conclusions. It also introduced certainty factors in order to incorporate probabilistic reasoning into an otherwise deterministic, decision-rule system [5]. Other systems that incorporated similar decision-rule technology as part of clinical information systems included the HELP system [6] at LDS Hospital in Salt Lake City and the Regenstrief Medical Record System at Indiana University [7]. More contemporary approaches have employed business rule management systems such as Drools to provide a maintenance and execution platform for systems employing production rules [8].

As additional institutions began to implement such technology, it became clear to researchers that a considerable amount of redundant effort was being mounted to encode the same or similar decision rules in formats that differed at least slightly from place to place. Moreover, decision rules encoded at one institution could not be used readily at another, thus inhibiting sharing of knowledge and increasing the cost of knowledge engineering. This underscored the need for a standard representation for encoding decision rules that would allow transfer of computable knowledge from one place to another without expending considerable effort to rewrite the units of knowledge and to link them to local data repositories. In addition, considering that both the procedural and the production-rule approach each offered advantages when implementing decision rules, it seemed that a hybrid of these approaches might be ideal. These lines of thought eventually culminated in efforts to compose standards for knowledge representation, and an early product of such efforts was the Arden Syntax.

9.4 A hybrid approach for knowledge transfer: Arden Syntax

In 1989, a consensus conference was held, bringing together workers in academia, industry and government with the goal of creating a standard for representing clinical logic in a shareable format. The eventual product of this effort, published as a standard in 1991 under the auspices of the American Society for Testing and Materials (ASTM), was the Arden Syntax for Medical Logic Systems [9]. Arden Syntax was moved under the auspices of another standards development organization, Health Level Seven International (HL7) in 1998, where it has subsequently evolved and been certified as an ANSI standard, culminating in the release of version 2.9 of the standard in 2012.

The unit of representation in the Arden Syntax is the medical logic module (MLM) [10]. Each MLM contains sufficient logic and references to data to make a single clinical decision. Each MLM is a procedure, in which the logical statements execute serially. However, each MLM also functions independently like the event-condition-action (ECA) variant of a production rule, with a separate trigger that, when satisfied by data, causes the inference engine to execute it and produce some action. Thus, this approach is considered a hybrid of the procedural and the production-rule forms of knowledge representation (Fig. 9.3).

A medical logic module is a text file consisting of English-language-like statements. Each MLM is organized into three labeled sections, called categories. Each category, in turn, has one or more attribute-value pairs known as slots that express in statement form clinical knowledge about the domain in question or knowledge about the MLM itself. The first category is the maintenance category. The MLM author uses this category to document the software engineering aspects of the MLM—who wrote it, when and where it was written, the version of Arden Syntax used, which version of this MLM this is, and so on. In order to facilitate this, the maintenance category contains the following slots: *title, mlmname, Arden Syntax version, version, institution, author, specialist, date and validation*. The statements in these slots are unstructured text.

The second category is the library category. The MLM author uses this category to describe the medical knowledge that underlies the logic of the MLM. In particular, the category is used to describe in narrative format the rationale behind the logic of the MLM, to explain how the MLM functions and to identify references to the biomedical literature and to other knowledge sources pertinent to the logic of the MLM. In order to facilitate this, the library category contains the following slots: *purpose, explanation, keywords, citations* and *links*. With the exception of the latter two slots, the other slots of this category have unstructured values. By contrast, the Arden Syntax documents a structure for citations and links. The library category allows the reader to discern at a glance the function of the MLM, without having to review its executable statements.

The third category, which contains the conditional logic actually executed by the CDSS, is the knowledge category. The slots in this category are *type* (with a fixed value of

```
maintenance:
  title:        Screen for positive troponin I;;
  filename:     troponin;;
  version:      1.40;;
  institution:  World-Famous Medical Center;;
  author:       Robert A. Jenders, MD, MS (jenders@ucla.edu);;
  specialist:   ;;
  date:         2013-04-30;;
  validation:   research;;

library:
  purpose:      Screen for evidence of recent myocardial infarction;;
  explanation:  Triggered by storage of troponin result.  Sends message
                  if result exceeds threshold;;
  keywords:     troponin; myocardial infarction;;
  citations:    ;;
knowledge:
  type:         data-driven;;
  data:
       troponin_storage := event {storage of troponin};

       /* get test result */
       tp := read last {select result from test_table where
            test_code = 'TROPONIN-I'};

       threshold := 1.5;

       /* email for research log */
       email_dest := destination {'email', 'name'= "jenders@ucla.edu"};

   ;;
  evoke:  troponin_storage;;
  logic:

    if (tp is not number) then conclude false;
    endif;

    if tp > threshold then conclude true;
    else conclude false;
    endif;

  ;;
  action:
    write "Patient may have suffered a myocardial infarction.  " ||
       "Troponin I = " || tp || " at " time of troponin
    at email_dest;

    ;;

  urgency:      50;;
end:
```

FIG. 9.3 Sample Arden Syntax MLM. A medical logic module consists of slots organized into 3 categories: maintenance, library and knowledge. Site-specific mappings—in this example, an event definition, a query string and a destination definition from a fictional organization—are enclosed by curly braces. Comments are delimited by /* */.

"data-driven"), *data*, *priority*, *evoke*, *logic*, *action* and *urgency*. The values of these slots are structured in order to facilitate execution by the computer. The *data slot* is used to represent all the data elements needed in the logic slot in order to render a medical decision. Typically these data elements are assigned to variables as the result of queries executed against a clinical database. In most cases, variables are very simple objects with two attributes (a value and a primary time) and no methods. The most recent version of the Arden Syntax provides a mechanism for building objects with multiple attributes. Because the developers of the Arden Syntax recognized that agreement on a common clinical database schema, query language and vocabulary would require many years, if it ever would occur at all, they introduced a construct known as the *curly braces* for the characters used to enclose it ('{}'). This provides a mechanism for the author to include institution-specific database mappings and links within an otherwise standard syntax. When an MLM is transferred from one institution to another, the statements in the curly braces may have to be adjusted to reflect the database mappings of the new host institution. Site-specific mappings enclosed in curly braces also are used to define events that are included as triggers in the evoke slot as well as delineation of destinations for messages from the CDS system.

Contributing additional structure to the knowledge category, the *evoke slot* identifies conditions, typically defined as constraints on data values, that state when the MLM should be executed. MLMs also may be called directly from other MLMs, through the action slot (described below), thus executing as a type of subroutine. The *logic slot* contains the IF-THEN statements and calculations that represent the deterministic reasoning over the available data. A number of operators are available to manipulate data in the logic slot. Among the more important ones are those that allow temporal reasoning. Arden Syntax offers a number of powerful operators for extraction of temporal information from data and reasoning over these times. This is especially important in medical reasoning, and these operators help facilitate the representation of clinical logic in this computable format.

The *action slot* specifies, like the RHS of a production rule, what is supposed to occur if the result of processing the logic slot returns a true value. Typically this involves sending a message to a clinician or writing a value to the database that can be used to trigger or process another MLM, thus facilitating forward chaining. The *priority slot* contains a numeric score that can be used in conflict resolution to establish the order of execution of MLMs that may be triggered at the same time. Finally, the *urgency slot* contains a numeric score that represents the clinical importance of the alert or reminder being encoded; the value in this slot can be used, for example, to decide what of several routes can be used to communicate an alert to a clinician, with faster routes being associated with higher values for urgency.

Arden Syntax has been adopted as the knowledge representation formalism for several large vendors of CDS software. It is used mainly in transaction-oriented clinical information systems, in which the acts of storage of discrete data elements, such as test results and visit information, represent individual events that can trigger execution of MLMs. In such systems, the Arden Syntax has been used to implement relatively simple alerts and reminders. Though capable of doing so, it has not been used, by and large, to represent the declarative knowledge of the typical clinical practice guideline. It has been adopted at a number of medical centers in the United States (US), principally customers of those vendors that have introduced it into their CDS systems, but it has seen some limited use outside the US, primarily in Europe.

One of the original design goals of the Arden Syntax was a simple yet powerful formalism that could both be written and validated by clinical domain experts lacking knowledge of specific information systems. Nevertheless, over time it became clear that it was programmers and knowledge engineers more than expert clinicians who were using the Syntax directly, and these workers demanded additional functionality to match constructs to which they were accustomed in other programming languages. Consequently, Arden Syntax has evolved in recent years to meet these needs, including adoption of a simple object data type, inclusion of iteration statements and most recently in v2.9 insertion of formal constructs for representing fuzzy logic [10,11]. In addition, in part to leverage tools for syntax-checking and translation to other formalisms, an alternative representation of the Arden Syntax in the Extensible Markup Language (XML) has been developed known as ArdenML. Linking Arden back to its origins in production rules, the utility of this approach has been demonstrated using the Drools business rule management system [8].

Arden Syntax offers a number of advantages as a hybrid formalism for knowledge representation in CDS systems. Its key advantage is that it is a standard. This facilitates sharing of computable knowledge by reducing the amount of revision that must be performed on each MLM in order for it to execute properly at an institution other than the one at which it was composed originally. It also facilitates development of tools for acquiring, debugging and maintaining knowledge encoded in this format. The hybrid nature of the Arden Syntax offers the best features of both the procedural and the production-rule formalisms: the control of flow available in a procedural representation with the modularity and separation of inferencing control from knowledge available in a production-rule representation.

Nevertheless, Arden Syntax has some disadvantages too, some of which are not specific to it but pertain to any formalism that might be used for knowledge sharing. One important challenge is the lack of standardized database mappings: the curly braces problem. Although the curly braces highlight those parts of the MLM that require attention as part of the process of knowledge transfer, thus helping to ensure that these mappings are addressed, the absence of standard mappings still requires time-intensive and potentially error-prone manual revision. Another potential disadvantage of Arden Syntax is its procedural code. Clinicians are more accustomed to viewing clinical knowledge in a declarative format, such as a decision tree or a narrative clinical practice guideline. The lack of familiarity with procedural code can make validation of the knowledge a challenge.

Indeed, although Arden Syntax has been used to encode clinical guidelines, a consensus has developed that it is best used for relatively simple single-step rules such as alerts and reminders. By contrast, under this consensus, workers believe that a declarative formalism that captures the specific features of guidelines, such as eligibility criteria (instead of transaction-based triggers) is more appropriate than procedural logic to express a clinical guideline in computable form. While this has led to the creation of a number of different formalisms [12], such as the Guideline Elements Model (GEM—a standard of ASTM for marking up narrative guideline content in a structured fashion) and the Guideline Interchange Format (GLIF), no widespread agreement has yet obtained regarding a standard formalism (see Chapter 16). Accordingly, workers in HL7 and other organizations have created standard components of an overall formalism as a decomposition of the problem. Approved as a distinct HL7 and ANSI standard in 2005, the Guideline Expression Language—Object-oriented (GELLO) represents a standard formalism intended to address these challenges.

9.5 Expression languages

The purpose of an expression language is to allow the knowledge engineer to build up statements that query data, logically manipulate them, provide for reasoning over them and facilitate calculations and other formulae involving them in a variety of applications. GELLO [13] was designed specifically to do this for the case of representing the logic in clinical guidelines, although it need not be restricted to this particular representation. GELLO is based on the Object Constraint Language (OCL), itself a standard of the Object Management Group (OMG), and can be used with any object-oriented data model. As a result, GELLO can be used with the standard HL7 Reference Information Model (RIM) to specify data to be extracted from clinical repositories and to manipulate those data, thus facilitating closer integration with other HL7 standards that use the RIM and taking advantage of the rich object model and relationships that this approach offers. An example of GELLO, including object references, queries, calculations and logical manipulations of data, is seen in Fig. 9.4.

GELLO was developed initially to represent the procedural component of the declarative guideline formalism GLIF. However, its generic nature allows it to be used in a number of applications. Moreover, GELLO addresses the curly braces challenge by facilitating the use of standard vocabularies and data models, thus enhancing the possibility of knowledge transfer. Accordingly, GELLO is a useful contribution in the effort to create a standard formalism for representing clinical guidelines as a common instance of deterministic reasoning represented in a decision rule. Nevertheless, although it remains an HL7 standard, because of its limited adoption, it serves more as an illustration of a stage in the evolution of computable knowledge representation that addresses database linkages, leading in turn to incorporation of more contemporary and widely adopted standard data models such as FHIR.

```
let lastTroponin: Observation = Observation→ select(code=
    ("SNOMED-CT", "102683006")).sortedBy(effectiveTime.high).last()

let threshold : PhysicalQuantity =
    Factory.PhysicalQuantity( "1.5, ng/dl")

let threshold_for_osteodystrophy : int = 70

let myocardial_infarction :Boolean =   if lastCreatinine <> null and
    lastCreatine.value.greaterThan(threshold)
  then
    true
  else
    false
  Endif

if myocardial_infarction then
  whatever action or message
else
  whatever action or message
endif
```

FIG. 9.4 Example of GELLO encoding a simple guideline. This guideline represents the same knowledge contained in the Arden Syntax MLM in Fig. 9.3. Because GELLO was created to extract data from clinical repositories, to manipulate those data and to reason over them, it does not have an explicit syntax for sending messages to clinicians. The GELLO code would be embedded in complete guideline representation or other application for use by the CDS system.

9.6 Standard data models for decision rules

A key design goal of efforts to create standards for representation of decision rules, including GELLO and the Arden Syntax, has been knowledge transfer: moving units of computable knowledge from one organization to another, ideally without needing to encode significant changes in order to be executed in the receiving system environment [14]. An important component of such standardization is use of a standard data model. Using this approach, a knowledge engineer could encode decision rules that process clinical data using standard references to these data. In this way, provided that an organization mapped its local data model or database schema to the standard, references to data could be automatically translated from the standard to its local instantiation with relatively little or no manual intervention, thus improving reliability and reducing the cost of knowledge engineering.

While the HL7 Reference Information Model has been discussed for this purpose, many workers regard it as overly high-level and insufficiently detailed for practical use to encode data references in decision rules. To help remedy these perceived challenges, a distinct HL7 standard, the Virtual Medical Record (vMR), was created specifically as a standard data model for clinical decision support [15]. The vMR is an object-oriented model with classes and their attributes that represent concepts and data important in clinical reasoning. Addressing the "curly braces problem" in part, workers have adapted the GELLO standard to use the vMR for standard data references. While experience in developing and working with the vMR provided key insights and requirements for a standard data model, like GELLO itself, the vMR, although used in practical projects such as immunization registries, did not enjoy wide uptake. Instead, reflecting its wide uptake for representation of clinical data, the standards community focuses on the use of the HL7 Fast Healthcare Interoperability Resources (FHIR) as a standard data model, which will advance the goal of knowledge sharing by reducing the manual changes required when knowledge is transferred between organizations using two different clinical data repositories.

9.6.1 HL7® FHIR® and data interoperability

In 2011 and 2012, HL7 began a project to modernize healthcare data standards called Fast Healthcare Interoperability Resources, or FHIR®, focused on existing web standards, implementability, and community, it represented a new direction for standards development and in the decade since its introduction has evolved into a full platform specification for healthcare data exchange with broad industry adoption throughout the world, and especially in the US where it is now regulated as part of ONC's 21st Century Cures Act Final Rule [16]. As of 2018, 82% of hospitals and 64% of clinics in the US are using a vendor product that exposes healthcare data using a FHIR API, and this trend is continuing [17]. A detailed description of FHIR is beyond the scope of this chapter (the reader can refer to Chapter 11 for a more detailed description for FHIR and its role in CDS), but from the perspective of Clinical Decision Support in general, and the sharing of Clinical Decision Support Rules in particular, FHIR forms a promising foundation for addressing the "curly-braces problem" by providing not only syntactic, but semantic interoperability both in theory and in practice.

As with any technology, the use of FHIR for Clinical Decision Support does have disadvantages. First, as with any large-scale specification, change management is an issue, with different versions of FHIR presenting sometimes significantly different versions of the data. To help address this, FHIR's maturity model [18] provides implementers a tool to assess the relative maturity of different areas of the specification. Second, as a platform specification designed for international usage, the base FHIR specification is intentionally silent on the precise semantics of many data elements. To enable semantic interoperability for data exchange, FHIR must be *profiled* to provide semantics and guidance. These profiles are grouped into *implementation guides* which in turn must be implemented and adopted. Third, and perhaps most relevant for CDS, despite the presence of standard and regulated implementation guides, different FHIR implementations sometimes exhibit differences in behavior that can significantly complicate usage, either because the specifications are unclear or incomplete, or due to unintentional variance. And fourth, because of the scope and complexity of healthcare, the number of data elements that must be standardized is large, and the adoption of FHIR, at least in the US, still covers only a fraction of the data elements even in a general-purpose clinical system. Of course, this is a problem of scale and would be present in any similar attempt to represent the entire spectrum of data in healthcare. In addition, several large-scale industry consortiums, known generally as FHIR Accelerators [19], are working on standardized data representation on everything from social determinants of health to payer and research data.

Early in the Clinical Quality Framework initiative, FHIR was recognized as a next generation standard that could provide many advantages for decision support and quality measurement. As a result, beginning during the development of FHIR DSTU2 and officially as of FHIR STU3, FHIR includes a Clinical Reasoning module that defines FHIR resources and guidance for the dual use cases of sharing and evaluation of clinical quality artifacts. The resources will be reviewed in more detail in a subsequent section.

9.7 Toward further standardization: Quality measures and Health e-Decisions

This push toward standardization manifested by standard formalisms for decision rules and standard data models is having an impact in other domains of clinical decision making. One such area pertains to quality indicators. Facing increasing demand to document value for money in the health care delivery system in the US, as discussed in Chapter 4, health care organizations are being required or incentivized to publish evaluations of their performance as reflected in measures that evaluate effectiveness, efficiency or other aspects of quality of care. Related to decision rules in that they result in conclusions regarding clinical care using decision logic and references to clinical data, quality indicators focus less on decision-making regarding the individual patient and more on assessing care across a population of patients. While these quality indicators can be encoded using extant formalisms such as the Arden Syntax, some workers in the field believe that a formalism that is less procedural, and more declarative in nature, corresponding to the typical narrative description of these indicators in scientific publications and regulations, would be more appropriate.

Nevertheless, as is the case with decision rules, the drive toward standardization in quality indicators is focused on improving knowledge transfer. If each health care organization required to use a quality indicator could execute it directly in its information systems to produce the desired quality report without the need for manual coding or revision, this would reduce the cost of knowledge engineering even as it enhanced the comparability of results across multiple organizations. A key effort in this regard is the HL7 Healthcare Quality Measure Format (HQMF), also known as the eMeasures formalism [20]. Encoded in XML, the HQMF allows the creator of a quality indicator to publish it in this standard, structured format so that it can be imported by health care organizations in order to create quality reports. Concern regarding the complexity of this standard has led to additional work, underway at the time of this writing, to create a second version of HQMF and to further work on the vMR to facilitate its use in quality measures.

Marking the further advancement of standardization with regard to decision rules and expressions is the Health e-Decisions project undertaken in the Standards & Interoperability Framework under the sponsorship of the US Office of the National Coordinator for Health IT (ONC). The S&I Framework is a collaborative community of participants from the public and private sectors that have the goal of facilitating the exchange of health information. The rationale for this work comes from the fact that, despite considerable effort to develop standards and promote their use in this domain during the last two decades, uptake of these standards has been incomplete at best. Accordingly, the goal is to couple an analysis of the need for standards in this area with the regulatory aspects of the Meaningful Use agenda in the US in the hope that this will promote use of knowledge representation standards that can further enhance knowledge sharing and reduce the cost of knowledge engineering.

Under the Health e-Decisions project, two key use cases were elaborated. Use Case 1 addressed CDS Artifact Sharing, identifying 3 key artifacts that are commonly used in the care of and decision making about patients—event-condition-action rules, order sets, and documentation templates. Not directly related to decision rules, Use Case 2 addressed a CDS Guidance Service, identifying the ideal characteristics of standardized access to knowledge-based CDS advice without necessarily specifying a standard formalism for knowledge transfer (see Chapter 29). While extant standards such as Arden Syntax and GELLO already can express ECA rules, additional effort was made to identify gaps in these and other standards in describing the key features of this construct. Although HL7 already had endorsed a standard to encode order sets used by physicians and other clinicians to produce orders, the S&I Framework strove to improve upon this standard as well. Finally, the specification for documentation templates broke new ground in specifying a standard for how to encode forms and templates for documenting clinical information. Overall, by standardizing the important characteristics of these different artifacts, the S&I Framework further promotes the standardization of decision rules that can enhance knowledge sharing.

9.7.1 Clinical Quality Framework

One of the key outcomes of the Health e-Decisions initiative was the recognition that the use of different standards for knowledge representation and sharing in the decision support and quality measurement domains resulted in duplicated effort for institutions consuming these artifacts. In March of 2014, the US Office of the National Coordinator for Health IT (ONC) and the Centers for Medicare and Medicaid Services (CMS) sponsored a public-private partnership initiative to identify, define, and harmonize electronic standards that promote integration between decision support and quality measurement in the areas of metadata, expression logic, and data modeling [21]. The initiative started by producing conceptual specifications for each of these areas, resulting in the HL7 Specification: Clinical Quality Common Metadata Conceptual Model, Release 1 [22], the HL7 Domain Analysis Model: Harmonization of Health Quality Artifact Reasoning and Expression Logic [23], and the HL7 Domain Analysis Model: Health Quality Improvement, Release 1, or QIDAM [24]. These conceptual models were then used to provide a foundation for harmonizing quality improvement artifact specifications. In 2017, sponsorship of the Clinical Quality Framework Initiative by ONC and CMS officially ended and the initiative was transferred to HL7 as a joint effort between the Clinical Decision Support (CDS) and Clinical Quality Information (CQI) work groups [25].

9.7.2 Clinical Quality Language

Building on lessons learned from Arden Syntax, GELLO, and Quality Data Model (QDM), the Clinical Quality Framework Initiative produced the Clinical Quality Language specification [26], which defines a representation for the expression of clinical logic that can be used across healthcare domains including decision support, quality measurement, and case reporting. Recognizing barriers to point-to-point knowledge sharing such as lack of tooling, complexity of implementation, and insufficient expressivity, the specification provides a solution for sharing logic by defining a syntax-independent, canonical representation for expression and query logic, called Expression Logical Model (ELM). The design of ELM is informed conceptually by the Health Quality Artifact Reasoning and Expression Logic domain analysis model and technically by compiler design best-practices, drawing especially from the domain of Database Management to ensure implementability and performance. In addition, the specification introduces a domain specific syntax, focused on clinical quality and targeted at measure and decision support artifact authors. This high-level syntax can then be rendered in the machine-friendly canonical representation provided by ELM. An important architectural pattern used in Database Management separates system concerns into three layers, (1) Conceptual, (2) Logical, and (3) Physical. The physical layer is concerned with how data is physically stored and accessed, the conceptual layer is concerned with how data is presented to users and applications, and the logical layer provides a level of indirection between the two. This architecture affords several important types of independence that isolate specific concerns to appropriate levels of the system. Two of the most important types of independence achieved by this separation are (1) Location independence, and (2) Model independence. Location independence refers to the fact that the logic is unaware of and isolated from the physical location of the data. This allows implementing systems flexibility in how data is provided to the logic when it is evaluated. Model independence refers to the fact that the logic specification is entirely separated from the model specification. As noted in the discussion on Arden Syntax, this separation is important, but also has disadvantages in that it means that logic encoded in this way makes assumptions about the data model, the "curly braces problem". We will revisit this in a subsequent section on data interoperability. Chapter 11 provides a more detailed discussion of this problem along with potential solutions.

At the conceptual (or authoring) level, CQL is organized into libraries consisting of any number of declarations. These declarations include expression definitions, function definitions, and terminology declarations that allow concepts used by the logic to be surfaced and mapped to standard terminologies. Libraries can reference any number of other libraries, allowing these declarations to be shared and organized.

CQL libraries specify data models with a *using* declaration, which identifies the name and version of a specific model. This model is described using a model information, or *modelinfo* file which is a simple, formal representation of the types available in the model. Authoring systems use this information to validate that CQL expressions are semantically correct (i.e. they reference known types in the model and the expressions and operations used within the logic are appropriate for the types of data being accessed).

CQL libraries also specify the *context*, which determines the focus of subsequent expressions in the library. For example, a typical library will specify the *Patient* as the focus, meaning that data accessed by expressions in that context will be limited to data related to the patient in context. This construct allows artifact authors to focus on expressions from a particular perspective, simplifying the logic.

Expressions of CQL are then made up of *literals* (values such as '5'), *operators* (such as '+' and '*'), *identifiers* (references to other declarations or data types in the model), and *functions* (such as 'First()' and 'Max()'). Every expression results in a value or 'null', indicating that information was missing, or the results of a computation were invalid (such as dividing by zero). Like other query languages (such as Structured Query Language (SQL) and XQuery), CQL uses 3-valued logic to deal with missing information.

CQL supports several categories of values, including the typical primitive values such as Booleans, integers, strings, dates, and times, as well as support for structured values and lists. Importantly, CQL also includes first-class support for intervals, including integer, decimal, and date, and time intervals.

To ensure sufficient expressivity and computational completeness, CQL includes several categories of operators and functions, including logical, arithmetic, comparison, date/time arithmetic, string manipulation, as well as interval and list operations.

All data access in CQL is performed through the *retrieve* operation, which allows authors to request data of a given type (selected from a model used by the library), and optionally filtered by a terminology to further refine the data returned. The result of a retrieve operation is any data of the requested type, scoped to the current context, and matching the specified terminology.

And finally, as a query language, CQL provides a *query* construct, similar to the SELECT.. FROM syntax of SQL, that allows authors to build sophisticated queries for accessing and shaping data from a clinical data source. An example of a CQL library illustrating common usage can be seen in Fig. 9.5. In addition, the datatypes and core expression language of CQL are aligned with FHIR through the FHIRPath specification [27].

FHIRPath is a path-based navigation and extraction language, similar to XPath and OCL. Operations are expressed in terms of the logical content of hierarchical data models and support traversal, selection, and filtering of data. FHIRPath is used throughout the FHIR

```
library OpioidCDSREC11 version '2.0.1'

using FHIR version '4.0.1'

include FHIRHelpers version '4.0.1' called FHIRHelpers
include OpioidCDSCommon version '2.0.1' called Common
include OpioidCDSRoutines version '2.0.1' called Routines

/*
**
** Recommendation #11
**   Clinicians should avoid prescribing opioid pain medication and benzodiazepines
**   concurrently whenever possible (recommendation category: A, evidence type: 3)
**
** When
**   Provider is prescribing an opioid analgesic with ambulatory misuse potential in the
outpatient setting
**   Provider is prescribing a benzodiazepine medication
**   Opioid review is useful for this patient:
**     Patient is 18 or over
**     Patient does not have findings indicating limited life expectancy
**     Patient does not have orders for therapies indicating end of life care
**     Patient is not undergoing active cancer treatment:
**       Patient has had at least 2 encounters within the past year with any diagnosis of cancer
**   Patient prescribed opioid analgesic with ambulatory misuse potential and benzodiazepine
medication concurrently
** Then
**   Recommend to avoid prescribing opioid pain medication and benzodiazepine concurrently
**     Will revise
**     Benefits outweigh risks, snooze 3 months
**     N/A - see comment, snooze 3 months
**
*/

// META: Plan Definition: http://fhir.org/guides/cdc/opioid-cds-r4/PlanDefinition/opioid-cds-11

parameter ContextPrescriptions List<MedicationRequest>

context Patient

define "Opioid Analgesic with Ambulatory Misuse Potential Prescriptions":
  Common."Is Opioid Analgesic with Ambulatory Misuse Potential?"( ContextPrescriptions )

define "Benzodiazepine Prescriptions":
  Common."Is Benzodiazepine?"( ContextPrescriptions )
```

FIG. 9.5 Example of CQL representing a decision support rule. This CQL library contains the logic used to implement recommendation #11 from the CDC Opioid Prescribing Guideline as a decision support rule intended to be invoked as part of a prescribing workflow to avoid co-prescribing opioids and benzodiazepines.

(Continued)

```
define "Patient Is Being Prescribed Opioid Analgesic with Ambulatory Misuse Potential":
 exists( "Opioid Analgesic with Ambulatory Misuse Potential Prescriptions" )

define "Patient Is Being Prescribed Benzodiazepine":
 exists( "Benzodiazepine Prescriptions" )

define "Is Recommendation Applicable?":
 "Inclusion Criteria"
  and not "Exclusion Criteria"

define "Inclusion Criteria":
 (
  (
   "Patient Is Being Prescribed Opioid Analgesic with Ambulatory Misuse Potential"
    and exists Common."Active Ambulatory Benzodiazepine Rx"
  )
   or (
    "Patient Is Being Prescribed Benzodiazepine"
     and exists Common."Active Ambulatory Opioid Rx"
   )
 )
  and Routines."Is Opioid Review Useful?"

define "Exclusion Criteria":
 Common."End of Life Assessment"

define "Get Indicator":
 if "Is Recommendation Applicable?"
  then 'warning'
 else null

define "Get Summary":
 if "Is Recommendation Applicable?"
  then 'Avoid prescribing opioid pain medication and benzodiazepine concurrently whenever
possible.'
 else null

define "Get Detail":
 if "Is Recommendation Applicable?"
  then
   if "Patient Is Being Prescribed Benzodiazepine"
    then 'The benzodiazepine prescription request is concurrent with an active opioid
prescription'
   else 'The opioid prescription request is concurrent with an active benzodiazepine prescription'
 else null
```

FIG. 9.5, CONT'D

specification to provide expressive capabilities such as formal description of search paths and invariants. FHIRPath is also used in CQL as the formal grammar for the core expression language (i.e. CQL is formally a superset of FHIRPath).

9.7.3 Representation of clinical decision support rules

The Clinical Quality Framework initiative expanded on the initial Clinical Decision Support Knowledge Artifact Specification (CDS KAS), modifying it to use CQL as the basis for logic expressions in CDS artifacts. At the same time, the Health Quality Measures Format (HQMF) was modified in the same way, allowing libraries of CQL to be shared between decision support and quality measurement. As has been noted, the CDS KAS drew heavily on lessons learned from Arden Syntax, focusing on four key areas of artifact representation, (1) Metadata such as identity, guidance, classification, and justification, (2) Expression logic, (3) Data model, and (4) Artifact structure. Both CDS KAS and HQMF were aligned with these components, especially the Metadata, Logic, and Data Model, using the conceptual models developed for each purpose. In CDS, the virtual Medical Record (vMR) was aligned with the QIDAM, and in quality measurement, the Quality Data Model (QDM) was similarly aligned.

At the same time, workers in the initiative used the CDS KAS, HQMF, and the conceptual Metadata model and QIDAM to express FHIR resources and guidance, resulting in the FHIR Clinical Reasoning Module.

The CDS KAS defines a generic artifact structure built up of components that can assembled to produce different types of artifacts. The initial specification focused on supporting Event-Condition-Action Rules, Documentation Templates, and Order Sets. In FHIR, this generic structure resulted in the definition of the PlanDefinition and ActivityDefinition resources. For Quality Measurement, a Measure resource was defined based on HQMF. In addition, a Library resource was introduced to support the representation of logic, in particular CQL libraries. The PlanDefinition, ActivityDefinition, and Measure resources can reference Library resources to provide the logic used by the artifacts. These resources, and others like them that have since been introduced are referred to generally as "Knowledge Resources", and are a special case of the more general "Conformance Resource" of FHIR. In addition, the conceptual Metadata model was used to provide consistent metadata elements on each of these resources, and a "Metadata Resource" pattern was introduced in FHIR that each of these resources follows.

Like the KAS artifact structure, the PlanDefinition resource is a flexible structure that can be used in a variety of different ways. The core of the resource is defined in the "action" element, which contains descriptive elements such as *title*, *description*, *textEquivalent*, and *code*, a *trigger* element that specifies the event that should trigger the action, a *condition* element that specifies whether the action is applicable, and a *definition* element that typically points to an ActivityDefinition, describing in detail the action to be performed. An example of a PlanDefinition resource is provided in Fig. 9.6.

Despite several pilots in the CQF initiative demonstrating feasibility and utility of the CDS Knowledge Artifact Specification, the specification did not receive broad adoption and has now largely been superseded by the FHIR Clinical Reasoning module. Many projects are

```xml
<PlanDefinition xmlns="http://hl7.org/fhir">
  <id value="opioidcds-11"/>
  <url value="http://fhir.org/guides/cdc/opioid-cds/PlanDefinition/opioidcds-11"/>
  <version value="2.0.1"/>
  <name value="CDC_opioid_11"/>
  <title value="PlanDefinition - CDC Opioid Prescribing Guideline Recommendation #11"/>
  <type>
    <coding>
      <system value="http://terminology.hl7.org/CodeSystem/plan-definition-type"/>
      <code value="eca-rule"/>
      <display value="ECA Rule"/>
    </coding>
  </type>
  <status value="draft"/>
  <experimental value="false"/>
  <date value="2018-03-19"/>
  <publisher value="Centers for Disease Control and Prevention (CDC)"/>
  <description value="Concurrently prescribing opioid medications with benzodiazepines
increases the risk of harm for the patient."/>
  <useContext>
    <code>
      <system value="http://terminology.hl7.org/CodeSystem/usage-context-type"/>
      <code value="focus"/>
      <display value="Clinical Focus"/>
    </code>
    <valueCodeableConcept>
      <coding>
        <system value="http://snomed.info/sct"/>
        <code value="182888003"/>
        <display value="Medication requested (situation)"/>
      </coding>
    </valueCodeableConcept>
  </useContext>
  <useContext>
    <code>
      <system value="http://terminology.hl7.org/CodeSystem/usage-context-type"/>
      <code value="focus"/>
      <display value="Clinical Focus"/>
    </code>
    <valueCodeableConcept>
```


FIG. 9.6 Example of a PlanDefinition representing a decision support rule. This PlanDefinition represents an ECARule that implements recommendation #11 from the CDC Opioid Prescribing Guideline as a decision support rule intended to be invoked as part of a prescribing workflow to avoid co-prescribing opioids and benzodiazepines. The PlanDefinition references the library defined in Fig. 9.5.

(Continued)

```xml
        <coding>
          <system value="http://snomed.info/sct"/>
          <code value="82423001"/>
          <display value="Chronic pain (finding)"/>
        </coding>
      </valueCodeableConcept>
    </useContext>
    <jurisdiction>
      <coding>
        <system value="urn:iso:std:iso:3166"/>
        <code value="US"/>
        <display value="United States of America"/>
      </coding>
    </jurisdiction>
    <purpose value="CDC's Guideline for Prescribing Opioids for Chronic Pain is intended to
improve communication between providers and patients about the risks and benefits of opioid
therapy for chronic pain, improve the safety and effectiveness of pain treatment, and reduce the
risks associated with long-term opioid therapy, including opioid use disorder and overdose. The
Guideline is not intended for patients who are in active cancer treatment, palliative care, or end-
of-life care."/>
    <usage value="Clinicians should avoid prescribing opioid pain medication and benzodiazepines
concurrently whenever possible."/>
    <copyright value="© CDC 2016+."/>
    <topic>
      <text value="Opioid Prescribing"/>
    </topic>
    <relatedArtifact>
      <type value="documentation"/>
      <display value="CDC guideline for prescribing opioids for chronic pain"/>
      <citation value="Dowell D, Haegerich TM, Chou R. CDC Guideline for Prescribing Opioids
for Chronic Pain — United States, 2016. MMWR Recomm Rep 2016;65(No. RR-1):1–49. DOI:
http://dx.doi.org/10.15585/mmwr.rr6501e1"/>
      <url value="https://www.cdc.gov/mmwr/volumes/65/rr/rr6501e1.htm"/>
    </relatedArtifact>
    <relatedArtifact>
      <type value="citation"/>
      <label value="212"/>
      <citation value="Park TW, Saitz R, Ganoczy D, Ilgen MA, Bohnert AS. Benzodiazepine
prescribing patterns and deaths from drug overdose among US veterans receiving opioid
analgesics: case-cohort study. BMJ 2015;350:h2698."/>
      <url value="http://www.ncbi.nlm.nih.gov/pubmed/26063215"/>
    </relatedArtifact>
    <library value="http://fhir.org/guides/cdc/opioid-cds/Library/OpioidCDSREC11"/>
    <action>
      <title value="Existing patient has concurrent opioid and benzodiazepine prescriptions."/>
```


FIG. 9.6, CONT'D

(Continued)

```
    <description value="Checking if the trigger prescription meets the inclusion criteria for
  recommendation #11 workflow."/>
    <documentation>
      <type value="documentation"/>
      <display value="CDC guideline for prescribing opioids for chronic pain"/>
      <url
  value="https://www.cdc.gov/mmwr/volumes/65/rr/rr6501e1.htm?CDC_AA_refVal=https%3A%
  2F%2Fwww.cdc.gov%2Fmmwr%2Fvolumes%2F65%2Frr%2Frr6501e1er.htm"/>
    </documentation>
    <documentation>
      <type value="documentation"/>
      <document>
        <extension url="http://hl7.org/fhir/StructureDefinition/cqf-strengthOfRecommendation">
          <valueCodeableConcept>
            <coding>
              <system value="http://terminology.hl7.org/CodeSystem/recommendation-strength"/>
              <code value="strong"/>
              <display value="Strong"/>
            </coding>
          </valueCodeableConcept>
        </extension>
        <extension url="http://hl7.org/fhir/StructureDefinition/cqf-qualityOfEvidence">
          <valueCodeableConcept>
            <coding>
              <system value="http://terminology.hl7.org/CodeSystem/evidence-quality"/>
              <code value="low"/>
              <display value="Low quality"/>
            </coding>
          </valueCodeableConcept>
        </extension>
      </document>
    </documentation>
    <documentation>
      <type value="justification"/>
      <display value="Risk increase in overdose death rate with current benzodiazepine
  prescription in VHA opioid users 2004-2009"/>
      <resource value="https://gps.health/coka/resources/Evidence/117"/>
    </documentation>
    <trigger>
      <type value="named-event"/>
      <name value="order-select"/>
    </trigger>
    <condition>
      <kind value="applicability"/>
      <expression>
```


FIG. 9.6, CONT'D

(Continued)

```
      <description value="Check whether the existing patient is using opioids concurrently with
benzodiazepines."/>
      <language value="text/cql-identifier"/>
      <expression value="Is Recommendation Applicable?"/>
    </expression>
  </condition>
  <groupingBehavior value="visual-group"/>
  <selectionBehavior value="exactly-one"/>
  <dynamicValue>
    <path value="action.description"/>
    <expression>
      <language value="text/cql-identifier"/>
      <expression value="Get Detail"/>
    </expression>
  </dynamicValue>
  <dynamicValue>
    <path value="action.title"/>
    <expression>
      <language value="text/cql-identifier"/>
      <expression value="Get Summary"/>
    </expression>
  </dynamicValue>
  <dynamicValue>
    <path value="action.extension"/>
    <expression>
      <language value="text/cql-identifier"/>
      <expression value="Get Indicator"/>
    </expression>
  </dynamicValue>
  <action>
    <description value="Will revise"/>
  </action>
  <action>
    <description value="Benefits outweigh risks, snooze 3 months"/>
  </action>
  <action>
    <description value="N/A - see comment, snooze 3 months"/>
  </action>
 </action>
</PlanDefinition>
```

FIG. 9.6, CONT'D

currently underway and in various stages of completion to continue testing and implementation of decision support rules using FHIR, including a project sponsored by ONC and the Centers for Disease Control and Prevention (CDC) focusing on improving processes for the development of standardized, shareable, computable decision support artifacts using the CDC Opioid Prescribing Guideline as a model case [28].

9.7.4 CDS Connect

Beginning in 2016, the US Agency for Healthcare Research and Quality (AHRQ) began a project called CDS Connect [29] to enable the clinical decision support community to identify, define, and share clinical decision support artifacts. The project is both a community of decision support authors, implementers, and other stakeholders, as well as a set of tools and technologies to support authoring, testing, evaluation, and implementation of clinical decision support artifacts. In particular, the CDS Connect repository provides both web-based and API access to a catalog of clinical decision support content that is vetted and maintained by the community and made freely available for use by clinical teams. In addition, the CDS Connect Authoring tool provides an easy-to-use, web-based authoring environment for Clinical Quality Language-based decision support. An example of an application that makes use of CQL content developed with the CDS Connect Authoring tool is the Pain Management Summary [30]. In addition, many of the current CQF community projects are either already published in CDS Connect, or on track to be published there [31]. As a platform and community, CDS Connect provides a valuable resource for sharing decision support rules.

9.7.5 Adapting clinical guidelines

Another related effort that highlights the use of standards and processes to create decision support rules, in 2018, the Centers for Disease Control and Prevention (CDC) launched an initiative called Adapting Clinical Guidelines for the Digital Age [32], focused on getting the agency's evidence-based guidance quickly, accurately, and consistently into patient care. The initiative consists of five workstreams, each focused on a different aspect of the guideline development process. CDS in particular is addressed by the Informatics Stream. Work from this effort, combined with lessons learned from various projects throughout the CQF community were combined in and published as the FHIR Clinical Guidelines implementation guide [33], which defines an approach and methodology for representing and implementing Clinical Practice Guidelines (CPGs). Decision support rules are an important delivery mechanism for CPGs, and the patterns established by the CDS KAS and FHIR Clinical Reasoning module continue to be refined and reflected in this work.

In addition, the World Health Organization is using FHIR-based decision support to represent and distribute the computable aspects of SMART Guidelines content [34], such as the WHO Antenatal Care SMART Guidelines, a FHIR implementation guide that conforms with the FHIR Clinical Guidelines implementation guide, using FHIR and CQL to represent and distribute event-condition-action decision support rules.

9.8 Future work

With increasing emphasis on patient safety and prevention of medical errors, coupled with increasing use of electronic health records, demand for computer-based CDS will grow. Responding to this demand will require leveraging the considerable investment in the creation of clinical practice guidelines to adapt them for use in CDS systems. A parallel trend is the emphasis on interoperability of clinical information systems. These trends will prompt convergence on a standard for representing decision rules in general and clinical practice guidelines in particular in a computable format, one component of which will be a standard expression language that can be executed in many different CDS systems with a minimum of adaptation. This will facilitate knowledge sharing by reducing the cost of knowledge engineering, which in turn will foster compliance with clinical practice guidelines and other evidence-based medicine, leading to an improvement in patient safety and clinical outcomes.

9.9 Conclusions

A decision rule is a representation of deterministic reasoning in which branching logic is used in combination with data to reach conclusions regarding diagnosis, treatment and other important clinical goals. One way to represent a decision rule in a computable format is through the sequential execution and explicit flow of control of a procedure. Another approach that separates the clinical knowledge from the inferencing mechanism and other control processes is the production rule, in which the knowledge of a domain is represented by a collection of modular IF-THEN expressions. Efforts to incorporate the advantages of these two approaches as well as create a standard formalism that would facilitate knowledge sharing led to the development of the HL7 standard Arden Syntax. In this formalism, knowledge is represented as modular procedures known as medical logic modules, which also can be triggered independently like a production rule. However, challenges with this approach, including nonstandard data mappings and a possibly inadequate data model, coupled with the need to implement the declarative knowledge of clinical practice guidelines, have led to pursuit of other approaches. Expression languages, such as the HL7 standards GELLO and CQL, have been developed in order to extract data from clinical repositories and manipulate those data, using standard data models and vocabularies, and thus address the challenge of creating an overall formalism to represent computable clinical practice guidelines. Further work in standardizing data models led to the creation of the HL7 Virtual Medical Record, which offers the promise of facilitating knowledge sharing still more by allowing standardized references to clinical data that can be mapped to local data repositories. Work in this domain has steadily evolved, with the popular FHIR standard now integrated with standards to facilitate data linkages. This drive toward CDS standards has been further manifest by creation of a standard for representing quality indicators (HQMF and CQL) and in the Health e-Decisions work of the S&I Framework, the CDS Connect project and in the Adapting Clinical Guidelines for the Digital Age effort.

With all of these initiatives we still have not yet seen a convergence within the USA, or a convergence with approaches elsewhere. Indeed, this domain is still in a considerable state of flux, and will be important to resolve if progress in widespread sharing and use of CDS are to occur.

References

[1] Miller RA. Medical diagnostic decision support systems—past, present and future: a threaded bibliography and brief commentary. J Am Med Inform Assoc 1994;1:8–27.

[2] Jackson P. Production systems. In: Introduction to expert systems. Wokingham: Addison-Wesley; 1990. p. 135–91.

[3] Papamarkos G, Poulovassilis A, Wood PT. Event-condition-action rule languages for the semantic web. In: Proc workshop on semantic web and databases, 29th annual international conference on very large data bases (VLDB'03). San Francisco, CA: Morgan Kaufmann; 2003.
[4] Shortliffe EH. Computer-based medical consultations: MYCIN. Artificial intelligence series, New York: Elsevier; 1976.
[5] Carter JH. Design and implementation issues. In: Clinical decision support systems: Theory and practice. New York: Springer; 1998. p. 169–97.
[6] Haug PJ, Gardner RM, Tate KE, et al. Decision support in medicine: examples from the HELP system. Comput Biomed Res 1994;27(5):396–418.
[7] McDonald CJ, Overhage JM, Tierney WM, et al. The Regenstrief Medical Record System: a quarter century experience. Int J Med Inform 1999;54(3):225–53.
[8] Jung CY, Sward KA, Haug PJ. Executing medical logic modules expressed in ArdenML using Drools. J Am Med Inform Assoc 2012;19(4):533–6.
[9] Pryor TA, Hripcsak G. The Arden Syntax for medical logic modules. Int J Clin Monit Comput 1993;10:29–224.
[10] Jenders RA, Adlassnig KP, Fehre K, Haug P. Evolution of the Arden Syntax: key technical issues from the standards development organization perspective. Artif Intell Med 2018;92:10–4. 27773563.
[11] Health Level Seven International (HL7). HL7 is an international standards development organization certified by the American National Standards Institute (ANSI). HL7 focuses on standards related to health care computing, including CDSS standards such as the Arden Syntax, GELLO and the Reference Information Model. Copies of these standards may be obtained at http://www.hl7.org.
[12] Peleg M, Tu S, Bury J, et al. Comparing computer-interpretable guideline models: a case-study approach. J Am Med Inform Assoc 2003;10(1):52–68.
[13] Sordo M, Boxwala AA, Ogunyemi O, Greenes RA. Description and status update on GELLO: a proposed standardized object-oriented expression language for clinical decision support. Medinfo 2004;11(Pt 1):164–8.
[14] Jenders RA, Corman R, Dasgupta B. Making the standard more standard: a data and query model for knowledge representation in the Arden Syntax. AMIA Annu Symp Proc 2003;323–30.
[15] Kawamoto K, Del Fiol G, Strasberg HR, et al. Multi-national, multi-institutional analysis of clinical decision support data needs to inform development of the HL7 Virtual Medical Record standard. AMIA Annu Symp Proc 2010;2010:377–81.
[16] ONC. Information blocking. Office of the National Coordinator for Health IT; 2021. https://www.healthit.gov/curesrule/final-rule-policy/2015-edition-cures-update. [Accessed 25 July 2021].
[17] Steve P, Wes B. Heat wave: The U.S. is poised to catch FHIR in 2019., 2018, https://www.healthit.gov/buzz-blog/interoperability/heat-wave-the-u-s-is-poised-to-catch-fhir-in-2019. [Accessed 25 July 2021].
[18] HL7. Fast healthcare interoperability resources (FHIR): Versioning and maturity. Health Level 7 International; 2019. http://hl7.org/fhir/versions.html#maturity. [Accessed 25 July 2021].
[19] HL7. FHIR accelerator home. Health Level 7 International; 2021. https://confluence.hl7.org/display/FA. [Accessed 20 December 2021].
[20] Chronaki C, Jaffe C, Dolin B. eMeasures: a standard format for health quality measures. Stud Health Technol Inform 2011;169:989–91.
[21] ONC. CQF charter and members. Office of the National Coordinator for Health IT; 2016. https://oncprojecttracking.healthit.gov/wiki/display/TechLabSC/CQF+Charter+and+Members. [Accessed 24 July 2021].
[22] HL7. HL7 specification: Clinical quality common metadata conceptual model, release 1. Health Level 7 International; 2015. http://www.hl7.org/implement/standards/product_brief.cfm?product_id=391. [Accessed 24 July 2021].
[23] HL7. HL7 domain analysis model: Harmonization of health quality artifact reasoning and expression logic. Health Level 7 International; 2014. http://www.hl7.org/implement/standards/product_brief.cfm?product_id=359. [Accessed 24 July 2021].
[24] HL7. HL7 domain analysis model: Health quality improvement, release 1. Health Level 7 International; 2014. www.hl7.org/implement/standards/product_brief.cfm?product_id=378. [Accessed 24 July 2021].
[25] HL7. Clinical quality framework. Health Level 7 International; 2018. https://confluence.hl7.org/display/CQIWC/Clinical+Quality+Framework. [Accessed 24 July 2021].
[26] HL7. Clinical quality language. Health Level 7 International; 2014. https://cql.hl7.org. [Accessed 24 July 2021].

[27] HL7. FHIRPath. Health Level 7 International; 2014. http://hl7.org/fhirpath/. [Accessed 25 July 2021].
[28] Alphora. CDC opioid prescribing guideline., 2019, http://build.fhir.org/ig/cqframework/opioid-cds/. [Accessed 25 July 2021].
[29] AHRQ. CDS Connect. Agency for Healthcare Research and Quality; 2017. https://cds.ahrq.gov/cdsconnect. [Accessed 25 July 2021].
[30] AHRQ. Factors to consider in managing chronic pain: A pain management summary. Agency for Healthcare Research and Quality; 2018. https://cds.ahrq.gov/cdsconnect/artifact/factors-consider-managing-chronic-pain-pain-management-summary. [Accessed 25 July 2021].
[31] CQF. HL7 clinical quality framework., 2019, https://github.com/cqframework/clinical_quality_language/wiki/Community-Projects#content. [Accessed 25 July 2021].
[32] CDC. Adapting clinical guidelines for the digital age. Centers for Disease Control and Prevention; 2018. https://www.cdc.gov/ddphss/clinical-guidelines/index.html. [Accessed 25 July 2021].
[33] CDS. FHIR clinical guidelines. HL7 Clinical Decision Support Work Group; 2021. https://hl7.org/fhir/uv/cpg. [Accessed 25 July 2021].
[34] WHO. SMART guidelines. World Health Organization; 2021. https://www.who.int/teams/digital-health-and-innovation/smart-guidelines. [Accessed 25 July 2021].

CHAPTER

10

Guidelines and workflow models

Mor Peleg[a] *and Peter Haug*[b,c]

[a]Department of Information Systems, University of Haifa, Haifa, Israel [b]Department of Biomedical Informatics, University of Utah, Salt Lake City, UT, United States [c]Advance Decision Support Group, Intermountain Healthcare, Salt Lake City, UT, United States

10.1 Introduction

Health care organizations strive to provide quality care to patients while spending resources effectively and efficiently. To support this goal, two main approaches have been developing over the past 40 years. One approach focuses on assuring the quality of care by delivering decision support to care providers based on clinical knowledge; the other focuses on supporting management of routine medical actions, such as ordering of laboratory tests and medications and scheduling patient visits. Interestingly, both approaches follow a process-based view for modeling clinical decision-making processes and task management. However, there is clear evidence that these approaches, when used individually, do not address important features that should be considered during care processes, such as integration of decision support and action management, support of flexibility of the care process, context awareness, personalization of treatment for patients' state and preferences, consideration of organizational issues during the care process, and management of comorbidities. A joint effort has been carried out in recent years by the heterogeneous community to address these issues.

10.1.1 Increasing and standardizing quality of care via clinical practice guidelines

In the last three decades, we have witnessed a movement toward evidence-based medicine (see Chapter 8), which seeks to base medical practice on evidence-based studies (such as clinical trials), employ outcome measures, and perform clinical audits. The influential 1999 report of the Institute of Medicine (IOM), *To Err is Human*, bolstered this movement by setting an agenda for reducing medical errors and improving patient safety through the design of a safer health system [1]. The report recommended making greater use of evidence-based approaches to health care and incorporation of information technology. The 2001 IOM report,

Clinical Decision Support and Beyond
https://doi.org/10.1016/B978-0-323-91200-6.00008-5

Crossing the Quality Chasm: a New Health System for the 21st Century [2], suggested ways to make scientific evidence more useful and accessible to clinicians and patients, such as authoring and dissemination of clinical practice guidelines (CPG). Although clinical guidelines have been used in health care since at least the early 1970s, the emphasis in the current health care agenda on safer, evidence-based medical practice has brought about a resurgence of interest in them.

Clinical guidelines are systematically developed statements to assist practitioners with decision making about appropriate health care for specific clinical circumstances [3]. Clinical guidelines aim to reduce errors, unjustified practice variation and wasteful commitment of resources, and to encourage best practices and accountability in medicine. Clinical guidelines are typically created by medical experts or panels convened by specialty organizations, who review the relevant studies, perform meta-analysis by contrasting and combining results from different studies (see Chapter 8) and, using a consensus-based process, compile a set of evidence-based recommendations. Their focus may be on screening, diagnosis, management, treatment, or referral of patients with specific clinical conditions. The recommendations are typically written as narrative text and tables, which point back to background material and evidence, ranking the strength of clinical validity, and the strength with which recommendations should be followed according to the guideline authors.

Although guidelines aim to be based on evidence, their usefulness is often limited by the tendency to construct them in a way that reflects the principles of care rather than in a way that reflects the flow of actual patient encounters, and thus they are sometimes difficult to apply. In order to solve this problem, clinical guidelines are sometimes portrayed as algorithms (flowcharts) to more directly specify for providers the recommended sequences of steps of data gathering, decision making, and actions (i.e., process flow) during patient encounters. The algorithms are based on the guidelines, but when evidence is unavailable the gaps are completed by the algorithm designers based on expert opinion. Clinical algorithms present the logic and flow of a care process for a single patient and focus on a single medical condition (e.g., diabetes, hypertension). They are often written for a single health professional role (e.g., physician, nurse). They typically assume a standard clinical setting in which timely access to relevant services can be expected.

By representing guideline knowledge in a formalism that enables computer-based execution and supports automatic inference, patient-specific advice can be delivered at the point of care, thus having the potential to have a larger impact on clinician behavior [4]. Such formalisms are known as Computer-Interpretable Guideline (CIG) modeling methodologies, or CIG formalisms [5].

The CIG formalisms that have been proposed or developed vary in the degree and ways in which they address integration with the actual clinical workflow and healthcare information systems (see also Chapter 20, which discusses CDS implementations in several academic medical centers and national programs). Yet, in all CIG formalisms, the focus is to represent the clinical knowledge that specifies the decision logic and the ordering of clinical actions that helps a healthcare professional to manage a single patient.

When healthcare organizations implement clinical guidelines as CIGs, they use the guidelines as templates in order to adapt them to their local settings and regulations, creating a *care pathway* [6]. Although they can be derived from guidelines, these care pathways basically differ from the typical guideline interpretation in that they utilize the resources available in the institution including relevant members of a multidisciplinary team and the diagnostic and therapeutic resources local to the care setting. Key goals generally include both the standardization of processes surrounding care and the implementation of the best clinical science as applied to a specific patient problem. The best CIG implementations reflect the different roles play by different caregivers, the resources available in a specific care setting, and potential transitions of care faced by the patient during any extended medical intervention.

10.1.2 Supporting management of routine medical actions via workflow systems

By their nature, CIG implementations invite the use of process (or workflow) management systems. Many definitions of workflow and workflow management systems exist in the literature. Unertl et al. [7] analyzed definitions of workflow in 127 papers and developed a conceptual framework of workflow-related terms. This framework features two levels. The *specific* level of their model is composed of the people performing actions (actors), the physical and virtual tools the actors are using (artifacts), specific details of the actions being performed (actions), characteristics that describe the actions (characteristics) and the end products of the actions (outcomes). The *pervasive* level consists of context, temporal factors, and aggregate factors. Context of physical workspace, virtual workspace, and organizational factors constrains and enables workflow. Temporality involves scheduling, temporal rhythms, and coordination of events. Aggregate factors are the relationship and interaction among different tasks and actors, including elements of coordination, cooperation, and conflict.

In his book *Workflow Management: Models, Methods and Systems* [8], van der Aalst explains that the main purpose of a workflow management system is the support of the definition, execution, registration and control of processes. Business process models convey the organizational knowledge about the different roles and departments in the organization, the tasks (actions) done by each participant, and the sequence of tasks that formulate an overall business process. Business processes (also referred to as workflows) are often scoped from the point of view of a customer, which in our case is a patient. In the health care domain, the workflow specifies the recommended sequence which a patient goes through during a visit in the health care institution. It is also possible to represent more complex workflow patterns, like split-joins, synchronizations, etc., which contributes to a higher expressivity of the manner in which the real workflow should be carried out. For example, the patient first checks in, then his vital signs are measured by a nurse, then a parallel set of activities can be carried out (e.g., two different laboratory tests of blood and urine), a physician sees him and uses the previous laboratory results to propose a diagnosis, he may be hospitalized, moved between wards, and finally released from a hospital.

To truly automate workflows that participate in patient care, the medical knowledge necessary for clinical decision making is required. This knowledge is care-provider oriented, and often captured in clinical guidelines and algorithms representing the principals of evidence-based care. Thus, workflows can be used to streamline and standardize routine tasks such as scheduling visits, ordering laboratory tests and medications, etc., while improving the collaboration and coordination of care among different organizational roles, and while using resources efficiently. They assist in the delivery of care as defined in evidence-based guidelines and clinical studies.

The definition and implementation of business processes is not unique to medicine; efforts in other industries date to the 1990s and earlier. The health care sector is one of the late adopters of business process management (BPM) technology [9]. One of the main reasons for the resistance that health care professionals have to the use of BPM technology has been the lack of flexibility seen in most workflow-based systems. Although production plants follow very routine processes, health care processes have numerous exceptions, and, due to the most important aim of patient safety, inflexible standard work processes often cannot be followed. The challenge of maintaining process adaptability and flexibility while employing process management methods has been the focus of much research in the BPM and Process Mining communities. The communities of researchers who target process support in the health care domain were originally known as the "ProHealth community" (https://sites.google.com/view/kr4hcprohealth2019/previous-events). This group's yearly meetings have been superseded by the International Workshop on Process-Oriented Data-Science for Healthcare, a collection of researchers with a focus on process mining (https://pm4health.com/4th-international-workshop-on-process-oriented-data-science-for-healthcare/). More recently, members of the Object Management Group (OMG) (https://www.omg.org/), an industry-sponsored standards organization, have developed a community with the goal of repurposing OMG's business process modeling standards for use in healthcare. Below, we further describe these standards and provide an example of their use in a workflow for diagnosing and treating pulmonary embolism.

Process management continues to be an area of active research. In 2020, De Ramón Fernández et al. [10] published a systematic literature review focused on Business Process Management (BPM) used to optimize clinical processes. They observed that these efforts are part of an ongoing exploration of techniques borrowed from the business community to improve quality and optimize efficiency in care processes. Other similar approaches used to improve clinical processes included Lean Manufacturing and Six Sigma. Lean is a production philosophy that focuses on redesigning processes to eliminate waste. It attempts to refine processes in accordance with scientific methods and targets those activities that most directly affect the product itself. Six Sigma focuses on minimizing process variability. The label, "Six Sigma", derives from the goal of achieving processes with a 99.99963% statistical probability of creating a product free of defects. The International Organization for Standardization (ISO) publishes a standard for methodologies used in Six Sigma process improvement (https://www.iso.org/standard/52901.html).

BPM directly addresses the modeling tools used to achieve process improvement. It emphasizes the use of information technologies as a key component in the continuous improvement of business processes. When fully realized, the BPM approach involves a process lifecycle beginning with process design and passing through the stages of modeling, execution, monitoring, and optimization.

In their review, De Ramón Fernández et al. focused on publications describing BPM methodologies applied in a clinical field with quantitative or qualitative outcome data. Most of the relevant articles explored redesign or standardization of processes using Business Process Management and Notation (BPMN) a standards-based authoring system developed by the OMG. The implementation of these processes led to partial or complete automation and typically provided the data needed to monitor the process and quantify impact through Key Process Indicators (KPIs). The study's findings suggested that "BPM methodology represents a novel approach for the field of health, with a very positive impact on the management and optimization of clinical processes." However, the authors also describe major challenges that hinder the implementation of BPM applications designed to standardize and optimize clinical processes. These include the orientation of typical hospital management teams toward functions and not toward processes, the unwillingness of some clinical roles to accept changes, the lack of trust needed to delegate tasks, the lack of communication between parts of the intended workflow, and others.

BPMN continues to be a focus of interest for developing medical CIGs. It is one of a collection of OMG standards created to support process management reaching across many industries. Its relevance in healthcare is being actively explored. Its history reflects the needs, felt in a variety of industries, to describe modeling paradigms that can be used to streamline and standardize key business processes. It is built upon experience with two earlier, industry-sponsored, approaches to workflow modeling.

10.1.3 The life-cycle of health care process and decision support systems

In this chapter, we follow a life-cycle approach, reviewing the contributions of the two communities addressing health care process and decision support, by considering the stage of the life-cycle that each achievement addresses: including knowledge acquisition, knowledge (process) modeling and specification, integration of guideline and workflow knowledge, validation and verification, enactment, process mining and improvement, and sharing of executable guideline knowledge (Fig. 10.1). Our approach of viewing the

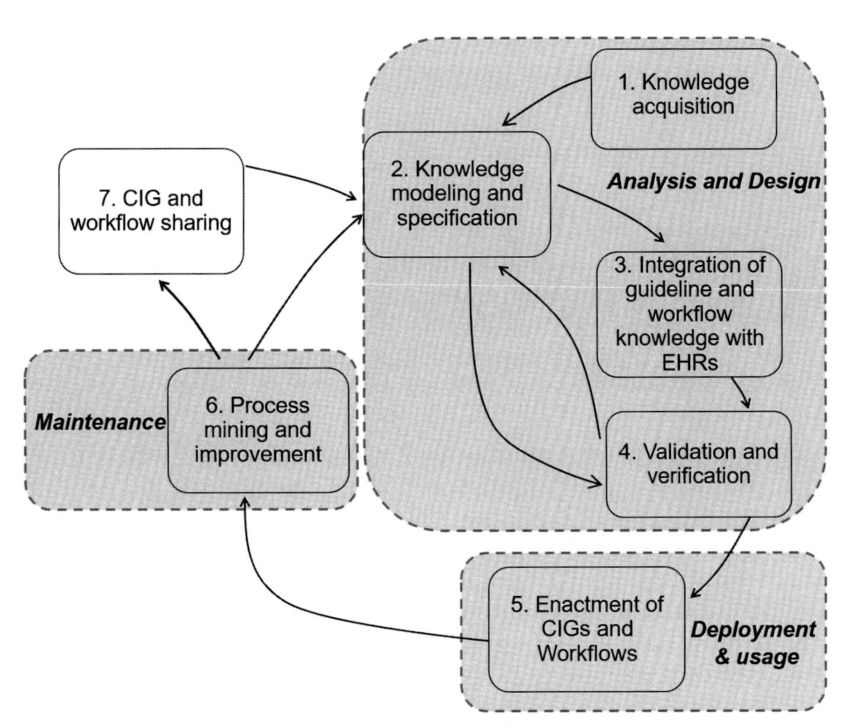

FIG. 10.1 The lifecycle of health care process and decision support systems.

achievements in terms of the stages of life-cycle support agrees and complements that of Gooch et al. [11] who provide an excellent review of implementation challenges for process-oriented health information systems and develop a conceptual model for implementing process-oriented health information systems. Their model presents the development of a process-oriented hospital information system as an iterative collaborative process that involves defining a clinical process model comprising formalized medical knowledge and organizational workflow. The generic model needs to be adapted to the local institutional setting. Before its execution, the process model needs to be verified, and strategies for helping the organization cope with the organizational changes required for implementing the process-oriented system need to be developed.

10.2 Supporting knowledge acquisition

The starting points for clinical-guideline-based decision support systems are typically narrative clinical guidelines. Unlike clinical trial protocols, which constrain clinical practice to clearly defined steps, narrative guideline documents contain a recommendation set that suggests options for optimal care. Because the nature of clinical guidelines is to suggest rather than impose a strict procedure for care, they are often written in a relaxed language that emphasizes the fact that the judgment of the clinician should determine the care process. However, the relaxed language used in narrative guidelines is not formal enough for computer processing, and the knowledge presented in a narrative guideline is thus often unclear, vague, incomplete, ambiguous, and even contradictory, which creates a problem in interpreting the guideline in order to computerize it.

10.2.1 The quality of narrative guidelines

Many approaches have been developed to improve the quality of narrative guidelines. Some approaches concentrate on the methodological quality of guideline development, that is, the nature of evidence and methodologies for aggregating research results of different studies that differ in patient population and settings, rather than on structuring the representation of guideline knowledge. For example, the 1992 IOM report *Guidelines for Clinical Practice: From development to use* [12] suggests eight attributes for assessing guideline quality. Four attributes relate to guideline content: validity, reliability and reproducibility, clinical applicability, and clinical flexibility. The other attributes relate to the process of guideline development or representation: clarity, multidisciplinary process, scheduled review, and documentation. The Australian Health Information Council (AHIC) suggests criteria that should be confirmed to ensure that a narrative guideline is reliable and valid (https://www.clinicalguidelines.gov.au/portal). These criteria include the validity of the knowledge source, which depends on systematic review of evidence and rating of levels of evidence, and the currency of the guideline (i.e., that the guideline is up to date). A variety of guideline assessment tools have been published, such as the Appraisal of Guidelines Research & Evaluation (AGREE) instrument (https://www.agreetrust.org/resource-centre/agree-ii/). These tools evaluate guidelines according to desirable attributes that can be mapped to the IOM attributes.

The reports of the IOM and the AHIC do not provide precise schemas for representing algorithmic guideline knowledge and for implementing guidelines. The GuideLine Implementability Appraisal (GLIA) (http://gem.med.yale.edu/glia) and its extensions complement the guideline quality appraisal instruments and addresses potential difficulties in implementation, including algorithmic control flow.

10.2.2 The types of knowledge contained in narrative guidelines

The Guidelines Elements Model (GEM) [13] is an XML-based knowledge model for guideline documents. GEM's 110 element types relate to a guideline's identity, developer, purpose, intended audience, method of development, target population, knowledge components, testing, and review plan. *Knowledge* components in guideline documents include tags for marking names of terms and their definitions and are used to structure guideline recommendations as conditional recommendations (decision rules) and imperative recommendations (clinical actions). Recommendations can be sequenced using a link element to represent guidelines that unfold over time. GEM is a standard of the American Society for Testing and Materials (ASTM) and is supported by many tools, available at the GEM Website (http://gem.med.yale.edu/default.htm), which include GEM-Cutter for marking up guidelines according to GEM elements, GEM-Q, GEM-COGS, and BRIDGE-Wiz, for assessing and improving the quality of marked-up guidelines, Extractor for review of recommendations, and GEM-Arden for translation of guidelines marked up in GEM into medical logic modules.

Another classification of guideline knowledge is evident in the constructs of CIG languages, discussed in Section 10.3. Such formalisms include constructs for decisions, goals, actions, and definitions used for data abstraction and interpretation.

10.2.3 From narrative to formal representations of guidelines

Narrative guidelines are written in a form that makes it extremely difficult to automatically convert them into their formal representation as CIGs. Several approaches have been developed to facilitate the transition from narrative to formal representations. These approaches, like GEM described above, mark up narrative text to indicate points at which phrases correspond to certain structural components of guidelines, according to markup ontologies. Structuring the narrative document is a step toward creating a computable implementation and can be used to link a formal representation to the narrative text. In addition, the mark-up process often can identify ambiguity that needs to be resolved, as well as areas where evidence is lacking and recommendations are not provided.

Section 4.2 of Ref. [5] provides a detailed review of how different methods support transformation of CPGs into CIGs, including (a) cognitive methods, (b) modeling methods and tools for gradual manual translation of CPGs into CIGs, and (c) semi-automated information extraction methodologies and tools for translation of CPGs into CIGs.

Who should be using the markup tools to transform narrative guidelines into computable specifications? A cognitive study has shown that different knowledge engineers/algorithm authors create dissimilar clinical algorithms using the same narrative clinical guideline as a starting point, depending in part on their degree of prior experience and knowledge of the domain [14]. In that study, the authors found that physicians who created algorithms tended to add organization and detail that were based on their knowledge, and which was not explicitly contained in the narrative guideline, while computer scientists tended to produce more consistent algorithms, but which reflected more literal interpretations of the narrative text. The algorithms of highest quality were created by teams involving clinicians and computer scientists. These results are in agreement with different methodologies for creating CIGs based on narrative guidelines, like the one of Shalom et al. [15] or the one of Peleg et al. [16]. According to Shalom et al., clinical experts in the guideline's domain and knowledge engineers first create an Ontology-Specific Consensus (OSC) regarding the semantics of the guideline to achieve shared understanding. Then the clinical editors, who need not be experts in the medical domain (e.g., residents or even interns), but have more knowledge in the use of the knowledge-specification tool, can mark up the clinical guideline, followed by gradual conversion into a formal representation by knowledge engineers. An evaluation study of the proposed methodology, using as a case study the Asbru language, discussed later in Section 10.3, and a markup tool, showed that, given an OSC, clinical editors with mark-up training can structure guideline knowledge with a high degree of completeness, where the main demand for correct structuring is training in the ontology's semantics.

GESHER and DELT/A are markup tools that have been used to gradually convert narrative guidelines into XML-based CIG models. GESHER and MEIDA are part of the Digital electronic Guideline Library framework (DeGeL) [17].The GESHER client-based module [17] assists modelers in marking up text, decomposing the actions embodied in the guideline into atomic actions and other sub-guidelines, and to define the control structure relating them (e.g., sequential, parallel, repeated application), based on the Asbru CIG language, including temporal patterns. Another tool, MEIDA, is used to search for vocabulary terms in controlled vocabularies and embed them in the guideline document.

The Document Exploration and Linking Tool (DELT/A) (http://ieg.ifs.tuwien.ac.at/projects/delta/) supports the transformation of narrative guidelines (in HTML format) into CIGs (in XML format) through links and macros. Links are used to show and connect related parts in HTML and XML markup. Macros combine multiple XML elements together with their attributes and can be used for simple construction of new XML documents. These macros are typical patterns of clinical guideline components (e.g., two mutually exclusive plans), which ease the implementation process.

Another approach for aiding the translation of narrative guidelines into CIGs involves the use of computer-interpretable design templates for representing guideline knowledge using clinical abstractions that are appropriate for particular guideline sub-domains. Peleg and Tu [18] developed such templates for screening and immunization guidelines and demonstrated their executability.

In the Protocure-II project, linguistic patterns have been developed, which can be used to formally represent the knowledge about medical actions contained in text [19]. Similar patterns are used in the LASSIE tool [20] to extract from narrative guidelines clinical actions that can be marked up using the DELT/A tool. Moreover, LASSIE supports "living guidelines" by automatically generating the mark-up and formalization of parts of the guideline and its CIG version that did not change from the previous version and highlighting the information extracted using information extraction methods from the new parts of the guideline.

Mulyar et al. [21] examined whether different CIG modeling languages support control-flow-based workflow patterns. The analysis shows that CIG languages support up to 22 of 43 different control-flow workflow patterns observed in industrial process models. It remains to be seen whether the additional patterns would be useful templates for defining the control flow of clinical processes.

Aigner, Kaizer, and Miksch describe information visualization techniques that could be used during different stages of the CIG lifecycle [22]. The visualization engines of DeGeL, AsbruView, PROforma (see Section 10.6), etc., present CIGs as process models. Other tools focus on presenting patients' data in multidimensional information space, in order to provide integrated views to create a comprehensive picture of a patient's status and its evolution over time. As reviewed in Ref. [22], visualization includes presentation of data over a time line (e.g., Gravi++, Interactive Parallel Bar Charts) and context-sensitive (i.e., guideline-specific) temporal abstraction of the data of individual or multiple patients (KNAVE-II, VISITORS [23], VIE-VENT). Finally, tools such as Tallis Tester, CareVis, and Guideline Overview Tool present patients' data in connection with clinical guidelines.

10.2.4 From narrative to formal representations of clinical workflows (an example)

Many examples exist illustrating the conversion of narrative clinical workflow descriptions to CIGs. Here, we illustrate one approach targeting CIGs modeled with the OMG business process modeling standards described above and further elaborated in Section 10.3.3. Intermountain Healthcare, a large health care provider in Utah, has created a procedure to develop clinical guidelines. These are termed Care Process Models (CPM). It has created over one hundred CPMs with the goal of defining the standard of care for various conditions. These are publicly available and consist of narrative descriptions of detailed care guidelines supported by graphical depictions of the associated workflow. In Fig. 10.2, we provide a snapshot of the title of the Venous Thromboembolism CPM (https://intermountainhealthcare.org/ckr-ext/Dcmnt?ncid=529597630)[a] and an image of one of several workflows provided as a simple graphical representation.

A number of specific illnesses are associated with venous thromboembolism including pulmonary embolism, the migration of blood clots from the lower extremities to the lungs. A key step in the workup of pulmonary embolism is to calculate the Revised Geneva Score (RGS, see Fig. 10.3). A simple, BPMN representation of the process that leads to the calculation of the revised Geneva score is presented in Fig. 10.4.

Section 10.3.3 elaborates the standard components of BPMN. Several standard components of a BPMN workflow are visible in Fig. 10.4. The activities decorated with small gears are *service tasks* designed to read data from an EHR. In a modern implementation of this process, standardized service calls would be used to enhance the interoperability of the application.

[a]Used with permission from Intermountain Healthcare.

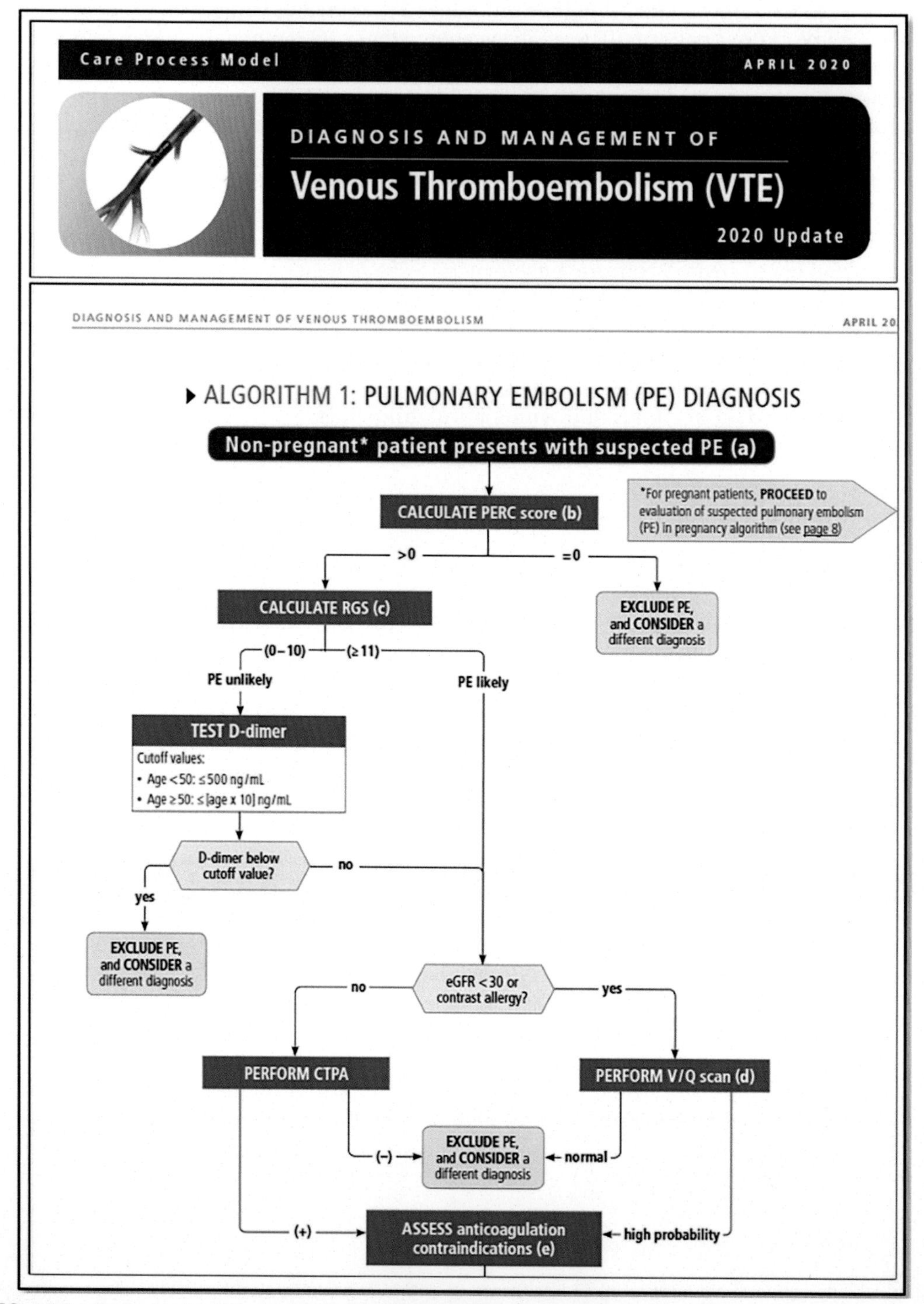

FIG. 10.2 Example of a clinical guideline developed at Intermountain Healthcare. These are called "Care Process Models" and include a narrative portion and algorithms depicted graphically. Care Process Models are developed by teams of clinicians and consist of detailed descriptions of care rendered in narrative and graphical process descriptions.

(c) Revised Geneva Score (RGS)	
Factor	**Points**
□ Age >65 years	1
□ Hemoptysis	2
□ Active malignant condition	2
□ Surgery or fracture within 1 month	2
□ Unilateral lower limb pain	3
□ Previous PE or DVT	3
□ Pain on lower-limb deep venous palpation and unilateral edema	4
□ Heart rate 75–94 beats/minute	3
□ Heart rate ≥95 beats/minute	5
ADD total points	□

FIG. 10.3 The scoring algorithm for the Revised Geneva Score.

These service calls are based on standards developed by Health Level 7 (HL7, http://www.hl7.org/) and follow the Fast Healthcare Interoperability Resource (FHIR, https://www.hl7.org/fhir/overview.html) standard (see Chapter 18). Each service call will return the specified data element if it is available in the health care computing system.

Following the arrows that indicate the process flow, the next step is a *human task* activity, decorated with a small human icon. In this case, a clinician will review the data collected, enter any missing data, and correct any erroneous data. The subsequent activity, decorated with a small table, is a *business rule task*, which performs the calculation. it invokes a service that implements the logic needed to direct workflow when two or more options are available. In this case, the RGS calculation determines whether the next step is a blood test (a low value from the D-Dimer test is effective in ruling out pulmonary embolism) or a *subprocess task* (decorated with a +) which will implement an embedded process leading to the ordering, execution of, and interpretation of the imaging exam needed to make the final diagnosis.

The service implementing the *business rule task* can be constructed using any of a variety of languages. However, the DMN specification, described above, provides a standards-based tool to express this logic. In this case, the model consists of a Decision Requirements Graph (Fig. 10.5) and two linked Decision Tables (Fig. 10.6). The DMN 1.3 standard specifies a procedure for converting this executable logic model into a service.

As with the other OMG standards, the Decision Modeling and Notation standard is focused on providing a language that will allow content experts to share their knowledge with process designers. Graphical representations of the structure of sets of decisions are combined with table-based logic to provide a shared lexicon.

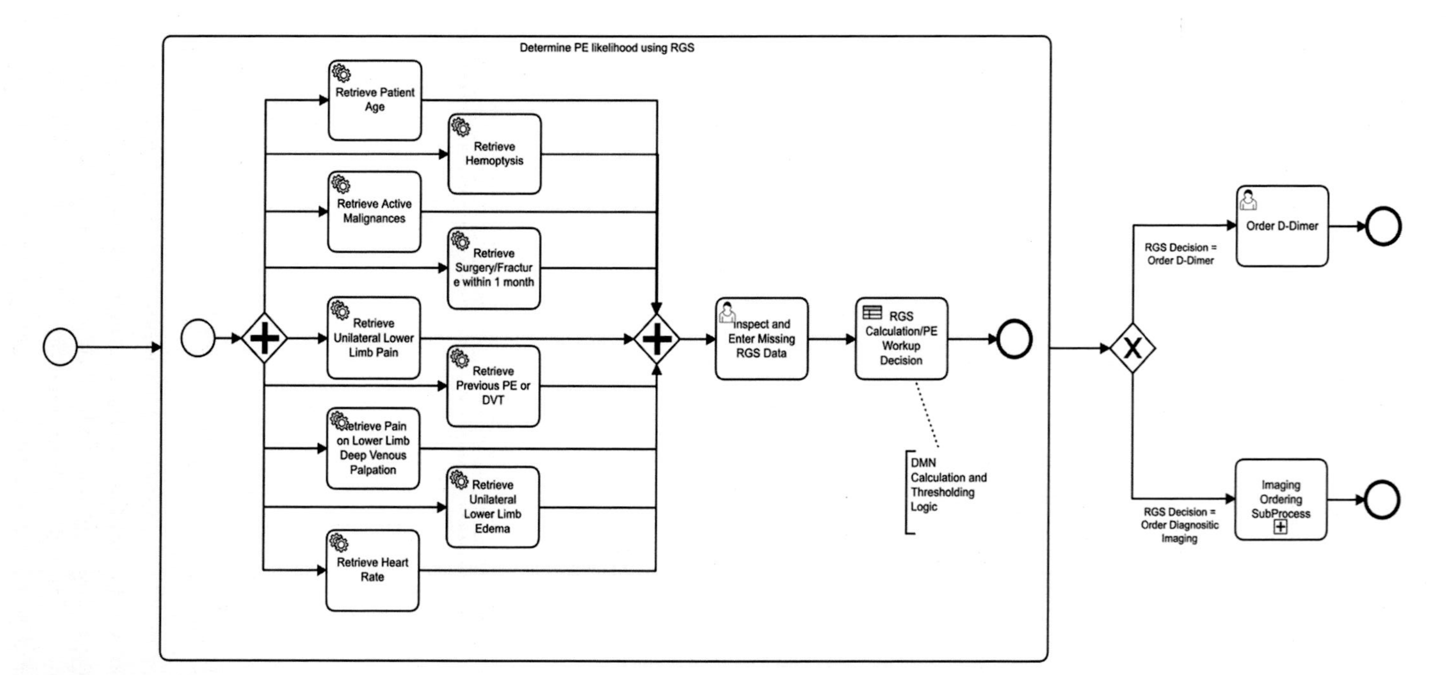

FIG. 10.4 BPMN process diagram for depicting the activities needed to calculate and interpret the RGS during the workup of a patient for pulmonary embolism. The workflow shown is encapsulated in a subprocess allowing it to be collapsed or expanded to visualize different parts of the larger model.

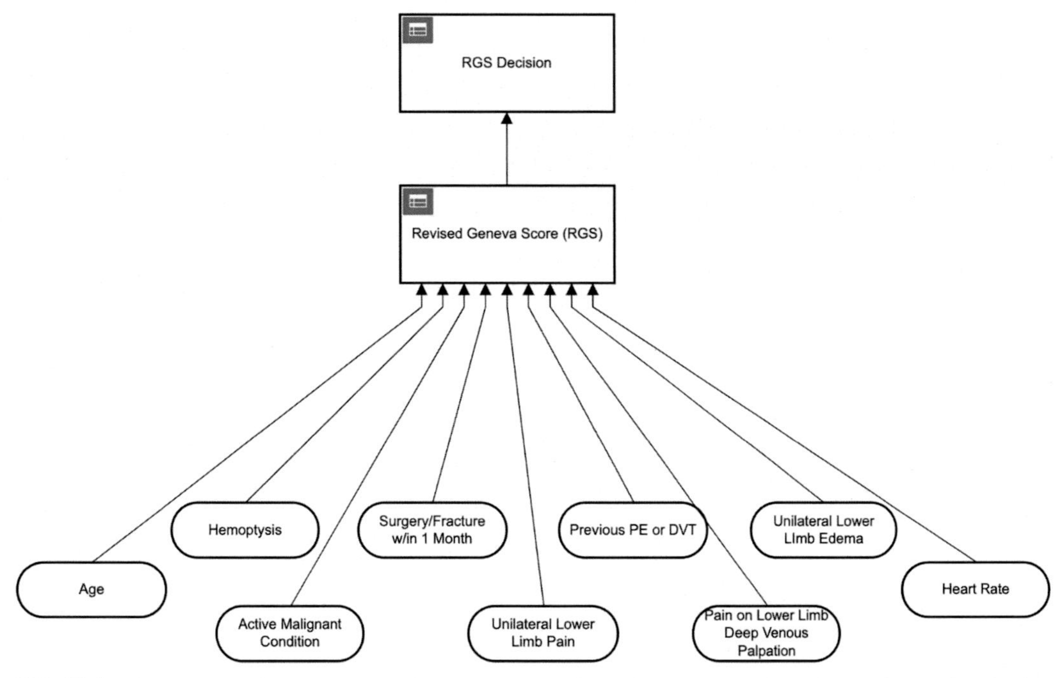

FIG. 10.5 An example of a Decision Requirements Graph. This artifact from the DMN standard identifies the data necessary for a decision and the logical constructs needed. In this example the logic for calculating a Revised Geneva Score will be provided in a decision table and an interpretation of the score will be provided in a second decision table. These are shown in Fig. 10.6.

Perhaps the most encouraging aspect of these OMG standards is their general availability. Their uptake in a variety of industries has led to a proliferation of useful tools with which to develop computable workflows and to execute these complex processes. This provides CIG developers in health care a broad choice among useful applications for authoring and implementing executable processes and the logic that directs them.

10.3 Formal methods for modeling and specifying CIGs

Specifying guideline knowledge formally as CIGs allows computer-based execution. During creation of the formal representation, ambiguities are removed, and areas are identified in which evidence is missing or for which no recommendations are given. Medical organizations that are adapting the recommendations into a care-flow process may fill in the gap with their opinions or leave the decision up to the end user. Many formalisms exist for specifying CIGs, each with its own motivations and features. Several papers have reviewed and compared formal methods for CIG specification [4,5,24,25]. In this section, we describe several well-known influential approaches from the research literature for formally representing guidelines as CIGs.

RGS Decision
tPEDecision
"Order D-Dimer","Oder Diagnostic Imaging"

	Inputs	**Outputs**	**ANNOTATIONS**
U	**Revised Geneva Score**	**RGS Decision**	**Description**
	Number	tPEDecision	
1	<11	"Order D-Dimer"	PE Unlikely with Low/ Intermediate RGS
2	>=11	"Order Diagnostic Imaging"	PE is Likely with High RGS

Revised Geneva Score
Number

	Inputs									**Outputs**	**ANNOTATIONS**
C+	**Age**	**Hemoptysis**	**Active Malignant Condition**	**Surgery/Fracture w/in 1 month**	**Unilateral Lower Limb Pain**	**Previous PE or DVT**	**Unilateral Lower Limb Edema**	**Pain on Lower Limb Deep Venous Palpation**	**Heart Rate**	**Revised Geneva Scorre**	**Description**
	tAge [0..120]	tPresence	tPresence	tPresence	tPresence	tPresence	tPresence	tPresence	tHeartrate	Number	
1	>65	-	-	-	-	-	-	-	-	1	Age > 65 worth 1 point
2	-	true	-	-	-	-	-	-	-	2	Hemoptysis Present
3	-	-	true	-	-	-	-	-	-	2	Malignant Condition Present
4	-	-	-	true	-	-	-	-	-	2	Surgery or Fracture Presetn within the Last Month
5	-	-	-	-	true	-	-	-	-	3	Lower Limb Pain Present
6	-	-	-	-	-	true	-	-	-	3	History of PE or DVT Present
7	-	-	-	-	-	-	true	true	-	4	Lower Limb Tendernes and Edema Present
8	-	-	-	-	-	-	-	-	[75..94]	3	Moderatly Elevated Heart Rate
9	-	-	-	-	-	-	-	-	>=95	5	Significantly Elevated Heart Rate

FIG. 10.6 Two examples of decision tables from the DMN logic for the Revised Geneva Score. The lower example calculates the score. Each row represents a rule; when a rule calculates true, the value in the Revised Geneva Score column is added to the overall score. The upper example uses the results of the lower example to categorize the status of the pulmonary embolism work up. It provides the result used by the BPMN in Fig. 10.4 to determine whether to continue the work up with a laboratory examination or an imaging examination.

10.3.1 Task-network models for CIGs

Many of the approaches have in common a process-flow-like model termed *Task-Network Model* (TNM) [24] – a hierarchical decomposition of guidelines into networks of component tasks that unfold over time (see Fig. 10.7). The task types vary in different TNMs, yet all of them support modeling of medical actions, decisions, and nested tasks. Following is a short review of some well-known TNMs that are currently in use, which highlights the distinguishing features of each methodology. The TNMs standards have influenced each other over the years in which they had been developed and maintained. Elkin et al. [26]

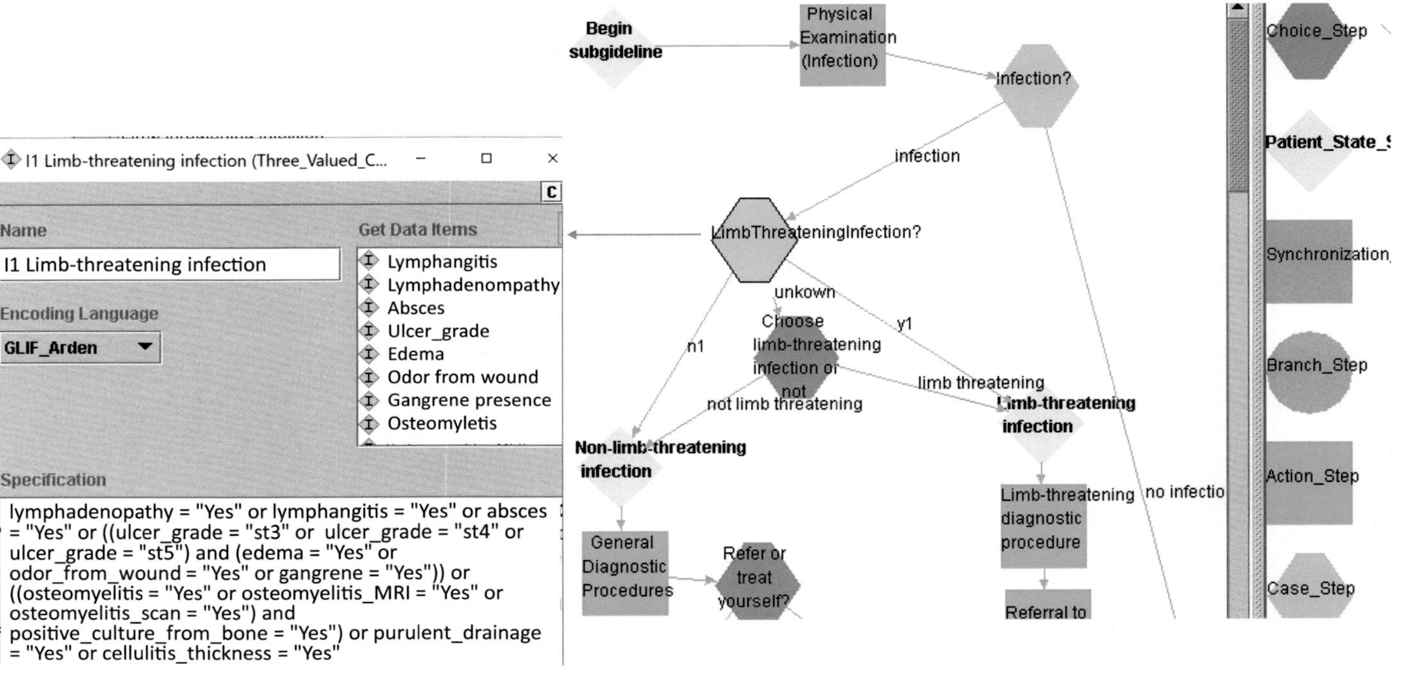

FIG. 10.7 Part of a diabetes foot management algorithm, dealing with infection (infection subguideline), encoded in GLIF3. The inset shows a formal specification of the decision criterion of the case step Limb Threatening Infection.

```
decision :: Additional_drug_choice ;
 caption :: 'Additional drug choice' ;
 choice_mode :: single ;
 support_mode :: symbolic ;

candidate :: ISA_beta_blocker ;
 argument :: excluding, ( Asthma = Yes or COPD = Yes ) ;
 argument :: excluding, ( STD_Heart_block = Yes ) ;
 argument :: excluding, ( Current_Rx = Ca_blocker_non_DHP ) ;
 argument :: for, ( Current_Rx = Thiazide_diuretic ) ;
 argument :: for, ( Current_Rx = Ca_blocker_DHP_long or Current_Rx =
Ca_blocker_DHP_short ) ;
 argument :: excluding, ( Current_Rx = Beta_blocker_ISA or Current_Rx =
Beta_blocker_non_ISA ) ;
 argument :: against, ( Type_1_Diabetes = Yesand Proteinuria <> None ) ;
 recommendation :: Netsupport( Additional_drug_choice, ISA_beta_blocker )
>= 1 ;
...
end decision .
```

FIG. 10.8 Part of the argumentation rule-set for selecting a second antihypertensive drug, in PROforma syntax (full example provided in http://www.openclinical.org/docs/ext/cigs/comparison/Hypertension_model_PROforma.txt).

presents the order in which these TNM formalisms have been initiated, starting with EON, followed by Asbru, PROforma, GLIF, and GLARE.

EON [27] and **GLIF** [28] have strongly influenced each other. In addition to including the generic tasks used by all TNMs, they use scenarios – partial specification of patient states allowing classification of a patient into an appropriate state within a CIG. **EON** uses a task-based approach to define decision support services that can be implemented using alternative techniques [27]. The decision-making task is supported by two classes of decision steps: simple if-then-else constructs and rule-in and rule-out criteria (that correspond to argumentation rules that confirm or refute a decision option, see Fig. 10.8)[b] as a way of setting qualitative preferences. Goals in EON are specified in a criteria language that uses patient data and abstractions based on classification hierarchies (e.g., disease hierarchies). Actions to be performed are represented as Management Diagrams (Activity Graphs) – networks of scenarios, actions, decisions, sub-guidelines, and branch and synchronization steps for modeling parallel paths. Data interpretation can be achieved using: (i) abstraction based on classification hierarchies, (ii) definition of terms referring to values of patient data items, and (iii) temporal abstractions. EON CIGs can be authored in Protégé-2000 (protege.stanford.edu) and executed by an execution engine that uses a temporal data mediator to support queries involving temporal abstractions and temporal relationships. A third component provides explanation services for other components.

Asbru [29] (see also http://www.asgaard.tuwien.ac.at/) represents guidelines as skeletal plans that can be hierarchically decomposed into (sub)plans or actions. The main emphasis is

[b] Argumentation rules originated in the PROforma formalism.

on representing not only the guideline's action prescriptions, but also the process and outcome intentions of the guideline and of its major subplans. Skeletal plans capture the essence of a procedure; when needed, they can go down to low-level action specification, but, when necessary, can leave enough room for execution-time flexibility in the achievement of particular intentions. Process intentions (for care-provider actions) and outcome intentions (for patient states) are specified as temporal patterns of actions or of external-world states that should be maintained, achieved, or avoided, during, or at the completion of a plan. The same temporal expression language is used for representing time-oriented actions, conditions, and intentions in a uniform fashion. The temporal expression language uses time annotations consisting of a time range (i.e., range of the start time, the end time, and the duration) and a time reference (i.e., point in time, or the time at which a plan changes state). Concepts can be abstracted from raw data using a domain-specific knowledge base, and may depend on context (e.g., pregnancy). Fig. 10.9 shows an example of an intention. Several tools support guideline authoring in Asbru: DELT/A (http://ieg.ifs.tuwien.ac.at/projects/delta/) and GESHER [17], mentioned in the previous section, which focus on easing the transition from narrative to formal representations via a mark-up stage, AsbruView (http://www.asgaard.tuwien.ac.at/tools/asbruview.html), which focuses on visualization and user interface for authoring, and CareVis (http://ieg.ifs.tuwien.ac.at/projects/carevis/), which provides multiple-coordinated views to cover different aspects of a complex underlying data structure of treatment plans and patient data. Asbru has been used in the implementation of CIGs as part of the MobiGuide project, mentioned in Section 10.4.

PRO*forma* [30] advocates the support of safe guideline-based decision support and patient management by combining logic programming and object-oriented modeling, and its syntax and semantics are formally defined. One aim of the PRO*forma* project is to explore the expressiveness of a deliberately minimal set of modeling constructs. PRO*forma* supports four tasks: actions, compound plans, decisions, and inquiries of patient data from a user. All tasks share attributes describing goals, control flow, preconditions, and post-conditions. An underlying premise is that the simple task ontology should make it easier to demonstrate soundness and to teach the language to encoders. PROforma's decisions are represented as argumentation rule-sets, where different candidate options are associated with arguments – conditions, which if true provide different degrees of support for that option: for, confirming, against, and excluding (see Fig. 10.8). This approach was later adopted by several CIG languages, including EON and GLIF3. A number of software components (e.g., Tallis (https://tallis.openclinical.net/), Deontics' Authoring Workbench and CDSS (https://deontics.com/products), and Arezzo (http://archive.cossac.org/proforma.html)) have been written to create, visualize, and enact PRO*forma* guidelines. A description of some of the projects may be found at http://archive.cossac.org/proformaInUse.html. A demonstrator system developed in collaboration with the American Association of Clinical Endocrinologists is presented in Ref. [31]. PRO*forma* is being used in the CAPABLE project mentioned in Section 10.9.

The Guideline Interchange Format version 3 (**GLIF**) [28] stresses the importance of sharing guidelines among different institutions and software systems, building on the most useful features of other CIG models, and incorporating standards. GLIF3 represents guidelines as clinical algorithms, similarly to EON's Management Diagrams. Its model of medical knowledge is used by action and decision steps to formally refer to patient data items, clinical concepts, and clinical knowledge. Patient data items are specified by a *medical concept*, the code for which is taken

```
<intentions>
 <intention type="intermediate-state" verb="maintain">
  <parameter-proposition parameter-name="blood-glucose">
   <value-description type="equal">
    <qualitative-constant value="HIGH"/>
   </value-description>
   <context>
    <context-ref name="GDM-Type-II"/>
   </context>
   <time-annotation>
    <time-range>
     <starting-shift>
      <earliest>
       <numerical-constant unit="week" value="24"/>
      </earliest>
      <latest>
       <numerical-constant unit="week" value="24"/>
      </latest>
     </starting-shift>
     <finishing-shift>
      <earliest>
       <qualitative-constant value="delivery"/>
      </earliest>
      <latest>
       <qualitative-constant value="delivery"/>
      </latest>
     </finishing-shift>
    </time-range>
    <time-point>
     <qualitative-constant value="conception"/>
    </time-point>
   </time-annotation>
  </parameter-proposition>
 </intention>
</intentions>
```

FIG. 10.9 A specification of an intention in Asbru: "In the context of GDM-Type-II, maintain blood glucose state at high level, starting week 24 and ending at delivery."

from a controlled clinical vocabulary and by a *data structure*, taken from a standard reference information model (RIM), such as the Observation, Medication, and Procedure classes of the Health Level 7 (HL7) RIM or the Observation, Medication_Statement, or Procedure resources of the HL7 FHIR (see Chapter 18). Clinical knowledge is expressed as relationships between medical concepts (e.g., contraindication relationships between a drug and a disease). GLIF3 has a formal language for expressing decision and eligibility criteria. This expression language was originally based on the Arden Syntax (see Chapter 12) and was replaced by an object-oriented language, called GELLO (https://wiki.hl7.org/index.php?title=Product_GELLO), which was accepted as an HL7 and ANSI standard (see Chapter 12), updated to Release 2 in 2010. Note that HL7's main current focus regarding an expression language is Clinical Quality Language (CQL https://cql.hl7.org/), which uses FHIR as a data model. GLIF3 is supported by two authoring and validation tools and two execution engines. A GLIF3 Editor and a GELLO Interpreter is currently maintained by the Australian company Medical Objects (https://www.medicalobjects.com/). Fig. 10.7 shows part of a GLIF3-CIG.

GLARE, see chapter by Terenziani et al. in Ref. [32], is an approach for modeling and executing clinical guidelines, which emphasizes management of temporal knowledge. The GLARE TNM has the following kinds of nodes: (i) atomic actions, which can be work actions, query actions, decisions, and conclusions, and (ii) composite actions, which are composed of actions that can be assembled in sequence, in parallel, iterated, or done in branching paths. In addition, temporal constraints can be defined between component actions. GLARE is supported by an authoring and validation tool, and by an execution engine, which supports temporal reasoning and a hypothetical reasoning facility that makes it possible to compare different paths in the guideline, by simulating what could happen if a certain choice was made. More recently META-GLARE [33] meta-system was developed which supports acquisition, representation and execution of CIGs in different formalisms.

10.3.2 Other CIG modeling methods

The Arden Syntax (see Chapter 12) is a standard of HL7 and ASTM suitable for representing individual decision rules in self-contained units called Medical Logic Modules (MLMs), which are usually implemented as event-driven single-step alerts or reminders. Arden Syntax was not meant to be used for encoding complex guidelines that involve multiple decisions or process flow sequences and there is no support to aid in human understanding of the way in which MLMs interact with one another; only if-then-else representation of decision rules is possible. Nonetheless, since rules can have action parts that invoke other rules, MLMs can be combined to represent computable guidelines.

Seroussi, Bouaud, and colleagues [34] proposed a method for representing guidelines halfway between formal knowledge representation and textual reading. Guideline knowledge is represented formally as decision trees. However, instead of automatically executing the decision tree, the user browses it as hypertext and flexibly interprets both patient data and guideline content, thus controlling the interpretation of the guideline knowledge in the specific context of a patient situation. The decision tree is built from clinical parameters that are identified in the guideline narrative and given labels chosen from standard classifications. All theoretically possible clinical situations are represented. This approach has been first applied to breast cancer therapeutic management with OncoDoc. Handling of chronic diseases, such as hypertension, involves considering the patient's therapeutic history (e.g., inadequate response to past treatment, or adverse effects, to select relevant patient-specific therapy among the recommendations). Therefore, the original knowledge base represented as a decision tree of clinical parameters was extended [34] by introducing a therapeutic level, structured along lines of therapy and levels of therapeutic intention. For each theoretical clinical situation, a range of pharmacological treatments is recommended. Matching patient's therapeutic history elements along with recommended therapies allows the system to rule out non-tolerated or non-efficient past treatments to finally select the best ones.

DESIREE [35] is the latest guideline-based DSS system by Seroussi, Bouaud, and colleagues for the breast cancer domain. DESIREE includes three types of decision-support capabilities: CIGs, an experience-base that allows learning from deviations from recommendations, and case-based reasoning. All three are integrated via an ontology formalized in Web Ontology Language (OWL). The ontology includes entities and properties related to the application domain, including lesion, tumor, side, histological type, and the e data model to represent clinical patient information, which is linked to an FHIR database. CIG knowledge is represented in Natural Rule Language (NRL). Notably, the THEN part of rules contains actions that build recommendations. A recommendation is composed of one or several orders, linked to clinical actions. Recommendations can be combined as refinements of each other or as complementary recommendations. Specific actions specify a level of evidence and level of expected conformance. The Rule Bases include generic rules describing common knowledge of the clinical domain and CPG-specific rules. Reasoning is provided via the EYE semantic reasoner that can handle ontology entity-property-value triples and which allows constructs like "not exist" or "for all". Rules are parsed against the ontology and are transformed into an intermediate XML representation similar to the Knowledge Artifacts specification proposed by HL7 organization that is independent of the source rule expression language (NRL) and of the rule execution language (N3 in our case). The XML format is then translated into N3 which is executed by the EYE reasoner to build the patient-specific recommendations.

10.3.3 Workflow formalisms

The BPMN standard has been widely adopted in the business world today and has been accepted by ISO as ISO/IEC 19510. Hence, the resulting BPMN is a modern standard for authoring complex, event-driven workflows. The BPMN graphical modeling environment supports workflow models that can be specified in greater or lesser detail. A fully specified process model can be executed via a standards-based execution engine to generate a computable workflow allowing the integration of tasks accomplished by humans with services provided by computers.

As a part of the growing effort to introduce formal, standards-based methods to healthcare, the OMG has sponsored a community of practice, the BPM+ Health Community (https://www.bpm-plus.org/). This group has authored and maintains the Field Guide to Shareable Clinical Pathways, as well as other resources (https://www.bpm-plus.org/healthcare-and-bpmn.htm). This publication introduces the medical informatics community to BPMN and other relevant standards, noting that they have proven their value in other industries. The community convenes regular meetings to encourage and demonstrate their use.

BPMN. A key component in the BPM process design phase is the use of a formal, graphical process description language which is used to describe the redesigned process. The original BPMN standard was focused exclusively on the design of process models; a model's largely graphical description would be turned over to software developers for implementation. However, industry groups in the OMG were interested in extending this graphical language to include an executable output that would simplify deployment of authored workflows. They subsequently extended BPMN to be fully executable. In BPMN version 2.0 (https://www.omg.org/spec/BPMN/2.0.1/), an adequately detailed BPMN model consists of both a graphical depiction of a business workflow and an XML-based executable workflow description that can be used to implement the process in a conformant execution engine.

Efforts to computerize processes have a long history in non-healthcare industries. In the 1990's, as computers became prevalent and businesses looked for ways to implement complex business processes, several industry groups attempted to standardize approaches to modeling business processes. Among these were the Workflow Management Coalition (WfMC) and the Business Process Management Institute (BPMI). In 1995, the WfMC published the Workflow Process Definition Language (WPDL). WPDL was soon to be re-expressed in XML and renamed the XML Process Definition Language (XPDL) (http://xpdl.org/). Its development represented an industry-driven standardization of a graphical modeling language for business processes.

XPDL. XPDL was the first widely recognized standard for workflow modeling and triggered a small number of initial efforts to explore the use of an industry sponsored process standard in medicine. In one such example, Huser V. et al. conducted initial testing of this workflow technology in healthcare [36]. They focused their efforts on a test of clinical decision logic against retrospective data and on the provision of a user-friendly representation of clinical logic. They reported success in testing their algorithms and indicated that the technology offered a user-friendly knowledge representation paradigm.

BPEL. The Business Process Management Institute (BPMI) was a second industrial group formed to develop a standard language for specifying executable business processes. The participants created the Web Services-Business Process Execution Language (WS-BPEL or BPEL) with the goal of defining execution semantics for business workflows. BPEL was adopted successfully by a number of large venders, notably IBM and Oracle. In 2005, BPMI merged with the OMG and created a task force to develop a next-generation business process management standard. The result was the Business Process Model and Notation (BPMN) standard released in 2006. Subsequently, the OMG iterated on this model, ultimately creating BPMN, version 2.0 which specified in detail both a graphical authoring environment for stateful processes and the execution semantics necessary to automate these processes. While both BPMN and BPEL provide an executable output, and there exists a translation from BPMN to BPEL,[c] there are constructs in BPMN that are difficult to simulate in BPEL (e.g., loops).

DMN. The success of BPMN in modeling and implementing workflows has led to a growing recognition of its deficiencies. The result has been an effort in the OMG to develop complementary standards designed to mitigate extend its usefulness. The first extension dealt with an ambiguous activity subtype in BPMN, the Business Rule Task. This activity is intended to invoke a business rules engine when complex logic is required to determine which path through the workflow to follow. In 2015, the OMG released a specification for Decision Modeling and Notation (DMN, https://www.omg.org/dmn/). This specification defines a modeling language for decision management and business rules. The standard combines a graphical authoring environment with logic specified in tables and a user-friendly expression language. The authoring environment was designed to be accessible to nonprogrammers. As with BPMN, the goal has been to create a knowledge authoring environment

[c]The published BPMN standard contains a guide to mapping BPMN models to WS-BPEL.

where subject-matter experts can collaborate with knowledge engineers to produce and maintain the complex logic underlying decisions and workflow.

PMML. For logic that cannot be expressed in tabular form, DMN provides an expression language crafted to be as readable as is reasonably possible: Friendly Enough Expression Language (FEEL). Support is also provided for Predictive Model Markup Language (PMML), a standard managed by the Data Mining Group (http://dmg.org/). PMML can be used to ease the integration of predictive algorithms generated through machine learning into the clinical workflow.

CMMN. A second companion standard was released by the OMG in 2014. Case Management Model and Notation (CMMN, https://www.omg.org/cmmn/) describes tools for use by practicing subject-matter experts in cases where a less structured workflow was needed. It provides access to tools and information needed within the context of a specific case while supporting the expert as she navigates a comparatively unpredictable sequence of activities.

The development of CMMN explicitly targeted the kind of flexibility needed by medical experts as they deal with individual cases. It can support key workflows and decision models while reserving overall direction of each case to the healthcare professional. However, adoption of CMMN has been slower than that of its sibling standards. The goal, to provide choice among workflows and decision processes to practicing experts, can, in many cases, be delivered through constructs in BPMN. CMMN's graphical authoring system has proven less intuitive to experts and process engineers alike.

10.3.3.1 The building blocks of BPMN

As indicated above, the BPMN standard specifies a largely graphical authoring language. Fig. 10.4 contains a brief example. Key components of BPMN include:

- *Events*: occurrences that trigger various behaviors in the workflow. These can be messages, temporal markers, errors, and others. A workflow may consume events or generate events for other processes.
- *Activities*: a unit of work that must be done. These include tasks, transactions, and invocations of various types of sub processes. Tasks can be assigned to users, business rule engines, or arbitrary services.
- *Gateways*: manage the forking of pathways through a process when alternative activities are available. They may test simple conditions to choose appropriate next steps.
- *Flows*: connections between events, activities, and gateways that direct the workflow, indicate the movement of information (such as message events and data), or provide a way to annotate flow elements.

These components can be combined to depict a care process in an intuitive way.

10.3.3.2 BPMN tools

A number of commercial and open-source tools both for authoring and for executing these workflows have been developed and continue to evolve. Proposed advantages to a defined, graphical representation include:

- *Familiar graphical authoring tools to develop workflows*: the standard dictates an authoring environment employing familiar graphical shapes and icons to illustrate process workflow using a flowchart motif. This approach facilitates regular review by content experts. A fully specified BPMN model will generate an executable framework for the targeted process.

- *Multiple vendors*: the presence of multiple sources for authoring and execution software precludes "vendor lock-in". If a business is dissatisfied with their current BPM system, they can export their process models and import them into another standards-based product.
- *Training*: the availability of qualified process engineers is encouraged by an abundance of accessible courses designed to teach the standard. A process engineer, once familiar with one modeling environment, can easily work in other systems that have embraced the standard.
- *Improvements in process execution*: well-designed process management tools improve the consistency and effectiveness of process execution. The result improves quality and reduces waste.

10.4 Integration of guidelines with workflow

In the 1990s, guideline modeling languages provided little support for integration with the organization's workflow and information systems. At that time, CIG languages supported modeling of guideline knowledge but, except for definitions of variables used in the encoding, they did not support data modeling intended to facilitate interfacing the guideline model with an EHR. These guideline modeling methodologies were very useful for implementing guidelines that require manual data entry or conducting a dialog of questions and answers. From the turn of the century, there has been a shift to support of the integration of CIGs into medical information systems. As a result, current CIG knowledge bases include definitions for mapping guideline knowledge to EHR schemas. Some of the CIG formalisms, including, EON and GLIF support part of this mapping by incorporating a patient information model as part of the CIG language. EON uses a virtual medical record (vMR) [37] based on the HL7 RIM, and GLIF uses a model based on the Medication, Observation, and Procedure classes of the HL7 RIM. This allows the decision criteria to refer to attributes of patient data classes, for example to address the dose and administration route of a certain class of medication. Note that the choice to rely on vMR or RIM, rather than FHIR, was simply due to the fact that historically, when EON and GLIF were developed, FHIR had not existed yet.

However, in order to query the relevant data from a specific EHR, mappings between the CIG's concepts and the EHR schema need to be defined. The Knowledge-Data Ontology Mapper (KDOM) [38] and MEIDA [17] provide architectures, mapping ontologies, and tools for defining such mappings. In KDOM, SQL queries can automatically be generated from the mappings. The mapping ontologies of KDOM can bridge the semantic gap between abstract CIG concepts and concrete EHR fields by allowing a guideline modeler to define logical combinations of data fields, define concept hierarchies, and specify temporal relationships between data items. The mappings can use the Global-as-View approach of data integration, where GLIF3's patient information model, which is a simplification of the HL7 RIM, is used as a connecting model; mappings are defined between the CIG concepts and the RIM model.

Database views following the RIM model are defined based on the EHRs fields. In this way, the mappings between the CIG and RIM can be reused when a guideline is implemented in an institution that uses a different EHR. Additionally, when a new guideline is implemented in the same institution, mappings between the EHR and RIM can be reused.

MEIDA [17] supports definitions of complex mappings, as in KDOM, and in addition supports unit conversion and disambiguation. However, the queries need to be defined by users and are not generated automatically from the mappings. MEIDA also offers terminology services for aiding in retrieving the correct vocabulary term that matches an EHR field, by utilizing several heuristics to cut down the number of irrelevant candidate terms. Basically, this can be done in three ways, (i) by selecting relevant terminologies (e.g., only LOINC terms are used for lab tests), (ii) string matching, and (iii) limiting the search based on measurement unit heuristics. Other recent approaches such as the one by Marcos et al. [39] propose to develop and use archetypes as a canonical data representation of the concepts used in a guideline, and show how to connect these archetypes to the target EHRs, using the LinkEHR tool and its capacity to generate both terminological and abstraction mappings based on logical expressions. Finally, some new service-oriented initiatives, like the Healthcare Services Specification Project (HSSP), created by the Object Management Group (OMG) and HL7, are being developed to leverage Service Oriented Architectures (SOA) as an interoperability solution (see Chapter 29), and they may be eventually used for supporting the data needs of CIG-based decision support systems.

The European FP7 MobiGuide project (https://sites.google.com/hevra.haifa.ac.il/mpeleg/research/mobiguide) [40] was a successful clinical decision support system for patients and their care providers that used Asbru as the CIG modeling language along with a multi-component SoA software system that included the Picard [41] Asbru enactment engine and a newer implementation of KDOM [35] for mapping medical knowledge concepts into patient data. The patient data was integrated via a Data Integrator [42] that was based on HL7's vMR standard. This integrated record allowed the complex MobiGuide systems, which included over 20 different software components to communicate with one another based on the most up to date representation of patients' data. The requirements of clinical data standards for developing interoperable knowledge-based DSS are discussed in Ref. [43].

A more recent emphasis in the HL7 community has been on the FHIR interoperability standard. While FHIR is an evolving standard, the more mature components have been recognized by the United States, Office of the National Coordinator for Health Information Technology (ONC) (https://www.healthit.gov/) and Centers for Medicare & Medicaid Services (CMS) (https://www.cms.gov/) as relevant for communication of healthcare data. A variety of FHIR data-interoperability models have been mandated for use by EHR vendors. This makes them attractive targets against which to develop CIGs.

Components of FHIR are designed to support service calls providing access to clinical data, allowing the ordering of tests, procedures, and medications, and a variety of other functions that support interaction with data stored in EHR's. In addition, HL7 has specified APIs for the Suitable Medical Apps, Reusable Technology (SMART) [44] standard. SMART provides support for standards- based user interfaces embedded in a EHR's user environment. The combination of HL7 standards with generally available process modeling tools can providing an environment where interoperable CIGs can be shared among the users of different EHRs with little effort required to customize an implementation for a new setting.

The FHIR CPG Implementation Guide [45] informs guideline modelers and developers of CIG-based DSSs how to develop FHIR computable knowledge artifacts. The modeling approach separates expressions of patient disease (clinical and physiological) processes that describe the state of a patient at a given point in time (Case) from expressions of clinician decision-making or care processes that describe what to do for a patient given their state (Plan) and how both are to be separated from expressions of how that care is to be delivered in a given setting at a given point in time (Workflow). The following main CPG content artifacts are used: (1) Recommendations are represented as FHIR Plan Definition largely derived from Event-Condition-Action Rule profile; (2) Guideline plans include strategies for relating, sequencing, or orchestrating individual (or groups of) recommendations and is represented via the FHIR Plan Definition; (3) CPGMetrics, which are patient level measurements or indicators of recommendation compliance and/or guideline adherence (corresponding to a process measure), reaching a stated goal or objective (corresponding to an end or intermediate outcome), or current status of a clinical activity; and (4) eCaseReport are intended to convey the set of data elements required to provide for more detailed outcomes research on the guideline topic itself as well as for a feedback loop for continuous improvement of the specified guideline.

Another aspect related to interoperability is the standardization of scientific evidence behind clinical recommendations. The goal of the FHIR Resources for Evidence-Based Medicine (EBM) Knowledge Assets project (EBMonFHIR) is to provide interoperability (standards for data exchange) for organizations and individuals producing, analyzing, synthesizing, disseminating and implementing clinical research (evidence) and recommendations for clinical care (clinical practice guidelines). This project created FHIR StructureDefinition Resources for Evidence, EvidenceVariable, Statistic, and OrderedDistribution FHIR resources [46]. The paper cited above demonstrate a computable expression of evidence with the results (summary effect estimate) of a meta-analysis of three randomized trials for the effect of remdesivir on 14-day mortality in patients with COVID-19 pneumonia. Examples of the integration of process interoperability standards and FHIR are beginning to appear including a combination of the OMG's Business Process Modeling and Notation (BPMN) standard in a prototype, pulmonary embolism diagnostic workflow, described in Section 10.2.4 [47].

Another level of integration support considers the workflow of activities that are taking place in the setting of the implementing institution and fits the guideline model within that workflow. This approach considers available resources, organizational roles that perform activities (e.g., a clinician ordering a prescription), care setting, and timing constraints. It helps in identifying the best way to implement a task in a given setting, taking into consideration the health care information system and environment factors.

The BPM community has recognized the low degree of flexibility of business processes in health care as one of the hurdles in adapting workflow technology to the health care domain. Much research in that community has addressed different ways of achieving flexibility in process-aware health care information systems. Approaches for addressing flexibility, adaptation to exceptions, and evolution of health care processes as medical knowledge grows, include supporting loosely specified process models, which can be refined during run-time according to pre-defined criteria and rules [9]. This need has been addressed by (a) adaptive process modeling and enactment environments such as ADEPT2 and its successors (https://www.uni-ulm.de/in/iui-dbis/forschung/abgeschlossene-projekte/adept2/), which enables instance-specific changes of a pre-specified model in a controlled, correct and secure manner, (b) case handling or Adaptive Case Management (ACM), technology used by LinkCare (http://www.linkcareapp.com/), which provides healthcare services designed to contribute to the change in the way care is delivered to chronic patients, and (c) declarative process modeling methods such as DECLARE (https://www.win.tue.nl/declare/).

In Ref. [48], Fox et al. highlight that the major challenges created by high levels of uncertainty, typical of clinical processes, may suggest ways in which advanced decision models could be integrated into business process models. For example, PROforma was designed to use an explicit model of argumentation-based decision making, which workflow systems currently lack and where AI and Knowledge Engineering techniques, used already for CIG development, could provide a way of enhancing BPM, including the management of temporal abstractions, or the ontological modeling of tasks and goals.

An alternative approach to integrating process and data is object-awareness which argues for full integration of processes with application data. According to Ref. [9], the modeling and execution of patient-related processes can be based on object behavior. Lenz suggests a document-based approach (called alpha-flow) to flexibly support inter-departmental health care processes [9]. Self-describing electronic documents serve as the unit for information interchange. By including process-related metadata in independent electronic documents, inter-institutional processes can be supported without the need to closely interconnect pre-existing IT-systems.

The approach of Terenziani [49] argues for specifying the workflow and the CIG models separately in their own formalisms which are suitable for different purposes but mapping both models onto a common system-internal representation, so that inferences requiring integration are performed on it.

Shvo and coauthors share their experience from the MobiGuide project regarding different strategies that may be used to integrate CIGs with organizational workflow systems: extending the CIG or BPM languages and their engines or creating an interplay between them. The latter approach was used by the authors and is demonstrated in Ref. [50].

10.5 CIG and workflow verification and exception-handling

The review of Peleg [5] on CIGs provides a section related to CIG verification. Below we provide further details on some of the prominent approaches.

The developers of Asbru were involved in the Protocure I and II projects (https://cordis.europa.eu/project/id/508794), which addressed the important topic of quality improvement of guidelines and protocols by integrating formal methods of software engineering in the lifecycle of guideline development and maintenance ("living guidelines"). In Protocure I, jaundice in newborns and diabetes protocols were formalized and verified against a set of properties. Verification helped detect ambiguities, incompleteness, inconsistencies, or redundancies, and checking concrete properties relevant for the selected protocols (e.g., when treating diabetes with insulin, a desirable property is to distribute the morning and evening insulin doses in a 2:1 ratio). In Protocure II, an original textual guideline was translated into an intermediate representation and then into Asbru. A semi-automatic translator was developed to convert the Asbru model into the specification format used by the Karlsruhe Interactive Verifier (KIV, http://homepages.inf.ed.ac.uk/wadler/realworld/kiv.html), which is an interactive theorem prover.

The team of Terenziani [51] proposed a way to integrate the GLARE CIG management system with the SPIN model-checker for verifying correctness of desired criteria and demonstrated their approach for an ischemic stroke guideline. Structural and medical validity, contextualization, applicability, and complex properties like paths of actions or sequences of patient's states can be checked with this approach. In Ref. [52], authors present *TNest*, a data-driven workflow modeling language for formalizing CIGs while verifying time-related properties, such as the *consistency* of the underlying temporal network or the *controllability*, i.e., the capability of executing a workflow for all possible durations of all tasks while satisfying all temporal constraints. In addition, GLARE has been extended to handle exceptions [53]. The description of the actions in the GLARE CIGs can be augmented to include the list of exceptions. Each exception is specified by listing its name, description, type (CIG exceptions that occur when an action fails or CIG-dependent patient exceptions that occur when foreseen patient states which were not part of the CIG specification occur), Plan, which contains a link to the plan to manage the exception, and modality which specifies what must be the interplay between the execution of the exception and the execution of the current CIG (i.e., concurrent, suspend, abort, abort&goto CIG's action, or ignore). GLARE's exception manager has been evaluated with real world examples in the context of severe trauma management.

Grando et al. [54] developed a state-based framework that enables specifying the goals of a guideline and linking them with recommended tasks that could satisfy the goals. Exceptions are linked with goals that manage them, which can be realized by tasks or plans. In addition, the authors developed design patterns for service assignment and delegation, which occur in collaborative work of health care teams and an exception manager for detecting and recovering from exceptions. They used a formal framework to specify design patterns and exceptions. They have proved, using Owicki–Gries Theory, that the proposed patterns satisfy the properties that characterize service assignment and delegation in terms of competence, responsibility and accountability in normal and abnormal (exceptional) scenarios. The proposed patterns were instantiated in an executable COGENT prototype, and were mapped into the Tallis tool that enacts PROforma language specifications of medical guidelines.

The BPM community has been working for many years on methods for verifying and testing business processes. Lately such methods have been adopted for the health care domain. As reviewed in Ref. [9], such methods include work by McCaull and colleagues who developed a multi-threaded model checker to reason about timed processes in careflows, sensitive to patient preferences and care team goals, using temporal logic extended with modalities of beliefs, desires and intentions. As described in Ref. [9], Osterweil and colleagues developed the Little-JIL process definition language and an integrated collection of tools supporting the precise definition, analysis, and execution of processes that coordinate the actions of humans, automated devices, and software systems for the delivery of health care. It is intended to support the continuous improvement of the health care delivery processes.

In the case of the OMG standards described above, much of the development of validation/verification techniques as occurred in the products of vendors who have adopted these standards. Examples include extensions of the BPMN authoring tools that support the tracing of workflows through their various activities. In these tools one or more tokens track the authored workflow under the control of the author. By following each of the relevant pathways implemented in the process model, an author can determine whether workflow components are access in the correct sequence and whether gateways and events are properly integrated to direct processes when choices exist. In some cases, the ability to inject data into a process and determine its effect on pathway choice is present.

An additional capability present in some authoring tools it the ability to do global simulation over the authored process. In this case, the tooling will automatically generate multiple traces through the possible pathways, exploring all available alternative choices at branching points. It will provide a summary of the available pathways and the usage to the author. This can identify path through the process that can be expected to be highly used and path that the author intended but that are never traversed in the model.

An evolving approach to evaluating process models comes from the realm of business process mining. This approach focusses attention on the data generated by existing unmanaged activities and can be used to critique newly designed process models by testing their conformance with the practices seen in the data. This kind of critique works either way: the newly-developed model may be evaluated in the context of the data, or the behavior of practitioners can be scrutinized by testing their conformance to the proposed process model. A further discussion of process mining is found in Section 10.7, below.

10.6 CIG and careflow enactment tools

As pointed out by Fox [55], decision making, workflow management and care planning and monitoring are important aspects of clinicians' work that can clearly benefit from computer support. All the CIG formalisms are supported by respective enactment tools (execution engines). PRO*forma* is the only CIG formalism that is supported by tools that are actively maintained and developed by a company; the Deontics Clinical Decision Support (CDS) System, which includes an authoring too, an enactment multiple web-based user interfaces (https://deontics.com/technology).

Isern and Moreno [56] analyze eight systems that allow the enactment of computerized clinical guidelines, represented using different languages, in a (semi) automatic fashion, including Arezzo, DeGeL, GLARE, GLEE, HeCase2, NewGuide, SAGE and SpEM. They describe the basic characteristics to be analyzed in a guideline execution engine, comparing 11 aspects of the different CIG engines, like the access to EHRs, the standards used, the approach followed (rule-based or event-based) the management of security issues, or the presence of coordination elements.

Another relevant approach that has been leveraged in the last five years is the enactment of CIGs by means of BPM tools, or more generally, Process Aware Information Systems (PAIS). Different PAIS like ADEPT2, AristaFlow, and YAWL have been used to cope with execution of health care processes. In Ref. [11] a deep review of implementation challenges for PAIS regarding the computerization of guidelines and care pathways is presented, highlighting the evolution toward the implementation of adaptive care pathways on the semantic web, incorporating formal, clinical, and organizational ontologies, and the use of WfMS.

A number of examples of the use of BPMN and DMN to model the clinical decision process have appeared. Sooter et al. [57] explore an approach to capturing a part of the knowledge embedded in the US Centers for Disease Control and Prevention (CDC) document titled "U.S. Selective Practice Recommendations for Contraceptive Use" in 2016. They used an authoring tool based on these standards to model the processes and decisions involved in the initiation of birth control. The effort captured the workflow and logic but illustrated the need for tools to manage terminologies and the data representations needed for each clinical scenario. The OMG has been exploring a potential specification to standardize to these representations called the Situational Data Modeling and Notation (SDMN) standard.

This model captures the process of initiating birth control but is has not been rendered fully executable. The pulmonary embolism workflow captured in the clinical guideline referenced in Fig. 10.2 has been modeled twice. The first effort created a functional BPMN-based application that was implemented in four emergency departments [58]. During a 2.5-year study, use of the electronic guideline was shown to increase the diagnostic yield of Computerized Tomography-Pulmonary Angiogram, the definitive test for this disease, from 10.0% to 21.5%.

The second effort extended the functioning pulmonary embolism application in a test environment by replacing the legacy data queries with HL7 FHIR data access services [47]. The user interface was replaced with a standards-based SMART UI. The goal of this effort was to demonstrate that a combination of standards could be used to construct an application capable of complete interoperability if integrated with a standards-compliance EHR.

10.7 Process mining and improvement

As discussed in Chapter 4, quality measurement and reporting feedback are important drivers for care improvement. As pointed out by Fox in his book "Safe and Sound" [59], although guidelines are based on evidence-based studies, this does not guarantee that the resulting clinical guideline will deliver safe and sound advice, nor does it guarantee that the guideline will contain all relevant evidence-based recommendations. Moreover, when a narrative guideline is translated into a CIG, decisions regarding interpretation of the narrative recommendations are done and details are added in order to obtain a recommended care process which is based not just on evidence-based studies. Therefore, to establish whether the resulting CIG is safe and sound, trials of the CIG-based CDS intervention should be conducted (see Section 10.10 for trial results).

When CIGs are not effective, it is important to differentiate between cases where the medical advice was followed by physicians and patients and cases where there was no adherence. As discussed by Quaglini [60], lack of compliance may be a result of lack of agreement on the part of physicians, for example, when they note relevant patient data relating to comorbid conditions that were not addressed by the clinical guideline. Another reason could be that the patients may find it hard to comply with treatments due to reasons such as complicated scheduling and troubling side effects. Compliance can be assessed against the guideline-recommended actions, or in a more relaxed way, against their intentions, as proposed by Shahar and supported by the Asbru guideline modeling language in which guideline intentions are specified as temporal patterns.

Assessing guideline compliance by mining log files or EHR data is a topic that has been intensely researched in the BPM community. This community has been developing (process) mining [61] techniques for process discovery, conformance testing (i.e., discovering deviations, which can be reported as a set of change operations), and process variant mining for discovering optimal process variants. Such techniques have been applied, for example, to assess compliance of Italian hospitals with guidelines for stroke patient care [62]. By considering the clinical outcomes achieved by actual process instances that may have deviated from the recommendations provided by a CIG, machine-learning techniques were used by Ghattas, Soffer, and Peleg to find the important patient groups, based on similarity of process paths and outcome, and provide semantic definitions for them based on contextual patient characteristics [9]. For each patient group, a path could be recommended, which has yielded desired outcomes in similar patients.

The evolution of process mining has led to a group of tools and practices that continue to provide insight into potential structures for CIGs in the medical domain. The substrate for this form of machine learning is an event log. This data set consists (minimally) of multiple rows each representing an instance of a particular process (called a case), an activity within that process, and a timestamp. Additional data elements such as duration of activities and resources used can augment this data set to allow additional analyses of the targeted processes.

The machine learning activity consists of extracting a process map from this data. This process might be represented in one of a number of forms such as a Petri net or a BPMN model. Three goals commonly motivate these efforts [63]. They include:

- *Process Discovery*: the goal is to discover a process model. The inferred model should be able to describe the sequences of activities observed in the event log data. The information generated may be used to explore, critique, and modify a healthcare process through a number of mechanisms. Implementing a new CIG would be one such approach.
- *Conformance Testing*: the event log is compared to an existing process model. Deviations from the model are analyzed as a part of auditing and compliance checking. Undesirable deviations can be reviewed and remedied through various interventions in the underlying workflow.
- *Process Enhancement*: in the case of process enhancement, an existing process is evaluated in the context of the event log. The anticipated result is improvement in the underlying model either in the form of repairs when the model fails to conform well to reality or extensions that seek to enhance the process often through evaluation in the context of an added perspective (improving efficiency, cost reduction, enhanced resource utilization, etc.).

Process mining in health care has been used in both the contexts of medical treatment and of organizational process management [64]. Clinical workflows have been studied across a broad range of medical fields although surgery and oncology have been the most prominent. A variety of process mining tools have been applied.

10.8 CIG-based decision support for multimorbidity patients

Many of the elderly patients suffer from multimorbidity. However, CPGs and their respective CIGs, focus on a single morbidity and refer to only some of the most common comorbidities. As a result, some unforeseen interactions may occur between the recommendations made by different CIGs when they are applied to multimorbidity patients. These interactions include redundant or conflicting recommendations, drug-drug interactions, drug-disease interactions, and temporal relationships between recommendations [65].

Many of the researchers who have been developing CIG-based decision-support systems have been developing methods to detect such interactions and mitigate them [65,66]. The approaches differ in their methodologies. The GoCom method developed by Kogan, Peleg, Tu, et al. [66] relies on modeling CIGs in PROforma and specifying metadata properties of plans and actions describing their goals in a standard way, while referring to the HL7 FHIR patient data model and to standard vocabularies, and in particular to the Veteran Affairs (VA)'s National Drug File Reference Terminology (NDF-RT). This allows a Controller algorithm to detect conflicting physiological effects of drugs (e.g., decreased platelet aggregation vs. increased platelet aggregation) and opposed start/stop actions that refer to drugs belonging to the same classification hierarchy (e.g., start aspirin vs. stop Nonsteroidal anti-inflammatory drug (NSAID), where aspirin is-a NSAID). When a goal is conflicted, the CIG's decision related to that goal is rerun to search for non-conflicted alternatives. The search for alternatives is generic and based on available actions meeting the specified clinical goals.

MitPlan [67], developed by Michalowski and Wilk uses planning methods to mitigate adverse interactions between multiple CIGs and to derive safe management plans. CIGs are represented as Actionable Graphs that are a form of task-network models, which can be easily derived from other representations, such as GLIF or PROforma. The CIGs and additional specific domain knowledge used for mitigation are automatically transformed into PDDL such that reasoning for detection interactions and strategies to mitigate them can be executed.

A multi-agent planning approach is used by Fdez-Olivares et al. [68] who propose a Multi-Agent Planning (MAP) framework. CIGs represented in Hierarchical Planning Description Domain Language (HPDL), which are Hierarchical Task Networks (HTNs). Possible plans, generated by agents corresponding to the individual CIGs, are evaluated using an objective function that considers plan cost and complexity assessed according to the patient's quantitative preferences to recommend the optimal plan.

Zamborlini et al. [69] represent CIG recommendations as stand-alone actions whose intended effects are state changes (e.g., administer PPI has the intention to decrease the risk of gastrointestinal bleeding from high to low). Logic-based reasoning is used to detect interactions: repetition, contradiction, alternative, side effect, repairable, and safety.

Piovesan et al. [70] describe a system that uses CIGs represented in GLARE and an OWL ontology of general medical knowledge integrating parts of SNOMED CT and ACT terminologies. The system employs temporal reasoning, cost-benefit analysis and model-based verification. d hierarchical graphs. The system supports clinicians in focusing on relevant parts of CIGs (where adverse interactions may occur), identifying alternative management options and testing these options in "what-if" analysis.

All of the approaches above have been tested but till now have not been used to provide decision support in real-world setting.

10.9 Support of patients as end-users via CIG and workflow models

With the growing trend of patient empowerment (see Chapter 24), the idea of having patients more informed and involved in their health care is receiving recognition among clinicians, patients, and researchers. When patients are actively taking part in shared clinician-patient decisions, the suitability of the proposed treatment and management plans to their daily life routine could be improved, hopefully improving adherence to recommended treatment, thus yielding better patient outcomes. Patients may be involved in accessing their patient health record, allowing access to their selected care providers, updating physicians (possibly via the CDS application interface) of important changes in their personal state, and interacting with the application to receive recommendations and alerts and answer dialog questions.

Klasnja and Pratt [71] mapped the space of mobile health interventions by looking at the features of mobile phones that are useful for health interventions (e.g., messaging, camera, sensors, internet access), and by identifying five basic intervention strategies that have been used in mobile-phone health applications across different health conditions. These include tracking health information, involving the healthcare team, leveraging social influence, increasing the accessibility of health information, and utilizing entertainment.

Below we review two mobile health decision-support systems, analyzing them from the perspective of these intervention strategies. The MobiGuide project [40], mentioned in Section 10.4, applied three of these strategies to engage patients and to keep them safe: tracking health information, involving the healthcare team, and increasing the accessibility of health information. Patients monitored their blood glucose, blood pressure, and ECG via mobile sensors that transmitted the measurements to the mobile phone, where they were analyzed by the local CDS running on the phone and interacting with the backend CDS that had access to the complete patient record and to the complete CIG knowledge base. Based on patterns in the sensor-based data, EHR data, and patients' self-reporting, the CDS (1) activated CIG plans regarding diet, exercise, and measurement schedule; (2) sent personalized reminders that considered patients' preferences regarding daily schedule; (3) provided feedback and education; and most importantly, (4) provided CIG-based patient-specific recommendations to patients (e.g., diet adjustments, changes to frequencies of monitoring, and what to do when experiencing atrial fibrillation symptoms, and when to see their doctor before the scheduled visit) – and to care providers (e.g., when to start insulin, when the ECG data needs to be inspected by healthcare professionals). The main benefit to patients was the fact that they stayed at home and were not required to go to the hospital for long monitoring sessions. Patients' feeling of safety increased. Patient compliance to the system's recommendations was 98% compliance for gestational diabetes patients and 0.5–0.82 compliance (depending on the recommendation) for atrial fibrillation patients [72].

This finding motivated researchers to explore additional ways to engage patients, considering the two additional strategies proposed by Klansja and Pratt: leveraging social influence and utilizing entertainment. In Ref. [73], Peleg et al. explored how behavioral economic principles could be applied to develop mobile health apps, leveraging all five strategies of Klansja and Pratt, but also founding the design of the strategies on behavioral theories. The behavioral theories used in that research included Prochaska's Trans-theoretical model of behavioral change, which argues that patients who are not ready for change should first be motivating via education that will allow them to see more pros than cons from the new behaviors that they are asked to develop (being engaged in their health care and compliant to therapy) and Abraham and Michie's taxonomy of behavior-change techniques.

A new research project that takes patient centrality to another level is the EU Horizon 2020 project CAncer PAtients Better Life Experience (CAPABLE). CAPABLE extends the knowledge-based decision support with data-based machine learning models to learn what is best to improve the quality of life of chronic patients. Topics explored include semantic data integration, machine learning to improve care processes, support for multimorbidity patients, and a Patient Coaching System that extends the work of Ref. [73]. This Coaching System uses Fogg's Behavioral Model [74] in conjunction with machine learning methods to find the most suitable moment [75] to deliver recommendations to patients – a time when patients are experiencing stress yet are not occupied with performing activities that require high cognitive load. The recommendations delivered by the Coaching System relate to non-medication evidence-based intervention that could improve mental wellbeing. These interventions come from the mindfulness, exercise, and positive psychology domains.

10.10 Discussion

Since 2011 the CIG and ProHealth communities have started interacting with each other, attending joint workshops and reaching an understanding of the promise and challenges involved in integrating the two approaches. The Knowledge Representation for HealthCare and the ProHealth workshops have been held together during the years 2012–2019 (https://sites.google.com/view/kr4hcprohealth2019/home). The workshops are a continuation of past CIG-centered guideline-specific meetings that started in 2000 (Boston 2000, Leipzig 2000, London 2001, Prague 2004) and continued annually since 2008 (https://sites.google.com/view/kr4hcprohealth2019/previous-events). The journey to integrating CIG-based decision support and workflow management has started, but true achievements still lie ahead of us.

Implementing clinical guidelines as decision support systems that provide patient-specific recommendations during clinical encounters increases the chances of affecting clinicians' behavior and achieving the benefits of guidelines. The road to achieving widespread use of such CDS systems is long and difficult. Along this road, we have known successes and failures, and many future directions can be taken to reach this goal.

10.10.1 Successes

A number of CIGs have been represented and implemented. Some of these systems have also been evaluated and shown to be effective and beneficial, as discussed below.

Twelve CDS applications have been implemented using PRO*forma* technology [76]. Quantitative trials have been carried out for seven of these systems. All seven have shown major positive effects on a variety of measures of quality and/or outcomes of care. The seven systems include: CAPSULE (assisting general practitioners in prescribing for common conditions), LISA (advising on dose adjustment in treatment of children with acute lymphoblastic leukemia), Retrogram (advice on the use of anti-retroviral therapy for HIV+ patients), a treatment planner for patients with type 2 diabetes and hypertension, RAGs (Helping GPs take the family history, assess risk and explain risk factors to patients), CADMIUM (combining conventional image processing with automated interpretation of images and diagnosis), and initial assessment of women referred to specialist breast clinics, and CREDO (http://archive.cossac.org/credo.html) decision-making and workflow management in the care of women at risk for with a proven diagnosis of breast cancer.

ATHENA [77] is an EON-based implementation of the JNC-VII hypertension guideline that has been successfully deployed as a CDS system at clinics in three medical centers of the US Department of Veterans Affairs. The ATHENA system has also been used to create a CDS application for opioid therapy. Analysis of the data collected during a clinical trial to test the impact of the ATHENA system is underway. Preliminary results indicate that, during the 15-month clinical trial, clinicians interacted with the advisory screen for 63% of patients eligible for guideline-based CDS. Use of the system remained high throughout the 15 months.

Two guidelines, management of diabetes-related foot disorders and post-coronary artery bypass surgery patient care planning, have been implemented in GLIF3, linked with an EHR, and executed using the GLEE engine in an educational setting [78].

Guidelines for diabetes, jaundice, neonatal care, and breast cancer have been implemented in the Asbru formalism as part of the ONCOCURE project (http://www.cvast.tuwien.ac.at/projects/OncoCure/). Guidelines for gestational diabetes and atrial fibrillation have been implemented in Asbru and evaluated by patients who had successfully used them for nine months in Spain and Italy, as part of the MobiGuide project (https://mpeleg.hevra.haifa.ac.il/research/mobiguide) [72].

The implementation of OncoDoc [34] for breast cancer management has been evaluated in a study that compared physicians' compliance and patient accrual in clinical trials before using OncoDoc, and just after its use (synchronous steps for a given patient). Compliance increased from 61.5% to 85%, and patient accrual increased by 50%. An uncontrolled before/after study has also been conducted in multidisciplinary staff meetings (MSMs) [79]. Compliance increased from 79% to 93%. Thus, the system has been adopted as a component of MSM decision processes and about 2000 decisions have been collected with a compliance rate stabilized at the same level. A multicenter randomized clinical trial involving 6 health institutions is currently being carried on to assess the actual impact of the system on MSM compliance.

GLARE is a domain-independent tool that has been applied to model clinical guidelines in different contexts, ranging from ischemic stroke to the management of harmful drinking and alcohol dependence in primary care. Currently, it is being used within the ROPHS (Report on the Piedmont Health System) project in order to train personnel in emergency medicine (and, specifically, in the treatment of severe trauma).

In the standardization arena, one success story has been the HL7 standardization of the GELLO guideline expression language that can be used for formally defining decision and eligibility criteria and patient states and the ongoing standardization of the vMR model for supporting representation and exchange of patient data for decision support. GELLO Release 2 is supported by an interpreter and engine by software from Medical-Objects Pty Ltd (https://www.medicalobjects.com/clinical-decision-support/).

These efforts have informed recent activities:

- HL7 is actively developing FHIR resources designed to support clinical workflow (https://www.hl7.org/fhir/workflow-module.html). This collection of models will help define the data that flows through medical information systems as CIGs are executed.
- HL7's GELLO has informed the Clinical Quality Language (CQL, https://cql.hl7.org/), which uses FHIR as a data model.

- The FHIR CPG Implementation Guide [45] is informed by the vast work that has been done in the CIG and Workflow area. It uses FHIR as the data model and CQL as the expression language.

10.10.2 Limitations

The process of seeking to create a standard computer-interpretable guideline model, while receiving wide support in HL7, has had many limitations, and has not yet been successful after more than a decade of effort. The field of computerized guidelines may be too immature for starting the standardization, while requirements and goals are still changing. Members of the community have disagreed about developing a full guideline model and instead are focusing on standardizing components of a guideline model. But disagreement still exists about which components should be included in such as a standard, and it is not yet clear that a complete CIG standard could be assembled from the component standards in the making. The CIG standard should be easy to use and should be supported by authoring, markup, and execution tools for it to be widely adopted. Standards would be more effective if consensus has been achieved and if users and industry are involved in the standard development process. Up to now, the standardization process has been driven by the developers of guideline modeling methodologies instead of other stakeholders, such as users, payers and vendors. There is no process for identifying urgent user needs.

Furthermore, CDS systems that use CIGs should not be seen as a silver bullet for the dissemination and implementation of clinical guidelines. Most of the literature that demonstrates the efficacy of computerized CDS focuses on alert- or reminder-based systems. Several recent evaluation studies show that evidence-based CDS systems may not necessarily have clinical impact [80,81]. Much more study is needed to understand the factors that make such an implementation successful.

10.10.3 Future research

Future and emerging research could concentrate on the following areas that constitute major limitations of widespread successful implementations of CIGs as well as opportunities for employing CDS systems in situations that have not been explored before:

- **Collaborative care processes**. Fox [55] has identified important areas of research that relate to shared care (see also Chapter 24). Sharing decisions and plans includes topics related to joint and distributed decision-making (where responsibility is shared), shared and distributed execution of plans in organizations, and managing distributed knowledge and data within and across organizations. The distribution of decision support could be between CDS systems belonging to different organizations who are cooperating in the care of a patient. Alternatively, the distribution could be related to the technology that the user of the CDS system is using and the data availability. An example might be use of a CDS application which operates on a mobile smartphone and draws data from a body area network of wearable sensors, while communicating with a fully-fledged web-based CDS system that also has access to a wide range of data sources (including hospital EHR data) that are all assembled into an integrated patient health record, as in the MobiGuide project (https://sites.google.com/hevra.haifa.ac.il/mpeleg/research/mobiguide).

- **Guideline and clinical workflow evolution and sharing**. The desire to share executable knowledge across implementing institutions, different countries and geographies is motivating the exploration of new formalisms that support such tasks. This endeavor is gaining the interest of guideline developers and publishers. Sharing of narrative guidelines is facilitated by electronic libraries of evidence-based narrative guidelines such as the ECRI Guidelines Trust (https://guidelines.ecri.org/), and the International Guideline Library (http://www.g-i-n.net/). However, with the variety of CIG formalisms and the fact that CIGs usually embed local adaptation of knowledge made by the implementing institution for which the CIG was developed, sharing a CIG is not a straightforward task. The Medinfo 2010 panel [82] discusses efforts made in this direction, highlighting technical, content-wise, and business model considerations. To facilitate sharing of CIGs, DeGel [17] and openclinical.org are setting up libraries of CIGs and executable components in different guideline representation formalisms. Interestingly, execution engines that could enact different TNM-based CIGs has been developed by Wang, as described in Ref. [56] and by Terenziani [52].
- **Knowledge portability**. A key feature of the BPMN standard is its design for a service-oriented environment. The RGS example shown in Section 10.2.1 illustrated the use of services to acquire needed data, but other services can also be invoked such as scheduling or messaging. Current efforts to implement BPM processes in healthcare environments where functionality based on HL7 standards is available show promise. Examples include prototypes of the pulmonary embolism diagnostic tool [47]. Here, a workflow designed to manage the diagnosis of pulmonary embolism is combined with data access and storage services expressed in the HL7 FHIR standards. This combination promises to ease the distribution of CIGs by (1) making them sharable among institutions that adopt the FHIR standards and (2) simplifying efforts to reconfigure parts of the underlying workflows for sites where the available resources and process choices do not match those found in the originating institution.
- **Acceptance of CIG-based decision support**. Evaluating the cost and impact of deployed CDS systems for guideline-based care and understanding the barriers and facilitators in the acceptance of such systems are areas of needed study.

10.11 Recommended resources

Institute of Medicine (2011). Clinical practice guidelines we can trust. Consensus report. Robin Graham, Michelle Mancher, Dianne Miller Wolman, Sheldon Greenfield, Earl Steinberg, editors. Washington (DC): National Academies Press (US); https://doi.org/10.17226/13058

This book was written by an expert committee appointed by the Institute of Medicine, which examined clinical guidelines, focusing on their development and implementation. The book discusses the strengths and limitations and how they can be used more effectively to benefit health care.

The ECRI Guidelines Trust https://guidelines.ecri.org/

A public resource for evidence-based clinical practice guidelines, initiated by the Agency for Healthcare Research and Quality (AHRQ), U.S. Department of Health and Human Services, and continued by ECRI Institute.

Mor Peleg (2013). Computer-interpretable Clinical Guidelines: a Methodological Review. *Journal of Biomedical Informatics*, 46(4), 744–763.

The paper reviews the literature on CIG-related methodologies since the inception of CIGs. The paper identifies eight themes that span the entire life-cycle of CIG development, usage, evolution and sharing.

Latoszek-Berendsen, a, Tange, H., van den Herik, H. J., & Hasman, a. (2010). From clinical practice guidelines to computer-interpretable guidelines. A literature overview. *Methods of information in medicine*, *49*(6), 550-70.

This literature review explains the characteristics of high-quality guidelines, and new advanced methods for guideline formalization, computerization, and implementation. The paper discusses the impact of guidelines on processes of care and the patient outcome and the reasons of low guideline adherence.

Guidelines International Network (G.I.N.) Web site https://g-i-n.net/home

Taken from the G.I.N. web site: G.I.N. is a major international initiative which seeks to improve the quality of health care by promoting systematic development of clinical practice guidelines and their applications into practice. The web site offers guideline resources such as a guideline library, development tools and resources, training material of guidelines, patient resources.

The Field Guide to Shareable Clinical Pathways Version 2.0. Web site: https://www.bpm-plus.org/healthcare-and-bpmn.htm

Acknowledgment

We would like to thank Dr. Arturo Gonzalez Ferrer for his contribution to a previous version of this chapter.

References

[1] Kohn LT, Corrigan JM, Donaldson MS. To err is human: Building a safer health system. Washington, DC: Committee on Quality of Health Care in America, Institute of Medicine, National Academy Press; 1999.

[2] Institute of Medicine. Crossing the quality chasm: A new health system for the 21st century. National Academy Press; 2001.

[3] Institute of Medicine. Clinical practice guidelines we can trust. Consensus report; 2013.

[4] Latoszek-Berendsen A, Tange H, van den Herik HJ, Hasman A. From clinical practice guidelines to computer-interpretable guidelines. A literature overview. Methods Inf Med 2010;49(6):550–70.

[5] Peleg M. Computer-interpretable clinical guidelines: a methodological review. J Biomed Inform 2013;46 (4):744–63.

[6] Bousquet J, Hellings PW, Agache I, Bedbrook A, Bachert C, Bergmann KC, et al. ARIA 2016: care pathways implementing emerging technologies for predictive medicine in rhinitis and asthma across the life cycle. Clin Transl Allergy 2016;6(1):47.

[7] Unertl KM, Novak LL, Johnson KB, Lorenzi NM. Traversing the many paths of workflow research: developing a conceptual framework of workflow terminology through a systematic literature review. J Am Med Inform Assoc 2008;17(3):265–73.

[8] van der Aalst W, van Hee K. Workflow management: Models, methods, and systems. MIT Press; 2004.

[9] Lenz R, Peleg M, Reichert M. Healthcare process support: achievements, challenges, current research. Int J Knowl-Based Organ Spec Issue Process Support Healthc 2012;2(4).

[10] De Ramón FA, Ruiz Fernández D, Sabuco GY. Business Process Management for optimizing clinical processes: a systematic literature review. Health Informatics J 2020;26(2):1305–20.
[11] Gooch P, Roudsari A. Computerization of workflows, guidelines, and care pathways: a review of implementation challenges for process-oriented health information systems. J Am Med Inform Assoc 2011;18(6):738–48.
[12] Field MJ, Lohr KN. Guidelines for clinical practice: From development to use. Washington, DC: Institute of Medicine, National Academy Press; 1992.
[13] Hajizadeh N, Kashyap N, George M, Shiffman R. GEM at 10: a decade's experience with the guideline elements model. In: Proceedings of AMIA symposium; 2011. p. 520–8.
[14] Patel VL, Allen VG, Arocha JF, Shortliffe EH. Representing a clinical guideline in GLIF: individual and collaborative expertise. J Am Med Inform Assoc 1998;5(5):467–83.
[15] Shalom E, Shahar Y, Taieb-Maimon M, Bar G, Yarkoni A, Young O, et al. A quantitative assessment of a methodology for collaborative specification and evaluation of clinical guidelines. J Biomed Inform 2008;41(6):889–903.
[16] Peleg M, Wang D, Fodor A, Keren S, Karnieli E. Lessons learned from adapting a generic narrative diabetic-foot guideline to an institutional decision-support system. In: Computer-based medical guidelines and protocols: A primer and current trends, studies in health technology and informatics. IOS Press; 2008. p. 243–52.
[17] Shahar Y. The "human Cli-Knowme" project: building a universal, formal, procedural and declarative clinical knowledge base, for the automation of therapy and research. In: Proceedings of the knowledge representation for health-care workshop (KR4HC); 2011. p. 1–22.
[18] Peleg M, Tu SW. Design patterns for clinical guidelines. Artif Intell Med 2009;47(1):1–24.
[19] Serban R, ten Teije A, van Harmelen F, Marcos M, Polo-Conde C. Extraction and use of linguistic patterns for modelling medical guidelines. Artif Intell Med 2007;39(2):137–49.
[20] Kaiser K, Miksch S. Versioning computer-interpretable guidelines: semi-automatic modeling of 'living guidelines' using an information extraction method. Artif Intell Med 2009;46(1):55–66.
[21] Mulyar N, van der Aalst WMP, Peleg M. A pattern-based analysis of clinical computer-interpretable guideline modeling languages. J Am Med Inform Assoc 2007;14(6):781–7.
[22] Aigner W, Kaiser K, Miksch S. Visualization methods to support guideline-based care management. In: ten Teije A, Miksch S, Lucas PJF, editors. Computer-based medical guidelines and protocols: A primer and current trends. Studies in health technology and informatics, vol. 139. IOS Press; 2008. p. 140–59 [chapter 8].
[23] Klimov D, Shahar Y, Taieb-Maimon M. Intelligent visualization and exploration of time-oriented data of multiple patients. Artif Intell Med 2010;49(1):11–31.
[24] Peleg M, Tu SW, Bury J, Ciccarese P, Fox J, Greenes RA, et al. Comparing computer-interpretable guideline models: a case-study approach. J Am Med Inform Assoc 2003;10(1):52–68.
[25] Riaño D, Peleg M, ten Teije A. Ten years of knowledge representation for health care (2009–2018): topics, trends, and challenges. Artif Intell Med 2019;100:101713.
[26] Elkin PL, Peleg M, Lacson R, Bernstam E, Tu S, Boxwala A, et al. Toward the standardization of electronic guidelines. MD Comput 2000;17(6).
[27] Tu SW, Musen MA. From guideline modeling to guideline execution: defining guideline-based decision-support services. In: Proceedings of AMIA symposium; 2000. p. 863–7.
[28] Boxwala AA, Peleg M, Tu S, Ogunyemi O, Zeng Q, Wang D, et al. GLIF3: a representation format for sharable computer-interpretable clinical practice guidelines. J Biomed Inform 2004;37(3):147–61.
[29] Shahar Y, Miksch S, Johnson P. The Asgaard project: a task-specific framework for the application and critiquing of time-oriented clinical guidelines. Artif Intell Med 1998;14(1–2):29–51.
[30] Sutton DR, Fox J. The syntax and semantics of the PROforma guideline modeling language. J Am Med Inform Assoc 2003;10(5):433–43.
[31] Peleg M, Fox J, Patkar V, Glasspool D, Chronakis I, South M, et al. A computer-interpretable version of the AACE, AME, ETA medical guidelines for clinical practice for the diagnosis and management of thyroid nodules. Endocr Pract 2014;20(4):352–9.
[32] Terenziani P, Montani S, Bottrighi A, Molino G, Torchio M. Applying artificial intelligence to clinical guidelines: the GLARE approach. Stud Health Technol Inform 2008;139:273–82.
[33] Bottrighi A, Terenziani P. META-GLARE: a meta-system for defining your own computer interpretable guideline system—architecture and acquisition. Artif Intell Med 2016;72:22–41.
[34] Seroussi B, Bouaud J, Chatellier G. Guideline-based modeling of therapeutic strategies in the special case of chronic diseases. Int J Med Inf 2005;74(2–4):89–99.

[35] Bouaud J, Pelayo S, Lamy J-B, Prebet C, Ngo C, Teixeira L, et al. Implementation of an ontological reasoning to support the guideline-based management of primary breast cancer patients in the DESIREE project. Artif Intell Med 2020;108:101922.
[36] Huser V, Rasmussen LV, Oberg R, Starren JB. Implementation of workflow engine technology to deliver basic clinical decision support functionality. BMC Med Res Methodol 2011;11(1):43.
[37] Johnson PD, Tu SW, Musen MA, Purves I. A virtual medical record for guideline-based decision support. Proc AMIA Symp 2001;294–8.
[38] Peleg M, Keren S, Denekamp Y. Mapping computerized clinical guidelines to electronic medical records: knowledge-data ontological mapper (KDOM). J Biomed Inform 2008;41(1):180–201.
[39] Marcos M, Maldonado J, Martínez-Salvador B, Moner D, Boscá D, Robles M. An archetype-based solution for the interoperability of computerised guidelines and electronic health records. In: Proceedings of artificial intelligence in medicine; 2011. p. 276–85.
[40] Peleg M, Shahar Y, Quaglini S, Fux A, García-Sáez G, Goldstein A, et al. MobiGuide: a personalized and patient-centric decision-support system and its evaluation in the atrial fibrillation and gestational diabetes domains. User Model User-Adapt Interact 2017;27(2):159–213.
[41] Shalom E, Shahar Y, Lunenfeld E. An architecture for a continuous, user-driven, and data-driven application of clinical guidelines and its evaluation. J Biomed Inform 2016;59:130–48.
[42] Marcos C, González-Ferrer A, Peleg M, Cavero C. Solving the interoperability challenge of a distributed complex patient guidance system: a data integrator based on HL7's Virtual Medical Record standard. JAMIA 2015;22(3):587–99.
[43] González-Ferrer A, Peleg M, Ferrer AG, Peleg M, González-Ferrer A, Peleg M. Understanding requirements of clinical data standards for developing interoperable knowledge-based DSS: a case study. Comput Stand Interfaces 2015;42(42):125–36.
[44] Mandel JC, Kreda DA, Mandl KD, Kohane IS, Ramoni RB. SMART on FHIR: a standards-based, interoperable apps platform for electronic health records. J Am Med Inform Assoc 2016;23(5):899–908.
[45] HL7. FHIR Clinical Guidelines [Internet]. Available from: https://build.fhir.org/ig/HL7/cqf-recommendations/.
[46] Alper BS, Dehnbostel J, Afzal M, Subbian V, Soares A, Kunnamo I, et al. Making science computable: developing code systems for statistics, study design, and risk of bias. J Biomed Inform 2021;115:103685.
[47] Haug PJ, Narus SP, Bledsoe J, Stanley H. Promoting national and international standards to build interoperable clinical applications. AMIA Annu Symp Proc 2018;2018:555–63.
[48] Fox J, Black E, Chronakis I, Dunlop R, Patkar V, South M, et al. From guidelines to careflows: modelling and supporting complex clinical processes. In: Studies in health technology and informatics, vol. 139. IOS Press; 2008. p. 44–62.
[49] Terenziani P. A hybrid multi-layered approach to the integration of workflow and clinical guideline approaches. In: Lecture notes in business information processing, vol. 43, Part 7; 2010. p. 539–44.
[50] Shabo Shvo A, Peleg M, Parimbelli E, Quaglini S, Napolitano C. Interplay between clinical guidelines and organizational workflow systems: experience from the mobiguide project. Methods Inf Med 2016;55(6):488–94.
[51] Bottrighi A, Giordano L, Molino G, Montani S, Terenziani P, Torchio M. Adopting model checking techniques for clinical guidelines verification. Artif Intell Med 2010;48(1):1–19.
[52] Combi C, Gambini M, Migliorini S, Posenato R. Modelling temporal, data-centric medical processes. In: Proceedings of ACM international health informatics symposium; 2012. p. 141–50.
[53] Leonardi G, Bottrighi A, Galliani G, Terenziani P, Messina A, Corte FD. Exceptions handling within GLARE clinical guideline framework. In: Proceedings of AMIA symposium; 2012.
[54] Grando MA, Peleg M, Cuggia M, Glasspool D. Patterns for collaborative work in health care teams. Artif Intell Med 2011;53(3):139–60.
[55] Fox J, Glasspool D, Patkar V, Austin M, Black L, South M, et al. Delivering clinical decision support services: there is nothing as practical as a good theory. J Biomed Inform 2010;43(5):831–43.
[56] Isern D, Moreno A. Computer-based execution of clinical guidelines: a review. Int J Med Inf 2008;77(12):787–808.
[57] Sooter LJ, Hasley S, Lario R, Rubin KS, Hasić F. Modeling a clinical pathway for contraception. Appl Clin Inform 2019;10(05):935–43.
[58] Bledsoe JR, Kelly C, Stevens SM, Woller SC, Haug P, Lloyd JF, et al. Electronic pulmonary embolism clinical decision support and effect on yield of computerized tomographic pulmonary angiography: ePE—A pragmatic prospective cohort study. J Am Coll Emerg Physicians Open 2021;2(4). [Internet] [cited 2021 Jul 21]. Available from: https://onlinelibrary.wiley.com/doi/10.1002/emp2.12488.
[59] Fox J, Das S. Safe and sound: Artificial intelligence in hazardous applications. AAAI Press; 2000.

[60] Quaglini S. Compliance with clinical practice guidelines. In: ten Teije A, Miksch S, Lucas PJF, editors. Computer-based medical guidelines and protocols: A primer and current trends. Studies in health technology and informatics, vol. 139. IOS Press; 2008. p. 160–79 [chapter 9].
[61] van der Aalst WMP. Process mining: Discovery, conformance and enhancement of business processes. Springer; 2011.
[62] Mans R, Schonenberg H, Leonardi G, Panzarasa S, Cavallini A, Quaglini S, et al. Process mining techniques: an application to stroke care. In: Studies in health technology and informatics. IOS Press; 2008. p. 573–8.
[63] Ghasemi M, Amyot D. Process mining in healthcare: a systematised literature review. Int J Electron Healthc 2016;9(1):60.
[64] Rojas E, Munoz-Gama J, Sepúlveda M, Capurro D. Process mining in healthcare: a literature review. J Biomed Inform 2016;61:224–36.
[65] O'Sullivan D, Woensel WV, Wilk S, Michalowski W, Abidi S, Carrier M, et al. Towards a framework for comparing functionalities of multimorbidity clinical decision support: a literature-based feature set and benchmark cases. In: AMIA Annual Symposium Proceedings. American Medical Informatics Association; 2021. p. 920–9.
[66] Kogan A, Peleg M, Tu SW, Allon R, Khaitov N, Hochberg I. Towards a goal-oriented methodology for clinical-guideline-based management recommendations for patients with multimorbidity: GoCom and its preliminary evaluation. J Biomed Inform 2020;112:103587.
[67] Michalowski M, Wilk S, Michalowsk W, Carrier I. MitPlan: a planning approach to mitigating concurrently applied clinical practice guidelines. In: Conference on artificial intelligence in medicine in Europe; 2019. p. 93–103.
[68] Fdez-Olivares J, Onaindia E, Castillo L, Jordán J, Cózar J. Personalized conciliation of clinical guidelines for comorbid patients through multi-agent planning. Artif Intell Med 2019;86:167–86.
[69] Zamborlini V, da Silveira M, Pruski C, ten Teije A, Geleijn E, van der Leeden M, et al. Analyzing interactions on combining multiple clinical guidelines. Artif Intell Med 2017;81:78–93.
[70] Piovesan L, Terenziani P, Molino G. GLARE-SSCPM: an intelligent system to support the treatment of comorbid GLARE-SSCPM: an intelligent system to support the treatment of comorbid patients. IEEE Intell Syst 2018;33 (6):37–46.
[71] Klasnja P, Pratt W. Healthcare in the pocket: mapping the space of mobile-phone health interventions. J Biomed Inf 2012;45(1):184–98.
[72] Peleg M, Shahar Y, Quaglini S, Broens T, Budasu R, Fung N, et al. Assessment of a personalized and distributed patient guidance system. Int J Med Inf 2017;101.
[73] Peleg M, Michalowski W, Wilk S, Parimbelli E, Bonaccio S, O'Sullivan D, et al. Ideating mobile health behavioral support for compliance to therapy for patients with chronic disease: a case study of atrial fibrillation management. J Med Syst 2018;42:234–48.
[74] Fogg BJ. Tiny habits: The small changes that change everything. Houghton Mifflin Harcourt; 2020.
[75] Lisowska A, Wilk S, Peleg M. Catching patient's attention at the right time to help them undergo behavioural change: Stress classification experiment from blood volume pulse. Porto, International Conference on Artificial Intelligence in Medicine. Cham: Springer; 2021. p. 72–82.
[76] Fox J, Patkar V, Thomson R. Decision support for health care: the PROforma evidence base. Inf Prim Care 2006;14 (1):49–54.
[77] Automating guidelines for clinical decision support: knowledge engineering and implementation. AMIA Annual Symposium Proceedings. American Medical Informatics Association; 2016. p. 1189–98.
[78] Peleg M, Shachak A, Wang D, Karnieli E. Using multi-perspective methodologies to study user interactions with the front-end of a guideline-based decision-support system for diabetic-foot care. Intl J Med Inf 2009;78 (7):482–93.
[79] Séroussi B, Bouaud J, Gligorov J, Uzan S. Supporting multidisciplinary staff meetings for guideline-based breast cancer management: a study with OncoDoc2. In: Proceedings of AMIA symposium; 2007. p. 656–60.
[80] Sittig D, Wright A, Osheroff J, Middleton B, Teich J, Ash J, et al. Grand challenges in clinical decision support. J Biomed Inform 2008;41(2):387–92.
[81] Bright TJ, Wong A, Dhurjati R, Bristow E, Bastian L, Coeytaux RR, et al. Effect of clinical decision-support systems: a systematic review. Ann Intern Med 2012;157(1):29–43.
[82] Peleg M, Fox J, Greenes R, Rafaeli S. Sharing guidelines knowledge: Can the dream come true? (panel presentation), 2010. Medinfo [Internet]. Available from: https://www.researchgate.net/profile/John-Fox-4/publication/267383922_Sharing_guidelines_knowledge_can_the_dream_come_true/links/54732cdc0cf24bc8ea19be57/Sharing-guidelines-knowledge-can-the-dream-come-true.pdf.

CHAPTER

11

Terminologies, ontologies and data models

Thomas A. Oniki[a], *Roberto A. Rocha*[b,c], *Lee Min Lau*[a], *Davide Sottara*[d], *and Stanley M. Huff*[e]

[a]3M Health Information Systems, Murray, UT, United States [b]Semedy Inc., Needham, MA, United States [c]Division of General Internal Medicine and Primary Care, Department of Medicine, Brigham and Women's Hospital, Harvard Medical School, Boston, MA, United States [d]Mayo Clinic, Rochester, MN, United States [e]Graphite Health, Murray, UT, United States

11.1 Introduction

11.1.1 Aims of the chapter

Why do we devote a chapter to ontologies and vocabularies in a book about Clinical Decision Support (CDS) and Knowledge-Enhanced Care? The reason is that, although this topic does not itself address CDS, the understanding, interpretation, sharing, and reuse of CDS cannot be advanced without common data models and terminologies for expressing them. This relates to both (i) the data about the patient being evaluated by CDS, as well as (ii) the clinical concepts in knowledge rules, probabilistic models, or other CDS artifact types that are used in the process of evaluating clinical data with respect to them.

We begin with some basic definitions.

- **Concept:** An item with a defined meaning.
- **Code:** An alpha, numeric, or alphanumeric value used to identify a concept (i.e., identifier). For example codes for gender might be "1" for male, "2" for female, and "3" for sex unknown.

Clinical Decision Support and Beyond
https://doi.org/10.1016/B978-0-323-91200-6.00027-9

- **Definition:**
  - **Formal:** An explicit expression of the meaning of a concept using computable means (e.g., relationships, description logics, conceptual graphs, etc.).
  - **Informal:** An unstructured narrative text describing the meaning of a concept (such as that seen in a dictionary).
- **Term:** A word or collection of words that describes or represents a concept. In the gender example above, the terms are "male" (for the concept identified by the code "1"), "female" (for the concept identified by the code "2"), and "sex unknown" (for the concept identified by the code "3"). Often, a concept can be described using multiple terms, e.g., "hypertension" and "high blood pressure." These are referred to as "**Synonyms**." Different code systems use different words for "term." "Label," "designation," and "description" are also used. A "**Thesaurus**" is a compendium of similar terms mapping to each other, ideally to a single concept. This can be used to create mappings of terms, for example, among different code systems.
- **Code System:** Loosely, any set of codes (and often their associated terms), whether organized or not. Code systems may vary in their organization, formality, and complexity. In its simplest form, a code system is a listing of codes and terms (a "**Code Set**"). A "**Glossary"** associates definitions with the codes/terms. "**Taxonomies**" and "**Partonomies**" are hierarchical arrangements of concepts based on parent-child "broader than/narrower than" or "whole/part of" relationships. "**Classifications**" are hierarchical arrangements designed to enable the fitting of individuals into categories, for the purpose of counting cases for statistical returns. "**Terminology**" is a general term referring to a particular code system, a type of code system, or the entire continuum of types of code systems. "**Nomenclature**" and "**Vocabulary**" are less formal terms, referring to "a set of specialized terms that facilitates precise communication by minimizing or eliminating ambiguity" [1]. A "**Controlled Vocabulary**" is "the set of individual terms in the vocabulary" [1]. "Controlled vocabulary" is also sometimes used more informally, to refer to any of the organizations of codes described in this chapter. Finally, a "**Structured Vocabulary**" or "**Reference Terminology**" is a more formal and structured code system which "relates terms to one another (with a set of relationships) and qualifies them (with a set of attributes) to promote precise and accurate interpretation" [1].
- **Ontology**: A formal system of organization of concepts depicting formal logical relationships between the concepts. A taxonomy is a specialized form of ontology. Ontologies enable applications to navigate the relationships and perform reasoning (drawing conclusions about individuals from the information contained in the relationships) and inference (discovering new information not stated explicitly by the existing relationships).
- A data structure-oriented "**data model**" or "**information model**" defines a data structure for *holding* information, rather than the information itself. For example, data models are created for "Observations" (or more specifically, for "Heart Rate Measurement Observations" or "Abdominal Tenderness Observations"), which define the structures by which instances of those observations are exchanged or stored.

Previous publications have described correct principles for developing and maintaining terminologies [2–11]. Other resources describe the details of how particular clinical terminologies like SNOMED International's Systematized Nomenclature of Medicine—Clinical Terms (SNOMED CT), the US National Library of Medicine (NLM) Unified Medical Language System (UMLS) Metathesaurus, the Logical Observation Identifiers Names and Codes (LOINC), and the NLM's RxNorm drug terminology have been developed and the conceptual content, relationships, and capabilities of those systems [12–24]. Much has been written about the structure, development, and maintenance of ontologies [25–27] and information models [25,28–37] in healthcare. In contrast, this chapter focuses specifically on how ontologies, vocabularies, and data models are essentially related to successful implementation of clinical decision support systems. We will discuss:

- Why standard coded data are essential for accurate and reliable execution of decision logic
- The semantic spectrum, including terminologies, ontologies, and data models
- Formalisms for representing these knowledge representations and implications of their combination
- The use of terminologies, ontologies, and data models in decision support
- The role of ontologies, vocabularies, and data models in the sharing of decision support logic

11.1.2 Referencing data in decision logic

CDS systems are highly dependent on clinical information to function. To be useful, the clinical information must contain sufficient detail and must be structured in a way that the CDS system understands. In proprietary decision support applications that are tightly bound to a particular clinical application, the representation of clinical information used in decision support may be the same as the representation used in the associated clinical application. However, when the decision support system is based on a more portable standard, such as Arden Syntax [38,39], Clinical Quality Language (CQL) [40], or GELLO [41,42], or the implemented system seeks to adapt and incorporate shared or commercially available interoperable content, the clinical information must be transformed from one format to another.

11.1.2.1 Facilitating portability among different information systems

As an example of decision logic, Arden Syntax (see Chapter 9) does not specify the format of a reference to data in an associated clinical application (although the most recent version of Arden does provide a mechanism to represent data using the Fast Healthcare Interoperability Resources [FHIR] standard). Rather, such references are implementation-specific and demarcated by curly braces within the code. Since these references to external data vary from one implementation to the next, the issue of dealing with references to external data has been dubbed "**The Curly Braces Problem.**" The most significant consequence of the curly braces problem is that CDS modules are not readily portable across different electronic health record (EHR) systems implementing standardized decision support languages. Implementers of the logic must fill in the curly braces with statements that will reference data from their local patient data store. The curly braces problem is not unique to the Arden Syntax. All decision support systems that strive to maintain portability across disparate clinical applications must address the issue of referencing external clinical information. In the worst case, a data reference in a given logic module would need to be mapped to a particular data element in every system where the logic was deployed (a many-to-many map from logic modules to local data elements). Further, if many logic rules reference the same data elements, the curly braces in each rule must repeat those mappings. In contrast, in the best case, the logic module is mapped to a shared common data model. Providing such a common model is a principal goal of HL7's FHIR [43] (a many-to-one map from logic modules to FHIR Resources). Each system that deploys the logic needs to map to FHIR Resources, but that effort can be undertaken once, and then utilized repeatedly, for a multitude of use cases and applications. Each decision support module does not need to perform the map to the EHR, and a many-to-many map for each data is avoided. Chapter 15 describes the role of FHIR and other relevant standards in the implementation and sharing of CDS.

Unless/until FHIR resources and FHIR profiles (artifacts which further constrain FHIR resources, see Section 11.4.4) are defined and adopted on a large scale, the mappings are not guaranteed to be predictable, preventing CDS logic from being universally portable.

Modern languages for CDS such as CQL try to further address this issue with stratification, decomposing the data access from the extraction of the information pertinent to the CDS, and the CDS logic itself.

11.1.3 The need for coded data

Coded data are required to address the curly braces problem and allow the structured specification of data. Coded data are ultimately required for accurate and reproducible execution of decision logic. Unstructured free text is too ambiguous and imprecise to support valid reasoning by computers; successful implementation of decision support relies on receiving coded data as input. Moreover, processing free text requires constant oversight and maintenance of the process since new text variations would have continued to arise over time.

Natural Language Processing (NLP) techniques continue to become better and more sophisticated, making free text processing more and more reliable. However, an objective of any NLP system should be to make the information in the text it processes reusable by a variety of downstream systems. Those systems need the information in a normalized form—a form provided by coded data.

11.1.3.1 Advantages of coded data

Coding of data has advantages other than just the execution of the logic. One advantage coded data facilitates is the ability to display different representations or forms to different users. For example, when appropriate, data can be displayed to nurses differently than to respiratory therapists. Use of codes also makes translation to different human languages easier. Most importantly, the rapid evolution of language and of medicine, combined with the inherent human variability in expressing ourselves makes maintenance of decision logic based upon free text very time and labor intensive and subject to errors and incompleteness, since text-based logic needs to reference all of the words that might be used to represent the needed concept rather than a code. Using coded data enables the use of terminologies that may already provide the needed sets of values that serve as valid and appropriate values for a data element (e.g., a set of coded values representing penicillins) and relationships (e.g., a "has-ingredient" relationship), and also provide regular updates, thus reducing the development and maintenance burden.

11.1.3.2 *Enabling a "Learning Health System" environment*

For over a decade, a "Learning Health System" (LHS) has been a vigorously pursued goal [44,45]. LHS developers and researchers design systems that facilitate discovery of new knowledge and then enable the rapid infusion of those discoveries back into the healthcare system—in clinical care, best practices, and policies—where they can improve care, reduce costs, and foment further discoveries, perpetuating a virtuous learning cycle. Development of an LHS has figured prominently into the US Federal Government's plans and priorities [46–48], and many Learning Health Systems have been developed [49–59].

The hallmark of a Learning Health System is a federated network of databases (EHRs, clinical trials databases, public health repositories, disease registries, repositories of environmental data, etc.) resulting in pooled data which are securely and immediately accessible to researchers, analysts, health care practitioners, patients, and other stakeholders. This wealth of data enables increased and more rapid discovery of new knowledge—not just from data collected during clinical trials, but from the data generated by patient care and a myriad of other sources.

Injecting the newfound knowledge back into clinical practice is the particular bailiwick of clinical decision support. A Learning Health System uses clinical decision support mechanisms such as computerized alerts, reminders, protocols, and the other mechanisms described in this chapter to infuse the new knowledge into practice quickly and in harmony with the practitioner's workflow. Clearly, the development and success of decision support in an LHS fundamentally depends on data in all these databases being interoperable. And central to interoperability is the use of coded data wherever practical in the patient record, to normalize semantics across the system. In addition, models of the type described in Section 11.2.1.3 are needed to provide the explicitly-defined semantic structures within which coded terminology is referenced.

11.1.3.3 *Coded data and ML/AI*

Despite the recognized advantages of high-quality coded data, most biomedical data are captured in unstructured forms. Examples include clinical notes, imaging data, time series data from biomedical devices, patient self-assessment questionnaires, social media feeds and data from consumer devices such as fitness trackers. In these situations, one must (i) identify the set of concepts scoped by the domain of interest, (ii) organize them into an appropriate arrangement, and (iii) recognize the instances/occurrences of those concepts in the unstructured data sets. Once recognized, concepts are codified and added as either data or metadata elements ("annotations").

Historically, the concept schemes have been defined by terminologists/ontologists, and then applied to data by analysts in the role of annotators. The commoditization of Artificial Intelligence (AI) and advances in scale and sophistication of Natural Language Processing (NLP) techniques have created an opportunity to automate these tasks. Problems (i) and (ii) are the subject of "Ontology Learning," a discipline that uses a combination of linguistic, statistical and logic-based approaches. Problem (iii), in contrast, is formally known as "Concept Recognition" or "Named Entity Recognition." Research in the field has led to promising but far from perfect results [60,61].

Like most applications of AI, adequacy is determined by the context of use; while likely not suitable for direct patient care, AI-driven codification can support and accelerate what would otherwise be a primarily manual effort. Notably, (semi-) automated codification is often used to prepare data sets that will be further used for machine learning, especially when the underlying algorithms require categorical input/output variables.

11.2 Introduction to the semantic spectrum: Ontologies, vocabularies/terminologies, and Data models

11.2.1 The semantic spectrum

Having asserted the benefits of coded and structured data to decision support, we will present an overview of the spectrum of knowledge representations involved. The notion of a "semantic spectrum" was first introduced by McGuinness and later refined by Obrst [62] and generally adopted on the Semantic Web to denote different approaches to the representation of semiotic resources, with different levels of formalism and precision. Although there is acknowledged overlap between terminologies, ontologies, and data models, we will use those groupings for the purposes of the discussion.

11.2.1.1 Terminologies

The constructs pertinent to the "Terminologies" section of the semantic spectrum are presented in the Aims of the Chapter, Section 11.1.1. Terminology formalisms are further described in Section 11.3.1.

11.2.1.2 Ontologies

Code systems become "ontologies" as they introduce formal logical concepts. They are based on the "is-A/kind of" relation with the addition of *necessary* and *necessarily true* facts and implications. For example, every possible—past, present and/or future—left arm *is-A* limb in an ontological sense. Ontologies can also define *sufficient* aspects of a concept, which are used to recognize individual instances.

A proper ontology attempts to address a particular domain of discourse, providing a common vernacular between all the parties involved in that domain. If the ontology possesses necessary and sufficient concept definitions expressed in logic-based formulas, new inferences that were not explicitly stated by the ontology may be made. Concepts can be *classified* to determine subsumption relationships, individuals can be *recognized* as instances of a concept, and logical inconsistencies can be *detected*. For example, given "Pneumonia is-A Inflammation that is situated in a Lung" and "Inflammation is-A Disorder" and "Lung is-A Organ that is part of some Respiratory System," one can infer that "Pneumonia is-A Disorder that is situated in some (Thing that is) part of the Respiratory System." Any other inference that excludes Pneumonia from affecting the Respiratory System would be detected as an inconsistency.

Special purpose languages—principally the W3C Web Ontology Language (OWL)—have been developed to express ontological knowledge. Expressing an ontology in such a language leverages tools such as editing tools (e.g., Protégé), databases (triple stores), and inference engines with query capabilities, and the use of a standard language helps provide *explicit, sharable* definitions.

11.2.1.3 Data models

As asserted in the Introduction, structure-oriented "data" or "information" models define data structures for *holding* information, rather than the information itself. For example, a model for "date of birth" is *not* a model of a Patient's "date of birth"—a necessary characteristic of all living beings—but is instead a model of a portion of a Patient *demographic record*, which may or may not include (data about) the date of birth.

Structure-oriented models include most object-oriented, frame-based, XML-based, and relational models. Such models are often realized as "schemas," and are effectively ad-hoc grammars that enable the description of the entities they represent, capturing necessary as well as *contingent* data. Models provide attributes (or "fields" or "slots") which are populated by data values. Because of those reasons described in Section 11.1.3, data values should be maintained as codes in terminologies or ontologies to the fullest extent possible. From now on, the term "terminology" will be used to refer to "terminology," "vocabulary," or "code system," as well as to "ontology." (Similarly, "terminologies" will refer to the various types of terminologies/vocabularies/code systems as well as to ontologies.)

Another example of a data model is a model for a medication order. In this case, a medication order is itself an artifact of an EHR (as opposed to an entity that exists in the "real world") and a medication order model is a model for the structure of that artifact in an EHR or other healthcare system or application. A simple medication order might be represented as follows:

MedicationOrder	
drug	Drug
dose	Decimal
route	DrugRoute
frequency	DrugFrequency

This model represents a MedicationOrder as a set of elements or attributes (the "drug," the "dose," the "route," and the "frequency"). The value of the "drug" element is a coded item whose value comes from the set of codes in a drug terminology (or the set of codes that are children of the "Drug" node in a more general-purpose terminology). The set of codes that are valid and appropriate values for a model element is called a "value set." The codes in a value set are drawn from a terminology. As mentioned in Section 11.1.3.1, some terminologies provide curated value sets as part of their content. When no value sets can be found ready-made in terminologies, the value sets must be specially composed for the purposes of the data model. (Often, the needed value sets are so specific to the use case that even if existing value sets can be found that approximately suit the use case, there is still tailoring that needs to be done.) Associating a model element to a value set is called "binding"; the value set is said to be "bound" to the model element.

The "dose" element in the model is a decimal number. The "route" element of the model is a coded item whose value comes from the set of codes that are children of the "DrugRoute" node in a terminology. And the "frequency" element of the model is a coded item whose value comes from the set of codes that are children of the "DrugFrequency" node in a terminology.

The model can then be used as a guide or template for creating instances of patient data. For example, an instance of a medication order for a particular patient could be represented as:

MedicationOrder	
drug	Penicillin VK
dose	500 mg
route	Oral
frequency	Every 6 h

The definition of MedicationOrder implies that there is a terminology that contains a Drug concept, where Drug has a computable relationship to other concepts that are drugs. A graphical representation of the drug hierarchy is shown in Fig. 11.1. Similar hierarchies would exist for DrugRoute and DrugFrequency.

Besides defining value sets for elements, the representation of the relationships between concepts in a terminology can also serve a second important function. The relationship information can be used to perform hierarchical inference. For example, if a decision support rule needs to evaluate whether Nafcillin is-A Antibiotic, this can be determined by querying the relationships and finding the one that asserts that Nafcillin is-A Penicillin, followed by a query that locates the relationship that asserts that Penicillin is-A Antibiotic.

The model and terminology combination may serve as either the logical model or the physical model for a database design, an API, or a messaging structure.

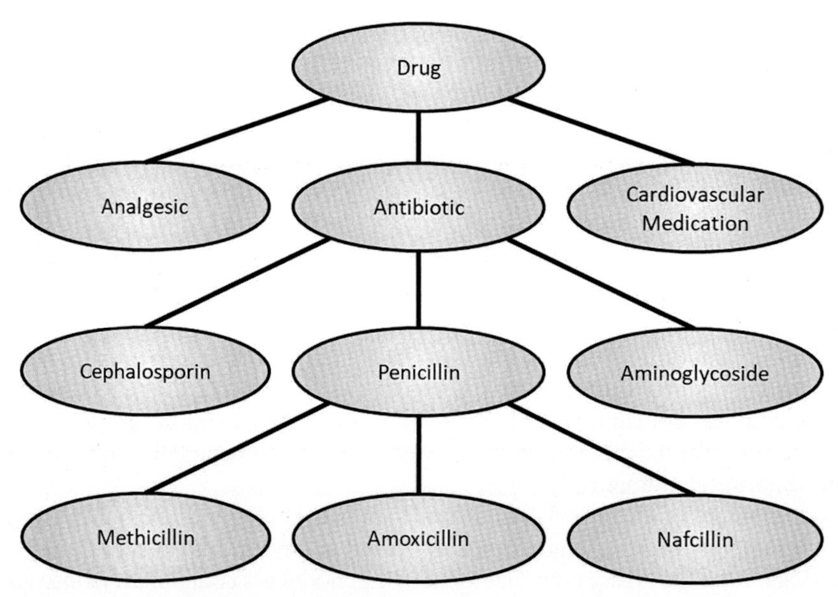

FIG. 11.1 An excerpt from a simple drug hierarchy.

11.3 Formalisms and use cases

11.3.1 Vocabulary formalisms

Several notable examples of vocabulary formalisms that illustrate the terminologies described in the Semantic Spectrum section (Section 11.2.1) are listed below:

Code set: Examples are the many "tables" in HL7 version 2, e.g., Table 0007 for "admission type" with values "A" (Accident), "C" (Elective), "E" (Emergency), etc. and codes for the fields of the United States federal Uniform Billing form (UB04), e.g., "Admit Type," with values "1" (Emergency), "2" (Urgent), "3" (Elective), etc. Code sets like these most often were created to simply document the valid values for fields in forms or messages.

Glossary: Examples are HL7 FHIR code systems such as "AccountStatus" or "ResponseType." They list codes, terms, and definitions. They are bound to attributes of FHIR resources. The definitions provide additional guidance (beyond what just the codes and terms provide) in selecting the appropriate codes.

Taxonomy and Partonomy: The Healthcare Provider Taxonomy is an example of a Taxonomy. Its codes are arranged in "is-A" hierarchies which facilitate navigation to the desired code and selection of the most appropriate granularity for the use case.

Thesaurus: Examples are the National Library of Medicine's Medical Subject Headings (MeSH) [63] and most things represented in the Simple Knowledge Organization System (SKOS) [64]. MeSH is a National Library of Medicine hierarchically arranged index to journal articles and books in the life sciences that allows users to search through and find desired articles and books. It organizes content under "subject headings," and contains multiple descriptions, explanatory text, and qualifiers (subheadings) for each entry.

Classification: ICD-10-CM [24] is the primary US example of a classification. ICD-10-CM is the US "clinical modification" of the World Health Organization's International Classification of Diseases, 10th Edition. Medical providers receive reimbursement from the US government for medical services based on their documentation of ICD-10-CM codes for diagnoses addressed. Many non-government healthcare payors also require the codes. It is a classification system, intended to enable placement of care into the appropriate classes associated with billing, as opposed to SNOMED CT, which is intended to provide sufficient detail for describing clinical care. There are 72,616 codes in the 2021 release.

Each ICD-10-CM code contains three to seven characters. The first digit is alphabetic and the second and third digits are numeric. A decimal point follows the third digit, and the fourth through seventh digits can be alphabetic or numeric. The digits dictate the hierarchy. A small portion of the hierarchy is shown below for illustration:

- J00-J99: Diseases of the respiratory system
  - J00-J06: Acute upper respiratory diseases
    - J01: Acute sinusitis
      - J01.0 Acute maxillary sinusitis
        - J01.00 Acute maxillary sinusitis unspecified
        - J01.01 Acute recurrent maxillary sinusitis

Classification schemes face limitations with overlaps and closures. For example, "Hypertension in Pregnancy" is logically a kind of both "Disease of Pregnancy" and

"Hypertension." ICD-10-CM assigns it to be a "Disease of Pregnancy," and, as a result, there is an "exclusion" under "Hypertension" for "Hypertension in pregnancy." ICD also introduces "not elsewhere classified" and "not otherwise specified" concepts to ensure that individual cases can always be classified within the scheme.

Structured vocabulary or Reference terminology: The creation of the Logical Observation Identifiers (LOINC) [21] code system began in 1994, initiated by the Regenstrief Institute. Its objective initially was to develop a set of codes for laboratory observations, but its scope now encompasses all clinical observations, including radiology, standardized survey instruments, and nursing assessment.

LOINC contains close to 100,000 codes (v2.69). Each LOINC code carries six attributes that uniquely define the concept:

1. Component: what is measured, evaluated, or observed (e.g., urea, creatinine, etc.)
2. Kind of property: characteristics of what is measured (e.g., mass, volume, time stamp, etc.)
3. Time aspect: interval of time over which the observation or measurement was made
4. System: context or specimen type within which the observation was made (e.g., blood, urine, etc.)
5. Type of scale: the scale of measure. The scale may be quantitative, ordinal, nominal or narrative
6. Type of method: procedure used to make the measurement or observation

LOINC is used to standardize observation identifiers and is intended for use in any data exchange and standardization efforts involving observations, e.g., lab reporting to state departments of health. It is recommended for use by the US Interoperability Standards Advisory (ISA) [65] and the US Core Data for Interoperability (USCDI) specification [66].

RxNorm [20] provides codes and normalized names for clinical drugs. RxNorm concepts are arranged in a "graph" of computable relationships to associate, for instance, clinical drugs with their ingredients, their dosage forms and their routes of administration, and branded drugs with their generic equivalents.

RxNorm is intended for use in the exchange of drug information between systems or applications using different drug vocabularies. In addition to providing codes and names, RxNorm also provides mappings to many commercial drug vocabularies to facilitate use with EHRs.

There are over 50,000 RxNorm codes. Because of the dynamic nature of drug development, RxNorm is released in full monthly updates and weekly additive updates. The NLM also provides an API through which its content may be programmatically accessed.

11.3.2 Ontology formalisms

To date, the de-facto standard for the expression of ontology knowledge in a sharable form is the Ontology Web Language (OWL). Other languages exist, but are usually limited to specific communities of practice. Examples include Common Logic, F(rame)-Logic, and Open Biomedical Ontology Language (OBO), as well as hybrid approaches [67].

Readers wanting a more complete introduction to OWL should refer to the *Introductory OWL Tutorial* by Horridge et al. [68] and the documentation on the W3C site [69]. Those interested in the underlying theory of description logics should refer to the *Description Logic*

Handbook [70]. Those interested in details of how SNOMED CT is implemented should consult the documentation and tutorials from SNOMED International, many of which are available on its website (http://www.snomed.org).

OWL is a language based on Description Logic, a subset of first order logic designed to balance expressivity and computational complexity. The key elements of OWL are *Classes*, *Properties* and *Individuals*.

- OWL Classes capture the notion of Universal Concept. An OWL Class is *extensional*: a Class is a Set of all (past, present and future) Individuals that are asserted or inferred to be *members* of that Class. The *intensional* counterpart of the Class is the Concept; Individual members are *instances of* the Class.
- OWL Individuals are *Particular Concepts* that exist at specific points in space and time.
  Individuals can be *recognized* as members of one or more Classes if their properties, asserted or inferred, satisfy the *sufficient* criteria that are part of a Class formal definition. Conversely, membership in a Class allows inference of additional properties of an Individual based on the *necessary* aspects of a Class definition.
- OWL Properties capture (binary) relationship concepts, such as "causes," "has medical condition," or "receives care from." OWL distinguishes between property types used in class definitions, e.g., "Patients are Person(s) that receive care from some Care Provider Organization," and properties asserted between individuals, e.g., "John receives care from Provider X." Property types are used in the formal definitions of Classes, to capture both necessary and sufficient aspects, while properties are used to *describe* individuals, and enable inferences.
- Note that the notion of "is-A" used semi-formally in this chapter maps to two distinct built-in property types: rdf:type, to assert membership of an Individual in a Class, and owl: subClassOf, to assert subsumption between two Classes.
  The key principle of an OWL ontology as a subsumption-based taxonomy can be formalized as follows: Given
  - X rdf:type owl:Individual (X is an "Individual")
  - C rdf:type owl:Class (C is a "Class")
  - S rdf:type owl:Class (S is a "Class")

  (X rdf:type C) AND (C owl:subClassOf S) IMPLIES (X rdf:type S)
  (In other words, if Individual X is a member of Class C AND Class C is a subclass of Class S, then Individual X is a member of Class S)

SNOMED CT [22] is a prime example of an ontology used in healthcare. SNOMED CT, maintained by SNOMED International, is a comprehensive vocabulary spanning many areas of medicine, e.g., procedures, findings, body structures, observables, drugs, etc. It is intended for the coding of information in electronic health records, research and clinical trials for individual patient care in an EHR, pooling of data for analytics, and exchange of information.

SNOMED CT contains more than 350,000 concepts and 800,000 terms (descriptions) that represent the concepts. It also contains 1.2 million relationships that represent computable associations between concepts according to a "concept model" that provides rules for combining concepts according to a formal ontology, making the content computable and logically sound. It is recommended for use by the ISA [65] and USCDI [66].

11.3.3 Data model formalisms

Many data model formalisms are used in healthcare. Two of the more pertinent to decision support will be described.

openEHR: The openEHR project [71] is operated by the openEHR Foundation, a not-for-profit organization supporting the research, development, and implementation of EHRs based on their modeling strategy. The openEHR specification shares a common heritage with the International Standardization Organization (ISO) 13606 standard [72]. Both ISO 13606 and openEHR adopt a two-level modeling approach. Level one consists of a general EHR reference model containing high-level constructs such as Composition, Section, Entry, and Data value. The reference model is then constrained at a second level by models called "archetypes [73]," which are formal definitions for finer-grained pieces of information such as a glucose measurement, a blood pressure measurement, an abdominal examination, and a discharge summary. (The value of greater specificity will be discussed later, in Section 11.4.4) Archetypes are expressed in the Archetype Definition Language (ADL) [30,73].

FHIR: HL7's FHIR standard [43] is currently the predominant standard addressing the exchange of patient data in a variety of settings. FHIR is a platform specification that builds on previous iterations of HL7 (version 2, version 3) but is different from those iterations in that:

- It supports exchange based on modern, web-based technologies (RESTful services) in addition to exchange by messaging and documents.
- It has a strong emphasis on implementation, providing concise documentation, a strong user community, and libraries, and a simple constraint/extension framework.

Its use is mandated by a number of US federal initiatives.

The FHIR standard defines "resources"—the common models of the data to be exchanged in healthcare. As of FHIR Release 4, 145 Resources have been defined. The FHIR specification expresses resources in a graphical structure on its website, but also provides expressions in XML, JSON, and other formats.

FHIR defines various categories of resources, including:

- "Foundation" resources (e.g., "Composition," "MessageHeader," "StructureDefinition") that support the FHIR data exchange infrastructure
- "Base" resources (e.g., "Encounter," "Organization," "Patient") that provide the basic contextual entities prevalent in healthcare data
- "Clinical" resources (e.g., "Condition," "Observation," "Procedure") that represent the types of clinical data commonly exchanged
- "Financial" resources (e.g., "Account," "Claim," "InsurancePlan") that represent the types of financial data commonly exchanged
- "Specialized" resources (e.g., "ResearchStudy," "Questionnaire," "MedicinalProduct") developed for specialized needs

Similar to the constraint strategy presented by the openEHR strategy on their reference model, FHIR allows constraint of its resources by means of "Profiles." The FHIR Shorthand language (FSH) [74] has been created for expressing such profiles.

11.3.3.1 Name value pair paradigm in model formalisms

As noted in the previous paragraphs, modeling formalisms provide attributes (or "fields" or "slots") which are populated by data values. Most modeling formalisms employ a "name-value-pair" (also known as "attribute-value-pair," "key-value-pair," or "field-value-pair") approach [75,76] wherein one attribute represents a "name," or a "question" and another attribute represents a "value," or the "answer" to the question. An example will illustrate. The HL7 FHIR standard contains a resource named "Observation" to represent clinical observations. The specification declares that an Observation has a "code" and a "value," among other attributes. An example instance of data that conforms to the Observation resource and that represents a heart rate of 66 beats per minute can be represented as follows:

Observation.code: LOINC 8867-4 ["Heart Rate"]
Observation.value: 66 beats/min

The "name" or "question" is "heart rate" and the "value" or "answer" is "66 beats per minute."

Another example is shown below:

Observation.code: LOINC 29544-4 ["Physical findings"]
Observation.value: SNOMED CT 43478001 ["Abdominal tenderness (finding)"]

The "name" or "question" is "physical finding" and the "value" or "answer" is "Abdominal tenderness." This paradigm allows the single Observation resource to be used to capture a wide variety of different data types, where the data type is specified by Observation.code and its value by Observation.value.

11.3.4 Theory of combining the formalisms

11.3.4.1 Semantic integrity in combining models with terminologies/ontologies

Simply put, terminology fills the slots in data models. Proper interpretation of models populated with terminology content, however, requires (i) semantically-consistent bindings of model elements to value sets and (ii) clear definition of relationships between elements. To illustrate, we return again to the "abdominal tenderness" example discussed previously:

Observation.code: LOINC 29544-4 ["Physical findings"]
Observation.value: SNOMED CT 43478001 ["Abdominal tenderness (finding)"]

In this model, in order for semantic integrity to be maintained, it is important that Observation.code be populated with a code that is truly the name of an observation (i.e., a "question"). If Observation.code were instead populated with an "answer," or with some other type of concept other than the name of an observation (e.g., a procedure type, a medication, etc.), a Decision Support application assuming a semantic type of "Observation type" in Observation.code may produce incorrect or nonsensical results.

Attributes in a model at least imply if not demand that they should be populated with coded values of certain semantic types. In addition to the Observation.code example, other examples are a "body location" attribute (which suggests that its value be drawn from a domain of body location codes), a "Procedure.code" attribute (which suggests that its value be drawn from a domain of codes that are names of procedures), or a "MedicationOrder.status" attributes (which suggests that its value be drawn from a domain of valid statuses of the life cycle of a medication order).

Relationships between model attributes are also implied or stated. Again referring to the Observation.code/Observation.value example, the two attributes are semantically related in that the Observation.value is assumed to be the "value" of or the "answer" to the Observation.code. If this assumption is violated by populating the attributes with codes that are not related in this way, a decision support application will not be able to produce valid, consistent results.

Returning to the example of "abdominal tenderness," another valid option (according to FHIR) is to populate Observation.code with a SNOMEC CT code, as follows:

Observation.code: SNOMED CT 43478001 ["Abdominal tenderness (finding)"]
Observation.value: SNOMED CT 52101004 ["Present (qualifier value)"]

In this case (valid according to FHIR), it might be argued that the Observation.code is not a "question," with Observation.value being the "answer." Very strictly speaking, a SNOMED CT "finding" is a physical condition (not an observation name) and "present" is qualifying information about the finding. Depending on how strictly a decision support application assumes the semantics of a model instance, this might result in logical inconsistencies.

Ensuring that only codes of a particular semantic type are allowed in a particular attribute requires careful curation of the sets of values (i.e., "value sets," or "domains") maintained for use with the attribute. Validation (either at design time or runtime) that an attribute is being populated with a code of a particular semantic type requires an infrastructure that stores metadata about attributes and codes (i.e., their semantic types) and then is able to programmatically check the semantic type associated with an attribute against the semantic type of a code being suggested for population of the attribute. Representing models and terminology using description logics and formal ontological knowledge representations assist the enablement of such infrastructure.

11.3.4.2 The value of ontological principles in data modeling

Ensuring the integrity of the bindings of terminology to models is one example of the importance of recognizing ontologic principles in data modeling and terminology. Ontologies deal with definitional, necessary aspects of classes and individuals, while Data Models focus more on contingent information *about* the Individuals that are members of the Classes. When ontologies are used in conjunction with data models, proper alignment of the ontologies and data models is critical if decision support is to correctly interpret the information conveyed by their combination. A data model might be viewed as a grammar which combines structural rules, built-in terminology, and bindings to vocabularies/ontologies and data types. Ideally in a data model, each element is ontologically committed to have exactly one interpretation, i.e., denote exactly one entity (class, individual, relationship, or combination thereof) that is defined in an ontology. Such a data model is known as a "semantic" data model. Theory, methods, and best practices for "Semantic Data Modeling" can be found in [77], but this section will focus only on a few key principles.

The violation of ontologic principles in the design of data models results in issues that are potentially harmful to decision support operation, as described below.

Construct Overload occurs when a data element may have multiple interpretations. As an example, as described in Section 11.3.3.1, the Observation Resource uses the name-value pair strategy, allowing the resource to be used for a wide variety of data, where the value of Observation.code dictates the type of data being represented. However, unconstrained use of the paradigm results in construct overload, which in turn results in difficulties for decision support applications attempting to interpret a resource that in one instance may represent a blood pressure measurement but in another instance may represent a urine color observation.

Likewise, since the permitted values of Observation.value are governed by the value of Observation.code, Observation.value is also susceptible to overload. It can hold data in a variety of data types (quantities, dates, boolean flags), depending on the value of Observation.code. Decision support must discern the data type of Observation.value based on the value of Observation.code before interpreting the data.

Construct overload reduces the number of models (because an element may have multiple interpretations), simplifying the implementation of interfaces and messaging systems. However, decision logic will usually need to disambiguate the overloaded elements before they can be processed properly.

Construct Excess occurs when a data element has no equivalent in the ontology. This occurs frequently when data constructs are created for technical purposes (consistency, ease of implementation, etc.) but have no analog in the ontology. For example, in a data model, a coded term, which could be represented in a very compact form, is sometimes wrapped in a complex data type (which has no analog in the ontology) with separate components for the code, the coding system, the version, the preferred term, and sometimes other components. Such data type structures are usually devised to satisfy technical constraints or make data organization more consistent. However, decision support must carry the overhead of parsing or deconstructing the data types to interpret the coded information and logic becomes more verbose and harder to understand.

Construct Redundancy occurs when ontologic concepts may be expressed in alternative ways. Pre-coordination and postcoordination (Section 11.4.2), the choice of information model vs. terminology (Section 11.4.3), and iso-semantic models (Section 11.4.5) all illustrate the susceptibility of systems to construct redundancy. Construct redundancy is often used to provide more direct access to data elements deemed more important, but it makes decision logic more complex, since alternative locations in the data must be processed.

Construct Deficiency: Some ontologic concepts that decision support is in need of processing may not be available in existing data models. Construct deficiency is usually compensated for with extension mechanisms which may be more difficult than the data models for decision support logic to interpret without additional technical solutions. For example, the FHIR extension mechanisms are used to convey information not inherently addressable by the Observation resource, but those mechanisms require a decision support system (or services or sub-systems it leverages) to support features above and beyond basic resource processing.

Clearly, a tradeoff exists between practicality on the one hand and the benefits of logical and ontological rigor on the other. When design decisions are made, the inertia of traditional systems, traditional modeling and development methods, and terminology and modeling standards that are not ontologically-based, combined with the availability of professionals familiar with those traditional artifacts and processes, often has outweighed the benefits of modeling, designing, and implementing with greater ontological rigor. As a consequence, historically, healthcare data models designed and optimized for data exchange tend to exhibit all these Construct issues, posing difficulties for decision support. Decision support designers and implementers must be aware of these issues and ensure the existence of compensatory mechanisms. Common approaches, treated in Section 11.4.4, involve the use of intermediary "logical" semantic data models and either (i) the remapping of the messaging data into logical data or (ii) the rewriting of expressions based on the logical model to expressions based on the underlying messaging model.

11.4 Implications

A number of issues that arise in combining models and terminology—especially in the face of solutions that do not completely follow ontological principles—and using the content in decision support will be described below.

11.4.1.1 Multiple name-value pair alternatives

Again we return to the abdominal tenderness example of Section 11.3.3.1. The representation in that section was as follows;

Observation.code: LOINC 29544-4 ["Physical findings"]
Observation.value: SNOMED CT 43478001 ["Abdominal tenderness (finding)"]

As has been noted, another valid population of the model was:

Observation.code: SNOMED CT 43478001 ["Abdominal tenderness (finding)"]
Observation.value: SNOMED CT 52101004 ["Present (qualifier value)"]

And still another valid population is:

Observation.code: LOINC 8694-2 ["Physical findings of Abdomen"]
Observation.value: SNOMED CT 247348008 ["Tenderness (finding)"]

All these combinations are valid according to the FHIR specification. The last example illustrates the problem of overlap between code systems—either a SNOMED CT code or a LOINC code is valid for use to express the same concept. And, taken together, the examples illustrate the problems of construct overload (a data element may have multiple interpretations) and construct redundancy (ontologic concepts may be expressed in alternative ways) described in Section 11.3.4.2.

11.4.2 Pre- and post-coordination

Consider the following representations for the concept "Ibuprofen, 200 mg oral tablet." (The drug terminology RxNorm is used in the examples. RxNorm codes are known as "RxCUIs.")

Representation 1:
Medication: RxCUI 310965 ["ibuprofen 200 MG Oral Tablet"]

Representation 2:
Medication: RxCUI 5640 ["ibuprofen"]
Dose:
value: 200
units: SNOMED CT 258684004 ["mg"]
Form: SNOMED CT 385055001 ["tablet" (a subtype of Oral dosage form)]

Both representations express the same information; however, they are structured quite differently. In the first representation, the substance, dose, and form are all expressed by a single concept code (RxCUI 310965), whereas in the second representation, several concepts (RxCUI 5640, SNOMED CT 258684004, and SNOMED CT 38505001) and the value "200" are compositionally combined to represent the same meaning. This example illustrates the alternative approaches of pre- and post-coordination. ISO/CD 17115 *Health informatics—Vocabulary of terminology* [78] defines *pre-coordinated concept representation* as "compositional concept representation within a formal system, with an equivalent single unique identifier," while *post-coordinated concept representation* is defined as "compositional concept representation using more than one concept from one or many formal systems, combined using mechanisms within or outside the formal systems." Thus, in the example shown above, representing the concept of *ibuprofen 200 MG Oral Tablet* with the single RxCUI 310965 is an example of pre-coordination while representing the same concept as a combination of multiple fields, codes, and values is an example of post-coordination.

Pre-coordinated representations are often easier to use—they translate more directly to human language. They can simplify data entry screens by requiring fewer fields for which the user needs to fill out or select a value, and decision logic only needs to reference a single item instead of multiple items. This is very convenient when the possible number of pre-coordinated concepts is relatively small and manageable, and for frequently used combinations, such as common prescribing forms of medications. Even if the list of possible combinations is large, modern search and type ahead technologies make selecting from a large list a surmountable challenge. In addition, pre-coordination avoids the problem of allowing pieces to be combined in ways that are incorrect or nonsensical.

However, when there are many pieces of information representing many dimensions, and the pieces can be combined in multiple ways, a pre-coordinated approach can lead to a combinatorial explosion of the number of concepts needed. For example, creating pre-coordinated concepts to record family histories of diseases (e.g., "family history of breast cancer," "family history of coronary artery disease," "family history of stroke") effectively doubles the number of disease concepts in a terminology. Extending this example, creating pre-coordinated concepts such as "maternal family history of breast cancer" or "paternal family history of coronary artery disease" causes an even greater combinatorial explosion of concepts.

Increasing specificity is another potential source of combinatorial explosion. For example, a concept for "Carcinoma of the breast" might be sufficient in some use cases, but other use cases might require further specificity, e.g., "infiltrating duct carcinoma of breast," or "infiltrating duct carcinoma of female breast." Exact location and laterality is a common type of specificity need that could lead to a proliferation of codes. "Carcinoma of the breast" might be further specified as "Carcinoma of the left breast," "Carcinoma of the right breast," "Carcinoma of the breast—lower, inner quadrant," Carcinoma of the breast—upper, outer quadrant," etc.

Post-coordination of concepts helps avoid the problem of combinatory explosion. A post-coordinated approach, rather than working with a set of preformed sentences, provides a dictionary of words or phrases from which an almost unlimited number of sentences can be generated. This allows the maintenance of only a parsimonious set of codes, with more atomic meanings.

While the flexibility of this approach can limit the representation of clinical information to manageable levels, it may also allow the creation of nonsensical statements. For example, representing a medication and its route in separate fields may allow a user to specify a route of "topical" for a medication that may only be administered intravenously. To prevent such situations, post-coordinated models require additional knowledge—metadata and rules that specify in a computable fashion which combinations of values are allowable under what circumstances.

The choice between pre- and post-coordinated approaches must take into consideration the nature of the concepts and the way they will be used. When the number of things that can be said is relatively small and well-constrained, pre-coordinated concepts may be the most useful. When pieces of information can be combined in many different ways, post-coordination should be considered.

11.4.3 Placing information in the terminology model or the information model

As shown in our examples, post-coordination of concepts may be performed by the combination of attributes and their values in the information model. However, by leveraging the functionality of modern terminology systems, post-coordination might also be performed completely in the terminology. Recall the representations of "ibuprofen 200 MG Oral Tablet" from the previous section:

Representation 1:
Medication: [RxCUI 310965, RxNorm, "ibuprofen 200 MG Oral Tablet"]
Representation 2:
Medication: [RxCUI 5640, RxNorm, "ibuprofen"]
Dose:
value: 200
units: [258684004, SNOMED-CT, "mg"]
Form: [385055001, SNOMED-CT, "tablet" (a subtype of Oral dosage form)]

In RxNorm, "ibuprofen 200 MG Oral Tablet" is expressed by RxCUI 310965, which possesses numerous RxNorm relationships and properties. A portion of those relationships and properties is shown in Fig. 11.2.

Hence, this single RxCUI, via its relationships and attributes, carries all the information expressed in Representation 2. Consequently, using the RxCUI 310965 in Representation 1 would enable a decision support application to query a terminology server containing RxNorm content to discover that the form of "ibuprofen 200 MG Oral Tablet" is "Oral Tablet," the strength is "200 MG," and the ingredient is "ibuprofen," without explicitly populating that information in the information model instance.

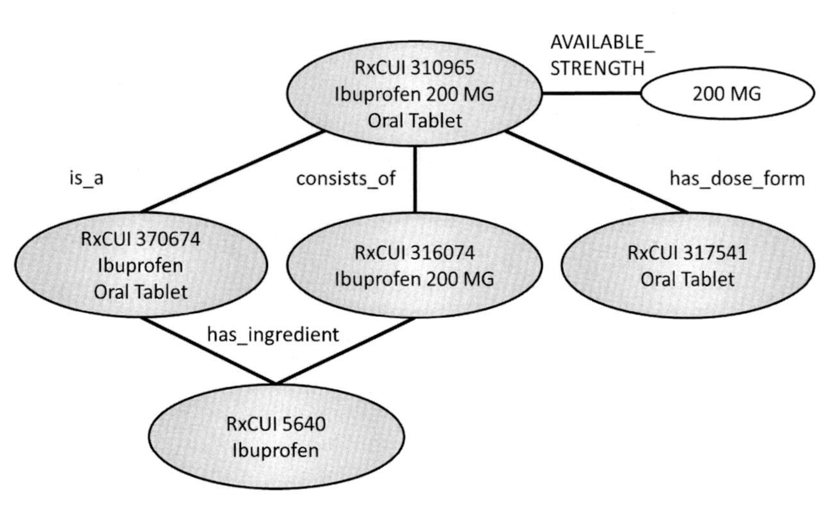

FIG. 11.2 Example relationships from RxNorm.

SNOMED CT is another notable example of a terminology that provides a wealth of relationships and properties in its terminology model, enabling postcoordination of information *in the terminology model* instead of in the information model. A SNOMED concept representing a "finding," for example, may carry with it (in the SNOMED terminology model) a finding site (e.g., "heart structure"), an observable entity it interprets (e.g., "heart sound"), and a finding method (e.g., "auscultation"). Similarly, a SNOMED procedure concept may carry a method (e.g., "excision"), a direct morphology (e.g., "aneurysm"), and a procedure site (e.g., "structure of the cardiovascular system"). When confronted with a SNOMED CT concept, "SNOMED-aware" terminology services would be able to navigate the SNOMED model to surface to the decision support any of this additional information about the concept.

The terminology model and information model ideally would be developed together, the developers making a clear delineation between the information provided by the information model and that provided by the terminology model. More often than not, though, the models are not developed together in a synchronized fashion; they are created separately, the creators making independent decisions on what information their models will contain. The result is that terminology models and information models often overlap, and decision support designers and implementers must often deal with overlapping representations. They must carefully evaluate how the information and terminology models at hand have been combined—areas in which the they overlap, or areas in which multiple alternate combinations might be possible—in order to ensure results are accurate and consistent.

As has been asserted in previous sections, agreeing upon common models would alleviate this difficulty. To this end, the "US Core" work sponsored by the US Office of the National Coordinator for Health IT (ONC) [79] actively works out the combination of HL7's FHIR specification with standard terminologies such as SNOMED CT and LOINC. The Intermountain Healthcare/Graphite Health group is creating Clinical Element Models (CEMs) that carefully address the association of the information model and the terminology model [34,35,80].

11.4.4 Need for explicit, specific models

The preceding sections illustrate different types of choices that exist in representing information in a combination of models and terminology—the choice between multiple name-value pair alternatives, the choice of pre-coordination vs. post-coordination, and the choice of placing information in the information model or the terminology model. The possibility that the same information might be represented in multiple ways (across systems or even within the same system), even using the same structure, is an illustration of Structural Overload and Structural Redundancy, as described in Section 11.3.4.2, and poses serious challenges to a decision support application. A solution to this problem was also presented in Section 11.3.4.2—create models and their paired terminology according to sound ontological principles. However, as noted in Section 11.3.4.2, often the inertia of legacy systems, processes, and standards cannot be overcome, and the decision support application is at the mercy of the existing (sub-optimal) knowledge representations.

Another solution to more exact and specific models, mentioned in Section 11.3.4.2, is to employ explicit logical models that are more specific than the typical name-value pair-based data models. Such logical models make decisions about the choices discussed, resulting in more semantically sound and exact definitions of healthcare data. They are used in conjunction with the higher level models (e.g., FHIR's "Observation" resource) to provide further specificity and constraint, thus overcoming the Construct Overload and Redundancy inherent in most modeling constructs. This "two-level" modeling is employed by ISO 13606/openEHR—via "archetypes" expressed in the Archetype Definition Language (ADL) [30,73], FHIR—which utilizes its constraint and extension mechanisms to create Profiles [34,43,80], Intermountain Healthcare/Graphite Health—via "Clinical Element Models" expressed in the Clinical Element Modeling Language (CEML) [32,33,35], and the Detailed Clinical Modeling efforts in the Netherlands [36,37]. Also—and critically significantly—the models definitively declare the coded values allowed in their attributes (i.e., they declare and standardize the allowed value sets). And those value sets are drawn from standard terminologies (LOINC, SNOMED CT, RxNorm, etc.).

Another alternative to these modeling paradigms is to use a language based on the internal representations used by Protégé, a language based in OWL, or an entirely new language [30,31,41,81]. These are not necessarily mutually exclusive options, because it is probably possible to translate among some of these languages.

11.4.5 Iso-semantic models

Even if and when common models are universally adopted, ontological principles are strictly followed, and/or more specific logical models are implemented, alternate models that express the same meanings in different ways will exist. Legacy systems will not immediately be refactored to use the common models. And different use cases (e.g., a user-facing form vs. a database structure vs. a messaging solution) will demand different structures and knowledge representations. These alternate models will make different choices about how to populate name and value in a name-value pair, whether to pre-coordinate or post-coordinate, and what to place in the information model vs. in the terminology. The alternate models represent the same meaning in different manners, i.e., they are "iso-semantic."

Consequently, a means of "translating" or "transforming" between the various members of an iso-semantic family of models is needed. For example, the translation between an instance of the pre-coordinated body location model and an instance of the post-coordinated model is given by:

instance of observation model where "code" = "fetal heart rate" =
instance of observation model where "code" = "heart rate" AND "subject" = "fetus"

Many commercial engines exist to perform this type of mapping. These tools provide graphical tools by which a user can create and maintain the mappings. AI and ML tools are also emerging that, after training, "deduce" the appropriate target structure given a source structure.

11.4.6 Protecting against changes in the terminology

Experience has led to another important requirement related to maintaining decision logic that is running in a production environment. Clinical terminologies are constantly changing; concepts, terms, and relationships are added, replaced, or retired with every release. For example, a particular drug or lab test may become obsolete and be replaced by a new test. Or the existence of two codes for the same concept might be detected, necessitating the retirement of one of the codes, even though decision support logic might be referencing it. When these situations occur, decision logic could cease to function correctly if it is unable to reference the new or remaining concept. The selection of a strategy to address these issues depends on how the decision logic references the terminology, and the division of labor between maintenance of the decision logic and the terminology. If the decision logic references a value set (or a class of concepts), and the terminology can be depended upon to make the correct changes within each release, then the decision logic could expect to continue to function appropriately. For example, if the decision logic references the drug class concept of "penicillins" in a drug terminology—using the parent-child relationship in the terminology to retrieve all specific drug concepts that is a penicillin, and can depend on the drug terminology to add any new penicillins and remove obsolete ones in each release, then the terminology updates can be applied every release without a negative impact on the decision logic.

However, it is more likely that value sets or grouping of concepts would need to be developed in the home institution specifically for the decision logic, when the needed relationships or groupings among concepts are absent, incomplete, or inaccurately built in an external terminology, even in the standards. For example, "is-a" relationships can be used to retrieve descendants of pneumonia in SNOMED CT release files. However, if the requirement is to identify pneumonia encoded across multiple terminologies, e.g., ICD-10-CM in addition to SNOMED CT, then a value set that explicitly enumerates all pertinent codes will likely need to be created and maintained. Therefore, terminology maintenance work will need to be performed with each new release, determining if updates need to be performed (i.e., new codes added to and obsolete codes deleted from the value set); however, the decision logic itself should not require changes.

Lastly, some situations would require an active management of dependencies between terminology and CDS interventions. That is, when a terminology change would affect the functioning of the decision logic, the knowledge engineers responsible for the affected decision logic should be notified before the terminology change is implemented. This is the case when the concept referenced by the decision logic is changed—such as when duplicated codes are recognized by a terminology so one of them is retired and the retired code is used in the decision logic.

A system or institution may implement just one of the above strategies, or some combination of the three strategies may be required. Regardless of the maintenance strategy selected, although a terminology change may not always trigger a change of the linked decision logic, it is important to confirm that the decision logic is still producing the expected results. Typically a database that contains rows showing what objects (queries, rules, data entry screens, reports, etc.) use a particular terminology concept or class of concepts is necessary. Services can be provided so that authoring tools can update the database as terminology items are included in decision logic, or a batch process can be run routinely to find concepts in decision logic and add or modify entries in the database.

11.4.7 Context

Over the years, the need to more effectively express the restrictions imposed by context on a variety of terminology entities has been recognized [82]. These contextual restrictions can help determine "defaults" and manage "exceptions" that arise in the clinical environment. For example, decision support applications need to leverage standard terminology services that take into account the desired language of designations (terms), returning the preferred designations for the concept codes in the supplied language [83]. In the UMLS Metathesaurus, the concept "Diabetes Mellitus, Insulin-Dependent" (CUI C0011854), can be expressed in English using the following designations (synonyms): "Diabetes Mellitus, Insulin-Dependent;" "Diabetes Mellitus, Brittle;" "Diabetes Mellitus, Juvenile-Onset;" "Diabetes Mellitus, Ketosis-Prone;" "Diabetes Mellitus, Sudden-Onset;" "Diabetes Mellitus, Type 1;" and "IDDM." While all these designations arguably convey the same meaning, the appropriateness of the designations "IDDM" and "Diabetes Mellitus, Ketosis-Prone" seems restricted to clinical providers that have familiarity with diabetes, while designations like "Diabetes Mellitus, Juvenile-Onset" and "Diabetes Mellitus, Insulin-Dependent" are more widely used by non-specialized providers. If "English" is supplied as the language context and "Endocrinologist" as the medical specialty context to a terminology service, the service would be able to return "IDDM" as the appropriate designation for concept C0011854 for a user that was an English-speaking endocrinologist.

If additional contextual details are taken into account, the returned designation can potentially be even more appropriate to the needs of the clinical user. It is important to clarify that the appropriateness of the returned designation does not influence the decision support system reasoning, but instead how results are communicated to the person who reads and acknowledges the results of decision logic.

11.4.7.1 Dimensions of context

Context dimensions can include characteristics of the patient (e.g., age, sex, clinical condition, care directive, language), the clinical provider (e.g., role, discipline, subspecialty, language), the clinical setting (e.g., inpatient vs. outpatient, type of unit or hospital), and the clinical system module (e.g., problem list, order entry, personal health record), among others. Designations, relationships, and concepts are the terminology entities most frequently affected by contextual restrictions. However, contextual restrictions can potentially apply to other terminology entities, including value sets and pick-lists, or even entire code systems. The representation of contextual restrictions within the terminology conveniently handles these nuances of meaning, enhancing the expressivity of information models and inference models [82,84]. The HL7 Context-Aware Knowledge Retrieval ("Infobutton") standard identifies similar context dimensions, representing an excellent first step toward the definition of a common set of context dimensions and their corresponding values [85].

11.4.7.2 Context and terminology services

Most clinical systems and terminology services currently in operation have some mechanism for handling contextual restrictions, at least for designations. The challenge, however, is that these restrictions are typically addressed in many different and inconsistent ways, making them cumbersome to create and maintain. An important consequence is the proliferation of overlapping domains and/or the adoption of pre-coordinated descriptors that eventually become unable to represent all the desired combinations of contextual restrictions.

When a mechanism to represent and apply contextual restrictions is not provided by a terminology service, clinical systems and decision support systems are forced to adopt a restrictive approach where all designations are identified as an explicit list in the source code logic, or the choices are pre-configured using application-specific dictionary tables.

More recent efforts dealing with terminologies and concept models have embraced OWL or related logic-based formalisms [84]. However, contextual constraints likely require additional components beyond what logic-based formalisms are designed to represent [82,86].

11.4.7.3 Context and individualized care

The ongoing progression toward individualized medicine and the dissemination of process improvement and continuous learning activities will increase the complexity of the clinical data. The increased level of detail needed to characterize individualized disease states and interventions requires ongoing extensions to terminologies and information models. However, an approach that includes the purposeful distribution of the more complex data semantics across different representation layers is required, with a clear demarcation of the details captured by terminology, models, and contextual dimensions [80]. An extended set of reusable contextual dimensions can be used to consistently qualify and constrain concepts and models, while enabling continuous evolution of the clinical domains.

11.5 Implementations of ontologies, terminologies, and models in CDS

11.5.1 Implementation of terminology to support CDS interventions

Terminology plays a very important role during all implementation phases of a CDS intervention, including deployments using stand-alone programs or more sophisticated decision support services. During the initial authoring phase, the terminology and the information models define and describe the clinical data that are available for CDS [87–89], effectively constraining how the decision logic can be implemented [90]. Also, during the authoring phase, the knowledge engineer should be able to invoke requests for new terminology items, particularly new classes and value sets that can help prevent the need to create direct references to lists of identifiers (codes) in the decision logic [91].

Once authoring is concluded, the testing phase enables the validation of the decision logic, including confirmation of how real data are represented in the system. During the testing phase, the most frequently used concepts are confirmed, along with the structural and semantic restrictions defined by the information model. Finally, during the deployment for production use, all dependent terminology content and models have to be transferred to the production environment, if not yet available, with particular attention to changes that might affect the performance and reliability of the system.

The steps above might also include the need to import and transform the decision logic, when the CDS intervention is obtained from an external collaborator, even when the logic is represented using standard formalisms. Similarly, when the deployment includes external services or relies on collaborators that do not share the same exact data definitions, extensive mappings and transformations might be required. Depending on the scope of the CDS intervention, extensions to the terminology might be necessary, particularly new concept classes used for inference. These new concepts and classes also need to be imported or exported depending on the desired deployment approach.

11.5.1.1 Authoring and browsing applications

Different terminology tools and services are needed during the initial authoring phase, and also in subsequent revisions of the CDS intervention. The knowledge engineer responsible for authoring the decision logic must be able to search, retrieve, and browse the terminology and the data definitions that are available in the EHR [92]. Once the necessary items are identified, the knowledge engineer needs to be able to reference these items using simple and reliable mechanisms, which can later be used to link (bind) the decision logic to the data definitions during testing and deployment.

Previous publications have described ways in which the author can find models and data definitions [93,94]. Besides finding and displaying models, there is a need to show the connection between models and the terminology concepts, classes, and value sets. For instance, in a medication order model, the author needs to be able see that the "ordered drug" refers to a hierarchy within the terminology that contains orderable drugs. The author should be able to navigate from the model into the terminology and review the kinds of drugs that are available, including the relationships that exist between drugs (e.g., ingredients contained by a combination preparation, usual route of administration for a preparation). Linking browsing of the terminology to the context of use of a code in an information model allows a knowledge engineer to understand how to formulate decision logic against the available data, or to recognize the need for additional detail in the information model, or additional content or relationships in the data dictionary.

Based on how the models are expressed and integrated with the target system where the decision logic will be ultimately deployed, one might need to run multiple searches against terminology and model repositories and, subsequently, copy the identifiers of the needed items into the decision logic. While this look-up and copy mechanism is relatively simple, it is also tedious and error prone. A complete and seamless integration is desirable when multiple CDS interventions need to be implemented and continuously maintained. More sophisticated authoring environments allow terminology and models to be exposed via services (or imported) along with the decision logic [89,92]. This level of integration prevents errors and enables the validation and tracking of dependencies between CDS interventions, terminology concepts and information models. Proper tracking of dependencies becomes critical in subsequent iterations, where the decision logic and/or the terminology and models need to be revised or extended.

11.5.1.2 Run-time terminology services

In the testing and deployment phases, the CDS interventions frequently rely on run-time services for accessing classes and relationships in the terminology. For example, a decision engine needs to be able to ask a terminology server whether a particular drug like Nafcillin is-A Penicillin, or if a Crush Syndrome problem code implies that the patient is in an Acute Renal Failure state. Answering questions about relationships between concepts and classifying findings into clinical states are essential terminology services for CDS. These run-time terminology services provide important "abstractions" that enable modular and reusable decision logic rules. Service consumers are freed from needing to understand the complexities of the service vendor's internal models, they can be oblivious to changes that the vendor may make to those models, and they avoid needing to make frequent changes to their applications as terminology content updates occur [95]. And if the service is an industry-wide standard, they are even freed from "vendor lock," able to migrate to another service vendor with relative ease. On the vendor side, the vendor can make internal changes as needed as long as they maintain the stable interface that consumers depend on.

Additionally, services enable continuous performance monitoring and optimization, taking into account terminology items (e.g., concepts, relationships, translations, context restriction) that are commonly used. And finally, standard services promote consistent sharing of standard codes, concepts, and relationships between developers of EHRs and decision support systems. For example, the adoption of standard terminology services leads to a consistent operational meaning for relationships like is-A, part-of, ingredient-of, etc. Standard services also enforce consistent use of terminology capabilities, such as translations between code systems, management of synonyms, hierarchical inferencing, concept classification, and translation between different human languages.

The first standard terminology services were created by the Object Management Group (OMG) and were called "TQS—Terminology Query Services" or "LQS—Lexicon Query Services" [96]. HL7 and OMG subsequently published a second and extended release of the Common Terminology Services (CTS) standard, known as CTS2 [83]. More recently, HL7 has created a new specification for terminology services within the context of the FHIR standard.

These standards provide the definition of Application Programming Interfaces (APIs) to terminology services and a platform-independent model that can be used to exchange terminology with other systems, and to download and import terminology content from reference sources (e.g., FHIR Terminology Service for VSAC Resources [97]). Decision support systems should be designed with terminology content and services as an integral part of their interoperability infrastructure.

11.5.1.3 Classifiers and reasoners

Akin to runtime terminology services, ontologic classifiers and reasonsers operate over relationships in coded content. However, they go beyond simply navigating existing stated relationships, but they can discover new knowledge implied by the relationships in ontologies. When codified data records reference concepts/classes defined in an ontology, linking the ontology to the data can add a variety of additional relationships, definitions, sub and super types—all of which can be used to add information which is not explicitly present in the data, but is either background or additional knowledge which can be used for reasoning.

In particular, ontologic knowledge can be used to categorize and situate the data more properly, allowing, for example, the triggering of more specific decision support logic. Using the additional knowledge to supplement the asserted data can further trigger additional expressions which would not match the data otherwise.

11.5.2 Referencing patient data based on models

As has been described, coded terminology alone does not suffice for specifying the information needed by decision support applications. Data or information models provide and define the structure within which terminology content is populated. Services then reference the models, which in turn contain the terminology, which is referenced by standard terminology services.

The hope is that, just as terminology is standardized, the models can also be standardized. For example, if a calculation for fever is based upon body temperature values having a unit of Fahrenheit, a value with the unit of Celsius would obviously not trigger the logic. To expect the decision logic to reference all the various representation forms in which the data might be found is impractical, because the possible ways that any particular kind of data can be represented are quite varied, and they can change over time without warning. Efficient decision support requires consistency of data [98].

It is also inefficient to put the burden of data conversion or terminology mapping onto the decision logic, slowing down performance and introducing potential errors. One approach, therefore, is to transform incoming data from their native form into a single canonical form when they are stored in the EHR or data warehouse. The CDS execution engine can then access the data in this canonical form, rather than needing to access each databases data differently. This transformation is accomplished by data integration software that is aware of the various canonical models for each kind of data and using a library of model mappings to convert from the inbound form of the data to the canonical models [33]. This sort of data model normalization function should be a part of all data integration programs, and part of the function of EHR database services.

Another option particularly suitable for independent CDS execution engines (not tightly coupled to the EHR or data warehouse) is to limit the scope of the CDS to widely used standard formats (e.g., Continuity of Care Document [99] or FHIR [43]) and to the standard terminologies that have been agreed upon for coded data. This option reduces the data normalization burden, but it limits the universe of clinical data available for CDS interventions to those included in the selected exchange standard and standard terminologies.

The hope, however, is that FHIR's ever-expanding scope, aided by a multitude of organizations' FHIR profiling efforts, will enlarge the variety of data FHIR, and thus CDS, can support. If these standardized models are specific enough (see Section 11.4.4), and/or are founded on sound ontological principles (see Section 11.3.4.2), the "curly braces" problem of Section 11.1.2.1 is resolved—all decision support modules can reference a "MedicationOrder" or a "Heart Rate Measurement" observation in the same manner. This is the hope of the FHIR services, augmented by the more specific models described in Section 11.4.4.

11.6 Sharing of decision logic

As has been noted, decision support is currently difficult to share between institutions and systems because of the "curly braces" problem (Section 11.1.2.1). To recap the assertions of previous sections, in order to promulgate the sharing of decision support language, the following are required:

- Standard terminologies
- Standard models that link to standard terminologies
- Standard services that reflect the standard models

The following sections list other support requirements that will realize the goal of sharing decision support logic.

11.6.1 Increased adoption of standards

Within the US, the Department of Health and Human Services (HHS) adopted specific terminologies starting with the **Health Insurance Portability and Accountability Act (HIPAA) of** 1996, namely: ICD (currently ICD-10-CM and ICD-10-PCS), CPT, HCPCS, CDT, and NDCs. As part of the 2009 American Recovery and Reinvestment Act (ARRA), the Health Information Technology for Economic and Clinical Health (HITECH) Act has the goal of using EHRs to promote patient safety and interoperability between and within healthcare systems. The initiatives outlined in the HITECH Act are known as "Meaningful Use"—now called Promoting Interoperability Program, and named additional standard terminologies including SNOMED CT, LOINC, RxNorm, and other code sets. The aforementioned Interoperability Standards Advisory and the US Core Data Initiative [65,66] have further standardized terminologies. As a result, most EHR vendors in the US are aware of the need to support these terminologies, although the degree of adoption and integration may differ, which may have an impact on how patient data is encoded and stored in the systems.

Additional standards—not just for terminologies, models, and services—will no doubt be needed. There are several groups working to build upon the FHIR standard, facilitating its use in building decision support and other applications, e.g., the SMART on FHIR project (funded by the ONC) [100,101], which standardizes the manner in which applications adhering to the FHIR standard are launched in an EHR, and the CDS Hooks standard [102], which works in conjunction with FHIR and SMART and standardizes the invocation of CDS within a clinician's workflow.

11.6.2 A repository for collecting and sharing clinical models

With a standard modeling language and standard terminologies in place, people can begin to produce a library of detailed clinical models coupled to standard terminologies. In order for the models to be shared and approved, there will need to be a common repository where the models can be stored and accessed. The repository will need to record mappings between different models that represent the same clinical data (families of iso-semantic models). Also, the repository will need to record meta-data about the models [103]. The meta data must include (but not be limited to):

- Creator
- Creation date/time
- Last updated date and time
- Status
- Name of approving body
- First date of clinical deployment
- Decision modules that reference the model

11.6.3 Selecting models for interoperability

A process for approving a single model from a family of iso-semantic models that will be used as the reference model for decision logic will need to be developed. Again, a first step in developing the process will be to define generally accepted characteristics of a good model.

11.7 Conclusions

Standard terminologies, ontologies, and information models are essential for interoperable sharing of data, applications, and computable decision support modules. There is much work to be done to realize the vision. A common misconception is that since standard terminologies are widely available now, terminology implementation is no longer an obstacle for CDS. Agreeing on the same standard terminologies to be referenced by the decision logic is indeed a good starting point. But next, identification of and agreement upon the concepts of interest is necessary, which are lamentably not straightforward tasks. For instance, an institution's decision support may involve "biomarkers." But what constitutes the members of this group (value set)? SNOMED CT has a "tumor marker measurement (procedure)" concept, but it is likely that some of the 70 inferred descendants of the concept are not suitable for the institution's decision support usage (e.g., urine pregnancy tests). Further, the SNOMED CT list is not the same as the list (of around 75 items) from the National Cancer Institute (NCI) [104]. Further, due to LOINC's specificity, there are over 900 lab tests that have "tumor marker" added to the "Related Names" column; and there is not yet a LOINC group for tumor markers. The groups (or high level concepts) that are intuitive to humans (biomarkers, statins, response to chemotherapy, etc.) often upon more detailed investigation prove to be less clear-cut. This is just one example of the "messy" details that are encountered when attempting to implement standards.

Creation of the standards and content that can support this vision of interoperability and, just as important, guidance on implementing the standards, is a challenging task, and greater than any one group can accomplish. The knowledge and science around modeling is evolving rapidly. Many groups are expending resources to achieve this goal of model-driven architecture and model-based interoperability. Everyone is invited to participate. A key factor in success will be the ability to work together. We will achieve the goal much faster, if we build on and reuse the work of others, and collaborate rather than compete.

References

[1] National Committee on Vital and Health Statistics. Report on uniform data standards for patient medical record information. 2000 [Internet] [cited 2021 September 25]. Available from: https://www.ncvhs.hhs.gov/wp-content/uploads/2014/08/hipaa000706.pdf.

[2] Cimino JJ. Desiderata for controlled medical vocabularies in the twenty-first century. Methods Inf Med 1998;37 (4-5):394–403.

[3] Cimino JJ. Controlled medical vocabulary construction: methods from the Canon Group [editorial]. J Am Med Inform Assoc 1994;1(3):296–7.

[4] Humphreys B, DAB L, Kingsland L, editors. Building the unified medical language system. In: Symposium on computer applications in medical care. IEEE Computer Society Press; 1989. p. 475–80.

[5] Rector A.L. Clinical terminology: why is it so hard? Methods Inf Med. 1999 [Internet] [cited 2021 August 27]; 38(4–5):239–52. Available from: https://pubmed.ncbi.nlm.nih.gov/10805008/.

[6] González Bernaldo de Quirós F., Otero C., Luna D. Terminology services: standard terminologies to control health vocabulary. Yearb Med Inform [Internet]. 2018 [cited 2021 August 27]; 27(1):227-233. Available from: https://pubmed.ncbi.nlm.nih.gov/29681027/ doi: https://doi.org/10.1055/s-0038-1641200.

[7] Chute CG. Clinical classification and terminology. J Am Med Inform Assoc 2000;7(03):298–303.

[8] Cimino JJ. High-quality, standard, controlled healthcare terminologies come of age. Methods Inf Med 2011;50 (02):101–4.

[9] Cornet R., Chute C.G. Health concept and knowledge management: twenty-five years of evolution. Yearb Med Inform 2016 [cited 2021 August 27]; Aug 2;Suppl 1(Suppl 1):S32-41. Available from: https://www.ncbi.nlm.nih.gov/pmc/articles/PMC5171511/ doi:10.15265/IYS-2016-s037.

[10] Chute CG, Cohn SP, Campbell JR. A framework for comprehensive health terminology systems in the United States: development guidelines, criteria for selection, and public policy implications. ANSI Healthcare Informatics Standards Board Vocabulary Working Group and the Computer-Based Patient Records Institute Working Group on Codes and Structures. J Am Med Inform Assoc 1998;5(6):503–10.

[11] Bodenreider O., Cornet R., Vreeman D.J. Recent developments in clinical terminologies—SNOMED CT, LOINC, and RxNorm. Yearb Med Inform [Internet]. 2018 [cited 2021 August 27]; 27(1):129-139. Available from: https://pubmed.ncbi.nlm.nih.gov/30157516/ doi:https://doi.org/10.1055/s-0038-1667077.

[12] Healthcare Services Platform Consortium. Solor: the simple healthcare terminology solution [Internet]. 2019 [cited 2021 August 18]; Available from: http://solor.io/.

[13] Stearns MQ, Price C, Spackman KA, Wang AY. SNOMED clinical terms: overview of the development process and project status. Proc AMIA Symp 2001;662–6.

[14] Wang AY, Sable JH, Spackman KA. The SNOMED clinical terms development process: refinement and analysis of content. Proc AMIA Symp 2002;845–9.

[15] Humphreys BL, Lindberg DA, Schoolman HM, Barnett GO. The Unified Medical Language System: an informatics research collaboration. J Am Med Inform Assoc 1998;5(1):1–11.

[16] Lindberg DA, Humphreys BL, McCray AT. The unified medical language system. Methods Inf Med 1993;32 (4):281–91.

[17] Forrey AW, McDonald CJ, DeMoor G, et al. Logical observation identifier names and codes (LOINC) database: a public use set of codes and names for electronic reporting of clinical laboratory test results. Clin Chem 1996;42 (1):81–90.

[18] Huff SM, Rocha RA, McDonald CJ, et al. Development of the LOINC (logical observation identifier names and codes) vocabulary. JAMIA 1998;5(3):276–92.
[19] McDonald CJ, Huff SM, Suico JG, et al. LOINC, a universal standard for identifying laboratory observations: a 5-year update. Clin Chem 2003;49(4):624–33.
[20] Nelson SJ, Brown SH, Erlbaum MS, et al. A semantic normal form for clinical drugs in the UMLS: early experiences with the VANDF. Proc AMIA Symp 2002;557–61.
[21] Regenstrief Institute, Inc., 2021 LOINC from Regenstrief. [Internet] [cited 2021 August 25]. Available from: https://loinc.org/.
[22] SNOMED International, 2021 The value of SNOMED CT [Internet]. [cited 2021 August 25]. Available from: https://www.snomed.org/snomed-ct/why-snomed-ct.
[23] National Library of Medicine. RxNorm [Internet]. 2021 [cited 2021 August 25]. Available from: https://www.nlm.nih.gov/research/umls/rxnorm/index.html.
[24] Centers for Disease Control and Prevention. International classification of diseases, tenth revision, clinical modification (ICD-10-CM) [Internet]. 2021 [cited 2021 August 25]. Available from: https://www.cdc.gov/nchs/icd/icd10cm.htm.
[25] Dimitrieski V, Petrović G, Kovačević A, Luković I, Fujita H. A survey on ontologies and ontology alignment approaches in healthcare. In: International conference on industrial, engineering and other applications of applied intelligent systems. Cham: Springer; 2016. p. 373–85.
[26] Rector AL, Qamar R, Marley T. Binding ontologies and coding systems to electronic health records and messages. Appl Ontology 2006;51–69.
[27] Lezcano L., Sicilia M.A., Rodríguez-Solano C. Integrating reasoning and clinical archetypes using OWL ontologies and SWRL rules. J Biomed Inform, 2011 [Internet] Apr [cited 2021 September 29];44(2):343-53. Available from: https://pubmed.ncbi.nlm.nih.gov/21118725/ doi:https://doi.org/10.1016/j.jbi.2010.11.005.
[28] Rector A.L., Nowlan W.A. (University of manchester). The GALEN representation and integration language (GRAIL) Kernel, version 1. In: The GALEN consortium for the EC. 1993.
[29] ISO/IEC 8825-2. Information technology—ASN.1 encoding rules: Specification of packed encoding rules (PER). Geneva, Switzerland: International Organization for Standardization; 1996.
[30] Beale T, Heard S. The archetype definition language version 2 (ADL2). Australia: The openEHR Foundation; 2006.
[31] W3C. OWL web ontology language semantics and abstract syntax., 2004, http://www.w3.org/TR/owl-semantics/.
[32] Coyle JF, Mori AR, Huff SM. Standards for detailed clinical models as the basis for medical data exchange and decision support. Int J Med Inform 2003;69(2–3):157–74.
[33] Oniki T.A., Zhuo N., Beebe C.E., Liu H., Coyle J.F., Parker C.G., Solbrig H.R., Marchant K., Kaggal V.C., Chute C.G., Huff S.M. Clinical element models in the SHARPn consortium. J Am Med Inform Assoc 2016; 23(2):248-56. [Internet] March [cited 2021 August 28]. Available from: https://pubmed.ncbi.nlm.nih.gov/26568604/ doi: https://doi.org/10.1093/jamia/ocv134.
[34] Lee J, Hulse NC, Wood GM, Oniki TA, Huff SM. Profiling fast healthcare interoperability resources (FHIR) of family health history based on the clinical element models. AMIA Annu Symp Proc 2017;2016:753–62.
[35] Oniki T.A., Coyle J.F., Parker C.G., Huff S.M. Lessons learned in detailed clinical modeling at Intermountain Healthcare. J Am Med Inform Assoc 2014 [Internet] [cited 2021 August 30]; 21(6):1076-1081. Available from: https://academic.oup.com/jamia/article/21/6/1076/786753 doi:https://doi.org/10.1136/amiajnl-2014-002875.
[36] Goossen W.T. Detailed clinical models: representing knowledge, data and semantics in healthcare information technology. Healthc Inform Res 2014 [Internet] [cited 2021 August 31]; 20(3):163-172. Available from: https://pubmed.ncbi.nlm.nih.gov/25152829/ doi:https://doi.org/10.4258/hir.2014.20.3.163.
[37] Goossen W., Goossen-Baremans A., van der Zel M. Detailed clinical models: a review. Healthc Inform Res [Internet]. 2010 [cited 2021 August 31]; 16(4):201-214. Available from: https://pubmed.ncbi.nlm.nih.gov/21818440/ doi:https://doi.org/10.4258/hir.2010.16.4.201.
[38] Health Level Seven. Arden syntax for medical logic systems. Ann Arbor, Michigan: Health Level Seven; 1999.
[39] Hripcsak G. Arden syntax for medical logic modules. MD Comput 1991;8(2):76. 78.
[40] HL7 International. Clinical quality language (CQL) [Internet]. 2014 [updated 2021 May 6; cited 2021 September 23]. Available from: https://cql.hl7.org/.

[41] Health Level Seven. GELLO: a common expression language, ANSI/HL7 V3 GELLO, R1-2005. Ann Arbor, Michigan: Health Level Seven; 2005.

[42] Sordo M, Ogunyemi O, Boxwala AA, Greenes RA. GELLO: an object-oriented query and expression language for clinical decision support. AMIA Annu Symp Proc 2003;1012.

[43] HL7 International. HL7 FHIR Release 4. 2019, [Internet] November 1 [updated 2019 November 1; cited 2021 August 4]. Available from: http://hl7.org/fhir/.

[44] McGinnis J.M., Aisner D., Olsen L.O. The learning healthcare system. Washington (DC), USA: National Academies Press; 2007 [Internet] [cited 2021 August 27]. 374 p. Available from: https://www.nap.edu/catalog/11903/the-learning-healthcaresystem-workshop-summary DOI: https://doi.org/10.17226/11903.

[45] Etheredge L.M. A rapid-learning health system. Health aff (Millwood) [Internet]. 2007 [cited 2021 August 10];007;26(2):w107-w118. Available from: https://pubmed.ncbi.nlm.nih.gov/17259191/ doi:https://doi.org/10.1377/hlthaff.26.2.w107.

[46] The Office of the National Coordinator for Health Information Technology. Connecting health and care for the nation: a ten year vision to achieve interoperable health IT infrastructure, 2014 [Internet] June 5 [cited 2021 August 9]. Available from: https://www.healthit.gov/sites/default/files/ONC10yearInteroperability ConceptPaper.pdf.

[47] The Office of the National Coordinator for Health Information Technology. Connecting health and care for the nation a shared nationwide interoperability roadmap [Internet]. 2015 October 6 [cited 2021 August 9]. Available from: https://www.healthit.gov/sites/default/files/hie-interoperability/nationwide-interoperability-roadmap-final-version-1.0.pdf.

[48] Interoperability Standards Priority (ISP) Task Force 2021. Report to the Health Information Technology Advisory Committee, 2021 [Internet] June 9 [cited 2021 August 9]. Available from: https://www.healthit.gov/sites/default/files/facas/2021-06-09_ISP_TF_2021_HITAC%20Recommendations_Report.pdf.

[49] Institute of Medicine. Roundtable on value and science-driven health care. Learning Healthcare System series of publications; 2010.

[50] National Research Council. Digital infrastructure for the learning health system: the foundation for continuous improvement in health and health care: workshop series summary. Washington, DC: The National Academies Press; 2011.

[51] Smith M., Saunders R., Stuckhardt L., M.G. JM, editors. Best care at lower cost: the path to continuously learning health care in America [Internet]. Washington (DC): National Academies Press (US); Committee on the Learning Health Care System in America: Institute of Medicine; May 10, 2013 [cited 2021 August 28]. Available from: https://pubmed.ncbi.nlm.nih.gov/24901184/ doi:10.17226/13444.

[52] Budrionis A., Bellika J.G. The Learning Healthcare System: Where are we now? A systematic review. J Biomed Inform 2016 [Internet] [cited 2021 August 30]; 64:87-92. Available from: https://pubmed.ncbi.nlm.nih.gov/27693565/ doi:https://doi.org/10.1016/j.jbi.2016.09.018.

[53] English M., Irimu G., Agweyu A., Gathara D., Oliwa J., Ayieko P., et al. Building learning health systems to accelerate research and improve outcomes of clinical care in low- and middle-income countries. PLoS Med [Internet]. 2016 April 12 [cited 2021 August 30]; 13(4):e1001991. Available from: https://pubmed.ncbi.nlm.nih.gov/27070913/ doi:10.1371/journal.pmed.1001991.

[54] Zurynski Y, Smith CL, Vedovi A, Ellis LA, Knaggs G, Meulenbroeks I, et al. Mapping the learning health system: a scoping review of current evidence. Sydney, Australia: Australian Institute of Health Innovation, and the NHRMC Partnership Centre for Health System Sustainability; 2020.

[55] Foley T., Horwitz L., Zahran R. Realising the potential of learning health systems. Newcastle: The Learning Healthcare Project; 2021 [Internet] [cited 2021 August 9]. Available from: https://learninghealthcareproject.org/wp-content/uploads/2021/05/LHS2021report.pdf.

[56] Platt J.E., Raj M., Wienroth M. An analysis of the learning health system in its first decade in practice: scoping review. J Med Internet Res 2020 [Internet] March 19 [cited 2021 August 30]; 22(3):e17026. Available from: https://pubmed.ncbi.nlm.nih.gov/32191214/ doi:https://doi.org/10.2196/17026.

[57] Foley T., Fairmichael F. The potential of learning healthcare systems. Newcastle: The Learning Healthcare Project; 2015 [Internet] [cited 2021 August 9]. Available from: http://www.learninghealthcareproject.org/LHS_Report_2015.pdf.

[58] Olsen L., Aisner D., M.G. JM, editors. Institute of Medicine (US). Roundtable on evidence-based medicine. the learning healthcare system: workshop summary [Internet]. Washington (DC): National Academies Press (US); 2007 [cited 2021 August 30]. Available from: https://pubmed.ncbi.nlm.nih.gov/21452449/ doi:10.17226/11903.

[59] Olsen L.A., Saunders R.S., M.G. JM, editors. Institute of Medicine (US). Patients charting the course: citizen engagement and the learning health system: workshop summary [Internet]. Washington (DC): National Academies Press (US); 2011 [cited 2021 August 30]. Available from: https://www.ncbi.nlm.nih.gov/books/NBK91496/ doi:10.17226/12848.
[60] M.N. Asim, M. Wasim, M.U.G. Khan, W. Mahmood, H.M. Abbasi, A survey of ontology learning techniques and applications, Database, vol 2018, 2018, bay101, doi:https://doi.org/10.1093/database/bay101.
[61] Song, HJ, Jo, BC, Park, CY, et al. Comparison of named entity recognition methodologies in biomedical documents. Biomed Eng Online 17, 158 (2018). doi:https://doi.org/10.1186/s12938-018-0573-6.
[62] Obrst L. Ontologies for semantically interoperable systems. In: Proceedings of the twelfth international conference on information and knowledge management (CIKM '03). New York, NY, USA: Association for Computing Machinery; 2003. p. 366–9. https://doi.org/10.1145/956863.956932.
[63] National Center for Biotechnology Information (NCBI). MeSH. 2021 [Internet]. [cited 2021 November 1]; Available from: https://www.ncbi.nlm.nih.gov/mesh/.
[64] W3C. SKOS simple knowledge organization system—home page. 2012 [Internet]. December 13, [cited 2021 November 1]; Available from: https://www.w3.org/2004/02/skos/.
[65] The Office of the National Coordinator for Health Information Technology. 2021 Interoperability standards advisory: reference edition 2021 [Internet]. [cited 2021 August 25]. Available from: https://www.healthit.gov/isa/sites/isa/files/inline-files/2021-ISA-Reference-Edition.pdf.
[66] The Office of the National Coordinator for Health Information Technology. United States core data for interoperability (USCDI), 2021 [Internet]. [cited 2021 August 25]. Available from: https://www.healthit.gov/isa/united-states-core-data-interoperability-uscdi.
[67] Lange C, et al. LoLa: a modular ontology of logics, languages, and translations. WoMO 2012.
[68] Horridge M, Knublauch H, Rector A, Stevens R, Wroe C. A practical guide to building OWL ontologies with the protege-OWL Plugin. 1st ed. University of Manchesteropen EHR Foundation; 2004.
[69] W3C. OWL 2 Web ontology language primer (2nd ed.). [Internet]. 2012 [cited 2021 November 1]; Available from: https://www.w3.org/TR/owl2-primer/.
[70] Baader F, Calvanese D, McGuinness DL, Nardi D, Patel-Schneider PF, editors. The description logic handbook. 2nd ed. Cambridge, UK: Cambridge University Press; 2007.
[71] openEHR, openEHR: Open industry specifications, models and software for e-health [Internet]. 2021 [cited 2021 August 4]. Available from: https://www.openehr.org/.
[72] International Standardization Organization (ISO), 2021. The ISO 13606 standard [Internet]. [cited 2021 August 4]. Available from: http://www.en13606.org/information.html.
[73] openEHR Specification Program, 2021. Archetype definition language 2 (ADL2) [Internet]. The openEHR Foundation. c2003-2021 [updated: 2021 March 9; cited 2021 September 25]. Available from: https://specifications.openehr.org/releases/AM/latest/ADL2.html.
[74] HL7 International, 2020. FHIR shorthand [Internet]. HL7 International—FHIR Infrastructure Group. c2020 [updated 2021 August 17; cited 2021 September 25]. Available from: https://build.fhir.org/ig/HL7/fhir-shorthand/The Clinical Data Interchange Standards Consortium. CDISC [Internet]. 2021 [cited 2021 August 4]. Available from: https://www.cdisc.org/.
[75] Nadkarni PM. QAV: querying entity-attribute-value metadata in a biomedical database. Comput Methods Programs Biomed 1997;53(2):93–103.
[76] Nadkarni PM, Marenco L, Chen R, Skoufos E, Shepherd G, Miller P. Organization of heterogeneous scientific data using the EAV/CR representation. J Am Med Inform Assoc 1999;6(6):478–93.
[77] Guizzardi G. Ontological foundations for structural conceptual models. Enschede: Telematica Instituut /CTIT, 2005. 416 p. (CTIT PhD Thesis Series; 05-74). (Telematica Instituut Fundamental Research Series; 015). "Ontological foundations for structural conceptual models." (2005).
[78] ISO/CD 17115. Health informatics—Vocabulary of terminology; 2007.
[79] HL7 International. US Core Implementation Guide: 4.0.0—STU4 Release, 2021. June 28 [Internet]. [cited 2021 August 30]. Available from: https://www.hl7.org/fhir/us/core/.
[80] Matney S.A., Heale B., Hasley S., Decker E., Frederiksen B., Davis N., et al. Lessons learned in creating interoperable fast healthcare interoperability resources profiles for large-scale public health programs. Appl Clin Inform 2019; 10(1):87-95. January [cited 2021 August 28]. Available from: https://pubmed.ncbi.nlm.nih.gov/30727002/ doi:https://doi.org/10.1055/s-0038-1677527.
[81] Noy NF, Crubezy M, Fergerson RW, et al. Protege-2000: an open-source ontology-development and knowledge-acquisition environment. AMIA Annu Symp Proc 2003;953.

[82] Schulz S, Rodrigues JM, Rector A, Chute CG. Interface terminologies, reference terminologies and aggregation terminologies: a strategy for better integration. Stud Health Technol Inform 2017;245:940–4.
[83] Object Management Group. Common Terminology Services 2—Version 1.2 [Internet]. April 2015 [cited 2021 August 27]. Available from: https://www.omg.org/spec/CTS2/.
[84] Rector AL. The interface between information, terminology, and inference models. Medinfo 2001;10(Pt 1):246–50.
[85] Del Fiol G., Huser V., Strasberg H.R., Maviglia S.M., Curtis C., Cimino J.J. Implementations of the HL7 Context-Aware Knowledge Retrieval ("Infobutton") Standard: challenges, strengths, limitations, and uptake. J Biomed Inform 2012 [cited 2021 August 27] [Internet] 45(4):726-35 Available from: https://pubmed.ncbi.nlm.nih.gov/22226933/ doi:https://doi.org/10.1016/j.jbi.2011.12.006.
[86] Rector AL. Defaults, context, and knowledge: alternatives for OWL-indexed knowledge bases. Pac Symp Biocomput 2004;226–37.
[87] Delvaux N., Vaes B., Aertgeerts B., Van de Velde S., Vander Stichele R., Nyberg P., Vermandere M. Coding systems for clinical decision support: theoretical and real-world comparative analysis. JMIR Form Res [Internet]. 2020 October 21 [cited 2021 August 28]; 4(10):e16094. Available from: https://pubmed.ncbi.nlm.nih.gov/33084593/ doi:https://doi.org/10.2196/16094.
[88] Lin Y, Staes CJ, Shields DE, Kandula V, Welch BM, Kawamoto K. Design, development, and initial evaluation of a terminology for clinical decision support and electronic clinical quality measurement. AMIA Annu Symp Proc 2015;2015:843–51.
[89] Lomotan E.A., Meadows G., Michaels M., Michel J.J., Miller K. To share is human! Advancing evidence into practice through a national repository of interoperable clinical decision support. Appl Clin Inform 2020; 11 (1):112-121. [Internet] January [cited 2021 August 28]; Available from: https://pubmed.ncbi.nlm.nih.gov/32052388/ doi:https://doi.org/10.1055/s-0040-1701253.
[90] Odigie E., Lacson R., Raja A., Osterbur D., Ip I., Schneider L., et al. Fast Healthcare Interoperability Resources, Clinical Quality Language, and Systematized Nomenclature of Medicine-Clinical Terms in Representing Clinical Evidence Logic Statements for the Use of Imaging Procedures: Descriptive Study. JMIR Med Inform [Internet]. 2019 May 13 [cited 2021 August 28]; 7(2):e13590. Available from: https://pubmed.ncbi.nlm.nih.gov/31094359/ doi:https://doi.org/10.2196/13590.
[91] Bodenreider O, Nguyen D, Chiang P, Chuang P, Madden M, Winnenburg R, McClure R, Emrick S, D'Souza I. The NLM value set authority center. Stud Health Technol Inform 2013;192:1224.
[92] Zhou L., Karipineni N., Lewis J., Maviglia S.M., Fairbanks A., Hongsermeier T., Middleton B., Rocha R.A. A study of diverse clinical decision support rule authoring environments and requirements for integration. BMC Med Inform Decis Mak 2012 [Internet]. [cited 2021 August 27] Nov 12;12:128. Available from: https://pubmed.ncbi.nlm.nih.gov/23145874/ doi:https://doi.org/10.1186/1472-6947-12-128.
[93] Huff SM, Rocha RA, Coyle JF, Narus SP. Integrating detailed clinical models into application development tools. Medinfo 2004;11(Pt 2):1058–62.
[94] Parker CG, Rocha RA, Campbell JR, Tu SW, Huff SM. Detailed clinical models for sharable, executable guidelines. Medinfo 2004;11(Pt 1):145–8.
[95] Peters L, Nguyen T, Bodenreider O. Terminology status APIs—mapping obsolete codes to current RxNorm, SNOMED CT, and LOINC concepts. Stud Health Technol Inform 2017;245:1333.
[96] Object Management Group. Lexicon Query Service. TC Document CORBAmed 98-03-22. Object Management Group; 1998.
[97] National Institutes of Health (NIH) National Library of Medicine. VSAC API Resources 2021 [Internet] July 22 [cited 2021 November 1]. Available from: https://www.nlm.nih.gov/vsac/support/usingvsac/vsacsvsapiv2.html.
[98] Tcheng JE, Bakken S, Bates DW, Bonner III H, Gandhi TK, Josephs M, Hamilton Lopez M, editors. Optimizing strategies for clinical decision support: summary of a meeting series. Washington, DC: National Academy of Medicine; 2017.
[99] Paterno MD, Goldberg HS, Simonaitis L, Dixon BE, Wright A, Rocha BH, Ramelson HZ, Middleton B. Using a service oriented architecture approach to clinical decision support: performance results from two CDS Consortium demonstrations. AMIA Annu Symp Proc 2012;2012:690–8.
[100] Mandel J.C., Kreda D.A., Mandl K.D., Kohane I.S., Ramoni R.B. SMART on FHIR: a standards-based, interoperable apps platform for electronic health records. J Am Med Inform Assoc [Internet]. 2016 [cited 2021 September 24];23(5):899-908. Available from: https://pubmed.ncbi.nlm.nih.gov/26911829/ doi:https://doi.org/10.1093/jamia/ocv189.

[101] Computational Health Informatics Program, Boston Children's Hospital. SMART [Internet]. Boston (MA): Computational Health Informatics Program, Boston Children's Hospital; 2019 [cited 2021 September 24]; Available from: https://smarthealthit.org/.

[102] HL7 & Boston Children's Hospital. CDS Hooks [Internet]. 2018 [cited 2021 November 1]. Available from: https://cds-hooks.org/.

[103] Alper B.S., Flynn A., Bray B.E., et al. Categorizing metadata to help mobilize computable biomedical knowledge. Learn Health Sys [Internet]. 2021 [cited 2021 October 30];e10271. Available from: https://onlinelibrary.wiley.com/doi/10.1002/lrh2.10271 doi:10.1002/lrh2.10271.

[104] NIH National Cancer Institute. Tumor markers in common use. [Internet]. May 11, 2021 [cited November 1, 2021]; Available from: https://www.cancer.gov/about-cancer/diagnosis-staging/diagnosis/tumor-markers-list.

CHAPTER

12

Grouped knowledge elements

Claude Nanjo[a] *and Aziz A. Boxwala*[b]

[a]Department of Biomedical Informatics, University of Utah, Salt Lake City, UT, United States [b]Elimu Informatics, Inc., La Jolla, CA, United States

12.1 Introduction

Two frequently occurring tasks in the clinical workflow where the health care provider and the computer communicate are during clinical documentation (both in structured data input and in report production) and in computerized-provider order entry (CPOE). In fact, newer CDS tools and applications often combine these elements of presenting data, requesting new data to be entered, and recommending actions based on the clinical care context.

The above kinds of tasks are facilitated by grouping of relevant knowledge elements, i.e., structured documentation templates and order sets respectively, to support care of specific conditions or care processes. Thus, we consider **Grouped Knowledge Elements (GKE) as a special class of CDS.** A structured documentation template is an organized collection of data items relevant to a particular clinical context (e.g., patient presenting with a headache) that can be used to collect or present codified information about a patient. An order set, similarly, is an organized collection of actions that can be ordered by a health care provider for the care of a patient in a specific clinical context (e.g., orders for a patient being admitted with stroke, post-operative orders).

The availability of data may be considered to be the most fundamental prerequisite to effective CDS of any kind, because analysis and guidance depend on it. Use of GKE for CDS both facilitates obtaining relevant data (e.g., in structured forms) and applying it to making relevant associations of data clear (e.g., in documentation templates and order sets).

Usefulness of CDS depends on the data being well-structured and unambiguous. Coding of data items is essential in order to understand and be able to manipulate them. Chapter 11 reviewed approaches to standardizing the terminology and information models used for data items. The latter includes the need for the data type of each item to be specified, including units for quantifiable data types or categorical values for dictionary/directory-based data

https://doi.org/10.1016/B978-0-323-91200-6.00012-7

types. Lack of ambiguity of these aspects of a data item is essential when data are communicated between the user and the computer.

As we have noted, GKEs may be considered as vehicles for providing clinical decision support. In a passive sense, they remind the health care professional about questions to ask the patient, specific observations to be made during an examination, tests to be ordered, or medications to be administered relevant to the particular clinical context. Documentation templates that display information can organize information in such a way as to play a mnemonic function, by facilitating identification of important data, recognition of trends, or making other associations. More actively, those who design the groupings of knowledge elements can use them to drive the behavior of the health care professional user. For example, in an order set, an intervention that is known to be more effective, per current evidence, can be made easier to order than alternatives by specifying that the intervention be selected by default [1]. Additionally, order items and documentation items can be dynamically presented based on the context, e.g., a documentation template can suggest asking about past history of rheumatic fever only if a murmur is noted as having been heard on auscultation. By anticipating needs for data entry or access, or for orders, such grouping not only provides CDS but also facilitates workflow, by eliminating extra steps that would otherwise be needed.

The specification of an order set's or a document template's structure and content is a form of *knowledge*. Standards continue to be developed for such specification to encourage the collection of higher quality, more interpretable, more comprehensive data, and to encourage reuse of document specifications, or parts thereof, where appropriate. This chapter will review those efforts, in terms of their degree of maturity and harmonization, and how they relate to clinical decision support.

The management of a collection of grouped knowledge element specifications has not received much attention in the clinical informatics and standards development communities. This is a knowledge management (KM) task, the purpose of which is to reconcile collections of document specifications in use in an enterprise, and encourage convergence on and reuse of specific ones that foster best practices and conformance with enterprise goals. Knowledge management for grouped knowledge elements is only recently being recognized as an important challenge, as KM systems begin to be introduced into health care enterprises to manage their knowledge content. Examples of approaches to curating documentation specifications, and supporting authors and editors in creating and updating them, are discussed in Chapter 28.

12.2 Clinical documentation

According to a report of the National Academy of Medicine (formerly Institute of Medicine (IOM)), "A learning healthcare system is designed to generate and apply the best evidence for the collaborative healthcare choices of each patient and provider; to drive the process of discovery as a natural outgrowth of patient care; and to ensure innovation, quality, safety, and value in healthcare" [2]. Thanks to a nation-wide focus in the United States (US) and other countries on improving health care quality, reducing costs and achieving better patient outcomes, adoption and use of electronic health record (EHR) systems is universal. However, to fully realize the potential of such systems, it is necessary to clearly identify the different roles of key clinical data and how such data are best integrated into clinical documentation at different stages in the clinical workflow, while specifying the roles care providers and patients play in relationship to an EHR [3]

According to the Health Level 7 (HL7) Clinical Document Architecture (CDA) definition, the purpose of clinical documentation is to capture clinical observations and services [4]. Under this definition, clinical documentation is not restricted to a medical record, but includes any patient-related clinical document that tracks and reports on operational activities carried out by health care providers, including a clinician interviewing or assessing a patient, recording observations about a patient's current health status, reviewing previous records, assessing laboratory test results, reading reports of studies, performing an examination, conducting a procedure, or consulting external sources of information. Among the key functions of clinical documentation, regardless of its type, are that it: a) communicates relevant clinical information between health care providers and health care teams; b) supports compliance with internal policies, quality improvement efforts, regulations, and law, and; c) provides inputs for other types of CDS such as alerts and reminders.

Many clinical documentation tool systems have been created to attempt to capture data in a structured format, while others allow the users to express the clinical data in a more narrative fashion. The structured data capture tools, by employing templates applicable to the clinical context, constrained data-entry fields, and validation rules, guide or restrict the user's data entry. This can result in data that are more usable for downstream applications [5] such as for measuring and reporting clinical quality [6], for clinical decision support [7,8], for consistent capture of validated patient assessment instruments such as the Glasgow Coma Score [9], and for serving as the basis for billing [10]. Computer-based clinical documentation systems also have an advantage in that the template can be dynamic, presenting elements for data capture that are pertinent to the clinical context, as in the case of Smart Forms (Fig. 12.1) in use at Partners HealthCare System, in Boston, MA. Smart Forms facilitate data entry in all forms (free text, structured, and coded), and provide dynamic, context-driven, actionable decision support at point of care, while ideally, if well designed, enhancing workflow and reducing the burden and redundancy of data entry [11]. The documentation system can determine the applicability of the data elements based on data elements in the patient's record, and on the data that already have been entered in the current template. For example, in a form in which the user is asked about presence of family history of breast cancer, a positive response could lead to invocation of a subsection in the form that requests details of the relatives affected, the age of onset of the disease, and whether breast cancer had been the cause of death in those relatives.

Although there are clear benefits of capturing structured data at the point of care in an electronic format, there always have been and continue to be significant challenges associated, not only with this task itself, but also with ensuring continued and maximized usefulness of the captured information. These challenges arise in part from the increased and tedious effort required to enter information into computer-based forms using a keyboard and mouse. Furthermore, by constraining the data to be entered, documentation templates may limit the ability to enter data from unusual or unexpected patient presentations [12].

The disparities in the structure and representation of data across existing systems make it more difficult to reuse data that are captured in a clinical document in other applications.

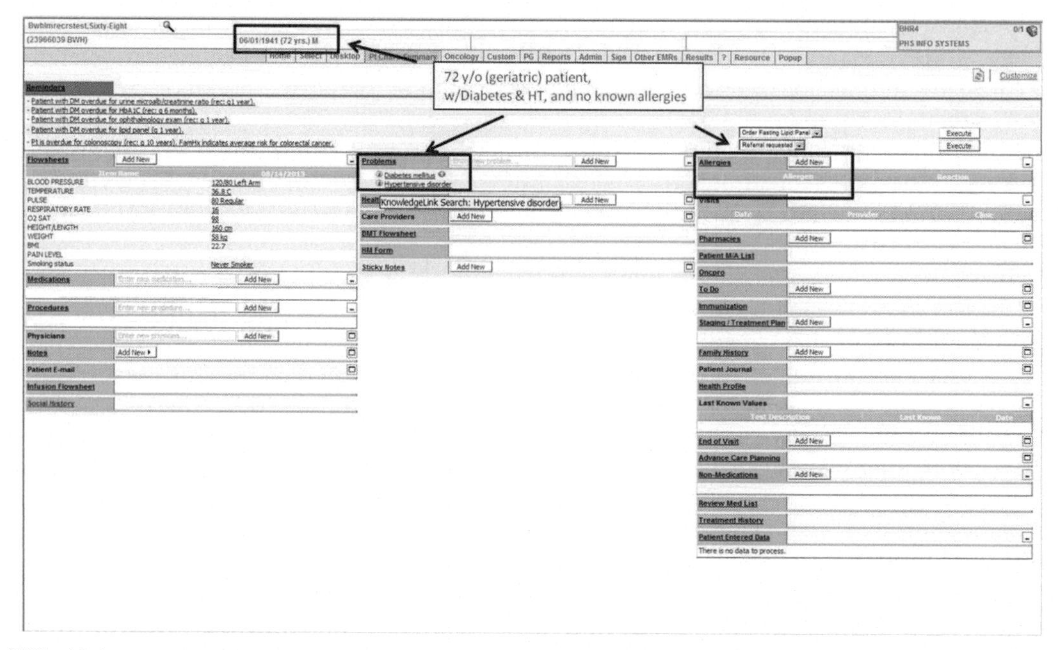

FIG. 12.1 A Smart Form for a primary care visit. It is primarily a documentation tool that incorporates several modalities of decision support such as reminders for overdue laboratory tests (top left portion of the form), infobuttons (in the problems section, center column of the form), and allowing physicians to add, edit, and delete coded and structured clinical information, such as medical problems, medications, allergies, vital signs, laboratory values, and health maintenance information, and to easily import that information into a visit note. *Courtesy of Partners Healthcare, Boston, MA.*

Previous efforts on CDA [4,13], Clinical Information Modeling Initiative or CIMI [14] and openEHR [15,16] have focused on developing and supporting standard-based technologies with a strong emphasis on semantic interoperability. The CDA is an ANSI-certified standard from HL7. Based on the HL7 Reference Information Model (RIM), the CDA specifies both syntax and full semantics of a clinical document based on persistence, stewardship, potential for authentication, context, wholeness, and human readability. A more detailed exposition of the CDA is provided later in the chapter. Similarly, openEHR, a European-led effort and ISO standard, is an open specification for management, storage, retrieval and exchange of heath data in EHRs. Clinical content is modeled in terms of two artefacts: archetypes and templates. Archetypes provide the means for defining reusable data element definitions binding clinical data models to terminologies (Fig. 12.2). Such definitions are independent from software implementations and infrastructure; they also are context-free, allowing for reusability in different settings and systems. Archetypes can be assembled into logical groups, or libraries of related data items. See Chapter 11 for further discussion of detailed clinical element models, and the various approaches beyond OpenEHR's archetypes.

Templates are used in OpenEHR to logically represent case-specific data sets, such as the data items making up a patient discharge summary, or a radiology report. Templates are built by reusing relevant elements from already defined archetypes. Templates must be consistent with the semantics of the archetypes from which they are built.

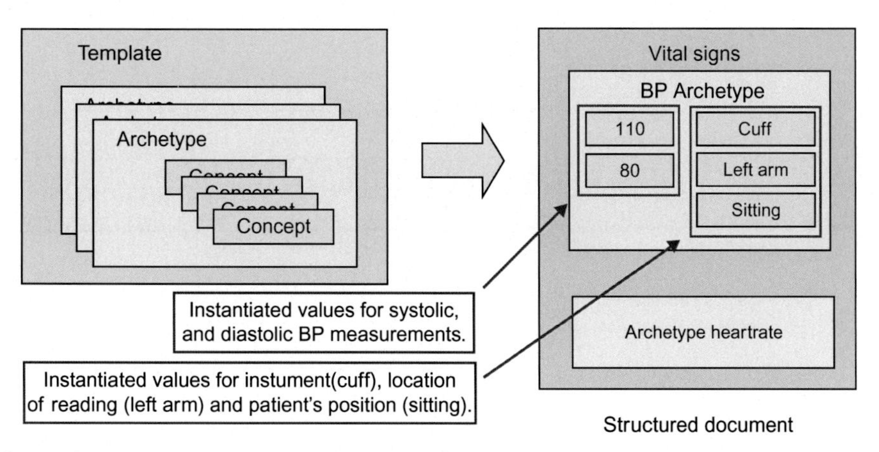

FIG. 12.2 A template with a simplified archetype for blood pressure. The BP archetype is an aggregation of five concepts: systolic and diastolic blood pressure values, instrument for measurement (cuff), location of reading (left arm), and patient's position (sitting). For a given BP measurement, the concepts would take on values and, for some concepts, these may be terminology codes (such as for the patient position). A template is constructed by grouping together different archetypes.

12.3 Order sets

Order sets, a critical feature of CPOE systems, are collections of orderable items aimed to facilitate the ordering process and improve the quality and consistency of care. Order sets organize and structure complex health care activities through structured groups of medical orders that provide mnemonic value, convenience, efficiency, and clinical decision support during the ordering process. A typical order set includes a variety of clinical activities to be performed, e.g., medications, treatment protocols, diagnostic procedures, laboratory tests, referrals and consultations, and nursing care. Order sets are process-specific as in the case of general admission and pre-operative order sets; condition-specific as in the case of diabetes, asthma, and hypertension; or more often a combination of process- and disease-specific, e.g., a post-operative order set for a specific procedure (Fig. 12.3), and a pancreatitis care pathway order set (Fig. 12.4).

As noted above, in addition to providing organization and structure for the ordering process, order sets also provide clinical decision support at the point of care [17,18]. Some of the ways CDS can be provided in order sets are by:

1. Facilitating selection of appropriate orders: Although order sets might include alternative interventions (e.g., beta-blockers, diuretics, calcium-channel blockers for a hypertension order set), the intervention that is more effective (or cost-effective) can be presented as a nudge in a manner that makes it the more likely choice by the provider [19]. Appropriate orders can also be suggested based on the patient's condition, e.g., selection of a diagnostic radiology exam [20].
2. Refining an order: The order set tool can suggest or constrain or calculate doses or frequencies of medication administration. The latter can be particularly useful in specific situations, such as when the dose is dependent on body weight or renal function.
3. Preventing unsafe ordering and offering safer alternatives: alerting physicians to a patient's allergies to a given medication, or prompting about potential adverse drug

ARTHROPLASTY (Hip/Knee) - Page 4 of 4 Active Pt: CLAUS, SANTA P

Medications continued
- [ ] Magnesium hydroxide (mom) 30 ML PO QD PRN:<R
- [ ] Bisacodyl (dulcolax) 10 mg pr qam prn:<r> start pod #3 if no stool

Anticoagulant
- [ ] Dalteparin sodium (fragmin) 2500 units sc x1 8 hrs after surgery
- [ ] Dalteparin sodium (fragmin) 5000 units sc qpm <di> start pod # 1
- [ ] Warfarin sodium 5 mg po x1 <fdi> If ordering warfarin sodium, please select this lab: [ ] PT-INR QAM x 4 days

PCA Medication [ ] Morphine sulfate pca <more> or [ ] Hydromorphone hcl pca <more>

Pain / Sleep Medication
- [ ] Oxycodone 5 mg/acetaminophen 325 mg (percocet 5 mg/325 mg) 1-2 tab po q4h prn:<r> <l>
- [ ] Propoxyphene nap/acetaminophen (darvocet n 100) 1 TAB PO Q4H PRN:<R<l>
- [ ] Acetaminophen (tylenol) 650 MG PO Q4H <DI> P
- [ ] Zolpidem tartrate (ambien) 5 mg po qhs <fdi> prn:<r>

Summary | Instructions

MAGNESIUM HYDROXIDE (MOM)
30 ML PO
QD
PRN: constipation

Prev Page | Next Page | Edit | OK | Cancel

FIG. 12.3 One screen of a multi-page post-operative order set for a patient having undergone arthroplasty of the hip or knee. This order set shows some of the medications being ordered; the order items included suggested doses. The provider must select the items from this page that he or she wishes to order for this patient. The provider can modify the parameters of some of the orders, e.g., the dose of magnesium hydroxide. *Courtesy of Partners Healthcare, Boston, MA.*

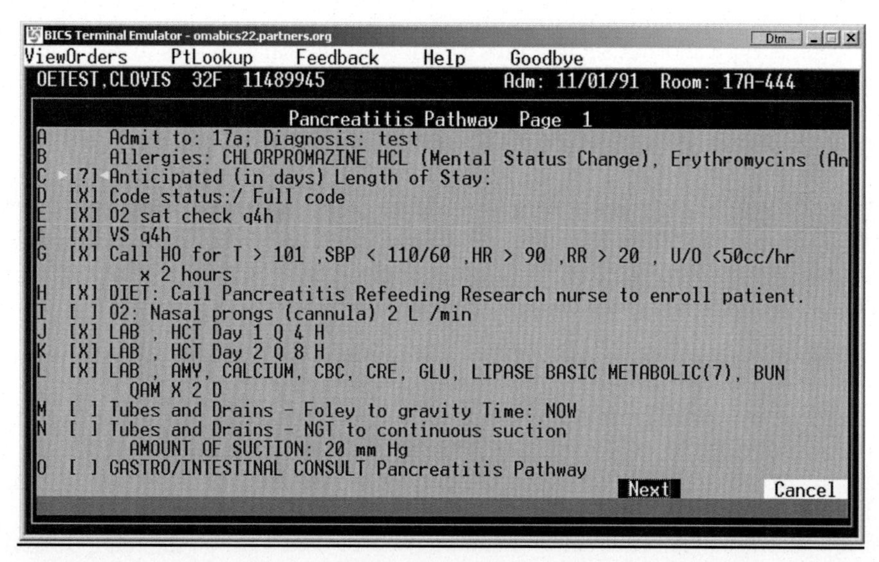

BICS Terminal Emulator - omabics22.partners.org

ViewOrders PtLookup Feedback Help Goodbye

OETEST,CLOVIS 32F 11489945 Adm: 11/01/91 Room: 17A-444

Pancreatitis Pathway Page 1

```
A       Admit to: 17a; Diagnosis: test
B       Allergies: CHLORPROMAZINE HCL (Mental Status Change), Erythromycins (An
C  [?]  Anticipated (in days) Length of Stay:
D  [X]  Code status:/ Full code
E  [X]  O2 sat check q4h
F  [X]  VS q4h
G  [X]  Call HO for T > 101 ,SBP < 110/60 ,HR > 90 ,RR > 20 , U/O <50cc/hr
          x 2 hours
H  [X]  DIET: Call Pancreatitis Refeeding Research nurse to enroll patient.
I  [ ]  O2: Nasal prongs (cannula) 2 L /min
J  [X]  LAB , HCT Day 1 Q 4 H
K  [X]  LAB , HCT Day 2 Q 8 H
L  [X]  LAB , AMY, CALCIUM, CBC, CRE, GLU, LIPASE BASIC METABOLIC(7), BUN
          QAM X 2 D
M  [ ]  Tubes and Drains - Foley to gravity Time: NOW
N  [ ]  Tubes and Drains - NGT to continuous suction
          AMOUNT OF SUCTION: 20 mm Hg
O  [ ]  GASTRO/INTESTINAL CONSULT Pancreatitis Pathway
```

Next Cancel

FIG. 12.4 One screen of a multi-page care pathway order set for a patient with pancreatitis, from an older version of the Partners predecessor system BICS. This order set addresses multiple aspects of care including vital sign monitoring (item F), diet (item H), laboratory tests (items J-L), incision care (tubes and drains, items M, N) and consultations by specialists (item O). The provider must select the items from this page that he or she wishes to order for this patient. *Courtesy of Partners Healthcare, Boston, MA.*

interactions between a medication a patient is currently taking and the medication that is about to be ordered.

4. Preventing omissions: The order set can suggest additional or corollary orders, e.g., recommend renal function tests when ordering vancomycin.

There are challenges associated with the development and maintenance of high-quality order sets. It has been shown that the effectiveness of order sets is dependent on careful design of the order sets [21]. Thus, it is highly desirable to be able to share order sets designed to promote best practice across organizations.

Since different EHR systems implement CPOE in different ways, order sets in use in the different CPOE systems are typically encoded in proprietary formats. This makes it difficult to import order sets that are shared or obtained from knowledge vendors, since they are typically not in a format that is suitable for integration into the CPOE systems. The underlying order items in different CPOE systems also may not be semantically aligned across systems. There are efforts to create a standard structure for order sets. The Clinical Decision Support Work Group in HL7 created a specification for order sets. The HL7 Order Set Publication, Release 1, a Draft Standards for Trial Use (DSTU), [22] is derived from the HL7 RIM [23]. The order set is made up of sections with each section containing order items. The latter are defined as subclasses of various Act classes from the RIM. Soon after, the Health eDecisions project sponsored by the United States (US) Office of the National Coordinator for Health Information Technology (ONC), Department of Health and Human Services, 2012) has produced a specification for CDS knowledge artifacts [24] that also can be used to represent order sets. With the rapid adoption of HL7's FHIR standards, a specification for order sets [25], based on the Health eDecision knowledge artifact specification, called the PlanDefinition resource was added to FHIR. For practical purposes, the PlanDefinition resource has replaced the other two standards. This specification is described later in Section 12.4.2.

12.4 Current standards for grouped knowledge elements

In this section, we describe important standards for grouped knowledge elements: the HL7 Clinical Document Architecture (CDA) and related templates, the HL7 FHIR Questionnaire and QuestionnaireResponse resources, and the HL7 FHIR PlanDefinition resource. The CDA is a specification for documents that contain health information previously collected. It enables patient data to be shared for a variety of purposes including care coordination, quality reporting and analysis [26], and clinical decision support [27]. FHIR Questionnaire and PlanDefinition are specifications for resources[a] for sharing knowledge in the form of templates for documentation and for order sets respectively. Associated with these two types of resources is the Clinical Quality Language (CQL)[28] that is used for specifying logical expressions for CDS (see Chapter 9 for details on CQL). These FHIR resources can be used for creating new health information. In other words, CDA documents and FHIR QuestionnaireResponse contain instances of patient data, and Questionnaire and PlanDefinition resources define the knowledge that is applied to patient data such as that contained in CDA documents.

[a] In FHIR, the unit of information that can be exchanged between two systems are called resources.

12.4.1 HL7 clinical document architecture

CDA is a specification for exchanging clinical information in the form of electronic documents [4]. A CDA document contains a header and a body. The header contains information about the source and author of the document, the subject of the document (e.g., the patient), and authentication information. The body contains the health information that is being shared. The CDA body provides a three-level model with increasing structure and semantics added to the body from Level 1 to Level 3. The contents of a document at Level 1 are largely narrative text such as a dictated reported or progress note. At Level 2, the document can be divided into a number of sections (e.g., History of Present Illness and Physical Examination). In a Level 3 document, the health information is codified. A discrete codified data element is referred to as an entry and specified as a Clinical Statement. Clinical Statements are common patterns used primarily for clinical messages. HL7 defines a Clinical Statement as "an expression of a discrete item of clinical (or clinically related) information that is recorded because of its relevance to the care of a patient. Clinical information is fractal in nature and therefore the extent and detail conveyed in a single statement may vary". Since the RIM is the basis of HL7 version 3 specifications, the goal of Clinical Statements and of Level 3 documents is to promote syntactic and semantic interoperability.

The CDA is intentionally designed to be flexible to allow its use in a variety of clinical documents ranging from progress notes to diagnostic examination reports to discharge summaries. For each of these applications, templates and implementation guides are used to constrain the contents of the body to achieve interoperability.

As part of the incentive programs for EHR adoption in the US (so-called "meaningful use" standards) [29] a number of templates for CDAs have been developed or refined. These templates include those for longitudinal care and transitions of care (e.g., continuity of care document, discharge summary), diagnostic reports (e.g., radiology reports), and quality reporting (namely, the Quality Reporting Document Architecture or QRDA templates). These templates specify how the documents will be populated with different types of Clinical Statements, their cardinality and optionality, and the value domains, in particular the sets of standard terminology codes that are considered valid.

12.4.2 FHIR questionnaire

To share documentation templates, FHIR provides a specification called the Questionnaire resource. These documentation templates are intended to support the input of data for many different clinical scenarios such as for risk assessment, soliciting patient-reported outcomes instruments, documenting ambulatory encounters, and charting in flowsheets during acute care encounters. Questionnaire resources can be exchanged in the popular Javascript Object Notation (JSON) and Extensible Markup Language (XML) formats. Documentation systems, such as EHRs and patient engagement applications, that support the FHIR Questionnaire standard can import Questionnaire resources and then use them as templates for structured input of data about a patient.

The Questionnaire resource contains attributes for specifying the metadata and the questions, i.e., the data elements to be documented. The metadata elements describe the provenance of the Questionnaire, attributions of authors, reviewers, publishers, licensing information, and codes indicating the applicability of the Questionnaire artifact in various scenarios. Together, these attributes can facilitate knowledge management by developers of the artifact and review for suitability of their use by consumers of the artifact.

The *item* attribute of the Questionnaire contains the questions. The questions can be grouped into sections. For example, encounter note templates may contain sections for History of Present Illness, Past Medical History, Review of Systems, and Physical Exam Findings. A question can be assigned an identifier, codes from standard terminology systems, and logic to skip questions. In addition, the artifact can specify for each whether a response is required, the data type of the response, other constraints on the response (e.g., minimum value or maximum value), default values, and codes from standard terminology systems to associate with a response. Fig. 12.5 shows an example of a Questionnaire resource.

When a Questionnaire is used to enter data about a patient, those responses are stored in a FHIR resource called QuestionnaireResponse. FHIR provides mechanisms to prepopulate QuestionnaireResponses with data from other FHIR resources. For example, if there is a question on whether a Patient has Asthma, the response for that may be obtained from the EHR by searching the Condition resources, which record a patient's diagnoses. When a QuestionnaireResponse is completed, the responses can be used to construct other FHIR resources and saved as those resources too. For example, if the user indicates the patient is taking Metformin tablets, that response can be saved in a MedicationStatement resource. For using Questionnaire and QuestionnaireResponse resources to obtain Patient Reported Outcomes, FHIR provides a detailed implementation guide (Health Level 7, 2019b).

12.4.3 FHIR PlanDefinition

In FHIR Release 3, two new resources were introduced to support the definition of declarative, sharable, and computable clinical decision-support rules and order sets using the FHIR constraint language—PlanDefinition and ActivityDefinition. The structure of the PlanDefinition resource is largely derived from the HL7 Knowledge Artifact Specification (KAS) Implementation Guide [24]. The purpose of KAS was to define a standard knowledge representation scheme for interoperable clinical decision support knowledge artifacts (Fig. 12.6). In HL7 KAS, the following constructs were provided:

- A metadata section that specifies provenance, attribution, and applicability information.
- An external data section that specifies the input requirements of the artifacts. For instance, the data needed for rule execution such as specific lab results or past medication orders.
- An expressions section that defines and names the logical expressions and formulae used in the artifact. These expressions are then referenced elsewhere in the artifact (e.g., in the condition section of the artifact).
- An events section that specifies the event when the artifact itself is triggered such as on opening the patient's chart.
- A conditions section that generally specifies the applicability requirements for the artifact (e.g., if patient age >= 18 years and patient is male).
- An action group/action section that specifies the 'consequent' of the rule or the orderables in an order set.
- A behaviors section that specifies how to select and configure items in an action group.

```
{
  "resourceType": "Questionnaire",
  "id": "phq-2-questionnaire",
  "version": "1.0",
  "title": "Patient Health Questionnaire (PHQ-2)",
  "status": "draft",
  "subjectType": [
    "Patient"
  ],
  "code": [
    {
      "system": "http://loinc.org",
      "code": "55757-9",
      "display": "Patient Health Questionnaire 2 item (PHQ-2) [Reported]"
    }
  ],
  "item": [
    {
      "linkId": "LittleInterest",
      "code": [
        {
          "system": "http://loinc.org",
          "code": "44250-9"
        }
      ],
      "text": "Little interest or pleasure in doing things",
      "type": "choice",
      "required": true,
      "answerValueSet": "http://loinc.org/vs/LL358-3"
    },
    {
      "linkId": "FeelingDown",
      "code": [
        {
          "system": "http://loinc.org",
          "code": "44255-8"
        }
      ],
      "text": "Feeling down, depressed, or hopeless",
      "type": "choice",
      "required": true,
      "answerValueSet": "http://loinc.org/vs/LL358-3"
    }
  ]
}
```


FIG. 12.5 A FHIR Questionnaire resource containing the PHQ-2 instrument in the JSON format. A limited number of the metadata fields are shown in this example, including version, title, and code. This artifact has two questions as seen within the item attribute.

Much like the KnowledgeDocument in KAS, PlanDefinition also supports the same constructs. PlanDefinition provides the overall structural framework for a group of knowledge artifacts and consists of three core attribute groupings:

1. The artifact's metadata such as the title, description, copyright information, author and publisher of the knowledge artifact. These metadata are similar to those of the Questionnaire resource.

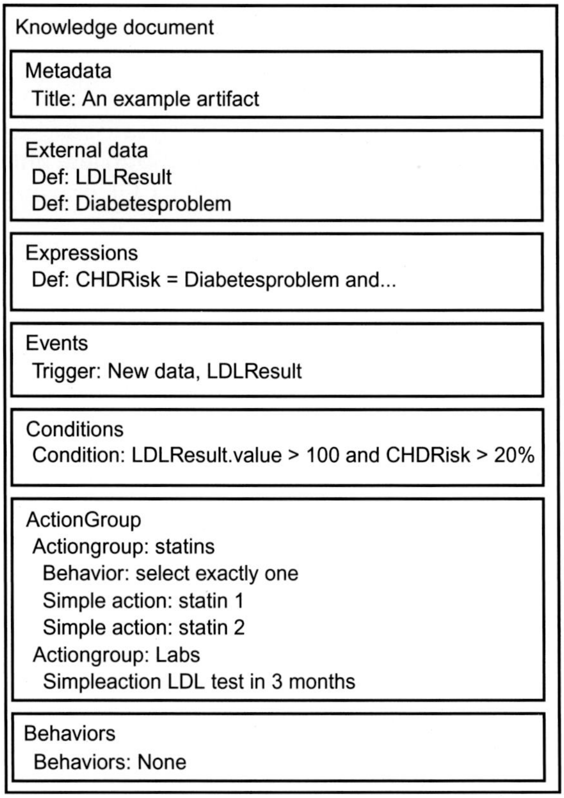

Knowledge document

Metadata
Title: An example artifact

External data
Def: LDLResult
Def: Diabetesproblem

Expressions
Def: CHDRisk = Diabetesproblem and...

Events
Trigger: New data, LDLResult

Conditions
Condition: LDLResult.value > 100 and CHDRisk > 20%

ActionGroup
Actiongroup: statins
Behavior: select exactly one
Simple action: statin 1
Simple action: statin 2
Actiongroup: Labs
Simpleaction LDL test in 3 months

Behaviors
Behaviors: None

FIG. 12.6 An example to provide a conceptual overview of how a knowledge artifact is represented in the KA schema. The major sections and their contents are shown. This rule is triggered when a new test result is available. The conditions of the test being above a specified value and the presence of risk of heart disease are evaluated and the actions specified in action groups are then executed. Conditions can be composed of expressions, and the expressions can refer to data from external sources.

2. The artifact's goals indicating the purpose of the knowledge artifact.
3. The artifact's actions which define the activities the artifact can perform in order to meet specified goals (when defined).

CDS knowledge artifacts are rarely simply static enumerations of activities to be performed in any clinical context (though, of course, they can be). Rather they typically define sets of activities that dynamically vary according to clinical context and patient (or population) characteristics. For instance, a CDS recommendation for a given patient may be tailored according to the patient's diagnoses, current active medications, relevant lab results, care venue, provider type, and step in the clinical workflow where the recommendation will be displayed.

In order to support such dynamic behaviors, PlanDefinition presents CDS knowledge artifact authors with a number of attributes that can be used to further elaborate when its actions should be considered:

- The PlanDefinition action.trigger attribute is typically used to specify when, in a clinical workflow, the action should be considered. For instance, it may specify that the action should be considered upon opening the patient's chart or when signing an order.
- The PlanDefinition action.condition attribute further refines when an action is applicable. For instance, it can be used to specify that the action should only be considered for persons 18 years of age or older with a hypertension diagnosis. Conditions are typically specified

using an expression language such as the Clinical Quality Language (CQL) and an information model such as FHIR.

- The PlanDefinition action.participant attribute can be used to specify which provider type or role is allowed to perform the action.

Each action in a plan definition can specify input data requirements (action.input) and output data requirements (action.output). These requirements may be specified in CQL or using a data structure called DataRequirement.

Logical expressions may be specified in the CQL syntax within the PlanDefinition resource itself or in reusable Library resources that are then referenced in the PlanDefinition. The CQL expression syntax has, among others, logical, set, temporal, and mathematical operators. Complex logical expressions can be decomposed into a set of smaller expressions to improve authoring and readability of the artifact. CQL supports coded terminology concepts as first-class elements of the language itself thus allowing knowledge artifacts to share common and formally coded semantics. CQL can be used to specify both the data requirements and conditional expressions of knowledge artifacts.

12.4.3.1 *Modeling rules*

PlanDefinition can be used to construct event-condition-action (ECA) rules. These rules are evaluated when the specified triggering event occurs. If the condition in the rule evaluates to true, the specified actions are performed. Chapter 9 describes in detail the use of PlanDefinition to model ECA rules.

12.4.3.2 *Modeling order sets*

PlanDefinition also supports the definition of order sets that can range from highly static, such as simple lists of sections and/or orderables to highly dynamic, where the actual content of the order set can adjust based on individual patient characteristics, provider characteristics, and clinical context. To indicate that a PlanDefinition is an order set, the type of the PlanDefinition is set to order-set as shown in the snippet:

```
"type": {
    "coding": [
    {
          "system": "http://terminology.hl7.org/CodeSystem/plan-
definition-type",
          "code": "order-set",
          "display": "Order Set"
    }
  ]
}
```

At its core, an order set is a collection of orderable items (e.g., Metoprolol 50 mg Tab 1 tablet taken by mouth twice daily) organized into sections (e.g., 'Medications', 'Lab Tests', 'Radiology'). To make order sets more dynamic, a knowledge author can make use of the following constructs:

- condition—to specify in which context an order set, section, or orderable is applicable

- dynamicValue—can be used to configure orderable and orderable sentence attribute values based on context.
- timing—when the order must be performed

Orderable items are typically defined using the ActivityDefinition resource. The kind of order must also be specified, such as MedicationRequest in the example below. The Activity-Definition resource specifies a number of attributes common to medication and non-medication orderables (Fig. 12.7).

The following example illustrates the definition of an order set section called "Medications" which may include a number of medication orderables, although only one is shown in the example below. Note the `groupingBehavior` of the action indicates that this action is a logical group (i.e., a section in the order set). The `selectionBehavior` of `at-most-one` indicates that only a single orderable may be chosen at a time in this section. In a CPOE system's interface, this selection behavior might be represented as radio buttons next to each orderable in the section.

```
"action": [
  {
  "action": [
    {
      "title": "Medications",
      "description": "Consider the following medications for
  stable patients to be initiated prior to the cardiology
  consultation.",
      "groupingBehavior": "logical-group",
      "selectionBehavior": "at-most-one"
      "action": [
        {
          "textEquivalent": "metoprolol tartrate 50 mg tablet 1
tablet oral 2 time daily",
  "definitionCanonical": "#metoprololTartrate50Prescription",
          "dynamicValue": [
            {
              "path": "status",
              "expression": {
                "language": "text/cql",
                "expression": "'draft'"
              }
            }
          ]
        }
      ]
    },
    ...
  ]
 }
]
```

```
"contained": [
  {
    "resourceType": "ActivityDefinition",
    "id": "metoprololTartrate50Prescription",
    "status": "draft",
    "kind": "MedicationRequest",
    "productReference": {
      "reference": "#metoprololTartrate50Medication"
    },
    "dosage": [
      {
        "text": "1 tablet by mouth 2 time daily",
        "timing": {
          "repeat": {
            "frequency": 2,
            "period": 1,
            "periodUnit": "d"
          }
        },
        "route": {
          "coding": [
            {
              "code": "26643006",
              "display": "Oral route (qualifier value)"
            }
          ],
          "text": "Oral route (qualifier value)"
        },
        "doseAndRate": [
          {
            "type": {
              "coding": [
                {
                  "system": "http://terminology.hl7.org/CodeSystem/dose-rate-
type",
                  "code": "ordered",
                  "display": "Ordered"
                }
              ]
            },
            "doseQuantity": {
              "value": 1,
              "unit": "{tbl}"
            }
          }
        ]
      }
    ],
    "dynamicValue": [
      {
        "path": "medicationRequest.dispenseRequest.quantity",
        "expression": {
          "language": "text/cql",
          "expression": "30"
        }
      }
    ]
  },
  {
    "resourceType": "Medication",
    "id": "metoprololTartrate50Medication",
    "code": {
```

FIG. 12.7 An example of an order item in an Order Set modeled as an ActivityDefinition. The item references a Medication resource also shown in the example.

(Continued)

```
        "coding": [
          {
            "system": "http://www.nlm.nih.gov/research/umls/rxnorm",
            "code": "866514"
          }
        ],
        "text": "metoprolol tartrate 50 MG Oral Tablet"
      },
      "doseForm": {
        "coding": [
          {
            "system": "http://snomed.info/sct",
            "code": "385055001",
            "display": "Tablet dose form"
          }
        ],
        "text": "Tablet dose form"
      }
    }
]
```

FIG. 12.7, CONT'D

PlanDefinition offers a number of other ways to refine how orderables are grouped and presented in an ordering interface. The `requiredBehavior` is used to indicate whether the action must be performed or may be overridden by the clinician after providing a reason. The `precheckBehavior` may be used to preselect orderables by default, such as for orders that may be preferred due to supporting evidence.

The full order set in JSON format is shown in Fig. 12.8.

12.5 Conclusions

This chapter presented a different perspective on CDS, as groups of knowledge elements. By grouping together elements such as order items or data elements relevant to a clinical context, it is possible to provide CDS to the health care provider in a natural and usable way that is part of the provider's usual workflow. The CDS might be passive and invoked on-demand, with the presence of the element in the group reminding the user of its applicability in that context. It may be more active, by dynamically modifying and presenting knowledge elements based on the user's selection and entries. If well designed, such use of CDS can also anticipate user needs and enhance workflow. New standards are evolving that will enable sharing of these grouped knowledge elements in the form of documentation templates, order sets, and reports.

```
{
  "resourceType": "PlanDefinition",
  "id": "example-os",
  "contained": [
    {
      "resourceType": "ActivityDefinition",
      "id": "metoprololTartrate50Prescription",
      "status": "draft",
      "kind": "MedicationRequest",
      "productReference": {
        "reference": "#metoprololTartrate50Medication"
      },
      "dosage": [
        {
          "text": "1 tablet oral 2 time daily",
          "timing": {
            "repeat": {
              "frequency": 2,
              "period": 1,
              "periodUnit": "d"
            }
          },
          "route": {
            "coding": [
              {
                "code": "26643006",
                "display": "Oral route (qualifier value)"
              }
            ],
            "text": "Oral route (qualifier value)"
          },
          "doseAndRate": [
            {
              "type": {
                "coding": [
                  {
                    "system": "http://terminology.hl7.org/CodeSystem/dose-rate-
type",
                    "code": "ordered",
                    "display": "Ordered"
                  }
                ]
              },
              "doseQuantity": {
                "value": 1,
                "unit": "{tbl}"
              }
            }
          ]
        }
      ],
      "dynamicValue": [
        {
          "path": "medicationRequest.dispenseRequest.quantity",
          "expression": {
            "language": "text/cql",
            "expression": "30"
          }
        }
      ]
    },
    {
      "resourceType": "Medication",
      "id": "metoprololTartrate50Medication",
      "code": {
        "coding": [
          {
            "system": "http://www.nlm.nih.gov/research/umls/rxnorm",
            "code": "866514"
```

FIG. 12.8 An order set expressed as a PlanDefinition in the JSON format.

(Continued)

```
          }
        ],
        "text": "metoprolol tartrate 50 MG Oral Tablet"
      },
      "doseForm": {
        "coding": [
          {
            "system": "http://snomed.info/sct",
            "code": "385055001",
            "display": "Tablet dose form"
          }
        ],
        "text": "Tablet dose form"
      }
    }
  ],
  "url": "http://example.com/plandefinition/orderset/simple-example-orderset",
  "identifier": [
    {
      "system": "urn:example.com.com:pd:os",
      "value": "OS123"
    }
  ],
  "version": "1.0",
  "name": "SimpleOrderSet",
  "title": "Example Order Set",
  "type": {
    "coding": [
      {
        "system": "http://terminology.hl7.org/CodeSystem/plan-definition-type",
        "code": "order-set",
        "display": "Order Set"
      }
    ]
  },
  "status": "active",
  "date": "2021-10-01",
  "publisher": "My Company",
  "description": "A simple order set to illustrate the capabilities of
PlanDefinition",
  "action": [
    {
      "action": [
        {
          "title": "Medications",
          "description": "Consider the following medications for stable
patients to be initiated prior to the cardiology consultation.",
          "groupingBehavior": "logical-group",
          "selectionBehavior": "at-most-one"

          "action": [
            {
              "textEquivalent": "metoprolol tartrate 50 mg tablet 1 tablet oral
2 time daily",
              "definitionCanonical": "#metoprololTartrate50Prescription",
              "dynamicValue": [
                {
                  "path": "status",
                  "expression": {
                    "language": "text/cql",
                    "expression": "'draft'"
                  }
                }
              ]
            }
          ]
        }
      ]
    }
  ]
}
```

FIG. 12.8, CONT'D

References

[1] Rubins D, Boxer R, Landman A, Wright A. Effect of default order set settings on telemetry ordering. J Am Med Inform Assoc 2019;26:1488–92. https://doi.org/10.1093/jamia/ocz137.
[2] Olsen LA, Aisner D, McGinnis JM, editors. Institute of Medicine (US) Roundtable on Evidence-Based Medicine. In: The Learning Healthcare System: Workshop Summary. Washington, DC: National Academies Press (US); 2007.
[3] Cusack CM, Hripcsak G, Bloomrosen M, Rosenbloom ST, Weaver CA, Wright A, et al. The future state of clinical data capture and documentation: a report from AMIA's 2011 Policy Meeting. J Am Med Inform Assoc 2013;20:134–40. https://doi.org/10.1136/amiajnl-2012-001093.
[4] Dolin RH, Alschuler L, Boyer S, Beebe C, Behlen FM, Biron PV, et al. HL7 Clinical Document Architecture, Release 2. J Am Med Inform Assoc 2006;13:30–9. https://doi.org/10.1197/jamia.M1888.
[5] Häyrinen K, Saranto K, Nykänen P. Definition, structure, content, use and impacts of electronic health records: a review of the research literature. Int J Med Inform 2008;77:291–304. https://doi.org/10.1016/j.ijmedinf.2007.09.001.
[6] Chan KS, Fowles JB, Weiner JP. Review: electronic health records and the reliability and validity of quality measures: a review of the literature. Med Care Res Rev 2010;67:503–27. https://doi.org/10.1177/1077558709359007.
[7] Kwok R, Dinh M, Dinh D, Chu M. Improving adherence to asthma clinical guidelines and discharge documentation from emergency departments: implementation of a dynamic and integrated electronic decision support system. Emerg Med Australas 2009;21:31–7. https://doi.org/10.1111/j.1742-6723.2008.01149.x.
[8] Schriger DL, Baraff LJ, Buller K, Shendrikar MA, Nagda S, Lin EJ, et al. Implementation of clinical guidelines via a computer charting system: effect on the care of febrile children less than three years of age. J Am Med Inform Assoc 2000;7:186–95.
[9] Teasdale G, Jennett B. Assessment of coma and impaired consciousness. A practical scale. Lancet 1974;2:81–4. https://doi.org/10.1016/s0140-6736(74)91639-0.
[10] Powers J, Gillett P, Goldblum K. Forms facilitating primary care documentation. Nurse Pract 2000;25:40–4. 49.
[11] Schnipper JL, Linder JA, Palchuk MB, Einbinder JS, Li Q, Postilnik A, et al. "Smart Forms" in an Electronic Medical Record: documentation-based clinical decision support to improve disease management. J Am Med Inform Assoc 2008;15:513–23. https://doi.org/10.1197/jamia.M2501.
[12] Rosenbloom ST, Denny JC, Xu H, Lorenzi N, Stead WW, Johnson KB. Data from clinical notes: a perspective on the tension between structure and flexible documentation. J Am Med Inform Assoc 2011;18:181–6. https://doi.org/10.1136/jamia.2010.007237.
[13] Dolin RH, Alschuler L, Beebe C, Biron PV, Boyer SL, Essin D, et al. The HL7 clinical document architecture. J Am Med Inform Assoc 2001;8:552–69.
[14] Oniki TA, Zhuo N, Beebe CE, Liu H, Coyle JF, Parker CG, et al. Clinical element models in the SHARPn consortium. J Am Med Inform Assoc 2016;23:248–56. https://doi.org/10.1093/jamia/ocv134.
[15] Garde S, Chen R, Leslie H, Beale T, McNicoll I, Heard S. Archetype-based knowledge management for semantic interoperability of electronic health records. Stud Health Technol Inform 2009;150:1007–11.
[16] International Standards Organization. Health informatics—Electronic health record communication—Part 1: Reference model; 2008.
[17] Bates DW, Teich JM, Lee J, Seger D, Kuperman GJ, Ma'Luf N, et al. The impact of computerized physician order entry on medication error prevention. J Am Med Inform Assoc 1999;6:313–21.
[18] Horsky J, Kuperman GJ, Patel VL. Comprehensive analysis of a medication dosing error related to CPOE. J Am Med Inform Assoc 2005;12:377–82. https://doi.org/10.1197/jamia.M1740.
[19] Jacobs BR, Hart KW, Rucker DW. Reduction in clinical variance using targeted design changes in computerized provider order entry (CPOE) order sets: impact on hospitalized children with acute asthma exacerbation. Appl Clin Inform 2012;3:52–63. https://doi.org/10.4338/ACI-2011-01-RA-0002.
[20] Ip IK, Schneider LI, Hanson R, Marchello D, Hultman P, Viera M, et al. Adoption and meaningful use of computerized physician order entry with an integrated clinical decision support system for radiology: ten-year analysis in an urban teaching hospital. J Am Coll Radiol 2012;9:129–36. https://doi.org/10.1016/j.jacr.2011.10.010.
[21] Leu MG, Morelli SA, Chung O-Y, Radford S. Systematic update of computerized physician order entry order sets to improve quality of care: a case study. Pediatrics 2013;131(Suppl 1):S60–7. https://doi.org/10.1542/peds.2012-1427g.
[22] Health Level 7. HL7 Version 3 Standard: Order Set Publication, Release 1. Ann Arbor, MI: Health Level 7; 2012.

[23] Health Level 7. HL7 Reference Information Model. Ann Arbor, MI: Health Level 7; 2012.
[24] Health Level 7. HL7 Implementation Guide: Clinical Decision Support Knowledge Artifact Implementations, Release 1. Ann Arbor, MI: Health Level 7; 2013.
[25] Health Level 7. FHIR v4.0.1 2019, http://hl7.org/fhir/R4/. [Accessed 25 September 2021].
[26] Velamuri S. QRDA—Technology overview and lessons learned. J Healthc Inf Manag 2010;24:41–8.
[27] Goldberg HS, Paterno MD, Rocha BH, Schaeffer M, Wright A, Erickson JL, et al. A highly scalable, interoperable clinical decision support service. J Am Med Inform Assoc 2013. https://doi.org/10.1136/amiajnl-2013-001990.
[28] Health Level 7. Clinical Quality Language (CQL) 2021., 2021, https://cql.hl7.org/. [Accessed 10 October 2021].
[29] Office of the National Coordinator for Health Information Technology (ONC), Department of Health and Human Services. Health information technology: standards, implementation specifications, and certification criteria for electronic health record technology, 2014 edition; revisions to the permanent certification program for health information technology. Final rule. Fed Regist 2012;77:54163–292.

CHAPTER

13

Infobuttons and point of care access to knowledge

Guilherme Del Fiol[a], Hong Yu[b], and James J. Cimino[c]

[a]Department of Biomedical Informatics, University of Utah Health, University of Utah, Salt Lake City, UT, United States [b]Center for Biomedical and Health Research in Data Sciences and Miner School of Computer & Information Sciences, University of Massachusetts, Lowell, MA, United States [c]Informatics Institute, The University of Alabama at Birmingham, Birmingham, AL, United States

13.1 Introduction

Much of this book deals with the use of computer algorithms and heuristics for providing clinical decision support (CDS). However, as described in Chapter 11, there is still a place for the use of archival or static knowledge resources (e.g., in the published literature, drug monographs, clinical guidelines, or evidence summaries) to support just-in-time decision making simply by educating the decision maker. That is, the clinician can make a better-informed decision by reading (or listening to or watching) relevant knowledge, which can then be incorporated into the clinician's cognitive processes in the context of a specific patient. The educational process used to train clinicians already overwhelms them with more knowledge than they can possibly record, let alone retain and recall when needed, and staying up to date after training is even more difficult. However, if the appropriate knowledge can be invoked at the time that the clinician needs it, then it has the potential to truly support clinical decision making as "just-in-time" information [1,2].

More and more, clinicians are making their decisions while using a computer. The need to make decisions may be triggered when a clinician receives new information about a patient and that information is often computer-based. The act of making decisions is often operationalized in the form of writing an order, and order writing is now largely a computer-based activity. Thus the clinician sitting in front of a computer presents an opportunity for CDS. First, the clinician is carrying out some limited set of activities, which suggests that the types of decision support needs that arise may be similarly limited. This makes automated solutions numerable, if not always tractable. Second, the clinician's task, discipline,

https://doi.org/10.1016/B978-0-323-91200-6.00017-6

and care setting as well as the specific patient information involved, can help to further narrow the prediction about the kind of decision support and specific kinds of knowledge that are most likely needed. Third, the user is already in front of the computer—the perfect place to retrieve and present knowledge resources that can address the need. Fourth, the information that triggers the request for decision support can be exploited not only to identify the need but to help get the answer.

In other words, a clinician using a clinical information system may be expected to have some typical, common CDS needs that can be suggested by what the clinician is doing and seeing on the system. A clever CDS capability can anticipate the needs and attempt to automatically satisfy them. For example, consider a nurse practitioner who is reviewing a urine culture result. When he sees that the organism is *Proteus mirabilis*, he might wonder several things: "How did this patient get Proteus mirabilis in her urine?", "Is this clinically significant?", "What is the best treatment for this?", "Are diagnostic studies of the urinary tract warranted?", etc. If the computer system presents these questions, and their answers from high-quality reputable knowledge resources, the nurse practitioner can learn exactly what he needs to know at the exact moment he needs to know it: just-in-time learning.

This sort of integration between clinical and knowledge systems [3] and architectures for their integration [4] have been envisioned for some time. Clearly, there are several challenges, not all of them technical, to realizing the above scenario. A number of developments in the past three decades, including better understanding of clinician information needs [5], more sophisticated controlled terminologies, wide adoption of electronic health record (EHR) systems [6], the advent of the World Wide Web (with all its attendant standards and resources), and wide adoption of knowledge resources [2] have facilitated the development of working solutions to the just-in-time education challenge.

Despite wide availability and adoption of knowledge resources, important barriers to the use of knowledge resources at the point-of-care (e.g., lack of time, lack of seamless access to resources, lack of knowledge about most relevant resources for a specific clinical question, and inability to identify specific clinical information needs) compromise the efficient and effective use of these resources. This chapter describes one CDS approach, called "infobuttons" that addresses this challenge. We first review what has been learned of clinician information needs. We then examine a variety of projects that have integrated health knowledge resources into clinical information systems. We focus on infobuttons, as one of those projects, and review its origins and evolution. We then describe how infobuttons have been implemented in a variety of settings, including related work on question answering, and the emerging strategies for managing them. Finally, we describe a widely adopted standard for integrating infobuttons into clinical systems along with open source software that can help health organizations and clinical system vendors to implement infobuttons.

The purpose of infobuttons, as well as other related approaches to point-of-care access to knowledge, is to provide methods for *automatically selecting and retrieving* information from appropriate knowledge resources, rather than methods for automatically executing inferences on knowledge bases, as is the case with many other CDS capabilities.

13.2 Understanding and addressing clinician information needs

13.2.1 Information needs in clinical practice

In a seminal study, Covell et al. found that primary care and specialist physicians in the outpatient clinical setting raised 2 questions for every 3 patients seen [7]. Of these questions, only 30% were answered during the patient visit, most commonly by another physician or health professional. While this study predated the Web and most electronic textbooks, a systematic review on clinicians' information needs including articles up to 2011 found 72 more recent studies that have had similar findings [5]. According to this systematic review, the median frequency of information needs across studies that use a method similar to Covell's was 0.6 needs per patient seen. Clinicians pursued 47% of these needs and, of those pursued, 80% were successfully met. Therefore, it can be estimated that over half of the information needs that clinicians raise in the context of patient care are not met. However, the rate of information needs may be even higher in the care of specific patient populations. For example, a study of clinicians' information needs in the aging population has shown a rate of over 4 questions per patient seen [8].

In order to better understand the types of clinical questions that commonly arise at the point-of-care, Ely et al. created a taxonomy with 64 question types using a set of 1396 questions observed at outpatient, primary care sites [9]. The three most common question types were "What is the drug of choice for condition X?" (11%); "What is the cause of symptom X?" (8%); and "What test is indicated in situation X?" (8%). A similar distribution was found across four other studies that applied the taxonomy to analyze clinical questions in other settings [10–13]. Interestingly, the frequency of question types follows a Pareto distribution, i.e., a relatively small percentage of question types accounted for most of the questions asked.

In a follow-up study, Ely et al. analyzed the questions that clinicians were unable to answer in the original taxonomy study [14]. The study revealed a set of recurrent reasons for each of these questions being difficult to answer. For example, a frequent pattern found in difficult questions consisted of a simple core question modified by a specific patient characteristic, such as co-morbidities. The authors proposed a set of recommendations for clinicians and knowledge authors to better address difficult questions. Clinicians were advised to (1) choose an appropriate resource for the question at hand; (2) rephrase the question to match the type of information in clinical resources; and (3) use more effective search terms. Knowledge authors were recommended to (1) provide explicit answers; and (2) provide practicable, actionable, and clinically-oriented information.

One of us (JJC) led a set of studies to examine the specific information needs that arise while clinicians are using clinical information systems. In an observational study, nurses and physicians were asked to think aloud as they used a clinical information system to review and enter patient data. Information needs arose most often while reviewing laboratory results and medication orders [15]. Subsequent analysis of the user interactions showed that fully half of the information needs were requests for health knowledge and, of these, 40% of the questions were medication-related [16]. Over half of health knowledge needs (55%), were not successfully resolved.

Many similar studies have been carried out since these initial explorations. Some have looked at specific settings or specific user types, and occasionally have looked across multiple settings and user types [17]. Results have consistently demonstrated that information needs are always underestimated and too often remain unresolved. Studies conducted in developing nations have demonstrated additional needs for that range across the spectrum from service delivery, through policy development, and program management [18].

13.2.2 Use and impact of online knowledge resources

A number of studies have examined how clinicians use online resources to help answer their questions. A detail review of these studies is provided in Chapter 11. In summary, research has shown that, despite the availability of online knowledge resources, clinicians still seldom use them. Educational interventions to increase uptake have had no or little impact [19]. Despite the apparent under-utilization of on-line knowledge resources, studies have shown that use of knowledge resources is associated with increased success in answering clinical questions, as well as positive impacts on clinician behavior and patient effects [20]. A wide range of barriers preclude more effective use of online resources at the point-of-care, such as doubt that an answer exists, lack of time, and uncertainty about where to look for information [21]. Chapter 11 provides a comprehensive discussion of those barriers.

In summary, the use of online health knowledge resources is a promising strategy to help clinicians meet their information needs, improving patient care decisions and patient outcomes. However, the lack of easy access to resources that can provide high quality and objective answers in a timely manner is still a critical barrier to the use of these online resources at the point of care. Therefore, interventions aimed at promoting the efficient and the effective use of online resources need to focus on lowering these barriers.

13.2.3 Understanding the context of information needs

Studies in areas such as medical informatics, anthropology, knowledge management, and pervasive computing have highlighted the role of context in predicting the nature of workers' information needs. In an ethnographic study of physicians' information needs, Forsythe et al. stated that understanding a question correctly requires interpretation in the light of the context in which it was expressed [22]. Khedr and Karmouch define context as "information about physical characteristics (such as location and network elements), the system (such as applications running and available services), and the user (such as privacy and presence)" and state "the environment becomes context-aware when it can capture, interpret, and reason about this information" [23]. Fischer and Ostwald propose that the context of the problem dictates the workers' information demands [24]. More specifically in the health care information retrieval domain, Lomax and Lowe stated "in the effort to characterize information seeking, it is important to accurately define the types of information clinicians may use as well as states of information need that trigger information search and retrieval" [25].

Research has been conducted as an attempt to understand the context in which those information needs arise while users are interacting with a computer. Pratt and Sim showed, with their Physician Information Customizer, that formal representation of information about users could help information retrieval systems identify articles of greatest interest to those users [26]. One of us (JJC) hypothesized that if computer systems are able to capture the context in which common information needs occur, such systems would be able to predict those information needs, automatically translating them into queries that can be executed by online resources [27]. One of us (GDF) demonstrated the importance of context in an XML-based order set model. In this model, context defines the care settings where an order set (and individual orders within the order set) can be used, the patients (in terms of age, gender, and clinical condition) that are eligible for this order set, and the providers who can use this order set to write orders [28].

Subsequent studies of specific attributes of the clinical context show that more accurate predictions of information needs can be made when additional factors are considered. These can range from specific concepts of interest, such as general laboratory tests versus laboratory tests measuring serum drug levels [29], all the way up to factors related to the health system in which a health worker is located [18]. Most recently, studies have used artificial intelligence technologies to automatically infer clinical diagnoses and assessments of patients during their visits [30,31]. In addition, specific knowledge of users' previous information seeking behavior may be one of the most reliable predictors of how the user will address current needs [32–34].

Studies that rely on usage logs to track users' information seeking processes are generally limited to capturing activity up to the point that a resource is invoked; tracking beyond that point requires relatively complex and robust proxy server functions. One exception has been tracking information seeking where the EHR itself is the information resource. Chen and colleagues used usage-mining techniques to identify typical paths taken by clinicians who were navigating through patients' EHR data. They found that knowledge about the user's starting point in the EHR and first traversal to a second point could be used to predict the most likely third point [35]. Yu and colleagues developed text-mining techniques to learn how physicians formulate clinical assessment in writing EHR notes to help infer patient's information needs [30,31].

13.2.4 History of linking clinical information systems to online resources

The integration of clinical information systems with online health knowledge resources has the potential to address the dual barriers of access and time constraints. Attempts exploring this approach date back to the late 1980s, before the advent of the Web. We summarize some of this work here, but the reader is directed to more extensive published reviews of the topic [36,37].

Because the National Library of Medicine's (NLM's) Medline database was one of the first and most prevalent online resources, Medline searches were typically the target of this initial work. Among the earliest such systems were Hepatopix [38] and Psychtopix [39]. These systems contained sets of topic-specific Medline search strategies that could be matched to "topics of interest" encountered in reports (liver biopsy reports and psychiatric records, respectively). Several variations of this approach automating bibliographic searches followed. Term Linker [40] Meta-1 Front End [41] allowed users to cut-and-paste a term in the EHR and transfer it to Medline for use in searching. Later systems, including the Interactive Query Workstation [42] and the Internet Gopher [43] allowed users to perform searches against a variety of resources besides Medline. The DeSyGNER system supported the integration of books, tutorials and simulation systems into a radiologist's clinical workstation [4].

The advent of the World Wide Web has contributed a great deal to reducing the barriers to accessing health information [44]. As a result, integrated systems became much easier to develop. Among the first was the MedWeaver system, which used a query formulator to translate a user's information request into a searchable form that could be passed to a retrieval manager. The retrieval manager, in turn, could access a variety of information resources and produce an integrated view of all information retrieved [45]. Integration of clinical information systems with Web-based retrieval systems followed soon after. One system, at Duke University, integrated Web-based clinical practice guidelines into a system for documenting well-child visits to a pediatric clinic [46]. The ActiveGuidelines system, at the Palo Alto Medical Clinic, integrated Web-based guidelines with a clinical information system and allow users to invoke guidelines based on relevant topics in a patient's electronic medical record (problems, medications, etc.) [47].

13.3 Infobuttons

13.3.1 History of infobutton development

Several integration efforts have taken the form of context-specific links to on-line resources, integrated into clinical systems. These links, called "infobuttons", not only invoke relevant resources, but anticipate information needs and initiate retrieval strategies to help the user navigate resources [48]. Infobutton research has included studies of information needs and their contexts. We focus here on large scale infobutton implementations within home-grown EHR systems in clinical settings at Columbia University and Intermountain Health Care, and briefly mention various approaches (some called infobuttons, some not) by other research groups. Since then, all these home-grown EHR systems have been replaced by commercial EHR products that provide native support for standards-based infobutton functionality that will be described below in this chapter.

13.3.1.1 Infobutton development at Columbia University

Infobutton development at Columbia can be traced back to an NLM-sponsored project to explore ways in which clinical data could be translated, using the UMLS, into MeSH terms to support automated bibliographic searches. Dubbed "the Medline Button", the system allowed users to select patient diagnoses and procedures, coded in ICD9-CM in a mainframe-based clinical information system, and use them to search a Medline database running on the same mainframe (see Fig. 13.1) [49].

With the creation of Web-based clinical and knowledge systems, the barriers to interfacing disparate systems were largely removed. We began to explore ways to link our new clinical information system to online resources, which led to the implementation of infobuttons in the New York Presbyterian Hospital's clinical information system, WebCIS (see Fig. 13.2).

The infobuttons were inserted into WebCIS in a variety of places, including applications for viewing laboratory results, microbiology culture results, microbiology antibiotic sensitivity

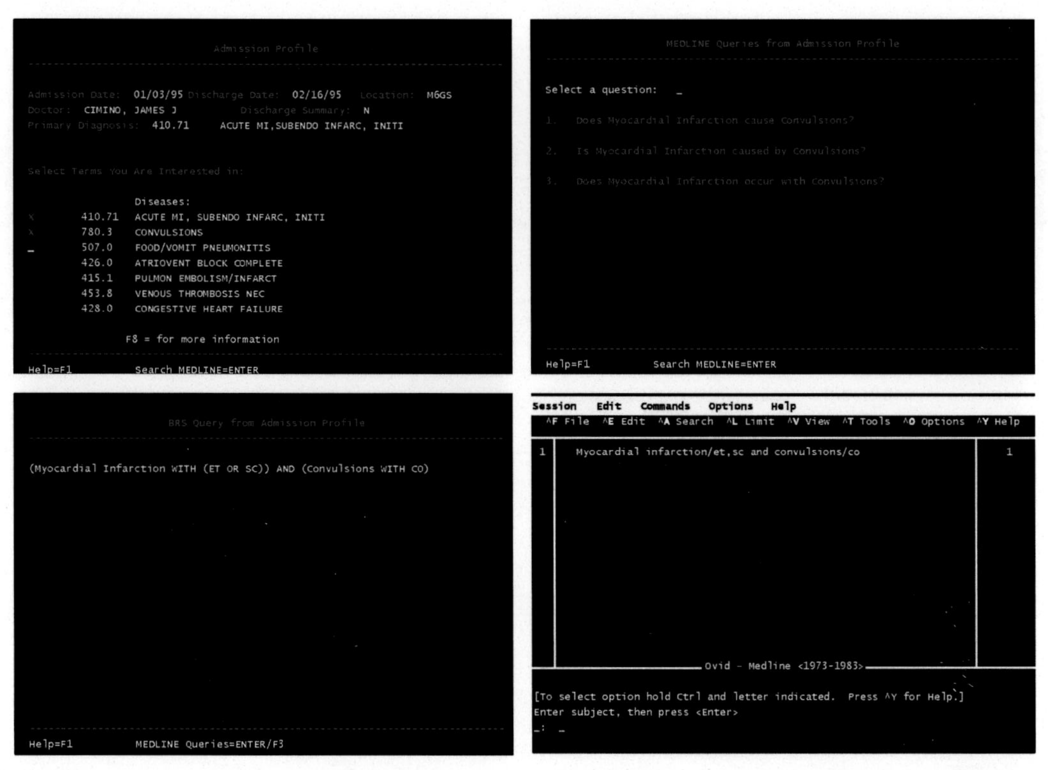

FIG. 13.1 Screen shots from the Medline Button. The top left screen shows patient diagnoses, coded in ICD9-CM, in the clinical information system. When the user selects two ICD9-CM diagnoses (in this case, "ACUTE MI, SUBENDOC INFARC, INITI" and "CONVULSIONS") and presses the F8 key, the Medline Button translates the diagnoses into MeSH terms ("Myocardial Infarction" and "Convulsions", in this case) and presents several possible questions of interest to the user (shown in the top right screen). When the user selects a question (in this case, question 2 "Is Myocardial Infarction caused by Convulsions?"), the system generates the Medline search strategy shown in the bottom left screen that, in turn, produces the search results shown in the bottom right screen (in this case, one article was returned). The user can then go on to review the citation and abstract (not shown).

results, and pharmacy orders (shown in Fig. 13.2). Analysis of log files showed that, depending on the context, users preferred infobuttons as much as nine to one over other available information resources [50].

13.3.1.2 Infobutton development at Intermountain Healthcare

In 2001, infobuttons were implemented within the medication list, problem list, and laboratory results modules of HELP2, Intermountain's home grown EHR system (Fig. 13.3) [51,52]. Infobuttons are placed next to each clinical concept (e.g., medication, problem) in these modules. When an infobutton is clicked, the user is presented with a list of questions about the concept of interest. The user can also select from a list of resources that cover the domain of the questions under consideration. When the user selects one of the questions, a search request is sent to the target resource, which then returns the search results.

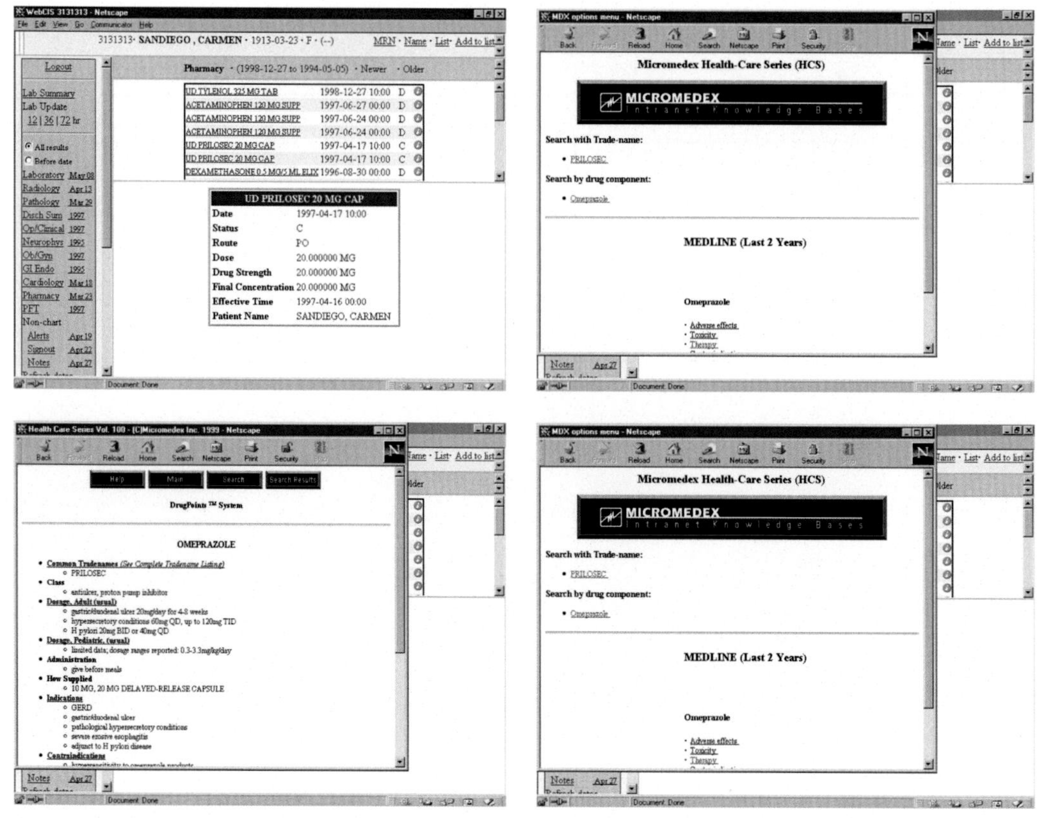

FIG. 13.2 Screen Shots of the Initial Infobutton Implementation at New York Presbyterian Hospital. The top left image shows a typical WebCIS screen, in this case a display of pharmacy orders; the infobuttons are the white-"i"-in-blue-circle icons, to the right of each medication. The top right image shows the result of clicking on an infobutton (in this case, the one to the right of "UD PRILOSEC 20 MG CAP"): a screen pops up with links to two resources, Micromedex and Medline (PubMed). Note that the infobutton has extracted the trade name "PRILOSEC" for use in searching Micromedex and has also used a terminologic knowledge base to recognize that the drug has the ingredient "Omeprazole", which is suitable for use in searching both resources. The bottom left image shows the result of clicking on the Micromedex "Omeprazole" link, while the bottom right image shows the result of clicking on the Medline "Adverse effects" link.

The HELP2 infobuttons used context information such as a concept of interest (e.g., a problem, medication, laboratory test), patient age and gender, and EHR task (e.g., order entry, medication list review). Using a terminology server, each concept of interest was translated into a suitable standard terminology, such as ICD-9-CM, LOINC [53], and the National Drug Codes (NDC). Moreover, a subtopic (e.g., diagnosis, treatment, prognosis, patient education) could be added to the request based on the user's selection.

13.3.1.3 Other infobutton development work

Several research groups investigated linking home grown EHR systems with knowledge resources. For example, MINDscape, at the University of Washington, integrated a digital

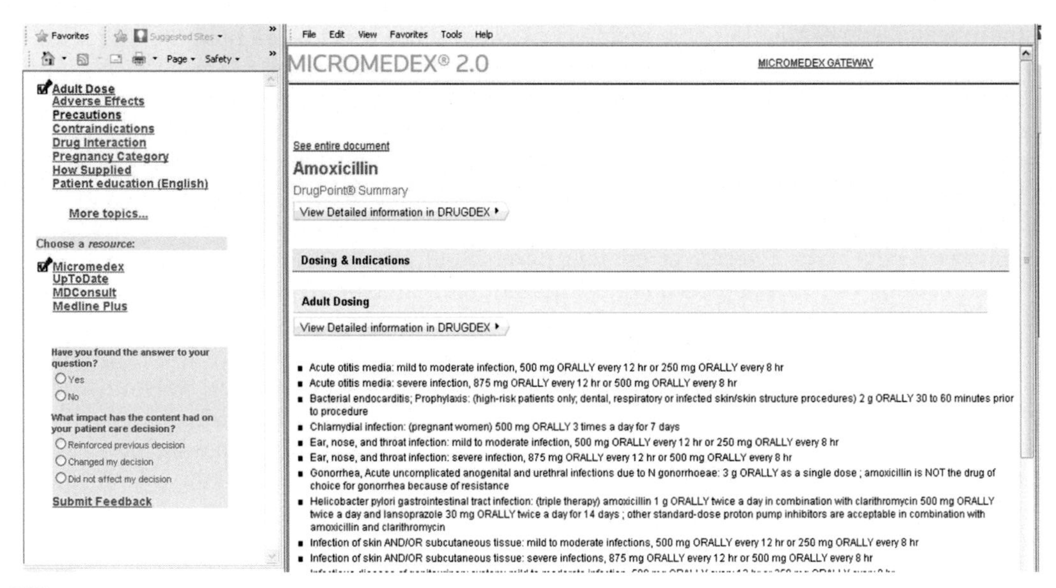

FIG. 13.3 An infobutton screen from Intermountain's HELP 2 system, showing a navigation panel (left) with links to relevant resources and content topics for the medication "amoxicillin."

library and electronic medical record. That system uses an "i" icon next to each term in the system's problem list to provide a link to a term-specific template that, in turn, provided links to a variety of resources. However, the links in MINDscape were all hard coded, creating problems with maintenance and with passing details about the user's context [54].

KnowledgeLink, at Partners HealthCare, embedded infobuttons within Partners' home grown EHR. An evaluation of that system showed that providers (physicians, nurse practitioners, and a wide variety of other medical staff) found answers to their medication questions in 74% percent of the KnowledgeLink sessions, supporting a previously made medical decision in 59% of the cases, and changing a medical decision in 25% of the cases [55].

A group at Vanderbilt University described a study of the Patient Care Provider Order Entry with Integrated Tactical Support (PC-POETS) component of their WizOrder system. PC-POETS provided links from order entry screens to on-line resources with searches for the item being ordered. The study showed that deployment of these links increased frequency of access to on-line resources by a ratio of over nine to one [56].

13.3.1.4 Infobutton software architecture

The experience with linking clinical systems to online resources has been consistently positive, but technical issues have constrained their deployment and scalability. For example, infobutton functionality was tightly coupled to the user interface of EHR systems at New York Presbyterian Hospital, Intermountain Healthcare, and the University of Washington with MINDscape [54]. As a result, custom programming was needed to enable infobuttons within different EHR modules and screens. In addition, custom programming for each knowledge resource was necessary to enable infobutton search requests. This approach was not scalable and inhibited experimentation with additional resources and queries.

A more scalable and maintainable architecture was to use service-oriented architectures to decouple the EHR systems from infobutton capabilities, resulting in a Web service component called *Infobutton Manager* [27]. In this service-oriented architecture, all the infobutton logic, including "mappings" from various contexts to clinical questions and connections to knowledge resources, is housed in an EHR-agnostic Infobutton Manager. To enable infobuttons, clinical systems only need to be able to build infobutton requests that are sent to the Infobutton Manager upon users' interaction with infobuttons. Infobutton requests are simple URLs with a series of context parameters about the patient, EHR task, user, and care setting. Fig. 13.4 depicts a typical Infobutton Manager architecture. Upon receiving an infobutton request, the Infobutton Manager (i) identifies relevant questions or topics for the request context; (ii) identifies resources that may have answers to those questions; (iii) builds custom resource-specific requests (also URLs) compliant with each resource application program interface (API); and (iv) returns to the user a list of hyperlinks for each relevant question/topic. Hyperlinks can be either *dynamic* search requests that include a list of search parameters or static links to a specific Web page.

The Infobutton Manager architecture has several advantages over previous infobutton approaches, which are typical advantages of service-oriented architectures for CDS [57]. *First*, an

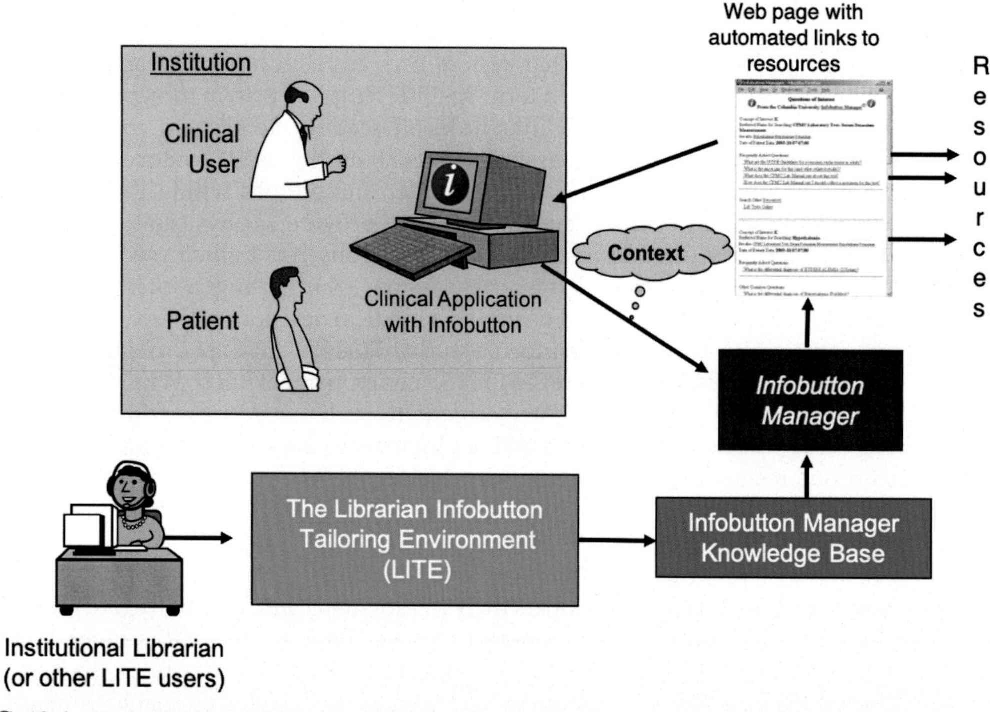

FIG. 13.4 Typical Infobutton Manager architecture. Upon user action, an infobutton request with the clinical context is sent to an Infobutton Manager, which uses its knowledge base to identify relevant questions or topics for the request context; builds resource-specific hyperlinks compliant with each resource application program interface (API); and returns to the user a Web page with a list of hyperlinks to a set of resources that address each relevant question/topic.

Infobutton Manager is agnostic of any EHR module, EHR product, or health care institution. As a result, a single Infobutton Manager can provide infobuttons to any module within an EHR, different EHR products, and multiple healthcare institutions. *Second*, since all infobutton logic is delegated to an Infobutton Manager, EHR developers do not need to become experts in infobuttons to implement infobutton functionality. *Third*, the logic within infobutton managers rely extensively on terminology inferences. Rather than embedding terminology knowledge within an infobutton application, infobutton managers delegate terminology inferences to terminology services, which can be maintained at a specific institution or externally. *Last*, the logic within an Infobutton Manager can be configured and shared among different sites.

13.3.1.5 Columbia University's infobutton manager

The Infobutton Manager (IM) approach at Columbia involved three design components [27]:

- Standardization of the set of context information that would be passed to the IM: user ID, user profession, user institution, patient ID, patient age, patient gender, clinical task being performed, and clinical data being reviewed
- An Infobutton Table containing questions that were determined through Columbia's empirical studies of clinician information needs; for each question, the developers identified a natural language version of the question (to display to users) and the URL for carrying out the search [58].
- A Context Table which matched context parameters passed to the IM; for each row that matched, a link to the corresponding resource in the Infobutton Table was assembled into a URL that displays the natural language question and contains the link to the resource; the question-resource-links were assembled into a Web page that was passed back to the user.

The IM architecture provided a great deal of flexibility of adding questions and resources. In one case, the chief medical officer of the hospital requested that a heparin administration guideline be added as an infobutton related to the laboratory display for partial thromboplastin time (PTT) results. Within 5 minutes, links were established for PTT results, as well as for heparin orders, with one guideline invoked for adult patients and a second guideline invoked for pediatric patients. These links were available immediately to the 4000 users of the system.

The Columbia IM had one additional, unique feature: it was able to translate the user's concept of interest to other, related concepts that might also relate to the user's information need. For example, if a user selected an infobutton next to a laboratory test, the IM could identify substances and conditions measured by the test, using Columbia's Medical Entities Dictionary (MED) [59]. Thus, in addition to providing links to laboratory test information (such as a lab manual), the IM could select questions about drug dosing (if the test measured a medication level) or disease diagnosis (if the test detected presence of a disease [48].

A number of other institutions have begun to take advantage of the IM by including links into their own systems. In one case, the New York Office of Mental Health has used the IM to provide drug information about the items in their patients' computerized medication lists. In another case, the developers of the Regenstrief Medical Record System, in Indianapolis, have added links to the IM for laboratory items, using LOINC codes as their controlled terminology, as well as links for medication-related items [60].

13.3.2 OpenInfobutton

OpenInfobutton (http://www.openinfobutton.org) is an open source suite of infobutton components that resulted initially from a project funded by the US Veterans Health Administration (VHA) Innovation Program that started in 2012 [61], and later funded through a variety of sources including grant funding and institutional support. Currently the project is hosted and coordinated by informatics researchers at the University of Utah Department of Biomedical Informatics and led by one of us (GDF). Collaborators and OpenInfobutton sites include the VHA, Intermountain Healthcare, Duke University, University of Alabama Birmingham, Regenstrief Institute, the ClinGen network, the EMERGE network, and the National Library of Medicine. OpenInfobutton has also been integrated with several homegrown and commercial EHR systems, including CPRS at the VHA; HELP2 at Intermountain Healthcare; Epic® at the University of Utah and Duke University; and Cerner® at Washington University St Louis, as well as healthcare apps integrated with EHR systems through the SMART on FHIR standard [62].

Overall, the architecture of OpenInfobutton is similar to Columbia's Infobutton Manager described above, except that infobutton capabilities are exposed to client applications through a RESTful Web service compliant with the *Health Level Seven (HL7) Context-Aware Knowledge Retrieval Standard*, also known as the *Infobutton Standard*, which is described below [63]. In addition, OpenInfobutton contains two other components: *Infobutton Responder* and a Web-based configuration tool called *LITE* (Fig. 13.5) [64,65].

The *Infobutton Responder* enables indexing of content that is not accessible through a search engine, such as institution-specific protocols. Once indexed in the Infobutton Responder, content becomes available to infobuttons for searching through a Web service compliant with the HL7 Infobutton Standard. Content in the Infobutton Responder is indexed according to the search parameters of the Infobutton Standard and using standard terminologies such as SNOMED-CT, ICD-10, LOINC, and RxNorm.

LITE is a configuration environment that allows individuals such as informaticians and medical librarians to configure Infobutton Manager logic and index Infobutton Responder content without any programming. LITE users can setup *resource profiles* that allow the

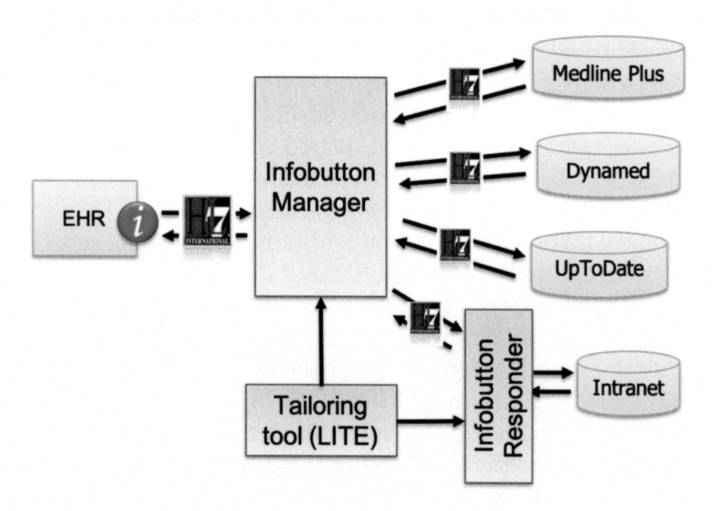

FIG. 13.5 OpenInfobutton's architecture with its core components: Infobutton Manager, Infobutton Responder, and authoring/tailoring environment (LITE). All communication with EHR systems and knowledge resources is done via HL7 Infobutton Standard.

Infobutton Manager to interact with various knowledge resources. A resource profile specifies the base URL for the resource's API; the contexts in which a knowledge resource is relevant; the questions or topics covered by the resource; the terminologies supported by the resource's search API; whether or not the resource is compliant with the HL7 Infobutton Standard; and for non-compliant resources, specify mappings between the Infobutton Standard and the resource's proprietary search parameters. Resource profiles can be shared with other OpenInfobutton sites through a cloud-based resource profile *store* (Fig. 13.6). Rather than configuring an entire OpenInfobutton instance from the ground up, OpenInfobutton sites can leverage the work done by other sites by downloading resource profiles of interest that are available in the profile store.

Similar to Columbia's Infobutton Manager, OpenInfobutton makes extensive use of *terminology inferences* leveraging the UMLS Terminology Services (UTS) and the RxNorm RESTful API as terminology services, as well as a cloud-based value set repository hosted within OpenInfobutton's GitHub repository. All three OpenInfobutton components (i.e., Infobutton Manager, Infobutton Responder, LITE) leverage terminology inferences such as concept translations (e.g., from SNOMED-CT to ICD-10 and vice-versa), hierarchical concept expansion using parent-child relationships, value set membership look up, and matching of search terms to terminology concepts. Since UTS and RxNorm are independently maintained and hosted by the National Library of Medicine, OpenInfobutton sites do not need to spend effort maintaining terminology content and services.

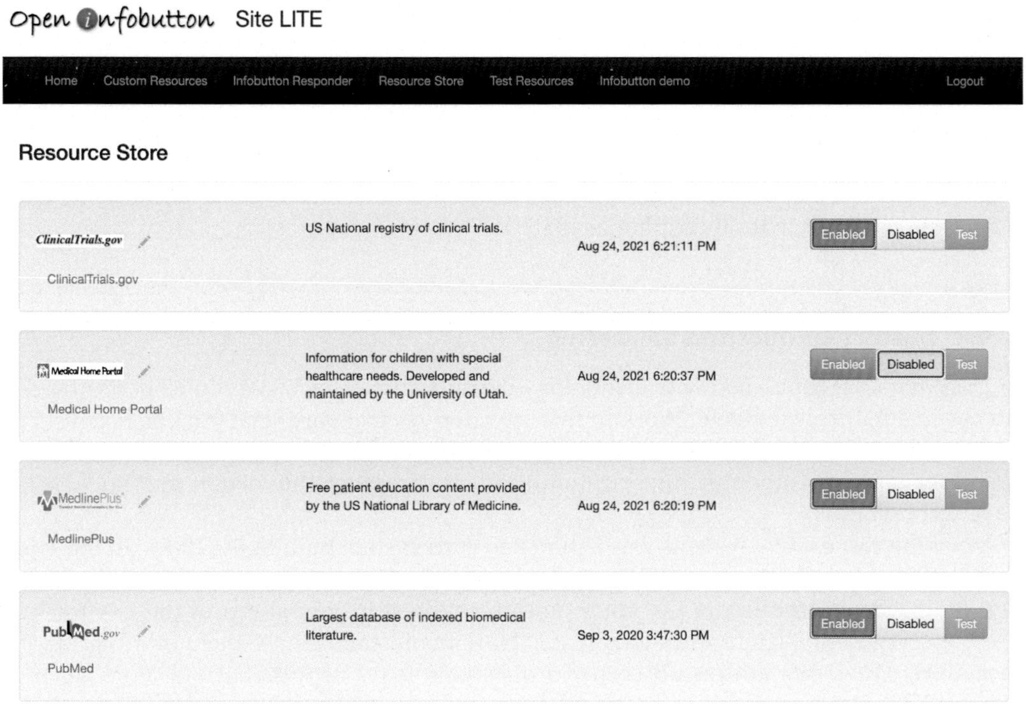

FIG. 13.6 Screen shot of LITE's resource profile store showing four resource profiles that have been downloaded to an institution's OpenInfobutton instance. Resource profiles can then be updated, enabled, and disabled as needed.

OpenInfobutton offers three deployment approaches with increasing levels of complexity and flexibility: *cloud-based* instance available at the University of Utah's Center for High Performance Computing, local installation using a *Docker Image*, and full local installation through local compilation of the *source code*. The cloud-based instance requires no software installation at the site, but provides little flexibility in terms of IT configuration and does not provide 24/7 technical support. This option is ideal for software development, testing, prototyping, and demonstration. The Docker Image is a software package that includes all OpenInfobutton components, including the compiled software and a MySQL database. The entire package can be installed through a few simple steps at a server owned by the site. This alternative provides a good trade-off between simplicity and flexibility, since the deployment is straightforward, but individual sites still have wide flexibility in configuring and controlling access to OpenInfobutton. A full local installation requires downloading and compiling the OpenInfobutton source, installing a relational database, running database creation scripts, and setting application properties. This approach is recommended only for sites that plan to make changes or extensions to the base source code. Detailed instructions are available in the OpenInfobutton Web site (http://www.openinfobutton.org) and the project's code repository (https://github.com/logicahealth/InfoButtons).

13.4 Question answering systems

In the field of artificial intelligence, the year 2011 will be remembered for Watson, an IBM question answering (QA) system that beat human champions to win Jeopardy [66]. Watson was soon hailed as a potential major impact on health care by answering providers' clinical questions and helping them with medical diagnoses and treatments [67]. During this same period, Apple's iOS application Siri and the Android app Iris have debuted, allowing users to converse naturally with their phones [68].

13.4.1 History of question answering

QA systems are concerned with technology that automatically answers questions posed by humans in a natural language. Work in this area derives from artificial intelligence research involving the fields of information retrieval and extraction, natural language understanding, processing and generation, summarization, human-computer interaction and spoken language technology.

One of the earliest QA systems was a punched card system built in the 1960s. In the 1990s, the Text REtrieval Conference (TREC), which provides infrastructure necessary for large-scale evaluation, introduced a QA track, and the earliest instantiations of the QA track focused on answering factoid questions (e.g., "How many calories are there in a Big Mac?"). Since 2003, TREC has addressed scenario questions (e.g., definitional questions such as, "What is X?") that require long and complex answers. In 2006 and 2007, TREC Genomics introduced passage retrieval for QA in the genomics domain [69].

13.4.2 Clinical question answering

Clinical QA is another approach to helping clinicians meet their patient care information needs. There have been numerous developments in QA. Cimino et al. [70] tagged clinical questions semantically to make them generic (for example, "Does aspirin cause ulcers?" became "Does <drug> cause <disease>?"). Rinaldi et al. [71] adapted an open-domain QA system to answer genomic questions (e.g., "Where was spontaneous apoptosis observed?"). The EpoCare project (Evidence at Point of Care) proposed a framework to provide physicians the best available medical information from both literature and clinical databases [72]. The CIQR (Context-Initiated Question Response) project [73] focuses on the analysis of the types of questions asked by clinicians when looking up references ([74]). The PERSIVAL (Personalized Retrieval and Summarization of Image, Video and Language Resource) system attempted to incorporate patient-specific information in seeking relevant and up-to-date evidence [75,76]. Essie is an information retrieval engine developed and used at the NLM that incorporates knowledge-based query expansion and heuristic ranking [77]. CQA-1.0 is designed as a clinical question answering system. It requires a user to enter a question in the PICO framework (Patient, Intervention, Comparison, and Outcome) and then provides semantic analysis at the document and text level, identifying PICO elements and documents that are of clinical relevance [78]. Sneiderman et al. integrated three systems (SemRep, Essie, and CQA-1.0) to achieve the best information retrieval system (which outperformed each of the three systems) in response to clinical questions [79]. Recent clinical QA development includes CliniQA [80] and MiPACQ [81]. AskHERMES is an online biomedical QA system that allows a user to enter an ad-hoc question and searches answers from MEDLINE articles and clinical guidelines [82]. Despite different approaches in QA, most QA systems share a common framework that we will describe below.

13.4.3 The QA framework

A typical QA system mainly comprises five components: question analysis, information retrieval, answer extraction, answer summarization and presentation, and a corpus from which the QA system draws answers. Fig. 13.7 shows the architecture of a QA system. A pre-processing step transforms a biomedical corpus, which can be biomedical literature and/or the World Wide Web. A question can be typed in or spoken. The information retrieval component returns either a ranked list of documents or paragraphs that contain the answer. Although most of existing QA systems are text-based, image and video can also be integrated. The answer extraction component further extracts exact answers from the documents/paragraphs returned from the information retrieval component. The summarization and answer presentation component removes redundant information, generates a coherent summary, and presents the summary to the user who posed the question. In the following, we will describe the state-of-the-art work for each QA component. We will also describe research advances in spoken QA and conclude with a real QA system and future work.

13.4.4 Question analysis

Automatically analyzing clinical questions is an important step for QA. Physicians often ask complex and verbose questions. There is a wealth of research proposing ways of

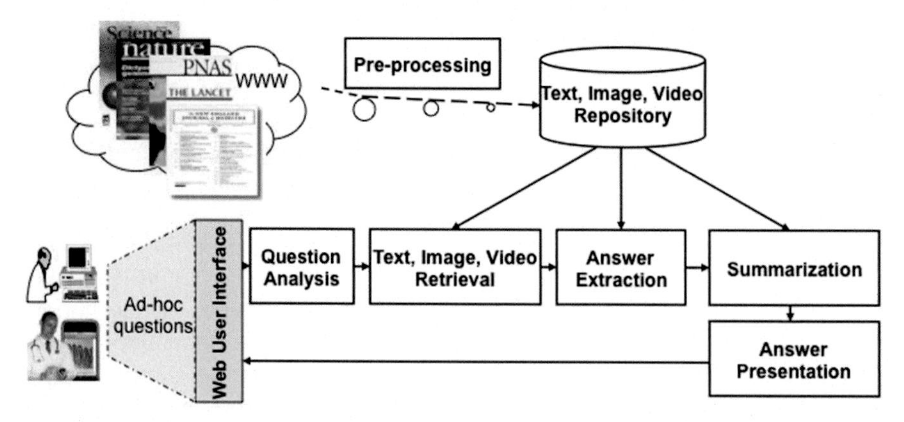

FIG. 13.7 QA system takes in a clinical question in natural language (either typed in or spoken) as input. *Question Analysis* identifies the information needs of the question. The processed output is sent to retrieve relevant documents or paragraphs from the text, image and video repository, which is a corpus of biomedical literature and online resources that are relevant to the clinical domain. *Answer extraction* extracts exact answers to the question. *Summarization* removes redundant information and formulates a coherent summary. *Answer Presentation* displays the answer to the user who posed the question.

categorizing such ad hoc questions, as stated in Section II. For example, Ely et al. manually mapped 1396 clinical questions to a set of 69 question types (e.g., "What is the cause of symptom X?" and "What is the dose of drug X?") and 63 medical topics (e.g., drug or cardiology) [9]. Cimino et al. defined a set of generic question types (e.g., "What is treatment for disease?") and then mapped ad hoc clinical questions to those types [70]. Such typologies offer different solutions for automated systems to overcome the wide range of variability in the forms that clinical questions may take. Other researchers have applied the popular PICO (Population, Intervention, Comparison, and Outcome) framework as a way of dealing with the variability in clinical questions [72,83,84]. Cao et al. developed supervised machine-learning classifiers to automatically assign predefined general categories (e.g., *etiology*, *procedure*, and *diagnosis*) to a clinical question [85]. They also explored both supervised and unsupervised approaches to automatically identify keywords that capture the main content of the question and used the keywords to identify relevant documents.

13.4.5 Information retrieval

Once a question is interpreted by a QA system, the next step is to identify documents or paragraphs that potentially answer the question. Most information retrieval approaches use keyword matching [72]; rely on other search engines such as PubMed [78] and Google [86]; apply the Vector Space Model [86], a model that is commonly used in information retrieval; or other probabilistic relevance models [82,87]. The QA approaches that go beyond the word-similarity models include the Minimal Logical Forms (MLFs) as a means to express the logical contents of sentences [71] and the CQA-1.0 framework [78] that structures both questions and documents using the PICO representation.

13.4.6 Information extraction

After documents or paragraphs are returned by information retrieval, the goal of information extraction is to identify further content that may incorporate an exact answer to the question. Most information extraction approaches first map free text to the clinical concepts, such as those in the UMLS [88]. In addition to concept mapping, Demner-Fushman and Lin proposed specific algorithms for identifying the PICO elements with either manually crafted rules or concepts identified by the UMLS; or a hybrid approach, which integrates rules, concepts, and a supervised machine-learning classifier [83].

The AskHERMES system uses a scoring function for measuring the similarity between a sentence (answer) and the question (S_S) [82]. The scoring function integrates both word-level and word-sequence-level similarity between a question and a sentence in the candidate answer passage, as shown in Eq. (13.1).

$$S_s = S_d \cdot TF_q \cdot UT_q \cdot \left(\frac{LCS}{\sqrt{L_q^2 + L_p^2}} \right), s \in d \tag{13.1}$$

S_d denotes question-document similarity based on the information retrieval function BM25 [87]; TF_q is the total number of query terms that appear in the sentence; UT_q is the unique number of query terms in the sentence; and *LCS* (longest common subsequence) is the similarity between the sentence and the whole question [89].

Incorporating the LCS score in the sentence score function can capture more detailed dependency information than mere bag-of-words or even bigrams. For example, given the question "How do I treat this man's herpes zoster?" the candidate answers represented by sentence 1 and sentence 2 below, have the same words and frequency that are matched against the extracted query terms ("treat", "herpes", and "zoster"), which means that an approach without LCS would rank the two answers the same. However, LCS assigns sentence 1 a value of 3 and sentence 2 a value of 2, giving sentence 1, which is the better answer for meeting the needs of the question, a higher ranking than sentence 2.

Sentence 1. Corticosteroids have been used to treat herpes zoster for much longer than the antiviral drugs, but the effect of corticosteroids on PHN does not appear to be consistent.
Sentence 2. A significant proportion of older subjects with herpes zoster develop post-herpetic neuralgia (PHN), a chronic condition that is difficult to treat.

Once the relevance score for each sentence is obtained, the score of a passage, S_p, is determined by the empirical metrics shown in Eq. (13.2).

$$S_p = \begin{cases} \max\left(S_{s1}^n\right) + \min\left(S_{s1}^n\right), & \max\left(S_{s1}^n\right) < 2 \times \min\left(S_{s1}^n\right) \\ \max\left(S_{s1}^n\right), & \text{otherwise} \end{cases} \tag{13.2}$$

where n is the number of sentences in this passage, $\max(S_s)$ is the maximum relevance score among all the sentences, and $\min(S_s)$ is the minimum relevance score among all the sentences.

13.4.7 Summarization and answer presentation

(1) Information retrieval and extraction components frequently return a list of sentences that come from different documents. Frequently, sentences convey redundant information. Summarization and answer presentation are components that aggregate answers, remove redundancy, generate a coherent summary, and present the summary as the output to the question.

(2) Few summarization approaches have been proposed in biomedical QA systems. AskHERMES implements a question-tailored approach [82]. As stated earlier, clinical questions are typically long and verbose and frequently relate to multiple topics. The automatic keyword extraction model effectively extracts content-rich keywords from ad hoc questions, and such keywords are then used to hierarchically structure the summarized answers and present such answers hierarchically.

(3) For example, in the question "How should I treat polymenorrhea in a 14-year-old girl?" the terms "treat", "polymenorrhea", "14-year-old" and "girl" are four important content terms, and an ideal answer would incorporate all four terms. However, in reality, most answer passages incorporate fewer content terms and sometimes contain only one of the four terms.

(4) The summarization system in the AskHERMES system is based on structural clustering using content-bearing terms as topic terms. An example output is shown in Fig. 13.8. Such summarization and presentation provides a user-friendly answer presentation interface that may help clinicians quickly and effectively browse answer clusters. AskHERMES employs the longest common substring ($LC_{Substring}$) to remove redundant information [90].

13.4.8 Spoken question answering

One challenge that clinical questions pose for QA systems is that they are typically long and complex [14]; busy clinicians rarely have the time to type questions of such length into

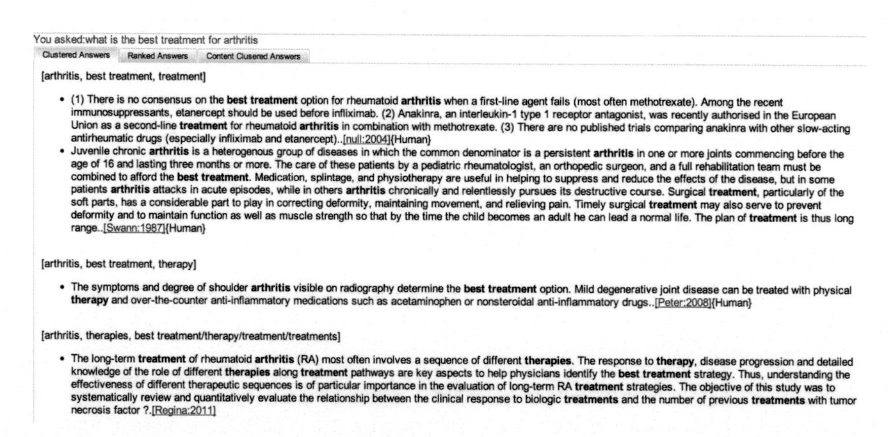

FIG. 13.8 AskHERMES's output for answering the question "What is the best treatment for arthritis?"

computers or portable devices, as QA systems traditionally required. Speech is a natural modality of interaction, providing an efficient way to address the aforementioned challenge in QA systems. The popularity of Siri and Iris is in large part due to their speech interface.

Work on automatic speech recognition (ASR) in the medical domain can be generally grouped into two categories: evaluation of multiple speech recognition software [91–93] and usability studies [94,95]. Liu et al. evaluated two state-of-the-art ASR systems, Nuance Dragon and SRI Decipher, and concluded that ASR systems need to be adapted to the clinical domain. Specifically, after domain adaptation, SRI Decipher reduces the error rate to 26.7%, in contrast to the 67.4% error rate of Nuance Med (version 10.1) [96]. The question is raised of whether such an error rate is robust to a spoken QA system. Miller et al. had physicians perform blind evaluations of results generated both by ASR transcripts of questions and gold standard transcripts of the same questions [97]. Their findings suggest that the medical domain differs enough from the open domain to require further improvement in ASR adapted for the clinical domain.

13.4.9 A fully-implemented biomedical QA system: AskHERMES

Although there are significant research developments in biomedical question answering, few systems are implemented and are available online. AskHERMES (http://www.askhermes.org) is a fully implemented online biomedical QA system. Fig. 13.8 shows an output of AskHERMES in response to the question "what is the best treatment for arthritis?".

13.4.10 Key challenges and future clinical QA

Key challenges for QA to be applicable in clinical setting are the ability to accurately formulate clinical information needs (generating questions), and efficiently access clinically relevant and high-quality resources in which answers exist (information retrieval or answer identification). The availability of large EHRs has made it possible to use computers to automatically generate patient-specific questions that may be relevant to clinicians. In addition to using structured data such as medication list and lab results, a promising direction is to train sophisticated deep neural networks on a large number of patients' longitudinal EHRs, and to infer a patient's clinical information needs from their own and other patients' experiences. The two main paradigms for answer identification are information-retrieval-based approaches and knowledge-based and hybrid approaches. Both paradigms can evolve to excel; however, none has the ability to synthesize and generate knowledge that is not explicitly expressed in the resources. Most current QA systems rely on the publicly available biomedical literature (i.e., MEDLINE), which is not designed for efficient and effective use in busy clinical settings. This is evidenced further by research in comparing AskHERMES with UpToDate, which found the performance of the two systems to be statistically equivalent. UpToDate is compiled by experts; while its content is manually curated by experts, it is limited by scope and timelineness. In contrast, AskHERMES automatically assembles information from the most up-to-date natural language texts. On the other hand, the texts from which AskHERMES extracted information are frequently not clinically relevant, and therefore, the non-relevancy leads to a lower performance [82].

Resources that can answer clinical questions include clinical guidelines, electronic medical textbooks and commercial databases, including UpToDate. Unfortunately, many of the aforementioned resources are not available for most QA developers, and therefore reduce the potential usefulness of any clinical QA system. In contrast, consumer-health-oriented QA benefits significantly from the WWW, which incorporates a wealth of information in consumer health [82].

13.5 Uptake, user satisfaction, and impact of infobuttons on clinician's decision making

Several studies have evaluated the use, satisfaction and clinical impact of infobuttons, particularly focused on home-grown EHR implementations at Columbia University, Intermountain Healthcare, and Partners Healthcare. Seventeen of these studies were summarized in a systematic review by Cook et al [98]. Overall, the systematic review found that infobuttons were used from 0.3 to 7.4 per month per potential user and clinicians found answers to their clinical questions in over 69% of infobutton sessions. Usage varied substantially between institutions, with one study showing steady increase in uptake over a four-year period [98]. A study looking at infobutton usage across four sites showed that most infobutton searches (70–84%) were conducted while prescribing or reviewing medications [99]. Studies looking at search efficiency and the effect of infobuttons on providers found that providers were able to answer 84–89% of their questions related to medications, leading to a high positive impact (decision enhancement or learning) in 56% of the sessions, and within a median of 35 seconds [55,100]. Despite evidence supporting the utility of infobuttons in efficiently answering most clinical questions, previous studies had important limitations, such as study outcomes limited to usage rates and self-reported measures, lack of patient outcomes, and most studies conducted at 3 sites with home-grown systems, which limits generalizability. Future studies are needed especially at sites using commercial EHR systems and using a wider range of outcomes, such as adherence to evidence-based guidelines, rate of adverse events, and cost.

13.6 The HL7 standard for context aware decision support

13.6.1 Motivation

The World Wide Web and the Internet enabled a stack of widely adopted standards for network communication (e.g., HTTP) and content representation (e.g., HTML, XML). These standards significantly facilitated the integration between EHR systems and Web-based knowledge resources. However, these protocols do not standardize communication at the semantic level. The lack of common information model and terminologies for representing clinical contexts required infobutton managers to establish custom, point-to-point communication with each knowledge resource. In addition, resources were unable to make inferences over the clinical context, leading to suboptimal content retrieval. To address this challenge, the HL7 Clinical Decision Support Work Group (CDS WG) developed the *HL7 Context Aware*

Knowledge Retrieval (Infobutton) Standard. The Infobutton Standard is composed of three specifications: a context information model (http://www.hl7.org/implement/standards/product_brief.cfm?product_id=22), a URL-based Implementation Guide (http://www.hl7.org/implement/standards/product_brief.cfm?product_id=22) and a Services-Oriented Architecture Implementation Guide (http://www.hl7.org/implement/standards/product_brief.cfm?product_id=283).

13.6.2 Context information model

The core element of the Infobutton Standard is a context information model. This model enables implementers to represent the clinical context in four dimensions: the *patient*, the *user*, the *care setting*, and the EHR *task* (e.g., medication order entry, laboratory results review). Each of these dimensions is composed of a set of attributes, such as patient *gender*, *age*, *language*, and a clinical *concept of interest*; *care setting type* (e.g., inpatient, intensive care, outpatient) and *organization*; and user *discipline*, *specialty*, and *language*. Each attribute is associated with a value set composed of codes drawn from standard terminologies, such as RxNorm, LOINC, ICD, and SNOMED-CT. Hence, HL7 compliant infobutton interactions consists of a set of context attributes that use standard codes as values.

13.6.3 Implementation guides

The Infobutton Standard also includes two specifications that specify how to implement infobuttons using RESTful Web services. The first implementation guide specifies a URL format for infobutton requests, in which attributes of the context information model are represented as URL parameter names and values. The second implementation guide specifies a standard format for infobutton responses, which is based on the Atom format, a widely adopted content syndication standard developed and maintained by the Internet Engineering Task Force (IETF). Fig. 13.9 provides an example of an Atom-based knowledge response to a request for patient education content on diabetes mellitus.

In a typical implementation using the RESTful protocol (Fig. 13.10), a clinical information system submits a knowledge request in URL format (Step 1) to an infobutton manager. The infobutton manager then submits independent knowledge requests to a set of relevant knowledge resources (Step 2) also as URLs. Each resource responds (Step 3) in the desired format specified in the knowledge request (e.g., HTML, XML). Finally, the infobutton manager aggregates the results from the multiple resources and sends an aggregate knowledge response back to the clinical information system (Step 4). Alternatively, a clinical information system can send an infobutton request directly to a knowledge resource, without using an infobutton manager.

13.6.4 Adoption

After its normative publication as an HL7/ANSI standard in 2010, the Infobutton Standard was rapidly adopted, especially by online knowledge resources such as UpToDate®, Micromedex®, Lexicomp®, MedlinePlus®, and Healthwise®. To assess the attitudes of

```
<feed>
    <title type="text">MedlinePlus</title>
    <subtitle type="text">Diabetes Mellitus</subtitle>
    <author>
        <name>National Library of Medicine</name>
        <uri>http://medlineplus.gov</uri>
    </author>
    <updated>2010-07-06T14:02:29Z</updated>
    <category>
        <v3:mainSearchCriteria>
            <v3:value code="250" codeSystem="2.16.840.1.113883.6.2" displayName="Diabetes
            Mellitus"/>
        </v3:mainSearchCriteria>
        <v3:informationRecipient typeCode="IRCP">
            <v3:patient classCode="PAT"/>
        </v3:informationRecipient>
    </category>
    <entry>
        <title>Diabetic diet</title>
        <link href="/diabeticdiet.html" rel="alternate" type="html" hreflang="en"/>
        <link href="/spanish/diabeticdiet.html" rel="alternate" type="html" hreflang="es"/>
        <updated>2011-15-02T00:00:00Z</updated>
        <summary type="html">If you have diabetes, your body cannot make or properly use
        insulin.  This leads to high blood glucose, or sugar, levels in your blood. Healthy
        eating helps to reduce your blood sugar.  It is a critical part of managing your
        diabetes, because controlling your blood sugar can prevent the complications of
        diabetes &lt;/a>.</summary>
        <content/>
    </entry>
    <entry>
        <title>Diabetic foot</title>
        <link href="/diabeticfoot.html" rel="alternate" type="html" hreflang="en"/>
        <link href="/spanish/diabeticfoot.html" rel="alternate" type="html" hreflang="es"/>
        <summary>If you have diabetes, your blood sugar levels are too high.  Over time,
        this can damage your nerves or blood vessels.  Nerve damage from diabetes can cause
        you to lose feeling in your feet.</summary>
        <category>
            <v3:mainSearchCriteria>
                <v3:value code="250.7" codeSystem="2.16.840.1.113883.6.2" displayName=
                "Diabetes with peripheral circulatory disorders"/>
            </v3:mainSearchCriteria>
        </category>
    </entry>
</feed>
```

FIG. 13.9 Atom-based knowledge response including two entries with patient education content for diabetes mellitus. Each entry contains metadata describing the entry (*category* tag), links to the full content in English and Spanish (*link* tag), and a summary of the content itself (*summary* tag).

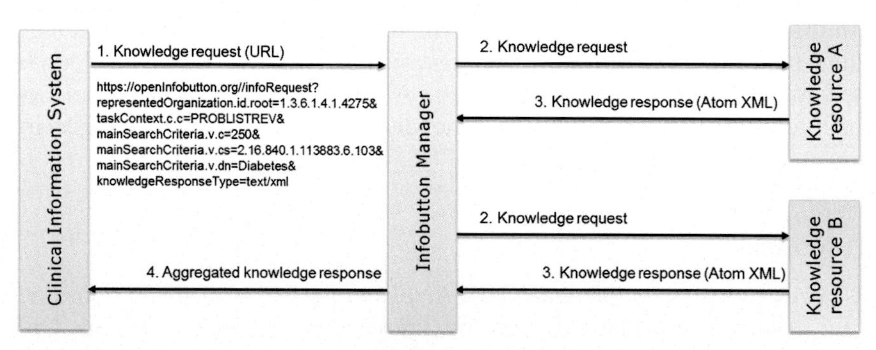

FIG. 13.10 Typical information flow in a RESTful implementation of the Infobutton standard.

implementers towards the HL7 Infobutton standard, in 2012 we conducted in-depth interviews with 17 health care organizations, knowledge publishers, and EHR vendors [63]. Overall, interviewees had a positive attitude towards the Infobutton standard and reported several strengths, such as simplicity, low cost implementation, and ability to leverage resources that are already available in the market and within health care organizations. At them time, most interviewees indicated slow adoption of the Infobutton standard among commercial EHR vendors as the main barrier to disseminating infobutton capabilities. Later in 2014, HL7-compliant infobuttons became a required CDS capability for EHR certification in the US Meaningful Use Program, both for provider reference and patient education resources. As a result, most EHR and personal health record (PHR) systems in the US know provide infobutton capabilities off-the-shelf. Notably, PHR systems such as Epic's MyChart® provide HL7-compliant infobuttons that link to patient education materials in resources such as MedlinePlus®, which is the largest patient education collection provided by the US government, with links to over 40,000 authoritative health information in 60 languages. MedlinePlus provides infobutton-access to its content through an HL7 Infobutton-compliant interface called MedlinePlus Connect [101]. In 2020, MedlinePlus Connect received 252 million infobutton requests from providers and patients through EHRs and patient portals worldwide [102].

13.6.5 Recent work

The Infobutton Standard was developed under the framework of the HL7 Version 3 standard, which was the most recent HL7 standard at that time. Since then, HL7 developed the FHIR standard, which is being rapidly adopted by health IT vendors and is progressively replacing HL7 Version 3. In addition, a generic CDS standard called CDS Hooks has been developed using FHIR as the data model (See Chapters 18 and 29 for details on CDS Hooks). Like the Infobutton Standard, both FHIR AND CDS Hooks use Web services technology as the foundation and are better positioned to support the next generation of the Infobtton Standard than HL7 Version 3. Thus, the HL7 CDS WG has explored the implementation of infobuttons using CDS Hooks and FHIR. The resulting work included one-to-one mappings between the Infobutton Standard and CDS Hooks as well as a prototype implementation using OpenInfobutton and a CDS Hooks sandbox. The prototype was successfully tested and integrated with different EHR systems at an HL7 FHIR Connectathon. Future work includes developing an implementation guide for infobutton implementation using CDS Hooks.

13.7 Ongoing and future research

13.7.1 Infobuttons for unrecognized information needs

Most infobutton implementations imply that the clinician recognizes an information need and initiates an infobutton session on-demand, typically by clicking on a hyperlink. However, at least equally important are information needs that are unrecognized by clinicians, such as the presence of a disease outbreak or evidence update that may be relevant for a particular patient.

To address unrecognized information needs, alternative infobutton implementations could trigger an infobutton session automatically upon certain user events in the EHR, such as opening the patient's record, selecting a medication for a prescription, or viewing a laboratory test result. For example, as a clinician in the emergency room types a chief complaint of "diarrhea", an infobutton request with the patient's postal code can be automatically submitted in the background to a registry of disease outbreaks. The registry would respond with potential outbreaks that are active within the patient's geographical area and that may be associated with the patient's chief complaint. When receiving a positive response, the EHR could display information as a non-interruptive "card" indicating the presence of a potentially relevant outbreak. We have tested this kind of approach in collaboration with the US Center for Disease Control and Prevention (CDC) as a scalable alternative for disseminating disease outbreak information to the point of decision-making. A similar approach can be adopted for other types of information needs that are typically unrecognized, such as a patient being eligible for a clinical trial, the availability of relevant evidence updates, or the availability of institution-specific protocols.

13.7.2 Machine learning techniques to improve search and retrieval

Most infobutton sessions rely on knowledge resources' search engines to retrieve content that is most relevant to a clinician's clinical question. Significant advances in general search engine technology, especially leveraging machine learning methods, have the potential to improve clinicians' search experience, whether through infobuttons or direct access to knowledge resources. One of the best examples of a widely used machine learning algorithm in the biomedical domain is PubMed's *Best Match* algorithm [34]. Best Match uses predictors such as previous searches, relevance score, publication date, and publication type to rank the most relevant articles for a given search. Several commercial search engines and knowledge resources also use similar machine learning techniques that leverage users' past search behavior to rank retrieved content. Other approaches attempt to identify scientifically sound or high impact studies. A study aiming to identify scientifically sound studies from PubMed found that compared with PubMed's Boolean Clinical Queries a deep learning approach had slightly lower sensitivity (98% vs 97%), but higher precision (34.6% vs 22.4%) [103]. In a different approach, a study aiming to identify studies with high clinical impact found that a machine learning classifier using bibliometric features (e.g., citation count), social media attention, journal impact factors, and citation metadata overperformed PubMed's former relevance sort algorithm on a gold standard of 502 high impact studies (top 20 precision = 34% vs. 11%) [104].

13.7.3 Knowledge summarization and visualization techniques

While infobuttons automate the selection and search of relevant knowledge resources, clinicians still need to scan the retrieved content to find answers to their clinical questions. This process may require substantial time and cognitive effort that are incompatible with a busy clinical workflow, especially when relevant content is buried within several paragraphs of text and/or scattered across multiple documents. Advances in semantic natural language processing and automatic knowledge summarization enable interesting approaches to address the problem described above. For example, Jonnalagadda et al. used a database of semantic predications extracted from Medline citations to automatically retrieve relevant sentences on the treatment of a particular condition. After ranking and aggregating the sentences, a knowledge summary is produced with statements about multiple treatment alternatives for a condition of interest [105]. In a simulated study with case vignettes, physicians perceived decision quality was significantly higher when accessing automatic, context-specific summaries of UpToDate content versus the full UpToDate content (16.6 vs. 14.4 on a 20-point scale) [106]. In another simulation with case vignettes, physicians were asked to review evidence from a preselected set of randomized controlled trials on a given treatment comparing a visual information display (Fig. 13.11) versus narrative article abstracts in PubMed. On a 9-point Likert scale, physicians significantly favored the interactive visual displays over narrative abstracts according to perceived efficiency (7.9 vs. 2.1), effectiveness (7.4 vs. 2.5), effort (7.5 vs. 2.5), user experience (7.5 vs. 2.5), and preference (8 vs. 2) [107]. Future studies are needed to investigate the use of automatic summarization and visualization techniques in clinical settings.

13.7.4 Infobuttons for specific knowledge domains and users

Many informatics researchers have explored additional infobutton-related issues and solutions. For example, infobuttons and infobutton-like links have been embedded in personal digital assistant- based systems to improve clinician screening for tobacco use and guideline-based tobacco cessation management [108]. Infobuttons have been extended to additional types of users, such as case managers working with HIV patients using a Continuity of Care Record (CCR)-based management system [109]. Prototypes have also been implemented with infobutton links to genomic knowledge resources such as ClinGen, Gene Reviews, and PharmGKB [110–112]; and links within i2b2 to resources for clinical researchers [113]. Possibly motivated by EHR certification requirements previously described, Infobuttons have also been used extensively to provide access to educational materials to patients within systems such as EHRs and patient portals [101,114–118].

13.8 Conclusions

The belief that better informed decisions lead to better patient outcomes is one of the underlying tenets of evidence-based medicine and, indeed, health care education. The integration of information resources into clinical information systems appears to be a viable approach to automated support of clinician decision making. By lowering barriers to information at the moment when it is needed, the hope is that clinicians will use the information to make better-informed decisions. While the impact on and quality of decisions are difficult to measure, we at least are seeing the increased access to information resources that we believe is a necessary (although not sufficient) step in the right direction.

The EHR Meaningful Use certification process in the United States has made a strong contribution expediting the wide adoption of infobutton capabilities across multiple EHR systems. This incentive has also facilitated worldwide adoption of infobuttons, since many of the EHR systems and knowledge resources that have implemented the HL7 Infobutton Standard are used in several locations worldwide. Additional work is needed to investigate the

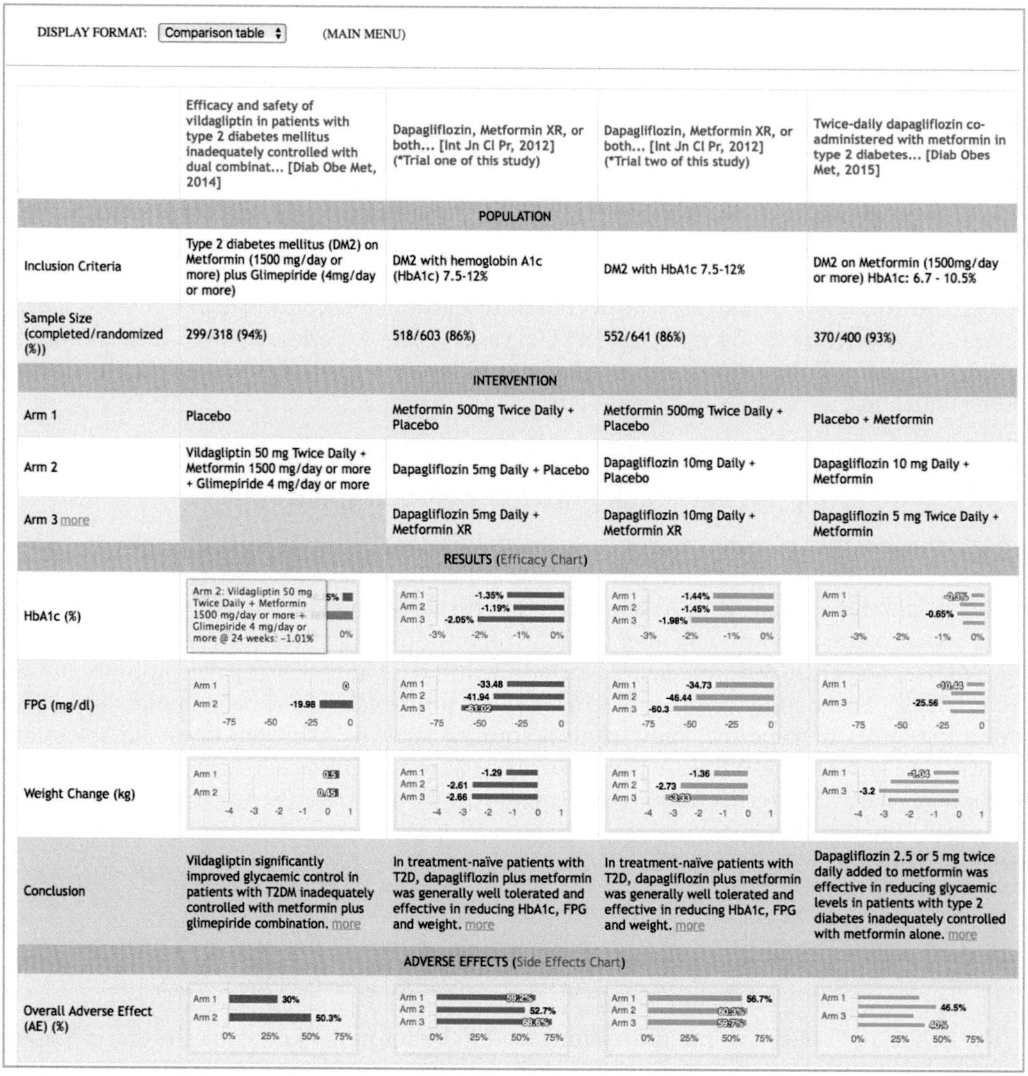

DISPLAY FORMAT: Comparison table (MAIN MENU)

	Efficacy and safety of vildagliptin in patients with type 2 diabetes mellitus inadequately controlled with dual combinat... [Diab Obe Met, 2014]	Dapagliflozin, Metformin XR, or both... [Int Jn Cl Pr, 2012] (*Trial one of this study)	Dapagliflozin, Metformin XR, or both... [Int Jn Cl Pr, 2012] (*Trial two of this study)	Twice-daily dapagliflozin co-administered with metformin in type 2 diabetes... [Diab Obes Met, 2015]
POPULATION				
Inclusion Criteria	Type 2 diabetes mellitus (DM2) on Metformin (1500 mg/day or more) plus Glimepiride (4mg/day or more)	DM2 with hemoglobin A1c (HbA1c) 7.5-12%	DM2 with HbA1c 7.5-12%	DM2 on Metformin (1500mg/day or more) HbA1c: 6.7 - 10.5%
Sample Size (completed/randomized (%))	299/318 (94%)	518/603 (86%)	552/641 (86%)	370/400 (93%)
INTERVENTION				
Arm 1	Placebo	Metformin 500mg Twice Daily + Placebo	Metformin 500mg Twice Daily + Placebo	Placebo + Metformin
Arm 2	Vildagliptin 50 mg Twice Daily + Metformin 1500 mg/day or more + Glimepiride 4 mg/day or more	Dapagliflozin 5mg Daily + Placebo	Dapagliflozin 10mg Daily + Placebo	Dapagliflozin 10 mg Daily + Metformin
Arm 3 more		Dapagliflozin 5mg Daily + Metformin XR	Dapagliflozin 10mg Daily + Metformin XR	Dapagliflozin 5 mg Twice Daily + Metformin
RESULTS (Efficacy Chart)				
HbA1c (%)	Arm 2: Vildagliptin 50 mg Twice Daily + Metformin 1500 mg/day or more + Glimepiride 4 mg/day or more @ 24 weeks: -1.01%	Arm 1 -1.35%; Arm 2 -1.19%; Arm 3 -2.05%	Arm 1 -1.44%; Arm 2 -1.45%; Arm 3 -1.98%	Arm 1; Arm 3 -0.65%
FPG (mg/dl)	Arm 1 0; Arm 2 -19.98	Arm 1 -33.48; Arm 2 -41.94; Arm 3 -61.02	Arm 1 -34.73; Arm 2 -46.44; Arm 3 -60.3	Arm 1; Arm 3 -25.56
Weight Change (kg)	Arm 1 0.5; Arm 2 0.45	Arm 1 -1.29; Arm 2 -2.61; Arm 3 -2.66	Arm 1 -1.36; Arm 2 -2.73; Arm 3 -3.33	Arm 1; Arm 3 -3.2
Conclusion	Vildagliptin significantly improved glycaemic control in patients with T2DM inadequately controlled with metformin plus glimepiride combination. more	In treatment-naïve patients with T2D, dapagliflozin plus metformin was generally well tolerated and effective in reducing HbA1c, FPG and weight. more	In treatment-naïve patients with T2D, dapagliflozin plus metformin was generally well tolerated and effective in reducing HbA1c, FPG and weight. more	Dapagliflozin 2.5 or 5 mg twice daily added to metformin was effective in reducing glycaemic levels in patients with type 2 diabetes inadequately controlled with metformin alone. more
ADVERSE EFFECTS (Side Effects Chart)				
Overall Adverse Effect (AE) (%)	Arm 1 30%; Arm 2 50.3%	Arm 1 59.2%; Arm 2 52.7%; Arm 3 68.6%	Arm 1 56.7%; Arm 2 60.3%; Arm 3 59.7%	Arm 1; Arm 3 46.5%; 40%

FIG. 13.11 Visual information display with four clinical trials comparing different treatments for diabetes mellitus. The display shows a summary of selected studies in a tabular format using the *PICO* (i.e., Population, Intervention, Comparison, Outcomes) framework. Study outcomes and adverse events are represented in bar graphs.

user experience, user satisfaction, implementation, efficiency, and effect of infobuttons on providers within commercial EHR implementations, as well as optimal methods to infobutton customization and to promote infobutton uptake. In addition, the increased adoption of the HL7 CDS Hooks standard opens up opportunities for innovative infobutton approaches that explore aspects such as more sophisticated inferences over the clinical context and different user experience approaches.

References

[1] Chueh H, Barnett GO. "Just-in-time" clinical information. Acad Med 1997;72(6):512–7.

[2] Aakre CA, Pencille LJ, Sorensen KJ, Shellum JL, Del Fiol G, Maggio LA, et al. Electronic knowledge resources and point-of-care learning: a scoping review. Acad Med 2018;93:S60-s7 (11S Association of American Medical Colleges Learn Serve Lead: Proceedings of the 57th Annual Research in Medical Education Sessions).

[3] Cimino JJ, Sengupta S. IAIMS and UMLS at Columbia-Presbyterian Medical Center. Med Decis Mak: Int J Soc Med Decis Mak 1991;11(4 Suppl):S89–93.

[4] Greenes RA. A "building block" approach to application development for education and decision support in radiology: implications for integrated clinical information systems environments. J Digit Imaging 1991;4 (4):213–25.

[5] Del Fiol G, Workman TE, Gorman PN. Clinical questions raised by clinicians at the point of care: a systematic review. JAMA Intern Med 2014;174(5):710–8.

[6] Henry J, Pylypchuk Y, Searcy T, Patel V. Adoption of electronic health record systems among US non-federal acute care hospitals: 2008–2015. ONC Data Brief 2016;35:1–9.

[7] Covell DG, Uman GC, Manning PR. Information needs in office practice: are they being met? Ann Intern Med 1985;103(4):596–9.

[8] Del Fiol G, Weber AI, Brunker CP, Weir CR. Clinical questions raised by providers in the care of older adults: a prospective observational study. BMJ Open 2014;4(7), e005315.

[9] Ely JW, Osheroff JA, Gorman PN, Ebell MH, Chambliss ML, Pifer EA, et al. A taxonomy of generic clinical questions: classification study. BMJ 2000;321(7258):429–32.

[10] Gorman PN, Helfand M. Information seeking in primary care: how physicians choose which clinical questions to pursue and which to leave unanswered. Med Decis Making 1995;15(2):113–9.

[11] González-González AI, Dawes M, Sánchez-Mateos J, Riesgo-Fuertes R, Escortell-Mayor E, Sanz-Cuesta T, et al. Information needs and information-seeking behavior of primary care physicians. Ann Fam Med 2007;5 (4):345–52.

[12] Graber MA, Randles BD, Monahan J, Ely JW, Jennissen C, Peters B, et al. What questions about patient care do physicians have during and after patient contact in the ED? The taxonomy of gaps in physician knowledge. Emerg Med J 2007;24(10):703–6.

[13] Ebell MH, Cervero R, Joaquin E. Questions asked by physicians as the basis for continuing education needs assessment. J Contin Educ Health Prof 2011;31(1):3–14.

[14] Ely JW, Osheroff JA, Maviglia SM, Rosenbaum ME. Patient-care questions that physicians are unable to answer. J Am Med Inform Assoc 2007;14(4):407–14.

[15] Currie LM, Graham M, Allen M, Bakken S, Patel V, Cimino JJ. Clinical information needs in context: an observational study of clinicians while using a clinical information system. AMIA Annu Symp Proc 2003;190–4.

[16] Allen M, Currie LM, Graham M, Bakken S, Patel VL, Cimino JJ. The classification of clinicians' information needs while using a clinical information system. AMIA Annu Symp Proc 2003;26–30.

[17] Collins SA, Currie LM, Bakken S, Cimino JJ. Information needs, Infobutton Manager use, and satisfaction by clinician type: a case study. J Am Med Inform Assoc 2009;16(1):140–2.

[18] D'Adamo M, Short Fabic M, Ohkubo S. Meeting the health information needs of health workers: what have we learned? J Health Commun 2012;17(Suppl. 2):23–9.

[19] Gagnon MP, Legare F, Labrecque M, Fremont P, Pluye P, Gagnon J, et al. Interventions for promoting information and communication technologies adoption in healthcare professionals. Cochrane Database Syst Rev 2009;1, CD006093.

[20] Maggio LA, Aakre CA, Del Fiol G, Shellum J, Cook DA. Impact of Clinicians' use of electronic knowledge resources on clinical and learning outcomes: systematic review and meta-analysis. J Med Internet Res 2019;21(7), e13315.

[21] Ely JW, Osheroff JA, Chambliss ML, Ebell MH, Rosenbaum ME. Answering physicians' clinical questions: obstacles and potential solutions. J Am Med Inform Assoc 2005;12(2):217–24.

[22] Forsythe DE, Buchanan BG, Osheroff JA, Miller RA. Expanding the concept of medical information: an observational study of physicians' information needs. Comput Biomed Res 1992;25(2):181–200.

[23] Khedr M, Karmouch A. Negotiating context information in context-aware systems. Intell Syst, IEEE 2004;19 (6):21–9.

[24] Fischer G, Otswald J. Knowledge management: problems, promises, realities, and challenges. Intell Syst, IEEE 2001;16(1):60–72.

[25] Lomax EC, Lowe HJ. Information needs research in the era of the digital medical library. Proc AMIA Symp 1998;658–62.
[26] Pratt W, Sim I. Physician's information customizer (PIC): using a shareable user model to filter the medical literature. Medinfo 1995;8(Pt 2):1447–51.
[27] Cimino JJ, Li J, Bakken S, Patel VL. Theoretical, empirical and practical approaches to resolving the unmet information needs of clinical information system users. Proc AMIA Symp 2002;170–4.
[28] Del Fiol G, Rocha RA, Bradshaw RL, Hulse NC, Roemer LK. An XML model that enables the development of complex order sets by clinical experts. IEEE Trans Inf Technol Biomed 2005;9(2):216–28.
[29] Cimino JJ. The contribution of observational studies and clinical context information for guiding the integration of infobuttons into clinical information systems. AMIA Ann Symp Proc/AMIA Symp 2009;2009:109–13.
[30] Hu B, Bajracharya A, Yu H. Generating medical assessments using a neural network model: algorithm development and validation. JMIR Med Inform 2020;8(1), e14971.
[31] Yang Z, Yu H, editors. Generating accurate electronic health assessment from medical graph. Proceedings of the conference on empirical methods in natural language processing conference on empirical methods in natural language processing. NIH Public Access; 2020.
[32] Del Fiol G, Haug PJ. Classification models for the prediction of clinicians' information needs. J Biomed Inform 2009;42(1):82–9.
[33] Del Fiol G, Haug PJ. Infobuttons and classification models: a method for the automatic selection of on-line information resources to fulfill clinicians' information needs. J Biomed Inform 2008;41(4):655–66.
[34] Fiorini N, Canese K, Starchenko G, Kireev E, Kim W, Miller V, et al. Best Match: new relevance search for PubMed. PLoS Biol 2018;16(8), e2005343.
[35] Chen ES, Bakken S, Currie LM, Patel VL, Cimino JJ. An automated approach to studying health resource and infobutton use. Stud Health Technol Inform 2006;122:273–8.
[36] Cimino JJ. Linking patient information systems to bibliographic resources. Methods Inf Med 1996;35(2):122–6.
[37] Stead WW, Miller RA, Musen MA, Hersh WR. Integration and beyond: linking information from disparate sources and into workflow. J Am Med Inform Assoc: JAMIA 2000;7(2):135–45.
[38] Powsner SM, Riely CA, Barwick KW, Morrow JS, Miller PL. Automated bibliographic retrieval based on current topics in hepatology: hepatopix. Comput Biomed Res, Int J 1989;22(6):552–64.
[39] Powsner SM, Miller PL. Automated online transition from the medical record to the psychiatric literature. Methods Inf Med 1992;31(3):169–74.
[40] Loonsk JW, Lively R, TinHan E, Litt H. Implementing the medical desktop: tools for the integration of independent information resources. In: Proceedings/the annual symposium on computer application [sic] in medical care symposium on computer applications in medical care; 1991. p. 574–7.
[41] Powsner SM, Miller PL. From patient reports to bibliographic retrieval: a Meta-1 front-end. In: Proceedings/the annual symposium on computer application [sic] in medical care symposium on computer applications in medical care; 1991. p. 526–30.
[42] Cimino C, Barnett GO, Hassan L, Blewett DR, Piggins JL. Interactive query workstation: standardizing access to computer-based medical resources. Comput Methods Programs Biomed 1991;35(4):293–9.
[43] Hales JW, Low RC, Fitzpatrick KT. Using the Internet Gopher Protocol to link a computerized patient record and distributed electronic resources. Proc Annu Symp Comput Appl Med Care 1993;621–5.
[44] Hersh W. "A world of knowledge at your fingertips": the promise, reality, and future directions of on-line information retrieval. Acad Med 1999;74(3):240–3.
[45] Detmer WM, Barnett GO, Hersh WR. MedWeaver: integrating decision support, literature searching, and Web exploration using the UMLS Metathesaurus. In: Proceedings: a conference of the American Medical Informatics Association/AMIA Annual Fall Symposium AMIA Fall Symposium; 1997. p. 490–4.
[46] Porcelli PJ, Lobach DF. Integration of clinical decision support with on-line encounter documentation for well child care at the point of care. In: Proceedings/AMIA Annual Symposium AMIA Symposium; 1999. p. 599–603.
[47] Tang PC, Young CY. ActiveGuidelines: integrating Web-based guidelines with computer-based patient records. In: Proceedings/AMIA Annual Symposium AMIA Symposium; 2000. p. 843–7.
[48] Cimino JJ, Elhanan G, Zeng Q. Supporting infobuttons with terminological knowledge. In: Proceedings: a conference of the American Medical Informatics Association / AMIA Annual Fall Symposium AMIA Fall Symposium; 1997. p. 528–32.
[49] Cimino JJ, Johnson SB, Aguirre A, Roderer N, Clayton PD. The MEDLINE button. In: Proceedings of the annual symposium on computer application in medical care. American Medical Informatics Association; 1992. p. 81.

[50] Cimino JJ, Li J, Graham M, Currie LM, Allen M, Bakken S, et al. Use of online resources while using a clinical information system. AMIA Annu Symp Proc 2003;175–9.
[51] Reichert JC, Glasgow M, Narus SP, Clayton PD. Using LOINC to link an EMR to the pertinent paragraph in a structured reference knowledge base. Proc AMIA Symp 2002;652–6.
[52] Del Fiol G, Rocha RA, Clayton PD. Infobuttons at Intermountain Healthcare: utilization and infrastructure. AMIA Annu Symp Proc 2006;180–4.
[53] Huff SM, Rocha RA, McDonald CJ, De Moor GJ, Fiers T, Bidgood Jr WD, et al. Development of the logical observation identifier names and codes (LOINC) vocabulary. J Am Med Inform Assoc: JAMIA 1998;5(3):276–92.
[54] Fuller SS, Ketchell DS, Tarczy-Hornoch P, Masuda D. Integrating knowledge resources at the point of care: opportunities for librarians. Bull Med Libr Assoc 1999;87(4):393–403.
[55] Maviglia SM, Yoon CS, Bates DW, Kuperman G. KnowledgeLink: impact of context-sensitive information retrieval on clinicians' information needs. J Am Med Inform Assoc 2006;13(1):67–73.
[56] Rosenbloom ST, Giuse NB, Jerome RN, Blackford JU. Providing evidence-based answers to complex clinical questions: evaluating the consistency of article selection. Acad Med 2005;80(1):109–14.
[57] Kawamoto K, Del Fiol G, Orton C, Lobach DF. System-agnostic clinical decision support services: benefits and challenges for scalable decision support. Open Med Inform J 2010;4:245–54.
[58] Cimino JJ, Li J, Allen M, Currie LM, Graham M, Janetzki V, et al. Practical considerations for exploiting the World Wide Web to create infobuttons. Stud Health Technol Inform 2004;107(Pt 1):277–81.
[59] Cimino JJ. Formal descriptions and adaptive mechanisms for changes in controlled medical vocabularies. Methods Inf Med 1996;35(3):202–10.
[60] McGowan JJ, Berner ES. Proposed curricular objectives to teach physicians competence in using the World Wide Web. Acad Med 2004;79(3):236–40.
[61] Del Fiol G, Curtis C, Cimino JJ, Iskander A, Kalluri AS, Jing X, et al. Disseminating context-specific access to online knowledge resources within electronic health record systems. Stud Health Technol Inform 2013;192:672–6.
[62] Mandel JC, Kreda DA, Mandl KD, Kohane IS, Ramoni RB. SMART on FHIR: a standards-based, interoperable apps platform for electronic health records. J Am Med Inform Assoc 2016;23(5):899–908.
[63] Del Fiol G, Huser V, Strasberg HR, Maviglia SM, Curtis C, Cimino JJ. Implementations of the HL7 context-aware knowledge retrieval ("Infobutton") standard: challenges, strengths, limitations, and uptake. J Biomed Inform 2012;45(4):726–35.
[64] Cimino JJ, Jing X, Del Fiol G. Meeting the electronic health record "meaningful use" criterion for the HL7 infobutton standard using OpenInfobutton and the Librarian Infobutton Tailoring Environment (LITE). AMIA Annu Symp Proc 2012;2012:112–20.
[65] Jing X, Cimino JJ, Del Fiol G. Usability and acceptance of the librarian infobutton tailoring environment: an open access online knowledge capture, management, and configuration tool for OpenInfobutton. J Med Internet Res 2015;17(11), e272.
[66] IBM, 2011 'Watson' Wins: 'Jeopardy' Computer Beats Ken Kennings, Brad Rutter 2011 [updated 2011/02/17/01:02:36 EST 2012/11/09/15:54:55. Available from: http://www.huffingtonpost.com/2011/02/17/ibm-watson-jeopardy-wins_n_824382.html http://www.huffingtonpost.com/2011/02/17/ibm-watson-jeopardy-wins_n_824382.html?view=print&comm_ref=false.
[67] Watson Meets Healthcare. Advances in clinical question answering: Watson meets healthcare. Organizers: John F Hurdle and Guergana Savova. Panelists: Marty Kohn, Rodney Nielsen, Dina Demner-Fushman, and Hong Yu; 2011.
[68] Siri (software). Wikipedia, the free encyclopedia; 2012.
[69] Hersh W, Cohen A, Yang JJ, Bhupatiraju RT, Roberts P, Hearst M, editors. TREC 2005 Genomics Track overview; 2005.
[70] Cimino JJ, Aguirre A, Johnson SB, Peng P. Generic queries for meeting clinical information needs. Bull Med Libr Assoc 1993;81(2):195–206.
[71] Rinaldi F, Dowdall J, Schneider G, Persidis A, editors. Answering questions in the genomics domain; 2004.
[72] Niu Y, Hirst G. Analysis of semantic classes in medical text for question answering; 2004.
[73] Mendonca EA, Cimino JJ, Johnson SB, Seol YH. Accessing heterogeneous souces of evidence to answer clinical questions. J Am Med Inform Assoc 2001;34:85–98.
[74] Chase HS, Kaufman DR, Johnson SB, Mendonca EA. Voice capture of medical residents' clinical information needs during an inpatient rotation. J Am Med Inform Assoc 2009;16(3):387–94.

[75] Elhadad N, McKeown K, Kaufman D, Jordan D. Facilitating physicians' access to information via tailored text summarization. AMIA Annu Symp Proc 2005;226–30.
[76] Elhadad N, McKeown K, editors. Towards generating patient specific summaries of medical articles; 2001.
[77] Ide NC, Loane RF, Demner-Fushman D. Essie: a concept-based search engine for structured biomedical text. J Am Med Inform Assoc 2007;14(3):253–63.
[78] Demner-Fushman D, Lin J. Answering clinical questions with knowledge-based and statistical techniques. Comput Linguist 2007;33(1):63–103.
[79] Sneiderman CA, Demner-Fushman D, Fiszman M, Ide NC, Rindflesch TC. Knowledge-based methods to help clinicians find answers in MEDLINE. J Am Med Inform Assoc 2007;14(6):772–80.
[80] Ni Y, Zhu H, Cai P, Zhang L, Qui Z, Cao F. CliniQA: highly reliable clinical question answering system. Stud Health Technol Inform 2012;180:215–9.
[81] Cairns BL, Nielsen RD, Masanz JJ, Martin JH, Palmer MS, Ward WH, et al. The MiPACQ clinical question answering system. AMIA Annu Symp Proc 2011;2011:171–80.
[82] Cao Y, Liu F, Simpson P, Antieau L, Bennett A, Cimino JJ, et al. AskHERMES: An online question answering system for complex clinical questions. J Biomed Inform 2011;44(2):277–88.
[83] Demner-Fushman D, Lin J. Answer extraction, semantic clustering, and extractive summarization for clincial question answering; 2006.
[84] Huang X, Lin J, Demner-Fushman D. Evaluation of PICO as a knowledge representation for clinical questions. AMIA Annu Symp Proc 2006;359–63.
[85] Cao YG, Ely J, Yu H, editors. Using weighted keywords to improve clinical question answering; 2009.
[86] Yu W, Yesupriya A, Wulf A, Qu J, Khoury MJ, Gwinn M. An open source infrastructure for managing knowledge and finding potential collaborators in a domain-specific subset of PubMed, with an example from human genome epidemiology. BMC Bioinform 2007;8(1):1–3.
[87] Robertson S, Zaragoza H, Taylor M, editors. Simple BM25 extension to multiple weighted fields; 2004.
[88] Humphrey B, Lindberg DAB, Schoolman HM, Barnett GO. The unified medical language system: an informatics research collaboration. JAMA 1998;5:1–11.
[89] Paterson M, Dancik V, editors. Longest common subsequences. Proceedings of 19th MFCS; 1994.
[90] Hirschberg DS. Algorithms for the longest common subsequence problem. J ACM (JACM) 1977;24(4):664–75.
[91] Zafar A, Overhage JM, McDonald CJ. Continuous speech recognition for clinicians. J Am Med Inform Assoc 1999;6(3):195–204.
[92] Zafar A, Mamlin B, Perkins S, Belsito AM, Overhage JM, McDonald CJ. A simple error classification system for understanding sources of error in automatic speech recognition and human transcription. Int J Med Inform 2004;73(9–10):719–30.
[93] Devine EG, Gaehde SA, Curtis AC. Comparative evaluation of three continuous speech recognition software packages in the generation of medical reports. J Am Med Inform Assoc 2000;7(5):462–8.
[94] Borowitz SM. Computer-based Speech Recognition as an Alternative to Medical Transcription. J Am Med Inform Assoc 2001;8(1):101–2.
[95] Havstam C, Buchholz M, Hartelius L. Speech recognition and dysarthria: a single subject study of two individuals with profound impairment of speech and motor control. Logoped Phoniatr Vocol 2003;28(2):81–90.
[96] Liu F, Tur G, Hakkani-Tür D, Yu H. Towards spoken clinical-question answering: evaluating and adapting automatic speech-recognition systems for spoken clinical questions. J Am Med Inform Assoc 2011;18(5):625–30.
[97] Miller T, Cimino JJ, Ravvaz K, Yu H. An investigation into the feasibility of spoken clinical question answering. AMIA Annu Symp Proc 2011;2011:954–9.
[98] Cook DA, Teixeira MT, Heale BSE, Cimino JJ, Del Fiol G. Context-sensitive decision support (infobuttons) in electronic health records: a systematic review. J Am Med Inform Assoc: JAMIA 2017;24(2):460–8.
[99] Cimino JJ, Overby CL, Devine EB, Hulse NC, Jing X, Maviglia SM, et al. Practical choices for infobutton customization: experience from four sites. AMIA Annu Symp Proc 2013;2013:236–45.
[100] Del Fiol G, Haug PJ, Cimino JJ, Narus SP, Norlin C, Mitchell JA. Effectiveness of topic-specific infobuttons: a randomized controlled trial. J Am Med Inform Assoc 2008;15(6):752–9.
[101] Burgess S, Dennis S, Lanka S, Miller N, Potvin J. MedlinePlus connect: linking health IT systems to consumer health information. IT Prof 2012;14(3):22–8.

[102] Jentsch J. MedlinePlus connect: 10 years of linking electronic health records to consumer health information. National Library of Medicine; 2021. updated 01/13/2021. Available from: https://nlmdirector.nlm.nih.gov/2021/01/13/medlineplus-connect-10-years-of-linking-electronic-health-records-to-consumer-health-information/.

[103] Del Fiol G, Michelson M, Iorio A, Cotoi C, Haynes RB. A deep learning method to automatically identify reports of scientifically rigorous clinical research from the biomedical literature: comparative analytic study. J Med Internet Res 2018;20(6), e10281.

[104] Bian J, Morid MA, Jonnalagadda S, Luo G, Del Fiol G. Automatic identification of high impact articles in PubMed to support clinical decision making. J Biomed Inform 2017;73:95–103.

[105] Jonnalagadda SR, Del Fiol G, Medlin R, Weir C, Fiszman M, Mostafa J, et al. Automatically extracting sentences from Medline citations to support clinicians' information needs. J Am Med Inform Assoc: JAMIA 2012.

[106] Del Fiol G, Mostafa J, Pu D, Medlin R, Slager S, Jonnalagadda SR, et al. Formative evaluation of a patient-specific clinical knowledge summarization tool. Int J Med Inform 2016;86:126–34.

[107] Bian J, Weir C, Unni P, Borbolla D, Reese T, Wan YJ, et al. Interactive visual displays for interpreting the results of clinical trials: formative evaluation with case vignettes. J Med Internet Res 2018;20(6), e10507.

[108] Bakken S, Roberts WD, Chen E, Dilone J, Lee NJ, Mendonca E, et al. PDA-based informatics strategies for tobacco use screening and smoking cessation management: a case study. Stud Health Technol Inform 2007;129(Pt 2):1447–51.

[109] Schnall R, Cimino JJ, Bakken S. Development of a prototype continuity of care record with context-specific links to meet the information needs of case managers for persons living with HIV. Int J Med Inform 2012;81(8):549–55.

[110] Heale BS, Overby CL, Del Fiol G, Rubinstein WS, Maglott DR, Nelson TH, et al. Integrating genomic resources with electronic health records using the HL7 Infobutton standard. Appl Clin Inform 2016;7(3):817–31.

[111] Crump J, Del Fiol G, Williams MS, Freimuth R. Prototype of a standards-based EHR and genetic test reporting tool coupled with HL7-compliant infobuttons. San Francisco, USA: AMIA Clinical Research Informatics Summit; 2018. p. 330–9.

[112] Overby CL, Rasmussen LV, Hartzler A, Connolly JJ, Peterson JF, Hedberg RE, et al. A template for authoring and adapting genomic medicine content in the eMERGE Infobutton project. AMIA Annu Symp Proc 2014;2014:944–53.

[113] Kennell Jr T, Dempsey DM, Cimino JJ. i3b3: Infobuttons for i2b2 as a mechanism for investigating the information needs of clinical researchers. AMIA Annu Symp Proc 2016;2016:696–704.

[114] Long J, Hulse NC, Tao C. Infobutton usage in Patient Portal MyHealth. AMIA Jt Summits Transl Sci Proc 2015;2015:112–6.

[115] Baorto DM, Cimino JJ. An "infobutton" for enabling patients to interpret on-line Pap smear reports. Proc AMIA Symp 2000;47–50.

[116] Borbolla D, Del Fiol G, Taliercio V, Otero C, Campos F, Martinez M, et al. Integrating personalized health information from MedlinePlus in a patient portal. Stud Health Technol Inform 2014;205:348–52.

[117] Tarver T, Jones DA, Adams M, Garcia A. The librarian's role in linking patients to their personal health data and contextual information. Med Ref Serv Q 2013;32(4):459–67.

[118] Ancker JS, Mauer E, Hauser D, Calman N. Expanding access to high-quality plain-language patient education information through context-specific hyperlinks. AMIA Annu Symp Proc 2016;2016:277–84.

CHAPTER

14

Information visualization and integration

Melanie C. Wright

College of Pharmacy, Idaho State University, Meridian, ID, United States

14.1 Introduction

Clinical decisions usually require consideration of information from a variety of information sources such as EHRs, patient monitoring devices, imaging and laboratory findings, and non-technological information sources (e.g., visual inspection, patient report). The effectiveness of clinical decision support is largely dependent on how well the decision support integrates into its broader context of use. Prior chapters have described the importance of associative grouping of information (Chapter 12) within the system in which the CDS is implemented and providing access to relevant supportive information at the point of decisions and actions (Chapter 13). These techniques help ensure that related information and actions are presented together, and to the right people at the right point in the clinical workflow, to make it faster and easier for users to make the right decisions and take the right actions.

Historically, individual devices or technologies were developed independently for specific purposes. For example, the electrocardiogram developed from efforts throughout the early 1900's to measure the difference in electrical potential between sites due to the heart's activation [1]. Plethysmography to measure lung volume and airway resistance also began prior to the 1900's, yet non-invasive measurement of oxygen saturation was not available commercially until the 1980's [2]. It has only been since the 1980's that device manufacturers have integrated different monitors (capnography, plethysmography, electrocardiogram, blood pressure monitors) into a single bedside monitoring device. Through meaningful use incentives in the US [3] and major advances in data storage and data transfer technologies, it is increasingly feasible to integrate patient information generated from a wide variety of sources for the purposes of both analysis and information presentation. And these capabilities are

Clinical Decision Support and Beyond
https://doi.org/10.1016/B978-0-323-91200-6.00032-2

now expanding from being available only in controlled clinical settings, into a wider variety of at-home, mobile, and wearable applications.

While advances in technology are generally perceived as positively impacting patient care, there are challenges. The quantity of information available now exceeds human capabilities to process that information within the time and manpower available to benefit patients. While artificial intelligence (AI) methods may be helpful in managing information overload, thus far, the use of AI approaches for these purposes is immature [4,5]. In order to make the best use of these technological advances, there is a need for an emphasis on user-centered *design* that takes advantage of both human and technological strengths. In the development of new technology, the primary emphasis is typically on technological innovations and new functionality. Attention to exactly how to implement new functionality, how users will interact with it, and how it will integrate with existing functions often is not considered until late in the process, and frequently follows the path of least (initial) resistance. That is, decisions regarding implementation or user interface design can partly be driven by how much work is required to achieve that design or implementation. The cost of early development effort (programming time, new hardware requirements) may not be appropriately balanced against future costs that are more difficult to quantify or are identified later in the process (the need to train users, time required for users to engage with new functionality, user errors, poor user adoption). Decisions regarding design or implementation may also be based on a limited view of the new functionality and its perceived value in isolation, without the broader understanding of how it fits into the full context of work for the user.

As an example, when we studied how critical care clinicians use information to make patient care decisions, at the top of the list of participant expressed needs was the ability to see dynamic quantitative data displayed over time (trends) and co-located on a time scale with other related data [6]. This format of information presentation makes it easy for clinicians to track a patients progress in relation to specific actions or to explore cause-effect relationships across different events or actions and a variety of physiologic or other patient responses to support diagnoses. The ability to visualize trends in quantitative data is not only a preference, but also has been shown to improve performance with respect to clinical decision-making [7–9]. However, clinicians also stated that fast information access was paramount, so much so that even though clinicians preferred to see data plotted as a trend, few would make additional "clicks" necessary to reformat data from a tabular presentation to plotted trends if it was required. Because users are unlikely to commit time to customize display formats to their needs and because it is difficult for developers to make key decisions such as what information to display together and what scales to use, despite the potential for patient outcome and healthcare cost benefits of trend display formats, solutions that are simpler to develop, such as tabular displays, persist.

CDS is, by nature, a human-technology partnership to support effective clinical decisions, thus, attention to the design of the human-user experience is critically important. User experience details matter. Trust in and subsequent adoption of recommendations generated by CDS can be critically influenced by whether relevant related information is easy to access at the right time. While some may consider *design* as a form of art or aesthetic, good user experience design results from the application of evidence-based design principles and well-established design methods. This does not mean that creativity and aesthetics are not also important. Creativity may be necessary to develop novel solutions to difficult problems.

In the implementation of CDS, efficiency and efficacy rank at the top of user priorities, ahead of aesthetic preferences, however, well-designed aesthetics can work hand-in-hand with principles that promote efficiency and efficacy such as visibility, consistency, appropriate visual emphasis, and visual balance. Bennett and Flach [10] espouse a design approach of *subtle science and exact art* [11] and expand the conventional dyad of user interface to user-interpreted meaning to include a third component of *ecology* which represents the context, work domain, or applied problem being addressed.

In this chapter, we draw on theoretical models of human cognition and visual perception, understanding of workflow and information needs in the context of clinical applications, and user-centered design methods to promote the design of effective information integration and visualization (see Fig. 14.1). Other chapters in this book cover theoretical and methodological frameworks for user-centered design (Chapter 19), methods appropriate for understanding clinical information needs (Chapter 6), and clinical evaluation (Chapter 21). In this chapter, we will expand on these ideas with respect to understanding human perception, information processing, and fundamental visual design principles that are particularly relevant to the design of information integration and visualization. We will first provide a grounding in how humans think, and see, which has led to the development of basic design principles that can guide the design of information integration and visualization. Then, we will expand on design methods that are particularly relevant to understanding information integration and visualization needs in CDS contexts. Finally, we will provide an overview of current problems and efforts to apply information integration and visualization for clinical decision support.

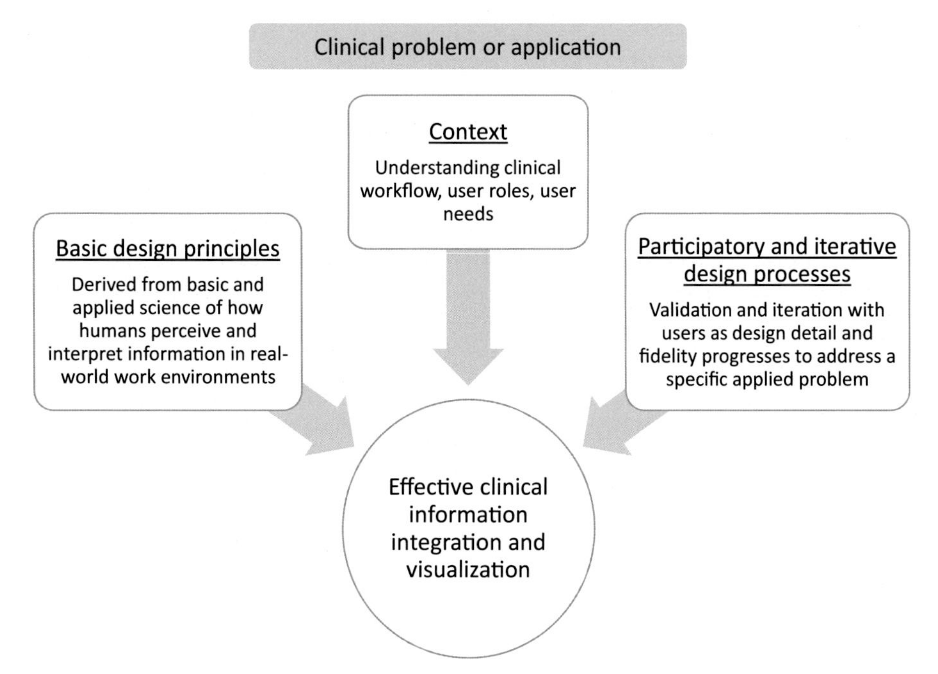

FIG. 14.1 Designing effective clinical information integration and visualization to address a clinical application. *Author's own work.*

14.2 How people think, and see

Chapter 3 presented multiple purposes of CDS such as answering questions, monitoring actions, supporting decisions, focusing attention, and enhancing visualization. There are a number of cognitive or information-processing theoretical models to help understand human perception, cognition and behavior in the context of CDS applications [12–14]. Chapter 19 covers theoretical frameworks of cognition and their implications across the broader context of healthcare teams and organizations. Here, we focus on aspects of theoretical models that provide insight into human perception and interpretation of visual information.

Wickens [13] developed a *human information processing* model from the input of environmental stimuli (e.g., a patient's visual appearance, verbal information from the patient or others, the display of information on a monitor) to a specific response or action. Key features of this model that are applicable to CDS design include:

(1) *Stimuli from the environment are first experienced by the human's short term, pre-attentive, sensory store.* While humans have a very large (veridical) capacity for visual and auditory input via this sensory store, it is rapidly decaying. Limitations of human attention greatly reduce the amount of sensory information that can subsequently be perceived and processed. This has important implications for ensuring that visual displays appropriately focus attention to critical information. This also has implications for the design of systems that may be able to draw on humans' expansive pre-attentive automated pattern recognition capabilities.
(2) *Information that is attended to is consciously perceived and processed by higher order cognitive centers.* Information may be perceived, recognized, detected, or categorized.
(3) *Decision and action are intricately tied to memory.* Working memory is engaged to interpret and process the information and long-term memory is engaged to place the information and cognitive processing in the context of one's experience and education.
(4) *The results of decisions or actions are monitored in a feedback loop*, further affecting perception and information processing.
(5) *Both attention and working memory are limited in humans* and thus pose limitations on human performance in the context of high information fast-paced work environments.

Human information processing can be *controlled*, which is the serial, deliberate, slow, and effortful processing of information that demands attention and working memory. Alternatively, information processing can be *automatic* [15,16]. Automatic processing is fast and fluid, demands little attention and working memory, and occurs largely outside of consciousness. This type of information processing develops with extensive consistent practice and is particularly useful when humans are required to manage multiple tasks concurrently.

Researchers have expanded this fundamental human information processing framework to more precise human-system performance constructs such as situation awareness, decision-making, and intuitive cognition. *Situation awareness* (SA) refers to an individual's (or team's) understanding of the current *dynamic* state of the work environment and is expanded on in detail in Chapter 19. SA is important because good decisions require an accurate and current understanding of relevant information in the environment. SA as a theoretical construct is central to the design of information integration and visualization because the choices that visual designers make regarding what information to group in a display and how to present that information are key to supporting the development of good SA [17].

Beyond having a good understanding of the dynamic information environment are the subsequent steps of decision and action. The study of human decision-making has generated interesting findings regarding how humans make decisions including their innate use of heuristics (rules-of-thumb) and how these can either be effective or problematic [16,18]. For designing information integration and visualization to support clinical decision-making in a dynamic time-critical environment, studies of decision-making in real world settings (naturalistic decision making [19–22]), provide important insight. The recognition-primed decision-making model [19] represents how individuals working in real-world dynamic work environments:

(1) experience a situation (and information available)
(2) judge it as familiar or unfamiliar
(3) develop expectancies and seek information to confirm expectancies
(4) mentally simulate courses of action
(5) continually reassess the situation in the process of taking actions

This model highlights the importance of designing displays that support recognition of visual patterns and comparisons between data and expectancies.

Drawing on these conceptual models of human perception and cognition, *intuitive cognition* has been described as a "quick grasp of meaningful gist of information based on experience or perceptual cues, without working memory or precise analysis" [23]. Supporting intuitive cognition is compelling for point of care systems where there is a need is to support fast and frugal decision-making and help clinicians quickly detect when it may be a good idea to slow down and look at a patient more carefully. Hegarty [24] asserts that visual displays enhance or "augment" cognition (see also [25]) by:

(1) providing an external representation which frees up working memory resources,
(2) organizing information by indexing it spatially, allowing for spatial relationships (e.g., proximity and grouping) to facilitate information search and integration,
(3) offloading cognition on to perception or "using vision to think"; for example, by mapping data to visual representations that allow patterns that are easily picked up by the visual system to emerge (see Fig. 14.2), and
(4) offloading cognition on to action, for example, by supporting the sorting of patients in a displayed list by features such as line of service or deterioration risk.

14.3 Visual design principles for CDS information integration and visualization

Research on human perception, cognition, and subsequent performance in a wide variety of basic and domain-specific applied human-computer tasks have led to the development of many design principles and guidelines [10,24,26–28]. In Table 14.1, we identify 21 principles that are relevant to the use of visualization and information integration in the design of CDS.

We start with foundational visual design principles. Because of the emphasis of clinicians on speed and simplicity (see Table 22.1, Ten Commandments for CDS), visual design choices

Hour	Heart rate	Respiratory rate
5:00	90	17
6:00	95	18
7:00	93	18
8:00	100	15
10:00	105	20
11:00	110	20
13:00	109	20
14:00	115	21
15:30	110	20
16:40	115	25
18:00	110	20
20:00	114	23
21:00	110	22
22:00	114	20

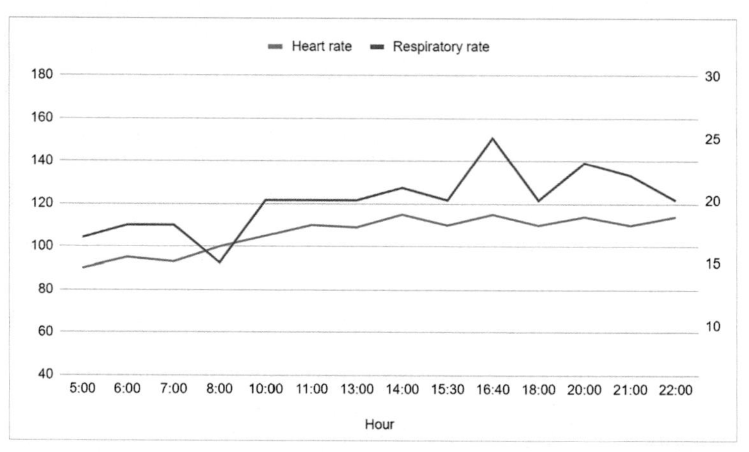

FIG. 14.2 A simple example of using vision think. It is more difficult to interpret change over time and the relationship between heart rate and respiratory rate by looking at the numbers in the table than it is to "see" those relationships in the plot. *Author's own work.*

TABLE 14.1 Principles to effective information integration and visualization for CDS [10,24,26–29].

Foundational visual design principles [29]:

1. Ensure that information has sufficient contrast to be readable by target users in the target setting.
2. Use colors judiciously. Over use of color can detract from content. About 5% of people have some form of color blindness. If meaning is conveyed by color, ensure that color differs not only by hue but also a significant difference in lightness or with redundant coding of meaning (e.g., an icon).
3. Ensure font style, size, and spacing are readable by expected user in the target setting. Ensure that symbols or icons are large enough and easily distinguishable. Simple icon designs are easier to distinguish than icons with fine details.
4. If shading or size is used to convey meaning, ensure that different levels are independently discriminable. That is, the range of intended users can differentiate relevant sizes or shading levels when contrasting alternatives are not visually present.
5. When a feature on a display is currently "in focus" or has interaction-capability, employ appropriate techniques to make that feature visually discriminable from "out of focus" or non-interaction display features.
6. Develop and follow a style guide (or a design system [30]) to ensure consistency of colors, font types and sizes, and methods of visual emphasis. Visual styles should be designed to make it easy for users to differentiate information into relevant categories such as headers, data, units, editable vs. non-editable, recent vs. old, normal vs. abnormal. Style guides also can help ensure that the number of different font styles, sizes, and colors are limited to reduce the risk of generating visually complex or cluttered designs that detract attention from meaningful information.

Task specificity:

7. Displays should be designed to support the specific use contexts and tasks for which they will be used [24]. Formal activities to understand and document use contexts and relevant tasks are recommended (see Section 14.3).
8. In the context of CDS, it may be necessary to develop variations or customization of designs for clinicians in different care roles or patients in different care settings. Where feasible, elicit and define unique needs by user roles and care settings and create "default" customizations instead of expecting working clinicians to personally customize the system to their needs.

TABLE 14.1 Principles to effective information integration and visualization for CDS [10,24,26–29]—cont'd

Effective and visually pleasing information organization:

9. Place the most important information in high visibility locations. In cultures where reading and writing progress from top to bottom and left to right, the upper and leftmost areas are prominent in the display space.
10. Create visually balanced information organizations with equal weighting of information content on the page (e.g., top and bottom; left and right).
11. Apply additional design principles such as regularity, predictability, symmetry, sequentiality, economy, and unity to ensure aesthetic and easy to use display layouts (see Galitz [26] for additional information).
12. Group together information that requires integration for decisions. Grouping can be achieved by placing in proximity or sequence and by using features such as headings and borders. Where appropriate, support temporal comparisons by visually aligning temporal scales (see Fig. 14.3).
13. Consistency in the placement of information supports learning and development of expectations that increase efficiency and error resilience. Consistency can be applied both with respect to the local or novel CDS visualization and should also be applied in the context of population expectations. For example, clinicians have come to expect patient demographics at the top and left of a screen and navigation controls sequenced across the top or along the left side of a page. Design guides are also useful for ensuring consistency across multiple applications and designers.

Visually emphasize relevant or meaningful information:

14. One overarching theme across many CDS applications is that clinicians are overloaded with information. Respect the limitations of human attentional capacity by presenting no more or less information than is needed for any given task.
15. Use visualization techniques to emphasize data (as opposed to non-data elements such as borders, frames, scale markings) in the context of presenting patient information to support decisions. One approach to this is maximizing the data-ink ratio [31] (see Fig. 14.4).
16. In clinical applications, errors due to lack of specificity or obscured details can have serious consequences. Balance the need for specificity with the need for emphasizing the most meaningful data. Techniques such as limiting the ability to obscure information (e.g., with movable windows), clearly annotating when information is "filtered," and minimizing (not removing) important but repetitive details support comprehensive information presentation while also making it easy to quickly see the most important information (see Fig. 14.5).
17. Another overarching theme of CDS applications is that new and urgent patient information is typically high priority. Emphasize new or urgent information by using high contrast, color, or annotations that draw attention. What makes information "new" or "urgent" may differ by clinician roles and care settings and should be determined for each use case using formal methods.
18. Where appropriate, minimize distraction by de-emphasizing patient data that conveys a normal or well state. Evaluate designs to ensure that de-emphasis of normal (e.g., using lower contrast colors) is not confused with designs that distinguish missing or dated information.

Effective use of graphics:

19. Match the graphic format to the data type. For example, use line graphs to visualize the relationship between two interval scale data types (such as time and heart rate) and bar graphs to evaluate relationships between nominal and interval scale data [28,31].
20. Choose natural mappings between graphic forms and meaning, and between spatial relationships and meanings [32]. For example, defined by gravity and human upright postures, humans perceive the vertical dimension with asymmetry, while the horizontal dimension is typically perceived to be symmetrical. In countries where writing and reading is left to right, time naturally maps from left (older) to right (most recent) and larger quantities are naturally mapped higher in the vertical dimension.
21. Choose graphics that are appropriate to relevant domain knowledge. For example, fishbone laboratory data shorthand or shared scales for a subset of physiology data may be easily interpreted in some settings but not in others.

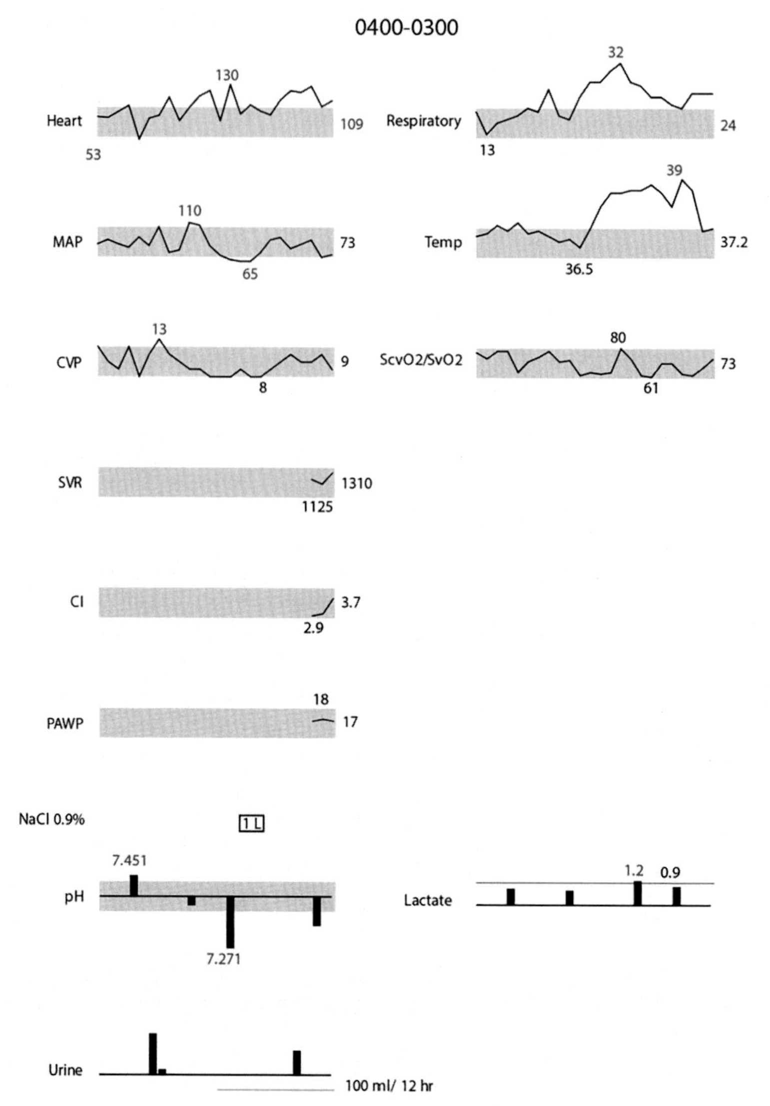

FIG. 14.3 In this integrated shock display, routinely available physiologic data are grouped in the upper section of the display (heart rate, mean arterial pressure, central venous pressure, respiratory rate, temperature, and oxygen saturation). Less frequently measured data are grouped in the center (systemic vascular resistance, cardiac index, and pulmonary artery wedge pressure). Relevant interventions and non-continuous data are grouped at the bottom of the page along the same timeline. The left column of data aligns (along a common timeline) information that is most relevant for integration to diagnose shock and determine the timing of shock onset. Data-ink ratio is kept to a minimum by labeling the 24-hour time scale that is relevant to all displays in one location and using spacing rather than grouping to separate plots. Gray shading of normal ranges minimizes the visibility of less relevant information (variation within normal ranges) and simple black on white supports high-contrast visibility of relevant information (data out of range and trending of data out of range). Red is used to indicate if minimum, maximum, and current values are out of the normal range. *Image courtesy of Thomas Reese (see also Reese TJ, Del Fiol G, Tonna JE, Kawamoto K, Segall N, Weir C, et al. Impact of integrated graphical display on expert and novice diagnostic performance in critical care. J Am Med Inform Assoc: JAMIA 2020;27:1287–92. https://doi.org/10.1093/jamia/ocaa086, adapted with permission).*

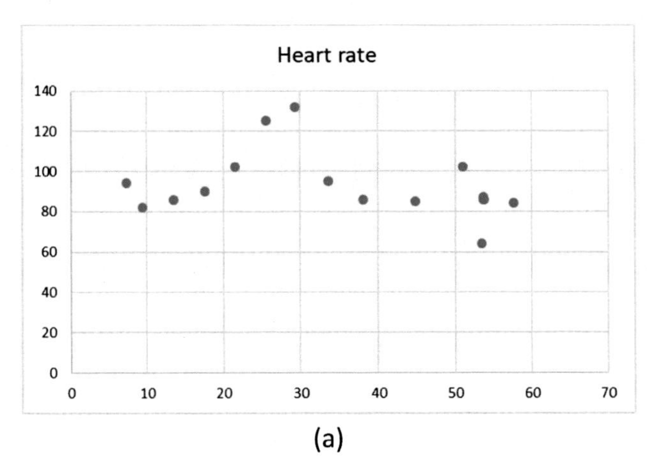

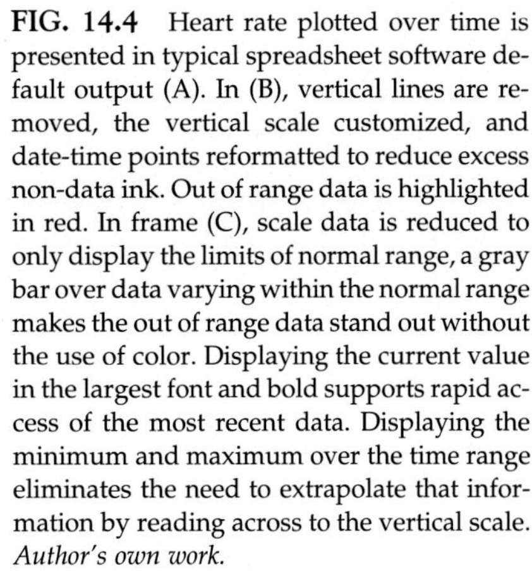

FIG. 14.4 Heart rate plotted over time is presented in typical spreadsheet software default output (A). In (B), vertical lines are removed, the vertical scale customized, and date-time points reformatted to reduce excess non-data ink. Out of range data is highlighted in red. In frame (C), scale data is reduced to only display the limits of normal range, a gray bar over data varying within the normal range makes the out of range data stand out without the use of color. Displaying the current value in the largest font and bold supports rapid access of the most recent data. Displaying the minimum and maximum over the time range eliminates the need to extrapolate that information by reading across to the vertical scale. *Author's own work.*

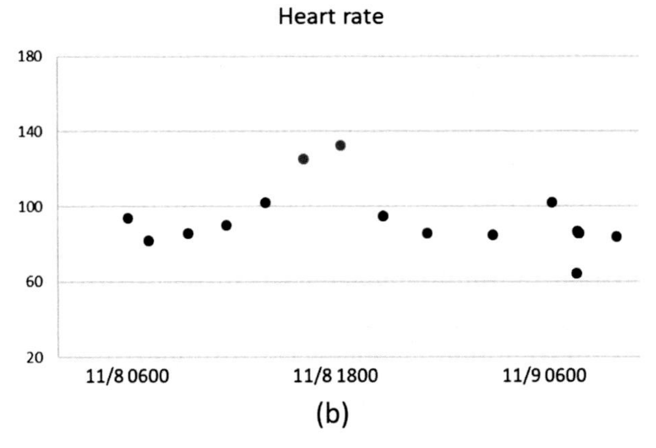

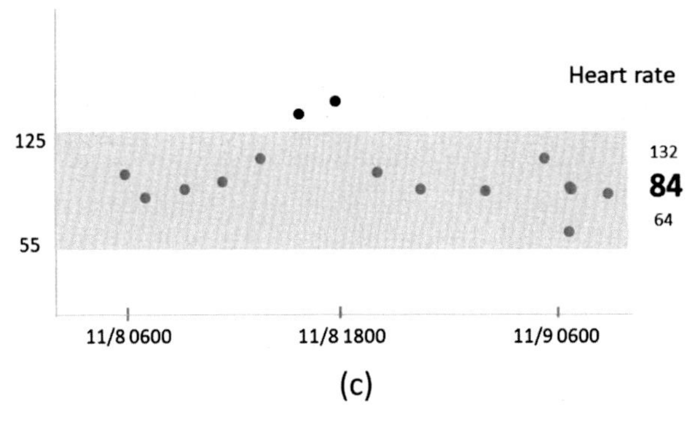

should favor rapid readability and interpretation over style. Even a perfectly reliable decision aid will fail if it is not seen, attended to, or its meaning mis-interpreted. These principles are expected to be widely known and integrated into good design practices of technology developers [10,24,26–29]. However, innovation often emerges from those who know the problem space well. For the design of CDS, this is often clinician users, clinical researchers, and

Blood Gases	Blood Gas pH Arterial	pCO2
3/16/2016 05:11 MDT	H 7.55	L 32 mmHG
Blood Gases	Total CO2	FiO2
3/16/2016 05:11 MDT	H 29 mEq/L	30
Chemistry-General	Sodium Level	Potassium Level
3/18/2016 02:25 MDT	L 131 mEq/L	L 3.4 mEq/L
3/17/2016 03:12 MDT	L 133 mEq/L	4.1 mEq/L
3/16/2016 18:00 MDT		3.7 mEq/L
3/16/2016 05:11 MDT	L 132 mEq/L	* C 2.9 mEq/L
Chemistry-General	BUN	Creatinine
3/18/2016 02:25 MDT	7 mg/dl	
3/17/2016 03:12 MDT	10 mg/dL	
3/16/2016 18:00 MDT	7 mg/dl	
Chemistry-General	Comment GFR	Calcium Total
3/16/2016 18:00 MDT	* *NOT VALUED*	L 7.3 mg/dL
3/16/2016 05:11 MDT		L 7.2 mg/dL
3/15/2016 05:11 MDT		L 7.2 mg/dL
Cardiac Isoenzymes	Troponin 1	
3/16/2016 05:11 MDT	* 0.03 ng/ML	
CBC	WBC Count	Red Blood Cell Count
3/17/2016 03:12 MDT	H 11.2 thou/cumm	L 2.99 million/mm3
3/16/2016 18:00 MDT	10.0 thou/comm	L 2.80 million/mm3
3/16/2016 05:11 MDT	9.8 thou/comm	L 2.94 million/mm3
CBC	MCHC	RDW
3/17/2016 03:12 MDT	33.1 gm/dL	
3/16/2016 18:00 MDT	32.4 gm/dL	
3/16/2016 05:11 MDT	33.7 gm/dL	
Differential	Diff Method	Neutrophil Percent
3/17/2016 03:12 MDT	AUTO	H 93.3 %
3/16/2016 18:00 MDT	AUTO	H 83.3 %
3/16/2016 05:11 MDT	AUTO	H 82.4 %

2016 MDT		
Blood Gases	Blood Gas pH Arterial	pCO2 (mmHG)
3/16 05:11	H 7.55	L 32
	Total CO2 (mEq/L)	FiO2
3/16 05:11	H 29	30
Chemistry-General	Sodium Level (mEq/L)	Potassium Level (mEq/L)
3/18 02:25	L 131	L 3.4
3/17 03:12	L 133	4.1
3/16 18:00		3.7
3/16 05:11	L 132	* C 2.9
	BUN (mg/dL)	Creatinine (mg/dL)
3/18 02:25	7	
3/17 03:12	10	
3/16 18:00	7	
	Comment GFR	Calcium Total (mg/dL)
3/16 18:00	* *NOT VALUED*	L 7.3
3/16 05:11		L 7.2
3/15 05:11		L 7.2
Cardiac Isoenzymes	Troponin 1 (ng/ML)	
3/16 05:11	* 0.03	
CBC	WBC Count (thou/mm3)	Red Blood Cell Count (million/mm3)
3/17 03:12	H 11.2	L 2.99
3/16 18:00	10	L 2.80
3/16 05:11	9.8	L 2.94
	MCHC (gm/dL)	RDW
3/17 03:12	33.1	
3/16 18:00	32.4	
3/16 05:11	33.7	
Differential	Diff Method	Neutrophil Percent
3/17 03:12	AUTO	H 93.3
3/16 18:00	AUTO	H 83.3
3/16 05:11	AUTO	H 82.4

FIG. 14.5 Reducing data-ink ratio speeds visual search (e.g., find the most recent red blood cell count). The table on the right contains requisite date and unit information, reduces repetitive framing, and emphasizes the most meaningful data (measured laboratory values and whether they fall outside of normal ranges). *Author's own work.*

technologists who may not have a strong grounding in visual design. Thus, we reiterate these principles here to ensure that failure to attend to these foundational principles does not doom otherwise good innovations in CDS.

For the remaining principles (covering task specificity, information organization, visual emphasis, and effective use of graphics) there are usually trade-offs to be balanced as they are applied. For example, information importance can vary by context. When does the principle of consistent information placement outweigh the principle of placing important information prominently? These trade-offs typically cannot be resolved through application of design principles alone; they require additional evaluation through methods such as those described in Section 14.3.

14.4 Framing problems and designing solutions

A number of methodologies or processes have been developed to generate effective design solutions. For example, TURF (see Chapter 19) advocates an approach that includes user analysis, functional analysis representational analysis, and task analysis. Different approaches for collecting and structuring relevant data primarily rely on variations of interviews, observations, or contextual inquiry (field interviews) [10,17,33–35]. There are differences between methodologies, often related to differences in the task or work environments in which they were developed, and some approaches may be more or less effective for different design problems. However, these are all user-centered design (UCD) approaches that have three key practices in common (Fig. 14.1):

(1) *Application of design principles* based on understanding the strengths, limitations, and expectations of people who will interact with the system.
(2) *A formal approach to understanding the user and the use environment* for the system being designed. This typically includes formal representation of goals, tasks, and relationships between information, people, and other aspects of the broader system and environment.
(3) Early engagement of representative users (from understanding the user and environment) through multiple stages of design that rely on *formal approaches to evaluation and iterative improvement* based on user feedback and objective metrics.

In a recent systematic review and meta-analysis, we extracted outcomes from 41 comparisons of novel critical care information display approaches to a baseline or alternative display approach. A majority of studies (73%) demonstrated improved accuracy, faster response, or lower workload with respect to clinical decisions associated with improved information display approaches. We were able to generate evidence in support of specific approaches to improving information presentation such as improving the representation of trend information or associative grouping. However, the characteristic that most clearly differentiated whether improved approaches were successful was *the degree to which these key UCD practices were applied in the design process* [36].

14.4.1 Knowing your user and contexts of use

Chapter 6 describes human-intensive techniques for extracting knowledge that is necessary for the development of clinical decision support. Beyond capturing the expert knowledge that can form the basis for decision support logic for generating decision support such as recommendations or risk scores, similar methods are necessary to make design decisions such as: to whom should the decision support be presented, on what device, or integrated with what additional data? While the emphasis of the methods described in Chapter 6 is on capturing and using expert *knowledge*, the emphasis for the purposes we describe is on the *context of use* such as competing work and attention demands and ranges of user characteristics such as roles and expertise. Methods described in Section 6.3.1, particularly contextual interview and ethnographic evaluation methods, are useful for these purposes (see also [33,35]). Investigators should consider the unique characteristics of different work environments and the experiences and perspectives of the individuals engaged in the data collection to select an effective approach.

As an example, in our effort to understand how information is used in critical care settings, we adapted a contextual interview approach by using an eye-tracker to record clinicians use of information while caring for patients. We then conducted interviews with the clinicians while they watched their recorded use of information technology. This provided the relevant context needed for rich recall of what information was being accessed and why, in the context of real patient care, but without the problem of disrupting high risk and high time-pressure patient care [6]. The Critical Decision Method, a formal approach to eliciting detailed recall of information cues and decision-making during critical incidents, is useful for understanding information needs for managing rare events which are unlikely to be seen in routine observation [37,38]. In adapting these methods, we have drawn on principles of collecting rich data in qualitative research methods and contextual design approaches [33,39] to ensure the rigor and validity of the approach.

There are also a number of techniques for mapping information captured from observations and interviews into design requirements or design innovations. Situation awareness-grounded methods focus on using interview data to identify goals, sub-goals, and the information needs to support those goals. TURF and ecological interface design (EID) methods focus on transforming the data into conceptual representations such as task, user, representation, and function (Chapter 19), functional and causal abstraction hierarchies [10,34], or, a variety of other relevant models such as collaboration, sequence, or decision point models [33].

In the context of our analysis of critical care information interview content, we applied a mix of data analytic techniques. We used contextual design interview interpretation methods [33] targeted toward understanding goals and contexts of clinician information use including: triggers or cues to seek out data, such as preparing for rounds; sources of data used—paper, electronic, or verbal; goals such as making a patient transfer decision; and characteristics of data engagement such as lookup of an anticipated lab value [6]. Because we wanted be open to exploring new ideas from the data, we also applied grounded theory open-coding qualitative analytic methods [39], and conducted visioning activities (re-engaging a small group of participants) to select specific problems and design approaches to pursue.

Independent of the methods that are selected for collecting and analyzing data in order to understand users, tasks, information requirements, and opportunities for information integration and visualization, it is important to ensure that the methods applied are executed with sufficient rigor, or validity. For example, how many interviews are sufficient to capture the information needs across an appropriate breadth of user roles or patient care contexts? Holtzblatt et al. [33] recommend three to four people in each role that is relevant to the project focus (consider also separate care settings or institutions with differing technologies or workflows). This results in about 6 people for a small focused project and 18–24 for a larger one. We have adopted the approach of relying on validity criteria for qualitative research to evaluate the rigor of our findings. For example, qualitative researchers rely on the concept of "saturation"—when no new insight is emerging from the data – to determine when they have captured data from a sufficient number of people and contexts. Table 14.2 lists the four primary criteria of validity espoused by Whittemore et al. [40], relevant assessment questions, and examples of how investigators or designers may meet these criteria relevant to CDS information integration and visualization. In addition to these primary criteria of validity, Whittemore et al. identify explicitness, vividness, creativity, thoroughness, congruence, and sensitivity as secondary criteria of validity.

TABLE 14.2 Primary criteria of validity in qualitative research [40].

Validity criteria	Assessment	Example approaches to meeting these criteria
Credibility	Do the results reflect the experience of intended users and contexts in a believable way?	Share findings with users and ask if they fit with their experience. Specifically solicit negative responses (i.e., what doesn't fit).
Authenticity	Do the findings demonstrate awareness of subtle differences in perspectives of different voices?	Justify the breadth of roles, care settings, and sites you have included in collecting the data. Ask other participants if you have missed any relevant roles or situations.
Criticality	Does the data collection and analysis process reveal evidence of critical appraisal?	Use a formal approach to data collection and analysis. Set and follow prescribed methods for data collection and analysis. Ensure that analysis includes exploration of alternative hypotheses or negative instances.
Integrity	Does the data collection and analysis process include checks for validity and transparency of biases and limitations?	Be transparent about potential biases of investigators. Are investigators vested in the outcomes of the data collection and analysis? Include investigators with different perspectives in the data collection and analysis and evaluate inter-investigator agreement. Be transparent about limitations in the breadth and depth of data collections and, thus, relevant roles and settings.

14.4.2 Participatory and contextual design methods

In a meta-analysis of the impact of difference UCD approaches on improvements in clinician performance, workload, or preferences with novel critical care information displays, the specific UCD activity that had the greatest impact was the iterative evaluation of designs with users [36]. Following design principles and understanding user needs is an important first step, but ultimately, the process of trying, testing, and fixing is critical to an effective solution. Participatory design refers to the process of engaging people who are users (or representative of users) in the design process. Including users as subject matter experts on the design or investigative team is an informal approach to participatory design. More formal approaches engage users in specific activities to answer specific design questions. Contextual design refers to grounding the design process and activities in the environment (or in a representation of the environment) in which they will be used. Examples of contextual approaches for CDS include: (1) generating design visualizations using real patient data and (2) evaluating visual search tasks in the visually and audibly noisy environment in which it will be used.

In our critical care analysis, we identified a need to support critical care information use through information integration that supported rapid review and prioritization of multiple patients, more effective representation of dynamic patient data, and more efficient and meaningful organization of patient information. To progress to the stage of generating design concepts, we required a more detailed understanding of, for example, "meaningful organization". What data, specifically, should be considered in any given associative grouping? This type of information may be found in literature, textbooks, or through more focused interviewing. Where answers are not available, different techniques may be used to answer these more targeted questions. Pickering et al. [41] asked clinicians to rank the value of different patient data when patients were admitted to the ICU (in context). Barwise et al. [42] conducted a survey to rank patient data information needs of rapid response teams. We employed card-sorting methods which not only rank specific data elements, but also provide information regarding clinicians' mental models of how specific data group together [43].

When you have sufficient understanding of the problem and clinical context to begin to put ideas to paper, a participatory design approach involves generating design artifacts (conceptual models, mockups, prototypes) and working closely with clinical users to identify strengths and weaknesses, innovate, and improve on design concepts (see Fig. 14.6) [44,45]. User-centered design with iterative participatory design stages draws on the strengths of both clinicians and designers in developing innovative solutions. Designers are tasked with creating design artifacts that make it easy for clinician users to interpret to a design concept, or to compare competing design concepts.

The nature of a participatory design interaction will vary depending on the design stage. Early interactions with clinician-users may include interviews to validate findings of user research activities including the problem definition, information requirements and ideas for addressing problems. Interviews may also be relevant for early design decisions such as validating the scope of the posed solution in terms of, for example, relevant roles, clinical settings, and data integration. Because the process is iterative, with opportunities to add new voices to the process or to correct early mis-understandings or mis-interpretations, small

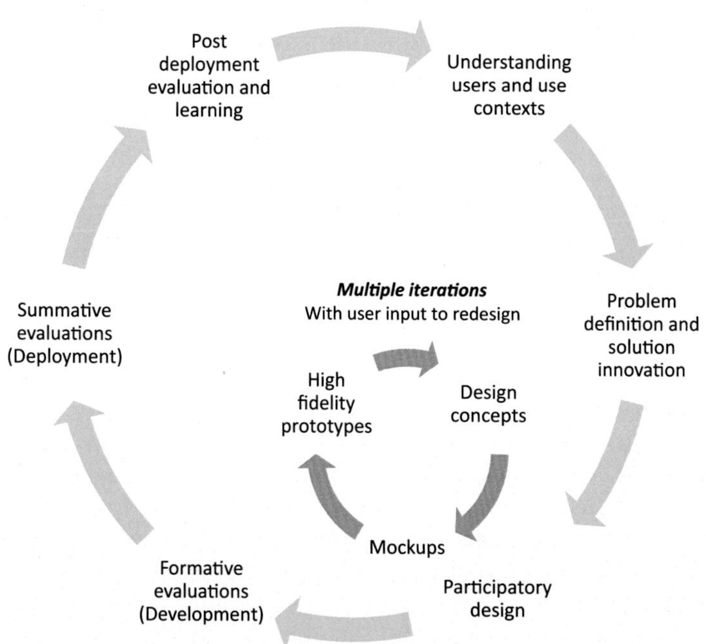

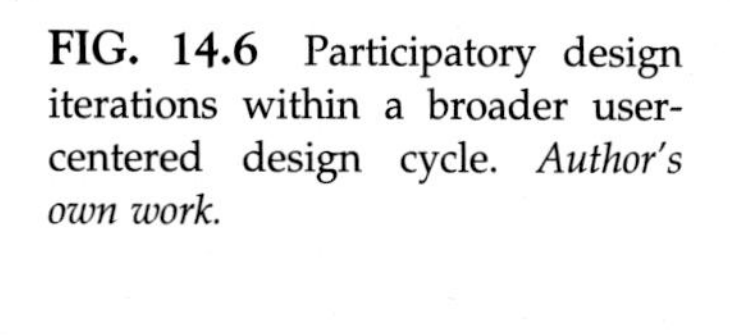

FIG. 14.6 Participatory design iterations within a broader user-centered design cycle. *Author's own work.*

sample sizes may be appropriate in these phases of design. However, principles of qualitative rigor, such as ensuring the inclusion of relevant perspectives and attention to limiting designer or investigator bias (see Table 14.1) remain relevant. For these early activities, we generally conduct individual or very small (maximum of 3 or 4 people) virtual group interviews. Convening a group can reduce the time spent in interviews and allow clinicians to react to input of others, but groups can also limit or bias the input of some participants.

As design concepts progress to more detailed design, mockups and design prototypes help clinicians to visualize the future design in context and allow them to react to competing design alternatives. We used individual participatory design interviews with progressively detailed design mockups to design a temporal data display widget, exploring different design alternatives for handling a variety of design decisions such as the appropriate time range to display, how to display reference ranges, and how to denote missing data [46]. To minimize the impact of designer bias in interpreting the opinions of clinicians, more than one investigator coded whether participants responded positively or negatively to specific design alternatives. To assess the reliability of any given finding, we calculated agreement between participants.

In the context of information integration and visualization, it is often a goal to ensure that visualization is understandable. If the visualization is intended to be intuitive, this may simply be evaluated by presenting examples and asking the participant to describe what the visualization represents and whether or not they found it easy to understand. If a visualization requires an introductory explanation, it may be sufficient to ask whether they believe the visualization is easy to understand and will be easy to use. Or, it may be appropriate to present the same visualization using different data to check that the participants interpretation of the meaning is correct. Early stages of participatory design may also involve asking participants to indicate which of different alternatives for presenting information they prefer, and why. In this case, it may be necessary to counter-balance presentation order of designs so that exposure to prior examples does not bias interpretations of later examples. Another goal of iterative participatory design may be to evaluate whether the information presented is sufficient for the clinical application. Initially, we may rely on clinicians' recall of their own experiences to evaluate whether all of the information necessary (and none that is unnecessary) is presented. However, in later design stages, it will be important to evaluate these questions in the context of a broadly representative set of realistic clinical decision scenarios. This will allow clinicians to more easily identify whether information is missing that is relevant to the specific clinical case.

Some specific design decisions may necessitate formal, objective, comparisons. For example, when narrowing a final set of icons, conducting a formal test of icon understanding and discriminability may be appropriate. When designs progress to more interactive prototypes, a variety of formative (intended to inform improvements) usability inspection methods such as cognitive walkthroughs, expert heuristic evaluations, or usability tests involving relevant user tasks may be appropriate [47,48]. Independent of the approaches used to evaluate the design, the goal of iterative participatory design is to draw on the findings from interacting with intended users to fix problems and improve on designs. The concept of saturation can also be applied to the participatory design process. When participants are no longer generating novel insight to improve upon the design and when designers have exhausted their leading design concepts, it is time to move to the next stage (e.g., from concepts to mockups to prototypes to development).

14.4.3 Measuring the user experience

Despite known design principles and methods to facilitate user involvement in the design process, there remains some "art" in the generation of design solutions and it is not known how users will interpret and respond to designs in practice. Participatory and contextual design processes rely largely on users' preferences and presumptions about how different design solutions will affect their clinical decision-making performance. These do not necessarily translate into optimal design solutions [49]. Ultimately, some form of objective testing of user experience is necessary. Objective usability testing typically involves asking users to perform relevant tasks (both representative and critical tasks) using a prototype system, perhaps in a simulated care setting using simulated patient data. In this controlled environment, relevant metrics such as accuracy of a clinical decision, time to make a response, errors, and subjective evaluations of usability [50], usefulness, or trust [48]. This type of evaluation is valuable for evaluating novel designs or to compare whether novel designs outperform conventional approaches in a way that justifies further development and evaluation in clinical settings [36,47,48].

Many CDS information integration and visualization design activities are evolutionary improvements to existing EMR or other clinical informatics systems. In these cases, it may be feasible to conduct objective evaluations in test environments or in a clinical pilot, before deploying new solutions more broadly. It is important that designers and developers have processes in place to implement necessary changes should testing identify problems, even if formative testing has already taken place. Due to the complexity and dynamic nature of clinical decision-making, clinical and ongoing post-deployment evaluation (see Chapter 21) also is necessary.

14.5 Progress in information integration and visualization in CDS

Applying information integration and visualization to enhance more conventional clinical patient information presentation has been shown to improve patient outcomes and clinical decisions [9,36,51–55]. Integrating related information and supporting visualization of data relationships provides an opportunity to support clinicians, and patients, in their decisions [56]. Ensuring that the right data is presented together, in the right way, not only reduces the time and effort required to search for key information, but can also help to ensure that critical information is not missed and can make it easy to identify information gaps. This type of decision support capitalizes on human strengths by making it easy for clinicians or patients to maintain a high level of situation awareness that allows them to make good judgments about what will happen in the future and respond appropriately [17]. Thus, we expect this approach to be especially useful for clinical decisions that are complex, requiring consideration of many variables, and that have high degrees of uncertainty or limited evidence-based guidance. Examples include monitoring and treating: (a) critically ill or rapidly deteriorating patients, (b) patients with interacting or uncertain diagnoses, and (c) patients with chronic problems such as diabetes that involve high inter- and intra-patient variability and many interacting management alternatives.

14.5.1 Integrated patient information displays

Early implementations of EMRs contained minimal structured data combined with scanned paper-based reports from external resources. These evolved to separate electronic systems pieced together with sometimes clunky integration. Current systems have higher levels of compatibility to support integrating data from multiple sources not only into a common platform, but feasibly, within a single display. With this expanded capability, there is a need and an opportunity to think about integrating patient data in new ways. As opposed to relying primarily on source-based information organization (e.g., ECG monitor, nursing observations, medication order entry, laboratory test results), researchers have advocated for approaches to information integration that are medical problem or clinical-concept oriented, to support clinical decision support [57,58]. Information needs of clinicians have been explored in broad ways [59], however, there is a need for research to understand and describe how specific patient data are inter-related and appropriate for co-presentation in the context of particular patient problems or clinical settings. We previously described efforts to capture information requirements in critical care, where there is a high volume and variety of patient data generated to inform care decisions [41–43].

Beyond understanding information needs and meaningful information grouping, researchers and developers have evaluated the impact of integrating related information from different sources on human-system performance and clinical outcomes. For example, researchers at the Mayo clinic identified a subset of high information value data from multiple sources including patient monitors, problem lists, and laboratory findings, and displayed the information grouped into meaningful categories (e.g., cardiovascular, respiratory, infections), with color coding to highlight urgent information (AWARE) [60,61]. They found that the integrated ICU information display decreased ICU length of stay [60] and reduced time spent in rounding [61].

Another example is an integrated anesthesia monitoring display (AlertWatch; [62]) that combines high value anesthesia-related information into a single page view (see Fig. 14.7). AlertWatch combines an anatomically-organized graphical display with conventional display presentations and color coding of chronic and acute patient data relevant to anesthesia monitoring. Kheterpal et al. [62] compared process and outcome metrics for cases in which clinicians used the system for less than 75% of the case duration to cases in which the system was used for 75% or more of the case duration. They observed positive impacts of system use on process measures related to maintaining appropriate blood pressure, ventilation, and fluid management.

Other integrated displays have been developed and evaluated using simulated clinical tasks. Examples include displays that integrate data from multiple infusion pumps into a tablet-based user interface [63] and displays that integrate data from infusion pumps, medication orders, ventilators, and monitors to present high-value nurse monitoring data in a single view [64–67]. With the integrated infusion pump application [63], nurses completed pump programming with fewer clicks and fewer errors compared to working with conventional separate pumps. In two different approaches to integrating critical care patient monitoring data (see Fig. 14.8), nurses identified patients at risk more quickly and made more accurate responses in comparison to conventional displays [64–67].

14.5.2 Dashboards and multi-patient integrated displays

Stephen Few [28] defines a *dashboard* as "a visual display of the most important information needed to achieve one or more objectives; consolidated and arranged on a single screen so the information can be monitored at a glance." He further asserts that key features of effective dashboards include: (a) customization to a specific user or application; (b) concise, clear,

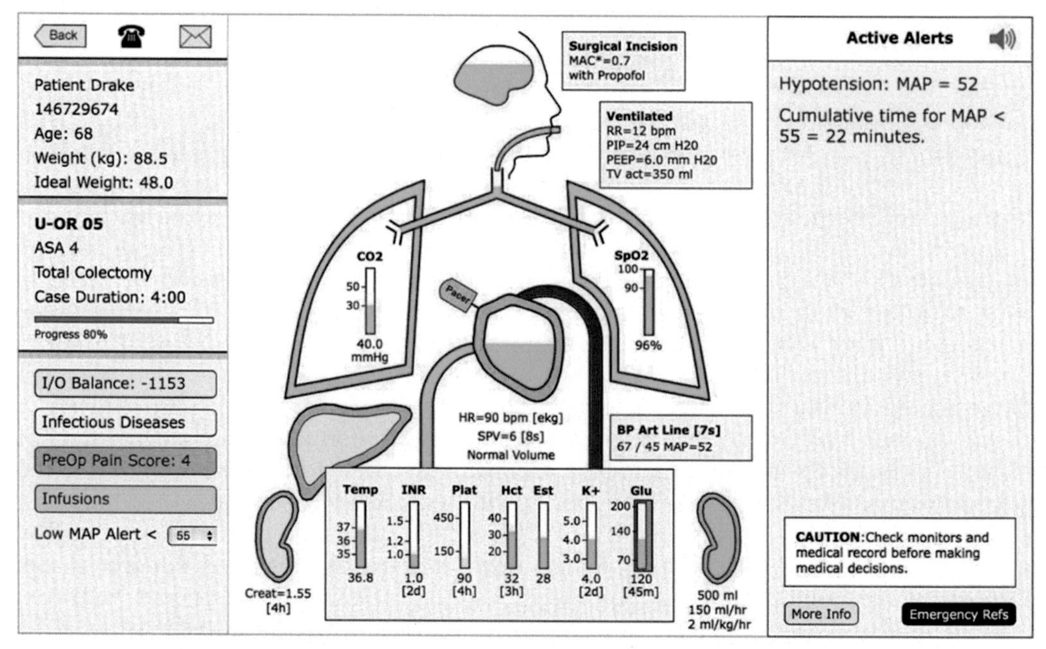

FIG. 14.7 AlertWatch integrated anesthesia display. *Reprinted with permission from Kheterpal S, Shanks A, Tremper KK. Impact of a novel multiparameter decision support system on intraoperative processes of care and postoperative outcomes. Anesthesiology 2018;128:272–82, Fig. 1.*

and intuitive display mechanisms (see Section 14.2); and (c) high level summaries. High level summaries provide an overview of specific concepts to convey rapid communication, at a glance. Few's approach to dashboard design limits the scope of the dashboard to information that can be displayed on a single page, without scrolling or other interaction. According to Few, dashboards are not defined by timing of data update. They may contain real-time (or near real-time) data or may be updated at any time interval that is appropriate for the specific application and data sources.

In clinical applications, the term dashboard has been applied more broadly to include data viewers that incorporate functionality such as customizing or filtering the data displayed or extend beyond a single page [53,68,69]. Dashboards or integrated information displays, have been implemented for a number of health care purposes including: (a) rapid assessment of patient status based on data integrated from multiple sources such as EHRs and ECG monitors [52,60,62], (b) multi-patient displays to support prioritization of attention or track compliance with specific prophylactic or safety practices [70–73], (c) support time and resource management in hospital settings [74–76], and (e) support population-level monitoring and prioritization of attention toward high risk patients [77,78].

Examples of multipatient integrated displays include several that have applied visualization techniques to make it easy to rapidly identify and respond to:

(a) which patients have the most urgent needs [70,73,79,80],
(b) potential problem areas with regard to hospital resource management [74–76], and
(c) gaps in compliance with patient safety or other important evidence-base practices [71,81,82].

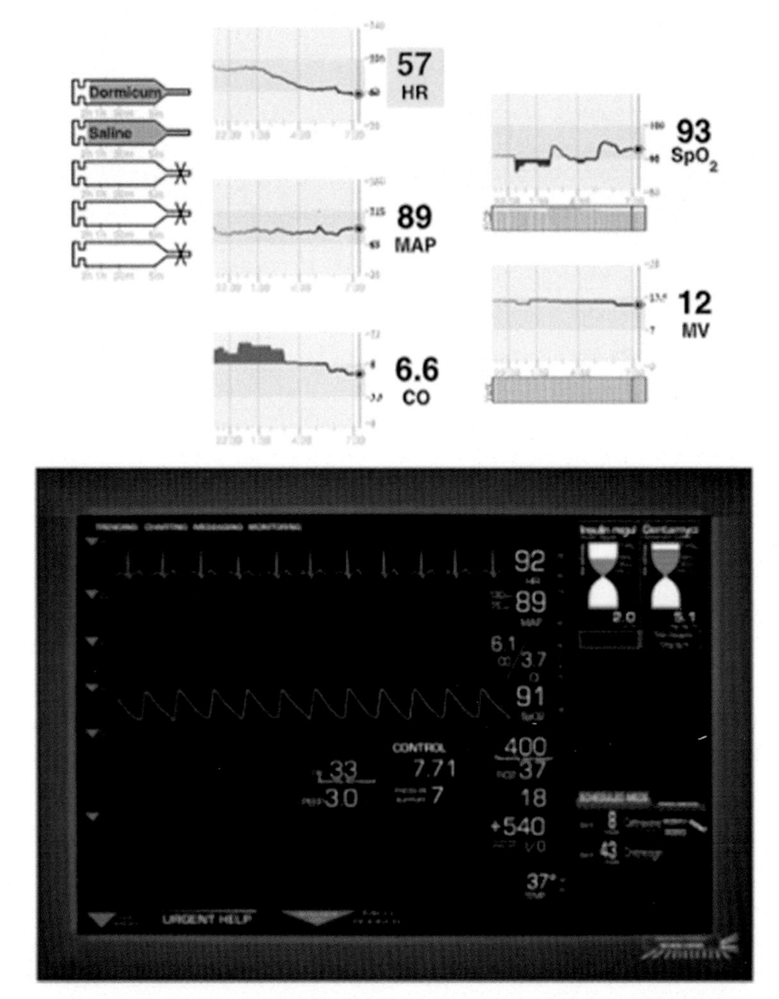

FIG. 14.8 Two unique approaches to integrating critical care patient monitoring information. *Reprinted with permission from (top) Görges M, Kück K, Koch SH, Agutter J, Westenskow DR. A far-view intensive care unit monitoring display enables faster triage. Dimens Crit Care Nurs 2011;30:206–17. Fig. 1. Reprinted with permission from (bottom) Koch SH, Weir C, Westenskow D, Gondan M, Agutter J, Haar M, et al. Evaluation of the effect of information integration in displays for ICU nurses on situation awareness and task completion time: a prospective randomized controlled study. Int J Med Inform 2013, Fig. 1.*

A common approach to support patient prioritization is to present multiple patients in a tabular view with rows representing patients and columns of data that may be relevant for sorting: location or line of service; patient data such as vital sign or laboratory metrics; or summarized risk scores such as a SOFA or early warning score (EWS)) [70,71,73] (see Fig. 14.9). Often, these systems implement stoplight analogy color coding with green (or gray), yellow (amber and/or orange), and red to represent different levels of acuity or risk to support rapid identification of patients requiring attention. Examples include systems that present: (a) general acuity or deterioration risk score based on vital signs, lab data, or other patient data [70,73], (b) predictions or acuity related to a specific clinical application such as glucose management [83,84], and (c) compliance with safety or evidence-based bundles to prevent

Bed	Patient name	Age	LOS		Spontaneous breathing trial			Prophylaxis		RASS					
					Vent	Screen	Trial	DVT	Stress ulcer	Target order	Score	Head of bed	Oral swab	Teeth	Hypopharyngeal Suction
9001	J, KM	72y	6 d	patient information	V	F		V	V	-4	-4	30			
9002	R, BA	80y	17 d	patient information	F			V	V	0	-2	45			
9003	T, PL	60y	34 d	patient information	V			V	V	-1	-1	30			
9004	S, JD	64y	7 d	patient information				V	V	0	-1	30	V	V	
9005	W, A	49y	30 d	patient information	V	F		V	V	-1	-3	30			
9006	B, JK	77y	11 d	patient information	V	P	P	V	V	-4	-4	30			
9007	P, KM	72y	6 d	patient information	V	F		V	V	-4	-4	30			
9008	M, BA	80y	17 d	patient information	V	F		V	V	0	-2	45			
9009	T, PL	60y	34 d	patient information	V			V	V	-1	-1	30			
9010	S, JD	64y	7 d	patient information				V	V	0	-1	30	V	V	
9011	W, A	49y	30 d	patient information	V	F		V	V	-1	-3	30			

FIG. 14.9 Example multi-patient tabular ventilator-associated pneumonia compliance dashboard. Color coding enhances visibility of tasks that require attention (yellow and red). *Author's adaptation, from Zaydfudim V, Dossett LA, Starmer JM, Arbogast PG, Feurer ID, Ray WA, et al. Implementation of a real-time compliance dashboard to help reduce SICU ventilator-associated pneumonia with the ventilator bundle. Arch Surg 2009;144:656–62. https://doi.org/10.1001/archsurg.2009.117.*

problems such as central-line associated blood stream infections or ventilator-associated pneumonia [71,81,82]. The selection of data to display in columns depends on the goals of each system. Frequently, displays are designed for persistent presentation, displayed on large overhead monitors on hospital units, or as screen savers, and minimize the use of patient identifying information to protect patient privacy. Evaluation of these types of displays has been linked to improved patient outcomes [51].

An example of a novel approach to integrated multipatient visualization is that of Nihon Kohden's 'CoMET display which presents cardiovascular and respiratory deterioration risk on a two-axis plot (see Fig. 14.10). The head of a "comet" (displaying patient identity as a bed number) is plotted at the scale position of the current score and risk is coded using both size of the dot and color [79]. A fading tail conveys the prior three hours of scores making it easy to visualize both rate and direction of change across two related deterioration risk scores. Using risk ratio (x-fold risk) as the measure presented capitalizes on the tendency of risk ratios to exhibit greater variance as risk levels increase, generating a natural emphasis on changes in patients at higher risk levels. Percentile data represented in gray scale provides an additional interpretation of level of risk in comparison to other patients. In a clinical trial of the CoMET display, the rate of septic shock remained constant on a control unit (medical ICU) at the same hospital, while a reduction in septic shock was demonstrated for an intervention unit using the novel display (surgical ICU) [79].

Some multi-patient integrated information systems are primarily intended as a separate information display (with action in response expected to occur external to the system), however, many systems also include functionality to support action in response to a problem such as the ability to drill-down to see additional information for a single patient or the ability to initiate a specific relevant order set. Direct link to a relevant action has been linked to increased perceptions of usefulness and engagement with these systems [85,86].

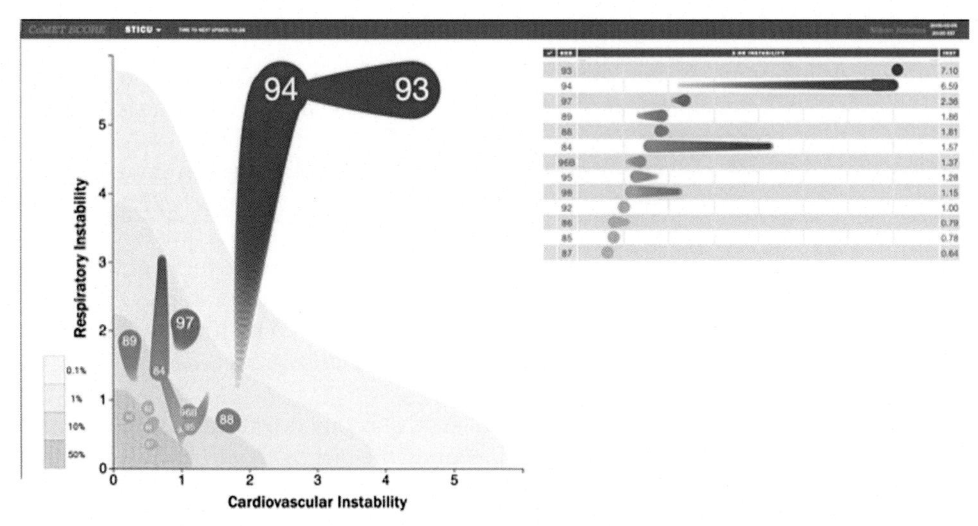

FIG. 14.10 Multipatient display with emerging visualization of changing cardiovascular and respiratory instability. *Figure provided courtesy of Randall Moorman, with permission from Nihon Kohden (see also Ruminski CM, Clark MT, Lake DE, Kitzmiller RR, Keim-Malpass J, Robertson MP, et al. Impact of predictive analytics based on continuous cardiorespiratory monitoring in a surgical and trauma intensive care unit. J Clin Monit Comput 2019;33:703–11. https://doi.org/10.1007/s10877-018-0194-4).*

14.6 Future challenges for CDS information integration and visualization

There has been significant progress in advancing our knowledge about how information integration and visualization can benefit clinical decision support. However, many challenges and areas for new exploration remain.

14.6.1 Shooting at a moving target

At the forefront of these is the significant challenge of complexity in human collaboration, information, patient, and data for supporting clinical decisions. Different patient conditions, clinical settings, and inter-disciplinary teams generate unique sets of information needs and priorities. And, as advancements are made through clinical research, information needs and priorities for clinical decisions change. For example, advancement of biosensors, point of care testing, and large data analytics will likely change practice in the diagnosis and treatment of sepsis [87,88]. Non-invasive patient monitoring via wearable devices is also likely to rapidly change the nature of low acuity hospital patient data (from being available at intermittent time increments ranging from minutes to hours to being near-continuous). As we work to develop information integration and visualization for AI models that predict conditions such as sepsis [89–91], our designs must remain open to the possibility that sources of diagnostic evidence and data rates will change. Participatory design activities engaging end users will draw on experiences of clinicians based on practices that may quickly become dated. Research and methods are needed that help to ensure that UCD methods keep pace with clinical research.

14.6.2 Integrating artificial intelligence

The value of AI in human-system performance is now plainly obvious in our daily lives. As described in Chapter 7, development of advanced AI models such as machine learning (ML) for predicting clinical outcomes is growing exponentially, yet the benefits of AI in the complex setting of health care are only beginning to be realized [4]. Research in automation in other industries has shown that, assuming the automation is not perfectly reliable and requires some human oversight, automation applied to information acquisition and information analysis (i.e., information integration and visualization) is superior to automation applied to decision-making or action-implementation [23,92–95]. As an example, AI to identify data inaccuracies or artifacts such that clinicians are not distracted by irrelevant information depicted in a trend display may be more valuable than AI that interprets data and suggests a specific action.

Research has established that humans tend to trust, and rely on, systems that are more reliable, transparent, and comprehensible [93,96]. Among the factors that influence human trust and reliance on AI, the factors with the greatest potential to control in health care delivery are system reliability and subsequent presentation (explanation and integration) of AI information. Moore and Swartout found that of 15 desired attributes of medical diagnostic systems, "never make an incorrect diagnosis" was ranked 14th most important while "explain their ... decisions" was most important [97].

Mueller et al. [96] reviewed 100's of articles including 37 evaluations of explainable AI systems involving human participants. They concluded that an explanation is *good* when it (1) contains appropriate detail, (2) is veridical (truthful), (3) is useful, (4) is clear, (5) is complete, (6) is observable, and (7) reveals boundaries. Mueller et al. further suggest that it may be most beneficial to offer a *sufficient explanation* which balances comprehensibility with level of detail. A sufficient explanation doesn't only describe how the AI works but also justifies its use. Why should one use it? Why not? And under what circumstances does it not work?

Research on information integration, visualization and explanation in AI healthcare applications is currently underway [98–100]. For example, Barda et al. [101] employed a UCD explanation framework and incorporated design feedback from focus groups with users to develop an explanation display for a pediatric intensive care unit 24-hour mortality prediction model (see Fig. 14.11). It remains to be seen whether and how detailed explanations such as this will be used in clinical practice. Diprose et al. [99] compared different approaches to explanation and identified trends toward physician preference for simple visualizations, explanations that are patient-specific, and explanations that confirm the clinical understanding.

Many questions remain regarding the integration of AI models for informing clinical decisions. Should an AI explanation replace other sources of patient data that may impact clinical decisions? Might there be cases where clinical decisions benefit from a display that integrates AI prediction model information, data that informs that model, and data that is not included in the AI model? Under what circumstances is it appropriate for AI models to provide interruptive push notifications as opposed to passive information visualizations? As the number of different models for different purposes expands, how do we integrate these multiple new sources of information? Might AI models be used to identify data that has little information value to help manage information overload problems?

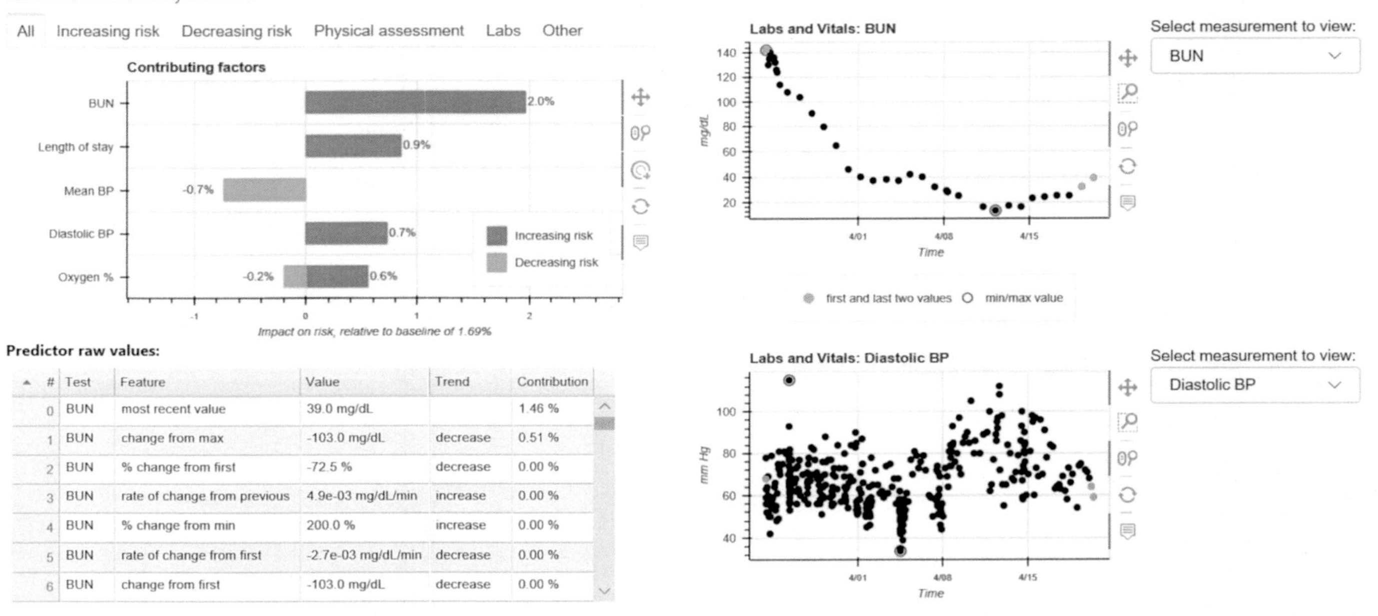

#	Test	Feature	Value	Trend	Contribution
0	BUN	most recent value	39.0 mg/dL		1.46 %
1	BUN	change from max	-103.0 mg/dL	decrease	0.51 %
2	BUN	% change from first	-72.5 %	decrease	0.00 %
3	BUN	rate of change from previous	4.9e-03 mg/dL/min	increase	0.00 %
4	BUN	% change from min	200.0 %	increase	0.00 %
5	BUN	rate of change from first	-2.7e-03 mg/dL/min	decrease	0.00 %
6	BUN	change from first	-103.0 mg/dL	decrease	0.00 %

FIG. 14.11 An example pediatric intensive care unit mortality risk prediction explanation display presents risk contributions of feature groups (upper left), detailed contributions of individual features (scrollable list, lower left) and temporal plots (right) that allow the clinician to view aligned trends for two selected variables [101]. *Figure reproduced under Creative Commons Attribution 4.0 (http://creativecommons.org/licenses/by/4.0), only the figure caption has been modified.*

14.6.3 An increasingly mobile and virtual world

The development of wearable devices and expanded use of virtual clinician-patient interactions systems also has important implications for information integration and visualization CDS. Designers will have to consider the implications of different team members and patients being in different locations. Is there a need to consider video and audio transmission integration with relevant patient information displays and clinical team and patient communication models?

Information presentation is now available in hardware formats ranging from wristwatches to wall displays. Users are accustomed to having critical information in a small lightweight package that can easily be carried with them. Responsive design is an approach to the design of web-based applications to generate designs that automatically adapt, in an effective way, to different screen formats such as mobile phones, tablets, laptops, and full-size display monitors [102]. As new information integration and visualization systems are designed for CDS, it is important to consider whether any given application can scale to smaller sizes such as smartphone displays and, if so, how the information presented should scale as the user opts for different display formats.

14.6.4 Promoting standard clinical data user-interface components

For every hour of clinical time, physicians spend about 2 hours on EHR-related tasks [103]. Information integration and visualization in CDS has made important advances over the past decade and many applications are now moving from research and development into commercial products and clinical practice. However, with expanding availability and complexity of data, it is unlikely that information overload, alarms, and other usability concerns will be significantly reduced. A quick scan through the pictorial examples presented in this chapter reveals a lack of consistency in clinical information design. This is partly because these examples are generated from academic medical centers and individual researchers relying on their own design creativity. Commercial interests motivate vendors to keep their design details, and their underlying design systems, closed to academic researchers and competing developers.

An important way to manage complexity is through consistency. In the 1980's, design elements and *widgets* such as icons, menus, and radio-buttons made their way into consistent use, supporting rapid development of UIs and ease of understanding by users. With the expansion of smartphones and other digital devices, there has been an explosion of user-interface innovations in display and interaction both to accommodate shortcomings (e.g., predictive text data entry for small devices) and to exploit technological advances (e.g., touch screen gestures). As a clinical example, it may be appropriate to specify a common symbol or indicator to highlight data that exceeds a threshold value and use the same indicator consistently across applications. A common standard for time scales, grid markings, spacing, line weights, current value position, and other components for a variety of trend display applications would likely make it easier for clinicians to interpret information as they switch between different clinical informatics applications.

Current clinical user interfaces do not sufficiently translate complex and voluminous patient information into standardized usable forms that support rapid interpretation of information. Just as clinical informatics practitioners have developed healthcare resource standards (e.g., Health Level Seven (HL7) Fast Healthcare Interoperability Resources (FHIR)), there is a need for open source clinical informatics design elements and best practices. A component library is a set of styles and components that can be used and shared [104]. A style guide is documentation that contains detailed descriptions of interface elements, color, typography, icons, and guidance for the application of specific design elements. The development of an open source component library and style guidance would allow developers and researchers to more rapidly generate designs that comply with basic visual design principles and support consistent interpretation of clinicians. Designers could then devote more time to tackling the challenges identified in preceding sections.

Acknowledgments

I am grateful for the contributions of Thomas Reese, Noa Segall, Guilherme Del Fiol, Brekk MacPherson, and Jonathan Mark for many years of collaboration that have informed the content presented here. Thanks also to Thomas Reese and Randall Moorman for providing figures and to Guilherme Del Fiol and Robert Greenes for their editorial guidance and input. This work was supported by the National Institute of General Medical Sciences, National Institute grant number R01GM137083.

References

[1] AlGhatrif M, Lindsay J. A brief review: history to understand fundamentals of electrocardiography. J Community Hosp Intern Med Perspect 2012;2. https://doi.org/10.3402/jchimp.v2i1.14383.
[2] Witt C. Vital signs are vital: the history of pulse oximetry. ACP Hospitalist 2014;January.
[3] Center for Medicare & Medicaid Services. Meaningful use objectives and measures for EPs, eligible hospitals, and CAHs for 2015 through 2017. vol. 42; 2016. CFR 495.22.
[4] Yu KH, Beam AL, Kohane IS. Artificial intelligence in healthcare. Nat Biomed Eng 2019;2:719–31. https://doi.org/10.1038/s41551-018-0305-z.
[5] Lynn LA. Artificial intelligence systems for complex decision-making in acute care medicine: a review. Patient Saf Surg 2019;13:6. https://doi.org/10.1186/s13037-019-0188-2.
[6] Wright MC, Dunbar S, Macpherson BC, Moretti EW, Del Fiol G, Bolte J, et al. Toward designing information display to support critical care: a qualitative contextual evaluation and visioning effort. Appl Clin Inform 2016;7:912–29. https://doi.org/10.4338/ACI-2016-03-RA-0033.
[7] Bauer DT, Guerlain S, Brown PJ. The design and evaluation of a graphical display for laboratory data. J Am Med Inform Assoc 2010;17:416–24. https://doi.org/10.1136/jamia.2009.000505.
[8] Segall N, Borbolla D, Fiol GD, Waller R, Reese T, Nesbitt P, et al. Trend displays to support critical care: a systematic review. In: 2017 IEEE international conference on healthcare informatics (ICHI); 2017. p. 305–13. https://doi.org/10.1109/ICHI.2017.85.
[9] Reese TJ, Del Fiol G, Tonna JE, Kawamoto K, Segall N, Weir C, et al. Impact of integrated graphical display on expert and novice diagnostic performance in critical care. J Am Med Inform Assoc : JAMIA 2020;27:1287–92. https://doi.org/10.1093/jamia/ocaa086.
[10] Bennett KB, Flach JM. Display and interface design: subtle science, exact art. Boca Raton: CRC Press; 2011.
[11] Rowling JK. Harry Potter and the Sorcerer's Stone. New York: Scholastic Press; 1997.
[12] Endsley MR. Toward a theory of situation awareness in dynamic systems. Hum Factors 1995;37:32–64.
[13] Wickens CD, Hollands JG, Banbury S, Parasuraman R. Engineering psychology and human performance. 4th ed. New York: Psychology Press; 2013.
[14] Card SK, Moran TP, Newell A. The psychology of human-computer interaction. Hillsdale, NJ: Lawrence Erlbaum Associates; 1983.
[15] Reason J. Human error. Cambridge, UK: Cambridge University Press; 1990.
[16] Kahneman D. Thinking, fast and slow. New York: Farrar, Strauss and Giroux; 2011.

[17] Endsley MR, Bolte B, Jones DG. Designing for situation awareness: an approach to human-centered design. London: Taylor and Francis; 2003.
[18] Tversky A, Kahneman D. Judgment under uncertainty: heuristics and biases. Science 1974;185:1124–31. https://doi.org/10.1126/science.185.4157.1124.
[19] Klein G. Sources of power: how people make decisions. Cambridge, MA: The MIT Press; 1998.
[20] Zsambok C, Klein G. Naturalistic decision making. New York: Psychology Press; 1997.
[21] Klein G. A naturalistic decision making perspective on studying intuitive decision making. J Appl Res Mem Cogn 2015;4:164–8. https://doi.org/10.1016/j.jarmac.2015.07.001.
[22] Hoffman R.R., Yates J.F.. Decision (?) making (?). IEEE Intell Syst n.d.;20:76–83.
[23] Patterson RE. Intuitive cognition and models of human-automation interaction. Hum Factors 2017;59:101–15. https://doi.org/10.1177/0018720816659796.
[24] Hegarty M. The cognitive science of visual-spatial displays: implications for design. Top Cogn Sci 2011;3:446–74. https://doi.org/10.1111/j.1756-8765.2011.01150.x.
[25] Card SK, Mackinlay JD, Shneiderman B. Readings in information visualization: using vision to think. San Francisco, CA: Morgan-Kaufmann; 1999.
[26] Galitz WO. The essential guide to user interface design: an introduction to GUI design principles and techniques. New York: John Wiley & Sons, Inc.; 2007.
[27] Schlatter T, Levinson D. Visual usability: principles and practices for designing digital applications. Boston: Elsevier; 2013.
[28] Few S. Information dashboard design. 2nd ed. Burlingame: Analytics Press; 2013.
[29] Accessibility Guidelines Working Group. Understanding WCAG 2.2., 2022, https://www.w3.org/WAI/WCAG22/Understanding/. [Accessed 9 February 2022].
[30] Vesselov S., Davis T.. Building design systems: unify user experiences through a shared design language. n.d.
[31] Tufte ER. The visual display of quantitative information. Cheshire, CT: Graphics Press; 2001.
[32] Tversky B. Visualizing thought. Top Cogn Sci 2011;3:499–535. https://doi.org/10.1111/j.1756-8765.2010.01113.x.
[33] Holtzblatt K, Wendell J, Wood S. Rapid contextual design. San Francisco, CA: Elsevier; 2005.
[34] Burns CM, Hajdukiewicz J. Ecological interface design. 1st ed. Boca Raton, FL: CRC Press; 2004.
[35] Wilson JR, Corlett N, editors. Evaluation of human work. 3rd ed. Boca Raton, FL: CRC Press; 2005.
[36] Wright MC, Borbolla D, Waller RG, Del Fiol G, Reese T, Nesbitt P, et al. Critical care information display approaches and design frameworks: A systematic review and meta-analysis. J Biomed Inform: X 2019. https://doi.org/10.1016/j.yjbinx.2019.100041.
[37] Hoffman RR, Crandall B, Shadbolt N. Use of the critical decision method to elicit expert knowledge: A case study in the methodology of cognitive task analysis. Human Factors Ergonom 1998;40:254–76.
[38] Crandall B, Getchell-Reiter K. Critical decision method: a technique for eliciting concrete assessment indicators from the "intuition" of NICU nurses. Adv Nursing Sci 1993;16:42–51.
[39] Charmaz K. Constructing grounded theory. Thousand Oaks, CA: Sage Publications Inc.; 2006.
[40] Whittemore R, Chase SK, Mandle CL. Validity in qualitative research. Qual Health Res 2001;11:522–37. https://doi.org/10.1177/104973201129119299.
[41] Pickering BW, Gajic O, Ahmed A, Herasevich V, Keegan MT. Data utilization for medical decision making at the time of patient admission to ICU. Crit Care Med 2013;41:1502–10. https://doi.org/10.1097/CCM.0b013e318287f0c0.
[42] Barwise A, Caples S, Jensen J, Pickering B, Herasevich V. Information needs for the rapid response team electronic clinical tool. BMC Med Inform Decis Mak 2017;17:142. https://doi.org/10.1186/s12911-017-0540-3.
[43] Reese T, Segall N, Nesbitt P, Del Fiol G, Waller R, MacPherson BC, et al. Patient information organization in the intensive care setting: expert knowledge elicitation with card sorting methods. J Am Med Inform Assoc 2018;25:1026–35. https://doi.org/10.1093/jamia/ocy045.
[44] Sanders E-N. From user-centered to participatory design approaches. In: Frascara J, editor. Design and the social sciences. Taylor & Francis; 2002.
[45] Muller MJ, Druin A. Participatory design: the third space in HCI. Particip Des 2010;4235:1–70.
[46] Reese T., Segall N., Del Fiol G., Tonna J.E., Kawamoto K., Weir C., et al. A novel approach to critical care information display: designing modular graphical components for presenting trend information. J Biomed Inform n.d.
[47] Nielsen J, Mack RL. Usability inspection methods. New York: John Wiley & Sons, Inc.; 1994.
[48] Tullis T, Albert B. Measuring the user experience. Boston: Elsevier Inc; 2008.

[49] Klein DE. When to ignore what users say. Ergon Des 2006;14:24–6. https://doi.org/10.1177/106480460601400106.

[50] Peres SC, Pham T, Phillips R. Validation of the system usability scale (SUS). In: Proceedings of the human factors and ergonomics society annual meeting, vol. 57. Santa Monica, CA: Human Factors and Ergonomics Society; 2013. p. 192–6.

[51] Waller RG, Wright MC, Segall N, Nesbitt P, Reese T, Borbolla D, et al. Novel displays of patient information in critical care settings: a systematic review. J Am Med Inform Assoc 2019;26:479–89. https://doi.org/10.1093/jamia/ocy193.

[52] Jung AD, Baker J, Droege CA, Nomellini V, Johannigman J, Holcomb JB, et al. Sooner is better: use of a real-time automated bedside dashboard improves sepsis care. J Surg Res 2018;231:373–9. https://doi.org/10.1016/j.jss.2018.05.078.

[53] Khairat SS, Dukkipati A, Lauria HA, Bice T, Travers D, Carson SS. The impact of visualization dashboards on quality of care and clinician satisfaction: integrative literature review. JMIR Hum Factors 2018;5, e22. https://doi.org/10.2196/humanfactors.9328.

[54] Kamaleswaran R, McGregor C. A review of visual representations of physiologic data. JMIR Med Inform 2016;4, e31. https://doi.org/10.2196/medinform.5186.

[55] Dixit RA, Hurst S, Adams KT, Boxley C, Lysen-Hendershot K, Bennett SS, et al. Rapid development of visualization dashboards to enhance situation awareness of COVID-19 telehealth initiatives at a multihospital healthcare system. J Am Med Inform Assoc 2020;27:1456–61. https://doi.org/10.1093/jamia/ocaa161.

[56] Powsner SM, Tufte ER. Graphical summary of patient status. Lancet 1994;344:386–9.

[57] Weed LL. Medical records that guide and teach. N Engl J Med 1968;278:593–600. https://doi.org/10.1056/NEJM196803142781105.

[58] Zeng Q, Cimino JJ, Zou KH. Providing concept-oriented views for clinical data using a knowledge-based system: an evaluation. J Am Med Inform Assoc 2002;9:294–305.

[59] Clarke MA, Belden JL, Koopman RJ, Steege LM, Moore JL, Canfield SM, et al. Information needs and information-seeking behaviour analysis of primary care physicians and nurses: a literature review. Health Info Libr J 2013;30:178–90. https://doi.org/10.1111/hir.12036.

[60] Olchanski N, Dziadzko MA, Tiong IC, Daniels CE, Peters SG, O'Horo JC, et al. Can a novel ICU data display positively affect patient outcomes and save lives? J Med Syst 2017;41:171. https://doi.org/10.1007/s10916-017-0810-8.

[61] Pickering BW, Dong Y, Ahmed A, Giri J, Kilickaya O, Gupta A, et al. The implementation of clinician designed, human-centered electronic medical record viewer in the intensive care unit: a pilot step-wedge cluster randomized trial. Int J Med Inform 2015;84:299–307. https://doi.org/10.1016/j.ijmedinf.2015.01.017.

[62] Kheterpal S, Shanks A, Tremper KK. Impact of a novel multiparameter decision support system on intraoperative processes of care and postoperative outcomes. Anesthesiology 2018;128:272–82. https://doi.org/10.1097/ALN.0000000000002023.

[63] Doesburg F, Cnossen F, Dieperink W, Bult W, de Smet AM, Touw DJ, et al. Improved usability of a multi-infusion setup using a centralized control interface: a task-based usability test. PLoS One 2017;12, e0183104. https://doi.org/10.1371/journal.pone.0183104.

[64] Koch SH, Weir C, Westenskow D, Gondan M, Agutter J, Haar M, et al. Evaluation of the effect of information integration in displays for ICU nurses on situation awareness and task completion time: a prospective randomized controlled study. Int J Med Inform 2013. https://doi.org/10.1016/j.ijmedinf.2012.10.002.

[65] Koch SH, Westenskow D, Weir C, Agutter J, Haar M, Görges M, et al. ICU nurses' evaluations of integrated information displays on user satisfaction and perceived mental workload. Stud Health Technol Inform 2012;180:383–7.

[66] Görges M, Kück K, Koch SH, Agutter J, Westenskow DR. A far-view intensive care unit monitoring display enables faster triage. Dimens Crit Care Nurs 2011;30:206–17. https://doi.org/10.1097/DCC.0b013e31821b7f08.

[67] Görges M, Westenskow DR, Markewitz BA. Evaluation of an integrated intensive care unit monitoring display by critical care fellow physicians. J Clin Monit Comput 2012;26:429–36. https://doi.org/10.1007/s10877-012-9370-0.

[68] Karami M, Langarizadeh M, Fatehi M. Evaluation of effective dashboards: key concepts and criteria. Open Med Inform J 2017;11:52–7. https://doi.org/10.2174/1874431101711010052.

[69] Dowding D, Randell R, Gardner P, Fitzpatrick G, Dykes P, Favela J, et al. Dashboards for improving patient care: review of the literature. Int J Med Inform 2015;84:87–100. https://doi.org/10.1016/j.ijmedinf.2014.10.001.

[70] Fletcher GS, Aaronson BA, White AA, Julka R. Effect of a real-time electronic dashboard on a rapid response system. J Med Syst 2017;42:5. https://doi.org/10.1007/s10916-017-0858-5.
[71] Zaydfudim V, Dossett LA, Starmer JM, Arbogast PG, Feurer ID, Ray WA, et al. Implementation of a real-time compliance dashboard to help reduce SICU ventilator-associated pneumonia with the ventilator bundle. Arch Surg 2009;144:656–62. https://doi.org/10.1001/archsurg.2009.117.
[72] Fuller TE, Pong DD, Piniella N, Pardo M, Bessa N, Yoon C, et al. Interactive digital health tools to engage patients and caregivers in discharge preparation: implementation study. J Med Internet Res 2020;22, e15573. https://doi.org/10.2196/15573.
[73] Bourdeaux CP, Thomas MJ, Gould TH, Malhotra G, Jarvstad A, Jones T, et al. Increasing compliance with low tidal volume ventilation in the ICU with two nudge-based interventions: evaluation through intervention time-series analyses. BMJ Open 2016;6, e010129. https://doi.org/10.1136/bmjopen-2015-010129.
[74] Jawa RS, Tharakan MA, Tsai C, Garcia VL, Vosswinkel JA, Rutigliano DN, et al. A reference guide to rapidly implementing an institutional dashboard for resource allocation and oversight during COVID-19 pandemic surge. JAMIA Open 2020;3:518–22. https://doi.org/10.1093/jamiaopen/ooaa054.
[75] Almasi S, Rabiei R, Moghaddasi H, Vahidi-Asl M. Emergency department quality dashboard; a systematic review of performance indicators, functionalities, and challenges. Arch Acad Emerg Med 2021;9:e47. https://doi.org/10.22037/aaem.v9i1.1230.
[76] Franklin A, Gantela S, Shifarraw S, Johnson TR, Robinson DJ, King BR, et al. Dashboard visualizations: supporting real-time throughput decision-making. J Biomed Inform 2017;71:211–21. https://doi.org/10.1016/j.jbi.2017.05.024.
[77] Ivanković D, Barbazza E, Bos V, Brito Fernandes Ó, Jamieson Gilmore K, Jansen T, et al. Features constituting actionable COVID-19 dashboards: descriptive assessment and expert appraisal of 158 public Web-based COVID-19 dashboards. J Med Internet Res 2021;23, e25682. https://doi.org/10.2196/25682.
[78] Nasir K, Javed Z, Khan SU, Jones SL, Andrieni J. Big data and digital solutions: laying the foundation for cardiovascular population management CME. Methodist Debakey Cardiovasc J 2020;16:272–82. https://doi.org/10.14797/mdcj-16-4-272.
[79] Ruminski CM, Clark MT, Lake DE, Kitzmiller RR, Keim-Malpass J, Robertson MP, et al. Impact of predictive analytics based on continuous cardiorespiratory monitoring in a surgical and trauma intensive care unit. J Clin Monit Comput 2019;33:703–11. https://doi.org/10.1007/s10877-018-0194-4.
[80] Van Eaton EG, McDonough K, Lober WB, Johnson EA, Pellegrini CA, Horvath KD. Safety of using a computerized rounding and sign-out system to reduce resident duty hours. Acad Med 2010;85:1189–95. https://doi.org/10.1097/ACM.0b013e3181e0116f.
[81] Pageler NM, Longhurst CA, Wood M, Cornfield DN, Suermondt J, Sharek PJ, et al. Use of electronic medical record-enhanced checklist and electronic dashboard to decrease CLABSIs. Pediatrics 2014;133:e738–46. https://doi.org/10.1542/peds.2013-2249.
[82] Shaw SJ, Jacobs B, Stockwell DC, Futterman C, Spaeder MC. Effect of a real-time pediatric ICU safety bundle dashboard on quality improvement measures. Jt Comm J Qual Patient Saf 2015;41:414–20. https://doi.org/10.1016/s1553-7250(15)41053-0.
[83] Lipton JA, Barendse RJ, Schinkel AFL, Akkerhuis KM, Simoons ML, Sijbrands EJG. Impact of an alerting clinical decision support system for glucose control on protocol compliance and glycemic control in the intensive cardiac care unit. Diabetes Technol Ther 2011;13:343–9. https://doi.org/10.1089/dia.2010.0100.
[84] Cox CE, Jones DM, Reagan W, Key MD, Chow V, McFarlin J, et al. Palliative care planner: a pilot study to evaluate acceptability and usability of an electronic health records system-integrated, needs-targeted app platform. Ann Am Thorac Soc 2018;15:59–68. https://doi.org/10.1513/AnnalsATS.201706-500OC.
[85] Phansalkar S, Zachariah M, Seidling HM, Mendes C, Volk L, Bates DW. Evaluation of medication alerts in electronic health records for compliance with human factors principles. J Am Med Inform Assoc 2014;21:e332–40. https://doi.org/10.1136/amiajnl-2013-002279.
[86] Curran RL, Kukhareva PV, Taft T, Weir CR, Reese TJ, Nanjo C, et al. Integrated displays to improve chronic disease management in ambulatory care: A SMART on FHIR application informed by mixed-methods user testing. J Am Med Inform Assoc 2020;27:1225–34. https://doi.org/10.1093/jamia/ocaa099.
[87] Pant A, Mackraj I, Govender T. Advances in sepsis diagnosis and management: a paradigm shift towards nanotechnology. J Biomed Sci 2021;28:6. https://doi.org/10.1186/s12929-020-00702-6.

[88] Alba-Patiño A, Vaquer A, Barón E, Russell SM, Borges M, de la Rica R. Micro- and nanosensors for detecting blood pathogens and biomarkers at different points of sepsis care. Microchim Acta 2022;189:74. https://doi.org/10.1007/s00604-022-05171-2.

[89] Wan Y., Del Fiol G., McFarland M., Wright M.. A protocol for a scoping review to identify trends in user interface approach used by integrated surveillance tools that detect patient deterioration and adverse events. BMJ Open n.d.

[90] Sendak MP, Ratliff W, Sarro D, Alderton E, Futoma J, Gao M, et al. Real-world integration of a sepsis deep learning technology into routine clinical care: implementation study. JMIR Med Inform 2020;8, e15182. https://doi.org/10.2196/15182.

[91] Vizcaychipi MP, Shovlin CL, McCarthy A, Howard A, Brown A, Hayes M, et al. Development and implementation of a COVID-19 near real-time traffic light system in an acute hospital setting. Emerg Med J 2020;37:630–6. https://doi.org/10.1136/emermed-2020-210199.

[92] Endsley MR. From here to autonomy. Hum Factors 2017;59:5–27. https://doi.org/10.1177/0018720816681350.

[93] Hoff KA, Bashir M. Trust in automation: integrating empirical evidence on factors that influence trust. Hum Factors 2015;57:407–34. https://doi.org/10.1177/0018720814547570.

[94] Parasuraman R, Sheridan TB, Wickens CD. A model of types and levels of human interaction with automation. IEEE Trans Syst Man Cybern 2000;30:286–97.

[95] Wright MC, Kaber DB. Effects of automation of information-processing functions on teamwork. Hum Factors 2005;47:50–66.

[96] Mueller ST, Hoffoman RR, Clancey W, Emrey A, Kleigh G. Explanation in human-AI Systems: a literature meta-review synopsis of key ideas and publications and bibliography for explainable AI. In: Defense Advanced Research Projects Agency; 2019.

[97] Moore JD, Swartout WR. ISI. Explanation in expert systems: a survey. Marina Del Ray: Information Sciences Institute; 1988. https://apps.dtic.mil/docs/citations/ADA206283;.

[98] Reeder B, Makic MBF, Morrow C, Ouellet J, Sutcliffe B, Rodrick D, et al. Design and evaluation of low-fidelity visual display prototypes for multiple hospital-acquired conditions. Comput Inform Nurs 2020;38:562–71. https://doi.org/10.1097/CIN.0000000000000668.

[99] Diprose WK, Buist N, Hua N, Thurier Q, Shand G, Robinson R. Physician understanding, explainability, and trust in a hypothetical machine learning risk calculator. J Am Med Inform Assoc 2020;27:592–600. https://doi.org/10.1093/jamia/ocz229.

[100] Van Belle V, Van Calster B. Visualizing risk prediction models. PLoS One 2015;10, e0132614. https://doi.org/10.1371/journal.pone.0132614.

[101] Barda AJ, Horvat CM, Hochheiser H. A qualitative research framework for the design of user-centered displays of explanations for machine learning model predictions in healthcare. BMC Med Inform Decis Mak 2020;20. https://doi.org/10.1186/s12911-020-01276-x.

[102] Marcotte E. Responsive web design. A list apart n.d. alistapart.com/article/responsive-web-design/.

[103] Sinsky C, Colligan L, Li L, et al. Allocation of physician time in ambulatory practice: a time and motion study in 4 specialties. Ann Intern Med 2016;165:753–60.

[104] Vassar M, Holzmann M. The retrospective chart review: important methodological considerations. J Educ Eval Health Prof 2013;10:12. https://doi.org/10.3352/jeehp.2013.10.12.

CHAPTER

15

The role of standards: What we can expect and when

Kensaku Kawamoto[a], *Guilherme Del Fiol*[a], *Bryn Rhodes*[b], *and Robert A. Greenes*[c]

[a]Department of Biomedical Informatics, University of Utah Health, University of Utah, Salt Lake City, UT, United States [b]Alphora, Orem, UT, United States [c]Biomedical Informatics, Arizona State University, Phoenix, AZ, United States

15.1 Introduction

Knowledge artifacts for clinical decision support (CDS) can be developed, stored, and accessed, and updated in a variety of ways. One-off encoding may be suitable for a very specific purpose. But if the goal is broader, to encourage adoption and widespread use, there is a need for accommodating the many different environments in which such adoption and use may occur, and to facilitate uptake by those responsible in such environments, as well as ongoing management and update of knowledge over time.

To accomplish that, a variety of approaches can be used that foster what have come to be known as the FAIR & T (Findable, Accessible, Interoperable, Reusable, and Trusted) principles for knowledge sharing and reuse [1]. The topic of this chapter is "Standards", but we include in the use of that term a range from knowledge representation and execution approaches to addressing aspects of FAIR & T, e.g., using predefined languages, formats, and templates for encoding knowledge, terminologies for encoding the data, methods for accessing the data, and methods for invocation of the knowledge resource at time of execution. Many of these approaches are discussed in previous chapters in this section of the book.

Thus, in the discussion below, we are using the term "Standards" to encompass those different goals, purposes, and methodologies.

Clinical Decision Support and Beyond
https://doi.org/10.1016/B978-0-323-91200-6.00007-3

15.2 The case for standards

As discussed in the preceding chapters of the book, there have been a variety of standards initiatives related to CDS, and additional initiatives are in progress. At this point, in order to put those initiatives in perspective, we reflect on what the principal reasons are for interest in such standards, whether the standards have had the desired impact, and what needs to be done to enhance CDS adoption and impact. In other words, we ask the question, "How will such standards help in achieving the goal of wide dissemination and adoption of CDS?"

15.3 CDS development with and without standards

A key advantage that standardization can provide is maximizing the *sharing* and *re-use* of CDS logic, capabilities and tools, as well as the *interoperability* of CDS components with clinical information systems. Let us consider first what happens in the absence of a standards-based approach. We will then reflect on how standards might help. The discussion to follow is summarized in Table 15.1.

TABLE 15.1 Tasks involved in deploying knowledge in operational settings.

Task	Usual practice	How standards can help
Knowledge generation and validation	Researchers publish findings	Findings accessible via external repositories, including both positive and negative results
Useful knowledge identified	Primary or literature research locally, based on need or driven by local champion	Identification by specialty bodies, or other authoritative groups; knowledge bases organized by domains, purposes, and other attributes to facilitate access; local efforts use external knowledge bases as starting point
Adoption in practice for some knowledge	Review to decide whether it can be utilized in local environment	Recommendation and prioritization by authoritative bodies
Encoding in computer-executable form	Knowledge engineering to make the knowledge unambiguous and interpretable	Standards-based representation
Local adaptation	Modification based on local practices and constraints	External knowledge as starting point for customization, tools for doing adaptation
Implementation, debugging, and operational use of CDS	Integration into an application	Ability to implement via execution engine; ability to provide modular CDS services; standard interfaces of information model to host data sources; standard mapping of results to actions to be carried out in host; standards-based invocation of CDS
Knowledge update	Changes identified and encoded	Updates received locally from external communal resource, reviewed for applicability
Incorporation of updates in applications	Applications using knowledge found and updated	Isolation of knowledge use, because of implementation via external CDS services or modules, facilitates update

The first task, before CDS development is actually undertaken, is the creation of knowledge to be ultimately used as a basis for the CDS. In previous chapters of this book, we discussed three main classes of methodologies—human-intensive, data-intensive, and literature-based (Chapters 6, 7, and 8 respectively)—utilized for the generation and validation of clinical knowledge for CDS, plus two somewhat hybrid areas of activity that draw on the others—namely, work in population health/management, and work on translational/precision medicine, moving the discoveries in genomics and related research to clinical application. In order to discover useful knowledge through such methodologies, specialized expertise, not only with respect to the medical domains being studied, but also in the application of the methodologies, is needed. Studies aimed at discovering knowledge, furthermore, are generally difficult and time-consuming to carry out. Typically, we learn about such results through their publication in the literature.

Having accomplished the goal of discovering or formulating (through the above methods) a valid, and possibly important, clinical relationship—that is, a unit of knowledge that might be applicable for CDS—then what? Typically (as we have cited earlier, and which seems to apply so clearly to this field) the technology transfer process characterized by Balas and Boren kicks in at this point—some of the knowledge derived as a result of these efforts makes its way into publication, although considerably less of it becomes adopted in practice (14% of the original findings), and the adoption process itself occurs over a protracted period of time (average of 17 years) [2]. For example, a decision to apply specific knowledge to practice in a particular setting may depend on the presence of a local subject matter expert or champion who believes that using that knowledge will offer benefits, and who is thus motivated to take on the work.

When it comes to CDS, it is likely, though undocumented, that even more attrition of potentially useful knowledge occurs, given that some of the discovered knowledge may not be amenable to expression in computer-executable form, or suitable application environments do not exist for its use, or data necessary for its execution are not accessible or practical to acquire.

If the assessment of usefulness of a knowledge resource makes it to this point, what typically happens next is that the knowledge is encoded and incorporated into CDS applications by healthcare organizations' clinical IT departments and/or by clinical IT vendors. However, this is not straightforward either. It has been shown that the process of encoding knowledge in executable form is not performed reliably, in that knowledge in the form of published results or narratives or even guidelines is subject to many ambiguities and differences in interpretation, resulting in often quite different renderings from the original.

Still further effort is required to adapt guidelines, or other knowledge, to account for local practices and policies, processes, workflow, available resources (e.g., laboratory tests, imaging technologies, or specialized surgical expertise), and other constraints (e.g., relating to financial, personnel, or time limitations).

Thus the general inertia in the technology transfer process is compounded in the case of CDS adoption by the burdens and costs of rendering the knowledge into executable form, and adapting it for use in various application settings and on various platforms—not to mention building and debugging the applications, and deploying them operationally. Additional burden is associated with maintaining the knowledge. Furthermore, deeper integration of knowledge into clinical applications may increase the effort required to modify instances of its use when an update becomes necessary.

The onus of these processes and the protracted time frame involved thus provide the major impetus for pursuing the development of standards. It would be highly advantageous if many of these processes could be done only once for a given item of knowledge, and if the knowledge would thenceforth be widely available, so that it could be accessed and used by anyone who wanted to implement or use corresponding CDS functionality. It would also be desirable if the whole sequence of processes could be accelerated so that the benefits of applying the knowledge could be realized more rapidly.

For these and other reasons, in health IT, the slow adoption rate may be even worse than in academic research. Also, when it comes to implementation, bad experiences, difficulties, or outright failures are rarely reported in the literature, and there is thus limited basis for developing implementation science in this area to make it easier to learn from experiences of the past.

15.4 Areas in need of standardization

Let us now examine in detail what needs to occur for these benefits to be realized. We begin with the knowledge generation and validation process. To reduce the time lag and the multiple points at which potentially useful discoveries get left by the wayside during the technology transfer process, it would be helpful if knowledge, once discovered, could be maintained in widely accessible repositories, as a starting point for use by other researchers, ideally in a standard format such as the emerging EBMonFHIR specification (see Chapter 8). This should include not only positive but negative results, and should include annotations regarding reasons for non-applicability when that is determined. The source of the knowledge, its authors and their credentials, its provenance, and other ways of assessing its quality should be annotated as well.

With such CDS knowledge repositories, the knowledge contained in them would ideally be in directly usable form, or in a form that could be readily adapted to such a form. A principal target for standardization in CDS is the specification of well-defined, unambiguous *representations of knowledge*, for example, for the logic expressions in alerts, reminders, and drug order validation and interaction checking rules, and for the decision rules and expressions that incorporate such logic, as discussed in Chapter 9, or for the depiction of the sequences of steps and process flow in clinical guidelines, which was the subject of Chapter 10. Other secondary targets for standardization in support of CDS include the *information models* for referencing data used by CDS and the *specifications for results* of CDS. Chapter 11 focused on efforts at standardization of the information model and the use of well-defined vocabularies and taxonomies for naming the concepts corresponding to the data elements. Methods for *accessing* knowledge or for *invoking* CDS are also potential targets for standardization. Chapter 13 discusses a standard (i.e., the Infobutton Standard) that supports the automatic retrieval of context-specific provider reference and patient education resources with clinical systems.

Many standardization efforts that bear on CDS beyond those covered in the previous chapters are also under way. Our omission of them does not suggest that those covered are more definitive. A review of some of the most prominent CDS standards is available in a journal article published in 2021 [3]. The primary standards covered in this review include the HL7 FHIR standard for representing and accessing clinical data; the HL7 Arden Syntax, Clinical Quality Language (CQL), and FHIR Clinical Reasoning standards for representing clinical knowledge; the HL7 SMART on FHIR standard for launching external applications from an EHR; and the HL7 CDS Hooks standard for leveraging external CDS Web services.

As should be apparent from the discussions in the preceding chapters, the state of the standardization process among the clinically relevant standards has historically been in considerable flux. More recently, there has been a convergence of the standards development and implementation communities around the use of FHIR-aligned standards, including the HL7 SMART on FHIR (for sharing of applications), CQL and FHIR Clinical Reasoning (for sharing of CDS logic), and CDS Hooks (for sharing of CDS services) standards. Of note, many of these standards have clear functional overlap with earlier generations of standards, or are explicitly derived from them. For example, the CQL standard was developed following close analysis of HL7 standards including the Arden Syntax and GELLO. A key benefit of the contemporary set of standards is that they are generally all aligned with FHIR and are able to be used in a complementary and coordinated fashion. For example, as discussed in the 2021 review of CDS standards (Strasberg, Rhodes, et al., 2021), CQL can be used to represent CDS logic within both CDS Hooks services and SMART on FHIR applications, the use of SMART on FHIR applications can be suggested by CDS Hooks services, and SMART on FHIR applications can use CDS Hooks services to encapsulate its executable clinical knowledge. Standardization efforts have also taken place with regard to the representation of grouped knowledge elements such as order sets and documentation templates as discussed in Chapter 12.

When the needed standards exist, the expectation is that the tasks of implementing CDS can be more easily carried out. Tasks that still need to be done locally would also be aided by standardization, partly because efforts could be confined to those tasks, and partly because methods and tools for modifying standard representations can be made available to aid in the tasks. Local adaptation can then be based on a well-defined starting point, so that dependencies could be tracked. In some deployment architectures, the implementation, debugging, and operational use of CDS could be localized to an external execution engine and to well-defined interfaces to the host system for invocation of CDS, data access, result communication, and mapping to actions. Whether relying on an external CDS execution engine or using a native execution engine (e.g., an EHR's own execution engine), knowledge management/update, and incorporation of updates in applications, could be done more easily by relying on a central source for authoritative knowledge, and updated based on tracked relationships to the knowledge sources.

15.5 Assessment of current state of CDS standards and needed future work

While there is still much work to be done, we have made significant strides in the development of CDS standards, and critically, their adoption across the healthcare community. Here, we provide our assessment of the current state of CDS standards. This assessment is based on a systematic review of the use of FHIR-aligned standards for CDS [4]. We also draw on our personal experience, which includes our service as current and former co-chairs of the

HL7 CDS Work Group, service on the U.S. Health IT Advisory Committee, and perhaps most importantly, our on-the-ground implementation of these CDS standards in clinical care settings. For example, through the University of Utah's ReImagine EHR initiative, several of the authors have implemented over ten FHIR-based CDS innovations, including in collaboration with corporate partners [5].

- **For many aspects of CDS, relevant and complementary standards are available, and they are gaining adoption by key stakeholders including EHR vendors.** These standards include, in particular, FHIR-aligned standards including the HL7 CQL and FHIR Clinical Reasoning standards for knowledge representation, the HL7 CDS Hooks standard for integrated CDS Web services at the point of care, and the HL7 SMART on FHIR standard for embedding applications—including CDS applications—into the EHR. There is high adoption of several of these standards among EHR vendors, including FHIR and SMART. Moreover, regulations based on the 21st Century Cures Act requires EHR vendor support for both FHIR and SMART [6]. CDS Hooks is also gaining increasing adoption by major EHR vendors, such as Epic® and Cerner®, and CQL is now used by the Centers for Medicare and Medicaid Services for the representation of electronic clinical quality measures. Thus, we are closer than ever before to achieving scalable CDS enabled by wide adoption of key standards.
- **Even where relevant standards exist, there are still important gaps.** While much can be accomplished with existing standards and their level of adoption, more work needs to be done. One key area of needed work is to extend FHIR beyond the base set, known as the US Core FHIR Implementation Guide [7], which is the basis of federal regulations and the core of EHR vendors' support for FHIR. In some cases, additional data may need to be added, such as *how much* and *how long* a patient has smoked to assess lung cancer risk, in addition to the currently required information on whether the patient smokes. In other cases, FHIR query parameters may need to be required—such as being able to query for data by date range, which may only be suggested for support but not required, leading to untenable query execution times. Further, there may need to be more standardization on how detailed clinical data elements are exchanged where there are multiple valid approaches allowed by the standard, such as to distinguish between different types of blood pressure measurements, each of which should be treated differently according to hypertension management guidelines (see Chapter 11). A key question will be how to ensure that these needed enhancements to the standards are developed and adopted in a timely manner, and whether the health IT industry will self-organize to enable these needed enhancements, or whether government intervention will be needed.
- **Lack of relevant standards or their adoption is a problem in some cases.** There are key areas of CDS that still lack either standards or community adoption. For example, order sets are a key type of CDS that is widely used to drive clinical care. However, despite significant standards development efforts in this area in HL7 [8], there is currently little adoption of standards in this area.
- **The availability of competing, overlapping standards is an important problem.** In many cases, the problem is not the lack of a relevant standard, but rather the availability of multiple competing standards with overlapping scope. For example, the Arden Syntax and CQL are both standards that can be used to represent CDS rule-based expressions

knowledge, and there are a multitude of competing standards available for representing patient data beyond FHIR. As another example, HL7 version 2 messages and Clinical Document Architecture (CDA) documents are still in widespread use in health systems. The increasing community adoption of FHIR-aligned approaches is helping with this problem, and the use of these standards can be coordinated. For example, CDA documents can be parsed into FHIR data resources, and the Arden Syntax now supports FHIR for data representation. However, as has always been the case, there remains the potential for overlapping standards to cause issues with market confusion and interoperability challenges.

Given this assessment of the present state, we discuss here what is needed for standards to enable the widespread adoption of effective CDS:

- **Clear business case for development and adoption of each CDS standard.** Ultimately, the implementation of CDS, as well as support for relevant CDS standards, requires an investment in resources. Thus, a clear business case needs to be developed and communicated for each stakeholder group that needs to be involved in the development and adoption of relevant CDS standards. In particular, if a CDS standard is not required by federal regulations, a business case should be clearly articulated for why an EHR system vendor should support a given standard. For example, the business case for an EHR system vendor to support an order set standard could be that a common standard would reduce integration costs with order set publishers, each of which may utilize a different knowledge representation format.
- **Standards that are easy to understand and to use.** When reviewing IT standards that are ubiquitously adopted, a common theme for many of these standards is that they are easy to understand and to use. Take, for example, SQL, XML Schemas, HTML, or FHIR. To the extent possible, future CDS standards should seek to make it easy to learn and use the standards.
- **Resources and tools to facilitate standards adoption.** Continuing the analogy just discussed, a common feature of standards such as SQL, XML Schemas, HTML, and FHIR is that there are a number of resources—many of them free—for facilitating adoption of those standards, including learning resources and programming tools. Making such resources and tools available for adopting CDS-related standards will be important.
- **Standards that are built upon a stack of universally adopted standards.** Ideally, industry-specific standards should leverage lower-level standards that have been widely adopted in other industries, such as XML, JSON, and RESTful Web services. This approach makes the standard easier to understand and use because implementers will be familiar with those foundational standards. In addition, widely adopted foundational standards generally have a wide range of implementation resources and tools with which implementers are also familiar, including robust open source packages and testing utilities.
- **Proactive engagement of the vendor community.** Ultimately, widespread adoption of CDS standards can only be achieved through commercial clinical information system vendors and commercial knowledge vendors. Thus, vendor engagement will be critical for influencing the outcome of future efforts to enable standards-based CDS.
- **Demonstration of the impact of interoperable CDS.** With the increasing adoption of key CDS standards, it is imperative that we collectively demonstrate the value of what those

standards have achieved, so that we can galvanize the community will to enhance these standards to enable the creation of even more value to key stakeholders including patients, providers, health systems, and health IT vendors. Already, we and others have begun to demonstrate such value through standards-based CDS solutions implemented in clinical care [4,5,9].

15.6 Beyond the standards—What is needed for widespread CDS adoption?

15.6.1 The vision for standards-enabled, scalable CDS

The premise underlying standardization efforts is that having standards such as those identified earlier would stimulate sharing and reuse *because* it would enable a number of developments and changes in procedures. The following are some of these potential capabilities and the benefits that could be derived from them.

- Collections of discovered knowledge of various types could be made widely available in the form of knowledge bases, ideally using standard representation formats such as EBMonFHIR. Evidence-based medicine repositories such as from the Cochrane Collaboration are examples of this. If the knowledge generation and validation have been done by authoritative, respected experts, this would obviate the need to rely entirely on local experts for carrying out or redoing such efforts in each institutional setting or for each vendor-based system. That is, sites could adopt a set of authoritative rules or guidelines instead of having to develop them locally.
- The management of the knowledge bases under the aegis of external content provider entities (commercial, professional society, government, consortial, or other) would relieve local sites or vendor systems from having to undertake this task.
- Knowledge could be flexibly provided in a variety of ways. For example, it could be made available for access by end users or systems when needed (e.g., through CDS Hooks services). Alternatively, knowledge content could be provided in standard format (e.g., Arden, CQL) for downloading and importing into local environments.
- If the knowledge has been encoded into executable form by knowledge engineers and software engineers supported via the external provider, this effort also would not need to be redone in each setting. The knowledge might still need to be translated or adapted to local platform-specific representations and interfaces, but even this can be reduced to the extent that standards-based interfaces to data and application services are supported in the local platforms.
- Beyond translation and interfacing to a host platform, local efforts could be confined to customization and adaptation to local requirements and constraints.
- Updates to knowledge can be coordinated by the provider of knowledge and communicated to users, and details of provenance and versioning can be maintained. Although the process of updating the instances in which the knowledge is used in local settings can still be difficult, at least part of the burden of creating and tracking updates can be borne by the provider of the knowledge.

- Given economies of scale that could be devoted to knowledge update, external knowledge bases can be kept more up-to-date and reliable than those that are developed or maintained locally.
- If suitable mechanisms exist for curation and management of the external knowledge bases (see Chapter 28), knowledge developed and created by local experts can also be uploaded and incorporated into those knowledge bases. This might serve to create a collaborative community for continuous knowledge base development and improvement.

15.6.2 Resources needed

In order to stimulate sharing and reuse of knowledge, it is insufficient to just define standards for knowledge representations, interfaces, and modes of access and invocation. What must also occur is the creation of artifacts that use them, as well as models for their integration into the life cycles of knowledge generation, knowledge management, and CDS method development/implementation. Three principal classes of artifacts are needed for standards-based CDS:

1. **Knowledge bases.** We have referred to external knowledge bases earlier and in Chapter 18, but have not discussed under what aegis they come about, what the business models are for their ongoing support, how they are structured, their mode of access or interface, or how they are maintained and updated. For example, the CDS Connect repository provides a public repository for standards-based CDS knowledge, EHR vendors increasingly have "app stores" for SMART on FHIR apps available for use with their systems, including many CDS apps, and knowledge publishers already provide CDS knowledge for order entry CDS capabilities such as drug-drug interaction checking [10].
2. **Tools for authoring and update.** Once we have an external standardized knowledge base of sufficient scale and utility for widespread use, it becomes feasible to invest in the development of a robust set of tools for authoring, review, editing, and publishing of knowledge of the types in the knowledge base. Further, to the extent that collaborative authoring and update are desirable, it is possible to provide other content management and collaboration capabilities that aid this process. For example, tools can be created to facilitate identification of similar knowledge to that which is being authored or modified, to provide a starting point for the work, and to detect potential inconsistencies, contradictions, redundancies, and gaps in the knowledge relating to a specific topic (see Chapter 28 for a detailed discussion of Clinical Knowledge Management). There have been efforts at large academic medical centers to build and use such tools (see Chapter 28). Moreover, there are efforts to develop such tools for use across institutions, such as in the CDS Connect effort, the OpenCDS initiative (www.opencds.org), and the LITE authoring environment of OpenInfobutton (see Chapter 13) [11,12].
3. **Tools for execution.** Once we have standardized knowledge in executable form, it is feasible to consider the development of execution tools that will operate on such knowledge, and that can be invoked by host environments. Certainly execution can be done in a host-specific way, in terms of its degree of integration with the host platform, its databases, and its applications. But independent "execution engines" can alternatively be developed. The use of an external execution engine can, for example, be used within CDS

Hooks services or SMART on FHIR apps to make use of FHIR Clinical Reasoning knowledge modules or CQL rule files. Examples of open-source tools available for such execution include OpenCDS and the CQL Evaluator [13].

15.7 How important are standards?

As compelling as the benefits of standardization appear to be for stimulation of knowledge sharing and reuse, we now pause to consider what evidence there is that this will make a substantial difference in the rate of dissemination and adoption of CDS. Although Chapters 1 and 20 have identified a number of barriers and areas of inertia that have impeded adoption of CDS, it is not immediately clear that the potential for access to external resources and reuse through standardization will have significant benefit in overcoming these barriers. Perhaps more important are other barriers that have little to do with standards. For example, the health care payment model in the U.S. still predominantly rewards care quantity over quality, such that there may be little financial incentive—and at times a financial disincentive—for health care systems to invest in quality improvement initiatives that reduce healthcare utilization. Another important barrier to CDS adoption is the difficulty involved in adapting successful demonstrations of CDS for use in settings with different operational environments, practice styles, organizational approaches, incentives, and constraints.

We have commented earlier (see Chapter 1) about the value of having a better scientific understanding of the human engineering and organizational strategies that are most effective. But would the availability of shared or reusable knowledge bases themselves play a significant role in accelerating CDS adoption?

In truth, the evidence for the benefits of sharing and reuse as a driver for widespread CDS adoption is still being established. With regard to relatively simple CDS, there are some good examples of where sharing and reuse in fact has occurred (see Chapter 11 for a detailed discussion of Knowledge Resources). Perhaps the best examples are "commodity" compendia such as those containing drug formulary or drug-drug interaction logic, which can be obtained from commercial knowledge vendors as well as some nonproprietary sources. These usually are provided in the form of simple tables that can be incorporated in relational databases in a host system and used in CPOE or other applications. Bibliographic databases (e.g., PubMed) and synthesized evidence resources (e.g., UpToDate®, Dynamed Plus®) are also valuable (e.g., for access via an infobutton manager), although they do not provide sources of executable knowledge. In these cases, however, it is worth noting that standards were not the driver of adoption—the driver was instead a combination of sizable knowledge bases and market demand for the underlying content. This observation points to the conclusion that whereas standards certainly have the potential to facilitate CDS adoption, they are by no means sufficient, and in some cases not even required.

Even for CDS content sharing sites sponsored by EHR system vendors, there often does not appear to be very robust peer-to-peer sharing, for the simple reason that while health care systems may have a strong incentive to download available, high-quality CDS content, they have little or no incentive to upload such content, especially if a competing health care system (e.g., in the same geographic region) could potentially make use of that content. Thus, while standards may facilitate *how* knowledge is shared, they do not in themselves address *why* a stakeholder organization would want to share such knowledge. Thus, it is critical that this motivation be explicitly identified and addressed. For example, the availability of a standard could motivate a knowledge vendor to produce more content using the standard due to the increased potential market size, and an EHR system vendor could in turn be motivated to support the standard due to the availability of more content for its customers. However, a typical health care organization will likely have very little incentive to share its knowledge content, with or without a standards-based framework that enables sharing.

Over the years, there have been several attempts at overcoming this fundamental challenge to standards-based knowledge sharing. These initiatives include the Institute for Medical Knowledge Implementation (IMKI), the Morningside Initiative, the CDS Consortium, OpenCDS, and CDS Connect. While these and other initiatives have shown that it is feasible to share CDS knowledge from a technical perspective, the challenge has remained the development of a self-sustaining business model for CDS knowledge sharing. To date, there is still limited evidence that there is a viable business model for creating and sharing CDS content other than for commercial content developers licensing content to a variety of customers, with sharing taking place simply as a mechanism to achieve economies of scale. Non-profit entities such as professional medical societies certainly have a desire to make their knowledge content (e.g., clinical practice guidelines) more widely adopted, but they often lack the financial incentives and accompanying resources to develop knowledge content in standard formats and to implement the sharing of executable CDS content to a significant extent. Moreover, a key challenge will be finding effective ways to harness the efforts at health care organizations to use and refine CDS knowledge resources. Potential strategies for enabling such knowledge sharing may include financial incentives (e.g., grant support for CDS content development at leading health care systems in exchange for CDS content uploading) or government mandate (e.g., requirement for CDS content developed at federal health care systems such as the Veterans Health Administration or Military Health System to be deposited to CDS knowledge repositories).

In summary, evidence demonstrating the value of shared, standards-based knowledge resources is still largely lacking. Nonetheless, it feels right to many experts that having such resources would be valuable, for reasons such as those listed earlier. We will further consider prospects for realizing those potential advantages in Section VI.

Note also that the main purpose for standards development to date in other areas has not been for sharing of external resources but for interoperability of systems, to facilitate the processing of transactions. This focus on transactions has made economic sense in that standards for messaging and transfer of data enable disparate systems and applications to communicate and cooperate as part of a value chain, whether for business or clinical purposes. When one considers sharing of external resources as a driver, the questions that come quickly to the top concern who owns, maintains, and takes responsibility for the resource, what the business model is for sharing and supporting the resource, and how it adds value to participants. Such questions need to be addressed in pursuing the goal of standardization of knowledge resources.

15.8 Vision for potential future impact of standards

Even though the role of standardization efforts in this realm is to define the representation of knowledge, interfaces to it, organizing schemas, and invocation methods—*not* to create the repositories, authoring/editing tools, and execution tools—nonetheless, we believe that by having such standards, efforts such as those we described earlier, devoted to developing the repositories and tools, are more likely to be undertaken. This is increasingly possible, given the ability to create self-contained Web services such as CDS Hooks services or API-invoked units of functionality such as SMART on FHIR applications as modular components.

A consequence of modularity is that the barriers to development of a component or entry into a marketplace are greatly reduced. Without standards, we have the usual lack of critical mass and focus that prevents forward movement. Having the repositories and tools, even in early forms, can stimulate refinement of them, and can facilitate explorations of new opportunities for creation of value, in terms of means for carrying out knowledge generation and validation activities, managing knowledge resources, and delivering knowledge content to users or applications that need it. This may involve new organizational and business models devoted to dissemination and reuse of shared knowledge content, involving modes of collaboration and commercial development not previously possible.

Will the creation of standards-based publicly available knowledge bases, seeded with significant initial content, and provision of open-source tools for authoring and editing, as is being done currently by several federal agencies, actually stimulate use? Will this combination of resources stimulate multi-party collaboration and refinement of the knowledge as well as of the tools? Will the existence of standards, knowledge bases, and tools prompt systems managers to provide interfaces and means of using such knowledge in their systems? Will clinical IT system vendors also provide such interfaces? Will demand stimulate the creation of a commercial marketplace of content, tool, and service providers that can provide added value to clinical systems? The answer to all these questions, increasingly, is yes. Thus, while we are still early in this evolution of the CDS landscape, it seems clear that we are increasingly realizing the vision of robust CDS deployed at scale through standards-based approaches.

References

[1] Alper BS, Flynn A, Bray BE, Conte ML, Eldredge C, Gold S, et al. Categorizing metadata to help mobilize computable biomedical knowledge. Wiley Online Library; Report No.: 2379-6146; 2021.

[2] Balas EA, Boren SA. Managing clinical knowledge for health care improvement. Yearb Med Inform 2000;1:65–70.

[3] Strasberg HR, Rhodes B, Del Fiol G, Jenders RA, Haug PJ, Kawamoto K. Contemporary clinical decision support standards using Health Level Seven International Fast Healthcare Interoperability Resources. J Am Med Inform Assoc 2021;28(8):1796–806.

[4] Taber P, Radloff C, Del Fiol G, Staes C, Kawamoto K. New standards for clinical decision support: a survey of the state of implementation. Yearb Med Inform 2021;30(1):159–71.

[5] Kawamoto K, Kukhareva PV, Weir C, Flynn MC, Nanjo CJ, Martin DK, et al. Establishing a multidisciplinary initiative for interoperable electronic health record innovations at an academic medical center. JAMIA Open 2021;4(3). ooab041.

[6] Office of the National Coordinator for Health Information Technology. Interoperability, information blocking, and the ONC health IT certification program. Fed Regist 2019;85.

[7] Health Level Seven International (HL7). HL7 FHIR US Core Implementation Guide, Standard for Trial Use Release 4; 2021.

[8] Health Level Seven International (HL7). HL7 Version 3 Standard: Order Set Publication; 2012.

[9] Kawamoto K, Kukhareva P, Shakib JH, Kramer H, Rodriguez S, Warner PB, et al. Association of an electronic health record add-on app for neonatal bilirubin management with physician efficiency and care quality. JAMA Netw Open 2019;2(11), e1915343.

[10] Lomotan EA, Meadows G, Michaels M, Michel JJ, Miller K. To share is human! Advancing evidence into practice through a national repository of interoperable clinical decision support. Appl Clin Inform 2020;11(1):112–21.

[11] Jing X, Cimino JJ, Del Fiol G. Usability and acceptance of the librarian infobutton tailoring environment: an open access online knowledge capture, management, and configuration tool for openInfobutton. J Med Internet Res 2015;17(11), e272.

[12] Cimino JJ, Jing X, Del Fiol G. Meeting the electronic health record "meaningful use" criterion for the HL7 infobutton standard using OpenInfobutton and the Librarian Infobutton Tailoring Environment (LITE). AMIA Annu Symp Proc 2012;2012:112–20.

[13] Dynamic Content Group. CQL Evaluator. 2021. Available from: https://github.com/dbcg/cql-evaluator.

CHAPTER

16

Population analytics and decision support

John Halamka and Paul Cerrato

Mayo Clinic Platform, Mayo Clinic, Rochester, MN, United States

In the forward to the second edition of this book, Jacob Reider, MD, then the chief medical officer for the Office of the National Coordinator for Health Information Technology, summarized clinical decision support's ultimate purpose in a few simple words: "Clinical decision support is how we distribute knowledge from where it *is* to where it *needs* to be." It can also be argued that population health analytics has the same purpose: At its most basic level, population health analytics is focused on collecting data about a group of individuals, obtaining actionable insights from that data, which in turn can then be distributed to where it needs to be, namely into the hands of providers, payors, and other stakeholders.

The U.S. Centers for Disease Control and Prevention (CDC) defines population health as "an interdisciplinary, customizable approach that allows health departments to connect practice to policy for change to happen locally" [1]. Duke University's School of Business defines population health analytics as "the act of applying quantitative methods and technology to reach advanced insight about a group" [2].

Since the topic of population health analytics covers many areas, this chapter will concentrate on the methods used to acquire, organize, secure and access population health data and the relevant analytic methods.

16.1 Population health data acquisition methods

In today's healthcare ecosystem, population health management and analysis are no longer on the wish list for most providers and payors. With ACOs and other value-based systems taking hold, they have become a necessity in order to survive and compete. The metrics currently being utilized to perform analytics include patient surveys, administrative enrollment and billing records, EHR reports, medical records from health plans and community health centers, and patient registries. For example, both the Hospital Consumer Assessment of

https://doi.org/10.1016/B978-0-323-91200-6.00020-6

Healthcare Providers and Systems (HCAHPS) and the Leapfrog Group rely on patient surveys to help evaluate providers' quality of care.

In recent years, most healthcare organizations have come to appreciate the value of population health metrics. Evidence to support that assertion comes from a variety of sources. A Health Catalyst survey of 101 healthcare leaders, for instance, found that more than 60% considered population health either very important or extremely important while only 16% rated it as somewhat important or not at all important [3]. According to Health Catalyst, the urgency to address this issue requires a strategic approach that includes "Integrated claims and clinical data sets to gain deep insights on patients and their risk, and flexible and powerful platforms to support sophisticated analytics needs, such as artificial intelligence (AI) and machine learning-powered tools to better manage their populations" [3].

Three of the most valuable data sources that can be used to inform population analytics are discussed below:

Patient surveys. For any population health analytics program to provide actionable insights, it must first create a solid foundation of data sources that stakeholders can rely on to generate those insights. Patient experience surveys have been an important source of data to help guide decisions on population health for many years, but these tools need to be updated in order for them to yield the best intelligence. Several associations, including the Federation of American Hospitals, the American Hospital Association, and the Catholic Health Association of the United State, are now calling for changes in the HCAHPS survey. The 28-question patient satisfaction survey is mandated by the Center for Medicare and Medicaid Services for all American hospitals. The call for modernization includes the recommendation that the survey devote more attention to metrics that evaluate patients' confidence in doctors and nurses, evaluate the efficiency and communication during hospital admission and discharge, and explore a patient's need to be heard and have input on their care [4].

The current version of the HCAHPS Survey (March 2021) [5], a portion of which is depicted in Fig. 16.1, provides several important metrics that can inform population health analytics, including 18 questions about communication with doctors and nurses, the responsiveness of hospitals staff, the cleanliness of the hospital, how quiet it is, pain management, and more. To make this survey more relevant, some critics believe it needs to include more open-ended questions and be more firmly grounded in research on what patients consider most important, rather than what administrators and clinicians perceive to be most important [6]. That concern is justified in light of studies that have found wide disagreement between physicians and patients [7].

Clinical data or patient registries. Like patient surveys, registries can serve as raw material to generate insights when incorporated into a comprehensive analytics program; however, they play a somewhat different role. By collecting details on specific treatments for a large group of patients as well as how patients responded, they provide a more objective source of data. As the American Medical Association points out: "A registry may focus on a disease or condition, a procedure, or a medical device. The registry defines a patient population, then recruits physicians and other health care professionals to submit data on a representative sample of those patients" [8]. In order to detect meaningful patterns in these disease registries, it's critical for registry staffers collecting patient data to perform the necessary quality checks to make sure it's correct, complete, and secure. EHR systems also provide registry functionality

HCAHPS Survey

SURVEY INSTRUCTIONS

- ☐ You should only fill out this survey if you were the patient during the hospital stay named in the cover letter. Do not fill out this survey if you were not the patient.
- ☐ Answer all the questions by checking the box to the left of your answer.
- ☐ You are sometimes told to skip over some questions in this survey. When this happens you will see an arrow with a note that tells you what question to answer next, like this:

  ☐ Yes
  ☑ No ➔ ***If No, Go to Question 1***

You may notice a number on the survey. This number is used to let us know if you returned your survey so we don't have to send you reminders.
Please note: Questions 1-29 in this survey are part of a national initiative to measure the quality of care in hospitals. OMB #0938-0981 (Expires November 30, 2021)

Please answer the questions in this survey about your stay at the hospital named on the cover letter. Do not include any other hospital stays in your answers.

YOUR CARE FROM NURSES

1. **During this hospital stay, how often did nurses treat you with courtesy and respect?**
   1☐ Never
   2☐ Sometimes
   3☐ Usually
   4☐ Always

2. **During this hospital stay, how often did nurses listen carefully to you?**
   1☐ Never
   2☐ Sometimes
   3☐ Usually
   4☐ Always

3. **During this hospital stay, how often did nurses explain things in a way you could understand?**
   1☐ Never
   2☐ Sometimes
   3☐ Usually
   4☐ Always

4. **During this hospital stay, after you pressed the call button, how often did you get help as soon as you wanted it?**
   1☐ Never
   2☐ Sometimes
   3☐ Usually
   4☐ Always
   9☐ I never pressed the call button

FIG. 16.1 Sample from the Hospital Consumer Assessment of Healthcare Providers and Systems: CAHPS Hospital Survey, March 2021. *Source: Hospital Consumer Assessment of Healthcare Providers and Systems: CAHPS Hospital Survey, March 2021.*

leveraging data already collected in the EHR; as such they do not necessarily require collection of new data as do the disease registries.

Unlike patient experience surveys, patient registries are a viable tool that serves several clinical decision support functions. For example they let clinicians and managers stratify risk in a large population of patients with Type 2 diabetes, allowing them to identify those most likely to experience exacerbations. Registries may also identify patients who are overdue for colorectal cancer screening, or who have yet to get the recommended vaccinations.

Designing a useful patient registry requires careful planning, the details of which are beyond the scope of this chapter; however, a few highlights are worth consideration for clinicians and researchers interested in making use of these resources to perform population health analysis [9]. You will want to know the types of questions that the data set is addressing, and whether they are focused on clinical issues, public health, or both. Similarly, was the registry designed to address questions that translate into measurable exposures and outcomes? How well defined is the population being measured? Has a comparison group been recruited? What's the size of the patient population contained in the data set? What exactly is the registry measuring: the natural history of a disease? The effectiveness and safety of a specific treatment protocol? Its cost effectiveness? Equally important: Is there any evidence of bias in collecting the data, and is it generalizable?

It is also important to familiarize oneself with the alphabet soup of acronyms used to describe the various types of data likely to be included in patient registries. Most clinicians are familiar with ICD (International Classification of Diseases), CPT (Current Procedural Terminology), and similar terms but may not be familiar with more technology-related terms, including SNOMED (Systemized Nomenclature of Medicine), RxNorm (the standardized nomenclature for clinical drugs), and LOINC (Logical Observation Identifiers Names and Codes), used to describe lab orders and results. Among the data elements you are likely to see in registries are obvious basics like contact information, patient's name, date of birth, and demographics. There may also be elements that describe various social determinants of health—which in the past have been undervalued as key factors in disease causation—biomarker test results like EGFR mutations in lung cancer, disease activity scores, unique device identifiers for medical devices, and alternative medicine treatments.

Electronic health records (EHRs). Since the inception of the HITECH act, the vast majority of U.S. healthcare providers have implemented EHR systems, which can serve as an invaluable source of data to improve public and population health outcomes. Because these digital tools share many common features and collect a common set of data elements, they can improve public health reporting and surveillance, help an organization's efforts in preventing disease, and expand communication between providers and public health officials.

In addition to the many structured data fields that provide details on an individual patient's diagnosis, treatment, and test results, EHR systems are also a source of unstructured data, including narrative notes that may include hidden treasure for researchers and clinicians interested in improving the health of a large patient population. With the appropriate use of natural language processing (NLP) software, it is often possible to derive actionable insights from these notes. In addition to NLP, there are a wide range of artificial intelligence and machine learning-enhanced algorithms now available to detect hidden patterns in EHR data, which in turn would be useful to investigators who are tasked with improving population health.

In addition, ML-enhanced algorithms rely of several modeling techniques that are now taking center stage as valuable adjuncts to clinicians' diagnostic and therapeutic skills. They include convolutional neural networks, random forest analysis, clustering, and gradient boosting, which are considered in more depth in Chapter 7.

16.2 Analytics methods

Over the decades, several traditional analytical techniques have been used to study population health data. More recently, AI- and machine learning-enhanced tools have been added to this tool kit. The Framingham heart health risk score has been used for many years to assess the likelihood of developing cardiovascular disease over a 10-year period. Because the scoring system can help predict the onset of heart disease, it can also serve as a useful tool in creating population-based preventive programs to reduce that risk. The tool requires patients to provide their age, gender, smoking status, total cholesterol, HDL cholesterol, systolic blood pressure, and whether they are taking antihypertensive medication. Because the Framingham risk scoring system is based on population data that is several years old, it may not be as useful in stratifying patients as it once was. Among the risk variables that have changed over the decades are the U.S. public's dietary habits and smoking status. The lack of this data in turn may skew the results.

The American Diabetes Association has developed its own risk scoring method to assess the likelihood of type 2 diabetes in the population. The tool takes into account age, gender, history of gestational diabetes, physical activity level, family history of diabetes, hypertension, height and weight. Another analytics methodology that has value in population health is the LACE Index. The acronym stands for length of stay, acuity of admission, Charlson comorbidity index (CCI), and number of emergency department visits in the preceding 6 months. It can be used to help predict 30-day hospital readmissions [10]. One study found that it had a c-statistic value of 0.628 for predicting 30-day readmissions. (For a more detailed discussion of analytic methods, including regression analysis, see Chapter 7.)

16.3 Addressing the problems associated with analytic tools

Predictive analytics tools have tremendous potential to detect hidden patterns in a population, but they can be easily misused. If you are a decision maker trying to determine the value of investing in a population analytics program, three of the most important issues to address are generalizability, data set shift, and accuracy. The study cited above that discussed the LACE Index, for instance, was performed in a Singapore population, so its application to other populations in other countries is questionable. Similarly, many machine learning-enhanced predictive algorithms have not been tested on more than one patient population. A recent analysis of 130 FDA approved AI-based products and services revealed that 93 did not report multi-site evaluation [11]. That's a significant shortcoming that can have far-reaching implications during clinical decision making.

Consider two contrasting examples [12]: Juan Banda and his colleagues developed a ML system to help identify patients most likely to have familial hypercholesterolemia [13]. To address the generalizability issue, developers tested their software on 2 completely separate data sets, namely, EHR data from Stanford Health Care patients and on patients from the Geisinger Healthcare System. Not every ML project has been this thorough in vetting its platform. John Zech, California Pacific Medical Center in San Francisco, and his associates, developed a convolutional neural network (CNN) to screen for pneumonia across 3 hospital systems using large data sets of X-ray images from the National Institutes of Health Clinical Center, Mount Sinai Hospital (MSH), and the Indiana University Network for Patient Care (IU) [14]. They found that, "When models were trained on pooled data from sites with different pneumonia prevalence, they performed better on new pooled data from these sites but not on external data," the result of overfitting. Similarly, if a CNN is developed to estimate who is at risk of being emergently admitted to a hospital, it may rely on data from past admissions of patients, namely, patients who have a specific set of symptoms, demographics, and so on. But in the real world, if those modeling features differ significantly from the characteristics of the hospital population one is trying to apply the neural network to—including bed availability, ethnic background, and insurance carriers—it's unlikely the results would be very useful.

Data set shift is another AI shortcoming that can interfere with any population analytics program. It occurs when the data collected during the development of an algorithm changes over time and is different from the data when the algorithm is eventually implemented. For example, the patient demographics used to create a model may no longer represent the patient population when the algorithm is put into clinical use. This happened when COVID 19 changed the demographic characteristics of patients, making the Epic sepsis prediction tool less effective. The Epic Sepsis Model (ESM) has been used on tens of thousands of inpatients to gauge their risk of developing this life-threatening complication. Part of the Epic EHR system, it is a model that the vendor has tested on over 400,000 patients in 3 health systems. However, investigators from the University of Michigan conducted a detailed analysis of the tool among over 27,600 patients and found it wanting. Andrew Wong and his associates found an area under the receiver operating characteristic curve (AURAC) of only 0.63. Their report states: "The ESM identified 183 of 2552 patients with sepsis (7%) who did not receive timely administration of antibiotics, highlighting the low sensitivity of the ESM in comparison with contemporary clinical practice. The ESM also did not identify 1709 patients with sepsis (67%) despite generating alerts for an ESM score of 6 or higher for 6971 of all 38,455 hospitalized patients (18%), thus creating a large burden of alert fatigue" [15]. Chapter 7 discusses technical approaches that can be used to address challenges related to data shift.

The accuracy of AI-enhanced analytical tools remains a concern for many in the healthcare community because there are too few prospective studies supporting them. The analysis of FDA approved devices mentioned above also found that among the 130 approved products and services, 126 of the approvals relied solely on retrospective studies and among the 54 high risk devices evaluated, none included prospective studies [11]. Those findings are consistent with a recent review we published in *NEJM Catalyst*, in which we pointed out that of the hundreds of AI enhanced algorithms reported on in the medical literature, only five were randomized controlled trials and 9 were non-RCT prospective in design [16] (See Table 16.1).

When evaluating the role of AI-based algorithms in population health, one also needs to consider the trustworthiness of these digital tools in the minds of the average clinician. Even when an algorithm is well supported by evidence, it will have little impact on practitioners in everyday practice if they do not understand the underlying technology. Machine learning based-algorithms rely on sophisticated mathematical calculations and an understanding of data science. Without that background, many clinicians will be skeptical about using them. There are several ways to address this explainability or "black box" issue. Chapter 7 describes potential technical approaches to address ML explainability issues.

TABLE 16.1 Randomized controlled trials and prospective studies on AI and machine learning.

Disease state	Findings	Reference
Randomized Controlled Trials		
Colorectal cancer	Colonoscopy combined with deep learning computer-assisted detection improved adenoma detection	[17]
Colorectal cancer	Neural network–assisted colonoscopy was more effective than unassisted colonoscopy in detecting adenomas	[18]
Upper gastrointestinal disease	Neural network–assisted esophagogastroduodenoscopy reduced blind spot rate	[19]
Colorectal cancer	Colonoscopy with artificial intelligence assistance increased adenoma detection rates, compared to standard colonoscopy	[20]
Childhood cataracts	CC-Cruiser, an artificial intelligence platform, was less accurate in detecting cataracts and making treatment decisions, compared to senior consultants.	[21]
Prospective Studies		
Diabetic retinopathy	Autonomous AI-enhanced algorithm proved effective for detecting mild retinopathy and diabetic macular edema	[22]
Diabetic retinopathy	Automated diabetic retinopathy grading system was at least as effective as manual grading in detecting moderate or worse disease	[23]
Diabetic retinopathy	AI-based grading of diabetic retinopathy was used to evaluate 193 patients; it judged 17 as having retinopathy, but only correctly identified 2 patients with true disease, compared to 15 false positive results.	[24]
Congenital cataracts	Convolutional neural network–based algorithm managed diagnosis, risk stratification, and treatment suggestions as accurately as ophthalmologists.	[25]
Breast cancer	Pathologists who were assisted by a deep learning algorithm demonstrated higher accuracy in detecting lymph node micrometastases, when compared to either the algorithm alone or unassisted pathologists	[26]
Colorectal cancer	Real time AI-enhanced ultra-magnifying colonoscopy improves differentiation of small polyps requiring resection from those not requiring resection.	[27]

Most scientific papers that explore the role of artificial intelligence and machine learning in medicine provide proof of concept support, retrospective analysis, or were conducted on relatively small data sets. The following list includes the small number of randomized controlled trials and prospective studies that bear closer examination by decision makers and clinicians.
Source: Halamka J, Cerrato, P. The digital reconstruction of health care. NEJM Catal 2020;1 (6). https://doi.org/10.1056/CAT.20.0082 (Used with permission of publisher).

Tutorials are available to simplify machine learning-related systems like neural networks, random forest modeling, clustering, and gradient boosting. Cerrato and Halamka provide a chapter on these machine learning techniques in a separate publication [28]. Similarly, *JAMA* [29] has created clinician friendly video tutorials designed to graphically illustrate how deep learning is used in medical image analysis and how such algorithms can be used to help detect lymph node metastases in breast cancer patients.

These resources require clinicians to take the initiative and learn a few basic AI concepts, but developers and vendors also have an obligation to make their products more transparent. One way to accomplish that goal is through saliency maps and generative adversarial networks. Using such techniques, it's possible to highlight the specific pixel grouping that a neural network has identified as a trouble spot, which the clinician can then view on a radiograph, for example. Alex DeGrave, with the University of Washington, and his colleagues [30], used this approach to help explain why an algorithm designed to detect COVID-19-related changes in chest X-rays made its recommendations. Amirata Ghrobani and associates [31] from Stanford University have taken a similar approach to help clinicians comprehend the echocardiography recommendations coming from a deep learning system. The researchers trained a convolutional neural network (CNN) on over 2.6 million echocardiogram images from more than 2800 patients and demonstrated it was capable of identifying enlarged left atria, left ventricular hypertrophy, and several other abnormalities. To open up the black box, Ghorbani et al presented readers with "biologically plausible regions of interest" in the echocardiograms they analyzed so they could see for themselves the reason for the interpretation that the model has arrived at. For instance, if the CNN said it had identified a structure such as a pacemaker lead, it highlighted the pixels it identifies as the lead. Similar clinician-friendly images are presented for a severely dilated left atrium and for left ventricular hypertrophy.

Another problem that interferes with population data analytics is the incomplete picture that usually surfaces when we rely too heavily on existing data sources like patient registries and EHR systems. Typically, these sources only contain data gleaned from episodic patient care, i.e. all the information collected while patients are being treated either in a hospital, clinic, or private office. Unfortunately, that information only tells a small part of the patient's story. The social context of patients' lives is probably more important to determining their health needs. That context, often referred to as the social determinants of health (SDOH), include their socioeconomic status, educational level, proximity to healthy food sources in their community, the safety of their neighborhood, their mode of transportation, and the availability of a secure internet connection to stay in touch with providers are equally important and need to be factored into any analysis. As Simukai Chigudu, Oxford Department of International Development, University of Oxford points out, health professionals are slowly beginning to realize that we cannot "remove health and illness from the social contexts in which they are produced" [32].

The Centers for Disease Control and Prevention (CDC) has numerous data sources to help incorporate SDOH into public health initiatives and medical practice. But as the agency points out, moving from data to action is the hard part. CDC has several programs designed to focus clinicians' attention on key social issues, including socioeconomic status, educational level, and work history. One initiative, for instance, zeros in on the role of EHRs. Its purpose is to support the incorporation and use of structured work information into health IT systems. How might this SDOH element inform a physician's differential diagnosis? Consider a patient with hypertension who doesn't respond to a low sodium diet or anti-hypertension medication. Awareness of the patient's 10-year history as a house painter might point the clinician in the direction of lead poisoning, a possible cause of hypertension. Similarly, a nurse practitioner may be at a loss to figure out why a patient with type 2 diabetes has recently seen a spike in her A1c levels. If an EHR system is linked to work history, when the nurse enters the reason for the clinic visit into the EHR field for chief complaint, this might trigger a pop up box that states that the patient works the night shift and that shift work can affect diabetes control. The system would then provide recommendations on diabetes management among shift workers. The same CDC program is also working on a work information data model, as well as national standards for vocabulary, system interoperability, and instructions for health IT system developers [33].

Ignoring the effect of SDOH on population health is part of a much larger problem facing the healthcare ecosystem. The larger issue is the wide variety of inequities that continue to impact clinical care. Conducting population health analytics using data sets and algorithms that do not take into account a population's racial and ethnic makeup, for instance, will likely generate misleading recommendations on how to manage that population. On the other hand, inappropriate use of race/ethnicity as predictors without a biologic reason can lead to racial discrimination. Many of these complex issues will be discussed at length in Chapter 25.

16.4 The Mayo Clinic approach to population data analysis

Mayo Clinic is addressing the equity issue in collaboration with Duke University School of Medicine and Optum/Change Healthcare by analyzing a data set that includes more than 35 billion healthcare events and over 15.7 billion insurance claims. The project's intent is to find such disparities so that they can be addressed. The analysis includes a review of economic vulnerability, education levels/gaps, race/ethnicity and household characteristics on about 125 million unique de-identified individuals.

Mayo Clinic has also taken steps to make certain any population health analysis it performs remains secure and that patients' privacy is maintained. In partnership with the data analytics firm nference, we have developed a de-identification approach that uses a protocol on EHR clinical notes that includes attention-based deep learning models, rule-based methods, and heuristics. Murugadoss et al. explain that "rule-based systems use pattern matching rules, regular expressions, and dictionary and public database look-ups to identify PII [personally identifiable information] elements" [34]. The problem with relying solely on such rules is they miss things, especially in an EHR's narrative notes, which often use non-standard expressions, including unusual spellings, typographic errors and the like. Such rules also require a great deal of time to manually create. Similarly, traditional machine learning based systems, which may rely on support vector machine or conditional random fields, have their shortcomings and tend to remain reliable across data sets.

The ensemble approach used at Mayo includes a next generation algorithm that incorporates natural language processing and machine learning. Upon detection of PHI, the system transforms detected identifiers into plausible, though fictional, surrogates to further obfuscate any leaked identifier (See Fig. 16.2). The project investigators evaluated the system with a publicly available dataset of 515 notes from the I2B2 2014 de-identification challenge and a

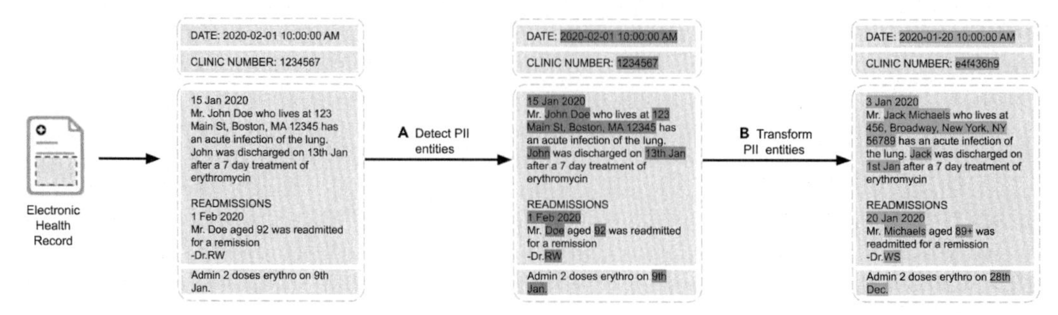

FIG. 16.2 Automated de-identification of electronic health records. Two steps in automated de-identification of EHRs: (A) Detecting PII entities and (B) transforming them by replacement with suitable surrogates. PII, personally identifiable information. *Source: Murugadoss K, Rajasekharan A, Malin B, et al. Building a best-in-class automated de-identification tool for electronic health records through ensemble learning. Patterns 2021. https://doi.org/10.1016/j.patter.2021.100255.*

dataset of 10,000 notes from Mayo Clinic. They compared that approach with other existing tools considered best-in-class. The results indicated a recall of 0.992 and 0.994 and a precision of 0.979 and 0.967 on the I2B2 and the Mayo Clinic data, respectively.

While this protocol has many advantages over older systems, it's only one component of a more comprehensive system used at Mayo to keep patient data private and secure. Experience has shown us that de-identified PHI, once released to the public, can sometimes be re-identified if a bad actor decides to compare these records to other publicly available data sets. There may be obscure variants within the data that humans can interpret as PHI but algorithms will not. For example, a computer algorithm expects phone numbers to be in the form of an area code, prefix, suffice i.e. (800) 555-1212. What if a phone number is manually recorded into a note as 80055 51212? A human might dial that number to re-identify the record. Further we expect dates to be in the form mm/dd/yyyy. What if a date of birth is manually typed into a note as 2104Febr (meaning 02/04/2021)? A de-identification algorithm might miss that.

With these risks in mind, Mayo Clinic is using a multi-layered defense referred to as data behind glass. The concept of data behind glass is that the de-identified data is stored in an encrypted container, always under control of Mayo Clinic Cloud. Authorized cloud sub-tenants can be granted access such that their tools can access the de-identified data for algorithm development, but no data can be taken out of the container. This prevents merging the data with other external data sources.

These security protocols are part of a larger system called the Mayo Clinic Platform (MCP), a coordinated portfolio approach to leverage emerging technologies, including AI, connected health care devices, and natural language processing to improve patient care. MCP projects span several domains, including virtual care, clinical data analytics, and remote diagnostic services. The recently published EAGLE trial demonstrates one of the tangible benefits of using this kind of platform approach to healthcare. The EAGLE trial combined a convolutional neural network with routine ECGs to detect low ejection fraction, a signpost for asymptomatic left ventricular systolic dysfunction [34].

The new algorithm, a joint effort between several of Mayo Clinic's clinical departments and Mayo Clinic Platform, included over 22,000 patients, divided into intervention and control groups and managed by 358 clinicians from 45 clinics and hospitals. The algorithm/ECG was used to evaluate patients in both groups, but only those clinicians allocated to the intervention arm had access to the AI results when deciding whether or not to order an echocardiogram. In the final analysis, 49.6% of patients whose physicians had access to the AI data underwent echocardiography, compared to only 38.1% (Odds ratio 1.63, $P < 0.001$). Xiaoxi Yao, with the Kern Center for the Science of Health Care Delivery, Mayo Clinic, and associates reported that "the intervention increased the diagnosis of low EF in the overall cohort (1.6% in the control arm versus 2.1% in the intervention arm) and among those who were identified as having a high likelihood of low EF." Using the AI tool enabled primary care physicians to increase the diagnosis of low EF overall by 32% when compared to the diagnosis rate among patients who received usual care. In absolute terms, for every 1000 patients screened, the AI system generated five new diagnoses of low EF compared to usual care.

Earlier research on the neural network used to create this AI tool has shown that it's supported by strong evidence. A growing number of thought leaders in medicine have criticized the rush to generate AI-based algorithms because many lack a solid scientific foundation required to justify their use in direct patient care. As we discussed earlier in the chapter, criticisms leveled at AI developers include concerns about algorithms derived from a dataset that is not validated with a second, external dataset, overreliance on retrospective analysis, lack of generalizability, and various types of bias, issues that we discuss in *The Digital Reconstruction of Healthcare* [28]. The EAGLE trial investigators addressed many of these concerns by testing their algorithm on more than one patient cohort. An earlier study used the tool on over 44,000 Mayo Clinic patients to train the convolutional neural network and then tested it again on an independent group of nearly 53,000 patients. And while this study was retrospective in design, other studies have confirmed the algorithm's value in clinical practice by using a prospective design. The most recent study by Yao et al. [35] was not only prospective in nature, it was also pragmatic, which reflects the real world in which clinicians practice. Traditional randomized controlled trials consume a lot of resources, take a long time to conduct, and usually include a long list of inclusion and exclusion criteria for patients to meet. The EAGLE trial, on the other hand, was performed among patients in everyday practice.

In addition to the EAGLE trial, MCP is developing a suite of digital tools designed to improve the ability of healthcare providers and payors to conduct more effective population analytics. Broadly speaking, these tools can be divided into four functional capabilities: Gather, Discover, Validate, and Deliver.

Gather's intent is to collect, harmonize, curate, and store patient data, allowing internal and external users to reduce the amount of time spent aggregating disparate data and improving data security and privacy.

Discover facilitates the development of algorithms and actionable insights. Products and services generated from this functionality have the potential to expand algorithm application to greater scope of use cases and can be used to improve the drug discovered process.

Validate will increase the speed with which an algorithm can reach the market by validating a proposed product.

Deliver refers to the delivery of insights into users' workflow, a critically important pain point for many algorithms as developers struggle to fit the tool into users' busy schedules. The goal is to free up providers from administrative work so they can spend more time in patient care. The combined algorithm/ECG tool used in the EAGLE trial is one such example.

As the quote from Dr. Reider at the beginning of the chapter pointed out, clinical decision support is how we distribute knowledge from where it *is* to where it *needs* to be. In the case of population health analytics, that knowledge is available from patient surveys, EHR records, clinical registries and a variety of other sources. But for these data sources to have a meaningful impact on population health, they must be derived from generalizable, unbiased data sets, incorporate social determinants of health, and respect patient privacy through the use of properly de-identified protocols. Our patients deserve nothing less.

References

[1] Centers for Disease Control and Prevention. What is population health? October 6, 2020, https://www.cdc.gov/pophealthtraining/whatis.html.
[2] Duke/Fuqua School of Business. Population health and the role of health analytics., Feb 13, 2020, https://www.fuqua.duke.edu/programs/msqm-health-analytics/articles/population-health-analytics/.
[3] Flaster A, Just E. Succeeding in population health management: Why the right tools matter. Health Catalyst; April 28, 2019. https://www.healthcatalyst.com/insights/population-health-management-solution-top-must-haves.
[4] McInturff, B, Roberts M. The executive summary of the HCAHPS redesign test. Federation of American Hospitals [Accessed 27 April 2021]. https://cdn720.s3.amazonaws.com/fah/documents/FAH_Executive_Summary_d2.pdf.
[5] Hospital Consumer Assessment of Healthcare Providers and Systems: CAHPS Hospital Survey., March 2021, https://www.hcahpsonline.org/globalassets/hcahps/survey-instruments/mail/qag-v16.0-materials/updated-materials/2021_survey-instruments_english_mail_updated.pdf. [Accessed 21 June 2021].
[6] Evans R, Berman S, Burlingame E, Fishkin S. It's time to take patient experience measurement and reporting to a new level: Next steps for modernizing and democratizing national patient surveys. Health Affairs Blog; March 16, 2020. https://www.healthaffairs.org/do/10.1377/hblog20200309.359946/full/.
[7] Sidorkiewicz S, Malmartel A, Prevost L, et al. Patient-physician agreement in reporting and prioritizing existing chronic conditions. Am Fam Med 2019;17:396–402.
[8] Staff News Writer. 5 things to know about clinical data registries. American Medical Association; Oct 15, 2014. https://www.ama-assn.org/practice-management/digital/5-things-know-about-clinical-data-registries.
[9] Gliklick RE, Leavy MB, Dreyer NA. Registries for evaluating patient outcomes: A user's guide. 4th ed; 2020. https://effectivehealthcare.ahrq.gov/products/registries-guide-4th-edition/users-guide.
[10] Low LL, Lee KH, Ong ME, et al. Predicting 30-day readmissions: performance of the LACE index compared with a regression model among general medicine patients in Singapore. Biomed Res Int 2015;2015:169870. https://www.ncbi.nlm.nih.gov/pmc/articles/PMC4670852/.
[11] Wu E, Wu K, Daneshjou R, et al. How medical AI devices are evaluated: limitations and recommendations from an analysis of FDA approvals. Nat Med 2021. https://doi.org/10.1038/s41591-021-01312-x. Published online April 5, 2021.
[12] Cerrato P, Halamka J. Reinventing clinical decision support. Boca Raton, FL: Taylor & Francis/HIMSS; 2020. p. 64.
[13] Banda J, Sarraju A, Abbasi F, et al. Finding missed cases of familial hypercholesterolemia in health systems using machine learning. Digit Med 2019;2:23. https://doi.org/10.1038/s41746-019-0101-5.
[14] Zech JR, Badgeley MA, Liu M, et al. Variable generalization performance of a deep learning model to detect pneumonia in chest radiographs: a cross-sectional study. PLOS Med 2018;15:e1002683. https://doi.org/10.1371/journal.pmed.1002683.

[15] Wong A, Otles E, Donnelly JP, et al. External validation of a widely implemented proprietary sepsis prediction model in hospitalized patients. JAMA Intern Med 2021;181(8):1065–70.
[16] Halamka J, Cerrato P. The digital reconstruction of health care. NEJM Catal 2020;1(6). https://doi.org/10.1056/CAT.20.0082.
[17] Wang P, et al. Lancet Gastroenterol Hepatol 2020;5(4):343–51. https://pubmed.ncbi.nlm.nih.gov/31981517/.
[18] Gong D, et al. Lancet Gastroenterol Hepatol 2020;5(4):352–61. https://pubmed.ncbi.nlm.nih.gov/31981518/.
[19] Wu L, et al. Gut 2019;68(12):2161–9. https://pubmed.ncbi.nlm.nih.gov/30858305/.
[20] Wang P, et al. Gut 2019;68(10):1813–9. https://pubmed.ncbi.nlm.nih.gov/30814121/.
[21] Lin H, et al. EClinicalMedicine 2019;52–9. https://doi.org/10.1016/j.eclinm.2019.03.001.
[22] Abràmoff M, et al. npj Digit Med 2018;1:39. https://doi.org/10.1038/s41746-018-0040-6. eCollection 2018 https://pubmed.ncbi.nlm.nih.gov/31304320/.
[23] Gulshan V, et al. JAMA Ophthalmol 2019. https://doi.org/10.1001/jamaophthalmol.2019.2004 [Epub ahead of print] https://pubmed.ncbi.nlm.nih.gov/31194246/.
[24] Kanagasingam Y, et al. JAMA New Open 2018;1(5):e182665. https://doi.org/10.1001/jamanetworkopen.2018.2665. https://www.ncbi.nlm.nih.gov/pmc/articles/PMC6324474/.
[25] Long E, et al. Nat Biomed Eng 2017;1:0024. https://doi.org/10.1038/s41551-016-0024.
[26] Steiner DF, et al. Am J Surg Pathol 2018;42(12):1636–46. https://doi.org/10.1097/PAS.0000000000001151. https://pubmed.ncbi.nlm.nih.gov/30312179/.
[27] Mori Y, et al. Ann Intern Med 2018;169:357–66. https://doi.org/10.7326/M18-0249. https://pubmed.ncbi.nlm.nih.gov/30105375/.
[28] Cerrato P, Halamka J. The digital reconstruction of healthcare. Boca Raton, FL: CRC Press; 2021.
[29] Livingston E. Machine learning for medical image analysis: How it works., Sept 18, 2018, https://www.youtube.com/watch?v=VKnoyiNxflk.
[30] DeGrave AJ, Janizek JD, Lee SI. AI for radiographic COVID-19 detection selects shortcuts over signal. Nat Mach Intell 2021;3:610–9.
[31] Ghorbani A, Ouyang D, Abid A, et al. Deep learning interpretation of echocardiograms. npj Digit Med 2020;3:10.
[32] Chigudu S. Book: an ironic guide to colonialism in global health. Lancet 2021;397:1874–975.
[33] Centers for Disease Control and Prevention. Electronic health records (EHRs) and patient work information., March 3, 2020, https://www.cdc.gov/niosh/topics/ehr/.
[34] Murugadoss K, Rajasekharan A, Malin B, et al. Building a best-in-class automated de-identification tool for electronic health records through ensemble learning. Patterns 2021. https://doi.org/10.1016/j.patter.2021.100255.
[35] Yao X, Rushlow D, Inselman JW, et al. Artificial intelligence–enabled electrocardiograms for identification of patients with low ejection fraction: a pragmatic, randomized clinical trial. Nat Med 2021;27:815–9.

CHAPTER

17

Expanded sources for precision medicine

Darren K. Johnson and Marc S. Williams

Department of Genomic Health, Geisinger, Danville, PA, United States

17.1 Introduction

If the 19th century was the Industrial Age and the 20th the Atomic Age, the 21st century may be remembered as the Age of Precision Medicine. While this may seem presumptuous given that we are only 2 decades into the century, the dramatic breakthroughs in genomics, informatics, and other enabling technologies culminating in the announcement by President Obama at the 2015 State of the Union address that called for investment in a large-scale precision medicine initiative, the All of Us program [1] would seem to make this assertion plausible. Despite the emergence of the concept of precision medicine into both research and clinic practice, there is still confusion about what the term precision medicine actually means.

17.1.1 What is precision medicine?

There are several terms that are sometimes used interchangeably, genomic medicine, personalized medicine, precision medicine, and precision health. Precision medicine is emerging as the most commonly used term, however there are distinctions that are useful to understand.

- **Genomic Medicine.** The National Human Genome Research Institute (NHGRI) defines genomic medicine as, "… an emerging medical discipline that involves using genomic information about an individual as part of their clinical care (e.g., for diagnostic or therapeutic decision-making) and the health outcomes and policy implications of that clinical use." [2] Genomic information broadly encompasses information derived from the DNA in the genome, including:

Clinical Decision Support and Beyond
https://doi.org/10.1016/B978-0-323-91200-6.00028-0

  - Single nucleotide variation (changes in single DNA bases that vary from individual to individual).
  - Structural variation (larger changes in DNA structure including insertion, duplication, or deletion of multiple DNA bases, rearrangements in the order of the bases such as a chromosomal inversion, and translocation of genetic material to another chromosomal location).
  - Somatic variation, where DNA changes occur in some cells but not others. This variation is a primary driver for cancer. Mosaicism refers to somatic DNA changes that occur before birth and affect only some tissues.
  - Epigenetics refers to modifications in DNA that alter the expression of genes but does not change the DNA sequence.
  - Genomic medicine also encompasses information obtained from family history. Family history captures information about shared genetic predispositions, but also includes the impact of shared environmental factors that can influence the expression of the genetic predispositions. Geneticists sometimes say that genetics loads the gun, but environment pulls the trigger. Obtaining a good family history can be very helpful in understanding individual risk for developing a condition.
- **Personalized Medicine.** Many definitions have been proposed however, a definition attributed to Stephen Pauker and Jerome Kassirer best encapsulates the key points encompassed by this term. "Personalized medicine is the practice of clinical decision-making such that the decisions made **maximize the outcomes that the patient most cares about and minimizes those that the patient fears the most**, on the basis of **as much knowledge about the individual's state as is available**." There are three key points captured by this definition. First is the focus on the outcomes of care. Second is the central role of the patient in defining what outcomes, positive or negative, are most important. Third, the words "genetic" or "genomic" do not appear only "... as much knowledge about the individual's state as is available." There are no assumptions that genetic or genomic information is superior to other information (since the definition was in fact proposed in 1987, well before genomic information was readily available). The discussion of personalized and precision medicine has been dominated by new technologies, with less attention paid to the critical role of the patient to define the desired outcomes of care from their perspective.
- **Precision Medicine.** Medicine as currently practiced is empiric and dependent on how much knowledge and experience the individual clinician has, leading to care that has high variability and sub-optimal outcomes. Christensen et al. [3] refer to this as "intuitive medicine" defined as "care for conditions that can be diagnosed only by their symptoms and only treated with therapies whose efficacy is uncertain." The same authors present the **vision of precision medicine** as "The provision of care for diseases that can be precisely diagnosed, whose causes are understood, and which consequently can be treated with rules-based therapies that are predictably effective." While some conflate genomic medicine with precision medicine, genomics is a subset of information used to inform precision care in conjunction with other information. **Precision health** encompasses precision medicine but extends it beyond the identification and treatment of disease, to emphasize the role of genomics and other technologies in prevention and health maintenance.

17.1.2 Is precision medicine a new idea?

While the use of genomic sequence variation in clinical care is relatively recent—dating to the discovery of chromosomal copy number variants' association with disease in the late 1950s, the underlying concepts of genetics, genomics, personalized, and precision medicine have been around for millennia.

Hippocrates of Kos (460–370 B.C.E.) stated that "it is more important to know what kind of person suffers from a disease than to know the disease a person suffers." This ancient statement encapsulates the essence of modern patient-centered personalized medicine. One of the first applications of personalized/precision medicine is the 1930 discovery of ABO blood groups by Karl Landsteiner that allowed safe transfusion of blood products for the first time. This early use of immune biomarkers has been greatly expanded to make possible the field of organ transplantation.

The first documented use of the word genetic was in 1831 by Carlyle who defined it as "pertaining to origins". The first use of it in a biological context was by Darwin in 1859 as "resulting from common origin". The modern sense of the word, "pertaining to genetics or genes", dates from 1908, while genome (a portmanteau word generated by combining genetic and chromosome) was first documented between 1925–1930 with genomic derived to mean the study of genomes.

However, the concept of inherited disorders (i.e., genetics) is much older. References to a genetic basis for some diseases date well into antiquity. The concept of genomic medicine stems from the second century C.E. The first documented application of individualized care based on presumed genetic information is documented in the Talmud (Yevamot 64b). Rabbi Judah the Prince (135–217 C.E.) ruled that if a woman's first two children died from blood loss after circumcision, the third son should not be circumcised. A second Rabbi disagreed and ruled that the third son may be circumcised; however, if this infant died, the fourth child should not be circumcised. There was agreement that abnormal bleeding was a hereditary trait with disagreement only regarding how many events were required to establish a pattern and therefore exempt a child from circumcision. Inheritance of certain conditions in royal lineages such as the Hapsburg lip and hemophilia were well documented although the underlying mechanism was unknown at that time. The first description of genetics in a modern medical context appears in the Croonian lectures presented by Sir Archibald Garrod in 1908 [4]. In these lectures Garrod articulated the principles of the practice of medical genetics through the lens of four inborn errors of metabolism: Pentosuria, Albinism, Cystinuria, and Alkaptonuria. Garrod was likely the first to recognize that these disorders followed patterns of inheritance first reported by Mendel and recognized the risk for recurrence in families.

17.1.3 Realizing the vision of precision medicine and health

It is evident that precision medicine and health are reliant on the generation, transformation, and intelligent use of disparate data elements. It is also known that this alone is not sufficient to drive the successful implementation of knowledge into care. Current healthcare systems as currently realized, are ill-equipped to deliver precision health at the individual level as treatment and prevention are based on studies in heterogeneous populations, that mask individual differences. This problem is exacerbated in the United States where, absent a national health service, there is high variability in application of even evidence-based population interventions. In 2007, the National Academy of Medicine (formerly the Institute of Medicine) published the first workshop summary defining the Learning Healthcare System [5], as "science, informatics, incentives, and culture are aligned for continuous improvement and innovation, with best practices seamlessly embedded in the delivery process and new knowledge captured as an integral by-product of the delivery experience." This approach contains the components needed to synthesize complex and disparate information and present this information to the clinician and patient at the time of clinical decision-making in a reliable and reproducible fashion.

Anticipating the emergence of genomics into practice and in recognition of the challenges identified by prior efforts, in 2015 the National Academy of Medicine published a workshop summary describing Genomics-Enabled Learning Health Care Systems [6]. While much of the focus was on the electronic health record (EHR) and data management (points that will be further explored below), on page 19 Friedman notes that a Learning Healthcare System is, "A health care system in which an infrastructure supports complete learning cycles that encompass both the analysis of data to produce results and the use of those results to develop changes in clinical practices is a system that will allow for optimal learning." This is a description of a system that supports a virtuous cycle that allows discoveries with sufficient evidence to be placed into clinical settings coupled with systems to capture relevant outcomes which can be examined for improvement opportunities that can be designed and implemented in clinical practice. These rapid improvement cycles can be iterated repeatedly until outcomes are optimized. Creation of a Genomic Learning Healthcare System is a central piece of the 2020 NHGRI strategic plan (Fig. 17.1).

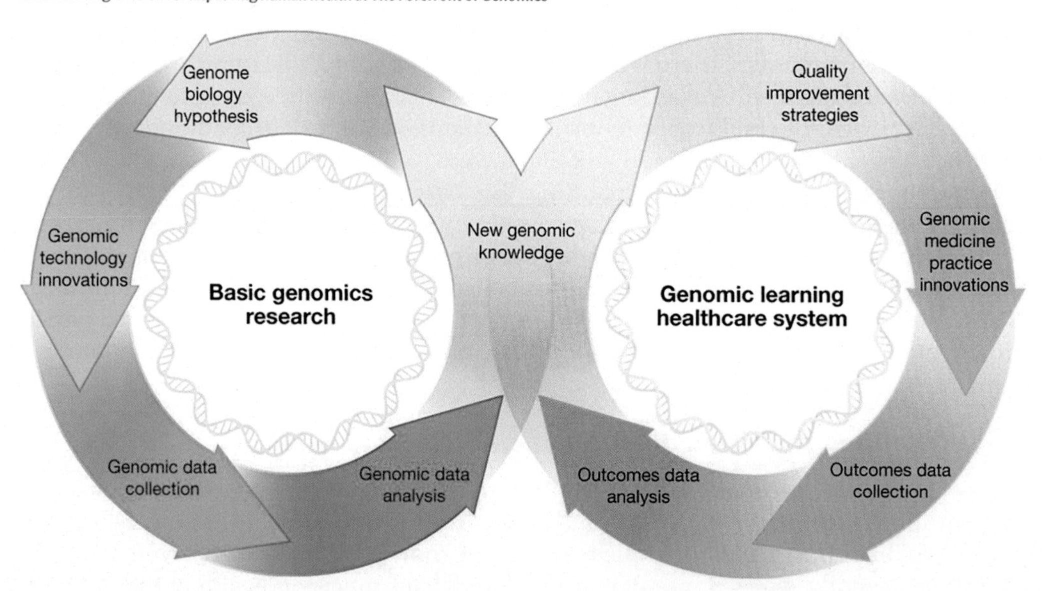

FIG. 17.1 Virtuous cycles in human genomics research and clinical care From: Strategic Vision for improving human health at The Forefront of Genomics *Credit: TBD.*

It is apparent that realizing the vision of precision health and the genomic learning healthcare system will require sophisticated informatics tools and comprehensive knowledge repositories intelligently interfacing with clinicians in the informatics ecosystem. In this chapter, we will examine the opportunities and challenges to developing and implementing information systems that can support the successful implementation of precision medicine in the clinic.

17.2 Current state of decision support in clinical genomics

17.2.1 Desiderata of CDS/EHR for clinical genomics

The effective use of sequencing information is hindered by inadequate laboratory reporting methods, insufficient numbers of trained professionals to interpret genetic information, and the overall complexity of genomic data. The rapid pace of change in genomics impacts our understanding of how that same genetic information should impact clinical care. Clinical Decision Support (CDS) tools utilizing genomic information can inform the clinical workflow and provide solutions, especially when integrated into the EHR [7]. Masys et al. [8] (#1–7) and Welch et al. [9] (#8–14) developed a set of implementation principles to help integrate genomic information into the EHR. The following are a summary of the desiderata.

1. **Maintain separation of primary molecular observations from the clinical interpretations of those data**
   - **(a)** Most biomolecular data for genomics are variants of unknown significance. Separation of the initial molecular observations allows for determination of significance in the future while also understanding that novel approaches to dynamic reporting of genomic information residing in knowledge bases are still being developed for implementation in the clinical environment.
2. **Support lossless data compression from primary molecular observations to clinically managed subsets**
   - **(a)** Healthcare institutions rarely store the complete raw reads of an individual's DNA sequence as most of the information is not used for clinical care, and the storage requirements are untenable. One solution for the portion of the data that is relevant to clinical care is to store this small subset of information in the EHR. However, given that more and more of the sequence will be useful to inform care in the future, access to the information through an ancillary system is desirable [10]. However, this still requires a significant amount of storage. Use of lossless compression can reduce the demand for storage while retaining information that could be clinically useful in the future. Genetic interpretation of these data is often reliant on a single change in one location of the genetic alphabet (a single nucleotide variant or SNV) thus any data compression must be able to reproduce an exact copy of the original sequence.
3. **Maintain linkage of molecular observations to the laboratory methods used to generate them**
   - **(a)** The methods used for measurement each have strengths and weaknesses which evolve as the technologies advance for reading and interpreting genomic sequences. Linking the observations to their initial laboratory methods allows data scientists to

better understand the pathway of the genomic data and clinicians to determine if the test method was adequate to detect the genetic change for a suspected condition. As an example, in a patient who presents with chorea, diagnosis of Huntington disease requires the ability to detect and quantify the number of trinucleotide repeats in the *HTT* gene. If the patient had prior exome sequencing, a review of the laboratory method would indicate whether the pathognomonic repeat expansion was able to be detected. With few exceptions, exome sequencing would not detect the repeat expansion, so clinical testing for Huntington disease is indicated.

4. Support compact representation of clinically actionable subsets for optimal performance
   (a) It is unreasonable to expect a query of an entire genome in real time for healthcare related decisions. Compact representations of clinically relevant information that can be queried in a timeframe consistent with an episode of care will allow clinical data to be displayed and utilized in real time within an EHR.

5. Simultaneously support human-viewable formats and machine-readable formats in order to facilitate implementation of decision support rules
   (a) Multiple molecular variations may contribute to disease risk and impact clinical outcomes. With rapidly evolving literature in genomics, there must be a way to build and implement automated methods to update databases and knowledge repositories that store and curate genetic variation data while supporting a human-readable format (ideally accompanied by CDS) that will present information to clinicians in a useable format with accompanying information that can guide care for conditions that they might not be familiar with.

6. Anticipate fundamental changes in the understanding of human variation
   (a) At present, our understanding of genetic variation is still quite limited. Of the estimated 20,000 genes in the human genome, we understand the function and associated diseases for fewer than one-third. Even at our current rate of discover of about 250 new gene-disease associations each year [11], with about 10–15,000 genes of unknown function, it will take 25–50 years of discovery if the pace doesn't change. Our knowledge of the variation in these genes is even less well developed. Any system developed to support the use of genomic information in clinical care must be flexible enough to accommodate new information rapidly.

7. Support both individual care and discovery science
   (a) This recommendation has been included in the 2020 NHGRI strategic plan referenced above and depicted in Fig. 17.1. Genetic discovery for a population merges the fields of clinical care and biomedical research. Every individual has some unique combination of DNA sequence variation that can contribute to the overall understanding of health and disease in the population. This type of learning system does raise issues about consent and individual privacy that are not well addressed by current practices. Successful planning for use of genomic information allows for use both in research, and in the personalized clinical care of an individual. Creation of the virtuous cycle accelerates discovery and facilitates implementation of new knowledge into clinical care.

8. CDS knowledge must have the potential to incorporate multiple genes and clinical information
   (a) Multiple genetic loci can often contribute to a specific phenotype or disease. To provide an accurate risk assessment or clinical interpretation, all relevant genetic

information along with phenotypic and other clinical data need to be considered. The volume and complexity of information requires computerization and associated decision support to optimize outcomes.

9. Keep CDS knowledge separate from the variant classification
  (a) CDS must have the ability to adapt as genetic variant interpretations change without requiring changes to the CDS architecture. By separating the CDS from variant classifications, the CDS knowledge can be utilized more effectively.

10. Have the capacity to support multiple EHR platforms with various data representations with minimal modification
  (a) Health information systems are each unique, storing and representing health information differently even when built to recommended standards. Ideally, CDS architecture would be based on generalizable standards that would preserve fidelity and reduce the need for customization when implemented in a system. Approaches such as SMART on FHIR [12] CDS Hooks [13] and Sync for Genes [14] are trying to achieve this objective.

11. Support a large number of gene variants while simplifying the CDS knowledge to the extent possible
  (a) With thousands of known disease-causing variants for specific genes, and anticipating that the knowledge around genes and variants will rapidly expand, variants with similar clinical impact should be classified in a similar way. This will allow one CDS tool to manage multiple variants. An example would be pathogenic variants in *BRCA1* associated with hereditary breast and ovarian cancer syndrome (HBOC). A single CDS rule (or, more accurately, one rule for men and another for women) would be able to accommodate existing and newly discovered variants that are curated as pathogenic. These CDS rules could also be used for other genes that have similar phenotypic consequences to *BRCA1*, such as *BRCA2* and *PALB2*. However, this approach is not always appropriate given that some genes are pleiotropic, that is variation of different types, or in different domains of the gene have different phenotypic consequences. Variation in the gene *LMNA* can cause cardiomyopathy, gray matter heterotopias, lipodystrophy, and other disorders depending on the type and location of the variation. CDS systems will have to be able to support these use cases and others that emerge.

12. Leverage current and developing CDS and genomic standards
  (a) There has been extensive research around the development and implementation of genomic standards. This area continues to mature but concerted efforts by standards organizations like HL7 [15] and organizations focused on genomic use cases, such as the Global Alliance for Genomics and Health (GA4GH) [16] are refining existing standards and filling gaps to accelerate the representation of genomic information and associated CDS in electronic health records and laboratory information systems.

13. Support a CDS knowledge base deployed at and developed by multiple organizations
  (a) One healthcare organization will be unable to shoulder the burden for creation of the entire CDS knowledge necessary for genomic medicine. This will require a system that allows CDS knowledge to be shared among institutions. Early examples of guideline and CDS repositories such as CPIC and CDS-KB (see knowledge repositories below) are emerging into use.

14. Access and transmit only the genomic information necessary for CDS
  (a) To satisfy both HIPAA and the processing capacity of CDS, only relevant genetic information should be accessed and sent to a CDS rules engine. This would be limited to relevant genes, molecular observations, variant interpretations, and associated phenotypic and clinical information.

These desiderata created a foundation for integrating genome sequencing information with CDS. They are not considered to be a final authoritative set and will need to be modified and augmented based on new knowledge. In 2021, the NHGRI hosted their thirteenth Genomic Medicine meeting titled "Developing a Clinical Genomic Informatics Research Agenda" (GMXIII) [17]. This meeting extended findings from a prior meeting also hosted by NHGRI focused on genomic CDS (GMVII) [18]. As part of both meetings, attendees were surveyed on these 14 desiderata and asked to rank their importance. In the 5 ½ years between the two meetings, some priorities of the desiderata were maintained. Two desiderata, both related to genomic CDS were deemed highly important, "CDS knowledge must have the potential to incorporate multiple genes and clinical information" and "CDS knowledge must have the capacity to support multiple EHR platforms with various data representations with minimal modification." Two CDS desiderata were deemed less important in 2021, "Support a CDS knowledge base deployed at and developed by multiple independent organizations" and "Simultaneously support human-viewable formats and machine-readable formats to facilitate implementation of decision support rules." The reasons for the change in importance are still being explored. The meeting also identified high priority topics that are related to the existing desiderata in some cases but extend the scope and propose new areas to explore. These topics include:

- Importance of assessing stakeholder preference and workflow
- Sustainability of resources
- Lack of methods for evaluation of innovation and implementation
- Impact of the consent and regulatory framework

Attendees also emphasized the need to address issues of equity and diversity, a topic that will be discussed later in this chapter.

These topics indicate that the emergence of precision medicine and health into the clinic has accelerated in the last 5 years emphasizing the need for research incorporating the tools of implementation science if genomic CDS is to be successfully implemented in the clinic.

17.2.2 Current state of CDS for precision medicine and health

- Pharmacogenomics CDS: Pharmacogenomics (PGx) is the influence of individualized genetic information on medication responses. The response could include therapeutic effects as well as adverse drug reactions. Pharmacogenomics CDS integrated into the EHR provides a more informative level of individualized care as clinicians are presented with personalized information about their patients to inform medication choice and dosage. This knowledge can help a clinician identify individuals at risk for specific medications or point the clinician towards a more efficacious medication. Commercial EHR vendors have not implemented pharmacogenomics, although some components to support implementation are appearing in some systems. This has resulted in a reliance on local customized solutions that have provided some evidence of effectiveness, but are not generalizable or, in most cases, sustainable without significant dedicated resources. To lower these barriers the NHGRI-Funded Implementing GeNomics In pracTicE (IGNITE) Network has made available a set of implementation guides to support pharmacogenomics [19].

Several institutions have implemented PGx CDS into their EHRs [20–26] Analysis of the impact of these implementations has identified recurring issues allowing for synthesis of recommendations that can be used to inform future implementations whether local, or through a generalizable approach tethered to vendor-based systems. These include the necessity for convening a multidisciplinary team that includes a full range of stakeholders and end users to develop and implement solutions, define and measure outcomes, and continuously improve the systems iteratively to optimize the health outcomes and user experience. One specialty does not have the breadth of expertise necessary to carry a full PGx project from development through implementation. Clinical domain experts should be invited to participate to identify the most significant medical problems, decrease clinical resistance, and inform the process and workflow which helps to determine the type of CDS tool that would be most appropriate. Alerts, while the most commonly implemented form of CDS, may not be the most effective, especially given the general dislike of alerts and reminders by clinicians and the risks associated with alert fatigue. Innovative approaches to CDS are needed and substantive engagement with end users is needed to inform the development and testing of novel approaches. As noted in the previous section, information contained in PGx reports must be represented as structured data yet retain human readability. These discrete data are critical for building CDS tools within the EHR. As with all genetic information, there must be a way to update the PGx CDS as our understanding of genomic information and PGx guidelines change.

PGx genes reported by laboratories are not all equal in clinical evidence. This requires curated knowledge repositories for PGx data (see PGRN and CPIC sections below) and clinical experts at the institution level to determine which CDS tools to support PGx should be configured, at which points the CDS should be triggered, and the information that should be presented to clinicians who may be less familiar with pharmacogenomics.

- Newborn screening (NBS): NBS is the only screening program primarily focused on genetic and inherited disease that is fully implemented in the United States and associated territories. With origins going back to the 1960s, it is one of the most successful public health screening programs in the United States [27]. Each year approximately 22,000 newborns (about 1 in 294) are identified with a disorder through the program, the majority of which are genetic [27]. The number of screened conditions has increased from 1 at its inception in 1963 to 81 in some states, with 75% (61/81) of these conditions recommended for inclusion on a uniform screening panel (RUSP) by a Federal advisory committee [28]. Screening methods encompass a variety of modalities including biomarker assays (T4, immunoreactive trypsinogen), *in vitro* cellular response (T-cell receptor excision circle), tandem mass spectrometry (metabolites for aminoacidopathies and fatty acid oxidation defects), genetic testing (*CFTR* for Cystic Fibrosis and *SMN1* for Spinal Muscular Atrophy), and point of care physiological assessments (pulse oximetry, Brainstem Auditory Evoked Potentials) [29].

The initial criteria for inclusion in a newborn screening program required onset in or shortly after the neonatal period (usually including a lack of overt signs of disease for a period of time) and specific effective treatment that prevented or significantly reduced morbidity and mortality [30]. However, the use of multiplex technologies such as tandem mass spectrometry and massively parallel sequencing has allowed identification of conditions with onset outside the neonatal period, expanding the potential reach of these programs. To fully realize the potential benefit of newborn screening requires not only identification of a potentially affected newborn, but effective and timely communication of this information to the newborn's parents, medical provider, and the healthcare system with rapid initiation of condition-specific interventions and referral to appropriate specialists for ongoing care. While the RUSP provides guidance at a national level, newborn screening programs are administered at the state and territorial level, meaning there is significant variability from state to state [31]. These conditions are rare and are, for the most part, unfamiliar to most primary care physicians. This combination of characteristics represents a significant opportunity for the use of CDS.

ACMG ACT Sheets [32]. Beginning in 2001, the American College of Medical Genetics and Genomics developed ACTion sheets (ACT Sheets) to support clinicians caring for a newborn with a positive newborn screening test. The ACT sheets consist of a narrative (human readable) description of the condition and the recommended actions to be taken based on the positive screening test. This represents an L1 (narrative, human readable) CDS artifact. ACT sheets also include a visual representation of the appropriate actions in a flow chart (an L2 CDS artifact). This is human readable but has the potential to be developed as a computable decision support artifact that could be used to trigger clinician actions such as an order set, referrals, etc. Lastly, the ACT sheets include a section that is customizable to allow inclusion of unique links to local and state resources relevant to the specific newborn screening program.

Newborn screening ACT sheets were included in the Agency for Healthcare Research and Quality's (AHRQ) 2013 release package and the 2015 priority list as a Normative Statement requirement type (SHALL) under Req-818 [33] and Req-2018 [34]. The AHRQ notes that these are available as downloadable PDFs but are not yet instantiated as computable CDS by EHR vendors. This reflects the current status of these artifacts. Some newborn screening laboratories distribute the relevant ACT sheet either electronically or as a physical copy with positive screening results, but these are not transferred directly into the EHR. Some institutions have promoted access to ACT sheets either by downloading them to a local knowledge repository or by using links in the EHR (such as infobuttons—see Chapter 16 for details) to access content on the ACMG site. ACT sheets have now been developed for conditions beyond newborn screening (See Knowledge Repositories below).

- Role of CDS in genetic testing for tailored disease risk prevention.

Genomic interventions, as with any new medical approach, will remain underutilized without a robust support structure. In the current practice of clinical care, only half of U.S. adults receive the recommended preventive care and fewer than half with a chronic disease receive guideline-recommended care [35]. Genomic interventions likely face greater barriers due to limited education and training for clinical personnel [36]. Utilizing CDS leverages the knowledge and condition-specific information for the stakeholders with relevant and pertinent clinical knowledge. For instance, a clinician may be alerted through a CDS that a patient has a pathogenic variant in *PCSK9* that confers risk for familial hypercholesterolemia (FH), a severe disorder of lipid metabolism. This information is clinically relevant, as an individual with FH has a 3-fold increased risk for the development of coronary artery disease compared to an individual without FH, even when controlled for the absolute level of LDL cholesterol [37]. CDS can provide this critical information to the clinician and present evidence-based cardiovascular disease and environmental preventive measures, specific to FH to be discussed with the patient. Condition-specific knowledge that could be presented through CDS could include:

- Differential risk compared with the general population: The genetic burden for coronary artery disease has been estimated at 40–60% [38]. There is a high positive predictive value when using pathogenic variants in major lipid-related genes [37] and lipid-related polymorphisms as a predictive factor for cardiovascular events [39]. As the number of cardiovascular disease variants increase, the risk increases as does the need for interventions.
- Treatment intervention: A recent clinical trial prescribing statins showed an absolute and relative benefit for individuals with the highest genetic risk of cardiovascular disease [39]. With research efforts expanding the understanding of underlying risk factors surrounding cardiovascular disease, clinical care can better identify appropriate treatment decisions in patients at risk. An example of this is the development of an entirely new class of drugs, PCSK9 inhibitors. These medications were developed based on discoveries about the protective effects of loss of function variants in *PCSK9* [40]. This provides new treatment options for patients with elevated LDL cholesterol and FH alike.
- Genetic testing for at risk family members: It's important to remember that single gene genetic disorders can also impact family members. This concept of the family as the patient is not readily appreciated by non-genetics professionals. In autosomal dominant conditions like FH, first degree relatives (i.e., parents, siblings, and children) have a 50% chance of carrying the same variant conferring increased risk of disease. If family members are tested (called cascade testing) gene carriers can be identified earlier in life, allowing condition-specific interventions of known effectiveness to be initiated [41]. CDS can be configured to present this information to clinicians and facilitate testing of at-risk relatives.

- Role of CDS in genetic testing to inform treatment

As evidence for improved outcomes using genomics in clinical care develops, extending CDS to genomic medicine can also inform clinical treatment at the point of care, as illustrated in the FH example in the previous section.

- A pharmacogenomic example which has been implemented in several systems, is a clinician-directed CDS alert for a patient with a recent vascular event being considered for treatment with clopidogrel discovered to have two loss of function *CYP2C19* variants. This means the patient is unable to convert clopidogrel from the pro-drug to the active agent and the prescription should be changed to an alternative P2Y12 inhibitor like prasugrel. Offering this recommendation automatically and within the clinician's workflow improves patient care [42].

- Genetic testing on tumors to direct treatment is becoming more routine in clinical care for cancer [43]. Increasing understanding of somatic tumor genetic profiles has led to universal screening protocols [44]. Many therapies can now target specific somatic variants. A robust CDS system built according to national standards and configured to represent evidence-based treatment guidelines could import results from tumor genetic testing and identify a genetic profile that could inform treatment. Presented within the EHR, this information could facilitate the provision of best care practices for the patient. Significant barriers and limitations for implementation of this process exist at this time and will be discussed below.
- Knowledge repositories

One major area of progress is in the availability of curated data sources and concerted efforts to transform these data into clinically useful knowledge. Through the efforts of the groups listed below and others, curated knowledge to support the use of genomic information in clinical care is more available than ever before. Unfortunately, these knowledge repositories are not integrated with laboratory information systems or EHRs requiring clinicians and laboratorians to disrupt their workflow to access the resources. Barriers and opportunities to improve the integration of the knowledge repositories within the informatics ecosystem will be discussed later in the chapter. What follows is a list of important genomic knowledge repositories with a brief description of their purpose. (see Chapter 11 for a broader discussion of knowledge resources for CDS beyond genomics).

- ACT sheets other than newborn screening [32]. Newborn screening ACT sheets are described above. Recognizing the utility of the ACT sheet, the ACMG has developed other ACT sheets to support care for other clinical genetics activities. These include: Carrier Screening, Diagnostic Testing for Duchenne Muscular Dystrophy and Fragile X syndrome, Family History collection to assess risk for hereditary colorectal cancer, Noninvasive Prenatal Screening for Chromosomal aneuploidy, Secondary Findings (the ACMG has recommendations for returning secondary or incidental findings from exome or genome sequencing that are medically important but not related to the test indication. Additional information is available at https://www.acmg.net/PDFLibrary/41436_2021_1171_OnlinePDF-1.pdf [45]), and Transition to Adult care for pediatric patients with genetic conditions. As with the Newborn Screening ACT sheets, these ACT sheets contain a narrative description of the condition and intervention (L1 artifact) and, a treatment flow chart where applicable (L2 artifact). To date, none of these has been integrated into either a laboratory information system, or EHR.
- ClinVar [13]. It is estimated that each of us carries between 3–10 million genetic variants in our genome. The vast majority of these have no impact on health and wellness. A smaller number contribute to disease risk in combination with other genetic changes (polygenic risk—see below). Only a fraction of the variants confers major risk of disease (e.g., a pathogenic variant in *BRCA1* which leads to substantial risk of developing breast, ovarian, and prostate cancer). When a variant is detected through genetic testing, interpretation of that variant is essential to understanding and conveying the consequences of the genetic variation to the patient. From the ClinVar website, "ClinVar is a freely accessible, public archive of reports of the relationships among human variations and phenotypes, with

supporting evidence. ClinVar thus facilitates access to and communication about the relationships asserted between human variation and observed health status, and the history of that interpretation." As of August 2021, almost 2000 submitters have deposited over 1.6 million interpreted variants in ClinVar, nearly 90% of which are accompanied by assertion criteria. To assist the use of the variant information, ClinVar reports the level of review supporting the assertion of clinical significance for each variant. These are visualized using a star system ranging from no stars (no assertion or assertion criteria submitted) to four stars (variant is explicitly included in a clinical practice guideline). To improve the number of variants with high quality assertions, expert curation panels have been established by groups such as ClinGen (see below). These panels develop criteria based on standard variant curation processes such as those published by the ACMG [46] but customized to incorporate gene-specific and variant-specific evidence which impact the curation. ClinVar data are accessible through the internet, and some research has been done to establish links to ClinVar data from the EHR using CDS Web services such as OpenInfobutton [47]. The lack of agreed upon standards for representation of genes, variants, and assertion has slowed the development of interfaces between ClinVar and EHRs and laboratory information systems. Groups such as the Global Alliance for Genomic Health (GA4GH) and the Health Level 7 (HL7) Clinical Genomics Working Group are working to develop and implement the necessary standards (see below).

- The Clinical Genome Resource (ClinGen) [13]. The success of ClinVar as a repository for knowledge about variant interpretation coupled with increased use of genetic testing in clinical settings identified significant gaps in available knowledge in several areas, most notably lack of a consistent approach to curate the association of genes with disease and the lack of collection of evidence of the effectiveness of interventions on outcomes of those at risk for a genetic disease based on presence of a gene variant. A meeting convened by NHGRI in 2011, "Characterizing and Displaying Genetic Variants for Clinical Action Workshop" [48,49] led to the funding of a cooperative agreement for ClinGen. From the ClinGen website, "ClinGen aims to create an authoritative central resource that defines the clinical relevance of genes and variants for use in precision medicine and research." The goals of ClinGen are to:

  - Share genomic and phenotypic data between clinicians, researchers, and patients through centralized and federated databases for clinical and research use.
  - Develop and implement standards to support clinical annotation and interpretation of genes and variants.
  - Develop data standards, software infrastructure and computational approaches to enable curation at scale and facilitate integration into healthcare delivery.
  - Enhance and accelerate expert review of the clinical relevance of genes and variants.
  - Disseminate and integrate ClinGen knowledge and resources to the broader community.

  Now in its third round of funding, ClinGen has developed standardized protocols for curation of gene-disease assertions, actionability of outcomes, impact of copy number variation, and other activities. ClinGen also supports expert panels involved in gene-disease curation, determination of actionability (for both children and adults) and variant curation expert panels that contribute 3-star variant assertions to ClinVar.

ClinGen and ClinVar goals are aligned and both projects play a critical role in the growing data sharing movement within the clinical genetics community.

During the first two phases of ClinGen the EHR Working Group explored using ClinGen resources, specifically the gene-disease validity curation, within the EHR as passive CDS [47]. While a test implementation in the developmental EHR was successfully piloted at Geisinger through use of OpenInfobutton, a full implementation was not achieved. Reasons for this included a lack of compatible standards for gene representation between ClinGen and the health record, challenges with integrating knowledge resources with the EHR through the application interfaces compliant with the HL7 Infobutton Standard implementing the open infobutton application interface (even though this was theoretically supported by the vendor system), and related issues [50,51]. Details about OpenInfobutton and the HL7 Infobutton Standard are available in Chapter 16.

- Online Mendelian Inheritance in Man (OMIM) [52]. OMIM may lay claim to being the oldest knowledge repository focused on genetics and genomics. It began in 1966 with the first publication of Mendelian Inheritance in Man, a hard cover print volume edited by Dr. Victor McKusick. OMIM, was created in 1985 through a collaboration between the National Library of Medicine and the William H. Welch Medical Library at Johns Hopkins with availability through the early internet in approximately 1987. In 1995, OMIM was developed for the World Wide Web by the National Center for Biotechnology Information (NCBI). OMIM is meant to be a comprehensive, authoritative compendium of human genes and associated phenotypes. The overviews in OMIM contain information on all known Mendelian disorders and over 16,000 genes. It is referenced and focuses on the relationship between phenotype and genotype.

As a knowledge repository OMIM has several strengths. Each entry is represented using a consistent summary template reducing searching, at least for experienced users. Each phenotype and gene are given a separate entry and are assigned stable, unique identifiers. Within a gene entry, associated diseases are included with links to the phenotype number and vice versa. As structured disease representations emerged, this was a relative weakness, as the OMIM number was not mapped to other code sets or ontologies. Over the last few years extensive efforts to correct this weakness have been implemented. For example, the entries in the Clinical Synopsis section are now mapped to several widely used controlled vocabularies including the Unified Medical Language System, Human Phenotype Ontology, International Classification of Diseases, and Orphanet [53]. All entries are versioned with updates provided daily through a large team of curators systematically searching the peer-reviewed biomedical literature. Within each entry are links to external resources such as ClinVar, the Exome Aggregation Consortium (ExAC), and the Genome Aggregation Database (gnomAD). OMIM also supports discovery research and collaboration through a feature called MIMmatch. With a free registration, a MIMmatch user will receive alerts about updates to selected genes and diseases of interest. Finally, download of OMIM data is facilitated by an Application Programming Interface (API) through a table linking OMIM genes to NCBI and Ensembl IDs (mim2gene.txt).

There are some weaknesses associated with the use of OMIM. While the entries are organized in a consistent structure, each section is organized chronologically, with new information added to existing content. This idiosyncratic approach is very useful for a user interested in the history of understanding of a gene or disorder but can hinder retrieval of information specific to contemporary diagnosis and treatment. OMIM cannot be accessed by EHR systems through standardized APIs and it is not HL7 infobutton compliant.

- Gene Reviews [54]. Gene Reviews dates to 1993. Gene Reviews "... provides clinically relevant and medically actionable information for inherited conditions in a standardized journal-style format, covering diagnosis, management, and genetic counseling for patients and their families." Each review is written by an expert or group of experts followed by peer review and editing. Version control is used, and periodic updates are required at a minimum of every 4–5 years, although these are expected more frequently if major new knowledge emerges. Currently Gene Reviews includes 806 chapters. These are comprehensive human readable documents, but do not contain structured data, thus are not computable.
- Genetic Test Reference (GTR®) [55]. The Genetic Test Reference supplanted the Gene Tests laboratory directory in 2013. From the website, "The overarching goal of the GTR is to advance the public health and research into the genetic basis of health and disease." To accomplish this, the GTR® relies on voluntary submission of information by genetic testing laboratories. Solicited information includes the test's purpose, methodology, validity, and evidence of clinical utility. Despite the reliance on voluntary submission, the site currently has information on more than 77,000 tests for over 10,000 conditions involving more than 18,000 genes. One weakness of the site is that there is no independent verification of the submitted information leading to reliance on the honor system. The search interface allows for multiple criteria to be used to rapidly restrict results to those that are most useful. GTR® is part of NCBI's Entrez system. Some structured data elements are accessible through APIs using E-utilities via web services or through a UNIX command line. This resource has the potential to be useful for CDS in the context of genetic test ordering. However, at least in the United States where the choice of testing modalities and laboratories is under the de facto control of health systems and payers through contracts and coverage agreements, the degree of customization needed to actually implement the GTR® in the EHR is prohibitive.

There are two knowledge repositories that are focused on support for pharmacogenomics.

- The Pharmacogenomics Knowledge Base (PharmGKB) [56]. PharmGKB is a publicly available, online knowledge base responsible for the aggregation, curation, integration, and dissemination of knowledge regarding the impact of human genetic variation on drug response. It encompasses clinically relevant information including dosing guidelines and drug labels, potentially clinically actionable gene-drug associations and genotype-phenotype relationships. The philosophy of curation is reflected in an adaptation of the Data, Information, Knowledge, Wisdom (DIKW) pyramid. The base of pyramid is aggregation of the medical literature relevant to PGx. Manual curation is used to extract PGx knowledge from the literature. The next level, Knowledge Annotation, Aggregation, and Integration, is achieved through a combination of variant annotation, combined with information from drug-centered pathways and presented as Very Important

Pharmacogenes (VIP) summaries. These are further curated for clinical use by synthesizing and grading the evidence available for clinical utility. The apex of the project is represented in the point of the pyramid which is focused on implementation. One such implementation resource is CPIC guidelines which will be discussed below. PharmGKB is currently the only genetic knowledgebase compliant with the HL7 Infobutton Standard lowering barriers for its use for passive clinical decision support [57].

- The Clinical Pharmacogenetics Implementation Consortium (CPIC) [58]. CPIC is a collection of evidence-based guidelines for the use of PGx information in clinical care. It was developed to address a significant barrier to the use of PGx information which is the difficulty in translating PGx results into appropriate prescribing decisions for drugs impacted by PGx information. The published guidelines use a standard format and standardized terminology, provide clinical recommendations accompanied by systematically graded evidence and are peer-reviewed. All published guidelines have an associated systematic evidence review. The guidelines are regularly updated as new information becomes available. Of particular relevance for this chapter, the CPIC informatics working group works with the guideline developers to ensure that the guidelines are adequately explicit for CDS. This proactively addresses a fundamental problem of translating practice guidelines into CDS products—use of ambiguous language that is imprecise and lacks clarity in written guidelines (i.e., not adequately explicit) [59]. As with the ACT sheets discussed above, in addition to the narrative description, all CPIC guidelines include a decision flow chart (L2 artifact) to guide care. The implementation strategy is described in detail [60].

To lower implementation barriers CPIC data are represented in a structured format accessible via a database or API available along with an implementation guide through GitHub [61]. In an AHRQ-funded project, the University of Maryland has implemented one CPIC guideline as a genomic CDS tool within their EHR [62]. Other institutions have also implemented CPIC guidelines either through local CDS rules, or by translating the guideline into the structure permitted by CDS rules engines provided through vendor-based EHR systems. Some companies providing PGx services are providing CDS based on CPIC guidelines to customers using APIs. Some attempts to integrate these into vendor-based EHR systems have begun. To date no generalized 'off the shelf' CDS system based on CPIC guidelines is available as a solution in an enterprise EHR. CPIC also partners with CDS-KB (below).

- Clinical Decision Support Knowledgebase (CDS-KB) [63]. As can be intuited from the preceding, there are numerous knowledgebases each with its own purpose and approach. This complicates the development and dissemination of CDS. Recognizing this as a barrier to implementation of precision medicine, two networks funded by the NHGRI, the electronic Medical Records and Genomics (eMERGE) and Implementing Genomics in Practice (IGNITE) collaborated to develop the CDS-KB. Its stated purpose is to support the sharing of CDS artifacts to serve as a collection of practical experiences and resources to enable more rapid translation and implementation of genomic and precision medicine. To further expedite implementation, CDS-KB collaborates with Open Clinical Decision Support "... a multi-institutional, collaborative effort to develop open-source, standards-based clinical decision support (CDS) tools and resources that can be widely adopted to enable CDS at scale." CDS-KB submitters have provided many CDS artifacts using a standardized format. Currently, artifacts consist of a narrative description (L1) and a workflow wireframe and pseudocode (L2), although it has the capability to represent fully specified knowledge representation (L3) and coded and implemented CDS rules (L4).

17.3 Challenges and opportunities

The preceding sections have introduced the reader to challenges and opportunities for CDS in precision medicine and health in the context of different use cases. This section will address some of the cross-cutting issues in more depth. A summary of issues with proposed short and long-term solutions from Walton et al. is presented in Table 17.1 [51].

TABLE 17.1 Challenges with proposed short and long term solutions to support use of genomic information in the EHR.

Challenge category	Challenge	Short-term solution	Long-term solution
Transmitting data from the laboratory to the EHR			
	Need for standardization of genetic phenotypes (i.e., disease, metabolizer status, gene-specific phenotype)	Create manual genetic phenotype mappings to each genetic testing laboratory for all genetic conditions imported	Engage the data standards and genomics communities to develop standards for genetic phenotypes
	Poorly defined role for the LIS in delivery of discrete genomic data	Bypass the LIS, sending discrete genomic information directly to the EHR	Engage LIS vendor to discuss future plans for genomic data
	Inadequate and discordant standards for genomics in the HL7 genomic report format and the Fast Healthcare Interoperability Resource (FHIR) molecular sequence resource	Develop a laboratory interface that maps JSON structured data to the HL7 genomic report format	Engage HL7 and genomics community to address deficiencies
Maintenance of clinical information and CDS in a rapidly changing field			
	Lack of publicly available resources for patient and provider facing information for genetic conditions	Develop and maintain internal resources for a limited number of conditions	Engage appropriate stakeholder groups to encourage development of standard resources for genomics
	Maintenance of CDS and clinical information requires technically competent individuals with training specific to the EHR	Create an external database that can be curated by content experts (geneticists, genetic counselors) with limited technical skills, and can then be used by the technical team to update the EHR	Work with EHR vendor to adopt an open API that will allow for third party specialized tools for maintenance of clinical information and CDS through nontechnical interfaces

Continued

TABLE 17.1 Challenges with proposed short and long term solutions to support use of genomic information in the EHR—cont'd

Challenge category	Challenge	Short-term solution	Long-term solution
	Lack of open APIs to access external Information resources and third-party tools in the EHR vendor deployment	Continue ongoing discussions with vendor regarding implementation of OpenInfobuttons	Encourage EHR vendor to implement a more mature adoption of CDS hooks, FHIR, and OpenInfobuttons
Disparate implementation needs for clinical geneticists (CGs) compared with primary care providers (PCPs)			
	PCPs are interested in genomics for population health and variant-guided health maintenance while CGs are interested in variants for diagnostic indications	Prioritize implementation for primary care/population health, delaying CG implementation for later stages in development, given the heightened complexity	Work with EHR vendor and CGs to develop a workflow specific to CG's needs
	CGs need access to variants of uncertain significance (VUS), while PCPs want variants with pathogenic interpretations only, as interpretation is outside of their training		
	PCPs favor a limited set of disorders with well-defined clinical recommendations CGs need to support rare disease where more limited knowledge and resources are available		
Differing perceptions/uses of discrete genomic data versus PDF of the genetic report			
	Inability to hide variants from public/patient view prior to clinical validation	Clinical expert to manually review all variant classifications prior to import into the EHR	Work with EHR/LIS vendor to develop variant queue that allows for clinical expert review before release to chart, Alternately develop external application for variant review
	CDS will fire from any variant classified by the laboratory as pathogenic		
	Risk of misinterpretation or inappropriate action from attributing a VUS to a patient	Do not import any variants that are not considered pathogenic	Study context sensitive delivery methods of variants based on variant pathogenicity and their impact or care
Classification and reclassification of variants			
	No process in place from an operational perspective to handle the global reclassification of variants	Develop thorough manual process for variant reclassification	Work with EHR vendor to develop automated workflows for reclassification

TABLE 17.1 Challenges with proposed short and long term solutions to support use of genomic information in the EHR—cont'd

Challenge category	Challenge	Short-term solution	Long-term solution
	Potential for discrete variant classification In the EHR to be discordant with the scanned PDF	Communicate with the laboratory after reclassification to consider issuing an updated report with changes	Work with EHR vendor to create a mark or note overlaying the scanned report that the variant has been reclassified by a provider in the system and allow for access to reclassification history
	The inability of EHR vendor solution to change the classification of all patients with a given variant, allowing for patients with the same variant to have discordant classifications	Manually track and change variant interpretations as needed	Engage stakeholders in clinical genomics and laboratory genomics communities to develop standard methodologies for reclassification of variants by clinical providers

Credit: Walton NA, Johnson DK, Person TN, Reynolds JC, Williams MS. Pilot implementation of clinican genomic data into the native electronic health record: challenges of scalability. ACI Open 2020;4:e162–e6. Creative Commons Attribution 4.0 International License. No changes were made in the table.

17.3.1 Adoption of clinical genomics standards

The lack of implemented standards to support the use of genomic information in the clinic is consistently identified as one of the most significant barriers to be overcome. The two areas most impacted by a lack of standards are representation of core genomic information, such as gene and variant names and variant classification and conventions for naming and delineating genetic disease and associated phenotypes. Standards within these larger categories including representation of patient-specific genomic data; representation of knowledge about genes, variants, disease, and phenotypes in a manner that reflects the most recent information in a rapidly advancing field; the definition of "genetic phenotypes" followed by the translation of these definitions into a computable format; the creation of standards-based interfaces to external knowledge resources; and the content and structure of information presented to the clinician, as informed by principles of user-centered design. The absence of these key components limits the ability to link knowledge contained in repositories such as those described above with CDS tools to support clinician access to point-of-care, just-in-time information relevant for the care of the patient and the ability to integrate this information and associated knowledge within other clinical applications that are critical to clinician workflow within the EHR. Some progress has been made using existing standards to support several use cases in genomic and precision medicine. (Table 17.2) While none have achieved wide dissemination or implementation (this will be discussed below), they do demonstrate the feasibility of approaches using currently available standards.

Three examples are presented, an early stand-alone effort, a multi-standard approach to the delivery of genetic information to the clinic, and an early example of an implementation using standards-based genomic capabilities within a vendor EHR system.

TABLE 17.2 Requirements, available standards, challenges, and resources to support clinician education in the electronic health record.

Requirements for clinical genomics implementation	Related standards and resources	Challenges	eMERGE/ClinGen efforts to overcome challenges
Storage of genomic data	Ancillary genomic systems Variant Call Format (VCF)	Inadequate ability of current EHRs to store detailed discrete genomic results Lack of consistent open source reference data structure that can robustly represent results	eMERGE XML provides an example of the content such standards should represent
Representation and exchange of patient genomic data in the EHR	HL7 v2 Clinical Genomic Implementation Guide	Need to represent heterogeneous result types (e.g., star alleles, diplotypes) Rapid evolution of data types and use cases related to clinical genomics Slow evolution of HL7 standards Low adoption of extant standards by EHR vendors and genetic testing laboratories	Interviews led by EHRI workgroup with eMERGE and CSER sites to understand intended use of genomic test reports and requirements for transferring reports and associated data from laboratories to sites Development of an XML standard capable of transmitting results within the eMERGE Network
	HL7 FHIR Genomic Reporting Implementation Guide		
	GA4GH Variant Representation		
	Specificatione		
	eMERGE XML standard		
			Interactions with HL7 to assist in incorporating the eMERGE XML standard into the FHIR standard
Representation and exchange of variant knowledge	ClinGen resource	Lack of resources with clinical genomics knowledge in computable format	eMERGE XML development and validation ClinGen resource: Variant Curation Working Groups
	GA4GH Variant Annotation model (in progress)		
	eMERGE XML standard		
	Monarch initiative (for ontology support)		

TABLE 17.2 Requirements, available standards, challenges, and resources to support clinician education in the electronic health record—cont'd

Requirements for clinical genomics implementation	Related standards and resources	Challenges	eMERGE/ClinGen efforts to overcome challenges
Clinical decision support (CDS)	HL7 Infobutton Standard, OpenInfobutton	Lack of EHR and laboratory support for representation of genetic data in standard formats	OpenInfobutton integration with ClinGen clinical genomic resources
	SMART on FHIR		
	CDS Hooks standard		CDSKB.org
		Lack of clinical genomic resources with knowledge accessible in computable, standards-compliant format	DocUBuild
			Use of ACMG genomic guideline ACT sheets to create genomic CDS
		Little experience with CDS for the use of genomic data in clinical care	Incorporation of CPIC Guidelines into ClinGen resource
		Lack of expert guidelines for clinical management of genomic findings to serve as the decision logic for CDS tools	ClinGen Actionability Working Group

eMERGE, Electronic Medical Records and Genomics Network; ClinGen, Clinical Genome Resource; XML, Extensible Markup Language; EHR, electronic health record; HL7, Health Level 7; FHIR, Fast Healthcare Interoperability Resources; GA4GH, Global Alliance for Genomics and Health; EHRI, Electronic Health Record Integration; CSER, Clinical Sequencing Exploratory Research; SMART, Substitutable Medical Applications, Reusable Technologies; ACMG, American College of Medical Genetics and Genomics; CPIC, Clinical Pharmacogenetics Implementation Consortium.

Credit: Williams MS, Taylor CO, Walton NA, Goehringer SR, Aronson S, Freimuth RR, et al. Genomic information for clinicians in the electronic health record: lessons learned from the clinical genome resource project and the electronic medical records and genomics network. Front Genet 2019;10:1059. Creative Commons Attribution 4.0 International License http://creativecommons.org/licenses/by/4.0/. No changes were made in the table.

- Gene Reviews was the first genetic knowledge repository to be linked to a production EHR—an effort lead by an author of this chapter (MSW) and editor of the book (Guilherme Del Fiol) [64]. The organization of the Gene Review was used to link to different parts of the review using the context awareness of infobuttons reducing the time needed to find relevant information within the review. Unfortunately, this was not widely adopted so access to this valuable repository is not integrated into clinician workflow like other e-knowledge resources.
- A project organized through the ClinGen EHR Working Group [65] developed a prototype standards-based genetic reporting application [57]. The purpose of this project was to demonstrate how a CDS tool constructed using currently available standards could lower the barriers to clinician access to genetic knowledge resources. The prototype utilized several current and emerging standards. The Genomic Medicine Assistant is a multi-gene sequencing panel report designed by Cutting et al. [66]. It was chosen as it is one of the only designs that was informed by clinician engagement and testing showed a favorable

usability rating. Gene and variant names included in the GMA are associated with infobuttons. These are supported through the HL7 Infobutton Standard and the Openinfobutton platform (see Chapter 16 for details). The former was chosen, as EHR certification under the US "Meaningful Use" program requires the implementation of infobutton functionality compliant with the HL7 Context-Aware Knowledge Retrieval ("Infobutton") Standard [67]. As noted in the previous section, most genomic knowledge resources are not HL7 infobutton compliant, therefore Openinfobutton was used to overcome this barrier. OpenInfobutton is described in detail in Chapter 16, but briefly, it is an HL7 compliant, open source software suite that includes an infobutton manager and a Web-based infobutton configuration application called LITE (Librarian Infobutton Tailoring Environment) [68]. For resources that are not HL7-compliant, (which is the case for most resources of relevance to genomics) the resource profile includes mappings between search parameters in the resource's API and the HL7 Infobutton Standard. Once a resource profile is configured in LITE, it can be downloaded using OpenInfobutton via LITE's resource profile store. The HL7 Fast Healthcare Interoperable Resources (HL7 FHIR) standard was used to create a set of data attributes extracted from the ClinGen data repository and made compatible with FHIR data types using standard terminologies where applicable and available creating a ClinGen FHIR resource. The approach was tested in a prototype EHR that was built with the relevant standards implemented. This was necessary as currently deployed EHR systems do not support all the recommended standards, and a developmental EHR 'sandbox' was not available at the time for testing.

- One EHR vendor, Epic Systems, has implemented a software solution called Genomic Indicators in their production EHR. As a proprietary system, the full extent to which existing standards are utilized is not fully documented, but standard nomenclatures for genes and variants, and use of the HL7 Genomic Standard are included. Studies describing early use of the system for pharmacogenomics [26,69] and broader genomic indications [69] are appearing in the literature. The latter article, by investigators at Penn, is of particular interest in that it reviews the current capability of the genomic indicators against the framework proposed by the ACMG, a specialty society committed to the use of genetic and genomic medicine in clinical care [70].

The implementation is presented in Fig. 17.2. The multidisciplinary team created a standardized process to create genetic documents that contain the defined data in human readable and, where appropriate, structured form. This process accommodates legacy reports (necessitating manual creation of the document), and future reports. While most reports generated at present will also require manual entry, the team did work with one outside laboratory vendor to develop an interface for computerized order entry of genetic tests through the EHR and a plan to develop an interface that will return the results as structured data into the EHR, specifically using the genomic indicator function. With the structured data represented as a genomic indicator, CDS rules can be triggered through native Epic alerts as appropriate. At the time of publication, this is live for one genomic medicine indication (Lynch syndrome) and one PGx rule (fluoropyrimidine dose adjustments in patients with dihydropyrimidine dehydrogenase deficiency identified on *DPYD* gene testing). Also noteworthy is CDS notification of both patients and clinicians, a novel strategy that will be discussed further below. The authors note the requirement for a substantial resource investment, given the heavy reliance on manual methods to bridge current gaps in technology and data transfer.

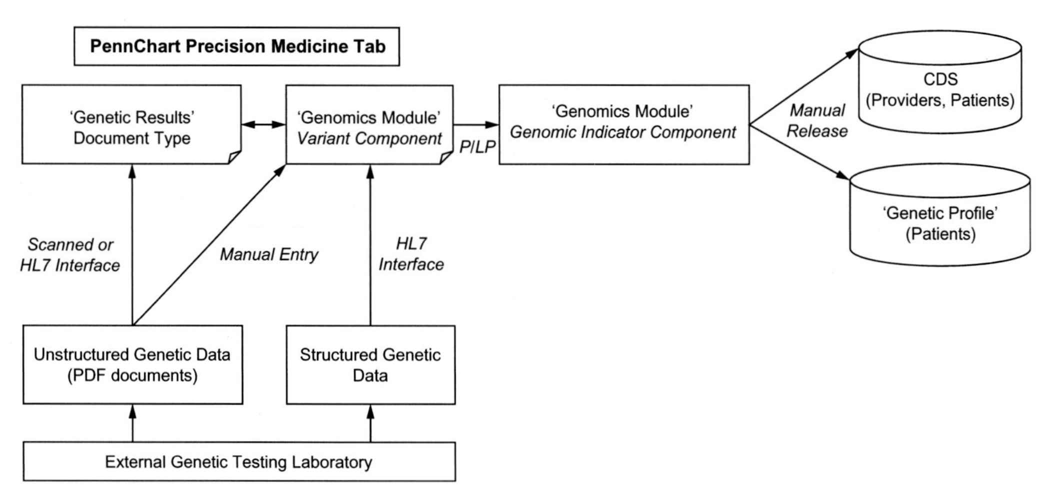

FIG. 17.2 Integration of genomic data into PennChart, the electronic health record system at Penn Medicine. HL7 = Health Level 7. P/LP = pathogenic/likely pathogenic variant. CDS = clinical decision support. *Credit: Lau-Min KS, Asher SB, Chen J, Domchek SM, Feldman M, Joffe S, et al. Real-world integration of genomic data into the electronic health record: the PennChart Genomics Initiative. Genet Med 2021;23(4):603–5. Creative Commons Attribution 4.0 International License http://creativecommons.org/licenses/by/4.0/. No changes were made to the figure.*

17.3.2 CDS logic and strategy

CDS has been studied for over 50 years, yet the improvement in outcomes attributable to the use of CDS has been suboptimal [71]. Common pitfalls (described in detail in Chapter 22—Evaluation of Clinical Decision Support) include workflow disruption, alert fatigue and inappropriate alerts, CDS reliance (also called user skill), dependence on computer literacy, system and content maintenance, operational impact of poor data quality and incorrect content, lack of transportability and interoperability, and financial challenges [72]. Greenes and colleagues proposed a cognitive framework for design, implementation, and evaluation of CDS interventions (Fig. 17.3) [71]. In this section, this framework will be used to explore the logic and strategy of CDS applied to precision medicine.

(a) Integration/Adaptation to Workflow. This is a critical issue for CDS that has been understudied in the context of precision medicine. At the aforementioned GMXIII meeting, one of the high-level themes was "Researching The Stakeholder Perspective: enablers and barriers that affect the integration of genomic-based clinical informatics resources in the healthcare system". Of the proposed areas that needed attention, research focused on the development, implementation, and maintenance of genomic-based workflows was highlighted. Within this research area specific topics included strategies to reduce burden on clinicians, engaging a broader range of stakeholders including the full health care delivery team and patients, and identifying strategies beyond alerts and reminders to improve the attention to and impact of CDS.

(b) Construction of CDS artifacts—structure and components. This has been discussed above, but the proposed framework makes it clear that stakeholder engagement and workflow analysis must be used to inform the construction of these artifacts. To date,

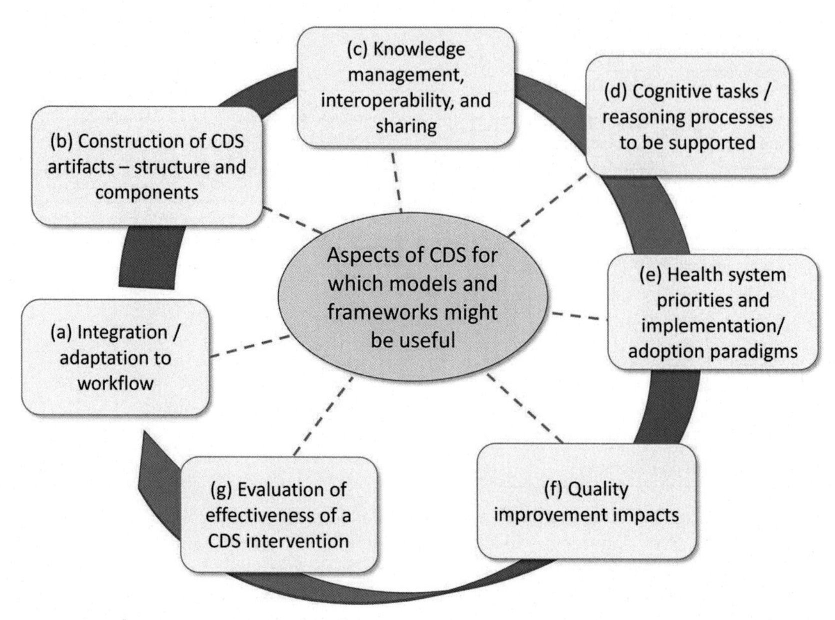

FIG. 17.3 Caption: Aspects of CDS for models and frameworks. Note that there is an implied need for an iterative cycle of attention to these aspects as a CDSS is developed, deployed, and refined. *Credit: TBD.*

most CDS artifacts have been developed by specialists and experts which could limit their applicability across the wider range of conditions and providers.

(c) Knowledge management, interoperability, and sharing. This has also been discussed above. One point that hasn't been made that is relevant to this component of the framework is the rapidly changing nature of genomic knowledge. Complex logic will be needed to ensure that the knowledge represented to clinicians is appropriately updated to include new findings. The need to regularly re-interpret genetic sequencing for those who have previously had testing for an indication where a diagnosis has not been made, or for new indications that arise in the future is another requirement. This will need to be reflected not only in content maintenance, but also in system updating and maintenance. Much of this work will need to take place within the knowledge repositories themselves, but the interfaces between the EHR and the knowledge repositories will need to support real time updates. Within the EHR ecosystem, as knowledge is updated, there is the possibility that advice given to a patient based on a prior interpretation, may need to be updated to reflect a new interpretation. This situation has been encountered in Geisinger's MyCode® Community Health Initiative (MyCode) and in Phase 3 of the eMERGE network. Variants curated as likely pathogenic/pathogenic that had been returned to participants in the course of these research studies were subsequently downgraded to variants of uncertain significance, meaning the condition specific recommendations were no longer relevant. Tracking of research participants allowed for recontact to provide updated information and counseling and to remove condition specific CDS that had been implemented. This functionality will need to be included in any clinical implementation.

(d) Cognitive tasks/reasoning processes to be supported. This component, similar to a) requires significant engagement with end users to understand clinical decision-making to be able to define the informational needs for each clinical decision impacted by genomic information and design CDS to best support the end user. This aspect is of particular importance with regards to risks associated with CDS reliance. There is ample literature supporting the relative lack of providers with adequate training in genetics and genomics. This coupled with an expanding number of applications of genomics in clinical care and a small number of providers with advanced training in genetics, increases the reliance on CDS. Bussone and colleagues have studied the impact of explanations on system reliance and trust [73]. Lack of explanation degrades trust in CDS, but providing explanations while building trust in the CDS, also increases reliance which can impact safety. This is an area that needs more research in general and is an area to be emphasized through the GMXIII recommendation for increased clinician engagement.

(e) Health care system priorities and implementation/adoption paradigms. With few exceptions, precision health hasn't been a major priority for most healthcare systems. This is especially true considering the impact of the COVID-19 pandemic on healthcare operations. As evidence of benefit accumulates, precision medicine implementation will be prioritized. The incorporation of standards and standardized approaches based on implementation science supported by increasing availability of genomic-enabled vendor systems will lower the barrier to incorporation of precision medicine in systems with less content expertise. A learning healthcare system model as proposed in the NHGRI strategic plan (Fig. 17.1) is another key paradigm to inform precision health strategy. This also is essential to the next two elements of the framework.

(f) Quality improvement impacts. Quality improvement has the goal of reducing unexplained variation in clinical practice through the use of evidence-based protocols and guidelines intended to improve health outcomes. At present, precision medicine has limited evidence for improved outcomes and few if any guidelines that could be implemented through CDS. However, as evidence of benefit accumulates for certain interventions, systems in the EHR facilitated by CDS must be built to deliver the evidence in a form that can be easily used by clinicians and to collect patient outcomes from transactional EHR data to ensure that the desired outcome goals are being achieved. Inherent in this is the need to continuously improve through iterative approaches (e.g., Plan-Do-Study-Act cycles) and, once the process is optimized, to monitor stability and update as new knowledge becomes available. Two key elements are the translation of evidence-based best practices into protocols that have the potential to be generalizable (e.g., ACMG ACT sheets discussed above) and definition and agreement on outcomes that can be measured within and across systems. This problem is illustrated by a recent systematic review by Sapp, et al. that examined disclosure practices and outcomes for medically actionable genomic secondary findings [74]. The investigators found substantial variability in the process of disclosure and in the definition and collection of outcomes. Attention must be paid to developing generalizable outcome measures that can facilitate comparison of different approaches to improve the quality of care. One example of outcomes harmonization was published through a collaboration between the eMERGE network and ClinGen [75].

(g) Evaluation of effectiveness of a CDS intervention. This component is related to the prior, however its intent is to focus on the effectiveness of a given CDS intervention to achieve the goals outlined in (f) above. This is another emerging area of research that will require some customization to meet the needs of precision health. One aspect of particular interest in this regard is the granularity inherent in precision health. Most CDS is applied to relatively large populations of patients, such as a recommendation for mammography for all women over the age of 40, or colonoscopy for all patients over the age of 50. Some CDS impacts smaller groups of patients such as medication allergies. However, precision health envisions treatment optimized at the individual level. This means new paradigms for the design (b) above) and the evaluation of CDS will be needed as this vision becomes reality. This was identified as a research priority in GMXIII.

Greenes et al. recognize that the proposed framework also implies an explicit feedback loop consistent with the concept of a learning system.

17.3.3 CDS architecture to overcome these challenges

Genomic data integrated with CDS has the potential to enhance the possibilities of precision health [76] including shortening the diagnostic odyssey, creating more precise phenotypic disease characterizations, targeted therapies at an individual level, and improving overall patient outcomes. Each potential advantage is associated with unique barriers to implementation in clinical practice. Static laboratory results presented as scanned documents, inadequate genetics workforce, training in genetics for non-geneticist clinicians, and national standards around representation of genomic information are areas where significant investment could help. Improvements in CDS architecture are also needed as illustrated by the following examples:

- Given the lack of trained personnel available, at least at the present time, it is unlikely that each institution will be able to address these issues through targeted recruitment of genetic clinicians. This requires innovative approaches to the design and implementation of CDS [77]. Instead of trying to master every area within genetics and its clinical applications, a service-oriented architecture (SOA) (see Chapter 29 for details on CDS architectures) will allow specific services to be built by experts in those areas. These services can be self-contained, and work within an organization, while supporting standards that allow it to be reusable among other institutions. Genomic applications will need to have standard core components yet be extensible and capable of evolving as new information is discovered. SOA can allow those applications to scale at a rapid pace [78].
- After obtaining the patient's genome sequence, the variant call format or genome variant format must be curated for its relevance to clinical care. Variants that may potentially impact a patient's health or medical treatment should be prioritized and compared against a genome variant knowledgebase like ClinVar or PharmGKB. A genome variant knowledgebase presents genomic and phenotypic data in a standardized format consumable by information systems and clinicians, researchers, and patients. These publicly available resources are still growing and being updated frequently, requiring systems or services that can revisit previous information.

- These annotated variants must be stored in a central repository, while readily allowing other services to access the information, such as the genomic ancillary system proposed above. The EHR, CDS, and the genome variant knowledge repository need the ability to read, write, and update genomic data. The phenotypic information that resides in the EHR can inform CDS for current patients while simultaneously informing researchers about specific phenotypic presentations of diseases consistent with the genomic learning health system (Fig. 17.1). Current EHRs use their own proprietary approach to collecting and storing information, but models like HL7, FHIR and other standards discussed above and in Chapters 12, 13, 14 and 18 must inform the architecture of the CDS.
- CDS within the EHR can accommodate diverse use cases of relevance to the clinical user [79]. These are typically formed from a CDS knowledge base which can include logic, decision rules, guidelines, algorithms, and expressions [42]. As explained in Chapter 29, using a SOA CDS allows specific CDS knowledge to be captured and contained within a specific service. This reduces the number of dependencies required to process this information and return a result based on the CDS knowledge. A separate controlling service can collect and centralize the relevant genomic information from the EHR, clinical genome database, and CDS knowledge base.

17.3.4 Integration with clinical workflow

Integrating CDS with clinician workflow is essential for success and this is no exception for CDS to support precision health. The need for research into precision health scenarios coupled with engagement with clinicians, patients, and other stakeholders to understand clinical workflows were key takeaways from the GMXIII meeting. It is essential to build CDS that supports optimized clinical care and outcomes as illustrated by the following scenarios.

- Diagnostic testing: Genetic testing based on clinical signs and symptoms, that is diagnostic testing, is still the most common application in clinical use [80]. The workflow for this type of testing is reasonably well understood for non-genetic tests so emulating successful approaches adapted for genetic and genomics to address barriers identified previously are likely to be successful. One difference from other tests is the potential for testing of at-risk relatives (cascade testing—described previously). It is likely that novel workflows that include innovations such as chatbots and technologies enabled by mobile devices will need to be developed.
- Diagnostic testing with pre-existing sequence: The future state of medicine will include a scenario where patients will have sequence data available for clinical care. The use of expert systems along with transactional EHR data can provide diagnostic aids within a typical clinical workflow. As we transition towards healthcare which assumes existing genomic data, diagnostics will focus on the interpretation of existing data within the EHR and proper phenotypic description of patients and diseases [81]. Collaborations like eMERGE and All of Us have built dozens of phenotypes for use in combination with genomic information [82].

  Genetic diagnosis at the point of care in real-time is novel, thus no research has been done to optimize its use. Diseases with a high genetic burden are present in both primary and specialty care, thus encompass a variety of different clinical workflows. Clinical experts in each field of disease should be interviewed about the patient trajectory to elucidate the optimal point within the clinical workflow where genomic CDS would be helpful while minimizing disruption. Each disease should be evaluated using Osheroff's "5 Rights of Clinical Decision Support"; the right information (what), at the right time (when), to the right person (whom), in the right format (how) and through the right channel (how).

- Pre-emptive screening: As sequencing becomes increasingly affordable, some programs are sequencing large numbers of patients without an indication. The availability of the sequences allows for screening these participants for genetic variants in genes that have definitive gene-disease associations and for which interventions such as enhanced surveillance, prophylactic surgery, and medication management are available. The Centers for Disease Control and Prevention's (CDC) Office of Public Health Genomics has identified three conditions, designated as Tier 1, defined as those having significant potential for positive impact on public health based on available evidence-based guidelines and recommendations [83]. These conditions include Hereditary Breast and Ovarian Cancer, Lynch syndrome (a cancer predisposition syndrome that increases risk for colorectal, endometrial, and other cancers), and Familial Hypercholesterolemia. MyCode has initiated screening for these conditions (and others). The follow-up study from the results disclosure has confirmed significant improvement in both process and intermediate outcomes known to predict health outcomes [84].

  To deliver the results and measure the outcomes required the development of a new clinical infrastructure. This was constructed using the principles of the Learning Healthcare System and was informed by extensive stakeholder engagement including patient-participants, primary care clinicians, specialists, laboratory services, ancillary service providers, and administrative leaders [85]. Information systems were essential to support the overall project, although given the barriers outlined above, the information system infrastructure necessitated extensive customization of existing resources, coupled with creation of local solutions. (Table 17.1) [51]

17.3.5 The role of different stakeholders

Genomic information is of great interest at both an individual and population level. Genetics are shared within geographical areas, influence behaviors, and predict traits of health such as cardiovascular diseases, cancer, diabetes, mental health disorders along with rare diseases. As genetic data become more prevalent, the responsibilities and roles of individuals have evolved. GM XIII emphasized the importance of engagement with a range of stakeholders to ensure that information systems are designed to support genomic medicine with attention to the need to not add additional work to already overburdened clinicians. The following brief descriptions illustrate the potential role different stakeholders will have in the delivery of precision health.

- Genetics professionals are likely to be the content experts for much of precision health. They will be expected to synthesize the information available in genomic knowledge resources with clinical information from the EHR and other sources and lead the development of best practices at the institutional level that are consistent with emerging evidence-based practice guidelines produced by the ACMG and other specialty organizations. There will be an expectation that they will lead the engagement with other clinicians at the organizational level to coordinate the delivery of care between primary care, specialists, and genetics professionals such that each clinician can deliver care most appropriate to their specialty. This will include an assessment of different workflows and resource requirements for different clinician types within each institution and development of systems to measure outcomes to support improvement activities. Recognizing that not all institutions will employ genetics professionals, the profession will also have the responsibility to develop resources to support clinicians in these institutions. The expansion of the ACMG ACT sheets beyond newborn screening (see knowledge repositories) is one example of how this might be accomplished. Adaptation of content in knowledge repositories such as ClinGen coupled with strategies for the content to be available with the EHR is another way genetics professionals can make vetted and up-to-date content available to other healthcare professionals [86].

- Laboratorians are at the interface between genetic testing and clinical care. Optimal interpretation of genetic tests necessitates the provision of sufficient clinical information to the laboratorian to identify the most relevant genes related to the clinical condition and facilitate the curation of variants within those genes to determine if any might be identified as likely to cause the patient's disease or, in the context of secondary findings or screening, have sufficient evidence of pathogenicity to warrant returning to the clinician. It is not unusual for this process to be iterative as a putative causal gene/variant may suggest the need for additional clinical information and vice versa. Providing information and links to knowledge repositories either within or associated with the laboratory report can provide passive CDS to the clinician receiving the report. There is also a need to communicate updated variant interpretations back to clinicians. At present there are no widely implemented solutions that address these workflows, but one proposed solution, GeneInsight, has been developed [87]. GeneInsight was implemented to support the results disclosure in eMERGE Phase 3, demonstrating that this approach was scalable and able to be implemented in multiple delivery systems [88].
- Non-genetics health professionals. As genomic medicine and precision health expands, health professionals without in depth training in genetics and genomics will become increasingly involved by the necessity of the expanding uses. Engagement with these professionals as stakeholders will be essential as illustrated by the following examples.
  - Specialists: Each specialty is expected to have a subset of conditions for which genomic information will be essential for the care of patients. Already tumor sequencing is informing precision oncology treatment with early evidence of improved outcomes with lower costs when implementation is informed by substantive engagement with specialist stakeholders [89]. For certain MyCode results, specialty-specific order sets were developed to reduce clinician effort and promote best practices for conditions such as cardiomyopathy, and several cancer predisposition syndromes. The MyCode program returns pathogenic variants in the gene *RYR1* which is associated with Malignant Hyperthermia Susceptibility (MHS), a condition that can result in significant morbidity and mortality when a triggering event leads to rhabdomyolysis and related metabolic dysfunction. A common triggering event for MHS is the use of certain

anesthetic agents and muscle relaxants prescribed by anesthesiologists. Engagement with anesthesiology led to the development of CDS directed at the anesthesiologists consisting of alerts tied to the patient's allergy list and order sets that supported the use of anesthetic agents unlikely to trigger an event, while ensuring that medications such as Dantrolene are available for use early in the course of an event to prevent a severe episode.

  - Primary Care Providers (PCPs): Effective use of PCPs has been shown to improve outcomes. PCPs will be essential stakeholders for genomic medicine and precision health. PCPs are the essential provider for the MyCode program. Extensive engagement with PCPs informed the most effective use of the EHR based on established patterns of use. The implementation design optimized the communication of results to the PCP while minimizing the impact on workflow. PCPs were also engaged in user-centered design activities to develop just in time CDS resources accessible from the EHR to provide actionable information specific to the result to the PCP at the point of care to guide clinical actions [90].
  - Pharmacists: Pharmacogenomics is emerging into clinical practice and, as is frequently the case with new disciplines, implementation takes different approaches. In some systems pharmacogenomics is considered within the purview of the ordering physician. However, other programs are implementing pharmacogenomics using a pharmacist-directed model [91]. Anticipating that pharmacogenomics may ultimately become the responsibility of pharmacists, the American College of Clinical Pharmacy has published a white paper exploring the integration of pharmacogenomics into clinical pharmacy practice [92].

- Patients: Germline genomic information remains essentially unchanged over a patient's lifetime. To get the maximum advantage from the information necessitates access to the information as the patient moves through the health care system. In the United States, the lack of a national health care delivery system leads to disorganized care. A major reason for this is the lack of interoperability of information systems that doesn't facilitate critical healthcare information to move with the patient. This is no different for genetic information. Recognizing that the patient is the only common actor in the delivery system, one potential solution is to explore the role of the patient as the source of access to the genomic information. This possibility was explored in a project at Geisinger that engaged patients to provide input and testing for a patient facing genomic test report accessible through the patient portal tethered to the EHR [93]. The reports were developed to support the care of patients and family members with rare genetic conditions (a diagnostic test use case) and were found to be effective and well-accepted by patients and care givers [94]. These reports were subsequently developed for pharmacogenomic information [95]. These promising early results suggest that patient facing CDS approaches can be used in conjunction with provider facing CDS to improve care.

Another key point of engagement is at the system level. At present most implementations of genomic medicine have involved well-resourced health care systems and academic medical centers. Another imperative from the GM XIII meeting was to engage with systems that are under resourced to develop implementation strategies that will allow these systems to initiate precision health initiatives. This will be essential to avoid exacerbation of existing health disparities (see below).

17.3.6 Implementation and dissemination

New discoveries with relevance to clinical care take some time to move into regular practice. Reliance on diffusion of knowledge has led to estimates of up to a 17-year lag from the time evidence of effectiveness is generated to routine use in clinical practice. This is an unacceptable delay that has a deleterious effect on health outcomes. To address this, a new discipline, Implementation Science, has emerged. Implementation Science is the scientific study of methods and strategies that facilitate the uptake of evidence-based practice and research into regular use by practitioners and policymakers. Recognition of its importance in precision health is the explicit inclusion of implementation science as a key component of the NHGRI 2020 Strategic Vision. Of the several bullets included in this section, two are of particular relevance to precision health, 1) Develop and assess strategies for implementing the use of genomic information in clinical care and, 2) Design and utilize genomic learning healthcare systems for knowledge generation and clinical care improvement (Fig. 17.1). Implementation has also been an underlying theme for the two NHGRI-sponsored informatics meeting. One goal of GM VII was to, "… define a prioritized genomic CDS implementation research agenda." Implementation was an overarching theme of GM XIII and a major conclusion of this meeting was that "Incorporating a broader range of methods that draw from implementation science … is needed to emphasize pragmatic approaches to research questions to inform the role of informatics in genomic medicine and precision health."

17.3.7 Health disparities

Health disparities are a major cause of morbidity and mortality that differentially impacts the underserved, disadvantaged, people of color, LGBTQ+, and differently abled. Chapter 25 addresses health disparities and CDS in detail, including issues related to inequities in access to healthcare technology and the "digital divide". Here we focus on issues related to lack of diverse representation in genomics research. Recent studies have demonstrated that care protocols, algorithms, and CDS tools can be inherently biased in that most were developed in settings that served people predominantly of northern European descent, meaning that the clinical questions to address and the data used to develop and test the algorithms were Euro-centric. While these issues will need to be accounted for in the development of CDS for precision health, there are some issues specific to genomics that are of particular relevance.

Inclusion of diverse populations, according to Bentley et al., [96] is critically important for two major reasons. The first is based on the principle of justice—individuals are more likely to benefit most from genomic research conducted in individuals with a similar ancestral background to them. The second represents a scientific imperative. Limiting scientific inquiry to currently well-represented populations will impede our knowledge of genomic variation with negative consequences. This has been shown time and again in variant interpretation, a fundamental process before genomic information can be used in clinical care. Many examples exist of inappropriate care being delivered based on flawed interpretation of genomic variation because the variant had not been studied in populations of non-European descent. The most notable example was that of the study of pathogenic variants in cardiomyopathy by Manrai, et al. [97]. In this study, the investigators found that multiple individuals of non-European ancestry (mostly African ancestry), had results returned as presumed pathogenic based on current interpretation. However, these interpretations were based on data obtained in European populations. When the variants were studied in African ancestry populations, the allele frequency was found to be much higher—nearly 15% in the case of one variant. These presumed pathogenic variants have been reclassified as benign based on interpretation that included the ancestrally relevant reference population. Using inaccurate information as the basis for CDS has the potential to cause harm. As Bentley et al. state, "Failure to fully engage diverse populations at all levels of genomic research perpetuates already considerable health disparities." [96]

Polygenic risk scores (PRS) are emerging into clinical practice both through traditional care delivery, but also through the increasing use of direct-to-consumer testing. A PRS is a number that the estimates the combined effect of many genetic variants on the risk for an individual to develop a complex disease (e.g., diabetes, coronary artery disease, etc.). In contrast to single gene, or Mendelian disorders where a single variant in a gene associated with a disease confers a very high risk for developing the disease, an elevated PRS generally increases risk for development of a complex disease by 2–4 fold, while a low score confers some protection, although to a lesser degree. PRS incorporate hundreds, thousands, and in some cases over a million variations each of which has a tiny contribution to the score. This genetic predisposition can be amplified by environmental factors further affecting the risk. PRS are determined by studying variation in a large population of individuals with a complex disease and comparing that variation to a large population without disease (Genome-wide Association Studies or GWAS). Relevant to this section, to date most PRS have been determined from populations of European ancestry. The applicability of the PRS to non-European ancestry populations is uncertain. Martin et al. [98] have identified this as a major scientific and ethical challenge given the huge differences in accuracy of PRS in European versus non-European populations. They state, "This disparity is an inescapable consequence of Eurocentric genome-wide association study biases. This highlights that—unlike clinical biomarkers and prescription drugs, which may individually work better in some populations but do not ubiquitously perform far better in European populations—clinical uses of PRS today would systematically afford greater improvement to European descent populations."

The NHGRI and other agencies are directing funding towards projects that engage diverse populations. Encouragingly, even modest increases of diversity in GWAS studies markedly improves the predictive accuracy of PRS in diverse populations. An explicit action item from the NHGRI GMXIII meeting is, "A need to ensure that the development and implementation of GBCITR [NB genomic-based clinical informatics tools and resources] is done in a manner that includes equitable representation from diverse and underserved populations." [17] This is also a point of emphasis in the NHGRI 2020 strategic vision [18]. In summary, development of PRS for precision medicine must account for systematic bias and its implementation must not exacerbate and ideally reduces health disparities.

17.4 Conclusion

Precision health has the potential to transform healthcare and improve outcomes. To realize this will require sophisticated information systems. This includes CDS systems that can present information relevant to individuals, incorporating clinical, laboratory, and genomic data coupled with awareness of the individual's goals for treatment, fulfilling the "5 Rights" of CDS. Such systems must be nimble in order to accommodate rapidly changing knowledge and recommendations, with awareness of inherent bias and diversity in order to reduce health care disparities.

References

[1] Collins FS, Varmus H. A new initiative on precision medicine. N Engl J Med 2015;372(9):793–5.

[2] Genomics and Medicine 2020 [cited 2021 8-30-2021]. Available from: https://www.genome.gov/health/Genomics-and-Medicine#:~:text=Genomic%20medicine%20is%20an%20emerging,implications%20of%20that%20clinical%20use.

[3] Christensen CMGJ, Hwang J. The Innovator's Prescription a disruptive solution for health care. New York: McGraw-Hill; 2009. p. 37.

[4] Weatherall DJ. The centenary of Garrod's Croonian lectures. Clin Med (Lond) 2008;8(3):309–11.

[5] Olsen LA AD, McGinnis JM, editors. Roundtable on Evidence-Based Medicine 2007. National Academies Press; 2007.

[6] Institute of Medicine. Genomics-Enabled Learning Health Care Systems: Gathering and Using Genomic Informatino to Improve Patient Care and Research: Workshop Summary. Washington DC: The National Academies Press; 2015.

[7] Overby CL, Kohane I, Kannry JL, Williams MS, Starren J, Bottinger E, et al. Opportunities for genomic clinical decision support interventions. Genet Med 2013;15(10):817–23.

[8] Masys DR, Jarvik GP, Abernethy NF, Anderson NR, Papanicolaou GJ, Paltoo DN, et al. Technical desiderata for the integration of genomic data into Electronic Health Records. J Biomed Inform 2012;45(3):419–22.

[9] Welch BM, Eilbeck K, Del Fiol G, Meyer LJ, Kawamoto K. Technical desiderata for the integration of genomic data with clinical decision support. J Biomed Inform 2014;51:3–7.

[10] Starren J, Williams MS, Bottinger EP. Crossing the omic chasm: a time for omic ancillary systems. JAMA 2013;309 (12):1237–8.

[11] Seaby EG, Rehm HL, O'Donnell-Luria A. Strategies to uplift novel mendelian gene discovery for improved clinical outcomes. Front Genet 2021;12, 674295.

[12] Smart on FHIR 2021 [Available from: https://smarthealthit.org/.

[13] CDS Hooks, 2022 [Available from: https://cds-hooks.org.

[14] Sync for Genes, 2022 [Available from: https://www.healthit.gov/topic/sync-genes.

[15] Health Level Seven International 2022;2022. [Available from: https://www.hl7.org/]; 2022.

[16] Global Alliance for Genomics and Health. [Available from: https://www.ga4gh.org/]. GA4GH 2022.

[17] GMXIII. Meeting Summary; 2021.

[18] Genomic Medicine Meeting VII, 2014.

[19] Duong BQ, Arwood MJ, Hicks JK, Beitelshees AL, Franchi F, Houder JT, et al. Development of customizable implementation guides to support clinical adoption of pharmacogenomics: experiences of the Implementing GeNomics In pracTicE (IGNITE) network. Pharmgenomics Pers Med 2020;13:217–26.

[20] Snyder SR, Mitropoulou C, Patrinos GP, Williams MS. Economic evaluation of pharmacogenomics: a value-based approach to pragmatic decision making in the face of complexity. Public Health Genomics 2014;17 (5–6):256–64.

[21] Herr TM, Bielinski SJ, Bottinger E, Brautbar A, Brilliant M, Chute CG, et al. Practical considerations in genomic decision support: The eMERGE experience. J Pathol Inform 2015;6:50.

[22] Hinderer M, Boerries M, Boeker M, Neumaier M, Loubal FP, Acker T, et al. Implementing pharmacogenomic clinical decision support into German hospitals. Stud Health Technol Inform 2018;247:870–4.

[23] Gill PS, Yu FB, Porter-Gill PA, Boyanton BL, Allen JC, Farrar JE, et al. Implementing pharmacogenomics testing: single center experience at Arkansas Children's hospital. J Pers Med 2021;11(5):394.

[24] Sissung TM, McKeeby JW, Patel J, Lertora JJ, Kumar P, Flegel WA, et al. Pharmacogenomics implementation at the National Institutes of Health Clinical Center. J Clin Pharmacol 2017;57(Suppl 10):S67–s77.

[25] Liu M, Vnencak-Jones CL, Roland BP, Gatto CL, Mathe JL, Just SL, et al. A tutorial for pharmacogenomics implementation through end-to-end clinical decision support based on ten years of experience from PREDICT. Clin Pharmacol Ther 2021;109(1):101–15.
[26] Caraballo PJ, Sutton JA, Giri J, Wright JA, Nicholson WT, Kullo IJ, et al. Integrating pharmacogenomics into the electronic health record by implementing genomic indicators. J Am Med Inform Assoc 2020;27(1):154–8.
[27] CDC, Ten great public health achievements—United States 2001–2010, 2011. Available from: http://www.cdc.gov/mmwr/preview/mmwrhtml/mm6019a5.htm.
[28] Advisory Committee on Heritable Disorders in Newborns and Children. Recommended Uniform Screening Panel; 2016.
[29] Fabie NAV, Pappas KB, Feldman GL. The current state of newborn screening in the United States. Pediatr Clin North Am 2019;66(2):369–86.
[30] Watson MS, et al. Newborn screening: toward a uniform screening panel and system. Genet Med 2006;8(Suppl. 1). 1s–252s.
[31] Conditions Screened by State, 2021 Available from: https://www.babysfirsttest.org/newborn-screening/states.
[32] ACMG. ACMG ACT sheets and algorithms. Bethesda; 2001.
[33] USHIK, 2013. Support appropriate newborn screening and follow-up. Available from: https://ushik.ahrq.gov/mdr/details/administeredItems/requirement/e6e34499-4bea-4f0c-a534-6d46610564fe?system=cehrf.
[34] USHIK, 2015. Support appropriate newborn screening and follow-up. Available from: https://ushik.ahrq.gov/mdr/details/administeredItems/requirement/3a9f1ecf-7109-45f7-92c9-1888215f059a?system=cehrf.
[35] McGlynn EA, Asch SM, Adams J, Keesey J, Hicks J, DeCristofaro A, et al. The quality of health care delivered to adults in the United States. N Engl J Med 2003;348(26):2635–45.
[36] Kawamoto K, Lobach DF, Willard HF, Ginsburg GS. A national clinical decision support infrastructure to enable the widespread and consistent practice of genomic and personalized medicine. BMC Med Inform Decis Mak 2009;9:17.
[37] Khera AV, Won HH, Peloso GM, Lawson KS, Bartz TM, Deng X, et al. Diagnostic yield and clinical utility of sequencing familial hypercholesterolemia genes in patients with severe hypercholesterolemia. J Am Coll Cardiol 2016;67(22):2578–89.
[38] Roberts R. Genetics of coronary artery disease. Circ Res 2014;114(12):1890–903.
[39] Mega JL, Stitziel NO, Smith JG, Chasman DI, Caulfield M, Devlin JJ, et al. Genetic risk, coronary heart disease events, and the clinical benefit of statin therapy: an analysis of primary and secondary prevention trials. Lancet 2015;385(9984):2264–71.
[40] Ogura M. PCSK9 inhibition in the management of familial hypercholesterolemia. J Cardiol 2018;71(1):1–7.
[41] Lee C, Rivera-Valerio M, Bangash H, Prokop L, Kullo IJ. New case detection by cascade testing in familial hypercholesterolemia: a systematic review of the literature. Circ Genom Precis Med 2019;12(11), e002723.
[42] Kawamoto K, Houlihan CA, Balas EA, Lobach DF. Improving clinical practice using clinical decision support systems: a systematic review of trials to identify features critical to success. BMJ 2005;330(7494):765.
[43] Forman A, Sotelo J. Tumor-based genetic testing and familial cancer risk. Cold Spring Harb Perspect Med 2020;10(8):a036590.
[44] de la Chapelle A, Hampel H. Clinical relevance of microsatellite instability in colorectal cancer. J Clin Oncol 2010;28(20):3380–7.
[45] Miller DT, Lee K, Gordon AS, Amendola LM, Adelman K, Bale SJ, et al. Recommendations for reporting of secondary findings in clinical exome and genome sequencing, 2021 update: a policy statement of the American College of Medical Genetics and Genomics (ACMG). Genet Med 2021;23(8):1391–8.
[46] Richards S, Aziz N, Bale S, Bick D, Das S, Gastier-Foster J, et al. Standards and guidelines for the interpretation of sequence variants: a joint consensus recommendation of the American College of Medical Genetics and Genomics and the Association for Molecular Pathology. Genet Med 2015;17(5):405–24.
[47] Heale BS, Overby CL, Del Fiol G, Rubinstein WS, Maglott DR, Nelson TH, et al. Integrating genomic resources with electronic health records using the HL7 infobutton standard. Appl Clin Inform 2016;7(3):817–31.
[48] Ramos EM, Din-Lovinescu C, Berg JS, Brooks LD, Duncanson A, Dunn M, et al. Characterizing genetic variants for clinical action. Am J Med Genet C Semin Med Genet 2014;166c(1):93–104.
[49] Institute NHGR, 2011. Characterizing and displaying genetic variants for clinical action workshop 2011. Available from: https://www.genome.gov/27546546/clinaction-agenda-powerpoints-and-videos.

[50] Williams MS, Taylor CO, Walton NA, Goehringer SR, Aronson S, Freimuth RR, et al. Genomic information for clinicians in the electronic health record: lessons learned from the clinical genome resource project and the electronic medical records and genomics network. Front Genet 2019;10:1059.
[51] Walton NA, Johnson DK, Person TN, Reynolds JC, Williams MS. Pilot implementation of clinican genomic data into the native electronic health record: challenges of scalability. ACI Open 2020;4, e162-e6.
[52] OMIM, 2021. Online Mendelian Inheritance in Man 2021. Available from: https://www.omim.org/.
[53] Amberger JS, Bocchini CA, Scott AF, Hamosh A. OMIM.org: leveraging knowledge across phenotype-gene relationships. Nucleic Acids Res 2019;47(D1):D1038–d43.
[54] Adam MP. Gene Reviews. Holly H Ardinger RAP, Stephanie E Wallace, editor, Gene Reviews. Seattle, WA: University of Washington; 2022.
[55] NCBI. Genetic Testing Registry 2021; 2021.
[56] PharmGKB, 2021. Available from: https://www.pharmgkb.org/.
[57] Crump JK, Del Fiol G, Williams MS, Freimuth RR. Prototype of a standards-based EHR and genetic test reporting tool coupled with HL7-compliant infobuttons. AMIA Jt Summits Transl Sci Proc 2018;2017:330–9.
[58] Clinical Pharmacogenetics Implementation Consortium, 2021. Available from: https://cpicpgx.org/.
[59] Grol R, Dalhuijsen J, Thomas S, Veld C, Rutten G, Mokkink H. Attributes of clinical guidelines that influence use of guidelines in general practice: observational study. BMJ 1998;317(7162):858–61.
[60] Hoffman JM, Dunnenberger HM, Kevin Hicks J, Caudle KE, Whirl Carrillo M, Freimuth RR, et al. Developing knowledge resources to support precision medicine: principles from the Clinical Pharmacogenetics Implementation Consortium (CPIC). J Am Med Inform Assoc 2016;23(4):796–801.
[61] CPIC Github, 2021. Available from: https://github.com/cpicpgx/cpic-data/wiki.
[62] Quality AfHRa. Electronic Health Record-linked Decision Support for Communicating Genomic Data, 2021. Available from: https://digital.ahrq.gov/ahrq-funded-projects/electronic-health-record-linked-decision-support-communicating-genomic-data.
[63] Clinical Decision Support Knowledgebase, 2021. Available from: https://cdskb.org/.
[64] Del Fiol G, Williams MS, Maram N, Rocha RA, Wood GM, Mitchell JA. Integrating genetic information resources with an EHR. AMIA Annu Symp Proc 2006;2006:904.
[65] Resource CG, 2021. EHR Working Group Aims 2021. Available from: https://clinicalgenome.org/working-groups/ehr/#:~:text=The%20EHR%20Working%20Group%20aims,health%20record%20and%20related%20systems.
[66] Cutting E, Banchero M, Beitelshees AL, Cimino JJ, Fiol GD, Gurses AP, et al. User-centered design of multi-gene sequencing panel reports for clinicians. J Biomed Inform 2016;63:1–10.
[67] Federal Register, 2012 Mar 7. 53 p.
[68] Cimino JJ, Jing X, Del Fiol G. Meeting the electronic health record "meaningful use" criterion for the HL7 infobutton standard using OpenInfobutton and the Librarian Infobutton Tailoring Environment (LITE). AMIA Annu Symp Proc 2012;2012:112–20.
[69] Lau-Min KS, Asher SB, Chen J, Domchek SM, Feldman M, Joffe S, et al. Real-world integration of genomic data into the electronic health record: the PennChart Genomics Initiative. Genet Med 2021;23(4):603–5.
[70] Grebe TA, Khushf G, Chen M, Bailey D, Brenman LM, Williams MS, et al. The interface of genomic information with the electronic health record: a points to consider statement of the American College of Medical Genetics and Genomics (ACMG). Genet Med 2020;22(9):1431–6.
[71] Greenes RA, Bates DW, Kawamoto K, Middleton B, Osheroff J, Shahar Y. Clinical decision support models and frameworks: seeking to address research issues underlying implementation successes and failures. J Biomed Inform 2018;78:134–43.
[72] Sutton RT, Pincock D, Baumgart DC, Sadowski DC, Fedorak RN, Kroeker KI. An overview of clinical decision support systems: benefits, risks, and strategies for success. NPJ Digit Med 2020;3:17.
[73] Bussone A, Stumpf S, O'Sullivan D. The role of explanations on trust and reliance in clinical decision support systems. In: International Conference on Healthcare Informatics; 2015. p. 160–9.
[74] Sapp JC, Facio FM, Cooper D, Lewis KL, Modlin E, van der Wees P, et al. A systematic literature review of disclosure practices and reported outcomes for medically actionable genomic secondary findings. Genet Med 2021;23(12):2260–9.
[75] Williams JL, Chung WK, Fedotov A, Kiryluk K, Weng C, Connolly JJ, et al. Harmonizing outcomes for genomic medicine: comparison of eMERGE outcomes to clingen outcome/intervention pairs. Healthcare (Basel) 2018;6(3):83.

[76] Welch BM, Loya SR, Eilbeck K, Kawamoto K. A proposed clinical decision support architecture capable of supporting whole genome sequence information. J Pers Med 2014;4(2):176–99.
[77] Evans JP, Wilhelmsen KC, Berg J, Schmitt CP, Krishnamurthy A, Fecho K, et al. A new framework and prototype solution for clinical decision support and research in genomics and other data-intensive fields of medicine. EGEMS (Wash DC) 2016;4(1):1198.
[78] Welch BM, Rodriguez-Loya S, Eilbeck K, Kawamoto K. Clinical decision support for whole genome sequence information leveraging a service-oriented architecture: a prototype. AMIA Annu Symp Proc 2014;2014:1188–97.
[79] Bates DW, Kuperman GJ, Wang S, Gandhi T, Kittler A, Volk L, et al. Ten commandments for effective clinical decision support: making the practice of evidence-based medicine a reality. J Am Med Inform Assoc 2003;10 (6):523–30.
[80] Roberts NJ, Vogelstein JT, Parmigiani G, Kinzler KW, Vogelstein B, Velculescu VE. The predictive capacity of personal genome sequencing. Sci Transl Med 2012;4(133), 133ra58.
[81] Katsanis SH, Katsanis N. Molecular genetic testing and the future of clinical genomics. Nat Rev Genet 2013;14 (6):415–26.
[82] Shang N, Liu C, Rasmussen LV, Ta CN, Caroll RJ, Benoit B, et al. Making work visible for electronic phenotype implementation: Lessons learned from the eMERGE network. J Biomed Inform 2019;99, 103293.
[83] Genomics Implementation Toolkit, 2021. Available from: https://www.cdc.gov/genomics/implementation/toolkit/tier1.htm.
[84] Buchanan AH, Lester Kirchner H, Schwartz MLB, Kelly MA, Schmidlen T, Jones LK, et al. Clinical outcomes of a genomic screening program for actionable genetic conditions. Genet Med 2020;22(11):1874–82.
[85] Williams MS, Buchanan AH, Davis FD, Faucett WA, Hallquist MLG, Leader JB, et al. Patient-centered precision health in a learning health care system: Geisinger's genomic medicine experience. Health Aff (Millwood) 2018;37 (5):757–64.
[86] Overby CL, Heale B, Aronson S, Cherry JM, Dwight S, Milosavljevic A, et al. Providing access to genomic variant knowledge in a healthcare setting: a vision for the ClinGen electronic health records workgroup. Clin Pharmacol Ther 2016;99(2):157–60.
[87] Aronson SJ, Clark EH, Babb LJ, Baxter S, Farwell LM, Funke BH, et al. The GeneInsight suite: a platform to support laboratory and provider use of DNA-based genetic testing. Hum Mutat 2011;32(5):532–6.
[88] eMERGE Consortium. Harmonizing Clinical Sequencing and Interpretation for the eMERGE III Network. Am J Hum Genet 2019;105(3):588–605.
[89] Haslem DS, Chakravarty I, Fulde G, Gilbert H, Tudor BP, Lin K, et al. Precision oncology in advanced cancer patients improves overall survival with lower weekly healthcare costs. Oncotarget 2018;9(15):12316–22.
[90] Williams JL, Rahm AK, Stuckey H, Green J, Feldman L, Zallen DT, et al. Enhancing genomic laboratory reports: A qualitative analysis of provider review. Am J Med Genet A 2016;170a(5):1134–41.
[91] Sperber NR, Carpenter JS, Cavallari LH, RM C-DH, Denny JC, et al. Challenges and strategies for implementing genomic services in diverse settings: experiences from the Implementing GeNomics In pracTicE (IGNITE) network. BMC Med Genomics 2017;10(1):35.
[92] Hicks JK, Aquilante CL, Dunnenberger HM, Gammal RS, Funk RS, Aitken SL, et al. Precision pharmacotherapy: integrating pharmacogenomics into clinical pharmacy practice. J Am Coll Clin Pharm 2019;2(3):303–13.
[93] Stuckey H, Williams JL, Fan AL, Rahm AK, Green J, Feldman L, et al. Enhancing genomic laboratory reports from the patients' view: a qualitative analysis. Am J Med Genet A 2015;167a(10):2238–43.
[94] Williams JL, Rahm AK, Zallen DT, Stuckey H, Fultz K, Fan AL, et al. Impact of a patient-facing enhanced genomic results report to improve understanding, engagement, and communication. J Genet Couns 2018;27 (2):358–69.
[95] Jones LK, Kulchak Rahm A, Gionfriddo MR, Williams JL, Fan AL, Pulk RA, et al. Developing pharmacogenomic reports: insights from patients and clinicians. Clin Transl Sci 2018;11(3):289–95.
[96] Bentley AR, Callier S, Rotimi CN. Diversity and inclusion in genomic research: why the uneven progress? J Community Genet 2017;8(4):255–66.
[97] Manrai AK, Funke BH, Rehm HL, Olesen MS, Maron BA, Szolovits P, et al. Genetic misdiagnoses and the potential for health disparities. N Engl J Med 2016;375(7):655–65.
[98] Martin AR, Kanai M, Kamatani Y, Okada Y, Neale BM, Daly MJ. Clinical use of current polygenic risk scores may exacerbate health disparities. Nat Genet 2019;51(4):584–91.

CHAPTER

18

Knowledge resources

Guilherme Del Fiol[a] and David A. Cook[b]

[a]Department of Biomedical Informatics, University of Utah Health, University of Utah, Salt Lake City, UT, United States [b]Office of Applied Scholarship and Education Science, Mayo Clinic College of Medicine and Science; and Division of General Internal Medicine, Mayo Clinic, Rochester, MN, United States

18.1 Introduction

Knowledge resources are repositories of medical knowledge developed and maintained by *knowledge publishers.* Knowledge resources serve as core components in the architecture of CDS systems (Fig. 18.1). A knowledge resource consists of *knowledge assets* (also known as knowledge artifacts), which serve as building blocks for CDS tools as well as knowledge for end user consumption [1]. Knowledge assets can be as simple as a narrative text (e.g., recommended antibiotics for community acquired pneumonia) or as complex as a tool that automatically scours the Web for best practices in community acquired pneumonia, uses a text analysis to identify salient recommendations, and reformats these recommendations to human-readable format. Knowledge assets can also refer to the components upon which other resources are built, such as vocabulary value sets (e.g., codes and controlled vocabularies that define a medical concept such as a disease, medication, or procedure), ontologies, order sets, decision rules, and machine learning algorithms (Table 18.1 provides examples of knowledge assets).

Knowledge assets contain information obtained using three primary types of *knowledge acquisition* methods: human-intensive acquisition, data driven acquisition, and evidence synthesis. Human-intensive methods involve manually eliciting knowledge from domain experts (e.g., interviews or expert-authored resources). Data driven methods (e.g., machine learning) automatically infer knowledge from data. Evidence synthesis methods extract medical evidence from the biomedical literature. Chapters 6, 7, and 8 provide details on each of these approaches respectively.

Knowledge assets can be represented using specific *knowledge representation* formats. These include human narrative language, decision tables, decision rules, and statistical models.

Clinical Decision Support and Beyond
https://doi.org/10.1016/B978-0-323-91200-6.00018-8

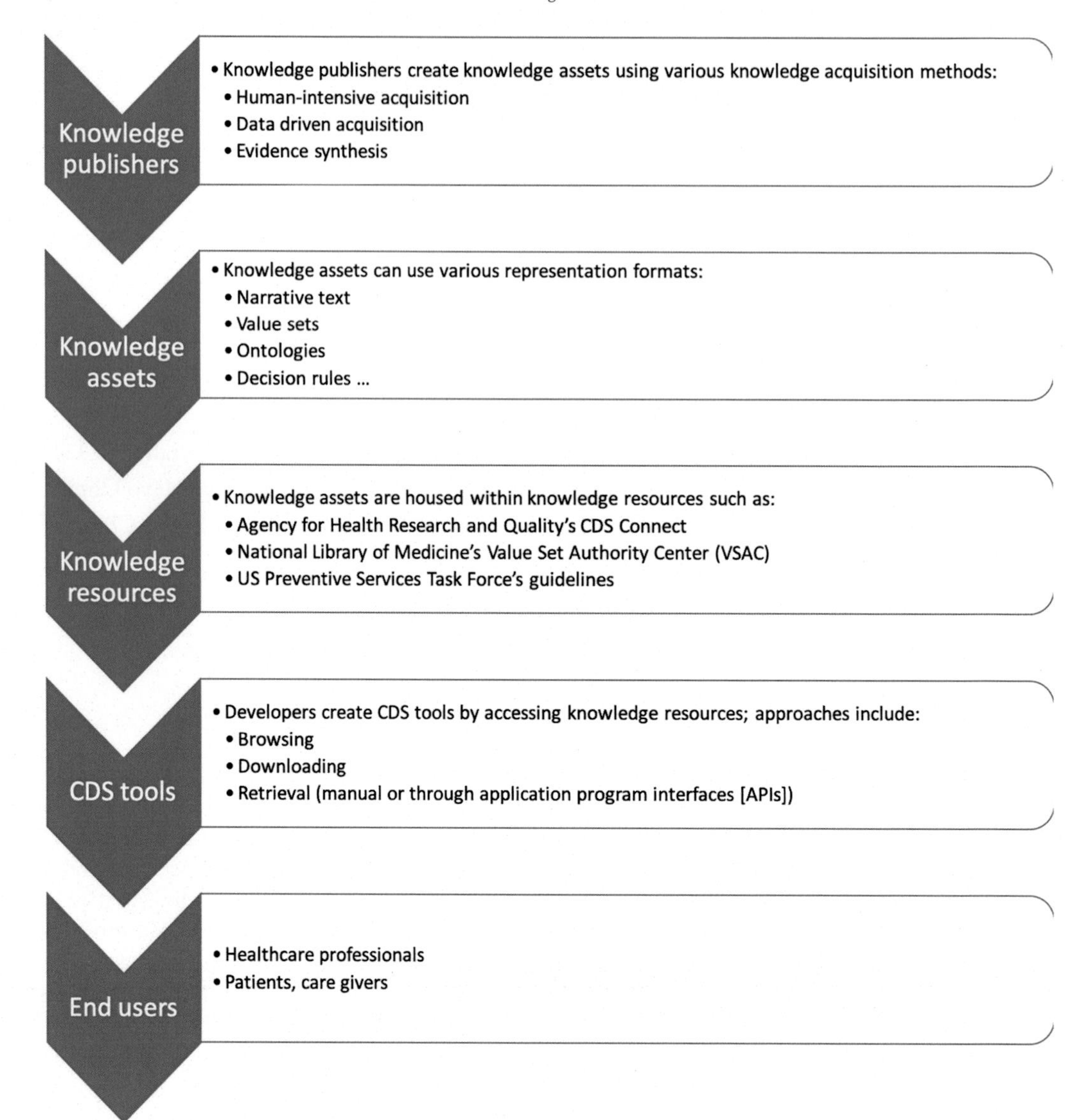

FIG. 18.1 Knowledge resources as building blocks for CDS systems.

Knowledge assets are made available to CDS developers, implementers, CDS tools, and end users through *knowledge access methods* such as search and retrieval.

In this chapter, we further consider the knowledge distribution architectures that make knowledge assets and knowledge resources available to developers, implementers, CDS tools, and users. We first describe the key components of a knowledge distribution architecture, with emphasis on components that are not covered in other chapters, i.e., knowledge publishers, knowledge resources and knowledge access methods; and then describe a general framework for centralized and decentralized architectures.

TABLE 18.1 Example of knowledge assets.

Type of knowledge asset	Subtype	Specific examples
Human readable	Clinical guidelines from national group	USPSTF guideline on prostate cancer screening AHA guideline on cardiac resuscitation CDC guidelines on COVID vaccination
	Systematic reviews	Cochrane review on screening for prostate cancer Meta-analysis for preparation of USPSTF recommendation
	Commercial knowledge summaries	UpToDate summary on diagnosis of prostate cancer DynaMed Plus summary on management of prostate cancer
	Locally-developed best practice summaries	Local guideline on prostate cancer screening
	Patient education materials	Medlineplus "Screening for prostate cancer"
Value set	NA	ICD-10 and SNOMED-CT codes covering different types of prostate cancer
Ontologies and vocabularies	Terminology resource	SNOMED-CT [2,3] concept hierarchy covering prostate cancer screening procedures
Rule-based logic for EHR alerts and reminders	Rule-based logic for preventive care reminders	Logic for an EHR reminder about prostate cancer screening
Order sets	NA	ProVation prostate cancer screening order set

18.2 Knowledge publishers

Knowledge publishers are entities—individuals, groups, organizations, commercial businesses, or government agencies—that produce, maintain, and distribute knowledge assets to CDS developers, providers, and consumers. Knowledge publishers can be classified by their funding/payment model (e.g., government, non-for-profit, commercial entities); or their target audience (e.g., internal, such as a healthcare organization publishing local protocols to guide care for their patients; or external, such as a government agency publishing guidelines for health practitioners).

In the United States, government agencies play an important role in providing health care guidance. For example, the Centers for Disease Control and Prevention (CDC) publishes up-to-date recommendations regarding disease identification, prevention and treatment for both providers and the public. Local, regional, and state agencies play similar roles. The breadth of these recommendations is fairly extensive. For example, a preliminary taxonomy of public health guidance for COVID-19 included 110 categories of recommendations for this topic alone [4]. Similarly, the Agency for Healthcare Research and Quality (AHRQ) and the National Institutes of Health (NIH) also publish knowledge assets; these typically take the form of narrative evidence synthesis, such as the AHRQ Evidence Based Practice Center (EPC) Reports [5]. The National Library of Medicine (NLM) publishes several types of knowledge assets. Some NLM assets are human-readable, such as patient education resources (e.g., MedlinePlus, Genetics Home Reference). Others serve as building blocks for CDS components, such as value sets through the Value set Authority Center (VSAC) [6], and terminology resources such as RxNorm [7] and the Unified Medical Language System [8]. As discussed in Chapter 5, government agencies in other countries play a similar role in creating and disseminating knowledge assets; these include the National Institute for Health and Care Excellence (NICE) in the United Kingdom [9] and the Canadian Task Force on Preventive Health Care, which is funded by the Public Health Agency of Canada [10]. Chapter 26 discusses the role of public health agencies in CDS.

Non-for-profit entities such as medical societies and national expert panels also play a significant role in the development and dissemination of knowledge assets that can be used for CDS. In the US, the US Preventive Services Task Force (USPSTF) is one of the core producers of evidence-based recommendations for primary care; these are implemented nationwide as rule-based preventive care CDS reminders, targeting both for providers and patients [11]. Likewise, the American College of Physicians (ACP) developed the Choosing Widely Guidelines [12], which have also been delivered through a variety of CDS tools such as alerts and order sets [13–15]. Medical specialty societies such as the American College of Cardiology (ACC) and the National Comprehensive Cancer Network (NCCN) also publish comprehensive libraries of clinical guidelines in narrative format, with recommendations that can be implemented as computable logic for delivery through CDS tools for delivery through CDS tools.

Lastly, commercial entities are playing an increasing role in the knowledge resource ecosystem. For example, drug knowledge publishers such as Medispan and First Data Bank provide various knowledge assets related to medications, including narrative drug monographs, terminologies, and logic for drug interaction alerts. Online textbooks such as UpToDate, Dynamed Plus, and Clinical Evidence have become tools used by clinicians worldwide for point-of-care decision support [16]. More recently, commercial knowledge vendors are broadening their scope from print resources to computable sources of knowledge. For example, Wolters Kluwer Health now publishes ProVation order sets; and Stanson Health has developed logic for CDS reminders and alerts, based on the Choosing Wisely recommendations from the ACP.

18.3 Knowledge assets

Knowledge assets consist of discrete pieces of information that can be used as building blocks for CDS or directly consumed by end users. This information is obtained using knowledge acquisition methods that can be classified as human-intensive approaches, such as cognitive task analysis, focus groups, expert panels, and evidence reviews (covered in Chapter 6); data driven approaches, such as data analytics and machine learning (covered in Chapter 7); and evidence synthesis approaches, such systematic reviews and meta-analyses (covered in Chapter 8).

Knowledge assets can represent knowledge using various knowledge representation formats. The most common is human-readable text in the form of publications such as systematic reviews, evidence statements, and clinical guidelines. End-users can of course discover and retrieve such assets on their own, but CDS tools such as Infobuttons can facilitate the process (see Chapter 13 for more information on Infobuttons). CDS developers can also use knowledge assets as building blocks for CDS tools, using representation formats such as decision tables (e.g., drug-drug interactions as rows and the drugs or drug classes that interact as columns; see Chapter 2), decision rules in the form of "if-then-else statements" (Chapter 9), computable guidelines and workflow models (Chapter 10), ontologies and value sets (Chapter 11), and grouped knowledge elements (e.g., order sets, documentation templates) (Chapter 12). Chapter 2 also provides a historical description of knowledge representation formats.

Knowledge assets can have interdependent and/or hierarchical relationships. One asset (or more) can work in parallel with another, or several component assets can serve as building blocks for a larger asset. For example, a decision rule that checks if a patient with new onset of hypertension is receiving recommended first-line anti-hypertensive therapy may refer to a value set with International Classification of Diseases (ICD) codes indicating a diagnosis of hypertension as well as a value set with RxNorm codes for first-line anti-hypertensive medications. The same decision rule could also refer to another decision rule that checks if the patient has a contraindication to receive any of the available first-line anti-hypertensive medications. As described in Chapter 28, such interdependencies can be managed through systematic knowledge management methods to ensure proper functioning of CDS tools.

18.4 Knowledge resources and knowledge distribution

Knowledge resources are collections of knowledge assets, and knowledge resources in turn provide building blocks for CDS tools. Three interrelated but distinct computer system architectures are involved in maintaining, distributing, and implementing knowledge assets. A knowledge management architecture is used to create and maintain knowledge assets (see Chapter 28). A CDS system architecture is used to implement CDS tools (e.g., infobuttons, alerts, or reminders; see Chapter 29). In between, there must be a system that distributes or disseminates assets from the knowledge resource to the CDS tools. A variety of software architectures can be used to provide this functionality; most can be classified as following either a centralized or decentralized approach.

18.4.1 Centralized knowledge architecture

Currently, most CDS implementations follow a centralized knowledge distribution approach (centralized architecture). In a centralized approach, knowledge assets are created, maintained, and retrieved directly within a CDS system, which is typically tightly coupled with a specific EHR. Knowledge assets are created and maintained directly by the EHR vendors or by the customers using embedded authoring tools. For example, most EHR products provide tools that enable healthcare organizations to develop various assets such as value sets, order sets, documentation templates, and rule logic for CDS reminders and alerts. Some of those assets are made available to different CDS tools that are embedded within the EHR itself. Centralized knowledge architectures allow tight control over the creation, maintenance and retrieval of knowledge assets. However, sharing and reusing knowledge assets across different EHR systems or organizations is limited.

18.4.2 Decentralized knowledge architecture

In a decentralized knowledge distribution architecture, CDS systems retrieve knowledge assets from external knowledge resources through various mechanisms, including manual retrieval and automated retrieval. In manual transfer, a CDS developer searches a knowledge resource portal (e.g., CDS Connect) to find relevant knowledge assets, downloads the asset in a sharable format (e.g., comma delimited file, Web Ontology Language [OWL], Clinical Quality Language, FHIR PlanDefinition resource) and imports the asset into the CDS tool for internal execution. In automated transfer, a CDS tool retrieves knowledge assets as-needed through Web services and application program interfaces (APIs). APIs are provided by most external knowledge resources. The timing of retrieval can vary depending on performance and functional requirements. For example, drug classes and value sets can be retrieved periodically (e.g., upon CDS initiation, daily) and cached for efficient internal execution; ontology inferences can be requested real-time upon CDS logic execution; and knowledge in human-readable format can be retrieved in real-time and on demand via Infobuttons (see Chapter 13) for end user access at the point of care.

Fig. 18.2 provides one example of such a decentralized knowledge architecture, illustrating how a preventive care reminder for hypertension screening in adults could be delivered at the point-of-care. Current USPSTF guidelines recommend screening for high blood pressure in individuals 18 years and older who do not have a previous diagnosis of hypertension [17]. A reminder to screen such patients might draw upon four independent knowledge resources: USPSTF guidelines, CDS Connect (AHRQ's repository of CDS logic), VSAC, and RxNorm. A human author might create rules to trigger a reminder based on USPSTF guidelines, and store the corresponding rule-based logic in CDS Connect. The reminder should not trigger for patients who have hypertension; this could be determined by checking the patient's record for hypertension-related diagnostic codes or current hypertension treatment. Diagnostic codes could be identified using value sets from VSAC (i.e., hypertension codes from ICD10

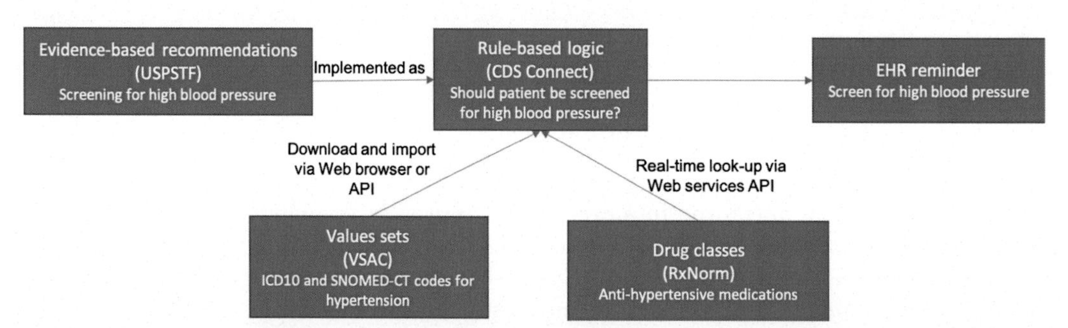

FIG. 18.2 Example of a hypothetical implementation of a decentralized CDS knowledge distribution architecture for preventive care reminders delivered in the EHR, including rule-based logic in CDS Connect based on evidence from USPSTF recommendations, and integrating data from value sets from VSAC and drug class relations from RxNorm.

and SNOMED-CT), which are downloaded as flat files and imported into CDS Connect. Hypertension treatment could be identified through drug class inferences using RxNorm's API. Drug class inferences can be computed real-time upon rule execution or retrieved periodically and processed internally to the CDS system. Using these rules, a patient who does not have hypertension could be flagged for screening as a preventive care reminder in the EHR.

A decentralized knowledge distribution architecture enables an ecosystem in which different people, publishers, and systems can each make distinct contributions to the authoring, maintenance and distribution of different types of knowledge assets. First, different knowledge publishers can focus on different types of knowledge assets based on the publisher's role, areas of expertise, and business model. For example, medical societies and guideline development entities can focus on the development of evidence-based recommendations; terminology development organizations can focus on assets such as value sets and ontologies; and CDS knowledge publishers can focus on the development of CDS logic such as decision rules. Second, different entities can focus on their main areas of expertise and interest, and rely on other entities for knowledge assets that are outside their expertise. Third, knowledge assets can be reused across multiple CDS interventions and healthcare organizations, reducing CDS development efforts. Last, each knowledge asset can be independently maintained and updated with minimal effort to incorporate new scientific evidence, regulatory issues, medications, and diagnostic codes; this contrasts with the centralized architecture, in which changes must typically be manually updated and propagated to all CDS tools and health care systems that use those knowledge assets.

Most CDS implementations still follow a centralized, non-reusable, non-scalable knowledge architecture, coupled with a specific EHR system. However, decentralized architectures are becoming more prevalent, especially with the increasing adoption of standards for CDS interoperability such as SMART on FHIR for third party EHR apps, Clinical Quality Language (CQL) for the representation of decision logic, and CDS Hooks for the integration of external CDS Web services (see Chapters 8 and 29) [18]. Such standards can work in concert to facilitate the implementation of decentralized knowledge architectures. For example, a SMART on FHIR app or a CDS Hooks Web service can execute CQL logic from CDS Connect, and the CQL logic can use value sets from VSAC. Still, several barriers need to be addressed to promote wide adoption of decentralized knowledge architectures, including management of knowledge content; implementation issues such as performance, technical support, reliability, security and privacy concerns; ongoing availability and interoperability of the various components; and funding models to support the long term sustainability of the various components.

18.5 Conclusion

With the advent of computers and the internet, biomedical knowledge resources in electronic format have become ubiquitous in healthcare. Knowledge resources are essential building blocks for today's CDS systems. Centralized knowledge distribution architectures, coupled within EHR systems, are still dominant. However, the rise of decentralized architectures is generating a marketplace of knowledge publishers that create, maintain, and share knowledge assets across CDS systems and healthcare organizations. Such decentralized architectures have the potential to enable the scalable development and dissemination of effective CDS tools [1,19].

References

[1] Lomotan EA, Meadows G, Michaels M, Michel JJ, Miller K. To share is human! Advancing evidence into practice through a national repository of interoperable clinical decision support. Appl Clin Inform 2020;11(1):112–21.

[2] Lee D, Cornet R, Lau F, De Keizer N. A survey of SNOMED CT implementations. J Biomed Inform 2013;46 (1):87–96.

[3] Al-Hablani B. The use of automated SNOMED CT clinical coding in clinical decision support systems for preventive care. Perspect Health Inf Manag 2017;14(Winter).

[4] Taber P, Staes CJ, Phengphoo S, Rocha E, Lam A, Del Fiol G, et al. Developing a sampling method and preliminary taxonomy for classifying COVID-19 public health guidance for healthcare organizations and the general public. J Biomed Inform 2021;120, 103852.

[5] Agency for Healthcare Research and Quality R, MD. Evidence-based practice center (EPC) reports, 2022. [updated 1/1/2022. Available from: https://www.ahrq.gov/research/findings/evidence-based-reports/index.html.

[6] Bodenreider O, Nguyen D, Chiang P, Chuang P, Madden M, Winnenburg R, et al. The NLM value set authority center. Stud Health Technol Inform 2013;192:1224.

[7] Bodenreider O, Cornet R, Vreeman DJ. Recent developments in clinical terminologies—SNOMED CT, LOINC, and RxNorm. Yearb Med Inform 2018;27(1):129–39.

[8] Amos L, Anderson D, Brody S, Ripple A, Humphreys BL. UMLS users and uses: a current overview. J Am Med Inform Assoc 2020;27(10):1606–11.

[9] (NICE) NIfHaCE. Guidance, NICE advice and quality standards United Kingdom, 2022. [updated 1/31/2022. Available from: https://www.nice.org.uk/guidance/published?ndt=Guidance&ndt=Quality%20standard.

[10] (CTFPHC) CTFoPHC. CTFPHC guidelines Canada, 2022. Available from: https://canadiantaskforce.ca/.

[11] (USPSTF) UPSTF. USPSTF recommendations United States, 2022. Available from: https://www.uspreventiveservicestaskforce.org/uspstf/.

[12] ACP) ACoP. Choosing Wisely United States, 2022. Available from: https://www.choosingwisely.org/getting-started/lists/.

[13] Gottheil S, Khemani E, Copley K, Keeney M, Kinney J, Chin-Yee I, et al. Reducing inappropriate ESR testing with computerized clinical decision support. BMJ Qual Improv Rep 2016;5(1). u211376.w4582.

[14] Chen D, Bhambhvani HP, Hom J, Mahoney M, Wintermark M, Sharp C, et al. Effect of electronic clinical decision support on imaging for the evaluation of acute low back pain in the ambulatory care setting. World Neurosurg 2020;134. e874-e7.

[15] Jenkins I, Doucet JJ, Clay B, Kopko P, Fipps D, Hemmen E, et al. Transfusing wisely: clinical decision support improves blood transfusion practices. Jt Comm J Qual Patient Saf 2017;43(8):389–95.

[16] Aakre CA, Pencille LJ, Sorensen KJ, Shellum JL, Del Fiol G, Maggio LA, et al. Electronic knowledge resources and point-of-care learning: a scoping review. Acad Med 2018;93. (11S Association of American Medical Colleges Learn Serve Lead: Proceedings of the 57th Annual Research in Medical Education Sessions): S60-s7.

[17] Force UPST. Screening for hypertension in adults: US preventive services task Force reaffirmation recommendation statement. JAMA 2021;325(16):1650–6.

[18] Strasberg HR, Rhodes B, Del Fiol G, Jenders RA, Haug PJ, Kawamoto K. Contemporary clinical decision support standards using health level seven international fast healthcare interoperability resources. J Am Med Infom Assoc: JAMIA 2021;28(8):1796–806.

[19] Richardson JE, Middleton B, Platt JE, Blumenfeld BH. Building and maintaining trust in clinical decision support: recommendations from the patient-centered CDS learning network. Learn Health Syst 2020;4(2), e10208.

SECTION IV

Adoption of clinical decision support and other modes of knowledge enhancement

CHAPTER

19

Cognitive considerations for health information technology in clinical team environments

Amy Franklin[a] *and Jiajie Zhang*[b]

[a]School of Biomedical Informatics, Center for Digital Health and Analytics, University of Texas Health Science Center at Houston, Houston, TX, United States [b]School of Biomedical Informatics, University of Texas Health Science Center at Houston, Houston, TX, United States

19.1 Introduction

Health information technology (HIT) has great potential to increase care quality, efficiency, and safety through its wide adoption and meaningful use. In the United States (US) in 2004, the national HIT Initiative started by President Bush and strengthened by President Obama with the $19 billion HITECH Act under the American Recovery and Reinvestment Act (ARRA) spurred the widespread adoption of electronic health record (EHR) systems. Nearly 20 years later, there have been many gains in the use of HIT along with continued challenges [1–5]. The cognitive, financial, security/privacy, social/cultural, and workforce hurdles have evolved with technological advances. For example, while hesitancy in the adoption of EHR systems has waned, similar concerns have arisen with the emergence of new forms of HIT such as adaptive CDS and repository systems support [6–8].

With the growth of HIT comes increased need to manage the volume, veracity, velocity, variability, variety, and value of new and compounding data which can strain the limits of human processing [9]. Cognitive support for HIT is intended to assist clinical problem solving and decision making such that the care for patients can be maximized along the six dimensions described by the National Academy of Medicine (formerly, Institute of Medicine) of quality (safe, effective, timely, efficient, equitable, and responsive) [10]. This chapter is devoted to exploring the methodologies of cognitive science as they are applied to more fully understanding the stresses of the clinical environment to aid in developing clinical decision

Clinical Decision Support and Beyond
https://doi.org/10.1016/B978-0-323-91200-6.00024-3

support (CDS) to meet these needs. Much of the stresses come from the nature of health care itself, the burdens of the information and knowledge explosion, the multiplicity of diagnostic and therapeutic choices available, the time pressures, and the fragmentation of care, which have led to demand for CDS in the first place. The need to better understand cognitive considerations is especially true for more complex care, when the patients themselves are more complicated, multiple participants are involved in the health care team, and often the environments themselves are stressful—such as in the emergency department, operating room, or critical care unit.

The National Center for Cognitive Informatics and Decision Making in Healthcare (http://www.sharpc.org), funded by the Strategic Health Advanced Research Projects (SHARP) grant program under the US Office of the National Coordinator for Health IT (ONC), characterizes the cognitive challenges for HIT as the gaps between HIT systems with good and poor cognitive support at three levels (Fig. 19.1). (a) At the **work domain level**, HIT systems with good cognitive support should have an explicit, unified, accurate, and comprehensive model that reflects the true ontology of the work domain, which provides a clear understanding of the care that is independent of how systems are implemented. What this means for HIT

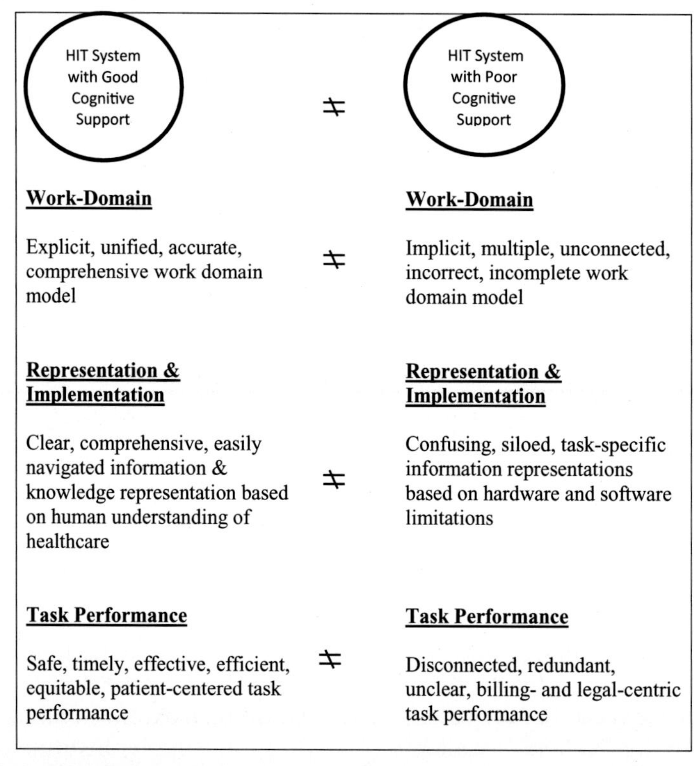

FIG. 19.1 Cognitive challenges for Health IT are characterized as the gaps at three levels between HIT systems that have good and poor cognitive support.

is that the systems should be developed with a work domain ontology for health care that reflects all the goals, needs and challenges of clinical care. Such a model should hold across sites regardless of the implementation (e.g., which EHR system is in place, or if providers are physicians or nurse practitioners). HIT systems with poor cognitive support typically suffer from having models of the work domain that are implicit, multiple, unconnected, disparate, incomplete, and often inaccurate. (b) At the **representation and implementation level**, HIT systems with good cognitive support are characterized as having clear, comprehensive, easy to navigate information and knowledge models optimized for human users. That is, the systems should be useful, usable, and satisfying for the end users, HIT systems with poor cognitive support usually have representations that are based on hardware and software features, which make them confusing, siloed, task-specific, difficult to use and learn, and hard to navigate, because they do not match human needs and expectations. (c) At the **level of task performance**, HIT systems with good cognitive support are characterized by having "built-in" safe, timely, effective, efficient, equitable, patient-centered task performance [10]. HIT systems with poor cognitive support often have disconnected, redundant, tedious, and unclear user models based on business and legal requirements that interfere with task performance. These gaps between good and poor systems highlight some of the issues the ONC named in their call for proposals for the SHARP programs. Strong cognitive support within a well-designed HIT system is built on appropriate models of how clinicians make decisions, provides information display and visualization to increase situation awareness, facilitates decision making under stress and time pressure, improves communication among clinicians, patients, and teams, and operates within highly usable systems.

19.2 Challenges for cognitive support in health care

19.2.1 Too much to find and buried too deep

Physicians need to perform life-critical tasks that require acquisition, processing, transmission, distribution, integration, search, and archiving of significant amounts of data in a distributed team environment in a timely manner. While HIT provides opportunities for support in these environments, there are also concerns regarding the impact of such technology on clinical performance. With the introduction of EHRs, increasingly augmented with further data from health information exchange, data comes at physicians in large volumes. Clinicians must not only manage potential information overload [11–13], they must also make efforts to ensure that the abundance of data in EHR systems and the unintended consequences of such technology do not lead to error [2,14–17].

With the increasing role of HIT and electronic data repositories in clinical settings, it is relevant to evaluate the role of technology in supporting (or impeding) clinical reasoning and decision-making [18]. Increases in information can lead to overload if, as Bawden [19] suggests, "information received becomes a hindrance rather than a help when the information is potentially useful." For example, in a survey of 229 general practitioners, Christensen and Grimsmo [20] found 37% of the group sometimes gave up searching for information simply because it was too time-consuming. Significant redundancy in data and the sheer volume of information make it difficult to both isolate individual pieces of data and also, at the same time, problematic to gain an appropriate overview of the patient's entire record. Kannampallil et al. [21] find that information seeking is challenged by both the cognitive limitations of clinicians, such as memory capacity and the aforementioned overload, as well as limitations imposed by technology. Intensive care physicians in this study [21] were observed iteratively swapping back and forth between electronic and paper resources (as well as within different segments of single sources such as the EHR) as they worked to find and re-find information. Implications from this additional cognitive burden include negative impacts on their ability to filter information for reasoning and decision making [22,23].

19.2.2 Unintended consequences

Some of the identified unintended consequences of HIT, particularly for computerized provider order entry and CDS, include changes to work and workflow (including increases in volume of effort), changes in roles and responsibilities, negative alterations to communication, new types of errors, and additional cognitive burdens such as alert fatigue and management of misleading content [2,14–17,24,25].

Similar consequences of poor EHR usability [2,26–28] are seen when the burden of tools such as CDS overshadow the benefits. For example, Kizzier-Carnahan and colleagues [29] estimate that an average ICU physician is met with 900 active and passive alerts per day. Studies exploring the frequency of alerts (as common as every other order [30]) and the rate of their dismissals (up to 95% [31]) demonstrate that difficulty of value amidst volume. Over reliance or automation bias may also occur as some providers become dependent on the monitoring provided by the system [32,33]. Management of these challenges include careful work-centered design of the systems and ongoing on monitoring of use. Sutton et al. [34] and Middleton et al. [35] provide overviews of the historical and maturing needs of CDSS.

19.2.3 Socio-technical design

19.2.3.1 Complex team environments

HIT with poor cognitive support disconnects tasks from the desired focus on patient-centered care by using representations that are not intuitive or are limited by the technology. Often such systems indicate a poor understanding of the work domain by their designers. Research on teamwork in complex environments has been going through a new and major challenge due to the explosive growth of information technology. The role of HIT is much more than the transformation of cognitive labor from people to machine. Information technology has become an inherent part of the complex work system, which includes passive artifacts, active agents, communication tools, workflow processes, and information and knowledge bases. Information technology also modifies the structures, processes, and outcomes of the complex work system. It not only changes how individuals and teams perceive, act, solve problems, reason, make decisions, communicate, and interact with other people but also determines these processes to a higher and higher degree.

Research on teamwork has been very active in several areas of social and behavioral sciences, such as industrial and organizational psychology, social psychology, organizational behavior, and management science. Research in these areas has focused on interactions among team members, individual mental structures and shared mental models, psychological processes and mechanisms, and influences of cognitive, personality, motivational, emotional, social, organizational, and cultural factors on team performance and dynamics [36–41]. The role of information technology on teamwork has been investigated in these areas [42,43], although the emphasis of these areas is on psychological and behavioral issues, not on information technology. Needs with respect to information technology were discussed in a panel at the Human Factors and Ergonomics Society Conference in 2012 focused on patient-centered communication, its role in patient outcomes, and the absence of attention to teamwork as managed by EHR systems [44]. In fact, in a systematic review of implemented systems, Van de Velde et al. [45] found that directing CDS to patients and including automatic display of CDS on shared screens has the potential to improve adherence. Additional support for teamwork, including patients, and the cognitive demands of group effort are required.

19.3 Developing cognitive support: Distributed cognition

The study of distributed cognition is a scientific discipline that is concerned with how cognitive activity is distributed across human minds, external cognitive artifacts, and groups of people, and how it is distributed across space and time [46–54]. In this view, people's cognitive behavior results from interactions with other people and with external cognitive artifacts (including information technology), and people's activities in concrete situations are guided, constrained, and, to some extent, determined by the physical, cultural, social, historical, and organizational contexts in which they are situated [54]. The unit of analysis for distributed cognition is an entire distributed system, composed of a group of people interacting with external cognitive artifacts, as depicted in Fig. 19.2. Such a distributed system (e.g., intensive care unit in a hospital or airplane cockpit) can have cognitive properties that differ radically from the cognitive properties of the components [55]. In general terms, the components of a distributed cognitive system are described as internal and external representations. Internal representations are the knowledge and structure in individuals' minds; and external representations are the knowledge and structure in the external environment [46,48].

19.3.1 Distributed cognition between individuals and artifacts

Many complex information processing tasks require the processing of information distributed across internal minds and the external artifacts [50]. External artifacts are defined as objects (e.g., buttons on a medical device), symbols (e.g., vital signs on a patient chart), tools (e.g., BMI calculator), and other entities that support or modify human cognitive behavior. It is the interwoven processing of internal and external information that generates much of a person's intelligence. Let us consider multiplying 965 by 273 using paper and pencil. The internal representations are the meanings of individual symbols (e.g., the numerical value of the arbitrary symbol "5" is the quantity five), the addition and multiplication tables, arithmetic procedures, etc., which must be retrieved from memory. The external representations are the shapes and positions of the symbols, the spatial relations of partial products, etc., which can be

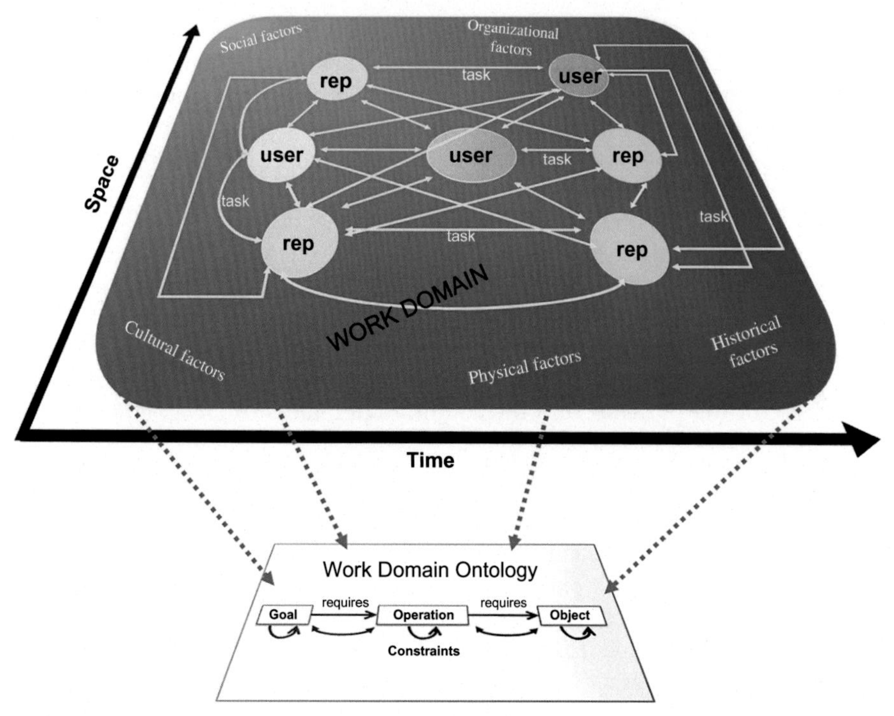

FIG. 19.2 The theoretical framework of distributed cognition. The upper part shows how cognition is distributed across users (internal representations) and representations (external representations), across space and time, and situated in social, cultural, physical, organizational, and historical backgrounds. The lower part is the abstract structure, called the work domain ontology, is an implementation-independent description of the work domain.

perceptually inspected from the environment. To perform the multiplication task, people need to process the information perceived from external representations and the information retrieved from internal representations in an interwoven, integrative, and dynamic manner.

19.3.2 The power of external representations

One important aspect emphasized by distributed cognition research is that external representations are more than inputs and stimuli to the internal mind. External representations have many non-trivial properties that empower human cognitive capability [50]. External representations make information displays and visualization (such as dashboards for the ED or ICU—see Chapter 14 for details) into powerful aids to human cognition due to the following features: they provide information that can be directly perceived and used, such that little effortful processing is needed to interpret and formulate the information explicitly [50,56]; they support perceptual operators that can recognize features easily and make inferences directly [57]; they stop time to make invisible and transient information visible and sustainable [58]; they provide short-term or long-term memory aids so that overall memory load can be reduced; they provide knowledge and skills that are unavailable from internal representations [59]; they anchor and structure cognitive behavior without conscious awareness [50,60]; and they change the nature of a task by generating more efficient action sequences [47].

19.3.3 Distributed cognition across individuals

Cognition can also be distributed across a group of individuals. For this type of distributed cognition, there are two different views. The reductionist view considers that the cognitive properties of a group can be entirely determined by the properties of individuals. In this view, to understand group behavior, all we need is to understand the properties of individuals. In contrast, the interactionist view considers that the interactions among the individuals can produce emergent group properties that cannot be reduced to the properties of the individuals. In this view, to study group behavior, we need to examine not only the properties of individuals but also the interactions among the individuals. Examples of emergent group properties include group affect [61], collective efficacy [62], and transactive memory systems [63].

One important issue in distributed cognition across a group of individuals is the group effectiveness problem [64]. A group of minds can be better than one (process gain), because in a group there are much more resources, task load and memory load are shared and distributed, errors are cross-checked, and so on. The performance of a group can also be worse than that of an individual (process loss), because in a group communication takes time, knowledge may not be shared, and different strategies may be used by different individuals. This phenomenon has been shown in a clinical environment where people work face-to-face, sharing tacit knowledge [52] and at a distance. It was also demonstrated empirically that whether two minds were better or worse than one mind depended on how the knowledge was distributed across the two minds [49]. The issue of group effectiveness is especially important in health care and has received some attention [53].

19.3.4 Cognitive work in a distributed system

From the distributed cognition perspective, cognitive work can be viewed in two different ways. From the individual user perspective, cognitive work is measured by the performance of the individuals in terms of time-on-task, success rate, error rate, etc. From the system perspective, cognitive work is measured by the performance of the distributed system composed of both users and technology. Information, knowledge, processes, and constraints can all be distributed across users and technology in various ways.

For a distributed system, there is an ontology of the work domain (see the lower part of Fig. 19.2) that the distributed system entails. The ontology of the work domain is the basic structure of the work that the system together with its human users will perform. It is an explicit, abstract, implementation-independent description of that work. The work domain ontology is composed of goals, operations, objects, and constraints. Correctly identifying the ontology of a work domain is essential for identifying the information needs for the design of user-centered information systems. More details of work domain ontologies are described in Section 19.4 about TURF (see also Table 19.1).

TABLE 19.1 The TURF framework and its components.

TURF Components	
User analysis	*Clinical roles* The process of identifying the types of users and the characteristics of each type of users Types of users: physicians at various levels (e.g., attending, fellow, resident, medical student); nurses with different responsibilities (e.g. charge nurse, floor nurse) User Characteristics: experience and knowledge of EHR, knowledge of computers, education background, cognitive capacities and limitations, perceptual variations, age related skills, cultural background, personality, etc.
Functional analysis	*Work domain ontology* The process of identifying the ontology of the work domain. The work domain ontology is the basic structure of the work that the system, together with its human users, will perform. It is an explicit, abstract, implementation-independent description of that work Components of the work domain ontology: goals (e.g., treating high glucose level in a pre-diabetic patient), operations (e.g., writing a medication prescription), objects (e.g., patient name, doctor's name, diagnosis, medication name, dosage, frequency, duration, route), and constraints (e.g., the relation between the operation "write a medication prescription" and the objects "Metformin" and "500mg")
Representational analysis	The process of evaluating the appropriateness of the representations for a given task performed by a specific type of user, such that the interaction between users and systems is in a direct interaction mode Examples of Representations: user interface objects such as icons, lists, tables, graphs, views, and windows Methods for representation analysis: isomorphic representations, affordance analysis, heuristic evaluations, etc.
Task analysis	The process of identifying the steps of carrying out an operation by using a specific representation, the relations among the steps, and the nature of each step (mental or physical) Methods for task analysis: key-stroke level modeling, user activity logs, workflow analysis, observations and interviews, etc.

Failure to consider the work across roles or across the system can be seen in implementations that do not appropriately support the shared work [65]. For example, in academic medicine, a CDS system may not support both the attending and resident information gathering approaches. Salwei et al. [66] found residents in their simulation study reported their pulmonary embolism diagnosis tool was incompatible with the resident workflow. CDS must strive to support the cognitive work of the teams in coordination rather than solely focus on the individual or task in isolation.

19.3.5 Organizational memory

Organizational memory is an important research topic for distributed cognition as well as for the field known as Computer-Supported Collaborative Work (CSCW) [67]. Organizational memory is the collection of knowledge embedded in individuals, artifacts, and processes in a team setting. It is the collective long-term memory of a team in a complex environment. It involves individuals, artifacts, organizational culture, organizational transformation, organizational structure, institution manuals, filing systems, databases, stories, etc. Its encoding, storage, organization, retrieval, and transmission are all potentially important factors for team performance.

Designing information systems' infrastructures for the capture of organizational memory and the distribution of this knowledge across a team requires not only an in-depth understanding of the numerous technical knowledge management activities, but also, more importantly and often omitted, an understanding and inclusion of the social, cultural, organizational, and cognitive aspects that not only occur within an individual or group of individuals but also occur across individuals and artificial agents. In general, any information system that supports organizational memory for a team or organization should have the following properties [68,69]:

- Provide a means for collaborative communication
- Capture informal knowledge
- Organize knowledge as searchable data
- Frame formal knowledge within context
- Increase search and retrieval capabilities
- Increase information sharing across team members
- Minimize repeated problem solving with routine tasks
- Decrease interruptions
- Redirect one-to-one to team communication patterns

19.3.6 Group decision making and technology

Groups have many functions: to communicate, share information, generate ideas, organize ideas, draft policies and procedures, collaborate on report writing, share a vision, build consensus, make decisions, etc. Nunamaker et al. [70] used an electronic meeting system for teams to demonstrate how information technology can affect team dynamics in significant ways. They considered the following factors for the design and evaluation of this system: group factors such as size, proximity, composition, and cohesiveness; task factors such as activities required to accomplish the task and task complexity; context factors such as organizational structure, time pressure, evaluative tone, and reward structure; and outcomes factors such as efficiency, effectiveness, and satisfaction. This system showed a number of positive effects on team performance: it increased many process gains and decreased many process losses that typically occur as a result of team behavior. The process gains that were increased included increases in information, synergy, objective evaluation, stimulation, and learning; and the process losses that were decreased included air-time fragmentation, attenuation blocking, concentration blocking, attention blocking, memory failure, conformance pressure, evaluation apprehension, socializing, domination, etc. The system has some drawbacks. Some process losses were increased, among them information overload, slower feedback, free riding, and incomplete use of information. Studies such as this one show that many of the properties of teams identified from behavioral studies of their operation without interactions with technology can be changed, eliminated, or transformed in systematic ways by introducing information technology. This also demonstrates that teams in a distributed system with technology do not behave in the same way as teams in a distributed system that is only composed of people, even if the tasks that are performed in the two systems are the same.

19.3.7 Group decision making in clinical contexts

Shared decision making involving the patient in conjunction with the clinical team is one area in which group decision making and HIT has received significant attention. Shared decision making (SDM) is defined as "...a formal process or tool that helps physicians and patients work together to choose the treatment option that best reflects both medical evidence and the individual patient's priorities and goals for his or her care [71]." Technology in the form of decision aids is explored in patient-physician decision making often with a focus on patient's access and ease of use [72], patient knowledge gain [73] and changes in patient's decisional conflict [74].

Cognitive interventions for other group contexts aim to support decision making by easing the burden of group meetings, increasing adherence to protocols or supporting individuals as components of the traditional team process [75]. Although care may be provided by a group, much of the literature on decision making in health care domains focuses on communication gaps and interventions [76], mutual agreement on treatment plans [77] or individual and/or role-based differences within a team [78].

Although not representing forms of group decision making per se, there are other ways that the power of the masses are being incorporated into HIT solutions, ranging from translation of SNOMED terms [79] to crowd-sourcing the identification of relationships between clinical problems and medications [80], and using the wisdom of the crowds to develop diagnostic decision support systems [81].

19.4 Building systems with distributed cognition in mind

The design of clinical decision support applications often focuses on the translation of an existing process to an EHR based interface. Review of prior implementations have led to guidelines highlighting the need for human factor principles to be incorporated into the designs through a focus on adherence to heuristics and identifying needed interactions (i.e., ability to filter, drill-down, manipulate the data [82]). These recommendations support the creation of the physical structure of the tool by shaping the visualizations (e.g. font size, color selection, item placement) and defining general needs (e.g. undo, feedback, help). Complementing these recommendations, approaches such as the TURF framework for evaluation of HIT provide a roadmap connecting cognitive principles to design through more granular content analysis.

19.4.1 TURF: A framework for HIT usability and cognitive support

TURF [83] is a cognitive framework originally developed for the evaluation, measurement, and design of EHR usability. However, its principles and methods can also be applied to address the cognitive factors for workflow and decision making in complex team environments. TURF stands for Task, User, Representation, and Function, which are the four core components of user-centered design. Table 19.1 describes these four components for the clinical environment. To develop good cognitive support for clinical care through HIT, we need to consider all four components. Different users have different needs, capabilities, and constraints. Therefore, HIT systems developed for different users should be customized. An ontology of work is a foundation for designing effective cognitive support. If the ontology of work is not correctly and completely identified, HIT systems developed may have overhead functions that are not essential for the work, interfere with the execution of the required work, and potentially induce errors. The representations, or user interfaces, of HIT systems should follow human-centered principles to maximize the effects of cognitive support and reduce usability problems. Also, tasks are sequences of activities that are carried out by specific users, using certain representations to achieve the goals in the work domain. Understanding the tasks is essential for optimizing the workflow and decision support. This four-part harmony builds cognitive support through good usability into HIT systems.

19.5 Developing tools to support cognition

The presentation of information to clinicians has the potential to profoundly influence their decision-making [84–86]. However, current systems often present information in a fragmented fashion, splitting a single patient record across screens and tables in different formats [87–89] Disjointed records, redundancy of information, and the sheer volume of data to be sifted through can prove challenging to users of HIT. The opportunities for support through technology and automated methods are being explored in data presentation or visualization and ways in which information can be clustered or aggregated to decrease human cognitive efforts (see Chapter 14 for a detailed discussion).

For example, medication reconciliation is a complex task requiring the consideration of multiple streams of disparate and potentially incomplete information. The physician must assimilate and reconcile all the patient's medications from all available sources (e.g., hospitalizations, specialists, over-the-counter medications) into a single active medication list. To facilitate this process, Plaisant and her research team [90] have created Twinlist, a collection of interface designs for this purpose. Features of Twinlist include the use of spatial layouts that highlight and separate the sources of medications as well as a multi-step animation that visually highlights the reconciliation process by indicating the movement of medications from each list as well as the drugs requiring management (Fig. 19.3).

Within this display, columns separate unique, similar (e.g., generic versus brand-name drugs, different timings of administration, varying dosages), and identical medications not requiring reconciliation. Each column is dedicated to a source of the lists (as seen above, showing intake versus hospital medication records). As a physician views the animation process, he or she can select between the medications that are similar and indicate to the system which medications (and their formulation, instructions, etc.) should be included in the final list. Visual indicators such as color coding highlight the medications that have been reconciled. Through this design, Twinlist seeks to improve the speed and accuracy of medication reconciliation by supporting the cognitive needs of the users.

19.5.1 Situation awareness

Situation Awareness (SA) is a theory that evolved out of the aviation industry and military decision making and is used to understand decision making and error. Endsley [91] describes situation awareness as *the perception of the elements in the environment in a volume of time and space, the comprehension of their meaning, and the projection of their status in the near future.* These components are linked to different levels of decision making, as they exist under the influence of task or system factors. Taken together, the distributed system and the situation awareness of individuals within it contribute to quality of decision-making. In health care, this framework has been applied and extended [92] to four levels: information perception, information comprehension, forecasting of future events, and choosing appropriate action (resolution) based on the first three levels. Understanding current circumstances "what's going on" while projecting to "what's next" is critical to this fourth level of decision making. Gaps in SA can lead to error; as Singh et al. [92] points out, there are many levels at which situation awareness may falter, and missing situational understanding can be propagated throughout the distributed team. Failures leading to compromised SA have focused on misperception, shortcuts in

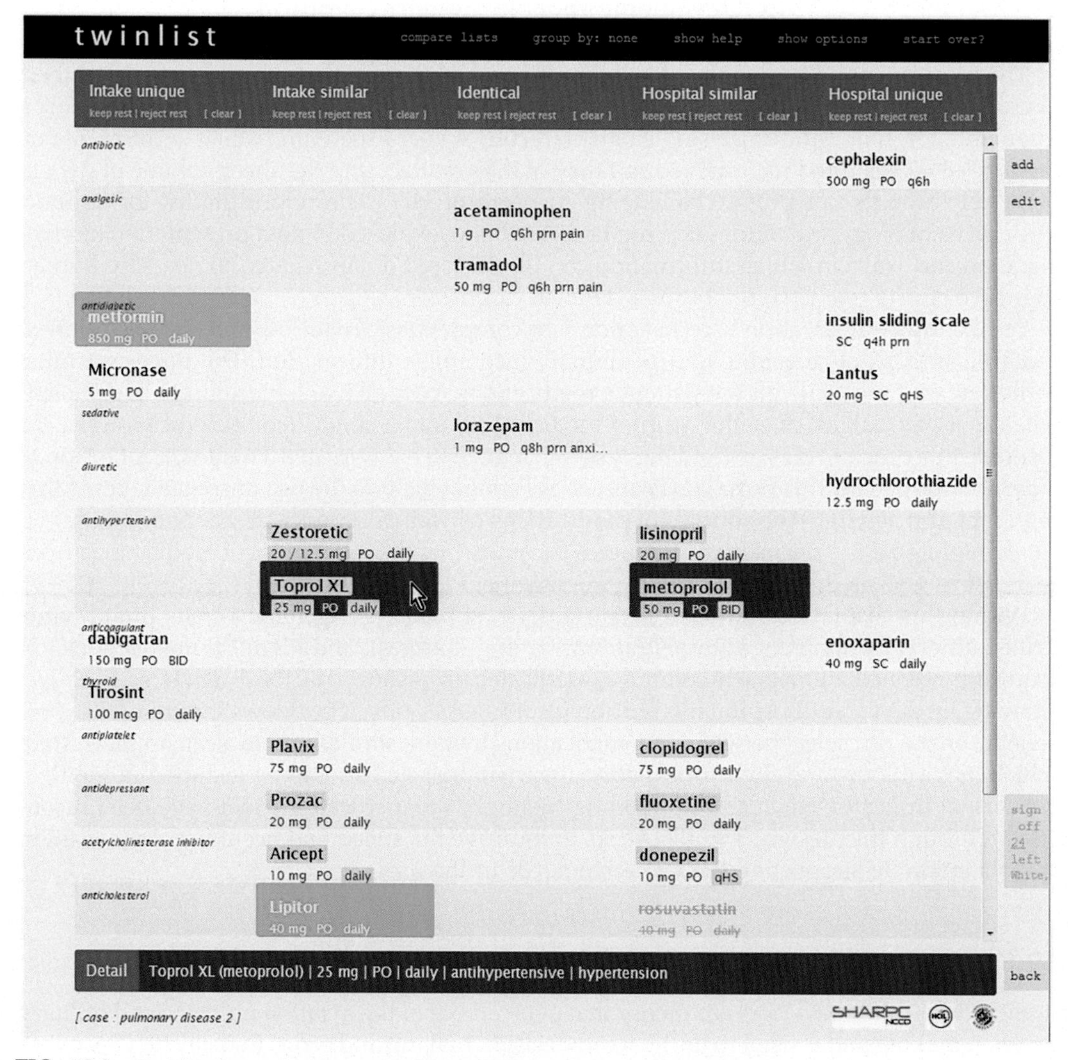

FIG. 19.3 Twinlist for medication reconciliation [90]. *Permission from C. Plaisant.*

reasoning and factors such as fatigue, stress or interrupted workflow [93,94]. Improving situation awareness as a means of improving safety and care has been embraced by anesthesiology [95] and in operating rooms [96], and SA has been deemed a critical human factors topic for patient safety by the World Health Organization [97].

Information dashboards are one way in which improving situation awareness has been approached in health care. Dashboards have become important business intelligence tools for many industries [98], and similar information systems arising from whiteboards and later electronic boards are seen in diverse settings from emergency care [99] to labor and delivery [100]. The intent of such systems includes facilitating group communication and team situation awareness [101], efficiency [102], general care, and administrative processes such as bed location monitoring [103]. One of the intents of such systems is to externalize some of the demands placed on clinicians to recall the details regarding all patients under their care.

For example, a dashboard in the Emergency Department might display a patient identifier, location or room of the patient, chief complaint, clinical team caring for the patient, status of treatment and testing as well as measures such as length of stay in the unit and the time elapsed from bed assignment to seeing a provider. The intent of these systems is to help clinicians manage patient care, maximize ER throughput, and make appropriate selections in their course of action based on needs at that time as shown in Fig. 19.4. Dashboards can also be used to highlight at-risk patients by indicating changes in vital signs, lab results or other critical values. Shifting demands in the environment such as forecasting overcrowding conditions or a need for diversion can also be improved through displays that increase the situation awareness of clinicians regarding bed availability, wait times, patient needs, staffing conditions, and trends such as patients leaving without being seen. Additionally, communication demands are eased through increased information available on such systems such as the status of labs or patient dispositions. Dashboards and other electronic information sources have proven to increase patient throughput [102,103] and improve individual task completion times [104].

19.5.2 Data aggregation

The development of such support systems has the potential to improve decision making, reduce memory burden, and improve quality of care. For example, one way quality may be improved is to bolster the performance of trainees so that they function more like experts. Patel et al. [105] suggest that the process of clinical comprehension differs between expert and novice clinicians with respect to selective filtering, pattern recognition, and accuracy of inferences generated. Experts use knowledge structures called *intermediate constructs* that represent clinically meaningful clusters of observations that lead toward specific diagnoses. In contrast, although non-experts may possess a large knowledge base, they tend to be less organized. Information aggregation *built into the system* can provide information to end users at an *intermediate* rather than raw level. The Psychiatric Clinical Knowledge Enhancement System (PSYCKES) [106] is one such example that presents information at the level of intermediate constructs through clusters of relevant information. This has been shown to lead to improved comprehension by residents when compared to current non-aggregated practice [107].

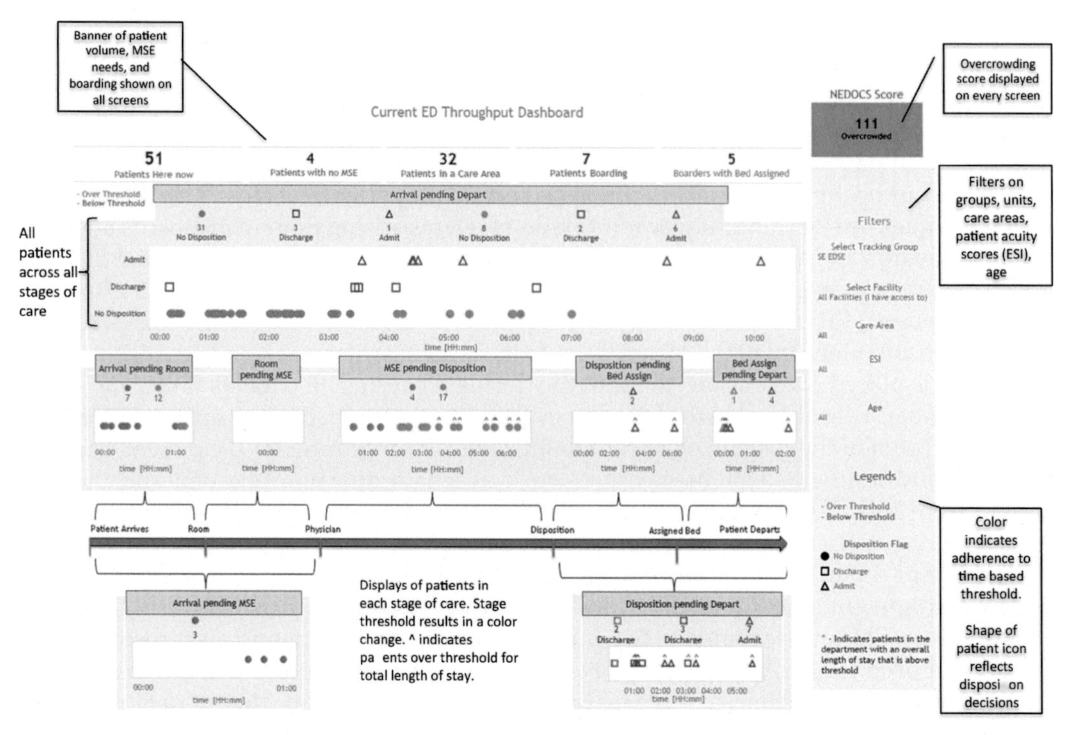

FIG. 19.4 Throughput dashboard to support identification of bottlenecks. Each symbol represents a patient's current stage in care and adherence to locally specified time-based thresholds. A drill-down function provides insight into the overall patient journey and details. *Reproduced with permissions from Franklin A, Gantela S, Shifarraw S, Johnson TR, Robinson DJ, King BR, Mehta AM, Maddow CL, Hoot NR, Nguyen V, Rubio A. Dashboard visualizations: supporting real-time throughput decision-making. J Biomed Inform 2017;71:211-21.*

Such aggregation and presentation of information to clinicians at the right time and in the right format underlies efforts in clinical summarization. Ranging for discharge summaries to daily progress notes, patient handoffs at change of shift, and oral case presentations, there are a variety of efforts to generate, automate, and standardize tools to improve information exchange while reducing data loss or gaps [108,109]

The *AORTIS* model of Feblowitz et al. [110] provides a conceptual model for the process of clinical summarization and provides a framework for various types of clinical summaries along with methods that could be employed to shift from human- to machine-generated summaries. AORTIS has five stages: Aggregation, Organization, Reduction and Transformation, Interpretation, and Synthesis. Methods such as this for data management and summarization have the potential to reduce the cognitive burden on clinicians and to improve safety. Fig. 19.5 illustrates an automated clinical summary built through AORTIS concepts. Annotations seen here indicate information types and additional drill-downs into other aspects of the patient record. In this snapshot, clinicians are provided with a synthesized and aggregated overview of patients' problems, history, and health over time. Like the Twinlist interface above, human cognition's abilities as well as limitations have been taken into consideration in the design of

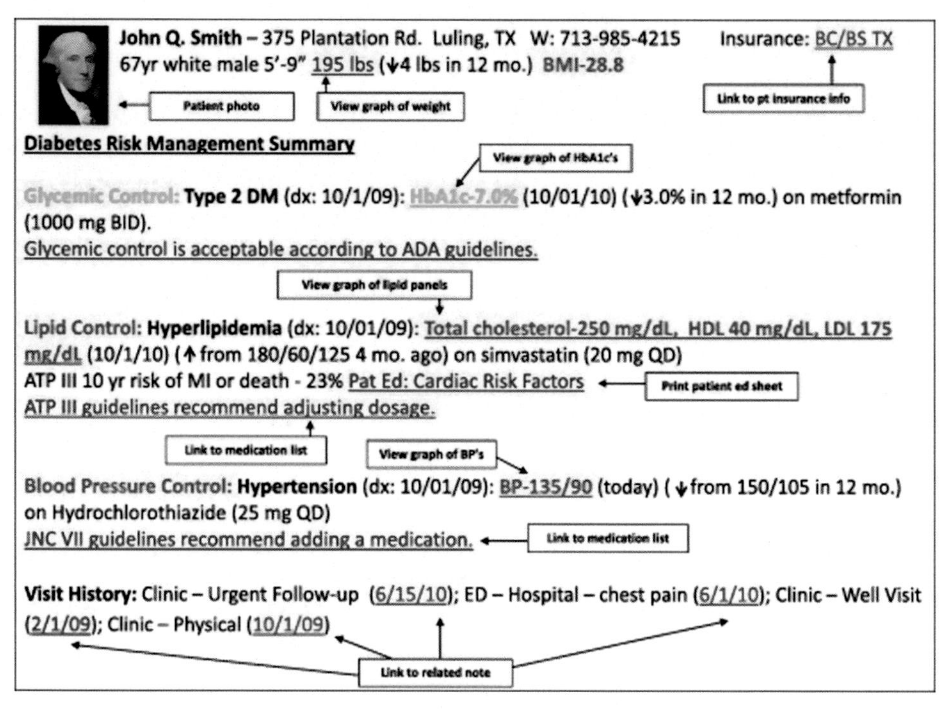

FIG. 19.5 AORTIS automated clinical summary [110].

this system. Relevant information is presented in a format that is easy to scan, color is used to increase recognition of abnormal values, and additional information is available with additional exploration of the system.

19.5.3 Visualization

At the heart of all these support tools is the need for information visualization. Better visualization is one way to improve the understanding of complex data and consequently increase the value of electronically available medical data [111]. The goal of visualization tools it to "provid[e] visual representations of datasets intended to help people carry out some task more effectively [112]." This includes both the cognitive activity of building mental models through visualization [113] and visualization's ability to "amplify cognition [114]." Rind et al. [115] provides an overview of visualization systems for exploring and querying EHRs. Chapter 14 provides an in-depth discussion of visualization approaches for CDS.

19.5.4 Automation, machine learning, and analytics

Cognitive support of clinical decision takes many forms from increasing awareness of information to aggregation of data beyond what human memory can hold. Automation in information acquisition, analysis, and prediction are other means of improving diagnostic performance [116–118]. Carayon et al. [119] evaluated CDS for the impact it had on clinic pathway selection in cases of pulmonary embolus. In addition to gains in time and reduction in workload, their human factors designed system improved accuracy in decision selection (a gain of 10% compared to related system.)

As machine learning becomes a more common as part of CDSS, for example to support early recognition of diseases such as COVID [120,121], sepsis [122], and stroke [123], these systems will be faced with similar hurdles in use and adoption of earlier EHR systems [124]. Clinical acceptance of these tools requires transparency, usability, and integration with workflow. Monitoring CDSS use [125] including the use of machine learning methods to uncover patterns in compliance and alert burden [126] face familiar adoption challenges as the data produced by the clinician becomes data within the system itself. Computational ethnography [127,128], including process mining and log analysis, will in the future support greater understanding of clinical process (as a proxy for clinical thinking) as EHR use patterns along with system interactions, such as response to CDS, become part of a learning, adaptive system.

19.6 Summary

Many health care activities occur in collaborative team environments involving clinicians, patients, and a variety of groups. Health information technology has been developed at a rapid rate and implemented on a large scale. The integration of HIT into the collaborative health care team environment has been fundamentally changing the way health care is delivered. HIT has potential to improve the quality, efficiency, and safety of care in significant ways. However, in order to fully achieve these potentials, we need to address the new cognitive challenges brought by the introduction of HIT into the health care systems.

This chapter has discussed the cognitive and usability factors for such complex team environments by exploring the challenges brought about by human cognitive limitations and the unintended consequences of technology.

Next, we presented both the challenges and opportunities in a distributed cognition framework. Distributed cognition considers a HIT-enabled health team environment as a cognitive system that is distributed between people and technology, among individuals across space and time, and situated in social, physical, and organizational backgrounds. To fully understand the impact of HIT on such a distributed system, we need to understand not just the impact HIT has on the behaviors of individuals but also the behaviors of interactions between individuals and technology and among individuals, and the emergent behaviors of the system as a whole.

The TURF framework was introduced as a methodology to analyze the cognitive factors in distributed cognitive systems through analyses of the users, functions (ontology of work domain), representations (user interfaces), and tasks (task sequences or workflow across multiple individuals or between people and technology). Through the TURF-based cognitive analyses, information systems can be designed to address the cognitive challenges in such distributed systems.

Finally, wide adoption of HIT, including EHR system, health information exchange systems, and other systems, provide opportunities for tools that support cognition through presentation and visualization of data in ways that support human processing, machine aggregation of information to not only manage the quantity of the data burden but also to provide cognitive support through pre-processing of information, and improvements of situation awareness for better decision making through dashboards and alert systems. The evolution of CDS to include machine learning and adaptive systems will be faced with new challenges to develop supports for users within distributed, complex systems.

References

[1] Southon G, Sauer C, Dampney K. Lessons from a failed information systems initiative: issues for complex organisations. Int J Med Inform 1999;55(1):33–46.

[2] Koppel R, Metlay JP, Cohen A, Abaluck B, Localio AR, Kimmel SE, Strom BL. Role of computerized physician order entry systems in facilitating medication errors. JAMA 2005;293(10):1197–203.

[3] Han YY, Carcillo JA, Venkataraman ST, Clark RS, Watson RS, Nguyen TC, Bayir H, Orr RA. Unexpected increased mortality after implementation of a commercially sold computerized physician order entry system. Pediatrics 2005;116(6):1506–12.

[4] Zhang J. Human-centered computing in health information systems. Part 1: analysis and design. J Biomed Inform 2005;38(1):1–3.

[5] Stead WW, Lin H. Committee on engaging the computer science research community in health care informatics. In: Computational technology for effective health care: immediate steps and strategic directions; 2009.

[6] Smith J. Setting the agenda: an informatics-led policy framework for adaptive CDS. J Am Med Inform Assoc 2020;27(12):1831–3.

[7] Petersen C, Smith J, Freimuth RR, Goodman KW, Jackson GP, Kannry J, Liu H, Madhavan S, Sittig DF, Wright A. Recommendations for the safe, effective use of adaptive CDS in the US healthcare system: an AMIA position paper. J Am Med Inform Assoc 2021;28(4):677–84.

[8] Lomotan EA, Meadows G, Michaels M, Michel JJ, Miller K. To share is human! Advancing evidence into practice through a national repository of interoperable clinical decision support. Appl Clin Inform 2020;11(01):112–21.

[9] Niculescu V. On the impact of high performance computing in big data analytics for medicine. Appl Med Inform 2020;42(1):9–18.

[10] Wolfe A. Institute of Medicine report: crossing the quality chasm: a new health care system for the 21st century. Pol Polit Nurs Practice 2001;2(3):233–5.

[11] Hall A, Walton G. Information overload within the health care system: a literature review. Health Inf Libr J 2004;21(2):102–8.

[12] Van Vleck TT, Wilcox A, Stetson PD, Johnson SB, Elhadad N. Content and structure of clinical problem lists: a corpus analysis. In: AMIA annual symposium proceedings, Vol. 2008. American Medical Informatics Association; 2008. p. 753.

[13] Singh H, Thomas EJ, Mani S, Sittig D, Arora H, Espadas D, Khan MM, Petersen LA. Timely follow-up of abnormal diagnostic imaging test results in an outpatient setting: are electronic medical records achieving their potential? Arch Intern Med 2009;169(17):1578–86.

[14] Ash JS, Berg M, Coiera E. Some unintended consequences of information technology in health care: the nature of patient care information system-related errors. J Am Med Inform Assoc 2004;11(2):104–12.

[15] Ash JS, Sittig DF, Poon EG, Guappone K, Campbell E, Dykstra RH. The extent and importance of unintended consequences related to computerized provider order entry. J Am Med Inform Assoc 2007;14(4):415–23.

[16] Sittig DF, Ash JS, Zhang J, Osheroff JA, Shabot MM. Lessons from "Unexpected increased mortality after implementation of a commercially sold computerized physician order entry system". Pediatrics 2006;118(2):797–801.

[17] Horsky J, Kuperman GJ, Patel VL. Comprehensive analysis of a medication dosing error related to CPOE. J Am Med Inform Assoc 2005;12(4):377–82.

[18] Patel VL, Kushniruk AW, Yang S, Yale JF. Impact of a computer-based patient record system on data collection, knowledge organization, and reasoning. J Am Med Inform Assoc 2000;7(6):569–85.

[19] Bawden D, Holtham C, Courtney N. Perspectives on information overload. In: Aslib proceedings. MCB UP Ltd.; 1999.

[20] Christensen T, Grimsmo A. Instant availability of patient records, but diminished availability of patient information: a multi-method study of GP's use of electronic patient records. BMC Med Inform Decision Making 2008;8(1):1–8.

[21] Kannampallil TG, Franklin A, Mishra R, Almoosa KF, Cohen T, Patel VL. Understanding the nature of information seeking behavior in critical care: implications for the design of health information technology. Artif Intell Med 2013;57(1):21–9.

[22] Patel VL, Yoskowitz NA, Arocha JF, Shortliffe EH. Cognitive and learning sciences in biomedical and health instructional design: a review with lessons for biomedical informatics education. J Biomed Inform 2009;42 (1):176–97.

[23] Patel VL, Kaufman DR. Medical informatics and the science of cognition. J Am Med Inform Assoc 1998;5 (6):493–502.

[24] Campbell EM, Sittig DF, Ash JS, Guappone KP, Dykstra RH. Types of unintended consequences related to computerized provider order entry. J Am Med Inform Assoc 2006;13(5):547–56.

[25] Ash JS, Sittig DF, Campbell EM, Guappone KP, Dykstra RH. Some unintended consequences of clinical decision support systems. In: AMIA annual symposium proceedings, Vol. 2007. American Medical Informatics Association; 2007. p. 26.

[26] Johnson CW. Why did that happen? Exploring the proliferation of barely usable software in healthcare systems. BMJ Qual Saf 2006;15(Suppl 1):i76–81.

[27] Karsh BT, Weinger MB, Abbott PA, Wears RL. Health information technology: fallacies and sober realities. J Am Med Inform Assoc 2010;17(6):617–23.

[28] Viitanen J, Hyppönen H, Lääveri T, Vänskä J, Reponen J, Winblad I. National questionnaire study on clinical ICT systems proofs: physicians suffer from poor usability. Int J Med Inform 2011;80(10):708–25.

[29] Kizzier-Carnahan V, Artis KA, Mohan V, Gold JA. Frequency of passive EHR alerts in the ICU: another form of alert fatigue? J Patient Saf 2019;15(3):246.

[30] Saiyed SM, Davis KR, Kaelber DC. Differences, opportunities, and strategies in drug alert optimization—experiences of two different integrated health care systems. Appl Clin Inform 2019;10(05):777–82.

[31] Moja L, Friz HP, Capobussi M, Kwag K, Banzi R, Ruggiero F, González-Lorenzo M, Liberati EG, Mangia M, Nyberg P, Kunnamo I. Effectiveness of a hospital-based computerized decision support system on clinician recommendations and patient outcomes: a randomized clinical trial. JAMA Netw Open 2019;2(12), e1917094.

[32] Goddard K, Roudsari A, Wyatt JC. Automation bias—a hidden issue for clinical decision support system use. In: International perspectives in health informatics; 2011. p. 17–22.

[33] Lyell D, Magrabi F, Raban MZ, Pont LG, Baysari MT, Day RO, Coiera E. Automation bias in electronic prescribing. BMC Med Inform Decision Making 2017;17(1):1.

[34] Sutton RT, Pincock D, Baumgart DC, Sadowski DC, Fedorak RN, Kroeker KI. An overview of clinical decision support systems: benefits, risks, and strategies for success. NPJ Digital Med 2020;3(1):1.

[35] Middleton B, Sittig DF, Wright A. Clinical decision support: a 25 year retrospective and a 25 year vision. Yearbook Med Inform 2016;25(S 01):S103–16.

[36] Middleton B, Bloomrosen M, Dente MA, Hashmat B, Koppel R, Overhage JM, Payne TH, Rosenbloom ST, Weaver C, Zhang J. Enhancing patient safety and quality of care by improving the usability of electronic health record systems: recommendations from AMIA. J Am Med Inform Assoc 2013;20(e1):e2–8.

[37] Mathieu JE, Heffner TS, Goodwin GF, Salas E, Cannon-Bowers JA. The influence of shared mental models on team process and performance. J Appl Psychol 2000;85(2):273.

[38] DeChurch LA, Mesmer-Magnus JR. The cognitive underpinnings of effective teamwork: a meta-analysis. J Appl Psychol 2010;95(1):32.

[39] Cooke NJ, Gorman JC, Myers CW, Duran JL. Interactive team cognition. Cogn Sci 2013;37(2):255–85.

[40] Salas EE, Fiore SM. Team cognition: understanding the factors that drive process and performance. American Psychological Association; 2004.

[41] Arrow H, McGrath JE, Berdahl JL. Small groups as complex systems: formation, coordination, development, and adaptation. Sage Publications; 2000.

[42] Bolstad CA, Endsley MR. Shared mental models and shared displays: an empirical evaluation of team performance. In: Proceedings of the human factors and ergonomics society annual meeting, vol. 43. Los Angeles, CA: SAGE Publications; 1999. p. 213–7. No. 3.

[43] Ho J, Intille SS. Using context-aware computing to reduce the perceived burden of interruptions from mobile devices. In: Proceedings of the SIGCHI conference on Human factors in computing systems; 2005. p. 909–18.
[44] Zachary W, Maulitz RC, Rosen MA, Cannon-Bowers J, Salas E. Clinical communications—human factors for the hidden network in medicine. In: Proceedings of the human factors and ergonomics society annual meeting, vol. 56. Los Angeles, CA: SAGE Publications; 2012. p. 850–4. No. 1.
[45] Van de Velde S, Heselmans A, Delvaux N, Brandt L, Marco-Ruiz L, Spitaels D, Cloetens H, Kortteisto T, Roshanov P, Kunnamo I, Aertgeerts B. A systematic review of trials evaluating success factors of interventions with computerised clinical decision support. Implement Sci 2018;13(1):1.
[46] Hutchins E. Cognition in the Wild. MIT press; 1995.
[47] Carroll JM, Long J, editors. Designing interaction: psychology at the human-computer interface. CUP Archive; 1991.
[48] Zhang J. The nature of external representations in problem solving. Cogn Sci 1997;21(2):179–217.
[49] Zhang J. A distributed representation approach to group problem solving. J Am Soc Inf Sci 1998;49(9):801–9.
[50] Zhang J, Norman DA. Representations in distributed cognitive tasks. Cogn Sci 1994;18(1):87–122.
[51] Hollan J, Hutchins E, Kirsh D. Distributed cognition: toward a new foundation for human-computer interaction research. ACM Trans Computer-Human Interact (TOCHI) 2000;7(2):174–96.
[52] Patel VL, Cytryn KN, Shortliffe EH, Safran C. The collaborative health care team: the role of individual and group expertise. Teach Learn Med 2000;12(3):117–32.
[53] Patel VL. Distributed and collaborative cognition in health care: implications for systems development. Artif Intell Med 1998;12.
[54] Clancey WJ. Situated cognition: on human knowledge and computer representations. Cambridge University Press; 1997.
[55] Hutchins E. How a cockpit remembers its speeds. Cogn Sci 1995;19(3):265–88.
[56] Gibson JJ. The ecological approach to visual perception. Boston: Houghton Mifling; 1979.
[57] Larkin JH, Simon HA. Why a diagram is (sometimes) worth ten thousand words. Cogn Sci 1987;11(1):65–100.
[58] Tweney RD. Stopping time: Faraday and the scientific creation of perceptual order; 1992.
[59] Reisberg D. External representations and the advantages of externalizinag one's thoughts. In: Proc. of the 9th Annual Conf. of the cognitive science society cognitive science society; 1987.
[60] Norman DA. The psychology of everyday things. Basic books; 1988.
[61] George JM. Personality, affect, and behavior in groups. J Appl Psychol 1990;75(2):107.
[62] Bandura A. Social cognitive theory of moral thought and action. Psychology Press; 2014.
[63] Wegner DM. Transactive memory: a contemporary analysis of the group mind. In: Theories of group behavior. New York, NY: Springer; 1987. p. 185–208.
[64] Foushee HC, Helmreich RL. Group interaction and flight crew performance. In: Human factors in aviation. Academic Press; 1988. p. 189–227.
[65] Carayon P, Hoonakker P. Human factors and usability for health information technology: old and new challenges. Yearbook Med Inform 2019;28(01):071–7.
[66] Salwei ME, Carayon P, Wiegmann D, Pulia MS, Patterson BW, Hoonakker PL. Usability barriers and facilitators of a human factors engineering-based clinical decision support technology for diagnosing pulmonary embolism. Int J Med Inform 2022;158, 104657.
[67] Baecker RM, editor. Readings in groupware and computer-supported cooperative work: assisting human-human collaboration. Morgan Kaufmann; 1993.
[68] Walsh JP, Ungson GR. Organizational memory. Acad Manag Rev 1991;16(1):57–91.
[69] Davenport TH, Jarvenpaa SL, Beers MC. Improving knowledge work processes. Sloan Manag Rev 1996;37:53–66.
[70] Nunamaker JF, Dennis AR, Valacich JS, Vogel D, George JF. Electronic meeting systems. Commun ACM 1991;34(7):40–61.
[71] Hill B, Proulx J, Zeng-Treitler Q. Exploring the use of large clinical data to inform patients for shared decision making. In: MEDINFO 2013. IOS Press; 2013. p. 851–5.
[72] Bass SB, Gordon TF, Ruzek SB, Wolak C, Ruggieri D, Mora G, Rovito MJ, Britto J, Parameswaran L, Abedin Z, Ward S. Developing a computer touch-screen interactive colorectal screening decision aid for a low-literacy African American population: lessons learned. Health Promot Pract 2013;14(4):589–98.

[73] Vlemmix F, Warendorf JK, Rosman AN, Kok M, Mol BW, Morris JM, Nassar N. Decision aids to improve informed decision-making in pregnancy care: a systematic review. BJOG Int J Obstet Gynaecol 2013;120(3):257–66.
[74] Stacey D, Légaré F, Lewis K, Barry MJ, Bennett CL, Eden KB, Holmes-Rovner M, Llewellyn-Thomas H, Lyddiatt A, Thomson R, Trevena L. Decision aids for people facing health treatment or screening decisions. Cochrane Database Syst Rev 2017;4.
[75] Kraemer KL, King JL. Computer-based systems for cooperative work and group decision making. ACM Comput Surv (CSUR) 1988;20(2):115–46.
[76] Abraham J, Kannampallil TG, Patel VL. Bridging gaps in handoffs: a continuity of care based approach. J Biomed Inform 2012;45(2):240–54.
[77] Have EC, Nap RE. Mutual agreement between providers in intensive care medicine on patient care after interdisciplinary rounds. J Intensive Care Med 2014;29(5):292–7.
[78] Kannampallil TG, Jones LK, Patel VL, Buchman TG, Franklin A. Comparing the information seeking strategies of residents, nurse practitioners, and physician assistants in critical care settings. J Am Med Inform Assoc 2014;21(e2):e249–56.
[79] Schulz S, Bernhardt-Melischnig J, Kreuzthaler M, Daumke P, Boeker M. Machine vs. human translation of SNOMED CT terms. In: MEDINFO 2013. IOS Press; 2013. p. 581–4.
[80] McCoy AB, Wright A, Laxmisan A, Ottosen MJ, McCoy JA, Butten D, Sittig DF. Development and evaluation of a crowdsourcing methodology for knowledge base construction: identifying relationships between clinical problems and medications. J Am Med Inform Assoc 2012;19(5):713–8.
[81] Hernández-Chan G, Rodríguez-González A, Alor-Hernández G, Gómez-Berbís JM, Mayer-Pujadas MA, Posada-Gómez R. Knowledge acquisition for medical diagnosis using collective intelligence. J Med Syst 2012;36(1):5–9.
[82] Miller K, Capan M, Weldon D, Noaiseh Y, Kowalski R, Kraft R, Schwartz S, Weintraub WS, Arnold R. The design of decisions: matching clinical decision support recommendations to Nielsen's design heuristics. Int J Med Inform 2018;117:19–25.
[83] Zhang J, Walji MF. TURF: toward a unified framework of EHR usability. J Biomed Inform 2011;44(6):1056–67.
[84] Dumont F. Inferential heuristics in clinical problem formulation: selective review of their strengths and weaknesses. Prof Psychol Res Pract 1993;24(2):196.
[85] Patel VL, Kaufman DR, Arocha JF. Emerging paradigms of cognition in medical decision-making. J Biomed Inform 2002;35(1):52–75.
[86] Mark DB. Decision-making in clinical medicine. In: Fauci AS, Braunwald E, Kasper DL, et al., editors. Harrison's principles of internal medicine. New York City: McGraw Hill; 2008. p. 16–7.
[87] Bourgeois FC, Olson KL, Mandl KD. Patients treated at multiple acute health care facilities: quantifying information fragmentation. Arch Intern Med 2010;170(22):1989–95.
[88] Alonso DL, Rose A, Plaisant C, Norman KL. Viewing personal history records: a comparison of tabular format and graphical presentation using LifeLines. Behav Inform Technol 1998;17(5):249–62.
[89] Pieczkiewicz DS, Finkelstein SM, Hertz MI. Design and evaluation of a web-based interactive visualization system for lung transplant home monitoring data. In: AMIA annual symposium proceedings, Vol. 2007. American Medical Informatics Association; 2007. p. 598.
[90] Plaisant C, Chao T, Wu J, Hettinger AZ, Herskovic JR, Johnson TR, Bernstam EV, Markowitz E, Powsner S, Shneiderman B. Twinlist: novel user interface designs for medication reconciliation. In: AMIA annual symposium proceedings, Vol. 2013. American Medical Informatics Association; 2013. p. 1150.
[91] Endsley MR. Toward a theory of situation awareness in dynamic systems. Hum Factors 1995;37(1):32–64.
[92] Singh H, Petersen LA, Thomas EJ. Understanding diagnostic errors in medicine: a lesson from aviation. BMJ Qual Saf 2006;15(3):159–64.
[93] Woodward S. Stop passing the buck—patient safety is nurses' problem too. Nurs Times 2010;106(31):25.
[94] Yule S, Flin R, Maran N, Rowley D, Youngson G, Paterson-Brown S. Surgeons' non-technical skills in the operating room: reliability testing of the NOTSS behavior rating system. World J Surg 2008;32(4):548–56.
[95] Fioratou E, Flin R, Glavin R, Patey R. Beyond monitoring: distributed situation awareness in anaesthesia. Br J Anaesth 2010;105(1):83–90.
[96] Parush A, Kramer C, Foster-Hunt T, Momtahan K, Hunter A, Sohmer B. Communication and team situation awareness in the OR: implications for augmentative information display. J Biomed Inform 2011;44(3):477–85.

[97] World Health Organization. Human Factors in Patient Safety. Review of Topics and Tools. Methods and Measures. Working Group of WHO Patient Safety; 2009.
[98] Few S. Information dashboard design. 1st ed. O'Reilly Media; 2006.
[99] Aronsky D, Jones I, Lanaghan K, Slovis CM. Supporting patient care in the emergency department with a computerized whiteboard system. J Am Med Inform Assoc 2008;15(2):184–94.
[100] Simms RA, Ping H, Yelland A, Beringer AJ, Fox R, Draycott TJ. Development of maternity dashboards across a UK health region; current practice, continuing problems. Eur J Obstet Gynecol Reprod Biol 2013;170(1): 119–24.
[101] France DJ, Levin S, Hemphill R, Chen K, Rickard D, Makowski R, Jones I, Aronsky D. Emergency physicians' behaviors and workload in the presence of an electronic whiteboard. Int J Med Inform 2005;74(10):827–37.
[102] Farley HL, Baumlin KM, Hamedani AG, Cheung DS, Edwards MR, Fuller DC, Genes N, Griffey RT, Kelly JJ, McClay JC, Nielson J. Quality and safety implications of emergency department information systems. Ann Emerg Med 2013;62(4):399–407.
[103] Franklin A, Gantela S, Shifarraw S, Johnson TR, Robinson DJ, King BR, Mehta AM, Maddow CL, Hoot NR, Nguyen V, Rubio A. Dashboard visualizations: supporting real-time throughput decision-making. J Biomed Inform 2017;71:211–21.
[104] Koch SH, Weir C, Westenskow D, Gondan M, Agutter J, Haar M, Liu D, Görges M, Staggers N. Evaluation of the effect of information integration in displays for ICU nurses on situation awareness and task completion time: a prospective randomized controlled study. Int J Med Inform 2013;82(8):665–75.
[105] Patel VL, Groen GJ. The general and specific nature of medical expertise: a critical look; 1991.
[106] Cohen T, Kaufman D, White T, Segal G, Staub AB, Patel V, Finnerty M. Cognitive evaluation of an innovative psychiatric clinical knowledge enhancement system. In: MEDINFO 2004. IOS Press; 2004. p. 1295–9.
[107] Dalai V, Gottipatti D, Kannampallil T, Cohen T. Characterizing the effects of a cognitive support system for psychiatric clinical comprehension. In: AMIA poster presentation; 2013.
[108] Stelfox HT, Perrier L, Straus SE, Ghali WA, Zygun D, Boiteau P, Zuege DJ. Identifying intensive care unit discharge planning tools: protocol for a scoping review. BMJ Open 2013;3(4), e002653.
[109] Abraham J, Kannampallil T, Patel VL. A systematic review of the literature on the evaluation of handoff tools: implications for research and practice. J Am Med Inform Assoc 2014;21(1):154–62.
[110] Feblowitz JC, Wright A, Singh H, Samal L, Sittig DF. Summarization of clinical information: a conceptual model. J Biomed Inform 2011;44(4):688–99.
[111] Chittaro L. Information visualization and its application to medicine. Artif Intell Med 2001;22(2):81–8.
[112] Munzner T. Visualization principles. In: Keynote at workshop on visualizing biological data (VIZBI 2011); 2011. http://www.cs.ubc.ca/~tmm/talks/vizbi11.pdf (cited 24 Jul 2012).
[113] Spence R. Information visualization: design for interaction. 2nd ed. Harlow, Essex, England: ACM Press Books Pearson Education Limited Edinburgh Gate; 2007.
[114] Card M. Readings in information visualization: using vision to think. Morgan Kaufmann; 1999.
[115] Rind A, Wang TD, Aigner W, Miksch S, Wongsuphasawat K, Plaisant C, Shneiderman B. Interactive information visualization to explore and query electronic health records. Found Trends Human-Comp Interact 2013;5 (3):207–98.
[116] Parasuraman R, Sheridan TB, Wickens CD. A model for types and levels of human interaction with automation. IEEE Trans Syst Man Cybernet-Part A: Syst Humans 2000;30(3):286–97.
[117] Henriksen K, Brady J. The pursuit of better diagnostic performance: a human factors perspective. BMJ Qual Saf 2013;22(Suppl 2):ii1–5.
[118] Bizzo BC, Almeida RR, Michalski MH, Alkasab TK. Artificial intelligence and clinical decision support for radiologists and referring providers. J Am Coll Radiol 2019;16(9):1351–6.
[119] Carayon P, Hoonakker P, Hundt AS, Salwei M, Wiegmann D, Brown RL, Kleinschmidt P, Novak C, Pulia M, Wang Y, Wirkus E. Application of human factors to improve usability of clinical decision support for diagnostic decision-making: a scenario-based simulation study. BMJ Qual Saf 2020;29(4):329–40.
[120] Karthikeyan A, Garg A, Vinod PK, Priyakumar UD. Machine learning based clinical decision support system for early COVID-19 mortality prediction. Front Public Health 2021;9.
[121] Subudhi S, Verma A, Patel AB, Hardin CC, Khandekar MJ, Lee H, McEvoy D, Stylianopoulos T, Munn LL, Dutta S, Jain RK. Comparing machine learning algorithms for predicting ICU admission and mortality in COVID-19. NPJ Digital Med 2021;4(1):1–7.

[122] van Doorn WP, Stassen PM, Borggreve HF, Schalkwijk MJ, Stoffers J, Bekers O, Meex SJ. A comparison of machine learning models versus clinical evaluation for mortality prediction in patients with sepsis. PLoS One 2021;16(1), e0245157.
[123] Sung SF, Hung LC, Hu YH. Developing a stroke alert trigger for clinical decision support at emergency triage using machine learning. Int J Med Inform 2021;152, 104505.
[124] Chew HS, Achananuparp P. Perceptions and needs of artificial intelligence in health care to increase adoption: scoping review. J Med Internet Res 2022;24(1), e32939.
[125] Yoshida E, Fei S, Bavuso K, Lagor C, Maviglia S. The value of monitoring clinical decision support interventions. Appl Clin Inform 2018;9(1):163–73. https://doi.org/10.1055/s-0038-1632397.
[126] Baron JM, Huang R, McEvoy D, Dighe AS. Use of machine learning to predict clinical decision support compliance, reduce alert burden, and evaluate duplicate laboratory test ordering alerts. JAMIA Open 2021;4(1), ooab006.
[127] Grando MA, Vellore V, Duncan BJ, Kaufman DR, Furniss SK, Doebbeling BN, Poterack KA, Miksch T, Helmers RA. Study of EHR-mediated workflows using ethnography and process mining methods. Health Inform J 2021;27(2). 14604582211008210.
[128] Zheng K, Padman R, Johnson MP, Diamond HS. An interface-driven analysis of user interactions with an electronic health records system. J Am Med Inform Assoc 2009;16(2):228–37.

CHAPTER

20

Governance and implementation

Richard Schreiber[a,b,c] *and John D. McGreevey III*[d,e]

[a]Penn State Health Holy Spirit Medical Center, Camp Hill, PA, United States [b]Geisinger Commonwealth School of Medicine, Scranton, PA, United States [c]Department of Health Policy and Management, Johns Hopkins Bloomberg School of Public Health, Baltimore, MD, United States [d]University of Pennsylvania Health System, Philadelphia, PA, United States [e]Perelman School of Medicine, University of Pennsylvania, Philadelphia, PA, United States

20.1 Introduction

Osler claimed, "to study the phenomena of disease without books is to sail an uncharted sea, while to study books without patients is not to go to sea at all" [1]. In today's electronic environment we contend that to see patients without clinical decision support, and to create clinical decision support in the absence of expert knowledge of clinical workflows risks traveling Osler's uncharted, tumultuous, and dangerous seas.

In this chapter we explore the overarching governance for clinical decision support (CDS), and guiding principles for its implementation [2]. The structures and principles apply to all healthcare delivery organizations, including small and large ambulatory, inpatient, rehabilitation, and nursing home facilities, and wherever clinicians see patients and make medical decisions daily. Clearly large organizations have more resources, both financial and personnel, and small organizations are constrained by a lack of breadth of subject matter experts. We describe general principles of CDS governance and implementation that are applicable to both large and small organizations, and include how smaller organizations can overcome disadvantages such as size, financial status, availability of technical and CDS personnel, and the type of electronic health record (EHR). We also reference that some of the newer CDS tools require sophisticated and advanced knowledge to implement and may thus be expensive for organizations to implement. Such realities may limit the ability for smaller institutions, or those serving disadvantaged populations, from being able to realize the benefits of these CDS tools without leadership to intentionally overcome inequalities in CDS access (see Section 20.24).

Clinical Decision Support and Beyond
https://doi.org/10.1016/B978-0-323-91200-6.00009-7

For purposes of this chapter, we define governance as the mechanism by which an organization evaluates requests for changes to electronic information systems and makes necessary changes consistent with organizational strategy, while at the same time preserving the stability and integrity of existing systems.

In healthcare we envision governance as a quartet: the confluence of leadership, management, communication, and accountability. Optimal governance is the intersection of these circles. See Fig. 20.1.

Leaders set the mission, goals, and tone of an organization's strategy. Management carries out that strategy through operations. In a healthcare organization, both leadership and management include physicians, nurses, pharmacy, and ancillary providers, as well as information technology (IT) experts. All must exercise effective communication up and down the chain of command, as well as between departments. Finally, all must be accountable not only to their own department (such as ensuring everyone knows their responsibilities and completes their assigned tasks on time) but to the organization as a whole (such as properly utilizing available resources consistent with the organization's mission). Good governance for information technology includes transparent documentation, a blueprint of committees and their roles and relationships, and links between this governance structure, leadership, and across administrative departments. Governance also includes a lifecycle of data gathering, consideration of the mission of the institution, goal orientation and goal setting (start with the end in mind), prioritization, decision-making, and an approval process. With this foundation, and plans on which everyone can rely, management can implement CDS effectively. After inauguration of the project or CDS, good governance requires reassessment and reconsideration.

To illustrate good CDS governance as shown in Fig. 20.1 with an example, consider implementing a complex CDS project such as an overhaul of order sets. Leadership guides the process in accordance with institutional goals and consonant with external requirements, while management is accountable to engage subject matter experts (SMEs) with the technical teams to deliver required content. In turn, the SMEs are accountable to the organization for delivery of order set content. Effective communication would include advance notice of the project, timely updates, and education once the project was complete. See also Section 20.9. Absence of any of these would constitute poor governance, falling somewhere outside the shaded area. See also Section 20.10.

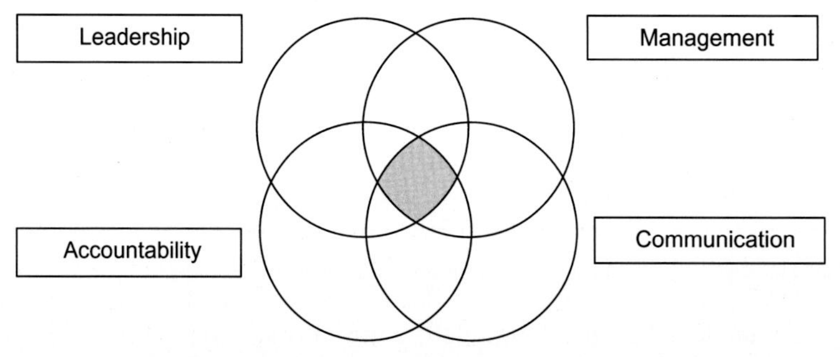

FIG. 20.1 The four circles represent leadership, management, accountability, and communication. Effective governance requires all elements represented by the shaded area in the middle.

20.1.1 Operational aspects of CDS governance

At every stage it is important to give end-users a voice [3]. After all, decisions and responsibility for medical care ultimately reside with the clinician who must use the tools available. Clinicians are the workflow experts; like all technology, clinical decision support must work for the benefit of clinician workflow, and not change it without a profound need to do so. Clinicians must feel heard and have a sense of ownership, which may enhance their buy-in for specific CDS initiatives and perhaps CDS efforts more generally. A commonly heard phrase from clinicians is "what's in it for me?" Effective governance will allow clinicians to participate in the decision-making processes, knowing the tools will work for them.

A critical aspect of governance is operational: how does the organization get tasks done? We will expand on this topic in the sections on implementation. See Section 20.14.1. Another aspect is the notion of oversight. Effective CDS governance provides assurance for the organization that changes are consistent with overall strategy, are affordable, and that personnel have the proper skill sets to ensure success. Simultaneously, good governance ensures that initiatives related to one another need to be linked to avoid duplication of effort, and to detect unintended consequences and undesirable precedents. Unrecognized hazards impair even well-intended projects.

20.2 Governance structures for CDS

Each organization will find its organizational structure unique to its needs. No one structure fits all circumstances. A few key tenets must be followed to ensure success.

1. The most important in our opinion is to have one or more physician champions, most commonly a Chief Medical Informatics or Information Officer (CMIO); there are many other abbreviations with nuances in responsibilities but overall similar functions [4] to lead the charge.
2. There must be an engaged CDS committee with key stakeholders. Larger institutions will certainly require a more complex set of committees.
3. A robust infrastructure is a prerequisite to support adequate CDS. This includes the right people making decisions, analysts who are skilled at CDS, having the technical architecture to support intra- and interoperability of servers, sufficient Wi-Fi access, high-speed internet, and other necessary functionalities.
4. There must be a comprehensive communication plan. Horstman's law of organizational communication says, "Say something seven times and half your folks will have heard it once" [5]. Organizations need several ways to communicate with all users of CDS tools.

Fig. 20.2 diagrams an archetypal functional governance structure. Informatics sits at the juncture of the medical staff, IT, and administration. One division within Informatics is CDS, which includes the entities on the left (indicated in yellow text). The CMIO leads these groups. Those on the right may also be appropriate areas for the CMIO to lead, or at least to have representation from the CMIO and others in informatics. Each of the groups requires representation from all stakeholders and subject matter experts as discussed in Section 20.3.

Members may serve on more than one committee; committees share their work with each other, and with the general informatics group, which in turn reports to the Medical Executive

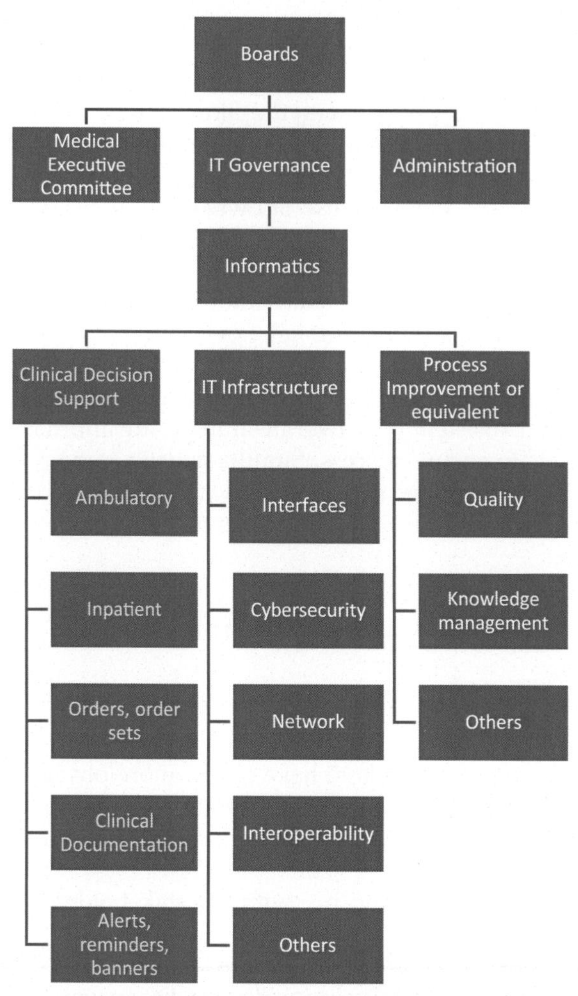

FIG. 20.2 Proposed general schema for a hierarchical clinical decision support (CDS) governance structure. Informatics includes the CMIO. Members of committees are shared at least one tier above and below.

Committee, IT governance councils, and administration. This schema may offer the advantages that hierarchical and consensus systems provide (see Section 20.4.1).

20.3 Participants in CDS governance

We recommend participation by diverse subject matter experts (SME), including clinician informaticists, clinician champions from all specialties, nurses, pharmacists, allied therapies, information technology experts, and others. See Table 20.1. Not all SMEs will be necessary for every initiative, so CDS governance benefits from flexibility and creative solutions for smooth operation [6]. Some organizations opt for regular, in-person meetings of all governance

TABLE 20.1 Key participants in clinical decision support governance. Not all institutions have the breadth and depth for all these roles.

Subject matter expert groups	Role as participant in governance
Clinicians	Define clinical goal of Clinical Decision Support
Advanced practitioners	May overlap with informatics specialists, if available
Nurses	
Pharmacists	
Physicians	
Informatics	Expertise regarding clinical decision support principles and knowledge management
Nursing	Arbiters between clinicians, technical staff, and administrators
Pharmacy	Broad cross-sectional knowledge of clinical and technical domains
Physician	
Information technology	Inform decision makers what is possible from a technical standpoint
Analysts	Subject matter experts for issues as they arise
Builders	
Data scientists	
Researchers	
Human factors engineers	
Optimization staff	
Administration	Define personnel, budgetary, and time resources
Legal staff	
Regulatory	
Safety officers	

From McGreevey JD, Mallozzi CP, Perkins RM, et al. Reducing alert burden in electronic health records: state of the art recommendations from four health systems. Appl Clin Inform 2020;11:1–12. Used with permission.

members (when public health conditions allow). Others opt for online sessions which are particularly suitable for geographically far-flung institutions. Ad hoc attendance is helpful to present specialty projects and advice. It is especially important for those with expertise in build and configuration to contribute to CDS decision-making to set expectations of what is possible, as well as to advise regarding time requirements for any given project.

Engaging stakeholders is difficult. Physicians, nurses, and ancillary staff must be allowed scheduled time away from their clinical duties to participate. This may involve additional compensation or other means of recognizing clinician work effort in the governance process. They must have a voice at the decision-making table to have ownership in the process. Likewise, information technology (IT) personnel must feel their participation is part of their daily duties, not an add-on task. One must appeal to IT management to support their employees, even if management itself is not involved in CDS and EHR decision-making. A top-down commitment to the process and necessary personnel is essential.

CDS involves medical decision-making. Ash, et al., [3] elaborated on the sociotechnical consequences of both computerized provider order entry (CPOE) [7] and CDS [3]. These include a feeling of loss of autonomy, a suggestion that professional skills are lacking, irritation, additional effort, especially the effort required to interact with interruptive alerts, and alert fatigue. Governance to deal with these concerns includes a recognition of the impact of CDS on clinicians' workflow. CDS developers must attend to CDS usability, user control and flexibility, CDS intrusiveness, and commit to proper training and support [8]. Engaging clinicians early in the process is key to overcoming these sociotechnical and emotional barriers.

20.3.1 Role of the chief medical informatics officer

The CMIO is and will remain the central agent of change management [9]. The roles of the CMIO have evolved over the years, including the title: Chief Health Information Officer, Chief Medical Information Officer, Chief Population Health Officer, and many others [4]. (For convenience, we will use CMIO as a convenient abbreviation to stand in for all these titles.) He or she, and along with the CMIO's staff and colleagues, are in the best position to understand clinician workflows, current medical knowledge, and effective clinical management strategies. The CMIO can judge whether clinicians and the institution are prepared for implementation of new or updated CDS. Clinicians who trust the CMIO will approach the CMIO when it appears the EHR or its CDS malfunctions or over-alerts, for example. The CMIO is the individual who can most effectively communicate these concerns to the IT staff, facilitate sometimes difficult conversations, and propose course corrections when necessary. Likewise, the CMIO can also translate technical decisions and system constraints for a clinician audience. When clinician requests diverge from health system strategy, CMIOs may need to decline clinician requests outright and communicate the rationales for those decisions to the clinicians involved.

A CMIO understands system performance and capabilities. Deep knowledge of workflows and care gaps enables the CMIO to address where CDS can benefit clinicians and the organization. Not all CMIOs have a budget and personnel reporting to him or her, but the CMIO can advise about necessary operational changes and allocate personnel and other resources appropriately. The CMIO has expertise to understand the impact of CDS, and equally to examine, remediate, and coordinate changes needed in workflows that may be necessary to enable the good clinical practices that CDS tools suggest.

20.3.2 Physician programmers

Another type of subject matter expert is the physician builder. Some institutions allow practicing clinicians to be content builders (while working with their analyst colleagues). An institution should have guidelines regarding what CDS artifacts builders are allowed to build, test, and move to production. Should their scope be limited to their specialty? Does their governance fall under the same guidelines as for any analyst, or do these builders represent a special skillset, and thus, should their activities be subject to some other type of oversight/governance?

20.4 Different governance structures

20.4.1 Structural forms

Organizations need to decide upon a governance structure which is consonant in large measure with institutional culture. One initial strategy may be a "CDS committee to rule them all," meaning one committee with all necessary participants, and see how it functions over time. If the organization is small enough, this approach may suffice, if there are sufficient subject matter experts willing to put in the time. There are many ways to assess the culture of an organization, which is beyond the scope of this discussion.

One can broadly consider governance models as either hierarchical or consensus. Most institutions will be somewhere in between, with a combination of both. The most successful design will depend on numerous factors within an institution. In our experience and observations, the structure will evolve over time. We offer several reasons for this evolution: the size of the organization; transformations as an institution grows; mergers with other institutions; development of specialty or centers of excellence; EHR transitions (see Section 20.23); and periodic reevaluations of an organization's structure as it evaluates its successes and failures.

One of the most difficult aspects of governance, but one which helps to shape its structure, is to formalize accountability. Physicians ought to take ownership of many CDS decisions but given busy clinical schedules and patient care responsibilities physicians rarely have time for committee work. Although it is often easiest to allow information technology (IT) services to take control of these decisions, CDS is not a technical endeavor as much as a clinical one. To encourage participation, consider protecting clinicians' time and/or compensating them for the extra work (see Section 20.3).

Another governance structural decision involves the level of autonomy or dependence that each operational group as depicted in Fig. 20.2 will have. Should each of these groups be empowered to make its own decisions regarding acceptance of requests? Or should they report to the central CDS committee for final approval or rejection? The former can lead to information silos, while the latter can be cumbersome. Either way, it is all too easy for governance to become fractured.

In either case, it is best to clearly delineate the roles of IT versus clinicians: IT brings to fruition content desired by and required for clinicians which they provide. Enforcement regarding how these CDS tools are actually used by end-users is beyond the scope of this chapter, but it is not the role of IT.

20.4.1.1 Hierarchical

In a hierarchical structure, decision-making resides in a centralized authority. A leader makes most major decisions after consideration of alternatives, as informed by informaticists, analysts, and appropriate subject matter experts [10]. Decision-making is thus "top down." Decisions can be made swiftly, as appropriate, but at the risk of disenfranchising important constituents in the process leading to a feeling of loss of autonomy for stakeholders.

20.4.1.2 Consensus

Organizations that rely heavily on consensus have more decentralized authority, with distributed decision-making. The advantage of this structure is that all parties have input into decision-making, which usually generates considerable discussion of alternatives, possible adverse consequences, and unanticipated results. The final decision may be a compromise of the original request. Well-guided consensus governance can result in broad acceptance. The main disadvantage is that it is a slow process, often leading to lengthy delays, stalled goals, and analysis paralysis. While it may seem that smaller, decentralized groups can make swifter decisions, if there are several teams working in silos this can lead to duplication of effort, and even decisions at cross purposes.

Consensus-based organizations can be challenging for CDS leaders given that everyone can be entitled to a say in the decision-making, requiring leadership that is tolerant and encouraging of such discussions while not allowing a group to get bogged down. It can be unclear in a consensus-based system who actually has authority to make decisions. Can a CDS governance chair decline a request? Or recommend that a request be paused so that the proposed CDS solution can be revised and resubmitted to the committee for review at a later date? Can a chair make a final decision on behalf of the committee if the committee is split for and against a proposed tool? Do others in the organization recognize the decision-making authority of the chair or not?

20.4.2 Reporting structures

Should the CDS committee act autonomously or report to another group? Many believe CDS governance should not exist within IT as IT is a technical, not clinical entity. In that case, should CDS governance report to the medical staff leadership, via the Medical Executive Committee or equivalent? This certainly gives ultimate strategic ownership of CDS processes to the medical staff, but CDS content may be of peripheral interest to most Medical Executive Committee members, aside from the informaticists and subject matter experts who may also sit within that group. Or, is the proper reporting connection to administration? If so, to whom specifically—the CMO and/or CNO, CEO, COO, CFO, or a chief quality officer? CDS governance reporting lines differ across organizations and evolve over time.

What is the relationship between CDS governance and quality assurance? (see also Chapter 4). Some organizations have separate leadership for informatics (such as a CMIO) and a quality officer (such as a Chief Quality Officer or Chief Health Informatics/Information Officer), and perhaps a Chief Population Health Officer. Who has the ultimate authority to insert, change, or remove CDS that some may consider to be a regulatory requirement or to satisfy quality metrics? In our view this intersection is precisely where informatics does its best work in recognizing the multiple stakeholders and fostering collaboration among them. Readers should appreciate the culture of their institution and build their governance in accordance with the size of their institution, their local perceptions and what works for them, and having considered the options and approaches in this chapter.

20.5 Structural and functional aspects of CDS governance

20.5.1 General considerations

Large organizations generally split CDS governance into two (inpatient and ambulatory) or several committees (e.g., inpatient surgery and perioperative; inpatient medical; inpatient other such as Women's health and pediatrics; with corresponding ambulatory committees). Further subdivisions are common: orders, order sets, and results; alerts and reminders; documentation; and perhaps others such as financial and billing. We conclude from our experience that bidirectional interaction is critical: each subcommittee must report up the structure to ensure full communication and coordination even as the central committee assigns tasks and delegates decision-making. Cross-seeding—having members of the subcommittee participate at the supervisory level, and vice versa—assists in the overall processes of both. Formalizing these interactions and reporting structures furthermore ensures transparency of the governance organization and process for end-users. It also roots out potential hazards, such as projects originating from different committees that have potential to collide, causing disruption or that depart from system defined CDS strategy.

Factors that contribute to successful governance for CDS include:

- Available personnel: Are sufficient subject matter experts available for all aspects of CDS, or will the organization need to rely on the electronic health record (EHR) vendor, contractors, or outside experts?
- Needs: Does the organization have the hardware, network, server, and other infrastructure to accomplish the goals of the CDS team?
- Financial resources
- Size of the organization: is it a single entity? Are there multiple sites? Is it an integrated delivery system (IDS)? Is the organization spread out geographically? To what extent can members of a geographically widespread organization operate semi-autonomously? Must they accept the CDS decisions and content of headquarters?
    - If the organization provides care across state lines, there may be legal and regulatory issues to address.

20.5.2 Cultural considerations

One must also consider the cultures of individual entities within a multisite, IDS, or academic medical complex:

- Do the entities serve differing socioeconomic and ethnic communities? If so, this may impact available resources suggested by CDS. For example, CDS which instructs patients regarding healthy fruits and vegetables for a diabetic diet will have little impact in food desert areas.
- Are any of the organizations religious-based, which may have different priorities and guidelines of care compared to non-denominational institutions? This, too, will have a profound impact on available medications, procedures, consent forms, and other CDS artifacts.
- Is the organization non-profit or for profit? Or mixed? This may significantly alter CDS or funding available for CDS development.
- Are there truly unique settings within the organization, such as a specialty hospital or clinic? If this is the case there will be significant differences in resource allocations. For example, a specialty orthopedic hospital will require more time from those familiar with orthopedic workflows.

20.5.3 Financial risk tolerance

Some CDS is extraordinarily expensive, in terms of actual cost, but also in resource and personnel allocation. A CDS project with low acquisition costs but which takes hundreds of hours of analyst time to configure, monitor, and maintain represents a significant risk to the organization as other projects may be delayed. Many organizations will screen CDS requests based on additional purchase and resource costs, allowing those under certain thresholds to move forward if otherwise acceptable and desirable. Those projects over these thresholds require deeper inspection. Some academic medical centers, particularly if well-endowed, may be more willing to be risk-tolerant to develop innovative, bleeding edge CDS. Other organizations, especially small practices and community hospitals do not have the financial or personnel resources to take such risks [6,9]. One must also consider ready-to-use commercial off-the-shelf product costs versus choosing homegrown or less expensive but customizable CDS, and ongoing costs of maintenance and optimization, as well as the impact of EHR version upgrades. Finally, one must consider the cost of failure: what if the CDS proves ineffective? Consider a multimillion-dollar tool which specialists begged for, but which it turns out they do not use or that does not deliver the expected benefits.

20.5.4 Differences between academic and community hospitals

The type of institution also influences governance approaches. One key question is who will receive the CDS? In teaching hospitals, residents and advanced practice providers enter most orders and hence are the targets of CDS messages. In community hospitals, seasoned attendings who enter orders feel they do not need to be alerted as much, especially for minor, moderate, or even severe but not contraindicated orders. The latter scenario tends to shift the alert burden to nurses, especially for duplicate or contradictory nursing orders, and pharmacists for significant medication alerts. One problem that is common to small ambulatory practices as well as large academic institutions is duplicate ordering. Duplicate orders, especially for laboratory tests, can be costly (monetarily and in terms of patient discomfort) and add work burden (for phlebotomists, nurses, laboratory staff). In both small and large institutions, CDS can help with this problem [11]. CDS serves to teach but also to control costs and steer clinicians toward standards of practice. Designing CDS that meets the needs of all types of users is complex and will entail compromises. Ultimately governance helps to balance all these opportunities and perils [3].

20.6 Prioritization of CDS activities

Good governance requires a carefully constructed evaluation and approval process. That process must be transparent to end users: they engage better if they know what to expect. A timeline helps to manage expectations: when might a project be reviewed and approved? How long will it take to implement a request? If a request is rejected, the requestor has a right to know why, so feedback is essential. Who gets to make these decisions?

20.6.1 CDS system safety

The foremost priority is that CDS structure in the EHR is safe. The Promoting Interoperability regulations from the Centers for Medicare and Medicaid Services require that certified healthcare organizations attest to an annual assessment of the Safety Assurance Factors for EHR Resilience (SAFER) Guides [12,13], starting in calendar year 2022 [14]. We advocate that institutions review and assess their processes according to the SAFER Guides, especially regarding CDS interventions in the sections on clinical content, the human-computer interface, workflow and communication, and measurement and monitoring, as a prerequisite before structuring a governance process for managing changes of the CDS.

20.6.2 Importance of explicit organizational leadership priorities

The next consideration is whether the health care organization's senior leadership has announced organizational priorities for a given cycle (e.g., 6 months, 1 year, 5 years). These goals are important from a CDS standpoint to serve as guideposts for CDS implementation work. Establishing clearly stated and unambiguous organizational priorities is where values, strategic and operational needs, financial pressures, existing investments in projects and departments, and social, political, and cultural forces all converge. Examples of goals could include "enhance medication prescribing safety" or "reduce factors associated with clinician burnout." Some organizations may avoid creating such a list of publicly announced priorities, as doing so may offend those working on initiatives ultimately deemed relatively lower priority or not prioritized at all. Some well-resourced organizations may take the view that they do not need to prioritize their goals, as they have the means to meet all of them simultaneously. Still, avoiding the difficult task of stating organizational priorities carries risk for the organization from a CDS standpoint. Without an explicitly stated and finite list of organizational priorities, CDS implementers will have no basis for evaluating the relative priorities when presented with dozens of implementation requests. Few CDS implementation teams would be well-resourced enough to approve all these requests, develop the CDS content thoughtfully and carefully, and deliver them all within a reasonable timeframe. In the absence of such organizational priorities, CDS implementation teams may be left in a position where they must assume that every request is a high priority, not having a basis to decide otherwise. This can lead to extensive resource (IT analyst, clinical informatics personnel expertise and time) consumption. It can create a hazardous situation in which the CDS team, inundated with new requests, lacks sufficient time to adequately consider and develop each new CDS artifact. Such conditions increase the likelihood that there will be shortcuts taken and abbreviated review of CDS tools performed, factors that could create hazards and unintended consequences when the CDS tools are ultimately deployed. In addition, devoting scarce CDS resources to every new request creates two major opportunity costs. First, what might the organization have been able to do better on high priority CDS initiatives had it focused its resources on those? Second, with all CDS resources always consumed, organizations miss a chance to foster creativity. IT teams that are fully subscribed will not have the ability to step back from day-to-day work and consider novel opportunities for CDS to solve health care organization challenges, including, for example, cross-institutional but less traditional approaches to interinstitutional, interoperable CDS enabled by standards such as SMART on FHIR. In a world of constrained resources, clearly stated organizational priorities will enable CDS teams to work more effectively and efficiently, and to act as good stewards of health care organization resources.

20.6.3 First level review of requests and projects

An initial question is often "what are our priority items?" First, we examine a high-level model for stratifying CDS requests. We envision the layers of this model much like a pyramid, with the basic, "must do" items at the bottom, and the "nice to have" items at the top. Fig. 20.3 illustrates a general prioritizing schema.

In this schema, level 1 patient safety issues (e.g., incorrect medication dosing, erroneous alerting, broken interfaces, issues found during a root cause analysis) as well as break-fixes (e.g., processes that fail, and no workaround exists) are the highest priority.

Regulatory and other external requirements (level 2) must be attended to in a timely way, sometimes immediately. These include laws and regulations, whether federal, state, or local. Policies and operations must be consonant with these necessities, as well as with Joint Commission and other accreditation bodies. Governance should allow for giving highest priority to address patient safety and break-fixes, assuming the organization does the difficult work of explicitly characterizing what should qualify as a significant patient safety issue. The next priority consideration is to address regulatory or licensing requirements. These changes require a smooth and efficient process that is transparent to end users, and quick. Governance for these issues may differ from lower priority requests, which usually require additional steps to avoid unintended consequences. In keeping with the need to promote transparency of the governance structure and process, organizations should document explicit circumstances (examples: regulatory body citation, pandemic conditions, natural disaster) under which it is permissible to bypass some levels of governance.

Next in line are high priority, mission-critical, time-sensitive items (level 3) that are affordable for the organization to pursue. High priority, mission sensitive, affordable, but not so time-sensitive requests (level 4) and "nice-to-haves" (level 5) are lower precedence. Governance for level 3, 4, and 5 projects should be equally transparent to and efficient for end users, but more deliberative to include considerations of cost, availability of personnel,

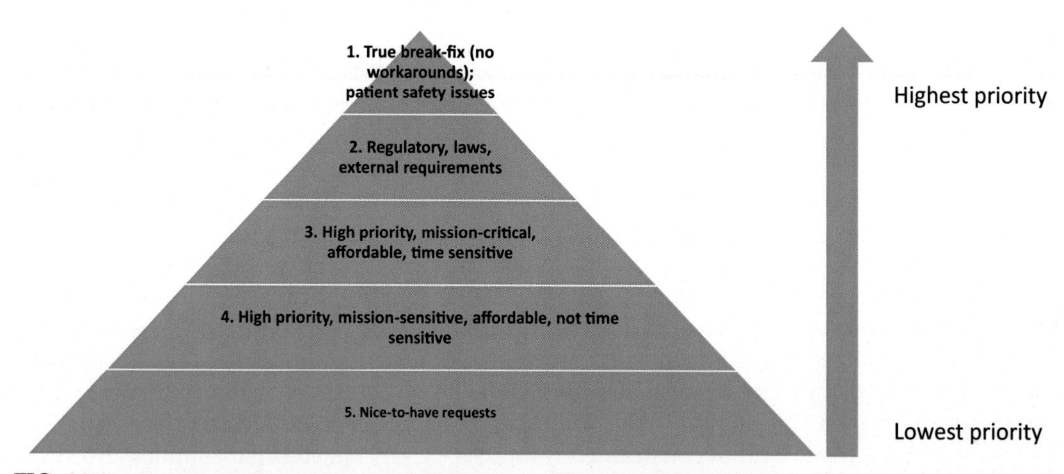

FIG. 20.3 General prioritizing schema for CDS requests. The most critical are level 1, at the apex of the pyramid, and must be addressed first. The size of the sections of the pyramid roughly denote the relative numbers of requests.

and institutional priorities matching the institution's mission. Despite the size of the sections in the figure, our experience is that there are usually far more "nice-to-haves" than all other requests.

20.6.4 CREATOR strategy

Our review of the CDS literature on prioritization of requests revealed few widely recognized guidelines for good CDS governance.

In the context of ideal alerting the authors developed 7 rules that we feel generally apply to CDS governance best practices [6]. The mnemonic CREATOR summarizes these principles:

- Consistent with organizational strategy and principles
- Relevant and timely
- Evaluable
- Actionable
- Transparent
- Overridable
- Referenced

20.6.4.1 Consistent with organizational strategy and principles

Clearly, the CDS governance structure must complement the institution's goals with respect to its EHR and overall financial abilities and quality goals. Setting expectations regarding what is affordable and consistent with the mission of the institution is critical to acknowledging clinicians' desires and attaining clinician buy-in and satisfaction.

While some strategies and principles will be well known and applicable broadly across the organization, others may present themselves day-to-day during CDS work and demand careful consideration. Institutions may be faced with the need to answer a multitude of specific CDS questions that help define and detail the CDS strategy. For example, does the institution allow personal order sets? Is personalization of order panels permissible? If the institution has certain clinical pathways, can individuals alter the orders, or should these pathway orders be locked down? Is it permissible for end users to block (or be exempt from) certain alerts, such as drug-drug interaction precautions, and if so, for how long? Should that be allowed generally, or specialty-by-specialty? For example, a cardiologist who frequently prescribes amiodarone and warfarin for patients with atrial fibrillation can be expected to be aware of the significant interaction between these two drugs, but what approach does the institution feel is safe for a clinician who is less familiar with the complexities of warfarin management who orders sulfamethoxazole/trimethoprim or a fluroquinolone for a patient already on warfarin?

20.6.4.2 Relevant and timely

As regulations change and medical practice evolves, the governance structure must be flexible and agile to respond promptly. This is also a reminder of the CDS 5 Rights [15] and specifically, that CDS tools should always be implemented at the optimal point in a particular workflow to the appropriate user, whenever possible (see also Implementation of CDS section that follows).

20.6.4.3 *Evaluable*

External regulations (such as Promoting Interoperability, including use of the SAFER guides [12,13]) and internal quality metrics evaluate the efficacy and safety of CDS. So, too, governing nursing and medical committees such as the Medical Executive Committee judge the governance structure itself by the CDS it produces and how well it operates. Simply put, is the CDS governance process itself performing optimally or is there room to improve?

20.6.4.4 *Actionable*

It goes almost without saying that CDS governance must have agency. The CDS governance must be empowered to evaluate, add, remove, modify, and restructure CDS, while considering input from various stakeholder groups. Actionable also refers to options for the clinician to take within the context of an interruptive alert (see also Implementation of CDS section that follows).

20.6.4.5 *Transparent*

Clinicians and other front-line workers often view information technology as a black box, impenetrably obscure and esoteric. This leads to misunderstanding, confusion, miscommunication, and misperceptions. CDS governance is more successful when users know what to expect, whether their project will be considered, how long something will take, and to be kept informed regarding progress. The adage "under promise but over deliver" applies to transparent CDS governance and good customer service.

20.6.4.6 *Overridable*

Regarding governance we intend this to mean that the CDS governance structure must have the power to say no. In this context, it may be policy that an institution will not endorse personal order sets. Perhaps a better term would be "overriding." CDS leadership often must make difficult choices to be able to pursue some CDS initiatives and not others.

20.6.4.7 *Referenced*

As with all medical interventions, CDS is a clinical endeavor which must be inherently evidence-based. CDS governance gains trust and greater compliance when guidance is from a solid source.

20.6.5 In-depth processes and review

The CREATOR framework serves as the groundwork for detailed prioritization. Fig. 20.4 illustrates a suggested flow diagram. An example is one organization which requires that a department leader submit all CDS requests, or if submitted by an individual, the department leader must approve the submission. If leadership has preapproved a project, work proceeds, recognizing the risks of unintended consequences if governance procedures are not followed.

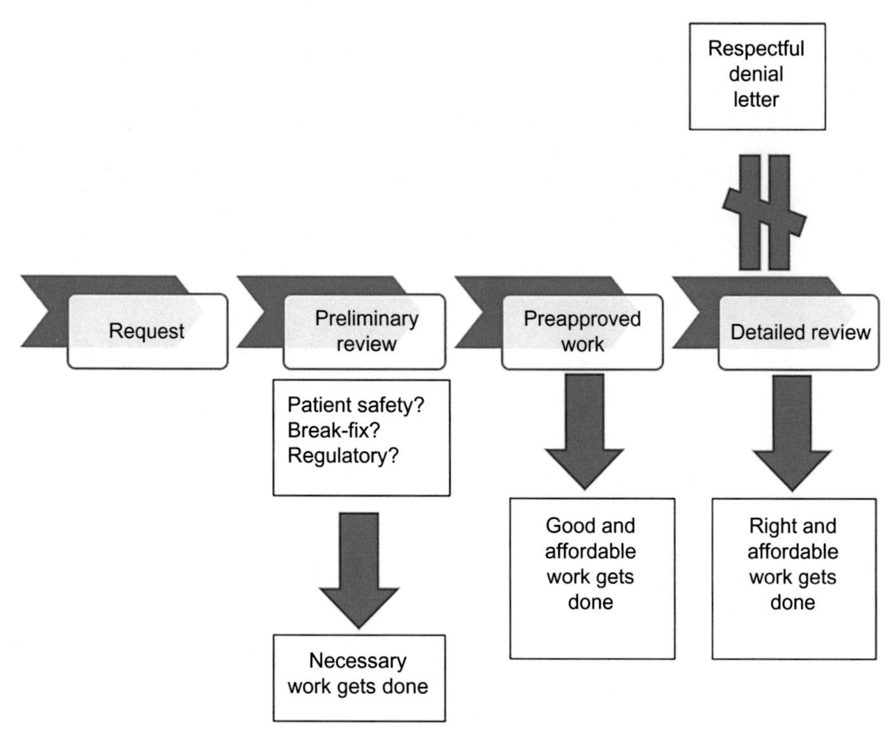

FIG. 20.4 Suggested flow diagram for CDS approval process.

More routine requests often involve an estimation of the cost of the request and the time it will take to complete it. If analysis reveals that a project comes under a certain time and personnel threshold, and the work is considered part of the organization's mission, the request is approved and work proceeds. If the project exceeds either threshold there may be a need for more detailed review for appropriateness by more senior leadership.

In contrast, other organizations may not require leadership approval for CDS requests to proceed through the evaluation process. In these organizations, it may be that all staff can submit ideas and requests as individuals. Sometimes even if driven by patient safety, break fix, or regulatory needs, there can be a collision in such organizations between well-intentioned requests placed by individuals and an organization's CDS strategy.

Once a request receives approval a key element in CDS governance is identifying an owner for each process, tool, or artifact. Ownership in this context means identifying a contact person who is also responsible for ensuring the CDS is up to date, consonant with current practice, and consistent with policy. Pharmacists or pharmacy informatics, if available, owns medication orders; nursing or nursing informatics owns nursing procedures, orders, and documentation. Alerts and other CDS ownership in other areas is difficult to assign, as alerts usually impact users across many disciplines. In the absence of a formalized structure, CDS ownership may ultimately fall to the person who proposes the CDS tool.

The effectiveness of CDS governance derives from the quality of the processes governance institutes. It is not just about decision-making, but includes the quality of the data inputs and defining duties of those involved in governance. Order set ownership is an especially vexing example. Usually the head of a department most impacted by the CDS, or his or her designee, retains ownership, such as the chief of the trauma service for an alert regarding use of volume expanders for hypovolemic shock. The chief of service can delegate this role. When CDS impacts a broad constituency, it may be necessary for the CMIO to assign an owner such as the chief of internal medicine for an insulin correction scale order set when an endocrinologist is not available. The CMIO may need to take ownership for a general alert such as one which directs the attending physician to cosign an admission order which an advanced practitioner or resident entered. When personnel change, there must be a process to hand off ownership and inform the new owner.

Another example of complex processes of governance and implementation is the management of problem and medication lists. Drug-problem, drug-drug interaction, duplicate drug, and test-condition CDS depend on reliable, accurate, and up to date problem and medication lists. Outdated lists, insufficiently specified or duplicated problems or medications, and other errors deprecate the quality of the CDS. Governance must address the ownership of these lists and more generally, the quality of the EHR data that serves as inputs to CDS tools. While some may consider EHR components such as problem lists and other data as outside the purview of CDS, we believe that careful attention to these elements is an integral part of CDS governance.

20.6.6 Scoring systems

There are many scoring systems to assess the clinical priority of CDS requests. In one model, an analyst, in concert with the requestor, assigns points to a request based on criticality, patient safety, affordability, resource demands (e.g., how many hours of build work?), number of people affected, complexity of the build (e.g., does the organization have the skill set to build the request or will the project require consultants?) and other factors. In other models, the CDS governance committee may determine these point assignments using a prioritization tool. For affordability, some organizations have thresholds of cost and time below which no administrative review is necessary, but above a certain level may require director or vice president level review; an even higher cost may require review by the CIO, CFO, COO or CEO. There is no scoring tool that the authors have found to be standard, nor one empirically known to be optimal. Even within a single organization, having stakeholder groups commit to using a common scoring tool for request evaluation can prove challenging. In turn, lack of a common scoring tool within an organization can lead to misunderstandings about the relative importance of requests and projects, especially if such projects require assessment by more than one stakeholder group.

20.7 Metrics, feedback, and anomalies

Best practice principles include upfront definition of metrics that will be used to evaluate CDS performance followed by constant reevaluation of CDS that has already been built. Good governance should address standards by which the institution will establish such metrics and perform such reviews. This is not as straightforward as it may sound. Optimization of CDS content is an iterative process which requires frequent review of CDS governance policies and procedures to maintain agility in the constantly changing environment of CDS. See Section 20.20.1 for more detailed discussion of implementation metrics.

No consensus exists regarding optimal metrics by which to evaluate CDS. Even the term "alert override" is ill-defined, and its measurement uncertain. In previous work [6] we listed 17 different alert metrics that are currently in use (Table 20.2). Moreover, a single metric may prove insufficient to fully characterize the impact of an alert or other CDS tool [16], in which case CDS developers may wish to consider tracking several metrics to provide a complete picture of a CDS tool across a range of dimensions.

It may be best to categorize CDS metrics as descriptive (e.g., alerts/100 orders, or alerts/visit), performance (e.g., sensitivity, specificity, and positive and negative predictive value), clinician response (e.g., override or acceptance rate, and clinician comments), and burden (e.g., number needed to alert, effectiveness, and efficiency of the CDS) [16].

These metrics apply mainly to alerts and reminders but may be applicable to other CDS tools such as order sets. How often do clinicians use a disease-specific order set, one which includes specific guideline-directed therapies that are known to influence clinical outcomes? Is there a way to lead a clinician to use these tools without substantially interfering with clinical care? It is then possible to compare outcomes between those clinicians who do versus those who do not use a particular CDS tool.

Some vendor products with which we are familiar include the ability to submit real-time, in-line comments to IT analysts or the organization's service desk regarding alerts and other processes. Even negative feedback, referred to as "cranky comments" [22] can be opportunities for improvement. In response, programs can be developed to effectively respond to those user-generated concerns such as the "Clickbusters" program [23]. Comments and complaints from users can be critical feedback for pointing out suspected errors, the need for

TABLE 20.2 Sociotechnical domains that may impede effective governance.

Domains	Issues	References
Physicians Other clinicians	Limited availability to participate Lose interest rapidly May not receive compensation	Authors' experience
EHR analysts	May not know system limitations early in project May be few knowledgeable analysts in health IT market for some applications Failure to recognize alert errors and anomalies	Authors' experience [17,18] [17,18]
Budget	Total cost of ownership not always clear Allotments may not be realistic	[19,20]
Administration	Impaired or interrupted institutional knowledge due to: Frequent hospital leadership turnover (avg.: 5.5 years) Frequent medical executive leadership turnover (often yearly)	[21]
Regulatory	Rapid cycle changes (e.g., 21st Century Cures Act)	Authors' experience

From McGreevey JD, Mallozzi CP, Perkins RM, et al. Reducing alert burden in electronic health records: state of the art recommendations from four health systems. Appl Clin Inform 2020;11:1–12. Used with permission.

improvement of CDS, and even suggestions for better CDS. Governance to organize and operationalize this feedback has been termed the creation of "tight loops" and is important as users dislike feeling that they lack input into the process. Still, for most organizations the resources needed to create and maintain feedback monitoring, management, and response programs are non-trivial.

Detecting CDS anomalies—quirks, errors, and misfirings, as well as alert logic errors in configurations—is an understudied area of CDS governance [17]. Some alerts stop firing, and often no one notices because one is not interrupted which many interpret as a good thing, even if it is due to "automation bias" [24]. Such misbehaving alerts may be due to changes in drug codes or formulary changes in the background, while the CDS was not updated [25,26]. Other causes include normal seasonal variation in alerts (such as for influenza vaccination) but where the alert was not adjusted for the date. Another example is an alert that was taken out of production for update but was not turned back on or transferred back into the production environment.

Governance should examine these CDS metrics and anomalies regularly, and attempt to review all CDS periodically. Regulations dominate this aspect, such as the Centers for Medicare and Medicaid Services requirement to review order sets periodically (although there is no specified interval) [27]. The burden for this work is enormous; many organizations struggle with identifying the personnel who should be responsible for doing these tasks on a routine basis as part of their jobs.

20.8 CDS governance models that work

Evaluating the success of CDS governance is subjective: each organization must design its governance structure as it sees fit (see Section 20.5). A lot depends on the style of the leadership, usually the CMIO.

If an organization is small, with a limited group of SMEs, the only option for governance is a small, centralized decision-making group. This group will frequently rely on outside consultants, or the vendor, for configuration advice. This model can be very efficient (see Section 20.5.1), but finances will usually constrain the possibilities of CDS projects. A common solution in these situations is to place the CDS governance groups in a spoke and wheel (or star) network with established medical staff and organizational units such as the medical executive, pharmacy and therapeutics, and nursing committees (see Fig. 20.5).

Although the technical considerations for CDS build may not be as apparent to these groups, at least members can impact the clinical decisions necessary for CDS in the EHR. The CMIO or IT usually drives the processes, making it clear that decisions rest with the clinicians. Balancing the right numbers of inputs from diverse clinicians and various steering committees is difficult; communication and consensus can be elusive.

In the best-case scenarios, the CMIO and CDS governance committee foster, encourage, and facilitate communication among the groups. Also, the CMIO can promote awareness of the activities of each group to avoid silos, and to share knowledge. Decision-making becomes more centralized, without losing the trust of those groups. As decisions become more efficient and coordinated, acceptance of change becomes the default.

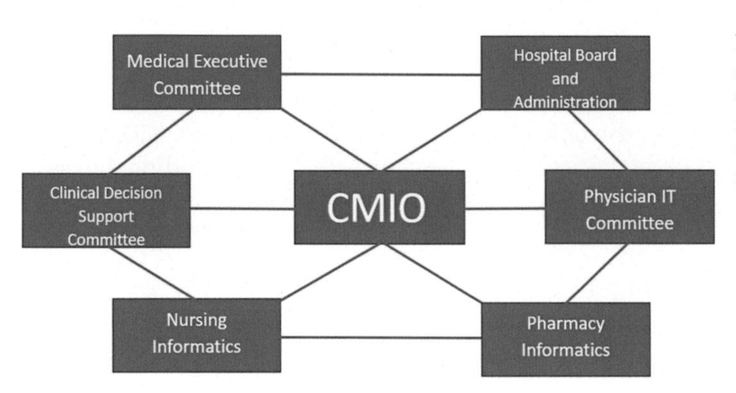

FIG. 20.5 A spoke and wheel, or star network diagram showing the CMIO as the coordinating leader for various informatics, administrative, and operational functions including the CDS governance committee.

20.9 CDS governance models that fail

In the worst-case scenarios, CDS decision-making bodies such as pharmacy, nursing, even the medical executive committee, become siloed, communication breaks down, change management disintegrates, and poor and unauthorized—even dangerous—workarounds abound. Often there is no identifiable accountability for changes. Leadership fails to maintain lines of communication, permits conflicting or duplicative processes, or fails to identify and hold management responsible to its governance policies.

Sometimes it is not a failure of IT or informatics governance. As noted in Section 20.3, few physicians can devote the time necessary to remain engaged. If the CDS governance structure depends on disengaged clinicians, processes will break down. It is important to include naysayers and highly critical, as well as highly supportive members, but the former are notoriously difficult to recruit and retain. It is particularly important to include all persons listed in Section 20.3, as well as physicians when system changes alter workflows [28]. But if these clinicians fail to come to the table, even when IT processes are intact, the clinicians will be unhappy.

Trust in the process is a key to its success, and lack of trust can cause the CDS development process to fray over time, especially in cases where requests for new CDS are revised by the CDS implementation team or in some cases, denied outright. In these situations, requestors may question the authority of the CDS process itself to make decisions on behalf of the organization and attempt to circumvent the governance process by contacting senior leaders directly to facilitate their desired work.

It may be difficult to keep up with regulatory and other legal changes, causing some organizations to bypass their governance to comply, even when that is not appropriate [29]. Equally problematic is what to do about governance during prolonged emergencies such as the SARS-CoV-19 pandemic. It became essential for organizations to enable rapid-cycle change, bypass their usual testing and control procedures and move items to production, despite the known risks [30]. See Section 20.14. Table 20.2 lists some of the sociotechnical domains that interfere with effective governance.

20.10 Governance evolution over time

In the authors' experience even successful CDS governance structures evolve over time as institutions learn better and more effective strategies. Many factors influence such changes.

"Every year, 1 out of every 6 hospital chief executive officers either retires or moves on to a new position" [31] and with it, strategic goals change [6,10]. Mergers and acquisitions, changes in demographics of the patients, and resource changes in the organization all impact CDS governance. On-again/off-again directives and regulatory changes cause CDS governance to alter its trajectories. And pandemics create upheaval at all levels, including the need to up-end all governance expectations to implement crucial new workflows.

20.11 Challenges to effective governance

One person's good governance is another's obstacle. We believe that a participatory, transparent, and relevant structure encourages good CDS development. There will always be disagreements on who gets to decide and the exact content of CDS artifacts, but if all stakeholder groups offer representation to the governance structure CDS tools are more likely to succeed. An absence of such well-defined decision-making as discussed above will lead to failure. If there is misalignment between leadership priorities, CDS governance, and the organization's mission, then management cannot maintain an organized approach to the declared priorities. Such lack of a clear mission causes frustration, mistrust, and even low-level chaos at times.

Transparency can go too far. Some end users get to know the analysts and specialists with whom they have worked. If they try to contact these analysts directly, it may interrupt the analyst's work and bypass the checks and balances of governance and proper procedures. This may seem efficient to the end users but in the long run is disruptive and chaotic. It also interferes with institutional priorities.

Lack of leadership interest—a mandatory participant in CDS governance (see Section 20.1)—is highly threatening, as discussed in Section 20.10. Having leadership support but without a corresponding availability of resources, including highly trained analysts, will also likely diminish the effectiveness of CDS governance. It is indeed difficult to solve the Goldilocks problem of what constitutes too much or too little governance. Too much can lead to the perception that the governance is a barrier to change, or an impenetrable black box. Too little risks failure as discussed above.

20.12 Governance of emerging models of CDS

Only a few institutions have incorporated third-party CDS applications and usually only in one or two specific use cases. To our knowledge the only institution with a number of such projects is University of Utah Health which has 10 instances of web-based CDS [32].

Adaptive CDS is that which "trains itself and adapts its algorithms based on new data" [33]. This is a new and rapidly evolving field that has many challenges, not the least of which is transparency and close attention to the risk of bias, especially in the development of the training data and its accuracy, and the applicability of that data to the real-world instances in which the artificial intelligence (AI) is used (see Chapters 7 and 16 for details) [34,35].

Governance must also be concerned with the development of new methods of CDS based on standards such as CDS Hooks, Clinical Quality Language, SMART on FHIR, and CDS machine learning, covered in Section 20.23 as well as Chapters 9, 15, and 29. The task for governance is to ensure that these technologies are compatible with the mission of the organization and with other CDS projects, affordable to implement, and that the organization has the technical and clinical subject matter experts to use these capabilities properly.

In our view it is unwise to pursue these novel methods separate from the main structure of CDS governance. The main risks include duplication of effort and the creation of silos, which complicates operations and communication. The challenge for CDS governance is to answer the serious questions that arise before instituting these applications. These include: Is there a clinical or operational problem that the novel form of CDS has been proven or is very likely to solve? Who controls and owns the data? Can these data be adequately de-identified, and not reidentified? Are the data secure during export to a third-party application? Are the data available to a third party for secondary use? Who is responsible for data integrity, data maintenance, and application upgrades? Who is responsible, financially and for the work, for interface updating when either party upgrades its software? [33].

20.13 Impact of crises on CDS governance

The recent Covid-19 pandemic threw the best-laid governance up in the air. Information about the novel viral infection flowed in daily. Detection methods and treatment plans changed overnight. New tests appeared; new dosing guidelines for old drugs came and went; new and unfamiliar medications came into daily use. Regulations and requirements for reporting changed equally as fast. All this created a need for rapid cycle decision-making and project approval. The cost included abandonment and/or suspension of other desired changes in the EHR and CDS. It was imperative to cut steps in decision-making deliberations and accelerate implementation. Accountability and reassessment faced challenges, too, as orders for therapies that did not work had to be removed, and treatment protocols changed. All these factors disrupted accepted traits of good governance—adequate testing, standardization, stabilization, optimization—which took a back seat to rapid-cycle innovation and operational efficiency [30,36]. Some might argue that the Covid-19 experience proves that much can be accomplished with minimal governance. However, it is important to recognize that overly simplified governance may create new risks, such as inadequate testing and stakeholder disenfranchisement.

What does this mean for CDS governance? One must recognize the need for flexibility and agility, and consider including plans in one's governance policies to deal with such exigencies. Perhaps there is a need for a governance model for usual times, and an emergency operations governance. What, then, is the threshold for activating this emergency model?

20.14 Implementation

20.14.1 The CDS package

While different types of CDS can be implemented independently, it is not uncommon for multiple CDS tools to be deployed in concert. This is the concept of the "CDS package" as described by Levick and Osheroff [37]. CDS developers should consider the entire workflow and operation surrounding a CDS intervention and account for it appropriately.

For example, to achieve optimal outcomes in patients with sepsis requires multiple steps in data gathering and clinical reasoning. A complete CDS package for sepsis would include an algorithm to determine patients at risk for sepsis, followed by alerting the appropriate clinicians when a patient meets those criteria. The algorithm should inform the clinician of the probability of sepsis, and the abnormalities that support that diagnosis. The clinician then needs immediate EHR access to notify a sepsis team (if one exists), and the ability to open an order set to place time-sensitive orders, arrange for any needed consultations, and consider transfer to an intensive care setting. This package of CDS tools may include: rule-based criteria or in some cases machine learning to detect the condition; timely notification; actionable alerts; order sets; communication tools; relevant informational material during the ordering process (e.g., renal function, culture reports, focused imaging results); drug-condition, drug-allergy, drug-drug interaction, and drug-dosing alerts; and ready access to documentation resources.

Finally, the CDS package for sepsis must include the ability to assess clinical outcomes (as discussed in Section 20.7). There may be a sepsis quality director who would benefit from a report showing how often the sepsis alert fired, for which criteria, and what actions clinicians took or did not take. In so doing, the quality director can evaluate in near real-time the performance of the suspected sepsis CDS program. All these elements are part of the CDS package.

20.15 Creating CDS that matches workflow

Others have written extensively on the importance of respecting the five rights of clinical decision support [15,37]. Clinical decision support functions best when it complements successful clinician workflows. Interrupting successful workflows risks alert fatigue and heightens clinician burden. Still, interruptive CDS can redirect erroneous or dangerous clinician workflows and thus prevent patient harm. To enable informaticists and CDS developers to determine the optimal type of CDS for a particular situation requires a complete understanding of current and desired workflows. See also Section 20.20.5.

Here we describe gradations of alerts and the varied ways they can intersect clinical workflows. In basic terms, there are two types of alerts: interruptive and non-interruptive. Interruptive alerts redirect the clinician's intended action while non-interruptive alerts allow the clinician's workflow to continue. Interruptive alerts that are purely informational are rarely, if ever desirable. As a general principle, interruptive alerts should be *actionable* [38].

Some alerts are "silent" because they are configured such that they are invisible to the clinician but still generate useful knowledge for the CDS development team. Silent alerts are useful for new alert testing to gain understanding of alert behavior in the real world. These data, including frequency of an event, give CDS developers valuable insights to inform revisions and other adjustments to alert and improve their quality, all before clinicians experience the alerts [6].

There are "hard" and "soft" stop alerts, but there is no universally accepted meaning of these terms, resulting in a continuum of definitions running from preventing clinicians from carrying out an intended action under any circumstances to permitting clinicians to proceed without requiring any acknowledgement. It may be useful to describe a more descriptive framework.

TABLE 20.3 Types of alerts, analogous to United States traffic lights.

Light	Traffic meaning	Alert type	Meaning	Example
Red	Stop	Absolute hard stop	Cannot proceed Must retract or go back	Order for isotretinoin for pregnant woman
Flashing red	Stop, look, go when safe	Overridable hard stop	Must enter data or make selection regarding reason to proceed, or go back	Order for a restricted antibiotic
Yellow	Slow down, caution, may need to stop	Cautionary	A warning usually requiring alert acknowledgement	Order for Clostridioides difficile assay in patient with solid stools or recently positive test
Flashing yellow	May proceed cautiously	Advisory	May proceed without need to acknowledge alert	Consider different antibiotic dose based on indication
Green	Safe to proceed	Informational	A banner, note to the side, or content within an order, to provide context	Display of a result pertinent to an order

A hard stop, like a red light, does not allow a user to proceed with a step in the clinical workflow under any circumstances (see Table 20.3). Perhaps the clearest example of an absolute hard stop alert is when a clinician attempts to prescribe isotretinoin, a well-known teratogen, for a pregnant patient. Arguably, there should never be any case in which isotretinoin is prescribed for a pregnant person. Such an alert would force the clinician to retract the isotretinoin order.

A soft stop allows a user to proceed with an intended clinical action, perhaps with modification, based on the alert's advice. However, there are gradations of soft stops. The most restrictive soft-stop alert, akin to a flashing red traffic signal, can be overridden but requires clinicians to enter a reason for their decision to override the alert. This might be a selection from a preset menu of override reasons, a free text description by the clinician about their rationale for overriding the alert, or even a requirement that the clinician confer with a relevant colleague and obtain the colleague's approval for proceeding with the intended action such as requiring an internist to enter the name of the infectious disease specialist who gave approval before prescribing a restricted antibiotic.

Less restrictive cautionary soft stops (or more precisely a "slow down and look" yellow traffic signal) may allow a clinician to proceed with an intended action without needing to enter data, make selections from within the alert, or document an authorized colleague's approval. The clinician can pause to undo an order or simply slow down momentarily to acknowledge and dismiss the alert. An example of a cautionary alert is one that advises a Clostridioides difficile test is not indicated unless the patient has liquid stool. If a clinician recognizes that a patient has actually been having solid stools, then the alert might prompt the clinician to stop and revoke the order. Otherwise, the clinician may simply click acknowledge or dismiss within the alert and proceed with ordering the test.

Even less restrictive alerts are advisory (like a flashing yellow traffic light). For example, an antibiotic ordering screen may urge the clinician to consider a different dose for a particular indication. If the clinician decides not to alter the dose, pharmacy might ultimately call the provider to discuss the chosen dose. Unlike a cautionary alert that requires acknowledgement, this type of alert does not have that requirement: the clinician may choose to order the originally intended antibiotic dose.

Finally, alerts that offer information to contextualize a situation are like a green traffic light. For example, a banner that displays a patient's most recent INR result could be informative but not suggestive in the process of ordering warfarin.

It is important to recognize that despite the framework proposed here, readers may encounter differing understandings of alert definitions in actual CDS practice. For example, some have advocated that a licensed clinician should never be prevented from taking an action that he or she deems clinically appropriate, arguing that the clinician is best positioned to weigh risks and benefits for a particular patient in a particular situation. In organizations where this belief is supported, then, the most restrictive alert that users would experience would be the equivalent of a soft stop as defined above, where individuals may be required or advised to enter a reason for a decision that deviates from the alert guidance before being able to proceed. See also Chapter 3 for details about CDS types, including alerts.

20.16 Key CDS guidance

Several principles inform CDS implementation. The original five rights of medication safety—the right medication, the right person, the right dose, the right time, and the right route became a model for the five rights of CDS: the right information, delivered to the right person, in the right format (e.g., alert, order set, banner etc.), delivered through the right channel (e.g., cell phone, electronic health record, email system), at the right time in workflow [15,39]. Bates and colleagues developed the 10 commandments of CDS: (1) speed is everything, (2) anticipate needs and deliver in real time, (3) fit into the user's workflow, (4) little things can make a big difference, (5) recognize that physicians will strongly resist stopping, (6) changing direction is easier than stopping, (7) simple interventions work best, (8) ask for additional information only when you really need it, (9) monitor impact, get feedback, and respond, and (10) manage and maintain your knowledge-based systems [2]. In some cases, guidance has been created for specific types of CDS. As referenced earlier, the CREATOR rules can guide the development of ideal alerts: Consistent with organizational strategy and principles, relevant and timely, evaluable, actionable, transparent, overridable, and referenced, if appropriate [6]. Likewise order sets have been the subject of evaluation, recommendations for development, and also scrutiny [40–43].

20.17 Facilitators and barriers in CDS implementation

As noted earlier in this chapter, a sound governance process lays the critical foundation for CDS implementation. See also Sections 20.1, 20.6–20.9, and 20.11–20.13. There are factors that can facilitate CDS implementation progress and others that can impair success.

20.17.1 Facilitators of CDS success

As CDS teams begin their work, they must recognize facilitators of and barriers to their success. Facilitators include strong support from leadership, dedicated and skilled staff

(IT builders, analysts, programmers, clinical informatics staff), a progressive organizational culture that is open to trying new methods of decision support, and, where available, ancillary resources such as quality improvement experts, researchers, and implementation scientists who can catalyze the evaluation and refinement of CDS tools, as well as help promote their impact publicly.

20.17.2 Barriers to CDS success

Barriers to successful CDS implementation are, not surprisingly, the absence of the above items so limited or no leadership support, the absence of technical and clinical informatics resources, and an organizational culture that is risk averse and reluctant to change. In addition, failure of clinicians (physicians, advanced practice providers, nurses, pharmacists, social workers) and subject matter experts to participate in the CDS development process, as noted earlier in this chapter, whether for lack of time or interest or support to do so or a combination, can hinder CDS implementations [6]. Additionally, limited participation by other key stakeholders, including those with expertise in regulatory, health information management, and quality and patient safety, can also impede CDS progress.

20.18 Role of vendors in CDS

Vendors, such as EHR, knowledge resource, and CDS vendors, can be partners in the CDS implementation process. At a fundamental level, as required to become certified health information technology, EHR vendors must provide basic capabilities in their systems that enable customers to build their own CDS (Office of the National Coordinator certified health information technology). Many vendor products have basic configurations, often referred to as community models that are ready "out-of-the-box." These products usually only have CDS that supports the requirements of Meaningful Use/Promoting Interoperability, such as allergy and drug duplication alerts. Vendors may supply the ability for customers to create alerts and reminders, order sets, and banners and iconography that can convey awareness of particular conditions to clinicians. However, the content and configuration of these tools is usually the responsibility of the customer to create and maintain. EHR vendors typically offer access to a customer network where customers can share experiences, challenges, and recommendations with one another [44]. And they may host in-person and virtual opportunities where exemplar customers who have developed novel uses of their software can demonstrate their techniques, describe their processes, and publicly showcase their success, an experience that not only benefits them as individuals and trailblazing institutions, but the vendor as well.

Knowledge resource and third-party CDS application vendors are playing an increasing role in CDS implementation. Notably, drug knowledge vendors play an important role in medication-related CDS, providing both narrative and computable logic for medication-related CDS tools such as alerts for drug-drug interactions, drug allergies, and duplicate therapy. As discussed in Chapter 18, these external resources provide capabilities for local customization that need to be addressed as a part of CDS implementation.

As described in Section 20.12 Governance of emerging models of CDS, recent interoperability standards such as SMART on FHIR and CDS Hooks (see Chapters 15 and 29 for technical details) are enabling a marketplace with a broad range of third-party vendors that provide tools such as CDS capabilities via Web services, CDS apps, patient-facing apps, and artificial intelligence algorithms. These third-party CDS tools often impose additional implementation complexity, such as interoperability with local terminologies, training, and long-term maintenance, which need to be addressed at the local level in collaboration with CDS vendors.

20.19 CDS costs and implementation strategy decisions

20.19.1 CDS costs

CDS development and implementation can be costly. There are software costs, for instance the EHR purchase, licensing, and maintenance charges [9]. As noted above, there may be complementary systems, such as drug knowledge vendor medication libraries, that also involve costs. There is the need to fund technical analysts who can build, maintain, and troubleshoot CDS artifacts and clinical informaticists to guide decisions about the most important CDS projects to pursue and the right way to pursue them [9]. Clinical subject matter experts and personnel from various clinical domains, quality and patient safety, regulatory affairs, and health information management are often needed to participate in the CDS development and review process, all of which are valuable and scarce resources for a health care organization as noted earlier in this chapter. See also Sections 20.1.1, 20.3, 20.7, 20.8, and 20.9.

20.19.2 Implementation strategy decisions

Faced with these costs, health care organizations must ask a fundamental question about how to pursue CDS: should they build, borrow, or buy? Building their own CDS tools requires the costly resources noted above but may yield a highly customized and successful CDS product that has the support of key stakeholders in the organization. At the same time, customized content requires a commitment, ideally, to a program that can inventory, review, and update CDS content on a regular basis, to assure clinical appropriateness as evidence and practices evolve.

In some cases, organizations may be able to borrow content developed elsewhere, such as a freely available CDS tool. An example is the neonatal bilirubin SMART on FHIR app that was developed at Boston Children's Hospital and then adapted for use, albeit with significant local enhancements and configuration work, at the University of Utah [45]. The Utah experience demonstrates that even if borrowing "free" content is pursued, organizations must recognize that there are in fact costs associated with this approach. There will be expertise and labor needed to retrofit the app or other content into one's own EHR and to continuously improve and maintain the tool. True plug and play functionality for CDS tools remains a goal but is not yet a reality.

Lastly, some organizations may decide to buy CDS content. Purchased CDS content offers the promise that someone else—experts in fact—will offload CDS work for a health care organization. And in fact, much of the CDS content available for purchase from CDS vendors may be rich and up to date. In addition, CDS content vendors may offer other benefits to customers, such as reporting and visualization tools that allow customers to view CDS performance and track it over time. Here too, while appealing in the sense that purchased content may obviate the need to reinvent the wheel with a local CDS development process, there are costs beyond the CDS content purchase price. As with SMART on FHIR apps and other tools, adapting purchased CDS content to a health care organization's local environment–its EHR, workflows, customs, equipment, medications, supplies, and personnel–can be a resource consuming activity.

20.20 Implementation principles

20.20.1 CDS leadership for implementation

As described by Levick and Osheroff [37], "Clinical leadership, including the CDS physician champion, must be able to articulate how the CDS program fits in and facilitates attaining the goals and priorities of the institution. High-level goals may include reducing medication errors or reducing re-admissions, decreasing cost of care or length of stay (LOS), attaining specified quality goals (e.g., disease management, core measures, or pay-for-performance targets). It is the role of the clinical leadership to communicate how specific CDS interventions can help achieve these goals in a win-win fashion for all stakeholders, and why they are necessary." In this critical phase of CDS change management, it is also important for health care organization leaders and CDS implementers to align on a CDS philosophy that will guide the work to come. As Ash and Hartzog observed [3], "Organizations with mature CDS programs tend to have philosophies that CDS should provide guardrails, should help the clinicians do the right thing, should make it easier to go down the right path, and CDS should be nearly invisible and remain in the background." However, even with a CDS philosophy, organization leaders and CDS implementers must have the humility to recognize that changing physician behavior can be a difficult and delicate task. "Organizations may seek to control physician behavior, but there are gentle ways of doing it that suggest and guide rather than mandate" [3].

20.20.2 Problem definition and characterization

During the CDS request intake process, problem definition is critical. What is the problem exactly and do stakeholders agree with how the problem is stated? Organizations need to do this well, as much as it can be tedious at times to crystallize the precise problem at hand. Many times requestors will have some idea of the problem without having been pressed to carefully consider and state it. During the CDS request intake process, requestors should bring data establishing the extent of the well-defined problem and its impact, to the extent possible. Here there must be latitude offered depending on the organization. In some organizations, data may be more anecdotal, for example a number of clinicians reporting a similar need or problem. In other organizations where data is more freely accessible, requestors may have a greater level of data at their disposal, including safety event data and hospital operations data

(LOS, average discharge time, patient satisfaction information). CDS leaders should consider having an intake form for new requests that ask some of the questions listed here. Notably, what has been tried to solve the problem to date?

20.20.3 Is CDS the right solution for a particular issue?

Once intake has been completed, CDS leaders must consider if a CDS solution is even appropriate for a particular request. In some cases, an education or workflow change may be a better solution to the problem than a new CDS tool. If a CDS solution should be pursued, what is the range of CDS options available to solve the problem? Importantly, when considering a CDS intervention, a number of principles must be followed as described earlier, notably the 10 Commandments and the 5 Rights of CDS [2,15]. For alert development, CDS leaders should consider the CREATOR rules [6].

20.20.4 Engaging stakeholders

20.20.4.1 Early phase engagement

Regardless of where a CDS tool concept originates—whether from front line users or from leadership—the most successful CDS tools are those designed with stakeholder involvement as noted earlier in this chapter. See also Sections 20.3, 20.8, and 20.9. In the case of CDS proposed by front line users, who are often in the best position to understand work as done and the clinical decision support needs, their request presents a natural opportunity for the CDS team to partner with them on creating effective tools. These requestors can often speak directly to the problem, what has been attempted to solve it to date, and the impact of the absence of this CDS. Such requestors will also be helpful in identifying other stakeholders who may be appropriate to include in the CDS development process. In contrast, CDS proposals that originate from leadership may require analysis first by the CDS team to determine the users likely to be impacted by the CDS. Once a list of user groups has been created, the CDS team should invite key representatives from those user groups into the CDS development process. In both cases, stakeholders should be respected as advisors to CDS development. Their knowledge will inform and complement the approach taken by the CDS team, which likewise has experience creating various types of CDS and institutional knowledge about what CDS is appropriate for a particular situation. As previously noted by Ash and Hartzog, "When a new CDS component is being designed, clinician users must be included in the interface and screen design discussions" [3].

20.20.4.2 Later phase engagement

Stakeholder engagement has benefits beyond the initial phases of CDS development as well. During the testing phase, stakeholders who have been invited to participate in development may also be willing to serve as testers for the tool(s) under development. And while stakeholders should be involved in testing and present a convenient group for doing so for CDS developers, it is important to remember that testing should also involve potential users of the CDS tool who are naïve to the CDS development process as well. Doing so avoids the risk of bias during the testing process, where involved stakeholders may be inclined to evaluate tools which they have helped develop more favorably than an independent testing group might. Stakeholder testers may likewise be less able to identify problems, errors, or other concerns with the CDS tool than independent testers given their close relationship to the tool itself. Still, as previously noted by Ash and Hartzog, including stakeholders in testing can increase user investment in the CDS being developed and advance broader acceptance of the CDS tool once deployed: "Not only can they provide valuable feedback, but their involvement will encourage their colleagues to accept the change more readily" [3].

20.20.4.3 Careful identification of stakeholders

Lastly, while direct users of the CDS may be the most obvious stakeholders to include in CDS development, it is important to think broadly to include as many representatives as possible. In many cases, stakeholders for CDS development may overlap with the stakeholders involved in CDS governance; key personnel to consider for both have been described [6]. See also Table 20.1. A CDS tool that prompts a provider to order an immunization, for example, will impact not only providers, but the pharmacists who must fulfill the vaccine order and the nurses who administer it. There may also be other stakeholders, for example, members of the quality improvement, regulatory, and patient safety teams, who have a vested interest in the CDS tool and its development.

20.20.5 Workflow

20.20.5.1 Participants in a sociotechnical system

CDS should be designed after the clinical workflow in which it will be deployed has been fully understood. In effect, the need to understand clinical workflows is one of the best examples of how the practice of clinical informatics, much to the surprise of some, has sometimes been described as being about technology 10% of the time, and about people 90% of the time. Any change to a clinical information system, whether CDS or otherwise, will occur within the context of a complex sociotechnical system, in this case, the health care organization [46,47]. Within these sociotechnical systems, a CDS intervention can reverberate with impacts on other parts of the system (people, tasks, technologies and tools, organizational policies and culture, and physical environment) and also be impacted by these same elements, leading to CDS intervention success or failure in achieving its goal [48]. As such, whenever a CDS intervention is being proposed, it is critical to take inventory of the sociotechnical system that will surround it.

Consider the downstream impacts of a physician deciding, based on a CDS intervention, to order or not order a lab test. This decision will impact phlebotomists and/or nurses, depending on how labs are drawn at that organization. The decision is likely to impact lab personnel, both on the front lines of processing tests and those in lab leadership who may have a vested interest in making sure the correct tests are ordered at the correct times, and in avoiding unnecessary test ordering. Depending on the medication involved, its indications, risks, and costs, others may also need to be identified as being impacted by the proposed CDS, such as pharmacy leaders charged with managing medication costs or perhaps antimicrobial stewardship leaders tasked with assuring appropriate use of antibiotics to limit the potential for antibiotic resistance. Understanding the full spectrum of individuals who may be affected by the CDS is essential to guarantee that the CDS is correctly considered and designed from the outset.

Finally, beyond the participants themselves, CDS developers must bear in mind other variations in the sociotechnical systems in which they implement CDS, all of which could impact the success and scalability of planned interventions, including:

- role differences (e.g., a nurse at site A is responsible for all blood draws whereas a nurse at site B does not perform blood draws; at site B a phlebotomist is available 24/7)
- physical differences in the clinical setting (e.g., long vs. short travel distances, ready access for all to a clinical unit vs. a locked unit with restricted access)
- equipment differences (e.g., phlebotomists with access to portable, point-of-care label printers at site A vs. only centralized, nursing station printers available at site B)
- and local cultural and practice differences (at site A physicians are always on site and are expected to personally place all necessary orders whereas at site B physicians may be transiently available and nurses are largely expected to place telephone orders on the physician's behalf)

20.20.5.2 Work processes

In addition to identifying the individuals impacted by a CDS intervention, CDS developers must also understand the work process in which an intervention will be deployed. How does a particular clinical task happen in a clinic or hospital ward? What are the steps involved? Is that process the same, with the same participants, at each site where the CDS intervention will appear, or are there variations that need to be catalogued and understood? Understanding how the work actually happens is critical to the success of a CDS intervention [49]. Too often, assumptions can be made about the work environment and processes that end up dooming a CDS intervention to failure if it is not designed to apply to each of these settings. An intervention that might succeed in one unit with one set of personnel and one work process may fail in another where the same task is done but with different personnel and/or different processes. For example, a unit might have phlebotomists in the morning but subsequent blood draws become the responsibility of nurses. There may also be policy or procedural expectations for blood draws that should be performed by physicians and other providers. For example, some sites may require that blood draws from indwelling venous catheters such as PICC lines only be performed by physicians, rather than by other members of the clinical team. This variation and complexity, while challenging to manage from a CDS build standpoint, is nonetheless a reality in many health care organizations that must be appreciated. In some cases, if the variability is too great, it may not be possible to create a CDS intervention that spans all eligible sites. At the very least, it may require a discussion with leadership that multiple, custom-made CDS interventions might be necessary to account for the level of variation in the organization, unless the variation can be reduced in a way that allows there to be a standard set of personnel and a consistent work process across the organization. Here too, the advice of Levick and Osheroff is incredibly salient for CDS developers, notably, not all problems and organizational priorities can be solved by CDS [37]. It is important for CDS developers to be selective.

20.20.5.3 *Workflow as guide to CDS development*

Understanding of workflow enables CDS team to gather baseline data and set success metrics for any CDS intervention, given that the entire end-to-end workflow in which the CDS will be deployed has been understood. It allows a CDS team to partner with operational stakeholders in the CDS intervention to define responsibilities for participants. Once the spectrum of participants is understood, the most appropriate individuals to experience the CDS intervention can be determined. As part of this determination, the notion of team-based care can help guide responsible party selections, where ideally each member of a clinical care team will be able to operate at the top of their clinical license. CDS developers, again, partnering with operational stakeholders, can develop "if-then" statements that acknowledge CDS as a series of steps that involve individuals collaborating to achieve a task. A full understanding of workflow can give CDS implementers insights into the organizational context of the organization as well that can have an impact on the success or failure of a CDS intervention. As Ash and Hartzog noted, "Many of the control, autonomy, and trust issues are related to the culture of the organization" [3]. A full understanding of workflow creates awareness among CDS developers about time and the need to be sensitive to it. As Levick and Osheroff noted, "The most effective CDS interventions will add value to the workflow by improving clinical quality and the delivery of knowledge and information at the point of care, or other points in the care delivery process without adversely affecting time required, or perhaps even improving it" [37]. Lastly, understanding the workflow will allow CDS developers to recognize situations in which CDS is appropriate and likewise situations in which non-CDS interventions might be warranted [37].

20.20.5.4 *Tools for identifying stakeholders and characterizing workflows*

In both prior editions of this book tools have been cited that may facilitate stakeholder identification and workflow characterization. Ash and Hartzog [3] cite stakeholder analysis as a first step and advise use of the Bridges implementers' guide. Levick and Osheroff [37] in their fig. 25–2 offer a tool for mapping out the entire CDS workflow process and applying the 5 rights of CDS to it. They encourage formal or informal workflow mapping and the use of standardized tools to do so successfully, whether by workflow process experts or interested parties without formal workflow training: "The workflow map should identify the various steps in the care process pertinent to the improvement target: who is performing each step, how each step will be accomplished, and opportunities to increase efficiency/effectiveness, and who is the target. The information and data that are being made available or being transferred at each step should also be identified. Review of the overall process may then identify opportunities where a CDS intervention could be designed and implemented" [37].

20.20.6 Testing and education

Prior to launch, all CDS tools require testing. CDS leaders must decide if there will be some level of testing that is standard for all CDS artifacts. In addition, is additional testing appropriate for particular interventions? Is there an opportunity to test the CDS intervention silently at first, so that it is invisible to end-users but supplies valuable information to CDS developers that can allow it to be revised if needed? Once a CDS tool has successfully passed the testing phase, decisions are needed about how to communicate about the intervention, if at all. Do all CDS tools at an organization have announcements when launched and associated education? Alternatively, do only certain CDS tools warrant such communication and education, being mindful of the risk of over-communicating to end users? Who are the end users who need communications and education? Who will develop the education, who will inspect and edit the education, and who will be responsible for education distribution and through what channel(s)?

20.20.7 CDS deployment

Launch day for a CDS tool can be one filled with both excitement and apprehension. There may be a lot of feedback, little feedback, or no feedback (the latter usually being the best indicator of early success) on the first day. Having some mechanism to gather end-user experiences and testimonials with the new CDS tool, even if anecdotally, can be useful in sharing good news with the development team, requestors, and leadership and also in pinpointing opportunities for improvement if there is less favorable initial feedback. Beyond the launch day, the team must begin the CDS intervention evaluation. Did the tool have its intended impact? Unintended consequences? Both? And what happens next in the lifespan of the CDS tool? For the evaluation of individual CDS tools and the CDS program as a whole, CDS leaders must assure that they have the analytics and reporting capabilities needed to provide visibility into the performance of CDS tools. As with any project that requires ongoing investment and leadership support, it is important to celebrate CDS achievements publicly. At the same time, a learning health system requires that CDS leaders also be transparent with findings and make adjustments when things do not go as planned. Each of these represents an opportunity to improve the next CDS tool.

20.21 CDS management

Over time, many health care organizations will accumulate a variety of CDS tools. To inventory, track, and assure the appropriateness of these tools with respect to current clinical practice, health care organizations should commit to ongoing review of their CDS libraries, one aspect of knowledge management which is reviewed in Chapter 28. A paradigm for monitoring, analysis, and optimization of alerts has been described [16]. In general, CDS management includes three components: content review, system performance, and optimization. Below, we discuss CDS optimization. Content review is covered in detail in Chapter 28 and CDS performance and evaluation is covered in Chapters 21 and 22.

20.21.1 CDS optimization and outcomes

To assure that CDS tools are built and functioning as well as they can be to enhance the end user experience and ensure improved outcomes requires optimization (see also Section 20.16). While metrics for optimal CDS performance remain largely undefined, it may become necessary for organizations to define locally what constitutes optimal or suboptimal performance. Optimization, like keeping content current and monitoring CDS system performance, can be a labor-intensive task. It may be that some organizations have the capability to optimize multiple types of CDS tools, such as order sets, alerts, and iconography simultaneously. Other organizations may choose to have a rotating focus, with order sets being the subject of optimization during one cycle and alerts the focus during a later cycle, for instance. Key questions that can be asked during optimization include: Is a particular alert firing to the right people and only the right people? Is there an opportunity to revise this order set's content because 80% of the time users place key one-off orders within 10 min of using the order set? Are there new EHR vendor capabilities that enable a new, fine-tuned, more desirable approach to providing clinical decision support than this existing CDS tool? Alternatively, optimization efforts can be guided by organizational goals rather than by analysis of individual CDS tools. For instance, an organizational goal to reduce interruptive alerts by 10% for nurses would be one way to begin an optimization process, with analysis to follow about which alerts that nurses experience offer the most opportunity to achieve the stated goal.

The tension in CDS outcomes analysis is that while it is relatively straightforward to assess process measures driven by CDS, it is far more difficult to gauge clinical success. The best data we have thus far come from process meta-analyses. A meta-analysis of 148 studies published in 2012 [50] showed odds ratios of adherence to CDS recommendations contained in alerts in the range of 1.42–1.72. Clinicians increased prescribing, test ordering, documentation, vaccination, and other parameters in response to CDS but the real question is whether, for example, increased alerting for clinicians to prescribe colonoscopy led to decreased incidence of colon cancer. A meta-analysis from 2020 of 122 trials showed modest, absolute improvements in desired care by 4.4–7.1% (mean 5.8%) [51]. CPOE is the structural foundation for CDS, and at one community hospital, LOS decreased as use of CPOE increased, independent of other secular trends affecting LOS [52]. In a follow up study decreases in CPOE led to a corresponding increase in LOS, followed by reversal of the trend when CPOE rates increased, consistent with cause and effect, but this was a correlational study so it is difficult to be certain [53]. Process measures remain, by and large, a proxy for clinical outcomes.

In addition to the approaches described earlier, organizations always have the option to approach optimization on an ad-hoc basis, where particular CDS tools are targeted for optimization when attention is called to them, for example, when a pharmacist reports that an alert is firing excessively for inpatient pharmacy staff and not creating any clinical value.

20.21.2 Enabling ongoing CDS optimization

The challenge for many organizations will be how to enable this important work. There are both structural and process aspects to consider.

20.21.2.1 Structural considerations

Structurally, does the organization have dedicated staff with the skillsets to perform content review, CDS performance evaluation, and optimization? Does the organization envision a separate team for this important work or is there an expectation that members of an existing CDS committee can and will take responsibility for these tasks? Organizations should recognize that effectively managing all of the CDS lifecycle tasks: request intake, evaluation, and decision-making, design, development, testing, deployment, evaluation, monitoring, content updates, and optimization for a large library of CDS tools can be an enormous undertaking. Most CDS committees are likely relatively spartan and unlikely to be capable of managing all these needs well. There may be a need for additional organization resources, especially when considering content review. Organizations should consider whether they have a structure that facilitates ongoing, prompt content review, both scheduled and ad hoc. Who, for instance, is on point for the organization to advise about cardiology-related content questions and is enabled to make decisions on behalf of the organization, recognizing that some organizations may be large, complex, and diverse in terms of geographic sites and cardiology practices?

Creating a network of specialty leads—an on-point content specialist from each clinical specialty or domain–can help organizations be ready and responsive when such content needs arise. Some organizations may have an informal network of specialty leads and may not support specialty lead time and effort for the role, expecting such individuals to advise about content questions on a voluntary basis, in addition to their existing clinical and perhaps teaching, administrative, and/or research responsibilities. Other organizations may intentionally design a specialty lead network and commit some level of either centralized or department-provided support, compensation, time, or other accommodation for this effort, recognizing that content review is real work that can require research, collaboration and consensus-building with specialty colleagues, meeting time with CDS technical and informatics teams, and valuable "soft" skills in areas such as diplomacy, advocacy, and communication. To the extent that specialty leads are also expected to serve as liaisons between their specialty and CDS teams, these individuals may also need to be effective educators for their colleagues about CDS tools, interpreters who can represent clinical needs to technical teams and convey technical capabilities to clinical colleagues, and leaders who can navigate difficult choices about functionality and workflow as well as disagreements among colleagues how best to proceed. See also Sections 20.2, 20.3, 20.5.1, 20.6.5, and 20.16.

Analytics tools represent another structural consideration. Does the organization have any analytics capabilities to monitor and evaluate CDS tools? If not, what are the opportunities to obtain these analytics capabilities that are essential to being able to monitor, maintain, and enhance a CDS system? Vendors may supply such tools in some cases. In other cases, organizations may turn to third-party vendor solutions. In others, organizations with sufficient capability may develop their own, homegrown analytics tools, including reports and dashboards.

20.21.2.2 Process considerations

In addition to these structural considerations for monitoring CDS tools, there are process considerations, too. As noted earlier, timing is a key issue. Each organization must determine a schedule to review its CDS artifacts. Medicare gives some guidance on this point, at least insofar as order sets are concerned, with a requirement that organizations review order sets periodically. This signals to organizations a need for a regular review process and encourages timely review but leaves decisions about what interval is most appropriate up to individual organizations. Organizations may thus create a schedule whereby order sets are reviewed typically every 2–3 years.

Monitoring CDS performance ideally should be a continuous practice. Optimization may be a bit more discretionary, suitable for an approach where optimization efforts are focused on a certain CDS tool, such as medication alerts, during a specified period of time, e.g., January to June, and then the next set of CDS tools is the focus for July–December. Administrative and/or project manager support for whatever content review process is established is key to making the content review program actually function and operate sustainably.

System performance monitoring requires consideration of timing. Will such monitoring happen at intervals, almost as spot checks, whether at defined intervals such as monthly or ad hoc? Or at the other extreme, will monitoring be continuous, with dedicated staff, akin to an energy utility operations center that monitors the power grid 24 h a day, 7 days a week?

The importance of defining optimization drivers and an optimization schedule were noted in Section 20.20.1. Once optimization opportunities have been identified, organizations can make decisions about whether each CDS tool set for optimization should be repaired, replaced, or retired individually or whether a strategic approach to a set of CDS tools makes sense, wherein a global approach to optimizing CDS could be employed. For instance, perhaps the optimization process will uncover opportunities to assure that every alert in an organization enables end users to take the alert-recommended action directly from the alert, a step that helps create consistency in the user experience across the CDS library. Indeed, such discoveries in the course of CDS optimization may create the basis for an organization to create a manual or style guide related to CDS tools, if it does not have one already.

20.22 CDS change management

For CDS implementations to be effective, a change management process must begin well before any tool is ever created. As noted earlier, a foundational step of every CDS program is alignment with organizational mission and priorities and endorsement of the CDS program by leadership.

A second key step in change management for CDS is committing to demonstrate value for those using CDS tools. Ash and Hartzog [3,54], and others [55,56] have observed: "Acceptance of decision support depends a great deal on the type of decision support, the reason for its use, how good it is, and the value the clinician places on it at any particular point in time." Additionally, Ash and Hartzog note that CDS acceptance depended on "perceived usefulness during the patient visit itself, the presence of facilitating conditions such as integration into workflow, and the availability of training opportunities, the CDS system's ease of use, and trust in the content of the CDS" [3].

A third step in CDS change management is assessing the readiness for change of the organization or at least, the population targeted for the CDS. While not every CDS tool developed in day-to-day practice requires such an assessment, it is advisable to assess change readiness for major CDS interventions. Major interventions could be considered those that will directly involve many clinicians throughout the organization, that represent a significant and/or challenging departure from existing workflows, or that result from regulatory requirements where CDS tools are used to fulfill a regulatory obligation. The Lewin Force Field model of change may be useful in these situations [3]. In these settings, a change calendar may also be beneficial to consider, to assure that the introduction of a new CDS tool does not overburden clinicians who may already be experiencing significant changes of other types, for example, opening of a new hospital, a recent transition to a new EHR, or workforce strains during a pandemic [57].

A fourth important step in managing CDS change is assuring user engagement, including users who may be skeptical about changing and the role of CDS. As noted previously, engaging with end users pays dividends. Involving end users can help those individuals feel invested in CDS changes [3]. These individuals can provide valuable insights into clinical workflows as a CDS intervention is considered and later, become testers of CDS tools that are developed. And involved end-users may become "CDS influencers" who can encourage their colleagues to accept the change more readily. During this phase, user expectations should be monitored and implementers should ask the question, "Do user expectations align with CDS implementer expectations?" A mismatch in these expectations can imperil the success of the planned CDS. Levick and Osheroff make a key point summarizing the value of end user involvement, noting "Clinical decision support is most successful when it is perceived as a team sport by the entire organization and done ***with*** stakeholders instead of ***to*** them" [37].

A fifth step in managing CDS change is assuring adequate testing, to assess if the intervention works technically and whether it is acceptable to end-users. As noted, stakeholders can be a source of testers for interventions. CDS interventions that do not pass testing initially can be reworked and improved until they ultimately can succeed in the testing process. A transparent and robust testing process serves as a public demonstration of the CDS program's commitment to developing quality products. When testing is curtailed, errors occur and when projects fall behind schedule, it is frequently the testing phase that is shortened, increasing the risk that CDS will malfunction.

The sixth step in CDS change management is communication and education. As noted earlier, the extent of communication and education required for any particular CDS tool may vary depending on its impact and degree to which it represents a departure from existing workflows. Of course, in an ideal state, CDS tools would be so well-developed that no training would be necessary at all. They would be completely self-explanatory when presenting to end-users for the first time. Still, for major CDS changes, it is important to recall the advice of Ash and Hartzog that training can be a way to enhance clinician understanding of the CDS tool, feelings of self-efficacy, and hopefully CDS adoption [3].

The seventh step in managing CDS change is reflecting on what has been done. This involves evaluating CDS tool performance, responding to user feedback, and making iterative improvements. Periodically, as previously noted, CDS implementation teams should report out about the performance of CDS tools and importantly, about the impact of CDS interventions. For example, did the CDS tools positively impact process or outcome measures for patients? Likewise did the CDS tools have a positive impact on process or outcome measures for the clinicians using the CDS? This reflection stage, if done well, demonstrates the CDS program's commitment to ongoing improvement and to being a core component of a learning health system. Moreover, strong performance in this stage can engender confidence among end-users in any future CDS tools that the CDS team may develop. These seven steps to effective CDS change management are outlined in Table 20.4.

TABLE 20.4 Seven steps to effective CDS change management.

1. Assure alignment with organizational mission and leadership endorsement
2. Commit to demonstrating value from CDS tools
3. Assess readiness to change
4. Assure user engagement
5. Commit to sufficient testing
6. Communicate and educate
7. Reflect on CDS tool performance and user reception

20.23 Impact of EHR-to-EHR transitions on CDS

While much of this chapter addresses day-to-day operations of a CDS implementation team, there is at least one special circumstance that is worth noting that can impact CDS. One is the impact of EHR-to-EHR transitions on CDS implementation. Even the most well-planned and supervised EHR transition can prove to be a major disruption and source of stress for a health care organization [9]. In the planning stage pre-transition, there is a need to account for the current state CDS tools and to anticipate how the decision support these tools provide to clinicians might be recreated in the destination EHR, usually without any firsthand experience at that point using or navigating the destination EHR [58]. For some CDS tools, there may be an easy 1:1 conversion, with the same functionality and look and feel preserved in the destination EHR. In other cases, there may be minor or major modifications to how the CDS tools function in the new system that need to be understood, both for technical build and end-user educational purposes. In yet other cases, the CDS functionality that exists in the current state may not exist at all in the destination EHR, which will prompt new questions about how some degree of decision support can be maintained in the future state, if at all. In addition, the analytics capabilities may differ from current to future state systems. The level to which an organization can configure CDS tools in the destination system may be more robust or more rudimentary than in the current EHR. Finally, from an educational perspective, CDS implementers must recognize that end-users not only learn how to use a new EHR and its CDS tools, but they simultaneously need to unlearn everything they are accustomed to using in the present state system. Moreover, there may be novel capabilities for CDS that the new EHR offers that will also require end-user training [9].

20.24 CDS frontiers

While so much of CDS implementation work can seem in the moment, it is important for CDS leaders to keep an eye on the frontier as well. For instance, while it may be of interest to understand how an organization performs in terms of CDS compared to other organizations, such an analysis is not possible today. There are no national benchmarks, for instance, by which to gauge one organization's performance. With time, however, it is possible that such benchmarking and comparisons may become possible, especially if the CDS community can coalesce around some standard metrics of CDS quality, performance, and impact. At such time, new questions and opportunities may arise, such as what is the state of the local marketplace in terms of CDS and where are there opportunities for one organization to differentiate itself? How does one organization's standing in that CDS marketplace impact its decisions about what CDS to develop, if it does at all? Does the landscape in that market, whether the only health care organization in town vs. one of many organizations in a highly competitive health care market, affect decisions about CDS development?

Some of the frontiers may be closer than we think. SMART on FHIR apps offer the hope that every health care organization can stop reinventing the wheel (see also Sections 20.6, 20.12, 20.15, 20.18, 20.19.2, and 20.21.2, as well as Chapter 15), each needing to develop CDS content on its own, a wasteful use of resources both for individual organizations and at a national level [59]. As noted earlier in this chapter, such SMART on FHIR apps can and have been built for use at one institution and deployed and used at another institution. While this technology holds promise, the reality is that it is not yet plug and play, with recipient institutions of SMART on FHIR apps still needing to do substantial work to configure the app to match the data and settings in the recipient health care organization. Ongoing initiatives such as the US Core Data for Interoperability (USCDI) [60] may help address this challenge (see Chapter 15 for details).

It is important to consider health equity in designing and implementing CDS. Anticipating and correcting for bias in machine learning is one aspect, while considering availability of CDS is another [61]. Do all EHRs have access to advanced CDS tools? Do those involved in CDS governance understand the new tools, and have the resources to implement them? How can we assure equitable access, whether nationally or globally, to increasingly sophisticated CDS tools that promote health and safety? These and other sociotechnical issues will be important considerations as the discipline of CDS informatics advances.

CDS Hooks also holds promise to offload the work of developing CDS content at each organization. CDS Hooks enables organizations to use cloud based CDS services, activating such services at key junctures in the clinical care process, such as at medication order entry. Minimum necessary patient/provider context data is sent to the CDS service and in return, the service sends back an informational statement, an alternative therapy suggestion, or even a link to additional knowledge resources, all of which can appear in-line within the native EHR that the clinician is using. While vendor support for CDS Hooks is increasing quickly, to date adoption of CDS Hooks appears to be quite limited [62]. Data security—given the need to send patient and clinician information outbound and to receive CDS service content inbound—has been one of the concerns raised that may be a factor limiting adoption. An additional concern is the degree to which the data send and receive process may potentially impair EHR system performance.

Other CDS is in its infancy or yet to be developed. These new tools will necessitate ongoing vigilance by CDS leaders. AI-enabled chatbots and digital voice assistants and virtual scribes for use with patients or clinicians, for example, might be considered an emerging form of clinical care and clinical decision support. Machine learning to filter alerts that clinicians will likely reject while maintaining very high sensitivity for alerts that will successfully result in the best desirable action are beginning to gain traction, but require more sociotechnical foundational work [61]. It remains to be seen what kind of oversight may be appropriate to assure that these tools abide by key CDS principles, do not exacerbate bias, health disparities, or clinician burden, and prove safe and effective [63].

References

[1] Osler SW. Remarks, on the dedication of the new building of the Boston medical library. Books Men 1901; CXLIV:60–1.

[2] Bates DW, Kuperman GJ, Wang S, et al. Ten commandments for effective clinical decision support: making the practice of evidence-based medicine a reality. J Am Med Inform Assoc 2003;10:523–30.

[3] Ash J, Hartzog T. Organizational and cultural change. In: Clinical decision support. Amsterdam: Academic Press (Elsevier); 2014. Chapter 23.

[4] Kannry J, Sengstack P, Thyvalikakath TP, et al. The chief clinical informatics officer (CCIO): AMIA task force report on CCIO knowledge, education, and skillset requirements. Appl Clin Inform 2016;7:143–76.

[5] Manager Tools. Horstman's law of project management, https://www.manager-tools.com/2009/01/horstman%E2%80%99s-law-project-management-part-1-hall-fame-guidance. [Accessed 25 November 2022].

[6] McGreevey JD, Mallozzi CP, Perkins RM, et al. Reducing alert burden in electronic health records: state of the art recommendations from four health systems. Appl Clin Inform 2020;11:1–12.

[7] Ash JS, Sittig DF, Dykstra RH, et al. Categorizing the unintended sociotechnical consequences of computerized provider order entry. Int J Med Inf 2007;76(Suppl. 1):S21–7.

[8] Ford E, Edelman N, Somers L, et al. Barriers and facilitators to the adoption of electronic clinical decision support systems: a qualitative interview study with UK general practitioners. BMC Med Inform Decis Mak 2021;21:193.

[9] Huang C, Koppel R, McGreevey JD, et al. Transitions from one electronic health record to another: challenges, pitfalls, and recommendations. Appl Clin Inform 2020;11:742–54.

[10] Wright A, Sittig DF, Ash JS, et al. Governance for clinical decision support: case studies and recommended practices from leading institutions. J Am Med Inform Assoc 2011;18:187–94.

[11] Ash JS, Sittig DF, Wright A, et al. Clinical decision support in small community practice settings: a case study. J Am Med Inform Assoc JAMIA 2011;18:879–82.

[12] Singh H, Sittig DF. A sociotechnical framework for safety-related electronic health record research reporting: the SAFER reporting framework. Ann Intern Med 2020;172:S92–100.

[13] Sittig DF, Ash JS, Singh H. The SAFER guides: empowering organizations to improve the safety and effectiveness of electronic health records. Am J Manag Care 2014;20:418–23.

[14] CMS.gov. Fiscal year (FY) 2022 Medicare hospital inpatient prospective payment system (IPPS) and long term care hospital (LTCH) rates final rule (CMS-1752-F)., 2021, https://www.cms.gov/newsroom/fact-sheets/fiscal-year-fy-2022-medicare-hospital-inpatient-prospective-payment-system-ipps-and-long-term-care-0. [Accessed 25 November 2022].

[15] Osheroff JA, Teich JM, Levick D, Saldana L, Velasco FT, Sittig DF, et al. Improving outcomes with clinical support: an Implementer's guide. 2nd ed. Boca Raton: CRC Press; 2012.

[16] McGreevey III JD, Wright A, Michalek C, Giannini R. Safe practices to reduce CPOE alert fatigue through monitoring, analysis, and optimization; 2021. https://d84vr99712pyz.cloudfront.net/p/pdf/hit-partnership/partnership_whitepaper_alertfatigue_final.pdf [Accessed 25 November 2022].

[17] Wright A, Ai A, Ash J, et al. Clinical decision support alert malfunctions: analysis and empirically derived taxonomy. J Am Med Inform Assoc 2018;25:496–506.

[18] Wright A, Ash JS, Aaron S, et al. Best practices for preventing malfunctions in rule-based clinical decision support alerts and reminders: results of a Delphi study. Int J Med Inf 2018;118:78–85.

[19] Koppel R. Is healthcare information technology based on evidence? Yearb Med Inform 2013;8:7–12.

[20] Koppel R, Lehmann CU. Implications of an emerging EHR monoculture for hospitals and healthcare systems. J Am Med Inform Assoc 2015;22:465–71.

[21] Khaliq AA, Thompson DM, Walston SL. Perceptions of hospital CEOs about the effects of CEO turnover. Hosp Top 2006;84:21–7.

[22] Aaron S, McEvoy DS, Ray S, et al. Cranky comments: detecting clinical decision support malfunctions through free-text override reasons. J Am Med Inform Assoc 2019;26:37–43.

[23] Govern P. Clickbusters program takes on EHR alert fatigue., 2020, https://news.vumc.org/2020/07/16/clickbusters-program-takes-on-ehr-alert-fatigue/. [Accessed 25 November 2022].

[24] Goddard K, Roudsari A, Wyatt JC. Automation bias: a systematic review of frequency, effect mediators, and mitigators. J Am Med Inform Assoc JAMIA 2012;19:121–7.

[25] Wright A, Hickman TT, McEvoy D, et al. Methods for detecting malfunctions in clinical decision support systems. Stud Health Technol Inform 2017;245:1385.

[26] Wright A, Aaron S, McCoy AB, et al. Algorithmic detection of Boolean logic errors in clinical decision support statements. Appl Clin Inform 2021;12:182–9.
[27] Department of Health and Human Services. Center for Medicaid and State Operations/Survey and Certification Group, https://www.cms.gov/Medicare/Provider-Enrollment-and-Certification/SurveyCertificationGenInfo/downloads/SCLetter09-10.pdf. [Accessed 25 November 2022].
[28] Leviss J. Anatomy of a preventable mistake: unrecognized workflow change in medication management. In: Leviss J, editor. Hit or miss: lessons learned from health information technology projects. Boca Raton: CRC Press Taylor & Francis Group; HIMSS; AMIA; 2019. p. 43–6.
[29] Wu E. Upgrading a hospital EHR for meaningful use. In: Leviss J, editor. Hit or miss: lessons learned from health information technology projects. Boca Raton: CRC Press Taylor & Francis Group; HIMSS; AMIA; 2019.
[30] Reeves JJ, Hollandsworth HM, Torriani FJ, et al. Rapid response to COVID-19: health informatics support for outbreak management in an academic health system. J Am Med Inform Assoc 2020;27:853–9.
[31] Khaliq AA, Walston SL, Thompson DM. Is chief executive officer turnover good for the hospital? Health Care Manag 2007;26:341–6.
[32] Kawamoto K, Kukhareva PV, Weir C, et al. Establishing a multidisciplinary initiative for interoperable electronic health record innovations at an academic medical center. JAMIA Open 2021;4:ooab041.
[33] Petersen C, Smith J, Freimuth RR, et al. Recommendations for the safe, effective use of adaptive CDS in the US healthcare system: an AMIA position paper. J Am Med Inform Assoc 2021;28:677–84.
[34] Obermeyer Z, Powers B, Vogeli C, et al. Dissecting racial bias in an algorithm used to manage the health of populations. Science 2019;366:447–53.
[35] Wong A, Otles E, Donnelly JP, et al. External validation of a widely implemented proprietary sepsis prediction model in hospitalized patients. JAMA Intern Med 2021;181:1065–70.
[36] Reeves JJ, Pageler NM, Wick EC, et al. The clinical information systems response to the COVID-19 pandemic. Yearb Med Inform 2021;30:105–25.
[37] Levick D, Osheroff J. A clinical decision support implementation guide: practical considerations. In: Greenes RA, editor. Clinical decision support. The road to broad adoption. Amsterdam: Academic Press (Elsevier); 2012.
[38] Schreiber R, Knapp J. Premature condemnation of clinical decision support as a useful tool for patient safety in computerized provider order entry. J Am Geriatr Soc 2009;57:1941–2.
[39] Agency for Healthcare Research and Quality. The five rights of medication administration: Section 2—Overview of CDS Five Rights, https://digital.ahrq.gov/ahrq-funded-projects/current-health-it-priorities/clinical-decision-support-cds/chapter-1-approaching-clinical-decision/section-2-overview-cds-five-rights. [Accessed 25 November 2022].
[40] Wells C, Loshak H. Standardized hospital order sets in acute care: a review of clinical evidence, cost-effectiveness, and guideline., 2019, https://www.ncbi.nlm.nih.gov/books/NBK546326/. [Accessed 25 November 2022].
[41] Institute for Safe Medication Practices. Guidelines for standard order sets., 2010, https://www.ismp.org/guidelines/standard-order-sets. [Accessed 25 November 2022].
[42] McGreevey III JD. Order sets in electronic health records: principles of good practice. Chest 2013;143:228–35.
[43] Li RC, Wang JK, Sharp C, et al. When order sets do not align with clinician workflow: assessing practice patterns in the electronic health record. BMJ Qual Saf 2019;28:987–96.
[44] Dinh A. Physician practice EHR resources. J AHIMA 2009;80:63–4.
[45] Kawamoto K, Kukhareva P, Shakib JH, et al. Association of an electronic health record add-on app for neonatal bilirubin management with physician efficiency and care quality. JAMA Netw Open 2019;2, e1915343.
[46] Carayon P, Schoofs Hundt A, Karsh B-T, et al. Work system design for patient safety: the SEIPS model. Qual Saf Health Care 2006;15(Suppl 1):i50–8.
[47] Carayon P, Wooldridge A, Hoonakker P, et al. SEIPS 3.0: human-centered design of the patient journey for patient safety. Appl Ergon 2020;84, 103033.
[48] Center for Quality and Productivity Improvement. SEIPS: the systems engineering initiative for patient safety, https://cqpi.wisc.edu/seips/. [Accessed 25 November 2022].
[49] Braithwaite J, Wears RL, Hollnagel E, editors. Resilient health care: reconciling work-as-imagined and work-as-done. London: CRC Press; 2019.
[50] Bright TJ, Wong A, Dhurjati R, et al. Effect of clinical decision-support systems: a systematic review. Ann Intern Med 2012;157:29.
[51] Kwan JL, Lo L, Ferguson J, et al. Computerised clinical decision support systems and absolute improvements in care: meta-analysis of controlled clinical trials. BMJ 2020;370, m3216.

[52] Schreiber R, Peters K, Shaha SH. Computerized provider order entry reduces length of stay in a community hospital. Appl Clin Inform 2014;5:685–98.
[53] Schreiber R, Shaha SH. Computerised provider order entry adoption rates favourably impact length of stay. J Innov Health Inform 2016;23:166.
[54] Ash JS, Sittig DF, Guappone KP, et al. Recommended practices for computerized clinical decision support and knowledge management in community settings: a qualitative study. BMC Med Inform Decis Mak 2012;12:6.
[55] Thornett A. Computerized decision support systems in general practice. Int J Inform Management 2001;21:39–47.
[56] Shibl R, Lawley M, Debuse J. Factors influencing decision support system acceptance. Dec Sup Sys 2013;54:953–61.
[57] Valusek JRS. The change calendar: a tool to prevent change fatigue. Jt Comm J Qual Patient Saf 2007;33:355–60.
[58] McGreevey III JD. Unexpected drawbacks of electronic order sets., 2016, https://psnet.ahrq.gov/web-mm/unexpected-drawbacks-electronic-order-sets. [Accessed 25 November 2022].
[59] Musen MA, Middleton B, Greenes RA. Clinical decision-support systems. In: Shortliffe EH, Cimino JJ, editors. Biomedical informatics: computer applications in health care and biomedicine. London: Springer; 2014. p. 670.
[60] Health IT.gov. United States core data for interoperabililty (USCDI), https://www.healthit.gov/isa/united-states-core-data-interoperability-uscdi. [Accessed 14 January 2022].
[61] Liu S, Kawamoto K, Del Fiol G, et al. The potential for leveraging machine learning to filter medication alerts. J Am Med Inform Assoc 2022. https://academic.oup.com/jamia/advance-article/doi/10.1093/jamia/ocab292/6497966. [Accessed 25 November 2022].
[62] Strasberg HR, Rhodes B, Del Fiol G, et al. Contemporary clinical decision support standards using Health Level Seven International Fast Healthcare Interoperability Resources. J Am Med Inform Assoc JAMIA 2021;28:1796–806.
[63] McGreevey III JD, Hanson CW, Koppel R. Clinical, legal, and ethical aspects of artificial intelligence-assisted conversational agents in health care. JAMA 2020;324:552–3.

CHAPTER

21

Managing the investment in clinical decision support

Tonya Hongsermeier[a] *and John Glaser*[b]

[a]Elimu Informatics, El Cerrito, CA, United States [b]Harvard Medical School, Acton, MA, United States

21.1 Introduction

Value-based healthcare purchasing continues to be a dominant force in reimbursement, and healthcare providers are looking to invest in systems that help them deliver tangibly cost-effective care. Investment in clinical decision support systems (CDSs) is influenced by the complex regulatory, technical, workflow, cultural and maintenance challenges and opportunities associated with their implementation.

We discuss some top-down incentive-based and regulatory drivers in this chapter from a U.S. perspective, but the point to recognize in this discussion is that, while such mechanisms may take various forms elsewhere, they are potentially important factors in the determination of priority and budget by organizations in adopting CDSs. In the U.S., the Medicare Access and CHIP Reauthorization Act of 2015, which resulted in the development of the Quality Payment Program, Merit Based Incentive Payments System (MIPS), and Alternative Payment Models (APMs), ultimately requires clinicians to effectively utilize Electronic Health Records (EHRs) and CDS enablers to successfully demonstrate high quality care and maximize reimbursement. These business drivers have refocused healthcare delivery organizations on building patient-centered, coordinated and collaborative approaches to care.

Business drivers are not the only force driving the advance of CDS. Artificial intelligence has introduced new methods and technologies that extend the power and capabilities of CDS [1] (see Chapters 7 and 16 for in depth descriptions of artificial intelligence methods and applications respectively).

The recent COVID19 Pandemic has forced healthcare organizations to become agile at aligning their IT infrastructure and CDS content with rapidly evolving requirements for virtual care and deployment of new innovations in diagnostics and therapeutics. Assessment

Clinical Decision Support and Beyond
https://doi.org/10.1016/B978-0-323-91200-6.00021-8

tools to screen patients for COVID19 had to be continuously updated as the exposure risk factors shifted from week to week. Computerized Provider Order entry (CPOE), embedded with order sets were seen as central to efforts to account for shifting supply chain problems and to guide evidence-based clinical management. Care pathway or health maintenance reminder functionality were key to identify high risk cohorts for prioritization of vaccination, track the vaccine status of patients and ensure the dosing sequence was timed correctly. Clinical documentation tools designed with decision support and quality measure reporting in mind assisted caregivers with screening and enabled researchers to better understand risk factors for severe disease as well as discover effective strategies for treatment. Our chronic disease population needed to be effectively managed remotely to mitigate their risk for COVID19 exposure. Population management systems with risk-assessment and disease management logic assisted physician-extenders and case-managers with diagnosis, secondary prevention and virtual care orchestration [2].

Developing an effective, pragmatic clinical decision support investment strategy, that responds to changes in business drivers, advances in technology and crises, is a competitive imperative [3,4].

Given the continuous advances in information systems, innovation, and the ever-changing business climate for providers, there is no such thing as a post-EHR implementation steady state. This conclusion requires that provider organizations establish management structures and processes that enable them to continuously prioritize decision support investment, develop and/or acquire the required clinical decision support solutions, orchestrate and update the knowledge expressed through clinical information systems and evaluate the impact of their strategies. The organization must be agile at designing well-orchestrated team workflows and aligning the key enabling clinical decision support solutions.

This chapter covers four areas of clinical knowledge management (CKM) from an institutional investment perspective, including management of clinical decision support knowledge as a component of clinical knowledge management, the boundaries of clinical knowledge management, key functions of clinical knowledge management, and the evolving "business case" for investing in clinical knowledge management. Other chapters in this book also cover closely related topics in clinical knowledge management, but from different perspectives. Specifically, Chapter 20 discusses CDS governance in general and Chapter 28 discusses technical infrastructure and tools for clinical knowledge management.

The organization of clinical knowledge management with regard to business alignment is reviewed including strategic objectives, governance, CDS impact on quality measurement and reporting, and approaches to insourcing and outsourcing of clinical decision support investment. Key IT strategies and considerations are examined including legacy systems, knowledge management tools and application foundations. The evaluation of the impact and value of clinical knowledge management is also discussed.

21.2 Clinical knowledge management

Investment in any set of organizational structures and processes that surround a significant information technology can benefit from a discussion of the concepts that will guide and frame that investment. For example, a discussion of the integration of an organization's applications should begin with attempts to answer the question, "What does integration mean to us?" The organization can develop very different strategies, e.g., single vendor or extensive use of application programming interfaces (APIs) or both, based on very different answers.

This section provides some concepts and context that should guide the organization's discussion of investment in clinical decision support.

21.2.1 Management of clinical decision support as a component of clinical knowledge management

Clinical Knowledge Management is essentially a framework for a "Learning Health Care Provider Organization" (see Fig. 21.1) [5]. Clinical decision support is a tactic that seeks to ensure that the caregiver (clinician or patient) has the right information necessary to document and deliver superior care. In Fig. 21.1, the "Care Framework" on the left illustrates a variety of care tasks that can be impacted by clinical decision support guidance. The "Learning Framework" on the right illustrates how data derived as a by-product of care delivery can be analyzed to develop new insights for how to improve care delivery with clinical decision support guidance as well as develop quality measurement reporting. As data are harvested from the Care Framework, CDS impacts are analyzed for insights that factor into governance decisions surrounding ongoing CDS curation. The CDS curation process is informed by a combination of governance prioritization, end-user feedback, and the maintenance demands of ensuring that the content is appropriately mapped to standard reference vocabularies such as LOINC, RxNorm, and SNOMED (see Chapter 11 for details on standard vocabularies). Further, CDS accuracy depends on correctly modeled value sets leveraging classification systems such as drug classes. Failure to maintain the modeling of correct data mapping and value sets is a frequent cause of degradation in the accuracy of clinical decision support as well as analytic tools.

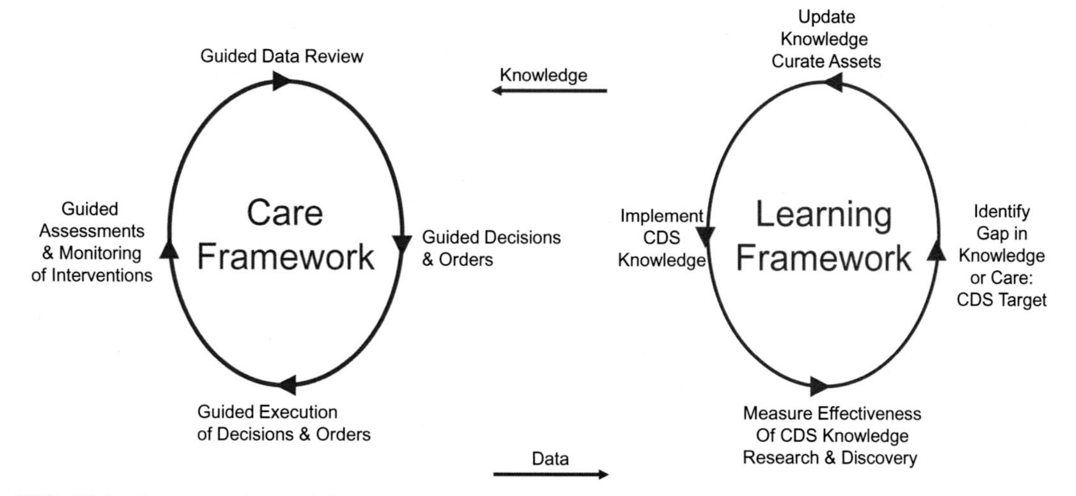

FIG. 21.1 Interdependence of the care framework and the learning framework.

A narrow organizational focus on the application of CDS knowledge in workflow may fail to consider equally important aspects of investment in knowledge discovery and knowledge asset management that help organizations become effective at learning and self-improvement. This narrow focus may also fail to consider other IT-based tactics for optimizing knowledge application. These tactics may include social media solutions, such as collaboration tools for governance, wikis for knowledge sharing and expertise location or end-user intelligence systems to optimize context-aware knowledge linking (see Chapter 13 for technical details on knowledge integration with EHR systems via context-aware "infobuttons").

The organization would be well served to step back and engage in an overall discussion of clinical knowledge management. Such a discussion would force consideration and creation of processes designed to identify the "best" CDS solutions, ensure CDS knowledge is maintained, align CDS with organizational business drivers, and broaden the focus to include a full range of IT-based and non-IT-based tactics.

While a more holistic view of clinical knowledge management is important, it can fall prey to various "traps," e.g., fuzzy boundaries, magical thinking about knowledge-based solutions, incomplete understanding of the scope of knowledge management processes and a complex business case. These issues are discussed in the following sections.

21.2.2 The boundaries of clinical knowledge management

Clinical knowledge management can have diffuse boundaries that encompass the entire organization. Translational research is a form of clinical knowledge management. Quality improvement is knowledge management. Training residents and allied health professionals is knowledge management. Training for managers on human resource issues and workflow expectations is knowledge management.

If clinical knowledge management is defined too broadly, it will be perceived (rightfully so) as too broad to be tractable and defying the ability to be managed by a common set of structures and processes. An organizational phenomenon that is too broad risks being seen as unmanageable and is hence dismissed from the management discussion. For example, no one in an organization proposes to be in charge of "decision making."

Boundaries can be defined in several ways, with each way being based on a different core concept, for example:

- Clinical goals. Knowledge management can focus on specific goals to improve clinical performance, e.g., reduce medication errors, reduce clinician burnout, or optimize management of congestive heart failure. IT-based and non-IT-based knowledge can be applied to prevention of errors or treatment of specific diseases for which there is a specific set of financial incentives, a high prevalence, or organizational focus on developing clinical excellence.
- Application. Knowledge management can address the broad array of knowledge that is contained in or expressed through specific applications, e.g., CPOE, care pathways, patient-specific disease status summaries, or clinical documentation.
- Knowledge implementation tactic. Knowledge management can focus on a specific implementation tactic, e.g., health maintenance reminders or predictive models, which might cut across applications and diseases.

An organization may pursue more than one concept. All of the concepts reflect "understandable" boundaries, i.e. you can explain them to a room full of practicing clinicians and they will "get it."

These concepts also supply a context. Knowledge management or decision support that has no context has no value. Achieving a clinical goal or improving the care of the chronically ill provides a reason for pursuing knowledge management.

21.2.3 The key functions of clinical knowledge management

Knowledge management, however an organization defines its boundaries, is essentially comprised of three key functions: knowledge application, knowledge asset management, and knowledge discovery. They are organized in a circle (Fig. 21.2) to emphasize that the knowledge management process is one of continuous learning and knowledge dissemination.

Knowledge application is the art of leveraging knowledge at the right places in workflow to achieve a strategic objective. Knowledge discovery is the process of analyzing data for the purpose of understanding performance, reporting, predicting, and/or harvesting new knowledge. Knowledge asset management is a set of processes for the acquisition, stewardship, curation, and deployment of knowledge.

Commercial EHR systems are typically designed to support knowledge application much more effectively than either discovery or asset management. For example, the tools for updating knowledge are often function-centric such that an editor for order sets is likely to be decoupled from an editor for value sets, drug classes, alerts or reminders. Thus, when an organization is attempting to build a diabetes management program, it must grapple with multiple disconnected editors to manage all the clinically relevant knowledge. The introduction of very targeted artificial intelligence algorithms is creating an additional set of disconnected knowledge. This non-integration of the knowledge curation tools can result in disconnects among the teams responsible for updating the content and/or measuring performance. Invisible dependencies, such as when the accuracy of a decision support

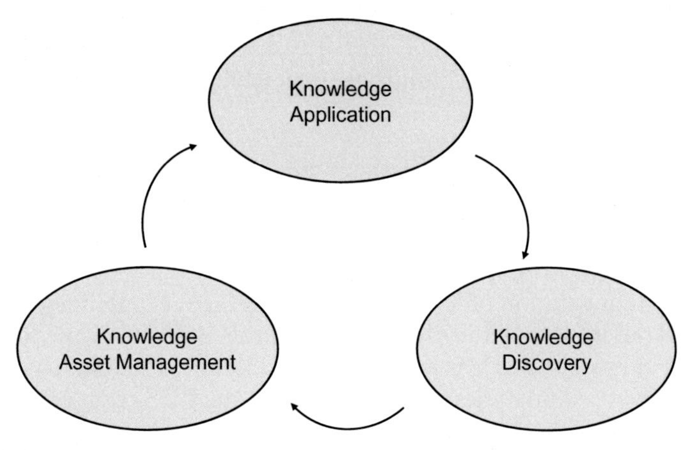

FIG. 21.2 Knowledge management core processes.

alert depends on well-curated medication and observation value sets, are a common problem. The relevant teams must track those dependencies carefully, often without the availability of native EHR dependency tracking functionality, and coordinate their changes with each other to avoid compromising the integrity of the decision support. Clinical decision support programs must encompass these three aspects of clinical knowledge management and focus on building governance structures that effectively align and integrate the various teams that have a role in this learning framework.

21.2.4 The business case for clinical knowledge management investment

The passage of the HITECH act in 2009 and multiple ensuing CMS regulations in the US emphasize IT functionality that removes interoperability barriers, enables quality measurement and reporting, and rewards quality performance. The business case rapidly evolved from answering the question of "Why invest?" to the question of "How do we allocate resource investment in clinical knowledge management such that it correlates with a tangible impact on business performance?"

Provider organizations are invariably confronted with tight budgets; capital budgets are constrained, and proposals to add expenses to operating budgets are subject to tough scrutiny. CKM requires a budget, and obtaining this budget requires that it compete effectively with other budget priorities.

The CKM business case faces several challenges:

- The term "knowledge management" is often too abstract and intangible for concrete, action-oriented provider organization managers. They may not fund it because the term "knowledge management" gets in the way; it doesn't mean anything to them.
- The knowledge management proponents may defend their case using terms such as "ontologies" or "semantics." These terms are incomprehensible to most managers, and generally managers will not support the funding of something that they don't understand.
- The organization may have no working experience with knowledge management; hence it is not sure how to organize the function or what clinical value will be realized. Sometimes the value of investment isn't tangible until consequences of poor management rise to the surface in the form of inaccurate CDS or poor outcomes. Managers are often quite conservative and hesitant to launch undertakings which they are unsure of their ability to manage.

A successful business case has several attributes:

- It links a proposal to an accepted organizational strategy or goal. For example, external business drivers such as MIPS and Accountable Care require knowledge-enriched clinical information systems.
- The creation or augmentation of knowledge management capabilities is often tightly linked to an overall investment in clinical information systems or medical care improvement. For example, CDS is an aspect of an overall acquisition of care coordination capabilities or advanced analytics applications, and the CDS costs are not presented separately. In this case, the knowledge management resources piggyback on the overall resource request, with the overall request being considered in light of organizational goals.

TABLE 21.1 Linkage of organizational goals to knowledge needs.

Organizational goal	Example knowledge need	Benefit
Medication safety	Geriatric medication dosing guidance in CPOE	Quality Payment Program compliance and incentives, reduced length of stay
Cost management	Radiology and medication order guidance in outpatient CPOE	Accountable care risk management
Patient wellness	Provider reminders	Increased reimbursement, increased care quality
Perioperative safety	Venous thromboembolism prevention protocols	Hospital accreditation, increased reimbursement
Disease management	Diabetes management protocols	Payor contract incentives, increased reimbursement

- Table 21.1 provides several examples of how knowledge management infrastructure can be explicitly aligned with business objectives to demonstrate a tangible gain. The current value-based purchasing climate means that a well-crafted clinical knowledge management proposal will tightly connect programs for CDS to both quality improvement and quality reporting programs. It is also just as important to describe the potential cost or risk to a provider organization of either not having the CDS knowledge or failing to adequately maintain it. Another CKM business driver on the horizon is the emergence of individualized medicine. Today, this exponential growth in the knowledge required to practice medicine is primarily impacting cancer care. However, in the coming years, molecular medicine will impact an ever-increasing percentage of clinical decisions, thus making it wholly unfeasible for clinicians to practice unless partnered with robust CDS solutions.
- The level of resources, such as staff, licensed content, and information systems, needed is deemed to be reasonable. Reasonableness is hard to empirically derive. Often organizations start with small numbers of staff and gradually increase effort, as they understand the nature of the challenge. Other times, benchmark data from other organizations provides guidance on needed resources. Regardless, the expense is deemed to be worth it.
- The business case describes the management structures, tools and processes needed to manage this knowledge. For example, who should make sure that our health maintenance reminders are kept current? How do we determine if our guidance on radiology procedure ordering is leading to reduced radiology costs? How do we correlate investment in clinical knowledge management with optimized operating margins from value-based reimbursement programs? Providing thoughtful answers to these questions helps to assure managers that the invested resources are likely to result in the desired gains.
- Lastly, the information technology infrastructure and content needed are defined. This infrastructure can include knowledge libraries, editors, content-lifecycle management systems, and collaboration tools. The tools proposed offer an evolutionary technology path that is robust and enduring.

21.3 Organization of the effort

Organization refers to structures and processes needed to manage the lifecycle of knowledge application, discovery and asset management.

This section will discuss objectives of organization, provide examples of organization structure and processes, and review implications for organizational design strategy.

21.3.1 Objectives of organization

CKM programs require governance structures, stewardship resources, and processes. The CKM team comprises the resources that continuously steward and update the CDS knowledge, support the governance activities, and direct the technical resources that manage the CDS content management systems. These structures and processes are intended to accomplish several objectives:

- Identify new types of knowledge that need to be incorporated into the organization's clinical information systems, e.g., the addition of a new Deep Vein Thrombosis Prevention intervention to the order entry, clinical documentation, and CDS system to reduce the incidence of this event and report on the relevant quality measure
- Ensure that CDS interventions are useful, impactful, and evidence-based through review of the literature and/or consensus-based decisions by appropriate clinical staff
- Ensure that existing knowledge is reviewed at an appropriate frequency to determine if "old" knowledge needs to be revised
- Examine the potential for CDS algorithm risk and develop approaches to mitigate that risk. This issue has become a growing concern with the potential bias of artificial intelligence algorithms (see Chapter 7 for a more detailed discussion).
- Ensure that CDS stewardship resources and tools are adequate to facilitate ongoing management engagement in CDS decision making, update existing CDS knowledge, and build new CDS interventions
- Recognizing the finiteness of information technology and clinical resources, provide direction on priorities for incorporating or modifying knowledge
- Educate the clinical staff on the rationale for introducing new CDS interventions
- Assess the impact of existing knowledge application tactics on provider decisions and practices to determine if the desired outcomes are being achieved
- Review strategies to improve the effectiveness of existing knowledge application tactics, e.g., does a computer-based intervention impede workflow, is it ignored, or does the application interface confuse rather than inform the user?
- Guide the efforts of information technology staff and/or the application vendor to ensure that appropriate specifications have been developed and testing performed.

Invariably, an organization will have several forums that pursue these objectives. The Pharmacy and Therapeutics Committee can be charged with managing all medication-centric knowledge for an inpatient clinical system. A Population Health Council may be convened by Accountable Care Organization leadership to develop decision support content to improve the health maintenance processes for chronic diseases such as diabetes, chronic obstructive pulmonary disease or congestive heart failure. A committee formed to reduce the costs of care operations may decide to examine ways of reducing inappropriate laboratory or radiology procedure utilization through CPOE. A committee that manages the evolution of an organization's clinical information systems may examine the systems to determine if there are "CDS knowledge gaps" that merit rectifying to meet pay-for-performance goals, e.g., inadequate CDS for sepsis prediction or venous thromboembolism prevention.

The result of assigning knowledge management tasks to a range of forums can lead to a complex maze of decision making. While each individual assignment may be the right assignment, the maze needs to be coordinated, conflicts may require resolution, and the resulting demands on the information technology staff will require prioritization. Further, it is not uncommon for leaders to direct IT staff in the deployment of clinical decision support interventions when they have not created the context for adoption by their staff. CDS typically is most helpful and effective when it reminds or assists a user to accomplish a task he/she already has a motivation to perform. Leadership participation in governance can inspire, educate and incentivize the adoption of guidance provided by the CDS. A well-crafted governance model with representation from informatics as well as leadership from clinical, operational and financial domains is an essential foundation for ensuring investment in CDS and CKM results in tangible improvements in clinical performance.

21.3.2 Examples of approaches

Several examples of approaches to organization are presented in the following sections. These examples are adapted from AMIA, 2005 [6].

21.3.2.1 Example 1

A Medical Information Systems Committee (MISC) is charged with overseeing the design and implementation of clinical information systems for the organization. The MISC is also responsible for ensuring that the clinical information systems conform to all regulations, Joint Commission requirements and the organization's policies.

The MISC has multi-stakeholder representation and reports to an Executive Medical Committee.

The MISC has a subcommittee that oversees the prioritization, design and development of CDS. This subcommittee receives requests from various task forces, committees and user groups. The subcommittee requests IT assessment of the costs and time required to fulfill the request. The subcommittee recommends priorities and forwards its recommendations to the MISC for approval.

21.3.2.2 Example 2

The Information Technology Strategy and Policy Committee (ITSPC) is responsible for strategic, policy and tactical decisions for all of the organization's information systems and information management. The Committee is composed of senior clinical, administrative, and IT leadership.

A Clinical Information Systems Committee reports to the ITSPC and is responsible for all patient care systems including CDS. The Clinical Information Systems Committee is responsible for reviewing all requests for decision support, identifying required resources, prioritizing requests and monitoring the effectiveness of existing decision support.

21.3.2.3 *Example 3*

The Clinical Systems Advisory Committee (CSAC) is responsible for providing direction and monitoring progress on the acquisition and implementation of clinical information systems. The CSAC members are operations, financial and practicing clinician leaders from across the organization. Requests for decision support are sent to the CSAC for review, and analysis of costs and effort and prioritization. Decision support requests that are approved are sent to a Clinical Data and Documentation Committee, a sub-committee of the Medical Executive Committee, to ensure that the requests conform to organizational policy and are supportive of organizational efforts to improve patient safety and medical care.

21.3.3 Clinical knowledge management organizations at Partners Healthcare, Lahey Health and Intermountain Healthcare

The previous examples center on the management of clinical decision support priorities and are aimed at ensuring that CDS activity is linked into, and fits with, other supporting activities such as the implementation of a clinical information system or medical policies. The following examples are drawn from larger health systems, some of which began their CKM journey in a context of internally developed clinical systems and subsequently have evolved their efforts in concert with migration to commercial EHR systems. Smaller entities have correspondingly fewer resources and must strategically prioritize their efforts accordingly.

At Partners Healthcare System (now called Mass General Brigham), a Clinical Knowledge Management Group was established in 2003 under the direction of Dr. Tonya Hongsermeier. This group grew and evolved over the years to serve enterprise-wide and site-specific CKM needs in an organization that had multiple internally developed EHRs in production. For example, a Clinical Content Committee (CCC) was created to direct and prioritize investment in the CDS components of an internally developed ambulatory EHR system. This committee's activities are supported by a dedicated ambulatory EHR CKM team that received proposals from clinical leaders and end users, analyzed these proposals, and facilitated the CCC's evaluation of such proposals for new CDS or changes to existing CDS. This CKM team maintained the content in the Ambulatory EHR with the support of solicited input from a variety of clinical discipline-centric expert panels focused on Primary Care, Pediatrics, Geriatrics, and the like which were chartered and sponsored by the CCC.

In 2013, Dr. Hongsermeier joined Lahey Health as Chief Medical Informatics Officer. In that capacity, she established a CKM model embedded in the EHR implementation team. This EHR-CKM team was governed by a Clinical Content Committee that was chartered with diverse, enterprise-wide leadership by the Medical Executive Committees of the system hospitals to steward clinical content in the EHR and ensure alignment with regulatory, compliance and clinical quality performance requirements. Fig. 21.3 is a schematic illustration of how the Clinical Content Committee was positioned vis-à-vis the overall EHR governance structure.

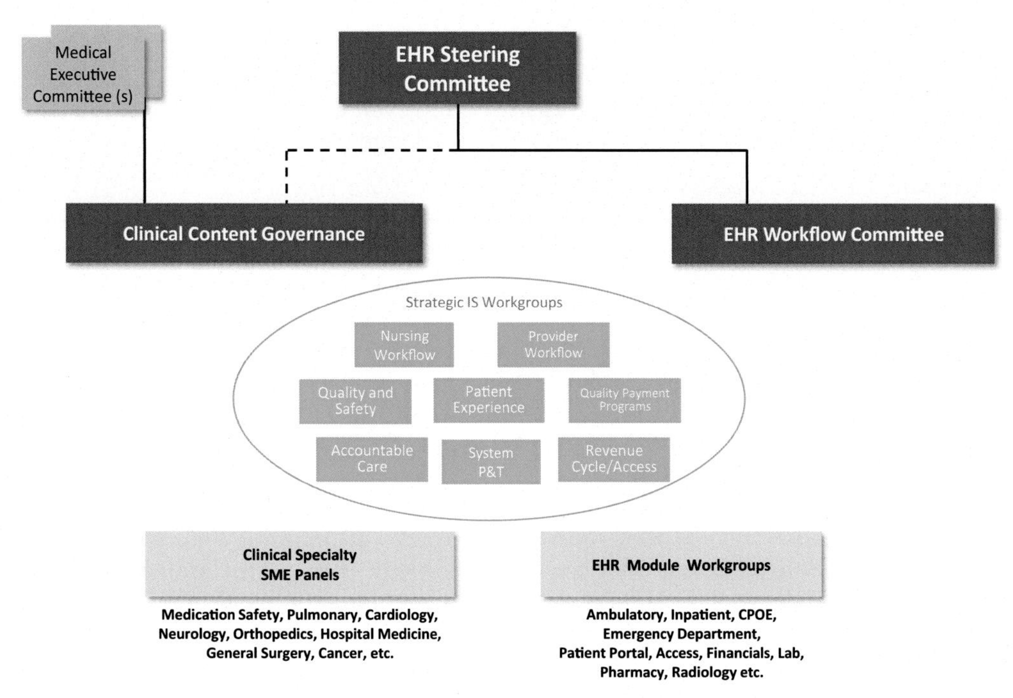

FIG. 21.3 Schematic illustration of CKM governance at Lahey Health.

The governance structure contained workgroups organized by the clinical user type (e.g., physicians or nurses), application, specialty, or operational domain level. Of note, every effort was made to move the decision making and design to the lowest level of stakeholder governance required for effective design stewardship of content. Assurance of strategic alignment, sound evidence-based design, stewardship and usability was accomplished by iteratively moving proposed CDS interventions through the relevant EHR workgroups of affected stakeholders and Clinical Content Committee.

At Intermountain Healthcare, there is a clinical leadership-driven governance model for CDS that explicitly links a focus on clinical performance targets to CDS initiatives [7,8]. There are multiple domain specific "Clinical Programs" that provide direction to the appropriate multidisciplinary workgroups and clinical information systems resources to develop CDS artifacts that help them meet their objectives. The Clinical Programs are headed by clinical staff who set goals, facilitate progress, and align resources. Within each Clinical Program, there are numerous multi-disciplinary, multi-stakeholder "Clinical Development Teams" that focus on specific clinical goals such as management of asthma, community-acquired pneumonia, congestive heart failure, diabetes and depression. These teams are tasked with developing a "Care Process Model" and providing direction to CDS implementer resources to integrate knowledge where appropriate into the clinical workflow systems. Once the Care Process Model is up and running, the Clinical Development Teams monitor progress and iteratively make improvements and updates where appropriate to achieve performance targets.

21.3.4 Observations on organization

As can be seen in the preceding examples, which represent a wide range of provider organizations, there is no single best way to organize. However, there are several commonalities and guidelines that can guide the organizing of knowledge management:

- *Understand the roles of oversight, stewardship, and stakeholder engagement.* In Fig. 21.4, the typical organizational components of a CKM governance model are outlined. These components include a steering committee that performs prioritization and resourcing decisions, a knowledge stewardship or CKM team that builds and maintains the CDS knowledge, expert panels representing the stakeholders who provide input on CDS design, and technical resources who manage the tools utilized by the knowledge stewardship team. This knowledge stewardship team, in some organizations, is referred to as a knowledge engineering team or CDS team. Regardless of the label or size of this resource pool, they typically combine clinical and technical expertise and sit at the center of the CDS maintenance process.
- *Leverage and evolve existing committees for prioritization and subject matter expert panel expertise.* For a hospital, an existing Pharmacy and Therapeutics (P&T) Committee could be asked to support CKM resources responsible for updating medication-centric knowledge. If the impacted stakeholders are part of a multi-hospital or ambulatory institution, system-wide P&T would need to be chartered. An existing committee devoted to improving cardiac care should be asked to oversee knowledge related to hypertension and congestive heart failure guidelines.

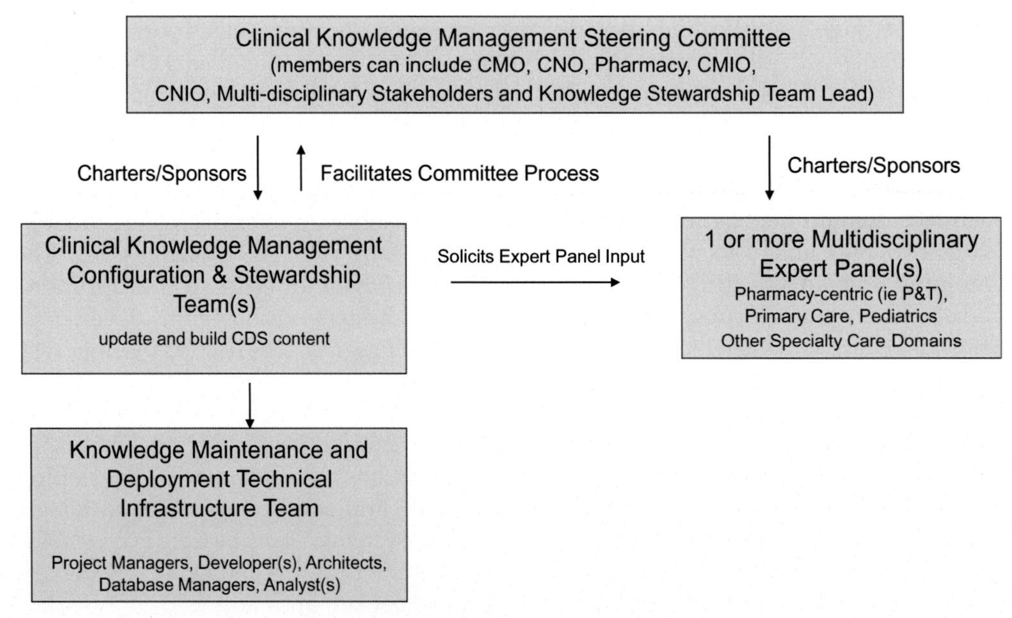

FIG. 21.4 Typical clinical knowledge management organization components.

- *Computer-based decision support is viewed as simply another tool available to the committee.* This tool may be new to them, and they may need time and education to become comfortable with understanding its strengths and weaknesses. As mentioned earlier, decision support is more effective when the target users already believe in the importance and validity of the guidance due to a combination of leadership, incentives and education. The use of existing care-oriented committees helps to address several critical aspects of knowledge management and medical decision support. First, the committees invariably possess the expertise and leadership necessary to determine the clinical utility of a specific decision support recommendation. While "anyone" can propose a specific set of decision support interventions, the experts and leaders must review, sponsor and approve it. The use of an existing, appropriate committee can help silence squabbles about who is "the expert" on specific decision support content. Second, decision support must be maintained. Content will need to be continuously updated by the CKM team and regularly reviewed by the appropriate expert panels. Some of that maintenance may be due to guideline changes. In most cases, the maintenance is necessary to ensure that the introduction of new medications, problem definitions, and diagnostic test types are incorporated into existing clinical content with data governance oversight. This lack of maintenance of is a frequent cause of errors in omission or commission with decision support. Oversight of this maintenance should be a formal responsibility of the committee. Third, education of clinicians must often occur to explain why the decision support was implemented. The committee can be given this responsibility. Fourth, the committee is in the best position to prioritize requests. For example, a patient safety committee will have the best organizational perspective on the major patient safety issues. Fifth, these committees are usually in the best position to "discover" new knowledge. This discovery can be based on the experiences of the organization or the review of the discoveries of others.
- *Examine committee composition.* Knowledge often spans domains. Leadership from multiple domains is often necessary and helpful to drive adoption. For example, there are obviously medication-centric rules that are of great interest to a committee focusing on cardiac care. To the degree that there is likely to be a significant set of knowledge that spans several committees, there should be cross-committee representation, e.g., a member of the P&T Committee on the Cardiac Care Committee. Often, this cross-committee representation is already in place; the boundary-spanning issues were present before the introduction of clinical information systems. Nonetheless, it can be useful to review committee composition and ensure that appropriate cross-representation is in place.
- *Cross-representation should not only account for clinical discipline, but overall perspective.* For example, it is important that clinicians representing the strategic concerns of the health system be balanced by those representing usability and efficiency concerns. Respected clinical champions can be those in management positions as well as the clinicians in a community practice who are greatly respected by their peers.
- *The addition of CKM team members from the IT department to these committees as either a member or liaison is essential.* These personnel can educate stakeholders on the capabilities and limitations of the clinical systems as well as draft and iterate proposed approaches to decision support design. They can update the committee on the status of proposed and released CDS interventions and assist with the vetting of proposed changes or additions. Regardless of organizational approach, these individuals can help the committee members

focus on the most feasible and effective informatics strategies to address a particular challenge, e.g., guidance at the time of ordering and the use of defaults and options for incorporating the knowledge into the workflow. They can assist with measurement of CDS intervention effectiveness and adoption. Furthermore, they can direct analysts, as they transform the clinical guidelines into proper CDS design specifications.

- *Ensure IT review and assessment.* CDS proposals must be examined for their impact on system performance, workflow, and productivity. The decision support technology will have limitations, some of which mean that some proposals cannot be practically implemented. The CKM and IT resource effort required to implement a new proposal must be understood. The staff that must "codify" and test the decision support will have a backlog that needs to be prioritized. Decision support can be a significant consumer of processing power; hence the machine performance of a specific decision support rule and the rules in aggregate must be monitored.
- *Define oversight group.* The actions of individual committees will often conflict. The conflict can center on:
- The definition of appropriate knowledge, e.g., different opinions on best practices such as between orthopedic surgeons and neurosurgeons on back pain management,
- Trade-offs between practicing best care and operational realities, e.g., the primary care physicians are so harried that additional health maintenance reminders will fall on deaf ears, support staff capacity increases may be required to assist with population health
- Prioritization of scarce organizational resources, e.g., budget limitations mean that some ideas can be implemented but not all ideas.

In addition to resolving conflicts, these individual committees must be coordinated. Coordination can be necessary for many reasons. For example, it may be the case that different committees independently embark on duplicative knowledge strategies (e.g., an inpatient Smoking Cessation team and an enterprise Chronic Obstructive Pulmonary Disease team both developing Smoking Cessation CDS). Different groups may be considering investments in redundant tools (e.g., different teams independently investing in analytic infrastructure).

Decision support must conform to the organization's medical policy and hence policy assurance must be determined. At times, the decision support idea may lead to a need to alter policy such as which staff can update a problem list. Decision support may also indicate the need to examine organizational roles, e.g., who should respond to an asynchronous panic lab value alert? This oversight committee must have members who can bridge into other important organizational groups, e.g., compliance, and have processes that enable it to turf some issues to those other forums.

An existing committee can be assigned the responsibility for overseeing knowledge management discussions and decisions. Many organizations have committees that have broad responsibility for care improvement, e.g., an integrated delivery system may have a Chief Medical Officer's forum.

In several of the examples cited earlier, this oversight group is one that has been formed to provide overall direction for the implementation and management of the organization's clinical information systems. The placing of decision support oversight responsibility with such a committee is common. This orientation is usually a reflection of the need for such committees during the implementation of major clinical information systems. These implementations are massive and complex undertakings, and a committee of senior leaders is necessary to ensure that progress is made. During implementation, CDS efforts will begin, and it is natural that decision support efforts become the purview of the committee.

However, CDS is a tool, and a natural evolution of tool oversight involves the transition from a tool-centric committee to a care-centric committee that has tools at its disposal, e.g., an Intermountain Clinical Program team.

As an example of this transition, many organizations had Internet Strategy Committees at the turn of the millennium. As understanding of the Internet increased, virtually all of these committees were disbanded, with responsibility for determining the best approaches to tool (the Internet) use being turned over to groups responsible for business performance.

21.4 Key IT strategies and considerations

Several chapters in this book have addressed specific aspects of the information technology and logic and data design of clinical decision support.

This section addresses three overall IT strategy considerations: legacy systems, tools and applications, and foundations. These considerations examine three critical aspects of defining and implementing the information technology infrastructure necessary for effective decision support.

21.4.1 Legacy systems

How can an organization address the challenge of implementing robust, content-enriched computer-based decision support while working within the constraints of legacy information systems investments? As an example, although the US HITECH act, Meaningful Use and Quality Payment Program incentives have spurred provider organizations to modernize their clinical information system infrastructure, there will still be constraints to contend with.

In pursuing the application of information technology to effect CDS, the organization will confront the reality of its clinical information system investments. In a large integrated delivery system, there may be several clinical information systems from multiple vendors. Each of these systems may have their own decision support technologies, and these technologies are likely to be of variable sophistication and utility. One need not be a large delivery system to face this challenge. A community hospital might find differing decision support capabilities in its laboratory, pharmacy and hospital information systems.

Replacing these investments may not be practical. The organization may not have enough capital or be culturally ready for migration and integration. Replacement can take years to implement, but the organization needs care improvements in the near term. Moreover, some clinical information systems work well in large hospitals but not in the small physician's practice; hence in a large health system there may be little prospect of finding one system that effectively addresses the needs of all constituents.

There is no easy answer to this challenge. Recent advances in API-centric architectures and substitutable applications can enable an organization to access CDS knowledge services outside their core clinical information system infrastructure, e.g., a cloud-based medication reconciliation service or a substitutable application for antibiotic selection, that effectively interoperate with heterogeneous applications. However, such approaches are still in their early phases of market penetration.

Faced with this problem, the organization can take several steps to make the most of its legacy investments.

a. *Define the content areas that are important to drive the business.* There are several content areas that can have a tangible effect on an organization's performance. For example, the Quality Payment Program identifies several quality measures for eligible providers and hospitals. In the hospital setting, critical quality performance topics include stroke management, hospital acquired infection prevention, and venous thromboembolism prophylaxis. In the outpatient setting, value-based reimbursement is aligned with quality performance measures for asthma, obesity prevention, smoking cessation, diabetes, cardiovascular disease and women's health management.

b. *Define the systems that will be the focus of applying decision support.* These systems are likely to include physician order entry, chart review, dashboards, clinical documentation, health maintenance systems, case management and the like.

c. *Evaluate the decision support capabilities of these applications.* It is important to evaluate, for example, what kind of medication decision support, order sets, templates, reminders, and reporting these applications support. This evaluation will lead to the development of the "lowest common denominator" of tools, in effect, establishing the limit to which decision support can be implemented across the enterprise with native EHR functionality. If it appears that the limitations of the legacy infrastructure are woefully inadequate for meeting the strategic goals, decision support and artificial intelligence vendors are available on the market in the form of specialized CDS application providers, substitutable application providers, and cloud-based service providers that can significantly augment the native capabilities of the legacy environment at much less than the cost of a new infrastructure purchase.

d. *Define CDS knowledge acquisition strategy* (see Chapter 18 for a description of the different kinds of knowledge resources that can serve as building blocks for CDS tools). In most health care delivery organizations, formal structures and resources are often lacking to undertake the process of transforming guidelines into the relevant CDS components and maintaining these artifacts. Most provider organizations are accustomed to licensing drug information as well as terminologies for problem list documentation and billing. The large content vendors offer a menu of prespecified content such as order sets, documentation templates and CDS rules. Some also offer tools for collaborative localization, update, and import into the EHR system, largely because very little of the licensed, importable content can be regarded as "plug-and-play." There are too many local considerations to account for, such as local operational process, data dictionaries and workforce composition, that determine CDS configuration, particularly in the hospital setting. Further, some offer CDS content embedded in an application system or cloud-based CDS service that can integrate with the EHR system. Typically, the cloud-based CDS services approach allows for less customization, but also outsources the content maintenance. The advent of personalized medicine and its dependence on complex genomic decision support content will make cloud-based CDS an imperative. The volume and complexity of such content will exceed the knowledge curation capacity of even the large provider organizations, not to mention the technical capabilities of most commonly used EHR vendor systems. Some EHR vendors, particularly for the ambulatory setting, are offering EHR and content-enriched CDS services as a complete package on a cloud-based platform. When licensing content for import and build into the native EHR system, the provider organization must bear the cost of localization and maintenance. With these considerations in mind, an organization must reconcile the cost of localization and maintenance with the value such investments create in clinical performance and usability. There is no easy answer, and as the CDS market evolves, most provider organizations will invest in an ever-evolving hybrid of home-grown, content license with localization, and CDS services strategies.

e. *Define strategies and resources needed to manage consistent knowledge across a heterogeneous set of applications and cultures*, e.g., applications across large academic health centers and small community hospitals. For example, if we have to implement a new provider reminder across six different applications in four different organizations within a single enterprise, how will we do that? How do we ensure that the logic is consistent across the organizations? Ensuring consistency and currency might require that a person at each organization, or for each relevant application, be tasked with implementing content. These individuals can be managed by a corporate person who ensures coordination.

f. *Develop/acquire an infrastructure for knowledge asset management* (see Chapter 28 for technical aspects of CKM infrastructure). The organization must be able to have a repository or library of the content and data definitions that it has implemented across the enterprise [9]. This library may be constrained to that content that has been determined to have significant value and/or must be consistent across all care settings. In the course of determining how to invest in knowledge management infrastructure, an organization must fully understand the comparative strengths and weaknesses of their legacy environment with respect to key functional capabilities. This assessment will lead to some form of the steps outlined above.

21.4.2 Knowledge management tools

Vendor systems are often designed with proprietary database design tools typically called "knowledge editors" which are used to build different content types such as value sets, rules, order sets, and documentation templates. Few vendor solutions offer functional support of other critical aspects of knowledge management such as governance, knowledge inventory, dependency tracking, propagation, versioning, knowledge vetting and design of complex cross-functional content such as disease management protocols. As highlighted earlier, the silo-ization of the different CDS content editors creates silo-ization of the content and presents a barrier to building integrated clinical program solutions. Hence, many clinical information systems remain undernourished and under-maintained from a knowledge perspective.

An inventory and library of decision support design specifications is a critical component of any knowledge asset management strategy. Collaboration tools are useful to support governance structures, subject matter expert review and validation of content. In the era of COVID19 and large integrated delivery systems, virtual team support has become more important than ever. Collaboration platforms have advanced significantly in recent years with advances in Web 2.0 standards [9]. They typically enable a combination of social interaction management, content life cycle management, and process management (see Chapter 28 for a more in depth description). Collaboration workspaces require dedicated resources to ensure they are deployed in a manner aligned with the strategic initiatives, support cross-disciplinary interaction, and are organized to facilitate stakeholder engagement. For example, a medication cost reduction panel and a geriatric panel may collaborate on safe, cost-effective pain management in the elderly. Content management systems are useful to support the scheduled maintenance, versioning, and overall life cycle management of contentsuch as the capture of critical metadata for CDS content such as author, business owner, purpose, subject matter expert validation, date of last update, schedule of next review, interdependencies with other content, and the like.

21.4.3 Foundations

The pursuit and progressive experience with knowledge-rich clinical information systems can lead the organization to begin to think of itself as implementing application foundations rather than strictly a set of clinical information system applications [4]. A foundation provides the broad ability to perform a never-ending series of application-leveraged small, medium, and occasionally large advances and improvements in organizational performance. Electronic health records are now increasingly regarded as platforms for innovation [10].

For example, a computerized provider order entry system is commonly regarded as a foundation to improve physician decision making. Once the system is implemented, the organization can introduce an unending series of decision-support rules and guides. These rules can address medication safety, ensure disease management referrals, critique the appropriateness of test and procedure orders, and facilitate the display to physicians of data relevant to a given order.

In effect applications become the foundation necessary to achieve the core goals of enabling ongoing delivery of new CDS and improving workflow. This view of applications as foundations has several ramifications.

Clearly, there will be a flurry of intense effort as the foundation is laid. Introduction of complex, multi-stakeholder clinical systems is difficult work that requires great skill and significant resources. But once the foundation is in place, there is an ongoing implementation of decision support and operationalization to achieve continuous innovation. In fact, implementation of a clinical information system never stops. Provider organizations are faced with continuously changing reference content, clinical guidelines, reimbursement rules, and regulations. Hence, organizational information system processes and management mechanisms must become agile to continuously innovate and iterate their implementations. This can imply that implementation teams do not disband but rather evolve their responsibilities from the team that installed the system and trained staff to the team that carries on ongoing optimization of decision support and workflow improvement.

The foundation must be able to evolve gracefully and support ongoing implementation. Tools that enable rule development, the safe addition of local modifications, incorporation of new data types and coding conventions, and efficient interoperability with other systems are essential. The foundation must function as a platform and be able to capitalize on new technologies and architectures with minimal disruption, and support growing organizational sophistication in applying the tools to improve care processes. In many ways, platforms that enable ongoing implementation and innovation are more important than the present functionality of the application. This emphasis affects the orientation of the application Request for Proposal (RFP) and the system selection criteria.

The RFP for an application generally centers on functionality. The RFP process for a foundation must be changed from this traditional focus to place a greater emphasis on tools, architectures, and core technologies. In addition, an implementation that never stops implies that using the RFP in an effort to fully define all functionality that will ever be needed will be misguided. It is important for an organization to be prepared to invest in ongoing iteration and operational transformation. Organizational strategic visions are often anchored to their legacy system capabilities and new horizons and opportunities may only become visible once the new platform foundation is implemented. Experience will be the teacher.

Assessing the return on investment (ROI) of a foundation during the process of deciding capital budgets is more difficult than determining the ROI of an application. Although it is essential to continue to evaluate the ROI, it is difficult to do, because the path of evolution is not always clear, and implementation is never-ending. The value realized is not only constitutive to the technology but highly dependent on the organization's cultural operational capability to take advantage of new capabilities and continuously transform themselves. In acquiring and implementing a foundation, the organization is investing in "an ability." It is difficult to assign an ROI to an ability be it technological, operational, or cultural. In a similar fashion, it is difficult to measure the ROI of effectively leadership, a well-educated workforce or having healthy capital reserves.

21.5 Evaluation of the impact and value of knowledge management

If the organization has identified decision support as a critical strategic enabler and has, as a result, committed resources to acquiring, implementing, and maintaining needed information systems and support resources, it will ask "Have our investments been effective? How much is it costing us to achieve our gains? Where must we focus our decision support efforts next?"

The evaluation of the impact and value of knowledge management must address three areas:

- The strength of alignment of the content to business goals and strategies
- Organizational performance relative to key measures
- The efficiency and effectiveness of the knowledge management function to enable rapid-cycle learning.

Evaluation does require that an organization has an approach to clinical data management and analysis. Assessing clinical performance and the impact of an intervention on that performance requires a set of well-defined data of known accuracy and timeliness. This approach must develop means to resolve issues that often plague the collection and management of necessary data.

Many health systems have poor access to clinical data for measurement and rely, instead, on billing and administrative data. The architecture of a typical transaction-oriented database is not optimized to support analysis. Further, the data that must be aggregated to enable deep analytics is typically located in many databases across an organization.

In the absence of a clinical data management and analysis strategy, those engaged in the process of understanding and reporting on clinical performance must often bear the cost and time delays of, for example, chart abstraction labor to collect clinical data, which consequently slows the translation of such insights into quality improvement.

21.5.1 Alignment

It is very useful for health care organizations to take a "begin with the end in mind" approach to decision support. In this way, business goals are linked to relevant measurement parameters and consequently, required decision support strategies.

Table 21.2 contains a sampling of the Ischemic Stroke eCQM measures to illustrate alignment among quality performance strategy, data sources for quality measurement, and clinical decision support components. Once the performance goals are identified and targeted, the measures are mapped to the necessary data sources of discrete data for measurement and necessary knowledge components to achieve performance improvement. Further, Fig. 21.5 illustrates how the CDS logic and data definitions that underlie a performance goal should be aligned with the quality measure logic and data. As one defines the CDS content and quality measure logic that informs a goal, such as ensuring that patients with Ischemic Stroke are discharged on anti-thrombotics, one can see the importance of definition alignment between the EHR care delivery system and the quality measure and reporting system.

Such goal, measurement, and decision support "tuples" are the centerpiece of alignment. Those measures can be complemented by measures that provide a form of overall assessment of alignment. For example, measures that might serve as complements include:

- Degree of knowledge asset coverage for key business-impact measures such as Quality Payment Program ((QPP), Joint Commission, National Quality Forum (NQF), and Pay-For-Performance Contracts. For example, in order to meet the Centers for Medicare and Medicaid Services (CMS) quality reporting requirements specified for congestive heart failure, one can measure the degree of CDS knowledge coverage by the clinical documentation elements, order sets, decision support rules, and reporting algorithms in production for inpatient, case management, and outpatient systems
- Application end-user satisfaction with clinical decision support. Are clinicians satisfied with decision support content? Has leadership incorporated stakeholder feedback into proposed interventions and sponsored effective adoption? Is the right balance achieved between quality improvement and workflow enhancement?

21.5.2 Performance

Table 21.2 also illustrates how decision support effectiveness can be measured in terms of direct impact on business performance. Effective knowledge management practices should result in better performance on key performance measures. Such measures can be translated into higher reimbursement on payer contracts or improved quality of care. Following are

TABLE 21.2 Clinical decision support intervention alignment with stroke quality measures.

Example eCQM measure	Measure description	Measurement data sources	Clinical knowledge for decision support
eCQM STK-2 or CMS104v9	Ischemic stroke patients prescribed or continuing to take antithrombotic therapy at hospital discharge	1. For antithrombotics, discharge Orders from discharge prescribing/ordering application 2. For indications: problem lists and/or discharge diagnoses for Ischemic Stroke 3. For patient contraindications, clinical documentation, allergies, laboratory data, problem list	1. Anticoagulation medication orders defaulted in discharge order sets when admission diagnosis is Ischemic Stroke 2. Documentation template (ideally embedded in order set) for contraindication capture 3. Decision support alert to notify clinician when qualifying patients have no discharge anticoagulation ordered and no contraindications are documented
eCQM STK-3 or CMS71v10	Ischemic stroke patients with atrial fibrillation/flutter who are prescribed or continuing to take anticoagulation therapy at hospital discharge	1. For anticoagulation medications: Discharge Orders from discharge prescribing/ordering application 2. For indications: problem lists and/or discharge diagnoses for Ischemic Stroke, Atrial Fibrillation and/or Atrial Flutter 3. For patient contraindications: orders (ie for comfort measures), clinical documentation, allergies, laboratory data, problem list	1. Anticoagulation medication orders defaulted in discharge order sets when admission diagnosis is Ischemic Stroke 2. Documentation template (ideally embedded in order set) for contraindication capture 3. Decision support alert to notify clinician when qualifying patients have no discharge anticoagulation ordered and no contraindications are documented
ECQM STK-5 or CMS72v9	Ischemic stroke patients administered antithrombotic therapy by the end of hospital day 2	1. For antithrombotics and timing, electronic medication administration record 2. For patient contraindications, orders (ie comfort measures) clinical documentation, allergies, laboratory data, problem list 3. For admission timing and destination, Admission/Discharge/Tranfer system	1. Ischemic stroke admission order set with antithrombotic options and instructions for timing 2. Monitoring alert that fires on open chart on day 2 if no antithrombotics are ordered and no contraindications are evident

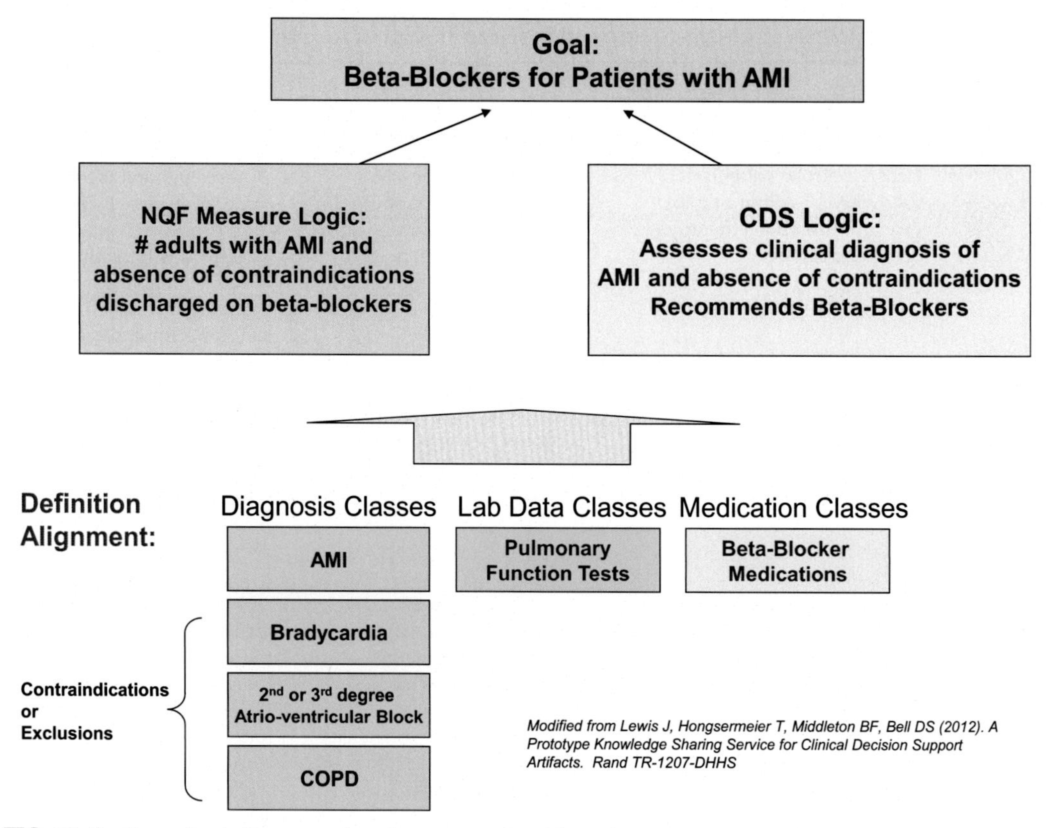

FIG. 21.5 Example of alignment of performance goals with quality measures and CDS logic.

examples of the kinds of performance measures that can be used to assess decision support effectiveness. Clinician acceptance of decision support recommendations is also a barometer. An organization should anticipate and accept some minimum override rate, because few decision support interventions are so specific that recommendations are always clinically correct. Conversely, if an override rate is too high, the decision support is probably overly sensitive and task interfering.

Examples of performance measures include:

- Quality Performance: HEDIS, QPP, Joint Commission, CMS, NQF, and Pay-For-Performance contracts measures
- Adverse Event Rate: Adverse drug events, bedsores, hospital-acquired infections, peri-operative venous thromboembolism, falls, confusion, etc.
- Compliance rate with decision support: Sensitivity and specificity analysis, override rates
- Patient Experience: Patient satisfaction scores and correlation with patient use of clinical decision support tools
- Malpractice: Insurance costs and trends in claims.

21.5.3 Knowledge management function and organizational learning

Keeping an inventory of decision support knowledge current with commonly accepted standards of practice can be a costly business. It means investing in a team that conducts ongoing literature review, localizes commercial content, and ensures that changes in the standard of practice are rapidly incorporated into the decision support content. As we have already noted, the advent of molecular medicine is increasing the speed of change in clinical knowledge, presenting new challenges for decision support maintenance. In addition, the knowledge engineering team must work closely with the quality improvement and analytics team that evaluates performance data to determine how decision support must change to achieve strategic objectives. They must work with end-users so that CDS is helpful and minimizes task interference. With each successive stage of decision support capability, health care performance becomes increasingly transparent. These CKM functions are critical to enabling a provider organization to become agile at self-improvement.

The organization will add to these costs of CKM the expenses of licensing fees, tools, and the sunk cost of clinical time spent on clinical decision support management.

The following lists some illustrative measures of CKM effectiveness:

- Coverage. Percentage of CDS assets with a clearly identified accountable CKM steward and business owner.
- Currency. Percentage of CDS assets on an explicit updating schedule (rather than waiting for a complaint) and/or percentage of CDS assets updated on time.
- Cycle time for content update. This cycle can be measured as the length of time it takes to convert an agreed-upon guideline into a decision support specification and then into production. This measure assumes there is a business cost to delayed alignment.
- Cycle time for content agreement. This measure evaluates broader organizational effectiveness in getting agreement on enterprise guidelines. For some organizations, depending on the complexity of the asset, this can take longer than converting the guideline into decision support.

Providers can measure CKM costs and assess the performance of the CKM function. While these measures are important, they do not lend themselves to a return on investment analysis. For example, it may not be clear if an increase in the CKM budget of 10% will be "worth it." How valuable is a reduction in content cycle time?

There is no easy answer to this challenge. While perhaps not much comfort, this challenge is confronted by many efforts to improve care quality and patient service experience.

21.6 Conclusions

Clinical decision support is a class of tactics for applying medical knowledge to achieve superior performance. An organization should devote strategic discussions to knowledge management overall to ensure that it has defined appropriate boundaries, understands the functions of knowledge management, and is able to prepare a business case that ensures necessary investments of organization resources.

Organization is a set of management structures and processes needed to ensure that an investment achieves desired organizational goals. Clinical decision support management structures and processes must achieve goals that include linkage to organizational strategies, prioritization of resources, and determination of the impact of clinical decision support. While there is some variation in the organizational approaches of different health care providers, common guidelines do emerge.

Clinical decision support implementation and management does require the consideration of key aspects of how an organization's clinical and business strategies drive the IT strategy. Specifically, this chapter discusses the application of clinical decision support across legacy systems, clinical decision support tools, knowledge acquisition and maintenance approaches, and the view of application systems as foundations.

Clinical decision support is utilized for one overarching goal—improving organizational performance. Achieving this goal requires leadership sponsorship, ensuring strategic alignment, measuring performance relative to goals and continuous improvement of the efficiency and effectiveness of the clinical knowledge management function.

References

[1] Davenport T, Hongsermeier T, McCord K. Using AI to improve electronic health records. Harvard Bus Rev 2018. https://hbr.org/2018/12/using-ai-to-improve-electronic-health-records.

[2] Ford D, Harvey JB, McElligott J, King K, Simpson KN, Valenta S, et al. Leveraging health system telehealth and informatics infrastructure to create a continuum of services for COVID-19 screening, testing, and treatment. J Am Med Inform Assoc 2020;27(12):1871–7.

[3] Sittig DF, Wright A, Osheroff JA, Middleton BF, Teich JM, Ash JS. Grand challenges in clinical decision support. J Biomed Inform 2008;41(2):387–92.

[4] Davenport T, Glaser J. Just-in-time delivery comes to knowledge management. Harvard Bus Rev 2002;80 (7):107–11.

[5] Lewis J, Hongsermeier T, Middleton BF, Bell DS. A prototype knowledge sharing service for clinical decision support artifacts. Technical Report: Rand TR-1207-DHHS. Available at: http://www.rand.org/pubs/technical_reports/TR1207.html.

[6] American Medical Informatics Association (AMIA). Comparison of eight governance models; 2005. Clinical Decision Support Governance Task Force.

[7] James BC, Savitz LA. How intermountain trimmed health care costs through robust quality improvement efforts. Health Aff (Millwood) 2011;30(6):1185–9.

[8] https://intermountainhealthcare.org/about/transforming-healthcare/hdi/clinical-management/ [Accessed 15 August 2021].

[9] Wright A, Bates DW, Middleton BF, Hongsermeier T, Kashyap V, Thomas SM. Creating and sharing clinical decision support content with web 2.0: issues and examples. J Biomed Inform 2009;42(2):334–46.

[10] Glaser J. It's time for a new kind of electronic health record. Harvard Bus Rev 2020. https://hbr.org/2020/06/its-time-for-a-new-kind-of-electronic-health-record.

CHAPTER

22

Evaluation of clinical decision support

Nicole M. Benson[a], *Hojjat Salmasian*[b], *and David W. Bates*[c]

[a]McLean Hospital, Belmont, MA, United States [b]Children's Hospital of Philadelphia, Philadelphia, PA, United States [c]Brigham and Women's Hospital, Boston, MA, United States

22.1 Clinical decision support adoption

Clinical decision support (CDS) systems provide a variety of different capabilities, for various purposes, as described in Chapter 1. Some CDS interprets patient data from the electronic health record (EHR), from an external source such as a clinical laboratory, or from data entered directly by a provider or patient, against a central repository of information to provide patient-specific guidance to clinicians through electronic tools such as alerts and reminders [1]. CDS is not limited to tools that make an explicit suggestion to the user (such as alerts and reminders); it also includes tools like condition-specific order sets, focused patient data summaries, note templates, and contextually relevant references or guidelines (e.g., through infobuttons) [2,3].

The earliest forms of clinical decision support can be traced to the 1960's and began as tools that aided in clinical diagnosis [4]. In one of the earliest iterations, patients presenting to Kaiser Permanente were asked Yes/No questions about their symptoms and a computer system generated a differential diagnosis [5]. The focus evolved such that management plans for various conditions were addressed as well [6,7]. Initially, systems such as these were designed independently of medical records or ordering systems, but later forms of CDS were designed to operate as part of or in conjunction with electronic health record (EHR) systems [4,8].

Rudimentary CDS within the computerized record began with clinical reminders, about important events or due dates for events, in the 1970s and evolved to include alerts about detected situations such as abnormal results or harmful interactions [9,10]. These integrated systems were advantageous because they allowed for more proactive methods of support. Incorporating active alerts, for example warning ordering clinicians about drug-drug interactions or inappropriate dosing of medications, reduced adverse patient events [10]. Early versions of computerized provider order entry (CPOE) developed in the 1990's included relatively little in the way of clinical decision support, as compared to today.

Clinical Decision Support and Beyond
https://doi.org/10.1016/B978-0-323-91200-6.00026-7

Over time, these systems have evolved to utilize a more proactive approach by integrating into routine clinician workflow, resulting in greater utilization of CDS tools. Today, decision support is embedded throughout the electronic record and can take a variety of forms, as outlined in Section 22.1.

With respect to CPOE, a key effort in the early versions aimed to anticipate clinician needs. For example, when a physician writes an order for digoxin, the order entry system could display any recent assessment of renal function, serum potassium, and digoxin levels. In addition, the system could suggest starting doses for medications. Regarding alerts within the order-entry system of CPOE, only a very small number were delivered initially. Specifically, systems could check for allergies to a few of the most important drug classes—such as penicillins and sulfonamides—and for a few of the most important drug-drug interactions—such as aspirin and warfarin. However, since then, the sophistication of alerts in CPOE has grown substantially. Some of the first additions were comprehensive rule sets for checking drug allergies and drug–drug interactions. The overall approach has been to layer on additional decision support gradually, to maximize the likelihood of clinician acceptance.

Another early feature was the "order set." Time-motion studies showed that submitting groups of orders was five times faster than writing them individually, which prompted developers to ensure that additional order sets were included as CPOE systems progressed [11]. Order sets not only increased the efficiency of the order entry process, but also could serve as a decision support tool that would guide users to perform certain actions together. One especially effective example is "corollary orders" [12], which come into play when you take one action that suggests you will likely want to take another—such as performing a pregnancy test when prescribing certain medications that are known to be teratogenic. There are now many such reminders integrated into CDS and their scope has expanded to many domains of healthcare quality—for instance, reminders may present the provider with the opportunity to prescribe heparin in the context of bed rest, which has increased compliance with recommended thrombotic prophylaxis [13]. A famous adage in CDS design has been to "make it easy for the users to do the right thing" and order sets offer the promise to achieve this goal.

Clinical information systems were created with the idea that a capable clinical information system with strong clinical decision support would be a linchpin of the next wave of quality and safety improvement in healthcare. As clinical information systems have evolved, standards have been developed to encode, store, and share CDS knowledge across healthcare systems. However, given that numerous standards exist, issues remain around harmonization of standards and adoption across systems [4]. Chapter 18 provides an in-depth discussion of the role of standards in the sharing of CDS knowledge and capabilities.

Over the years, many studies have been performed to assess the impact of clinical decision support on a wide range of outcomes, including safety, quality, costs, satisfaction, and provider time. In this chapter, the results of these studies are reviewed. Finally, the generalizable lessons learned across the studies are discussed, as are the next set of frontiers.

22.2 Evaluation methods and reporting of CDS

Leveraging the EHR and EHR-enabled approaches (e.g., machine learning), robust evaluations of CDS interventions can be conducted [14].

22.2.1 Study types

There are several advantages to using the EHR for studying processes and outcomes. First, widespread adoption of EHRs has allowed for data collection on large populations of patients that can be used in observational studies [15]. Second, though randomized controlled trials (RCT) are considered the gold standard for comparing interventions, these can be expensive and time-consuming to conduct. With the EHR, it is possible to conduct trials quickly and in real-world conditions. Pletcher et al. describe how RCTs can be embedded within EHRs so interventions can be tested quickly. Issues remain with respect to optimal cohort identification in real-time, intervention delivery, and outcome identification (both intended and unintended consequences) [16]. However, when evaluating any EHR-based implementation, a critical component to consider is the local culture or environment, including clinician attitudes, training, and support with respect to interventions [17].

22.2.2 Implementing and reporting of CDS studies

Kawamoto and McDonald describe several recommendations for implementing and reporting on various types of CDS studies. When designing a study, they recommend that developers ensure that the necessary EHR data are available to inform the CDS, carefully consider the optimal users and workflows, minimize burden on users, and select decision rules that are consistent with local culture and practice. Further, they advocate for choosing meaningful evaluation measures (including financial implications as appropriate), and reporting the study within the context of the local environment [1]. They, and others, also promote standardization and transparency of interventions, allowing them to be more easily translated across sites [14,18].

22.2.3 Emerging areas

As machine learning (ML) and artificial intelligence (AI) evolve, there is a desire to utilize these techniques in healthcare, though adoption has been slow. There are several limiting factors for implementing ML- or AI-based interventions. First, because these types of interventions are created in a local environment, they require external validation to ensure model performance remains stable across datasets and populations. Second, quantifying model performance, including uncertainty of a model, and how interventions were implemented is critical, but not yet standardized [19]. AI and ML-based interventions have great potential to improve healthcare and provide individualized recommendations for patients and clinicians, however, they must be implemented thoughtfully for optimal benefit. A detailed description of ML development and validation methods is provided in Chapter 7.

22.3 Clinical decision support effectiveness

Since their inception, CDS interventions have demonstrated improved compliance with clinical guidelines by providers and have sometimes been associated with a reduction in adverse event rates [20]. When CDS recommendations are integrated into the clinical workflow and provided at the time of clinical decision making, up to 90% have been shown to significantly improve clinical practice, most commonly improving preventive care practice, improving medication prescribing, and reducing adverse events [21,22]. Notably, although these systems are effective, the most common improvements are generally related to process of care measures rather than clinical outcomes [23]. In addition, the marginal level of improvement is often modest—approximately 5–15%—and substantial opportunities to improve generally remain.

22.3.1 Medication-related decision support

22.3.1.1 Medication appropriateness

One of the simplest ways that CDS can improve medication prescribing is to suggest patient-appropriate dosing for ordered medications. These types of CDS can reduce variability in how medications are dosed and reduce errors in prescribing, in some studies reducing errors up to 42–75% [24].

In theory, clinical decision support tools could also help improve care by incorporating indication-based prescribing. Using the EHR and CDS tools to provide only the medications that are appropriate for a given indication during the ordering or pharmacy verification process is a way to reduce medication overuse and can help reduce medication errors, particularly for medications with similar names. This type of tool can also help ordering providers identify the optimal medication for a given indication when they are presented with only the appropriate medications to choose from [25,26].

Similarly, antibiotic stewardship program interventions use a variety of mechanisms to guide ordering prescribers and improve adherence to guideline concordant therapies [27,28]. One study by Goss et al. implemented an EHR-based, non-interruptive intervention that allowed ordering providers to access antibiotic recommendations and guidelines when prescribing based on the diagnosis entered. The intervention resulted in an increase in medication appropriateness to more than 70%, with the highest improvements seen for pneumonia and pyelonephritis [28].

Another way to implement indication-based prescribing is to prompt clinicians to provide the indication or corresponding problem when ordering a medication. Falck et al. implemented CDS that requested ordering providers to select a problem based on the medication being ordered; if the problem was not already in the problem list, providers could add it to the patient's problem list right from the order screen. Accurate problems were chosen almost 60% of the time and less than 5% of inaccurate problems were added to the problem list via the order screen [29]. Galanter et al. implemented a similar intervention that prompted ordering clinicians to select an indication for medications being ordered if there was not already an appropriate problem on the problem list. They found indication-based prescribing to have varying degrees of effectiveness in reducing medication overuse, but notably, they also found that implementation of indication-based prescribing resulted in a decrease in wrong-patient medication errors [30].

Two main challenges in implementing indication-based prescribing are that reliable ways to capture and encode indications are lacking, and that medication-indication knowledge bases are limited, inconsistent, and have data granularity issues [31].

22.3.1.2 *Adverse drug events*

A key goal of clinical decision support is to reduce the frequency of adverse drug events. The *Adverse Drug Event Prevention Study* at Brigham and Women's Hospital (BWH) was one of the first studies to describe the epidemiology of preventable adverse drug events (ADEs) and potential ADEs in hospitalized patients [32]. A key finding was that 60% of serious medication errors—i.e., those that harmed someone or had the potential to do so—occurred at the prescription or transcription stages and, as such, were potentially preventable with CPOE and CDS. Furthermore, many of the serious prescribing errors appeared to be related to lack of access to clinical knowledge or data [33]. Intriguingly, for many ADEs, the key piece of data needed to prevent the ADE was already in the EHR, but not accessed or properly considered. With these considerations, initial decision support efforts focused on preventing ADEs.

Phase 2 of the *ADE Prevention Study* tested two interventions: CPOE and a team-based intervention targeted at the administration and dispensing stages of the process. The team-based intervention had no effect. CPOE, however, reduced the serious medication error rate by 55%, despite minimal decision support consisting mainly of default dosages and limited drug-allergy and drug-drug interaction checking [34]. Subsequently, much more decision support has been added across CPOE systems, so this clearly represents a conservative estimate of its effect. As features were added over time, including comprehensive drug-allergy modules and comprehensive drug-drug interaction checking, the cumulative impact of these changes was studied in another report that examined all medication errors as the primary outcome [35]. In an interrupted time-series analysis with samples at one-year intervals, the overall medication error rate fell 83% simply with computerization of prescribing.

Two of the most important types of decision support that have developed include special medication dosing for patients with renal impairment [36] and for geriatric patients [37]. One example of a specialized renal dosing application, *Nephros* at BWH, utilized existing data for the patient's age, gender, and most recent serum creatinine, coupled with weight information which was entered by the clinician. For drugs that need to be renally dosed, the computer used the Cockcroft-Gault equation to estimate a patient's creatinine clearance based on latest available lab results, and then suggested a drug dose appropriate for the patient's level of renal function (see Fig. 22.1). In a controlled trial, the appropriate dose was selected 67% of the time in the intervention group compared to 54% in the control group. In addition, dosing frequency was more often appropriate—59% in the intervention group, compared to 35% in the control group. Moreover, length of stay was shorter by one half-day for patients in the intervention group [36]. Over time and with changes in the electronic health record, the utility of CDS for renal insufficiency has declined [38]. Recent data show that clinicians frequently override CDS that is aimed at renal-based dose adjustment, although less than 3% of these overrides may be appropriate [39].

Geriatric patients were another common target for CDS; these patients often receive initial doses that are too high and CDS was designed to improve medication dosing in this group of patients. In a 2005 study, investigators helped an expert panel develop initial dosing recommendations for psychoactive medications, for which the application then suggested an appropriate default starting dosage (see Fig. 22.2). Key results were that patients more often got the recommended dosage (29% vs. 19%), had a lower rate of tenfold overdose (2.8% vs. 5%), and were less likely to fall (2.8 vs. 6.4 falls per 1000 patient-days) [37]. There was no difference in the frequency of mental status change or in length of stay. A systematic review published in 2019

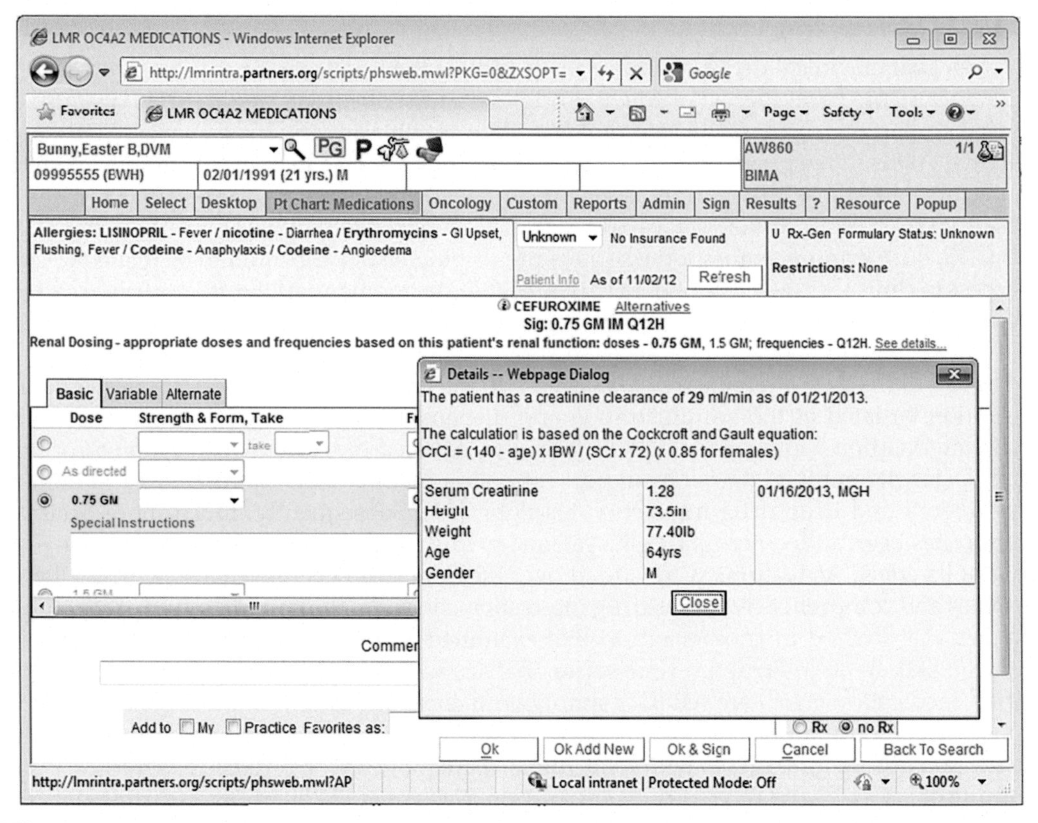

FIG. 22.1 Renal dose adjustment example. *Notes*: This screen shows a renal dose adjustment for the medication cefuroxime. The dose has been reduced given the patient's impaired renal function. The details dialog box shows the calculation of the patient's creatinine clearance.

examined CDS in geriatric prescribing and demonstrated that CDS tools routinely facilitate a decrease in the number of potentially inappropriate prescriptions and an increase in the appropriate discontinuation of potentially inappropriate medications [40].

In addition to larger interventions, several drug-specific guidelines have also been implemented with success in different systems. For example, BWH researchers implemented the Centers for Disease Control and Prevention's guidelines for prescribing vancomycin in the CPOE system in response to cases of vancomycin overuse [41]. After implementation of the guideline, the number of vancomycin-days per provider decreased. Notably, this effect was seen more in terms of shortening of the courses of vancomycin than decreasing the number of times vancomycin was started. Guidelines have been implemented for many other medications, including human growth hormone and several high-cost or high-potency antibiotics.

An array of medication-related decision support functions have also been implemented in the outpatient setting, starting with relatively simple suggestions such as drug-allergy and drug-drug interaction checking, and then adding a variety of more sophisticated decision

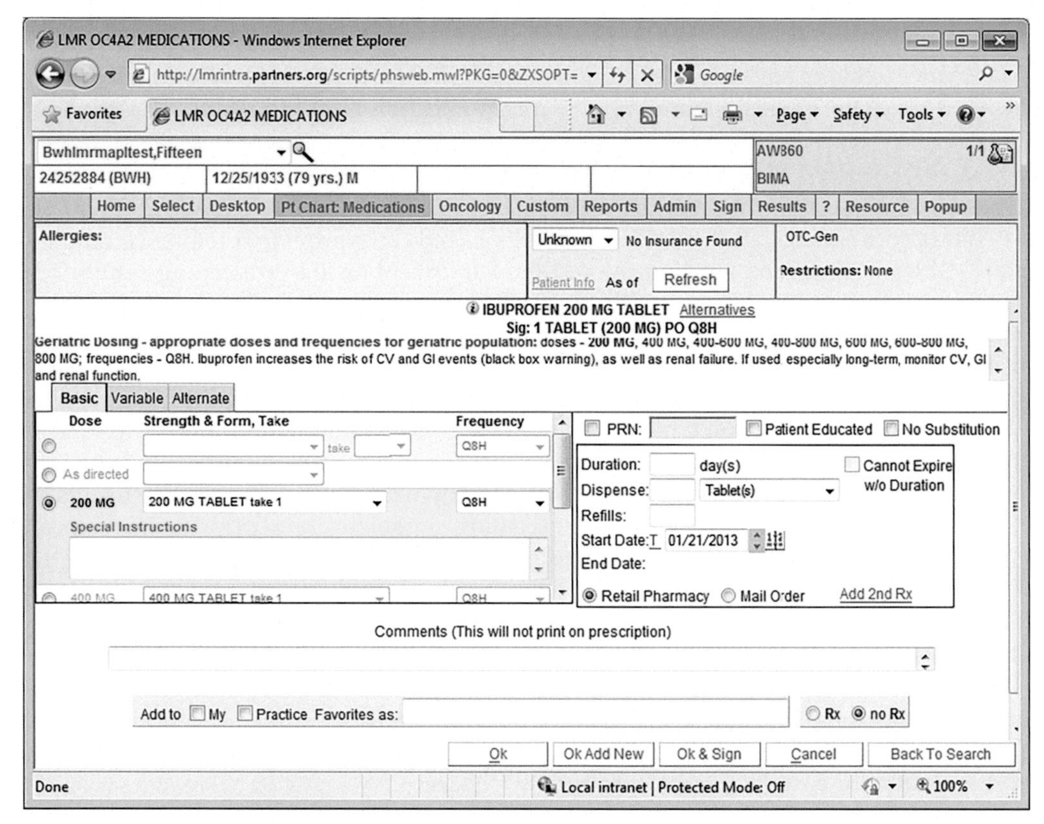

FIG. 22.2 Geriatric dose adjustment example. *Notes*: The geriatric dosing alert (shown above the ordering information) tells the provider that higher doses of ibuprofen may trigger cardiovascular and gastrointestinal events as well as renal failure in geriatric patients, and suggests increased monitoring to prevent these adverse events.

support, including drug-pregnancy checks, drug-age checking, and drug-disease checking. In one study of tiered drug alerts [42], a total of 18,115 drug alerts were presented to physicians over a six-month period. In this study, 71% were deemed less important, and presented in a non-interruptive fashion, while 29% were interruptive. Of the interruptive alerts, 67% were accepted, which compares very favorably to some other reports, which have found acceptance levels of only 10 to 30% [43,44]. Some of the keys to this success were the fact that the designers were highly selective in determining which alerts to display, iterating to identify alerts with high override rates, and using the interruptive approach only for truly critical alerts. Knowledge bases derived largely from studies conducted at the BWH have recently been made available [45,46].

Another evaluation focused on drug-allergy alerts found that 80% of the drug-allergy alerts were overridden [47]. However, only 10% of alerts were triggered by an exact match between the drug prescribed and the allergy listed. On close evaluation, all the overrides appeared clinically justifiable. To address this, a group of recommendations was developed to fine-tune the specificity of warnings, thereby increasing the utility of the allergy alerting system.

22.3.2 Laboratory interventions

Many interventions have also been implemented in an attempt to improve the appropriateness of use of the clinical laboratory and communication around results.

22.3.2.1 *Notification of abnormal results*

One of the simplest ways to alert providers to abnormal findings is to highlight them during result reporting (e.g., show abnormal results in red) and provide a reference range for normal values to clinicians viewing the results. For critical or life-threatening results, real-time notification is needed to ensure the successful communication of these results [48].

Another key issue is follow-up of abnormal results, which is often suboptimal [49–51]. Several studies have suggested that about a third of abnormal test results, even for tests such as Pap smears and mammograms, do not receive appropriate follow-up. With the transition from paper records to electronic health records, the documentation and follow-up rates for results has increased, with patient notifications increasing from 66% to 80% [52]. In an effort to help address this issue, Dr. Eric Poon led the development of a tool to organize results and make it easier for clinicians to handle, by aggregating, organizing, and prioritizing them, and then making it easy for providers to generate letters [53].

22.3.2.2 *Display of charges*

The concept behind this group of interventions is that if providers are exposed to the estimated cost (or charge) of the laboratory tests when they are ordering them, they would be more "cost conscious" and less likely to order tests that are low value. In a randomized controlled trial evaluating the impact of charge display, one study presented ordering providers with the charge for each test and the cumulative charges for a session were shown using a "cash register" function [54]. Physicians liked seeing the charges, but no statistically significant impact on ordering practices was observed. The intervention group, however, showed a beneficial trend of 4.5% fewer tests performed, with this figure being greater for tests that are ordered infrequently, which tend to be more expensive. The annual estimated cost reduction benefit in the intervention group was $1.7 million. Since then, several other studies have explored ways in which displaying the cost of laboratory tests, radiology studies, or medications in the EHR may impact their overuse. Results of a 2013 RCT and a large quasi-experimental study published in 2018, to name a few, have been consistent: they also showed a modest decrease in utilization of diagnostics and therapeutics [55,56].

22.3.2.3 *Redundant tests*

A common way to study redundant tests is to focus on tests that are expected to be routinely repeated—such as measuring the hemoglobin A1c (HbA1c) level in diabetic patients or the lipid panel in patients with hyperlipidemia. One study demonstrating that 28% of 12 target tests were ordered earlier than a test-specific predefined minimum interval [57], thus showing substantial unnecessary utilization, which amounted to an estimated $930,000 per year in charges. Alerts for these tests were implemented and studied in a randomized controlled trial. Even though the tests considered to be redundant by the interval criteria were performed only 24% of the time in the intervention group versus 51% of the time in the control group, the savings realized were a mere $35,000 versus a prior projection of $436,000. The reasons for the difference were multifactorial. First, only 44% of the redundant tests performed were ordered through use of the computer (many were being sent to the laboratory without an order). Second, 31% of the reminders were overridden. Third, half of the electronic orders were placed through order sets and the software development team had elected to exempt order sets from the intervention without realizing the full impact this would have. There are at least two take-away messages from this experience: first, it is important to design "closed-loop" systems to ensure inclusion of all points of order entry, particularly when many orders are not entered electronically. Second, it is important to include in the screening those orders that are in order sets, which may be especially prone to redundancy. More recent studies show that redundant testing continues to be an issue; for instance, while the American Diabetes Association (ADA) recommends testing HbA1c level once every 90 days in patients with controlled diabetes and twice as frequently in those with uncontrolled diabetes, analysis of data across 119,000 patients over 15 years at one hospital system shows that in practice, clinicians often re-test HbA1c much more frequently [58]. Subsequent studies, including a study by Bellodi et al. have demonstrated that implementing CDS that provides information on appropriate timing of repeat tests decreased the number of tests ordered by 16.4% and decreased costs by 16.5% [59].

22.3.2.4 *Therapeutic drug monitoring*

In another evaluation, antiepileptic drug level testing was targeted. Prior work demonstrated that only 26–29% (depending on the drug) of antiepileptic drug level tests performed in the hospital appeared to be appropriate [60]. Subsequently, an intervention in which guidelines for drug level testing were displayed at the time of ordering resulted in a 19.5% decrease in the use of these levels, despite a 19.3% increase in overall test volume during the study period [61]; there was also a major decrease in the proportion of tests that appeared to be inappropriate.

In the outpatient setting, there have been several studies of reminders to monitor for specific medications such as renal monitoring in long-term users of non-steroidal anti-inflammatory drugs. One study by Lau et al. demonstrated that CDS implementation was associated with an increase in laboratory monitoring compliance for safety in patients with diabetes, leading to an increase in monitoring of hepatic function and renal function [62]. In the ELMO study, a cluster randomized, open-label trial, providers were presented with evidence-based order sets for their patients with chronic conditions. Implementation of this CDS increased the proportion of appropriate tests ordered [63].

22.3.2.5 *Tests pending at discharge*

In a 2005 study, researchers at BWH found that 41% of patients leaving the hospital had one or more test results pending at discharge and, of these results, 9.4% were potentially actionable; however, physicians were unaware of these actionable results 61.6% of the time [64]. In order to reduce this problem, BWH implemented a tracking system for pending tests, which alerted both ordering providers and primary care physicians of results of tests that came in after discharge, and the system has been viewed favorably by providers [65]. A systematic review examining tests pending at discharge revealed that the implementation of automated notifications improved awareness of these pending tests [66].

22.3.3 Radiology interventions

22.3.3.1 Appropriateness of ordered studies

In the inpatient setting, nearly all imaging studies have been ordered electronically since the initial implementation of CPOE. When providers order an imaging study, they are asked to enter coded historical findings and the clinical question they would like answered with the study. This information is then compared to a knowledge base, and providers are given feedback about the appropriateness of studies they plan to order. If the system determines that a study is unlikely to provide useful information, it attempts to suggest a more informative alternative. These alternatives have proven key to acceptance: in a trial of decision support for abdominal radiographs [67], providers were unlikely to heed suggestions to cancel inappropriate orders, even if the examinations were virtually certain to provide no useful information. However, when alternative views or studies were offered, suggestions were much more likely to be accepted, though still only about half the time [67].

Other attempts to improve CDS efficacy for appropriateness-based use cases have incorporated local best practices to align with clinician workflow. For example, one study by Raja et al. examined the implementation of a CDS alert that incorporates patient characteristics of a patient presenting with renal colic (e.g., age, or history of uncomplicated nephrolithiasis) to suggest alternative studies for those who are less likely to have findings on imaging. After implementation of this type of alert, CT orders for patients determined to be lower risk decreased from 24% to 15% [68]. Future methods could also include incorporating elements found in clinical documentation to inform CDS. For example, Rousseau et al. found that, at the time of imaging order entry, unstructured clinical notes contained high-impact attributes (both positive and exclusionary) that could inform CDS alerts and guide imaging decision making [69].

Early studies suggested that displaying the charges for radiographs had no impact on the overall level of utilization, however, more recent studies have found that displaying the cost of imaging during the order process can result in a reduction in image orders [54,56].

22.3.3.2 Critical result tracking

A more recent focus in the area of radiology has been communication of abnormal test results in the inpatient setting. Most hospitals have had longstanding processes in place to alert providers to the most critical abnormal results (generally with a phone call or page). More recently, BWH has developed and implemented a system called ANCR (Alert Notification of Critical Radiology Results) [70] which provides and tracks alerts and manages escalation for three levels of abnormality color-coded as Red, Orange or Yellow. The Red category includes new or unexpected findings that are potentially immediately life-threatening, such as tension pneumothorax, ischemic bowel, or intracerebral hemorrhage), which require immediate interruptive notification of the ordering physician, covering physician, or other care team member who can initiate the appropriate clinical action for the patient. The Orange category includes new or unexpected findings that could result in mortality or significant morbidity if not appropriately treated urgently (within 2–3 days), such as an intra-abdominal abscess or impending pathological hip fracture, where notification must be within 3 h of discovery of findings. The Yellow category includes new or unexpected findings that could result in mortality or significant morbidity if not appropriately treated, but are not immediately life-threatening or urgent, such as a nodule on a chest X-ray or a solid renal mass on an ultrasound examination. Notification must be within 3 days of discovery of findings.

The ANCR system ensures that all alerts are seen by the responsible provider, and escalates notifications as needed according to a severity-specific schedule, resulting in a significant increase in communication for critical alerts [71]. Analysis of data collected by ANCR has signified the role of various information sources in elucidating diagnostic process errors [72]. This led to the development of a new system called RADAR (short for Result Alert and Development of Automated Resolution) which was highly adopted and led to the identification of several incidental radiology findings and more timely execution of closed-loop communication and follow-up [73,74].

22.3.3.3 *Radiology decision support*

One way to ensure CDS functions for radiology is to use electronic ordering and mapping of all key historical factors and indications for radiographs, so that providers request studies using controlled vocabularies. Indications for procedure requests can be filtered for appropriateness and checking for redundancy can also be done. In one study, an evaluation was done of patients who underwent abdominal imaging for abnormal liver function tests, and found that of all modalities evaluated, CT scan had the highest yield, and that unexpected new findings that appeared to be clinically important were found in a higher proportion of patients than anticipated [75].

22.3.4 Patient portals

Associated with many electronic health records are patient portals that allow patients to view the contents of their medical record and communicate with their providers. With increasing adoption of patient portals, there is an increasing interest in designing patient-facing CDS tools that are delivered through the portals as well.

The Prepare for Care study was a large, prospective, randomized trial of a package of personal health record interventions for patients designed to improve the quality of care, which included modules from screening, medication history documentation and diabetes management [76]. Patients in the intervention arm were more likely to have their diabetes regimens altered [77], had more accurate medication histories [78], and were more likely to receive some, but not all, indicated preventive services [79]. Otte-Trojel et al. reviewed several studies examining patient portals and found that, in general, use of patient portals can result in improvement in clinical outcomes and patient experience [80]. Disparities in access and use remain a key challenge for patient portals [81–83].

Although patient portals started as—and continue to be mainly used as—a tool used in the outpatient setting, recent studies have demonstrated notable impacts in using them in the inpatient settings as well; interestingly, though the data are mixed, inpatient portals generally have been found to be less susceptible to disparities [83–85].

22.3.5 Problem lists

A common issue across electronic health records is maintenance of clinical problem lists as they are frequently inaccurate and incomplete, particularly if the problem list is manually

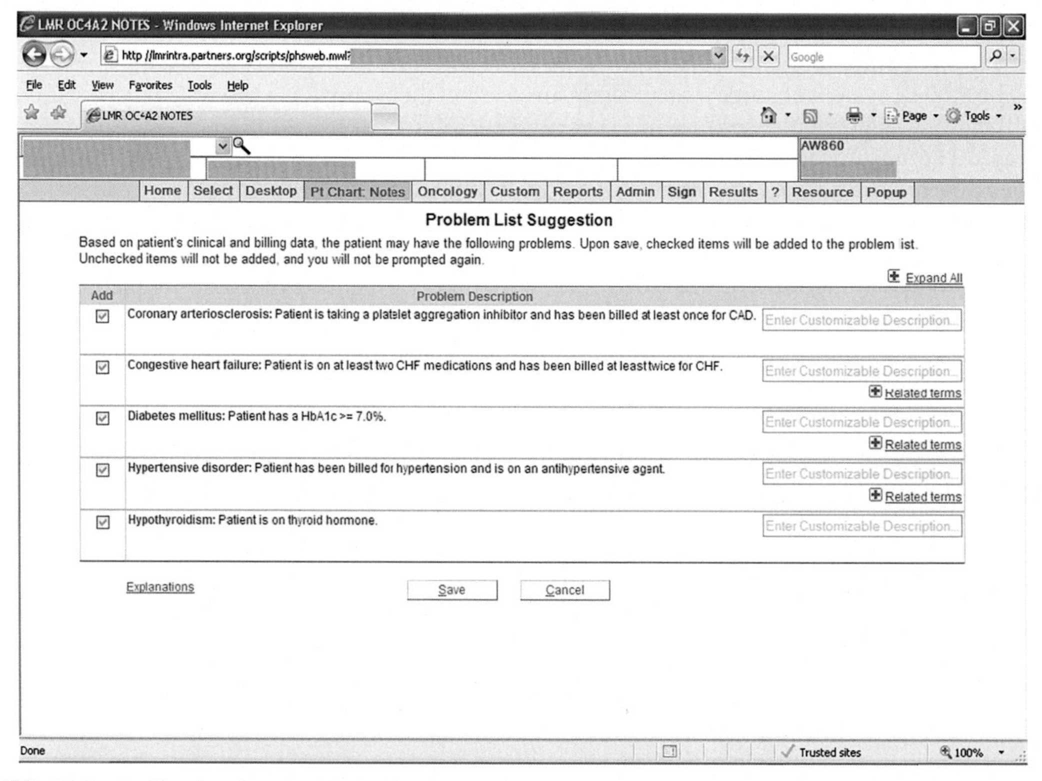

FIG. 22.3 Problem list alert example. *Notes*: This problem list alert tells the provider that his patient is likely to have several diagnoses which are not currently present on the problem list. Explanations are provided for each alert, and if the provider accepts them, the problems will be added to the patient's problem list.

maintained [86]. This is especially problematic because a wide range of CDS tools rely on the problem list [10]. In one multi-site study, completeness of the problem list ranged from 60.2% to 99.4% across 10 sites [86] and another study revealed a range of completeness of 72.9–93.5% depending on condition studied [87]. In one study from a large academic medical center, only 62% of patients with diabetes had diabetes on their problem list [88]. To address this, researchers developed a series of inference rules to identify patient problems and then showed an alert to providers when their patients were very likely to have a problem which was not found on the problem list (see Fig. 22.3) [89]. In a randomized trial, the intervention led to a three-fold increase in documentation of problems.

22.3.6 Cost-effectiveness of CDS

In addition to the many individual studies described earlier, the cost-effectiveness of computerized physician order entry and real-time clinical decision support have been assessed [90,91]. One study at BWH from 1992 to 2003, revealed that BWH spent $11.8 million to develop, implement and maintain its CPOE system. Over the same period, though, the system saved $28.5 million, yielding a significant overall return on investment. The most important driver of this cost savings was the success of renal dosing guidance. Second was an improvement in nursing efficiency and time utilization, followed by specific drug guidance and ADE prevention.

In addition, the cost-effectiveness of implementing the outpatient electronic health record has been evaluated [92]. Wang estimated the net financial benefit over costs for a primary care provider over a five-year period at $86,400 per provider, mostly from savings related to lower drug expenditures, more appropriate utilization of radiology tests, better capture of charges, and a decrease in the rate of billing errors. These data suggest that the greatest savings from CPOE adoption come in the form of recommendations for lower cost drugs as well as alerts which identify potentially inappropriate radiology and laboratory test orders, especially in settings where more appropriate usage affects physician reimbursement.

Other studies have demonstrated similar findings, showing that healthcare expenditures can be reduced with CDS, however, not always to a degree that offsets the costs of implementation [91]. Overall, several systematic reviews have concluded that cost results are mixed and most CDS trials do not incorporate cost outcomes [93–96].

22.4 Provider responses to clinical decision support

22.4.1 Impact of alerts on providers

Several studies have examined the alerts providers encounter and the impact these alerts have on their behavior. One study assessed the number of alerts and time associated with responding to alerts for primary care physicians. They found that ordering providers received 56 alerts per day which translated to 49 min of processing time [97]. Other studies have found that providers encounter 0.79 alerts/visit [21]. Due to the high number of alerts and varied effectiveness of alert implementation, clinicians have been found to ignore up to 96% of alerts [1,98].

The high rate of alert override is a sign of CDS fatigue which occurs when clinicians are tired of frequent alerts and may be desensitized to alerts and clinical reminders. This is most likely to occur when there are high rates of false-positive alerts, when alerts are presented at the wrong time during clinical workflow, or when alerts are not relevant to the ordering provider [21]. Studies examining these alert overrides have found that, in many instances, the override is clinically justifiable (e.g., alerts were irrelevant) [98]. This suggests that for CDS to be most effective, alerts must be actively evaluated and subsequently updated when no longer relevant. One example where this was done successfully is a study by Kho et al. who examined the impact of inpatient isolation guidance for multidrug-resistant organisms when a full-time expert was maintaining the alert. In this study, providers were compliant with the CDS alerts 95% of the time, an increase from 33% prior to implementation [99].

22.4.2 Satisfaction with CDS

Satisfaction with various CDS implementations has been formally assessed. In one early study in 1996 [100], physicians and nurses were quite satisfied with CPOE overall, including the embedded decision support, although internists were more satisfied than surgeons. Satisfaction was highly correlated with the user's perception of the CPOE system's impact on productivity, ease of use, and speed, and was less strongly associated with features directed at improving the quality of care. This suggests that decision support must be fast to be tolerated and confirms that users may not perceive the need to improve quality even if it is present. Encouragingly, in a separate study, a majority of providers studied believed that EHRs and CDS would increase quality of care, access, and communication [101]. Other studies, have identified factors associated with provider satisfaction with CDS, including that the information presented be reliable, relevant, and actionable [102,103].

22.4.3 Optimal alert presentation

Ideally, clinical decision support alerts should be integrated into the ordering provider's workflow, including showing an alert when opening a patient's record or ordering a medication. For example, when a provider orders lithium, the recommended medication monitoring labs could be presented to the provider with a suggested action (e.g., order now, order for 6 months from now) [1]. Further, use of a CDS system that automatically alerts the user rather than requiring a user to activate it increases the performance of the alert [104].

Another aspect of CDS is to consider the optimal recipient of the alert. One review examining factors associated with successful CDS implementation found that targeting different healthcare provider roles improved compliance with CDS alerts [105]. For example, immunization reminders that are presented to nurses and physicians or implemented as standing immunization orders for eligible patients increased the rates of immunization when compared to reminders presented to physicians alone [106]. Identifying alternative recipients for CDS alerts, when indicated and able to be done safely, reduces the overall burden of alerts on ordering providers and can mitigate alert fatigue.

22.5 Clinical decision support measures and utilization

The impact of CDS interventions is commonly examined through an evaluation of utilization by providers and their responses to the intervention. For example, how often the alert is accepted rather than ignored is a common way to evaluate how useful the alert is to clinicians. The reasons for alert override are also important when considering utility of an alert [107]. When an alert is overridden the majority of the time by providers, with or without a valid reason, a reevaluation of the utility of the alert is required [98,108].

The optimal types of alerts are useful to the ordering clinician and presented at the right time, reducing unnecessary alerts to mitigate alert fatigue. Incorporating human factors principles into CDS design can help with successful alert development and deployment [109]. First, total number of alerts should be minimized, and any active alerts should prompt users for corrective action (if needed) in a seamless manner, through as few additional steps as possible. False alarms or outdated alerts should be minimized. Alerts should have a prioritization level and should be placed within the ordering process or located on the computer screen based on the order of importance [21,109].

22.5.1 Reminders

A key type of CDS is a reminder system, which alerts clinicians to guideline-based screening, and preventive and monitoring interventions. However, the optimal way to present reminders and how frequently is unknown. Reminders have been successful in increasing screening and/or treatment for preventive care, including reminders for influenza vaccination and cancer screening [110,111]. In a randomized trial of preventive care reminders by Sequist, there were significant improvements in recommended care for diabetes and coronary artery disease when physicians were shown reminders; however, some reminders were more effective than others [112]. Other studies have demonstrated success by repeating reminders until an ordering provider responds [99].

22.5.2 Alert fatigue

With the implementation of health information technology and CDS, there have been many unanticipated outcomes. A study of expert panelists identified several unintended consequences from implementation of CDS including difficulty maintaining currency of CDS content, incorrect content informing CDS, and alteration of workflows (e.g., eliminating individuals as gatekeepers, increasing work for ordering providers) [113,114].

One particularly notable unintended consequence of CDS has been alert fatigue, or the overwhelming presence of alerts such that they become ineffective. Some of the earliest studies to describe alert fatigue are from 2007 [113]. However, at the same time, studies emerged describing potential solutions for alert fatigue and ways to improve the efficacy of CDS. Potential solutions include implementing minimal and only necessary alerts, applying human factors when designing alerts, and engaging clinical stakeholders (e.g., physicians) into the design of CDS [115,116]. These remain some of the best ways to address and mitigate alert fatigue.

Recent reviews on contributors to alert fatigue suggest that many problematic alerts and resulting override processes stem from not following the above suggested solutions, including not incorporating stakeholders or human factors when designing alerts [117]. In a review examining the appropriateness of alert overrides, inappropriate overrides were common, particularly within certain categories of alerts (e.g., drug-drug interaction, renal related alerts), however, many alert overrides even within these categories were classified as appropriate. These findings provide several areas for improvement. First, presenting ordering clinicians with a clear rationale for the alert to reduce unnecessary overrides. Second, ensuring that CDS tools are current and updated to avoid alerts that lead to clinically appropriate overrides [118]. Other recent reviews offer potential solutions for how to optimize CDS including engaging a multidisciplinary committee in combination with other approaches to ensure CDS alerts remain relevant and up to date [119].

22.5.3 Malfunctions

An emerging essential metric to follow when using CDS systems is the number of times the CDS malfunctions, or the number of times that the intervention does not function as expected [120]. In a survey of Chief Medical Information Officers, 93% reported experiencing at least one CDS malfunction during their tenure, with two thirds reporting malfunctions each year [121]. Malfunctions can occur from a variety of causes, compiled by Wright et al. The most common cause they identified was an error in the way a CDS rule was built, even if the design of the rule was correct. Other causes included rules that were designed incorrectly, system upgrades, and changes in medication or laboratory codes causing the CDS to stop firing [121]. Ensuring a robust monitoring strategy of CDS is critical to mitigate any malfunctions.

22.6 Limitations of clinical decision support

Over the years, there have been several success stories with clinical decision support, but there have also been probably just as many, if not more, failures. At BWH, efforts to support documentation for diabetes and congestive heart failure failed, because the tools were sufficiently complex that clinicians would not use them. Reminders and guidelines have had little or no impact, especially when clinicians were not fully convinced that the message being delivered was correct. For instance, a suggestion not to use intravenous ketorolac was routinely ignored, because physicians believed it was more effective than the alternatives suggested, despite lack of supporting evidence.

Another study examined provider responses to an alert encouraging the use of beta-blockers in patients with myocardial infarction. They found that the alert was ignored initially, however, over time and with education from cardiology, providers began to respond. These experiences suggest that, even when presented with CDS and accompanying evidence, providers may not comply with alerts until they have integrated the suggested workflow into their clinical practice [1].

Another limitation of CDS is that it relies on information available in the electronic health record. It is important to consider the culture and documentation style within an institution when designing CDS. For example, will all the data needed to trigger an alert be contained in structured and accessible fields? Understanding the frequency and reliability of data needed for alerts is critical for successful CDS implementation.

The true impact of CDS is difficult to measure. Evaluations of many CDS implementations demonstrate a change in process measures, but are unable to demonstrate changes in clinical outcomes [23]. This may be because the clinical outcomes are far removed from the CDS intervention, making it difficult to attribute outcome to the specific CDS intervention.

A few studies have examined unintended consequences from CDS and CPOE [122]. One study by Ash et al. found that implementation of CDS around chest x-ray ordering resulted in discontinuation in the human monitoring of orders, which led to an increase in chest x-ray orders [113]. As discussed previously, outdated information or inappropriate information can cause unintended consequences as well (e.g., alert fatigue).

Electronic health record vendors are also critical to successful implementation of CDS. Alerts are created within the constraints of the existing EHR systems. For example, nearly all vendors allow medication-specific decision support, however, some may not allow patient-specific decision support. In one academic health system, alerts from a prominent EHR system were evaluated. The study examined appropriateness and overrides for renal medication-related clinical decision support, finding that the alerts were frequently inappropriately presented and overridden [38]. Another study examined frequency and appropriateness of alerts, showing wide variation for these metrics across different EHR vendors [123].

22.7 Keys to succeeding with CDS

Building on the original guidance to optimize CDS by Wright et al., Kawamoto et al. describe several principles for pragmatic use of CDS. They advocate for adding CDS only if the recipients want it, integrating CDS in a way that is least disruptive to workflow, and ensuring that benefits observed from the CDS implementation outweigh the costs [21]. A summary of the lessons learned over the years has been published [124] as "ten commandments" for clinical decision support" (see Table 22.1).

TABLE 22.1 Ten commandments for clinical decision support.

1. Speed is everything: A routine goal is sub-second screen flips, since providers will not tolerate much longer than that, and minimizing the number of screens used is also important.
2. Anticipate needs and deliver in real time: One example is showing the potassium lab values when a drug that lowers potassium is prescribed.
3. Fit into the user's workflow: If a suggestion seems to come out of left field or at a time when the user is focused on another issue, it is much less likely to be heeded.
4. Little things can make a big difference: In the prototypical example, the decision regarding how a default is set can have an enormous impact on the frequency that a provider will choose a specific action. Generally, it is good informatics practice to set the default to an action that is most likely to be correct.
5. Physicians resist stopping: Here, the point is that, when you suggest that a physician not take an action, but fail to provide an alternative the initial action is likely to be continued even if it is virtually certain to have little or no yield.
6. Changing direction is fine: The corollary to 5 is that when one does suggest a superior clinical alternative, physicians tend to accept the recommendation.
7. Simple interventions work best: Here, the point is that the level of success has been highest for straightforward guidelines and much less for more complex guidelines, nearly all of which have required substantial adaptation before they could be computerized.
8. Ask for additional information only when you really need it: Implementation of many guidelines or pieces of clinical decision support has required some information, such as the weight for renal dosing, which was not already available. Although clinicians eventually supplied the weight in most instances, even getting this small piece of clinical information routinely required an effort, which seemed completely disproportionate. Getting multiple pieces of data would undoubtedly prove even harder.
9. Monitor impact, get feedback, and respond: For most pieces of clinical decision support implemented, at least some additional changes are required. Failure to make multiple incremental changes can result in lack of benefit, and even promote errors [125].
10. Manage and maintain your knowledge-based systems: This is related to the preceding tenet, but it is useful to routinely track how often each piece of decision support is triggered, and try to ensure that there is an "owner" for each rule, and that each will get periodic follow-up to make sure it still applies.

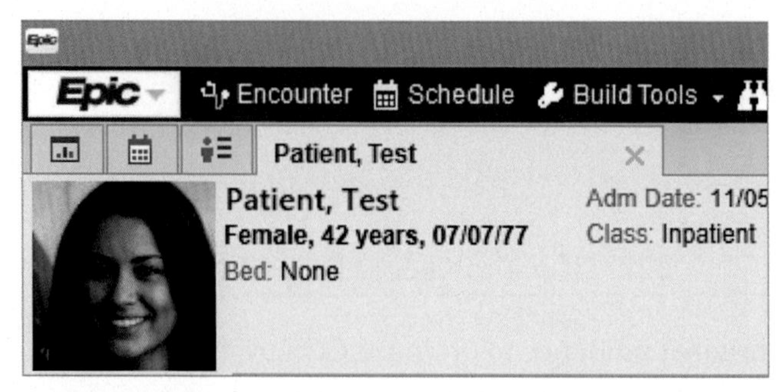

FIG. 22.4 Example of a patient photograph in the epic electronic health record. *Notes*: Demographic data and the face image are not from a real human; the image was generated using Style Generative Adversarial Network (owned by Nvidia). *The screenshot is used with permission from Epic Systems Corporation, © 2021 Epic Systems Corporation.*

In addition, it has recently become clear that leveraging some of the techniques of behavioral economics such as "nudges" can help improve performance. Data from the BEARI trial which was focused on decreasing unnecessary antibiotic usage in acute respiratory infections showed that comparisons to peers around performance in real time was more effective than simply presenting the evidence [126]. Further, utilizing elements from the patient chart in subtle and unobtrusive, yet helpful, ways is critical for the next phase of CDS. Recently studies have emerged demonstrating that use of patient photos in the EHR (see Fig. 22.4), a mechanism to remind ordering providers of the record they are in, can decrease wrong patient orders [127].

Overall, the best performance is generally realized if the "ten commandments" are followed, and what has been learned from behavioral economics is implemented as well. There is still a great deal of opportunity for improvement in the way most of the clinical decision support in place today is being implemented.

References

[1] Kawamoto K, McDonald CJ. Designing, conducting, and reporting clinical decision support studies: recommendations and call to action. Ann Intern Med 2020;172(11 Suppl):S101–9.
[2] Centers for Medicare & Medicaid Services EHR Incentive Program. Clinical decision support: More than just 'alerts' Tipsheet, July 2014. Available from https://www.healthit.gov/sites/default/files/clinicaldecisionsupport_tipsheet.pdf. [Retrieved 25 June 2021].
[3] Cook DA, Teixeira MT, Heale BS, Cimino JJ, Del Fiol G. Context-sensitive decision support (infobuttons) in electronic health records: a systematic review. J Am Med Inform Assoc 2017;24(2):460–8.
[4] Wright A, Sittig DF. A four-phase model of the evolution of clinical decision support architectures. Int J Med Inform 2008;77(10):641–9.
[5] Collen MF, Rubin L, Neyman J, Dantzig GB, Baer RM, Siegelaub AB. Automated multiphasic screening and diagnosis. Am J Public Health Nations Health 1964;54:741–50.
[6] Bleich HL. Computer evaluation of acid-base disorders. J Clin Invest 1969;48(9):1689–96.
[7] Shortliffe EH, Davis R, Axline SG, Buchanan BG, Green CC, Cohen SN. Computer-based consultations in clinical therapeutics: explanation and rule acquisition capabilities of the MYCIN system. Comput Biomed Res 1975;8(4):303–20.
[8] Middleton B, Sittig DF, Wright A. Clinical decision support: a 25 year retrospective and a 25 year vision. Yearbook of Medical Informatics 2016;Suppl. 1:S103–16.

[9] McDonald CJ, Murray R, Jeris D, Bhargava B, Seeger J, Blevins L. A computer-based record and clinical monitoring system for ambulatory care. Am J Public Health 1977;67(3):240–5.
[10] Sutton RT, Pincock D, Baumgart DC, Sadowski DC, Fedorak RN, Kroeker KI. An overview of clinical decision support systems: benefits, risks, and strategies for success. NPJ Digit Med 2020;3:17.
[11] Bates DW, Boyle DL, Teich JM. Impact of computerized physician order entry on physician time. In: Proc Annu Symp Comput Appl Med Care; 1994. p. 996.
[12] Overhage JM, Tierney WM, Zhou XH, McDonald CJ. A randomized trial of "corollary orders" to prevent errors of omission. J Am Med Inform Assoc 1997;4(5):364–75.
[13] Teich JM, Merchia PR, Schmiz JL, Kuperman GJ, Spurr CD, Bates DW. Effects of computerized physician order entry on prescribing practices. Arch Intern Med 2000;160(18):2741–7.
[14] Auerbach A, Bates DW. Introduction: improvement and measurement in the era of electronic health records. Ann Intern Med 2020;172(11 Suppl):S69–72.
[15] Callahan A, Shah NH, Chen JH. Research and reporting considerations for observational studies using electronic health record data. Ann Intern Med 2020;172(11 Suppl):S79–84.
[16] Pletcher MJ, Flaherman V, Najafi N, Patel S, Rushakoff RJ, Hoffman A, Robinson A, Cucina RJ, McCulloch CE, Gonzales R, Auerbach A. Randomized controlled trials of electronic health record interventions: design, conduct, and reporting considerations. Ann Intern Med 2020;172(11 Suppl):S85–91.
[17] Haynes RB, Del Fiol G, Michelson M, Iorio A. Context and approach in reporting evaluations of electronic health record-based implementation projects. Ann Intern Med 2020;172(11 Suppl):S73–8.
[18] Wright A, McCoy AB, Choudhry NK. Recommendations for the conduct and reporting of research involving flexible electronic health record-based interventions. Ann Intern Med 2020;172(11 Suppl):S110–5.
[19] Bates DW, Auerbach A, Schulam P, Wright A, Saria S. Reporting and implementing interventions involving machine learning and artificial intelligence. Ann Intern Med 2020;172(11 Suppl):S137–44.
[20] Kaushal R, Shojania KG, Bates DW. Effects of computerized physician order entry and clinical decision support systems on medication safety: a systematic review. Arch Intern Med 2003;163(12):1409–16.
[21] Kawamanto K, Flynn MC, Kukhareva P, ElHalta D, Hess R, Gregory T, Walls C, Wigren AM, Borbolla D, Bray BE, Parsons MH, Clayson BL, Briley MS, Stipelman CH, Taylor D, King CS, Del Fiol G, Reese TJ, Weir CR, Taft T, Strong MB. A pragmatic guide to establishing clinical decision support governance and addressing decision support fatigue: a case study. AMIA Annu Symp Proc 2018;2018:624–33.
[22] Kawamoto K, Houlihan CA, Balas EA, Lobach DF. Improving clinical practice using clinical decision support systems: a systematic review of trials to identify features critical to success. BMJ 2005;330(7494):765.
[23] Bright TJ, Wong A, Dhurjati R, Bristow E, Bastian L, Coeytaux RR, Samsa G, Hasselblad V, Williams JW, Musty MD, Wing L, Kendrick AS, Sanders GD, Lobach D. Effect of clinical decision-support systems: a systematic review. Ann Intern Med 2012;157(1):29–43.
[24] Kuperman GJ, Bobb A, Payne TH, Avery AJ, Gandhi TK, Burns G, Classen DC, Bates DW. Medication-related clinical decision support in computerized provider order entry systems: a review. J Am Med Inform Assoc 2007;14(1):29–40.
[25] Grissinger M. Is an indication-based prescribing system in our future? P T 2019;44(5):232–66.
[26] Schiff GD, Seoane-Vazquez E, Wright A. Incorporating indications into medication ordering—time to enter the age of reason. N Engl J Med 2016;375(4):306–9.
[27] Davey P, Marwick CA, Scott CL, Charani E, McNeil K, Brown E, Gould IM, Ramsay CR, Michie S. Interventions to improve antibiotic prescribing practices for hospital inpatients. Cochrane Database Syst Rev 2017;2: CD003543.
[28] Goss FR, Bookman K, Barron M, Bickley D, Landgren B, Kroehl M, Williamson K, Zane R, Wiler J. Improved antibiotic prescribing using indication-based clinical decision support in the emergency department. J Am Coll Emerg Phys Open 2020;1(3):214–21.
[29] Falck S, Adimadhyam S, Meltzer DO, Walton SM, Galanter WL. A trial of indication based prescribing of antihypertensive medications during computerized order entry to improve problem list documentation. Int J Med Inform 2013;82(10):996–1003.
[30] Galanter W, Falck S, Burns M, Laragh M, Lambert BL. Indication-based prescribing prevents wrong-patient medication errors in computerized provider order entry (CPOE). J Am Med Inform Assoc 2013;20(3):477–81.
[31] Salmasian H, Tran TH, Chase HS, Friedman C. Medication-indication knowledge bases: a systematic review and critical appraisal. J Am Med Inform Assoc 2015;22(6):1261–70.

[32] Bates DW, Cullen DJ, Laird N, Petersen LA, Small SD, Servi D, Laffel G, Sweitzer BJ, Shea BF, Hallisey R, et al. Incidence of adverse drug events and potential adverse drug events. Implications for prevention. ADE prevention study group. JAMA 1995;274(1):29–34.
[33] Leape LL, Bates DW, Cullen DJ, Cooper J, Demonaco HJ, Gallivan T, Hallisey R, Ives J, Laird N, Laffel G, et al. Systems analysis of adverse drug events. ADE prevention study group. JAMA 1995;274(1):35–43.
[34] Bates DW, Leape LL, Cullen DJ, Laird N, Petersen LA, Teich JM, Burdick E, Hickey M, Kleefield S, Shea B, Vander Vliet M, Seger DL. Effect of computerized physician order entry and a team intervention on prevention of serious medication errors. JAMA 1998;280(15):1311–6.
[35] Bates DW, Teich JM, Lee J, Seger D, Kuperman GJ, Ma'Luf N, Boyle D, Leape L. The impact of computerized physician order entry on medication error prevention. J Am Med Inform Assoc 1999;6(4):313–21.
[36] Chertow GM, Lee J, Kuperman GJ, Burdick E, Horsky J, Seger DL, Lee R, Mekala A, Song J, Komaroff AL, Bates DW. Guided medication dosing for inpatients with renal insufficiency. JAMA 2001;286(22):2839–44.
[37] Peterson JF, Kuperman GJ, Shek C, Patel M, Avorn J, Bates DW. Guided prescription of psychotropic medications for geriatric inpatients. Arch Intern Med 2005;165(7):802–7.
[38] Shah SN, Amato MG, Garlo KG, Seger DL, Bates DW. Renal medication-related clinical decision support (CDS) alerts and overrides in the inpatient setting following implementation of a commercial electronic health record: implications for designing more effective alerts. J Am Med Inform Assoc 2021;28(6):1081–7. https://doi.org/10.1093/jamia/ocaa222.
[39] Nanji KC, Seger DL, Slight SP, Amato MG, Beeler PE, Her QL, Dalleur O, Eguale T, Wong A, Silvers ER, Swerdloff M, Hussain ST, Maniam N, Fiskio JM, Dykes PC, Bates DW. Medication-related clinical decision support alert overrides in inpatients. J Am Med Inform Assoc 2018;25(5):476–81.
[40] Monteiro L, Maricoto T, Solha I, Ribeiro-Vaz I, Martins C, Monteiro-Soares M. Reducing potentially inappropriate prescriptions for older patients using computerized decision support tools: systematic review. J Med Internet Res 2019;21(11), e15385.
[41] Shojania KG, Yokoe D, Platt R, Fiskio J, Ma'luf N, Bates DW. Reducing vancomycin use utilizing a computer guideline: results of a randomized controlled trial. J Am Med Inform Assoc 1998;5(6):554–62.
[42] Shah NR, Seger AC, Seger DL, Fiskio JM, Kuperman GJ, Blumenfeld B, Recklet EG, Bates DW, Gandhi TK. Improving acceptance of computerized prescribing alerts in ambulatory care. J Am Med Inform Assoc 2006;13(1):5–11.
[43] Payne TH, Nichol WP, Hoey P, Savarino J. Characteristics and override rates of order checks in a practitioner order entry system. Proc AMIA Symp 2002;602–6.
[44] Weingart SN, Toth M, Sands DZ, Aronson MD, Davis RB, Phillips RS. Physicians' decisions to override computerized drug alerts in primary care. Arch Intern Med 2003;163(21):2625–31.
[45] Phansalkar S, Desai AA, Bell D, Yoshida E, Doole J, Czochanski M, Middleton B, Bates DW. High-priority drug-drug interactions for use in electronic health records. J Am Med Inform Assoc 2012;19(5):735–43.
[46] Phansalkar S, van der Sijs H, Tucker AD, Desai AA, Bell DS, Teich JM, Middleton B, Bates DW. Drug-drug interactions that should be non-interruptive in order to reduce alert fatigue in electronic health records. J Am Med Inform Assoc 2013;20(3):489–93.
[47] Hsieh TC, Kuperman GJ, Jaggi T, Hojnowski-Diaz P, Fiskio J, Williams DH, Bates DW, Gandhi TK. Characteristics and consequences of drug allergy alert overrides in a computerized physician order entry system. J Am Med Inform Assoc 2004;11(6):482–91.
[48] Lacson R, Prevedello LM, Andriole KP, O'Connor SD, Roy C, Gandhi T, Dalal AK, Sato L, Khorasani R. Four-year impact of an alert notification system on closed-loop communication of critical test results. AJR Am J Roentgenol 2014;203(5):933–8.
[49] Litchfield I, Bentham L, Hill A, McManus RJ, Lilford R, Greenfield S. Routine failures in the process for blood testing and the communication of results to patients in primary care in the UK: a qualitative exploration of patient and provider perspectives. BMJ Qual Saf 2015;24(11):681–90.
[50] Maillet E, Pare G, Currie LM, Raymond L, Ortiz de Guinea A, Trudel MC, Marsan J. Laboratory testing in primary care: a systematic review of health IT impacts. Int J Med Inform 2018;116:52–69.
[51] Poon EG, Haas JS, Louise Puopolo A, Gandhi TK, Burdick E, Bates DW, Brennan TA. Communication factors in the follow-up of abnormal mammograms. J Gen Intern Med 2004;19(4):316–23.
[52] Elder NC, McEwen TR, Flach J, Gallimore J, Pallerla H. The management of test results in primary care: does an electronic medical record make a difference? Fam Med 2010;42(5):327–33.

[53] Poon EG, Wang SJ, Gandhi TK, Bates DW, Kuperman GJ. Design and implementation of a comprehensive outpatient results manager. J Biomed Inform 2003;36(1–2):80–91.

[54] Bates DW, Kuperman GJ, Jha A, Teich JM, Orav EJ, Ma'luf N, Onderdonk A, Pugatch R, Wybenga D, Winkelman J, Brennan TA, Komaroff AL, Tanasijevic MJ. Does the computerized display of charges affect inpatient ancillary test utilization? Arch Intern Med 1997;157(21):2501–8.

[55] Feldman LS, Shihab HM, Thiemann D, Yeh HC, Ardolino M, Mandell S, Brotman DJ. Impact of providing fee data on laboratory test ordering: a controlled clinical trial. JAMA Intern Med 2013;173(10):903–8.

[56] Silvestri MT, Xu X, Long T, Bongiovanni T, Bernstein SL, Chaudhry SI, Silvestri JI, Stolar M, Greene EJ, Dziura JD, Gross CP, Krumholz HM. Impact of cost display on ordering patterns for hospital laboratory and imaging services. J Gen Intern Med 2018;33(8):1268–75.

[57] Bates DW, Boyle DL, Rittenberg E, Kuperman GJ, Ma'Luf N, Menkin V, Winkelman JW, Tanasijevic MJ. What proportion of common diagnostic tests appear redundant? Am J Med 1998;104(4):361–8.

[58] Pivovarov R, Albers DJ, Hripcsak G, Sepulveda JL, Elhadad N. Temporal trends of hemoglobin A1c testing. J Am Med Inform Assoc 2014;21(6):1038–44.

[59] Bellodi E, Vagnoni E, Bonvento B, Lamma E. Economic and organizational impact of a clinical decision support system on laboratory test ordering. BMC Med Inform Decis Mak 2017;17(1):179.

[60] Schoenenberger RA, Tanasijevic MJ, Jha A, Bates DW. Appropriateness of antiepileptic drug level monitoring. JAMA 1995;274(20):1622–6.

[61] Chen P, Tanasijevic MJ, Schoenenberger RA, Fiskio J, Kuperman GJ, Bates DW. A computer-based intervention for improving the appropriateness of antiepileptic drug level monitoring. Am J Clin Pathol 2003;119(3):432–8.

[62] Lau B, Overby CL, Wirtz HS, Devine EB. The association between use of a clinical decision support tool and adherence to monitoring for medication-laboratory guidelines in the ambulatory setting. Appl Clin Inform 2013;4(4):476–98.

[63] Delvaux N, Piessens V, Burghgraeve T, Mamouris P, Vaes B, Stichele RV, Cloetens H, Thomas J, Ramaekers D, Sutter A, Aertgeerts B. Clinical decision support improves the appropriateness of laboratory test ordering in primary care without increasing diagnostic error: the ELMO cluster randomized trial. Implement Sci 2020;15(1):100.

[64] Roy CL, Poon EG, Karson AS, Ladak-Merchant Z, Johnson RE, Maviglia SM, Gandhi TK. Patient safety concerns arising from test results that return after hospital discharge. Ann Intern Med 2005;143(2):121–8.

[65] Dalal AK, Schnipper JL, Poon EG, Williams DH, Rossi-Roh K, Macleay A, Liang CL, Nolido N, Budris J, Bates DW, Roy CL. Design and implementation of an automated email notification system for results of tests pending at discharge. J Am Med Inform Assoc 2012;19(4):523–8.

[66] Whitehead NS, Williams L, Meleth S, Kennedy S, Epner P, Singh H, et al. Interventions to improve follow-up of laboratory test results pending at discharge: a systematic review. J Hosp Med 2018. https://doi.org/10.12788/jhm.2944.

[67] Harpole LH, Khorasani R, Fiskio J, Kuperman GJ, Bates DW. Automated evidence-based critiquing of orders for abdominal radiographs: impact on utilization and appropriateness. J Am Med Inform Assoc 1997;4(6):511–21.

[68] Raja AS, Pourjabbar S, Ip IK, Baugh CW, Sodickson AD, O'Leary M, Khorasani R. Impact of a health information technology-enabled appropriate use criterion on utilization of emergency department CT for renal colic. AJR Am J Roentgenol 2019;212(1):142–5.

[69] Rousseau JF, Ip IK, Raja AS, Schuur JD, Khorasani R. Can emergency department provider notes help to achieve more dynamic clinical decision support? J Am Coll Emerg Phys Open 2020;1(6):1269–77.

[70] Anthony SG, Prevedello LM, Damiano MM, Gandhi TK, Doubilet PM, Seltzer SE, Khorasani R. Impact of a 4-year quality improvement initiative to improve communication of critical imaging test results. Radiology 2011;259(3):802–7.

[71] Lacson R, O'Connor SD, Sahni VA, Roy C, Dalal A, Desai S, Khorasani R. Impact of an electronic alert notification system embedded in radiologists' workflow on closed-loop communication of critical results: a time series analysis. BMJ Qual Saf 2016;25(7):518–24.

[72] Cochon L, Lacson R, Wang A, Kapoor N, Ip IK, Desai S, Kachalia A, Dennerlein J, Benneyan J, Khorasani R. Assessing information sources to elucidate diagnostic process errors in radiologic imaging—a human factors framework. J Am Med Inform Assoc 2018;25(11):1507–15.

[73] Desai S, Kapoor N, Hammer MM, Levie A, Sivashanker K, Lacson R, Khorasani R. RADAR: a closed-loop quality improvement initiative leveraging a safety net model for incidental pulmonary nodule management. Jt Comm J Qual Patient Saf 2021;47(5):275–81.
[74] Hammer MM, Kapoor N, Desai SP, Sivashanker KS, Lacson R, Demers JP, Khorasani R. Adoption of a closed-loop communication tool to establish and execute a collaborative follow-up plan for incidental pulmonary nodules. AJR Am J Roentgenol 2019;1–5.
[75] Rothschild JM, Khorasani R, Silverman SG, Hanson RW, Fiskio JM, Bates DW. Abdominal cross-sectional imaging for inpatients with abnormal liver function test results: yield and usefulness. Arch Intern Med 2001;161 (4):583–8.
[76] Wald JS, Businger A, Gandhi TK, Grant RW, Poon EG, Schnipper JL, Volk LA, Middleton B. Implementing practice-linked pre-visit electronic journals in primary care: patient and physician use and satisfaction. J Am Med Inform Assoc 2010;17(5):502–6.
[77] Grant RW, Wald JS, Schnipper JL, Gandhi TK, Poon EG, Orav EJ, Williams DH, Volk LA, Middleton B. Practice-linked online personal health records for type 2 diabetes mellitus: a randomized controlled trial. Arch Intern Med 2008;168(16):1776–82.
[78] Schnipper JL, Gandhi TK, Wald JS, Grant RW, Poon EG, Volk LA, Businger A, Williams DH, Siteman E, Buckel L, Middleton B. Effects of an online personal health record on medication accuracy and safety: a cluster-randomized trial. J Am Med Inform Assoc 2012;19(5):728–34.
[79] Wright A, Poon EG, Wald J, Feblowitz J, Pang JE, Schnipper JL, Grant RW, Gandhi TK, Volk LA, Bloom A, Williams DH, Gardner K, Epstein M, Nelson L, Businger A, Li Q, Bates DW, Middleton B. Randomized controlled trial of health maintenance reminders provided directly to patients through an electronic PHR. J Gen Intern Med 2012;27(1):85–92.
[80] Otte-Trojel T, de Bont A, Rundall TG, van de Klundert J. How outcomes are achieved through patient portals: a realist review. J Am Med Inform Assoc 2014;21(4):751–7.
[81] Anthony DL, Campos-Castillo C, Lim PS. Who isn't using patient portals and why? Evidence and implications from a national sample of US adults. Health Aff (Millwood) 2018;37(12):1948–54.
[82] Sinha S, Garriga M, Naik N, McSteen BW, Odisho AY, Lin A, Hong JC. Disparities in electronic health record patient portal enrollment among oncology patients. JAMA Oncol 2021;7(6):935–7.
[83] Walker DM, Hefner JL, Fareed N, Huerta TR, McAlearney AS. Exploring the digital divide: age and race disparities in use of an inpatient portal. Telemed J E Health 2020;26(5):603–13.
[84] Grossman LV, Masterson Creber RM, Ancker JS, Ryan B, Polubriaginof F, Qian M, Alarcon I, Restaino S, Bakken S, Hripcsak G, Vawdrey DK. Technology access, technical assistance, and disparities in inpatient portal use. Appl Clin Inform 2019;10(1):40–50.
[85] Masterson Creber RM, Grossman LV, Ryan B, Qian M, Polubriaginof FCG, Restaino S, Bakken S, Hripcsak G, Vawdrey DK. Engaging hospitalized patients with personalized health information: a randomized trial of an inpatient portal. J Am Med Inform Assoc 2019;26(2):115–23.
[86] Wright A, McCoy AB, Hickman TT, Hilaire DS, Borbolla D, Bowes 3rd WA, Dixon WG, Dorr DA, Krall M, Malholtra S, Bates DW, Sittig DF. Problem list completeness in electronic health records: a multi-site study and assessment of success factors. Int J Med Inform 2015;84(10):784–90.
[87] Wang EC, Wright A. Characterizing outpatient problem list completeness and duplications in the electronic health record. J Am Med Inform Assoc 2020;27(8):1190–7.
[88] Wright A, Pang J, Feblowitz JC, Maloney FL, Wilcox AR, Ramelson HZ, Schneider LI, Bates DW. A method and knowledge base for automated inference of patient problems from structured data in an electronic medical record. J Am Med Inform Assoc 2011;18(6):859–67.
[89] Wright A, Pang J, Feblowitz JC, Maloney FL, Wilcox AR, McLoughlin KS, Ramelson H, Schneider L, Bates DW. Improving completeness of electronic problem lists through clinical decision support: a randomized, controlled trial. J Am Med Inform Assoc 2012;19(4):555–61.
[90] Kaushal R, Jha AK, Franz C, Glaser J, Shetty KD, Jaggi T, Middleton B, Kuperman GJ, Khorasani R, Tanasijevic M, Bates DW, Brigham, C. W. G. Women's Hospital. Return on investment for a computerized physician order entry system. J Am Med Inform Assoc 2006;13(3):261–6.
[91] Lewkowicz D, Wohlbrandt A, Boettinger E. Economic impact of clinical decision support interventions based on electronic health records. BMC Health Serv Res 2020;20(1):871.

[92] Wang SJ, Middleton B, Prosser LA, Bardon CG, Spurr CD, Carchidi PJ, Kittler AF, Goldszer RC, Fairchild DG, Sussman AJ, Kuperman GJ, Bates DW. A cost-benefit analysis of electronic medical records in primary care. Am J Med 2003;114(5):397–403.

[93] Fillmore CL, Bray BE, Kawamoto K. Systematic review of clinical decision support interventions with potential for inpatient cost reduction. BMC Med Inform Decis Mak 2013;13:135.

[94] Jacob V, Thota AB, Chattopadhyay SK, Njie GJ, Proia KK, Hopkins DP, Ross MN, Pronk NP, Clymer JM. Cost and economic benefit of clinical decision support systems for cardiovascular disease prevention: a community guide systematic review. J Am Med Inform Assoc 2017;24(3):669–76.

[95] Mackintosh N, Terblanche M, Maharaj R, Xyrichis A, Franklin K, Keddie J, Larkins E, Maslen A, Skinner J, Newman S, De Sousa Magalhaes JH, Sandall J. Telemedicine with clinical decision support for critical care: a systematic review. Syst Rev 2016;5(1):176.

[96] Main C, Moxham T, Wyatt JC, Kay J, Anderson R, Stein K. Computerised decision support systems in order communication for diagnostic, screening or monitoring test ordering: systematic reviews of the effects and cost-effectiveness of systems. Health Technol Assess 2010;14(48):1–227.

[97] Murphy DR, Reis B, Sittig DF, Singh H. Notifications received by primary care practitioners in electronic health records: a taxonomy and time analysis. Am J Med 2012;125(2):209. e201–207.

[98] McCoy AB, Thomas EJ, Krousel-Wood M, Sittig DF. Clinical decision support alert appropriateness: a review and proposal for improvement. Ochsner J 2014;14(2):195–202.

[99] Kho A, Dexter P, Warvel J, Commiskey M, Wilson S, McDonald CJ. Computerized reminders to improve isolation rates of patients with drug-resistant infections: Design and preliminary results. In: AMIA Annu Symp Proc; 2005. p. 390–4.

[100] Lee F, Teich JM, Spurr CD, Bates DW. Implementation of physician order entry: user satisfaction and self-reported usage patterns. J Am Med Inform Assoc 1996;3(1):42–55.

[101] Pizziferri L, Kittler AF, Volk LA, Honour MM, Gupta S, Wang S, Wang T, Lippincott M, Li Q, Bates DW. Primary care physician time utilization before and after implementation of an electronic health record: a time-motion study. J Biomed Inform 2005;38(3):176–88.

[102] Kim J, Chae YM, Kim S, Ho SH, Kim HH, Park CB. A study on user satisfaction regarding the clinical decision support system (CDSS) for medication. Healthc Inform Res 2012;18(1):35–43.

[103] Lee KA, Felix W, Jackson G. Concordance, decision impact, and satisfaction for a computerized clinical decision support system in treatment of lung cancer patients. Prevention, Early Detection, Epidemiology, Tobacco Control Annnals of Oncology 2019;30(Supplement 2):ii19. https://doi.org/10.1093/annonc/mdz070.010.

[104] Garg AX, Adhikari NK, McDonald H, Rosas-Arellano MP, Devereaux PJ, Beyene J, Sam J, Haynes RB. Effects of computerized clinical decision support systems on practitioner performance and patient outcomes: a systematic review. JAMA 2005;293(10):1223–38.

[105] Van de Velde S, Heselmans A, Delvaux N, Brandt L, Marco-Ruiz L, Spitaels D, Cloetens H, Kortteisto T, Roshanov P, Kunnamo I, Aertgeerts B, Vandvik PO, Flottorp S. A systematic review of trials evaluating success factors of interventions with computerised clinical decision support. Implement Sci 2018;13(1):114.

[106] Dexter PR, Perkins SM, Maharry KS, Jones K, McDonald CJ. Inpatient computer-based standing orders vs physician reminders to increase influenza and pneumococcal vaccination rates: a randomized trial. JAMA 2004;292(19):2366–71.

[107] Wright A, McEvoy DS, Aaron S, McCoy AB, Amato MG, Kim H, Ai A, Cimino JJ, Desai BR, El-Kareh R, Galanter W, Longhurst CA, Malhotra S, Radecki RP, Samal L, Schreiber R, Shelov E, Sirajuddin AM, Sittig DF. Structured override reasons for drug-drug interaction alerts in electronic health records. J Am Med Inform Assoc 2019;26(10):934–42.

[108] Nanji KC, Slight SP, Seger DL, Cho I, Fiskio JM, Redden LM, Volk LA, Bates DW. Overrides of medication-related clinical decision support alerts in outpatients. J Am Med Inform Assoc 2014;21(3):487–91.

[109] Phansalkar S, Edworthy J, Hellier E, Seger DL, Schedlbauer A, Avery AJ, Bates DW. A review of human factors principles for the design and implementation of medication safety alerts in clinical information systems. J Am Med Inform Assoc 2010;17(5):493–501.

[110] Baron RC, Melillo S, Rimer BK, Coates RJ, Kerner J, Habarta N, Chattopadhyay S, Sabatino SA, Elder R, Leeks KJ, S. Task Force on Community Preventive. Intervention to increase recommendation and delivery of screening for breast, cervical, and colorectal cancers by healthcare providers a systematic review of provider reminders. Am J Prev Med 2010;38(1):110–7.

[111] Stockwell MS, Catallozzi M, Camargo S, Ramakrishnan R, Holleran S, Findley SE, Kukafka R, Hofstetter AM, Fernandez N, Vawdrey DK. Registry-linked electronic influenza vaccine provider reminders: a cluster-crossover trial. Pediatrics 2015;135(1):e75–82.
[112] Sequist TD, Gandhi TK, Karson AS, Fiskio JM, Bugbee D, Sperling M, Cook EF, Orav EJ, Fairchild DG, Bates DW. A randomized trial of electronic clinical reminders to improve quality of care for diabetes and coronary artery disease. J Am Med Inform Assoc 2005;12(4):431–7.
[113] Ash JS, Sittig DF, Campbell EM, Guappone KP, Dykstra RH. Some unintended consequences of clinical decision support systems. In: AMIA Annu Symp Proc; 2007. p. 26–30.
[114] Coiera E, Ash J, Berg M. The unintended consequences of health information technology revisited. In: Yearbook of Medical Informatics, vol. 1; 2016. p. 163–9.
[115] Magid SK, Pancoast PE, Fields T, Bradley DG, Williams RB. Employing clinical decision support to attain our strategic goal: the safe care of the surgical patient. J Healthc Inf Manag 2007;21(2):18–25.
[116] Xie M, Johnson K. Applying human factors research to alert-fatigue in e-prescribing. AMIA Annu Symp Proc 2007;1161.
[117] Olakotan OO, Mohd Yusof M. The appropriateness of clinical decision support systems alerts in supporting clinical workflows: a systematic review. Health Informatics J 2021;27(2). 14604582211007536.
[118] Poly TN, Islam MM, Yang HC, Li YJ. Appropriateness of overridden alerts in computerized physician order entry: systematic review. JMIR Med Inform 2020;8(7), e15653.
[119] Van Dort BA, Zheng WY, Sundar V, Baysari MT. Optimizing clinical decision support alerts in electronic medical records: a systematic review of reported strategies adopted by hospitals. J Am Med Inform Assoc 2021;28 (1):177–83.
[120] Wright A, Ai A, Ash J, Wiesen JF, Hickman T-TT, Aaron S, McEvoy D, Borkowsky S, Dissanayake PI, Embi P, Galanter W, Harper J, Kassakian SZ, Ramoni R, Schreiber R, Sirajuddin A, Bates DW, Sittig DF. Clinical decision support alert malfunctions: analysis and empirically derived taxonomy. J Am Med Inform Assoc 2017;25 (5):496–506.
[121] Wright A, Hickman TT, McEvoy D, Aaron S, Ai A, Andersen JM, Hussain S, Ramoni R, Fiskio J, Sittig DF, Bates DW. Analysis of clinical decision support system malfunctions: a case series and survey. J Am Med Inform Assoc 2016;23(6):1068–76.
[122] Bloomrosen M, Starren J, Lorenzi NM, Ash JS, Patel VL, Shortliffe EH. Anticipating and addressing the unintended consequences of health IT and policy: a report from the AMIA 2009 health policy meeting. J Am Med Inform Assoc 2011;18(1):82–90.
[123] Shah SN, Seger DL, Fiskio JM, Horn JR, Bates DW. Comparison of medication alerts from two commercial applications in the USA. Drug Saf 2021;44(6):661–8.
[124] Bates DW, Kuperman GJ, Wang S, Gandhi T, Kittler A, Volk L, Spurr C, Khorasani R, Tanasijevic M, Middleton B. Ten commandments for effective clinical decision support: making the practice of evidence-based medicine a reality. J Am Med Inform Assoc 2003;10(6):523–30.
[125] Koppel R, Metlay JP, Cohen A, Abaluck B, Localio AR, Kimmel SE. Role of computerized physician order entry systems in facilitating medication errors. JAMA 2005;293(10):1197–203.
[126] Meeker D, Linder JA, Fox CR, Friedberg MW, Persell SD, Goldstein NJ, Knight TK, Hay JW, Doctor JN. Effect of behavioral interventions on inappropriate antibiotic prescribing among primary care practices: a randomized clinical trial. JAMA 2016;315(6):562–70.
[127] Salmasian H, Blanchfield BB, Joyce K, Centeio K, Schiff GB, Wright A, Baugh CW, Schuur JD, Bates DW, Adelman JS, Landman AB. Association of display of patient photographs in the electronic health record with wrong-patient order entry errors. JAMA Netw Open 2020;3(11), e2019652.

CHAPTER

23

Legal and regulatory issues related to the use of clinical software in healthcare delivery

Steven Brown[a,b] *and Apurva Desai*[b]

[a]Department of Biomedical Informatics, Vanderbilt University Medical Center, Nashville, TN, United States [b]Office of Knowledge Based Systems, Department of Veterans Affairs/VHA/OHI/CIDMO, Washington, DC, United States

23.1 Basic legal standards

23.1.1 Legal issues related to using embedded and free-standing decision support software in clinical settings

Several decades ago, in an article in the Annals of Internal Medicine, Miller et al. discussed the question of legal liability for injuries resulting from use of computer software in health care [1]. That discussion followed an earlier, more general series of articles addressing the broader issue of liability for software-related injuries [2–6]. In the ensuing years, hospitals have increased their reliance on automated patient record systems and automated medical devices, and clinicians have increasingly embraced clinical decision support (CDS) software to assist with diagnosis and treatment. American courts have begun to elucidate the conditions under which vendors, care-providing institutions, or clinicians (e.g., physicians and nurses) might be liable for harm to patients arising from the use of computer software [7].

The issue of liability arising from the medical use of software programs continues to attract widespread coverage in both the legal and biomedical literature. Most commentators have explored possible theories of liability, and have concluded that the tort system offers injured plaintiffs—in this case, patients—the best chance of remedies [1,8,9]. American tort law distinguishes between intentional and unintentional injuries. For intentional injuries, the plaintiff must show only that the defendant has intentionally caused the plaintiff's injuries in order to recover damages. If the defendant has unintentionally caused the plaintiff's injuries, tort

https://doi.org/10.1016/B978-0-323-91200-6.00001-2

law can apply several different standards to determine whether the defendant should be considered legally responsible, and hence liable for damages. American tort law at present relies on two major standards of liability: negligence and strict liability [10–12].

The principle of negligence holds that defendants are liable for unintentionally causing a plaintiff's injuries where the defendant caused such injuries due to wrongful or unreasonable conduct. In the medical context, most jurisdictions hold that a defendant's conduct is negligent when it diverges from the customary treatment or medical practice that would be followed by the profession. Most jurisdictions also consider national custom rather than the way most doctors in a particular locality would treat a particular condition. So, in most jurisdictions, a plaintiff in a medical malpractice case would need to show that the standard of care in the medical community was not being followed by the physicians involved [1].

A minority of U.S. jurisdictions rely on a different rule in the medical context. Following the case *Helling v. Carey*, these jurisdictions may consider physicians negligent even if they followed national custom, if a "reasonably prudent" physician would have followed another treatment that might have averted the patient's injuries [13]. Some jurisdictions thus may impose liability even if the physician's treatment adhered to general custom, on the grounds that "a whole calling may have unduly lagged behind" with respect to adopting new treatments and methods of care [14]. In some jurisdictions, the physician would be considered negligent if the provided care diverged from the care that a reasonably prudent physician would have exercised under the circumstances. In *Helling*, a court found ophthalmologists negligent when they failed to administer a glaucoma test to a patient under 40 despite a general medical custom to administer such tests only to patients over 40 [15]. However, many have criticized *Helling's* holding as inefficient for the healthcare profession as a whole and argued that custom should prevail as the general standard [16]. Moreover, in a jurisdiction that followed *Helling*, the plaintiff generally would need expert testimony to establish that a reasonably prudent physician would have followed the non-customary practice [17]. Thus, if the liability standard is negligence, a defendant would be held responsible for a plaintiff's unintentional injuries if the defendant's conduct caused such injuries and if the defendant's conduct either diverged from customary practice or fell below an objectively "reasonable" level of care.

The principle of strict liability, on the other hand, does not consider whether the defendant was exercising precautions or following customary practice. Instead, it merely requires proof that the defendant's conduct was the direct cause of the plaintiff's injuries. Originally, strict liability applied to situations where the defendant was engaged in an inherently hazardous activity. Even if the defendant was exceptionally cautious in conducting such activities, courts allowed anyone injured as a result to recover. The rationale was that such activities carried an inherent risk of harm and that the defendant, in choosing to conduct such an activity, should bear its ultimate costs [18]. Compensation of the injured is thus a primary goal of strict liability [1].

In the past several decades, the principle of strict liability has been extended to situations where a manufacturer's product ends up harming the consumer. The Restatement of Torts (Third) defines a product as follows:

> Tangible personal property distributed commercially for use or consumption. Other items, such as real property and electricity, are products when the context of their distribution and use is sufficiently analogous to the distribution and use of tangible personal property that it is appropriate to apply the rules stated in this Restatement [19].

Under the present doctrine of products liability, a plaintiff harmed by a seller or manufacturer's product could recover for such injuries if they were caused by manufacturing, design, or warning defects in the product, regardless of the care the seller or manufacturer used in manufacturing, designing, and marketing the product. As presently defined, a manufacturing defect means that the product did not comply with the manufacturer's own design standards. A design defect means that the product has been designed in such a way that it carries unreasonable risks for the consumer. A warning defect means that, absent warning labels on the product about its intended use and possible hazards that would not be obvious to the consumer, the product may be unreasonably dangerous [20]. Rationales for strict liability for harms caused by products include:

- That consumers have imperfect information and cannot adequately assess a product's safety on their own
- That manufacturers use their market power to rely on standard form contracts protecting themselves from liability
- That manufacturers are the most preventable parties, whereas preventing consumers from experiencing accidents would be difficult
- That, if the accident is not preventable, manufacturers should be held liable because they can spread the risk of the product through pricing mechanisms.

In recent decades, products liability has been extended to a range of situations on the grounds that it represents the only way to protect consumers from otherwise unaccountable manufacturers or sellers. It has been extended to apply not only to the product's original manufacturer, but to anyone involved in the stream of commerce who is selling or distributing the particular product [21]. Based on the extension of the doctrine, some commentators have argued that the strict liability standard used in products liability cases should govern courts' disposition of claims from software-related injuries. Based on this approach, software vendors would be held liable for all injuries caused by malfunctions, irrespective of whether the vendor exercised due care when developing the product. Commentators, however, have not addressed whether hospitals using such software should also be strictly liable for these harms [1,12]. In the software context, products liability likely would center on a design defects theory, as opposed to manufacturing or warning defects. Software, as a more intangible product, is more easily and perfectly replicated than tangible goods, making it unlikely that the software a buyer received from a vendor would have any unique differences from the source code (unless, as is often the case with EHR implementations, the manufacturer inserted potentially faulty end-user-specific customizations prior to distribution of the software to the local customer). Hence, manufacturing defects theories generally may not apply. Similarly, problems arising from software are unlikely to be the sort where a warning would have allowed a user to avert potential harms, making warning defects theories inapplicable. Instead, outside of the customizations mentioned above, injuries arising from the use of software are most likely to relate to design defects such as programming errors or miscalculations that relate to how the software was created and how it functions.

Whether negligence or products liability is the applicable liability standard has enormous implications in the context of clinical software. Some commentators have suggested that products liability should be the general liability standard applied in this context. Although courts appear unlikely to adopt such a sweeping rule, it could have significant effects on whether manufacturers of CDS systems would enter the marketplace [1]. Vendors, hospitals, and physicians are likely to alter their perceptions of the necessary institutional precautions, possible costs, and relative benefits of clinical software based on whether exercising care and/or following customary practice is enough to absolve them of responsibility for patients' unintentional injuries.

The authors of this chapter argue that the use of clinical software is too varied for either negligence or products liability to be the appropriate standard for all possible situations where the use of such software ultimately harms the patient. The authors contend, first, that products liability as a standard, if applicable, should apply only to vendors, and not to hospitals. Second, products liability should apply to vendors only in certain situations. Software-related injuries likely to arise in a medical context are diverse, and thus strict products liability, though appropriate in some situations, cannot be universally applied. Instead, the applicability of strict liability may be a function of the extent to which the software relied upon is automated as a "closed loop" that provides little opportunity for human intervention. For example, cardiologists (and their patients) now rely upon embedded computer software to perform arrhythmia detection within implantable cardiac pacemakers of various sorts. In such settings, when the software malfunctions, holding a vendor responsible is most likely to be appropriate, since the pacemaker represents a closed loop system not easily inspected or interrupted by clinicians in most situations. In such settings, embedded software in a device is considered to be an integral part of the medical device, and such devices can be considered products. In contrast, when physicians use computer software as a diagnostic aid, strict liability is less apposite, because the physicians are using the software to enhance what is ultimately their own judgment and professional responsibility regarding diagnosis. We note that, while it is too soon to tell, the growing trend toward explanation-free Artificial Intelligence/Machine Learning in clinical decision support may make strict liability more appropriate in the future. Moreover, vendors of clinical software and its users—hospitals and physicians—may face different liability standards based on their relative ability to prevent accidental injuries, and the desirability of distributing such costs.

In order to illustrate the range of legal issues associated with clinical software, this chapter therefore considers two broad scenarios: medical device software and CDS software. The authors conclude that although products liability may be an appropriate means of holding vendors responsible for defects in medical device software, it should not be applied to most decision support situations. We conjecture that partially effective legal recourse via product liability may have led to increased regulation as an alternative.

23.1.1.1 *Software used in medical devices*

Hypothetical examples of cases in which flaws in software (or the operation of software) could produce catastrophic errors include (1) a defect in the embedded software that regulates cardiac pacemaker function, (2) software errors ("glitches" or "bugs") that misreport life-critical serum chemistry test results from a laboratory system to an electronic medical record system, and (3) a programming error in an electronic prescribing system that alters intended doses in prescriptions written by doctors in a manner not readily apparent to the physicians. The US Food and Drug Administration (FDA) estimates that software flaws in medical devices were responsible for approximately 7% of all medical device recalls in the US from the early to mid-1990s, and that the figure is likely to rise: of the 10,000 categories of medical devices available in the US, approximately half rely on embedded software to function [22].

The most serious examples of medical device software-related injuries to patients to date have involved radiation treatment devices. In two separate cases, the software design features of these devices led to multiple severe patient injuries, including some deaths. In the late 1980s, programming errors in the Therac-25, a then leading-edge radiation therapy device, caused two patients' deaths and severely injured another when a software bug caused the device to ignore technicians' corrections to dosage inputs [22,23]. The device administered up to 15 times the estimated fatal dosage of radiation. Later, the FDA prohibited a Missouri company, Multidata Systems, from manufacturing or distributing radiation treatment devices, after numerous patients in a Panama facility died of radiation overexposure attributed to software malfunctions in Multidata's device [22,24].

Such device-related software malfunctions create several possible bases for liability. First, the vendors of such devices could be held liable for producing defective software. Second, technicians or physicians using these devices to treat patients could be liable for their patients' resulting injuries. In this scenario, absent egregious behavior by the technicians or physicians, vendors are much more likely to bear responsibility. The authors conclude that products liability may be an appropriate standard for vendors in this situation.

23.1.1.2 Vendors' responsibilities for devices containing embedded software

Patients could try to hold vendors responsible for injuries arising from software defects by claiming that the vendor was negligent in developing the device's software. However, vendors are unlikely to face substantial liability if the applicable tort standard is negligence, even in a seemingly egregious situation like the Therac-25 case. Although shortcomings in the vendor's software demonstrably caused serious harm to patients, plaintiffs would still have a difficult time showing that the vendor failed to take sufficient precautions or diverged from the industry's general safety practices. Even with extensive testing and software debugging, the precise software flaw in the Therac-25, which prevented modification of the technician's instructions if the instructions were entered at a particularly rapid rate and a particular modification was made, would not necessarily have been apparent. Absent evidence of very poor programming practices or a vendor's failure to test the software, plaintiffs may find it difficult to hold vendors responsible for their injuries. Moreover, if an institution's technicians have made even seemingly trivial modifications to the equipment or to software, vendors may be able to avoid or mitigate liability on the grounds that the institution contributed to the negligence or that it constituted an intervening cause of the accident that should break the chain of liability.

A hospital using a defective device containing embedded software might sue the vendor based on a contractual theory of breach of an implied or express warranty that the device would perform as intended. However, as several commentators have noted, vendors are unlikely to be held liable under contractual theories, since vendors generally include warranty disclaimers in their contracts [25,26].

Given these challenges, commentators have turned to products liability as a possible theory that would hold vendors legally responsible for the harmful consequences of software malfunctions [27]. Although no court has ever applied products liability standards to computer software or online sites [7], products liability standards appear to be well suited to redress injuries arising from medical device software defects and to provide sufficient incentives for efficient vendor behavior. As outlined earlier, a plaintiff would need to show that the device was a product, that the product was defective for products liability purposes, and that the product caused the plaintiff's injuries [1].

Pursuing such a theory first would require plaintiffs to demonstrate that the computer software in a device-defect case should be considered a good rather than a service. The Restatement of Torts (Third), excerpted earlier, distinguishes between products and services and defines products as "tangible personal property distributed commercially for use or consumption [28]." Irrespective of whether the software is incorporated within the medical device or sold separately for the purpose of operating the device, such software would appear to meet the definition of a product for the purposes of products liability. Indeed, the Ninth Circuit hypothesized as much in its decision in *Winter v. G.P. Putnam's Sons* [29]. There, the court opined that defective computer software that diverged from its intended functions, like a defective technical tool that a user relied upon, might be considered a product. Similarly, in C.G. Bryant v. Tri-County Electric Membership Corporation, electricity was considered a product for the purposes of products liability, because the irregularities in the utility company's electricity supply caused the plaintiff's sawmill to burn down [30].

Some have questioned classifying software as a product because its programming aspects are seen to represent more of a service [28], but medical device malfunctions caused by defective software would appear to fall squarely in the category of products. Software defects in medical devices should be encompassed within products liability, because the physical manifestations of the software program have caused harm. The software used to operate medical devices has no function outside of the device; its sole intended use is within a physical product for medical care. This distinction is significant because it is not true of all clinical software; for example, CDS software may be used in a less tangible way by a clinician to choose among alternative therapies.

Assuming the relevant software in a case of a defective device is considered a product, a products liability suit would most likely proceed under a theory of design defects, as discussed earlier. Currently, American jurisdictions are split among three approaches to assessing whether there is a design defect. The first approach, based on the Restatement (Third) of Torts, allows a plaintiff to show that a product is defective through three further theories. First, the plaintiff can present a reasonable alternative design (RAD) for the product, illustrating that the RAD would have reduced the foreseeable risks of harm and that the omission of the RAD prevented the product from being reasonably safe [31]. Alternately, a plaintiff can show that the product is defective through §3 of the Restatement based on a *res ipsa loquitor* theory, or the idea that the product can be inferred to have caused the plaintiff's harm because the injury is of a kind that would not ordinarily result absent a defective product, and the specific harm is one that might come from such a defect. Finally, the plaintiff could show that the product in question has a manifestly unreasonable design [32]. Examples include exploding cigars—a product with such a low utility to consumers and such a high level of risk that the product, from a risk-utility standpoint, should never have been marketed.

The second approach, based on §402A of the Restatement (Second) of Torts, requires a plaintiff to show that a product is unreasonably dangerous based on one of two further theories: that the magnitude of risks in the product's design outweighs the product's utility, or that the product is manifestly defective because the plaintiff's injury is not the type of harm that could have occurred without a serious flaw in the product [20]. This is similar to the *res ipsa* theory in the Restatement (Third).

Finally, the third approach, or the consumer expectations test, allows plaintiffs to show a design defect through two scenarios (derived from [20,33,34]). If the product is of such type that an average consumer could have definite expectations regarding how a product was supposed to function, and the plaintiff's accident arose in that context, the issue of whether the product conformed to reasonable consumer expectations is left to the jury [20,35]. If, instead, the plaintiff's accident is of a type beyond the knowledge or expectation of an ordinary consumer, the plaintiff must rely upon expert testimony to provide a risk-benefit analysis similar to those relied upon under the approaches of the Second and Third Restatements [36,37].

Whatever the jurisdiction, arguments in medical software defects cases seem likely to center on some version of a risk-benefit analysis and the issue of whether the plaintiff's injury manifestly suggests a product defect (*res ipsa*). These are the predominant aspects of the first two approaches, and medical software, as a highly technical product, is likely to fall outside an ordinary consumer's knowledge. The Therac-25 case, for example, might have been well suited to a *res ipsa* theory, since an excessive dose of radiation is not an injury patients are likely to suffer save device malfunctioning, and the evidence had ruled out technician error. Cases where the device has a less severe flaw—for example, if the software programming failed to adequately prevent known side effects of a particular device-related intervention—would likely require a risk-utility test or, in some jurisdictions, proof of a reasonable alternative design. Plaintiffs would find this route far more difficult. Beyond the expense of hiring expert witnesses capable of proposing an alternative product or sufficiently analyzing all the risks and relative benefits of the device, the likelihood of pinpointing the precise error or omission in the software code would be low as well as time-intensive. The code is generally proprietary information, although plaintiffs might acquire it in discovery. Given the complexity and magnitude of most software, however, and the fact that errors in only a few lines of software can be responsible for catastrophic malfunctions, a plaintiff might be required to review hundreds of thousands of lines of code in order to find a few lines that could have been altered to prevent the malfunction.

The use of software-based devices in clinical settings may also raise difficult questions about causation [1]. In many cases, technicians may not have followed the vendor's instructions precisely. For example, in Panama, the technicians operating Multidata's radiation device used five lead shields around patients rather than four; because Multidata's software had trouble identifying the five shields, the technicians identified the five shields as one large shield, causing the machine to drastically miscalculate the appropriate radiation dosage. Vendors may be able to point to this sort of behavior to limit their own liability. Multidata has relied on a misuse defense, arguing that the technicians' modifications could not have been reasonably foreseen by the vendor at the time the product was sold [38]. Depending on the type of modifications made by users in hospitals, and the foreseeability of such modifications at the time of sale, such a defense may prevail.

Plaintiffs realistically may recover only for injuries caused by the most blatant and severe software malfunctions; however, the normative case for applying products liability standards in this scenario remains strong. The scenario corresponds well with the general rationales for applying products liability. Because the software code is embedded in the device and opaque to the user, neither doctors nor patients can assess whether the device is safe. Indeed, doctors and patients may not even be able to examine the software's content, since such information is proprietary. Users generally depend upon the vendor's operating manual or technical support for assistance. Additionally, vendors of software-operated devices have almost universally relied upon contractual language that disclaims warranties, eliminating contractual liability. Furthermore, vendors of software-operated medical devices appear to be the most preventable parties, since they have the greatest knowledge of software content, can most easily test devices for problems, and could adjust the code, unlike users. Finally, although some software bugs are inevitable given the complexity and length of code required for most devices, vendors, rather than patients or doctors, should be liable because they can include the costs of non-preventable accidents in the device price.

One thing to note with respect to vendor liability, the Supreme Court in *Riegel v. Medtronic, Inc.*, 552 U.S. 312, 317-18 (2008), held that in cases where the device is subject to the FDA's oversight under the Medical Device Amendments of 1976 (MDA), MDA's pre-emption clause bars common-law claims challenging the safety or effectiveness of a medical device marketed in a form that received premarket approval from the FDA [39]. The effect of the preemption of state tort claims may cause, as some commentators claim, a shift in litigation risk to physicians or other stakeholders in the healthcare industry, to the extent that those systems (or software such as CDS or EHRs) utilize the FDA's premarket approval process and should be monitored by practitioners and vendors alike.

Technicians and physicians' responsibilities while using devices containing embedded software

Patients could also potentially bring a case against the technicians or physicians who operated a malfunctioning device containing embedded software, as well as against the clinical facility where the device was used. The question of whether hospitals and clinics can be liable for defective devices implanted or used in surgery has been raised with increasing frequency in the context of medical recalls on defective pacemakers, breast implants, and Teflon jaw implants. Courts in this context have thus considered whether hospitals can be considered part of the distribution chain of the device in question for purposes of products liability. With a few exceptions, courts generally have found that hospitals are not subject to strict liability in such situations [40]. The main rationale has been that even where hospitals have charged patients markup for the devices, hospitals are healthcare service providers rather than product distributors. Even though a product may be used as part of patient care, it is being used because it is essential to providing a course of treatment [41]. Even courts that have considered hospitals as potential distributors have declined to apply products liability standards to them on public policy grounds, concluding that making hospitals accountable for thoroughly testing all medical products used in a clinical setting would be unreasonable and would detract from hospitals' primary mission of providing patient care [42]. Thus, although hospitals can be strictly liable for defective products sold in their gift shops, in general, they will not be liable for defects in products used as part of medical care. The more a particular device seems inseparable from the service of assisting in patient care, the less likely it is that any court would apply a strict liability standard to the hospital [21,43]. Thus, it is particularly unlikely that a hospital would be strictly liable for harms arising from malfunctioning software in a radiation device, or similar software-based errors.

Hospitals, technicians, and physicians could still be held responsible for injuries arising from malfunctioning devices under negligence. As mentioned earlier, jurisdictions differ on the standard used to evaluate negligence. In most jurisdictions, the question would be whether the care provided fell below the standard of care customary in the medical profession. In a few jurisdictions—the ones that follow *Helling*—the question would be whether a "reasonable physician or technician" would have acted differently. In either case, if all relevant precautions were followed in using the device, it would be difficult to prove negligence. However, in the Multidata case, the technicians diverged from instructions without authorization in operating the device and failed to adequately observe patients receiving the treatment. A Panamanian court ultimately convicted the technicians of involuntary manslaughter for criminally negligent behavior [38]. However, physicians and hospitals can still be liable under certain situations see for example, *Lamb v. Candler Gen. Hosp.*, 413 S.E.2d 720, 721-22 (Ga. 1992) ("It is well recognized that a hospital may be liable in ordinary negligence for furnishing defective equipment for use by physicians and surgeons in treating patients."); and *Berg v. United States*, 806 F.2d 978, 983 (10th Cir. 1986) (upholding a verdict for the plaintiff whose injuries were caused in part by a lack of adequate testing and maintenance of equipment and a lack of adequate training of technicians) [44,45].

In conclusion, in situations involving software-based malfunctions in a clinical context, it seems appropriate to hold vendors accountable through products liability. The application of such a standard in this context comports with the general purposes behind products liability, because in this type of situation the vendor is uniquely situated to control possible malfunctions. Hospitals should not be held liable under the same standard, because technicians and physicians have little to no control over how the product functions as a program. Moreover, hospitals are using the devices only as part of patient care; it is difficult to compare hospitals to product distributors. Finally, as case law suggests, holding hospitals strictly liable for such devices would have extremely detrimental effects for patient care. It is appropriate to hold hospitals liable in situations where technicians or physicians fail to adequately supervise patients receiving treatment when device malfunctions occur.

23.1.1.3 CDS software used by licensed practitioners during medical practice

The previous analysis suggested that products liability is an appropriate standard to apply to vendors who sell malfunctioning software-containing medical devices when injury to patients occurs. In this section, we discuss whether CDS software should be governed by the same standard. CDS software enhances practitioners' abilities to collect, manage, and draw inferences from patient-related data and from general biomedical information.

We describe two possible scenarios for liability involving its use: reliance upon erroneous clinical advice provided by such software, and failure to use CDS software when its use would have prevented improper treatment of the patient. Again, two possible classes of defendants are considered: vendors who produce the CDS software, and physicians—and their hospitals—that rely on CDS software in treating the patient.

We argue that CDS software should not be governed by a products liability standard for either class of defendants, because the decision support context materially differs from defective software-operated devices. Scenarios arising from the use of CDS software appear more complex, and thus appear to require different sets of liability rules. Vendors should not, and are unlikely to be, held liable under a products liability standard. However, if vendors' CDS software provides erroneous advice to its users, vendors may be considered negligent. Questions arise as to whether a vendor would or should be held to the same standard of care as a physician might be in a situation where the CDS software dispensed erroneous advice. Additionally, if a doctor relied upon CDS software that provided erroneous information, it is likely that the doctor might be considered negligent.

Vendors' responsibilities for erroneous information provided by CDS software

Vendors are unlikely to face strict liability in the event that their software dispenses erroneous advice to licensed clinicians, ultimately causing harm to patients. CDS software is unlikely to be considered a product for purposes of products liability. Most courts have considered computer software to be a "good" within the meaning of the Uniform Commercial Code (UCC) [46,47], and thus governed by the UCC. If, however, the vendor has provided institution-specific programming and tailored the software to the hospital's particular needs, courts are more likely to consider the contract to involve a service [48]. The UCC usually sustains warranty disclaimers provided that they are obvious from the contract [49,50]. Patients are unlikely to prevail under third-party beneficiary theories (i.e., patients injured as "bystanders" to a contract between the vendor and the hospital), as there is no privity of contract (i.e., mutual or successive relationship to the same rights) between patients and the hospital on the one hand, and the hospital and vendor on the other [51]. Even if there were such a connection, vendors' disclaimers would still limit, if not eradicate, patients' ability to recover under breach of contract theories. Second, patients and/or hospitals might sue vendors for injuries arising from software defects through the torts system, by claiming that the vendor was negligent in developing the device's software. To prevail on this claim, plaintiffs would need to demonstrate that the vendor owed them a general duty of care, that the vendor breached this duty by failing to take adequate precautions to check the software code or by employing programming practices that were below the customary level of vendors' practices, that the vendor's negligence with the software proximately caused the plaintiff's injuries, and that the plaintiff's damages are recoverable within the tort system.

Decisions relating to whether something is a product for purposes of products liability have long distinguished between harms arising from the functionality of a thing and the ideas it contains. The former is generally classified as a product, the latter is not. For example, in *Winter v. G.P. Putnam's Sons*, the Ninth Circuit distinguished between things that graphically illustrated technical information, like a compass, and things like books identifying poisonous and edible mushrooms, which it considered more like instructions on how to use a technical device [29]. The latter was not considered a product in the case, which found that the ideas and expression in a text or other work could not be considered a product because of their intangible properties. Furthermore, in *James v. Meow Media*, a Kentucky court found that the violent images and ideas contained in video games and other media could not be considered products in a case where these ideas were alleged to have inspired a school shooting. That court suggested that the ideas and images had no tangible expression or physical manifestation in of themselves; the tangible actor was the school shooter, not the video games [52]. Finally, comments to §19(a) of the Third Restatement have approved of such decisions, again distinguishing between information and the tangible medium in which it appeared [53].

Thus, CDS software generally can be distinguished from software within medical devices. Medical devices can be classified as products because their tangible effects, such as targeted radiation, are the source of patients' injuries and, as discussed below, are subject to significant regulatory oversight. CDS software, however, is not generally marketed as intended to replace the judgment or the functions of a physician; instead, manufacturers assert that it merely augments the physician's existing knowledge by providing further information [1]. It is true that CDS software may also involve record entries or other data that might provide the physician with erroneous information if the software were to malfunction; however, unlike medical devices used for patient treatment, such information is easily verified by a physician, and should be verified in the course of treatment. Even patient record information remains closer to an idea than to something capable of tangible expression.

Moreover, to include CDS software within products liability would be inconsistent with the purposes of imposing strict liability. Products liability would find a vendor of CDS software liable for errors in information provided to doctors even if the doctor should have known that the program's information was blatantly false; for example, when the computer-suggested regimen included prescribing a medication wholly and obviously unrelated to the patient's illness. Doctors, as possessors of specialized expertise, should be incentivized to take the utmost care when treating patients. Unlike possible accidents involving the use of medical devices to perform treatment, CDS software represents what game theorists call a sequential care situation: even if the software provides an erroneous diagnosis, the doctor subsequently acting upon the information possesses specialized knowledge of the patient and of general medical conditions and is therefore in the best position to evaluate and reject faulty or inapposite advice. This is, however, not to say that vendors of CDS software should be able to avoid liability on the grounds that the attending physician, rather than the software's erroneous information, was the proximate cause of the patient's injuries. Instead, both the vendor and physician should be held accountable. The issue remains to be resolved and as noted elsewhere, the growth in clinical algorithms and machine learning may cause clinicians to rely on them in lieu of their own judgment, without truly understanding the basis of the software's conclusion ("that machine-learning algorithms, even when they make an accurate prediction, may not be able to explain the basis of their predictions in terms intelligible to a human being.") making it harder to avoid shifting more of the liability onto the vendor [54].

Thus, though negligence appears to be the appropriate standard with respect to vendor liability for non-device-related CDS software, the question remains whether a vendor would be held to the same or higher standard as a physician might be in such a situation. Vendors might face a different general standard given the nature and purposes of CDS software and the fact that vendors are not necessarily physicians. In jurisdictions that generally look to medical custom, vendors of CDS software might be considered negligent if their software failed to dispense the advice or information that a reasonably prudent physician would provide. However, such jurisdictions could also hold vendors to the standard of custom within the decision support field, looking to whether in general, competitors' software would have provided the appropriate advice or information. Alternately, jurisdictions might define custom in this area as what is the best available, current disseminated knowledge in the medical field, on the grounds that CDS software compiles this knowledge and intends to provide physicians with a broader range of knowledge than the unaided physician might possess. It is difficult to anticipate how a court would resolve the matter. It seems clear, however, that difficulties will arise in defining the standard of care to which CDS software vendors should be held. At the very least, CDS software must at least meet the standard of care to which the unaided physician would be held. But to hold vendors merely to this standard may be at odds with the purpose of CDS software.

One possible exception to the proposed rule that vendors should be subject to a negligence standard may arise in situations where a patient (or family member, in the case of a child) relies on purchased medical software installed on or accessed by a home-based personal computer (PC) to diagnose the patient's symptoms and to provide advice as to whether additional medical treatment is necessary. In such a scenario, licensed practitioners would be bypassed. The patient might inappropriately rely upon the software to select appropriate therapy; for example, the PC-based software product might suggest taking two acetaminophen tablets for a headache that actually was caused by meningococcal meningitis, a treatable and rapidly progressive, often fatal infection. One might argue that products liability should apply in this situation, drawing an analogy to the case of the pilot relying upon a misleading aeronautical chart [55] or alternatively, due to patients' lack of professional education or experience, stricter regulatory oversight by the appropriate regulatory agenc(ies).

This scenario suggests the difficulties in making legal distinctions between products and services. Purchased medical software is being sold as a product, but it is intended to replace the service of qualified medical advice. Applying products liability would hold the software maker liable for any harms arising from advice given by the software. Applying a negligence standard would involve the same difficulties in defining the standard of care as described earlier; however, the negligence standard should then at least hold the software to the same standard of care as the medical profession as a whole. We argue that a negligence standard is more appropriate in this situation, and that the proper standard should be the same as whatever standard applies to CDS software in the previous example.

Whereas aeronautical charts rely on exact knowledge to provide crucial in-flight information, diagnostic software used by the consumer ideally is based on the best available knowledge prevailing in the medical profession. A negligence standard should be sufficient to incentivize vendors of such software to keep abreast of changing medical knowledge with respect to the intended CDS software function, for example, to incorporate progress in new approaches to diagnosis in new releases of a CDS software package.

Practitioners' and healthcare Facilities' responsibilities for erroneous information provided by CDS software

Given that products liability is unlikely to apply to hospitals whose physicians or technicians use medical devices with software malfunctions, it is even less likely to apply to hospitals using CDS software. Decision support software is even further integrated into the process of providing patients with a course of treatment. Even the small number of jurisdictions that hold hospitals strictly liable for defective implants or pacemakers would be unlikely to extend the rule to the use of CDS software, because such software is incidental to the service of providing medical advice, and is not passed on to the consumer the way an implanted device or treatment might be [18].

Instead, physicians (and the hospitals that employ them) may be considered negligent if they fail to question erroneous advice given by CDS software and proceed to provide improper care to the patient. Again, the standard for determining negligence would vary depending on the jurisdiction, and would either be custom or the care provided by a "reasonably prudent physician" [1,13–17,56]. Liability in such cases would depend on the care provider's actions in the particular case. In this respect, the use of CDS software would not expose the care provider to any additional grounds for liability. The care provider is ultimately accountable for the care given and held to the same standard of care irrespective of whether CDS software was used. This is consistent with the aims of CDS software: when functioning properly, it can help enhance diagnostic abilities and prevent misdiagnosis or other errors. However, decisions in treatment are ultimately left to the care provider, and the care provider should be considered responsible for these decisions.

Finally, any eventual lawsuit involving erroneous advice dispensed by CDS software is likely to involve joint and several liability. Joint and several liability holds multiple defendants—in this scenario, the vendor, hospital or clinic, and physician—responsible for the ultimate injuries suffered by the patient. It would mean that any liability assigned to one of the defendants would be shared by all, and one defendant could compel the others to contribute to any damages awarded to the patient. Joint and several liability could apply in this situation because the ultimate harm to patients is an "indivisible harm"—meaning that without both the software vendor and the attending physician's negligence, harm to the patient would not have occurred [57]. In other words, the software vendor's erroneous advice was harmful only because the attending physician failed to correct it, or the attending physician may have recommended a course of treatment only because it was recommended by the CDS software.

Failure to use CDS software to prevent medical errors

The previous section has highlighted possible areas of legal liability when physicians rely on CDS software. However, it is also possible that licensed practitioners could be considered negligent if they failed to use CDS software to avert medical errors. A clinician can be considered negligent due to omissions in care as well as for overt actions. As mentioned earlier, most jurisdictions look to whether the physician followed the national standard of care when determining whether the physician's conduct was negligent. If the use of CDS software became the national norm in medical practice, a physician who failed to use such a program, to the detriment of a patient, might thus be liable for the patient's injuries. But at present, the use of CDS software does not appear to have uniformly reached the level of custom across all clinical specialties.

In those jurisdictions that follow *Helling* and look to a more general reasonableness standard (other than custom), clinicians could be found liable for failing to use CDS software to avert an error, if evidence were introduced that a reasonably prudent physician would have done so in the case at hand. This would require expert testimony and would likely be left to a jury to assess [53]. Thus, if reliable CDS software were available and its use might have prevented the patient from injuries caused by the chosen treatment, even if such software is not used by a majority of physicians, it is possible that a particular physician might be found liable for having failed to use that technology if this failure caused the patient's injury [1].

Some precedent exists for this particular scenario. A Washington court found that a physician's failure to consult available literature, such as Medline, went against general notions of "good basic medicine" and constituted negligence in a case where consulting such literature might have led to a proper diagnosis:

> Finally, we address the government's concern expressed at oral argument as to "how a doctor ought to know that he doesn't know" whether there is information that need be disclosed. To justify ignorance of this type of risk would insulate the medical profession beyond what is legally acceptable. Here, there is expert testimony ... that it would be "just good basic medicine" to conduct a literature search or contact specialists in response to a direct question to a physician such as the one posed here. With the demands of their profession, no one can expect doctors to have all material information stored in their minds. We do not decide the extent to which a literature search must reach. Some limits are appropriate. This may best be defined by reference to what is material and reasonably available. As we have stated, a risk is not material unless expert testimony can establish its existence, nature, and likelihood of occurrence ... A literature search will thus put a physician on notice of these risks [56].

As this quote demonstrates, even in jurisdictions that define medical malpractice based on a general reasonableness standard, courts faced with this issue will ultimately weigh difficult questions about what sort of knowledge doctors should be expected to possess. The court distinguished between the doctor's own knowledge and knowledge from other readily available sources. In this case, CDS software could have been one of many possible sources of the type of knowledge that would have avoided the patient's injuries. Instances where the use of CDS software would have been the only means of preventing the patient's injuries are less common, but it has recently been demonstrated for certain intensive care unit protocols that computer-based advice/adjustments are superior to purely human-mediated attempts to follow carefully defined protocols for ventilator management and hemodynamic monitoring.

A search of a legal database in early 2013 revealed more than a hundred cases during the preceding decade that involved clinical decision support systems. Nevertheless, the cases pertained to contract law (whether a vendor had delivered the contracted functionality to a purchaser) and intellectual property issues (invention, legal ownership, and protection of algorithms, data, and rules underpinning a decision support system). The search uncovered no specific cases involving successful malpractice litigation regarding use of, or failure to use, a clinical decision support system in healthcare settings in the US. However, as outlined below, later studies found that with the widespread deployment of EHR systems, malpractice suits focusing on inherent defects of the software or in its implementation or use have started to occur.

In conclusion, this section discussed how tort law might treat CDS software use during patient care, and how different levels of liability may affect the incentives of the vendors and healthcare providers who develop and use it. Whether products liability should apply to vendors who develop and sell such software should depend upon whether the software is used in an automated medical device (closed-system context) or whether the licensed practitioner can make an independent decision regarding the adequacy of the program's advice before treating the patient (open-system context). In any event, products liability should not apply to hospitals using any such software.

23.1.1.4 *Malpractice cases involving EHRs*

With the federal government's push for EHR adoption through the HITECH Act, a part of the American Recovery and Reinvestment Act of 2009 (ARRA), there has been considerable growth in the deployment and use of EHRs by physician practices and hospitals in the last two decades (see further discussion below, in "HITECH Act and MU Certification") fueling an increase in plaintiff's malpractice claims specifically targeting the EHR software itself.

In a 2010 study, "Medical Malpractice Liability in the Age of Electronic Health Record," the authors noted that their "analysis is based on a review of the limited available literature on the liability implications of EHRs and a much larger body of literature on the effects of EHRs on quality of care and the role of clinical practice guidelines in malpractice litigation" [58]. The study focused on four core functions of an EHR: documentation of clinical findings, recording of test and imaging results, computerized provider order entry, and clinical-decision support. The authors further divided the cases chronologically into phases, beginning with the initial rollout (when facilities often have a hybrid operating environment mixing paper and electronic) to post complete migration or implementation. During each phase, the authors found a varying assortment of risks that needed to be managed (Table 23.1).

TABLE 23.1 Medical Malpractice Risks in the Age of the Electronic Health Record.

Initial deployment	• Migration from paper to electronic increases chances of gaps in patient documentation • Failure to meet a "reasonable provider" standard in configuration of the EHR during roll-out, increasing the potential liability of clinicians • Inadequate training on the new systems increases likelihood of user error, increasing risks to patient safety • Errors by users could result in missing or incorrect data records • Inconsistent use of the EHR by users could lead to gaps in documentation and communication • Systemwide bugs or failures in the EHR could result in adverse effects on clinical care that could lead to patient injuries and medical claims
Maturing in production	• Increased use of email to communicate multiples the number of patient encounters which could increase the potential of claims if the clinical advice is provided without thorough review of the patient's clinical record • More extensive documentation of clinical decisions and activities could increase the range of discoverable information, including metadata • Clinicians may succumb to the temptation to merely "copy and paste" the patient's medical history into new records, risking perpetuating past mistakes and failing to catch new clinical information • Information overload may result in clinicians missing key information • Clinicians choosing not to follow clinical practice guidelines could strengthen plaintiffs' malpractice claims
With widespread adoption of EHRs and HIEs	• Better access to greater clinical knowledge via the EHR may impose new legal duties to act on the information in the provision of care • Widespread use of clinical decision support could establish standards of care that otherwise may be up for debate • Rise of HIEs may increase clinicians' duty to investigate and review patient information created by other clinicians • Failure to adopt and use new technologies in itself may constitute a deviation from the standard of care and open clinicians to liability

And in 2014, a study of malpractice claims found that in a 2012–13 study by CRICO of 147 cases in which EHRs were found to be a contributing factor to the plaintiff's injuries [59]. Computer systems that don't "talk" to each other, test results that aren't routed properly, and mistakes caused by faulty data entry or copying and pasting were among the EHR-related problems found in the claims, which represented $61 million in direct payments and legal expenses. The top five categories of claims were:

- Incorrect information in the EHR—20%
- Hybrid health records/EHR conversion issues—16%
- System failure—electronic routing of data—12%
- System failure—unable to access data—10%
- Pre-populating/copy & paste—10%

Subsequently, in a later study published by the Doctors Company in 2017, further supported these findings, noting that over a 9-year period, about 1.1% of the malpractice claims were based on problems with the EHRs themselves (Fig. 23.1) [60].

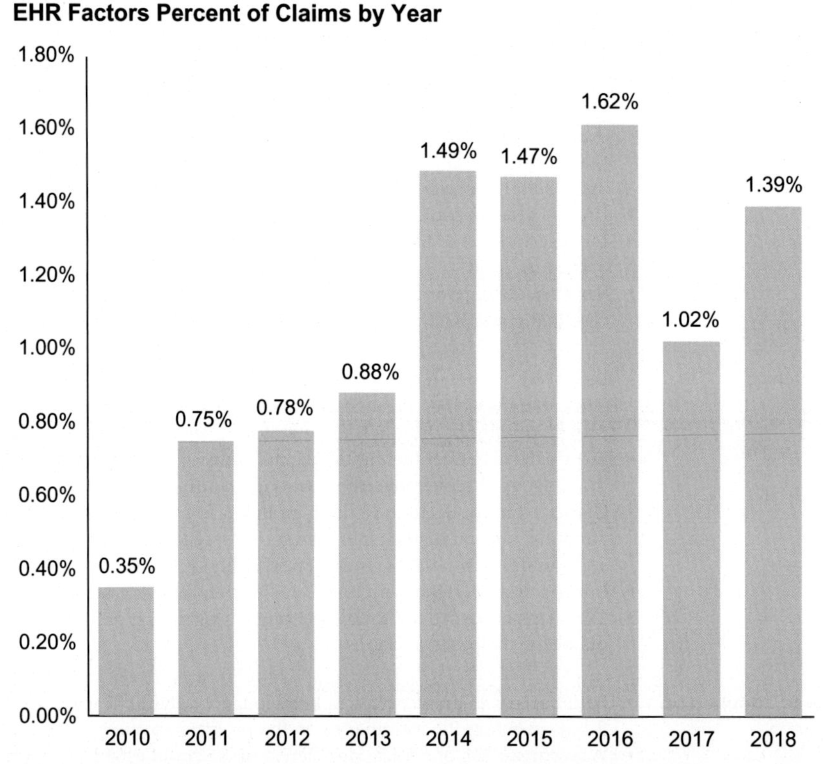

FIG. 23.1 Breakdown of EHR factors by percent of claims by year. *Reprinted with the permission of The Doctors Company 2022.*

These then further broke down into eight distinct types of claims (Table 23.2).

TABLE 23.2 Breakdown of claims by top technology and design issues.

Top system technology and design issues	Claim count	Percent
Other	30	14%
Electronic systems/technology failure (EHR)	26	12%
Lack of or failure of EHR alert or alarm	15	7%
Fragmented record	14	6%
Failure/lack of electronic routing of data	10	5%
Insufficient scope/area for documentation in EHR	8	4%
Lack of integration/incompatible systems	5	2%
Failure to ensure information security	1	0%
Grand total	**104**[a]	**48%**

[a] *Note that the percentages are of the total number of EHR claims (n=216).*
Reprinted with the permission of The Doctors Company 2022.

As part of its study, Mangalmurti et al. had hypothesized that the use of an EHR could affect (both positively and negatively), the course of malpractice cases due to increased documentation of acts or inactions on the part of clinicians and hospitals. This appears to have come true in a few cases and as EHRs are more prevalent, courts appear ready to impose responsibilities on hospitals with respect to their EHR implementations.

In the case of *Richter v. Presbyterian Healthcare Services,* Timothy Richter, as personal representative for his wife, brought a wrongful death action against Presbyterian Healthcare Services (PHS) and Regional Lab Corporation d/b/a Tri-Core Laboratories (TriCore), alleging negligent delivery of Mrs. Richter's laboratory test results in 2001 [61]. Plaintiff also brought a medical malpractice action against the two physicians treating Mrs. Richter at the time of her death, alleging medical negligence in her treatment. The district court granted some motions for summary judgment in favor of Plaintiff and some in favor of PHS and TriCore and granted a partial directed verdict in favor of Dr. Winterkorn, one of Mrs. Richter's treating physicians. Mr. Richter appealed the district court's rulings. Relevant to our discussion is the discussion by the appellate court with respect to the importance of the hospital's records re: Mrs. Richter's test results. The Court found that the district court had not erred in finding that question of whether TriCore's delays in delivery of the test results rose to the level of professional negligence (thus requiring expert testimony) or ordinary negligence (which didn't) was clearly a claim of professional negligence, thus requiring expert testimony, the same conclusion did not apply to PHS and its duty to have a complete record for each patient. Instead, it stated that "*hospitals do have a clearly established duty to maintain their patients' medical charts in good order*, and that duty includes posting completed lab tests as received." (emphasis added).

The court in *Presbyterian* cited with approval the court in *Johnson v. Hillcrest Health Center, Inc.*, and the function and the importance of the patient chart [62]. Specifically—"The obvious purpose of the charting requirement is to provide a record to assist the physician in properly treating the patient. Physicians depend on the reliability and trustworthiness of the chart. As far as a hospital is concerned, there is no more important record than the chart for indicating the diagnosis, the condition, and the treatment required for patients. In our view, no degree of knowledge or skill is required other than that possessed by the average person to conclude that the applicable standard of care required the hospital to include completed lab tests and lab reports in the patient's chart to aid the doctor in diagnosing and treating the patient—regardless of whether lab tests are made available on the computer." (Id. ¶ 15).

In response, PHS raised 4 grounds to support its request for summary judgment in its favor: (1) In the absence of results reflecting critical values under the Clinical Laboratory Improvement Amendments (CLIA), 42C.F.R. § 493.1291(g) (2012), it had no duty to give any notice to Mrs. Richter's physicians; (2) PHS fulfilled whatever duty it had by making the test results available to the physicians by computer access on the morning of April 23, 2001, and by placing the chart copy in the patient's chart when it was delivered by TriCore Laboratory; (3) its Rule 11-406(A) NMRA evidence of its habitual handling of lab results created a presumption that it followed that habit in this case; and (4) Mrs. Richter's physicians had their own duty to follow up on the test results. The court agreed to the 1st and found that the 4th basis (intervening negligence) did not absolve PHS, only implicated a need for comparative analysis if the physicians are found to have completed malpractice. The court rejected both the 2nd and 3rd grounds based on the facts and how physicians used the computer system at the time Mrs. Richter was a patient, thus overturning and remanding to the district court.

Thus as hospitals migrate to EHRs and rely on them to contain the patient's complete record, the legal obligation to ensure the EHR is properly configured and contains all information is likely to impose a significant liability on them.

Another area where courts have held hospitals liable for malpractice has been found with respect to the need to preserve and provide access to metadata from the hospital EHR to plaintiff as reasonable application of state statute re: Civil Procedure and Discovery, finding that hospitals are required to maintain audit trails under federal and state law.

In *Vargas v. Lee*, a medical malpractice suit, the court found that denial of plaintiff's motion to compel defendant hospital to produce the audit trail of plaintiff's records was error because CPLR art. 3101(a) was to be interpreted liberally and plaintiff demonstrated that the part of the audit trail at issue was reasonably likely to yield evidence relevant to the negligence allegations regarding plaintiff's post-operative care and the request was limited to the period immediately following the injured plaintiff's surgery; further, defendant failed to demonstrate that the requested disclosure was improper or otherwise unwarranted, and allegations that producing the audit trail would be time consuming were conclusory and unsubstantiated [63]. In this case, the plaintiff sought to recover damages for medical malpractice, the plaintiffs appeal from an order of the Supreme Court, Kings County (Gloria M. Dabiri, J.), dated August 1, 2016. The order, insofar as appealed from, denied that branch of their renewed motion which was to compel the defendant Wyckoff Heights Medical Center to produce the audit trail of the patient records of the plaintiff Jose Vargas for the period of May 1, 2012, through May 17, 2012.

On review the Court found that CPLR 3101 provides that "[t]here shall be full disclosure of all matter material and necessary in the prosecution or defense of an action, regardless of the burden of proof" (CPLR 3101 [a] and that "the Court of Appeals has emphasized that "[t]he words, 'material and necessary', are ... to be interpreted liberally to require disclosure, upon request, of any facts bearing on the controversy which will assist preparation for trial" and thus "in this context, "[i]f there is any possibility that the information is sought in good faith for possible use as evidence-in-chief or for cross-examination or in rebuttal, it should be considered [matter] 'material' in the action" citing (*Shutt v Pooley*, 43 AD2d at 60). The court found that the plaintiffs met their burden and the Supreme Court had erred in ruling against them. That "The plaintiffs demonstrated, and Wyckoff does not dispute, that an audit trail generally shows the sequence of events related to the use of a patient's electronic medical records; i.e., who accessed the records, when and where the records were accessed, and changes made to the records" (see *Gilbert v Highland Hosp.*, 52 Misc 3d 555, 557, 31 NYS3d 397 [Sup Ct, Monroe County 2016]).

> As argued by the plaintiffs, the requested audit trail was relevant to the allegations of negligence that underlie this medical malpractice action in that the audit trail would provide, or was reasonably likely to lead to, information bearing directly on the post-operative care that was provided to the injured plaintiff. Moreover, the plaintiffs' request was limited to the period immediately following the injured plaintiff's surgery. The plaintiffs further demonstrated that such disclosure was also needed to assist preparation for trial by enabling their counsel to ascertain whether the patient records that were eventually provided to them were complete and unaltered.

This position was later followed in *Heinrich v. State of New York*, involving the administrator of an estate who sought the production, pursuant to CPLR 3101(a), of the audit trails in the decedent's electronic medical record (EMR) in conjunction with the allegations of medical malpractice that led to their death [64]. The Court held even though the liability phase of discovery had completed, where the request is narrow enough to limit the burden on defendant and had been timely filed, there was a clear showing that the EMR and audit trail contained information that would be useful to claimant and should be granted.

Thus, while the increased deployment and use of EHRs may improve the delivery of clinical care, a result that has not been fully documented by independent studies, pervasive use of EHRs may change the standard of care and make the failure to adopt (or properly use) an EHR system a deviation from the standard of care and give rise to its own cause of action.

23.1.2 Responsibility for CDS software at the institutional level and potential governmental regulation

Clinical software systems are defined as algorithmic programs, related knowledge bases, and embedded interfaces, that directly contribute to the delivery of health care, in contrast with inventory or accounting functions. Clinical software systems are ubiquitous in medium- to large-sized healthcare facilities, although CDS systems represent only a small minority of such systems. Healthcare practitioners, clinical facilities, industry, and regulatory agencies share an obligation to manage clinical software systems responsibly using a common framework [65]. The previous section of this chapter reviewing legal issues related to CDS systems indicated that use of clinical software systems does not often cause substantial harm to patients. However, concerns for safety at both the individual practitioner and institution levels must be addressed. Portions of the following discussion are reproduced with permission from the Annals of Internal Medicine [65].

23.1.2.1 *Complexity of institutional clinical software environments*

Because clinical software systems are diverse and complex, determination of their safety is difficult. In an ACMI Distinguished Lecture in the late 1990s, Dr. Clement McDonald, at that time at the University of Indiana, estimated that every large academic medical center in the US had at least three dozen major software systems installed. Such systems serve a variety of purposes, including (among others) billing, inventory control, scheduling, compliance, electronic mail, message handling/routing, ADT (admission, discharge, and transfer) patient census functions, various laboratory functions, radiology image capture and retrieval, pharmacy ordering and dispensing, electronic patient chart/electronic health record functions, computerized provider order entry (CPOE), electronic textbook/reference functions, and clinical decision support.

Thousands of clinical software products compete in the commercial marketplace. A large number of "home-grown" systems exist, and variable-quality biomedically oriented World Wide Web sites and more recently, mobile apps, proliferate in an uncontrolled manner. Most overall institutional installations are one-of-a-kind. Significant functional changes occur when a software product is integrated locally into a clinical information management infrastructure. Upgrades and maintenance also increase the variety and complexity of clinical systems. If there are six possible vendors (including the institution itself for "home-grown" products) for each of the presumed 36 major systems at a large institution, and three possible versions of each software package (most recent release, recent past release, and institutionally customized older release), then the number of potential configurations at an institution would be 18 to the 36th power (six vendors times three configurations for each of 36 systems). This impossibly large number of system configurations in an institution's environment is further complicated by the observation that variability in interactions between clinical software programs and individual users may cause unpredictable outcomes not related to software malfunctions.

Because of high local variability in both system configurations and usage patterns, it is not possible for a centralized monitoring agency such as the FDA to monitor local software environments for safety at all institutions in the US. Just as responsibility for monitoring of safety for human subjects' research was delegated by Congress to be shared among the FDA, National Institutes of Health (NIH), and local Institutional Review Boards (IRBs), monitoring of clinical software for patient safety is arguably only possible at the local level [65].

23.2 IMDRF

In recognition of the dangers and complexities of health-related software, regulators in the US and abroad continued efforts have continued efforts to design a uniform regulatory format that balanced the needs to minimize the risk of harm to patients with the costs of regulatory compliance on manufacturers. One such group, the International Medical Device Regulators Forum (IMDRF), http://www.imdrf.org/, is a voluntary group of medical device regulators from around the world who have come together to build on the strong foundational work of the Global Harmonization Task Force on Medical Devices (GHTF) which seeks to accelerate international medical device regulatory harmonization and convergence.

IMDRF was establish in October 2011, when representatives from the medical device regulatory authorities of Australia, Brazil, Canada, China, European Union, Japan and the United States, as well as the World Health Organization (WHO) met in Ottawa to address the establishment and operation of this new Forum. The USA is one of the managing members and FDA represents the US at the IMDRF.

Through a set of four documents (2013–17), the IMDRF laid out a proposed comprehensive, interlocking regulatory pathway (Fig. 23.2) designed to achieve the goal of "a common and converged understanding of clinical evaluation and principles for demonstrating the safety, effectiveness and performance of SaMD" [66–69].

23.2.1 Software as a medical device (SaMD)

In 2013 the IMDRF released a white paper entitled "Software as a Medical Device (SaMD): Key Definitions" [70]. In this white paper IMDRF adopted the position that software did not need to be embodied in a physical device to be considered a medical device.

In their definition "'Medical device' means any instrument, apparatus, implement, machine, appliance, implant, reagent for in vitro use, ***software***, material or other similar or related article, intended by the manufacturer to be used, alone or in combination, for human beings, for one or more of the specific medical purpose(s)." The IMDRF white paper then defined covered medical purposes to include diagnosis, prevention, monitoring, treatment or alleviation of disease or injury as two of seven classes of medical purposes served by devices.

23.2.1.1 Examples of software considered to be SaMD

Per IMDRF guidance:

- Software with a medical purpose that operates on a general-purpose computing platform, i.e., a computing platform that does not have a medical purpose, is considered SaMD. For example, software that is intended

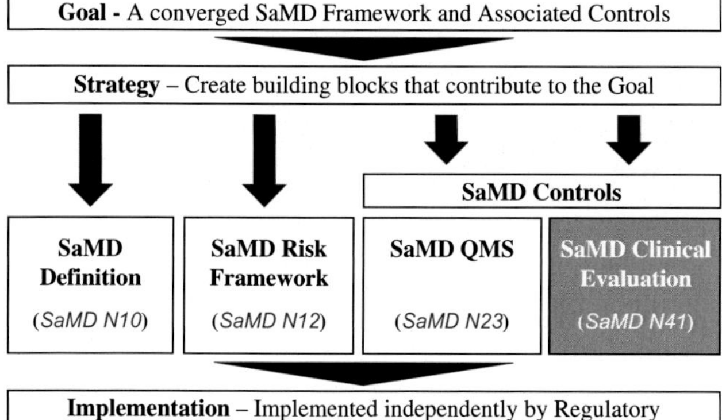

FIG. 23.2 IMDRF proposed regulatory pathway, IMDRF No 41 at 6.

for diagnosis of a condition using the tri-axial accelerometer that operates on the embedded processor on a consumer digital camera is considered a SaMD.

- Software that is connected to a hardware medical device but is not needed by that hardware medical device to achieve its intended medical purpose is SaMD and not an accessory to the hardware medical device. For example, software that allows a commercially available smartphone to view images for diagnostic purposes obtained from a magnetic resonance imaging (MRI) medical device is SaMD and not an accessory to MRI medical device.
- The SaMD definition notes states that "SaMD is capable of running on general purpose (nonmedical purpose) computing platforms." SaMD running on these general-purpose computing platforms could be located in a hardware medical device, for example, software that performs image post-processing for the purpose of aiding in the detection of breast cancer (CAD - computer-aided detection software) running on a general purpose computing platform located in the image-acquisition hardware medical device is SaMD.
- The SaMD definition notes states that "SaMD may be interfaced with other medical devices, including hardware medical devices and other SaMD software, as well as general purpose software." Software that provides parameters that become the input for a different hardware medical device or other SaMD is SaMD. For example, treatment planning software that supplies information used in a linear accelerator is SaMD.

23.2.1.2 *Examples of software not considered to be SaMD*

In contrast, per IMDRF guidance the following would not be considered SaMD:

- The SaMD definition states "SaMD is defined as software intended to be used for one or more medical purposes that perform these purposes without being part of a hardware medical device". Examples of software that are considered "part of" include software used to "drive or control" the motors and the pumping of medication in an infusion pump; or software used in closed loop control in an implantable pacemaker or other types of hardware medical devices. These types of software, sometimes referred to as "embedded software", "firmware", or "micro-code" are, not SaMD.
- Software required by a hardware medical device to perform the hardware's medical device intended use is not SaMD even if/when sold separately from the hardware medical device.
- Software that relies on data from a medical device, but does not have a medical purpose, e.g., software that encrypts data for transmission from a medical device is not SaMD.
- Software that enables clinical communication and workflow including patient registration, scheduling visits, voice calling, video calling is not SaMD.
- Software that monitors performance or proper functioning of a device for the purpose of servicing the device, e.g., software that monitors X-Ray tube performance to anticipate the need for replacement; or software that integrates and analyzes laboratory quality control data to identify increased random errors or trends in calibration on IVDs is not SaMD.

23.2.2 IMDRF risk categorization framework

The IMDRF risk categorization framework relies on two key factors of the intended use of a SaMD. The first key factor is the significance of the information provided by the SaMD to the healthcare decision. Significance to healthcare decision is determined by the presence or absence of a human intermediary and the time frame in which resulting actions may occur. Highly significant information is used to "treat or diagnose" a condition without human intervention in the immediate or near term. In general, high significance SaMD would effect treatment or diagnosis via connection to other hardware or software systems. The intermediate significance level, to "drive clinical management" is assigned to SaMD whose outputs will be used as an aid to a healthcare provider in determining next steps in treatment or diagnosis. The lowest significance level, to "inform clinical management," is applied to SaMD outputs that will not trigger an immediate or near term action.

The second key factor in the IMDRF regulation framework relates to the healthcare situation or condition in which the SaMD is being applied. A "Critical" situation or condition is one in which life is threatened without major therapeutic intervention. A "Serious" situation or condition is one in which intervention is necessary to mitigate long term consequences or unnecessary interventions. Serious situations are generally not time critical. Finally the "Non-serious" healthcare situation or condition category is applied when mitigation of long-term consequences and major interventions are not required.

The IMDRF framework combines the factors of significance to treatment/diagnosis and the risk of the healthcare circumstance into a single category for a SaMD. For example, Category IV, the most consequential for SaMD is applied only when the outputs are used to treat or diagnose a critical clinical situation without human intervention. A complete table of the possible combinations of significance and impact in the IRMDF framework is shown below.

	Significance of information provided by SaMD to healthcare decision		
State of healthcare situation or condition	**Treat or diagnose**	**Drive clinical management**	**Inform clinical management**
Critical	IV	III	II
Serious	III	II	I
Non-serious	II	I	I

23.2.2.1 Examples of SaMD for each risk category

Per IMDRF:

Category IV: "SaMD that performs diagnostic image analysis for making treatment decisions in patients with acute stroke, i.e., where fast and accurate differentiation between ischemic and hemorrhagic stroke is crucial to choose early initialization of brain-saving intravenous thrombolytic therapy or interventional revascularization."
Category III: "SaMD that uses data from individuals for predicting risk score in high-risk population for developing preventive intervention strategies for colorectal cancer."
Category II: "SaMD that uses data from individuals for predicting risk score for developing stroke or heart disease for creating prevention or interventional strategies."
Category I: "SaMD that uses data from individuals for predicting risk score (functionality) in healthy populations for developing the risk (medical purpose) of migraine (non-serious condition)."

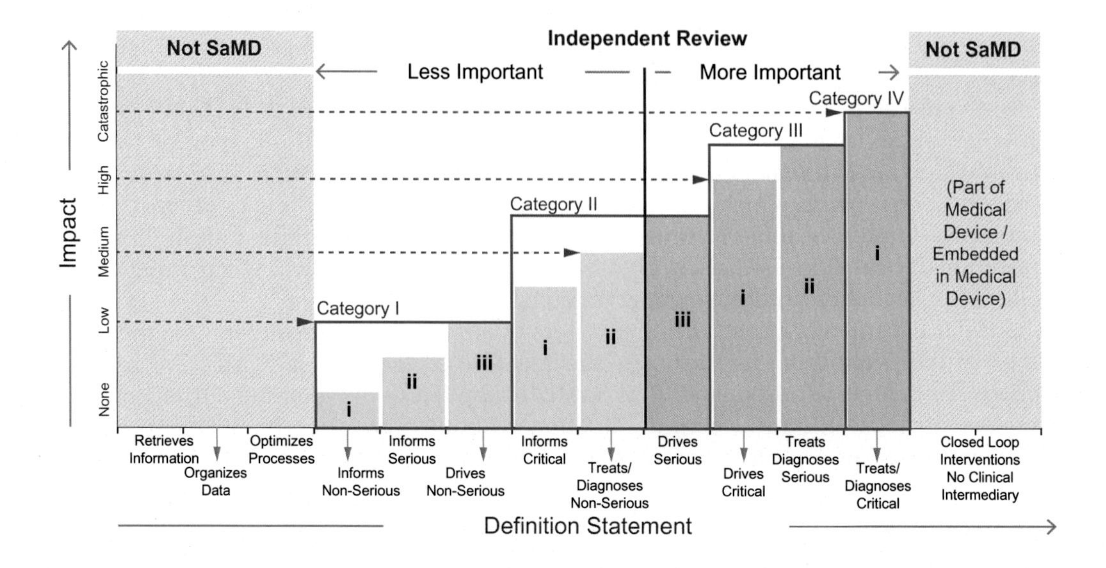

23.3 Legislation and regulation in the United States

Under Article I of the US Constitution, the Congress was vested with the sole power to pass legislation. Through these statutes, Congress apportions funds; establish areas of responsibility for federal agencies, and direct agencies and their sub-divisions to execute specific tasks. An example is the Federal Food, Drug, and Cosmetic Act of 1938 (FD&C Act), Public Law 75-717 (June 25, 1938) [71], which gave authority to the U.S. Food and Drug Administration to oversee the safety of food, drugs, medical devices, and cosmetics within the United States.

Often in drafting in these laws, Congress makes an explicit delegation to the agencies to establish the appropriate rules and regulations necessary to achieve the goals outlined in the statutory provisions. This allows Congress to rely on the agencies, which often have significant amount of expertise and can "fill in" the technical details of programs that Congress created (or modified) by statute. Agencies are authorized to draft and issue rules and regulations that are then binding on the sections of the public subject to the statute's provisions. The regulations issued pursuant to this authority carry the force and effect of law and can have substantial implications for both policy implementation and day-to-day operations. Given the considerable power inherent in this delegation, Congress separately established a uniform framework for all agencies to follow for the exercise of rulemaking authority, under the Administrative Procedure Act (APA), Public Law 79-404 (June 11, 1946) [72]. Under the APA, an agency must not take action that goes beyond its statutory authority or violates the Constitution. Agencies must follow an open public process when they issue regulations, according to the APA.

Like other US government executive branch agencies, The FDA's authority comes from legislation and regulations. The mission of the United States Food and Drug Administration within the Department of Health and Human Services is to protect "the public health by ensuring the safety, efficacy, and security of human and veterinary drugs, biological products, and medical devices" [73].

The FDA's primary empowering legislation is the Federal Food, Drug, and Cosmetic (FFD&C) Act of 1938 [71,74]. This legislation was enacted after a (then) legally marketed toxic antibiotic elixir killed 107 people [75]. The FFD&C Act has been extended by amendments over the years. For example, The Medical Device Amendments of 1976 applied safety and effectiveness safeguards to new medical devices after a Senate inquiry determined that numerous injuries and deaths had been caused by faulty medical devices [76].

One of the primary ways that FDA meets its responsibility is by restricting access to the marketplace for regulated products (including devices) via pre-market approval and post-market monitoring. The FDA Premarket approval (PMA) processes are used to evaluate the safety and effectiveness of devices that "support or sustain human life, are of substantial importance in preventing impairment of human health, or which present a potential, unreasonable risk of illness or injury" [77]. Premarket approval regulations of "Class III" devices are located in Title 21 Code of Federal Regulations (CFR) Part 814, Premarket Approval of Medical Devices [78].

In the following sections we will specifically address legislation, regulations and agency guidance that impact Clinical Decision Support.

23.4 FDA and CDS software regulation

23.4.1 Past FDA regulation of clinical software systems

Through its mandate from Congress to safeguard the public, the FDA has regulated marketing and use of medical devices. Section 201(h) of the 1976 Federal Food, Drug, and Cosmetic Act defines a medical device as any "instrument, apparatus, implement, machine, contrivance, implant, in vitro reagent, or other similar or related article, including any component, part, or accessory, which is ... intended for use in the diagnosis of disease or other conditions, or in the cure, mitigation, treatment, or prevention of disease ... or intended to affect the structure or any function of the body" [76]. In the past, FDA representatives have stated that clinical software programs, whether associated with biomedical devices or stand-alone, are "contrivances," and therefore fall within the FDA's realm of responsibility [65].

The FDA regulates medical devices that are commercial products used in patient care, devices used in the preparation or distribution of clinical biological materials (such as blood products), and experimental devices used in research involving human subjects [65]. Commercial vendors of specified types of medical devices must register as manufacturers with the FDA and list their devices as products with the FDA. Upon listing, the FDA classifies medical devices by categories. In its regulation of classified medical devices, the FDA usually takes one of three courses of action [65]. First, the FDA can "exempt" specific devices, or categories of devices, deemed to pose minimal risk. Second, the FDA employs the so-called 510(k) process—premarket notification—for non-exempt systems [79]. Through the 510(k) process, manufacturers attempt to demonstrate that their devices are equivalent, in purpose and function, to low-risk (FDA Class I or Class II) devices previously approved by the FDA (or to devices marketed before 1976). Such devices can be cleared by the FDA directly. Finally, the FDA requires premarket approval (PMA) for higher-risk (FDA Class III) products and for products with new (unclassified) designs invented after 1976. Through the premarket approval process, a manufacturer provides evidence to the FDA that a product performs its stated functions safely and effectively [77]. Premarket approval is especially important for those products that pose significant potential clinical risk. The processes of 510(k) premarket notification and premarket approval typically take a few to many months to complete, and may involve numerous iterations [65].

Exemption can take place in two ways: a device can be exempt from registration, and thus not subject to 510(k) requirements at all; or, a category of listed (classified) devices may be specifically exempted from certain regulatory requirements [65]. Whenever a nonexempt product is modified substantially (as defined by FDA guidelines), the vendor must reapply to the FDA for new clearance through the 510(k) or premarket approval mechanisms.

In mid-1996, the FDA called for new discussions on the regulation of software programs as medical devices [58]. Previously, a 1989 draft policy [80], never adopted formally, served as guidance for FDA conduct. The draft policy recommended that the FDA exempt from regulation "generic" software (e.g., content-free spreadsheet and database programs), educational systems merely providing information, and systems that generated advice for clinician users in a manner that they could easily override. In response to the 1996 FDA announcement, a consortium of health information-related organizations developed and published in 1997 a set of recommendations for public and private actions that were intended to accomplish responsible monitoring and regulation of clinical software systems [65,81]. The list of 1997 consortium members included the American Medical Informatics Association (AMIA), the Center for Healthcare Information Management (CHIM); the Computer-based Patient Record Institute (CPRI), the Medical Library Association (MLA), the Association of Academic Health Sciences Libraries (AAHSL), the American Health Information Management Association (AHIMA), the American Nurses Association (ANA), and the American College of Physicians (ACP). Not all boards of directors of all consortium members formally endorsed the consortium recommendations [65]. Dr. Reed Gardner at the University of Utah subsequently obtained NIH grant funding to develop prototypic models of the Software Oversight Committees (SOCs) at each of four separate healthcare centers that had advanced information systems installed (see [82] for a discussion of early SOC activity at Brigham and Women's Hospital in Boston).

In July 2011, the FDA published "Draft Guidance for Industry and Food and Drug Administration Staff; Mobile Medical Applications; Availability" in the Federal Register [83]. The draft guidance made two major points. First, it announced FDA's intention to apply its authority to specific types of mobile applications referred to as "mobile medical apps." The draft guidance narrowly defines mobile medical apps as any "mobile app that meets the definition of "device" in section 201(h) of the Federal Food, Drug, and Cosmetic Act (FD&C Act) (21 U.S.C. 321); 1 and either: Is used as an accessory to a regulated medical device or Transforms a mobile platform into a regulated medical device." The draft guidance asked for comments regarding device extensions that aid in use of the connected device, extend the intended use of the connected device or create a new intended use for the device.

The second major point of the guidance addressed stand-alone clinical decision support. FDA's guidance on mobile applications explicitly excludes stand-alone (i.e., not device-based) CDS. Despite this restriction, the document clearly states FDA's intent to issue additional guidance on stand-alone CDS and solicits comments regarding risk classification factors, criteria for assuring safety and effectiveness, and controls CDS manufacturers should implement to reduce risk and costs of regulatory compliance.

In September 2011 the FDA held a public workshop on their Mobile Medical Application draft guidance [84]. The workshop included presentations and a panel discussion addressing stand-alone clinical decision support and its potential regulation. During this workshop the FDA broadly defined CDS as software that uses an individual's information from various sources (electronically or manually entered) and converts this information into new information that is intended to support a clinical decision [85]. The FDA cited several methods that could be applied to convert information including using algorithms, formulae, database lookups and rules and associations. Factors deemed "relevant" included the impact of the CDS on subject health, the degree of acceptance of the CDS in clinical practice and the ability to easily identify erroneous outputs. Despite the public notice and workshop of 2011, the FDA has not published draft guidance on stand-alone CDS as of May 2013.

23.4.2 FDASIA

In 2012 Congress passed the Food and Drug Administration Safety and Innovation Act ("FDASIA") [Public Law 112-144] which included provisions that directed the FDA, Federal Communications Commission ("FCC"), and Office of the National Coordinator for Health Information Technology ("ONC") to work together to create recommendations for a "risk-based regulatory framework pertaining to health information technology." [See §618 of FDASIA]. Pursuant to §618, the three agencies issued the FDASIA Health IT Report ("the Health IT Report") in April 2014, wherein they outlined recommendations and a proposed strategy for balancing innovation and patient safety in health IT.

In the Health IT Report the agencies proposed a limited, narrowly-tailored approach that primarily relies on ONC-coordinated activities and private sector capabilities is prudent. The three Agencies also recommended that no new or additional areas of FDA oversight are needed. The Report divided health information technology ("health IT") into three components: (1) administrative health IT functions, (2) health management health IT functions, and (3) medical device health IT functions. Under this framework, administrative health IT, which was defined to include such functions as billing and claims processing, practice and inventory management, and scheduling were deemed to pose limited or no risk to patient safety and, thus, did not require additional oversight. The second category—health management IT—was defined "to include, but are not limited to, health information and data exchange, data capture and encounter documentation, electronic access to clinical results, most clinical decision support, medication management, electronic communication and coordination, provider order entry, knowledge management, and patient identification and matching" [Health IT Report at 3.]. While some risks were noted with this segment, the potential benefits were deemed to be greater than the risks which the report found to be "low."

[Id.] Oversight of this type of health IT was to be split into two areas—that portion of health management IT that manifested itself in medical devices would continue to fall under existing FDA oversight and new powers/oversight would not be pursued. For the remainder, the Agencies proposed a strategy and set of recommendations that focused primarily on the creation of a risk-based framework for health management health IT functionalities.

The report outlined four key priority areas that the Agencies felt would both more fully realize the benefits of health IT while managing the associated risks: (I) Promote the Use of Quality Management Principles; (II) Identify, Develop, and Adopt Standards and Best Practices; (III) Leverage Conformity Assessment Tools; and (IV) Create an Environment of Learning and Continual Improvement. The Agencies proposed a risk-based model which incorporated a public/private partnership approach which would include the creation of a Health IT Safety Center. This public-private entity would be created by ONC, in collaboration with FDA, FCC, and the Agency for Healthcare Research and Quality (AHRQ), with involvement of other Federal agencies, and other health IT stakeholders. This new Health IT Safety Center would "convene stakeholders in order to focus on activities that promote health IT as an integral part of patient safety with the ultimate goal of assisting in the creation of a sustainable, integrated health IT learning system that avoids regulatory duplication and leverages and complements existing and ongoing efforts." Participation would be fully voluntary and the "proposed framework and priority areas contained in this report are not binding, do not create new requirements or expectations for affected parties, and do not create or confer any rights for or on any person.", so no safe harbor would accrue to vendors or healthcare facilities who choose to participate.

Highlighting the need for a risk-based approach to regulatory oversight, the FDASIA Health IT Report indicated that the FDA did not intend to focus its oversight on CDS programs because most are likely to pose a low risk of patient harm. The agencies were mindful, however, to explain that such low-risk CDS programs are "not intended to replace clinicians' judgment, but rather to assist clinicians in making timely, informed, higher quality decisions." Even then, there was a strong reluctance to exert more direct oversight and regulation on clinical decision support software/applications, with agencies making the claim that clinical staff (unlike non-clinical, e.g., patients/civilians) could still independently analyze the underlying basis for the CDS recommendations/suggestions. This seems to ignore the fact that as technology progresses, CDS becomes more "black-boxed" in nature and forces even clinicians to accept the results of CDS and EHRs blindly. Nowhere is this more evident than in relation to AI in health IT.

23.4.3 21st Century Cures Act

The 21st Century Cures Act of 2016, comprised of 4 major divisions and 18 titles, has significant impact on Clinical Decision Support and its regulation [86]. One of the legislation's main themes is enhancing health through discovery, development, and delivery of innovations.

Division 1 Title III (Development) subtitle F (Medical Device Innovations) section 3060 "Clarifying Medical Software Regulation" is the most extensive legislation addressing Clinical Decision Support in the US. As we have noted, the FDA until this point issued and retracted various guidance regarding CDS regulation For example the FDA Policy for the Regulation of Computer Products of 1987 was retracted in 2005 [87,88].

The express purpose of Section 3060 was to bring clarity to FDA's regulation of medical software that may or may not include aspects of clinical decision support. It includes separate sections on what clearly ..is.. subject to regulation and what is clearly ..not.. subject to regulation. Section 3060 specifically includes for regulation devices that are designed to acquire, process or analyze medical images, or signals from an in vitro diagnostic system or some other signal acquisition system.

Section 3060 then specifically excludes from regulation various types of Health IT that are not deemed to be clinical decision support. Examples of HIT that are excluded outright from FDA regulation include systems that support administrative operations (e.g., claims), promote healthy lifestyles in general, or transfer and display lab test results or other device data without interpretation or analysis. HIT that provides the electronic version of a paper medical record for healthcare professionals or their direct supervisees is also excluded, however the stipulation of audience is of considerable practical importance.

After excluding HIT that is not considered to be CDS, Congress uses the balance of § 3060 to describe specific types of medical software that includes CDS may or may not be excluded from regulatory oversight. The underlying principle applied in the section is to exclude systems in which an informed healthcare provider is in between the HIT system and the patient. Both conditions, "informed" and "healthcare provider" are relevant to application of this law. Systems that allow a healthcare provider to review the basis of the recommendation are excluded from regulation. In "The Demise of the Greek Oracle Model for Medical Diagnostic systems," Miller and Masarie note that "explanation is a difficult problem" because of the multistep and multipath complexity of the diagnostic process [89].

23.4.3.1 *CDS excluded from regulation under the Cures Act*

The initial type of excluded medical software CDS under §3060 provides medical information such as peer-reviewed clinical studies and clinical guidelines that may (or may not) apply to an individual patient's case. Further extending the "informed practitioner" reasoning, Congress excluded CDS systems that support or provide "recommendations to a healthcare professional about prevention, diagnosis, or treatment of a disease or condition" IF the receiving healthcare professional can "independently review the basis for such recommendations that such software presents." The distinction is of even greater importance today than when described by Miller in 1990 because of a recent blossoming in the use of technologies that cannot provide a basis for their outputs (e.g., many machine learning algorithms) [90].

23.4.3.2 *FDA draft guidance on CDS software regulation*

In September of 2019 the FDA published in the Federal Register "Clinical Decision Support Software Draft Guidance for Industry and FDA Staff" for public comment [91]. The stated purpose of the guidance was to clarify FDA's risk-based regulatory approach to CDS software functions by providing practical explanations and examples. At first glance, it appears that the FDA sought to balance its approach, which was based on the IMDRF frameworks, with the restrictions placed on its authority by the Cures Act.

As noted above, the two key determinants impacting the regulatory classification of CDS software in the Cures Act are: (1) if the intended user is a healthcare provider;

and (2) If the user is able to independently review the rationale behind any recommendation proffered. The 2019 draft guidance summarizes the impact of these factors in tabular fashion:

Is the intended user an HCP? [part of criteria (3) and (4)]	Can the user independently review the basis?[a] [part of criterion (4)]	Is it device CDS?
Yes	Yes	No, it is Non-Device CDS because it meets all of section 520(o)(l)(E) criteria
	No	Yes, it is Device CDS
No, it is a patient or caregiver	Yes	Yes, it is Device CDS
	No	Yes, it is Device CDS

[a] *"Can the user independently review the basis?" asks whether the function is intended for the purpose of enabling the user to independently review the basis for the recommendations so that it is not the intent that user rely primarily on any such recommendation [part of criterion (4)].*

23.4.3.3 *Best practices for AI/ML based CDS*

The 21st Century Cures Act and related regulation raises important challenges for AI/ML based CDS. In particular, most AI and ML methods are not able to provide a human understandable basis for their outputs [92]. Thus, innovations based on these methods are potentially subject to regulation prior to distribution in the marketplace. As a service to CDS developers and implementors, both the American Medical Informatics Association (AMIA) and a consortium of international device regulators have published guidance on related best practices.

In January of 2021 AMIA published a position paper entitled "Recommendations for the safe, effective use of adaptive CDS in the US healthcare system" [93]. The authors defined adaptive CDS from an artificial intelligence perspective as "CDS that can learn and change performance over time, incorporating new clinical evidence, new data types and data sources, and new methods for interpreting data." Several of the papers recommendations focused on metrics of transparency for FDA precertification and lifecycle regulatory approach covering topics such as semantics and provenance of training datasets, data set preprocessing, data set representativeness and rigor in software engineering practices. The AMIA authors pursued a second line of recommendations regarding communications standards. Examples where standards for communication were supported included expected performance parameters and limitations, intended use, testing and optimization processes.

In October of 2021, the FDA, Health Canada, and the United Kingdom's Medicines and Healthcare products Regulatory Agency (MHRA) published "Good Machine Learning Practice for Medical Device Development: Guiding Principles" [94]. A summary table from the guidance is reproduced in Table 23.3.

The three countries' device regulatory agencies were clearly influenced by the AMIA position paper, for example by advocating for representative data sets, clear user instructions, and good software engineering practices. The regulatory agency authors included additional principles in areas such as human involvement in multidisciplinary teams and human-AI team performance. We strongly suggest that AI/ML CDS developers read, understand

TABLE 23.3 Good Machine Learning Practice for Medical Device Development: Guiding Principles.

Multi-disciplinary expertise is leveraged throughout the total product life cycle	Good software engineering and security practices are implemented
Clinical study participants and data sets are representative of the intended patient population	Training data sets are independent of test sets
Selected reference datasets are based upon best available methods	Model design is tailored to the available data and reflects the intended use of the device
Focus is placed on the performance of the human-AI team	Testing demonstrates device performance during clinically relevant conditions
Users are provided clear, essential information	Deployed models are monitored for performance and re-training risks are managed

and follow these best practice guidelines in ways commensurate with the scale and maturity of their innovation.

23.4.3.4 Who should seek FDA premarket approval

CDS developers should carefully evaluate their product's functionality as a primary consideration. Two important factors are the intended user and if the basis for the CDS output is reviewable. Any CDS directed toward patients is categorized as "device CDS" that may require FDA review. CDS directed toward practitioners that does not allow the practitioner to independently review the basis for the CDS output is also categorized as "device CDS." Only practitioner directed CDS that allows independent decision basis review is considered "not device CDS." A third aspect of functionality to be considered is IMDRF risk category. Applications with high IMDRF risk categories (III or IV) require more careful premarket evaluation than lower risk categories (I–II).

The intended deployment of a "device CDS" product also has bearing on the need for FDA evaluation. Developers who qualify as Manufacturers that intend to market or widely distribute their CDS are clearly subject to FDA oversight. Certain developers may not be considered manufacturers by FDA and would therefore not be subject to oversight. Examples shared in 2019 guidance include: licensed practitioners that develop device CDS for use solely in their own professional practice or group practice. Similarly excluded are persons who manufacture device CDS solely for use in research, teaching, or analysis and do not introduce such devices into commercial distribution. The guidance unfortunately does not address clinical entities larger than group practices.

The FDA encourages developers and manufacturers to contact them to determine what regulatory requirements apply. An overview of mechanisms to request FDA feedback or meetings was published in January 2021 [95] and is available on the FDA Digital Health Center of Excellence website (https://www.fda.gov/medical-devices/digital-health-center-excellence).

23.5 21st Century Cures and other HHS agencies

The 21st Century Cures Act and subsequent regulations directly addressed criteria for subjecting CDS to regulatory oversight. The Act also addressed an important complimentary topic—the useful and free sharing of clinical data among HIT systems.

Semantic interoperability refers to the sharing of data between systems in a manner that allows the receiving system to use the data as if it were generated natively—i.e., no loss of meaning from either a human data use or algorithmic data use perspective. It is our opinion that clinical decision support is a compelling algorithmic use for interoperable clinical data.

23.5.1 API

Division 1 Title IV Section 4002 requires as a condition for HIT certification that the developer "has published application programming interfaces and allows health information from such technology to be accessed, exchanged, and used without special effort." At the time the legislation was authored, HL7 Fast Healthcare Interoperability Resources (FHIR) was the standard API technology to be implemented. Section 4002 further established that the expected scope for API-based data exchange to be "access to all data elements of a patient's electronic health record to the extent permissible under applicable privacy laws."

23.5.2 Interoperability

The 21st Century Cures Act established the Health Information Technology Advisory Committee (HITAC) (section 3002) to make recommendations to the Office of the National Coordinator (ONC) for Health Information Technology regarding policies, standards, implementation specifications, and certification criteria to advance the electronic access, exchange, and use of health information via regulatory actions. HITAC recommendations to ONC included that "all data elements" referred to in 21 century cures be operationally defined as those data elements contained in the United States Core Data for Interoperability (USCDI) [96].

23.5.2.1 Defining interoperability

Under the 21st Century Cures Act, Congress amended §300jj of the Public Health Service Act (§3000 of the Public Health Service Act (42 U.S.C. 300jj)) to include a definition of interoperability with respect to Health IT:

> INTEROPERABILITY.—The term 'interoperability', with respect to health information technology, means such health information technology that—
>
> (A) enables the secure exchange of electronic health information with, and use of electronic health information from, other health information technology without special effort on the part of the user;
> (B) allows for complete access, exchange, and use of all electronically accessible health information for authorized use under applicable State or Federal law; and
> (C) does not constitute information blocking as defined in section 3022(a).

Section 4003 of the 21st Century Cures sections mandated ONC actions promoting the establishment of a "Trusted Exchange Framework" to ensure appropriate and secure exchange of clinical information.

23.5.3 Information blocking under §4004

In May of 2020, ONC released the 21st Century Cures Act: Interoperability, Information Blocking, and the ONC Health IT Certification Program ("Final Rule") [45C.F.R. Part 171]. Under the Final Rule, ONC establishes three categories or "actors" subject to its provisions: Health Care Providers (HCPs); Health Information Networks (HINs) or Exchanges (HIEs); and Health IT Developers of Certified HealthIT. The provisions went into effect on April 5, 2021.

Under its provisions, actors would be guilty of "information blocking" if they undertook any actions determined to be likely to interfere with, prevent, or materially discourage access to, exchange of, or use of Electronic Health Information ("EHI"). EHI is currently defined as all the data elements defined in USCDI v2 such as allergies, problems, labs and procedures [97].

Information blocking is defined as any practice "[E]xcept as required by law or covered by an exception set forth in subpart B or subpart C of this part, is likely to interfere with access, exchange, or use of electronic health information" [§171.103(a)(1)].

The Final Rule sets out different types of prohibited actions depending on Actor class:

- For HCPs, they would be guilty of information blocking if they implement a practice the HCP "knows that such practice is unreasonable and is likely to interfere with access, exchange, or use of electronic health information." [§171.103(a)(3)].
- For HIN/HIE or Health IT Developers, scope of what qualifies as "information blocking" under the Final Rule is defined as anything the developer, network or exchange "knows, or should know, that such practice is likely to interfere with access, exchange, or use of electronic health information" [§171.103(a)(2)].

The Final Rule also included a range of five exceptions (§171.200 *et seq.*), practices by any or all of the Actors, that would be treated as an exception to the list of actions prohibited under §§171.103(a)(1)-(3):

- Preventing harm (§ 171.201)—practices that are reasonable and necessary to prevent harm to a patient or another person, provided all conditions set for in rule are met.
- Privacy (§ 171.202)—practices designed to protect the privacy of EHI, pursuant to specific, relevant privacy laws (e.g., HIPAA)
- Security (§ 171.203)—practices designed to protect the security of EHI, which are specifically tailored to the security risk and implemented in a consistent and non-discriminatory manner
- Infeasibility (§ 171.204)—when an actor is permitted to not provide EHI where it can be shown that fulfilling the request is objectively infeasible.
- Health IT performance (§ 171.205)—practices which are implemented to address Health IT performance, maintenance, security issues or to prevent harm to a patient.

The legislation also established penalties of up to $ one million per violation for entities that deliberately impede the access and exchange of clinical data—a practice referred to as "Information Blocking" (Section 4004). Taken together, §§3002, 4002-4 of the 21st Century Cures made significant progress toward free nation-wide flow of useable and useful electronic data for multiple purposes including CDS.

The 21st Century Cures Act final rule added 4 new certification criteria, known as the 2015 Edition Cures Update [98]. § 170.315(b)(10) Electronic Health Information (EHI) Export Focuses on the ability to export the electronic health information to support patient EHI access requests and exporting an entire patient population to transition to another health IT system. § 170.315(g)(10) Standardized API for Patient and Population Services Requires the use of the HL7® Fast Healthcare Interoperability Resources (FHIR®) Release 4 standard for single and multiple patients data.

23.6 HiTECH act and MU certification

Congress enacted the HITECH Act in 2009 with the purpose of promoting the adoption and "'meaningful use' of EHRs—that is, their use by providers to achieve significant improvements in care." The HITECH Act's goal was to incentivize healthcare providers to implement EHRs through Medicaid and Medicare payments totaling $27 billion through 2019. Under the act, in order to receive these payments, clinicians and hospitals had to show clear improvements in "healthcare processes and outcomes," by meeting the objectives set by the Secretary of Health and Human Services.

The resulting Medicare and Medicaid EHR Incentive Programs, more broadly known as "Meaningful Use" had major impact on the use of EHRs and CDS in the US. The Meaningful Use program provided fiscal incentives and disincentives for individual healthcare professionals and hospitals to use certified EHRs, including decision support, in specific and measurable ways. In 2018 the incentive programs were renamed "Promoting Interoperability Programs."

23.6.1 Legal overview

Meaningful Use grew out of the Health Information Technology for Economic and Clinical Health (HITECH) Act, a part of the American Recovery and Reinvestment Act of 2009 (ARRA) [99]. The HITECH Act created Title XXX of the Public Health Service Act entitled "Health Information Technology and Quality." This federal legislation authorizes the US Department of Health & Human Services (HHS) to "establish programs to improve healthcare quality, safety, and efficiency through the promotion of health IT, including electronic health records and private and secure electronic health information exchange."

In order to establish the Congressionally authorized EHR incentive programs, the HHS Centers for Medicare & Medicaid Services (CMS) published the Medicare and Medicaid EHR Incentive Programs proposed rule in the Federal Register (75 FR 1844) in January 2010 [100]. The proposed rule invited public comment on a definition for Stage 1 meaningful use of Certified EHR Technology and regulations associated with the incentive payments. The "Medicare and Medicaid EHR Incentive Programs final rule" (75 FR 44314) was published on July 28, 2010 [101]. On the same day, the ONC published a related final rule "Health Information Technology: Initial Set of Standards, Implementation Specifications, and Certification Criteria for Electronic Health Record Technology" (75 FR 54163) [102]. As a whole, these laws and regulations established financial incentives for healthcare providers and hospitals to "meaningfully use" certified EHRs, defined certification criteria and defined "meaningfully use."

23.6.2 Incentives and penalties

Meaningful Use provided financial incentives for professionals and hospitals to use EHRs. Financial incentives begin as additional payments from Medicare and Medicaid. Potential maximal payments decrease over time and eventually disappear, to be replaced by payment penalties for the non-compliant. Eligible professionals starting in 2012 could receive up to $44,000 over 5 years through the Medicare incentives and up to $63,750 over 6 years through the Medicaid incentives. Hospital payments started at $2 Million.

Beginning in 2015, Medicare payment penalties were implemented on non-compliant eligible professionals [103]. In the first year, Medicare reduced physician fees for all covered professional services performed by non-compliant eligible providers by 1%. In subsequent years penalties increased by 1% per year to a maximum of 5% in 2019. The numbers of EPs receiving payment adjustments were 257,000 in 2015, 209,000 in 2016 and 171,000 in 2017 [104]. A hardship exemption application process was created to avoid Medicare payment adjustments.

23.6.3 Policy

Meaningful Use was designed to address four overarching health policy priorities established by the HIT Policy Committee (HITPC), an HHS federal advisory committee: (1) Improve quality, safety and efficiency and reduce health disparities; (2) Engage patients and their families in health care; (3) Improve care coordination; and (4) Improve population and public health. The HITPC identified care goals to meet each health policy priority.

The HITPC identified five care goals to improve quality, safety and efficiency while reducing health disparities: (1) Provide access to comprehensive patient health data for a patient's healthcare team; (2) Use evidence-based order sets and computerized provider order entry (CPOE); (3) Apply clinical decision support at the point of care; (4) Generate lists of patients who need care and use them to reach out to those patients; and (5) Report information for quality improvement and public reporting. HHS policy makers elected to pursue these four priorities and related care goals via an incremental approach with three stages enacted over time.

Stage 1: Capture health information electronically in a structured format; use that information to track key clinical conditions and coordinate care; Implement clinical decision support tools; use EHRs to engage patients and families and report clinical quality measures and public health information.
Stage 2: Encourage the use of health IT for continuous quality improvement at the point of care and the exchange of information in the most structured format possible.
Stage 3: Promote improvements in quality, safety and efficiency leading to improved health outcomes by focusing on decision support, patient access to self-management tools, access to comprehensive patient data and improving population health.

23.6.4 MU and CDS

Clinical decision support is one of the central pillars of Meaningful Use. CDS is defined in 75 FR 1844 as "health information technology functionality that builds upon the foundation of an EHR to provide persons involved in care processes with general and person-specific information, intelligently filtered and organized, at appropriate times, to enhance health and health care." The Meaningful Use CDS definition builds upon the AMIA CDS Roadmap definition by adding specific reference to EHRs as a required foundation. (AMIA definition: "Clinical decision support (CDS) provides clinicians, staff, patients or other individuals with knowledge and person-specific information, intelligently filtered or presented at appropriate times, to enhance health and health care") [105].

Stage 1 Meaningful use objectives and measures refer to several different types of CDS interventions as classified by Osheroff et al. [106]. In somewhat simplified form these include:

- CPOE—More than 30% of all unique patients have at least one medication order entered using CPOE
- Drug-drug and drug-allergy interaction checks—More than 80% of all unique patients have at least one entry or an indication that no problems are known
- Implementation of one clinical decision support rule
- Patient-specific education resources identified with an EHR are provided to more than 10% of unique patients.

Stage 2 Meaningful Use objectives and measures increase CDS related requirements [107] include:

- CPOE—Use CPOE for more than 60% of medication orders, 30% of laboratory orders, and 30% of radiology orders.
- Drug-drug and drug-allergy interaction checks are continuously implemented
- Implementation of five clinical decision support interventions related to four or more clinical quality measures
- Patient-specific education resources identified with EHR are provided to more than 10% of unique patients.

Stage 3 Meaningful Use: Stage 3 criteria relating to decision support were largely unchanged from Stage 2. Order entry targets remained at 60% for medication orders and 30% for laboratory and radiology orders. Requirements for drug-drug and drug-allergy checks, CDS rule implementation and provision of patient-specific education were also essentially unchanged in Stage 3 [108].

23.6.5 MU impact

In order to understand the impact of the EHR incentive programs on CDS utilization, we will first examine national data gathered before the incentive programs were implemented in 2011.

DesRoches et al. conducted a survey of 2758 ambulatory care physicians' use of EHRs in 2007–08 [109]. They found that 4% of ambulatory care physicians had a fully functional EHR system including CDS components (drug warnings, abnormal lab result identification, guideline-based reminders), and order entry management. An additional 13% had a basic system with order entry limited to medications and no CDS.

In 2009 Jha et al. published results of an EHR use survey sent to all US acute care general medical and surgical member hospitals in 2008 [110]. They found that only 1.5% of non-federal hospitals had a comprehensive EHR system implemented across all major clinical units. The percentages of hospitals fully implementing various types of CDS were reported as: guidelines (17%), reminders (23%), drug-allergy alerts (46%), drug-drug interaction alerts (45%), drug-lab interaction alerts (34%), drug dose support (31%), lab CPOE (20%), and medication CPOE (17%).

Meaningful Use Stage 1 had a major impact on the use of EHRs and CDS in the United States. In 2011, 833 hospitals and 57,652 professionals attested for meaningful use. In 2012 an additional 1726 hospitals and 132,395 professionals were added. According to CMS data presented to the HITPC in April,2013, over 77% of eligible hospitals and approximately 53% of EPs are meaningful users of EHRs according to Stage 12,011 standards. All meaningful users were required to implement one CDS rule. Over 80% of professionals and hospitals met the CPOE for medications requirement and over 90% met the medication allergy list requirement.

By 2017 79.7% of office based physicians had adopted a certified EHR and 86% adopted any EHR [111]. Hospital adoption of certified EHRs in 2017 was 96% [112].

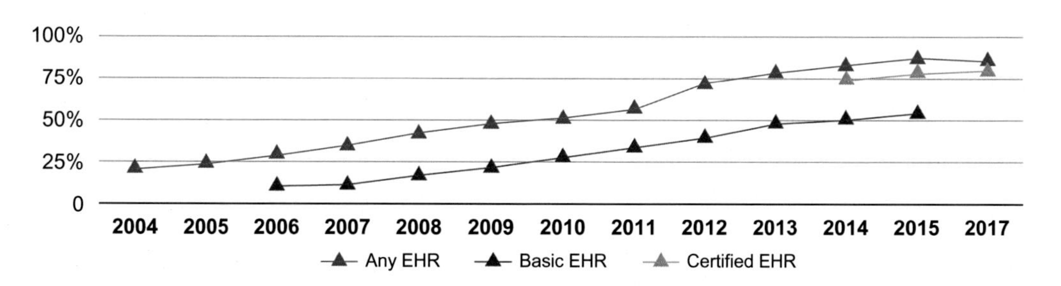

In April 2018 the Meaningful Use Incentive Program was renamed Promoting Interoperability Program with increased focus on interoperability and increasing patient access to health information. Meaningful Use 3 metrics from 2015 remain mostly in place with a small number of modifications in the 2015 Edition Cures Update [98] [see above]. Payment adjustments continue to be levied but are now included as a component of funding calculations. Eligible professionals participating in the Medicaid Promoting Interoperability Program are required to report on any six eCQMs and at least one outcome measure in 2021 [113]. Hospitals participating in the Medicare promoting interoperability program are required to use certified EHRs and report on four self-selected eCQMs in 2021.

23.7 Conclusion

The legal and regulatory environment for health information technology has changed dramatically since the first version of this chapter was written in 2007. Rapid legal and regulatory change has both driven and been driven by a massive uptick in EHR adoption. Fewer than 5%

of ambulatory care physicians and hospitals had fully functional EHR systems deployed in 2008. The HITECH act of 2009 and Meaningful Use incentive programs made EHR use nearly ubiquitous. With rising adoption, malpractice litigation involving health IT has become more common place. The IMDRF health IT risk framework was a significant international advance in prioritizing implementations for safety enhancing oversight. The FDA's role in premarket oversight for safety and efficacy was significantly clarified by the 21st Century Cures Act and related regulations and guidance. In its implementation of 21st Century Cures legislation, FDA adopted the IMDRF risk categorization framework as a method for targeting its regulatory efforts. These efforts are extremely timely given the upward trajectory of AI and ML algorithms in the provision of clinical decision support. It merits mention that the AI/ML boom has been fueled in part by the availability of large electronic data sets for training resulting from Meaningful Use driven EHR adoption. The transition of regulatory focus from EHR use to clinical information access and meaningful interoperability speaks to the success of past efforts. While many challenges still await us, the progress made in the past 15 years has been remarkable.

As noted earlier, this chapter has detailed the legal and regulatory environment from a US perspective. It is not feasible to include detail about efforts in other nations, because such an environment tends to be nation-specific. However, this chapter can serve to illustrate the kinds of issues that must be considered and ways to address them that are likely to be needed in any environment.

Acknowledgments

Randolph A. Miller co-authored the preceding 2012 version of this Chapter; substantial material has been retained in the current version. Sarah M. Harris, J.D. (nee Sarah M. Miller) co-authored the preceding 2007 version of this Chapter; some material has been retained in the current version. This chapter is based in part on earlier collaborations of Randolph A. Miller with Kenneth Schaffner, MD, PhD, and Alan Meisel, JD, at the University of Pittsburgh in the 1980s [1] and with Reed Gardner, PhD, at the University of Utah and others in the 1990s [65,81]. Permission has been granted by the Annals of Internal Medicine to reproduce herein portions of copyrighted materials previously published in that journal [65,81].

References

[1] Miller RA, Schaffner KF, Meisel A. Ethical and legal issues related to the use of computer programs in clinical medicine. Ann Intern Med 1985;102:529–36.
[2] Gemignani MC. Products liability and software, 8 Rutgers Computer & Tech. L.J. 173; 1981. p. 196–9.
[3] Note: Strict products liability and computer software: Caveat vendor, 4 Computer/L.J. 373; 1983.
[4] Note: Negligence: Liability for defective software, 33 Okla. L. Rev. 848, 855; 1980.
[5] Note: Computer software and strict products liability, 20 San Diego L. Rev. 439; 1983.
[6] Note: Easing plaintiffs' burden of proving negligence for computer malfunction, 69 Iowa L. Rev. 241; 1983.
[7] Rustad ML. Punitive damages in cyberspace: where in the world is the consumer? Chapman Law Rev 2004;7 (39).
[8] American Law Institute. Restatement (Second) of Torts. St. Paul, MN: American Law Institute Publishers; 1965.
[9] Uniform Commercial Code. Chicago: National Conference of commissioners of, Uniform State Laws; 1962.
[10] Keeton WP. Prosser and Keeton on torts. 5th ed. St. Paul, MN: West Publishing Co.; 1984.
[11] Epstein RA. Modern products liability law. Westport, CT: Quorum Books; 1980.
[12] American Law Institute. Restatement (Third) of Torts: products liability. St. Paul, MN: American Law Institute Publishers; 1998.

[13] Peters PG. The quiet demise of deference to custom: malpractice law at the millennium, 57 Wash. & Lee L Rev; 2000. p. 163.
[14] Hooper TJ. 60 F. 2d 737 (2d Cir. Wash.); 1932.
[15] Helling v. Carey. 83 Wn. 2d 514 (Wash.); 1974.
[16] Epstein RA. The path to the T.J. Hooper: the theory and history of custom in the law of tort. J Legal Stud 1992;21 (1).
[17] Peters, 171–172, 185–187; n.d.
[18] Rylands v. Fletcher, L. R. 3 E. & I. App. 330 (H. L. 1868); n.d.
[19] Restatement (Third) of Torts, x19; n.d.
[20] Restatement (Second) of Torts, x402A; n.d.
[21] Silverhart v. Mount Zion Hospital. 20 Cal. App. 3d 1022, 98 Cal. Rptr. 187; 1971.
[22] Gage D, McCormick J. Can software kill? eWeek.com; 2004. Available at: http://www.eweek.com/article2/0,1759,1544225,00.asp. [Accessed 8 March 2004].
[23] Leveson NJ, Turner CS. An investigation of the Therac-25 accidents. IEEE Computer 1993;26(7):18–41.
[24] American Health Line. FDA blocks Missouri company from making radiation devices; 2003.
[25] Dahm LL. Restatement (Second) of Torts Section 324A: an innovative theory of recovery for patients injured through use or misuse of health care information systems. John Marshall J Comput Inf Law 1995;14(2):91–2.
[26] Gable JK. An overview of the legal liabilities facing manufacturers of medical information. Quinnipiac Health Law J 2001;5:127. 140–141.
[27] Dahm, Notes 3–6; n.d.
[28] Gable, 146–147; n.d.
[29] Winter v. G. P. Putnam's Sons. 938 F.2d 1033, 1036, Ninth Circuit, July 12, 1991; 1991.
[30] Bryant v. Tri-County Elec. Membership Corp. 844 F. Supp. 347, 349 (KY); 1994.
[31] Restatement (Third) of Torts, x2(b); n.d.
[32] Restatement (Third) of Torts, x2(e); n.d.
[33] Barker v. Lull Engineering Co. 20 Cal. 3d 413 (CA); 1978.
[34] Green v. Smith & Nephew AHP, Inc. WI 109 (WI); 2001.
[35] Heaton v. Ford Motor Co. 248 Ore. 467 (OR); 1967.
[36] Soule v. General Motors Corp. 8 Cal. 4th 548 (CA); 1994.
[37] Vautour v. Body Masters Sports Indus. 147 N.H. 150 (NH); 2001.
[38] McCormick J, Steinert-Threlkeld T. Panama technicians found guilty. Baseline; 2004. Available at: https://www.baselinemag.com/business-intelligence/Panama-Technicians-Found-Guilty/.
[39] Riegel v. Medtronic, Inc. vol. 312; 2008.
[40] The main exception was Missouri, whose courts permitted the application of products liability to hospitals in this situation. However, the Missouri legislature disagreed, and subsequent Missouri court decisions have concluded that the Missouri legislature's intent was to statutorily bar the extension of such liability. See, for example, Budding v. SSM Healthcare Sys, 19 S.W.3d 678 (MO 2000); n.d.
[41] Ayyash v. Henry Ford Health Systems, Mich. App., 210 Mich. App. 142, 533 N.W.2d 353 (1995) (strict liability not applied to hospital for defective jaw implants); St. Mary Medical Center, Inc. v. Casko, 639 N.E.2d 312, 315 (Ind. 1994) (strict liability not applied to hospital for defective pacemaker); Hoff v. Zimmer, 746 F. Supp. 872 (W. D. Wis. 1990) (strict liability not applied to hospital for failure of prosthetic hip); Easterly v. HSP of Texas, Inc., 772S.W.2d 211 (Tex. Ct. App. 1989) (strict liability not applied to hospital supplying epidural kit with defective needle); Hector v. Cedars-Sinai Medical Center, 180 Cal. App. 3d 493, 225, Cal. Rptr. 595 (1986); n.d.
[42] Parker v. St. Vincent Hosp. NMCA 70, 122 NM 39, 46, 919 P.2d 1104, 1111 (Ct., App. 1996); 1996.
[43] Magrine v. Krasnica. 94 N.J. Super 228 [227 A.2d 539]; 1967.
[44] Lamb v. Candler Gen. Hosp. vol. 413; 1992.
[45] Berg v. United States. vol. 806; 1986.
[46] Advent Sys. v. Unisys Corp. 925 F.2d 670, 675–676 (3d Cir.); 1991.
[47] RRX Indus. v. Lab-Con, Inc. 772 F.2d 543, 546 (9th Cir.); 1985.
[48] Micro-Managers, Inc. v. Gregory. 434 N.W.2d 97, 100 (Wis. Ct. App.); 1988.
[49] Dahm, 91–92 and note 77; n.d.
[50] Uniform Commercial Code 2-316(1); n.d.
[51] Dahm, 93; n.d.

[52] James v. Meow Media. 90 F. Supp. 2d 798, 810–1 (United States District Court for the Western District of Kentucky); 2000.
[53] Restatement (Third) of Torts, x19(a) Comment d; n.d.
[54] Evans B, Ossorio P. The challenge of regulating clinical decision support software after 21st Century Cures. Am J Law Med 2018;44:237–51. https://doi.org/10.1177/0098858818789418.
[55] Aetna Casualty & Surety Co. v. Jeppesen & Co. 642 F.2d 339 (9th Cir.); 1981.
[56] Harbeson v. Parke Davis, Inc. 746 F.2d 517, 525 (9th Cir.); 1984.
[57] Kingston v. Chicago & N.W. R. Co. 191 Wis. 610, 211 N.W. 913; 1927.
[58] Mangalmurti SS, Murtagh L, Mello MM. Medical malpractice liability in the age of electronic health records. N Engl J Med 2010;363:2060–7. https://doi.org/10.1056/NEJMhle1005210.
[59] Ruder D.B. Malpractice claims analysis confirms risks in EHRs; n.d. 5.
[60] Ranum D. Electronic health records continue to lead to medical malpractice suits; n.d. https://www.thedoctors.com/articles/electronic-health-records-continue-to-lead-to-medical-malpractice-suits/. [Accessed 14 December 2021].
[61] Richter v. Presbyterian Healthcare Servs. vol. 326; 2014.
[62] Johnson v. Hillcrest Health Ctr., Inc. vol. 70; 2003.
[63] Vargas v. Lee. vol. 170; 2019.
[64] Heinrich v. State of New York; 2021.
[65] Miller RA, Gardner RM. Summary recommendations for responsible monitoring and regulation of clinical software systems. Ann Intern Med 1997;127(9):842–5.
[66] 2013—Software as a Medical Device (SaMD) Key definitio.pdf; n.d.
[67] IMDRF.pdf; n.d.
[68] 2017—Software as a Medical Device (SAMD) Clinical Eval.pdf; n.d.
[69] imdrf-tech-151002-samd-qms.pdf; n.d.
[70] Software as a Medical Device (SaMD): Key definitions; 2013. p. 9.
[71] Act of June 25, 1938 (Federal Food, Drug, and Cosmetic Act), Public Law 75-717, 52 STAT 1040, which prohibited the movement in interstate commerce of adulterated and misbranded food, drugs, devices, and cosmetics; n.d. https://catalog.archives.gov/id/299847. [Accessed 5 January 2022].
[72] Administrative Procedure Act (P.L. 79-404); n.d. https://www.justice.gov/jmd/ls/administrative-procedure-act-pl-79-404. [Accessed 13 January 2022].
[73] What we do | FDA; n.d. https://www.fda.gov/about-fda/what-we-do#mission. [Accessed 30 December 2021].
[74] Laws enforced by FDA | FDA; n.d. https://www.fda.gov/regulatory-information/laws-enforced-fda. [Accessed 30 December 2021].
[75] The accidental poison that founded the modern FDA. The Atlantic; n.d. https://www.theatlantic.com/technology/archive/2018/01/the-accidental-poison-that-founded-the-modern-fda/550574/. [Accessed 5 January 2022].
[76] Public Law 94-295. Medical Device Amendments to the Federal Food, Drug, and, Cosmetic Act (passed on May 28, 1976); 1976.
[77] Premarket approval (PMA) | FDA; n.d. https://www.fda.gov/medical-devices/premarket-submissions-selecting-and-preparing-correct-submission/premarket-approval-pma. [Accessed 30 December 2021].
[78] eCFR:: 21 CFR Part 814—Premarket approval of medical devices; n.d. https://www.ecfr.gov/current/title-21/chapter-I/subchapter-H/part-814. [Accessed 5 January 2022].
[79] Premarket notification 510(k) | FDA; n.d. https://www.fda.gov/medical-devices/premarket-submissions-selecting-and-preparing-correct-submission/premarket-notification-510k. [Accessed 13 January 2022].
[80] Young FE. Validation of medical software: present policy of the Food and Drug Administration. Ann Intern Med 1987;106:628–9.
[81] Miller RA, Gardner RM. Recommendations for responsible monitoring and regulation of clinical software systems. J Am Med Inform Assoc 1997;4(6):442–57.
[82] Abookire S, Martin MT, Teich JM, Kuperman GJ, Bates DW. An institution-based process to ensure clinical software quality. Proc AMIA Symp 1999;461–5.
[83] Draft guidance for industry and Food and Drug Administration staff; mobile medical applications; availability. Fed Regist July 21, 2011;76:43689.
[84] Food and Drug Administration. vol. 2013; 2011.

[85] K. Meier. Food and Drug Administration, vol. 2013; 2011.
[86] PLAW-114publ255.pdf; n.d.
[87] Parasidis—2018—CLINICAL DECISION SUPPORT ELEMENTS OF A SENSIBLE.pdf; n.d.
[88] FDA policy for the regulation of computer products; 1987.
[89] Miller RA, Masarie FE. The demise of the "Greek Oracle" model for medical diagnostic systems. Methods Inf Med 1990;29:1–2.
[90] Matheny ME, Whicher D, Thadaney Israni S. Artificial intelligence in health care: a report from the National Academy of Medicine. JAMA 2020;323:509–10. https://doi.org/10.1001/jama.2019.21579.
[91] Clinical decision support software draft guidance for industry and Food and Drug Administration staff; n.d. https://www.fda.gov/regulatory-information/search-fda-guidance-documents/clinical-decision-support-software. [Accessed 7 July 2021].
[92] Ghassemi M, Oakden-Rayner L, Beam AL. The false hope of current approaches to explainable artificial intelligence in health care. Lancet Digit Health 2021;3:e745–50. https://doi.org/10.1016/S2589-7500(21)00208-9.
[93] Petersen C, Smith J, Freimuth RR, Goodman KW, Jackson GP, Kannry J, et al. Recommendations for the safe, effective use of adaptive CDS in the US healthcare system: an AMIA position paper. J Am Med Inform Assoc 2021;28:677–84. https://doi.org/10.1093/jamia/ocaa319.
[94] US FDA, Health Canada, Medicines and Healthcare Regulatory Agency. Good Machine Learning Practice for Medical Device Development: Guiding Principles. Guiding Principals—GMLP; n.d. https://www.fda.gov/media/153486/download. [Accessed 13 January 2022].
[95] Requests for feedback and meetings for medical device submissions: the Q-Submission program | FDA; n.d. https://www.fda.gov/regulatory-information/search-fda-guidance-documents/requests-feedback-and-meetings-medical-device-submissions-q-submission-program. [Accessed 5 January 2022].
[96] USCDI A The USCDI is a standardized set of health data classes and constituent data elements for nationwide, interoperable health information exchange; n.d. 18.
[97] United States Core Data for Interoperability (USCDI) | Interoperability Standards Advisory (ISA); n.d. https://www.healthit.gov/isa/united-states-core-data-interoperability-uscdi#uscdi-v2. [Accessed 28 January 2022].
[98] 2015 edition cures update; n.d. https://www.healthit.gov/curesrule/final-rule-policy/2015-edition-cures-update. [Accessed 26 September 2021].
[99] American Recovery and Reinvestment Act of 2009::P.L. 111-5, as signed by the President on February 17, 2009: law, explanation and analysis. Chicago, IL: CCH; 2009. p. 678.
[100] Health information technology: initial set of standards, implementation specifications, and certification criteria for electronic health record technology. Interim final rule. Fed Regist Jan 13, 2010;75:2013.
[101] Medicare and Medicaid programs; electronic health record incentive program. Final rule. Fed Regist Jul 28, 2010;75:44313.
[102] Health information technology: initial set of standards, implementation specifications, and certification criteria for electronic health record technology. Final rule. Fed Regist Jul 28, 2010;75:44589.
[103] 2017 Medicare Electronic Health Record (EHR) incentive program payment adjustment fact sheet for eligible professionals; n.d. 3.
[104] 171K providers subject to meaningful use payment adjustments; n.d. https://ehrintelligence.com/news/171k-providers-subject-to-meaningful-use-payment-adjustments. [Accessed 13 January 2022].
[105] Osheroff JA, et al. A roadmap for national action on clinical decision support. J Am Med Inform Assoc Mar–Apr, 2007;14:141.
[106] Osheroff JA, Pifer E, Teich J, Sittig D, Jenders R. Improving outcomes with clinical decision support: an implementer's guide. Chicago: Healthcare Information Management Systems Society; 2005.
[107] Medicare and Medicaid programs; electronic health record incentive program—stage 2. Final rule. Fed Regist Sep 4, 2012;77:53967.
[108] Meaningful use stage 3 final rule; 2015.
[109] C.M. DesRoches et al., Electronic health records in ambulatory care—a national survey of physicians. N Engl J Med 359, 50 (Jul 3, 2008). n.d.
[110] Jha AK, DesRoches CM, Campbell EG, Donelan K, Rao SR, Ferris TG, et al. Use of electronic health records in U.S. hospitals. N Engl J Med 2009;360:1628–38. https://doi.org/10.1056/NEJMsa0900592.
[111] Office-based physician electronic health record adoption | HealthIT.gov; n.d. https://www.healthit.gov/data/quickstats/office-based-physician-electronic-health-record-adoption. [Accessed 14 December 2021].

[112] Non-federal acute care hospital electronic health record adoption | HealthIT.gov; n.d. https://www.healthit.gov/data/quickstats/non-federal-acute-care-hospital-electronic-health-record-adoption. [Accessed 13 January 2022].

[113] 2021 program requirements | CMS; n.d. https://www.cms.gov/regulations-guidance/promoting-interoperability/2021-program-requirements. [Accessed 13 January 2022].

CHAPTER

24

The promise of patient-directed decision support

Jessica S. Ancker[a] *and Meghan Reading Turchioe*[b]

[a]Department of Biomedical Informatics, Vanderbilt University Medical Center, Nashville, TN, United States [b]Columbia University School of Nursing, New York, NY, United States

24.1 Introduction

The informatics community has long recognized the imperative for clinical decision support to assist providers in making timely, evidence-based decisions about the care of patients. But supporting patients in their own decisions has not attracted the same amount of attention. As a result, some of the most interesting work in this domain is occurring outside of informatics. Bringing informatics awareness to this important work could strengthen the field by ensuring that the tools patients need for decisions are grounded in the best evidence, developed with attention to patient needs and requirements, and (where appropriate) integrated into clinical systems, teams, and workflow.

24.2 Some definitions

Clinical decision support (CDS) is generally designed to provide timely information to clinicians, not to patients. When CDS focuses on outcomes important to patients and addresses patient-specific clinical and demographic characteristics, it can be considered *patient-centered*. However, in this chapter, we discuss decision support designed specifically to support patients and their caregivers, with or without their care teams, in health-related decisions and actions. For simplicity, we call this type *patient-directed decision support*.

Patient-directed decision support represents only a small fraction of consumer health information technology. Other important consumer technologies out of scope for this chapter are those designed to provide care (e.g., technology-supported cognitive behavioral therapy

Clinical Decision Support and Beyond
https://doi.org/10.1016/B978-0-323-91200-6.00004-8

or physical therapy), teach skills (e.g., meditation and yoga apps), provide access to medical data (e.g., patient portals), or facilitate patient-provider communication (e.g., telemedicine, secure messaging). We also do not focus on technologies that simply provide patient education because information alone is not sufficient to support patient decision-making. As we discuss below, information about options must be accompanied by methods of eliciting patient values and preferences, guidance about how to apply values and preference to decisions, and support for patient-provider shared decision making discussions. Instead, we focus on systems and tools designed to support patients in making either one-time decisions about disease and preventive care or recurring decisions about disease self-management.

24.2.1 Decision support for challenging one-time decisions

Although all medical decisions are likely to be challenging for patients, the case of the preference-sensitive medical decision is perhaps the most challenging. In a preference-sensitive decision, there is no accepted "best" option; several options are recognized as having some efficacy but have different risk-benefit profiles. Patient education is critical, but not sufficient. Patient preferences and values are also key to making the choice.

An excellent example is the case of colorectal cancer screening. Some form of colorectal cancer screening is strongly recommended for all adults beginning at age 50 [1], so whether to get screened is not considered a preference-sensitive choice. However, the choice of screening modality is preference-sensitive because several options are recognized as effective. Patients who value a highly sensitive screening option that permits immediate surgical treatment if needed might opt for colonoscopy. Other patients who want to avoid the discomforts and risks of the procedure, and who are willing to accept lower sensitivity, can opt for more frequent testing with fecal immunochemical tests or fecal occult blood tests. All of these options are recognized as reasonable choices [1].

Research and policy on how to support patients in decisions such as these has emphasized the importance of *shared decision making* between patient and provider, moving away from a traditional paternalistic model [2–6]. Shared decision making depends in part on a mutual understanding of the options, their advantages and disadvantages; this typically requires some education for the patient. Education, however, is not sufficient to support shared and informed decisions. In addition, and perhaps more importantly, shared decision making depends upon eliciting the patient's values and preferences about those options, and empowering patients to incorporate these values and preferences into discussions with providers [7]. The *decision aid* is a form of patient-directed decision support designed to support these one-time decisions [8–10]. According to the International Patient Decision Aid Standards (IPDAS), a decision aid must describe the health condition and its natural course without intervention, describe all options in an unbiased and easily understood way (including the option of taking no action), explain the details of each option including its potential harms and benefits, provide opportunities to reflect on values and priorities, and provide structured guidance about how to make the decision and discuss it with their provider [8]. Critically, decision aids must not steer patients toward one of the available options. (Because personal testimonials influence patient choices so strongly [11], they are not recommended in decision aids [12].) A decision aid designed in accordance with these criteria can be used to educate, engage, and support patients in the shared decision-making process. Documentation of a shared decision making encounter, using an evidence-based decision aid, is now required by CMS for a few preference-sensitive decisions, including initiating lung cancer screening with low-dose computed tomography [13] and placing implantable automatic defibrillators [14].

Decision aids are effective in improving patient understanding of their options and making it easier to make a decision. Randomized trials have demonstrated that decision aids help people be more knowledgeable, informed, clear about values, accurate in judging risks, and overall active in decision-making compared to usual care [15,16]. Decision aids have been developed and demonstrated effective for colorectal cancer screening [17], genetic testing [18], breast cancer screening and prophylaxis [19–21], lung cancer screening [22], prostate cancer screening [23], cancer treatment choices [24], left ventricular assist device placement [25], and many other decisions [26]. Among the outcomes assessed in decision aid research are knowledge, decisional conflict (how difficult it is to make a decision), decisional regret (after the decision), and self-efficacy [27].

Decision aids must also be developed to meet the needs of their potential users to ensure they are usable and useful. IPDAS has established a model for systematically developing decision aids with user input that involves the following general steps: (1) defining the scope, purpose, and target audience, (2) establishing a steering group of patients and experts, (3) conducting preliminary user-centered design activities, (4) prototyping, and (5) performing several iterative rounds of alpha and beta testing with patients and clinicians, followed by review and redesign guided by the steering group [28].

A recent case study illustrates the design process. Patients with atrial fibrillation (AF) face a preference-sensitive decision about whether to undergo a minimally invasive surgery called radiofrequency ablation. The decision is preference-sensitive, because current evidence regarding outcomes after ablation is mixed [29,30], and the decision depends on a patient's goals of care and risk tolerance. In interviews with AF patients and clinicians, both sets of stakeholders agreed that many patients did not fully understand the risks, benefits, or alternative options. These stakeholders expressed a preference for a web decision aid to be used both before and after a consultation with a doctor. Stakeholders suggested visualizations and videos in addition to text. This information on patients' and clinicians' lived experiences and preferences for consuming information directly informed decision aid prototypes.

24.2.2 Supporting ongoing disease management and self-management decisions

Longer-lasting medical conditions, including chronic diseases, involve ongoing self-management decisions about medications, disease-management activities (such as glucose monitoring in diabetes or self-weighing in heart failure), and daily behaviors such as exercise and diet. Patients can benefit from decision support designed for these routine daily or hourly decisions.

Simple reminders (through text, phone, and other modalities) are effective in promoting a variety of health-related activities, including helping patients take medications on time and make appointments for needed care [31–34]. Reminders are generally effective in situations when patients have already made a general decision (for example, to begin a course of medication or to get vaccinated) and need nudges to operationalize their decision effectively.

Situations involving making more complex self-management decisions typically require more information exchange and education than is possible through a reminder system. For example, in diabetes, apps and programs for patients generally provide a suite of supportive features which may include medication reminders, blood glucose self-measurement and tracking, hemoglobin A1c upload, general education materials for the patient to access as needed, and specific guidance about how the patient should interpret their own data and incorporate it into decisions [35]. Several of these also include patient-provider communication through secure messaging, so that the patient can ask for additional support as needed, and social networking opportunities with other patients. Evidence about effectiveness of these combination patient-directed systems is somewhat mixed, with some associated with improvements in hemoglobin A1c when paired with support from a clinician or study staff member [35].

Instruments for supporting self-management decisions can also leverage patient-reported outcomes (PROs), which are outcomes reported by patients themselves on the basis of their lived experience of disease or treatment, rather than being recorded by or interpreted by a clinician [36–38]. Regularly collecting PROs and providing them to clinicians is associated with better health-related quality of life in cancer and even with improved survival [38,39]. PRO collection is not itself a form of patient decision support, but can be made so if the results are presented back to patients with decision guidance. In a recent project, a cancer center developed an electronic system to give patients management guidance for the 10 days after ambulatory cancer surgery [40]. Patients were invited to report PROs pertinent to their surgery, and an interpretive report was immediately provided back to the patient. PRO values that were within the range expected post-surgery (i.e., localized moderate pain) were labeled with a reassuring message that the patient's recovery was "on track." However, a value that fell outside the normal range (such as one indicating possible infection) was delivered to the patient with an instruction to seek care immediately; a message was also automatically delivered to the covering nurse to contact the patient. This system is designed to support patients through day-to-day decisions during their recovery, encouraging patients who need care to seek it, while reassuring patients whose recovery is progressing as expected. A randomized trial is under way to assess the impact on post-surgical outcomes such as emergency admission [41]. A similar automated symptom management system for cancer patients in chemotherapy significantly reduced symptom severity and duration [42,43].

24.2.3 Challenges to be addressed

Technology offers tremendous potential for one-time and recurring patient decision support, but patient decision support faces critical challenges before it will be universally effective. Future research and innovation could help address some of these, while policy changes might be needed to address others.

24.2.3.1 Technology access and literacy

One set of concerns centers on technology access, broadband access, and digital literacy [44]. Although smartphones are common even among disadvantaged segments of the public, there remain up to 15% of Americans without smartphones; these are disproportionately from populations with high disease burden, including older individuals and people with low income [45]. Broadband access also varies widely in different settings across the country [44]. Any informatics intervention must address this technology access disparity to avoid worsening health equity [46].

24.2.3.2 *Patient engagement in decision making*

A second set of concerns is about patient desire for—and self-efficacy for—engagement in health-related decision making. Special design considerations are usually needed to support decision-making around disease management in vulnerable communities. Most tools focused on disease self-management to date have been developed with input and feedback from "data enthusiasts," who are excited to learn from their personal health data and use it to inform their decision-making about daily behaviors. Not all real-world patients are so enthusiastic about their personal health data, however. One patient with diabetes said that looking at personal health data in the patient portal was difficult: "You get reminded you're a sick person." [47] The high prevalence of chronic conditions and risk of worse outcomes among individuals in vulnerable communities means they stand to benefit the most from tools that support chronic disease management—but are least likely to engage with these types of tools [48]. One study with low-income, Latino adults living with type 2 diabetes identified key differences in the ways the participants engaged in disease management; there was a greater reliance on clinicians to initially suggest tracking health information, provide guidelines about how often and for how long, interpret the data, and recommend behavior changes [49]. Participants reported difficulty interpreting their health data and low confidence making decisions about the appropriate next steps. When the participants were shown design options for a mobile application to self-manage diabetes, they gravitated toward options that rewarded them for collecting health data but disliked options that required them to analyze, evaluate, and make decisions themselves based on the data. In general, they strongly preferred having their care team conduct these tasks [50]. Overall, individuals with lower desire and self-efficacy to manage a chronic illness using technology may need additional context, support, and guidance [49].

In discussing patient engagement with decisions, it is worth noting that patients are of course deeply embedded in their own social networks and information networks [51–53]. A full discussion of the influence of peer networks, social media, and news media on individual health-related decisions is beyond the scope of this chapter. Yet it is inescapable, particularly in context of the COVID-19 pandemic, that patient decisions are strongly informed by peers, family, social media, and news media. Decision aids and shared decision-making conversations between patient and provider may need to directly address patient perceptions derived from these other sources of information.

24.2.3.3 *Health literacy and numeracy*

A third important set of challenges centers on ensuring information can be understood, interpreted, and applied by people who need it. Perhaps one-quarter to one-third of the population has limited health literacy—the set of skills and knowledge needed to access, interpret, and apply information [54]. An even larger proportion have low numeracy, the skills needed to access, interpret, and apply health-related numbers [55–57]. Any material for patients, including decision support, needs to be appropriate for low-literacy and low-numeracy audiences [58]. Unfortunately, existing decision aids are often written with the assumption of a high level of literacy [59], making it unlikely they will benefit less well-educated individuals. In addition, individuals with low numeracy tend to have less self-efficacy for shared decision making, probably because of their awareness of challenges understanding quantitative information [60]. A number of excellent references are available for decision aid designers seeking to develop materials that are easy to read [58,61].

Information developed by and for expert users such as physicians and researchers must be transformed before being useful for patients. For example, the Patient-Reported Outcome Measurement Information System (PROMIS), a standardized set of PROs, are reported as T-scores, a metric largely meaningless to the average patient – or provider [62]. However, research has shown that most people find scores on a scale of 1–10 much more approachable [63], suggesting that a simple transform can improve comprehension.

Even relatively simple wording changes can transform jargon syntax to actionable decision guidance. In one study, getting rid of the standard "take 2 pills twice a day" wording and replacing it with "take 2 pills in the morning and 2 in the evening" nearly doubled the number of patients who could correctly take their medications [64,65].

Visualizations have potential for conveying both probabilistic information such as risks [20,66] and quantitative information such as lab values, medication doses, and PROs [67]. Icon arrays (fields of stick figures or other icons representing the numerator and denominator of a risk) have demonstrated efficacy in conveying information about probabilities to low numeracy individuals [68]. As a result, icon arrays showing risks and benefits have become common in decision aids. Illustrations of dosing instruments sharply increase the proportion of parents who can correctly administer pediatric medication [69,70].

However, for more complex concepts such as time trends, trade-offs, or covariate relationships, the inherent complexity of the concepts can make it difficult to develop effective graphics. For example, survival curves can guide decision making about therapeutic options, but are very challenging for many patients. Guided practice exercises can help patients make sense of them [71], yet there may be limited applicability of this finding to clinical practice, where there is unlikely to be time to train patients. Because up to 40% of the U.S. population may have inadequate graph literacy, it cannot be assumed that data graphics are always better than numbers [72].

In cases where precise understanding of quantities is unimportant, patient decision making may be improved by replacing quantitative data graphics with visual analogies. Some studies have used weather icons (stormy, sunny) to represent mental health status [73], or images of full or depleted batteries to represent self-reported energy levels [74]. Such graphics can convey a gist of "more" or "less", "better" and "worse" without requiring patients to extract the gist from the numbers themselves.

The choice of graphics appears to be highly context- and population-dependent. One sample of hospitalized patients with heart failure and mild to moderate cognitive impairment demonstrated better comprehension of visual analogies than of line graphs or text [73]. However, a national online sample assessing the same visualizations had the highest level of comprehension for the line graphs [75].

Many patients, especially those with challenges reading because of low literacy, vision limitations, or other barriers, could potentially benefit from voice-based conversational agents.

These interactive voice tools are accessible even with limited dexterity or other issues that prohibit or limit the use of screen interfaces. Smart speakers in the home offer other voice-based opportunities for reminders and daily instruction. Interactive voice solutions have been used in many use cases including medication adherence [76], physical activity [77], alcohol consumption reduction [78], chronic disease and mental health management [79].

24.2.3.4 *Provider and patient workflows*

Finally, patient decision support will become most effective when it is fully integrated into both clinicians' workflow and patients' illness-related work [26]. Libraries of high-quality decision aids are already available (e.g., https://decisionaid.ohri.ca/cochinvent.php) and commercial solutions are also available (e.g., https://www.healthwise.org/specialpages/imdf.aspx). However, decision support will be most effective when it is delivered to patients at the moment when they need it, and in a way that facilitates shared decision-making between patient and provider. The right timing and delivery strategy may be nuanced and context-dependent. In one study, providers had low self-efficacy about conducting shared decision-making with patients, suggesting the need for implementation to focus on this skill set [80]. Another study of a decision aid for individuals with type 2 diabetes showed that patients preferred to read decision aids at home, rather than with their provider during a visit [81]. Delivering decision aids to patients prior to specific visits could therefore help them prepare for the visit, while reducing the encounter time burden on providers, who report that time limitations are a major barrier to shared decision making and the use of decision aids [82]. However, conversely, provider attitudes toward the decision aid influence the patient's interest and willingness to use it, suggesting that providing the aid *after* an orientation visit could be even more effective [81]. Furthermore, determining which visit will require the use of a decision aid might be a challenge that could create an advance-planning burden on providers. Fully integrating decision aids into clinical workflow requires attention to these preferences by patients, as well as factors at the organizational level (leadership, culture, resources, and workflows) and systems level (policies, guidelines, incentives, and education) [16,26,83].

Patient-directed decision support should also reduce data-input and workflow burdens on patients. Initially interested patients may lose motivation after being required to repeatedly input information for self-tracking or decision guidance [84,85]. In one high-profile trial, nearly one-third of patients stopped using the fitness app they were assigned to; recorded complaints were that it was difficult to use and "tedious." [85] These problems may become less salient with the increasing use of passive or automatic data capture through activity trackers, QR scanning, and integration with home medical devices such as blood glucose monitors, scales, and blood pressure monitors. Self-report data is still important in patient-directed decision support, however. One way to reduce the burden of self-report data may be through techniques such as ecological momentary assessment (EMA), the use of very short instruments delivered at randomly selected times to invite participants to report behavior, thoughts, or mood in real time. EMA has greater ecological validity than retrospective questionnaires and is generally simpler for patients to complete [86].

To fully incorporate patient-directed decision support into clinical workflow, EHR integration will be required [87]. Massachusetts General Hospital has been a leader in allowing providers to order decision aids through their EHR [26]. Patients with active patient portal accounts receive the decision aid in online form, and those without are mailed a DVD and a booklet [26]. In a subsequent initiative, patients were given order sheets to order their own decision aids. During a 3-month trial period, providers ordered decision aids for 28 patients, but an additional 280 patients ordered their own decision aids [26]. This finding suggests that expecting providers to take on the entire responsibility for supplying decision aids may result in considerable unmet need among their patients.

Fuller EHR integration, through native functionality or technology such as SMART on FHIR [88], could allow automatic data capture from the EHR and patient portal, integration of patient-generated health data (PGHD) and other information contributions and requests from the patient, and layering onto existing clinical decision support [89]. However, this area is still emerging. In a recent review, only 5% of articles about SMART on FHIR implementations involved patient-facing apps, and some of these focused narrowly on patient education rather than more comprehensive decision support [90]. One example of a shared decision-making FHIR app is a lung cancer decision tool developed at the University of Michigan and the Department of Veterans Affairs, integrated into the EHR and displayed for potentially eligible patients to support providers in conducting these conversations [88]. The promise of integrated technologies should be considered in conjunction with their challenges. In a recent paper, integrating a SMART on FHIR decision aid for uterine fibroids into a commercial EHR required an 18-month timeline and committed involvement from a clinical champion, a health informaticist, and the EHR software developer team [91]. Additional research and innovation may help to simplify the process to make it more accessible for a wider range of medical centers.

Implementation also requires provider training in shared decision-making skills and fostering of an organizational culture receptive to shared decision-making [26]. Cultural change in healthcare organizations, however, may also require restructuring incentives to ensure that providers are incentivized to conduct shared decision-making. As mentioned above, the Center for Medicaid and Medicare Services has recently required documented shared decision-making encounters for decisions with significant risks and benefits, including low-dose computed tomography lung cancer screening [92] and implantable automatic defibrillators [93]. Policies such as these are a step toward ensuring that providers have adequate incentives and resources to work together with patients to make these decisions.

24.3 Technology innovation

Applying other technological innovations to patient-directed decision support is of great interest. For example, conversational agents offer the promise of communication in a form that patients find natural. Symptom-checker chatbots became popular during the COVID-19 pandemic as patients overwhelmed existing phone lines; for common questions such as whether to seek COVID testing, simple decision tree infrastructures were sufficient [94]. More sophisticated artificial-intelligence-based conversational agents are under active development to provide education (e.g., answers to sensitive questions that patients may be reluctant to ask face-to-face), elicit medical history, get help with smoking cessation and other health coaching, and manage chronic disease [95]. These sorts of AI-based conversational agents would seem to have great promise in assisting with the types of one-time and recurring health decisions discussed in this chapter, which integrate factual education about options, elicitation and clarification of values and preferences, and guidance for patient-provider conversations.

Smart devices, sensors, and other internet-of-things (IOT) technologies also have the potential to support health-related decision making. Some clearly health-related foci to date have been to provide methods for patients to accomplish otherwise challenging activities (such as allowing them to substitute voice commands for typed search) and to monitor for safety problems (e.g., allowing patients to share camera access with a family member, or route alerts about potential falls to a caregiver) [96]. By integrating principles from shared decision making and the behavioral sciences, these technologies will have increased applications to the types of health-related decisions we discuss here.

24.4 Conclusion

Although informatics has classically focused on facilitating provider decision making, the field also has the potential to transform and support patient decision making. Technological solutions for patients include decision aids for one-time decisions and disease management systems for recurring disease-related decisions. All solutions for patients must take into account technology access and literacy, self-efficacy for decision making, health literacy and numeracy, and implementation into workflow. With increasing numbers of informaticists developing innovations in this space, patient-directed decision support may become as ubiquitous as clinical decision support.

References

[1] Lin JSPL, Henrikson NB, Bean SI, Blasi PR. Screening for colorectal cancer: an evidence update for the U.S. Preventive Services Task Force 2021, https://www.ncbi.nlm.nih.gov/books/NBK570913/.

[2] Agency for Healthcare Research and Quality. Shared decision making: the SHARE approach, http://www.ahrq.gov/professionals/education/curriculum-tools/shareddecisionmaking/index.html.

[3] Elwyn G, Frosch D, Thomson R, et al. Shared decision making: a model for clinical practice. J Gen Intern Med 2012;27(10):1361–7. https://doi.org/10.1007/s11606-012-2077-6.

[4] Kriston L, Scholl I, Holzel L, Simon D, Loh A, Harter M. The 9-item shared decision making questionnaire (SDM-Q-9). Development and psychometric properties in a primary care sample. Patient Educ Couns 2010;80(1):94–9. https://doi.org/10.1016/j.pec.2009.09.034.

[5] Edwards A, Elwyn G, Wood F, Atwell C, Prior L, Houston H. Shared decision making and risk communication in practice: a qualitative study of GPs' experiences. Br J Gen Pract 2005;55(510):6–13.

[6] Sheridan SL, Harris RP, Woolf SH, for the Shared Decision-Making Workgroup of the US Preventive Services Task Force. Shared decision making about screening and chemoprevention: a suggested approach from the US preventive services task force. Am J Prev Med 2004;26(1):56–66.

[7] Berry ABL, Lim C, Hartzler AL, et al. Creating conditions for patients' values to emerge in clinical conversations: perspectives of health care team members. DIS (Des Interact Syst Conf) 2017;2017:1165–74. https://doi.org/10.1145/3064663.3064669.

[8] Elwyn G, O'Connor A, Stacey D, et al. Developing a quality criteria framework for patient decision aids: online international Delphi consensus process. BMJ 2006;333(7565):417. https://doi.org/10.1136/bmj.38926.629329.AE.

[9] O'Connor A, Stacey D, Entwistle V, et al. Decision aids for people facing health treatment or screening decisions. Cochrane Database Syst Rev 2007;2.

[10] Stacey D, Legare F, Col NF, et al. Decision aids for people facing health treatment or screening decisions. Cochrane Database Syst Rev 2014;1.

[11] Ubel PA, Jepson C, Baron J. The inclusion of patient testimonials in decision aids: effects on treatment choices. Med Decis Making 2001;21(1):60–8. https://doi.org/10.1177/0272989x0102100108.

[12] Shaffer VA, Brodney S, Gavaruzzi T, et al. Do personal stories make patient decision aids more effective? An update from the international patient decision aids standards. Med Decis Making 2021. https://doi.org/10.1177/0272989x211011100. 272989x211011100.

[13] Center for Medicare and Medicaid Services, n.d.. Screening for lung cancer with low dose computed tomography (LDCT) decision memo. https://www.cms.gov/medicare-coverage-database/view/ncacal-decision-memo.aspx?proposed=N&NCAId=274.

[14] Center for Medicare and Medicaid Services. National coverage determination: implantable automatic defibrillators, https://www.cms.gov/medicare-coverage-database/view/ncd.aspx?NCDId=110&ncdver=4&NCAId=39.

[15] Stacey D, Légaré F, Lewis K, et al. Decision aids for people facing health treatment or screening decisions. Cochrane Database Syst Rev 2017;4(4):Cd001431. https://doi.org/10.1002/14651858.CD001431.pub5.

[16] Elwyn G, Frosch DL, Kobrin S. Implementing shared decision-making: consider all the consequences. Implement Sci 2016;11:114. https://doi.org/10.1186/s13012-016-0480-9.

[17] Miller DP, Spangler JG, Case LD, Goff DC, Singh S, Pignone MP. Effectiveness of a web-based colorectal cancer screening patient decision aid: a randomized controlled trial in a mixed-literacy population. Am J Prev Med 2011;40(6):608–15.

[18] Green MJ, Peterson SK, Baker MW, et al. Effect of a computer-based decision aid on knowledge, perceptions, and intentions about genetic testing for breast cancer susceptibility: a randomized controlled trial. JAMA 2004;292:442–52.

[19] Metcalfe K, Poll A, O'Connor A, et al. Development and testing of a decision aid for breast cancer prevention for women with a BRCA1 or BRCA2 mutation. Clin Genet 2007;72:208–17.

[20] Zikmund-Fisher B, Ubel P, Smith D, et al. Communicating side effect risks in a tamoxifen prophylaxis decision aid: the debiasing influence of pictographs. Patient Educ Couns 2008;73. https://doi.org/10.1016/j.pec.2008.05.010.

[21] Fagerlin A, Zikmund-Fisher BJ, Smith DM, et al. Women's decisions regarding tamoxifen for breast cancer prevention: responses to a tailored decision aid. Breast Cancer Res Treat 2010;119(3):613–20.

[22] Volk RJ, Linder SK, Leal VB, et al. Feasibility of a patient decision aid about lung cancer screening with low-dose computed tomography. Prev Med 2014;62:60–3. https://doi.org/10.1016/j.ypmed.2014.02.006.

[23] Allen JD, Othus MK, Hart Jr A, et al. A randomized trial of a computer-tailored decision aid to improve prostate cancer screening decisions: results from the take the wheel trial. Cancer Epidemiol Biomark Prevent: Publ Am Assoc Cancer Res Cosponsored Am Soc Prevent Oncol 2010;19(9):2172–86. https://doi.org/10.1158/1055-9965.epi-09-0410.

[24] Sawka AM, Straus S, Gafni A, et al. A usability study of a computerized decision aid to help patients with, early stage papillary thyroid carcinoma in, decision-making on adjuvant radioactive iodine treatment. Patient Educ Couns 2011;84. https://doi.org/10.1016/j.pec.2010.07.038.

[25] Allen LA, McIlvennan CK, Thompson JS, et al. Effectiveness of an intervention supporting shared decision making for destination therapy left ventricular assist device: the DECIDE-LVAD randomized clinical trial. JAMA Intern Med 2018;178(4):520–9. https://doi.org/10.1001/jamainternmed.2017.8713.

[26] Sepucha KR, Simmons LH, Barry MJ, Edgman-Levitan S, Licurse AM, Chaguturu SK. Ten years, forty decision aids, and thousands of patient uses: shared decision making at Massachusetts General Hospital. Health Aff (Millwood) 2016;35(4):630–6. https://doi.org/10.1377/hlthaff.2015.1376.

[27] Patient Decision Aids: Evaluation Measures, n.d.. https://decisionaid.ohri.ca/eval.html.

[28] H VRL-T. 2012 Update of the IPDAS Collaboration Background Document: Using a Systematic Development Process. In: H VRL-T, editor. Updatee of the IPDAS Collaboration Background Document; 2012. p. 2012. http://ipdas.ohri.ca/resources.html.

[29] Packer DL, Mark DB, Robb RA, et al. Effect of catheter ablation vs antiarrhythmic drug therapy on mortality, stroke, bleeding, and cardiac arrest among patients with atrial fibrillation: the CABANA randomized clinical trial. JAMA 2019;321(13):1261–74. https://doi.org/10.1001/jama.2019.0693.

[30] Schnabel RB, Pecen L, Rzayeva N, et al. Symptom burden of atrial fibrillation and its relation to interventions and outcome in Europe. J Am Heart Assoc 2018;7(11). https://doi.org/10.1161/jaha.117.007559.
[31] Schwebel FJ, Larimer ME. Using text message reminders in health care services: a narrative literature review. Internet Interv 2018;13:82–104. https://doi.org/10.1016/j.invent.2018.06.002.
[32] Jacobson Vann JC, Jacobson RM, Coyne-Beasley T, Asafu-Adjei JK, Szilagyi PG. Patient reminder and recall interventions to improve immunization rates. Cochrane Batabase System Rev 2018;1(1):Cd003941. https://doi.org/10.1002/14651858.CD003941.pub3.
[33] Eaton C, Comer M, Pruette C, Psoter K, Riekert K. Text messaging adherence intervention for adolescents and young adults with chronic kidney disease: pilot randomized controlled trial and stakeholder interviews. J Med Internet Res 2020;22(8):e19861. https://doi.org/10.2196/19861.
[34] Pandey A, Choudhry N. AREST MI: adherence effects of a comprehensive reminder system for post-myocardial infarction secondary prevention. J Am Coll Cardiol 2015;65(10_Supplement):A1384. https://doi.org/10.1016/S0735-1097(15)61384-5.
[35] Veazie S, Winchell K, Gilbert J, et al. Rapid evidence review of mobile applications for self-management of diabetes. J Gen Intern Med 2018;33(7):1167–76. https://doi.org/10.1007/s11606-018-4410-1.
[36] Basch E. New frontiers in patient-reported outcomes: adverse event reporting, comparative effectiveness, and quality assessment. Annu Rev Med 2014;65:307–17. https://doi.org/10.1146/annurev-med-010713-141500.
[37] Basch E, Reeve BB, Mitchell SA, et al. Development of the National Cancer Institute's patient-reported outcomes version of the common terminology criteria for adverse events (PRO-CTCAE). J Natl Cancer Inst 2014;106(9). https://doi.org/10.1093/jnci/dju244.
[38] Basch E, Deal AM, Kris MG, et al. Symptom monitoring with patient-reported outcomes during routine cancer treatment: A randomized controlled trial. J Clin Oncol 2016;34(6):557–65. https://doi.org/10.1200/JCO.2015.63.0830.
[39] Basch E, Deal AM, Dueck AC, et al. Overall survival results of a trial assessing patient-reported outcomes for symptom monitoring during routine cancer treatment. JAMA 2017;318(2):197–8. https://doi.org/10.1001/jama.2017.7156.
[40] Ancker JS, Stabile C, Carter J, et al. Informing, reassuring, or alarming? Balancing patient needs in the development of a postsurgical symptom reporting system in cancer. In: Proceedings of the American medical informatics association annual symposium; 2018. p. 166–74.
[41] Stabile C, Temple LK, Ancker JS, et al. Ambulatory cancer care electronic symptom self-reporting (ACCESS): a randomized controlled trial. BMJ Open 2019;9(9), e030863. https://doi.org/10.1136/bmjopen-2019-030863.
[42] Kolb NA, Smith AG, Singleton JR, et al. Chemotherapy-related neuropathic symptom management: a randomized trial of an automated symptom-monitoring system paired with nurse practitioner follow-up. Support Care Cancer 2018;26(5):1607–15. https://doi.org/10.1007/s00520-017-3970-7.
[43] Mooney KH, Beck SL, Wong B, et al. Automated home monitoring and management of patient-reported symptoms during chemotherapy: results of the symptom care at home RCT. Cancer Med 2017;6(3):537–46. https://doi.org/10.1002/cam4.1002.
[44] Benda NC, Veinot TC, Sieck CJ, Ancker JS. Broadband internet access is a social determinant of health! Am J Public Health 2020;110(8):1123–5. https://doi.org/10.2105/ajph.2020.305784.
[45] Pew Research Center, n.d.. Mobile fact sheet. Pew Research Center. https://www.pewinternet.org/fact-sheet/mobile/.
[46] Veinot TC, Mitchell H, Ancker JS. Good intentions are not enough: how informatics interventions can worsen inequality. J Am Med Inform Assoc 2018;25(8):1080–8. https://doi.org/10.1093/jamia/ocy052.
[47] Ancker JS, Witteman HO, Hafeez B, Provencher T, Van de Graaf M, Wei E. "You get reminded you're a sick person": personal data tracking and patients with multiple chronic conditions. J Med Intern Res 2015;17(8): e202. https://doi.org/10.2196/jmir.4209.
[48] Régnier F, Chauvel L. Digital inequalities in the use of self-tracking diet and fitness apps: interview study on the influence of social, economic, and cultural factors. JMIR Mhealth Uhealth 2018;6(4), e101. https://doi.org/10.2196/mhealth.9189.
[49] Reading Turchioe M, Burgermaster M, Mitchell EG, Desai PM, Mamykina L. Adapting the stage-based model of personal informatics for low-resource communities in the context of type 2 diabetes. J Biomed Inform 2020;110:103572. https://doi.org/10.1016/j.jbi.2020.103572.
[50] Turchioe MR, Heitkemper EM, Lor M, Burgermaster M, Mamykina L. Designing for engagement with self-monitoring: a user-centered approach with low-income, Latino adults with type 2 diabetes. Int J Med Inform 2019;130:103941. https://doi.org/10.1016/j.ijmedinf.2019.08.001.

[51] Ancker JS, Carpenter K, Greene P, et al. Peer-to-peer communication, cancer prevention, and the internet. J Health Commun 2009;14:38–46. https://doi.org/10.1080/10810730902806760.

[52] Piltch-Loeb R, Savoia E, Goldberg B, et al. Examining the effect of information channel on COVID-19 vaccine acceptance. PLOS ONE 2021;16(5), e0251095. https://doi.org/10.1371/journal.pone.0251095.

[53] Christakis NA, Fowler JH. The collective dynamics of smoking in a large social network. N Engl J Med 2008;358 (21):2249–58. https://doi.org/10.1056/NEJMsa0706154.

[54] Parker RM, Ratzan SC, Lurie N. Health literacy: a policy challenge for advancing high-quality health care. Health Aff 2003;22(4):147–53.

[55] Ancker JS, Kaufman D. Rethinking health numeracy: a multidisciplinary literature review. J Am Med Inform Assoc: JAMIA 2007;14(6):713–21. https://doi.org/10.1197/jamia.M2464.

[56] Zikmund-Fisher B, Smith D, Ubel P, Fagerlin A. Validation of the subjective numeracy scale (SNS): effects of low numeracy on comprehension of risk communications and utility elicitations. Med Decis Making 2007;27. https://doi.org/10.1177/0272989x07303824.

[57] Lipkus IM, Samsa G, Rimer BK. General performance on a numeracy scale among highly educated samples. Med Decis Making 2001;21:37–44.

[58] DeWalt DA, Broucksou KA, Hawk V, et al. Developing and testing the health literacy universal precautions toolkit. Nurs Outlook 2011;59(2):85–94. https://doi.org/10.1016/j.outlook.2010.12.002.

[59] McCaffery KJ, Holmes-Rovner M, Smith SK, et al. Addressing health literacy in patient decision aids. BMC Med Inform Decis Making 2013;13(S2):S10.

[60] Galesic M, Garcia-Retamero R. Do low-numeracy people avoid shared decision making? Health Psychol 2011;30 (3):336–41. https://doi.org/10.1037/a0022723.

[61] Baur C, Prue C. The CDC clear communication index is a new evidence-based tool to prepare and review health information. Health Promotion Practice 2014;15(5):629–37. https://doi.org/10.1177/1524839914538969.

[62] Gigerenzer G, Gaissmaier W, Kurz-Milcke E, Schwartz LM, Woloshin S. Helping doctors and patients make sense of health statistics. Psychol Sci Public Interest 2007;8(2):53–96. https://doi.org/10.1111/j.1539-6053.2008.00033.x.

[63] Garcia-Retamero R, Cokely ET. Designing visual aids that promote risk literacy: a systematic review of health research and evidence-based design heuristics. Hum Factors 2017;59(4):582–627. https://doi.org/10.1177/0018720817690634.

[64] Davis TC, Federman AD, Bass PF, et al. Improving patient understanding of prescription drug label instructions. J Gen Intern Med 2009;24(1):57–62. https://doi.org/10.1007/s11606-008-0833-4.

[65] Sharko M, Sharma MM, Benda NC, Chan M, Wilsterman E, Liu LG, et al. Strategies to optimize comprehension of numerical medication instructions: a systematic review and concept map. Patient Educ Couns 2022;105 (7):1888–903. https://doi.org/10.1016/j.pec.2022.01.018.

[66] Ancker JS, Senathirajah Y, Kukafka R, Starren JB. Design features of graphs in health risk communication: a systematic review. J Am Med Inform Assoc 2006;13(6):608–18. https://doi.org/10.1197/jamia.M2115.

[67] Zikmund-Fisher BJ, Scherer AM, Witteman HO, et al. Graphics help patients distinguish between urgent and non-urgent deviations in laboratory test results. J Am Med Inform Assoc 2016;24(3):520–8. https://doi.org/10.1093/jamia/ocw169.

[68] Galesic M, Garcia-Retamero R, Gigerenzer G. Using icon arrays to communicate medical risks: overcoming low numeracy. Health Psychol 2009;28(2):210–6.

[69] Yin HS, Dreyer BP, van Schaick L, Foltin GL, Dinglas C, Mendelsohn AL. Randomized controlled trial of a pictogram-based intervention to reduce liquid medication dosing errors and improve adherence among caregivers of young children. Arch Pediatr Adolesc Med 2008;162(9):814–22. https://doi.org/10.1001/archpedi.162.9.814.

[70] Yin HS, Mendelsohn AL, Fierman A, van Schaick L, Bazan IS, Dreyer BP. Use of a pictographic diagram to decrease parent dosing errors with infant acetaminophen: a health literacy perspective. Acad Pediatr 2011;11 (1):50–7. https://doi.org/10.1016/j.acap.2010.12.007.

[71] Armstrong K, Fitzgerald G, Schwartz JS, Ubel PA. Using survival curve comparisons to inform patient decision making: can a practice exercise improve understanding? J Gen Intern Med 2001;16:482–5.

[72] Galesic M, Garcia-Retamero R. Graph literacy: a cross-cultural comparison. Med Decis Making 2011;31 (3):444–57. https://doi.org/10.1177/0272989x10373805.

[73] Reading Turchioe M, Grossman LV, Myers AC, Baik D, Goyal P, Masterson Creber RM. Visual analogies, not graphs, increase patients' comprehension of changes in their health status. J Am Med Inform Assoc 2020;27 (5):677–89. https://doi.org/10.1093/jamia/ocz217.

[74] Arcia A, Suero-Tejeda N, Bales ME, et al. Sometimes more is more: iterative participatory design of infographics for engagement of community members with varying levels of health literacy. J Am Med Inform Assoc: JAMIA 2016;23(1):174–83. https://doi.org/10.1093/jamia/ocv079.

[75] Mangal S., Reading Turchioe M., Park L., et al., n.d. Know your audience: comprehension of health information varies by visual format [under review].

[76] Tsoli S, Sutton S, Kassavou A. Interactive voice response interventions targeting behaviour change: a systematic literature review with meta-analysis and meta-regression. BMJ Open 2018;8(2), e018974. https://doi.org/10.1136/bmjopen-2017-018974.

[77] Migneault JP, Dedier JJ, Wright JA, et al. A culturally adapted telecommunication system to improve physical activity, diet quality, and medication adherence among hypertensive African-Americans: a randomized controlled trial. Ann Behav Med 2012;43(1):62–73. https://doi.org/10.1007/s12160-011-9319-4.

[78] Rose GL, Badger GJ, Skelly JM, MacLean CD, Ferraro TA, Helzer JE. A randomized controlled trial of brief intervention by interactive voice response. Alcohol Alcohol 2017;52(3):335–43. https://doi.org/10.1093/alcalc/agw102.

[79] Bérubé C, Schachner T, Keller R, et al. Voice-based conversational agents for the prevention and Management of Chronic and Mental Health Conditions: systematic literature review. J Med Internet Res 2021;23(3), e25933. https://doi.org/10.2196/25933.

[80] Reese TJ, Schlechter CR, Kramer H, et al. Implementing lung cancer screening in primary care: needs assessment and implementation strategy design. Transl Behav Med 2021. https://doi.org/10.1093/tbm/ibab115.

[81] Tong WT, Ng CJ, Lee YK, Lee PY. What will make patients use a patient decision aid? A qualitative study on patients' perspectives on implementation barriers and facilitators. J Eval Clin Pract 2020;26(3):755–64. https://doi.org/10.1111/jep.13161.

[82] Légaré F, Ratté S, Gravel K, Graham ID. Barriers and facilitators to implementing shared decision-making in clinical practice: update of a systematic review of health professionals' perceptions. Patient Educ Couns 2008;73(3):526–35. https://doi.org/10.1016/j.pec.2008.07.018.

[83] Scholl I, LaRussa A, Hahlweg P, Kobrin S, Elwyn G. Organizational- and system-level characteristics that influence implementation of shared decision-making and strategies to address them—a scoping review. Implement Sci 2018;13(1):40. https://doi.org/10.1186/s13012-018-0731-z.

[84] Ancker JS, Witteman HO, Hafeez B, Provencher T, Wei E. The invisible work of personal health information management among people with multiple chronic conditions: qualitative interview study among patients and providers. J Med Internet Res 2015;17(6), e137. https://doi.org/10.2196/jmir.4381.

[85] Laing BY, Mangione CM, Tseng C-H, et al. Effectiveness of a smartphone application for weight loss compared with usual care in overweight primary care patients: a randomized, controlled trial smartphone application for weight loss in overweight primary care patients. Ann Intern Med 2014;161(10_Supplement):S5–S12. https://doi.org/10.7326/m13-3005.

[86] Shiffman S, Stone AA, Hufford MR. Ecological momentary assessment. Annu Rev Clin Psychol 2008;4:1–32. https://doi.org/10.1146/annurev.clinpsy.3.022806.091415.

[87] Marcial LH, Richardson JE, Lasater B, et al. The imperative for patient-centered clinical decision support. EGEMS (Washington, DC) 2018;6(1):12. https://doi.org/10.5334/egems.259.

[88] Kawamoto K, Kukhareva PV, Weir C, et al. Establishing a multidisciplinary initiative for interoperable electronic health record innovations at an academic medical center. JAMIA Open 2021;4(3):ooab041. https://doi.org/10.1093/jamiaopen/ooab041.

[89] Bloomfield RA, Polo-Wood F, Mandel JC, Mandl KD. Opening the Duke electronic health record to apps: implementing SMART on FHIR. Int J Med Inform 2017;99:1–10. https://doi.org/10.1016/j.ijmedinf.2016.12.005.

[90] Taber P, Radloff C, Del Fiol G, Staes C, Kawamoto K. New standards for clinical decision support: a survey of the state of implementation. Yearb Med Inform 2021;30(1):159–71. https://doi.org/10.1055/s-0041-1726502.

[91] Scalia P, Ahmad F, Schubbe D, et al. Integrating option grid patient decision aids in the epic electronic health record: case study at 5 health systems. J Med Internet Res 2021;23(5), e22766. https://doi.org/10.2196/22766.

[92] Anonymous. Decision Memo for Screening for Lung Cancer with Low Dose Computed Tomography (LDCT) (CAG-00439N); 2015.
[93] Anonymous. National Coverage Determination (NCD) for Implantable Automatic Defibrillators (20.4); 2019.
[94] Lerman R. The chatbot will see you now: health-care chatbots boom but still can't replace doctors. Wash Post.
[95] Schachner T, Keller R, Wangenheim FV. Artificial intelligence-based conversational agents for chronic conditions: systematic literature review. J Med Internet Res 2020;22(9):e20701. https://doi.org/10.2196/20701.
[96] Choi YK, Thompson HJ, Demiris G. Use of an internet-of-things Smart home system for healthy aging in older adults in residential settings: pilot feasibility study. JMIR Aging 2020;3(2), e21964. https://doi.org/10.2196/21964.

CHAPTER

25

Clinical decision support and health disparities

Jorge A. Rodriguez[a,b] *and Lipika Samal*[a,b]

[a]Division of General Internal Medicine and Primary Care, Brigham and Women's Hospital, Boston, MA, United States [b]Harvard Medical School, Boston, MA, United States

25.1 Introduction

Health equity is critical to delivering effective care. However, marginalized populations have faced persistent health disparities and barriers to their care. Health information technologies, like clinical decision support (CDS), represent a potential tool to achieve health equity. In this chapter, we will (1) define health disparities and health equity, (2) present how CDS can be a tool for equity, (3) reveal barriers to broader implementation in underserved settings, and (4) highlight future efforts to apply CDS for health inequity, including artificial intelligence. We acknowledge that the application of CDS to address health disparities is an understudied area with great potential.

25.2 Health disparities

A health disparity is defined as "a health difference that adversely affects [marginalized populations]" [1]. Marginalized populations are those that "have systematically experienced greater obstacles to health based on their racial or ethnic group; religion; socioeconomic status; gender; age; mental health; cognitive, sensory, or physical disability; sexual orientation or gender identity; geographic location; or other characteristics historically linked to discrimination or exclusion" [2]. Additionally, health equity is "the opportunity to attain one's full health potential without barriers due to avoidable or preventable differences among groups of people, whether those groups of people are defined socially, economically, demographically, or geographically" [3].

Clinical Decision Support and Beyond
https://doi.org/10.1016/B978-0-323-91200-6.00016-4

There have been multiple critical moments in the development of health disparities research [4]. The Heckler Report, in 1985, revealed gaps in health status across race and ethnicity. This report suggested that race and ethnicity may contribute to health outcomes. In the report, *Unequal Treatment: Confronting Racial and Ethnic Disparities in Health Care*, the Institute of Medicine examined the disparities that arise in healthcare and provided recommendations to address these existing gaps [5]. In 2020, the coronavirus disease 2019 (COVID-19) pandemic revealed striking disparities in healthcare access and outcomes experienced by marginalized populations [6].

Marginalized populations experience greater burden in access to care, quality of care, health outcomes, patient engagement, and chronic disease management. Minority populations are more likely to have diabetes than White populations. Latinos, for example, are disproportionately affected by type II diabetes with an age-adjusted incidence rate of 9.7 per 1000 persons, nearly double the rate for non-Latino Whites [7]. Black populations have a higher risk of death from diabetes, heart disease, and cancer [8]. Similar patterns extend to multiple health outcomes, including infant mortality, maternal health, and overall health status. These disparities were made strikingly clear during the COVID-19 pandemic. People from minority populations had higher rates of infection, hospitalization, and death from the coronavirus compared with Whites [6,9,10]. In fact, in 2020, while life expectancy fell for the entire United States population, it was greater for Black and Hispanic populations [11]. While race and ethnicity are often the focus of health disparities, they exist across a broad set of factors. For example, low-income individuals and lesbian, gay, bisexual, and transgender (LGBT) populations report worse health status [12].

In terms of quality of care, marginalized populations often receive worse effectiveness of care and preventative measures. Black patients are less likely to have their blood pressure controlled and receive an influenza and pneumococcal vaccination [13]. Additionally, people with low income, low literacy or those lacking health insurance are less likely to have recommended cancer screenings for cervical and colorectal cancer [14]. Even in treatment, minority populations are less likely to receive appropriate care. Black patients with atrial fibrillation, an irregular heart rhythm which increases the risk of stroke, are less likely to receive newer and more effective blood thinners for stroke prevention [15].

In addition to healthcare considerations, health disparities are driven by multiple interrelated genetic, cultural, behavioral, environmental, economic, and social factors (Fig. 25.1). Particular emphasis is placed on factors that exists outside of the healthcare system, namely social determinants of health (SDOH). SDOH are "the conditions in the environments where people are born, live, learn, work, play, worship, and age that affect a wide range of health, functioning, and quality-of-life outcomes and risks" [16]. For example, people living with food insecurity have worse diabetes control [17].

There are also structural barriers to health equity, like systemic racism and clinician bias, which contribute to existing health disparities [18,19]. Systemic racism refers to "processes of racism that are embedded in laws (local, state, and federal), policies, and practices of society and its institutions that provide advantages to racial groups deemed as superior, while differentially oppressing, disadvantaging, or otherwise neglecting racial groups viewed as inferior" [18]. Clinician bias can also impact care delivery with data suggesting that clinicians have implicit bias that advantages certain populations over others [20].

Domains of Influence (Over the Lifecourse)		Levels of Influence*			
		Individual	Interpersonal	Community	Societal
	Biological	Biological Vulnerability and Mechanisms	Caregiver–Child Interaction Family Microbiome	Community Illness Exposure Herd Immunity	Sanitation Immunization Pathogen Exposure
	Behavioral	Health Behaviors Coping Strategies	Family Functioning School/Work Functioning	Community Functioning	Policies and Laws
	Physical/Built Environment	Personal Environment	Household Environment School/Work Environment	Community Environment Community Resources	Societal Structure
	Sociocultural Environment	Sociodemographics Limited English Cultural Identity Response to Discrimination	Social Networks Family/Peer Norms Interpersonal Discrimination	Community Norms Local Structural Discrimination	Social Norms Societal Structural Discrimination
	Health Care System	Insurance Coverage Health Literacy Treatment Preferences	Patient–Clinician Relationship Medical Decision-Making	Availability of Services Safety Net Services	Quality of Care Health Care Policies
Health Outcomes		Individual Health	Family/ Organizational Health	Community Health	Population Health

FIG. 25.1 Factors contributing to health disparities. *From National Institute on Minority Health and Health Disparities (2018).*

Thus, clinical teams must consider SDOH, structural barriers along with expanding clinical knowledge as part of their in their effort to deliver equitable care. This can create a challenge, which could be addressed by clinical decision support.

25.3 Clinical decision support and health disparities

Clinical decision support offers an opportunity to narrow disparities in care [21]. Specifically, CDS can improve implementation and adherence to clinical guidelines and practices, increase screening for preventable diseases and chronic disease complications, promote patient safety, reduce adverse outcomes, decrease redundant testing, and minimize clinician bias [22–24]. CDS could decrease disparities by raising clinician awareness of clinical gaps, especially among marginalized populations [25]. More rapid retrieval of information could support clinician efficiency and allow for more time for shared decision making and trust building, which are critical for equitable care. Though more disparities focused research is needed, there are examples of CDS leading to more equitable care [26].

Early work provided evidence that CDS could help improve health disparities. A systematic review by Milley and Kukafka from 2010, found that CDS interventions demonstrated improvements in care quality in under-resourced settings [27]. Samal et al. used primary care visit data from the National Ambulatory Medical Care Survey to examine the association between implementing CDS and racial and ethnic disparities in blood pressure control [28]. They found that among physicians who used CDS, there were no disparities in blood pressure control between White and Black Patients.

Further studies have shown that CDS implementation can lead to improvements in chronic disease management and quality of care [29,30]. A diabetes dashboard that included CDS was implemented in two urban safety net clinics serving primarily Latino patients [31]. The dashboard offered providers clinical alerts, reminders related to diabetes care, treatment plan templates, and educational resources. Welch et al. found that twice as many patients in the intervention arm achieved improved glycemic control. Additionally, patients reported significant improvement in social and diabetes-related distress. These findings are consistent with work that CDS can improve referral to diabetes education among underserved populations [32]. Ganju et al. further showed that CDS help attenuate disparities in complications from diabetes [33]. Their implementation of CDS to address racial disparities in limb amputation found that CDS adoption led to narrowing of the gap in amputation rates between Black and White patients. As it relates to quality, in a large study conducted at an academic internal medicine ambulatory practice, CDS was used to address quality gaps in care between Black and White patients [25]. The intervention included electronic clinical reminders, EHR-integrated decision support tools to promote adherence to clinical guidelines, and regular provider feedback on quality disparities. Of seven quality metrics that had known disparities, two improved (prescriptions of antiplatelet therapy for patients with coronary heart disease and colorectal cancer screening) demonstrating that CDS can support an equitable healthcare system.

CDS has also promoted equity in cancer screening. The implementation of CDS in two outpatient Federally Qualified Health Centers narrowed disparities in breast cancer screening [34]. A specialized software generated an individualized breast cancer risk and CDS for appropriate monitoring. CDS was provided to primary care clinicians at the time of the encounter. The use of this CDS led to increased mammography screening among women of racial and ethnic minority groups who had a high breast cancer risk. Similarly, Baker et al. demonstrated that using CDS as part of a multifaceted intervention to increase colorectal cancer screening led to increased screening rates in an underserved community that included people who had low socioeconomic status, low health literacy, and limited English proficiency [35].

The use of CDS in attenuating preventative disease disparities has shown promise as well. Electronic alerts focused on routine immunization were implemented in a largely Black and urban pediatric patient population in Philadelphia. These led to increased routine immunization opportunities and follow through immunizations. CDS has also shown to increase infectious disease preventative testing in underserved areas. Chak et al. aimed to increase screening for Hepatitis B among Asian and Pacific Islander patients using a CDS alert [36]. They found that in the alert group 119 ($n = 1484$) patients were screened compared to 48 ($n = 1503$) in the control group. Similar findings have been reported for using CDS to increase HIV and Hepatitis C screening [37,38]. Patient-facing CDS interventions have also demonstrated opportunities to address cancer screening disparities. For example, serial text messaging to patients increased colon cancer screening rates by 17.7% points in an urban clinic serving primarily Black patients [39].

25.3.1 Limitation of CDS for health disparities

CDS, of course, is not a panacea, there is also evidence the CDS does not lead to decreased disparities. For example, Mishuris et al. used data from the National Ambulatory and National Hospital Ambulatory Medical Care Surveys to demonstrate that CDS did not improve cancer screening disparities [40]. We discuss further limitations below.

25.4 CDS implementation for underserved populations

25.4.1 Clinical decision support diffusion disparities

The impact of CDS on health disparities depends on equitable diffusion of these technologies. EHR and CDS uptake has been mixed in underserved areas due to limited technology infrastructure and availability of new innovations [27,41,42]. Progress has been made in terms of basic EHR adoption across a variety of clinical settings, 80.5% of hospitals have adopted at least a basic EHR system. However, only 37.5% of hospitals adopted advanced EHR functions. Further, critical access hospitals, which care for rural populations, have been found to be less likely to adopt advanced functions often implemented in CDS, like using EHR data to identify high-risk patients, monitor adherence to guidelines, and identify gaps in care for specific patient populations [43]. Conversely, there has been little difference in uptake among safety-net and non-safety-net hospitals creating an environment where CDS can support health equity.

25.4.2 Clinician decision support implementation considerations for health equity

As CDS continues to diffuse in underserved communities, the success of future implementation will depend on applying an equity lens as part of the development, implementation, and evaluation process (Table 25.1) [23,44]. Particular attention should be placed on the sustainability of CDS tools to support the health of communities [45].

25.5 CDS, artificial intelligence, and health equity

Clinical decision support is increasingly augmented by artificial intelligence (AI) and holds promise to achieve health equity [46]. The National Academy of Medicine specifically highlights health equity in their report, *Artificial Intelligence in Health Care: The Hope, The Hype, The Promise, The Peril* [47]. However, the potential for AI to perpetuate existing disparities through data and algorithmic bias tempers this potential [48]. For example, Obermeyer et al. found that an algorithm that help guide care allocation assigned the same level of risk to Black

TABLE 25.1 Health equity consideration for clinical decision support.

- Aim to address recognized health disparities from the start
- Use available technology to maximize reach to underserved populations (e.g., text messaging)
- Include stakeholders from relevant communities in CDS development and implementation
- Leverage multiple types of data to promote the most current and relevant evidence-based guidelines
- Ensure the guideline is derived from studies that included marginalized populations
- Link to services that can support recommended intervention (e.g., refer to local food pantry)
- Make recommendations actionable and beneficial to all patients
- Test CDS in a variety of settings including with a diverse cohort of patients
- Design CDS tools that support sustainable use in underserved settings.
- Align with existing quality and safety initiatives

patients compared to White patients, even though the Black patients were sicker [49]. This occurred due to inappropriately determining the level of need based on health care costs. Similar concerns have been raised by the use of race in a variety of CDS tools and algorithms [50]. If equity is not central to the future development of AI-powered CDS, similar patterns may arise.

25.6 Conclusion

CDS provides an effective tool to minimize many of the drivers of health disparities in healthcare. There is promising evidence that these tools can serve to promote equity across the healthcare spectrum. However, future iterations require a focus on equity from the start and inclusive and sustainable development, implementation, and evaluation practices.

References

[1] Duran DG, Pérez-Stable EJ. Novel approaches to advance minority health and health disparities research. Am J Public Health 2019;109(S1):S8–10.
[2] Healthy People 2030 Questions & Answers | health.gov., 2021. [Internet]. [cited 2021 Oct 2]. Available from: https://health.gov/our-work/national-health-initiatives/healthy-people/healthy-people-2030/questions-answers#q9.
[3] Social determinants of health., 2021. [Internet]. [cited 2021 Oct 2]. Available from: https://www.who.int/westernpacific/health-topics/social-determinants-of-health.
[4] Dankwa-Mullan I, Perez-Stable EJ, Gardner K, Zhang X, Rosario AM, editors. The science of health disparities research. Hoboken, NJ: Wiley-Blackwell; 2021.
[5] Institute of Medicine. Unequal treatment: confronting racial and ethnic disparities in health care. Washington, DC: The National Academies Press; 2003. [Internet]. Available from: https://www.nap.edu/catalog/12875/unequal-treatment-confronting-racial-and-ethnic-disparities-in-health-care.
[6] Lopez III L, Hart III LH, Katz MH. Racial and ethnic health disparities related to COVID-19. JAMA 2021; 325(8):719–20.
[7] National Diabetes Statistics Report. CDC, 2020. [Internet]. [cited 2021 Oct 2]. Available from: https://www.cdc.gov/diabetes/data/statistics-report/index.html.
[8] KFF. 2019 Key facts on health and health care by race and ethnicity—Conclusion—8878-02, 2019. [Internet]. [cited 2021 Oct 2]. Available from: https://www.kff.org/report-section/key-facts-on-health-and-health-care-by-race-and-ethnicity-conclusion/.
[9] Rubin-Miller L, Alban C. COVID-19 racial disparities in testing, infection, hospitalization, and death: analysis of epic patient data. KFF; 2020. Sep 16 SSP, [Internet] [cited 2021 Oct 2]. Available from: https://www.kff.org/coronavirus-covid-19/issue-brief/covid-19-racial-disparities-testing-infection-hospitalization-death-analysis-epic-patient-data/.
[10] CDC. Cases, data, and surveillance. Centers for Disease Control and Prevention; 2020. [Internet]. [cited 2021 Oct 2]. Available from: https://www.cdc.gov/coronavirus/2019-ncov/covid-data/investigations-discovery/hospitalization-death-by-race-ethnicity.html.
[11] Arias E, Betzaida T-V, Ahmad F, Kochanek K. Provisional life expectancy estimates for 2020. National Center for Health Statistics (U.S.); 2021. [Internet]. [cited 2021 Oct 2]. Available from: https://stacks.cdc.gov/view/cdc/107201.
[12] KFF. Disparities in health and health care: 5 key questions and answers. KFF; 2021. [Internet]. [cited 2021 Oct 2]. Available from: https://www.kff.org/racial-equity-and-health-policy/issue-brief/disparities-in-health-and-health-care-5-key-question-and-answers/.
[13] 2019 National healthcare quality and disparities report disparities measures; 2017. p. 139.

[14] Benavidez GA. Disparities in meeting USPSTF breast, cervical, and colorectal cancer screening guidelines among women in the United States. Prev Chronic Dis 2021;18. [Internet]. [cited 2021 Oct 2]. Available from: https://www.cdc.gov/pcd/issues/2021/20_0315.htm.
[15] Essien UR, Kim N, Hausmann LRM, Mor MK, Good CB, Magnani JW, et al. Disparities in anticoagulant therapy initiation for incident atrial fibrillation by race/ethnicity among patients in the veterans health administration system. JAMA Netw Open 2021;4(7), e2114234.
[16] Social Determinants of Health. Healthy people 2030 | health.gov, 2021. [Internet]. [cited 2021 Oct 2]. Available from: https://health.gov/healthypeople/objectives-and-data/social-determinants-health.
[17] Berkowitz SA, Karter AJ, Corbie-Smith G, Seligman HK, Ackroyd SA, Barnard LS, et al. Food insecurity, food "deserts," and glycemic control in patients with diabetes: a longitudinal analysis. Diabetes Care 2018; 41(6):1188–95.
[18] Williams DR, Lawrence JA, Davis BA. Racism and health: evidence and needed research. Annu Rev Public Health 2019;40(1):105–25.
[19] Evans MK, Rosenbaum L, Malina D, Morrissey S, Rubin EJ. Diagnosing and treating systemic racism. N Engl J Med 2020;383(3):274–6.
[20] Hall WJ, Chapman MV, Lee KM, Merino YM, Thomas TW, Payne BK, et al. Implicit racial/ethnic bias among health care professionals and its influence on health care outcomes: a systematic review. Am J Public Health 2015;105(12):e60–76.
[21] Zhang X, Hailu B, Tabor DC, Gold R, Sayre MH, Sim I, et al. Role of health information technology in addressing health disparities: patient, clinician, and system perspectives. Med Care 2019;57:S115.
[22] Pérez-Stable EJ, Jean-Francois B, Aklin CF. Leveraging advances in technology to promote health equity. Med Care 2019;57(Suppl 2):S101–3.
[23] Tcheng JE, National Academy of Medicine (U.S.), editors. Optimizing strategies for clinical decision support: summary of a meeting series. Washington, DC: National Academy of Medicine; 2017. p. 1 [The learning health system series.].
[24] Grasso C, Goldhammer H, Thompson J, Keuroghlian AS. Optimizing gender-affirming medical care through anatomical inventories, clinical decision support, and population health management in electronic health record systems. J Am Med Inform Assoc 2021. ocab080.
[25] Jean-Jacques M, Persell SD, Thompson JA, Hasnain-Wynia R, Baker DW. Changes in disparities following the implementation of a health information technology-supported quality improvement initiative. J Gen Intern Med 2012;27(1):71–7.
[26] Implementing Clinical Decision Support Systems | CDC | DHDSP., 2021. [Internet]. [cited 2021 Oct 3]. Available from: https://www.cdc.gov/dhdsp/pubs/guides/best-practices/clinical-decision-support.htm.
[27] Millery M, Kukafka R. Health information technology and quality of health care: strategies for reducing disparities in underresourced settings. Med Care Res Rev 2010;67(5_suppl). 268S–298S.
[28] Samal L, Lipsitz SR, Hicks LS. Impact of electronic health records on racial and ethnic disparities in blood pressure control at US primary care visits. Arch Intern Med 2012;172(1):75–6.
[29] Sequist TD, Adams A, Zhang F, Ross-Degnan D, Ayanian JZ. Effect of quality improvement on racial disparities in diabetes care. Arch Intern Med 2006;166(6):675–81.
[30] Hicks LS, Sequist TD, Ayanian JZ, Shaykevich S, Fairchild DG, Orav EJ, et al. Impact of computerized decision support on blood pressure management and control: a randomized controlled trial. J Gen Intern Med 2008; 23(4):429–41.
[31] Welch G, Zagarins SE, Santiago-Kelly P, Rodriguez Z, Bursell S-E, Rosal MC, et al. An internet-based diabetes management platform improves team care and outcomes in an urban latino population. Dia Care 2015. dc141412.
[32] James TL. Improving referrals to diabetes self-management education in medically underserved adults. Diabetes Spectr 2021;34(1):20–6.
[33] Ganju KK, Atasoy H, McCullough J, Greenwood B. The role of decision support systems in attenuating racial biases in healthcare delivery. Manag Sci 2020;66(11):5171–81.
[34] Schwartz C, Chukwudozie IB, Tejeda S, Vijayasiri G, Abraham I, Remo M, et al. Association of population screening for breast cancer risk with use of mammography among women in medically underserved racial and ethnic minority groups. JAMA Netw Open 2021;4(9), e2123751.
[35] Baker DW, Brown T, Buchanan DR, Weil J, Balsley K, Ranalli L, et al. Comparative effectiveness of a multifaceted intervention to improve adherence to annual colorectal cancer screening in community health centers: a randomized clinical trial. JAMA Intern Med 2014;174(8):1235–41.

[36] Chak E, Taefi A, Li C-S, Chen MS, Harris AM, MacDonald S, et al. Electronic medical alerts increase screening for chronic hepatitis B: a randomized, double-blind, controlled trial. Cancer Epidemiol Biomarkers Prev 2018; 27(11):1352–7.

[37] Avery AK, Del Toro M, Caron A. Increases in HIV screening in primary care clinics through an electronic reminder: an interrupted time series. BMJ Qual Saf 2014;23(3):250–6.

[38] Tapp H, Ludden T, Shade L, Thomas J, Mohanan S, Leonard M. Electronic medical record alert activation increase hepatitis C and HIV screening rates in primary care practices within a large healthcare system. Prev Med Rep 2020;17, 101036.

[39] Huf SW, Asch DA, Volpp KG, Reitz C, Mehta SJ. Text messaging and opt-out mailed outreach in colorectal cancer screening: a randomized clinical trial. J Gen Intern Med 2021;36(7):1958–64.

[40] Mishuris RG, Linder JA. Racial differences in cancer screening with electronic health records and electronic preventive care reminders. J Am Med Inform Assoc 2014;21(e2):e264–9.

[41] Jha AK, DesRoches CM, Shields AE, Miralles PD, Zheng J, Rosenbaum S, et al. Evidence of an emerging digital divide among hospitals that care for the poor: a central policy question is whether health information technology investments prompted by the 2009 federal stimulus law will help close the gap. Health Aff 2009;28(Suppl 1): w1160–70.

[42] Shields AE, Shin P, Leu MG, Levy DE, Betancourt RM, Hawkins D, et al. Adoption of health information technology in community health centers: results of a national survey. Health Aff (Millwood) 2007;26(5):1373–83.

[43] Adler-Milstein J, Holmgren AJ, Kralovec P, Worzala C, Searcy T, Patel V. Electronic health record adoption in US hospitals: the emergence of a digital "advanced use" divide. J Am Med Inform Assoc 2017;24(6):1142–8.

[44] López L, Green AR, Tan-McGrory A, King R, Betancourt JR. Bridging the digital divide in health care: the role of health information technology in addressing racial and ethnic disparities. Jt Comm J Qual Patient Saf 2011; 37(10):437–45.

[45] Green LA, Potworowski G, Day A, May-Gentile R, Vibbert D, Maki B, et al. Sustaining "meaningful use" of health information technology in low-resource practices. Ann Fam Med 2015;13(1):17–22.

[46] Clark CR, Wilkins CH, Rodriguez JA, Preininger AM, Harris J, DesAutels S, et al. Health care equity in the use of advanced analytics and artificial intelligence technologies in primary care. J Gen Intern Med 2021;36(10):3188–93.

[47] Matheny M, Thadaney IS, Ahmed M, Whicher M. AI in health care: the hope, the hype, the promise, the peril. National Academy of Medicine; 2019.

[48] Gianfrancesco MA, Tamang S, Yazdany J, Schmajuk G. Potential biases in machine learning algorithms using electronic health record data. JAMA Intern Med 2018;178(11):1544–7.

[49] Obermeyer Z, Powers B, Vogeli C, Mullainathan S. Dissecting racial bias in an algorithm used to manage the health of populations. Science 2019;366(6464):447–53.

[50] Vyas DA, Eisenstein LG, Jones DS. Hidden in plain sight—reconsidering the use of race correction in clinical algorithms. N Engl J Med 2020;383(9):874–82.

CHAPTER

26

Population health management

Guilherme Del Fiol

Department of Biomedical Informatics, University of Utah Health, University of Utah, Salt Lake City, UT, United States

26.1 Introduction

Much of this book deals with the use of clinical decision support (CDS) tools at the point of care delivery. However, as described in Chapters 16 (Population Analytics) and 27 (CDS for Public Health), there are critical needs and increased opportunities for improving health at the population level, which are in part motivated by the proposed goals in the Quintuple Aim for Healthcare Improvement: (i) improving population health; (ii) enhancing the care experience; (iii) reducing costs [1]; (iv) improving the work life of the health care workforce, addressing growing concerns about burnout among healthcare professionals [2]; and (v) advancing health equity [3]. Another important driver towards a focus on population health, especially in the United States, is the emergence of new models of care such as value-based care [4] and accountable care organizations [5], in which healthcare systems are financially rewarded for improving population-level healthcare processes and outcomes rather than for providing higher volumes of care. More recently, the COVID-19 pandemic motivated the rapid implementation of technology innovations such as telehealth [6–8], conversational agents [9,10], and at-home testing [11]. A common pattern across these innovations is the shift of patient care from physical care settings to patients' homes and their communities, and from individual patients to patient populations. As proposed by Horn et al. [12], similar approaches could be adopted for a wide range of preventive services such as screening for colorectal cancer [13], hypertension, and diabetes. Such approaches are also valuable for long-term protocol-driven chronic disease management, using tools such as patient-interactive and sensor-enabled data capture, self-care, and other connected-care capabilities [14,15].

Population health management (PHM) is closely aligned with the Quintuple Aim and new models of care in that it is "...a proactive, organized, and cost-effective approach to prevention that utilizes newer technologies to help reduce morbidity while improving the health status, health service use, and personal productivity of individuals in defined populations" [16].

Clinical Decision Support and Beyond
https://doi.org/10.1016/B978-0-323-91200-6.00023-1

As described above, PHM can also be used for chronic disease management and has the potential to reduce provider burden by shifting healthcare interventions outside the scope of face-to-face patient encounters and from individual providers to the healthcare system. Last, PHM has the potential to promote health equity by proactively identifying individuals suffering from health inequities and proactively offering affordable and accessible interventions that are adapted to their background and needs.

26.2 Scope of this chapter

There is an extensive body of literature describing different aspects of PHM, from conceptual frameworks [17] to specific PHM interventions [18–20] and broader PHM programs [16,18,21,22]. In this chapter, we focus on the components of PHM and a few examples of PHM interventions. This chapter also has synergy with other chapters in the book. Notably, Chapter 16 covers population analytics approaches, which are critical for PHM, and Chapter 27 covers CDS in the context of public health, which can benefit from PHM.

26.3 Components of population health management

A PHM program relies on a robust and scalable technical infrastructure with several components (Fig. 26.1). At the core of a PHM platform is a set of patient *data sources*, in some cases integrated into a centralized enterprise data warehouse (EDW), coupled with *population analytics* tools. Examples of data sources include electronic health records (EHRs), healthcare claims, external registries such as immunization registries, and health information exchange (HIE). Population analytics tools support a variety of functions such as quality measurement (with focus on the Quintuple Aim), identification of *priority cohorts* for specific health interventions (e.g., patients who are overdue for colorectal cancer screening), and *risk stratification*

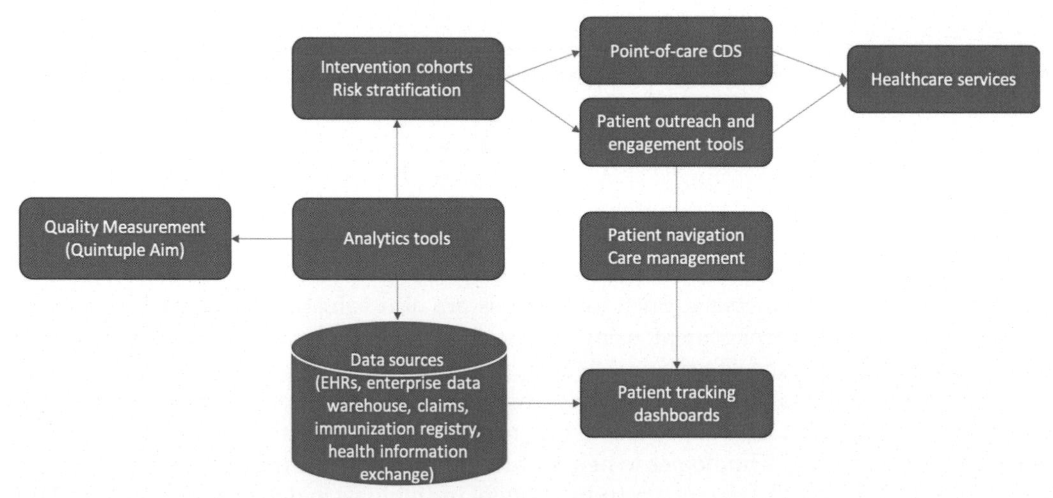

FIG. 26.1 Components of a population health management platform.

(e.g., patients with high risk of hospitalization) for risk-tailored interventions. Once priority cohorts are identified, a variety of electronic tools can be used for *patient outreach, engagement* and connection with relevant *healthcare services*, including text messaging, e-mail, conversational agents, and patient portals. These tools can be supplemented with human interventions such as *patient navigators* and *care managers* who reach out to patients to help address barriers (e.g., social, transportation, financial), logistics, and to facilitate a connection to relevant services. In addition, *point of care* CDS tools such as preventive care reminders can be used opportunistically to ensure that patients in a target cohort are offered relevant interventions within a specific healthcare encounter. Last, PHM *dashboards* support the management of a specific PHM intervention, including identification of care gaps, tracking of individual patient status, review of specific patient data, and launching of patient outreach efforts.

Several of these components are described in detail in other chapters of this book. Patient registries, data warehouses, analytics tools, and risk stratification are described in Chapter 16. Quality measurement and its role in improving provider performance is covered in Chapter 4. Point-of-care CDS tools such as alerts and reminders are covered in Chapters 1, 9, and 20. In the subsections below, we focus on concepts that are not covered in other chapters, i.e., patient outreach tools such as e-mail, text messaging, and conversational agents (chatbots); and patient tracking dashboards.

26.3.1 Automated patient outreach tools

Unlike reactive, clinic-level approaches that require a clinic visit, patient-level interventions identify specific groups outside the scope of patient encounters to proactively connect those individuals with relevant healthcare services. Patient outreach tools are an essential component of such interventions in that they allow healthcare systems to reach out to patients who have gaps in care offering access to needed services. Patient outreach approaches include electronic tools such as text messaging, e-mail, and conversational agents; paper-based tools such as patient letters; and human-driven approaches such as patient navigation.

26.3.1.1 E-mail, text messaging, and letters

Automated patient outreach tools such as text messaging and e-mail are available in most EHR systems and their patient portals [23]. These tools are being increasingly used, especially among large integrated delivery networks and academic medical centers. Paper-based letters are also often used, both as a fallback approach for patients who do not have access to electronic tools and for patients who prefer to use multiple methods of communication. A common use of such tools is to automatically remind patients about preventive care for which they are due, such as breast cancer screening, colorectal cancer screening, and immunizations.

26.3.1.2 Conversational agents

Conversational agents are automated, scripted, interactive agents designed to mimic human interaction. They are increasingly popular in various health contexts as they can be easily accessed through a Web browser via smartphones, tablets, laptops or desktops. Conversational agents have many advantages for engaging patients including providing scripted education interactively, chunking information into small segments that are easier to process, and allowing for choice in the amount of information received. Thus, a conversational agent interaction can be tailored based on individual level factors rather than targeting population groups. Conversational agents are accessible to the vast majority of U.S. adults—76% of individuals with low socioeconomic status (i.e., households with annual incomes less than $30,000) own a smart phone [24–26]. Delivery of health services through conversational agents has been successfully tested in various health domains [27], with demonstrated benefits in lowering rates of depression and anxiety [28–30], and improving self-adherence to asthma, diabetes, and pain management care [31–33]. Conversational agents have also been used in the COVID-19 pandemic to help answer questions, screen for testing and vaccination eligibility, and to schedule testing and vaccination [10,34].

26.3.1.3 Patient navigation

Coupled with technology-based patient outreach, patient navigators can be a powerful intervention to help connect patients with healthcare and social services. They can be especially useful when engaging with groups from historically marginalized minorities, low socioeconomic status and rural/frontier locations who often have less awareness, access, and use of prevention services [35,36]. Practical advice from patient navigators (e.g., social workers, community health workers, care managers) can educate and overcome hesitancy and engagement barriers such as logistics, transportation, and expense, and of critical importance, healthcare providers welcome the use of these approaches with their patients [37–39]. Therefore, patient navigation can be a critical component of PHM programs that aim to advance health equity.

26.3.2 Patient outreach tracking dashboards

A patient outreach tracking dashboard enables PHM staff members to launch patient outreach activities, such as text messaging campaigns, to target cohorts; track the status of each patient in the cohort, including responses to outreach messages that require human attention; and track human outreach activities such as phone calls.

26.4 Case studies

Studies have shown that PHM approaches using outreach methods such as patient reminders are effective tools in increasing the rate of completion of several types of preventive care, including immunizations [40] and cancer screening [41–44]. Examples of such PHM approaches are provided below.

26.4.1 Colorectal cancer screening

In this section, we illustrate the application of the PHM components described above in the context of the Colorectal Cancer Control Program (CRCCP) sponsored by the US Center for Disease Control and Prevention (CDC) [13]. According to the US Preventive Services Task

Force guidelines for CRC screening, individuals 45 and older should receive colorectal cancer screening, which can be implemented as a Fecal Immunochemical Test (FIT) every year, FIT-DNA test every three years, or colonoscopy every 10 years [45]. Despite its benefits, the rates of CRC screening in the US are relatively low, with large disparities among individuals with lower access to care. For example, CDC data from 2008 shows that the CRC screening rate in the US was 63%, but as low as 35% among those who are uninsured [46].

The overall goal of the CRCCP program is to use population-based approaches to increase the rate of colorectal cancer (CRC) screening, especially among individuals with low access to care. The program starts with using analytics tools to identify the target population and assess baseline quality measurements, including process measures (e.g., rate of ordered FIT tests for which patients collect a sample and send to the lab, rate of completed colonoscopy referrals) and outcome measures (e.g., CRC screening rates). Further analyses are completed to identify cohorts that could be targets for specific interventions. Next, multi-level interventions are implemented for different cohorts. For example, a clinic-level intervention reminds clinical staff to schedule or carry out CRC screening for patients who are coming due or are overdue. A patient-level intervention sends electronic reminders (e.g., text messaging, e-mail) offering an at-home FIT test to be mailed to patients who do not have an upcoming clinic appointment and are eligible for a FIT test. Patients who do not mail a sample back to the clinic or the laboratory also receive a reminder. For patients who need a colonoscopy, a patient navigator, assisted by a patient outreach tracking dashboard, reaches out to patients to schedule the procedure and help address any barriers. Last, providers and clinical staff receive provider and clinic-level reports showing CRC screening rates among their patient population compared with their peers.

Similar PHM programs with multi-level interventions have been implemented with demonstrated benefits through randomized controlled trials. A meta-analysis of studies evaluating such programs found strong evidence supporting the use of PHM interventions such as patient outreach for FIT testing and patient navigation to increase CRC screening rates. Further increase was observed when those interventions were combined [43].

26.4.2 PHM for chronic disease management

PHM programs have also been used successfully to support chronic disease management using strategies such as protocol-driven care, risk stratification, care management, and self-care [14,15,47,48]. Common approaches involve data analytics methods to identify individuals who could benefit from PHM interventions; computable logic to stratify patients according to risk; and protocol-based logic to assign individuals within each stratum to specific interventions.

A systematic review of 232 randomized controlled trials identified studies assessing the effect of PHM programs on diabetes, heart failure, asthma, chronic obstructive pulmonary disease, and cancer [15]. PHM interventions included self-monitoring with feedback and action plans, education, lifestyle interventions (e.g., tobacco cessation, exercise), medication reminders for increase compliance, and step-up care as needed for those at higher risk. The benefits of interventions with a single self-management component were not consistent and showed mixed-results. Education-based interventions were effective for diabetes, but not for other conditions. Active self-monitoring and feedback was the most common intervention component, resulting in consistent benefits for patients with diabetes and heart failure, but mostly no benefits for COPD and asthma. Remote consultations were often provided as a step-up care approach to help patients with poor outcomes or at elevated risk. Benefits were consistent especially in heart failure management, with reduced hospitalization and mortality. A systematic review assessed the effect of multi-component interventions (patient education, health care provider education, patient-tailored self-monitoring, protocol-based blood glucose goals, use of blood glucose data to modify behavior, feedback to patients, and interactive shared-decision making) and found that programs that included at least 6 out of 7 components had the strongest benefit in HbA1C reduction [49].

26.4.3 Genetic testing for hereditary cancers

Several studies support individualizing risk-stratified cancer screening, with selective application of specific screening interventions best suited to the individual [50–58]. Early identification of individuals at high risk is critical to initiating personalized cancer prevention and treatment to reduce disparities in morbidity and mortality. For example, the US Preventive Services Task Force recommends tailoring breast cancer screening based on a woman's individual risk profile [57] and the US Multi-Society Task Force on Colorectal Cancer recommends tailoring colorectal cancer screening based on hereditary cancer risk [58]. Despite recommendations from evidence-based guidelines [57,58], the adoption of risk-stratified cancer screening is still low. Primary care providers have reported significant barriers to risk-stratified screening including low self-efficacy, lack of time, and lack of access to services such as genetic counseling [59–62]. On the other hand, PHM is a promising approach to identify and offer genetic testing to eligible individuals based on their family health history.

To address the challenges above, our informatics group at the University of Utah has developed GARDE (Genetic Cancer Risk Detector), a PHM platform that identifies patients who are eligible for genetic testing of hereditary cancers based on meet evidence-based criteria set by the National Comprehensive Cancer Network (NCCN). GARDE uses an open architecture, external to any EHR system, and exposed through Web services compliant with the Health Level Seven (HL7) Fast Healthcare Interoperability Resources (FHIR) and CDS Hooks standards [63,64]. Through bulk data analytics processes, GARDE retrieves data (e.g., demographics, family health history) from the EHR; runs algorithms to identify patients who meet NCCN criteria for genetic testing of breast, ovarian and colorectal cancer; and uses conversational agents for patient outreach, education, connection to genetic testing, and explanation of test results [65,66].

A PHM program implemented with GARDE is being evaluated in the BRIDGE study, a multi-site randomized controlled trial comparing conversational agent outreach with usual care (human-based patient outreach and education) [67]. The conversational agent provides tailored patient education about heredity; explains the benefits, risks, logistics, and costs of genetic testing; and offers patients the option to receive an at-home genetic test kit through the mail [68]. Patients self-collect their sample and mail it to the laboratory. Once results are returned to the clinic, patients with a negative result are offered access to another conversational agent that explains the implications of the results. Patients with a positive test for a pathogenic mutation receive a phone call to schedule a genetic counseling appointment. Results of the BRIDGE trial are expected towards the end of 2022.

26.4.4 PHM to advance health equity

Populations from historically marginalized minorities, low socioeconomic status, and rural/frontier areas suffer profound health inequities across a wide variety of diseases and conditions. For example, mortality attributed to tobacco use accounted for nearly half of total male mortality among individuals with low socioeconomic status in North America, Poland, and the United Kingdom, twice as much as individuals from higher social economic strata [69]. Profound disparities were also found in the COVID-19 pandemic [70]. Several factors contribute to these disparities, including low health literacy and lack of access to high quality healthcare.

To advance health equity, the Huntsman Center for Health Outcomes and Population Equity (HOPE), in collaboration with our informatics group at the University of Utah, is conducting several studies investigating the effect of PHM approaches using interventions such as text messaging, conversational agents, and patient navigation to connect patients to services such as tobacco cessation treatment (quitlines) [71], colorectal cancer screening (funded by the CDC's CRCCP program described above), HPV immunization, and COVID-19 testing and vaccination [72]. The interventions target patients who seek care at Community Health Centers across Utah and other states in the Mountain West region. Fig. 26.2 shows an example of a text messaging intervention offering patients access to COVID-19 vaccination and at-home testing. Messages aim to reach specific groups based

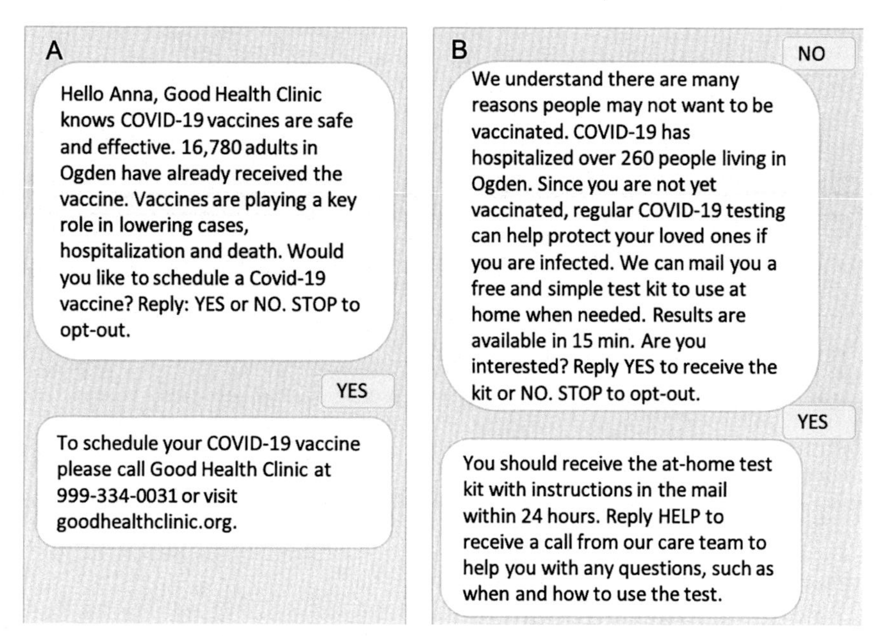

FIG. 26.2 Text messaging offering access to COVID-19 vaccination (A) and at-home testing (B).

on factors such as eligibility to receive the vaccine according to data from the EHR and the Utah state immunization registry. Messages are sent on behalf of the patient's clinic and include up-to-date data about the pandemic at the patient's location. Patients can reply with a single-touch response to be connected to services. For patients who are interested in vaccination or testing, a patient navigator calls to address questions, barriers, and concerns.

26.5 Conclusion

PHM is a promising approach to help achieve the Quintuple Aim of healthcare: (i) improving population health through population-level interventions; (ii) enhancing the care experience by shifting healthcare from the clinic to the patient's home; (iii) reducing costs by focusing on health promotion and prevention; (iv) improving the work life of the health care workforce by reducing clinic workload; and (v) advancing health equity by maximizing reach through a combination of digital and human-based patient outreach interventions. Healthcare systems are increasingly using PHM approaches, especially after the COVID-19 pandemic. PHM is expected to experience substantial growth with novel digital health technologies, such as sensors, phone apps, conversational agents, and virtual reality; artificial intelligence; and new data sources (e.g., social determinants of health, exposure to air pollution, family habits). These innovations have the potential to raise PHM to the next level through advances in (i) health analytics methods for cohort identification, risk stratification, and intervention assignment to individuals; (ii) real-time tailoring of interventions to each individual based on self-monitoring and context-aware data (e.g., geolocation, activity); and (iii) research identifying optimal interventions in multi-component programs for each condition and patient.

References

[1] Berwick DM, Nolan TW, Whittington J. The triple aim: care, health, and cost. Health Aff (Millwood) 2008; 27(3):759–69.
[2] Bodenheimer T, Sinsky C. From triple to quadruple aim: care of the patient requires care of the provider. Ann Fam Med 2014;12(6):573–6.
[3] Nundy S, Cooper LA, Mate KS. The quintuple aim for health care improvement: a new imperative to advance health equity. JAMA 2022;327(6):521–2.
[4] Porter ME. What is value in health care? N Engl J Med 2010;363(26):2477–81.
[5] McClellan M, McKethan AN, Lewis JL, Roski J, Fisher ES. A national strategy to put accountable care into practice. Health Aff (Millwood) 2010;29(5):982–90.
[6] Monaghesh E, Hajizadeh A. The role of telehealth during COVID-19 outbreak: a systematic review based on current evidence. BMC Public Health 2020;20(1):1193.
[7] Doraiswamy S, Abraham A, Mamtani R, Cheema S. Use of telehealth during the COVID-19 pandemic: scoping review. J Med Internet Res 2020;22(12), e24087.
[8] Wosik J, Fudim M, Cameron B, Gellad ZF, Cho A, Phinney D, et al. Telehealth transformation: COVID-19 and the rise of virtual care. J Am Med Inform Assoc 2020;27(6):957–62.
[9] Miner AS, Laranjo L, Kocaballi AB. Chatbots in the fight against the COVID-19 pandemic. npj Digital Medicine 2020;3(1):65.
[10] Amiri P, Karahanna E. Chatbot use cases in the Covid-19 public health response. J Am Med Inform Assoc 2022; 29(5):1000–10.

[11] Thoumi A, Kaalund K, Phillips E, Mason G, Harris Walker G, Carrillo G, et al. Redressing systemic inequities: five lessons for community-based COVID-19 testing from the RADx-UP initiative. Health Affairs Forefront 2022; May 20. https://www.healthaffairs.org/do/10.1377/forefront.20220519.204170/.
[12] Horn DM, Haas JS. Covid-19 and the mandate to redefine preventive care. N Engl J Med 2020;383(16):1505–7.
[13] Joseph DA, DeGroff AS, Hayes NS, Wong FL, Plescia M. The colorectal cancer control program: partnering to increase population level screening. Gastrointest Endosc 2011;73(3):429–34.
[14] Allegrante JP, Wells MT, Peterson JC. Interventions to support behavioral self-management of chronic diseases. Annu Rev Public Health 2019;40:127–46.
[15] Hanlon P, Daines L, Campbell C, McKinstry B, Weller D, Pinnock H. Telehealth interventions to support self-management of long-term conditions: a systematic metareview of diabetes, heart failure, asthma, chronic obstructive pulmonary disease, and cancer. J Med Internet Res 2017;19(5), e172.
[16] Steenkamer BM, Drewes HW, Heijink R, Baan CA, Struijs JN. Defining population health management: a scoping review of the literature. Popul Health Manag 2017;20(1):74–85.
[17] Starfield B. Basic concepts in population health and health care. J Epidemiol Community Health 2001; 55(7):452–4.
[18] Loeppke R, Nicholson S, Taitel M, Sweeney M, Haufle V, Kessler RC. The impact of an integrated population health enhancement and disease management program on employee health risk, health conditions, and productivity. Popul Health Manag 2008;11(6):287–96.
[19] Kaspin LC, Gorman KM, Miller RM. Systematic review of employer-sponsored wellness strategies and their economic and health-related outcomes. Popul Health Manag 2013;16(1):14–21.
[20] Mehta SJ, Jensen CD, Quinn VP, Schottinger JE, Zauber AG, Meester R, et al. Race/ethnicity and adoption of a population health management approach to colorectal cancer screening in a community-based healthcare system. J Gen Intern Med 2016;31(11):1323–30.
[21] Hassett MJ, Uno H, Cronin AM, Carroll NM, Hornbrook MC, Ritzwoller D. Detecting lung and colorectal cancer recurrence using structured clinical/administrative data to enable outcomes research and population health management. Med Care 2017;55(12):e88–98.
[22] Clarke JL, Bourn S, Skoufalos A, Beck EH, Castillo DJ. An innovative approach to health care delivery for patients with chronic conditions. Popul Health Manag 2017;20(1):23–30.
[23] Han H-R, Gleason KT, Sun C-A, Miller HN, Kang SJ, Chow S, et al. Using patient portals to improve patient outcomes: systematic review. JMIR Hum Factors 2019;6(4), e15038.
[24] Anderson M. Mobile technology and home broadband. 2020 Retrieved from https://www.pewresearch.org/internet/2019/06/13/mobile-technology-and-home-broadband-2019/. Contract No.: 05/14/2020.
[25] PEW Research Center. Mobile fact sheet. Retrieved from: https://www.pewresearch.org/internet/fact-sheet/mobile/.
[26] PEW Research Center. Internet/Broadband fact sheet. Retrieved from https://www.pewresearch.org/internet/fact-sheet/internet-broadband/.
[27] Kocaballi AB, Berkovsky S, Quiroz JC, Laranjo L, Tong HL, Rezazadegan D, et al. The personalization of conversational agents in health care: systematic review. J Med Internet Res 2019;21(11), e15360.
[28] Inkster B, Sarda S, Subramanian V. An empathy-driven, conversational artificial intelligence agent (Wysa) for digital mental well-being: real-world data evaluation mixed-methods study. JMIR Mhealth Uhealth 2018;6 (11), e12106.
[29] Fitzpatrick KK, Darcy A, Vierhile M. Delivering cognitive behavior therapy to young adults with symptoms of depression and anxiety using a fully automated conversational agent (woebot): a randomized controlled trial. JMIR Ment Health 2017;4(2), e19.
[30] Fulmer R, Joerin A, Gentile B, Lakerink L, Rauws M. Using psychological artificial intelligence (tess) to relieve symptoms of depression and anxiety: randomized controlled trial. JMIR Ment Health 2018;5(4), e64.
[31] Harper R., Nicholl P., McTear M., Wallace J., Black L., Kearney P., editors. Automated phone capture of diabetes patients readings with consultant monitoring via the Web. 15th Annual IEEE international conference and workshop on the engineering of computer based systems (ecbs 2008); 2008 31 March–4 April 2008.
[32] Levin E, Levin A. Evaluation of spoken dialogue technology for real-time health data collection. J Med Internet Res 2006;8(4), e30.
[33] Rhee H, Allen J, Mammen J, Swift M. Mobile phone-based asthma self-management aid for adolescents (mASMAA): a feasibility study. Patient Prefer Adherence 2014;8:63–72.
[34] Almalki M, Azeez F. Health chatbots for fighting COVID-19: a scoping review. Acta Inform Med 2020;28 (4):241–7.

[35] Ali MK, Bullard KM, Imperatore G, Benoit SR, Rolka DB, Albright AL, et al. Reach and use of diabetes prevention services in the United States, 2016–2017. JAMA Netw Open 2019;2(5), e193160-e.
[36] Ariel-Donges AH, Gordon EL, Dixon BN, Eastman AJ, Bauman V, Ross KM, et al. Rural/urban disparities in access to the National Diabetes Prevention Program. Transl Behav Med 2019;10(6):1554–8.
[37] McBrien KA, Ivers N, Barnieh L, Bailey JJ, Lorenzetti DL, Nicholas D, et al. Patient navigators for people with chronic disease: a systematic review. PLoS One 2018;13(2), e0191980-e.
[38] Peart A, Lewis V, Brown T, Russell G. Patient navigators facilitating access to primary care: a scoping review. BMJ Open 2018;8(3), e019252.
[39] Valaitis RK, Carter N, Lam A, Nicholl J, Feather J, Cleghorn L. Implementation and maintenance of patient navigation programs linking primary care with community-based health and social services: a scoping literature review. BMC Health Serv Res 2017;17(1):116.
[40] Szilagyi PG, Bordley C, Vann JC, Chelminski A, Kraus RM, Margolis PA, et al. Effect of patient reminder/recall interventions on immunization rates—a review. JAMA 2000;284(14):1820–7.
[41] Issaka RB, Avila P, Whitaker E, Bent S, Somsouk M. Population health interventions to improve colorectal cancer screening by fecal immunochemical tests: a systematic review. Prev Med 2019;118:113–21.
[42] Jager M, Demb J, Asghar A, Selby K, Mello EM, Heskett KM, et al. Mailed outreach is superior to usual care alone for colorectal cancer screening in the USA: a systematic review and meta-analysis. Dig Dis Sci 2019;64(9):2489–96.
[43] Dougherty MK, Brenner AT, Crockett SD, Gupta S, Wheeler SB, Coker-Schwimmer M, et al. Evaluation of interventions intended to increase colorectal cancer screening rates in the United States: a systematic review and meta-analysis. JAMA Intern Med 2018;178(12):1645–58.
[44] Payne TH, Galvin M, Taplin SH, Austin B, Savarino J, Wagner EH. Practicing population-based care in an HMO: evaluation after 18 months. HMO Pract 1995;9(3):101–6.
[45] Force UPST. Screening for colorectal cancer: US preventive services task force recommendation statement. JAMA 2021;325(19):1965–77.
[46] Centers for Disease Control and Prevention (CDC). Vital signs: colorectal cancer screening among adults aged 50–75 years—United States, 2008. MMWR Morb Mortal Wkly Rep 2010;59(26):808–12.
[47] Mendu ML, Ahmed S, Maron JK, Rao SK, Chaguturu SK, May MF, et al. Development of an electronic health record-based chronic kidney disease registry to promote population health management. BMC Nephrol 2019; 20(1):72.
[48] Schmittdiel JA, Gopalan A, Lin MW, Banerjee S, Chau CV, Adams AS. Population health management for diabetes: health care system-level approaches for improving quality and addressing disparities. Curr Diab Rep 2017;17(5):31.
[49] Greenwood DA, Young HM, Quinn CC. Telehealth remote monitoring systematic review: structured self-monitoring of blood glucose and impact on A1C. J Diabetes Sci Technol 2014;8(2):378–89.
[50] Heidinger O, Heidrich J, Batzler WU, Krieg V, Weigel S, Heindel W, et al. Digital mammography screening in Germany: impact of age and histological subtype on program sensitivity. Breast 2015;24(3):191–6.
[51] Armstrong AC, Evans GD. Management of women at high risk of breast cancer. BMJ 2014;348, g2756.
[52] Walter LC, Schonberg MA. Screening mammography in older women: a review. JAMA 2014;311(13):1336–47.
[53] Wang AT, Vachon CM, Brandt KR, Ghosh K. Breast density and breast cancer risk: a practical review. Mayo Clin Proc 2014;89(4):548–57.
[54] Drukteinis JS, Mooney BP, Flowers CI, Gatenby RA. Beyond mammography: new frontiers in breast cancer screening. Am J Med 2013;126(6):472–9.
[55] Cairns SR, Scholefield JH, Steele RJ, Dunlop MG, Thomas HJ, Evans GD, et al. Guidelines for colorectal cancer screening and surveillance in moderate and high risk groups (update from 2002). Gut 2010;59(5):666–89.
[56] Ong MS, Mandl KD. National expenditure for false-positive mammograms and breast cancer overdiagnoses estimated at $4 billion a year. Health Aff (Millwood) 2015;34(4):576–83.
[57] Force USPST, Owens DK, Davidson KW, Krist AH, Barry MJ, Cabana M, et al. Risk assessment, genetic counseling, and genetic testing for BRCA-related cancer: US preventive services task force recommendation statement. JAMA 2019;322(7):652–65.
[58] Rex DK, Boland CR, Dominitz JA, Giardiello FM, Johnson DA, Kaltenbach T, et al. Colorectal cancer screening: recommendations for physicians and patients from the U.S. multi-society task force on colorectal cancer. Gastroenterology 2017;153(1):307–23.

[59] Gramling R, Nash J, Siren K, Eaton C, Culpepper L. Family physician self-efficacy with screening for inherited cancer risk. Ann Fam Med 2004;2(2):130–2.

[60] Taber P, Ghani P, Schiffman JD, Kohlmann W, Hess R, Chidambaram V, et al. Physicians' strategies for using family history data: having the data is not the same as using the data. JAMIA Open 2020;3(3):378–85.

[61] Birmingham WC, Agarwal N, Kohlmann W, Aspinwall LG, Wang M, Bishoff J, et al. Patient and provider attitudes toward genomic testing for prostate cancer susceptibility: a mixed method study. BMC Health Serv Res 2013;13:279.

[62] Ahmed S, Hayward J, Ahmed M. Primary care professionals' perceptions of using a short family history questionnaire. Fam Pract 2016;33(6):704–8.

[63] Kawamoto K, Lobach DF. Proposal for fulfilling strategic objectives of the U.S. Roadmap for national action on clinical decision support through a service-oriented architecture leveraging HL7 services. J Am Med Inform Assoc 2007;14(2):146–55.

[64] Strasberg HR, Rhodes B, Del Fiol G, Jenders RA, Haug PJ, Kawamoto K. Contemporary clinical decision support standards using Health Level Seven International Fast Healthcare Interoperability Resources. J Am Med Inform Assoc 2021;28(8):1796–806.

[65] Del Fiol G, Kohlmann W, Bradshaw RL, Weir CR, Flynn M, Hess R, et al. Standards-based clinical decision support platform to manage patients who meet guideline-based criteria for genetic evaluation of familial cancer. JCO Clin Cancer Inform 2020;4:1–9.

[66] Bradshaw RL, Kawamoto K, Kaphingst KA, Kohlmann WK, Hess R, Flynn MC, et al. GARDE: a standards-based clinical decision support platform for identifying population health management cohorts. J Am Med Inform Assoc 2022;29(5):928–36.

[67] Kaphingst KA, Kohlmann W, Chambers RL, Goodman MS, Bradshaw R, Chan PA, et al. Comparing models of delivery for cancer genetics services among patients receiving primary care who meet criteria for genetic evaluation in two healthcare systems: BRIDGE randomized controlled trial. BMC Health Serv Res 2021;21(1):542.

[68] Chavez-Yenter D, Kimball KE, Kohlmann W, Lorenz Chambers R, Bradshaw R, Espinel WF, et al. Patient interactions with an automated conversational agent delivering pretest genetics education: descriptive study. J Med Internet Res 2021;23(11):e29447.

[69] Jha P, Peto R, Zatonski W, Boreham J, Jarvis MJ, Lopez AD. Social inequalities in male mortality, and in male mortality from smoking: indirect estimation from national death rates in England and Wales, Poland, and North America. The Lancet 2006;368(9533):367–70.

[70] Adhikari S, Pantaleo NP, Feldman JM, Ogedegbe O, Thorpe L, Troxel AB. Assessment of Community-Level Disparities in Coronavirus Disease 2019 (COVID-19) Infections and Deaths in Large US Metropolitan Areas. JAMA Netw Open 2020;3(7), e2016938-e.

[71] Fernandez ME, Schlechter CR, Del Fiol G, Gibson B, Kawamoto K, Siaperas T, et al. QuitSMART Utah: an implementation study protocol for a cluster-randomized, multi-level Sequential Multiple Assignment Randomized Trial to increase Reach and Impact of tobacco cessation treatment in Community Health Centers. Implement Sci 2020;15(1):9.

[72] Schlechter CR, Del Fiol G, Lam CY, Fernandez ME, Greene T, Yack M, et al. Application of community—engaged dissemination and implementation science to improve health equity. Prev Med Rep 2021;24, 101620.

CHAPTER

27

CDS for public health

Leslie A. Lenert

Medical University of South Carolina, Charleston, SC, United States

27.1 Introduction

In contrast to many other fields of healthcare, decision support for public health is both a tool to improve the practice of public health professionals, within their own computer systems *and* a tool to implement public health policy, in external organizations. Public health implements policy through decision support systems designed to influence the healthcare system in ways that support public health goals and objectives (for example, increasing uptake of recommended immunizations). While in many areas, decision support technologies for direct support of public health practice are undeveloped, the use of decision support as a means to change clinical providers' behaviors is technically mature and widely disseminated in the United States and represents an exemplar how decision support technologies might be employed in other aspects of the health system. In this chapter, we will review decision support technologies for public health from both the internal and external use perspectives.

27.2 Decision support for public health operations

To understand the role of decision support for public health practice, one must first understand the types of decisions that public health practitioners make and where computer decision support could be helpful. These activities fall into three broad categories: Assessment, Policy Development, and Assurance [1]. *Assessment* refers to the task of monitoring, detecting, and responding, through investigations, to acute and chronic public health issues. A central role of public health is to identify new health threats, especially threats from emerging infectious diseases, but also threats from pollution, and climate change, and to characterize those threats. The role of decision support in the Assessment task is to facilitate recognition and immediate mitigation of public health threats. *Policy Development* is the second broad area of services and is how public health responds to identified threats to the health of the public.

Clinical Decision Support and Beyond
https://doi.org/10.1016/B978-0-323-91200-6.00015-2

Through health policies, laws, and planning, public health officials seek to control, mitigate, and eventually eliminate threats. The development of effective policies often requires evaluation of the risks and benefits of particular actions. As a result, simulation methods play an important role in decision support for policy development. *Assurance* is the task of creating and maintaining social and technical conditions that allow the successful implementation of policies that promote public health. This includes addressing issues of equitable opportunities to achieve good health, building a workforce to perform assessment and policy development, and continuously assessing and improving methodologies. Decision support for Assurance related tasks strengthens systems, as opposed to having a direct impact. Examples of this type of decision support are described below.

27.3 Decision support for assessment operations

Assessment operations for public health collect data on the incidence (new cases) and prevalence (population levels of existing cases) of infectious and non-infectious health threats to a population. The most common role for decision support systems in Assessment operations is the detection of a new infectious disease outbreak or health threat. For most infectious disease-related threats and many others, there is a background incidence of infection (the attack rate of new cases) and prevalence of the disease (the percentage of the population with the disease of interest). Outbreak events occur when there is a special source of the disease that results in *clusters* of *incident cases*. Decision support systems for Assessment operate by enhancing the detection of new clusters of disease and distinguishing those from background events.

There are three approaches to distinguish background events from new clusters: (1) time series analysis of the rate of events to detect upticks in the rate, (2) spatial analysis of events based on geocoding or zip codes to detect spatial clustering of events, and (3) subtyping of species and strain of the causative agents to find disparate events linked by a common causative organism. One type of Assessment system, called a syndromic surveillance system, monitors the observed prevalence of patients' chief complaints in the healthcare system from electronic medical records. An example of this system is CDC's Biosense system [2] and its successors. These systems use time series analysis to distinguish increases in rates of different chief complaint-related symptoms (10 or 12 clusters of symptoms) seen at a particular hospital or in a region from the background and seasonal variation. Geospatial analysis is more typically applied to find outbreaks of chronic illnesses such as cancer and lung disease to find clusters of patients who share a common environmental exposure, such as a toxin in the water system [3]. Decision support systems for detection of spatially related outbreaks test the null hypothesis of equal distribution of disease events across an area, adjusted for populations [4].

Detection of clusters of events linked by common biology is a particularly important technique in foodborne illness, where a contamination problem in a food producer can result in widely disseminated infections that are imperceptible from background rates in any given region but when linked by the genetic markers of that identify a single organism as the cause. Decision support for the identification of outbreaks using this method requires a centralized database with species subtypes, using whole gene sequencing of the organism (and previously by western blotting methods), of background infections. Pulsenet is an example of this type of system [5]. When a new infection or a local outbreak is identified, public health officials consult the database and attempt to match the organism to those from other background infections to identify larger-scale patterns [6]. Bioinformatics techniques that measure the similarity of gene sequences are particularly important in this task for this type of decision support [7].

27.4 Decision support for policy development operations

Public health intervenes in society primarily through the development and implementation of policies, regulations, and laws. This kind of intervention can focus on *urgent threats* such as outbreaks and pandemics. Actions undertaken in response to urgent threats include the development of and implementation of quarantine regulations for the infected, contact tracing of the exposed, design of and recommendations for social distancing regulations such as issuing of mask requirements and school closures, prioritizing of populations for vaccinations, and development and implementation of policies to mitigate social impacts of a pandemic (homeless sheltering, social isolation interventions, etc.) The role of decision support in policy development is to help public health professionals develop a realistic estimate of the positive benefits and the costs and risks of policies for use by decision-makers. In each of the above threats to health, federal, state, and local governments have a variety of options that could be applied as laws or regulations that have different impacts on the perceptions of individual freedom and on the functioning of society. The role of computerized decision support in policy development is the accurate simulation of the baseline impacts of public health threats and the forecasting of the impact of specific mitigation measures.

There are numerous examples of simulation modeling tools that provide decision support for public health policy development. An important class of models that forecasts the public health decision-makers are mathematical models that forecast the impact of social distancing policies on the potential course of an outbreak or pandemic. An example of this is work by Alagoz and colleagues on timing and adherence to social distancing measures [8] and of Naimark and colleagues specifically on the effect of school closures in transmission in Ontario Canada [9]. "Locally informed" (tailored) simulation studies on hospital capacity can play an important role in city and count decision making on specific social distancing measures. For example, Weissman and colleagues, used simulation modeling to estimate surges in clinical demand using Monte Carlo simulation methods [10]. This in turn was used to plan to staff and to make recommendations for social distancing.

Other models, often based on process simulation techniques, help policymakers prepare for disasters by developing effective systems of response in advance of disasters and pandemics. One of the nation's tools for response to pandemics and other large-scale disasters is the Strategic National Stockpile of materials. This includes medicals for bacterial and viral threats, intravenous fluids, and general medical materials such as face masks and other personal protective equipment and medical devices such as ventilators. There were numerous questions raised and debates about the adequacy of preparations for a pandemic in the design of the stockpile and decision-making on its release during the COVID-19 pandemic [11].

Better use of decision support systems in planning the content and disbursement of items from the stockpile may have mitigated these issues. Ajrawat and coauthors [12] describe a decision support system to aid policymakers in the development of the stockpile and policies on the release of its contents. Blood product ensuring blood product availability for certain types of public health disasters (hurricanes, mass casualty events) can be challenging. Simonetti and coauthors describe a tool for helping policymakers develop policies and programs for inter-regional transfers of blood products in response to disasters that increase demand for these due to injuries [13].

27.5 Decision support for public health assurance operations

Decision support for Assurance operations might be described as "meta" level decision support. Public Health Assurance often takes the form of toolkits, surveys, and maturity models that help governments characterize their health services infrastructure and identify gaps in policies, services, and infrastructure. Computer data systems for health and disease surveillance are integral to the assurance task, and, as described in Assessment, measure compliance with regulations or counts of diseases, deaths, and other adverse outcomes. This sets the stage for quality improvement efforts that are part of Assurance.

Focused decision support systems for Assurance operations are an emerging area in public health informatics. Rios-Zertuche and coauthors describe one of these systems, which uses survey methods to broadly collect data from health facilities, rolls up this data into the district and central levels, and then helps public health officials explore gaps through the stratified sampling of specific facilities across a health system [14].

Another example of decision support software that specifically addresses Assurance issues, is the Forecasting Immunization Test Suite (FITS) software package from the National Institute of Standards and Technology [15]. This software tests the decision support systems integrated into Immunization Information Systems (IIS). IIS systems have integrated software that provides recommendations on vaccines administration, particularly scheduling and prioritization of vaccinations (described below). IIS systems are implemented at a state level with wide variability in the logic used to drive their recommendations [16]. FITS has a library of standardized cases to test the logic in IIS recommendations to ensure that the appropriate recommendations are made by these systems, ensuring the quality of the diverse IIS national enterprise with its differing systems, rule sets, and platforms providing advice. By requiring states to have systems that are certified by FITS as part of funding requirements, the Federal government can exert influence on technical quality standards on state IIS systems.

27.6 Public health decision support systems designed to influence clinical care providers

Public health is unique in that the majority of efforts for decision support are not directed at public health practitioners; instead, most efforts are focused on influencing the healthcare system to undertake required actions to improve public health and well-being. Public health does this both to obtain data from the health system and to change provider behaviors to be more consistent with the public health policies and goals. While laws and administrative rules provide a framework for mandates to the healthcare system, the absence of enforcement mechanisms can limit the effectiveness of regulation. For example, only a small portion of notifiable infectious diseases (diseases where the law requires healthcare provider reporting of the disease) are reported [17]. Without an effective means to compel providers to provide the information needed, public health has largely adopted a strategy of two-way communication with providers to promote compliance with public health regulations. Decision support is at the center of two-way public health to clinical care system communication and helps create virtuous cycles of information sharing. In this section, we will discuss evolving systems in three areas: prescription drug monitoring registries, immunization Information systems, and electronic reporting of public health notifiable conditions.

27.7 Prescription drug monitoring programs

One of the most common models for decision support in public health is to serve as a centralized database where critical information for treatment is aggregated across providers. This architecture combines decision support for providers with a database used for policy decision-making. Often aggregation occurs by policy mandate. For example, a state may require the pharmacies dispensing drugs to report all narcotic and other sedative drug prescriptions to its Prescription Drug Monitoring Program (PDMP) database [18]. By aggregating data on clinical activities such as opioid prescription, policymakers can assess the impact of policies on health, identify potential health disparities, and identify providers who may not be complying with policies. And, by opening access to databases to support decision making for individual patients, public health provides decision support for clinical care. For example, PDMP resources track prescriptions for opioid drugs and other controlled substances through reporting requirements from pharmacies. This allows regulators to identify providers who are writing excessive numbers of opioid prescriptions and investigate whether the providers are breaking the law by dispensing them inappropriately. Some providers, for example, cancer specialists and pain control physicians, may appropriately dispense opioids to large numbers of patients. Forty of 50 states have some requirement for reporting of opioid or other controlled substance prescribing to a PDMP. In general, PDMP systems have been highly effective tools in the mitigation of prescription opioid medications as a cause of opioid overdoses [19,20].

In addition, by opening PDMP to physicians for decision support, PDMP systems allow providers to identify patients obtaining opioid drugs from other physicians and assess the risk-to-benefit ratio of any new prescription. With the opioid crisis, some states have begun to require healthcare providers to check opioid databases before issuing opioid prescriptions, however, there is wide variability across states and in the specific conditions [21]. Furthermore, there are no specific mechanisms to audit providers to ensure that they check these PDMP databases before prescriptions, though self-auditing has been studied [22]. Some PDMP vendors now offer direct integration of PDMP checks with electronic health record systems, eliminating the need for providers to log onto a second system and look up a patient's prescriptions. The Government Accounting Organization, in a report of PMDP effectiveness, estimated that the average time for a provider to log in and look up patients in a PDMP was 2–3 min, whereas direct integration with the EHR reduced this time 2–20 s [19].

PDMP's also have begun to add decision support tools to reporting opioid and other controlled substance prescriptions. At the simplest level, these tools calculate the total dose of opioids being prescribed by converting the dose of different types of medications to an estimated dose in "morphine equivalents" (one of the more common opioid drugs prescribed for pain). This allows providers to rapidly determine if the total dose of opioids places the patient at high risk of accidental overdose (patients receiving more than 50 mg equivalents of morphine a day have been shown to be at high risk) [23]. More advanced tools are in development that use standards-based approaches to decision support, such as CDS Hooks (see Chapter 15 for details on the role of standards in CDS), to implement CDC guidelines for chronic opioid prescribing, offering patient advice based on those guidelines [24,25].

27.8 Immunization information systems

Another common area for state-level collection of data for decision support is immunizations. Driven by both state laws [26] and Federal Meaningful Use of Electronic Health Record system regulations [27], Certified EHR systems must be able to submit reports of vaccination, using a standardized message architecture, to public health authorities. Medicare and Medicaid regulations also require the submission of vaccination data for certification for payment. While not all public health authorities can receive and process these electronic immunization messages at a state level, most can [16]. In some states, IIS systems have not had the technical capabilities to receive automated reports of vaccinations from providers, limiting the effectiveness of their IIS systems. Overall, the accuracy of reporting mechanisms is high across states with more than 80% of states and jurisdictions which participate in testing reporting accurate processing of data from EHR messages. However, there are issues in mapping to the data models inside of IISs. In tests of queries and reporting of the data from IIS, only about 50% of states have completely accurate reporting or minor differences. This may limit the usefulness of IIS data to practitioners.

Similar to PDMP systems, Immunization registries support public health operations by providing important data on immunization rates in the population, gaps in immunizations, and support for the distribution of vaccines for public health policymakers. Opening decision state immunization registries reporting tools to physicians provides critical information for decision support on what vaccines have been administered and what vaccines may be needed. In addition, a CDC recommended component of IIS systems is vaccine forecasting—that is determining which vaccine a patient might require next and when it should be administered. This is particularly difficult for children in the process of receiving multiple immunizations early in life, where some vaccine combinations are not compatible with each other, and there is a need for delaying dosing [28,29].

The basic design of decision support systems that offer recommendations for vaccination in IISs dates from the 1990s [30,31]. IIS use decision tables to represent facts about the patient and vaccinations received. Decision tables represent facts as a series of discrete variables linked to deterministic policies [32]. Facts are aggregated through a series of tables to build deeper and more meaningful conclusions. Tables are linked in ad hoc "neighborhood" models (conceptual entity-relationship models) that explain and constrain relationships between variables in tables [33]. Several proprietary implementations of reasoning systems using this table data model exist. Some are directly integrated within the IIS system [34]. Others offer platform-independent decision support accessible by web services approaches [35,36].

CDC's CDSi (Clinical Decision Support *immunizations*) [37] program illustrates how national public health agencies can contribute to state/jurisdictional level resources for decision support for vaccination. CDC's role focuses on, in addition to funding, knowledge representation and generation for use by the implementers of immunization registries in a platform-independent way. The CDC CDSi program has compiled and maintains a comprehensive set of decision tables, in spreadsheet form, linked to the latest recommendations for immunizations for use by IIS developers in their software implementations, as well as training materials and test cases. The components of the CDSi are shown in Fig. 27.1.

The design of the CDSi is both complex and robust. Tables are applied sequentially to split vaccines into antigen components, assess the timing of administration of antigens to determine if the regimen is complete, compute next dose recommendations which may include

The CDSi resources translate clinical recommendations into technical information

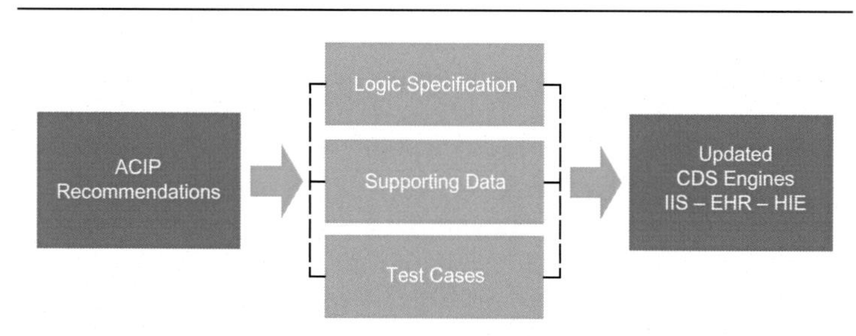

The complete suite of CDSi resources comprises:

Logic Specification	Supporting Data	Test Cases
– Vocabulary	– Excel Format	– Excel Format
– Business Rules	– XML Format	
– Decision Tables	– Release Notes	
– Processing Definitions		
– Domain Model		
Training Materials	– Brochure – Quick Guides	– Practice Exercise – Quiz – Videos

FIG. 27.1 CDSi conceptual model. *Source: CDSi-miniGuide.pdf.*

either combination of antigens in multivalent vaccines products or single antigen vaccines. Two types of recommendations are generated: (1) routine vaccinations for persons who have received prior vaccinations in a timely way; and (2) "Catchup" vaccinations that are needed to determine if a series of vaccinations need to be replicated and to prioritize among potential sequences. The advantages of this approach include externalization of knowledge in the form of spreadsheets that are both relatively easy to understand and maintain.

The Immunization Calculation Engine (ICE) [38] is an open-source software platform that implements the table models from CDSi and automatically evaluates a patient's medical history against recommendations. Medical records need to be summarized in the Virtual Medical Record (VMR [39,40]), a variant of Continuity of Care Document format, to be read into this system, which then generates recommendations for next dose and catch-up vaccination. Unfortunately, the VMR format has not been widely adopted by EHR vendors as a standard for summarizing medical histories, limiting the impact of this work. A Smart-on-FHIR middleware "adaptor" that converted FHIR resources to a VMR formatted CCD could extend the value of investments in the ICE, similar to what is described below for electronic case reporting.

There are several other technical disadvantages inherent in the CDSi model of knowledge representation. Discrete logic used to model the timing, doses, and interactions is somewhat simplistic. For example, time intervals between doses are modeled as either "acceptable/unacceptable intervals" and "Unacceptable/Absolute minimum/Minimum interval". There is no formal temporal model [41] which may result in simplistic recommendations. In addition, the logic for individualization of dosing of vaccines, when there are gaps, does not consider individuals' preferences. Priorities are determined by policymakers and do not incorporate patients' (or their parents') preferences. This has resulted in growing calls for use of shared decision-making methods in vaccine administration and the incorporation of such methods into CDSi [42,43]. The decision table format is not tied to specific tools for an explanation of its computations. If a provider does not understand the recommendation from the CDSi, there is no way to generate an automated explanation of that recommendation [44].

Last, but importantly, there is not a standardized approach for integrating the recommendations computed by CDSi back into an EHR. Recommendations might be visualized in a linked but separate browser window or paper recommendations, as described by Wilkinson and coauthors for Human Papilloma Virus vaccinations [45]; however, CDSi recommendations have not been automatically translated into order sets or other interventions inside the EHR that would facilitate providers acting on them, and as a result, this limits their effectiveness [46]. That said, in qualitative interviews, health systems leaders interviewed about the impact of IIS in Minnesota, believed CDSi features were valuable [47]. But, until there are widely adopted standardized approaches for integration of results back into EHRs, such as CDS Hooks or the Immunization-Recommendation FHIR resource [48], the impact of CDSi on actual clinical decision making may be limited. Once these technologies become more available, further research will be needed to confirm the impact of CDSi on outcomes.

The CDSi program is an extremely innovative and adaptive approach that takes patient-level decision support to many different IIS implementations in a platform-independent way. As such, it is a model for how government agencies can support advanced decision support concepts in clinical practice by building, documenting, and maintaining platform-independent models for decision support.

27.9 Electronic case reporting (eCR)

eCR is an example of a decision support system for public health that is designed to improve the quality of reporting from healthcare providers to the public health system. It currently does not provide advice on the management of infectious diseases but is designed to do this. Currently, its primary function is to improve the completeness, timeliness, and accuracy of clinical reports to public health of patients with public health notifiable conditions. This system, while rapidly evolving, is also becoming a model for how clinical decision support for critical issues might be implemented at a national level.

Case reporting for public health notifiable conditions is a fundamental task that spans clinical care and public enterprise [49]. Reporting of specified conditions is required by state law. Most states require reporting between 45 and 50 diseases when diagnosed. The specific diseases and conditions (types of test results and the timing of test results) for reporting are different in each state (definitions of what diseases are reportable and timing of reporting). Reporting allows public health to track and investigate new reports of contagious diseases. Reports contain detailed but summarized clinical information and have typically been submitted on paper forms or postcards. Disease reporting forms are supposed to be completed by physicians or healthcare providers; however, widespread under-reporting occurs [50] and reporting is often not timely or complete, despite state law requirements [51]. Decision support in this context helps providers understand when a case is reportable to public health and what types of data need to be included, as well as improves the timeliness of reporting [51,52]. While substantial progress has been made on the automated transmission of laboratory data to public health for notifiable conditions in the Meaningful Use of electronic health records era [53], it is only recently, with the SAR-COV-2019 pandemic that substantial progress has been made in electronic case reporting.

27.9.1 Why is electronic case reporting complex?

Many factors make it difficult to apply decision support techniques to improve public health reporting of notifiable conditions. Public health reporting requires providers to submit data that may not have a direct benefit to their patients and without compensation for the time and effort required in reporting. Making this a simpler task by pre-populating forms with clinical data from an electronic health record improves the rate of submission of case reports. This was first shown by Lazarus and coauthors in 2009 [51] in a demonstration project. Dixon and coauthors advanced the principle by repopulating data from all care sites participating in a regional health information exchange [52]. While offering providers help in completing forms seems like a simple idea, the rules for reporting vary substantially by state, and matching the timing of submission to clinical processes can be difficult. Some states require reporting of negative results for some conditions but in most cases, reports are required only when there is a new "diagnosis," as defined by state regulations. In addition, these reports contain personally identified data and are considered highly sensitive. Public health is reluctant to share these data except where authorized by state legislation. The basic workflow for eCR is shown in Fig. 27.2.

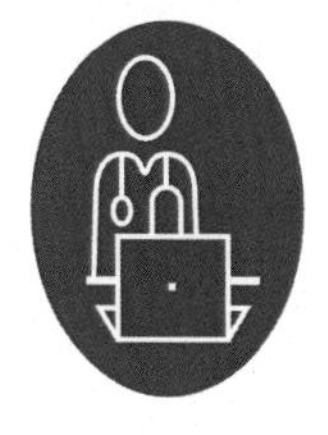

Patient is diagnosed with a reportable condition, such as COVID-19
Healthcare provider enters patient's information into the electronic health record (EHR)
Data in the EHR automatically triggers a case report that is validated and sent to the appropriate public health agencies if it meets reportability criteria
The public health agency receives the case report in real time and a reponse about reportability is sent back to the provider
State or local health department reaches out to patient for contact tracing, services, or other public health action

FIG. 27.2 eCR workflow. *Source: The figure appears in Digital Resources 2021. https://www.cdc.gov/ecr/digital-resources.html [Accessed 27 December 2021].*

Achieving automation of this workflow is complex because there are numerous competing groups with potentially different agendas for case reporting as well as state laws and regulations for reporting. Healthcare provider organizations are the focus of this work but want the minimum impact of regulation on clinical workflows. The CDC provides funding for state public health data systems and wants a rapid collection of data into a complete national picture of the impact of a disease, particularly in the setting of a pandemic. Individual states' public health departments have legacy data systems, largely supported by private vendors, that may or may not be compatible with different approaches to deliver eCR. States also have public health laboratories that provide testing services for many of the more rare public health notifiable conditions. State Health Departments and Laboratories are organized at a national level in competing/collaborating organizations (the Association State and Territorial Health Officers and the Association of Public Health Laboratories (APHL). These organizations receive funding from CDC at a national level to distribute to states for eCR and other public health programs. Integration into electronic health records of decision support systems for this area requires the cooperation of electronic health record (EHR) vendors with public health data system vendors. This is driven by Federal regulations created by other government agencies (Center for Medicare and Medicaid Services (CMS) and Office of the National Coordinator for Health Information Technology (ONC). CMS can (and has created) regulations for healthcare providers with incentive payments for reporting in various ways to public health. This incentivizes provider organizations to use a system that has the required functionality. ONC creates policies for certification of EHR technology produced by vendors to ensure that the systems that providers buy will perform the functions required by CMS. In addition to policy, whatever EHR and public health data systems need to do must be translated into national standards that EHR and public health software vendors can support. This is usually done by consensus in the Health Level 7 (HL7) organization (see Chapter 15 for details on the role of standards in CDS). This is a complex ecosystem with multiple competing interests, which made progress difficult even though the right approach has never been in question.

27.9.2 Digital bridge as a model for progress

In 2016 when the Robert Wood Johnson Foundation in collaboration with other charitable foundations, clinical partners (Kaiser Permanente, Intermountain Healthcare, and the American Medical Association), major EHR vendors (Epic and Cerner), ASTHO, APHL, National Association of City and County Health Organization (NACCHO), and with government participation (CMS, ONC, CDC), along with facilitating organizations (Deloitte Consulting and Public Health Informatics Institute) came together to work on a strategy for eCR that included decision support called Digital Bridge [54]. What was unique about this effort was its long term non-governmental funding for collaborations to produce the system architecture, standards, and working implementations for eCR and other public health reporting issues, with stable funding and support independent of CDC, giving state and local public health authorities an equal voice. The design of the national eCR system created by the Digital Bridge Project is shown in Fig. 27.3.

The design is based on a push operation with the transmission of data to public health of a potential case for investigation and confirmation. This is because many state laws do not allow public health departments to have detailed clinical data on an individual unless that individual is suspected of having a disease of public health significance (infection, toxins, etc.). EHRs use a set of condition-specific trigger codes that capture events such as results of laboratory tests, orders for laboratory tests, new diagnoses, and pharmaceuticals given primarily for public health notifiable diseases (for example, penicillin given in the formulation, dose

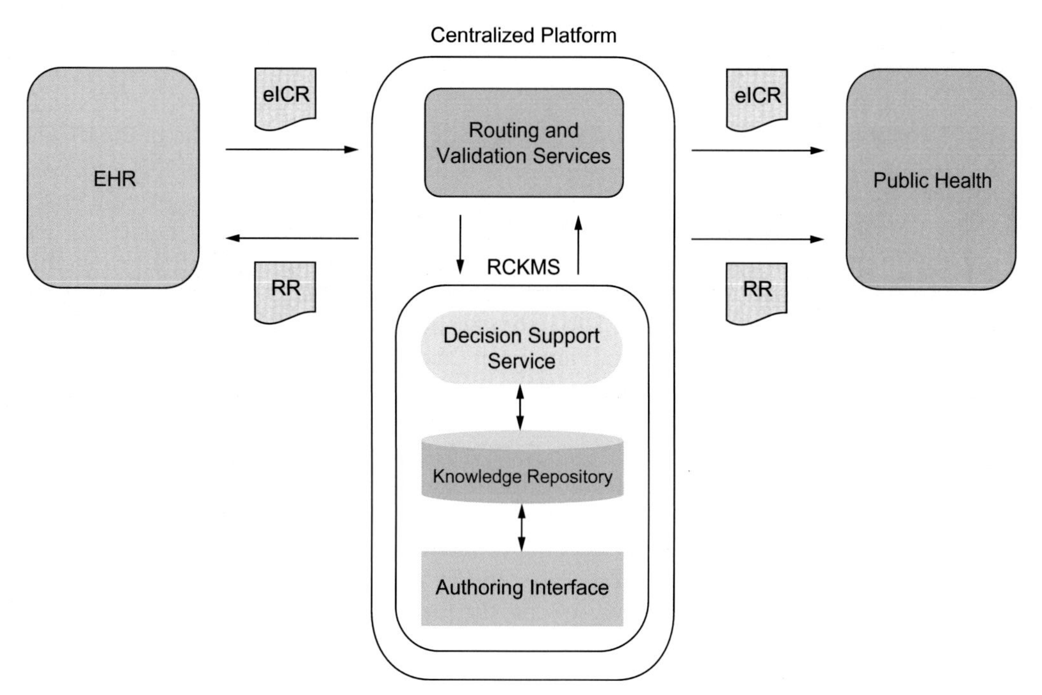

FIG. 27.3 Conceptual architecture of eCR reporting system. *Source: The figure appears in About RCKMS. https://www.rckms.org/about-rckms/ [Accessed 27 December 2021].*

and amounts usually used to treat syphilis), to trigger the creation of an HL7 Continuity of Care Document (CCD), called the eICR prepopulated with required the required clinical content for case reporting. The list of trigger codes and their relationship to notifiable conditions is maintained by CSTE. The value sets include International Classification of Disease (ICD) version 10 codes, Systematic Nomenclature of Medicine (SNOMED) codes (for suspected or new diagnoses, problems, complaints), Logical Observation, Identifier, and Name Codes (LOINC) for ordered laboratory tests, and National Drug Codes (NDC) for ordered or administered drugs. EHR vendors subscribe to this list and help clinical providers keep their list of trigger codes up to date.

Inside their EHRs, vendors use proprietary methods to bring partially completed forms to the provider's attention for submission when triggered based on the HL7 CCD case reporting standard. The data in the form are converted to the eCR CCD standard [55] and then submitted to a national eCR resource, an intermediary (jointly controlled by ASTHO and CDC) that reviews the content to determine if the case needs to be submitted to public health authorities. EHRs receive back, via the eCR Response Return CDC standard [55], the assessment from the Reportable Conditions Knowledge Management System (RCKMS) of reportability and the final determination of reportability from the public health authority. These are returned by Direct messaging to the EHR. There are three key innovations in the process, namely: the completion of draft reporting forms by healthcare providers based on sensitive but not specific triggers in the EHR (with the majority of data automatically extracted from the EHR), use of an intermediary using decision support methodologies to electronic screen submissions at a national level before routing positive or suspicious ones to state public health authorities, and provision of timely feedback to the clinical system on the reportability of cases, creating a two-way communications loop.

The initial implementation of the case report triggering and CCD process was EHR vendor-specific. Later a FHIR-based application was developed to move the creation of the CCD for the eCR to an external piece of middleware software [56]. The architecture of this app, which sits between the RCKMS and the EHR is shown in Fig. 27.4. The app is triggered using the SMART on FHIR decision support standard [48], and then retrieves FHIR resources using the application programming interface (API) of the EHR to get data required to evaluate the triggers, retrieve other data, and then generate the eCR CCD form. The app is envisioned to subscribe, using FHIR standards [57], to national sources of trigger codes, being automatically and remotely configurable. The advantage of this approach is that it requires much less infrastructure within an EHR for reporting. The disadvantage of the FHIR based approach is that it requires advanced FHIR APIs for the EHR and SMART on FHIR decision support web services triggering. Depending on how often the SMART on FHIR app is triggered, it could be both intrusive for the user and computationally inefficient for the EHR system.

Routing eCR documents to a central national resource of the RCKMS create a hub and spoke architecture, where EHR vendors route to one internet address, and the national resource then re-routes proven or possible cases to the appropriate public health authorities. This may be two or more separate jurisdictions: the jurisdiction where the patient was diagnosed and the jurisdiction where the patient resides. Potential case documents are sent via Direct Messaging or APHL's AIMS (APHL Informatics Messaging System) standard. EHR vendors do not need to know the DIRECT or AIM protocol addresses of various agencies or have data use agreements in place with these facilities, providing some return on their

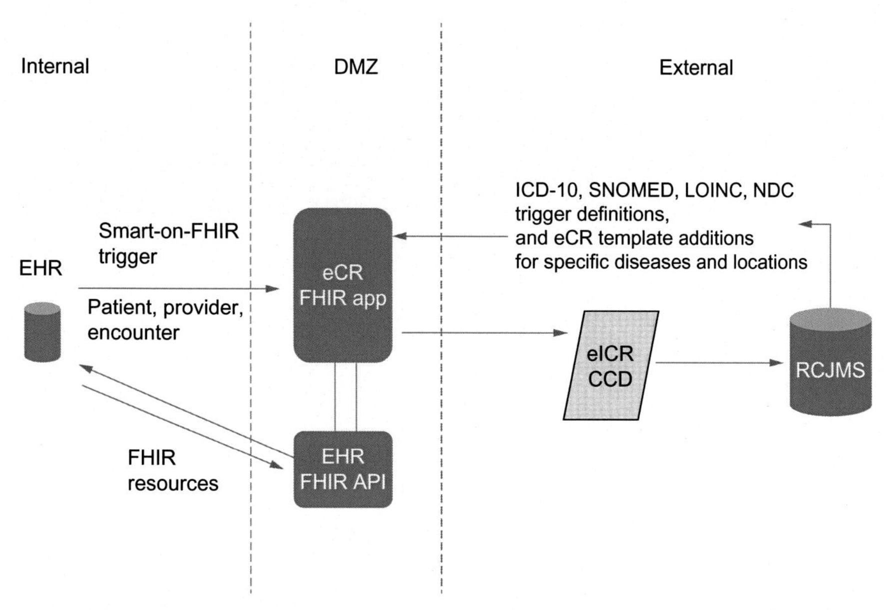

FIG. 27.4 FHIR app integration with eCR reporting. Definitions for triggers and the creation of templates occur in the FHIR app, which receives data from the EHR through FHIR queries of data resources.

investments in infrastructure. Messages can be sent through HIEs to the central resource, where additional data may be added in future versions.

At the central resource, the eCR case reporting document is parsed and evaluated against requirements for reporting called the RCKMS (for Reportable Conditions Knowledge Management System) (Fig. 27.5) [58]. Inside the RCKMS, each jurisdiction works to create its templates for evaluation of cases from pre-existing baseline or standardized definitions for eCR, varying test results, timing, and other details as their laws require. Results are classified into three categories: not reportable, possibly reportable, and definitely reportable. Agencies in the jurisdictions of healthcare treatment and the residence of the patient are forwarded the eCR CCDs from possibly and definitely reportable results, using either the Direct Messaging or AIMS protocol. Providers are also notified of the result of analysis, using a unique HL7 CCD transmission standard (eCR Result Return [55]), creating a feedback loop to improve the quality of reporting. The Result Return document is designed to also be able to deliver links to trusted advice on treatment of the disorder and potentially forms for collection and transmission of additional data not in the eCR to public health. At the state public health agency, eCR reports are integrated into the state case management system, and when confirmed, forwarded after anonymization to CDC to its National Notifiable Disease Surveillance System.

27.9.3 Progress on eCR to date

In 2017, work was begun on gaining approval for an HL7 standard for eICR (electronic-Initial Case-Report form) [55], a variant of the CCD architecture dedicated to the eCR and

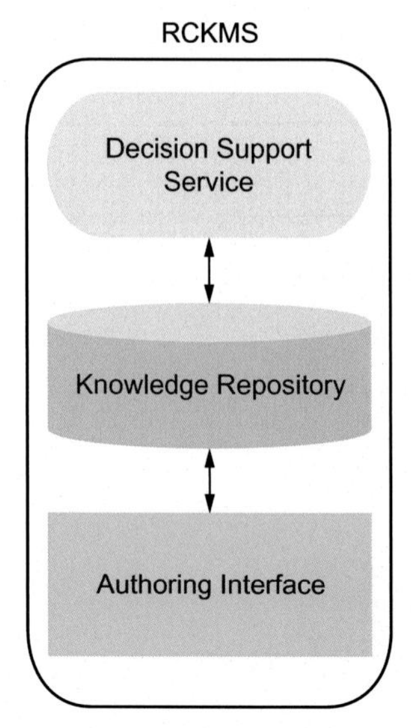

FIG. 27.5 Components of the RCKMS system as illustrated in [59]. *Source: About RCKMS. https://www.rckms.org/about-rckms/ [Accessed 27 December 2021].*

its content. In 2018, the first demonstrations of the architecture were launched, including the authoring of reporting requirements for states using RCKMS [60]. In 2019, the results of the evaluation of these demonstration projects at four sites were mixed. Using qualitative and quantitative analysis, evaluators found substantial variability and issues for sites in implementation, particularly focused on the lack of LOINC coding of clinical data and the use of other non-standard codes within EHRs. The greatest success was seen in Utah with its work with Intermountain Health, where a strong technical foundation preceded the pilot study. In other test sites, there was less than complete success [61].

However, in 2020, with the COVID-19 Pandemic, eCR work accelerated focused on SARS-COV 2 2019 case reporting [62,63]. Work was rapid because of the urgency and burdens posed by reporting, but also because it was simpler with limited numbers of ICD-10, SNOMED, and LOINC codes for diagnosis and tests, and absence of need for drug treatment triggers (as there were no treatments initially). As the project moved to operations, CDC renamed the project, eCR Now program [63]. Emergency funding [64] expanded the capacity of systems used by APHL for routing messages and by CSTE for RCKMS for evaluating message content. This same funding allowed state health departments to add the capability for receipt of messages. This resulted in extremely rapid uptake of the standard. By December of 2021, all 50 states and 8 territorial public health jurisdictions could receive SARS-COV 2 eCR messages and more than 10,000 healthcare sites across the United States were using this approach to report COVID-19 infections (see Fig. 27.6) [65].

In October of 2021, based in part on the degree of progress seen in uptake of eCR, as well as policy advocacy grounded in the urgency of the pandemic [62], CMS has introduced draft regulations to require eCR reporting as one of four required public health reporting standards by the end of 2022 [66], making wide adoption extremely likely. Further, the EHR manufacturer Cerner has committed to making an eCR FHIR application for general reporting [67] and

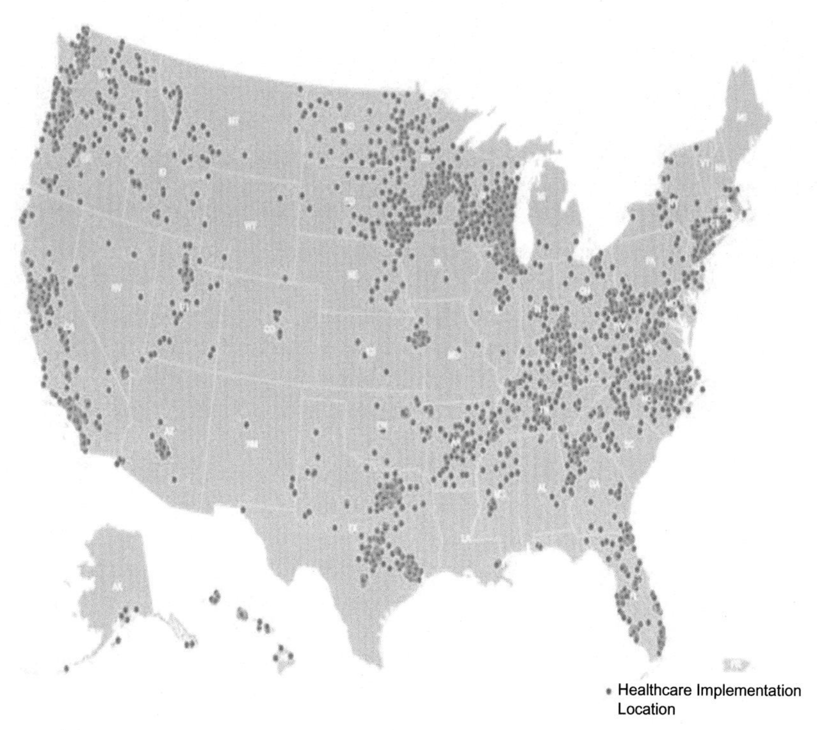

FIG. 27.6 Map of the distribution of healthcare facilities reporting data using eICR standards to public health as of December 2021. *Figure from CDC website: CDC. Healthcare Facilities in Production for COVID-19 Electronic Case Reporting 2021. https://www.cdc.gov/coronavirus/2019-ncov/hcp/electronic-case-reporting/hcfacilities-map.html [Accessed 21 December 2021].*

releasing initial versions in the Spring of 2022. The Digital Bridge group has called for the expansion of the FHIR app into a general "public health API" [68] for data access. CDC is also developing prototypes to expand the eCR FHIR application into a general public health and research API for EHRs in the MedMorph program [69]. Overall, this eCR program could represent one of the most successful and sophisticated applications of decision support technology at a national level. However, validation data (comparing data quality to curated EHR resources) and evaluation data (sensitivity, specificity of eCR methods versus gold standards) is lacking at the time of writing and further research is needed on whether this system produces timely and accurate (relative to curated data sources such as the National Covid Cohort Collaborative [70]). In addition, validation of the user experience and impact on clinical workflow has not been performed. This type of system might be expected to trigger alert fatigue responses [71], particularly if prone to false-positive interrupts for reporting. Further work on its clinical effects at scale, with triggers for multiple diseases, is needed.

27.10 Summary

Decision support systems are used by public health practitioners in internal operations but are also, and possibly more successfully, used as a tool to influence and collect data from the healthcare system. Public Health decision support systems excel in the role of influencing the clinical care system, driven by strong policy advocacy and adoption of standards with single programs. In the past few years, in partial response to the opioid overdose epidemic and the COVID-19 pandemic, decision support systems for public health have become some of the most mature and developed examples of cross platforms for decision support. Starting from a basis of using regulation to aggregate data across health systems (PDMP and IIS systems) for decision support and then adding robust methods for calculation of advice for individuals in IIS systems and with recent innovations in standards-based approaches for triggering data collection and forwarding (eCR), public health, with the support policymakers, EHR vendors, and clinical users, public health is leading the way in many aspects of clinical decision support, despite chronic underfunding and fragmentation of efforts across federal, state and local government entities. Government laws and policies as well as external support by non-governmental organizations that fund coordination of activities in collaboratives such as Digital Bridge have been critical to the successes seen.

Successes are somewhat limited by the program-specific nature of implementations for decision support methods: PDMP, IIS, and eCR programs use different platforms for knowledge representation and interventions. The eCR program, and its successor, the Public Health API, potentially lays a strong foundation for integrating decision support across programs. Efforts to reuse and integrate program-specific clinical decision support tools are critical to future success.

References

[1] Thacker SB, Qualters JR, Lee LM, Centers for Disease Control and Prevention. Public health surveillance in the United States: evolution and challenges. MMWR Suppl 2012;61:3–9.

[2] Gould DW, Walker D, Yoon PW. The evolution of BioSense: lessons learned and future directions. Public Health Rep 2017;132:7S–11S.

[3] Musa GJ, Chiang P-H, Sylk T, Bavley R, Keating W, Lakew B, et al. Use of GIS mapping as a public health tool-from cholera to cancer. Health Serv Insights 2013;6:111–6.

[4] Sahar L, Foster SL, Sherman RL, Henry KA, Goldberg DW, Stinchcomb DG, et al. GIScience and cancer: state of the art and trends for cancer surveillance and epidemiology. Cancer 2019;125:2544–60.

[5] Kubota KA, Wolfgang WJ, Baker DJ, Boxrud D, Turner L, Trees E, et al. PulseNet and the changing paradigm of laboratory-based surveillance for foodborne diseases. Public Health Rep 2019;134:22S–8S.

[6] Tolar B, Joseph LA, Schroeder MN, Stroika S, Ribot EM, Hise KB, et al. An overview of PulseNet USA databases. Foodborne Pathog Dis 2019;16:457–62.

[7] Ribot EM, Freeman M, Hise KB, Gerner-Smidt P. PulseNet: entering the age of next-generation sequencing. Foodborne Pathog Dis 2019;16:451–6.

[8] Alagoz O, Sethi AK, Patterson BW, Churpek M, Safdar N. Effect of timing of and adherence to social distancing measures on COVID-19 burden in the United States: a simulation modeling approach. Ann Intern Med 2021;174:50–7.

[9] Naimark D, Mishra S, Barrett K, Khan YA, Mac S, Ximenes R, et al. Simulation-based estimation of SARS-CoV-2 infections associated with school closures and community-based nonpharmaceutical interventions in Ontario, Canada. JAMA Netw Open 2021;4, e213793.

[10] Weissman GE, Crane-Droesch A, Chivers C, Luong T, Hanish A, Levy MZ, et al. Locally informed simulation to predict hospital capacity needs during the COVID-19 pandemic. Ann Intern Med 2020;173:21–8.

[11] New GAO. Report identifies early missteps in pandemic response, reinforces need for continued federal action., 2021, https://coronavirus.house.gov/news/press-releases/new-gao-report-identifies-early-missteps-pandemic-response-reinforces-need. [Accessed 30 December 2021].

[12] Ajrawat K, Fintzy A, Miles J, Shaffer C, Barbera J, Gralla E, et al. Decision support tool for Strategic National Stockpile (SNS) supply chain policies. In: 2018 systems and information engineering design symposium (SIEDS); 2018. p. 82–7.

[13] Simonetti A, Ezzeldin H, Walderhaug M, Anderson SA, Forshee RA. An inter-regional US blood supply simulation model to evaluate blood availability to support planning for emergency preparedness and medical countermeasures. Disaster Med Public Health Prep 2018;12:201–10.

[14] Rios-Zertuche D, Gonzalez-Marmol A, Millán-Velasco F, Schwarzbauer K, Tristao I. Implementing electronic decision-support tools to strengthen healthcare network data-driven decision-making. Arch Publ Health 2020;78. https://doi.org/10.1186/s13690-020-00413-2.

[15] NIST. Software tool improves your Doctor's vaccination advice., 2020, https://www.nist.gov/news-events/news/2020/12/nist-software-tool-improves-your-doctors-vaccination-advice. [Accessed 20 November 2021].

[16] Abbott EK, Coyle R, Dayton A, Kurilo MB. Measurement and improvement as a model to strengthen immunization information systems and overcome data gaps. Int J Med Inform 2021;148:104412.

[17] Revere D, Hills RH, Dixon BE, Gibson PJ, Grannis SJ. Notifiable condition reporting practices: implications for public health agency participation in a health information exchange. BMC Public Health 2017;17:247.

[18] Castillo-Carniglia A, González-Santa Cruz A, Cerdá M, Delcher C, Shev AB, Wintemute GJ, et al. Changes in opioid prescribing after implementation of mandatory registration and proactive reports within California's prescription drug monitoring program. Drug Alcohol Depend 2021;218:108405.

[19] Raths D. GAO report: PDMPs valuable but better EHR integration needed., 2020, https://www.hcinnovationgroup.com/clinical-it/prescription-drug-monitoring-program-pdmp/article/21156859/gao-report-pdmps-valuable-but-better-ehr-integration-needed. [Accessed 29 November 2021].

[20] Mauri AI, Townsend TN, Haffajee RL. The association of state opioid misuse prevention policies with patient- and provider-related outcomes: a scoping review. Milbank Q 2020;98:57–105.

[21] When Are Prescribers Required to Use Prescription Drug Monitoring Programs? n.d. https://www.pewtrusts.org/en/research-and-analysis/data-visualizations/2018/when-are-prescribers-required-to-use-prescription-drug-monitoring-programs [Accessed 24 November 2021].

[22] Brown WC, Whitted K. Provider prescription drug monitoring program utilization and self-auditing-a pilot study. J Dr Nurs Pract 2020;13:142–7.

[23] Dowell D, Haegerich TM, Chou R. CDC guideline for prescribing opioids for chronic pain—United States, 2016. MMWR Recomm Rep 2016;65:1–49.

[24] Spithoff S, Mathieson S, Sullivan F, Guan Q, Sud A, Hum S, et al. Clinical decision support systems for opioid prescribing for chronic non-cancer pain in primary care: a scoping review. J Am Board Fam Med 2020;33:529–40.

[25] Sinha S, Jensen M, Mullin S, Elkin PL. Safe opioid prescription: a SMART on FHIR approach to clinical decision support. Online J Public Health Inform 2017;9, e193.

[26] Martin DW, Lowery NE, Brand B, Gold R, Horlick G. Immunization information systems: a decade of progress in law and policy. J Public Health Manag Pract 2015;21:296–303.

[27] Kempe A, Hurley LP, Cardemil CV, Allison MA, Crane LA, Brtnikova M, et al. Use of immunization information systems in primary care. Am J Prev Med 2017;52:173–82.

[28] IIS Functional Standards v4.0., 2021, https://www.cdc.gov/vaccines/programs/iis/functional-standards/func-stds-v4-0.html. [Accessed 20 November 2021].

[29] Bacci JL, Hansen R, Ree C, Reynolds MJ, Stergachis A, Odegard PS. The effects of vaccination forecasts and value-based payment on adult immunizations by community pharmacists. Vaccine 2019;37:152–9.

[30] Yasnoff WA, Miller PL. Decision support and expert systems in public health. Health Inform 2003;494–512. https://doi.org/10.1007/0-387-22745-8_23.

[31] Miller PL, Frawley SJ, Sayward FG, Yasnoff WA, Duncan L, Fleming DW. Combining tabular, rule-based, and procedural knowledge in computer-based guidelines for childhood immunization. Comput Biomed Res 1997;30:211–31.

[32] Shiffman RN, Greenes RA. Improving clinical guidelines with logic and decision-table techniques: application to hepatitis immunization recommendations. Med Decis Making 1994;14:245–54.

[33] Clinical Decision Support for Immunization (CDSI): Logic Specification for Recommendations. Centers for Disease Control and Prevention, National Center for Immunization and Respiratory Disease; 2021.

[34] Muscoplat MH, Rajamani S. Immunization information system and informatics to promote immunizations: perspective from Minnesota immunization information connection. Biomed Inform Insights 2017;9. 1178222616688893.

[35] Zhu VJ, Grannis SJ, Rosenman MB, Downs SM. Implementing broad scale childhood immunization decision support as a web service. AMIA Annu Symp Proc 2009;2009:745–9.
[36] Cunningham RM, Sahni LC, Kerr GB, King LL, Bunker NA, Boom JA. The Texas Children's Hospital immunization forecaster: conceptualization to implementation. Am J Public Health 2014;104:e65–71.
[37] IIS: Clinical Decision Support (CDSi)., 2021, https://www.cdc.gov/vaccines/programs/iis/cdsi.html. [Accessed 20 November 2021].
[38] Arzt NH. Clinical decision support for immunizations (CDSi): a comprehensive, collaborative strategy. Biomed Inform Insights 2016;8:1–13.
[39] HL7 Version 3 Standard: Virtual medical record for clinical decision support (vMR-CDS) XML specification, release 1 n.d. https://www.hl7.org/implement/standards/product_brief.cfm?product_id=342 [Accessed 7 December 2021].
[40] Kawamoto K, Del Fiol G, Strasberg HR, Hulse N, Curtis C, Cimino JJ, et al. Multi-national, multi-institutional analysis of clinical decision support data needs to inform development of the HL7 virtual medical record standard. In: AMIA Annu Symp Proc, vol. 2010; 2010. p. 377–81.
[41] Shahar Y. Dimension of time in illness: an objective view. Ann Intern Med 2000;132:45–53.
[42] Shen AK, Michel JJ, Langford AT, Sobczyk EA. Shared clinical decision-making on vaccines: out of sight, out of mind. J Am Med Inform Assoc 2021;28:2523–5.
[43] Kempe A, Lindley MC, O'Leary ST, Crane LA, Cataldi JR, Brtnikova M, et al. Shared clinical decision-making recommendations for adult immunization: what do physicians think? J Gen Intern Med 2021;36:2283–91.
[44] Holzinger A, Langs G, Denk H, Zatloukal K, Müller H. Causability and explainability of artificial intelligence in medicine. Wiley Interdiscip Rev Data Min Knowl Discov 2019;9, e1312.
[45] Wilkinson TA, Dixon BE, Xiao S, Tu W, Lindsay B, Sheley M, et al. Physician clinical decision support system prompts and administration of subsequent doses of HPV vaccine: a randomized clinical trial. Vaccine 2019;37:4414–8.
[46] Dixon BE, Kasting ML, Wilson S, Kulkarni A, Zimet GD, Downs SM. Health care providers' perceptions of use and influence of clinical decision support reminders: qualitative study following a randomized trial to improve HPV vaccination rates. BMC Med Inform Decis Mak 2017;17:119.
[47] Rajamani S, Bieringer A, Sowunmi S, Muscoplat M. Stakeholder use and feedback on vaccination history and clinical decision support for immunizations offered by public health. In: AMIA Annu Symp Proc, vol. 2017; 2017. p. 1450–7.
[48] Strasberg HR, Rhodes B, Del Fiol G, Jenders RA, Haug PJ, Kawamoto K. Contemporary clinical decision support standards using health level seven international fast healthcare interoperability resources. J Am Med Inform Assoc 2021. https://doi.org/10.1093/jamia/ocab070.
[49] Lee LM, Thacker SB, St. Louis ME. Principles and practice of public health surveillance. Oxford University Press; 2010.
[50] Overhage JM, Grannis S, McDonald CJ. A comparison of the completeness and timeliness of automated electronic laboratory reporting and spontaneous reporting of notifiable conditions. Am J Public Health 2008;98:344–50.
[51] Lazarus R, Klompas M, Campion FX, McNabb SJN, Hou X, Daniel J, et al. Electronic support for public health: validated case finding and reporting for notifiable diseases using electronic medical data. J Am Med Inform Assoc 2009;16:18–24.
[52] Dixon BE, Grannis SJ, Gibson J. Enhancing provider reporting of notifiable diseases using HIE-enabled decision support. OJPHI 2019;11. https://doi.org/10.5210/ojphi.v11i1.9706.
[53] Gluskin RT, Mavinkurve M, Varma JK. Government leadership in addressing public health priorities: strides and delays in electronic laboratory reporting in the United States. Am J Public Health 2014;104:e16–21.
[54] Cooney MA, Iademarco MF, Huang M, MacKenzie WR, Davidson AJ. The public health community platform, electronic case reporting, and the digital bridge. J Public Health Manag Pract 2018;24:185–9.
[55] HL7 Public Health Work Group. http://www (web archive link, 26 December 2021). hl7.org/Special/committees/pher/index.cfm (web archive link, 26 December 2021). Narrative text reportablility response communication (RR) n.d. http://hl7.org/fhir/us/ecr/2018Sep/design-considerations.html#fhir-design-considerations [Accessed 26 December 2021].
[56] eCR Now FHIR App n.d. https://ecr.aimsplatform.org/ecr-now-fhir-app [Accessed 20 December 2021].
[57] Subscription—FHIR v4.0.1 n.d. https://www.hl7.org/fhir/subscription.html [Accessed 24 November 2019].

[58] RCKMS n.d. https://www.rckms.org/ [Accessed 26 December 2021].
[59] About RCKMS n.d. https://www.rckms.org/about-rckms/ [accessed 27 December 2021].
[60] LaVell C, Krishan S, Lichtenstein M, Minami M. Reportable conditions knowledge management system (RCKMS): real-life implementation progress and our future. In: 2019 CSTE annual conference. CSTE; 2019.
[61] digital-bridge-ecr-evaluation-report-12-32019.pdf n.d.
[62] EHRIntelligence. Pew calls for electronic case reporting for potential pandemics n.d. https://ehrintelligence.com/news/pew-calls-for-electronic-case-reporting-for-potential-pandemics [Accessed 21 December 2021].
[63] CDC. COVID-19 electronic case reporting for healthcare providers 2021, https://www.cdc.gov/coronavirus/2019-ncov/hcp/electronic-case-reporting.html. [Accessed 21 December 2021].
[64] COVID-19 Funding., 2021, https://www.cdc.gov/cpr/readiness/funding-covid.htm. [Accessed 27 December 2021].
[65] CDC. Healthcare facilities in production for COVID-19 electronic case reporting., 2021, https://www.cdc.gov/coronavirus/2019-ncov/hcp/electronic-case-reporting/hcfacilities-map.html. [Accessed 21 December 2021].
[66] Myers E. Federal agencies align to promote public health reporting—health IT buzz., 2021, https://www.healthit.gov/buzz-blog/health-it/federal-agencies-align-to-promote-public-health-reporting. [Accessed 21 December 2021].
[67] EHRIntelligence. EHR Vendor Cerner Develops National eCR Public Health Solution n.d. https://ehrintelligence.com/news/ehr-vendor-cerner-develops-national-ecr-public-health-solution [Accessed 21 December 2021].
[68] Drafted by the Digital Bridge Public Health API Workgroup Workgroup Chair – Walter Suarez. Public Health API White Paper Version 1.0 n.d.
[69] Michaels M, Syed S, Lober WB. Blueprint for aligned data exchange for research and public health. J Am Med Inform Assoc 2021;28:2702–6.
[70] Haendel MA, Chute CG, Gersing K. The national COVID cohort collaborative (N3C): rationale, design, infrastructure, and deployment. J Am Med Inform Assoc 2020. https://doi.org/10.1093/jamia/ocaa196.
[71] Bryant AD, Fletcher GS, Payne TH. Drug interaction alert override rates in the meaningful use era: no evidence of progress. Appl Clin Inform 2014;5:802–13.

SECTION V

The journey to a knowledge-enhanced health and health care system

CHAPTER

28

Clinical knowledge management program

Roberto A. Rocha[a,b], *Saverio M. Maviglia*[a,b], and *Beatriz H. Rocha*[b,c]

[a]Semedy Inc., Needham, MA, United States [b]Division of General Internal Medicine and Primary Care, Department of Medicine, Brigham and Women's Hospital, Harvard Medical School, Boston, MA, United States [c]Wolters Kluwer Health, Waltham, MA, United States

28.1 Introduction and program overview

We discuss in this chapter a comprehensive model for a program to foster the systematic, sustainable, and scalable acquisition, management, and adaptation of knowledge assets for different types of computer-based clinical decision support (CDS). Such a program, by its very nature, must leverage collective knowledge and take into account local and institutional settings, requirements, and priorities. A successful program enables healthcare institutions to effectively utilize knowledge-driven computer systems to promote continuous learning, overcome patient safety and quality challenges, and embrace new care delivery models and scientific advances.

We explain the program motivation, requirements, and implementation, taking into account needs from an institutional or enterprise perspective. We also describe essential process and infrastructure details, including critical aspects for the effective integration with clinical information systems. Given variations in size and differing capabilities of institutions for establishing such a program, part of our discussion is about how collaboration, and collective national or public-private initiatives can contribute to certain aspects of the program. Related topics, such as CDS Implementation and Governance and Managing the Investment in CDS are discussed in more detail in Chapters 20 and 21 respectively.

Clinical Decision Support and Beyond
https://doi.org/10.1016/B978-0-323-91200-6.00003-6

28.1.1 Motivation and opportunities

Healthcare is an industry characterized by very complex processes and extensive reliance on decision practices guided by a constantly evolving body of knowledge [1]. The complexity is largely the result of process fragmentation and limited information exchange; factors that can be significantly improved with the adoption of modern communication technologies and interoperability standards [2]. However, even with integrated and streamlined processes supported by efficient information exchange mechanisms, the delivery of safe and high-quality care also depends on clinical decisions that are supported by the latest evidence, while taking into account available local resources and preferences [3]. In addition, productivity is a very important goal that is not always emphasized. Improving the productivity of the healthcare industry "knowledge workers" is essential to mitigate ever-increasing costs and demand for services [4].

28.1.1.1 Knowledge-driven processes

CDS has long been recognized as an effective mechanism to help clinicians make better decisions [5,6]. Clinical systems currently in use offer a variety of features that can be customized to provide useful CDS to care providers and patients [7], as reviewed throughout this book. Despite the benefits offered by these CDS features, several important challenges associated with the creation and ongoing maintenance of computer-interpretable knowledge remain essentially unchanged [8]. Different phases of the knowledge engineering lifecycle have specific challenges, but the overarching problems include costly and inefficient ongoing maintenance, limited extensibility, and inability to achieve proper domain coverage. While several knowledge engineering problems can be mitigated with well-trained personnel equipped with efficient tools and processes, proper domain coverage can only be achieved with distributed and collaborative activities. Collaboration in this case implies cooperation across teams and institutions, with the adoption of common tools, processes, and standards, and taking into account reasonable restrictions imposed by institutional interests.

28.1.1.2 Process fragmentation

While ongoing efforts attempt to ensure continuity of care by transmitting and reconciling clinical data across settings and institutions [9], the same is not true regarding knowledge assets and CDS. The need for continuous and consistent CDS "following" the patient is a critical requirement for Accountable Care Organizations (ACOs), along with pay-for-value reimbursement models, where emphasis on management of the entire patient, and on wellness, disease prevention, and more effective engagement of the patient in disease management are essential [10]. Even though clinical systems implemented at different institutions might have very similar CDS features, they are frequently not configured in the same way. As a result, there is no guarantee that CDS interventions triggered in one clinical setting will be confirmed or re-enacted in subsequent settings, even when the same triggering elements are present. Without continuity and consistency across settings and institutions, CDS interventions have decreased effectiveness as tools for disseminating the best available evidence and reducing unwarranted care process variability [11].

28.1.1.3 Knowledge maintainability

Despite the increasing demand for computer-interpretable clinical knowledge [12,13], the initial creation and the ongoing testing and maintenance of knowledge assets remain largely manual and resource intensive processes. The authoring of a knowledge asset intended for computer-based implementation typically requires detailed review of the evidence and confirmation of the availability of patient data to trigger and/or execute the proposed CDS intervention. Knowledge assets are frequently defined using associations to other assets, creating a complex web of dependencies that must be tracked and managed. A knowledge asset can easily become inoperative when the triggering data are no longer available (e.g., medication is removed from the formulary, diagnostic test is superseded by a newer test), or when other assets that it depends on are modified or retired [14–16].

The availability of collections of knowledge assets created and maintained by commercial vendors, or open-source consortia reduces some of the knowledge engineering hurdles [17,18]. A reasonable amount of customization is still frequently needed before knowledge assets can be deployed and used within a specific clinical setting [19,20]. Customizations include simple actions to remove or change knowledge assets in support of specific local CDS interventions. However, extensive editing is needed when more complex "localizations" are required, such as the inclusion of contextual constraints to improve the specificity of a CDS intervention, or to adapt to workflow and staffing requirements, or when knowledge assets are integrated with assets from other sources to implement more sophisticated interventions [21]. When extensive local editing is needed, the implementation of third-party knowledge assets becomes difficult and costly, particularly if subsequent releases override local customizations. These difficulties are well recognized, and some knowledge vendors now provide tools that enable each customer to create its own local customizations [22]. Unfortunately, existing vendor tools are not able to support more complex CDS interventions that integrate assets from multiple sources.

28.1.1.4 Constraints and preferences

The need to contextualize available clinical evidence is an important activity that is frequently overlooked [23,24]. Computer-interpretable knowledge is commonly engineered without proper separation of clinical evidence from contextual constraints introduced during local vetting and deployment of CDS interventions. Contextual constraints represent known characteristics, limitations, and preferences of patients, care providers, and care settings where CDS interventions are occurring. Particularly in the case of patients, a complex set of characteristics may need to be considered, including age group, gender, and co-morbidities, as well as genes, lifestyle, diet, and environment, among others [25,26]. Additional constraints might include features and characteristics of the clinical system being used, or the particular type of knowledge assets being deployed. The explicit recognition of contextual constraints separate from the evidence greatly improves the consistency and reliability of CDS interventions, while promoting reuse and modularity [27,28].

28.1.1.5 Liability and oversight

As discussed further in Chapter 23, potential risks to patients, along with liability concerns, arise when knowledge assets do not accurately represent clinical evidence, or when the evidence is weak, outdated, or difficult to generalize to a well-defined patient population [29]. For example, knowledge assets available for a given domain may fail to include evidence for frequently occurring conditions, resulting in an incomplete set of CDS recommendations (false-negative results, or "failure to warn"). Similarly, knowledge assets might include a large number of conditions without proper stratification, creating an excessive number of recommendations (false-positive results, "excessive warnings"). These situations can mislead decision makers and potentially cause harm to patients [30], as well as the well-known problem of "alert fatigue" [31], sometimes causing providers to turn off alerts and reminders altogether.

It is important to recognize that problems affecting the accuracy, completeness, and currency of knowledge assets can be introduced at different stages of the knowledge engineering lifecycle, as well as during asset deployments within clinical systems. Therefore, the development and implementation of clinical knowledge assets should always follow a systematic process where knowledge design and modeling decisions are vetted by domain experts and documented by knowledge engineers. Similarly, once knowledge assets are deployed, they should be extensively tested and validated, and monitoring mechanisms should be used to continuously gather utilization details and metrics [32]. Periodic reviews should be performed to confirm projected utilization patterns and analyze unexpected deviations [33,34].

28.1.1.6 Personalized decisions

The continual growth of biomedical knowledge creates an ongoing demand not only for new computer-interpretable knowledge, but also for ongoing refinements to existing knowledge assets [35]. In some cases, such as the release of a new medication, or the replacement of a diagnostic test, changes are relatively simple and require only minimal modifications. However, advancements toward more stratified and personalized clinical decisions, e.g., through pharmacogenomics, greatly increase the number of discrete conditions that need to be considered [36]. For example, indications for new "targeted" therapeutic interventions are significantly more intricate and specialized [37,38]. Under these more complex conditions, the resulting decision-making process has to take into account a much larger number of parameters (e.g., individualized diagnoses, larger number of diagnostic tests, targeted therapies) [39], not only increasing the overall complexity of new knowledge assets, but also invalidating the more generic evidence represented in existing assets. Other important challenges include the lack of comprehensive reference resources about genomic information (Chapter 17 covers this topic in detail) curated for clinical application [40] and the high percentage of inconsistencies present in existing curated sources [41]. These challenges highlight once again the need for collaborative knowledge engineering activities.

28.1.1.7 Consumer empowerment

Shared care planning and decision-making give patients the opportunity to express their preferences and to choose treatment options in collaboration with their providers [42]. Currently, a typical knowledge engineering process considers the development of CDS for patients as an additional and separate effort. However, as patients gain more access to their electronic records and become active decision makers, knowledge assets must evolve and become shared resources for patients and providers. CDS interventions will have to reflect preferences and utility expressed by patients, and also the benefits covered by the patient's health plan [11]. Consistently communicating clinical evidence in a clear and concise manner is critical for enabling patients as informed decision makers [43,44].

28.1.1.8 Regulatory incentives

In the US, key objectives of the Health Information Technology for Economic and Clinical Health Act (HITECH) [45] fostered institutions to implement a knowledge management program. For instance, several "Meaningful Use" objectives outlined the implementation of CDS interventions [9]. More recently, the 21st Century Cures Act [46] brought forward the need for safety and effectiveness oversight of CDS interventions. Responding to the mandate for oversight, the FDA has issued preliminary guidance related to CDS software [47–49]. The draft guidance emphasizes regulatory activities using a risk-based approach. The proposed approach adopts a framework that characterizes "Software as a Medical Device" (SaMD) and requires the implementation of a quality management system that augments the software engineering process with medical device quality practices [50]. These quality practices are derived from the ISO 9000 family of standards. Since 2015, the ISO 9001 standard recognizes "knowledge" as a resource that must be effectively managed, establishing a direct connection to knowledge management [51]. Consequently, organizations involved with the development and commercialization of software systems with complex embedded CDS interventions are expected to implement a quality management program that integrates software engineering and knowledge management activities [52]. This requirement is particularly relevant for organizations involved with the development of mobile applications ("apps") designed to interoperate with different types of clinical systems [53].

28.1.2 Requirements

28.1.2.1 Historical perspective

In the early years of computer based CDS, end-user functionality was the principal objective, and long-term knowledge maintainability was often a secondary concern. Most early CDS interventions were funded as research projects, so when the project funding expired, so did the resources to maintain the knowledge assets. In many cases, lack of availability of domain experts to ensure ongoing oversight was also a problem, particularly when the "initial experts" were junior faculty members that ended up pursuing other interests. The aftermath was a CDS intervention without clear ownership and governance, resulting in ad hoc, reactive, and partial maintenance efforts to deal only with serious problems or errors [33].

Another reason for knowledge maintainability to be relegated to second-class status has been that user experience, and therefore acceptability of the CDS intervention, is impacted directly by the user interface, but indirectly by the knowledge assets. The argument sometimes made was that the required knowledge assets for an intervention prototype did not have to be fully developed until the intervention was first "proven" beneficial. More often than not, by the time the CDS intervention was validated, the focus of the business owners, particularly those with research funding, would have shifted to another priority.

It is also important to recognize that the knowledge assets used by early clinical systems were relatively simple, sparse, and non-overlapping, feeding the misperception that the effort to build and maintain such assets was relatively small compared to the implementation as a whole. Notable exceptions were the early "expert systems," which commonly required large, unwieldy, and difficult-to-test repositories of interconnected knowledge assets [54]. The need for robust tools and processes to manage the complexity and size of the expert systems' "knowledge bases" was widely recognized [55]. Unfortunately, more ubiquitous CDS interventions with simpler logic were often embedded by software developers directly into either the source code or data tables of the hosting clinical system, and designed and specified using different types of documents (e.g., spreadsheets, narrative documents, screenshots). A knowledge engineering process based on independent documents grouped into folders cannot adequately support versioning, quality assurance, or cross-referencing to other related or dependent knowledge assets. Similarly, a document-centric approach is not adequate to capture the interactions between knowledge engineers and subject matter experts, or to reliably reveal how software developers ultimately implement the decision logic in the hosting clinical system.

Eventually, early adopters and pioneers of CDS interventions became the first to experience the "rich man's problem" of having the latest and most advanced interventions running on top of numerous large silos of overlapping and inadequately maintained knowledge assets [8]. From this situation emerged the understanding that a sound knowledge engineering process must respect the key principles that define a complete lifecycle for knowledge assets, along with the realization that the knowledge lifecycle is independent of the software lifecycle governing the evolution of clinical systems hosting CDS interventions [21,32,56]. In response to the growing number and diversity of CDS interventions, a few institutions have assigned the responsibility for the long-term maintenance of knowledge assets to dedicated knowledge engineering teams [57,58].

28.1.2.2 Goals and objectives

The overarching goal of a clinical knowledge management program at an institutional or enterprise level is to establish an efficient and reliable knowledge engineering process, while ensuring that knowledge assets accurately and completely represent the latest clinical evidence without compromising long-term maintainability. Another import goal for a comprehensive program is to be able to handle multiple types of "computer-accessible" and "computer-interpretable" knowledge assets, i.e., the ability to create and manage knowledge assets for a wide-range of CDS interventions [8,34]. In order to achieve these goals, the clinical knowledge management program has to fulfill important operational and technical objectives. For example, operational objectives include (a) enabling direct and ongoing communication among knowledge engineers, domains experts, and end-users ("knowledge workers"); (b) providing accessible and easy-to-use methods for browsing and inspecting the available knowledge assets; and (c) ensuring ongoing integration and effective use of reference knowledge sources. Similarly, important technical objectives include (a) consolidation and standardization of authoring and review processes, (b) representation and management of dependencies among assets, and (c) implementation of automated testing and validation processes.

Taking into account the goals and objectives just outlined, the overarching requirements for a clinical knowledge management program include (a) knowledge assets used in CDS interventions must represent the latest clinical evidence with sufficient coverage, while taking into account local best practices and available resources; (b) knowledge assets must be easily accessible to all stakeholders, including detailed information about provenance, change history, and ongoing utilization; (c) knowledge assets must be developed and maintained using efficient, reliable, and systematic knowledge engineering processes; and (d) knowledge engineering processes must enable continuous collaboration with domain experts and stakeholders, including formal review, testing, and approval phases.

28.1.2.3 Governance and structure

The governance model for a clinical knowledge management program should rely on a clear and well-articulated vision that communicates to institution leaders and key stakeholders the importance and benefits of the program [59–62]. The program vision must be aligned with the core business strategies of the institution, and also emphasize government incentives and regulations that create opportunities for the deployment of CDS interventions. Along with a clear vision statement, specific examples can be used to portray the clinical knowledge management program as an efficient model for knowledge translation [63]. Several areas provide excellent examples and opportunities for knowledge-driven interventions, including quality assurance, patient safety, disease management, and protocol-based care, among others. These examples also suggest the need for continuous evaluation and improvement, which, in turn, require a long-term commitment to knowledge maintenance and evolution [64]. In addition, the clinical knowledge management program must be conceived with direct stakeholder participation, not only during the initial planning phases, but, more importantly, during implementation and execution. Selected clinical leaders must be identified as program champions, providing not only the necessary accountability, but also assisting with the identification and recruiting of clinical domain experts, and with the definition of meaningful incentives and methods of recognition for active participation [65]. See also Chapters 20 and 21 for additional details regarding organizational culture, commitment, framework, and business drivers needed for such a program.

28.1.2.4 Organization

A clinical knowledge management program typically includes a steering group that is responsible for the institution's CDS strategy. The CDS strategy should be derived from institutional goals defined by the clinical leadership, taking into account safety, quality, affordability, and efficiency objectives, as indicated above. The steering group also defines the relative priority of the knowledge management activities, with particular attention to competing activities and availability of resources. Depending on the characteristics of the institution, the steering group might also be responsible for the analysis and approval of CDS interventions requested by different stakeholders, including researchers that may want to conduct experiments requiring new CDS interventions. The composition of the steering group should be representative of the various stakeholders involved with CDS, including practicing clinicians, IT professionals, informaticians, and knowledge engineers.

The lack of integration among clinical systems within an institution frequently results in different knowledge management steering groups, typically one for each implemented system. Lack of integration requires ongoing communication and coordination among steering groups, which invariably become the responsibility of the knowledge engineering team. Lack of integration also prevents the implementation of more sophisticated CDS interventions, particularly those requiring continuity during and after care setting transitions.

A clinical knowledge management program must also include panels of domain experts that are consulted during different phases of the knowledge engineering process. Panel members should be practicing clinicians recognized by their peers as experts in a particular clinical domain. Effective domain experts usually have extensive practical experience with local clinical systems, particularly from an end-user perspective, as well as good understanding of CDS interventions. Domain experts are essential to help interpret the latest clinical evidence, and to guide local customizations that improve the relevance and utility of CDS interventions. Expert panels should be established as independent consultative groups that primarily take into account the needs of providers and patients. Steering group decisions should always take into account the applicable recommendations made by the experts panels.

In addition to the steering group and the expert panels, a clinical knowledge management program must also include a knowledge engineering team. Depending on the number of implemented CDS interventions, the knowledge engineering team may include specialized knowledge engineers that have mastered tools and processes applicable to similar intervention across clinical systems. The knowledge engineering team typically combines various degrees of technical and clinical expertise relevant to the different modalities of CDS [8]. The team is frequently involved not only with the creation and management of knowledge assets, but also with design and implementation of knowledge engineering processes and supporting software tools [21]. Given this broad range of responsibilities, the ideal team should include a combination of clinically trained knowledge engineers, informaticians, business analysts, software engineers, and project managers.

The size of the knowledge engineering team frequently determines how many knowledge assets the team is able to create and manage, since processes are commonly not integrated, and tools often lack important features [21]. Without integrative tools and processes, the size of the knowledge engineering team has to grow as the number of assets increase, which is not sustainable. Instead, the knowledge engineering team's growth should be proportional to the interdependencies among knowledge assets.

The scale of the clinical knowledge management program is defined by several factors, particularly the size of the institution, the number and complexity of the CDS interventions, and the degree of integration and sophistication of the clinical systems. Institutions using integrated clinical systems are better equipped to implement a larger number of CDS interventions, when compared to institutions of similar size that have disparate systems. However, most clinical systems still lack important features that significantly limit the efficiency and reliability of knowledge engineering activities [21,66].

28.1.2.5 Resilience and agility

Parallel to the constant evolution of clinical evidence and the development of best practices, clinical systems evolve and progressively incorporate new features and technologies that enable the implementation of novel CDS interventions. For example, the integration of new data types, such as genomic biomarkers and data streams from medical devices, can enable more complete and timely interventions. Therefore, the ability to evolve and adapt to new opportunities and requirements is another important characteristic of a clinical knowledge management program. Resilience and agility are essential to endure and benefit from frequent changes and adaptations, suggesting that these important characteristics need to be incorporated in all components of the clinical knowledge management program [67].

28.1.3 Implementation

CDS interventions must be continuously aligned with clinical business drivers, particularly care delivery safety, quality, and effectiveness. Several best practices and activities need to be considered during the implementation of a CDS intervention [33].

28.1.3.1 Resources and roles

Taking into account the broad set of requirements outlined above, before a clinical knowledge management program can be established, a dedicated multidisciplinary team has to be assembled. As expected, traditional management roles are responsible for governance and planning, and the ongoing coordination of projects and tasks. However, more specialized roles directly involved with the knowledge translation process are of critical importance for the success of the program [63]. These specialized roles include knowledge engineer, terminology (ontology) engineer, and knowledge modeler.

Knowledge and terminology engineers are responsible for the creation and maintenance of knowledge assets using available editing tools and relying on frequent interactions with domain experts. On the other hand, the knowledge modelers are responsible for the definition of models to represent the knowledge and terminology assets, taking into account not only the characteristics of the CDS interventions and the availability of reference sources, but also configurable features provided by the available software tools and the imperative of ongoing and long-term maintenance.

Individuals assigned to knowledge translation roles must have a strong process orientation and excellent analytical and communication skills. The individuals need to have sufficient understanding of the clinical domain to effectively communicate with domain experts and clinicians (end-users). During the knowledge acquisition and representation phases it is critical for the knowledge engineers to have a complete understanding of the scope and intent of each CDS intervention. This understanding requires frequent interactions with domain experts, when intervention details and requirements are discussed and documented. A more experienced domain expert is also critical during these phases, helping knowledge translation engineers to properly frame questions for the other domain experts. Also, during this phase, it is critical for knowledge translation engineers to communicate back to domain experts regarding how the implementation can occur, taking into account features and limitations of the target clinical system.

28.1.3.2 Clinical and technical skills

It is very difficult to recruit candidates that already have experience with the creation and maintenance of knowledge assets, as a result of which institutions should anticipate the need for a long initial period of 'on-the-job' training. Recruiting is frequently a lengthy process that attempts to identify candidates with the appropriate combination of clinical and technical skills. Qualified candidates commonly have formal clinical training (e.g., physicians, nurses, pharmacists) with experience as a practicing clinician, plus practical experience with clinical systems and information technology; formal training in biomedical informatics is highly desirable, particularly for knowledge modelers, but in general not required. Another possible alternative is to seek candidates with formal technical training (e.g., engineers, computer scientists, etc.), but with extensive practical experience in a healthcare setting, including development and implementation of clinical systems and direct contact with practicing clinicians.

Depending on the background and experience of the individual, but also on type of knowledge asset and the modality of CDS, the period of on-the-job training can last between 6 and 18 months. Resource managers and other senior members of the team are responsible for the training program, with incremental exposure to different knowledge engineering processes and software tools. The expectation after the training period is a proper understanding of the knowledge management principles, along with necessary skills to create and maintain specific types of knowledge assets. In addition, institutions with established clinical knowledge management programs frequently have partnerships with local academic programs to help raise awareness and train future generations of knowledge engineers.

Retention of talent is also a challenge given the growing emphasis and incentives to implement CDS interventions [68]. Healthcare institutions, vendors of clinical systems, and also vendors of reference knowledge content are constantly seeking specialized individuals for clinical knowledge management positions. In response to these challenges, institutions with established clinical knowledge management programs have created specialized job families for the more critical roles. Job families give individuals the opportunity to continue advancing their professional careers with higher degrees of specialization and autonomy, while at the same time helping mentor and train others. Finally, given the highly specialized nature of these roles, a differentiated salary structure is commonly implemented within each job family.

28.1.3.3 Technology framework

The technology framework to support a clinical knowledge management program has to include configurable components that enable the implementation of distributed and collaborative processes, and the creation and maintenance of an ever-growing collection of knowledge assets. The framework implementation has to take into account the open-ended nature of the knowledge translation activities, where new CDS interventions requiring novel types of knowledge assets are commonly identified. Similarly, the framework implementation also needs to enable open integration with multiple knowledge reference sources (inbound) and knowledge consumers (outbound).

The implementation of distributed and collaborative processes has to take into account the entire lifecycle of knowledge assets and the multiple roles of individuals that participate at each phase. The discrete processes implemented for each lifecycle phase should minimize the need for manual and repetitive steps, while generating new revisions that record modifications implemented by knowledge engineers and relevant contributions from domain experts and other participants. The integrity of knowledge assets should be verified at each lifecycle phase transition using configurable rules. These configurable rules should be designed to confirm compliance with structural and semantic constraints, and to also verify if dependencies on other assets continue to be valid. The expectation is that knowledge engineers will be asynchronously creating and editing different knowledge assets—the platform has to ensure that changes to a given asset do not invalidate other assets that depend on it for their own definition. The processes themselves also have to be configurable, allowing discrete steps and interactions to evolve as new models and rules are created and deployed.

The technical framework should also allow flexible and extensible knowledge representation and storage, without compromising the modularity and compositional nature of knowledge assets. For example, using an object-oriented design, each type of knowledge asset corresponds to an object class, and each class is defined by a series of attributes or properties. A common set of metadata properties should be assigned to each type of knowledge asset, enabling consistent behaviors related to lifecycle evolution and provenance documentation. Similarly, unique combinations of properties should be used to explicitly define structural and semantic characteristics of each type of knowledge asset, ideally using logic-based formalisms that simplify the implementation of arbitrarily complex validation rules. The same formalism should be used to represent dependencies and associations among knowledge assets, mappings to external reference sources, as well as applicable contextual constraints. The compositional structure defined using logic-based constructs is critical for enabling automated and introspective validation, ensuring long-term consistency and maintainability [32,69]. Finally, it is also important to highlight that the knowledge representation and storage should be optimized for knowledge asset curation and not transactional use. The detailed representation for knowledge curation frequently needs to be streamlined and optimized before assets can be efficiently utilized by transactional systems.

28.1.3.4 *Software platform*

The software platform used to implement the proposed technical framework has to take into account some important requirements. Software tools should be implemented as a complete "workbench" that integrates functions supporting multiple asset types and all lifecycle phases, while enabling productive collaborations between the various participants. Each software feature should be implemented as a modular component, maximizing the opportunity for reuse and consistency. Also, as explained above, the software platform should be highly configurable, enabling the definition and deployment of multiple concurrent lifecycles, asset types, and validation rules, as well as easy integration with reference sources and consumers. The configurability of the software platform must enable the definition of new processes, models, properties, rules, and integration profiles without requiring changes to the platform itself. In an effort to support open integration with other tools and platforms, the configurable features should, whenever possible, be compatible with available interoperability standards. Finally, the software platform should include utilization monitoring and analytical capabilities to support the ongoing evaluation and refinement of knowledge engineering processes.

28.1.3.5 *Challenges*

Several challenges have to be mitigated during the implementation of a clinical knowledge management program. From a business perspective, an important risk to be considered is the lack of clinical ownership and stewardship, particularly if CDS interventions are perceived primarily as research or IT-oriented activities. Another important business challenge is related to inadequate funding. While recognizing the significant implementation costs of a clinical knowledge management program, most institutions fail to recognize the potential liability created by outdated or incorrect knowledge assets, or the cost of not having CDS interventions [70]. The proper allocation of resources is yet another significant business challenge for the implementation of a new program. Institutions are frequently implementing multiple parallel programs at the same time, and the highly specialized resources needed for the knowledge translation activities are typically very limited. As a result, the clinical knowledge management program might be perceived as a competitor with other activities, forcing key stakeholders and domain experts to commit very limited amounts of time, compromising the overall quality and sustainability of the program [71–73].

From an operational perspective, institutions are frequently eager to deploy new CDS interventions without considering long-term maintenance implications. An important risk is the lack of a well-defined process for documenting decisions and implementation assumptions. Without proper documentation, subsequent changes to the same knowledge assets become very time-consuming and are perceived by domain experts as redundant and wasteful. Similarly, the accessibility and integrity of the knowledge assets might also be compromised if the CDS interventions are implemented as software programs that are indistinguishable from other portions of the hosting clinical system. When poor implementation choices are made, the knowledge engineering team is unable to properly test and maintain the knowledge assets without direct support from software developers. Another common operational challenge is underestimating the cost and commitment required for maintaining knowledge assets over a long period of time, particularly within institutions using poorly integrated clinical systems and/or with large number of CDS interventions [73].

From a technical perspective, an important challenge is the utilization of software tools that are unable to recognize and manage dependencies among knowledge assets, or that are unable to implement and manage critical lifecycle phases when assets are created, retired, or superseded. Without efficient and integrated tools, knowledge engineers are forced to rely on generic productivity tools to keep track of lifecycle changes and dependencies. Reliance on generic productivity tools (e.g., spreadsheets) or content management platforms (e.g., wikis) ultimately increases the overall knowledge engineering burden, introducing inconsistencies and errors that can only be identified and resolved with extensive manual effort. Another common technical limitation is the lack of utilization monitoring capabilities, directly compromising the ability of the knowledge engineering team to identify and track utilization patterns for different CDS interventions. The availability of utilization data is critical for measuring success of the interventions and also for detecting discrepancies created by unanticipated changes affecting the clinical system and the deployed knowledge assets.

28.2 Knowledge engineering process

28.2.1 Knowledge lifecycle

Boxwala et al. have described four levels of detail (1 narrative, 2 semi-structured, 3 structured, 4 executable) in which knowledge about a CDS intervention may be documented or conveyed [74,75]. The purpose of the knowledge engineering lifecycle is to iteratively translate a knowledge artifact through these levels until it is executable within a clinical information system. To do so with minimal error and with maximal assurance of effectiveness, the knowledge engineering lifecycle should include the successive states or phases of authorization, specification, implementation, testing, monitoring, and evaluation, as depicted in

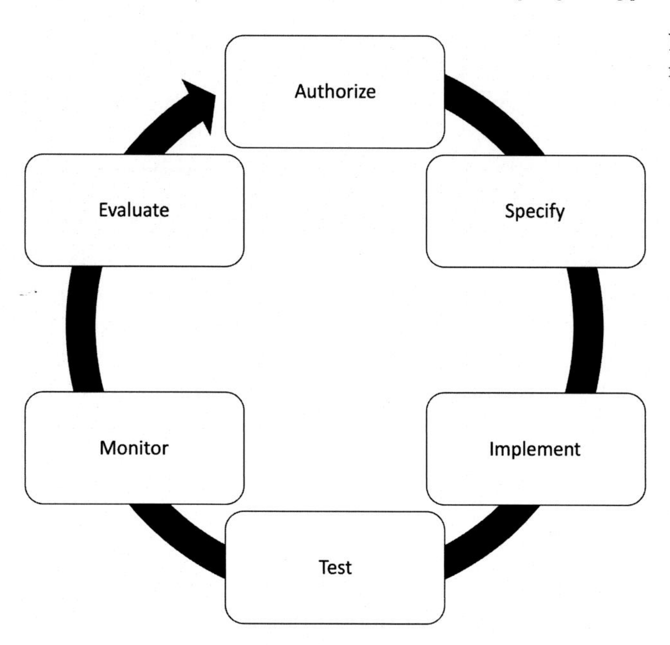

FIG. 28.1 Phases of the knowledge engineering lifecycle.

Fig. 28.1. Each of these phases, and the transitions between them, impose requirements for a clinical knowledge management program and its supporting software infrastructure. They also present opportunities for iterative process refinements which can simultaneously increase efficiency and promote quality.

In the discussion to follow, we will use Partners HealthCare System, in Boston, MA ("Partners"), where the authors previously worked, and now called Mass General Brigham, as an example to illustrate many of the key concepts. In 2013, Partners began to replace its suite of mostly homegrown clinical information systems with a single integrated commercial electronic health record system (EHR). Over the subsequent 4 years, the authors were involved with cataloging, translating, and re-implementing within the new EHR as much legacy CDS as was deemed necessary, desirable, and possible. With new clinical workflows introduced by the new EHR, some legacy CDS interventions were no longer relevant; or new CDS interventions, not previously available, were required. When important CDS interventions could not be easily migrated within the time and resource constraints imposed by the project, mitigation measures were developed and implemented. New governance structures for the project had to be established, and then evolved as the project transitioned from implementation to post-go-live operational status. Though all of these activities focused on alerts and reminders (about 500, as of the completion of the transition project in 2017), the discussion to follow applies equally to other types of CDS interventions, such as care pathways and protocols, patient registries, and context-specific delivery of reference information (infobuttons).

28.2.1.1 Authorization

The first phase of the knowledge engineering lifecycle is authorization. This is the decision to either build a new knowledge asset, or to edit an existing one. The asset can be a single concept or rule, or a multi-step chronic disease treatment protocol; and in this phase, it is typically summarized at a level 1 or 2. As indicated above, a recognized group within the institution that has the authority to commit resources to a project or task must be responsible for authorizing the knowledge engineering activity. Due to limitations in resources and time, this phase must include mechanisms to prioritize among approved tasks.

Multiple governance models can be used, depending on the size, scale, and type of institution. Though a comprehensive review is beyond the scope of this chapter, a common approach which was followed at Partners is to have a clinical content steering committee at the top that communicates vision, chooses strategy, sets priorities, sponsors investment in resources and infrastructure, sanctions editorial policies and procedures, defines relationships to other institutional groups, and resolves disputes [57,76]. This steering committee is usually comprised of executive clinical leaders, such as the chief medical officer, the chief nursing officer, the chief medical informatics officer, and directors of ancillary services.

Beneath this steering committee, the bulk of the decision-making about actual knowledge assets occurs within one or more subject matter panels staffed by clinicians of various relevant domains. These panels may meet regularly or ad hoc, in person or virtually. Their deliberations are supported, and their decisions executed by a team of knowledge engineers, who facilitate the panels' decision-making activities, design and propose the knowledge assets, craft editorial policies and procedures, and implement CDS interventions in the appropriate clinical systems. All these activities are supported by a set of tools and processes implemented using a configurable KM software platform.

28.2.1.2 Specification

The next lifecycle phase is specification, which is the actual design and authoring of knowledge assets. This phase of the knowledge engineering lifecycle requires extensive collaboration between knowledge engineers and domain experts. Collaboration is a recursive, multidisciplinary, goal-oriented, creative, and consensus-driven process undertaken by two or more people [76,77]. In terms of requirements, the KM software platform must allow communication, sharing, and decision making to occur, concentrated or distributed through space and time, among a dynamically changing group of contributors, mediated by a smaller and more static group of knowledge engineers. The flow of information among multiple contributors must be accurate, efficient, transparent, and auditable. The output of this phase is a design specification at level 3 or platform-independent level 4. The output includes all required terminology concepts, mappings between terms, decision rules, order sets, and documentation templates required to implement the intervention in any clinical information system. Ideally, the specification includes annotations with decisions, their rationales, and extensive metadata to support search and analytics.

28.2.1.3 Implementation

The process of implementing a design specification into a working CDS intervention (i.e., translating it into platform-specific level 4 artifacts) is what happens in the implementation phase of the lifecycle. At Partners, the knowledge translation process utilized dedicated tools ("workbenches"), editing guidelines ("building instructions"), and validation and testing procedures ("quality assurance") analogous to other manufacturing processes. The output of this phase is an operational CDS intervention within a testing environment. Ideally, a separate build specification is also produced, which documents any deviations from the design specification, again with reasons and rich metadata.

The distinction between the design specification and the build specification is sometimes under-appreciated, and there may be pressure to merge the two phases. However, these phases should remain separate for a number of reasons. First, without an explicit specification phase, discussions and decisions are rarely documented of why an intervention was built the way it was. Second, separating specification from implementation to the extent that the resulting design is agnostic to the clinical information system in which it is ultimately implemented, promotes more novel, interoperable, and reusable solutions. Ideally, the two activities are led by employees with different strengths: specification by those with deep informatics experience, and implementation by those with extensive application-specific training.

28.2.1.4 Testing

Testing is universally appreciated as vital to ensure quality, but commonly sacrificed when there are resource and time constraints, despite the higher cost of remediating errors after delivery compared to identifying and correcting them up front. The same testing processes utilized in software engineering should be followed when engineering knowledge. The process includes the creation of detailed positive and negative test cases, authored by someone other than the engineer responsible for the implementation of the intervention, and executed by someone other than the author of the test cases. Further, a CDS intervention must never be released until it passes all tests, no matter how small the intervention, or how minor the change. Lastly, any intervention which fails after release despite passing all its tests should be considered an opportunity to create new test cases, in order to make the testing phase more robust and to help prevent regression errors.

28.2.1.5 Monitoring

The knowledge engineering lifecycle does not end when knowledge assets are created and CDS interventions are implemented and released for use. Instead, assets and interventions must constantly be monitored to proactively detect malfunctions [14], similar to post-marketing surveillance of pharmaceutical agents [78,79]. For example, for CDS interventions involving event-action rules, monitoring is done through measurement of alert volume, overall and stratified by site, recipient role, and recipient activity. Monitoring can occur before or after an alert is activated. Pre-activation monitoring is the measurement of how often an alert fires in silent mode. Not all clinical information systems have this functionality, but when available, pre-activation monitoring not only extends testing, but also helps to establish a baseline alert burden, and a baseline or control acceptance rate, which is needed in the next lifecycle phase (evaluation) to know the true effectiveness of an intervention. After the alert is unsilenced, post-activation monitoring confirms stability and allows measurement of acceptance rate change from the silent baseline. Post-activation monitoring can occur ad hoc, intermittently, or even continuously.

At Partners, the policy was that all CDS was actively monitored for 2 weeks following un-silencing, and then again whenever there was a user report of a problem with an alert, or before/after significant clinical information system events, such as go-lives or system upgrades. Partners also implemented a continuous passive monitoring system in which automated change detection algorithms alerted knowledge engineers whenever a significant change from baseline was detected, either in frequency or pattern [80]. The aim of continuous monitoring is to detect CDS malfunctions before they are noticed by end users, or which may never get reported, such as an alert that stops firing. In the first year, continuous monitoring at Partners flagged 128 instances of alert pattern changes – 10% were true positives where the CDS had to be revised, 65% were true positives but no action was required (the change was caused by an external factor), and 25% were false positives which were often opportunities to refine the automated detection algorithm [80].

28.2.1.6 Evaluation

Detection of malfunction is necessary but not sufficient to have effective CDS interventions. Even well-functioning CDS may not accomplish the original desired effect that was the reason that the intervention was authorized in the first place. While monitoring answers the question "Is the CDS behaving as expected?", evaluation answers the question "Is the CDS solving the problem it was intended to solve?" Each question requires different data and outcome measures, analyzed with different study designs and statistical methods, presented to different audiences, and resulting in different next steps. For alerts and reminders, the easiest but crudest way to assess effectiveness is to measure override rates. Though a reasonable and simple way to start, override rates can be tricky to measure, since a user may reverse their decision in a subsequent action, whether or not they originally override the alert. One should also discount the baseline or control rate of performing the correct action (hence the importance of pre-activation monitoring). More sophisticated calculations such as "number needed to remind" and other process-oriented measures can be presented on a CDS quality dashboard, similar to how clinician dashboards show clinical effectiveness statistics about patients and patient panels [81]. Under active research are quantitative models of alert fatigue that may be used to dynamically upgrade or downgrade the intrusiveness of CDS interventions based on the predicted receptivity of the receiver [82,83], or the probability of an adverse outcome [84]. Given that the purpose of CDS is to deliver a message meant to guide decisions and/or change behaviors, the problem of increasing specificity and impact is very similar to what advertisers and marketers face, and many of the techniques and strategies developed and perfected by these sectors have yet to be tested in clinical systems. Finally, evaluation methodologies of CDS interventions other than alerts and reminders, including artificial intelligence and machine learning interventions [85], and adaptive CDS [49], have yet to be standardized and validated.

28.2.1.7 Optimizing the knowledge engineering lifecycle

Whether documented or not, an institution's collection of knowledge assets follows a lifecycle that includes many if not all of the phases described above. Undocumented lifecycle processes are more likely to be chaotic, meaning that the asset's journey through the lifecycle is unpredictable. Undocumented lifecycle processes cannot be systematically improved. Therefore, the first step to optimize one's lifecycle process is to comprehensively describe it. Once that is done, then the processes within each phase can be iteratively modified to extract efficiency and quality improvement benefits.

At Partners, this was achieved over 3 years. First, the addition of an explicit design step prior to implementation reduced the error rate from over 30% to about 10%. Adding pre-activation monitoring after testing reduced the error rate further, to under 3%. Together, these

TABLE 28.1 Knowledge lifecycle phase durations (median business days).

Lifecycle phase	2015	2016	2017
Specify	22	12	5
Implement	42	25	14
Test	19	24	15
Monitor	24	27	6
Total	107	88	40

changes resulted in an 80% reduction in support tickets as a percentage of released CDS, and anecdotally the tickets were easier to evaluate and simpler to rectify due to the increased documentation introduced in all of the lifecycle phases.

Throughput also increased (see Table 28.1). The number of business days between start of design (or build, prior to implementing a formal design phase) to activation went from a median of 107 days in 2015, to 88 days in 2016, and down to only 40 days in 2017. These improvements were observed for new CDS interventions, enhancements, and revisions due to errors. And they occurred despite diminishing resources devoted to CDS implementation over this time span.

Furthermore, as a formal and robust evaluation phase was gradually implemented near the end of this period, the conversation began to change from CDS quantity to quality, with more emphasis placed on turning off or revising ineffective CDS than on implementing new interventions [86–92].

28.2.2 Example at a national level

The National Institute for Clinical Excellence (NICE) was launched in 1999, initially to decide which medicines and treatments were cost-effective and therefore would be universally provided by the National Health Service (NHS) [93]. Its scope quickly expanded to develop evidence-based clinical guidelines, ultimately whole pathways, covering the entire timeline of all aspects of care, from screening, to diagnosis, through pharmaceutical, surgical, and behavioral interventions. In 2005, the organization was rebranded as the National Institute for Health and Care Excellence but retained the same acronym.

NICE developed a systematic methodology for creating these artifacts, and over time optimized its appraisal process, from a 54-week timetable to a 39-week one [94]. NICE produces a variety of artifact types. A guidance artifact ($n = 1745$) is a list of graded recommendations following the PICO methodology (Population, Intervention, Comparator, Outcome), and includes assessments of quality of evidence, patient preference and values, and assessments of equity, acceptability, and feasibility. Pathways are visual representations of recommendations on a topic. An advice artifact ($n = 397$) is a critical appraisal of evidence, but without a formal recommendation. All artifacts are linked to primary supporting literature, to other artifacts, and to related tools and resources. They are regularly reviewed and updated and have well documented change histories.

The governance structure of NICE includes a board which oversees financials of the organization; an executive team reports to the board and which develops strategy, objectives, business plans, policies and procedures, and creates orders for artifacts; and a guidance executive team which meets weekly to consider and approve artifacts for publication [94].

Artifacts (specifically clinical guidelines) are developed by appointed independent advisory committees comprised of healthcare professionals as well as lay members. The committees are supported by a team of project managers, information specialists, systematic reviewers and health economists. Finally, registered stakeholders such as organizations that have an interest in the guideline topic or represent people whose practice or care may be directly affected by the guideline, are consulted prior to publication [93].

NICE artifacts are published on an open web portal [95], and searchable based on title or keyword, date of last update, and type of artifact. Artifacts are presented as HTML pages or downloadable PDFs. NICE is committed to transparency, so in addition to publishing knowledge artifacts also makes available to the general public its working documents, such as charter, principles, strategy, and business plans. Its meetings are open to the public, and it welcomes participation by the general public at different levels, from stakeholders, to consultants, to committee members.

NICE is an example of a mature national knowledge management program that implements many of the requirements and principles described in this chapter. However, the knowledge assets created by NICE are not computable CDS interventions per se, but artifacts that would support the creation and implementation of such interventions. NICE artifacts are specified at level 2, so significant effort is still required to engineer level 4 (operational) CDS interventions from them.

28.3 Software infrastructure

28.3.1 General capabilities

Considering again the enterprise-specific needs for knowledge management, the software infrastructure of a clinical knowledge management program must support several key capabilities, including collaborative authoring and review of knowledge assets, integration and effective use of reference knowledge sources, and publication and distribution of knowledge assets for subsequent deployment as part of CDS interventions. While open source and commercially available solutions might provide one or some of these capabilities, there is still a significant effort to integrate each component into a complete software infrastructure for clinical knowledge management.

Taking into account the cyclic nature of the knowledge engineering activities, the architecture of the software infrastructure should enable the implementation of a comprehensive suite of tools and services that can be integrated with 'inbound' (*upstream*) knowledge sources and 'outbound' (*downstream*) knowledge consumers, as depicted in Fig. 28.2.

A few essential capabilities must be consistently implemented by all software infrastructure components. These essential capabilities include the combined management of different types of knowledge assets, from terminology concepts [96,97] to complex collections of computer-executable rules that implement clinical protocols and pathways [98,99]. The

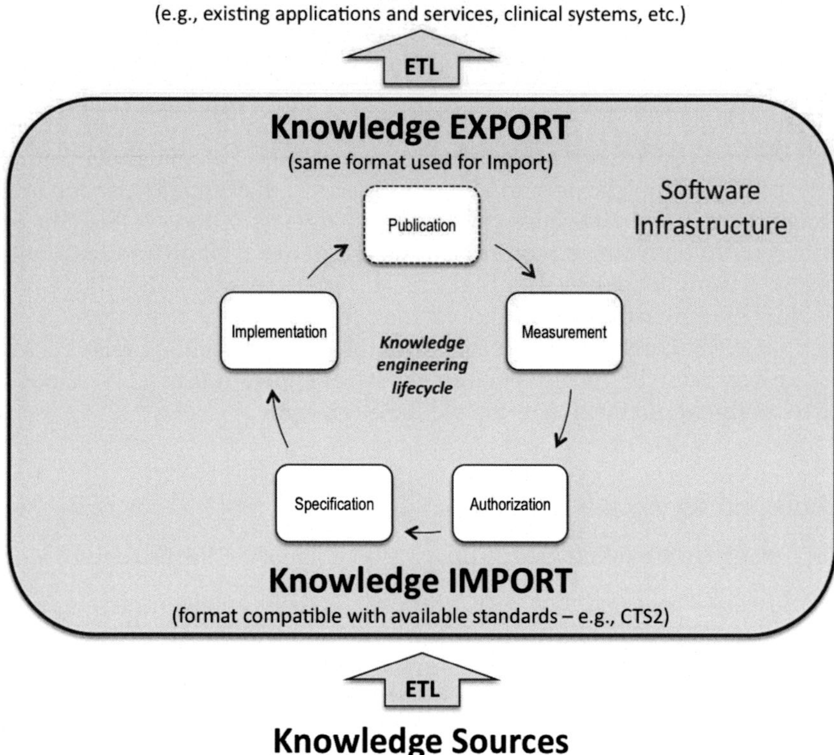

FIG. 28.2 Integration of the knowledge management software infrastructure with "inbound" knowledge sources and "outbound" knowledge consumers. Publication phase added to the lifecycle, highlighting the need to track exported assets.

combined management of different types of assets with the same software infrastructure is justified by the need to support complete lifecycles and to effectively manage dependencies among assets. The management of complete lifecycles provides consistent engagement of all stakeholders, resulting in well-defined decisions with proper documentation. The management of dependencies prevents inconsistencies and integrity violations caused by ongoing modifications and extensions to the different types of knowledge assets, ultimately preventing content errors and malfunctions.

Another essential capability is the representation and management of knowledge assets from multiple sources. For example, knowledge assets defined by standard reference sources that are required for clinical system integration and interoperability, or commercially available knowledge assets that simplify the deployment of commonly needed CDS interventions. The software infrastructure must be able to manage a wide variety of knowledge sources with their specific namespaces and identifiers, and also preserve the original characteristics of each source [100]. The effective management of namespaces enables local customizations and extensions, which, in turn, need to later be reconciled when new versions of reference and commercial sources become available [96]. The use of independent, non-proprietary, and durable identifiers, ideally in a locally defined namespace, is critical for versioning and the long-term maintenance of knowledge assets, and also for preserving the intellectual property of locally authored assets.

Finally, another essential capability is the automation of activities designed to test and confirm the integrity and consistency of knowledge assets. These automated validation features must not be limited to syntactic or structural verifications; they should also include as much confirmation of the meaning and intent of the knowledge assets as possible. The implementation of these validation features requires expressive representation and reasoning models, particularly those using logic-based formalisms [101]. Automated validation is vital for the overall reliability of knowledge assets and the long-term sustainability of the knowledge engineering activities. Similarly, a software infrastructure with reasoning capabilities can enable the progressive expansion of the number and variety of knowledge assets without always requiring the recruitment of additional knowledge engineers.

28.3.2 Tools and services

The main functions of the knowledge management software infrastructure include:

(a) *Representation and storage*—define how different types of knowledge assets are stored and versioned.
(b) *Authoring*—integrated knowledge assets creation and editing.
(c) *Collaboration*—lifecycle-driven activities involving collaboration among knowledge engineers, domain experts, and other stakeholders.
(d) *Open communication*—making knowledge assets available for searching and viewing.
(e) *Integration*—services used to search, retrieve, display, import, and export knowledge assets.
(f) *Utilization analytics*—services to track and analyze the utilization of the software infrastructure and the deployed knowledge assets.

The features and requirements for each function are described below in more detail.

28.3.2.1 Representation and storage

Despite the need to support multiple types of knowledge assets and their corresponding lifecycles, essential features can be implemented using a limited set of metadata properties that are common to all assets. A common core model includes properties to identify each knowledge asset, define and track its provenance and revisions, assign to it one of multiple designations, and establish associations with other assets, as depicted in Fig. 28.3. The common set of properties also enables the implementation of generic services to store, retrieve, search, and display knowledge assets. The implementation of common features and services contributes to the overall consistency and extensibility of the knowledge management platform. Contributors and stakeholders involved with the knowledge engineering lifecycle also benefit from having the same set of features available in all tools, and with consistent behavior across knowledge asset types.

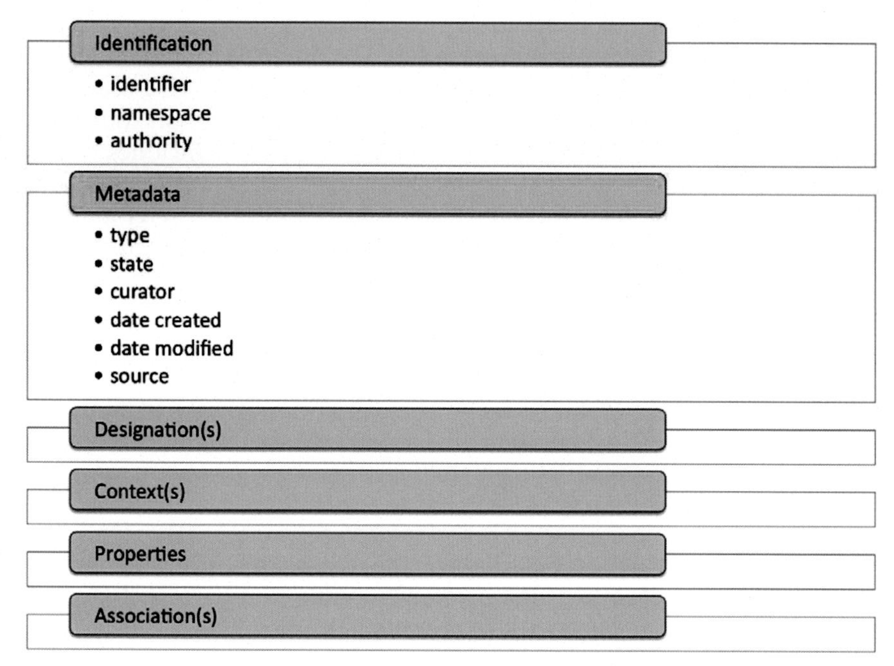

FIG. 28.3 Common core model to represent all types of knowledge assets.

In addition to a common "core" model with metadata properties, another important implementation detail is the mechanism to extend this core model to represent the specific characteristics of each new type of knowledge asset. This mechanism in effect creates a hierarchy of knowledge asset types, with progressive composition of more complex asset types through inheritance and extension. The implementation of a common mechanism to extend asset types and create new types again contributes to the overall consistency of the knowledge management platform. Knowledge engineers involved with the design and modeling of knowledge types benefit from having a consistent method for creating new types. Similarly, the implementation of automated validation rules can be stratified across types, enabling rules created for generic types to be inherited by other more specific types.

In terms of storage, a knowledge repository provides the storage infrastructure for all types of knowledge assets, including the basic services to create, read, update and delete assets. The knowledge repository is also used to store models, queries, and rules, which are essential for ensuring the validity and consistency of all knowledge assets. Given the diverse nature of the knowledge assets, the repository must store and manage different content types (e.g., text, binary types, XML). The storage model must be able to represent each discrete state of a complete knowledge asset lifecycle, and also the complete history of revisions implemented over time.

As indicated above, the repository storage model includes a common set of metadata properties that are shared by all asset types, and also a configurable set of extended properties that are used to define discrete knowledge asset types. Ideally, the modeling and implementation of new asset types does not require changes to the knowledge repository structure, or modifications to the basic repository services. The repository model also enables the consistent representation of dependencies among entities. Definitional dependencies (references) are created when the range of values for a property of a given knowledge asset type is defined as another asset type—for example: the "*medication*" asset type includes a property "*medication ingredient*," which is defined as having a range that includes assets of type "*ingredient*." In this situation, a change to an ingredient may affect medications defined with that ingredient, which in turn may affect other knowledge assets that refer to these medications. The repository services must implement validation rules that ensure the integrity of these types of dependencies.

Other important characteristics of the repository include the ability to associate knowledge assets using ad hoc relationships, and the ability to constrain these associations to be valid only when specific conditions are present. The simple associations create a mechanism to establish "semantic links" between knowledge assets [102,103], extensible to all types of assets. The associations constrained by specific conditions enable dynamic links that depend on context of use [104]. Ideally, context variables should also be defined as knowledge asset types.

28.3.2.2 Authoring

The main users of authoring tools are knowledge engineers responsible for the creation and ongoing maintenance of knowledge assets. The authoring tools include generic and easy-to-use features that allow knowledge engineers to efficiently and consistently manage large collections of knowledge assets of multiple types, including configurable dashboards and actions like changing lifecycle states and updating metadata. The authoring tools also include specialized editors that enable the creation of different types of assets, but without compromising the overall consistency of the underlying lifecycle or knowledge engineering process—each editor provides the same consistent set of features for searching, linking, and validating knowledge assets. The authoring tools also offer different forms of presentation for each type of knowledge asset, along with generic views that simplify the visualization of metadata and semantic interdependencies.

Given the different types of knowledge assets, the authoring tools must provide multiple configurable mechanisms to consistently validate knowledge assets, particularly during critical transitions between lifecycle states. As indicated above, the validation process has to be configurable and extensible, enabling knowledge engineers to progressively refine underlying models and associated validation rules. A particularly important form of validation must ensure the referential integrity of knowledge assets, where dependencies are confirmed before knowledge assets can transition to the next lifecycle state. In fact, if changes to a given asset compromise the integrity of another asset, these changes must not become active unless the affected dependencies are resolved. This is particularly important when knowledge assets are inactivated and superseded by other assets. Taking into account the highly compositional nature of clinical knowledge assets, it is virtually impossible for a knowledge engineer to effectively author and maintain large collections of assets, while keeping track of all relevant dependencies.

Access control is another important requirement related to the authoring tools. Access control mechanisms are configured to limit access to specific tools, to specific knowledge asset types, and to specific instances of each asset type. The most common scenario, where multiple knowledge engineers are able to create and manage multiple types of knowledge assets, is for a given knowledge engineer to have access to most features of the authoring tools but have editing rights to only a subset of knowledge asset types. For example, a knowledge engineer might be able to create and manage medications, ingredients, order sets, and drug-drug interaction rules, but not able to create documentation templates or health monitoring reminders. Also, in order to avoid problems created by concurrent editing, particularly the need for a complex check-out/check-in mechanism, a knowledge asset created by a given knowledge engineer is not editable by another knowledge engineer. However, the ownership of any knowledge asset should be easily transferable from one knowledge engineer to another if necessary.

All information related to authoring and maintenance of knowledge assets has to be easily accessible to the assigned knowledge engineer, including the original request to create or modify the asset, along with the task assignment and expected timeline. The knowledge engineer is also able to exchange information with other knowledge engineers, particularly when the completion of an authoring task depends on other assets still under development. In addition to providing messages and direct access to relevant information, the authoring tools also allow knowledge engineers to "subscribe" to specific knowledge assets managed by others. The subscription feature generates notifications describing changes to knowledge assets. These notifications enable knowledge engineers to proactively track the evolution of assets they might need to complete their own assignments, or to have the opportunity to re-validate assets that have dependencies on other assets that have recently changed.

28.3.2.3 Collaboration

As explained in previous sections, a successful knowledge management program requires the ongoing participation of clinical stakeholders and the continuous support and contributions from domain experts. The collaboration tools implement a variety of functions that enable knowledge engineers, domain experts, and other stakeholders to effectively work together. The main characteristic of the collaboration tools is that they enable synchronous and asynchronous activities that guide and endorse the ongoing creation and maintenance of knowledge assets. Collaboration tools provide mechanisms to organize, carry out, and document interactions and communications that would otherwise occur during meetings, or via exchanged messages and documents [77].

The notion of virtual "workspaces" is central to collaboration tools for knowledge engineering. A collaboration workspace enables the implementation and execution of essential knowledge engineering activities. Each workspace allows multiple participants with complementary roles: knowledge engineers acting as organizers and facilitators, domain experts with responsibility for reviews and approvals, and other stakeholders contributing with comments and external validation. The activities within a workspace cover the entire lifecycle of a knowledge asset, from design and implementation to long-term maintenance. The workspace offers asynchronous and synchronous types of collaboration and provides seamless access to knowledge assets, reference knowledge sources, and other supporting documentation, including prior collaboration records.

Asynchronous collaboration features are particularly effective in supporting review and vetting processes that require the expertise of busy clinicians [77]. Using a configurable workspace, knowledge engineers prepare groups of knowledge assets for review. Once the preparations are completed, notifications with specific links to the applicable workspaces are sent to domain experts requesting their participation. Experts respond and use the links to access collaboration workspaces at their convenience. Throughout the process, knowledge engineers are able to monitor the progress of the activities, while also contributing to discussions and adding new supporting evidence as needed. Knowledge engineers can also rely on voting and scoring features to obtain final approval for the clinical deployment of knowledge assets. Once the review and vetting process is concluded, knowledge engineers summarize conclusions and decisions with ample supporting evidence provided by a rich audit trail generated from the interactions with domain experts and stakeholders. Efficient access to the most qualified domain experts justifies the substantial effort required from the knowledge engineer during all phases of collaboration.

The concept of distributed "workplans" is also an important feature of collaboration tools for knowledge engineering. Workplans enable concurrent and interdependent knowledge engineering activities by efficiently allocating resources and tracking tasks and deliverables. The lifecycles of the identified knowledge assets are used to create dynamic workplans that take into account dependencies among assets, along with the expected effort and complexity of authoring and review tasks. Collaboration features integrated with the authoring tools, including simple publish and subscribe functions, enable knowledge engineers to promptly communicate their progress and needs, which, in turn, update the applicable workplans for optimal coordination.

28.3.2.4 Open communication

All knowledge assets should be accessible to clinicians within the institution, along with the documentation of the knowledge engineering process. Accessibility to assets and process records establishes a transparent and open knowledge management program that welcomes contributions and comments from clinicians. These contributions frequently provide insights on how the CDS interventions are being used, including implementation details and conditions that might have been previously overlooked. A simple and effective way to enable open access is a web-based portal available to all users of the institution's intranet. Depending on the goals of the knowledge management program, the institution might also decide to make the knowledge content accessible via the Internet to outside audiences.

The main features of the knowledge portal include searching and viewing knowledge assets. The search function must be simple to use, starting with an initial retrieval based on words or partial strings that retrieve the knowledge assets by name (designation). Once the initial set of knowledge assets is retrieved, the portal also provides effective mechanisms to select and filter the results using characteristics common to all knowledge assets, such as type, state, namespace, review date, and other metadata properties. Users searching and browsing a large collection of knowledge assets will depend on selection and filtering features to find what they are looking for. For example, a user looking for clinical reminders and order sets for diabetes management can start by simply searching for "diabetes" and, once the search results are displayed, select the knowledge asset types of "clinical reminder" and "order set" to filter the results.

After locating the desired knowledge assets, the portal enables users to retrieve the asset and display all details available. The detailed display includes all defined properties, along with provenance information, links to/from other assets, and associated design, review, and vetting process records. In an effort to provide a more user-friendly display, the portal allows users to see assets in formats similar to how they are presented within clinical systems. Finally, the portal user is also able to "subscribe" to any knowledge asset, and also provide feedback to the asset curator. With the subscription, a notification is sent to the portal user whenever modifications to the asset are made. The ability to contact the knowledge engineer ("curator") responsible for each asset also provides an efficient mechanism to ask questions and communicate concerns, again contributing to an open and transparent process.

28.3.2.5 Integration

The successful deployment of the knowledge management software infrastructure requires extensive integration with systems that consume or produce knowledge assets. The integration requires services and tools to search, retrieve, display, import, and export knowledge assets, ideally implemented in compliance with available standards. The search, retrieval, and display services should enable external systems and applications to obtain easy access to knowledge assets, preventing the need to maintain local copies that can become outdated. A subset of the features available via the web-based portal should be exposed as read-only web-services, given the need to embed links (URLs) to knowledge assets within documents and other web portals. It is also important to recognize that these web-services should not be used by applications with a high volume of transactions, since the knowledge management software infrastructure is optimized for asset creation and management, and not for transactional deployment.

Another very important integration feature is the ability to import (upload) reference knowledge assets from external (third-party) sources. While external sources might also expose their assets using web-services, a complete integration usually requires the representation of reference assets using the same models adopted by the knowledge management software infrastructure. Without common representational models, the ongoing management of dependencies between local and reference assets becomes more difficult, making it virtually impossible to guarantee that local extensions to the reference assets are correctly implemented and preserved with subsequent updates. When multiple representational models and formats do exist, preference should be given to a standard model that can represent multiple types of knowledge assets. For example, OMG's CTS2 Platform Independent Model (PIM) [105,106].

Similarly, the ability to export (download) and distribute knowledge assets is yet another very important integration feature. The export process should allow efficient selection of specific types of knowledge assets, as well as arbitrary collections that might include multiple types of assets. The process should also enable the selection of assets with specific lifecycle states, but warnings should be issued if state and type restrictions violate the integrity of assets included in the collection for export. Ideally, the export format should comply with an open standard model that can be processed by consuming applications, and also easily transformed into other formats as needed.

28.3.2.6 Utilization analytics

Effective management of the knowledge engineering activities and CDS interventions requires mechanisms to track and analyze the utilization of the software infrastructure and the deployed knowledge assets. Utilization tracking services allow continuous monitoring of all activities enabled by the knowledge management software infrastructure, essentially covering the entire knowledge asset lifecycle from creation to review and vetting, and to subsequent deployment. Monitoring services need to capture essential details about different lifecycle activities, along with information from users and knowledge assets involved. These utilization services should not be confused with detailed audit logs that record each and every transaction of the knowledge management software infrastructure. Instead, utilization services should provide configurable methods to monitor specific activities with different levels of detail, including events that might not result in transactions against the knowledge repository.

The ability to monitor CDS transactions is also a very important aspect of utilization analytics. Similar utilization services can be used if clinical systems responsible for the interventions are able to call external services. Otherwise, a mechanism to capture utilization details has to be established within each clinical system. In the latter case, tracking data generated by the clinical system has to be periodically exported and uploaded into the knowledge management software infrastructure. Details captured by the monitoring process include information about the specific CDS intervention, particularly the triggering mechanism and the user response, if applicable. The monitoring process also captures the knowledge assets used in the intervention, along with limited details about the user, clinical system, and setting, but details about the patient are not needed. Instead, a reference to the actual transaction is sufficient, enabling subsequent data retrieval and aggregation if necessary.

The utilization monitoring data are analyzed and presented to those involved with the knowledge engineering processes, and also to those responsible for the knowledge management program. Utilization metrics should be established for various activities, including thresholds that trigger more detailed reviews to explain unexpected variations, and to justify new editing cycles to update knowledge assets. Process metrics for knowledge engineering activities are defined whenever a new type of knowledge asset is created, taking into account interdependencies with other assets and the overall complexity of the new type. Utilization metrics for CDS interventions are defined during the review and vetting stages, taking into account the intent and scope of the interventions, and leveraging insights and expectations of participating domain experts. As indicated above, the knowledge management software infrastructure should have all the data required to generate process metrics for knowledge engineering activities. However, the analysis of CDS interventions might require data from clinical systems and should be performed within a clinical data warehouse environment.

28.4 Integration with clinical systems

28.4.1 Knowledge engineering activities

The integration of knowledge engineering activities with clinical systems, including EHRs, can occur in different ways. See also Chapter 29, which discusses a variety of new and evolving implementation architectures. A first scenario requires the implementation of tools and features to establish a complete knowledge management software infrastructure within the clinical system. The basic assumption supporting this first scenario is that different phases of the knowledge engineering lifecycle are managed using the clinical system software, i.e., the clinical system is able to produce and maintain its own knowledge assets.

This first scenario remains the most common approach, where clinical systems attempt to incorporate tools and features necessary to create and maintain their own knowledge assets [107]. While this "tethered" approach may seem to provide the best level of integration, the reality is that most clinical systems rely on distinct configuration tools to create and maintain different types of knowledge assets. Each configuration tool is frequently implemented from narrowly defined requirements that neglect very important phases of the knowledge asset lifecycle, forcing the adoption of "external" ancillary tools that lead to inefficient and inconsistent processes. Also, disparate configuration tools frequently fail to proactively manage interdependencies among knowledge assets, leading to configuration errors that can only be detected with rigorous regression testing procedures. Moreover, embedded configuration tools typically mix the definition of knowledge assets with attributes describing how assets will be deployed and presented at a given implementation site, compromising subsequent reuse and increasing the overall maintenance burden, while limiting the ability to share and exchange with other institutions. Another limitation is that this approach fails to account for external assets imported or accessed real-time via mechanisms such as web services.

A second and more suitable scenario redirects the integration to mechanisms that enable collections of knowledge assets to be periodically uploaded from external sources. The main assumption supporting this second scenario is that the knowledge engineering lifecycle is managed by an independent software infrastructure, making the clinical system a "consumer" and not a "producer" of knowledge assets. As indicated above, the knowledge management activities being considered here include creation and management of different types of knowledge assets, from terminology concepts to complex rules and protocols. In this second scenario, limited adjustments to knowledge assets obtained from external sources might be necessary and should be supported by clinical systems.

This second integration scenario has gained popularity, as institutions using clinical systems recognize the complexity of processes required to create and maintain knowledge assets. At the same time, a growing number of knowledge content vendors are offering more sophisticated collections of clinical knowledge assets [108]. Clinical systems have also implemented proprietary integration methods to allow periodic uploads of knowledge assets produced by content vendors, particularly order sets, care plans, and clinical documentation templates. In addition to periodic uploads, another option is to utilize software services to update and maintain knowledge assets obtained from external sources. While this last approach provides a very efficient integration mechanism, it also requires more sophisticated implementation standards that are not limited to a small number of knowledge asset types. The option to deliver CDS as a service will be discussed later, along with the execution platform requirements.

28.4.2 Integration of knowledge assets

While recognizing the second scenario as the preferred method of integration with clinical systems, several important requirements for the clinical system also need to be considered. As described above, the core premise of the second scenario is that all knowledge engineering activities occur outside the clinical system, ideally using a complete knowledge management software platform. Knowledge assets considered ready for deployment are uploaded into a "central" configuration environment of the clinical system using standard services and interfaces. This central environment provides intuitive and user-friendly tools to efficiently complete the configuration and deployment of the uploaded knowledge assets, including the necessary "bindings" to internal models used for triggering events, data entry screens and controls, as well as messaging and auditing functions. Eventual adjustments made to knowledge assets within the central environment must be easily traceable. Modified items should be available for export using a non-proprietary format, and each adjustment reflected in the original knowledge assets that were uploaded.

The central environment also has to provide a reliable mechanism to selectively distribute relevant knowledge assets to other environments (e.g., testing, training, production). The distribution process takes into account the interdependencies between knowledge assets and ideally does not require downtime of the "target" environment. The central environment also has the ability to track and report all editing and distribution activities performed. Similarly, each clinical system environment receiving knowledge assets has the ability to selectively use assets, taking into account identified utilization constraints. Each environment is also able to independently track and report on the utilization of knowledge assets. Finally, each environment is able to group and export knowledge assets being used for subsequent analysis.

28.4.3 Clinical decision support interventions

28.4.3.1 Executable knowledge assets

Several steps are required for the conversion of knowledge assets from a "declarative" to an "executable" form. The first and most important step is the translation of declarative knowledge statements into machine executable expressions. The translation process includes clear definitions of logic and data requirements, ideally using standard data (information) models and concepts from reference terminologies. Executable expressions need to be explicit, avoiding generic statements such as "if the patient has a severe acute disease." The notion of "severe acute disease" has to be explicitly defined and compatible with data available in the clinical system. Executable expressions also need to cover all possible variations of a given clinical scenario. If covering all possible variations is not feasible, logic gaps have to again be explicit, enabling a graceful "way out" (e.g., message indicating that the scenario in question cannot be analyzed). Finally, executable knowledge assets need to explicitly handle common data idiosyncrasies, including missing, incomplete, or contradictory clinical data. These data statements should be expressed using the same standard information models used for all other data statements, with proper consistency across data types, and using concepts drawn from the same standard reference terminologies.

For example, the 2020 Global Hypertension Practice Guidelines from the International Society of Hypertension classifies blood pressure (BP) based on office measurements as "high-normal" if BP is 130–139/85–89 mmHg, and as "hypertension" if BP > 140/90 mmHg [109]. In order to translate this narrative knowledge statement into executable expressions (rules), the following questions related to the clinical data have to be answered:

(a) What is the data model used by the clinical system to represent vital signs (e.g., Are systolic blood pressure and diastolic blood pressure stored as a single result or two separated results)?

(b) How does the data model used by the clinical system map into the standard model used by the executable rule?
(c) How does the dictionary identifier used by the clinical system for a blood pressure measurement map to the reference terminology code used by the executable rule?

Regarding the actual rule logic, the following questions also need to be answered:

(a) How does the rule handle a blood pressure measurement that does not fall within the defined intervals, or are additional rules needed to handle these exceptions?
(b) If only the systolic measurement is available (i.e., diastolic pressure cannot be measured), should the execution proceed, or an error be issued instead?

Another very important step that is commonly underestimated when transforming knowledge assets from declarative to executable format is the amount of effort required for quality assurance. The quality assurance process confirms if the declarative and executable formats are equivalent (i.e., represent the same knowledge statements), and also validates the accuracy of results produced by the executable rules. A comprehensive set of positive and negative test scenarios must be implemented, validating every possible execution path and exposing all known data idiosyncrasies.

Overall, the transformation process has to be intuitive and systematic, relying on software tools that clearly expose interdependencies with other rules and with the underlying standard data model and reference terminologies. The quality assurance process must also be systematic, again relying on software tools that enable the effective creation, maintenance, and review of complete sets of testing scenarios.

28.4.3.2 Execution platform requirements

During the execution of a CDS intervention, several requirements have to be taken into account, independent of the types of knowledge assets involved. The most important requirement is reliability of the execution process. The execution software platform (engine) has to guarantee that no data are lost and that all transactions are completed. The reliability requirement also applies to all data inputs and outputs. The second most important requirement is performance of the execution process. The execution software platform has to be able to process large and complex sets of rules in sub-second time, even when these rules evoke very large amounts of clinical data. Since data fetching from the clinical system is a known performance "bottleneck," the execution software platform must be able to identify and request all clinical data prior to execution. Other important execution platform requirements include scalability, enabling progressive expansion of the number of CDS interventions, and the ability to update executable knowledge assets without requiring downtime.

Efficient deployment and management of the execution software platform also depends on several interoperability requirements. Given the opportunity to use external execution platforms via software services, an important requirement is compliance with the HL7 "CDS Hooks" standard [110]. The CDS Hooks standard defines interfaces for an execution software platform to communicate with a clinical system, greatly simplifying the integration and utilization of external execution engines [111]. Another relevant interoperability requirement is the adoption of a standard model for representing data statements and for exchanging data with clinical systems during rule execution. The emerging HL7 standard known as the "Fast

Healthcare Interoperability Resources" (FHIR) fulfills this requirement; see also Chapter 15. The ONC has selected FHIR Release 4.0.1 as the foundational standard to support data exchange [112]. Considering that all clinical systems in the US will have to adhere in time with this standard, it can be considered the "default" mechanism for exchanging data between knowledge execution platforms and clinical systems, with minimal additional integration effort. Finally, despite ongoing standardization efforts, current standards provide limited options for the representation and communication of the output produced by a knowledge execution process, particularly detailed messages and recommendations that need to be interpreted and acknowledged by clinical systems. For instance, CDS Hooks provides a mechanism to represent knowledge execution output, but does not support detailed messages and recommendations, despite the fact that it includes the ability to recommend actions.

28.4.3.3 Execution platform Integration

Many clinical information systems already include at least a basic execution engine and are able to implement multiple types of CDS interventions. However, recent efforts, including some already showcasing the newer standards mentioned above, have demonstrated the important advantages of using external execution platforms integrated with clinical systems using software services [113–115]. The opportunity to distribute and consequently reduce the knowledge engineering costs for each institution is a very important advantage, particularly considering the imminent need for a significantly larger number of complex CDS interventions, as discussed earlier. Another critical advantage is the ability to ensure proper continuity of CDS interventions, particularly when a patient transitions from a care setting supported by a given clinical system to another setting supported by a different system, including, for example, a patient-controlled health record. Similar to the notion of having "patient data follow the patient" [116], the deployment using an external execution engine enables the CDS interventions to also follow the patient.

Several requirements must be considered when an external execution software platform (engine) is integrated with a clinical system. A first requirement is the ability to consistently trigger (initiate) a CDS intervention and call the external execution engine immediately following any data transaction (e.g., storage and/or retrieval of clinical observations, orders, documents) or workflow step, including triggering logic resulting from lapse of predefined time intervals or from other interventions. In essence, all data in a clinical system should be accessible to create new triggering logic statements, including data resulting from new data-entry forms and workflows defined using customization tools available in the clinical system. Taking into account the standardization requirements presented above, all data available for building triggering logic must be uniformly structured and coded, irrespective of how it was generated or stored. Similarly, all clinical system events should also be accessible for building new triggering logic statements, including events initiated by end-users and those associated with automated data transactions. Finally, the clinical system has to provide extensible built-in functions and operators for creating logic statements used as CDS triggers.

A second requirement is the ability of a clinical system to retrieve and send to an external engine all data required for the execution of a CDS intervention. As indicated above, ideally the data fetching process assembles all required data prior to the activation of the external engine. Also, data sent to the external engine should be structured using a standard data model and codified using reference terminologies. Given that data fetching can have a significant impact on performance of the CDS intervention, software routines and services responsible for data selection, retrieval, and translation must be highly optimized for efficient execution.

A third requirement is the ability to effectively communicate results of clinical decision interventions to end-users of a clinical system. Whenever applicable, users must be able to promptly carry out or override actions suggested by a CDS intervention. In addition, synchronous and asynchronous methods of communication supported by the clinical system should be accessible to the CDS interventions, including configurable 'push' and 'pull' methods integrated with the clinical system user interface and workflow processes. Similarly, "actionable" components of the communication should be able to evoke any function or workflow step supported by a clinical system, including the initiation, modification, or interruption of actions and orders. It is also very important to have communication methods configured according to the intervention intent and context (e.g., user role, activity, location, patient age group). Whenever possible, these actionable elements should be presented as "inline" changes that can be conveniently recognized and manipulated by end-users using functions available in the clinical system. For example, a suggestion for a different medication dose that takes into account the patient age group is pre-selected and presented in the medication-ordering screen; this new dose is highlighted in a different color or format to make it easily identifiable. In the case of synchronous communication methods, the intervention might generate interactive dialogs that enable end-users to justify in some detail the reason(s) suggested actions were ignored or overridden, and to provide comments and suggestions regarding clinical relevance and appropriateness of the intervention.

A fourth requirement is the ability to ensure successful completion of all interactions involving an external engine with proper levels of performance, accuracy, reliability, and scalability [117]. This requirement includes the ability to recover from failed and/or interrupted requests without compromising the intended outcome of the CDS intervention. Similarly, this requirement also includes the ability to "rollback" previous interventions when the original triggering conditions change due to updated preliminary or invalid results, or as the result of unexpected errors affecting the clinical data and/or triggering events [118].

A fifth requirement, anticipating the need for complex CDS interventions that might require the use of multiple external execution engines, is the ability of a clinical system to coordinate and synchronize interactions with different engines. This requirement is particularly important if a given CDS intervention combines synchronous and asynchronous invocation methods, along with streamlined access to contextually relevant information. In this case, it might be more appropriate to have all coordination and synchronization activities under the responsibility of yet another external engine that is specifically configured to orchestrate the execution of very complex interventions [119].

A sixth and final requirement is the ability to continuously record and monitor the triggering, generation, and communication of all CDS interventions, along with actions taken by end-users. All triggering events and all resulting communications and actions should generate auditable records with sufficient detail to allow events to be re-enacted at a later time. In particular, all "patient states" and "execution states" inferred or updated during triggering or generation of each intervention should be stored and maintained as retrievable data in the clinical system [120]. Examples of patient states are "impaired liver function," "recent antimicrobial use," and "received 1st dose of Hepatitis A vaccine." Examples of execution states are "enrolled in the nosocomial pneumonia treatment protocol" and "eligible for 2nd dose of Hepatitis A vaccine." Given the importance of patient and execution states to the integrity and continuity of a CDS intervention, a clinical system must represent each assertion using prevailing interoperability standards, enabling them to be communicated to other systems.

28.5 Future directions

28.5.1 Continuous clinical decision support interventions

Opportunities for more comprehensive and continuous forms of CDS have been identified in several clinical domains. These opportunities are frequently portrayed as highly relevant and specific interventions and, consequently, require extensive details describing a single chronic clinical condition or multiple comorbid conditions, the characteristics and preference of the enrolled patient, the care delivery context across multiple sequential encounters, and also characteristics and decision support needs of care providers involved with each encounter [121]. Advanced continuous interventions unfolded across multiple sequential encounters also need to take into account different communication channels that provide direct and ubiquitous access to patients and providers, and the increasing availability of relevant data obtained from activities of daily living [122].

An important barrier for continuous CDS imposed by existing clinical systems is the inability to support "stateful" interventions. Despite several demonstrations confirming the benefits of knowledge-driven care workflows, clinical systems lack important features that prevent the implementation of longitudinal CDS interventions, particularly clinical practice guidelines [123,124]. These limitations reduce the utility of CDS interventions as mechanisms to implement care redesign and workflow reengineering efforts, despite clear benefits for continuous quality improvement and continuous learning [64,125]. These same limitations also restrict the implementation of CDS interventions beyond organizational boundaries, reducing the usefulness CDS for complex patients that must rely on care delivered by multiple providers and settings [126]. The reduced effectiveness of CDS interventions might be explained, at least partially, by these same limitations [127]. Refer to Chapter 10 for a description of guideline-based systems.

Several pieces of the infrastructure necessary to overcome known limitations to continuous CDS are being actively developed [110,128]. Given the common expectation that the number and complexity of CDS interventions will continue to increase, clinical system vendors are implementing mechanisms to enable the deployment of "composite" CDS interventions, using combinations of internal CDS features with external capabilities provided by apps and cloud-based services [53]. The implications for future architectures of clinical systems are discussed in Chapter 29.

28.5.2 Novel clinical decision support methodologies

Novel knowledge acquisition and reasoning methodologies, particularly those involving machine learning algorithms (see Chapter 7 for additional insights about machine learning models), have gained attention in recent years despite the limited number of rigorous evaluation studies [129,130]. The ability to ingest and process large quantities of data obtained from myriad sources to identify relevant patterns is an essential feature to enable individualized medicine [37,131]. Within the context of CDS interventions, the ability to learn and adapt as new data patterns are identified is also extremely important given the need to support knowledge evolution and intervention maintainability.

The reliable and sustainable adoption of these novel methodologies implies new goals for a knowledge management program. In practical terms, the focus of the knowledge management lifecycle expands from traditional knowledge engineering to include the engineering of large, annotated datasets [132]. These datasets are essential for algorithm training and validation and their development must follow principled and reproducible processes [49,133]. Knowledge management processes can be applied with relatively simple modifications, essentially fulfilling the same process requirements outlined in previous sections of this chapter. The same underpinnings required for assembling comparable and consistent knowledge assets are directly applicable to dataset engineering, including the integration of reference terminologies and ontologies, and the adoption of descriptive information models [134].

A recent publication outlined guiding principles to enable safe, effective, and high-quality application of machine learning methodologies [135]. The proposed principles highlight the iterative and data-driven nature of the machine learning model development and the need for a complete lifecycle to systematically engineer representative datasets. The proposed lifecycle should include methods to consistently acquire and normalize relevant data sources using reference standards (i.e., dataset engineering), plus methods to test, validate, and monitor performance initially and over time. This publication corroborates the applicability of knowledge management principles and processes to the successful adoption of these novel methodologies.

28.5.3 Advanced curation and maintenance tools

The ultimate goal of a clinical knowledge management program is to evolve from a "cottage industry," where each collection of knowledge assets is extensively customized by highly specialized knowledge engineers, to a state-of-the-art factory that efficiently produces high-quality and interoperable assets that can be assembled into highly complex but manageable structures. This evolution requires a deep understanding of the knowledge engineering lifecycle, respect for asset dependencies, and robust mechanisms for governance, collaboration, and automation. Ideally, each of these activities is supported by an advanced software infrastructure that streamlines and optimizes the work of knowledge engineers and subject matter experts. For example, an error or defect identified in a knowledge asset becomes an opportunity to configure additional validation rules, which can subsequently prevent the error from recurring. Similarly, dynamic and knowledge-driven processes can be implemented, taking into account explicit structural and semantic properties of the knowledge assets. Such an infrastructure fosters and accelerates institutional learning and opens the possibility for a new kind of CDS, targeted at the curators of knowledge assets instead of the asset consumers.

28.6 Conclusions

We discussed in this chapter the motivation and detailed requirements for a clinical knowledge management program to author and curate knowledge assets for a variety of CDS interventions. Despite important advancements during the past decade, sustainable programs capable of managing multiple modalities of CDS interventions remain elusive. Even with the availability of new methodologies and CDS paradigms, renewed efforts are still needed toward known groups of challenges: (a) biomedical knowledge is constantly evolving and generally not represented in a computable format. The effort to produce computable assets is significant, requiring specialized tools that are costly to implement and skilled knowledge engineering resources that are difficult to train and retain. Opportunities to effectively collaborate outside the context of time-limited research efforts are uncommon, resulting in unnecessary redundancy and costly variability; (b) most commercially available knowledge assets and CDS interventions in widespread use are distributed and implemented using proprietary methods. Candidate standards for knowledge representation and CDS functional interoperability are constantly evolving, compromising understanding and limiting widespread adoption. Furthermore, clinical systems frequently provide only limited support for CDS-focused interoperability standards; (c) CDS interventions commonly have to be customized to accommodate local needs and resources given the large variety of healthcare organizations and activities. Local customizations require maintenance when knowledge assets and interventions are updated. The effective management of a growing number of independent CDS interventions obtained from different sources again requires specialized tools and skilled resources.

Most requirements and strategies presented in this chapter apply to healthcare institutions of any size, including smaller institutions not involved with educational and research activities. These same requirements and strategies also apply to software vendors of any size, specifically those involved with the development of systems, services, or apps with knowledge-driven features. Within the US, ongoing incentives for the widespread adoption of clinical information systems with CDS features create the opportunity for disseminating knowledge management processes that enable the efficient and sustainable production of knowledge assets. While organizations involved with the development of new CDS interventions need to implement a complete knowledge engineering lifecycle, organizations focused on the implementation of interventions developed by others can concentrate their activities on the implementation and monitoring phases of the lifecycle. Unfortunately, the inability to efficiently integrate, customize and deploy knowledge assets developed by others remain as critical limitations that can perhaps be characterized as the CDS "last mile." The limited adoption of interoperability standards compromises the integration of knowledge assets with existing clinical systems. In addition to efficient integration, healthcare organizations also need knowledge management tools that enable coordinated intervention tailoring and monitoring. These limitations continue to prevent a thriving ecosystem of CDS interventions.

The need for a principled approach to design and implement CDS interventions is as important as ever, particularly considering the opportunity to implement new interventions targeting patients and other non-clinical stakeholders. The evolutionary transition of clinical knowledge engineering "powerhouses" from large academic medical centers to commercial ventures is almost complete. The transition process has remained focused on a subset of all CDS interventions ever attempted, giving priority to interventions perceived as relevant due to quality incentives, ease of implementation, and/or regulatory requirements. This transition reaffirms not only the need for cost-effective and sustainable knowledge engineering processes, but also the opportunity to extend CDS interventions to a broader cohort of healthcare institutions. Coincidentally, a brand-new cycle of exploration and development focused on new CDS paradigms is getting started, reflecting the excitement with clinical knowledge acquisition and reasoning applying machine learning and artificial intelligence methodologies [136].

Acknowledgments

We would like to thank all those that have contributed to the knowledge management programs at Intermountain Healthcare and Partners Healthcare, and also colleagues and customers at Semedy and Wolters Kluwer Health.

References

[1] Byyny RL. The data deluge: the information explosion in medicine and science. Pharos Alpha Omega Alpha Honor Med Soc 2012;75:2–5.
[2] Basole RC, Rouse WB. Complexity of service value networks: conceptualization and empirical investigation. IBM Syst J 2008;47:53–70.
[3] Bohmer R. Fixing health care on the front lines; 2010. http://hbr.org/product/fixing-health-care-on-the-front-lines/an/R1004D-PDF-ENG. [Accessed 28 October 2021].
[4] Drucker PF. Knowledge-worker productivity: the biggest challenge. Calif Manage Rev 1999;41(2):79–94.
[5] McDonald CJ, Overhage JM, Tierney WM, Abernathy GR, Dexter PR. The promise of computerized feedback systems for diabetes care. Ann Intern Med 1996;124:170–4.
[6] Berner E. Clinical decision support systems: state of the art. Rockville, Maryland: Agency for Healthcare Research and Quality; June 2009.
[7] Wright A, Sittig DF, Ash JS, Sharma S, Pang JE, Middleton B. Clinical decision support capabilities of commercially-available clinical information systems. J Am Med Inform Assoc 2009;16:637–44.
[8] Clayton PD, Hripcsak G. Decision support in healthcare. Int J Biomed Comput 1995;39:59–66.
[9] Blumenthal D, Tavenner M. The "meaningful use" regulation for electronic health records. N Engl J Med 2010;363:501–4.
[10] Berwick DM. Making good on ACOs' promise- -the final rule for the Medicare shared savings program. N Engl J Med 2011;365:1753–6.
[11] Clancy CM, Cronin K. Evidence-based decision making: global evidence, local decisions. Health Aff (Millwood) 2005;24:151–62.
[12] Fineberg HV. Shattuck lecture. A successful and sustainable health system—how to get there from here. N Engl J Med 2012;366:1020–7.
[13] Friedman CP, Wong AK, Blumenthal D. Achieving a nationwide learning health system. Sci Transl Med 2010;2:57cm29.
[14] Wright A, Hickman TT, McEvoy D, Aaron S, Ai A, Andersen JM, et al. Analysis of clinical decision support system malfunctions: a case series and survey. J Am Med Inform Assoc 2016;23(6):1068–76.
[15] Kassakian SZ, Yackel TR, Gorman PN, Dorr DA. Clinical decisions support malfunctions in a commercial electronic health record. Appl Clin Inform 2017;8(3):910–23.
[16] Lyell D, Magrabi F, Coiera E. Reduced verification of medication alerts increases prescribing errors. Appl Clin Inform 2019;10(1):66–76.
[17] Middleton B. The clinical decision support consortium. Stud Health Technol Inform 2009;150:26–30.
[18] Lomotan EA, Meadows G, Michaels M, Michel JJ, Miller K. To share is human! Advancing evidence into practice through a National Repository of interoperable clinical decision support. Appl Clin Inform 2020; 11(1):112–21.
[19] Kuperman GJ, Reichley RM, Bailey TC. Using commercial knowledge bases for clinical decision support: opportunities, hurdles, and recommendations. J Am Med Inform Assoc 2006;13:369–71.
[20] Tiwari R, Tsapepas DS, Powell JT, Martin ST. Enhancements in healthcare information technology systems: customizing vendor-supplied clinical decision support for a high-risk patient population. J Am Med Inform Assoc 2013;20:377–80.

[21] Zhou L, Karipineni N, Lewis J, Maviglia SM, Fairbanks A, Hongsermeier T, et al. A study of diverse clinical decision support rule authoring environments and requirements for integration. BMC Med Inform Decis Mak 2012;12:128.
[22] Bubp JL, Park MA, Kapusnik-Uner J, Dang T, Matuszewski K, Ly D, et al. Successful deployment of drug-disease interaction clinical decision support across multiple Kaiser Permanente regions. J Am Med Inform Assoc 2019;26(10):905–10.
[23] Sordo M, Rocha BH, Morales AA, Maviglia SM, Oglio ED, Fairbanks A, et al. Modeling decision support rule interactions in a clinical setting. Stud Health Technol Inform 2013;192:908–12.
[24] Maviglia SM, Yoon CS, Bates DW, Kuperman G. KnowledgeLink: impact of context-sensitive information retrieval on clinicians' information needs. J Am Med Inform Assoc 2006;13:67–73.
[25] Scheuner MT, Sieverding P, Shekelle PG. Delivery of genomic medicine for common chronic adult diseases: a systematic review. JAMA 2008;299:1320–34.
[26] Hamburg MA, Collins FS. The path to personalized medicine. N Engl J Med 2010;363:301–4.
[27] Musen MA. Dimensions of knowledge sharing and reuse. Comput Biomed Res 1992;25:435–67.
[28] Pincus Z, Musen MA. Contextualizing heterogeneous data for integration and inference. AMIA Annu Symp Proc 2003;514–8.
[29] Kesselheim AS, Cresswell K, Phansalkar S, Bates DW, Sheikh A. Clinical decision support systems could be modified to reduce 'alert fatigue' while still minimizing the risk of litigation. Health Aff (Millwood) 2011;30:2310–7.
[30] Bates DW. Clinical decision support and the law: the big picture. Journal of Health Law & Policies 2012;5:319–24.
[31] Ash JS, Sittig DF, Campbell EM, Guappone KP, Dykstra RH. Some unintended consequences of clinical decision support systems. AMIA Annu Symp Proc 2007;26–30.
[32] Batarseh FA, Gonzalez AJ. Validation of knowledge-based systems: a reassessment of the field. Artif Intell Rev 2013;1–16.
[33] Bates DW, Kuperman GJ, Wang S, Gandhi T, Kittler A, Volk L, et al. Ten commandments for effective clinical decision support: making the practice of evidence-based medicine a reality. J Am Med Inform Assoc 2003;10:523–30.
[34] Wyatt JC. Decision support systems. J R Soc Med 2000;93:629–33.
[35] Meric-Bernstam F, Farhangfar C, Mendelsohn J, Mills GB. Building a personalized medicine infrastructure at a major cancer center. J Clin Oncol 2013;31:1849–57.
[36] Abdullah-Koolmees H, van Keulen AM, Nijenhuis M, Deneer VHM. Pharmacogenetics guidelines: overview and comparison of the DPWG, CPIC, CPNDS, and RNPGx guidelines. Front Pharmacol 2021;11:595219.
[37] Feero WG, Guttmacher AE, Collins FS. Genomic medicine- -an updated primer. N Engl J Med 2010;362:2001–11.
[38] Trusheim MR, Berndt ER, Douglas FL. Stratified medicine: strategic and economic implications of combining drugs and clinical biomarkers. Nat Rev Drug Discov 2007;6:287–93.
[39] Kohane IS, Masys DR, Altman RB. The incidentalome: a threat to genomic medicine. JAMA 2006;296:212–5.
[40] Feero WG, Guttmacher AE, Collins FS. The genome gets personal—almost. JAMA 2008;299:1351–2.
[41] Shugg T., Pasternak A.L., London B., Luzum JA. Prevalence and types of inconsistencies in clinical pharmacogenetic recommendations among major U.S. sources. NPJ Genom Med 2020;5:48.
[42] Bates DW, Bitton A. The future of health information technology in the patient-centered medical home. Health Aff (Millwood) 2010;29:614–21.
[43] Sepucha K, Mulley AG. A perspective on the patient's role in treatment decisions. Med Care Res Rev 2009;66:53S–74S.
[44] Epstein RM, Alper BS, Quill TE. Communicating evidence for participatory decision making. JAMA 2004;291:2359–66.
[45] Health Information Technology for Economic and Clinical Health (HITECH) Act. Office of the National Coordinator for Health Information Technology (ONC); 2009. https://www.healthit.gov/sites/default/files/hitech_act_excerpt_from_arra_with_index.pdf. [Accessed 28 October 2021].
[46] 114th Congress. 21st Century Cures Act. Public Law; 2016. p. 1033–344. 130 Stat.
[47] US Food & Drug Administration. Clinical decision support software—draft guidance for industry and food and drug administration staff., 2019, September 27, https://www.fda.gov/regulatory-information/search-fda-guidance-documents/policy-device-software-functions-and-mobile-medical-applications. [Accessed 28 October 2021].

[48] US Food & Drug Administration. Proposed regulatory framework for modifications to artificial intelligence/machine learning (AI/ML)-based software as a medical device (SaMD)., 2021, https://www.fda.gov/media/122535/download. [Accessed 28 October 2021].
[49] Petersen C, Smith J, Freimuth RR, Goodman KW, Jackson GP, Kannry J, et al. Recommendations for the safe, effective use of adaptive CDS in the US healthcare system: an AMIA position paper. J Am Med Inform Assoc 2021;28(4):677–84.
[50] IMDRF SaMD Working Group. Software as a medical device (SaMD): key definitions; 2013. p. 1–9.
[51] Maximo EZ, Pereira R, Malvestiti R, Souza JA. ISO 30401: the standardization of knowledge. Int J Dev Res 2020;10(06):37155–9.
[52] Vasconcelos JB, Kimble C, Carreteiro P, Rocha A. The application of knowledge management to software evolution. Int J Inform Manag 2017;38(1):1499–506.
[53] Barker W, Johnson C. The ecosystem of apps and software integrated with certified health information technology. J Am Med Inform Assoc 2021;28(11):2379–84.
[54] Berner ES, Webster GD, Shugerman AA, Jackson JR, Algina J, Baker AL, et al. Performance of four computer-based diagnostic systems. N Engl J Med 1994;330:1792–6.
[55] Warner HR, Sorenson DK, Bouhaddou O. Knowledge engineering in health informatics. New York: Springer; 1997.
[56] Fox J, Thomson R. Clinical decision support systems: a discussion of quality, safety and legal liability issues. In: Proceedings of the AMIA symposium; 2002. p. 265–9.
[57] Davenport TH, Glaser J. Just-in-time delivery comes to knowledge management. Harv Bus Rev 2002;80:107–11 [126].
[58] Rocha R, Bradshaw RL, Hulse NC, Rocha BHSC. The clinical knowledge management infrastructure of Intermountain Healthcare. In: Greenes RA, editor. Clinical decision support: the road ahead. Burlington: Academic Press; 2006.
[59] Gray JA. Where's the chief knowledge officer? To manage the most precious resource of all. BMJ 1998;317 (7162):832.
[60] Hibble A, Kanka D, Pencheon D, Pooles F. Guidelines in general practice: the new tower of babel? BMJ 1998;317 (7162):862–3.
[61] Ash JS, Sittig DF, Guappone KP, Dykstra RH, Richardson J, Wright A, et al. Recommended practices for computerized clinical decision support and knowledge management in community settings: a qualitative study. BMC Med Inform Decis Mak 2012;12:6.
[62] Wright A, Sittig DF, Ash JS, Bates DW, Feblowitz J, Fraser G, et al. Governance for clinical decision support: case studies and recommended practices from leading institutions. J Am Med Inform Assoc 2011;18:187–94.
[63] Davis D, Evans M, Jadad A, Perrier L, Rath D, Ryan D, et al. The case for knowledge translation: shortening the journey from evidence to effect. BMJ 2003;327:33–5.
[64] Clemmer TP, Spuhler VJ, Berwick DM, Nolan TW. Cooperation: the foundation of improvement. Ann Intern Med 1998;128:1004–9.
[65] Plsek PE, Wilson T. Complexity, leadership, and management in healthcare organisations. BMJ 2001;323:746–9.
[66] Stefanelli M. Knowledge and process management in health care organizations. Methods Inf Med 2004;43:525–35.
[67] Fischer G, Ostwald J. Knowledge management: problems, promises, realities, and challenges. IEEE Intell Syst 2001;16:60–72.
[68] Haesli A, Boxall P. When knowledge management meets HR strategy: an exploration of personalization-retention and codification-recruitment configurations. Int J of Human Resour Manag 2005;16:1955–75.
[69] McGuinness DL. Configuration. In: Franz Baader DC, McGuinness D, Nardi D, Patel-Schneider P, editors. Description logic handbook: theory, implementation and applications. Cambridge University Press; 2003.
[70] Synnot A, Bragge P, Lunny C, Menon D, Clavisi O, Pattuwage L, et al. The currency, completeness and quality of systematic reviews of acute management of moderate to severe traumatic brain injury: a comprehensive evidence map. PLoS ONE 2018;13(6), e0198676.
[71] Helmons PJ, Suijkerbuijk BO, Nannan Panday PV, Kosterink JG. Drug-drug interaction checking assisted by clinical decision support: a return on investment analysis. J Am Med Inform Assoc 2015;22(4):764–72.
[72] Herwig C, Garcia-Aponte OF, Golabgir A, Rathore AS. Knowledge management in the QbD paradigm: manufacturing of biotech therapeutics. Trends Biotechnol 2015;33(7):381–7.

[73] Sutton RT, Pincock D, Baumgart DC, Sadowski DC, Fedorak RN, Kroeker KI. An overview of clinical decision support systems: benefits, risks, and strategies for success. NPJ Digit Med 2020;3:17.

[74] Boxwala AA, Rocha BH, Maviglia S, Kashyap V, Meltzer S, Kim J, et al. A multi-layered framework for disseminating knowledge for computer-based decision support. J Am Med Inform Assoc 2011;18(Suppl. 1):i132–9.

[75] CIRD. Clinical Decision Support Consortium (CDSC). Clinical Informatics Research and Development (CIRD). Online: AHRQ; 2013. https://digital.ahrq.gov/ahrq-funded-projects/clinical-decision-support-consortium. [Accessed 28 October 2021].

[76] Hongsermeier T, Kashyap V, Sordo M. Knowledge management infrastructure: evolution at Partners Healthcare System. In: Greenes RA, editor. Clinical decision support: the road ahead. Burlington: Academic Press; 2006.

[77] Collins SA, Bavuso K, Zuccotti G, Rocha RA. Lessons learned for collaborative clinical content development. Appl Clin Inform 2013;4:304–16.

[78] Sordo M, Colecchi J, Dubey A, Dubey AK, Gainer V, Murphy SN. STROBE-based methodology for detection of adverse events across multiple communities. AMIA Annu Symp Proc 2008;1144.

[79] Brownstein JS, Murphy SN, Goldfine AB, Grant RW, Sordo M, Gainer V, et al. Rapid identification of myocardial infarction risk associated with diabetes medications using electronic medical records. Diabetes Care 2010;33:526–31.

[80] Yoshida E, Fei S, Bavuso K, Lagor C, Maviglia S. The value of monitoring clinical decision support interventions. Appl Clin Inform 2018;9(1):163–73.

[81] Einbinder JS. AHRQ CDS TEP CDSC updates—dashboards demonstration. Online: AHRQ; 2010. http://healthit.ahrq.gov/sites/default/files/docs/page/2-2010_0201_TEP_Demos_Dashboards.pdf. [Accessed 28 October 2021].

[82] Coleman JJ, Van Der Sijs H, Haefeli WE, Slight SP, McDowell SE, Seidling HM, et al. On the alert: future priorities for alerts in clinical decision support for computerized physician order entry identified from a European workshop. BMC Med Inform Decis Mak 2013;13:111.

[83] Poly TN, Islam MM, Muhtar MS, Yang HC, Nguyen PAA, Li YJ. Machine learning approach to reduce alert fatigue using a disease medication-related clinical decision support system: model development and validation. JMIR Med Inform 2020;8(11), e19489.

[84] Chazard E, Beuscart JB, Rochoy M, Dalleur O, Decaudin B, Odou P, et al. Statistically prioritized and contextualized clinical decision support systems, the future of adverse drug events prevention? Stud Health Technol Inform 2020;270:683–7.

[85] Bates DW, Auerbach A, Schulam P, Wright A, Saria S. Reporting and implementing interventions involving machine learning and artificial intelligence. Ann Intern Med 2020;172(11 Suppl):S137–44.

[86] Kane-Gill SL, O'Connor MF, Rothschild JM, Selby NM, McLean B, Bonafide CP, et al. Technologic distractions (part 1): summary of approaches to manage alert quantity with intent to reduce alert fatigue and suggestions for alert fatigue metrics. Crit Care Med 2017;45(9):1481–8.

[87] Saiyed SM, Davis KR, Kaelber DC. Differences, opportunities, and strategies in drug alert optimization-experiences of two different integrated health care systems. Appl Clin Inform 2019;10(5):777–82.

[88] McGreevey J.D. 3rd, Mallozzi C.P., Perkins R.M., Shelov E., Schreiber R.. Reducing alert burden in electronic health records: state of the art recommendations from four health systems. Appl Clin Inform 2020; 11(1):1–12.

[89] Chaparro JD, Hussain C, Lee JA, Hehmeyer J, Nguyen M, Hoffman J. Reducing interruptive alert burden using quality improvement methodology. Appl Clin Inform 2020;11(1):46–58.

[90] Baron JM, Huang R, McEvoy D, Dighe AS. Use of machine learning to predict clinical decision support compliance, reduce alert burden, and evaluate duplicate laboratory test ordering alerts. JAMIA Open 2021;4(1): ooab006.

[91] Shah SN, Seger DL, Fiskio JM, Horn JR, Bates DW. Comparison of medication alerts from two commercial applications in the USA. Drug Saf 2021;44(6):661–8.

[92] Van Dort BA, Zheng WY, Sundar V, Baysari MT. Optimizing clinical decision support alerts in electronic medical records: a systematic review of reported strategies adopted by hospitals. J Am Med Inform Assoc 2021;28 (1):177–83.

[93] National Institute for Health and Care Excellence (NICE)., 2021, https://www.nice.org.uk/. [Accessed 4 September 2021].

[94] Garbi M. National Institute for health and care excellence clinical guidelines development principles and processes. Heart 2021;107:949–53.

[95] National Institute for Health and Care Excellence (NICE). Published guidance., 2021, https://www.nice.org.uk/guidance/published. [Accessed 4 September 2021].

[96] Rocha RA, Huff SM, Haug PJ, Warner HR. Designing a controlled medical vocabulary server: the VOSER project. Comput Biomed Res 1994;27:472–507.

[97] Cimino JJ. Terminology tools: state of the art and practical lessons. Methods Inf Med 2001;40:298–306.

[98] Maviglia SM, Zielstorff RD, Paterno M, Teich JM, Bates DW, Kuperman GJ. Automating complex guidelines for chronic disease: lessons learned. J Am Med Inform Assoc 2003;10:154–65.

[99] Morris AH. Computerized protocols and bedside decision support. Crit Care Clin 1999;15:523–45 [vi].

[100] Hole WT, Carlsen BA, Tuttle MS, Srinivasan S, Lipow SS, Olson NE, et al. Achieving "source transparency" in the UMLS Metathesaurus. Stud Health Technol Inform 2004;107:371–5.

[101] Kifer M. Rules and ontologies in F-logic. Springer Berlin Heidelberg: Reasoning Web; 2005.

[102] Cimino JJ, Clayton PD, Hripcsak G, Johnson SB. Knowledge-based approaches to the maintenance of a large controlled medical terminology. J Am Med Inform Assoc 1994;1:35–50.

[103] Rector A.L., Rogers J., Roberts A., Wroe C.. Scale and context: issues in ontologies to link health- and bioinformatics. Proceedings of the AMIA Symposium 2002;642–6.

[104] Del Fiol G, Rocha RA, Bradshaw RL, Hulse NC, Roemer LK. An XML model that enables the development of complex order sets by clinical experts. IEEE Trans Inf Technol Biomed 2005;9:216–28.

[105] Tao C, Pathak J, Solbrig HR, Wei WQ, Chute CG. Terminology representation guidelines for biomedical ontologies in the semantic web notations. J Biomed Inform 2013;46:128–38.

[106] Object Management Group, Inc (OMG). Documents associated with common terminology services 2 (CTS2) version 1.0; 2013. http://www.omg.org/spec/CTS2/1.0/. [Accessed 28 October 2021].

[107] Sittig DF, Wright A, Meltzer S, Simonaitis L, Evans RS, Nichol WP, et al. Comparison of clinical knowledge management capabilities of commercially-available and leading internally-developed electronic health records. BMC Med Inform Decis Mak 2011;11:13.

[108] Fung KW, Kapusnik-Uner J, Cunningham J, Higby-Baker S, Bodenreider O. Comparison of three commercial knowledge bases for detection of drug-drug interactions in clinical decision support. J Am Med Inform Assoc 2017;24(4):806–12.

[109] Unger T, Borghi C, Charchar F, Khan NA, Poulter NR, Prabhakaran D, et al. International society of hypertension global hypertension practice guidelines. Hypertension 2020;75:1334–57.

[110] Strasberg HR, Rhodes B, Del Fiol G, Jenders RA, Haug PJ, Kawamoto K. Contemporary clinical decision support standards using health level seven international fast healthcare interoperability resources. J Am Med Inform Assoc 2021;28(8):1796–806.

[111] Kawamoto K, Jacobs J, Welch BM, Huser V, Paterno MD, Del Fiol G, et al. Clinical information system services and capabilities desired for scalable, standards-based, service-oriented decision support: consensus assessment of the health level 7 clinical decision support work group. In: AMIA annual symposium proceedings; 2012. p. 446–55.

[112] ONC. 21st century cures act: interoperability, information blocking, and the ONC health IT certification program. Fed Regist 2020;85:25642–961.

[113] Goldberg HS, Paterno MD, Rocha BH, Schaeffer M, Wright A, Erickson JL, et al. A highly scalable, interoperable clinical decision support service. J Am Med Inform Assoc 2014;21(e1):e55–62.

[114] Kawamoto K, Del Fiol G, Orton C, Lobach DF. System-agnostic clinical decision support services: benefits and challenges for scalable decision support. Open Med Inform J 2010;4:245–54.

[115] Open Clinical Decision Support (OpenCDS). OpenCDS community; 2013. http://www.opencds.org/. [Accessed 28 October 2021].

[116] Minnesota e-Health Initiative. Conference call update on the hitech act and minnesota e-health activities; 2009. https://docplayer.net/19240109-Minnesota-e-health-initiative-conference-call-update-on-the-hitech-act-thursday-september-17-2009-4-00-p-m-to-4-45-p-m.html. [Accessed 28 October 2021].

[117] Sun C, El Khoury E, Aiello M. Transaction management in service-oriented systems: requirements and a proposal. IEEE Trans Serv Comput 2011;4:167–80.

[118] Kuperman GJ, Hiltz FL, Teich JM. Advanced alerting features: displaying new relevant data and retracting alerts. In: Proceedings of the AMIA annual fall symposium; 1997. p. 243–7. 9357625. PMCID: PMC2233581.

[119] Alexandrou D, Mentzas G. Research challenges for achieving healthcare business process interoperability. In: International conference on eHealth, telemedicine, and social medicine. eTELEMED '09, 1–7 February. IEEE; 2009. p. 58–65.

[120] Wang D, Peleg M, Tu SW, Boxwala AA, Greenes RA, Patel VL, et al. Representation primitives, process models and patient data in computer-interpretable clinical practice guidelines: a literature review of guideline representation models. Int J Med Inform 2002;68:59–70.

[121] Douthit BJ, Staes CJ, Del Fiol G, Richesson RL. A thematic analysis to examine the feasibility of EHR-based clinical decision support for implementing *Choosing Wisely®* guidelines. JAMIA Open 2021;4(2):ooab031.

[122] Haque A, Milstein A, Fei-Fei L. Illuminating the dark spaces of healthcare with ambient intelligence. Nature 2020;585(7824):193–202.

[123] Peleg M. Computer-interpretable clinical guidelines: a methodological review. J Biomed Inform 2013;46:744–63.

[124] Nabhan C, Feinberg BA. Clinical pathways in oncology: software solutions. JCO Clin Cancer Inform 2017;1:1–5.

[125] Fraser SW, Greenhalgh T. Coping with complexity: educating for capability. BMJ 2001;323:799–803.

[126] Zayas-Cabán T, Haque SN, Kemper N. Identifying opportunities for workflow automation in health care: lessons learned from other industries. Appl Clin Inform 2021;12(3):686–97.

[127] Kwan JL, Lo L, Ferguson J, Goldberg H, Diaz-Martinez JP, Tomlinson G, et al. Computerised clinical decision support systems and absolute improvements in care: meta-analysis of controlled clinical trials. BMJ 2020;370: m3216.

[128] Kawamoto K, Lobach DF, Willard HF, Ginsburg GS. A national clinical decision support infrastructure to enable the widespread and consistent practice of genomic and personalized medicine. BMC Med Inform Decis Mak 2009;9:17.

[129] Topol EJ. High-performance medicine: the convergence of human and artificial intelligence. Nat Med 2019;25 (1):44–56.

[130] Verma AA, Murray J, Greiner R, Cohen JP, Shojania KG, Ghassemi M, et al. Implementing machine learning in medicine. CMAJ 2021;193:E1351–7. https://doi.org/10.1503/cmaj.202434.

[131] Shah P, Kendall F, Khozin S, Goosen R, Hu J, Laramie J, et al. Artificial intelligence and machine learning in clinical development: a translational perspective. NPJ Digit Med 2019;2:69.

[132] Cohen JP, Cao T, Viviano JD, Huang CW, Fralick M, Ghassemi M, et al. Problems in the deployment of machine-learned models in health care. CMAJ 2021. https://doi.org/10.1503/cmaj.202066. early-released August 30, 2021.

[133] Lyell D, Coiera E, Chen J, Shah P, Magrabi F. How machine learning is embedded to support clinician decision making: an analysis of FDA-approved medical devices. BMJ Health Care Inform 2021;28(1), e100301.

[134] Chute CG, Ullman-Cullere M, Wood GM, Lin SM, He M, Pathak J. Some experiences and opportunities for big data in translational research. Genet Med 2013;15(10):802–9.

[135] Good Machine Learning Practice for Medical Device Development: Guiding Principles., 2021, https://www.fda.gov/medical-devices/software-medical-device-samd/good-machine-learning-practice-medical-device-development-guiding-principles. [Accessed 28 October 2021].

[136] Kashyap S, Morse KE, Patel B, Shah NH. A survey of extant organizational and computational setups for deploying predictive models in health systems. J Am Med Inform Assoc 2021;28(11):2445–50.

CHAPTER

29

Integration of knowledge resources into applications to enable CDS: Architectural considerations

Preston Lee[a], *Robert A. Greenes*[b], *Kensaku Kawamoto*[c], *and Emory A. Fry*[d]

[a]BDR Solutions, Silver Spring, MD, United States [b]Biomedical Informatics, Arizona State University, Phoenix, AZ, United States [c]University of Utah Health, University of Utah, Salt Lake City, UT, United States [d]Cognitive Medical Systems, San Diego, CA, United States

29.1 Introduction

Much of this textbook focuses on either the supply-side concerns of knowledge generation and representation, *or* the delivery of that content via clinical decision support (CDS) and other means. In this chapter, we examine in detail the bridge between the authoring of clinical knowledge and its operationalization in production care contexts. In other words, how knowledge resources can be integrated with clinical information systems (CISs)—electronic health record (EHR) and other systems—to enable CDS.

While such integration can be relatively straightforward for a single knowledge resource in a specific clinical information system, the core challenge remains that even "standards-compliant" knowledge resources and clinical information systems are quite diverse. Consequently, there is no single knowledge integration architecture that can address all combinations of circumstances and deployed systems. However, there are several architectural patterns for knowledge integration that can, taken together, enable the effective integration of knowledge resources into running applications. This topic is of particular importance in furthering the vision of true "plug-and-play" interoperability of both structured knowledge content and executable services across vendor platforms in heterogeneous data contexts. When we refer to "knowledge resources" here, we are referring to both the clinical knowledge processing methods and the content it operates on, unless specifically stated otherwise.

Clinical Decision Support and Beyond
https://doi.org/10.1016/B978-0-323-91200-6.00014-0

Since the second edition of this book in 2014 the CDS standards community has established several new technical standards closely aligned with new U.S. regulations on Fast Healthcare Interoperability Resources (FHIR)-based data access, thus drawing the focus of implementers. Health Level 7 (HL7)—the primary body establishing health IT standards both in the U.S. and globally—has made FHIR the centerpiece of the organization. Building on FHIR and web-based REST (**Re**presentational **S**tate **T**ransfer) principles, HL7 published the first CDS Hooks specification in 2018.[a] CDS Hooks has already seen implementation by several commercial vendors and, for those already familiar with FHIR, can provide a generally straightforward path for "hooking" certain classes of CDS functions into the user experiences of EHRs and other systems. It also supports the launching of SMART-on-FHIR applications.

Outside of HL7, the Object Management Group (OMG) is working towards formal health IT knowledge representation and run-time standards from a different angle. OMG, well known for the Unified Modeling Language (UML) specification often seen lingering on engineering whiteboards, envisions its Business Process Model and Notation™ (BPMN), Case Management Model and Notation™ (CMMN), and Decision Model and Notation™ (DMN) standards implemented in healthcare for both knowledge representational and *run-time* use—the later the subject of much debate. This trio of standards, as applied to healthcare, has been named "BPM+Health." The group driving it has released a free "Field Guide to Shareable Clinical Pathways"[b] towards this end.

The revisions to this chapter continue to describe the CDS knowledge integration architectures from the second revision, but with a stronger focus on the newer approaches that are heavily driven by the standards community and the desire of implementers for simpler means of establishing plug-and-play compatibility. In the U.S., FHIR-based data representation and methods are incrementally becoming public policy, already included in requirements coming from the Centers for Medicare and Medicaid Services (CMS), Office of the National Coordinator for Health IT (ONC), and other agencies under the Department of Health and Human Resources (HHS) umbrella. The COVID-19 pandemic amplified the call for interoperable data and detailed FHIR implementation guides for solving real-world problems. Regardless of any personal or organizational critiques of FHIR, the fact that it is in U.S. regulations signals that it is here to stay long term.

Being a platform-level information representation specification, FHIR is also intended to cover payment and claim-related use cases. The HL7 Da Vinci FHIR accelerator program[c] is rapidly developing concrete guidance for implementers in that space. The community can expect to see SMART-on-FHIR, CDS Hooks and other FHIR-based interface technologies to stay aligned with the needs of policy makers. Beyond the United States, there is also significant interest in, and increasing adoption of, FHIR and related standards internationally.

The appropriateness of a given architecture for a particular organization depends on many factors, including the existing clinical information system infrastructure and the type of CDS capability involved (e.g., real-time vs. non-real-time applications). In this chapter we outline various approaches, with special attention being placed on knowledge integration

[a] https://cds-hooks.hl7.org.

[b] https://www.bpm-plus.org/healthcare-and-bpmn.htm.

[c] http://www.hl7.org/about/davinci/.

architectures aligned with updated trends in the IT landscape, such as service-oriented architectures, cloud-based computing, app-based software ecosystems, and rethinking of EHRs as a *component* of health systems architecture as opposed to the platform constraining it. We also discuss how CDS architectures must align with larger changes in the health care industry as a whole, as shifts in health care reimbursement models towards value-based care increasingly require continuity of care across multiple organizations and health IT systems centered around patients and populations rather than individual care episodes at various care settings.

29.2 Generic system architectures, examples, and their pros and cons

Ideally, as in the conceptual model presented in Fig. 29.1, CDS implementation is a loosely coupled integration of core user interface systems and externalized decision model/execution engines.

The externalization of the execution engine as a separate component enables modular implementation of the methodologies of logical condition expression evaluation, Bayesian inference, database search, or other types of decision support. In effect, such externalization allows for incremental improvement of clinical user experiences without the burden of massive, monolithic user applications in which CDS is "baked in" to the source code. The externalized execution engine can be within a local computer, server, or non-local service—the methods of invocation and communication differ in each such situation. In one such architectural approach, these externalized tidbits of CDS (and other) functions accessed over a network, typically via APIs, are called *microservices*.

Beyond the execution engine, a second required component is the knowledge base of content that must be interpreted. Depending on the model, this could be the actual logic expressions, probability matrices, or database tables, for example, needed by that model. Both the

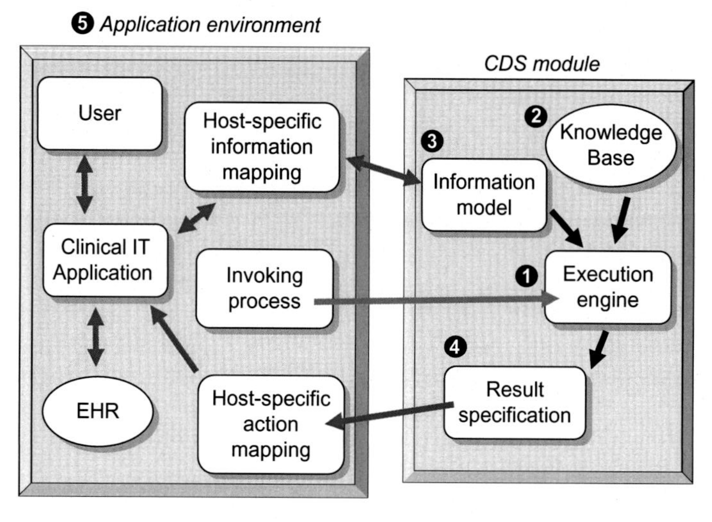

FIG. 29.1 Conceptual model of CDS as integrated but architecturally externalized capabilities supplementing clinical user application experiences.

decision model and the knowledge content can be separated from the application that uses it if appropriate interfaces are established.

Other components include the information model, a specification of how to refer to the data elements used in the CDS (both for patient-specific data and for clinical elements referred to in the knowledge model such as lab result threshold specification, diagnosis names or categories, or medication names or categories), and a result specification, defining the nature of the output as a consequence of executing CDS. A final component, at least conceptually, is the application environment. The constituents of this conceptual model can be architected in a variety of ways. Historically, there has been a gradual but consistent evolution from tightly coupled, singular tools to loosely coupled, distributed solutions. Coupling refers to how interconnected an application or its subsystems are. Tight coupling means that components have detailed knowledge of each other's internal operations and will often call these methods directly. Loose coupling implies a degree of functional and technical abstraction that hides implementation details behind generalized, ideally standardized APIs. A brief discussion of these implementation architectures follows.

29.2.1 Range of architectural approaches

The simplest way to implement a single instance of CDS, such as evaluating a logic condition, is to embed it right into a program's source code. For example, to create a simple heart disease risk calculator, one could simply code the desired mathematical operations directly into the application.

Alternatively, if a large number of rules needs to be evaluated in different circumstances, a CDS system might build the logic interpreter directly in an application but maintain the logic itself in a separate knowledge base. This might be done, for example, for a reminder system that evaluates a corpus of rules to determine what, if any, reminders should be sent to the clinician.

A more decoupled approach might be when an application supports alternative, dynamically selected inference or execution methods. For example, at times it might require a rule interpreter for deterministic rules, but in other circumstances utilize a fuzzy logic interpreter or one supporting an argumentation model. Such systems separate both the inference method (execution engine) and the knowledge base from the application framework in which the CDS is intended to operate.

There are many examples to illustrate each of these variations. Some diagnostic systems, as discussed in Chapter 1, were used in a stand-alone fashion. But even in the early days, systems such as MYCIN [1], its successor eMYCIN [2], Internist [3], and DXplain [4] were architected in such a way that they used external knowledge bases (a collection of production rules for MYCIN, disease and findings tables with associated numerical weights and rankings for Internist and DXplain) as part of their designs. RMRS [5], HELP [6], and BICS [7] all implemented, either initially or in subsequent iterations, logic interpreters operating on external knowledge bases.

Nevertheless, these early CDS systems are still representative of tightly coupled implementations. Despite their modularity and potentially elegant separation of concerns, the technical abstraction that their components or modules realized fell short of the generalized, and ideally, standardized APIs required of a loosely coupled architecture. For dedicated functions such as a specialty calculator or simplistic flow control where domain logic rarely changes, a tightly coupled approach is both appropriate and efficient. But when CDS systems become more complex, such as when the corpus of input knowledge grows and/or changes rapidly, the cost of maintaining, modifying, reusing, and regression testing tightly coupled applications becomes impractical. Reflecting evolving software engineering practices and realities of sustainable enterprise architecture, modern CDS applications typically favor loosely-coupled design principles.

29.2.2 Service oriented architecture

Service Oriented Architecture (SOA) represents a potentially more loosely-coupled architecture that has become increasingly popular since the late 1990s. SAGE [8], EON [9], SEBASTIAN [10], and numerous other implementations were early pioneers of this approach. Two important functional characteristics of SOA are (a) modularity and (b) remote invocation. *Modularity* refers to the decomposition of complex system behavior into a series of generalized services. A well-designed service hides the complexity, idiosyncrasies, and internal contracts of a typical business process behind an interface that describes its public behavior in simpler, functional terms. Such abstraction not only enables legacy functionality to be encapsulated and exposed, extending the life expectancy of those systems, but also allows for greater system flexibility, making complex business processes potentially easier to orchestrate and manage than those that rely on embedded functionality.

Remote Invocation, the second functional characteristic, aligns well with the heterogeneous computing environments often found in clinical settings. Since required resources might be distributed across numerous machines, different services, and potentially multiple networks, the ability of a SOA design to remotely invoke and orchestrate required service behaviors can deliver significant improvements in scalability, reuse, and performance.

Nevertheless, SOA designs demand careful attention to several implementation details if they are to be successfully deployed. First, business or clinical functionality must be generalized and well abstracted if a service is to prove reusable and have the broadest possible applicability. SOA implementations can be expensive to implement and if services are not well designed, the anticipated return on investment may not materialize. Slapping a web service on top of an existing functional capability is not a strategy likely to meet with success. Second, service invocation is often a synchronous, blocking transaction in which participant process threads are engaged and not available until the transaction completes. Services, while perhaps loosely coupled from a technical perspective, are still tightly coupled functionally and this can have significant implications on scalability as transaction loads increase. In general, SOA is best applied to command and control architectures that must vertically integrate functional components or where transactional processes require commit and rollback capabilities.

29.2.3 Event-driven architecture

Complementing SOA is Event-Driven Architecture (EDA). While SOA is typically referenced in a synchronous request-response context, EDA typifies asynchronous "publish/subscribe" and "producer/consumer" patterns. In publish/subscribe models, business or clinical events are broadcast by a publisher to a *multitude* of unknown subscribers. In similar producer/consumer use cases, messages are queued for a *single* consumer (or "worker") to accept and process. Either way, implementations vary in terms of a guaranteed delivery, order enforcement, ability to query, and other qualities. EDA focuses on the triggers that drive on-demand responses to rapidly changing environmental conditions. Its components are loosely coupled in that they share only the semantics of the messages exchanged, and this separation lends itself well to the type of adaptive behavior CDS systems seek to achieve. EDA, and related technologies such as Complex Event Processing, are ideal for applications that demand interdependence of process components that cross organizational boundaries, or that must operate at a scale difficult to achieve using only request/response patterns. They are ideal for reacting to unexpected episodic events and for triggering effective responses in real-time.

In practice, EDA and SOA complement each other so well that they are often deployed together in what are called "SOA 2.0" or "event-driven SOA" (ED-SOA) architectures, a largely academic distinction. The take-home message is that modern enterprises usually employ several architecture styles concurrently to address diverse functional requirements. Often, these combinatory deployments combine REST-based web applications and APIs for request/response use with some form of event streaming platform or Enterprise Service Bus (ESB) for real-time messaging, query, routing, job processing and transform functions. Indeed, as the scale and complexity of their processes increase, organizations utilize many recognized enterprise integration patterns [11] simultaneously. Event streaming platforms and more traditional ESBs are well suited to supporting these patterns using either SOA or EDA as appropriate. Their popularity with many different business domains is well warranted.

SOA, EDA, or architectures leveraging both, are proven, effective, and scale well within the scope of an established organization or business domain managed using hierarchical, vertically aligned command and control strategies. Nevertheless, the globalization of business and the rapidity of both technological and cultural change create challenges for even the most effective business-oriented ED-SOA designs.

29.2.4 Agent architecture

As an autonomous entity that can adapt to and interact with its environment, an agent is characterized by being (a) proactive and displaying goal-oriented behavior, (b) reactive and able to respond to events in its environment, (c) autonomous and able to control its behavior independently of other components, and (d) social in that it can interact with other agents to accomplish its goals [12]. The critical distinguishing feature of an agent-based architecture is that, unlike SOA or EDA components whose state is managed by an external control entity, an agent manages state internally without external choreography. It can reason about that state to make decisions about how to control its behavior accordingly.

While often associated with AI platforms and not a dominant paradigm in healthcare, agent-based architectures are well suited for non-deterministic uses where service actors are largely indifferent to each other and strict adherence to a prescribed order of activities is not required. As compared to SOA and EDA, agent architectures are typically not used as an enterprise-wide integration pattern.

29.3 Exemplar FHIR-aligned system architectures

The architectural patterns described above can be seen in FHIR-aligned system architectures that are increasingly coming into routine use. As these approaches are described in detail in Chapter 9, only a brief summary is provided here. A detailed review of these standards is also available in a recent publication [13].

Currently, the most widely adopted among these standards in the United States is SMART-on-FHIR. This SOA approach enables a Web-based application, including CDS applications, to be embedded into CISs such as EHR systems and to read data from, and potentially write data to, CIS systems using FHIR. These SMART-on-FHIR architectures, as well as other FHIR-aligned system architectures, generally use FHIR APIs that conform to the requirements of the US Core Data for Interoperability. Support for SMART-on-FHIR and the US Core Data for Interoperability are now required as a part of US regulations of certified EHR systems.

An ED-SOA approach that is increasing in adoption is CDS Hooks. In this approach, a CIS event triggers the invocation of an external CDS service, which can retrieve data from the CIS's FHIR server. CDS Hooks services can then return guidance and recommendation in the user's workflow. Of note, CDS Hooks messages may provide a link to open a SMART-on-FHIR application.

FHIR Clinical Reasoning and Clinical Quality Language (CQL) allow encapsulation of clinical knowledge in a standardized, FHIR-compatible form. These knowledge resources can then be used by CDS execution engines in a variety of contexts, including in execution engines tightly coupled within CIS systems or external execution engines independent of the CIS.

Of note, these different exemplar architectures can overlap. For example, SMART-on-FHIR applications can be invoked by CDS Hooks services; SMART-on-FHIR applications can conduct its CDS computations using CDS Hooks, FHIR Clinical Reasoning, or CQL; and CDS Hooks services can also use FHIR Clinical Reasoning and CQL to enable its CDS computations.

29.4 Role of the CIS architecture

In considering the system architecture for CDS, it is important to acknowledge that a CDS knowledge integration architecture is one component of a larger CIS architecture. Thus, we consider here trends in the architectures used in CISs, and in particular EHR systems. Of course, every CIS system is different, such that making general statements about their architectures is potentially fraught with error. However, we provide here general trends in the industry that we believe are accurate on the whole.

29.4.1 Trend to more open, component-based, service-oriented architecture

Historically, many EHR systems—as with any large IT system—tended to utilize a monolithic architecture. Traditional perspectives have continued to view EHR systems as the central foundation upon which a health IT architecture is built, much to the benefit of EHR vendors. Since publication of the second edition of this book, a slow but gradual dethroning has begun to occur: reframing EHR systems not *as* the foundation of health IT, but a primary player *within* the foundation of health IT and a critical component of a connected ED-SOA-esque architecture. While EHR vendors remain incentivized to keep the center of gravity around their existing solutions, the trajectory of connected care across providers, patient devices, out-of-band app stores, patient data access regulations, and other factors requires architects to accept that EHRs cannot, and should not, continue the monolithic tradition.

CMS' Interoperability and Patient Access final rule, ONC's Interoperability and Information Blocking Final Regulation, Trusted Exchange Framework Common Agreement (TEFCA), and broader 21st Century Cures Act have cemented that EHR systems will become more open—at least in terms of data access—over time, and specifically using FHIR. This API-based approach is driving a resurgence towards component-based, service-oriented architecture *outside* of EHR walled gardens. While this trend is certainly not global, many factors have contributed to moving vendors towards this direction, including:

- **General IT trend**. The IT industry continues to move in this direction. For example, some EHR system vendors' move towards cloud-based computing and mobile-first access closely mirrors the move of many businesses to cloud-based and mobile computing. Also, as new vendors enter the marketplace, and with less need to support legacy applications developed in the 1970s and 1980s, it is easier for new companies to adopt modern software architectures that have decisively moved away from monolithic or EHR-centric architectures.
- **Market consolidation**. Many of the larger EHR system vendors have grown by acquiring other EHR system vendors. As a result, a single EHR system vendor may in fact be supporting multiple distinct EHR systems. Use of a more open, service-oriented architecture can be a way to deal with this need for greater interoperability *within* a vendor's own product lines.
- **U.S. Regulations**. There is certainly debate over the United States HITECH Act of 2009 [14,15] incentives for EHR adoption and Meaningful Use [16] factors. For example, interoperability could have been better served by investing Meaningful Use funds more deliberately over 5–10 years rather than in 2–3 years. However, these regulations undeniably pushed health systems and EHR solutions towards greater interoperability. Meaningful Use, and now the 21st Century Cures Act, are major factors for EHR systems increasingly opening access to data. As TEFCA begins rolling out in 2022, EHRs will face additional pressure to support a higher level of "network of networks" interoperability. As noted earlier, other countries are also increasingly exploring the use of FHIR and related standards to promote interoperability.
- **Rise of an app culture**. Mobile devices and the need to integrate data from multiple data sources in an enterprise have driven some of the larger health care organizations to either develop or obtain apps that provide new ways of visualizing and managing their data and workflows, especially as required by transitions and coordination of care. Many EHR vendors provide such capabilities or provide APIs into their systems to enable such third-party apps. This trend has been accelerated by vendors such as Apple. iPhone's iOS can now use FHIR to extract data from disparate health system sources onto patient users' *local* mobile device(s). These aggregated, longitudinal views in patients' hands are thus not visible in totality to any given health system. This patient-centric paradigm of care

coordination breaks conventional assumptions of master data management (MDM) and is thus extremely complicated. Health systems have not yet resolved how trusted exchange of health information can optimally occur across parties both claiming to be authoritative for certain records and at the behest of the patient. Major efforts are ongoing in this area.

Regardless of the true underlying reasons, by virtue of regulatory factors alone, the EHR system vendors *must* comply with increasingly open API requirements necessary for component-based, service-oriented architectures to flourish. Thus, SMART-on-FHIR and other applications capable of leveraging those APIs can be granted access to data without the exclusive blessing of the EHR vendor. Seeing the inherent business opportunity, EHR vendors have generally pivoted to also charging for API use and supported cloud-based hosting environments. Also, many vendors now provide platforms for clients and knowledge vendors to develop custom CDS capabilities. For example, popular EHR vendors now allow the invocation of web-based SMART-on-FHIR CDS systems from within their EHR systems, which are then presented as a part of the native EHR system's user interface.

29.4.2 Implications for knowledge integration

The obvious implication of these trends in EHR system architectures is the increasing possibilities for integrating knowledge resources into EHR systems. In particular, through the HL7 SMART-on-FHIR standard widely supported by EHR vendors, the HL7 CDS Hooks standard starting to gain adoption among EHR vendors, and the US Core Data for Interoperability (USCDI) and its implementation in FHIR. At the same time, vendor support for such standards varies, and even when a given resource is supported (e.g., ability to retrieve clinical data elements using an API), the subtle idiosyncrasies of implementation still limit true plug-and-play interoperability. In any event, the ultimate goal is for standards-based interface specifications—not vendor-specific implementations—to form the basis of a standardized health IT ecosystem in which CDS can be seamlessly integrated in a scalable manner.

29.5 Scaling considerations for CDS and knowledge-enhanced health/healthcare

Depending on one's perspective, "scalability" has numerous interpretations, and consequently, many implications. This section will examine several such considerations, highlighting implementation issues that, regardless of deployment architecture or philosophy, can impact an organization's ability to encapsulate and disseminate domain knowledge in an effective manner. For example, the discipline of artificial intelligence recognizes that optimal knowledge representation and management require (a) standardized representations of domain data, (b) standardized vocabularies to unambiguously define and declare domain concepts, and (c) an acceptable description logic syntax with which to encode domain knowledge. Clinical decision support is similarly constrained by these [17] and other requirements; organizations must carefully consider the strength and weakness of the technology they choose to deploy when addressing these requirements.

29.5.1 Rule syntax and execution

As previously discussed, the most straightforward approach for CDS would use conventional programming approaches that embed inference and control logic into discrete procedural steps within an application [18]. Such embedded representations can be readily developed using the technologies and programmer skill sets available since the 1970s. Coupling control and inference logic, however, makes reusing and maintaining large collections of knowledge modules more difficult and ties CDS performance and scalability to the run-time characteristics of the containing application. Fortunately, while some early systems did this, even early CDS research tended to already separate inference engines and knowledge bases at least conceptually.

Production Rules found popularity in separating domain knowledge from the control logic required by the application [19]. Encapsulation of inference logic inside a production rule allowed for the development of specialized run-time engines that could be scaled horizontally to meet performance requirements, and ultimately provided for the evolution of domain-specific languages (DSL) largely abstracted from implementation concerns [20]. One early DSL, Arden Syntax, organizes domain knowledge into Medical Logic Modules (MLMs) that incorporate the notion of an activation event used to direct module execution similar to the concept of a trigger commonly found in modern production rule implementations [21]. Unfortunately, the original Arden Syntax had a somewhat simplistic object model where the implementation specifics for the data used by a rule evaluation were under-specified and deferred to the implementer [22], as further discussed in Chapter 9: often referred to as "the curly braces problem." Another HL7 standard, GELLO, sought to improve data bindings by providing a standard query and expression language based on the Object Management Group (OMG) Object Constraint Language (OCL) and was designed to leverage standard information models such as the HL7 Reference Information Model (RIM). While capable of encapsulating decision logic, both languages relied on the application framework to ensure that triggering events correctly invoke the appropriate rule. As the number of rules or facts to be evaluated increases, such approaches rapidly impose performance constraints. Ultimately, both Arden Syntax and GELLO experienced limited adoption within the health care vendor community.

The need to deploy efficient rule execution capabilities that are decoupled from application logic, that can process enormous quantities of clinical data and execute specific algorithms from a repository of potentially thousands of knowledge modules is not unique to health care. Several efficient approaches have been developed to meet similar requirements for the larger business community. The most popular, Rete [23], is an efficient pattern-matching algorithm used in several production rule systems promising orders of magnitude better performance and scalability than brute-force approaches that must re-evaluate all existing rules when new facts are introduced.

The tradeoff with these optimized algorithms is that all employ proprietary languages tightly coupled to their specific run-time implementations with little to no interoperability between systems. Recognizing the need for a standard rule syntax to facilitate knowledge encoding and exchange, the knowledge management community is sponsoring several competing attempts to develop a true lingua franca. Rule Interchange Format (RIF)[d] is a W3C recommendation designed to complement its stable of semantic web standards. Another major initiative is RuleML (Rule Markup Language),[e] a rule modeling and metalogic language also designed to represent domain knowledge abstractly and facilitate translation to run-time syntaxes. To date, no standards-oriented approach has gained consensus adoption. See also Chapter 9, which discusses these activities.

Since the second edition, the CQL[f] specification has been created as a DSL specifically aiming to bring interoperability to decision support logic and clinical quality measures. Being easy for developers to understand, accommodating of clinical informatics concepts such as "value sets," directly based on FHIR R4, and supporting multiple paradigms of run-time execution, it has gained significant recognition as well as attention of EHR vendors. As of 2022, portions of CQL are recognized as an ANSI normative standard. See Chapter 9 for details on CQL and rule-based CDS approaches.

29.5.2 Data model

As FHIR has gained in adoption, it has become the most important data model for CDS as well as health IT generally. In particular, the US Core FHIR Implementation Guide, which implements the requirements of the US Core Data for Interoperability, fulfills US federal regulations and is widely adopted. In addition, there are ongoing efforts to develop a stable logical model for CDS aligned with FHIR, so that CDS knowledge resources can be written against a more stable and CDS-optimized data model rather than to different versions of FHIR as the standard evolves. Other data models that have been used for CDS historically include the HL7 Reference Information Model, [g] HL7 Clinical Document Architecture (CDA), [h] and HL7 Virtual Medical Record (vMR) [24,25]. Details on relevant data models for CDS are provided in Chapter 11.

29.5.3 Semantics and terminology services

Implementation strategies for vocabulary and terminology services, used to unambiguously define domain concepts and facilitate interoperability, can also significantly impact CDS performance and scalability. Health care enjoys a wealth of vocabularies, developed initially within clinical communities with domain-specific requirements; SNOMED, LOINC, CPT, RxNorm, and ICD-10, to name just a few, already define hundreds of thousands of concepts. Nevertheless, current terminologies are highly variable in both conceptual granularity and their comprehensiveness. This discordance creates several challenges for CDS, since it is

[d] http://www.w3.org/2005/rules/wiki/RIF_Working_Group.

[e] http://ruleml.org.

[f] https://cql.hl7.org.

[g] http://www.hl7.org/implement/standards/product_brief.cfm?product_id=77.

[h] http://www.hl7.org/implement/standards/product_brief.cfm?product_id=7.

not uncommon to store data in one system using a vocabulary (e.g., diagnoses using ICD-10) for which there is no precise, corresponding concept in the vocabulary (e.g., SNOMED) used by another system encoding similar data. When aggregated for the purpose of CDS analysis, data and terminology differences must optimally be resolved (translated) before a single rule can reason over the collective. Similarly, when rules are used to analyze data of varying conceptual granularity, the system must determine the conceptual equivalence of the data to ensure all relevant evaluations are performed.

In today's typical health care organization, terminology services are most often employed in, and even synonymous with, vocabulary mapping or translation. Mapping is the process of associating concepts or terms from one coding system to those in another and defining their degree of equivalence in accordance with a documented rationale and for a given purpose. There are several mapping subtypes:

- One-to-one: One concept is mapped between both the source and target system. For example, a concept in SNOMED is *precisely the same* as its mapped concept in ICD 10.
- One-to-many: One concept in the source system is mapped to multiple concepts in the target system that may or may not be exactly equivalent—in other words, there may only be a "best" match. Such one-to-many mapping concerns cannot easily be adjudicated without potentially affecting the conceptual analysis of the data [26].
- Many-to-one. Multiple concepts in the source vocabulary are mapped to one concept in the target. An example is the pre-coordinated and post-coordinated versions (see below) of many clinical concepts in SNOMED that might need to be mapped, for example, to each other or to ICD-10.
- Many-to-many. Multiple concepts are mapped to multiple concepts in the source and target system.

The traditional ("static") mapping technique of linking different source and target items is, by and large, the approach utilized by the 3M HDD [27] and by the National Library of Medicine Unified Medical Language System (UMLS) [28]. While functional, static mapping struggles with the volume of clinical terms introduced each year. An alternative approach relies on the gradual evolution of current medical terminologies into ontologies that can be maintained and resolved dynamically in different environments using semantic reasoning. Regardless of mechanism, terminology services play a crucial role in any scalable, real-world CDS architecture. See Chapter 11 for deeper discussion.

As with many clinical informatics concepts, FHIR-specific terminology services aim to satisfy the most common types of concept queries needed by application developers at run-time: notably term lookups, finding mappings to other code systems, expanding value sets, and testing if a term is "subsumed" by another concept. Such operations are critical to DSLs such as CQL that require terminology services support at run-time. A comprehensive description of terminologies in CDS is provided in Chapter 11.

29.5.4 Localization

A significant impediment to the acceptance and adoption of CDS relates to the sensitivity and specificity of the knowledge representation itself. Evidence-based guidelines, by their nature, summarize best clinical practice for general populations. There may not be an optimal guideline for an individual patient with unusual characteristics or additional comorbidities. Even if CDS artifacts could capture the patient-specific nuances that distinguish exemplary from adequate care, they are not guaranteed to be appropriate for a given run-time context. Computerized practice guidelines, being consensus-driven, make idealized assumptions regarding physician capabilities, patient preferences, normal reference ranges, clinical acuity, diagnostic services, and treatment capabilities. In practice, there are countless exceptions to these assumptions, and it is impractical to accommodate all combinatorial possibilities explicitly.

Several attempts at creating clinical guideline engines that can adapt at run-time to accommodate individualized circumstances have been reported. Some of these projects interpret process flow as plans in the AI sense where a sequence of actions aims to achieve a defined goal. In [29], case-based reasoning techniques are used to define a default plan suitable for a class of problems. This plan may then be refined using proper planning techniques [30] to match the identified context.

In the approach proposed by Bottrighi et al. [31] the guideline plan is not modified dynamically to accommodate exceptional situations. Instead, the CDS system performs compliance checks to determine if the actions that are directed by the user are compatible with its original specifications. The criteria for determining such compatibility are explicitly defined as adjunct rule sets that evaluate whether alternative actions might be justified. Notice that the criteria to detect the anomalies and the logic to deal with them are not part of the guideline but are defined in modules that can be composed as needed. Whenever the execution flow deviates from planned or expected behavior, a violation is logged and then evaluated. If a violation can be justified, the user's actions are permitted, and the system compensates for the perceived plan violation. If the action cannot be justified, the conformance violation is reported, and the system may optionally cancel the guideline's execution.

29.5.5 State management

An important implementation decision affecting scalability is whether the CDS architecture provides stateless or stateful services. A stateless system will typically present one or more interfaces for requesting a specific rule evaluation, or alternatively, for presenting data for evaluation by one or more knowledge modules.

A stateless CDS service maintains no history or knowledge of prior rule evaluations, treating each request as an independent transaction. Stateless designs typically require fewer resources to deploy than stateful designs, can be used simultaneously by multiple clients, are often easily scaled, and if they fail, can quickly be restarted. As a consequence, however, the data required for a rule evaluation must either be delivered with each request or retrieved from an appropriate source by the knowledge service after invocation. If a series of rule evaluations require similar inputs, the system may have to query repeatedly for the same data, resulting in significant transactional loads unless appropriate compensating mechanisms are implemented, such as caching or rule chaining. To address this issue, some stateless services group and chain evaluations so that the same information only needs to be queried and evaluated once. Unless such performance optimization mechanisms are put into place, stateless CDS designs may face performance constraints as the complexity of the logical or statistical operations or the volume of data to be analyzed increases.

Many applications, particularly clinical guideline systems that attempt to combine rule evaluation with process orchestration, incorporate stateful components in their architecture. A stateful service maintains, typically in memory, an internal data structure, the state of which is potentially modified by each invocation. Subsequent calls, even with identical parameters, can yield different results because of this accumulated invocation history.

Whether a stateless or stateful design is most appropriate for a given application depends ultimately on understanding their respective trade-offs. The architectures described briefly above each have distinct advantages and disadvantages in terms of complexity, performance, and resource requirements. Choosing an optimal design will require careful analysis and prioritization. Nevertheless, the diversity of implemented design patterns and the breath of requirements being addressed reflect the community's maturing understanding of how best to implement scalable, sustainable clinical decision support capabilities.

29.5.6 Agent-based approaches

The Architecture for Real-Time Application of Knowledge Artifacts [32] blends the capabilities of conventional rule-based systems with the paradigms of event-driven architecture and agent-based support services. Unlike conventional CDS architectural paradigms, ARTAKA attempts to reorient the preconception of CDS away from the comfort of discrete functional components operating on a deterministic, stateful process. Instead, ARTAKA defines CDS as a contextual *stream of possible future and preferred alternate past states* driven by continuous event processing (CEP) of data concerning a particular subject; generally a patient or user in a non-predetermined situation. By largely removing the concept of explicit request/response invocation and introducing the concept of alternative states for purposes of user experience orchestration and other purposes, the ARTAKA paradigm more strongly resembles agent-based simulation architecture used in control systems design than those historically used in clinical informatics.

29.5.7 Human tasks

Delivering high quality health care is inherently a complex activity requiring human and non-human resource coordination, task management, and state monitoring. Traditional SOA designs, supported by production rule engines, workflow servers, and business activity monitoring tools, are known to orchestrate non-human services and interactions reliably. Specifications such as WS-Human Task[i] and BPEL4People[j] are attempts to standardize the human-interactions with software processes. While BPEL, WSDL and other non-REST-oriented specifications have largely faded in enthusiasm in favor of ED-SOAs and microservice-based architectures friendlier to mobile and web devices, the implementation

[i] http://docs.oasis-open.org/ns/bpel4people/ws-humantask/200803.

[j] https://www.oasis-open.org/committees/bpel4people/charter.php.

of these standards into commercial SOA products from major industry vendors is testimony to their value, extensibility and interoperability.

The OASIS WS-BPEL Extension For People Technical Committee proposed to layer human interaction capabilities on top of the existing Business Process Execution Language (BPEL) standard. It specified a BPEL4People extension defining "People Activity" as a sub-process that can be invoked from within BPEL using standardized interfaces and provides other extensions to facilitate resource assignments, task properties, and role-based process behavior. The related OASIS WS-Human Task specification defines human tasks, including their properties, a set of allowable operations, and a coordination protocol to organize their behavior. It defines a standalone state engine for managing task lifecycle and client interfaces for applications operating on the task. This separation of *process* and *task* engines ensures they can be deployed, managed, and scaled independently.

Human Task is of particular interest to health care because it provides the vocabulary and conceptual framework necessary to describe task assignments, care transitions, and task process state. The reader is referred to the full specification for a comprehensive discussion of the Human Task state engine diagram illustrated in Fig. 29.2, but three specific concepts are illustrative of the important services a Human Task server might provide for SOA-enabled CDS.

- *Assignment:* A human task must have people assigned to it. Human Task defines several roles, three of which, *Task Stakeholder, Actual Owner*, and *Potential Owner*, will be discussed here briefly. A stakeholder is the person ultimately responsible for a specific task and its

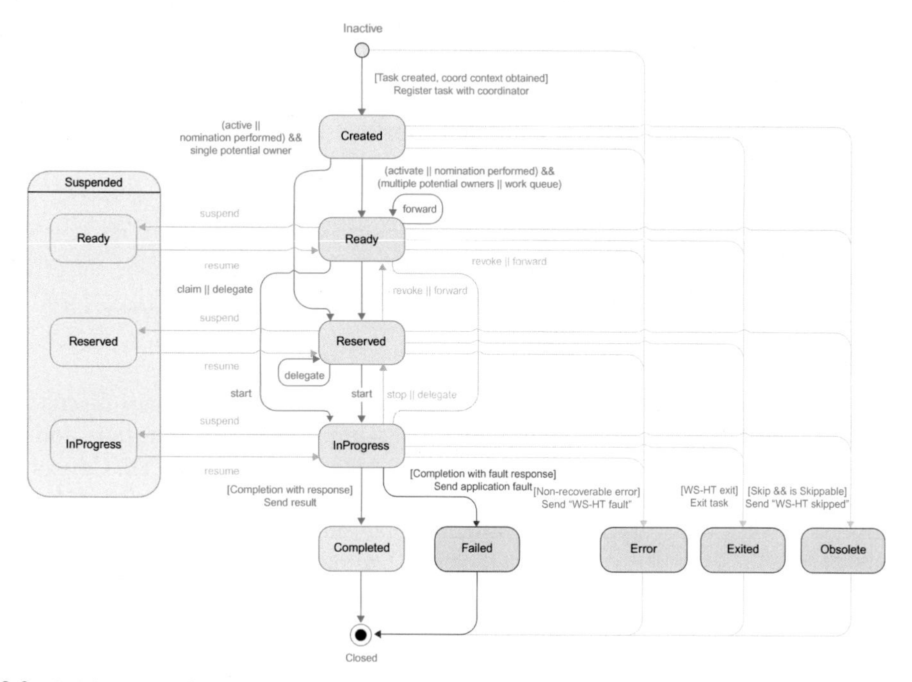

FIG. 29.2 WS-human task state engine.

outcome. This role MUST be assigned for each task—its occupant has the right to monitor, influence, modify, and administrate the progress of a task. In the setting of a teaching hospital, think *Attending Physician.* Stakeholders may or may not be the task's actual owner—the person ultimately assigned the responsibility of performing the task. Using the teaching hospital analogy again, think *Resident*. Task assignments might be done explicitly to a specific person, or to a group of Potential Owners that can lay claim to a task based on availability and assume the role of owner.

- *Delegation*: Depending on how a specific task is configured, its owner may be allowed to delegate a task to another user, making the new user the actual owner of the task. There are two basic delegation types. Revocable delegations allow the delegating owner to maintain control by keeping the task in his or her work list, effectively giving the person the right to observe task progress or to revoke the delegation. Irrevocable delegations remove the task from the delegating owner's work list preventing the person from continued task oversight.
- *Notifications:* Notifications are simple human interactions used to communicate a noteworthy event, for example a new medication order or a task assignment. They are also used during escalation events to notify a user when a task is overdue. A notification recipient can be explicitly provided or be assigned following a query to organizational model.

The engine in Fig. 29.2 illustrates the states (created, ready, reserved, in progress, completed, suspended), which, when combined with other WS-Human Task semantics, provide the basic tools to define and trace how clinical task responsibility/accountability are created, assigned, and transferred. All these are critical dimensions in patient care coordination activities, and when not executed or monitored appropriately, are a significant source of adverse events, sub-optimal outcomes, and unnecessary expense. The authors anticipate that that as the scope of standardization for CDS evolves to include full clinical workflow orchestration and human interaction, robust implementation options will be increasingly important in deploying scalable and interoperable platforms.

29.6 Other issues to be considered

29.6.1 Legal issues

Since Chapter 23 is dedicated to legal and regulatory issues, we will not discuss them at length here. However, we note that legal issues are an important consideration, and one that needs to be directly addressed. For example, in the AHRQ-sponsored CDS Consortium research project completed in 2013, issues of data sharing, intellectual property, accountability, and liability required a dedicated and significant legal effort before CDS sharing could even begin in earnest [33].

Similarly, the diverse nature terminology licensing is particularly problematic. For example, not all countries use SNOMED. The ones that do may have different licensing requirements. In the United States, access to SNOMED and many other accepted terminologies is wrapped in the Unified Medical Language System (UMLS) license that does not allow for organization-wide acceptance and requires annual reporting at the individual level. As a second example, the LOINC license is very permissive but does not permit certain competitive use. Some other popular terminologies are proprietary, others are abandoned IP. These idiosyncrasies can make the redistribution of works dependent on them legally murky. The irony of widely accepted terminologies not following widely accepted licensing models is most often ignored.

It should be noted, however, that legal issues can represent more than simply a barrier to CDS knowledge sharing, because they can also be a potentially critical enabler. There have been discussions, for example, of whether compliance with a federally approved clinical practice guideline could be used as an officially sanctioned defense in a medical malpractice lawsuit. If such a policy were adopted as federal law, then it could provide an enormous impetus for widespread adoption of CDS that supports compliance with clinical practice guidelines.

29.6.2 Reusability and standards

As with legal issues, standards have been discussed at length in this book, particularly Chapter 23. Therefore, we will not dwell in detail on the issue of standards here. However, we will note that with standards, the cost of knowledge integration can be diminished—*if* there are enough deployments required in new environments, and *if* the standard approach is well designed, easy to understand, and easy to use.

Consider, for example, Fig. 29.3, which provides a conceptual framework for the impact of standards on the effort required to deploy knowledge in a new environment. The default is the use of a non-standard, existing approach—say, the use of a SOA approach to CDS using custom APIs. The effort level in this case rises linearly with the number of deployments in new environments. The two other lines represent the effort required to use a standard approach—say, the use of a SOA approach to CDS using standard APIs. In both cases, the initial effort required will be higher, because there is a need for the knowledge producers and the CIS vendors to conform to a new standard rather than to use their existing APIs. However, the use of standards will reduce the amount of incremental effort required for each

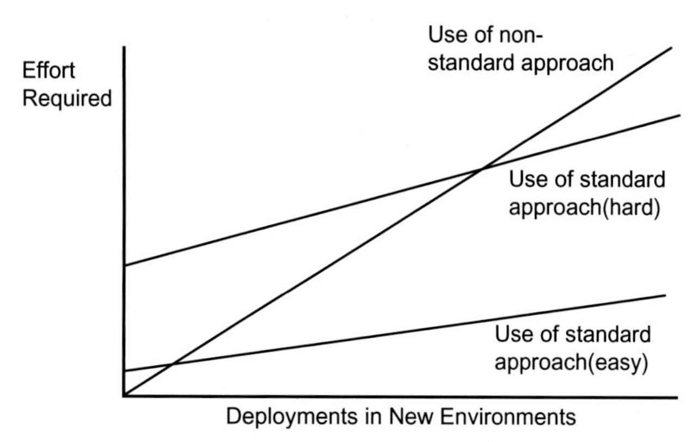

FIG. 29.3 Impact of standards on knowledge integration effort.

new deployment once support for the standards is in place. As shown in the figure, the two variables that impact when it makes sense for stakeholders to support the standard are (i) how many deployments will be made, and (ii) how easy it is to understand and comply with those standards. The need from a standards perspective is that they be as easy as possible to understand and implement, so that the break-even point for standards compliance can be reached as quickly as possible.

Despite the resurgent interest in CDS and the enormous capital investments currently underway, CDS as a fundamental infrastructure capability continues to struggle with typical implementation concerns. As with any early-stage deployment of knowledge management infrastructure, numerous inter-dependent business processes remain poorly understood and incompletely implemented. Clinical processes, architectures, domain standards, and behaviors remain hotly contested as participants initially focus on requirements they believe to be singularly important to preserve. Such turmoil hampers the implementation of CDS that benefits significantly from standardized, predictable processes and business requirements.

29.7 Conclusions

As with the general IT industry, the software architectures being used in health IT are continuing to evolve towards more open, component-based, service-oriented architectures scaled with event-driven integrations. In particular, HL7 SMART-on-FHIR, and increasingly HL7 CDS Hooks, is gaining adoption among EHR vendors. This shift provides new opportunities for the optimal integration of knowledge resources into applications. At the same time, fundamental issues remain, such as the medicolegal implications of the approach used and the appropriate role of standards and licensing.

Other factors to consider are forces benefiting from patient data being moved outside the governance of a health system, such as cloud repository vendors, SaaS applications, payer or regional aggregators, mobile device applications, federated PHI alliances, and consumer technology giants. The movement to a patient-centered health record bank is advocated by some [34] but has not yet taken hold, perhaps for lack of a sustainable business model and the shift of control that this portends, which is still at odds with interests of health care organizations and EHR-based technologies. At present, also, there is no effective, *bi-directional* integration of EHRs with personal health records (PHRs) except when tethered to health care organization or payer-based patient portals. However, the need for integrated longitudinal patient records is building, stimulated in large part by the movement towards value-based care and their need to shift focus to wellness, disease prevention, and improved care coordination across venues of care.

As this shift inevitably occurs, both the capabilities of aggregate data and longitudinal records fed by multiple input sources (EHRs, PHRs, wearable devices etc.) will evolve. Looking forward, this may usher in a "golden age" for CDS. Moreover, longitudinal records aggregated across health IT systems may allow more effective CDS, and "big data" may enable new CDS capabilities when coupled with machine learning, as discussed in Chapter 7.

Given the rise of a significant "app" culture in all industries, and so-called "mHealth" initiatives in health care, we should expect more CDS to be delivered by specialized apps. Either used directly by providers and patients, e.g., on mobile devices, but also on desktops that access data from both the EHR and other source to provide information, education, advice, visualization, and other capabilities in innovative ways. These will increasingly leverage the APIs and SOA capabilities provided by EHRs and other health IT resources described above. As discrete apps become increasingly sophisticated and are used for more health and care purposes, we can expect further change in the nature of health IT architectures, moving towards increasing reliance on interoperable information, broad libraries of heterogeneous services, and cloud-based access to comprehensive data with sophisticated access controls.

The trends we have identified in this chapter are taking hold in many industries. Their adoption in health care is slower, for many reasons including legacy investment, which has in fact been paradoxically stimulated in the US by government policy emphasizing rapid adoption of EHRs. It is unclear whether and how rapidly new health care finance models and policies will bring about change, but the trends we describe will inevitably occur eventually. The purpose of this chapter is thus to look forward, while at the same time recognizing today's realities, and to seek to stimulate engagement of readers in initiatives that will help bring about this new future.

The future contribution of CDS to quality of care and patient outcomes remains uncertain and its near-term impact should be viewed with healthy skepticism. The practical effectiveness of existing CDS has been evaluated several times in recent years. While many researchers agree that CDS can result in significant changes in practitioner performance, the final effectiveness varies according to the specific application, with some activities (e.g., drug prescription) being more suitable candidates for CDS than others. The impact on patient outcomes is considerably less clear with fewer studies identifying positive benefits [35], and with the multiplicity of other downstream factors that affect outcome, making studies of outcome impact difficult. Moreover, many CDS studies utilize sample sizes that are too small to reliably detect improvements in outcomes, because of which, improved outcomes might not be able to be detected even if such improvements were present [36]. Such cautionary findings underscore the need for a considered, iterative approach to CDS implementation. Health care is a community poised for making large capital investments in information technology; it should consider carefully whether it truly has the conceptual refinement, the organizational maturity, and the necessary standards to implement large-scale CDS deployments without considerable risk. Given the social-political-economic environment, health care has an unprecedented opportunity to redefine its technological foundation; it should remain cognizant that considerable organizational introspection and process reengineering are needed to implement any technology successfully.

Nevertheless, given the successful application of inference and workflow technology in other communities, there should be considerable optimism that this most complex of human endeavors can indeed be decomposed into more manageable and predictable system component and interfaces. It is also true that early and seemingly irreconcilable differences often prove less important once the reengineering effort matures. If the experiences, standards and technology capabilities that found success in business markets at large can be harmonized with health care concerns, the full integration of CDS into the very fabric of our health care delivery system may be closer than one might otherwise expect.

References

[1] Shortliffe EH, Davis R, Axline SG, Buchanan BG, Green CC, Cohen SN. Computer-based consultations in clinical therapeutics: explanation and rule acquisition capabilities of the MYCIN system. Comput Biomed Res 1975;8 (4):303–20.

[2] van Melle W, Shortliffe EH, Buchanan BG. EMYCIN: a knowledge engineer's tool for constructing rule-based expert systems. In: Buchanan BG, Shortliffe EH, editors. Rule-based expert systems: the MYCIN experiments of the Stanford heuristic programming project. Reading, MA: Addison Wesley; 1984. p. 302–13.

[3] Miller PL. Critiquing anesthetic management: the "ATTENDING" computer system. Anesthesiology 1983; 58(4):362.

[4] Barnett GO, Cimino JJ, Hupp JA, Hoffer EP. An evolving diagnostic decision-support system. JAMA 1987;258:67–74.

[5] McDonald CJ, Murray R, Jeris D, Bhargava B, Seeger J, Blevins L. A computer-based record and clinical monitoring system for ambulatory care. Am J Public Health 1977;67(3):240.

[6] Kuperman GL, Maack BB, Bauer K, Gardner RM. The impact of the HELP computer system on the LDS Hospital paper medical record. Top Health Rec Manage 1991;12:1–9.

[7] Bates DW, Teich JM, Lee J, Seger D, Kuperman GJ, Ma'Luf N, Boyle D, Leape L. The impact of computerized physician order entry on medication error prevention. J Am Med Inform Assoc 1999;6(4):313–21.

[8] Tu SW, Campbell JR, Glasgow J, Nyman MA, McClure R, McClay J, Parker C, Hirabak K, Welds T, Mansfield J, Musen M, Abarbanel RM. The SAGE guideline model: achievements and overview. J Am Med Inform Assoc 2007;14(5):589–98.

[9] Musen MA, Tu SW, Das AK, Shahar Y. EON: A component-based approach to automation of protocol-directed therapy. J Am Med Inform Assoc 1996;3(6):367–88.

[10] Kawamoto K, Lobach DF. Design, implementation, use, and preliminary evaluation of SEBASTIAN, a standards-based web service for clinical decision support. AMIA Annu Symp Proc 2005;2005:380–4.

[11] Fowler M. Patterns of enterprise application architecture. Chicago, IL: Addison-Wesley Longman Publishing Co., Inc; 2002.

[12] Wooldridge M, Jennings NR. Intelligent agents: theory and practice. Knowl Eng Rev 1995;10(2):115–52.

[13] Strasberg HR, Rhodes B, Del Fiol G, Jenders RA, Haug PJ, Kawamoto K. Contemporary clinical decision support standards using health Level seven international fast healthcare interoperability resources. J Am Med Inform Assoc 2021;28(8):1796–806.

[14] Burde H. Health law the hitech act-an overview. Virtual Mentor 2011;13(3):172–5.

[15] Pipersburgh J. The push to increase the use of EHR technology by hospitals and physicians in the United States through the HITECH act and the Medicare incentive program. J Health Care Finance 2011;38(2):54–78.

[16] Murphy J. The journey to meaningful use of electronic health records. Nurs Econ 2010;28(4):283.

[17] Parker CG, Rocha RA, Campbell JR, Tu SW, Huff SM. Detailed clinical models for sharable, executable guidelines. Medinfo 2004;11(Pt 1):145–8.

[18] Sherman EH, Hripcsak G, Starren J, Jenders RA, Clayton P. Using intermediate states to improve the ability of the Arden syntax to implement care plans and reuse knowledge. In: Proceedings of the annual symposium on computer application in medical care (p. 238). American Medical Informatics Association; 1995.

[19] Quinlan JR. Generating production rules from decision trees. In: IJCAI (vol. 87, pp. 304-307); 1987, August.

[20] Greenes RA. Why clinical decision support is hard to do. In: Annual Symposium Proceedings. AMIA; 2006. p. 1169–70.

[21] Pryor TA, Hripcsak G. The Arden syntax for medical logic modules. Int J Clin Monit Comput 1993;10(4):215–24.

[22] Jenders RA, Sujansky W, Broverman CA, Chadwick M. Towards improved knowledge sharing: assessment of the HL7 reference information model to support medical logic module queries. In: Proceedings of the AMIA annual fall symposium (p. 308). American Medical Informatics Association; 1997.

[23] Forgy CL. Rete: a fast algorithm for the many pattern/many object pattern match problem. Artif Intell 1982;19 (1):17–37.

[24] Kawamoto K, Del Fiol G, Strasberg HR, Hulse N, Curtis C, Cimino JJ, Rocha BH, Maviglia S, Fry E, Scherpbier HJ, Huser V, Redington PK, Vawdrey DK, Dufour J, Price M, Weber JH, White T, Hughes KS, McClay JC, Wood C, Eckert K, Bolte S, Shields D, Tattam PR, Scott P, Liu Z, McIntyre AK. Multi-national, multi-institutional analysis of clinical decision support data needs to inform development of the HL7 virtual medical record standard. In: AMIA annual symposium proceedings (vol. 2010, p. 377). American Medical Informatics Association; 2010.

[25] Kawamoto K, Del Fiol G, Orton C, Lobach DF. System-agnostic clinical decision support services: benefits and challenges for scalable decision support. Open Med Inform J 2010;4:245–54.
[26] Doerr M. Semantic problems of thesaurus mapping. J Digit Inf 2001;1(8):1–27.
[27] Che C, Monson K, Poon KB, Shakib SC, Lau LM. Managing vocabulary mapping services. Am Med Inform Assoc 2005;2005:916.
[28] McCray AT, Nelson SJ. The representation of meaning in the UMLS. Methods Inf Med 1995;34(1–2):193–201.
[29] Montani S. Case-based reasoning for managing noncompliance with clinical guidelines. Comput Intell 2009;25 (3):196–213.
[30] Anselma L, Montani S. Planning: supporting and optimizing clinical guidelines execution. In: Computer-based medical guidelines and protocols: a primer and current trends; 2008. p. 101–20.
[31] Bottrighi A, Chesani F, Mello P, Molino G, Montali M, Montani S, Storari S, Terenziani P, Torchio M. A hybrid approach to clinical guideline and to basic medical knowledge conformance. In: Artificial intelligence in medicine. Berlin, Heidelberg: Springer; 2009. p. 91–5.
[32] Lee P. Automated injection of curated knowledge into real-time clinical systems CDS architecture for the 21st century; 2018.
[33] Hongsermeier T, Maviglia S, Tsurikova L, Bogaty D, Rocha RA, Goldberg H, Meltzer S, Middleton B. A legal framework to enable sharing of clinical decision support knowledge and services across institutional boundaries. In: AMIA annual symposium proceedings (vol. 2011, p. 925). American Medical Informatics Association; 2011.
[34] Yasnoff WA, Sweeney L, Shortliffe EH. Putting health IT on the path to success. JAMA 2013;309(10):989–90.
[35] Jaspers MW, Smeulers M, Vermeulen H, Peute LW. Effects of clinical decision-support systems on practitioner performance and patient outcomes: a synthesis of high-quality systematic review findings. J Am Med Inform Assoc 2011;18(3):327–34.
[36] Kawamoto K, McDonald CJ. Designing, conducting, and reporting clinical decision support studies: recommendations and call to action. Ann Intern Med 2020;172(11 Suppl):S101–s9.

CHAPTER

30

Getting to knowledge-enhanced health and healthcare

Robert A. Greenes[a] *and Guilherme Del Fiol*[b]

[a]Biomedical Informatics, Arizona State University, Phoenix, AZ, United States [b]Department of Biomedical Informatics, University of Utah Health, University of Utah, Salt Lake City, UT, United States

30.1 Where we are now

The subtitle of the Second Edition of this book, published in 2014, was "The Road to Broad Adoption." After decades of work, the field had reached a stage in which progress toward adoption of clinical decision support (CDS) was beginning to move from very slow and haphazard, over several decades, as described in the First Edition, in 2007, to a stage where underpinnings were beginning to be put in place that would enable more widespread adoption and use. As we noted in Chapter 1 and elaborated on in Chapter 3, these included progress in standards and interoperability and new architectural patterns and methods for integrating computer-based CDS with host systems, organizational efforts to build implementation skills and teams, tools for knowledge management, and evaluation studies that demonstrated value of CDS.

It is quite amazing that during that same short period of time, a number of other factors that had just begun to take shape were also growing into major forces stimulating interest in and need for CDS. These included the *rise of new or expanded capabilities* such as precision medicine, patient/user engagement, aggregation of data for population health, onset of an "app culture," and interoperability and standards development and adoption.

In addition to new capabilities, we saw the rise of *new demands, stimuli, and incentives* for CDS brought about by the factors such as broad EHR adoption, national programs such as the US Meaningful Use requirements, value-driven healthcare delivery and financing models, and quality monitoring and reporting.

All these rising capabilities and demands have laid a foundation for continued progress and expansion of efforts since that time. Some of these have continued to expand, and other new emphases have evolved over the period, especially remarkable shifts during the COVID-19

Clinical Decision Support and Beyond
https://doi.org/10.1016/B978-0-323-91200-6.00005-X

TABLE 30.1 Current drivers for adoption of CDS and knowledge-enhanced health and healthcare.

Policy and societal drivers
• Focus on wellness, fitness, and prevention, and more proactive management of disease • Health finance and health system transformation • Top-down national initiatives, incentives, and regulations • The Learning Health System movements • The Quintuple Aim
Science and technology enablers
• Precision medicine • Apps and services platforms and architectures • Mobile health • Telehealth • At-home diagnostic testing, biosensors, and monitoring • Population health data and analytics • Cognitive support and visualization and workflow enhancement

pandemic, toward innovations and expansions of capabilities such as telehealth, home monitoring, and home testing. The final row of Table 1.3, expanded here as Table 30.1, identifies key current drivers.

Our introduction, in this Third Edition, of the "*and Beyond*" aspects of knowledge-enhanced health and healthcare—including not only CDS but other ways to incorporate knowledge and guidance into activities of both patients/users and providers—is based on these trends. We have discussed this in various chapters.

The expanded foci have resulted in 10 completely new chapters in this edition, and substantial revision of most others. This is testimony to the fact that we are now entering a period of, not just broader adoption, but qualitatively broader kinds of use of knowledge to include new methods and approaches to aiding the process of care, new ways of integrating those methods into the care process, including wellness and disease prevention settings, expanded data sets on which to base them, and new and more comprehensive realms in which to apply them.

As we did in the previous edition, we use this chapter to venture into uncharted territory, by discussing how the various forces may evolve, and some of the requirements that will need to be met to capitalize on these enabling trends and capabilities. We hope that this will stimulate discussion and action and that the suggestions contained herein will be helpful to readers engaged in this field to and expand on the ideas presented here and further push the envelope.

30.2 Impediments still with us

Over the past six decades, the pursuit of CDS has mostly been stimulated by three main kinds of interests: (a) the intellectual challenge of understanding and improving on the cognitive processes and information base of the human; (b) the moral and ethical imperative to address important issues in patient safety, health care quality, and access to health care; and

(c) business and policy reasons relating to allocation of limited resources and control of costs of an increasingly expensive health care system.

Until the most recent two decades, innovation and experimentation in the use of CDS has been largely carried out by academic medical center-based researchers and developers addressing local needs and with goals of error prevention and quality improvement. They tended to be largely ad hoc, and rarely led to large-scale deployment and dissemination outside of the entities for which they were developed, although the approaches used in some cases did spawn some of the commercial systems existing today.

Business reasons for implementation of CDS have also been somewhat opportunistic and locally driven, although frequently tied to changes in health care financing and reimbursement models, efforts to shift care from hospitals to office or home, introduction of managed care, and approaches to curbing overutilization by requiring preapproval/prior authorizations for high-cost procedures, referrals, or medications. CDS was introduced in those situations as a means of coping with and addressing government or payer regulations and restrictions, as a defensive measure by health care organizations and providers to ward off such intrusions, and as a means of achieving efficiencies. As a result, business-oriented uses of CDS tended also to be implemented in institution-specific fashion.

Because such responses by institutions have largely been either local and opportunistic, academically driven, or reactive and defensive, not as a result of top-down policy and a coordinated set of standards, it is not surprising that CDS—in the most prevalent forms of logic rules, order sets, and documentation templates—has been implemented in a manner that is highly dependent on local needs, constraints, and preferences. As a consequence of the individualized nature of the implementations, with setting-specific adaptations and customizations, and the proprietary incompatible platforms in which they have been done, there has been considerable difficulty and little perceived benefit to sharing of CDS knowledge and experience.

30.3 Need for new mechanisms

As we venture into speculation on the future, we recognize the continued presence of obstacles and challenges. Motivations have become more coordinated and integrated into policy over the past decade, stimulated by factors such as we have listed in Table 30.1, such as national initiatives for EHR adoption in developed nations [1,2], and as an example of a further specific driver in the US, the requirements for Meaningful Use of such systems [3]. However, the opportunistic and diverse modes of implementation, and the impediments of them, are still with us. If we are to greatly expand our capabilities, as we need to, just incrementally trying to do more the way we have been will not be the way to do it. It simply won't scale as the complexity and extent of CDS demand increase.

Technical advances that we have reviewed in previous chapters are beginning to make the process easier, including computer technologies and systems architectures, in Chapter 29; development of some of the important standards needed for data and knowledge representation and communication, in several chapters of Section III; and increased understanding of organizational strategies to encourage CDS use, as discussed from several perspectives in

Section IV. Yet, adoption of EHRs has largely been through legacy systems, some of which are 20–30 years old, with proprietary environments, data models, and CDS capabilities. In the US, although well-intentioned, the HITECH Act of 2009 and the incentives for rapid increase in EHR adoption throughout the nation, in fact greatly expanded the uptake of legacy systems, as some of the older dominant EHR vendors grabbed significant shares of the market. Thus, despite progress in EHR adoption, we are still saddled perversely with a high degree of dependence on proprietary systems and incompatibilities.

The burden of knowledge management and CDS governance, well beyond the scope of all but the largest institutions, and even then exceedingly complex and costly, is particularly troublesome, as discussed in Chapters 20 and 28. This all but demands multi-stakeholder participation in a more robust, scalable approach that can share and coordinate the tool developments needed, and establish an ongoing process of making high-quality computable knowledge resources broadly available.

Eric Topol's 2013 book "The Creative Destruction of Medicine" [4] was apt in its characterization of how significant the forces are that are leading to major transformation of our health care system, which continue to gather momentum, although often in fits and starts. Many of the forces he identifies overlap with the factors we list in Table 30.1. Topol's focus is on describing the forces, and not on how the transformation will come about. In fact, there is very little written about that, but it is clear that at least part of what will be needed is a significant rethinking of the digital infrastructure required to support it to achieve truly patient-centered care, a focus on lifetime health and wellness, and coordination of care processes across venues of care.

We do not yet have the IT infrastructure and framework to support these goals. The changes will need to rely on much greater integration of data for a patient, the ability to aggregate and harness the power of population data, with more powerful analytics for population management, much more availability of point-of-care knowledge for CDS, much greater interoperability and ability to create workflows and processes across venues of care, and much better tracking of care processes and outcomes, and quality assessment.

Table 30.2 lists 13 desiderata for IT capabilities for the health care systems of the future that are beginning to shape.

TABLE 30.2 Digital infrastructure desiderata for future health care systems.

Feature	Description	Importance
1. Ontology of health/health care processes	A framework for describing the entire spectrum of care processes, actions, and settings	Context of health and health care activity can be used in conjunction with CDS to target appropriate knowledge and advice
2. Ontologies of clinical problems/diseases, findings, actions, status	Characterization of problems/diseases and their attributes, and diagnostic, treatment, and other actions (e.g., by clinical ontologies such as SNOMED-CT and ICD-11), as well as current trajectory for both problems and findings	These descriptors provide the ability to characterize a given patient's health/disease state at a point in time

TABLE 30.2 Digital infrastructure desiderata for future health care systems—cont'd

Feature	Description	Importance
3. Models of context-specific cognitive processes and activities	Models of thought processes, intentions, and actions by healthcare providers and patients	Association models can be used to discover patterns of clinical problems, observations, and actions performed, and possible relations among them, in terms of the sequence of events that occurred (e.g., medication change as a result of an elevated lab test result, or diagnosis made based on a set of findings or response to treatment). Cognitive models can also be used to anticipate or suggest potential decisions and activities for a given context
4. A universally adopted robust clinical information model	Data model with rich enough sets of attributes and constraints to maximize interoperability (e.g., as represented in FHIR profiles such as US Core)	Need for consistency and richness of access to clinical data for decision support, and for quality measurement
5. Longitudinal individual lifetime care record	A method for effectively integrating data on individual patients over their lifetime, and across venues of care, whether done explicitly by harvesting data from various primary sources into a central repository, or virtually on demand, through federated architectures	Continuity of care, lifetime wellness goals and CDS&B to support these require a care record that is more comprehensive than individual institutional records. Can avoid some of the redundancy of HIE and need to reconcile CCD documents, etc., if we have a single source record that is continually updated with audit trails of all transactions
6. Population data resources	Methods for obtaining data from different sources, including genomics, personal data and biosensors, harvesting from records by NLP, imaging features, etc. Methods of normalizing data, aggregating data across patient populations, data mining, and predictive modeling	More refined population management capabilities, predictive analytics, and immediate access to data on similar patients during care can provide important new types of CDS&B
7. Privacy and role-based access	Methods for assuring privacy of individual data and managing role-based access, needs to be bolstered by regulations and enforcement mechanisms	Necessary in order to obtain confidence of the public and necessary protections for the benefits of lifetime records and big data to be achieved
8. Methods of organizing available knowledge and CDS	Methods to encapsulate CDS, e.g., as SOA services, and tag them by descriptors such as in the ontologies of (1) and (2) above	Will provide a framework for knowledge management, update, and selection for CDS based on precise context and setting. Also will enable identification of gaps, discrepancies, and conflicting knowledge, and focus on priorities for addressing them

Continued

TABLE 30.2 Digital infrastructure desiderata for future health care systems—cont'd

Feature	Description	Importance
9. Reusable methods/apps	Methods/paradigms to create innovations over EHR platforms including advanced visualization, summarization, trend analysis, analytics, and other decision aids, in interoperable or portable form. SMART-on-FHIR and CDS Hooks are emerging as robust approaches for providing such functionality	Availability will stimulate an app industry that can significantly expand the innovation capacity, fostering more receptivity of health systems and EHR vendors to incorporate them. This will further enable the ability to meet the needs of lifetime care and coordination across venues, as will be needed by future health systems
10. Infrastructure for patient-facing digital health tools	Methods/infrastructure to integrate 3rd party digital health tools (e.g., sensors, self-care, shared-decision making) with EHRs and the clinical workflow	Allow healthcare systems and providers to use novel digital health tools as an approach to maximize patient outreach, self-care, home monitoring, telehealth, shared-decision making, etc.
11. Approaches to promote health equity	Methods/infrastructure that maximize reach to groups from historically marginalized minorities, low socioeconomic status, and rural/frontier areas	Technology advances often contribute to creating or widening a digital gap. Need for CDS approaches that are deliberately designed to advance health equity
12. Sharable repositories of best-available knowledge, in unambiguous form, amenable to use in CDS	Ideally, each item is systematically annotated in terms of how it was developed; source or EBM review, consensus, or peer review process to derive it; transparency in terms of performance, external validation (e.g., for machine learning models), clinical benefit, continuous updates, and measures to prevent bias; responsible party or sponsor; date of creation and last update; standards and conventions used; and situations (context and setting such as from capabilities (1) and (2) above) to which it applies	Rediscovery or individual compilation of such repositories is beyond the scope of even the largest enterprise. A communal process, possibly public-private, possibly commercial, is needed to achieve the scale and continual update required
13. Orchestration services	Coordination and management of the infrastructure capabilities (1)–(9) above. This could be in the form of third-party services to which individuals and entities can subscribe	The above capabilities need to be overseen by entities that can set priorities, identify and allocate resources, and manage the process on behalf of their clients

30.4 Orchestration

The infrastructure, tools, and resources identified above are both daunting to individual efforts and require concerted action that has not yet become organized to any significant degree, although they are beginning to be available in various forms and levels of maturity. Addressing the gap between current capabilities and those needed for future widespread knowledge-enhanced health and healthcare, in our opinion, requires, in particular, the last item (13) in the list in Table 30.2—a set of **orchestration capabilities**.

An early call to action on coordinated efforts was voiced in a June 2006 white paper produced on contract from the US Office of the National Coordinator for Health IT ONC) to the American Medical Informatics Association [5], which outlined a proposed *Roadmap for National Action on Clinical Decision Support for the U.S.* In that white paper, a number of steps were proposed to create an environment conducive to the general goal we address. In the intervening years, what we observe is that the needs have increased, recognition of them has grown, incentives such as Meaningful Use have been attempted, and standards and interoperability initiatives have been promoted. We have seen specific initiatives spring up, including activities in the HL7 CDS Work Group,[a] an initiative known as OpenClinical [6] to track and collate diverse activities going on related to CDS, collaborative projects such as the Morningside Initiative/SHARPC 2B project [7], the CDS Consortium [8], and OpenCDS [9]. We have seen the establishment under ONC sponsorship of the S&I Framework Initiatives [10] as public-private activities; and the work of the National Library of Medicine (NLM), which launched the Value Set Authority Center and the NLM Value Set Authority Center (VSAC) [11], which "provides downloadable access to all official versions of vocabulary value sets contained in the 2014 Clinical Quality Measures. The value sets provide lists of codes and display names from standard vocabularies used to define the clinical concepts (e.g. diabetes, clinical visit) used in the quality measures." The intent is for this site and other related sites to be repositories of computable resources, such as quality measure definitions and decision rules.

Also occurring have been national efforts in countries with single-payer systems (see Chapter 5), international standards organizations and collaborations, voluntary consortia, and public-private workgroups focusing on specific challenges of standards and interoperability. Noteworthy among the latter, as of this writing, are the CDS Connect[b] platform, sponsored by the US Agency for Healthcare Research and Quality (AHRQ)[c]; the SMART App Gallery[d]; and Logica,[e] a public-private partnership involving healthcare organizations and EHR vendors, which provides sandbox environments to develop and test the interoperability of CDS apps and services.

Despite the value of such initiatives, no overarching effort has been established to align the various activities and create a sustainable framework for moving the whole effort forward and

[a] https://confluence.hl7.org/display/CDS/WorkGroup+Home, last accessed 6/8/2022.

[b] https://cds.ahrq.gov/cdsconnect, last accessed 6/22/2022.

[c] https://www.ahrq.gov/, last accessed 6/22/2022.

[d] https://apps.smarthealthit.org/apps/featured, last accessed 6/22/2022.

[e] https://www.logicahealth.org/, last accessed 6/22/2022.

to create a fully standards-based infrastructure model or pattern for managing shared knowledge, updating it, and facilitating its incorporation and adaptation for use in local settings.

We expect to see continued maturation and expansion of such orchestration capabilities over the coming decade, most likely in the form of services to which individual practices and larger organizations can subscribe. Personal health records and interactions with them to manage one's health and healthcare could also use such services (perhaps evolving as expansions of health/healthcare data tracking services now available to consumers).

This evolution is currently being aided by various initiatives at the national level, through enabling efforts such as the US 21st Century CURES Act,[f] referred to in Chapter 3, section 3.1, various consortia, and a few commercial offerings. The CURES Act, for example, includes legislation that governs relevant topics for CDS such as prohibiting information blocking by EHR systems; and requiring health care systems to provide patients with access to their data through mechanisms such as third-party patient apps. Increasing focus on initiatives that can provide funding is needed to agree on priorities, demonstrate benefits, align motivations, and overcome barriers. Also, since such capabilities will ultimately need to become self-sustaining, attention needs to be given to prospects for commercial (or non-commercial) viability, e.g., through service fees to clients.

To summarize the preceding, we believe that for the deployment of CDS to progress at other than the glacial speed that has occurred to date, the communities of interest—the stakeholders invested in delivering safe, high-quality, cost-effective care—need to proactively organize entities, public and/or private to provide coordinating roles in the evolution of CDS capabilities and tools, processes, and knowledge resources that they require, and the services to orchestrate them.

We believe that the ideal architecture achieved through orchestration of services and resources needs to evolve through an iterative process involving recurring cycles of pilot projects, feedback, refinement, and end-to-end implementation, as depicted in Fig. 30.1. Funding mechanisms such as from national healthcare agencies, e.g., by the US ONC and AHRQ, cited earlier, are already undertaking such efforts; we expect to see this kind of activity as well as efforts by public/private stakeholder bodies grow and begin to build momentum. Goals could be to initiate and/or fund projects by institutions or consortia that would serve as appropriate testbeds. It would probably be best for these projects to be of limited duration. If they are successful, they will provide feedback for improvement of evolving infrastructure, resources, and tools available to all. It may also be necessary to provide funds aimed specifically at technology transfer, in order to get successful projects to the point where they are self-sustaining at local or consortial sites, and for refining the process of adoption of the approach at other sites. Ultimately the goal would be for further replication and adoption of established approaches to be supported through the commercial marketplace. This might also need to be stimulated through a series of small grants to business.

[f]https://www.federalregister.gov/documents/2020/05/01/2020-07419/21st-century-cures-act-interoperability-information-blocking-and-the-onc-health-it-c, last accessed 6/4/2022.

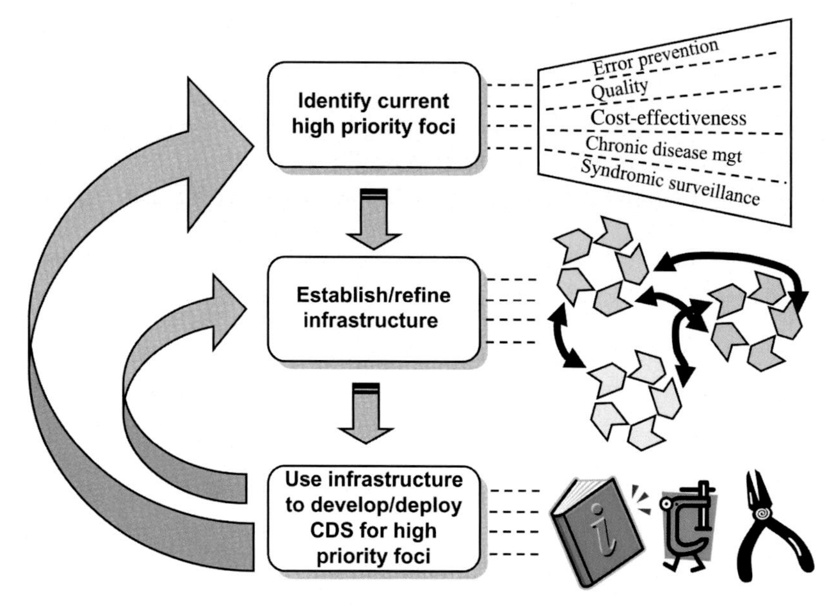

FIG. 30.1 Priorities for CDS&B are likely to fall into specific categories, such as the five areas listed as examples (top right). The three interrelated lifecycle processes involved in generation of knowledge, knowledge management, and incorporation into operational settings require infrastructure for supporting them (middle right; these are the three interlocking lifecycle processes discussed in Chapter 3). The result of applying these lifecycle processes to the priority areas will be knowledge bases and authoring and implementation tools (lower right), which get tried out in selected testbeds. The whole process iterates as we learn more about how to create infrastructure to support it, and as priorities change.

30.5 A possible paradigm for future CDS&B

In this penultimate Section, we would like to switch gears and venture into somewhat uncharted territory by speculating on how we can move the needle and stimulate broader CDS&B adoption and use.

As we review the range of opportunities and the huge amount of new activity occurring in many realms that bear on CDS&B, we are struck by the lack of scalability of current approaches. This is mainly due to the need to essentially handcraft the interfaces between the host Application Environment (AE) and CDS&B modules in our idealized 5-component model of CDS&B presented in Chapter 1, Fig. 1.3, and discussed further in the introduction of Chapter 3. When, where, and how a CDS&B module is invoked by the AE, and how the returned actions and recommendations are to be communicated and utilized in the AE, in terms of insertion into workflow and interaction with the user, typically need to be customized, tested, and refined—this is, in fact, where much of the effort tends to be expended.

Can we do anything to reduce this effort by developing patterns of implementation, and by automating the process of implementation, to facilitate incorporation of CDS&B into AEs at scale? We believe so, and we illustrate our proposed approach with some ideas, based in part on the ACTS Future Vision [12] presented in the introduction to Chapter 1.

30.5.1 Context and state as an organizing framework

Consider a new paradigm for invoking CDS&B, relying on the increasing ability of systems to be aware of setting, situation, activity, and other ambient conditions, which we collectively refer to here as user "**context**." While precise definition of each of the attributes of context is difficult because they overlap, some examples should make the idea clear. Who is the user of the system, what kind of system, where using it, for what activity, focusing on what individual (including self), for what condition or question, initiated by what? Most of these parameters are readily available or obtainable from the system environment and application being carried out. Patient/consumer data can also be continually augmented by smart devices and sensors and the data they continually collect.

We can consider the patient/consumer as also having a health/disease "**state**," defined by all current data and their trajectories—for example, most recent HbA1c% result for diabetes assessment has a value, an indicator of its normal or abnormal status (e.g., by color), date/time relative to current date/time, a time interval since a prior result (if any), and an indicator of qualitative change since that prior result. For example:

HbA1c 9.5% *2w / D 6m* ↗ *(where arrow* ***colors*** *indicate change in normal/abnormal status and* ***angle*** *indicates degree of change since prior)*

Any event that occurs, such as a new data entry (from any source including setting update), a preset time event, or a user action creates a state update. Any state update could be used as a basis for invoking a CDS&B module.

Think of the myriad rules we now have. Think of every node on a computer-interpretable guideline. Think of an order set, a documentation template, a piece of knowledge retrieved by an infobutton manager or generated by a predictive model. Ideally, each of these has either an explicit or implicit set of "eligibility criteria" for when they are relevant to consider. These include characteristics of the patient, and the setting and activities of a user, for which the CDS intervention is appropriate, i.e., an invoking state.

The targeting of particular CDS resources could be made highly explicit if we were able to continually have access to context data about the user, e.g., patient, nurse, doctor, pharmacist, what they are doing, e.g., checking or entering an order, communicating with a patient by phone, going on rounds, and details about the patient, e.g., demographics, problem list, medications, preferences and trajectory of current findings.

We thus call for an expanded set of ontologies of these attributes as the first two key components of shared knowledge infrastructure in Table 30.2. Attributes of *user context*, including intended activities, and *patient clinical state* could in fact be the basis for a semantic framework for organizing CDS knowledge components. Table 30.3 gives shows some example dimensions.

Another aspect that would be helpful is to develop a set of context-specific and state-specific associations, as described in row (3) of Table 30.2. For example, relations among clusters of conditions, observations, and actions (e.g., treatments) can be discovered by combinations of semantic modeling and data mining. Associations can be assembled by using context tuples of type (<subject>|<predicate>|<object>) derived from the literature based

TABLE 30.3 Some possible dimensions for characterizing user context and subject clinical state and needs.

User context	User intention	Subject state
• Who –role, expertise/focus	• Identify/diagnose	• Conditions
• What – type of encounter	• Validate	• Data with trends/ changes
• Where – setting	• Decide	• Reason for encounter
• When – what is going on	• Implement (e.g., carry out a treatment)	• Alerted observations
• Why – what triggered it	• Monitor	• Pending or due or suggested actions
• How – kind of interaction	• Explain	
• For what – purpose, goal	• Document	
• Activity – at precisely what point of action/ system interaction		

User intention is an elaboration of the user purpose/goal dimension.

on semantic tools such as SemRep from the US National Library of Medicine (NLM) [13,14]. By filtering to clinically related concepts, one can have tables with relative frequencies of predicate relations of types such as A causes B, A is a finding of B, A is treated by B, or A is a side effect of B. Loupe [15] is a tool that has been developed to find associations among clinical data elements in an EHR, with labeling of potential relation types. Another way to augment and boost clinical relevance of the relations to be proposed is to create context tuples by mining of clinical records. Colicchio et al. [16] have done some excellent initial work in formally representing patient care context.

Expanding on these ideas, we could use the ontologies describing axes of context and setting, in conjunction with the collections of relevant cognitive thought processes, intentions, and actions to anticipate needs of a user. In a specific context and for a specific patient state, if a user were to focus on particular set of clinical parameters (e.g., current conditions, observations, and actions), the context tuples could be used to anticipate and propose potential notes to be made and decisions that could be made and actions that could be performed.

A side benefit of the assembly of such resources is that we could, over time, identify situations where there are no knowledge resources, which could help focus attention on those that are important to discover. In situations where there are multiple alternative approaches, we could develop methods to use them in combination or as alternative "opinions." We could harness population database for situations where no artifacts exist, or to add to those results the experience of patients maximally similar to the current patient ("patients like mine") in this precise situation, e.g., in terms of which medication was more likely to be beneficial. See [17] for example.

What strikes us here is the similarity of this kind of organizing framework for knowledge-enhanced care to a technology we are already very familiar with—the GPS navigator. GPS is able to monitor where we are in the physical world, and can be set for different modes (e.g., walking vs. driving or sailing), and can be used in a passive mode giving us awareness of our immediate surroundings, information about resources that are available (e.g., restaurants, service stations, and ATMs), and what lies ahead in the direction we are going (e.g., next exits, traffic jams, accidents, and weather conditions). In a directive mode, we can give it a destination, and it develops a plan and helps to keep us on course or to get back on course.

We actually introduced this idea in the First Edition of this book in 2007, but we now have technology to make it work in health and healthcare at scale. If we have a contextual and situational awareness monitoring capability, why can't we create something analogous to GPS in the non-physical world of health and health care management? We could call this a PGS—Personal Guidance System. We know how to build GPS navigators to be very user-friendly, and they are widely adopted and used. Why not build something like this for health care? Thus, we believe that the PGS and the semantic modeling of context and situation applying to health and health care can become a highly effective, scalable framework for organizing and delivering CDS&B.

We can easily obtain context and situation on a continual basis, if an individual opts into it and if we set up our systems to allow it. The user profile, job role and specialty, if you are a provider, roles and restrictions, the applications you are using, the physical location, your immediate prior history, and the patient you are interacting with, or whose record you are viewing and his/her problems and data could be readily used to define context and situation.

We could thus set about refining these contexts and situations, organizing our knowledge artifacts, and identifying where resources exist, where conflicts may occur, and where gaps exist, as a basis for further development of knowledge resources. We can fill in gaps with big data analytics. In my (RG) own laboratory, we are beginning to explore the idea of a PGS framework for organizing available knowledge and integrating it with human activities.

Looking at the future interaction of health/healthcare participants (patients and providers) with computer and communication systems, we can imagine environments that are always proactive in monitoring health status or patients, analyzing the trajectory of findings/status information, anticipating possible needs, interacting with patients or the care team to put those needs forward, and making it easy to review, decide on, and enact necessary actions. Of particular interest are the opportunities to enable the Learning Health System [18] and Quintuple Aim [19]. While it is tempting to add capabilities to an EHR platform, for example, it is necessary to do so in ways that don't overcomplicate the interface or contribute to provider "burnout." Thus, this is one of the specific outcomes to be avoided by the Quintuple Aim.

To drive progress in adopting CDS&B, the above ideas appear to us to be very promising. A number of other ideas that have been proposed or implemented as pilots are also attractive. In fact, we believe that real-life implementation of such ideas for how health and healthcare can work better—ideas that will capture imagination and generate excitement—is necessary for widespread progress. The US Agency for Healthcare Research and Quality (AHRQ) ACTS Framework [12] introduced in Chapter 1 was an attempt by a group of stakeholders to do exactly that. In the remainder of this section, we would like to illustrate some of these ideas.

30.5.2 Examples

We have experimented with various user interaction ideas, such as those suggested below, as examples of what can be done. Consider the hypothetical case presented in Chapter 1 and reproduce below for convenience (adapted from the ACTS Framework, along with examples of CDS&B ideas for the future.[g]

Briefly, the ACTS example scenario considers Mae, a 63-year-old female who has been followed by the same primary care physician (PCP) for 10 years. She has been treated for hypertension and osteoarthritis. Recently, the PCP had to discontinue a nonsteroidal anti-inflammatory drug (NSAID) for her osteoarthritis pain due to progression of stage 3 chronic kidney disease (CKD). Acetaminophen has been ineffective, as has a recent trial of tramadol. The PCP elected to start her on low doses of opiates (oxycodone, 5 mg every 4 hours). This was initially effective in that it reduced her pain, improved her function, and enhanced her overall quality of life. However, Mae told her PCP after several months that it wasn't working as well as it had at the start of treatment.

30.5.2.1 From the patient perspective

For management of hypertension, Mae is enrolled in a population health management program for chronic diseases. As a part of the program, a risk algorithm based on machine-learning suggests an individualized plan that includes self-monitoring, physical activity guided by an individualized virtual coach, and electronic reminders to help with medication adherence. Mae wears a device at home that measures her blood pressure, heart rate, sleep patterns, physical activities, diet and medication intake. The device reminds Mae to take her medications. A phone app associated with the device presents Mae with her measures over time and provides protocol-based feedback. The virtual coach includes motivational components to help ensure adherence and adjusts Mae's physical activity plan according to her level of fitness and achievement of goals. Recommended physical activities are adjusted frequently based on factors such as Mae's preferences, weather forecast, air pollution levels, and proximity to parks. The app also allows Mae to schedule physical activities with her friends and find groups that engage in physical activity with similar levels, goals, and interest as hers.

Mae's device and app continuously send data to her electronic health record (EHR). Mae's state is updated whenever important changes occur, leading to re-computation of machine learning algorithms that detect adverse outcomes and assesses response to treatment. Mae's data are also monitored by her care team and healthcare system as described in the sections below.

30.5.2.2 From provider/care team perspective

Whenever Mae has an encounter with the healthcare system, new data and actions are logged, and are tagged with state parameters, such as interval since prior data or action, change in direction, magnitude, and normality status for a result, or change for a medication such as added, discontinued, or dose modified. These data help provide a context for the provider in preparing for and engaging in the encounter. The encounter may be in-person, virtual, or asynchronous (such as review of results based on an alert).

[g] https://digital.ahrq.gov/acts, last accessed 10/22/2022.

Mae is similar to many patients with multiple problems, several of which are chronic, each with various treatments, and perhaps managed by different members of the care team. Let's assume that, in addition to the problems described in the case scenario above, Mae also has diabetes, hypertension, and depression. For a visit, the provider has limited time available to review prior data and actions, talk with and examine the patient, discuss findings and options, decide on actions, document the encounter, and enter orders. To meet this set of needs, a more dynamic EHR interactive environment is provided, akin to a dashboard with the ability to invoke a set of specialized views to facilitate:

- focusing on pertinent data and trends,
- identifying potentially important relations among data,
- anticipating and offering possible diagnostic considerations, actions to be performed, and statements to include in notes.

(a) **Summary** screen display showing timeline plus the three columns for Conditions, Observations, and Actions.
(b) Detail view of the **patient timeline (journey)** showing a summary of events over time.
(c) In this example, when specific items are selected from any columns by the user ☑ to view, **a sociated conditions, observations and actions** in the record are highlighted (in yellow), and inferred secondary associations (with the yellow items) are included (in gray). Additionally, decision support in the form of a warning icon for the BEERS criteria regarding potential association between Amitriptyline and low sodium is overlayed.
(d) **Trend view** showing selected observations, lab results, and medications over time. Ranges can be highlighted to show possible associations.
(e) **Context-aware menu** of possible relevant user actions, based on where the mouse is hovering.

Fig. 30.2A provides an example of what such a summary view could look like:

- An initial display shows three columns:
  (1) active conditions (problems, including reason for encounter, exposures, risks, and actions due to be performed)
  (2) observations (history of present illness, symptoms, findings, results) with time interval relative to now, direction and normal/abnormal status change, and time interval since prior
  (3) actions current and pending (tests, procedures, referrals, medications, immunizations, education/training, and app/device provisioning.

The summary data in the three columns can be filtered in a number of ways, for example, by relevance of conditions to the provider's focus, recency of last notation, urgency, or implication of the data as a result of the presence of specific observations or the administration of particular treatments.

Special tools are available to view:

- **The patient's journey (timeline)** for all encounters/events or filterable for selected encounter types, including prior visits and notes, laboratory test results, diagnostic imaging, and medication prescriptions. In Fig. 30.2B, a timeline is labeled by icons indicating type of event. The timeline is collapsible/expandible (left arrow), scalable, e.g., in weeks, months, years, logarithmic (most recent in greater detail); or focused on clusters of activity (episodes of care). Hovering over an event icon, the relevant EHR notes/reports can be accessed, as well as trend plots.

(A)

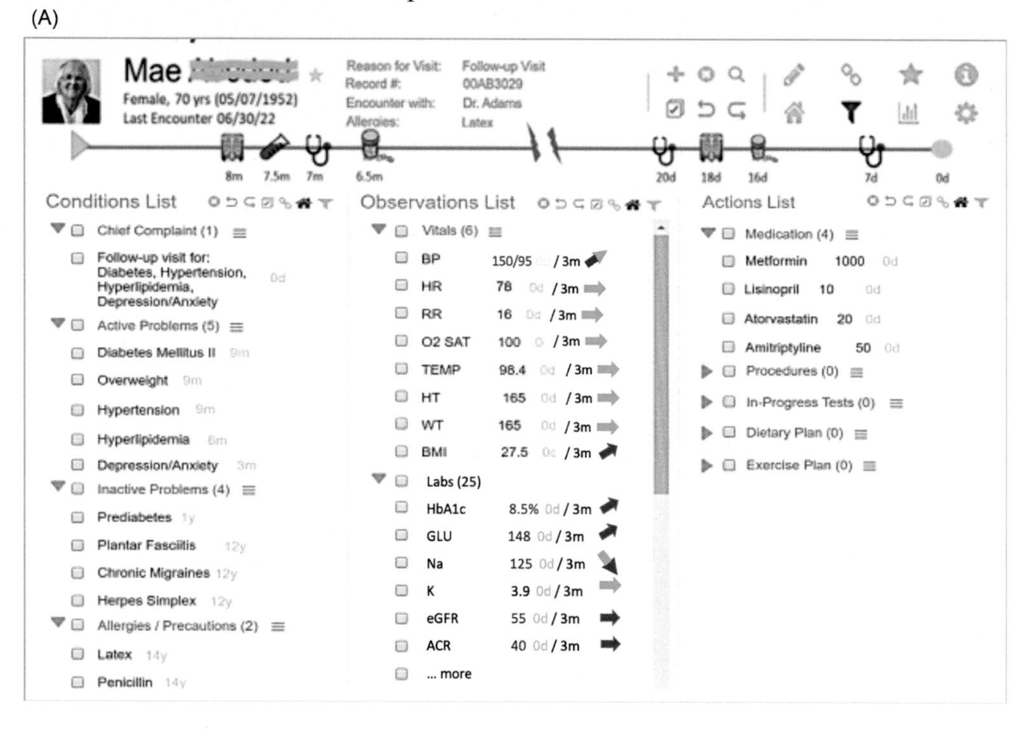

(B)

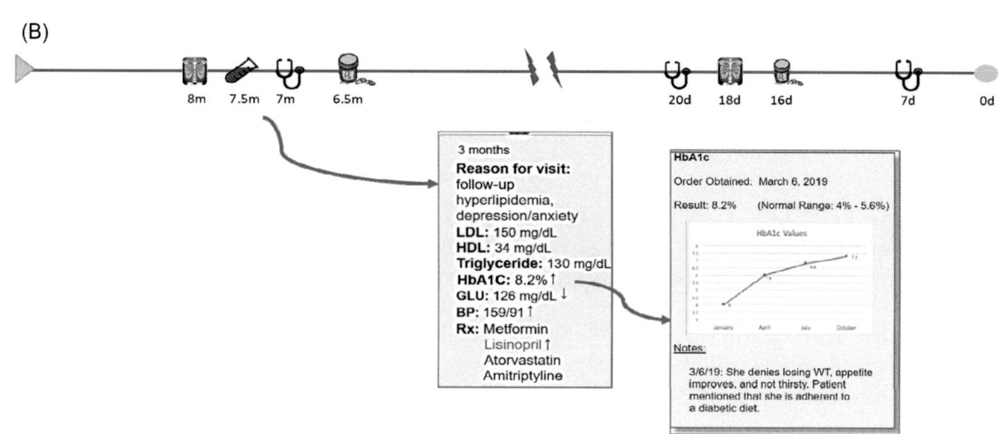

FIG. 30.2 A potential patient visit summary to support visualization of complex relationships and facilitate cognition and planning. (A) Summary view showing timeline and three-column view of conditions, observations, and actions. (B) Patient timeline (journey) showing events and providing access to notes/reports.

(Continued)

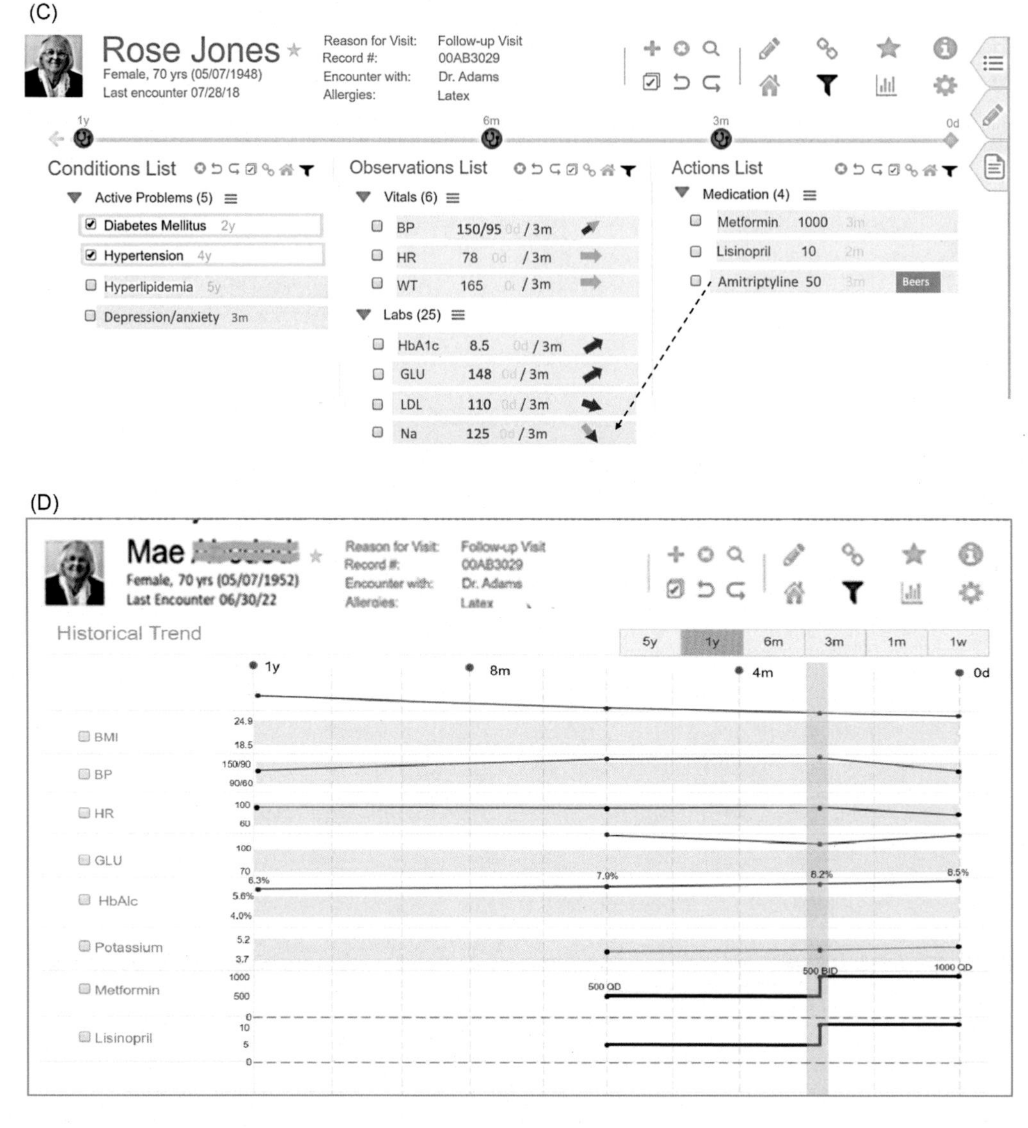

FIG. 30.2, CONT'D (C) Selected items by user from summary, showing inferred associations of related items. (D) Trend view of selected items, showing changes and associations among related items over time.

- **Associations among the data fields.** Fig. 30.2C illustrates an interaction in which, a provider selects certain elements from the summary in (a) (e.g., the conditions of hypertension, obesity, and diabetes), and elements relevant to these are highlighted, including observations and medications relevant to the conditions or the selected observations. In the example, medication links can show intended vs. undesirable associations (such as adverse reactions or likely interactions). This capability is provided by the use of tools such as those described in Table 30.2 row 3.

- **A trend view showing the timeline for selected conditions, observations, and actions** (procedures, medications, etc.), again with similar ways of adjusting time scale. Fig. 30.2D shows quantitative observations as graphs against a normal range; events as point actions or continuing bar, if resulting in a status change (e.g., post-surgery or radiation); and medications as relative dose levels. Clusters of entries in a particular time interval can be highlighted by the system or by the user with vertically bars, calling attention to possible causal associations.

Knowledge support can also be provided in terms of a continually updated menu (e.g., following mouse movement, perhaps in combination with eye-tracking):

- **Floating context bar. Context-aware suggestions for views, notes and actions** can be instantly available. As the user hovers over any part of a display, an intention model (based on the capabilities in Table 30.2, row 3) can be used to provide a floating context bar that offers these options dynamically. This might also be aided by eye-tracking which enables the mouse to quickly jump to a position where the user is looking, when clicked. For example, looking at an upward-trending lab result, the context logic might analyze the presence of relevant medications during the interval, and suggest the possibility of revising the treatment, as well as the phrasing for a statement to be added to the progress note.

Using the tools above, along with data collected from Mae's wearable devices and phone app, Mae's care team continuously monitors Mae's and other patients' status. The care team uses a population-level dashboard to help prioritize care and identify those who need attention (Fig. 30.3). For example, if Mae's hypertension shows signs of deterioration, a machine learning algorithm may flag her record for clinical review and schedule a telehealth visit. Prior to the visit, her provider reviews her case using the visualization timeline described above, where they identify a reduction in physical activity, reduction in medication adherence, and an upward trend in her blood pressure measurements in the last 2 months. A member of Mae's care team calls her to talk about her status. Mae reports that she is not tolerating her hypertension medication well; side effects are limiting some of her daily activities, especially exercising.

During a telehealth visit, Mae's provider uses a patients-like-mine analytic algorithm (see [17]) that suggests alternate treatment approaches that are most likely to help Mae reach her target blood pressure goal based on factors such as prior treatments, medication adherence levels, blood pressure trends, physical activity levels, body mass index, social determinants of health, co-morbidities and Mae's genetic profile. Mae's provider discusses those options with her through an interactive decision-making tool that is shared via Mae's computer screen, and helps visualize individualized risks (e.g., specific side effects) and likelihood of reaching treatment goals under different treatment scenarios (Fig. 30.4).

When Mae develops stage 3 chronic kidney disease, a population analytics algorithm helps her PCP choose the next best option to treat her arthritis based on her data profile. The algorithm reveals that patients with a clinical, genetic, social, and behavioral profile similar to Mae did not benefit from acetaminophen or tramadol, so those options are discarded upfront. Other alternatives are presented, along with a visual depiction of individualized risks and likelihood of alleviating her pain symptoms.

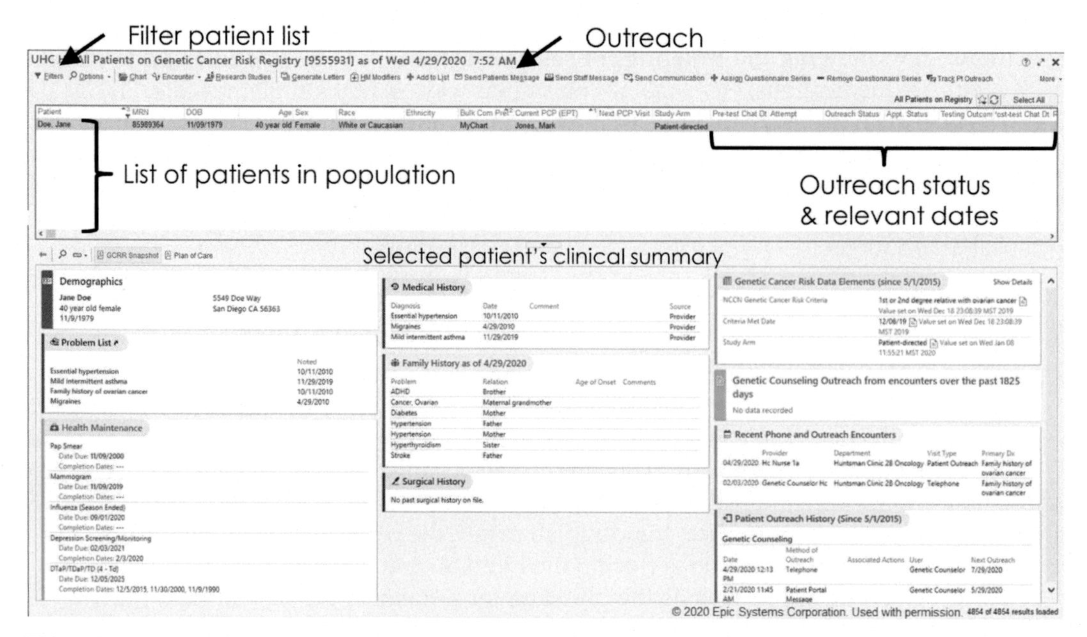

FIG. 30.3 Population health management dashboard for care teams to track patients in a specific cohort, prioritize care, and conduct patient outreach. Main display features include: (1) a list of patients in the cohort is displayed along with key data elements with (2) the ability to filter and sort; (3) status of specific outreach activities for each patient; (4) relevant data groups for a specific selected patient; and (5) outreach functions for contacting (e.g., via text messaging, e-mail, letter) individual patients or multiple patients in bulk.

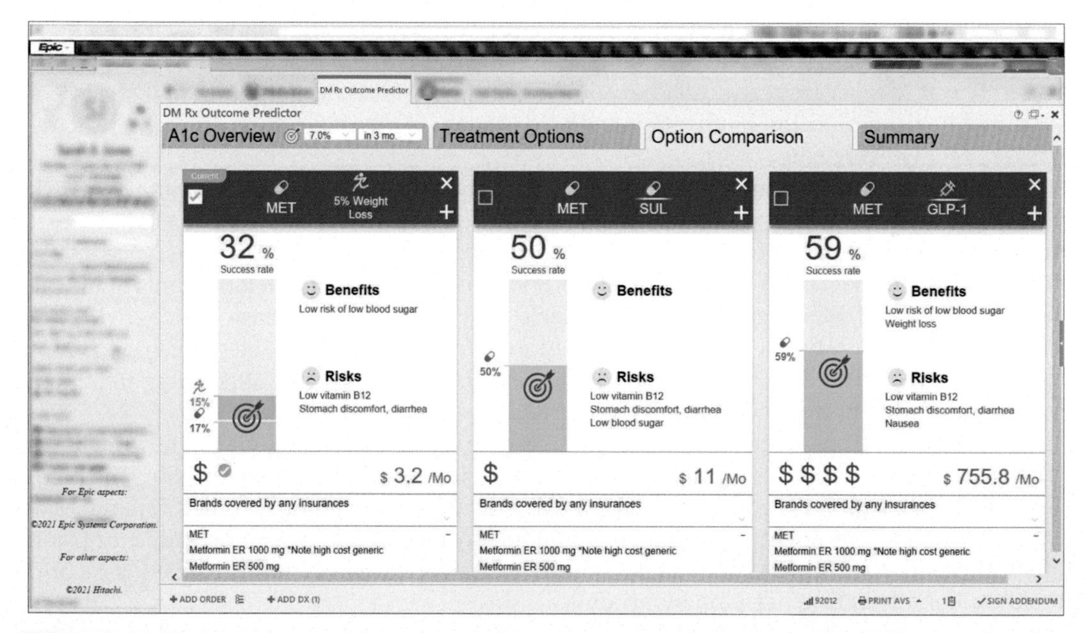

FIG. 30.4 Shared-decision making tool (developed by the Reimagine EHR group at the University of Utah and Hitachi) to help select optimal diabetes treatment [20]. Three top alternatives to achieve a specified HbA1c goal are displayed based on a "patients-like-mine" machine learning algorithm [21]. Each alternative includes a likelihood of treatment success, a list of benefits and risks, and the cost for the patient.

30.5.2.3 From healthcare system/public health perspective

The healthcare system within which Mae receives her care leverages a large variety of data including patient, social determinants of health, healthcare system (e.g., available services, financial data), community (e.g., access to services, green areas, healthy foods), exposures (e.g., air pollution, allergens) to continuously monitor patients and identify cohorts that may benefit from targeted interventions. For example, assume Mae lives by herself in an area with difficult access to supermarkets that offer fresh vegetables and fruits. The healthcare system uses digital outreach tools such as text messaging and conversational agents to offer Mae and other similar patients a service that delivers a box of fruits and vegetables weekly at their homes. Since Mae lives in an area with seasonal high air pollution, the healthcare system also offers her access to air purifiers with HEPA-certified filters. Following the vision of a learning health system [18], the healthcare system continuously monitors the effectiveness of these interventions from the patient and healthcare utilization perspectives, and iteratively refines interventions and target cohorts using a data-driven approach. Likewise, public health agencies such as county and state departments of health use data from patients like Mae to conduct health surveillance and plan population-level interventions. At the national-level, ubiquitous access to similar, de-identified data, enables a nationwide learning health system [22].

30.6 Looking ahead: Epilogue as prologue

We are poised at a point where the need to accelerate efforts for CDS&B adoption is great, but where ill-conceived or inadequately founded efforts could contribute more to chaos than to benefit. We are already overwhelmed by knowledge, so just having many varieties of it deployed in the form of CDS&B is no guarantee that patient safety, quality, cost-effectiveness, or other objectives will be achieved. In fact, sorting through and reconciling conflicting knowledge may be particularly frustrating.

As we seek to accomplish approaches to sharing the results of knowledge generation and knowledge management required for the preceding, we also need to continue an active process of experimentation to learn how to best deploy CDS&B for maximum benefit and acceptability by users. Thus, we need to lower the barriers for this process. By considering CDS&B as an external capability, we are also shifting the paradigm from a built-in functionality of a clinical system to an added value that can be incorporated into clinical applications in a variety of ways. This opens up the process not only to initiative and experimentation but also to business opportunities, by creating niches for content, software, and services that would otherwise not be there.

Thus, there are many reasons for moving in the general direction outlined. The road has progressed from a bumpy one that was initially largely unpaved, to the current point, since our Second Edition, in which some of the pavement has been done, which can now allow our speed to accelerate. With the new "&B" capabilities now feasible since our Second Edition, progress on the broader goal of knowledge-enhanced health and healthcare can be expected to accelerate.

Efforts need to also focus on how to enable the incorporation of CDS&B into health and healthcare to be done at scale. Organizing our collective effort appears to be the only feasible path for enabling us to cope with the many opportunities and challenges, particularly in the context of a health care system that is itself undergoing transformation. It is encouraging to see efforts to do this mounting in various nations, both in standards efforts, national health care infrastructure development, EHR adoption, and professional and public calls to action as well as collective efforts such as public/private consortia.

So, as in our previous two editions, we do hope that this Epilogue will indeed be a Prologue and that we are able to move toward a truly knowledge-enhanced health and healthcare system for the benefit of all of us.

References

[1] NHS. National Programme for IT in the NHS. Connecting for Health. National Health Service; 2006. http://www.connectingforhealth.nhs.uk/.
[2] Pipersburgh J. The push to increase the use of EHR technology by hospitals and physicians in the United States through the HITECH Act and the Medicare incentive program. J Health Care Finance 2011;38:54–78.
[3] Murphy J. The journey to meaningful use of electronic health records. Nurs Econ 2010;28:283–6.
[4] Topol EJ. The creative destruction of medicine: How the digital revolution will create better health care. Paperbak, Basic Books; 2013.
[5] Osheroff JA, Teich JM, Middleton B, Steen EB, Wright A, Detmer DE. A roadmap for national action on clinical decision support. J Am Med Inform Assoc 2007;14:141–5.
[6] OpenClinical. Open Clinical: Knowledge Management for Health Care., 2013, http://www.openclinical.org/home.html.
[7] Greenes R, Bloomrosen M, Brown-Connolly NE, Curtis C, Detmer DE, Enberg R, et al. The morningside initiative: collaborative development of a knowledge repository to accelerate adoption of clinical decision support. Open Med Inform J 2010;4:278–90. https://doi.org/10.2174/1874431101004010278.
[8] Middleton B. The clinical decision support consortium. Stud Health Technol Inform 2009;150:26–30.
[9] OpenCDS. Open Clinical Decision Support (OpenCDS) Tools and Resources., 2015, http://www.opencds.org/. [Accessed 21 March 2015].
[10] S&I. S&I framework., 2013, http://www.siframework.org/.
[11] NLM. Value Set Authority Center., 2013, https://vsac.nlm.nih.gov/.
[12] AHRQ Evidence-Based Care Transformation Support (ACTS). Digital Healthcare Research n.d. https://digital.ahrq.gov/acts [Accessed 23 April 2022].
[13] Kilicoglu H, Rosemblat G, Fiszman M, Shin D. Broad-coverage biomedical relation extraction with SemRep. BMC Bioinform 2020;21:188. https://doi.org/10.1186/s12859-020-3517-7.
[14] Liu Y, Bill R, Fiszman M, Rindflesch T, Pedersen T, Melton GB, et al. Using SemRep to label semantic relations extracted from clinical text. AMIA Annu Symp Proc 2012;2012:587–95.
[15] Boxwala A. Less is more: context-relevant views of EHRs. Elimu Informatics, Inc.; 2018. https://reporter.nih.gov/project-details/9465578.
[16] Colicchio TK, Dissanayake PI, Cimino JJ. Formal representation of patients' care context data: the path to improving the electronic health record. J Am Med Inform Assoc 2020;27:1648–57. https://doi.org/10.1093/jamia/ocaa134.
[17] Li P, Yates SN, Lovely JK, Larson DW. Patient-like-mine: A real time, visual analytics tool for clinical decision support. In: 2015 IEEE international conference on big data (big data); 2015. p. 2865–7. https://doi.org/10.1109/BigData.2015.7364104.
[18] McGinnis JM, Fineberg HV, Dzau VJ. Advancing the learning health system. N Engl J Med 2021;385:1–5. https://doi.org/10.1056/NEJMp2103872.
[19] Coleman K, Wagner E, Schaefer J, Reid R, LeRoy L. Redefining primary care for the 21st century. vol. 16. Rockville, MD: Agency for Healthcare Research and Quality; 2016. p. 1–20.
[20] Kawamoto K, Kukhareva PV, Weir C, Flynn MC, Nanjo CJ, Martin DK, et al. Establishing a multidisciplinary initiative for interoperable electronic health record innovations at an academic medical center. JAMIA Open 2021;4:ooab041. https://doi.org/10.1093/jamiaopen/ooab041.

[21] Tarumi S, Takeuchi W, Qi R, Ning X, Ruppert L, Ban H, et al. Predicting pharmacotherapeutic outcomes for type 2 diabetes: an evaluation of three approaches to leveraging electronic health record data from multiple sources. J Biomed Inform 2022;129:104001. https://doi.org/10.1016/j.jbi.2022.104001.
[22] Friedman CP, Wong AK, Blumenthal D. Achieving a nationwide learning health system. Sci Transl Med 2010;2:57cm29. https://doi.org/10.1126/scitranslmed.3001456.

Index

Note: Page numbers followed by *f* indicate figures, *t* indicate tables, and *b* indicate boxes.

D

H

I

M

N

Q

T

Printed in the United States
by Baker & Taylor Publisher Services